Greek Alphabet

Alpha	A	α	Iota	I	ι	Rho		
Beta	B	β	Kappa	K	κ	Sigma	Σ	σ
Gamma	Γ	γ	Lambda	Λ	λ	Tau	T	τ
Delta	Δ	δ	Mu	M	μ	Upsilon	Υ	υ
Epsilon	E	ε	Nu	N	ν	Phi	Φ	ϕ
Zeta	Z	ζ	Xi	Ξ	ξ	Chi	X	χ
Eta	H	η	Omicron	O	o	Psi	Ψ	ψ
Theta	Θ	θ	Pi	Π	π	Omega	Ω	ω

Conversion Table for Units

Length

meter (SI unit)	m	
centimeter	cm	$= 10^{-2}$ m
ångström	Å	$= 10^{-10}$ m
micron	μ	$= 10^{-6}$ m

Volume

cubic meter (SI unit)	m^3	
liter	L	$= dm^3 = 10^{-3}$ m^3

Mass

kilogram (SI unit)	kg	
gram	g	$= 10^{-3}$ kg

Energy

joule (SI unit)	J	
erg	erg	$= 10^{-7}$ J
rydberg	Ry	$= 2.179\ 87 \times 10^{-18}$ J
electron volt	eV	$= 1.602\ 18 \times 10^{-19}$ J
inverse centimeter	cm^{-1}	$= 1.986\ 45 \times 10^{-23}$ J
calorie (thermochemical)	cal	$= 4.184$ J
liter atmosphere	l atm	$= 101.325$ J

Pressure

pascal (SI unit)	Pa	
atmosphere	atm	$= 101325$ Pa
bar	bar	$= 10^5$ Pa
torr	Torr	$= 133.322$ Pa
pounds per square inch	psi	$= 6.894\ 757 \times 10^3$ Pa

Power

watt (SI unit)	W	
horsepower	hp	$= 745.7$ W

Angle

radian (SI unit)	rad	
degree	°	$= \dfrac{2\pi}{360}$ rad $= \left(\dfrac{1}{57.295\ 78}\right)$ rad

Electrical dipole moment

C m (SI unit)		
debye	D	$= 3.335\ 64 \times 10^{-30}$ C m

Physical Chemistry
for the Life Sciences

Thomas Engel
University of Washington

Gary Drobny
University of Washington

Philip Reid
University of Washington

PEARSON

Prentice Hall

Upper Saddle River, NJ 07458

Library of Congress Cataloging-in-Publication Data

Engel, Thomas
   Physical chemistry for the life sciences / Thomas Engel, Gary Drobny, Philip Reid.
      p. cm.
   Includes bibliographical references and index.
   ISBN-13: 978-0-8053-8277-8
   ISBN-10: 0-8053-8277-1
   1. Physical biochemistry. 2. Chemistry, Physical and theoretical. 3. Life sciences.
I. Drobny, Gary. II. Reid, Philip. III. Title.
   QP517 .P49E54 2008
   572' .43—dc22

                                                    2007014794

Editor-in-Chief, Science: Nicole Folchetti	Art Studio: Argosy
Acquisitions Editor: Jeff Howard	AV Project Manager: Connie Long
Associate Editor: Carol Dupont	Director of Logistics, Operations, and Vendor Relations: Barbara Kittle
Editorial Assistant: Laurie Varites	Senior Operations Supervisor: Alan Fischer
Senior Managing Editor: Kathleen Schiaparelli	Director, Image Resource Center: Melinda Patelli
Production Editor: Emily Bush, Carlisle Publishing Services	Manager, Rights and Permissions: Zina Arabia
Composition: Carlisle Publishing Services	Interior Image Specialist: Beth Brenzel
Director of Design: Christy Mahon	Image Permission Coordinator: Annette Linder
Art Director: Kenny Beck	Photo Researcher: Kristin Piljay
Cover and Interior Design: Hespenheide Design	Cover Image: Ribbon structure/Greg Williams

© 2008 Pearson Education, Inc.
Pearson Prentice Hall
Pearson Education, Inc.
Upper Saddle River, NJ 07458

Printed in the United States of America
10 9 8 7 6 5 4 3 2 1

ISBN-10:   0-8053-8277-1
ISBN-13: 9780-8053-8277-8

Pearson Education LTD., *London*
Pearson Education Australia PTY, Limited, *Sydney*
Pearson Education Singapore, Pte. Ltd.
Pearson Education North Asia Ltd., *Hong Kong*
Pearson Education Canada, Ltd., *Toronto*
Pearson Educación de Mexico, S.A. de C.VB.
Pearson Education—Japan, *Tokyo*
Pearson Education Malaysia, Pte. Ltd.

This book is dedicated to my parents, Walter and Juliane, who were my first teachers, and to my cherished family, Esther and Alex, with whom I am still learning.

—Thomas Engel

This book is dedicated to my family: Annika, Joshua, and Elizabeth

—Gary Drobny

This book is dedicated to my friends for their faith in me.

—Philip Reid

Brief Contents

Contents

About the Authors

Thomas Engel has taught chemistry at the University of Washington for more than 20 years, where he is Professor Emeritus of Chemistry. Professor Engel received his bachelor's and master's degrees in chemistry from the Johns Hopkins University, and his Ph.D. in chemistry from the University of Chicago. He then spent 11 years as a researcher in Germany and Switzerland, in which time he received the Dr. rer. nat. habil. degree from the Ludwig Maximilians University in Munich. In 1980, he left the IBM research laboratory in Zurich to become a faculty member at the University of Washington.

Professor Engel's research interests are in the area of surface chemistry, and he has published more than 80 articles and book chapters in this field. He has received the Surface Chemistry of Colloids Award from the American Chemical Society and a Senior Humboldt Research Award from the Alexander von Humboldt Foundation. When not writing, he is likely to be hiking, sea kayaking, or cross-country skiing.

Gary Drobny has taught chemistry at the University of Washington since he joined the chemistry faculty in 1985. Professor Drobny received his bachelor's degree in chemistry from San Francisco State University in 1976, and his Ph.D. in chemistry in 1981 from the University of California at Berkeley.

Professor Drobny's interests are in the areas of solution and solid-state nuclear magnetic resonance, protein-nucleic acid recognition, biomaterials, biomineralization, and structural studies of proteins at biomaterial interfaces. He has published more than 120 articles in these fields.

Philip Reid has taught chemistry at the University of Washington since he joined the chemistry faculty in 1995. Professor Reid received his bachelor's degree from the University of Puget Sound in 1986, and his Ph.D. in chemistry from the University of California at Berkeley in 1992. He performed postdoctoral research at the University of Minnesota, Twin Cities, campus before moving to Washington.

Professor Reid's research interests are in the areas of atmospheric chemistry, condensed-phase reaction dynamics, and nonlinear optical materials. He has published more than 90 articles in these fields. Professor Reid is the recipient of a CAREER award from the National Science Foundation, is a Cottrell Scholar of the Research Corporation, and is a Sloan fellow. He received the Distinguished Teaching Award from the University of Washington in 2005 for his contributions to undergraduate education.

Preface

This book grew out of our experience in teaching physical chemistry to undergraduate students majoring in chemistry, biochemistry, and the biological sciences. The following objectives, illustrated with brief examples, outline the distinctive features of this book:

- **Focus on teaching core concepts.**   The central principles of physical chemistry are explored by focusing on core ideas, and then extending these ideas to a variety of problems. For example, the Gibbs energy, bioenergetics, and chemical equilibrium are at the heart of thermodynamics and are explored in depth in this text. Similarly, a very good understanding of quantum mechanics can be obtained from a few basic systems: the particle in a box, the harmonic oscillator, and the hydrogen atom. Therefore, care is taken to fully explain and develop these key systems in order to provide a solid foundation for the student. A similar approach has been taken in other areas of physical chemistry. The goal is to build a solid foundation of student understanding rather than cover a wide variety of topics in modest detail.

- **Illustrate the relevance of physical chemistry to the world around us.**   Many students struggle to connect physical chemistry concepts to the world around them. To address this issue, example problems and specific topics are tied together to help the student develop this connection. Biological membranes and the energetics of ion transport are discussed in a chapter focused on bioenergetics. Fuel cells, refrigerators, and heat pumps are discussed in connection with the second law of thermodynamics. Glycolysis, the Krebs cycle, and the electron transport chain are discussed in a chapter on biochemical equilibria.

- **Demonstrate the importance of quantum mechanics in the biological sciences.** Many everyday phenomena cannot be understood without quantum mechanics. The particle-in-a-box model is used to explain why metals conduct electricity and why valence electrons rather than core electrons are important in chemical bond formation. The real-world applications of quantum mechanics are in chemical spectroscopy. In-depth discussions of structural determinations of biomolecules using multidimensional NMR, the use of Raman spectroscopy to image living cells with chemical specificity, the use of fluorescence spectroscopy to sequence the human genome, and the use fluorescence resonance energy transfer (FRET) as a spectroscopic ruler to measure donor–acceptor distances show the student the importance of having a solid foundation in quantum mechanics.

- **Present exciting new science in the field of physical chemistry.**   Physical chemistry lies at the forefront of many emerging areas of modern chemical research. Examples discussed in this text include the use of atomic force microscopy to obtain nanometer-scale structural information about biological systems *in situ* and in real time, the use of single-molecule spectroscopy to understand kinetics at a molecular level, the use of FRET to determine the magnitude of the structural change introduced by substrate binding to an enzyme, and the use of multidimensional NMR to determine biomolecular structures in solution.

- **Use Web-based simulations to illustrate the concepts being explored and avoid math overload.**   Mathematics is central to physical chemistry; however, the mathematics can distract the student from "seeing" the underlying concepts. To circumvent this problem, Web-based simulations have been incorporated as end-of-chapter problems throughout the book so that the student can focus on the science and avoid a math overload. These Web-based simulations can also be used by instructors during lectures. More than 50 such Web-based problems are available on the course Web site. An important feature is that each problem has been designed as an assignable exercise with a printable answer sheet that the student can submit to the instructor. The course Web site also includes a graphing routine with a curve-fitting capability, which allows students to print and submit graphical data.
- **Show that learning problem-solving skills is an essential part of physical chemistry.**   Many example problems are worked through in each chapter. The end-of-chapter problems cover a range of difficulties suitable for students at all levels. Conceptual questions at the end of each chapter ensure that students learn to express their ideas in the language of science.
- **Use color to make learning physical chemistry more interesting.**   Color is used to enhance both the pedagogy and content of the text. For example, four-color images are used to enhance the understanding of biochemical cycles, to display atomic and molecular orbitals both quantitatively and attractively, and to make complex images such as multidimensional NMR spectra understandable.

This text contains more material than can be covered in a one- or two-semester course, and this is entirely intentional. Effective use of the text does not require one to proceed sequentially through the chapters, or to include all sections. Many sections are self-contained so that they can be readily omitted if they do not serve the needs of the instructor. The text is constructed to be flexible to your needs, not the other way around. We welcome the comments of both students and instructors on how the material was used and on how the presentation can be improved.

Thomas Engel
University of Washington

Gary Drobny
University of Washington

Philip Reid
University of Washington

Acknowledgments

Many individuals have helped us to bring the text into its current form. Students have provided us with feedback directly and through the questions they have asked, which has helped us to understand how they learn. We especially thank our colleagues at the University of Washington including Rachel Klevit, Mickey Schurr, and Gabriele Varani who contributed valuable advice about biological NMR spectroscopy and transport. The help of Nicholas Breen, Gil Goobes, and Dirk Stueber, who reviewed Chapter 20, is also gratefully acknowledged. The help of William Parson and Ronald Stenkamp in critically reading Chapter 21 has been invaluable. Our own approach to thermodynamics and statistical thermodynamics has been influenced by the excellent textbooks of Lennard Nash and Gilbert Castellan. The biologically oriented physical chemistry texts by Eisenberg and Crothers and Tinoco, Sauer, Wang, and Puglisi influenced some of our approaches to topics in transport and spectroscopy. We are also fortunate to have access to some end-of-chapter problems that were written by Joseph Noggle and Gilbert Castellan in their physical chemistry textbooks. The reviewers, who are listed separately, have made many suggestions for improvement, for which we are very grateful. All those involved in the production process have helped to make this book a reality through their efforts. Special thanks are due to Jim Smith, who helped us initiate this project, and to Katie Conley and Jeff Howard who have guided the production process.

Reviewers

Alexander Angerhofer
University of Florida

Jochen Autschbach
State University of New York, Buffalo

Dor Ben-Amotz
Purdue University

Sunney Chan
California Institute of Technology

David Chen
University of British Columbia

Phillip Geissler
University of California Berkeley

Javier B. Giorgi
University of Ottawa

Martina Kaledin
Kennesaw State University

Kathleen Knierim
University of Louisiana, Lafayette

Krzysztof Kuczera
University of Kansas

Challa V. Kumar
University of Connecticut

Alexander D. Li
Washington State University

Bob Pecora
Stanford University

Glenn Penner
University of Guelph

Jacob Petrich
Iowa State University

Donald E. Sands
University of Kentucky

Caroline M. Taylor
Michigan Technological University

Andrew Teplyakov
University of Delaware

Engel/Reid's *Physical Chemistry*, 1/e Reviewers

Ludwik Adamowicz
University of Arizona

Daniel Akins
City College of New York

Peter Armentrout
University of Utah

Joseph BelBruno
Dartmouth College

Eric Bittner
University of Houston

Juliana Boerio-Goates
Brigham Young University

Alexandre Brolo
University of Victoria

Alexander Burin
Tulane University

Laurie Butler
University of Chicago

Ronald Christensen
Bowdoin College

Jeffrey Cina
University of Oregon

Robert Continetti
University of California, San Diego

Susan Crawford
California State University, Sacramento

Ernest Davidson
University of Washington

H. Floyd Davis
Cornell University

Jimmie Doll
Brown University

D. James Donaldson
University of Toronto

Robert Donnelly
Auburn University

Doug Doren
University of Delaware

Bogdan Dragnea
Indiana University

Cecil Dybowski
University of Delaware

Donald Fitts
University of Pennsylvania

Patrick Fleming
San Jose State University

Edward Grant
Purdue University

Arthur Halpern
Indiana State University

Ian Hamilton
Wilfrid Laurier University

Cynthia Hartzell
Northern Arizona University

Rigoberto Hernandez
Georgia Institute of Technology

Ming-Ju Huang
Jackson State University

Ronald Imbihl
University of Hannover

George Kaminski
Central Michigan University

Katherine Kantardjieff
California State University, Fullerton

Chul-Hyun Kim
California State University, Hayward

Keith Kuwata
Macalester College

Kimberly Lawler-Sagarin
Elmhurst College

Katja Lindenberg
University of California, San Diego

Lawrence Lohr
University of Michigan

John Lowe
Penn State University

Peter Lykos
Illinois Institute of Technology

Peter Macdonald
University of Toronto, Mississauga

David Micha
University of Florida

David Nesbitt
University of Colorado

Daniel Neumark
University of California, Berkeley

Simon North
Texas A&M University

Maria Pacheco
Buffalo State College

Robert Pecora
Stanford University

Lee Pedersen
University of North Caroline, Chapel Hill

Jacob Petrich
Iowa State University

Vitaly Rassolov
University of South Carolina

David Ritter
Southeast Missouri State University

Peter Rossky
University of Texas, Austin

Marc Roussel
University of Lethbridge

Ken Roussland
University of Puget Sound

George Schatz
Northwestern University

Robert Schurko
University of Windsor

Roseanne J. Sension
University of Michigan

Alexa Serfis
Saint Louis University

Robert Wofford
Wake Forest University

Michael Trenary
University of Illinois, Chicago

Carl Trindle
University of Virginia

Michael Tubergen
Kent State University

Tom Tuttle
Brandeis University

James Valentini
Columbia University

Carol Venanzi
New Jersey Institute of Technology

Michael Wagner
George Washington University

Robert Walker
University of Maryland

Gary Washington
United States Military Academy, West Point

Charles Watkins
University of Alabama at Birmingham

Rand Watson
Texas A&M University

Mark Young
University of Iowa

Problem Solvers

Alexander Angerhofer
University of Florida

Krzysztof Kuczera
University of Kansas

Donald Sands
University of Kentucky

Dirk Steuber
University of Washington

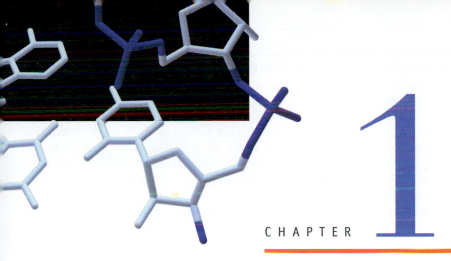

CHAPTER 1

Fundamental Concepts of Thermodynamics

Thermodynamics provides a description of matter on a macroscopic scale. In this approach, matter is described in terms of bulk properties such as pressure, density, volume, and temperature. The basic terms employed in thermodynamics, such as system, surroundings, intensive and extensive variables, adiabatic and diathermal walls, equilibrium, temperature, and thermometry, are discussed in this chapter. The usefulness of equations of state, which relate the state variables of pressure, volume, and temperature, is also discussed for real and ideal gases.

1.1 What Is Thermodynamics and Why Is It Useful?

Thermodynamics is the branch of science that describes the behavior of matter and the transformation between different forms of energy on a **macroscopic scale** (i.e., the human scale and larger). Thermodynamics describes a system in terms of its bulk properties. Only a few such bulk property variables are needed to describe the system, and these variables are generally obtained via measurements. A thermodynamic description of matter does not make reference to its structure and behavior at the microscopic level. For example, 1 mol of gaseous water at a sufficiently low density is completely described by two of the three **macroscopic variables** of pressure, volume, and temperature. By contrast, the **microscopic scale** refers to dimensions on the order of the size of molecules. At the microscopic level, water is a dipolar triatomic molecule, H_2O, with a bond angle of 104.5° that forms a network of hydrogen bonds.

Given that the microscopic nature of matter is becoming increasingly well understood using theories such as quantum mechanics, why is a macroscopic science like thermodynamics relevant today? The need to approach problems from a macroscopic point of view may seem debatable. Indeed, an argument exists for describing physical problems from a microscopic point of view using quantum or classical mechanics, then deriving macroscopic properties statistically. Such a strategy, commonly called the "bottom-up" approach, is often justifiable in a field such as chemistry where nature is frequently investigated at the molecular level, but in many fields of engineering and biology, nature is not viewed exclusively in detail at the molecular level. In these cases, a "top-down" strategy is followed wherein macroscopic properties are investigated without reference to the underlying microscopic composition or mechanics of the system. Even if an engineer or biologist desires in principle a knowledge of the fundamental molecular properties that underlie a given problem at hand, the system under study may be so complex and poorly characterized at the atomic or molecular level that the only practical approach to understanding material properties or biological function frequently is to study the system's macroscopic properties. As such, thermodynamics is a method of choice for understanding the energetics of complex systems. It is no surprise that in his seminal text *Bioenergetics: The Molecular Basis of Biological Energy Transformations,* Albert L. Lehninger stated, "The proper study of biology should really *begin* with the theme of energy and its transformations."

The usefulness of thermodynamics can be further illustrated by describing three applications of thermodynamics to biological and nonbiological problems:

- You have built a plant to synthesize ammonia gas (NH_3) from N_2 and H_2. You find that the yield is insufficient to make the process profitable and decide to try to improve the ammonia yield by changing the temperature and/or the pressure. However, you do not know whether to increase or decrease the values of these variables. As you can conclude after reading Chapter 6, the ammonia yield will be higher at equilibrium if the temperature is decreased and the pressure is increased.
- You are developing a drug that inhibits a bacterial enzyme. You want to determine the binding affinity and the number of binding sites that the drug has on the enzyme. Also, if the drug binds to more than one site on the enzyme, you want to determine if binding occurs to each site independently or if binding to one site enhances binding to other sites. As we will explain in Chapter 11, thermodynamics affords an efficient method for determining all of this information.
- As a biochemist interested in the energetics of membrane transport, you want to determine the energy associated with transport of sodium and potassium ions across a cell membrane. As will be shown in Chapter 10, the amount of ATP that must be hydrolyzed to effect active transport of sodium and/or potassium ions across a cell membrane, and the pH gradient across the inner mitochondrial membrane required to synthesize a certain amount of ATP, can be calculated using thermodynamic principles.

In this book, you will learn first about thermodynamics and then about statistical thermodynamics. **Statistical thermodynamics** (Chapters 22 and 23) uses atomic and molecular properties to calculate the macroscopic properties of matter. For example, statistical thermodynamics can show that liquid water is the stable form of aggregation at a pressure of 1 bar and a temperature of 90°C, whereas gaseous water is the stable form at 1 bar and 110°C. Using statistical thermodynamics, macroscopic properties are calculated from underlying molecular properties.

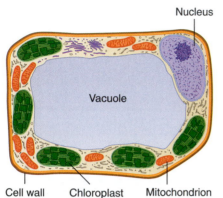

Nucleus

Vacuole

Cell wall   Chloroplast   Mitochondrion

Plant cell

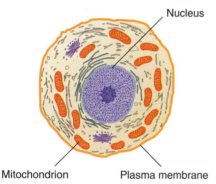

Nucleus

Mitochondrion   Plasma membrane

Animal cell

FIGURE 1.1
Animal and plant cells are open systems. The contents of the animal cell include the cytosol fluid and the numerous organelles (e.g., nucleus and mitochondria) that are separated from the surroundings by a lipid-rich plasma membrane. The plasma membrane acts as a boundary layer that can transmit energy and is selectively permeable to ions and various metabolites. A plant cell is surrounded by a cell wall that similarly encases the cytosol and organelles including chloroplasts, which are the sites of photosynthesis. (Source: Adapted from W.M. Becker, L.J. Kleinsmith, and J. Hardin, *The World of the Cell,* 6th ed. (San Francisco, CA: Pearson Benjamin Cummings), p. 2. Copyright 2006. Reprinted by permission of Pearson Education, Inc.)

1.2 Basic Definitions Needed to Describe Thermodynamic Systems

A thermodynamic **system** consists of all the materials involved in the process under study. This material could be the contents of an open beaker containing reagents; a solution consisting of a macromolecular solute, buffer components, and ligands; or the total of the electrolytes, the cytosol, and organelles distributed within a cell membrane. In thermodynamics, the rest of the universe is referred to as the **surroundings.** If a system can exchange matter with its surroundings, it is called an **open system;** if not, it is a **closed system.** Living cells are open systems (see Figure 1.1). Both open and closed systems can exchange energy with their surroundings. Systems that can exchange neither matter nor energy with their surroundings are called **isolated systems.** The contents of a perfectly insulating and sealed thermos vessel provide an example of an isolated system.

The interface between the system and its surroundings is called the **boundary,** which often takes the form of a **wall.** The boundaries determine if energy and matter can be transferred between the system and the surroundings and lead to the distinction between open, closed, and isolated systems. Consider the Earth's oceans as a system, with the rest of the universe being the surroundings. The system–surroundings boundary consists of the solid–liquid interface between the continents and the ocean floor and the water–air interface at the ocean surface. Or consider a living cell as a system whose contents are separated from the surrounding extracellular matrix by a cell membrane, a boundary layer that is composed of various lipids, integral cell membrane proteins, etc. In each case, energy and matter can be exchanged between the system and surroundings by transport through the boundary layer.

The exchange of energy and matter across the boundary between system and surroundings is central to the important concept of **equilibrium.** The system and surroundings can be in equilibrium with respect to one or more of several different **system variables** such as pressure (P), temperature (T), and concentration. **Thermodynamic equilibrium** refers to a condition in which equilibrium exists with respect to P, T, and concentration. Equilibrium exists with respect to a single variable only if that variable does not change with time, and if it has the same value in all parts of the system and surroundings. For example, the interior of a soap bubble[1] (the system) and the surroundings (the room) are in equilibrium with respect to P because the movable wall of the bubble can reach a position where P on both sides of the wall is the same, and because P has the same value throughout the system and surroundings. Equilibrium with respect to concentration exists only if transport of all species across the boundary in both directions is possible. If the boundary is a movable wall that is not permeable to all species, equilibrium can exist with respect to P, but not with respect to concentration. Because N_2 and O_2 cannot diffuse through the (idealized) bubble, the system and surroundings are in equilibrium with respect to P, but not to concentration. Equilibrium with respect to temperature is a special case that is discussed next.

Temperature is an abstract quantity that can only be measured indirectly, for example, by measuring the volume of mercury confined to a narrow capillary, the electromotive force generated at the junction of two dissimilar metals, or the electrical resistance of a platinum wire. At the microscopic level, temperature is related to the mean kinetic energy of molecules. Although each of us has a sense of a "temperature scale" based on the qualitative descriptors *hot* and *cold,* we need a more quantitative and transferable measure of temperature that is not grounded in individual experience. To make this discussion more concrete, we consider a dilute gas under conditions in which the ideal gas law of Equation (1.1) describes the relationship among P, T, and the molar density $\rho = n/V$ with sufficient accuracy:

$$P = \rho RT \tag{1.1}$$

where R is a proportionality constant that will be discussed together with the ideal gas law in Section 1.4.

In thermodynamics, **temperature** is the property of a system that determines if the system is in thermal equilibrium with other systems or the surroundings. Equation (1.1) can be rewritten as follows:

$$T = \frac{P}{\rho R} \tag{1.2}$$

This equation shows that for ideal gas systems having the same molar density, a pressure gauge can be used to compare the systems and determine which of T_1 or T_2 is greater. **Thermal equilibrium** between systems exists if $P_1 = P_2$ for gaseous systems with the same molar density.

We use the concepts of temperature and thermal equilibrium to characterize the boundary between a system and its surroundings. Consider the two systems with rigid walls shown in Figure 1.2a. Each system has the same molar density and is equipped

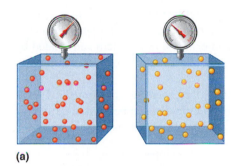

(a)

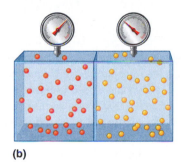

(b)

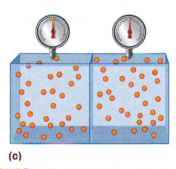

(c)

FIGURE 1.2
(a) Two separated systems with rigid walls and the same molar density have different temperatures. **(b)** The two systems are brought together so that the walls are in intimate contact. Even after a long time has passed, the pressure in each system is unchanged because the walls are adiabatic. **(c)** As in part (b), the two systems are brought together so that the walls are in intimate contact. After a sufficient time has passed, the pressures are equal because the walls are diathermal.

---

[1] For this example, the surface tension of the bubble is assumed to be so small that it can be set equal to zero. This is in keeping with the thermodynamic tradition of weightless pistons and frictionless pulleys.

with a pressure gauge. If we bring the two systems into direct contact, two limiting behaviors can be observed based on the properties of the walls of each system. If neither pressure gauge changes, as in Figure 1.2b, we refer to the walls as being **adiabatic.** Because $P_1 \neq P_2$, the systems are not in thermal equilibrium and, therefore, have different temperatures. An example of a system surrounded by adiabatic walls is coffee in a Styrofoam cup with a Styrofoam lid.[2] Experience shows that it is not possible to bring two systems enclosed by adiabatic walls into thermal equilibrium by bringing them into contact, because adiabatic walls insulate against the transfer of "heat." If you push a Styrofoam cup containing hot coffee against one containing ice water, they will not reach the same temperature. Rely on your experience at this point regarding the meaning of heat; a thermodynamic definition will be given in Chapter 2.

The second limiting case is shown in Figure 1.2c. In bringing the systems into intimate contact, both pressures change and reach the same value after some time. We conclude that the systems have the same temperature, $T_1 = T_2$, and say that they are in thermal equilibrium. We refer to the walls as being **diathermal**. Two systems in contact separated by diathermal walls reach thermal equilibrium because diathermal walls conduct heat. Hot coffee stored in a copper cup is an example of a system surrounded by diathermal walls. Because the walls are diathermal, the coffee will quickly reach room temperature.

The **zeroth law of thermodynamics** generalizes the experiment illustrated in Figure 1.2 and asserts the existence of an objective temperature that can be used to define the condition of thermal equilibrium. The formal statement of this law is as follows:

> Two systems that are separately in thermal equilibrium with a third system are also in thermal equilibrium with one another.

There are four laws of thermodynamics, all of which are generalizations from experience rather than mathematical theorems. They have been rigorously tested in more than a century of experimentation, and no violations of these laws have been found. The unfortunate name assigned to the "zeroth" law is due to the fact that it was formulated after the first law of thermodynamics, but logically precedes it. The zeroth law tells us that we can determine if two systems are in thermal equilibrium without bringing them into contact. Imagine the third system to be a thermometer, which is defined more precisely in the next section. The third system can be used to compare the temperatures of the other two systems; if they have the same temperature, they will be in thermal equilibrium if placed in contact.

1.3 Thermometry

The discussion of thermal equilibrium requires only that a device exist, called a **thermometer,** that can measure relative hotness or coldness. However, scientific work requires a quantitative temperature scale. For any useful thermometer, the empirical temperature, t, must be a single-valued, continuous, and monotonic function of some thermometric system property designated by x. Examples of thermometric properties are the volume of a liquid, the electrical resistance of a metal or semiconductor, and the electromotive force generated at the junction of two dissimilar metals. The simplest case that one can imagine is if the empirical temperature, t, is linearly related to the value of the thermometric property, x:

$$t(x) = a + bx \tag{1.3}$$

----

[2] In this discussion, Styrofoam is assumed to be a perfect insulator.

Equation (1.3) defines a **temperature scale** in terms of a specific thermometric property, once the constants a and b are fixed. Constant a determines the zero of the temperature scale because $t(0) = a$, and constant b determines the size of a unit of temperature, called a degree.

One of the first practical thermometers was the mercury-in-glass thermometer. It utilizes the thermometric property that the volume of mercury increases monotonically over the temperature range in which it is in the liquid state (between $-38.8°$ and $356.7°C$). In 1745, Carolus Linnaeus gave this thermometer a standardized scale by arbitrarily assigning the values 0 and 100 to the freezing and boiling points of water, respectively. This chosen interval was divided into 100 equal degrees, and the same size degree is used outside of the interval. Because there are 100 degrees between the two calibration points, it is called the **centigrade scale.**

The centigrade scale has been superseded by the **Celsius scale,** which is in widespread use today. The Celsius scale (denoted in units of $°C$) is similar to the centigrade scale. However, rather than being determined by two fixed points, the Celsius scale is determined by one fixed reference point at which ice, liquid water, and gaseous water are in equilibrium. This point is called the triple point (see Section 7.2) and is assigned the value $0.01°C$. On the Celsius scale, the boiling point of water at a pressure of 1 atmosphere is $99.975°C$. The size of the degree is chosen to be the same as on the centigrade scale.

Although the Celsius scale is used widely throughout the world today, the numerical values for this temperature scale are completely arbitrary. It would be preferable to have a temperature scale derived directly from physical principles. There is such a scale, called the **thermodynamic temperature scale** or **absolute temperature scale.** For such a scale, the temperature is independent of the substance used in the thermometer, and the constant a in Equation (1.3) is zero. The **gas thermometer** is a practical thermometer with which the absolute temperature can be measured. The thermometric property is the temperature dependence of P for a dilute gas at constant V. The gas thermometer provides the international standard for thermometry at very low temperatures. At intermediate temperatures, the electrical resistance of platinum wire is the standard, and at higher temperatures, the radiated energy emitted from glowing silver is the standard.

How is the gas thermometer used to measure the thermodynamic temperature? Measurements carried out by Jacques Charles in the 19th century demonstrated that the pressure exerted by a fixed amount of gas at constant V varies linearly with temperature on the Celsius scale as shown in Figure 1.3. At the time of Boyle's experiments, temperatures below $-30°C$ were not attainable in the laboratory. However, the P versus T data can be extrapolated to the limiting T value at which $P \rightarrow 0$. It is found that these straight lines obtained for different values of V intersect at a common point on the T axis that lies near $-273°C$.

The data show that at constant V, the thermometric property P varies with temperature as

$$P = c + dt \tag{1.4}$$

where t is the temperature on the Celsius scale, and c and d are experimentally obtained proportionality constants.

Figure 1.3 shows that all lines intersect at a single point, even for different gases. This suggests a unique reference point for temperature, rather than the two reference points used in constructing the centigrade scale. The value zero is given to the temperature at which $P \rightarrow 0$. However, this choice is not sufficient to define the temperature scale, because the size of the degree is undefined. By convention, the size of the degree on the absolute temperature scale is set equal to the size of the degree on the Celsius scale, because the Celsius scale was in widespread use at the time the absolute temperature scale was formulated. With these two choices, the absolute and Celsius temperature scales are related by Equation (1.5). The scale measured by the ideal gas

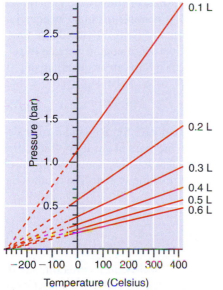

FIGURE 1.3
The pressure exerted by 5.00×10^{-3} mol of a dilute gas is shown as a function of the temperature measured on the Celsius scale for different fixed volumes. The dashed portion indicates that the data are extrapolated to lower temperatures than could be achieved experimentally by Charles.

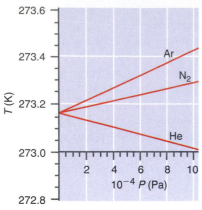

FIGURE 1.4

The temperature measured in a gas thermometer defined by Equation (1.6) is independent of the gas used only in the limit that $P \rightarrow 0$.

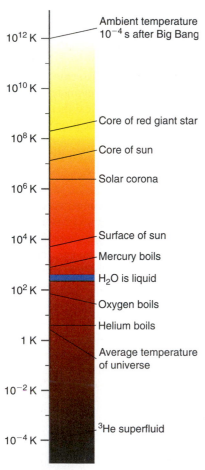

FIGURE 1.5

The absolute temperature is shown on a logarithmic scale together with the temperature of a number of physical phenomena.

thermometer is the absolute temperature scale used in thermodynamics. The unit of temperature on this scale is called the **Kelvin,** abbreviated K (without a degree symbol):

$$T(\text{K}) = T(^{\circ}\text{C}) + 273.15 \tag{1.5}$$

Using the triple point of water as a reference, the absolute temperature, $T(\text{K})$, measured by an ideal gas thermometer is given by

$$T(\text{K}) = 273.16 \frac{P}{P_{tp}} \tag{1.6}$$

where P_{tp} is the pressure corresponding to the triple point of water. On this scale, the volume of an ideal gas is directly proportional to its temperature; if the temperature is reduced to half of its initial value, V is also reduced to half of its initial value. In practice, deviations from the ideal gas law that occur for real gases must be taken into account when using a gas thermometer. If data were obtained from a gas thermometer using He, Ar, and N_2 for a temperature very near T_{tp}, they would exhibit the behavior shown in Figure 1.4. We see that the temperature only becomes independent of P and of the gas used in the thermometer if the data are extrapolated to zero pressure. It is in this limit that the gas thermometer provides a measure of the thermodynamic temperature. Because 1 bar = 10^5 Pa, gas-independent T values are only obtained below $P \sim 0.01$ bar. The absolute temperature scale is shown in Figure 1.5 on a logarithmic scale together with associated physical phenomena.

1.4 Equations of State and the Ideal Gas Law

Macroscopic models in which the system is described by a set of variables are based on experience. It is particularly useful to formulate an **equation of state,** which relates the state variables. Using the absolute temperature scale, it is possible to obtain an equation of state for an ideal gas from experiments. If the pressure of He is measured as a function of the volume for different values of temperature, the set of nonintersecting hyperbolas shown in Figure 1.6 is obtained. The curves in this figure can be quantitatively fit by the functional form

$$PV = \alpha T \tag{1.7}$$

where T is the absolute temperature as defined by Equation (1.6), allowing α to be determined, which is found to be directly proportional to the mass of gas used. It is useful to separate out this dependence by writing $\alpha = nR$, where n is the number of moles of the gas, and R is a constant that is independent of the size of the system. The result is the ideal gas equation of state:

$$PV = NkT = nRT \tag{1.8}$$

where the proportionality constants k and R are called the **Boltzmann constant** and the **ideal gas constant,** respectively; N is the number of molecules; and n is the number of moles of the gas. The equation of state given in Equation (1.8) is known as the **ideal gas law.** Because the four variables are related through the equation of state, knowledge of any three of these variables is sufficient to completely describe the ideal gas. Note that the total number of moles—not the number of moles of the individual gas components of a gas mixture—appears in the ideal gas law.

Of these four variables, P and T are independent of the amount of gas, whereas V and n are proportional to the amount of gas. A variable that is independent of the size of the system (for example, P and T) is referred to as an **intensive variable,** and one that is proportional to the size of the system (for example, V) is referred to as an **extensive variable.** Equation (1.8) can be written in terms of intensive variables exclusively:

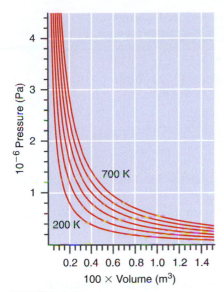

FIGURE 1.6
Illustration of the relationship between pressure and volume of 0.010 mol of He for fixed values of temperature, which differ by 100 K.

$$P = \rho RT \qquad (1.8a)$$

For a fixed number of moles, the ideal gas equation of state has only two independent intensive variables: any two of P, T, and ρ.

EXAMPLE PROBLEM 1.1

An average-sized person has about 5.0 liters of blood, which circulates through the system in about 1 minute. Each liter of blood can carry 0.20 L of oxygen, measured at $T = 273$ K and $P = 1.0$ atm. Calculate the number of moles of oxygen that the blood can transport from the lungs each minute if its full oxygen-carrying potential is used. Calculate also the number of oxygen molecules transported from the lungs per minute in the blood.

Solution

Use Equation (1.8) to calculate the number of moles of oxygen:

$$n = \frac{PV}{RT} = \frac{1.0 \text{ atm} \times 0.20 \text{ L}}{0.08206 \text{ L atm mol}^{-1}\text{K}^{-1} \times 273\text{K}} = 0.0089 \text{ mol}$$

The number of moles of oxygen circulated per minute is given by

$$= 5.0 \text{ L blood} \times 0.0089 \text{ mol O}_2 \text{ per L blood per minute}$$

$$= 0.045 \text{ mol O}_2 \text{ per minute}$$

The number of oxygen molecules circulated per minute is given by

$$= 0.045 \text{ mol min}^{-1} \times 6.022 \times 10^{23} \text{ molecules mol}^{-1}$$

$$= 2.7 \times 10^{22} \text{ molecules min}^{-1}$$

For an ideal gas mixture

$$PV = \sum_i n_i RT \qquad (1.9)$$

because the gas molecules do not interact with one another. Equation (1.9) can be rewritten in the form

$$P = \sum_i \frac{n_i RT}{V} = \sum_i P_i = P_1 + P_2 + P_3 + \dots \qquad (1.10)$$

In Equation (1.10), P_i is the **partial pressure** of each gas. This equation states that each ideal gas exerts a pressure that is independent of the other gases in the mixture. We also have

$$\frac{P_i}{P} = \frac{\dfrac{n_i RT}{V}}{\displaystyle\sum_i \dfrac{n_i RT}{V}} = \frac{\dfrac{n_i RT}{V}}{\dfrac{nRT}{V}} = \frac{n_i}{n} = x_i \qquad (1.11)$$

which relates the partial pressure of a component in the mixture, P_i, with its **mole fraction,** $x_i = n_i/n$, and the total pressure, P.

In the SI system of units, pressure is measured in Pascal (Pa) units, where 1 Pa = 1 N/m². The volume is measured in cubic meters, and the temperature is measured in kelvin. However, other units of pressure are frequently used, and these units are related to the Pascal as indicated in Table 1.1. In this table, numbers that are not exact have been given to five significant figures. The other commonly used unit of volume is the liter (L), where $1 \text{ m}^3 = 10^3 \text{ L}$ and $1 \text{ L} = 1 \text{ dm}^3 = 10^{-3} \text{ m}^3$.

TABLE 1.1   Units of Pressure and Conversion Factors

Unit of Pressure	Symbol	Numerical Value
Pascal	Pa	$1\ N\ m^{-2} = 1\ kg\ m^{-1}\ s^{-2}$
Atmosphere	atm	$1\ atm = 101{,}325\ Pa$ (exactly)
Bar	bar	$1\ bar = 10^5\ Pa$
Torr or millimeters of Hg	Torr	$1\ Torr = 101{,}325/760 = 133.32\ Pa$
Pounds per square inch	psi	$1\ psi = 6{,}894.8\ Pa$

TABLE 1.2   The Ideal Gas Constant, R, in Various Units

$R = 8.314\ J\ K^{-1}\ mol^{-1}$

$R = 8.314\ Pa\ m^3\ K^{-1}\ mol^{-1}$

$R = 8.314 \times 10^{-2}\ L\ bar\ K^{-1}\ mol^{-1}$

$R = 8.206 \times 10^{-2}\ L\ atm\ K^{-1}\ mol^{-1}$

$R = 62.36\ L\ Torr\ K^{-1}\ mol^{-1}$

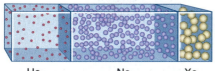

He	Ne	Xe
2.00 L	3.00 L	1.00 L
1.50 bar	2.50 bar	1.00 bar

In the SI system, the constant R that appears in the ideal gas law has the value $8.314\ J\ K^{-1}\ mol^{-1}$, where the Joule (J) is the unit of energy in the SI system. To simplify calculations for other units of pressure and volume, values of the constant R with different combinations of units are given in Table 1.2.

EXAMPLE PROBLEM 1.2

Consider the composite system, which is held at 298 K, shown in the following figure to the left. Assuming ideal gas behavior, calculate the total pressure, and the partial pressure of each component if the barriers separating the compartments are removed. Assume that the volume of the barriers is negligible.

Solution

The number of moles of He, Ne, and Xe is given by

$$n_{He} = \frac{PV}{RT} = \frac{1.50\ bar \times 2.00\ L}{8.314 \times 10^{-2}\ L\ bar\ K^{-1}\ mol^{-1} \times 298\ K} = 0.121\ mol$$

$$n_{Ne} = \frac{PV}{RT} = \frac{2.50\ bar \times 3.00\ L}{8.314 \times 10^{-2}\ L\ bar\ K^{-1}\ mol^{-1} \times 298\ K} = 0.303\ mol$$

$$n_{Xe} = \frac{PV}{RT} = \frac{1.00\ bar \times 1.00\ L}{8.314 \times 10^{-2}\ L\ bar\ K^{-1}\ mol^{-1} \times 298\ K} = 0.0403\ mol$$

$$n = n_{He} + n_{Ne} + n_{Xe} = 0.464$$

The mole fractions are

$$x_{He} = \frac{n_{He}}{n} = \frac{0.121}{0.464} = 0.261$$

$$x_{Ne} = \frac{n_{Ne}}{n} = \frac{0.303}{0.464} = 0.653$$

$$x_{Xe} = \frac{n_{Xe}}{n} = \frac{0.0403}{0.464} = 0.0860$$

The total pressure is given by

$$P = \frac{(n_{He} + n_{Ne} + n_{Xe})RT}{V} = \frac{0.464\ mol \times 8.3145 \times 10^{-2}\ L\ bar\ K^{-1}\ mol^{-1} \times 298\ K}{6.00\ L}$$

$$= 1.92\ bar$$

The partial pressures are given by

$$P_{He} = x_{He}P = 0.261 \times 1.92\ bar = 0.501\ bar$$

$$P_{Ne} = x_{Ne}P = 0.653 \times 1.92\ bar = 1.25\ bar$$

$$P_{Xe} = x_{Xe}P = 0.0860 \times 1.92\ bar = 0.165\ bar$$

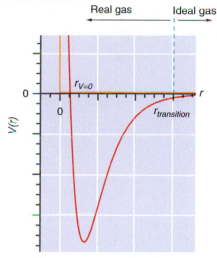

FIGURE 1.7
The potential energy of interaction of two molecules or atoms is shown as a function of their separation, r. The orange curve shows the potential energy function for an ideal gas. The dashed blue line indicates an approximate r value below which a more nearly exact equation of state than the ideal gas law should be used. $V(r) = 0$ at $r = r_{V=0}$ and as $r \to \infty$.

1.5 A Brief Introduction to Real Gases

The ideal gas law provides a first look at the usefulness of describing a system in terms of macroscopic parameters. However, we should also emphasize the downside of not taking the microscopic nature of the system into account. For example, the ideal gas law only holds for gases at low densities. Experiments show that Equation (1.8) is accurate to higher values of pressure and lower values of temperature for He than for NH_3. Why is this the case? Because we need to take *non*ideal gas behavior into account in the following chapters, we introduce in this section an equation of state that is valid for higher densities.

An ideal gas is described by two assumptions: the atoms or molecules of an ideal gas do not interact with one another, and the atoms or molecules can be treated as point masses. These assumptions have a limited range of validity, which can be discussed using the potential energy function typical for a real gas, as shown in Figure 1.7. This figure shows the potential energy of interaction of two gas molecules as a function of the distance between them. The intermolecular potential can be divided into regions in which the potential energy is essentially zero ($r > r_{transition}$), negative (attractive interaction) ($r_{transition} > r > r_{V=0}$), and positive (repulsive interaction) ($r < r_{V=0}$). The distance $r_{transition}$ is not uniquely defined and depends on the energy of the molecule. It can be estimated from the relation $|V(r_{transition})| \approx kT$

As the density is increased from very low values, molecules approach one another to within a few molecular diameters and experience a long-range attractive van der Waals force due to time-fluctuating dipole moments in each molecule. This strength of the attractive interaction is proportional to the polarizability of the electron charge in a molecule and is, therefore, substance dependent. In the attractive region, P is lower than that calculated using the ideal gas law. This is the case because the attractive interaction brings the atoms or molecules closer than they would be if they did not interact. At sufficiently high densities, the atoms or molecules experience a short-range repulsive interaction due to the overlap of the electron charge distributions. Because of this interaction, P is higher than that calculated using the ideal gas law. We see that for a real gas, P can be either greater or less than the ideal gas value. Note that the potential becomes repulsive for a value of r greater than zero. As a consequence, the volume of a gas well above its boiling temperature approaches a finite limiting value as P increases. By contrast, the ideal gas law predicts that $V \to 0$ as $P \to \infty$.

Given the potential energy function depicted in Figure 1.7, under what conditions is the ideal gas equation of state valid? A real gas behaves ideally only at low densities for which $r > r_{transition}$, and the value of $r_{transition}$ is substance dependent. At this point we introduce a real gas equation of state, because it will be used in the next few chapters. The **van der Waals equation of state** takes both the finite size of molecules and the attractive potential into account. It has the form

$$P = \frac{nRT}{V - nb} - \frac{n^2 a}{V^2} \qquad (1.12)$$

This equation of state has two parameters that are substance dependent and must be experimentally determined. Parameters b and a take the finite size of the molecules and the strength of the attractive interaction into account, respectively. (Values of a and b for selected gases are listed in Table 1.3, see Appendix B, Data Tables.) The van der Waals equation of state is more accurate in calculating the relationship between P, V, and T for gases than the ideal gas law because a and b have been optimized using experimental results. However, there are other more accurate equations of state that are valid over a wider range than the van der Waals equation. Such equations of state include up to 16 adjustable substance-dependent parameters.

EXAMPLE PROBLEM 1.3

Van der Waals parameters are generally tabulated with either of two sets of units:

1. $Pa\ m^6\ mol^{-2}$ or $bar\ dm^6\ mol^{-2}$

2. $m^3\ mol^{-1}$ or $dm^3\ mol^{-1}$

Determine the conversion factor to convert one system of units to the other. Note that $1\ dm^3 = 10^{-3}\ m^3 = 1\ L$.

Solution

$$Pa\ m^6\ mol^{-2} \times \frac{bar}{10^5\ Pa} \times \frac{10^6\ dm^6}{m^6} = 10\ bar\ dm^6\ mol^{-2}$$

$$m^3\ mol^{-1} \times \frac{10^3\ dm^3}{m^3} = 10^3\ dm^3\ mol^{-1}$$

In Example Problem 1.4, a comparison is made of the molar volume for N_2 calculated at low and high pressures, using the ideal gas and van der Waals equations of state.

EXAMPLE PROBLEM 1.4

a. Calculate the pressure exerted by N_2 at 300. K for molar volumes of 250. and 0.100 L using the ideal gas and the van der Waals equations of state. The values of parameters a and b for N_2 are 1.370 bar dm^6 mol^{-2} and 0.0387 dm^3 mol^{-1}, respectively.

b. Compare the results of your calculations at the two pressures. If P calculated using the van der Waals equation of state is greater than those calculated with the ideal gas law, we can conclude that the repulsive interaction of the N_2 molecules outweighs the attractive interaction for the calculated value of the density. A similar statement can be made regarding the attractive interaction. Is the attractive or repulsive interaction greater for N_2 at 300. K and $V_m = 0.100$ L?

Solution

a. The pressures calculated from the ideal gas equation of state are

$$P = \frac{nRT}{V} = \frac{1\ mol \times 8.314 \times 10^{-2}\ L\ bar\ mol^{-1}K^{-1} \times 300.\ K}{250.\ L} = 9.98 \times 10^{-2}\ bar$$

$$P = \frac{nRT}{V} = \frac{1\ mol \times 8.314 \times 10^{-2}\ L\ bar\ mol^{-1}K^{-1} \times 300.\ K}{0.100\ L} = 249\ bar$$

The pressures calculated from the van der Waals equation of state are

$$P = \frac{nRT}{V - nb} - \frac{n^2 a}{V^2}$$

$$= \frac{1\ mol \times 8.314 \times 10^{-2}\ L\ bar\ mol^{-1}K^{-1} \times 300.\ K}{250.\ L - 1\ mol \times 0.0387\ dm^3\ mol^{-1}} - \frac{(1\ mol)^2 \times 1.370\ bar\ dm^6\ mol^{-2}}{(250.\ L)^2}$$

$$= 9.98 \times 10^{-2}\ bar$$

$$P = \frac{1\ mol \times 8.314 \times 10^{-2}\ L\ bar\ mol^{-1}K^{-1} \times 300K}{0.100\ L - 1\ mol \times 0.0387\ dm^3\ mol^{-1}} - \frac{(1\ mol)^2 \times 1.370\ bar\ dm^6\ mol^{-2}}{(0.100\ L)^2}$$

$$= 270.\ bar$$

b. Note that the result using the van der Waals equation of state is identical with that for the ideal gas law for $V_m = 250.$ L, but that the van der Waals result calculated for $V_m = 0.100$ L deviates from the ideal gas law result. Because $P_{real} > P_{ideal}$, we conclude that the repulsive interaction is more important than the attractive interaction for this specific value of molar volume and temperature.

Vocabulary

absolute temperature scale	extensive variable	mole fraction	thermodynamic equilibrium
adiabatic	gas thermometer	open system	thermodynamic temperature
Boltzmann constant	ideal gas constant	partial pressure	scale
boundary	ideal gas law	statistical thermodynamics	thermometer
Celsius scale	intensive variable	surroundings	van der Waals equation of
centigrade scale	isolated system	system	state
closed system	Kelvin scale	system variables	wall
diathermal	macroscopic scale	temperature	zeroth law of
equation of state	macroscopic variables	temperature scale	thermodynamics
equilibrium	microscopic scale	thermal equilibrium	

Questions on Concepts

Q1.1 The location of the boundary between the system and the surroundings is a choice that must be made by the thermodynamicist. Consider a beaker of boiling water in an airtight room. Is the system open or closed if you place the boundary just outside the liquid water? Is the system open or closed if you place the boundary just inside the walls of the room?

Q1.2 Real walls are never totally adiabatic. Order the following walls in increasing order with respect to their being diathermal: 1-cm-thick concrete, 1-cm-thick vacuum, 1-cm-thick copper, 1-cm-thick cork.

Q1.3 Why is the possibility of exchange of matter or energy appropriate to the variable of interest a necessary condition for equilibrium between two systems?

Q1.4 At sufficiently high temperatures, the van der Waals equation has the form $P \approx RT/(V_m - b)$. Note that the attractive part of the potential has no influence in this expression. Justify this behavior using the potential energy diagram of Figure 1.7.

Q1.5 Parameter a in the van der Waals equation is greater for H_2O than for He. What does this say about the form of the potential function in Figure 1.7 for the two gases?

Problems

Problem numbers in **RED** indicate that the solution to the problem is given in the *Student Solutions Manual*.

P1.1 A sealed flask with a capacity of 1.00 dm^3 contains 5.00 g of ethane. The flask is so weak that it will burst if the pressure exceeds 1.00×10^6 Pa. At what temperature will the pressure of the gas exceed the bursting pressure?

P1.2 Consider a gas mixture in a 2.00-dm^3 flask at 27.0°C. For each of the following mixtures, calculate the partial pressure of each gas, the total pressure, and the composition of the mixture in mole percent:

    a. 1.00 g H_2 and 1.00 g O_2

    b. 1.00 g N_2 and 1.00 g O_2

    c. 1.00 g CH_4 and 1.00 g NH_3

P1.3 Approximately how many oxygen molecules arrive each second at the mitochondrion of an active person? The following data are available: oxygen consumption is about 40. mL of O_2 per minute per kilogram of body weight, measured at $T = 300.$ K and $P = 1.0$ atm. An adult with a body weight of 64 kg has about 1×10^{12} cells. Each cell contains about 800. mitochondria.

P1.4 In a normal breath, about 0.5 L of air at 1.0 atm and 293 K is inhaled. About 25.0% of the oxygen in air is absorbed by the lungs and passes into the bloodstream. For a respiration rate of 18 breaths per minute, how many moles of oxygen per minute are absorbed by the body? Assume the mole fraction of oxygen in air is 0.21. Compare this result with Example Problem 1.1.

P1.5 Suppose that you measured the product PV of 1 mol of a dilute gas and found that $PV = 22.98$ L atm at 0.00°C and 31.18 L atm at 100°C. Assume that the ideal gas law is valid, with $T = t(°C) + a$, and that the value of R is not known. Determine R and a from the measurements provided.

P1.6 Devise a temperature scale, abbreviated G, for which the magnitude of the ideal gas constant is $1.00 \, J \, G^{-1} \, mol^{-1}$.

P1.7 A rigid vessel of volume $0.500 \, m^3$ containing H_2 at $20.5°C$ and a pressure of $611 \times 10^3 \, Pa$ is connected to a second rigid vessel of volume $0.750 \, m^3$ containing Ar at $31.2°C$ at a pressure of $433 \times 10^3 \, Pa$. A valve separating the two vessels is opened and both are cooled to a temperature of $14.5°C$. What is the final pressure in the vessels?

P1.8 In normal respiration, an adult exhales about 500. L of air per hour. The exhaled air is saturated with water vapor at body temperature $T = 310. \, K$. At this temperature water vapor in equilibrium with liquid water has a pressure of $P = 0.062 \, atm$. Assume water vapor behaves ideally under these conditions. What mass of water vapor is exhaled in an hour?

P1.9 At $T = 293 \, K$ and at 50.% relative humidity, the pressure of water vapor in equilibrium with liquid water is $0.0115 \, atm$. Using the information in Problem P1.8, determine what mass of water is inhaled per hour and the net loss of water through respiration per hour.

P1.10 A compressed cylinder of gas contains $1.50 \times 10^3 \, g$ of N_2 gas at a pressure of $2.00 \times 10^7 \, Pa$ and a temperature of $17.1°C$. What volume of gas has been released into the atmosphere if the final pressure in the cylinder is $1.80 \times 10^5 \, Pa$? Assume ideal behavior and that the gas temperature is unchanged.

P1.11 As a result of photosynthesis, 1.0 kg of carbon is fixed per square meter of forest. Assuming air is 0.046% CO_2 by weight, what volume of air is required to provide 1.0 kg of fixed carbon? Assume $T = 298 \, K$ and $P = 1.00 \, atm$. Also assume that air is approximately 20.% oxygen and 80.% nitrogen by weight.

P1.12 A balloon filled with 10.50 L of Ar at $18.0°C$ and 1 atm rises to a height in the atmosphere where the pressure is 248 Torr and the temperature is $-30.5°C$. What is the final volume of the balloon?

P1.13 One liter of fully oxygenated blood can carry 0.20 L of O_2 measured at $T = 273 \, K$ and $P = 1.00 \, atm$. Calculate the number of moles of O_2 carried per liter of blood. Hemoglobin, the oxygen transport protein in blood, has four oxygen-binding sites. How many hemoglobin molecules are required to transport the O_2 in 1.0 L of fully oxygenated blood?

P1.14 Myoglobin is a protein that stores oxygen in the tissues. Unlike hemoglobin, which has four oxygen-binding sites, myoglobin has only a single oxygen-binding site. How many myoglobin molecules are required to transport the oxygen absorbed by the blood in Problem 1.13?

P1.15 Consider a 20.0-L sample of moist air at $60.°C$ and 1 atm in which the partial pressure of water vapor is 0.120 atm. Assume that dry air has the composition 78.0 mol % N_2, 21.0 mol % O_2, and 1.00 mol % Ar.

  a. What are the mole percentages of each of the gases in the sample?

  b. The percent relative humidity is defined as $\%RH = P_{H_2O}/P^*_{H_2O}$ where P_{H_2O} is the partial pressure

of water in the sample and $P^*_{H_2O} = 0.197 \, atm$ is the equilibrium vapor pressure of water at $60.°C$. The gas is compressed at $60.°C$ until the relative humidity is 100%. What volume does the mixture contain now?

  c. What fraction of the water will be condensed if the total pressure of the mixture is isothermally increased to 200. atm?

P1.16 A mixture of $2.50 \times 10^{-3} \, g$ of O_2, $3.51 \times 10^{-3} \, mol$ of N_2, and 4.67×10^{20} molecules of CO is placed into a vessel of volume 3.50 L at $5.20°C$.

  a. Calculate the total pressure in the vessel.

  b. Calculate the mole fractions and partial pressures of each gas.

P1.17 Carbon monoxide (CO) competes with oxygen for binding sites on the transport protein hemoglobin. CO can be poisonous if inhaled in large quantities. A safe level of CO in air is 50. parts per million (ppm). When the CO level increases to 800. ppm, dizziness, nausea, and unconsciousness occur, followed by death. Assuming the partial pressure of oxygen in air at sea level is 0.20 atm, what ratio of O_2 to CO is fatal?

P1.18 A normal adult inhales 0.500 L of air at $T = 293 \, K$ and 1.00 atm. To explore the surface of the moon, an astronaut requires a 25.0-L breathing tank containing air at a pressure of 200. atm. How many breaths can the astronaut take from this tank?

P1.19 Liquid N_2 has a density of $875.4 \, kg \, m^{-3}$ at its normal boiling point. What volume does a balloon occupy at $18.5°C$ and a pressure of 1.00 atm if $2.00 \times 10^{-3} \, L$ of liquid N_2 is injected into it?

P1.20 Yeast and other organisms can convert glucose ($C_6H_{12}O_6$) to ethanol (CH_3CH_2OH) by a process called alcoholic fermentation. The net reaction is

$$C_6H_{12}O_6(s) \rightarrow 2C_2H_5OH(l) + 2CO_2(g)$$

Calculate the mass of glucose required to produce 1.0 L of CO_2 measured at $P = 1.00 \, atm$ and $T = 300. \, K$.

P1.21 A sample of propane (C_3H_8) is placed in a closed vessel together with an amount of O_2 that is 3.00 times the amount needed to completely oxidize the propane to CO_2 and H_2O at constant temperature. Calculate the mole fraction of each component in the resulting mixture after oxidation assuming that the H_2O is present as a gas.

P1.22 Calculate the volume of all gases evolved by the complete oxidation of 0.25 g of the amino acid alanine (NH_2CHCH_3COOH) if the products are liquid water, nitrogen gas, and carbon dioxide gas and the total pressure is 1.00 atm and $T = 310. \, K$.

P1.23 A gas sample is known to be a mixture of ethane and butane. A bulb having a 200.0-cm^3 capacity is filled with the gas to a pressure of $100.0 \times 10^3 \, Pa$ at $20.0°C$. If the weight of the gas in the bulb is 0.3846 g, what is the mole percent of butane in the mixture?

P1.24 A glass bulb of volume 0.136 L contains 0.7031 g of gas at 759.0 Torr and $99.5°C$. What is the molar mass of the gas?

P1.25 The total pressure of a mixture of oxygen and hydrogen is 1.00 atm. The mixture is ignited and the water is removed. The remaining gas is pure hydrogen and exerts a pressure of 0.400 atm when measured at the same values of T and V as the original mixture. What was the composition of the original mixture in mole percent?

P1.26 The photosynthetic formation of glucose in spinach leaves via the Calvin cycle involves the fixation of carbon dioxide with ribulose 1-5 diphosphate $C_5H_8P_2O_{11}^{4-}(aq)$ to form 3-phosphoglycerate $C_3H_4PO_7^{3-}(aq)$:

$$C_5H_8P_2O_{11}^{4-}(aq) + H_2O(l) + CO_2(g)$$
$$\rightarrow 2C_3H_4PO_7^{3-}(aq) + 2H^+(aq)$$

If 1.00 L of carbon dioxide at $T = 273$ K and $P = 1.00$ atm is fixed by this reaction, what mass of 3-phosphoglycerate is formed?

P1.27 Calculate the pressure exerted by Ar for a molar volume of 1.42 L mol^{-1} at 300. K using the van der Waals equation of state. The van der Waals parameters a and b for Ar are 1.355 bar dm^6 mol^{-2} and 0.0320 dm^3 mol^{-1}, respectively. Is the attractive or repulsive portion of the potential dominant under these conditions?

P1.28 Calculate the pressure exerted by benzene for a molar volume of 1.42 L at 790. K using the Redlich–Kwong equation of state:

$$P = \frac{RT}{V_m - b} - \frac{a}{\sqrt{T}}\frac{1}{V_m(V_m + b)} = \frac{nRT}{V - nb} - \frac{n^2 a}{\sqrt{T}}\frac{1}{V(V + nb)}$$

The Redlich–Kwong parameters a and b for benzene are 452.0 bar dm^6 mol^{-2} K$^{1/2}$ and 0.08271 dm^3 mol^{-1}, respectively. Is the attractive or repulsive portion of the potential dominant under these conditions?

P1.29 When Julius Caesar expired, his last exhalation had a volume of 500. cm^3 and contained 1.00 mol % argon. Assume that $T = 300$. K and $P = 1.00$ atm at the location of his demise. Assume further that T and P currently have the same values throughout the Earth's atmosphere. If all of his exhaled Ar atoms are now uniformly distributed throughout the atmosphere (which for our calculation is taken to have a thickness of 1.00 km), how many inhalations of 500. cm^3 must we make to inhale one of the Ar atoms exhaled in Caesar's last breath? Assume the radius of the Earth to be 6.37×10^6 m. (*Hint:* Calculate the number of Ar atoms in the atmosphere in the simplified geometry of a plane of area equal to that of the Earth's surface and a height equal to the thickness of the atmosphere. See Problem P1.30 for the dependence of the barometric pressure on the height above the Earth's surface.)

P1.30 The barometric pressure falls off with height above sea level in the Earth's atmosphere as $P_i = P_i^0 e^{-M_i g/RT}$ where

P_i is the partial pressure at the height z, P_i^0 is the partial pressure of component i at sea level, g is the acceleration of gravity, R is the gas constant, T is the absolute temperature, and M_i is the molecular mass of the gas. Consider an atmosphere that has the composition $x_{N_2} = 0.600$ and $x_{CO_2} = 0.400$ and that $T = 300$. K. Near sea level, the total pressure is 1.00 bar. Calculate the mole fractions of the two components at a height of 50.0 km. Why is the composition different from its value at sea level?

P1.31 Assume that air has a mean molar mass of 28.9 g mol^{-1} and that the atmosphere has a uniform temperature of 25.0°C. Calculate the barometric pressure at Denver, for which $z = 1600$. m. Use the information contained in Problem P1.30.

P1.32 A mixture of oxygen and hydrogen is analyzed by passing it over hot copper oxide and through a drying tube. Hydrogen reduces the CuO according to the reaction CuO + $H_2 \rightarrow$ Cu + H_2O, and oxygen reoxidizes the copper formed according to Cu + 1/2 $O_2 \rightarrow$ CuO. At 25°C and 750. Torr, 100.0 cm^3 of the mixture yields 84.5 cm^3 of dry oxygen measured at 25°C and 750. Torr after passage over CuO and the drying agent. What was the original composition of the mixture?

P1.33 Aerobic cells metabolize glucose in the respiratory system. This reaction proceeds according to the overall reaction

$$6O_2(g) + C_6H_{12}O_6(s) \rightarrow 6CO_2(g) + 6H_2O(l)$$

Calculate the volume of oxygen required at STP to metabolize 0.010 kg of glucose ($C_6H_{12}O_6$). STP refers to standard temperature and pressure, that is, $T = 273$ K and $P = 1.00$ atm. Assume oxygen behaves ideally at STP.

P1.34 Consider the oxidation of the amino acid glycine (NH_2CH_2COOH) to produce water, carbon dioxide, and urea (NH_2CONH_2):

$$NH_2CH_2COOH(s) + 3O_2(g)$$
$$\rightarrow NH_2CONH_2(s) + 3CO_2(g) + 3H_2O(l)$$

Calculate the volume of carbon dioxide evolved at $P = 1.00$ atm and $T = 310$. K from the oxidation of 0.0100 g of glycine.

P1.35 An initial step in the biosynthesis of glucose ($C_6H_{12}O_6$) is the carboxylation of pyruvic acid ($CH_3COCOOH$) to form oxaloacetic acid ($HOOCCOCH_2COOH$):

$$CH_3COCOOH(s) + CO_2(g) \rightarrow HOOCCOCH_2COOH(s)$$

If you knew nothing else about the intervening reactions involved in glucose biosynthesis other than that no further carboxylations occur, what volume of CO_2 is required to produce 0.50 g of glucose? Assume $P = 1$ atm and $T = 310$. K.

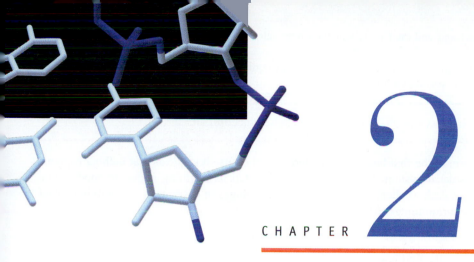

CHAPTER

2

Heat, Work, Internal Energy, Enthalpy, and the First Law of Thermodynamics

In this chapter, the internal energy, U, is introduced. The first law of thermodynamics relates ΔU to the heat (q) and work (w) that flows across the boundary between the system and the surroundings. Other important concepts introduced include the heat capacity, the difference between state and path functions, and reversible versus irreversible processes. The enthalpy, H, is introduced as a form of energy that can be directly measured by the heat flow in a constant pressure process. We also show how q, w, ΔU, and ΔH can be calculated for processes involving ideal gases and biologically relevant compounds.

2.1 The Internal Energy and the First Law of Thermodynamics

In this section, we focus on the change in energy of the system and surroundings during a thermodynamic process such as an expansion or compression of a gas. In thermodynamics, we are interested in the internal energy of the system, as opposed to the energy associated with the system relative to a particular frame of reference. For example, a spinning container of gas has a kinetic energy relative to a stationary observer. However, the internal energy of the gas is defined relative to a coordinate system fixed on the container. Viewed at a microscopic level, the internal energy can take on a number of forms such as

- the kinetic energy of the molecules;
- the potential energy of the constituents of the system; for example, a crystal consisting of dipolar molecules will experience a change in its potential energy as an electric field is applied to the system;
- the internal energy stored in the form of molecular vibrations and rotations; and
- the internal energy stored in the form of chemical bonds that can be released through a chemical reaction.

The total of all of these forms of energy for the system of interest is given the symbol U and is called the **internal energy.**

The **first law of thermodynamics** is based on our experience that energy can be neither created nor destroyed, if both the system and the surroundings are taken into account.

This law can be formulated in a number of equivalent forms. Our initial formulation of this law is stated as follows:

> The internal energy, U, of an isolated system is constant.

This form of the first law looks uninteresting, because it suggests that nothing happens in an isolated system. How can the first law tell us anything about thermodynamic processes such as chemical reactions? When changes in U occur in a system in contact with its surroundings, ΔU_{total} is given by

$$\Delta U_{total} = \Delta U_{system} + \Delta U_{surroundings} = 0 \tag{2.1}$$

Therefore, the first law becomes

$$\Delta U_{system} = -\Delta U_{surroundings} \tag{2.2}$$

That is, for any decrease of U_{system}, $U_{surroundings}$ must increase by exactly the same amount. For example, if a gas (the system) is cooled, and the surroundings is also a gas, the temperature of the surroundings must increase.

How can the energy of a system be changed? There are many ways to alter U, several of which are discussed in this chapter. Experience has shown that all changes in a closed system in which no chemical reactions or phase changes occur can be classified as heat, work, or a combination of both. Therefore, the internal energy of such a system can only be changed by the flow of heat or work across the boundary between the system and surroundings. For example, U for a gas can be increased by heating it in a flame or by doing compression work on it. This important recognition leads to a second and more useful formulation of the first law:

$$\Delta U = q + w \tag{2.3}$$

where q and w designate heat and work, respectively. We use ΔU without a subscript to indicate the change in internal energy of the system. What do we mean by heat and work? In the following two sections, we define these important concepts and distinguish between them.

The symbol Δ is used to indicate a change that occurs as a result of an arbitrary process. The simplest processes are those in which only one of P, V, or T changes. A constant temperature process is referred to as **isothermal**, and the corresponding terms for constant P and V are **isobaric** and **isochoric**, respectively.

2.2 Work

Work in thermodynamics is defined as any quantity of energy that "flows" across the boundary between the system and surroundings that can be used to change the height of a mass in the surroundings. An example is shown in Figure 2.1. We define the system as the gas inside the adiabatic cylinder and piston. Everything else shown in the figure is in the surroundings. As the gas is compressed, the height of the mass in the surroundings is lowered and the initial and final volumes are defined by the mechanical stops indicated in the figure.

Consider the system and surroundings before and after the process shown in Figure 2.1, and note that the height of the mass in the surroundings has changed. It is this change that distinguishes work from heat. Work has several important characteristics:

- Work is transitory in that it only appears during a change in state of the system and surroundings. Only energy, and not work, is associated with the initial and final states of the systems.

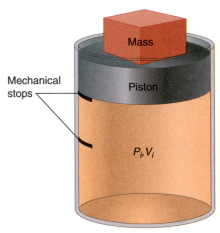

Initial state

Final state

FIGURE 2.1
A system is shown in which compression work is being done on a gas. The walls are adiabatic.

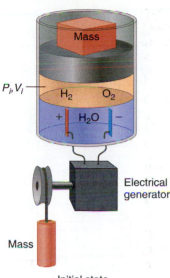

P_i, V_i — H$_2$ O$_2$
+ H$_2$O −

Electrical generator

Mass

Initial state

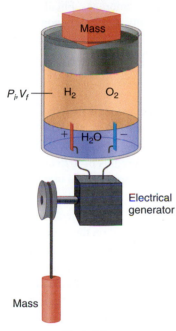

P_i, V_f — H$_2$ O$_2$
+ H$_2$O −

Electrical generator

Mass

Final state

FIGURE 2.2
Current produced by a generator is used to electrolyze water and thereby do work on the system as shown by the lowered mass linked to the generator. The gas produced in this process does P–V work on the surroundings, as shown by the raised mass on the piston.

- The net effect of work is to change U of the system and surroundings in accordance with the first law. If the only change in the surroundings is that a mass has been raised or lowered, work has flowed between the system and the surroundings.
- The quantity of work can be calculated from the change in potential energy of the mass, $E_{potential} = mgh$, where g is the gravitational acceleration and h is the change in the height of the mass, m.
- The sign convention for work is as follows: if the height of the mass in the surroundings is lowered, w is positive; if the height is raised, w is negative. In short, $w > 0$ if $\Delta U > 0$. It is common usage to say that if w is positive, work is done on the system by the surroundings. If w is negative, work is done by the system on the surroundings.

How much work is done in the process shown in Figure 2.1? Using a definition from physics, work is done when an object subject to a force, **F**, is moved through a distance, $d\mathbf{l}$, according to the path integral

$$w = \int \mathbf{F} \cdot d\mathbf{l} \qquad (2.4)$$

Using the definition of pressure as the force per unit area, the work done in moving the mass is given by

$$w = \int \mathbf{F} \cdot d\mathbf{l} = -\iint P_{external} \, dA dl = -\int P_{external} \, dV \qquad (2.5)$$

The minus sign appears because of our sign convention for work. Note that the pressure that appears in this expression is the external pressure, $P_{external}$, which need not equal the system pressure, P.

An example of another important kind of work, namely, electrical work, is shown in Figure 2.2. In this figure the contents of the cylinder is the system. Electrical current flows through a conductive aqueous solution, and water undergoes electrolysis to produce H$_2$ and O$_2$ gas. The current is produced by a generator, like that used to power a light on a bicycle through the mechanical work of pedaling. As current flows, the mass that drives the generator is lowered. In this case, the surroundings do the electrical work on the system. As a result, some of the liquid water is transformed to H$_2$ and O$_2$. From electrostatics, the work done in transporting a charge, q, through an electrical potential difference, ϕ is

$$w_{electrical} = q\phi \qquad (2.6)$$

For a constant current, I, that flows for a time, t, $q = It$. Therefore,

$$w_{electrical} = I\phi t \qquad (2.7)$$

The system also does work on the surroundings through the increase in the volume of the gas phase at constant pressure. The total work done is

$$w = w_{PV} + w_{electrical} = I\phi t - \int P_{external} \, dV = I\phi t - P_{external} \int dV = I\phi t - P_{external}(V_f - V_i) \qquad (2.8)$$

Other forms of work include the work of expanding a surface, such as a soap bubble, against the surface tension. Table 2.1 shows the expressions for work for five different cases. Each of these different types of work poses a requirement on the walls separating the system and surroundings. To be able to carry out the first three types of work, the walls must be movable, whereas for electrical work, they must be conductive.

The manner in which living systems accomplish work will be treated in detail in Chapter 7. Briefly, work accomplished by living cells includes mechanical work, electrical work, and osmotic work. The mathematical form of mechanical work is given by the work integral expression in Equation (2.4). Mechanical work includes contractile

TABLE 2.1   Types of Work

Types of Work	Variables	Equation for Work	Conventional Units
Volume expansion	Pressure (P), volume (V)	$w = -\int P_{external}\, dV$	Pa m^3 = J
Stretching	Tension (γ), length (l)	$w = -\int \gamma\, dl$	N m = J
Surface expansion	Surface tension (γ), area (σ)	$w = -\iint \gamma\, d\sigma$	(N m^{-1}) (m^2) = J
Electrical	Electrical potential (ϕ), electrical charge (q)	$w = \int \phi\, dq$	V C = J
Transport (Osmotic)	Temperature (T), activity (a). The activity of a species in solution is proportional to the concentration, as will be discussed in Section 9.11.	$w = \int RT d \ln a$	J

motion of muscle tissue, heart motions, and constriction of arteries. Electrical work includes movement of ions across the electrical potential in cell membranes. Finally, work is required to transport material against concentration gradients, a phenomenon called active transport or osmotic work.

Several examples of work calculations are given in Example Problem 2.1.

EXAMPLE PROBLEM 2.1

a. Calculate the work involved when a human adult exhales. Assume the adult lungs expel 0.50 L of air per exhalation against an atmospheric pressure of 1.00 atm.

b. A water bubble is expanded from a radius of 1.00 cm to a radius of 3.25 cm. The surface tension of water is 71.99 N m^{-1}. How much work is done in the process?

c. A current of 3.20 A is passed through a heating coil for 30.0 s. The electrical potential across the resistor is 14.5 V. Calculate the work done on the coil.

d. An aqueous solution of glucose is divided into two chambers by a membrane that is permeable to the passage of glucose. In one chamber the concentration of glucose is 0.010 M and in the other chamber the concentration is 0.050 M. Calculate the work required to move 2 g of glucose from the 0.010-M chamber to the 0.050-M chamber.

Solution

a. $w = -\int P_{external}\, dV = -P_{external}(V_f - V_i)$

$= -(1\text{ atm}) \times \dfrac{1.01 \times 10^5\text{ Pa}}{\text{atm}} \times 0.50\text{ L} \times \dfrac{10^{-3}\text{ m}^3}{\text{L}} = -50.\text{ J}$

b. A factor of 2 is included in the calculation below because a bubble has an inner and an outer surface:

$w = -\iint \gamma d\sigma = 2\gamma 4\pi(r_f^2 - r_i^2)$

$= -8\pi \times 71.99\text{ Nm}^{-1}(3.25^2\text{ cm}^2 - 1.00^2\text{ cm}^2) \times \dfrac{10^{-4}\text{ m}^2}{\text{cm}^2}$

$= -1.73\text{ J}$

c. $w = \int \phi dq = \phi q = I\phi t = 3.20\text{ A} \times 14.5\text{ V} \times 30.0\text{ s} = 1.39\text{ kJ}$

d. The molecular weight of glucose is 180 g mol^{-1}. Therefore, we have

$$n = \frac{2.00 \text{ g}}{180 \text{ g mol}^{-1}} = 1.11 \times 10^{-2} \text{ mol}$$

We set the activity equal to the concentration because the solution is dilute.

$$w = nRT \int_{c_1 = 0.010\, M}^{c_2 = 0.050\, M} d \ln c = nRT \ln \frac{c_2}{c_1}$$

$$= 0.011 \text{ mol} \times 310. \text{ K} \times 8.314 \text{ J K}^{-1} \text{ mol}^{-1} \times \ln 5.0 = 46 \text{ J}$$

2.3 Heat

Heat is defined in thermodynamics as the quantity of energy that flows across the boundary between the system and surroundings because of a temperature difference between the system and the surroundings. Just as for work, several important characteristics of heat are of importance:[1]

- Heat is transitory, in that it only appears during a change in state of the system and surroundings. Only energy, and not heat, is associated with the initial and final states of the system and the surroundings.
- The net effect of heat is to change the internal energy of the system and surroundings in accordance with the first law. If the only change in the surroundings is a change in temperature of a reservoir, heat has flowed between system and surroundings. The quantity of heat that has flowed is directly proportional to the change in temperature of the reservoir.
- The sign convention for heat is as follows: if the temperature of the surroundings is lowered, q is positive; if it is raised, q is negative. It is common usage to say that if q is positive, heat is withdrawn from the surroundings and deposited in the system. If q is negative, heat is withdrawn from the system and deposited in the surroundings.

Defining the surroundings as the rest of the universe is impractical, because it is not realistic to search through the whole universe to see if a mass has been raised or lowered and if the temperature of a reservoir has changed. Experience shows that in general only those parts of the universe close to the system interact with the system. Experiments can be constructed to ensure that this is the case, as shown in Figure 2.3. Imagine that you are interested in an exothermic chemical reaction that is carried out in a rigid sealed container with diathermal walls. You define the system as consisting solely of the reactant and product mixture. The vessel containing the system is immersed in an inner water bath separated from an outer water bath by a container with rigid diathermal walls. During the reaction, heat flows out of the system ($q < 0$), and the temperature of the inner water bath increases to T_f. Using an electrical heater, the temperature of the outer water bath is increased so that, at all times, $T_{outer} = T_{inner}$. Because of this condition, no heat flows across the boundary between the two water baths, and because the container enclosing the inner water bath is rigid, no work flows across this boundary. Therefore, $\Delta U = q + w = 0 + 0 = 0$ *for the composite system* made up of the inner water bath and everything within it. Therefore, this composite system is an isolated system that does not

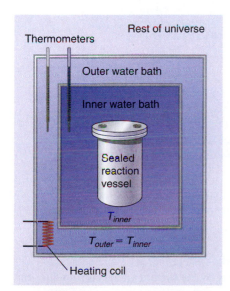

FIGURE 2.3
An isolated composite system is created in which the surroundings of the system of interest are limited in extent. The walls surrounding the inner water bath are rigid.

---

[1]*Heat* is perhaps the most misused term in thermodynamics as discussed by Robert Romer [*American Journal of Physics, 69* (2001), 107–109]. In common usage, it is incorrectly referred to as a substance as in the phrase "Close the door; you're letting the heat out!" An equally inappropriate term is heat capacity (discussed in Section 2.4), because it implies that materials have the capacity to hold heat, rather than the capacity to store energy. We use the terms *heat flow* or *heat transfer* to emphasize the transitory nature of heat. However, you should not think of heat as a fluid or a substance.

interact with the rest of the universe. To determine q and w for the reactant and product mixture, we need to examine only the composite system and can disregard the rest of the universe.

EXAMPLE PROBLEM 2.2

A heating coil is immersed in a 100-g sample of liquid H_2O at 100°C in an open insulated beaker on a laboratory bench at 1 bar pressure. In this process, 10% of the liquid is converted to the gaseous form at a pressure of 1 bar. A current of 2.00 A flows through the heater from a 12.0-V battery for 1.00×10^3 s to effect the transformation. The densities of liquid and gaseous water under these conditions are 997 and 0.590 kg m^{-3}, respectively.

a. It is often useful to replace a real process with a model that exhibits the important features of the process. Design a model system and surroundings, like those shown in Figures 2.1 and 2.2, that would allow you to measure the heat and work associated with this transformation. For the model system, define the system and surroundings as well as the boundary between them.

b. Calculate q and w for the process.

c. How can you define the system for the open insulated beaker on the laboratory bench such that the work is properly described?

Solution

a. The model system is shown in the following figure. The cylinder walls and the piston form adiabatic walls. The external pressure is held constant by a suitable weight.

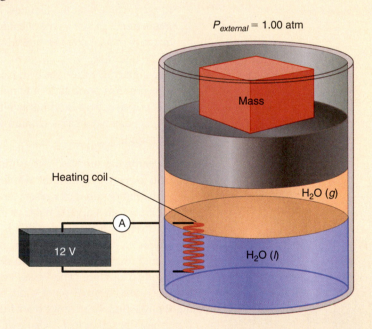

b. In the system shown in part (a), the heat input to the liquid water can be equated with the work done on the heating coil. Therefore,

$$q = I\phi t = 2.00 \text{ A} \times 12.0 \text{ V} \times 1.00 \times 10^3 \text{ s} = 24.0 \text{ kJ}$$

As the liquid is vaporized, the volume of the system increases at a constant external pressure. Therefore, the work done by the system on the surroundings is

$$w = -P_{external}(V_f - V_i) = -10 \text{ Pa} \times \left(\frac{10.0 \times 10^{-3} \text{ kg}}{0.590 \text{ kg m}^{-3}} + \frac{90.0 \times 10^{-3} \text{ kg}}{997 \text{ kg m}^{-3}} - \frac{100.0 \times 10^{-3} \text{ kg}}{997 \text{ kg m}^{-3}} \right)$$

$$= -1.70 \text{ kJ}$$

Note that the electrical work done on the heating coil is much larger than the $P-V$ work done in the expansion.

c. Define the system as the liquid in the beaker and the volume containing only molecules of H_2O in the gas phase. This volume will consist of disconnected volume elements dispersed in the air above the laboratory bench.

2.4 Heat Capacity

Heat flow can be quantified in terms of the easily measured electrical work done on a heating coil immersed in a liquid or a gas, or wrapped around a solid sample, $w = I\phi t$. The response of a single-phase system of constant composition to heat input is an increase in T. This is not the case if the system undergoes a phase change, such as the vaporization of a liquid. For example, the temperature of an equilibrium mixture of liquid and gas at the boiling point remains constant as heat flows into the system. However, the mass of the gas phase increases at the expense of the liquid phase.

The response of the system to heat flow is described by a very important thermodynamic property called the **heat capacity.** The heat capacity is a material-dependent property defined by the relation

$$C = \lim_{\Delta T \to 0} \frac{q}{T_f - T_i} = \frac{\mathchar'26\mkern-10mu d q}{dT} \tag{2.9}$$

where C is in the SI unit of J K^{-1}. It is an extensive quantity that, for example, doubles as the mass of the system is doubled. Often, the molar heat capacity, C_m, is used in calculations and it is an intensive quantity with the units of J K^{-1} mol^{-1}. Experimentally, the heat capacity of fluids is measured by immersing a heating coil in the fluid and equating the electrical work done on the coil with the heat flow into the sample. For solids, the heating coil is wrapped around the solid. In both cases, the experimental results must be corrected for heat losses to the surroundings. The significance of the notation $\mathchar'26\mkern-10mu d q$ for an incremental amount of heat is explained in the next section.

The value of the heat capacity depends on the experimental conditions under which it is determined. The most common conditions are constant volume or constant pressure, for which the heat capacity is denoted C_V and C_P, respectively. Values of $C_{P,m}$ at 298.15 K for pure substances are tabulated in Tables 2.2 and 2.3 (see Appendix B, Data Tables), and formulas for calculating $C_{P,m}$ at other temperatures for gases and solids are listed in Tables 2.4 and 2.5, respectively (see Appendix B, Data Tables).

An example of how $C_{P,m}$ depends on T is illustrated in Figure 2.4 for Cl_2. To make the functional form of $C_{P,m}(T)$ understandable, we briefly discuss the relative magnitudes of $C_{P,m}$ in the solid, liquid, and gaseous phases using a microscopic model. A solid can be thought of as a set of interconnected harmonic oscillators, and heat uptake leads to the excitations of the collective vibrations of the solid. At very low temperatures, these vibrations cannot be activated, because the spacing of the vibrational energy levels is large compared to kT. As a consequence, energy cannot be taken up by the solid. Hence, $C_{P,m}$ approaches zero as T approaches zero. For the solid, $C_{P,m}$ rises rapidly with T because the thermal energy available as T increases is sufficient to activate the vibrations of the solid. The heat capacity increases discontinuously as the solid melts to form a liquid. This is the case because the liquid retains all of the local vibrational modes of the solid, and more modes with low frequencies become available on melting. Therefore, the heat capacity of the liquid is greater than that of the solid. As the liquid vaporizes, the local vibrational modes present in the liquid are converted to translations that cannot take up as much energy as vibrations. Therefore, $C_{P,m}$ decreases discontinuously at the vaporization temperature. The heat capacity in the gaseous state increases slowly with temperature as the vibrational modes of the molecule are activated. These changes in $C_{P,m}$ can be calculated for a specific substance using a microscopic model and statistical thermodynamics, as will be discussed in detail in Chapter 23.

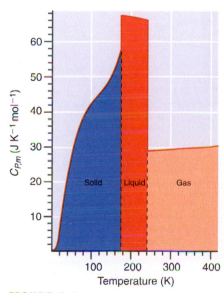

FIGURE 2.4

The variation of $C_{P,m}$ with temperature is shown for Cl_2.

Once the heat capacity of a variety of different substances has been determined, we have a convenient way to quantify heat flow. For example, at constant pressure, the heat flow between the system and surroundings can be written as follows:

$$q_P = \int_{T_{sys,i}}^{T_{sys,f}} C_P^{system}(T)\,dT = -\int_{T_{surr,i}}^{T_{surr,f}} C_P^{surroundings}(T)\,dT \qquad (2.10)$$

By measuring the temperature change of a thermal reservoir in the surroundings at constant pressure, q_P can be determined. In Equation (2.10), the heat flow at constant pressure has been expressed both from the perspective of the system and from the perspective of the surroundings. A similar equation can be written for a constant volume process. Water is a convenient choice of material for a heat bath in experiments because C_P is nearly constant at the value 4.18 J g^{-1} K^{-1} or 75.3 J mol^{-1} K^{-1} over the range from 0° to 100°C.

EXAMPLE PROBLEM 2.3

The volume of a system consisting of an ideal gas decreases at constant pressure. As a result, the temperature of a 1.50-kg water bath in the surroundings increases by 14.2°C. Calculate q_p for the system.

Solution

$$q_P = -\int_{T_{surr,i}}^{T_{surr,f}} C_P^{surroundings}(T)\,dT = -C_P^{surroundings}\Delta T$$

$$= -1.50 \text{ kg} \times 4.18 \text{ J g}^{-1}\text{K}^{-1} \times 14.2 \text{ K} = -89.1 \text{ kJ}$$

How are C_P and C_V related for a gas? Consider the processes shown in Figure 2.5 in which a fixed amount of heat flows from the surroundings into a gas. In the constant

FIGURE 2.5
Not all of the heat flow into a system can be used to increase ΔU in a constant pressure process, because the system does work on the surroundings as it expands. However, no work is done for constant volume heating.

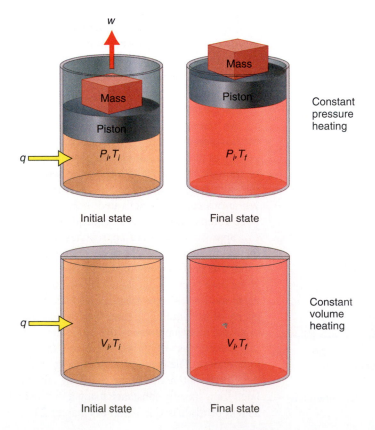

pressure process, the gas expands as its temperature increases. Therefore, the system does work on the surroundings. As a consequence, not all of the heat flow into the system can be used to increase ΔU. No such work occurs for the corresponding constant volume process, so all of the heat flow into the system can be used to increase ΔU. Therefore, $dT_p < dT_V$ for the same heat flow dq. For this reason, $C_P > C_V$ for gases.

The same argument applies to liquids and solids as long as V increases with T. Nearly all substances follow this behavior, although notable exceptions occur such as liquid water between 4° and 0°C for which the volume increases as T decreases. However, because ΔV on heating is much smaller than for a gas, the difference between C_P and C_V for a liquid or solid is much smaller than that for a gas.

The preceding remarks about the difference between C_P and C_V have been qualitative in nature. However, the following quantitative relationship, which will be proved in Chapter 3, holds for an ideal gas:

$$C_P - C_V = nR \text{ or } C_{P,m} - C_{V,m} = R \qquad (2.11)$$

2.5 State Functions and Path Functions

An alternate statement of the first law is that ΔU is independent of the path between the initial and final states and depends only on the initial and final states. We validate this statement for kinetic energy with the following equation. (Note that the argument can also be extended to the other forms of energy listed in Section 2.1.) Consider a single molecule in the system. Imagine that the molecule of mass m initially has the speed v_1. We now change its speed incrementally following the sequence $v_1 \rightarrow v_2 \rightarrow v_3 \rightarrow v_4$. The change in the kinetic energy along this sequence is given by

$$\Delta E_{kinetic} = \left(\frac{1}{2}m(v_2)^2 - \frac{1}{2}m(v_1)^2\right) + \left(\frac{1}{2}m(v_3)^2 - \frac{1}{2}m(v_2)^2\right)$$

$$+ \left(\frac{1}{2}m(v_4)^2 - \frac{1}{2}m(v_3)^2\right)$$

$$= \left(\frac{1}{2}m(v_4)^2 - \frac{1}{2}m(v_1)^2\right) \qquad (2.12)$$

Even though v_2 and v_3 can take on any arbitrary values, they still do not influence the result. We conclude that the change in the kinetic energy depends only on the initial and final speed and that it is independent of the path between these values. Our conclusion remains the same if we increase the number of speed increments in the interval between v_1 and v_4 to an arbitrarily large number. Because this conclusion holds for all molecules in the system, it also holds for ΔU.

This example supports the conclusion that ΔU depends only on the final and initial states and not on the path connecting these states. Any function that satisfies this condition is called a **state function,** because it depends only on the state of the system and not the path taken to reach the state. This property can be expressed in a mathematical form. Any state function, for example, U, must satisfy the equation

$$\Delta U = \int_i^f dU = U_f - U_i \qquad (2.13)$$

This equation states that in order for ΔU to depend only on the initial and final states characterized here by i and f, respectively, the value of the integral must be independent of the path. If this is the case, then U can be expressed as an infinitesimal quantity, dU, that when integrated, depends only on the initial and final states. The quantity dU is called an **exact differential.** We defer a discussion of exact differentials to Chapter 3.

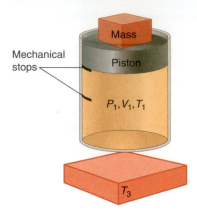

Mechanical stops

Mass

Piston

P_1, V_1, T_1

T_3

Initial state

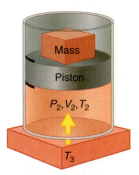

Mass

Piston

P_2, V_2, T_2

T_3

Intermediate state

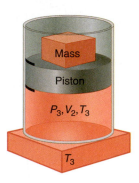

Mass

Piston

P_3, V_2, T_3

T_3

Final state

FIGURE 2.6

A system consisting of an ideal gas is contained in a piston and cylinder assembly. The gas in the initial state V_1, T_1 is compressed to an intermediate state, whereby the temperature increases to the value T_2. It is then brought into contact with a thermal reservoir at T_3, leading to a further rise in temperature. The final state is V_2, T_3.

It is useful to define a cyclic integral, denoted by the symbol $\oint$, as applying to a **cyclic path** such that the initial and final states are identical. For U or any other state function,

$$\oint dU = U_f - U_f = 0 \qquad (2.14)$$

because the initial and final states are the same in a cyclic process.

The state of a single-phase system at fixed composition is characterized by any two of the three variables P, T, and V. The same is true of U. Therefore, for a system of fixed mass, U can be written in any of the three forms $U(V,T)$, $U(P,T)$, or $U(P,V)$. Imagine that a gas of fixed mass characterized by V_1 and T_1 is confined in a piston and cylinder system that is isolated from the surroundings. There is a thermal reservoir in the surroundings at a temperature $T_3 < T_1$. We first compress the system from an initial volume V_1 to a final volume V_2 using a constant external pressure $P_{external}$. The work is given by

$$w = -\int_{V_i}^{V_f} P_{external}\, dV = -P_{external}\int_{V_i}^{V_f} dV = -P_{external}(V_f - V_i) = -P_{external}\Delta V \quad (2.15)$$

Because the height of the mass in the surroundings is lower after the compression (see Figure 2.6), w is positive and U increases. Because the system consists of a uniform single phase, U is a monotonic function of T, and T also increases. The change in volume, ΔV, has been chosen such that the temperature of the system, T_2, after the compression satisfies the inequality $T_1 < T_2 < T_3$.

We next let an additional amount of heat q flow between the system and surroundings at constant V by bringing the system into contact with the reservoir at temperature T_3. The final values of V and T after these two steps are V_2 and T_3.

This two-step process is repeated for different values of the mass. In each case the system is in the same final state characterized by the variables V_2 and T_3. The sequence of steps that takes the system from the initial state V_1, T_1 to the final state V_2, T_3 is referred to as a **path.** By changing the mass, a set of different paths is generated, all of which originate from the state V_1, T_1, and end in the state V_2, T_3. According to the first law, ΔU for this two-step process is

$$\Delta U = U(T_3, V_2) - U(T_1, V_1) = q + w \qquad (2.16)$$

Because ΔU is a state function, its value for the two-step process just described is the same for each of the different values of the mass.

Are q and w also state functions? For this process,

$$w = -P_{external}\Delta V \qquad (2.17)$$

and $P_{external}$ is different for each value of the mass, or for each path. Therefore, w is also different for each path; one can take one path from V_1, T_1 to V_2, T_3 and a different path from V_2, T_3 back to V_1, T_1. Because the work is different along these paths, the cyclic integral of work is not equal to zero. Therefore, w is not a state function.

Using the first law to calculate q for each of the paths, we obtain the result

$$q = \Delta U - w = \Delta U + P_{external}\Delta V \qquad (2.18)$$

Because ΔU is the same for each path, and w is different for each path, we conclude that q is also different for each path. Just as for work, the cyclic integral of heat is not equal to zero. Therefore, neither q nor w is a state function; they are instead called **path functions.**

Because both q and w are path functions, there are no exact differentials for work and heat. Incremental amounts of these quantities are denoted by $\text{đ}q$ and $\text{đ}w$, rather than dq and dw, to emphasize the fact that incremental amounts of work and heat are not exact differentials and so the differential notations $\text{đ}q$ and $\text{đ}w$, are meant to indicate that the criterion of exactness is not fulfilled and thus the heat and work integrals are path dependent. This is an important result that can be expressed mathematically:

$$\Delta q \neq \int_i^f \! dq \neq q_f - q_i \quad \text{and} \quad \Delta w \neq \int_i^f \! dw \neq w_f - w_i \tag{2.19}$$

In fact, there are no such quantities as Δq, q_f, q_i and Δw, w_f, w_i. This is an important result. One cannot refer to the work or heat possessed by a system, because these concepts have no meaning. After a process involving the transfer of heat and work between the system and surroundings is completed, the system and surroundings possess internal energy, but neither possesses heat or work.

Although the preceding discussion may appear pedantic on first reading, it is important to use the terms *work* and *heat* in a way that reflects the fact that they are not state functions. Examples of systems of interest are batteries, fuel cells, refrigerators, internal combustion engines, and living cells. In each case, the utility of these systems is that work and/or heat flows between the system and surroundings. For example, in a refrigerator, electrical energy is used to extract heat from the inside of the device and release it to the surroundings. One can speak of the refrigerator as having the capacity to extract heat, but it would be wrong to speak of it as having heat. In living cells, chemical energy stored in the bonds of the six-carbon-atom molecule glucose ($C_6H_{12}O_6$) is released when glucose is broken down into lactic acid ($CH_3(CHOH)COOH$):

$$C_6H_{12}O_6 \rightarrow 2CH_3(CHOH)COOH$$

The chemical energy made available by the breakdown (i.e., the metabolism) of glucose to 2 three-carbon molecules of lactate is not given off as heat. Instead, cells use the energy made available from glucose metabolism for protein synthesis, transport of metabolites through cell membranes, and other forms of work. To couple energy-producing reactions with reactions that require energy, cells use a molecule called ATP. The metabolism of glucose is enzymatically coupled to the phosphorylation of adenosine diphosphate (ADP) to produce adenosine triphosphate (ATP):

$$C_6H_{12}O_6 + 2HPO_4^{2-} + 2ADP^{3-} \rightarrow 2CH_3(CHOH)COO^- + 2ATP^{4-} + 2H_2O$$

When the need arises for energy to accomplish a cellular reaction, ATP is hydrolyzed back to ADP, releasing the necessary energy. However, it is incorrect to say that the chemical bonds in glucose contain work, although the energy liberated by glucose metabolism furnishes the work required to phosphorylate ADP. It is similarly incorrect to say that the chemical bonds in ATP store work, although the hydrolysis of ATP eventually furnishes the work to drive various cellular processes.

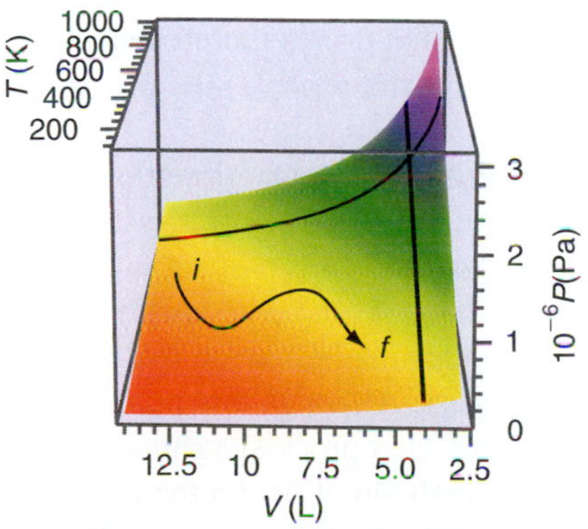

2.6 Equilibrium, Change, and Reversibility

Thermodynamics can only be applied to systems in internal equilibrium, and a requirement for equilibrium is that the overall rate of change of all processes such as diffusion or a chemical reaction be zero. How do we reconcile these statements with our calculations of q, w, and ΔU associated with processes in which there is a macroscopic change in the system? To answer this question, it is important to distinguish between the system and surroundings each being in internal equilibrium, and the system and surroundings being in equilibrium with one another.

We first discuss the issue of internal equilibrium. Consider a system made up of an ideal gas, which satisfies the equation of state $P = nRT/V$. All combinations of P, V, and T consistent with this equation of state form a surface in P–V–T space as shown in Figure 2.7. All points on the surface correspond to equilibrium states of the system. Points that are not on the surface do not correspond to any physically realizable state of the system. Nonequilibrium situations cannot be represented on such a plot, because P, V, and T do not have unique values for a system that is not in internal equilibrium.

FIGURE 2.7

All combinations of pressure, volume, and temperature consistent with 1 mol of an ideal gas lie on the colored surface. All combinations of pressure and volume consistent with $T = 800$ K and all combinations of pressure and temperature consistent with a volume of 4.0 L are shown as black curves that lie in the P–V–T surface. The third curve corresponds to a path between an initial state i and a final state f that is neither a constant temperature nor a constant volume path.

Pulley

1 kg

1 kg

FIGURE 2.8
Two masses of exactly 1 kg each are connected by a wire of zero mass running over a frictionless pulley. The system is in mechanical equilibrium and the masses are stationary.

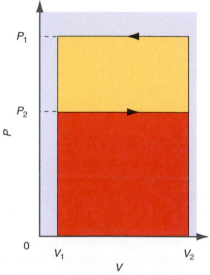

FIGURE 2.9
The work for each step and the total work can be obtained from an indicator diagram. For the compression step, w is given by the total area in red and yellow; for the expansion step, w is given by the red area. The arrows indicate the direction of change in V in the two steps. The sign of w is opposite for these two processes. The total work in the cycle is the yellow area.

Next, consider a process in which the system changes from an initial state characterized by P_i, V_i, and T_i to a final state characterized by P_f, V_f, and T_f as shown in Figure 2.7. If the rate of change of the macroscopic variables is negligibly small, the system passes through a succession of states of internal equilibrium as it goes from the initial to the final state. Such a process is called a **quasi-static process,** and internal equilibrium is maintained in this process. If the rate of change is sufficiently large, the rates of diffusion and intermolecular collisions may not be high enough to maintain the system in a state of internal equilibrium. However, thermodynamic calculations for such a process are still valid as long as it is meaningful to assign a single value for the macroscopic variables P, V, T, and concentration to the system undergoing change. For rapid changes, local fluctuations may occur in the values of the macroscopic variables throughout the system. If these fluctuations are small, quantities such as work and heat can still be calculated using the mean values of the macroscopic variables. The same considerations hold for the surroundings.

The concept of a quasi-static process allows one to visualize a process in which the system undergoes a major change in terms of a directed path along a sequence of states in which the system and surroundings are in internal equilibrium. We next distinguish between two very important classes of quasi-static processes, namely, reversible and irreversible processes. It is useful to consider the mechanical system shown in Figure 2.8 when discussing reversible and irreversible processes. Because the two masses have the same value, the net force acting on each end of the wire is zero, and the masses will not move. If an additional mass is placed on either of the two masses, the system is no longer in mechanical equilibrium and the masses will move. In the limit that the incremental mass approaches zero, the velocity at which the initial masses move approaches zero. In this case, one refers to the process as being **reversible,** meaning that the direction of the process can be reversed by placing the infinitesimal mass on the other side of the pulley.

Reversibility in a chemical system can be illustrated by liquid water in equilibrium with gaseous water surrounded by a thermal reservoir. The system and surroundings are both at temperature T. An infinitesimally small increase in T results in a small increase in the amount of water in the gaseous phase, and a small decrease in the liquid phase. An equally small decrease in the temperature has the opposite effect. Therefore, fluctuations in T give rise to corresponding fluctuations in the composition of the system. If an infinitesimal opposing change in the variable that drives the process (temperature in this case) causes a reversal in the direction of the process, the process is reversible.

If an infinitesimal change in the driving variable does not change the direction of the process, the process is considered **irreversible**. For example, in the chemical system just discussed, if a large stepwise temperature increase is induced using a heat pulse, the amount of water in the gas phase increases abruptly. In this case, the composition of the system cannot be returned to its initial value by an infinitesimal temperature decrease. This relationship is characteristic of an irreversible process. Although any process that takes place at a rapid rate in the real world is irreversible, real processes can approach reversibility in the appropriate limit. For example, by small variations in the electrical potential in an electrochemical cell, the conversion of reactants to products can be carried out nearly reversibly.

2.7 Comparing Work for Reversible and Irreversible Processes

We concluded in Section 2.5 that w is not a state function and that the work associated with a process is path dependent. This statement can be put on a quantitative footing by comparing the work associated with the reversible and irreversible expansion and compression of an ideal gas. This process is discussed next and illustrated in Figure 2.9.

Consider the following irreversible process, meaning that the internal and external pressures are not equal. A quantity of an ideal gas is confined in a cylinder with a weightless movable piston. The walls of the system are diathermal, allowing heat to flow between the system and surroundings. Therefore, the process is isothermal at the temperature of the surroundings, T. The system is initially defined by the variables T, P_1, and V_1. The position of the piston is determined by $P_{external} = P_1$, which can be changed by adding or removing weights from the piston. Because they are moved horizontally, no work is done in adding or removing the weights. The gas is first expanded at constant temperature by decreasing $P_{external}$ abruptly to the value P_2 (weights are removed), where $P_2 < P_1$. A sufficient amount of heat flows into the system through the diathermal walls to keep the temperature at the constant value T. The system is now in the state defined by T, P_2, and V_2, where $V_2 > V_1$. The system is then returned to its original state in an **isothermal process** by increasing $P_{external}$ abruptly to its original value P_1 (weights are added). Heat flows out of the system into the surroundings in this step. The system has been restored to its original state and, because this is a cyclic process, $\Delta U = 0$. Are q_{total} and w_{total} also zero for the cyclic process? The total work associated with this cyclic process is given by the sum of the work for each individual step:

$$w_{total} = \sum_i -P_{external,i}\Delta V_i = w_{expansion} + w_{compression} = -P_2(V_2 - V_1) - P_1(V_1 - V_2)$$
$$= -(P_2 - P_1) \times (V_2 - V_1) > 0 \text{ because } P_2 < P_1 \text{ and } V_2 > V_1 \quad (2.20)$$

The relationship between P and V for the process under consideration is shown graphically in Figure 2.9, in what is called an **indicator diagram**. An indicator diagram is useful because the work done in the expansion and contraction steps can be evaluated from the appropriate area in the figure, which is equivalent to evaluating the integral $w = -\int P_{external}\, dV$. Note that the work done in the expansion is negative because $\Delta V > 0$, and that done in the compression is positive because $\Delta V < 0$. Because $P_2 < P_1$, the magnitude of the work done in the compression process is more than that done in the expansion process and $w_{total} > 0$. What can one say about q_{total}? The first law states that because $\Delta U = q_{total} + w_{total} = 0$, $q_{total} < 0$.

The same cyclical process is carried out in a reversible cycle. A necessary condition for reversibility is that $P = P_{external}$ at every step of the cycle. This means that P changes during the expansion and compression steps. The work associated with the expansion is

$$w_{expansion} = -\int P_{external}\, dV = -\int P\, dV = -nRT \int \frac{dV}{V} = -nRT \ln \frac{V_2}{V_1} \quad (2.21)$$

This work is shown schematically as the red area in the indicator diagram of Figure 2.10.

If this process is reversed and the compression work is calculated, the following result is obtained:

$$w_{compression} = -nRT \ln \frac{V_1}{V_2} \quad (2.22)$$

which indicates that the magnitudes of the work in the forward and reverse processes are equal. The total work done in this cyclical process is given by

$$w = w_{expansion} + w_{compression} = -nRT \ln \frac{V_2}{V_1} - nRT \ln \frac{V_1}{V_2}$$
$$= -nRT \ln \frac{V_2}{V_1} + nRT \ln \frac{V_2}{V_1} = 0 \quad (2.23)$$

Therefore, the work done in a reversible isothermal cycle is zero. Because $\Delta U = q + w$ is a state function, $q = -w = 0$ for this reversible isothermal process. Looking at the heights of the weights in the surroundings at the end of the process, we find that they are the same as at the beginning of the process. As will be shown in Chapter 5, q and w are not equal to zero for a reversible cyclical process if T is not constant.

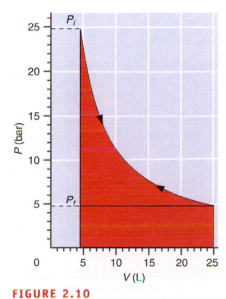

FIGURE 2.10

Indicator diagram for a reversible process. Unlike Figure 2.9, the areas under the P–V curves are the same in the forward and reverse directions.

EXAMPLE PROBLEM 2.4

In this example, 2.00 mol of an ideal gas undergoes isothermal expansion along three different paths: (1) reversible expansion from $P_i = 25.0$ bar and $V_i = 4.50$ L to $P_f = 4.50$ bar; (2) a single-step expansion against a constant external pressure of 4.50 bar; and (3) a two-step expansion consisting initially of an expansion against a constant external pressure of 11.0 bar until $P = P_{external}$, followed by an expansion against a constant external pressure of 4.50 bar until $P = P_{external}$.

Calculate the work for each of these processes. For which of the irreversible processes is the magnitude of the work greater?

Solution

The processes are depicted in the following indicator diagram:

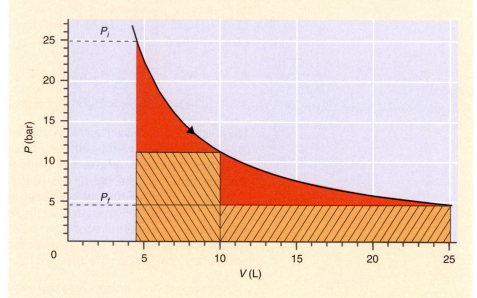

We first calculate the constant temperature at which the process is carried out, the final volume, and the intermediate volume in the two-step expansion:

$$T = \frac{P_i V_i}{nR} = \frac{25.0 \text{ bar} \times 4.50 \text{ L}}{8.314 \times 10^{-2} \text{ L bar mol}^{-1} \text{ K}^{-1} \times 2.00 \text{ mol}} = 677 \text{ K}$$

$$V_f = \frac{nRT}{P_f} = \frac{8.314 \times 10^{-2} \text{ L bar mol}^{-1} \text{ K}^{-1} \times 2.00 \text{ mol} \times 677 \text{ K}}{4.50 \text{ bar}} = 25.0 \text{ L}$$

$$V_{int} = \frac{nRT}{P_{int}} = \frac{8.314 \times 10^{-2} \text{ L bar mol}^{-1} \text{ K}^{-1} \times 2.00 \text{ mol} \times 677 \text{ K}}{11.0 \text{ bar}} = 10.2 \text{ L}$$

The work of the reversible process is given by

$$w = -nRT_1 \ln \frac{V_f}{V_i}$$

$$= -2.00 \text{ mol} \times 8.314 \text{ J mol}^{-1} \text{ K}^{-1} \times 677 \text{ K} \times \ln \frac{25.0 \text{ L}}{4.50 \text{ L}} = -19.3 \times 10^3 \text{ J}$$

We next calculate the work of the single-step and two-step irreversible processes:

$$w_{single} = -P_{external} \Delta V = -4.50 \text{ bar} \times \frac{10^5 \text{ Pa}}{\text{bar}} \times (25.00 \text{ L} - 4.50 \text{ L}) \times \frac{10^{-3} \text{ m}^3}{\text{L}}$$

$$= -9.23 \times 10^3 \text{ J}$$

$$w_{two\text{-}step} = -P_{external}\Delta V = -11.0\text{ bar} \times \frac{10^5\text{ Pa}}{\text{bar}} \times (10.2\text{ L} - 4.50\text{ L}) \times \frac{10^{-3}\text{ m}^3}{\text{L}}$$

$$-4.50\text{ bar} \times \frac{10^5\text{ Pa}}{\text{bar}} \times (25.00\text{ L} - 10.2\text{ L}) \times \frac{10^{-3}\text{ m}^3}{\text{L}}$$

$$= -12.9 \times 10^3\text{ J}$$

The magnitude of the work is greater for the two-step process than for the single-step process, but less than that for the reversible process.

Example Problem 2.4 suggests that the magnitude of the work for the multistep expansion process increases with the number of steps. This is indeed the case, as shown in Figure 2.11. Imagine that the number of steps n increases indefinitely. As n increases, the pressure difference $P_{external} - P$ for each individual step decreases. In the limit that $n \to \infty$, the pressure difference $P_{external} - P \to 0$, and the total area of the rectangles in the indicator diagram approaches the area under the reversible curve. Therefore, the irreversible process becomes reversible and the value of the work equals that of the reversible process.

By contrast, the magnitude of the irreversible compression work exceeds that of the reversible process for finite values of n and becomes equal to that of the reversible process as $n \to \infty$. The difference between the expansion and compression processes results from the requirement that $P_{external} < P$ at the beginning of each expansion step, whereas $P_{external} > P$ at the beginning of each compression step.

On the basis of these calculations for the reversible and irreversible cycles, we introduce another criterion to distinguish between reversible and irreversible processes. Suppose that a system undergoes a change through one or a number of individual steps,

FIGURE 2.11

The work done in an expansion (red area) is compared with the work done in a multistep series of irreversible expansion processes at constant pressure (yellow area) in the top panel. The bottom panel shows analogous results for the compression, where the area under the black curve is the reversible compression work. Note that the total work done in the irreversible expansion and compression processes approaches that of the reversible process as the number of steps becomes large.

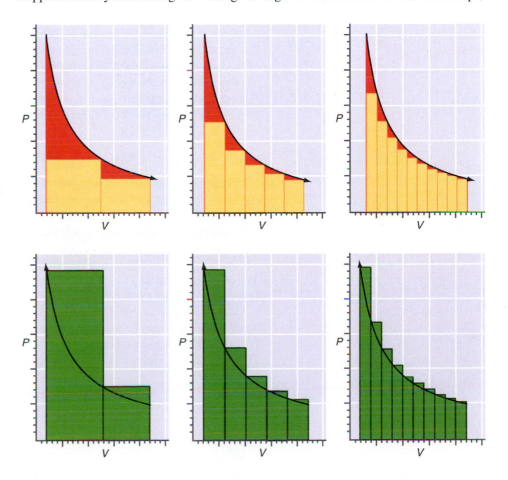

and that the system is restored to its initial state by following the same steps in reverse order. The system is restored to its initial state because the process is cyclical. If the surroundings are also returned to their original state (all masses at the same height and all reservoirs at their original temperatures), the process is reversible. If the surroundings are not restored to their original state, the process is irreversible.

One is often interested in extracting work from a system. For example, it is the expansion of the fuel–air mixture in an automobile engine on ignition that provides the torque that eventually drives the wheels. Is the capacity to do work similar for reversible and irreversible processes? This question is answered using the indicator diagrams of Figures 2.9 and 2.10 for the specific case of isothermal expansion work, noting that the work can be calculated from the area under the P–V curve. We compare the work for expansion from V_1 to V_2 in a single stage at constant pressure to that for the reversible case. For the single-stage expansion, the constant external pressure is given by $P_{external} = nRT/V_2$. However, if the expansion is carried out reversibly, the system pressure is always greater than this value. By comparing the areas in the indicator diagrams of Figure 2.11, it is seen that

$$|w_{reversible}| \geq |w_{irreversible}| \qquad (2.24)$$

By contrast, for the compression step,

$$|w_{reversible}| \leq |w_{irreversible}| \qquad (2.25)$$

The magnitude of the reversible work is the lower bound for the compression work and the upper bound for the expansion work. This result for the expansion work can be generalized to an important statement that holds for all forms of work: *the maximum work that can be extracted from a process between the same initial and final states is that obtained under reversible conditions.*

Although the preceding statement is true, it suggests that it would be optimal to operate an automobile engine under conditions in which the pressure inside the cylinders differs only infinitesimally from the external atmospheric pressure. This is clearly not possible. A practical engine must generate torque on the drive shaft, and this can only occur if the cylinder pressure is appreciably greater than the external pressure. Similarly, a battery is only useful if one can extract a sizable rather than an infinitesimal current. To operate such devices under useful irreversible conditions, the work output is less than the theoretically possible limit set by the reversible process.

Biochemical pathways are also interesting examples of optimized work production. For example, in some cells, glucose is broken down into ethanol (C_2H_5OH) by alcoholic fermentation, a process consisting of a sequence of reversible and irreversible chemical reactions. The net reaction is

$$C_6H_{12}O_6\ (s) \rightarrow 2C_2H_5OH\ (l) + 2CO_2\ (g)$$

In cells, fermentation is coupled enzymatically to the phosphorylation of ADP to ATP. As discussed earlier, ATP is a compound used by cells to couple energy-producing processes to processes that require energy to proceed. For every mole of glucose that is broken down, sufficient energy is recovered to enable the phosphorylation of two moles of ADP:

$$C_6H_{12}O_6(s) + 2HPO_4^{2-} + 2ADP^{3-} + 2H^+ \rightarrow 2C_2H_5OH(l) + 2CO_2(g) + 2ATP^{4-}$$

The efficiency with which energy is recovered from the conversion of glucose to ethanol in the form of work used to produce ATP is about 30%.

Suppose this reaction were carried out irreversibly and in a single step. The only work produced would be volume work due to the expansion of the system when 2 mol of carbon dioxide gas are produced from a single mole of solid glucose. Assuming $T = 298$ K and a negligible contribution from the volume of ethanol, the volume of the carbon dioxide is

$$V = \frac{2\ \text{mol} \times 0.0821\ \text{L} \times \text{atm} \times \text{mol}^{-1} \times \text{K}^{-1} \times 298\ \text{K}}{1\ \text{atm}} = 48.9\ \text{L}$$

and the work done by this expansion is

$$w = -P_{ext}\Delta V = -1\text{atm} \times 48.9\ \text{L} \times 101.325\text{J} \times \text{L}^{-1} \times \text{atm}^{-1} = 4.96\ \text{kJ}$$

The vast majority of the energy released from the breaking of the chemical bonds of glucose is dissipated as heat. Only about 3% of the total energy released by this irreversible reaction is recovered as work.

2.8 Determining ΔU and Introducing Enthalpy, a New State Function

How can the ΔU for a thermodynamic process be measured? This will be the topic of Chapter 4, in which calorimetry is discussed. However, this topic is briefly discussed here to enable you to carry out calculations on ideal gas systems in the end-of-chapter problems. The first law states that $\Delta U = q + w$. Imagine that the process is carried out under constant volume conditions and that nonexpansion work is not possible. Because under these conditions, $w = -\int P_{external} \, dV = 0$, and

$$\Delta U = q_V \tag{2.26}$$

Equation (2.26) states that ΔU can be experimentally determined by measuring the heat flow between the system and surroundings in a constant volume process.

What does the first law look like under reversible and constant pressure conditions? We write

$$dU = dq_P - P_{external} \, dV = dq_P - P \, dV \tag{2.27}$$

and integrate this expression between the initial and final states:

$$\int_i^f dU = U_f - U_i = \int dq_P - \int P \, dV = q_P - (P_f V_f - P_i V_i) \tag{2.28}$$

To evaluate the integral involving P, we must know the functional relationship $P(V)$, which in this case is $P_i = P_f = P$. Rearranging the last equation, we obtain

$$(U_f + P_f V_f) - (U_i + P_i V_i) = q_P \tag{2.29}$$

Because P, V, and U are all state functions, $U + PV$ is a state function. This new state function is called **enthalpy** and is given the symbol H. [A more rigorous demonstration that H is a state function can be given by invoking the first law for U and applying the criterion of Equation (3.5) to the product PV.]

$$H \equiv U + PV \tag{2.30}$$

As is the case for U, H has the units of energy, and it is an extensive property. As shown in Equation (2.29), ΔH for a process involving only P–V work can be determined by measuring the heat flow between the system and surroundings at constant pressure:

$$\Delta H = q_P \tag{2.31}$$

This equation is the constant pressure analogue of Equation (2.26). Because chemical reactions are much more frequently carried out at constant P than constant V, the energy change measured experimentally by monitoring the heat flow is ΔH rather than ΔU.

2.9 Calculating *q*, *w*, ΔU, and ΔH

In this section, we discuss how ΔU and ΔH, as well as q and w, can be calculated from the initial and final state variables if the path between the initial and final state is known. The problems at the end of this chapter ask you to calculate q, w, ΔU, and ΔH for simple and multistep processes. Because an equation of state is often needed to carry out such calculations, the system will generally be an ideal gas. Using an ideal gas as a surrogate for more complex systems has the significant advantage of simplifying the mathematics, allowing one to concentrate on the process rather than the manipulation of equations and the evaluation of integrals.

What does one need to know to calculate ΔU ? The following discussion is restricted to processes that do not involve chemical reactions or changes in phase. Because U is a state function, ΔU is independent of the path between the initial and final states. To describe a fixed amount of an ideal gas (i.e., n is constant), the values of two of the variables P, V, and T must be known. Is this also true for ΔU for processes involving ideal gases? To answer this question, consider the expansion of an ideal gas from an initial state V_1, T_1 to a final state V_2, T_2. We first assume that U is a function of both V and T. Is this assumption correct? Because ideal gas atoms or molecules do not interact with one another, U will not depend on the distance between the atoms or molecules. Therefore, U is not a function of V, and we conclude that ΔU must be a function of T only for an ideal gas, $\Delta U = \Delta U(T)$.

We also know that for a temperature range over which C_V is constant,

$$\Delta U = q_V = C_V (T_f - T_i) \tag{2.32}$$

Is this equation only valid for constant V? Because U is a function of T only for an ideal gas, Equation (2.32) is also valid for processes involving ideal gases in which V is not constant. Therefore, if one knows C_V, T_1, and T_2, ΔU can be calculated, regardless of the path between the initial and final states.

How many variables are required to define ΔH for an ideal gas? We write

$$\Delta H = \Delta U(T) + \Delta(PV) = \Delta U(T) + \Delta(nRT) = \Delta H(T) \tag{2.33}$$

which indicates that ΔH is also a function of T only for an ideal gas. In analogy to Equation (2.32),

2.1 Heat Capacity

$$\Delta H = q_P = C_P(T_f - T_i) \tag{2.34}$$

Because ΔH is a function of T only for an ideal gas, Equation (2.34) holds for all processes involving ideal gases, whether P is constant or not, as long as it is reasonable to assume that C_P is constant. Therefore, if the initial and final temperatures are known or can be calculated, and if C_V and C_P are known, ΔU and ΔH can be calculated *regardless of the path* for processes involving ideal gases using Equations (2.32) and (2.34), as long as no chemical reactions or phase changes occur. Because U and H are state functions, the previous statement is true for both reversible and irreversible processes. Recall that for an ideal gas $C_P - C_V = nR$, so that if either C_V or C_P is known, the other can be readily determined.

We next note that the first law links q, w, and ΔU. If any two of these quantities can be calculated, the first law can be used to calculate the third. In calculating work, often only expansion work takes place. In this case one always proceeds from the equation

$$w = -\int P_{external} \, dV \tag{2.35}$$

2.2 Reversible Isothermal Compression of an Ideal Gas

2.3 Reversible Isobaric Compression and Expansion of an Ideal Gas

2.4 Isochoric Heating and Cooling of an Ideal Gas

2.5 Reversible Cyclic Processes

This integral can only be evaluated if the functional relationship between $P_{external}$ and V is known. A frequently encountered case is $P_{external}$ = constant, such that

$$w = -P_{external}(V_f - V_i) \tag{2.36}$$

Because $P_{external} \neq P$, the work considered in Equation (2.36) is for an irreversible process.

A second frequently encountered case is that the system and external pressure differ only by an infinitesimal amount. In this case, it is sufficiently accurate to write $P_{external} = P$, and the process is reversible:

$$w = -\int \frac{nRT}{V} \, dV \tag{2.37}$$

This integral can only be evaluated if T is known as a function of V. The most commonly encountered case is an isothermal process, in which T is constant. As demonstrated in Section 2.2, for this case

$$w = -nRT \int \frac{dV}{V} = -nRT \ln \frac{V_f}{V_i} \qquad (2.38)$$

In solving thermodynamic problems, it is very helpful to understand the process thoroughly before starting the calculation, because it is often possible to obtain the value of one or more of q, w, and ΔU and ΔH without a calculation. For example, $\Delta U = \Delta H = 0$ for an isothermal process because ΔU and ΔH depend only on T. For an adiabatic process, $q = 0$ by definition. If only expansion work is possible, $w = 0$ for a constant volume process. These guidelines are illustrated in the following three example problems.

EXAMPLE PROBLEM 2.5

A system containing 2.50 mol of an ideal gas for which $C_{V,m} = 20.79$ J mol^{-1} K^{-1} is taken through the cycle in the following diagram in the direction indicated by the arrows. The curved path corresponds to $PV = nRT$, where $T = T_1 = T_3$.

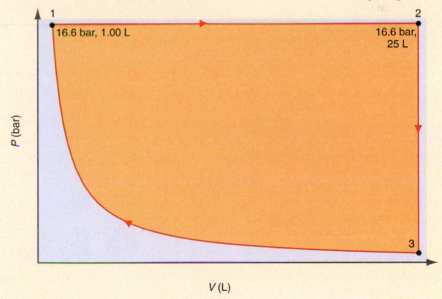

a. Calculate q, w, and ΔU and ΔH for each segment and for the cycle.

b. Calculate q, w, and ΔU and ΔH for each segment and for the cycle in which the direction of each process is reversed.

Solution

We begin by asking whether we can evaluate q, w, ΔU, or ΔH for any of the segments without any calculations. Because the path between states 1 and 3 is isothermal, ΔU and ΔH are zero for this segment. Therefore, from the first law, $q_{3 \rightarrow 1} = -w_{3 \rightarrow 1}$. For this reason, we only need to calculate one of these two quantities. Because $\Delta V = 0$ along the path between states 2 and 3, $w_{2 \rightarrow 3} = 0$. Therefore, $\Delta U_{2 \rightarrow 3} = q_{2 \rightarrow 3}$. Again, we only need to calculate one of these two quantities. Because the total process is cyclic, the change in any state function is zero. Therefore, $\Delta U = \Delta H = 0$ for the cycle, no matter which direction is chosen. We now deal with each segment individually.

Segment 1→2

The values of n, P_1 and V_1, and P_2 and V_2 are known. Therefore, T_1 and T_2 can be calculated using the ideal gas law. We use these temperatures to calculate ΔU as follows:

$$\Delta U_{1 \to 2} = nC_{V,m}(T_2 - T_1) = \frac{nC_{V,m}}{nR}(P_2 V_2 - P_1 V_1)$$

$$= \frac{2.50 \text{ mol} \times 20.79 \text{ J mol}^{-1}\text{K}^{-1}}{2.50 \text{ mol} \times 0.08314 \text{ L bar K}^{-1}\text{mol}^{-1}} \times (16.6 \text{ bar} \times 25.0 \text{ L} - 16.6 \text{ bar} \times 1.00 \text{ L})$$

$$= 99.6 \text{ kJ}$$

The process takes place at constant pressure, so

$$w = -P_{external}(V_2 - V_1) = -16.6 \text{ bar} \times \frac{10^5 \text{ N m}^{-2}}{\text{bar}} \times (25.0 \times 10^{-3}\text{m}^3 - 1.00 \times 10^{-3}\text{m}^3)$$

$$= -39.8 \text{ kJ}$$

Using the first law

$$q = \Delta U - w = 99.6 \text{ kJ} + 39.8 \text{ kJ} \approx 139.4 \text{ kJ}$$

We next calculate T_2, T_3, and then $\Delta H_{1 \to 2}$:

$$T_2 = \frac{P_2 V_2}{nR} = \frac{16.6 \text{ bar} \times 25.0 \text{ L}}{2.50 \text{ mol} \times 0.08314 \text{ L bar K}^{-1} \text{ mol}^{-1}} = 2.00 \times 10^3 \text{ K}$$

We next calculate $T_3 = T_1$

$$T_1 = \frac{P_1 V_1}{nR} = \frac{16.6 \text{ bar} \times 1.00 \text{ L}}{2.50 \text{ mol} \times 0.08314 \text{ L bar mol}^{-1} \text{ K}^{-1}} = 79.9 \text{ K}$$

$$\Delta H_{1 \to 2} = \Delta U_{1 \to 2} + \Delta(PV) = \Delta U_{1 \to 2} + nR(T_2 - T_1)$$

$$= 99.6 \times 10^3 \text{ J} + 2.5 \text{ mol} \times 8.314 \text{ J mol}^{-1}\text{K}^{-1} \times (2000 \text{ K} - 79.9 \text{ K}) = 139.4 \text{kJ}$$

Segment 2→3

As noted in the earlier calculations, $w = 0$, and

$$\Delta U_{2 \to 3} = q_{2 \to 3} = C_V(T_3 - T_2)$$

$$= 2.50 \text{ mol} \times 20.79 \text{ J mol}^{-1}\text{K}^{-1}(79.9 \text{ K} - 2000 \text{ K})$$

$$= -99.6 \text{ kJ}$$

The numerical result is equal in magnitude, but opposite in sign to $\Delta U_{1 \to 2}$ because $T_3 = T_1$. For the same reason, $\Delta H_{2 \to 3} = -\Delta H_{1 \to 2}$.

Segment 3→1

For this segment, $\Delta U_{3 \to 1} = 0$ and $\Delta H_{3 \to 1} = 0$ as noted earlier and $w_{3 \to 1} = -q_{3 \to 1}$. Because this is a reversible isothermal compression,

$$w_{3 \to 1} = -nRT \ln \frac{V_1}{V_3} = -2.50 \text{ mol} \times 8.314 \text{ J mol}^{-1} \text{ K}^{-1} \times 79.9 \text{ K} \times \ln \frac{1.00 \times 10^{-3}\text{m}^3}{25.0 \times 10^{-3}\text{m}^3}$$

$$= 5.35 \text{ kJ}$$

The results for the individual segments and for the cycle in the indicated direction are given in the following table. If the cycle is traversed in the reverse fashion, the magnitudes of all quantities in the table remain the same, but all signs change.

Path	q (kJ)	w (kJ)	ΔU (kJ)	ΔH (kJ)
1→2	139.4	−39.8	99.6	139.4
2→3	−99.6	0	−99.6	−139.4
3→1	−5.35	5.35	0	0
Cycle	34.5	−34.5	0	0

EXAMPLE PROBLEM 2.6

In this example, 2.50 mol of an ideal gas with $C_{V,m} = 12.47$ J mol^{-1} K^{-1} is expanded adiabatically against a constant external pressure of 1.00 bar. The initial temperature and pressure of the gas are 325 K and 2.50 bar, respectively. The final pressure is 1.25 bar. Calculate the final temperature, q, w, ΔU, and ΔH.

Solution

Because the process is adiabatic, $q = 0$, and $\Delta U = w$. Therefore,

$$\Delta U = nC_{v,m}(T_f - T_i) = -P_{external}(V_f - V_i)$$

Using the ideal gas law,

$$nC_{v,m}(T_f - T_i) = -nRP_{external}\left(\frac{T_f}{P_f} - \frac{T_i}{P_i}\right)$$

$$T_f\left(nC_{v,m} + \frac{nRP_{external}}{P_f}\right) = T_i\left(nC_{v,m} + \frac{nRP_{external}}{P_i}\right)$$

$$T_f = T_i\left(\frac{C_{v,m} + \dfrac{RP_{external}}{P_i}}{C_{v,m} + \dfrac{RP_{external}}{P_f}}\right)$$

$$= 325\text{ K} \times \left(\frac{12.47\text{ J mol}^{-1}\text{ K}^{-1} + \dfrac{8.314\text{ J mol}^{-1}\text{ K}^{-1} \times 1.00\text{ bar}}{2.50\text{ bar}}}{12.47\text{ J mol}^{-1}\text{ K}^{-1} + \dfrac{8.314\text{ J mol}^{-1}\text{ K}^{-1} \times 1.00\text{ bar}}{1.25\text{ bar}}}\right) = 268\text{ K}$$

We calculate $\Delta U = w$ from

$$\Delta U = (nC_{v,m}T_f - T_i) = 2.5\text{ mol} \times 12.47\text{ J mol}^{-1}\text{ K}^{-1} \times (268\text{ K} - 325\text{ K}) = -1.78\text{ kJ}$$

Because the temperature falls in the expansion, the internal energy decreases:

$$\Delta H = \Delta U + \Delta(PV) = \Delta U + nR(T_2 - T_1)$$

$$= -1.78 \times 10^3\text{ J} + 2.5\text{ mol} \times 8.314\text{ J mol}^{-1}\text{ K}^{-1} \times (268\text{ K} - 325\text{ K}) = -2.96\text{ kJ}$$

EXAMPLE PROBLEM 2.7

An athlete exercising at a moderate rate for 30 minutes performs about 920. kJ of work. During the first hour of exercise, fats provide about 50.% of the energy consumed doing work. During the second and third hours of exercise fats provide 70.% and 80.% of the energy consumed, respectively. Suppose an athlete exercises for 3 hours at the rate of work just stated. How many grams of fat will be metabolized to provide this work? Assume the energy derived from metabolism of fat is about 34 kJ g^{-1}. Assume the energy produced from metabolism of fat is converted to work at 50.% efficiency.

Solution

The rate of work is $\dfrac{w}{\Delta t} = \dfrac{920\text{ kJ}}{30.\text{ min}} \times 60.\text{ min h}^{-1} = 1.8 \times 10^3\text{ kJ h}^{-1}$

Energy provided by fat: $0.50 \times 1.8 \times 10^3\text{ kJ} + 0.70 \times 1.8 \times 10^3\text{ kJ} + 0.80 \times 1.8 \times 10^3\text{ kJ} = 3.6 \times 10^3\text{kJ}$

Mass of fat metabolized during 3 hours of moderate work:

$$\frac{3.6 \times 10^3\text{ kJ}}{34\text{ kJ g}^{-1}} \times 2 = 2.12 \times 10^2\text{ g}$$

2.6 Reversible Adiabatic Heating and Cooling of an Ideal Gas

2.10 The Reversible Adiabatic Expansion and Compression of an Ideal Gas

The adiabatic expansion and compression of gases are important meteorological processes. For example, the cooling of a cloud as it moves upward in the atmosphere can be modeled as an adiabatic process because the heat transfer between the cloud and the rest of the atmosphere is slow on the timescale of its upward motion.

Consider the adiabatic expansion of an ideal gas. Because $q = 0$, the first law takes the form

$$\Delta U = w \text{ or } C_v dT = -P_{external} dV \tag{2.39}$$

For a reversible adiabatic process, $P = P_{external}$, and

$$C_V dT = -nRT \frac{dV}{V} \text{ or, equivalently, } C_V \frac{dT}{T} = -nR \frac{dV}{V} \tag{2.40}$$

Integrating both sides of this equation between the initial and final states,

$$\int_{T_i}^{T_f} C_V \frac{dT}{T} = -nR \int_{V_i}^{V_f} \frac{dV}{V} \tag{2.41}$$

If C_V is constant over the temperature interval $T_f - T_i$, then

$$C_V \ln \frac{T_f}{T_i} = -nR \ln \frac{V_f}{V_i} \tag{2.42}$$

Because $C_P - C_V = nR$ for an ideal gas, Equation (2.42) can be written in the form

$$\ln\left(\frac{T_f}{T_i}\right) = -(\gamma - 1) \ln\left(\frac{V_f}{V_i}\right) \text{ or, equivalently, } \frac{T_f}{T_i} = \left(\frac{V_f}{V_i}\right)^{1-\gamma} \tag{2.43}$$

where $\gamma = C_{P,m}/C_{V,m}$. Substituting $T_f/T_i = P_f V_f/P_i V_i$ in the previous equation, we obtain

$$P_i V_i^{\gamma} = P_f V_f^{\gamma} \tag{2.44}$$

for the adiabatic reversible expansion or compression of an ideal gas. Note that our derivation is only applicable to a reversible process, because we have assumed that $P = P_{external}$.

Reversible adiabatic compression of a gas leads to heating, and reversible adiabatic expansion leads to cooling, as can be concluded from Figure 2.12. Two systems containing 1 mol of N_2 gas have the same volume at $P = 1$ atm. Under isothermal conditions, heat flows out of the system as it is compressed to $P > 1$ atm, and heat flows into the system as it is expanded to $P < 1$ atm to keep T constant. No heat flows into or out of the system under adiabatic conditions. Note in Figure 2.12 that in a reversible adiabatic compression originating at 1 atm, $P_{adiabatic} > P_{isothermal}$ for all $P > 1$ atm. Therefore, this value of P must correspond to a value of T for which $T > T_{isothermal}$. Similarly, in a reversible adiabatic expansion originating at 1 atm, $P_{adiabatic} < P_{isothermal}$ for all $P < 1$ atm. Therefore, this value of P must correspond to a value of T for which $T < T_{isothermal}$.

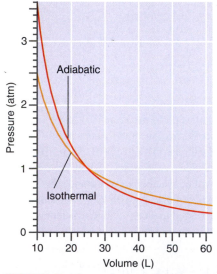

FIGURE 2.12
Two systems containing 1 mol of N_2 have the same P and V values at 1 atm. The red curve corresponds to reversible expansion and compression about $P = 1$ atm under adiabatic conditions. The yellow curve corresponds to reversible expansion and compression about $P = 1$ atm under isothermal conditions.

EXAMPLE PROBLEM 2.8

A cloud mass moving across the ocean at an altitude of 2000. m encounters a coastal mountain range. As it rises to a height of 3500. m to pass over the mountains, it undergoes an adiabatic expansion. The pressure at 2000. and 3500. m is 0.802 and 0.602 atm, respectively. If the initial temperature of the cloud mass is 288 K, what is the cloud temperature as it passes over the mountains? Assume that $C_{P,m}$ for air is 28.86 J K^{-1} mol^{-1} and that air obeys the ideal gas law. If you are on the mountain, should you expect rain or snow?

Solution

$$\ln\left(\frac{T_f}{T_i}\right) = -(\gamma - 1)\ln\left(\frac{V_f}{V_i}\right) = -(\gamma - 1)\ln\left(\frac{T_f}{T_i}\frac{P_i}{P_f}\right)$$

$$= -(\gamma - 1)\ln\left(\frac{T_f}{T_i}\right) - (\gamma - 1)\ln\left(\frac{P_i}{P_f}\right)$$

$$= -\frac{(\gamma - 1)}{\gamma}\ln\left(\frac{P_i}{P_f}\right) = -\frac{\left(\dfrac{C_{P,m}}{C_{P,m} - R} - 1\right)}{\dfrac{C_{P,m}}{C_{P,m} - R}}\ln\left(\frac{P_i}{P_f}\right)$$

$$= -\frac{\left(\dfrac{28.86\ \text{J K}^{-1}\ \text{mol}^{-1}}{28.86\ \text{J K}^{-1}\ \text{mol}^{-1} - 8.314\ \text{J K}^{-1}\ \text{mol}^{-1}} - 1\right)}{\dfrac{28.86\ \text{J K}^{-1}\ \text{mol}^{-1}}{28.86\ \text{J K}^{-1}\ \text{mol}^{-1} - 8.314\ \text{J K}^{-1}\ \text{mol}^{-1}}} \times \ln\left(\frac{0.802\ \text{atm}}{0.602\ \text{atm}}\right) = -0.0826$$

$$T_f = 0.9207 T_i = 265\ \text{K}$$

You can expect snow.

Vocabulary

cyclic path	indicator diagram	isothermal	quasi-static process
enthalpy	internal energy	isothermal process	reversible process
first law of thermodynamics	irreversible process	path	state function
heat	isobaric	path function	work
heat capacity	isochoric		

Questions on Concepts

Q2.1 Electrical current is passed through a resistor immersed in a liquid in an adiabatic container. The temperature of the liquid is varied by 1°C. The system consists solely of the liquid. Does heat or work flow across the boundary between the system and surroundings? Justify your answer.

Q2.2 Explain how a mass of water in the surroundings can be used to determine q for a process. Calculate q if the temperature of a 1.00-kg water bath in the surroundings increases by 1.25°C. Assume that the surroundings are at a constant pressure.

Q2.3 Explain the relationship between the terms *exact differential* and *state function*.

Q2.4 Why is it incorrect to speak of the heat or work associated with a system?

Q2.5 Two ideal gas systems undergo reversible expansion starting from the same P and V. At the end of the expansion, the two systems have the same volume. The pressure in the system that has undergone adiabatic expansion is lower than in the system that has undergone isothermal expansion. Explain this result without using equations.

Q2.6 A cup of water at 278 K (the system) is placed in a microwave oven and the oven is turned on for 1 minute during which it begins to boil. Which of q, w, and ΔU are positive, negative, or zero?

Q2.7 What is wrong with the following statement?: *because the well-insulated house stored a lot of heat, the temperature didn't fall much when the furnace failed.* Rewrite the sentence to convey the same information in a correct way.

Q2.8 What is wrong with the following statement?: *burns caused by steam at 100°C can be more severe than those caused by water at 100°C because steam contains more heat than water.* Rewrite the sentence to convey the same information in a correct way.

Q2.9 Describe how reversible and irreversible expansions differ by discussing the degree to which equilibrium is maintained between the system and the surroundings.

Q2.10 A chemical reaction occurs in a constant volume enclosure separated from the surroundings by diathermal walls. Can you say whether the temperature of the surroundings increases, decreases, or remains the same in this process? Explain.

Problems

Problem numbers in **RED** indicate that the solution to the problem is given in the *Student Solutions Manual*.

P2.1   3.00 moles of an ideal gas at 27.0°C expand isothermally from an initial volume of 20.0 dm^3 to a final volume of 60.0 dm^3. Calculate w for this process (a) for expansion against a constant external pressure of 1.00×10^5 Pa and (b) for a reversible expansion.

P2.2   A major league pitcher throws a baseball at a speed of 150. km/h. If the baseball weighs 220. g and its heat capacity is 2.0 J g^{-1} K^{-1}, calculate the temperature rise of the ball when it is stopped by the catcher's mitt. Assume no heat is transferred to the catcher's mitt. Assume also that the catcher's arm does not recoil when he or she catches the ball.

P2.3   3.00 moles of an ideal gas are compressed isothermally from 60.0 to 20.0 L using a constant external pressure of 5.00 atm. Calculate q, w, ΔU, and ΔH.

P2.4   A system consisting of 57.5 g of liquid water at 298 K is heated using an immersion heater at a constant pressure of 1.00 bar. If a current of 1.50 A passes through the 10.0-ohm resistor for 150 s, what is the final temperature of the water? The heat capacity for water can be found in Appendix B.

P2.5   Using the results from Problem P1.4, determine the average heat evolved by the oxidation of foodstuffs in an average adult per hour per kilogram of body weight. Assume the weight of an average adult is 70. kg. State any assumptions you make. Assume 420. kJ of heat are evolved per mole of oxygen consumed as a result of the oxidation of foodstuffs.

P2.6   Suppose an adult body were encased in a thermally insulating barrier. If as a result of this barrier all the heat evolved by metabolism of foodstuffs were retained by the body, what would the temperature of the body reach after 3 hours? Assume the heat capacity of the body is 4.18 J g^{-1}K^{-1}. Use the results of Problem P2.5 in your solution.

P2.7   For 1.00 mol of an ideal gas, $P_{external} = P = 200. \times 10^3$ Pa. The temperature is changed from 100.°C to 25.0°C, and $C_{V,m} = 3/2R$. Calculate q, w, ΔU, and ΔH.

P2.8   Consider the isothermal expansion of 5.25 mol of an ideal gas at 450. K from an initial pressure of 15.0 bar to a final pressure of 3.50 bar. Describe the process that will result in the greatest amount of work being done by the system with $P_{external} \geq 3.50$ bar and calculate w. Describe the process that will result in the least amount of work being done by the system with $P_{external} \geq 3.50$ bar and calculate w. What is the least amount of work done without restrictions on the external pressure?

P2.9   A hiker caught in a thunderstorm loses heat when her clothing becomes wet. She is packing emergency rations that, if completely metabolized, will release 30. kJ of heat per gram of rations consumed. How much rations must the hiker consume to avoid a reduction in body temperature of 4.0 K as a result of heat loss? Assume the heat capacity of the body equals that of water. Assume the hiker weighs 55 kg. State any additional assumptions.

P2.10   A muscle fiber contracts by 2.0 cm and in doing so lifts a weight. Calculate the work performed by the fiber and the weight lifted. Assume the muscle fiber obeys Hooke's law with a constant of 800. N m^{-1}.

P2.11   Calculate ΔH and ΔU for the transformation of 1.00 mol of an ideal gas from 27.0°C and 1.00 atm to 327°C and 17.0 atm if

$$C_{P,m} = 20.9 + 0.042 \frac{T}{K} \text{ in units of J K}^{-1}\text{mol}^{-1}$$

P2.12   Calculate w for the adiabatic expansion of 1.00 mol of an ideal gas at an initial pressure of 2.00 bar from an initial temperature of 450. K to a final temperature of 300. K. Write an expression for the work done in the isothermal reversible expansion of the gas at 300. K from an initial pressure of 2.00 bar. What value of the final pressure would give the same value of w as the first part of this problem? Assume that $C_{P,m} = 5/2R$.

P2.13   In the adiabatic expansion of 1.00 mol of an ideal gas from an initial temperature of 25.0°C, the work done on the surroundings is 1200. J. If $C_{V,m} = 3/2R$, calculate q, w, ΔU, and ΔH.

P2.14   According to a story told by Lord Kelvin, one day when walking down from Chamonix to commence a tour of Mt. Blanc, "whom should I meet walking up (the trail) but (James) Joule, with a long thermometer in his hand, and a carriage with a lady in it not far off. He told me he had been married since we parted from Oxford, and he was going to try for (the measurement of the) elevation of temperature in waterfalls." Suppose Joule encountered a waterfall 30. m in height. Calculate the temperature difference between the top and bottom of this waterfall.

P2.15   An ideal gas undergoes an expansion from the initial state described by P_i, V_i, T to a final state described by P_f, V_f, T in (a) a process at the constant external pressure P_f and (b) in a reversible process. Derive expressions for the largest mass that can be lifted through a height h in the surroundings in these processes.

P2.16   An automobile tire contains air at $320. \times 10^3$ Pa at 20.0°C. The stem valve is removed and the air is allowed to expand adiabatically against the constant external pressure of $100. \times 10^3$ Pa until $P = P_{external}$. For air, $C_{V,m} = 5/2R$. Calculate the final temperature. Assume ideal gas behavior.

P2.17   Count Rumford observed that using cannon-boring machinery, a single horse could heat 11.6 kg of water ($T = 273$ K) to $T = 355$ K. in 2.5 hours. Assuming the same rate of work, how high could a horse raise a 150.-kg weight in 1 minute? Assume the heat capacity of water is 4.18 kJ K^{-1} kg^{-1}.

P2.18   Count Rumford also observed that nine burning candles generate heat at the same rate that a single horse-driven cannon-boring piece of equipment generates heat. James Watt observed that a single horse can raise a 330.-lb. weight 100. feet in 1 minute. Using the observations of Watt and Rumford, determine the rate at which a candle generates heat. (*Note:* 1.00 m = 3.281 ft.)

P2.19   3.50 moles of an ideal gas are expanded from 450. K and an initial pressure of 5.00 bar to a final pressure of 1.00 bar, and $C_{P,m} = 5/2R$. Calculate w for the following two cases:

  a. The expansion is isothermal and reversible.

  b. The expansion is adiabatic and reversible.

Without resorting to equations, explain why the result for part (b) is greater than or less than the result for part (a).

P2.20   An ideal gas described by T_i = 300. K, P_i = 1.00 bar, and V_i = 10.0 L is heated at constant volume until P = 10.0 bar. It then undergoes a reversible isothermal expansion until P = 1.00 bar. It is then restored to its original state by the extraction of heat at constant pressure. Depict this closed-cycle process in a P–V diagram. Calculate w for each step and for the total process. What values for w would you calculate if the cycle were traversed in the opposite direction?

P2.21   3.00 mols of an ideal gas with $C_{V,m} = 3/2R$ initially at a temperature T_i = 298 K and P_i = 1.00 bar are enclosed in an adiabatic piston and cylinder assembly. The gas is compressed by placing a 625-kg mass on the piston of diameter 20.0 cm. Calculate the work done in this process and the distance that the piston travels. Assume that the mass of the piston is negligible.

P2.22   A bottle at 21.0°C contains an ideal gas at a pressure of 126.4×10^3 Pa. The rubber stopper closing the bottle is removed. The gas expands adiabatically against $P_{external} = 101.9 \times 10^3$ Pa, and some gas is expelled from the bottle in the process. When $P = P_{external}$, the stopper is quickly replaced. The gas remaining in the bottle slowly warms up to 21.0°C. What is the final pressure in the bottle for a monatomic gas, for which $C_{V,m} = 3/2R$, and a diatomic gas, for which $C_{V,m} = 5/2R$?

P2.23   A pellet of Zn of mass 10.0 g is dropped into a flask containing dilute H_2SO_4 at a pressure of P = 1.00 bar and temperature of T = 298 K. What is the reaction that occurs? Calculate w for the process.

P2.24   One mole of an ideal gas for which $C_{V,m}$ = 20.8 J K^{-1} mol^{-1} is heated from an initial temperature of 0.00°C to a final temperature of 275°C at constant volume. Calculate q, w, ΔU, and ΔH for this process.

P2.25   One mole of an ideal gas, for which $C_{V,m}$ = 3/2R, initially at 20.0°C and 1.00×10^6 Pa undergoes a two-stage transformation. For each of the stages described in the following list, calculate the final pressure, as well as q, w, ΔU, and ΔH. Also calculate q, w, ΔU, and ΔH for the complete process.

  a. The gas is expanded isothermally and reversibly until the volume doubles.

  b. Beginning at the end of the first stage, the temperature is raised to 80.0°C at constant volume.

P2.26   One mole of an ideal gas, for which $C_{V,m}$ = 3/2R, initially at 298 K and 1.00×10^5 Pa undergoes a reversible adiabatic compression. At the end of the process, the pressure is 1.00×10^6 Pa. Calculate the final temperature of the gas. Calculate q, w, ΔU, and ΔH for this process.

P2.27   The temperature of 1 mol of an ideal gas increases from 18.0° to 55.1°C as the gas is compressed adiabatically. Calculate q, w, ΔU, and ΔH for this process assuming that $C_{V,m} = 3/2R$.

P2.28   A 1.00-mol sample of an ideal gas for which $C_{V,m} = 3/2R$ undergoes the following two-step process: (1) From an initial state of the gas described by T = 28.0°C and P = 2.00 $\times 10^4$ Pa, the gas undergoes an isothermal expansion against a constant external pressure of 1.00×10^4 Pa until the volume has doubled. (2) Subsequently, the gas is cooled at constant volume. The temperature falls to −40.5°C. Calculate q, w, ΔU, and ΔH for each step and for the overall process.

P2.29   A cylindrical vessel with rigid adiabatic walls is separated into two parts by a frictionless adiabatic piston. Each part contains 50.0 L of an ideal monatomic gas with $C_{V,m} = 3/2R$. Initially, T_i = 298 K and P_i = 1.00 bar in each part. Heat is slowly introduced into the left part using an electrical heater until the piston has moved sufficiently to the right to result in a final pressure P_f = 7.50 bar in the right part. Consider the compression of the gas in the right part to be a reversible process.

  a. Calculate the work done on the right part in this process and the final temperature in the right part.

  b. Calculate the final temperature in the left part and the amount of heat that flowed into this part.

P2.30   A vessel containing 1.00 mol of an ideal gas with P_i = 1.00 bar and $C_{P,m} = 5/2R$ is in thermal contact with a water bath. Treat the vessel, gas, and water bath as being in thermal equilibrium, initially at 298 K, and as separated by adiabatic walls from the rest of the universe. The vessel, gas, and water bath have an average heat capacity of C_P = 7500. J K^{-1}. The gas is compressed reversibly to P_f = 10.5 bar. What is the temperature of the system after thermal equilibrium has been established?

P2.31   DNA can be modeled as an elastic rod that can be twisted or bent. Suppose a DNA molecule of length L is bent such that it lies on the arc of a circle of radius R_c. The reversible work involved in bending DNA without twisting is $w_{bend} = \dfrac{BL}{2R_c^2}$ where B is the bending force constant. The DNA in a nucleosome particle is about 680 Å in length. Nucleosomal DNA is bent around a protein complex called the histone octamer into a circle of radius 55 Å. Calculate the reversible work involved in bending the DNA around the histone octamer if the force constant $B = 2.00 \times 10^{-28}$ J m^{-1}.

P2.32   Compare the energy of DNA bending calculated in Problem 2.31 to the thermal energy $k_B T$, where k_B is Boltzmann's constant. Assume T = 310. K. Propose a source for the excess energy required to bend the DNA in Problem 2.31.

P2.33   The reversible work involved in twisting a short DNA molecule depends quadratically on ϕ/L, the angle of twist per unit length, where the twist angle is expressed in units of radians. The expression for the reversible work is: $w_{twist} = \dfrac{CL}{2}\left(\dfrac{\phi}{L}\right)^2$. A DNA oligomer 20 base pairs in length undergoes a twisting deformation of 36 degrees in order to bind to a protein. Calculate the reversible work involved in this twisting deformation. Assume $C = 2.5 \times 10^{-28}$ J m. Assume also that each base pair is 3.4 Å in length (1 Å $= 10^{-10}$ m).

P2.34 The formalism of Young's modulus is sometimes used to calculate the reversible work involved in extending or compressing an elastic material. Assume a force F is applied to an elastic rod of cross-sectional area A_0 and length L_0. As a result of this force, the rod changes in length by ΔL. Young's modulus E is defined as

$$E = \frac{\text{tensile stress}}{\text{tensile strain}} = \frac{F/A_0}{\Delta L/L_0} = \frac{FL_0}{A_0 \Delta L}$$

a. Derive Hooke's law from the Young's modulus expression just given.

b. Using your result from part (a), show that the reversible work involved in changing by ΔL the length L_0 of an elastic cylinder of cross-sectional area A_0

is $w = \dfrac{1}{2}\left(\dfrac{\Delta L}{L_0}\right)^2 EA_0 L_0$.

P2.35 The heat capacity of solid lead oxide is given by

$$C_{P,m} = 44.35 + 1.47 \times 10^{-3}\, \frac{T}{K} \quad \text{in units of J K}^{-1}\,\text{mol}^{-1}$$

Calculate the change in enthalpy of 1 mol of PbO(s) if it is cooled from 500 to 300 K at constant pressure.

P2.36 Consider the adiabatic expansion of 0.500 mol of an ideal monatomic gas with $C_{V,m} = 3/2R$. The initial state is described by $P = 3.25$ bar and $T = 300.$ K.

a. Calculate the final temperature if the gas undergoes a reversible adiabatic expansion to a final pressure of $P = 1.00$ bar.

b. Calculate the final temperature if the same gas undergoes an adiabatic expansion against an external pressure of $P = 1.00$ bar to a final pressure of $P = 1.00$ bar.

c. Explain the difference in your results for parts (a) and (b).

P2.37 The relationship between Young's modulus and the bending force constant for a deformable cylinder is $B = EI$, where, $I = \pi R^4/4$ and R is the radius of the cylinder.

a. Calculate the Young's modulus associated with a DNA of radius 10. Å (1 Å $= 10^{-10}$ m). Assume the value of B given in Problem 2.31.

b. Suppose a DNA molecule 100 base pairs in length is extended by 10. Å. Calculate the reversible work assuming the DNA can be treated as a deformable rod.

c. Compare this work to the thermal energy. Assume $T = 310.$ K.

P2.38 The Young's modulus of muscle fiber is approximately 3.12×10^7 Pa. If a muscle fiber 2.00 cm in length and 0.100 cm in diameter is suspended with a weight M hanging at its end, calculate the weight M required to extend the length of the fiber by 10%.

P2.39 An ideal gas undergoes a single-stage expansion against a constant external pressure $P_{external}$ at constant temperature from T, P_i, V_i to T, P_f, V_f.

a. What is the largest mass m that can be lifted through the height h in this expansion?

b. The system is restored to its initial state in a single-state compression. What is the smallest mass m that must fall through the height h to restore the system to its initial state?

c. If $h = 10.0$ cm, $P_i = 1.00 \times 10^6$ Pa, $P_f = 0.500 \times 10^6$ Pa, $T = 300.$ K, and $n = 1.00$ mol, calculate the values of the masses in parts (a) and (b).

P2.40 Calculate q, w, ΔU, and ΔH if 1.00 mol of an ideal gas with $C_{V,m} = 3/2R$ undergoes a reversible adiabatic expansion from an initial volume $V_i = 5.25$ m^3 to a final volume $V_f = 25.5$ m^3. The initial temperature is 300. K.

P2.41 A nearly flat bicycle tire becomes noticeably warmer after it has been pumped up. Approximate this process as a reversible adiabatic compression. Assume the initial pressure and temperature of the air before it is put in the tire to be $P_i = 1.00$ bar and $T_i = 298$ K, respectively. The final volume of the air in the tire is $V_f = 1.00$ L and the final pressure is $P_f = 5.00$ bar. Calculate the final temperature of the air in the tire. Assume that $C_{V,m} = 5/2R$

P2.42 One mole of an ideal gas with $C_{V,m} = 3/2R$ is expanded adiabatically against a constant external pressure of 1.00 bar. The initial temperature and pressure are $T_i = 300.$ K and $P_i = 25.0$ bar, respectively. The final pressure is $P_f = 1.00$ bar. Calculate q, w, ΔU, and ΔH for the process.

P2.43 One mole of N_2 in a state defined by $T_i = 300.$ K and $V_i = 2.50$ L undergoes an isothermal reversible expansion until $V_f = 23.0$ L. Calculate w assuming (a) that the gas is described by the ideal gas law and (b) that the gas is described by the van der Waals equation of state. What is the percent error in using the ideal gas law instead of the van der Waals equation? The van der Waals parameters for N_2 are listed in Table 1.3.

P2.44 One mole of an ideal gas, for which $C_{V,m} = 3/2R$, is subjected to two successive changes in state: (1) From 25.0°C and 100. $\times 10^3$ Pa, the gas is expanded isothermally against a constant pressure of 20.0 $\times 10^3$ Pa to twice the initial volume. (2) At the end of the previous process, the gas is cooled at constant volume from 25.0° to -25.0°C. Calculate q, w, ΔU, and ΔH for each of the stages. Also calculate q, w, ΔU, and ΔH for the complete process.

P2.45 The adhesion of leukocytes (white blood cells) to target cells is a crucial aspect of the body's immune system. The force required to detach leukocytes from a substrate can be measured with an atomic force microscope (AFM). The work of de-adhesion is proportional to the cell–substrate contact area A_c, which in turn is related to Young's modulus, the force F applied to the cell by the AFM, and the radius R of the cell by $A_c = \pi \times \left(\dfrac{RF}{E}\right)^{2/3}$. Assuming $R = 5.00 \times 10^{-6}$ m and $F = 2.00 \times 10^{-7}$ N, calculate the change in contact area when Young's modulus is reduced from 1.40 to 0.30 kPa. By what amount does the work of de-adhesion change when Young's modulus is changed from 1.40 to 0.30 kPa? Explain this effect.

Web-Based Simulations, Animations, and Problems

W2.1   A simulation is carried out in which an ideal gas is heated under constant pressure or constant volume conditions. The quantities ΔV (or ΔP), w, ΔU, and ΔT are determined as a function of the heat input. The heat taken up by the gas under constant P or V is calculated and compared with ΔU and ΔH.

W2.2   The reversible isothermal compression and expansion of an ideal gas is simulated for different values of T. The work, w, is calculated from the T and V values obtained in the simulation. The heat, q, and the number of moles of gas in the system are calculated from the results.

W2.3   The reversible isobaric compression and expansion of an ideal gas is simulated for different values of pressure as heat flows to/from the surroundings. The quantities q, w, and ΔU are calculated from the ΔT and ΔV values obtained in the simulation.

W2.4   The isochoric heating and cooling of an ideal gas are simulated for different values of volume. The number of moles of gas and ΔU are calculated from the constant V value and from the T and P values obtained in the simulation.

W2.5   Reversible cyclic processes are simulated in which the cycle is either rectangular or triangular on a P–V plot. For each segment and for the cycle, ΔU, q, and w are determined. For a given cycle type, the ratio of work done on the surroundings to the heat absorbed from the surroundings is determined for different P and V values.

W2.6   The reversible adiabatic heating and cooling of an ideal gas are simulated for different values of the initial temperature. The quantities $\gamma = C_{P,m}/C_{V,m}$ and $C_{P,m}$ and $C_{V,m}$ are determined from the P, V values of the simulation, and ΔU and ΔH are calculated from the V, T, and P values obtained in the simulation.

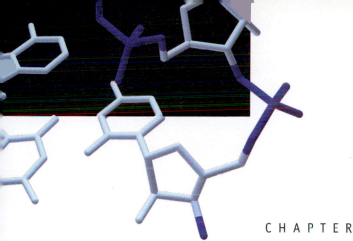

3

The Importance of State Functions: Internal Energy and Enthalpy

The mathematical properties of state functions are utilized to express the infinitesimal quantities dU and dH as exact differentials. By doing so, expressions can be derived that relate the change of U with T and V and the change in H with T and P to experimentally accessible quantities such as the heat capacity and the coefficient of thermal expansion. For most substances, the dependence of U and H on T is much greater than the dependence on P and V, so that it is generally sufficient to regard U and H as functions of T only. An important exception is the isenthalpic expansion of real gases, which is commercially used in the liquefaction of gases such as N_2, O_2, He, and Ar.

3.1 The Mathematical Properties of State Functions

In Chapter 2 we demonstrated that U and H are state functions and that w and q are path functions. We also discussed how to calculate changes in these quantities for an ideal gas. In this chapter, the path independence of state functions is exploited to derive relationships with which ΔU and ΔH can be calculated as functions of P, V, and T for real gases, liquids, and solids. In doing so, we develop the formal aspects of thermodynamics and show that the formal structure of thermodynamics provides a powerful aid in linking theory and experiment. However, before these topics are discussed, the mathematical properties of state functions need to be outlined.

The thermodynamic state functions of interest here are defined by two variables from the set P, V, and T. In formulating changes in state functions, we will make extensive use of partial derivatives, which are reviewed in the Math Supplement (Appendix A). The following discussion does not apply to path functions such as w and q because a functional relationship such as Equation (3.1) does not exist for path-dependent functions. Consider 1 mol of an ideal gas for which

$$P = f(V,T) = \frac{RT}{V} \tag{3.1}$$

Note that P can be written as a function of the two variables V and T. The change in P resulting from a change in V or T is proportional to the following **partial derivatives:**

$$\left(\frac{\partial P}{\partial V}\right)_T = \lim_{\Delta V \to 0} \frac{P(V + \Delta V, T) - P(V,T)}{\Delta V} = -\frac{RT}{V^2}$$

$$\left(\frac{\partial P}{\partial T}\right)_V = \lim_{\Delta T \to 0} \frac{P(V,T + \Delta T) - P(V,T)}{\Delta T} = \frac{R}{V} \tag{3.2}$$

The subscript T in $(\partial P/\partial V)_T$ indicates that T is being held constant in the differentiation with respect to V. The partial derivatives in Equation (3.2) allow one to determine how a function changes when the variables change. For example, what is the change in P if the values of T and V both change? In this case, P changes to $P + dP$ where

$$dP = \left(\frac{\partial P}{\partial T}\right)_V dT + \left(\frac{\partial P}{\partial V}\right)_T dV \tag{3.3}$$

Consider the following practical illustration of Equation (3.3). You are on a hill and have determined your altitude above sea level. How much will the altitude (denoted z) change if you move a small distance east (denoted by x) and north (denoted by y)? The change in z as you move east is the slope of the hill in that direction, $(\partial z/\partial x)_y$, multiplied by the distance that you move. A similar expression can be written for the change in altitude as you move north. Therefore, the total change in altitude is the sum of these two changes or

$$dz = \left(\frac{\partial z}{\partial x}\right)_y dx + \left(\frac{\partial z}{\partial y}\right)_x dy$$

These changes in the height z as you move first along the x direction and then along the y direction are illustrated in Figure 3.1. Because the slope of the hill is a function of x and y, this expression for dz is only valid for small changes in dx and dy. Otherwise, higher order derivatives need to be considered.

Second or higher derivatives with respect to either variable can also be taken. The mixed second partial derivatives are of particular interest. Consider the mixed partial derivatives of P:

$$\left(\frac{\partial}{\partial T}\left(\frac{\partial P}{\partial V}\right)_T\right)_V = \frac{\partial^2 P}{\partial T \partial V} = \left(\partial\left[\frac{\partial\left[\frac{RT}{V}\right]}{\partial V}\right]_T\middle/ \partial T\right)_V = \left(\partial\left[-\frac{RT}{V^2}\right]\middle/ \partial T\right)_V = -\frac{R}{V^2}$$

$$\left(\frac{\partial}{\partial V}\left(\frac{\partial P}{\partial T}\right)_V\right)_T = \frac{\partial^2 P}{\partial T \partial V} = \left(\partial\left[\frac{\partial\left[\frac{RT}{V}\right]}{\partial T}\right]_V\middle/ \partial V\right)_T = \left(\partial\left[\frac{R}{V}\right]\middle/ \partial V\right)_T = -\frac{R}{V^2} \tag{3.4}$$

For all state functions f and for our specific case of P, the order in which the function is differentiated does not affect the outcome. For this reason,

$$\left(\frac{\partial}{\partial T}\left(\frac{\partial f(V,T)}{\partial V}\right)_T\right)_V = \left(\frac{\partial}{\partial V}\left(\frac{\partial f(V,T)}{\partial T}\right)_V\right)_T \tag{3.5}$$

Because Equation (3.5) is only satisfied by state functions f, it can be used to determine if a function f is a state function. If f is a state function, one can write $\Delta f = \int_i^f df = f_{final} - f_{initial}$. This equation states that f can be expressed as an infinitesimal quantity, df, that when integrated depends only on the initial and final states; df is called an **exact differential**. An example of a state function and its exact differential is U and $dU = \mathchar'26\mkern-12mu dq - P_{external}\, dV$.

Two other important results from differential calculus will be used frequently. Consider a function, $z = f(x,y)$, which can be rearranged to $x = g(y,z)$ or $y = h(x,z)$. For example, if $P = nRT/V$, then $V = nRT/P$ and $T = PV/nR$. In this case

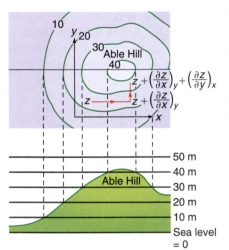

FIGURE 3.1
Side (bottom panel) and on top (top panel) views of a hill are shown. The top view includes topographical contours. Starting at the point labeled z on the hill, you first move in the positive x direction and then along the y direction. If dx and dy are sufficiently small, the change in height dz is given by $dz = \left(\frac{\partial z}{\partial x}\right)_y dx + \left(\frac{\partial z}{\partial y}\right)_x dy$.

$$\left(\frac{\partial x}{\partial y}\right)_z = \frac{1}{\left(\dfrac{\partial y}{\partial x}\right)_z} \tag{3.6}$$

The **cyclic rule** will also be used:

$$\left(\frac{\partial x}{\partial y}\right)_z \left(\frac{\partial y}{\partial z}\right)_x \left(\frac{\partial z}{\partial x}\right)_y = -1 \tag{3.7}$$

Equations (3.6) and (3.7) can be used to reformulate Equation (3.3):

$$dP = \left(\frac{\partial P}{\partial T}\right)_V dT + \left(\frac{\partial P}{\partial V}\right)_T dV$$

Suppose this expression needs to be evaluated for a specific substance, such as N_2 gas. What quantities must be measured in the laboratory to obtain numerical values for $(\partial P/\partial T)_V$ and $(\partial P/\partial V)_T$? Using Equations (3.6) and (3.7),

$$\left(\frac{\partial P}{\partial V}\right)_T \left(\frac{\partial V}{\partial T}\right)_P \left(\frac{\partial T}{\partial P}\right)_V = -1$$

$$\left(\frac{\partial P}{\partial T}\right)_V = -\left(\frac{\partial P}{\partial V}\right)_T \left(\frac{\partial V}{\partial T}\right)_P = -\frac{\left(\dfrac{\partial V}{\partial T}\right)_P}{\left(\dfrac{\partial V}{\partial P}\right)_T} = \frac{\beta}{\kappa} \quad \text{and}$$

$$\left(\frac{\partial P}{\partial V}\right)_T = -\frac{1}{\kappa V} \tag{3.8}$$

where β and κ are the readily measured **volumetric thermal expansion coefficient** and the **isothermal compressibility**, respectively, defined by

$$\beta = \frac{1}{V}\left(\frac{\partial V}{\partial T}\right)_P \quad \text{and} \quad \kappa = -\frac{1}{V}\left(\frac{\partial V}{\partial P}\right)_T \tag{3.9}$$

Both $(\partial V/\partial T)_P$ and $(\partial V/\partial P)_T$ can be measured by determining the change in volume of the system when the pressure or temperature is varied, while keeping the second variable constant.

The minus sign in the equation for κ is chosen so that values of the isothermal compressibility are positive. For small changes in T and P, the differentials dT and dP can be replaced by the differences, ΔT and ΔP. In this case, Equations such as (3.9) can be written in the more compact form $V(T_2) = V(T_1)(1 + \beta[T_2 - T_1])$ and $V(P_2) = V(P_1)(1 - \kappa[P_2 - P_1])$. Values for β and κ for selected solids and liquids are shown in Tables 3.1 and 3.2, respectively. Note that β is generally much larger for solids than for liquids.

Equation (3.8) is an example of how seemingly abstract partial derivatives can be directly linked to experimentally determined quantities using the mathematical properties of state functions. Using the definitions of β and κ, Equation (3.3) can be written in the form

$$dp = \frac{\beta}{\kappa} dT - \frac{1}{\kappa V} dV \tag{3.10}$$

which can be integrated to give

$$\Delta P = \int_{T_i}^{T_f} \frac{\beta}{\kappa} dT - \int_{V_i}^{V_f} \frac{1}{\kappa V} dV \approx \frac{\beta}{\kappa}(T_f - T_i) - \frac{1}{\kappa}\ln\frac{V_f}{V_i} \tag{3.11}$$

TABLE 3.1   Volumetric Thermal Expansion Coefficient for Solids and Liquids at 298 K

Element	$10^6 \beta$ (K^{-1})	Element or Compound	$10^4 \beta$ (K^{-1})
Ag(s)	57.6	Hg(l)	1.81
Al(s)	69.3	CCl_4(l)	11.4
Au(s)	42.6	CH_3COCH_3(l)	14.6
Cu(s)	49.5	CH_3OH(l)	14.9
Fe(s)	36.9	C_2H_5OH(l)	11.2
Mg(s)	78.3	$C_6H_5CH_3$(l)	10.5
Si(s)	7.5	C_6H_6(l)	11.4
W(s)	13.8	H_2O(l)	2.04
Zn(s)	90.6	H_2O(s)	1.66

Sources: W. Benenson, J. W. Harris, H. Stocker, and H. Lutz, *Handbook of Physics,* Springer, New York, 2002; D. R. Lide, Ed., *Handbook of Chemistry and Physics,* 83rd ed., CRC Press, Boca Raton, FL, 2002; R. Blachnik, Ed., *D'Ans Lax Taschenbuch für Chemiker und Physiker,* 4th ed., Springer, Berlin, 1998.

TABLE 3.2   Isothermal Compressibility at 298 K

Substance	$10^6 \kappa$/bar^{-1}	Substance	$10^6 \kappa$/bar^{-1}
Al(s)	1.33	Br_2(l)	64
SiO_2(s)	2.57	C_2H_5OH(l)	110
Ni(s)	0.513	C_6H_5OH(l)	61
TiO_2(s)	0.56	C_6H_6(l)	94
Na(s)	13.4	CCl_4(l)	103
Cu(s)	0.702	CH_3COCH_3(l)	126
C(graphite)	0.156	CH_3OH(l)	120
Mn(s)	0.716	CS_2(l)	92.7
Co(s)	0.525	H_2O(l)	45.9
Au(s)	0.563	Hg(l)	3.91
Pb(s)	2.37	$SiCl_4$(l)	165
Fe(s)	0.56	$TiCl_4$(l)	89
Ge(s)	1.38		

Sources: W. Benenson, J. W. Harris, H. Stocker, and H. Lutz, *Handbook of Physics,* Springer, New York, 2002; D. R. Lide, Ed., *Handbook of Chemistry and Physics,* 83rd ed., CRC Press, Boca Raton, FL, 2002; R. Blachnik, Ed., *D'Ans Lax Taschenbuch für Chemiker und Physiker,* 4th ed., Springer, Berlin, 1998.

The second expression in Equation (3.11) holds if ΔT and ΔV are small enough that β and κ are constant over the range of integration. Example Problem 3.1 shows a useful application of this equation.

EXAMPLE PROBLEM 3.1

In conducting an experiment, you have inadvertently arrived at the end of the range of an ethanol-in-glass thermometer so that the entire volume of the glass capillary is filled. By how much will the pressure in the glass capillary increase if the temperature is increased by another 10.0°C? $\beta_{glass} = 2.00 \times 10^{-5}$ (°C)$^{-1}$, $\beta_{ethanol} = 11.2 \times 10^{-4}$(°C)$^{-1}$, and $\kappa_{ethanol} = 11.0 \times 10^{-5}(bar)^{-1}$. Do you think that the thermometer will survive your experiment?

Solution

Using Equation (3.11),

$$\Delta P = \int \frac{\beta_{ethanol}}{\kappa} dT - \int \frac{1}{\kappa V} dV \approx \frac{\beta_{ethanol}}{\kappa} \Delta T - \frac{1}{\kappa} \ln \frac{V_f}{V_i}$$

$$= \frac{\beta_{ethanol}}{\kappa} \Delta T - \frac{1}{\kappa} \ln \frac{V_i (1 + \beta_{glass} \Delta T)}{V_i} \approx \frac{\beta_{ethanol}}{\kappa} \Delta T - \frac{1}{\kappa} \frac{V_i \beta_{glass} \Delta T}{V_i}$$

$$= \frac{(\beta_{ethanol} - \beta_{glass})}{\kappa} \Delta T$$

$$= \frac{(11.2 - 0.200) \times 10^{-4}(°C)^{-1}}{11.0 \times 10^{-5}(bar)^{-1}} \times 10.0 \,°C = 100. \, bar$$

In this calculation, we have used the relations $V(T_2) = V(T_1)(1 + \beta[T_2 - T_1])$ and $\ln(1 + x) \approx x$ if $x \ll 1$. The glass is unlikely to withstand such a large increase in pressure.

3.2 The Dependence of *U* on *V* and *T*

In this section, the fact that dU is an exact differential is used to establish how U varies with T and V. For a given amount of a pure substance or a mixture of fixed composition, U is determined by any two of the three variables P, V, and T. One could choose other combinations of variables to discuss changes in U. However, the following discussion will demonstrate that it is particularly convenient to choose the variables T and V. Because U is a state function, an infinitesimal change in U can be written as

$$dU = \left(\frac{\partial U}{\partial T}\right)_V dT + \left(\frac{\partial U}{\partial V}\right)_T dV \tag{3.12}$$

This expression says that if the state variables change from T,V to $T + dT, V + dV$, the change in U, dU can be determined in the following way. We determine the slopes of $U(T,V)$ with respect to T and V and evaluate them at T,V. Next, these slopes are multiplied by the increments dT and dV, respectively, and the two terms are added. As long as dT and dV are infinitesimal quantities, higher order derivatives can be neglected.

How can numerical values for $(\partial U/\partial T)_V$ and $(\partial U/\partial V)_T$ be obtained? In the following, we only consider $P-V$ work. Combining Equation (3.12) and the differential expression of the first law,

$$đq - P_{external} dV = \left(\frac{\partial U}{\partial T}\right)_V dT + \left(\frac{\partial U}{\partial V}\right)_T dV \tag{3.13}$$

The symbol $đq$ is used for an infinitesimal amount of heat as a reminder that heat is not a state function. We first consider processes at constant volume for which $dV = 0$, so that Equation (3.13) becomes

$$đq_V = \left(\frac{\partial U}{\partial T}\right)_V dT \tag{3.14}$$

Note that in the previous equation, $đq_V$ is the product of a state function and an exact differential. Therefore, $đq_V$ behaves like a state function, but only because the path (constant V) is specified. The quantity $đq$ is *not* a state function.

Although the quantity $(\partial U/\partial T)_V$ looks very abstract, it can be readily measured. For example, imagine immersing a container with rigid diathermal walls in a water bath,

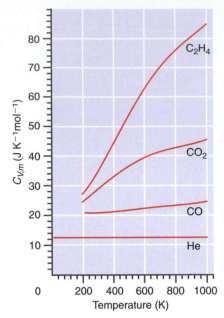

FIGURE 3.2

Molar heat capacities $C_{V,m}$ are shown for a number of gases. Atoms have only translational degrees of freedom and, therefore, have comparatively low values for $C_{V,m}$ that are independent of temperature. Molecules with vibrational degrees of freedom have higher values of $C_{V,m}$ at temperatures sufficiently high to activate the vibrations.

where the contents of the container are the system. A process such as a chemical reaction is carried out in the container and the heat flow to the surroundings is measured. If heat flow, dq_V, occurs, a temperature increase or decrease, dT, is observed in the system and the water bath surroundings. Both of these quantities can be measured. Their ratio, dq_V / dT, is a special form of the heat capacity discussed in Section 2.4:

$$\frac{dq_V}{dT} = \left(\frac{\partial U}{\partial T} \right)_V = C_V \tag{3.15}$$

where dq_V / dT corresponds to a constant volume path and is called the **heat capacity at constant volume.**

The quantity C_V is extensive and depends on the size of the system, whereas $C_{V,m} = C_V/n$ is an intensive quantity. Experiments show that $C_{V,m}$ has different numerical values for different substances under the same conditions. Observations show that $C_{V,m}$ is always positive for a single-phase, pure substance or for a mixture of fixed composition, as long as no chemical reactions or phase changes take place in the system. For processes subject to these constraints, U increases monotonically with T. Figure 3.2 shows $C_{V,m}$ as a function of temperature for several gaseous atoms and molecules.

Before continuing with our discussion about how to experimentally determine changes in U with T at constant V for systems of pure substances or for mixtures, let's study heat capacities in a little more detail by looking at them in terms of a microscopic model.

In thermodynamics, the origin for the substance dependence of $C_{V,m}$ is not a matter of inquiry, because thermodynamics is not concerned with the microscopic structure of the system. To obtain numerical results in thermodynamics, system-dependent properties such as $C_{V,m}$ are obtained from experiment or theory. A microscopic model is required to explain why $C_{V,m}$ for a particular substance has its measured value. For example, why is $C_{V,m}$ smaller for gaseous He than for gaseous methanol? To increase the temperature by an amount dT for a system containing helium, the translational energy of the atoms is increased. By contrast, to arrive at the same increase dT in a system containing methanol, the rotational, vibrational, and translational energies of the molecules are all increased simultaneously, because all of these degrees of freedom have the same temperature. Therefore, for a given temperature increment dT, more energy must be added to a mole of methanol than to a mole of He. For this reason, $C_{V,m}$ is larger for gaseous methanol than for gaseous He.

The preceding explanation is valid for high T, but at low T, additional considerations apply. Far less energy is required to excite molecular rotations than molecular vibrations. Therefore, the rotational degrees of freedom contribute to $C_{V,m}$ for molecules even at low temperatures, making $C_{V,m}$ greater for CO than for He at 300 K. However, vibrational degrees of freedom are only excited at higher T. This leads to the additional gradual increase in $C_{V,m}$ observed for CO and CO_2 relative to He for $T > 300$ K. Because the number of vibrational degrees of freedom increases with the number of atoms in the molecule, $C_{V,m}$ is larger for a polyatomic molecule such as C_2H_4 than for CO and CO_2 at a given temperature.

With the definition of C_V, we now have a way to experimentally determine changes in U with T at constant V for systems of pure substances or for mixtures of constant composition in the absence of chemical reactions or phase changes. After C_V has been determined as a function of T as discussed in Section 2.4, the integral is numerically evaluated:

$$\Delta U_V = \int_{T_1}^{T_2} C_V \, dT = n \int_{T_1}^{T_2} C_{V,m} \, dT \tag{3.16}$$

Over a limited temperature range, $C_{V,m}$ can often be regarded as a constant. If this is the case, Equation (3.16) simplifies to

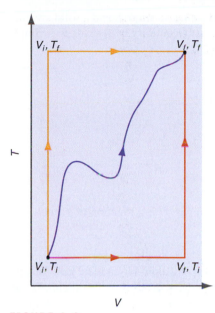

FIGURE 3.3
Because U is a state function, all paths connecting V_i, T_i and V_f, T_f are equally valid in calculating ΔU. Therefore, a specification of the path is irrelevant.

$$\Delta U_V = \int_{T_1}^{T_2} C_V \, dT = C_V \Delta T = nC_{V,m}\Delta T \qquad (3.17)$$

which can be written in a different form to explicitly relate q_V and ΔU:

$$\int_i^f \mathit{d}q_V = \int_i^f \left(\frac{\partial U}{\partial T}\right)_V dT \qquad \text{or} \qquad q_V = \Delta U \qquad (3.18)$$

Although $\mathit{d}q$ is not an exact differential, the integral has a unique value if the path is defined, as it is in this case (constant volume). Equation (3.18) shows that ΔU for an arbitrary process in a closed system in which only $P-V$ work occurs can be determined by measuring q under constant volume conditions. As will be discussed in Chapter 4, the technique of bomb calorimetry uses this approach to determine ΔU for chemical reactions.

Next consider the dependence of U on V at constant T, or $(\partial U / \partial V)_T$. This quantity has the units of J/m^3 = (J/m)/m^2 = kg m s^{-2}/m^2 = force/area = pressure, and is called the **internal pressure.** To explicitly evaluate the internal pressure for different substances, a result will be used that is derived in the discussion of the second law of thermodynamics in Section 5.3

$$\left(\frac{\partial U}{\partial V}\right)_T = T\left(\frac{\partial P}{\partial T}\right)_V - P \qquad (3.19)$$

Using this equation, the total differential of the internal energy can be written as

$$dU = dU_V + dU_T = C_V \, dT + \left[T\left(\frac{\partial P}{\partial T}\right)_V - P\right]dV \qquad (3.20)$$

In this equation, the symbols dU_V and dU_T have been used, where the subscript indicates which variable is constant. Equation (3.20) is an important result that applies to systems containing gases, liquids, or solids in a single phase (or mixed phases at a constant composition) if no chemical reactions or phase changes occur. The advantage of writing dU in the form given by Equation (3.20) over that in Equation (3.12) is that $(\partial U / \partial V)_T$ can be evaluated in terms of the system variables P, V, and T and their derivatives, all of which are experimentally accessible.

Once $(\partial U / \partial V)_T$ and $(\partial U / \partial T)_V$ are known, these quantities can be used to determine dU. Because U is a state function, the path taken between the initial and final states is not important. Three different paths are shown in Figure 3.3, and dU is the same for these and any other paths connecting V_i, T_i and V_f, T_f. To simplify the calculation, the path chosen consists of two segments, in which only one of the variables changes in a given path segment. An example of such a path is $V_i, T_i \rightarrow V_f, T_i \rightarrow V_f, T_f$. Because T is constant in the first segment,

$$dU = dU_T = \left[T\left(\frac{\partial P}{\partial T}\right)_V - P\right]dV$$

Because V is constant in the second segment, $dU = dU_V = C_V \, dT$. Finally, the total change in dU is the sum of the changes in the two segments.

3.3 Does the Internal Energy Depend More Strongly on *V* or *T*?

Chapter 2 demonstrated that U is a function of T alone for an ideal gas. However, this statement is not true for real gases, liquids, and solids for which the change in U with V must be considered. In this section, we ask if the temperature or the volume dependence of U is most important in determining ΔU for a process of interest. To answer

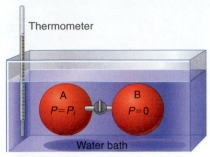

FIGURE 3.4
Schematic depiction of the Joule experiment to determine $(\partial U/\partial V)_T$. Two spherical vessels, A and B, are separated by a valve. Both vessels are immersed in a water bath, the temperature of which is monitored. The initial pressure in each vessel is indicated.

this question, systems consisting of an ideal gas, a real gas, a liquid, and a solid are considered separately. Example Problem 3.2 shows that Equation (3.19) leads to a simple result for a system consisting of an ideal gas.

EXAMPLE PROBLEM 3.2

Evaluate $(\partial U/\partial V)_T$ for an ideal gas and modify Equation (3.20) accordingly for the specific case of an ideal gas.

Solution

$$T\left(\frac{\partial P}{\partial T}\right)_V - P = T\left(\frac{\partial[nRT/V]}{\partial T}\right)_V - P = \frac{nRT}{V} - P = 0$$

Therefore, $dU = C_V\,dT$, showing that for an ideal gas, U is a function of T only.

Example Problem 3.2 shows that U is a function of T only for an ideal gas. Specifically, U is not a function of V. This result is understandable in terms of the potential function of Figure 1.7. Because ideal gas molecules do not attract or repel one another, no energy is required to change their average distance of separation (that is, to increase or decrease V):

$$\Delta U = \int_{T_i}^{T_f} C_V(T)dT \qquad (3.21)$$

Recall that because U is a function of T only, Equation (3.21) holds for an ideal gas even if V is not constant.

Next consider the variation of U with T and V for a real gas. The experimental determination of $(\partial U/\partial V)_T$ was carried out by James Joule using an apparatus consisting of two glass flasks separated by a stopcock, all of which were immersed in a water bath. An idealized view of the experiment is shown in Figure 3.4. As a valve between the volumes is opened, a gas initially in volume A expands to completely fill the volume A + B. In interpreting the results of this experiment, it is important to understand where the boundary between the system and surroundings lies. Here, the decision was made to place the system boundary so that it includes all of the gas. Initially, the boundary lies totally within V_A, but it moves during the expansion so that it continues to include all gas molecules. With this choice, the volume of the system changes from V_A before the expansion to $V_A + V_B$ after the expansion has taken place.

The first law of thermodynamics [Equation (3.13)] states that

$$đq - P_{external}\,dV = \left(\frac{\partial U}{\partial T}\right)_V dT + \left(\frac{\partial U}{\partial V}\right)_T dV$$

However, all of the gas is contained in the system; therefore, $P_{external} = 0$ because a vacuum cannot exert a pressure. Therefore Equation (3.13) becomes

$$đq = \left(\frac{\partial U}{\partial T}\right)_V dT + \left(\frac{\partial U}{\partial V}\right)_T dV \qquad (3.22)$$

To within experimental accuracy, Joule found that $dT_{surroundings} = 0$. Because the water bath and the system are in thermal equilibrium, $dT = dT_{surroundings} = 0$. With this observation, Joule concluded that $đq = 0$. Therefore, Equation (3.22) becomes

$$\left(\frac{\partial U}{\partial V}\right)_T dV = 0 \qquad (3.23)$$

Because $dV \neq 0$, Joule concluded that $(\partial U/\partial V)_T = 0$. Joule's experiment was not definitive because the experimental sensitivity was limited, as shown in Example Problem 3.3.

EXAMPLE PROBLEM 3.3

In Joule's experiment to determine $(\partial U/\partial V)_T$, the heat capacities of the gas and the water bath surroundings were related by $C_{surroundings}/C_{system} \approx 1000$. If the precision with which the temperature of the surroundings could be measured is $\pm 0.006°C$, what is the minimum detectable change in the temperature of the gas?

Solution

View the experimental apparatus of Figure 3.4 as two interacting systems in a rigid adiabatic enclosure. The first is the volume within vessels A and B, and the second is the water bath and the vessels. Because the two interacting systems are isolated from the rest of the universe,

$$q = C_{water\ bath}\ \Delta T_{water\ bath} + C_{gas}\ \Delta T_{gas} = 0$$

$$\Delta T_{gas} = -\frac{C_{water\ bath}}{C_{gas}}\ \Delta T_{water\ bath} = -1000. \times (\pm 0.006°C) = \mp 6°C$$

In this calculation, ΔT_{gas} is the temperature change that the expanded gas undergoes to reach thermal equilibrium with the water bath, which is the negative of the temperature change during the expansion.

Because the minimum detectable value of ΔT_{gas} is rather large, this apparatus is clearly not suited for measuring small changes in the temperature of the gas in an expansion.

More sensitive experiments that were carried out by Joule in collaboration with William Thomson (Lord Kelvin) demonstrate that $(\partial U/\partial V)_T$ is small, but nonzero for real gases.

Example Problem 3.2 has shown that $(\partial U/\partial V)_T = 0$ for an ideal gas. We next calculate $(\partial U/\partial V)_T$ and $\Delta U_T = \int_{V_{m,i}}^{V_{m,f}} (\partial U/\partial V)_T\ dV_m$ for a real gas, in which the van der Waals equation of state is used to describe the gas, as illustrated in Example Problem 3.4.

EXAMPLE PROBLEM 3.4

For a gas described by the van der Waals equation of state, $P = RT/(V_m - b) - a/V_m^2$. Use this equation to complete these tasks:

a. Calculate $(\partial U/\partial V)_T$ using $(\partial U/\partial V)_T = T(\partial P/\partial T)_V - P$.

b. Derive an expression for the change in internal energy, $\Delta U_T = \int_{V_{m,i}}^{V_{m,f}} (\partial U/\partial V)_T\ dV_m$, in compressing a van der Waals gas from an initial molar volume $V_{m,i}$ to a final molar volume $V_{m,f}$ at constant temperature.

Solution

a. $T\left(\dfrac{\partial P}{\partial T}\right)_V - P = T\left(\dfrac{\partial\left[\dfrac{RT}{V_m - b} - \dfrac{a}{V_m^2}\right]}{\partial T}\right)_V - P = \dfrac{RT}{V_m - b} - P$

$\quad = \dfrac{RT}{V_m - b} - \dfrac{RT}{V_m - b} + \dfrac{a}{V_m^2} = \dfrac{a}{V_m^2}$

b. $\Delta U_{T,m} = \displaystyle\int_{V_{m,i}}^{V_{m,f}} \left(\dfrac{\partial U_m}{\partial V}\right)_T dV_m = \displaystyle\int_{V_{m,i}}^{V_{m,f}} \dfrac{a}{V_m^2}\ dV_m = a\left(\dfrac{1}{V_{m,i}} - \dfrac{1}{V_{m,f}}\right)$

Note that $\Delta U_{T,m}$ is zero if the attractive part of the intermolecular potential is zero.

Example Problem 3.4 demonstrates that $(\partial U/\partial V)_T \neq 0$ for a real gas, and that $\Delta U_{T,m}$ can be calculated if the equation of state of the real gas is known. This allows the relative importance of $\Delta U_{T,m} = \int_{V_{m,i}}^{V_{m,f}} (\partial U_m/\partial V)_T dV_m$ and $\Delta U_{V,m} = \int_{T_i}^{T_f} C_{V,m} dT$ to be determined in a process in which both T and V change, as shown in Example Problem 3.5.

EXAMPLE PROBLEM 3.5

A sample of N_2 gas undergoes a change from an initial state described by $T = 200.$ K and $P_i = 5.00$ bar to a final state described by $T = 400.$ K and $P_f = 20.0$ bar. Treat N_2 as a van der Waals gas with the parameters $a = 0.137$ Pa m^6 mol^{-2} and $b = 3.87 \times 10^{-5}$ m^3 mol^{-1}. We use the path $N_2(g, T = 200.$ K, $P = 5.00$ bar$) \rightarrow$ $N_2(g, T = 200.$ K, $P = 20.0$ bar$) \rightarrow N_2(g, T = 400.$ K, $P = 20.0$ bar$)$, keeping in mind that all paths will give the same answer for ΔU for the overall process.

a. Calculate $\Delta U_{T,m} = \int_{V_{m,i}}^{V_{m,f}} (\partial U_m/\partial V)_T dV_m$ using the result of Example Problem 3.4. Note that $V_{m,i} = 3.28 \times 10^{-3}$ m^3 mol^{-1} and $V_{m,f} = 7.88 \times 10^{-4}$ m^3 mol^{-1} at 200 K, as calculated using the van der Waals equation of state.

b. Calculate $\Delta U_{V,m} = \int_{T_i}^{T_f} C_{V,m} dT$ using the following relationship for $C_{V,m}$ in this temperature range:

$$\frac{C_{V,m}}{J\ K^{-1}\ mol^{-1}} = 22.50 - 1.187 \times 10^{-2} \frac{T}{K} + 2.3968 \times 10^{-5} \frac{T^2}{K^2} - 1.0176 \times 10^{-8} \frac{T^3}{K^3}$$

The ratios T^n/K^n ensure that $C_{V,m}$ has the correct dimension.

c. Compare the two contributions to ΔU_m. Can $\Delta U_{T,m}$ be neglected relative to $\Delta U_{V,m}$?

Solution

a. Using the result of Example Problem 3.4,

$$\Delta U_{T,m} = a\left(\frac{1}{V_{m,i}} - \frac{1}{V_{m,f}}\right) = 0.137\ \text{Pa m}^6\ \text{mol}^{-2}$$

$$\times \left(\frac{1}{3.28 \times 10^{-3}\ \text{m}^3\ \text{mol}^{-1}} - \frac{1}{7.88 \times 10^{-4}\ \text{m}^3\ \text{mol}^{-1}}\right) = -132\ \text{J}$$

$$\Delta U_{V,m} = \int_{T_i}^{T_f} C_{V,m}\, dT = \int_{200.}^{400.} \left(\begin{array}{l} 22.50 - 1.187 \times 10^{-2} \dfrac{T}{K} + 2.3968 \times 10^{-5} \dfrac{T^2}{K^2} \\ -1.0176 \times 10^{-8} \dfrac{T^3}{K^3} \end{array}\right)$$

b. $$\times\, d\left(\frac{T}{K}\right) \text{J mol}^{-1}$$

$$= (4.50 - 0.712 + 0.447 - 0.0610)\ \text{kJ mol}^{-1} = 4.17\ \text{kJ mol}^{-1}$$

c. $\Delta U_{T,m}$ is 3.1% of $\Delta U_{V,m}$ for this case. In this example, and for most processes, $\Delta U_{T,m}$ can be neglected relative to $\Delta U_{V,m}$ for real gases.

The calculations in Example Problems 3.4 and 3.5 show that to a good approximation $\Delta U_{T,m} = \int_{V_{m,i}}^{V_{m,f}} (\partial U_m/\partial V)_T dV_m \approx 0$ for real gases under most conditions. Therefore, it is sufficiently accurate to consider U as a function of T only [$U = U(T)$] for real gases in processes that do not involve unusually high gas densities.

Having discussed ideal and real gases, what can be said about the relative magnitude of $\Delta U_{T,m} = \int_{V_{m,i}}^{V_{m,f}} (\partial U_m/\partial V)_T dV_m$ and $\Delta U_{V,m} = \int_{T_i}^{T_f} C_{V,m} dT$ for processes involving liquids and solids? From experiments, it is known that the density of liquids and solids varies only slightly with the external pressure over the range in which these two forms of matter are stable. This conclusion is not valid for extremely high pressure conditions such as those in the interior of planets and stars. However, it is safe to say that dV for a solid or liquid is very small in most processes. Therefore,

$$\Delta U_T^{solid,liq} = \int_{V_1}^{V_2} \left(\frac{\partial U}{\partial V}\right)_T dV \approx \left(\frac{\partial U}{\partial V}\right)_T \Delta V \approx 0 \qquad (3.24)$$

because $\Delta V \approx 0$. This result is valid even if $(\partial U/\partial V)_T$ is large.

The conclusion that can be drawn from this section is as follows: Under most conditions encountered by chemists in the laboratory, U can be regarded as a function of T alone for all substances. The following equations give a good approximation even if V is not constant in the process under consideration:

$$U(T_f, V_f) - U(T_i, V_i) = \Delta U = \int_{T_i}^{T_f} C_V \, dT = n\int_{T_i}^{T_f} C_{V,m} \, dT \qquad (3.25)$$

Note that Equation (3.25) is only applicable to a process in which there is no change in the phase of the system, such as vaporization or fusion, and in which there are no chemical reactions. Changes in U that arise from these processes will be discussed in Chapters 4 and 6.

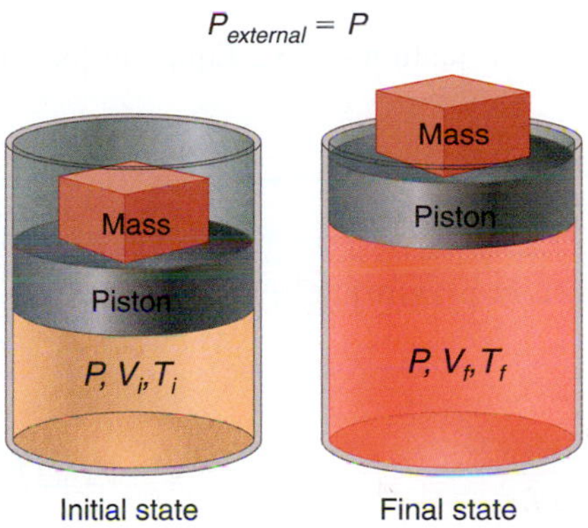

3.4 The Variation of Enthalpy with Temperature at Constant Pressure

As for U, H can be defined as a function of any two of the three variables P, V, and T. It was convenient to choose U to be a function of T and V because this choice led to the identity $\Delta U = q_v$. Using similar reasoning, we choose H to be a function of T and P. How does H vary with P and T? The variation of H with T at constant P is discussed next, and a discussion of the variation of H with P at constant T is deferred to Section 3.6.

Consider the constant pressure process shown schematically in Figure 3.5. For this process defined by $P = P_{external}$,

$$dU = đq_P - P \, dV \qquad (3.26)$$

Although the integral of $đq$ is in general path dependent, it has a unique value in this case because the path is specified, namely, $P = P_{external} = $ constant. Integrating both sides of Equation (3.26),

$$\int_i^f dU = \int_i^f đq_P - \int_i^f P \, dV \quad \text{or} \quad U_f - U_i = q_P - P(V_f - V_i) \qquad (3.27)$$

Because $P = P_f = P_i$, this equation can be rewritten as

$$(U_f + P_f V_f) - (U_i + P_i V_i) = q_P \quad \text{or} \quad \boxed{\Delta H = q_P} \qquad (3.28)$$

The preceding equation shows that the value of ΔH can be determined for an arbitrary process at constant P in a closed system in which only $P-V$ work occurs by simply measuring q_P, the heat transferred between the system and surroundings in a constant pressure process. Note the similarity between Equations (3.28) and (3.18). For an arbitrary process in a closed system in which there is no work other than $P-V$ work, $\Delta U = q_V$ if the process takes place at constant V, and $\Delta H = q_P$ if the process takes place at constant P. These two equations are the basis for the fundamental experimental techniques of bomb calorimetry and constant pressure calorimetry discussed in Chapter 4.

$P_{external} = P$

Mass
Piston
P, V_i, T_i

Mass
Piston
P, V_f, T_f

Initial state   Final state

FIGURE 3.5
The initial and final states are shown for an undefined process that takes place at constant pressure.

A useful application of Equation (3.28) is in experimentally determining the ΔH and ΔU of fusion and vaporization for a given substance. Fusion (solid $\rightarrow$ liquid) and vaporization (liquid $\rightarrow$ gas) occur at a constant temperature if the system is held at a constant pressure and heat flows across the system–surroundings boundary. In both of these phase transitions, attractive interactions between the molecules of the system must be overcome. Therefore, $q > 0$ in both cases and $C_P \rightarrow \infty$. Because $\Delta H = q_P$, ΔH_{fusion} and $\Delta H_{vaporization}$ can be determined by measuring the heat needed to effect the transition at constant pressure. Because $\Delta H = \Delta U + \Delta(PV)$, at constant P,

$$\Delta U_{vaporization} = \Delta H_{vaporization} - P\Delta V_{vaporization} > 0 \tag{3.29}$$

The change in volume on vaporization is $\Delta V_{vaporization} = V_{gas} - V_{liquid} \gg 0$; therefore, $\Delta U_{vaporization} < \Delta H_{vaporization}$. An analogous expression to Equation (3.29) can be written relating ΔU_{fusion} and ΔH_{fusion}. Note that ΔV_{fusion} is much smaller than $\Delta V_{vaporization}$ and can be either positive or negative. Therefore, $\Delta U_{fusion} \approx \Delta H_{fusion}$. The thermodynamics of fusion and vaporization will be discussed in more detail in Chapter 7.

Because H is a state function, dH is an exact differential, allowing us to link $(\partial H/\partial T)_P$ to a measurable quantity. In analogy to the preceding discussion for dU, dH is written in the form

$$dH = \left(\frac{\partial H}{\partial T}\right)_P dT + \left(\frac{\partial H}{\partial P}\right)_T dP \tag{3.30}$$

Because $dP = 0$ at constant P, and $dH = đq_P$ from Equation (3.28), Equation (3.30) becomes

$$đq_P = \left(\frac{\partial H}{\partial T}\right)_P dT \tag{3.31}$$

Equation (3.31) allows the **heat capacity at constant pressure, C_P**, to be defined in a fashion analogous to C_V in Equation (3.15):

$$C_P \equiv \frac{đq_P}{dT} = \left(\frac{\partial H}{\partial T}\right)_P \tag{3.32}$$

Although this equation looks abstract, C_P is a readily measurable quantity. To measure it, one need only measure the heat flow to or from the surroundings for a constant pressure process together with the resulting temperature change in the limit in which dT and $đq$ approach zero and form the ratio $\lim_{dT \rightarrow 0} (đq/dT)_P$.

As was the case for C_V, C_P is an extensive property of the system and varies from substance to substance. The temperature dependence of C_P must be known in order to calculate the change in H with T. In general, for a constant pressure process in which there is no change in the phase of the system and no chemical reactions,

$$\Delta H_P = \int_{T_i}^{T_f} C_P(T)\,dT = n\int_{T_i}^{T_f} C_{P,m}(T)\,dT \tag{3.33}$$

If the temperature interval is small enough, it can usually be assumed that C_P is constant. In that case,

$$\Delta H_P = C_P\Delta T = nC_{P,m}\Delta T \tag{3.34}$$

Enthalpy changes that arise from chemical reactions and changes in phase cannot be calculated using Equations (3.33) and (3.34), and the calculation of ΔH for these processes will be discussed in Chapters 4 and 6.

EXAMPLE PROBLEM 3.6

A 143.0-g sample of C(s) in the form of graphite is heated from 300. to 600. K at a constant pressure. Over this temperature range, $C_{P,m}$ has been determined to be

$$\frac{C_{P,m}}{\text{J K}^{-1}\,\text{mol}^{-1}} = -12.19 + 0.1126\frac{T}{\text{K}} - 1.947 \times 10^{-4}\frac{T^2}{\text{K}^2} + 1.919 \times 10^{-7}\frac{T^3}{\text{K}^3}$$
$$-7.800 \times 10^{-11}\frac{T^4}{\text{K}^4}$$

Calculate ΔH and q_P. How large is the relative error in ΔH if you neglect the temperature-dependent terms in $C_{P,m}$ and assume that $C_{P,m}$ maintains its value at 300. K throughout the temperature interval?

Solution

$$\Delta H = \frac{m}{M}\int_{T_i}^{T_f} C_{P,m}(T)\,dT$$

$$= \frac{143.0\text{ g}}{12.00\text{ g mol}^{-1}}\frac{\text{J}}{\text{mol}}\int_{300.}^{600.}\left(\begin{array}{c} -12.19 + 0.1126\dfrac{T}{\text{K}} - 1.947 \times 10^{-4}\dfrac{T^2}{\text{K}^2} + 1.919 \\ \times 10^{-7}\dfrac{T^3}{\text{K}^3} - 7.800 \times 10^{-11}\dfrac{T^4}{\text{K}^4} \end{array}\right) d\frac{T}{\text{K}}$$

$$= \frac{143.0}{12.00} \times \left[\begin{array}{c} -12.19\dfrac{T}{\text{K}} + 0.0563\dfrac{T^2}{\text{K}^2} - 6.49 \times 10^{-5}\dfrac{T^3}{\text{K}^3} + 4.798 \\ \times 10^{-8}\dfrac{T^4}{\text{K}^4} - 1.56 \times 10^{-11}\dfrac{T^5}{\text{K}^5} \end{array}\right]_{300}^{600}\text{J} = 46.85\text{ kJ}$$

From Equation (3.28), $\Delta H = q_P$.

If we had assumed $C_{P,m} = 8.617$ J mol^{-1} K^{-1}, which is the calculated value at 300. K, $\Delta H = 143.0$ g/12.00 g mol^{-1} × 8.617 J K^{-1} mol^{-1} × [600 K − 300 K] = 30.81 kJ. The relative error is (30.81 kJ − 46.85 kJ)/46.85 kJ = −34%. In this case, it is not reasonable to assume that $C_{P,m}$ is independent of temperature.

3.5 How Are C_P and C_V Related?

To this point, two separate heat capacities, C_P and C_V, have been defined. Are these quantities related? To answer this question, the differential form of the first law is written as

$$dq = C_V\,dT + \left(\frac{\partial U}{\partial V}\right)_T dV + P_{external}\,dV \qquad (3.35)$$

Consider a process that proceeds at constant pressure for which $P = P_{external}$. In this case, Equation (3.35) becomes

$$dq_P = C_V\,dT + \left(\frac{\partial U}{\partial V}\right)_T dV + P\,dV \qquad (3.36)$$

Because $dq_P = C_P\,dT$,

$$C_P = C_V + \left(\frac{\partial U}{\partial V}\right)_T\left(\frac{\partial V}{\partial T}\right)_P + P\left(\frac{\partial V}{\partial T}\right)_P = C_V + \left[\left(\frac{\partial U}{\partial V}\right)_T + P\right]\left(\frac{\partial V}{\partial T}\right)_P$$

$$= C_V + T\left(\frac{\partial P}{\partial T}\right)_V\left(\frac{\partial V}{\partial T}\right)_P \qquad (3.37)$$

To obtain Equation (3.37), both sides of Equation (3.36) have been divided by dT, and the ratio dV/dT has been converted to a partial derivative at constant P. Equation 3.19 has been used in the last step. Using Equation (3.9) and the cyclic rule, one can simplify Equation (3.37) to

$$C_P = C_V + TV\frac{\beta^2}{\kappa} \text{ or } C_{P,m} = C_{V,m} + TV_m\frac{\beta^2}{\kappa} \tag{3.38}$$

Equation (3.38) provides another example of the usefulness of the formal theory of thermodynamics in linking seemingly abstract partial derivatives with experimentally available data. The difference between C_P and C_V can be determined at a given temperature knowing only the molar volume, the coefficient for thermal expansion, and the isothermal compressibility.

Equation (3.38) is next applied to ideal and real gases and to liquids and solids, in the absence of phase changes and chemical reactions. Because β and κ are always positive for real and ideal gases, $C_P - C_V > 0$ for these substances. First, $C_P - C_V$ is calculated for an ideal gas, and then it is calculated for liquids and solids. For an ideal gas, $(\partial U/\partial V)_T = 0$ as shown in Example Problem 3.2 and $P(\partial V/\partial T)_P = P(nR/P) = nR$ so that Equation (3.37) becomes

$$C_P - C_V = nR \tag{3.39}$$

This result was stated without derivation in Section 2.4. The partial derivative $(\partial V/\partial T)_P = V\beta$ is much smaller for liquids and solids than for gases. Therefore, generally

$$C_V \gg \left[\left(\frac{\partial U}{\partial V}\right)_T + P\right]\left(\frac{\partial V}{\partial T}\right)_P \tag{3.40}$$

so that $C_P \approx C_V$ for a liquid or a solid. As shown earlier in Example Problem 3.1 it is not feasible to carry out heating experiments for liquids and solids at constant volume because of the large pressure increase that occurs. Therefore, tabulated heat capacities for liquids and solids list $C_{P,m}$ rather than $C_{V,m}$.

3.6 The Variation of Enthalpy with Pressure at Constant Temperature

In the previous section, we learned how H changes with T at constant P. To calculate how H changes as both P and T change, $(\partial H/\partial P)_T$ must be calculated. The partial derivative $(\partial H/\partial P)_T$ is less straightforward to determine in an experiment than $(\partial H/\partial T)_P$. As will be seen, for many processes involving changes in both P and T, $(\partial H/\partial T)_P dT \gg (\partial H/\partial P)_T dP$ and the pressure dependence of H can be neglected relative to its temperature dependence. However, the knowledge that $(\partial H/\partial P)_T$ is not zero is crucial for understanding the operation of a refrigerator and the liquefaction of gases. The following discussion is applicable to gases, liquids, and solids.

Given the definition $H = U + PV$, we begin by writing dH as

$$dH = dU + P\,dV + V\,dP \tag{3.41}$$

Substituting the differential forms of dU and dH,

$$C_P\,dT + \left(\frac{\partial H}{\partial P}\right)_T dP = C_V\,dT + \left(\frac{\partial U}{\partial V}\right)_T dV + P\,dV + V\,dP$$

$$= C_V\,dT + \left[\left(\frac{\partial U}{\partial V}\right)_T + P\right]dV + V\,dP \tag{3.42}$$

For isothermal processes, $dT = 0$, and Equation (3.42) can be rearranged to

$$\left(\frac{\partial H}{\partial P}\right)_T = \left[\left(\frac{\partial U}{\partial V}\right)_T + P\right]\left(\frac{\partial V}{\partial P}\right)_T + V \tag{3.43}$$

Using Equation (3.19) for $(\partial U/\partial V)_T$,

$$\left(\frac{\partial H}{\partial P}\right)_T = T\left(\frac{\partial P}{\partial T}\right)_V \left(\frac{\partial V}{\partial P}\right)_T + V$$

$$= V - T\left(\frac{\partial V}{\partial T}\right)_P \tag{3.44}$$

The second formulation of Equation (3.44) is obtained through application of the cyclic rule [Equation (3.7)]. This equation is applicable to all systems containing pure substances or mixtures at a fixed composition, provided that no phase changes or chemical reactions take place. The quantity $(\partial H/\partial P)_T$ is evaluated for an ideal gas in Example Problem 3.7.

EXAMPLE PROBLEM 3.7

Evaluate $(\partial H/\partial P)_T$ for an ideal gas.

Solution

$(\partial P/\partial T)_V = (\partial[nRT/V]/\partial T)_V = nR/V$ and $(\partial V/\partial P)_T = RT(d[nRT/P]/dP)_T = -nRT/P^2$ for an ideal gas. Therefore,

$$\left(\frac{\partial H}{\partial P}\right)_T = T\left(\frac{\partial P}{\partial T}\right)_V \left(\frac{\partial V}{\partial P}\right)_T + V = T\frac{nR}{V}\left(-\frac{nRT}{P^2}\right) + V = -\frac{nRT}{P}\frac{nRT}{nRT} + V = 0$$

This result could have been derived directly from the definition $H = U + PV$. For an ideal gas, $U = U(T)$ only and $PV = nRT$. Therefore, $H = H(T)$ only for an ideal gas and $(\partial H/\partial P)_T = 0$.

Because Example Problem 3.7 shows that H is a function of T only for an ideal gas,

$$\Delta H = \int_{T_i}^{T_f} C_P\,(T)\,dT = n\int_{T_i}^{T_f} C_{P,m}\,(T)\,dT \tag{3.45}$$

for an ideal gas. Because H is a function of T only, Equation (3.45) holds for an ideal gas even if P is not constant. This result is also understandable in terms of the potential function of Figure 1.7. Because ideal gas molecules do not attract or repel one another, no energy is required to change their average distance of separation (that is, no increase or decrease in P).

Equation (3.44) in its general form is next applied to several types of systems. As shown in Example Problem 3.7, $(\partial H/\partial P)_T = 0$ for an ideal gas. For liquids and solids, the first term in Equation (3.44), $T(\partial P/\partial T)_V(\partial V/\partial P)_T$, is usually much smaller than V. This is the case because $(\partial V/\partial P)_T$ is very small, which is consistent with our experience that liquids and solids are difficult to compress. Equation (3.44) establishes that for liquids and solids, $(\partial H/\partial P)_T \approx V$ to a good approximation, and dH can be written as

$$dH \approx C_P\,dT + V\,dP \tag{3.46}$$

for systems that consist only of liquids or solids.

EXAMPLE PROBLEM 3.8

Calculate the change in enthalpy when 124 g of liquid methanol initially at 1.00 bar and 298 K undergoes a change of state to 2.50 bar and 425 K. The density of liquid methanol under these conditions is 0.791 g cm^{-3}, and $C_{P,m}$ for liquid methanol is 81.1 J K^{-1} mol^{-1}.

Solution

Because H is a state function, any path between the initial and final states will give the same ΔH. We choose the path methanol (l, 1.00 bar, 298 K) $\rightarrow$ methanol (*l*, 1.00 bar, 425 K) → methanol (*l*, 2.50 bar, 425 K). The first step is isothermal, and the second step is isobaric. The total change in *H* is

$$\Delta H = n \int_{T_i}^{T_f} C_{P,m}\, dT + \int_{P_i}^{P_f} V\, dP \approx n\, C_{P,m}\, (T_f - T_i) + V(P_f - P_i)$$

$$= 81.1 \text{ J K}^{-1} \text{ mol}^{-1} \times \frac{124 \text{ g}}{32.04 \text{ g mol}^1} \times (425 \text{ K} - 298 \text{ K})$$

$$+ \frac{124 \text{g}}{0.791 \text{ g cm}^{-3}} \times \frac{10^{-6} \text{ m}^3}{\text{cm}^3} \times (2.50 \text{ bar} - 1.00 \text{ bar}) \times \frac{10^5 \text{ Pa}}{\text{bar}}$$

$$= 39.9 \times 10^3 \text{ J} + 23.5 \text{ J} \approx 39.9 \text{ kJ}$$

Note that the contribution to ΔH from the change in *T* is far greater than that from the change in *P*.

Example Problem 3.8 shows that because molar volumes of liquids and solids are small, *H* changes much more rapidly with *T* than with *P*. Under most conditions, *H* can be assumed to be a function of *T* only for solids and liquids. Exceptions to this rule are encountered in geophysical or astrophysical applications, for which extremely large pressure changes can occur.

The conclusion that can be drawn from this section is the following: under most conditions encountered by chemists in the laboratory, *H* can be regarded as a function of *T* alone for liquids and solids. It is a good approximation to write

$$H\,(T_f, P_f) - H\,(T_i, P_i) = \Delta H = \int_{T_1}^{T_2} C_P\, dT = n \int_{T_1}^{T_2} C_{P,m}\, dT \tag{3.47}$$

even if *P* is not constant in the process under consideration.

Note that Equation (3.47) is only applicable to a process in which there is no change in the phase of the system, such as vaporization or fusion, and in which there are no chemical reactions. Changes in *H* that arise from chemical reactions and changes in phase will be discussed in Chapters 4 and 6.

Having dealt with solids, liquids, and ideal gases, we are left with real gases. For real gases, $(\partial H / \partial P)_T$ and $(\partial U / \partial V)_T$ are small, but still have a considerable effect on the properties of the gases on expansion or compression. Conventional technology for the liquefaction of gases and for the operation of refrigerators is based on the fact that $(\partial H / \partial P)_T$ and $(\partial U / \partial V)_T$ are not zero for real gases. To derive a formula that will be useful in calculating $(\partial H / \partial P)_T$ for a real gas, the Joule–Thomson experiment is discussed first in the next section.

3.7 The Joule–Thomson Experiment

If the valve on a cylinder of compressed N_2 at 298 K is opened fully, it will become covered with frost, demonstrating that the temperature of the valve is lowered below the freezing point of H_2O. A similar experiment with a cylinder of H_2 leads to a considerable increase in temperature and, potentially, an explosion. How can these effects be understood? To explain them, we discuss the **Joule–Thomson experiment.**

The Joule–Thomson experiment shown in Figure 3.6 can be viewed as an improved version of the Joule experiment because it allows $(\partial U / \partial V)_T$ to be measured with a much higher sensitivity than in the Joule experiment. In this experiment, gas flows from the high-pressure cylinder on the left to the low-pressure cylinder on the right through a porous plug in an insulated pipe. The pistons move to keep the pressure unchanged in each region until all the gas has been transferred to the region to the right of the porous

In the Joule–Thomson experiment, a gas is forced through a porous plug using a piston and cylinder mechanism. The pistons move to maintain a constant pressure in each region. There is an appreciable pressure drop across the plug, and the temperature change of the gas is measured. The upper and lower figures show the initial and final states, respectively. As shown in the text, if the piston and cylinder assembly forms an adiabatic wall between the system (the gases on both sides of the plug) and the surroundings, the expansion is isenthalpic.

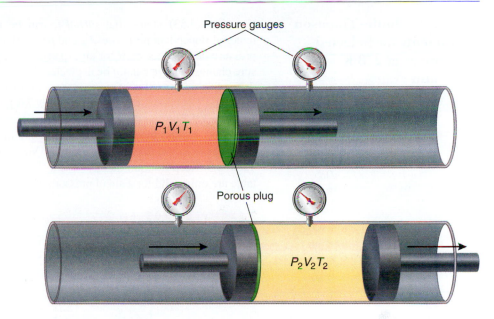

plug. If N_2 is used in the expansion process ($P_1 > P_2$), it is found that $T_2 < T_1$; in other words, the gas is cooled as it expands. What is the origin of this effect? Consider an amount of gas equal to the initial volume V_1 as it passes through the apparatus from left to right. The total work in this expansion process is the sum of the work performed on each side of the plug separately by the moving pistons:

$$w = w_{left} + w_{right} = -\int_{V_1}^{0} P_1 \, dV - \int_{0}^{V_2} P_2 \, dV = -P_2 V_2 + P_1 V_1 \tag{3.48}$$

Because the pipe is insulated, $q = 0$, and

$$\Delta U = U_2 - U_1 = w = -P_2 V_2 + P_1 V_1 \tag{3.49}$$

This equation can be rearranged to

$$U_2 + P_2 V_2 = U_1 + P_1 V_1 \text{ or } H_2 = H_1 \tag{3.50}$$

Note that the enthalpy is constant in the expansion, that is, the expansion is **isenthalpic.** For the conditions of the experiment using N_2, both dT and dP are negative, so $(\partial T/\partial P)_H > 0$. The experimentally determined limiting ratio of ΔT to ΔP at constant enthalpy is known as the **Joule–Thomson coefficient:**

$$\mu_{J-T} = \lim_{\Delta P \to 0} \left(\frac{\Delta T}{\Delta P} \right)_H = \left(\frac{\partial T}{\partial P} \right)_H \tag{3.51}$$

If μ_{J-T} is positive, the conditions are such that the attractive part of the potential dominates, and if μ_{J-T} is negative, the repulsive part of the potential dominates. Using experimentally determined values of μ_{J-T}, $(\partial H/\partial P)_T$ can be calculated. For an isenthalpic process,

$$dH = C_P \, dT + \left(\frac{\partial H}{\partial P} \right)_T dP = 0 \tag{3.52}$$

Dividing through by dP and making the condition $dH = 0$ explicit,

$$C_P \left(\frac{\partial T}{\partial P} \right)_H + \left(\frac{\partial H}{\partial P} \right)_T = 0$$

$$\text{giving } \left(\frac{\partial H}{\partial P} \right)_T = -C_P \mu_{J-T} \tag{3.53}$$

TABLE 3.3 Joule–Thomson Coefficients for Selected Substances at 273 K and 1 atm

Gas	μ_{J-T} (K/MPa)
Ar	3.66
C_6H_{14}	−0.39
CH_4	4.38
CO_2	10.9
H_2	−0.34
He	−0.62
N_2	2.15
Ne	−0.30
NH_3	28.2
O_2	2.69

Source: P. J. Linstrom and W. G. Mallard, Eds., *NIST Chemistry Webbook: NIST Standard Reference Database Number 69,* National Institute of Standards and Technology, Gaithersburg, MD, retrieved from *http://webbook.nist.gov.*

Equation (3.53) states that $(\partial H/\partial P)_T$ can be calculated using the measurement of material-dependent properties C_P and μ_{J-T}. Because μ_{J-T} is not zero for a real gas, the pressure dependence of H for an expansion or compression process for which the pressure change is large cannot be neglected. Note that $(\partial H/\partial P)_T$ can be positive or negative, depending on the value of μ_{J-T} at the P and T of interest.

If μ_{J-T} is known from experiment, $(\partial U/\partial V)_T$ can be calculated as shown in Example Problem 3.9. This has the advantage that a calculation of $(\partial U/\partial V)_T$ based on measurements of C_P, μ_{J-T}, and the isothermal compressibility κ is much more accurate than a measurement based on the Joule experiment. Values of μ_{J-T} are shown for selected gases in Table 3.3. Keep in mind that μ_{J-T} is a function of P and ΔP, so the values listed in the table are only valid for a small pressure decrease originating at 1 atm pressure.

EXAMPLE PROBLEM 3.9

Using Equation (3.43), $(\partial H/\partial P)_T = [(\partial U/\partial V)_T + P](\partial V/\partial P)_T + V$, and Equation (3.19), $(\partial U/\partial V)_T = T(\partial P/\partial T)_V - P$, derive an expression giving $(\partial H/\partial P)_T$ entirely in terms of measurable quantities for a gas.

Solution

$$\left(\frac{\partial H}{\partial P}\right)_T = \left[\left(\frac{\partial U}{\partial V}\right)_T + P\right]\left(\frac{\partial V}{\partial P}\right)_T + V$$

$$= \left[T\left(\frac{\partial P}{\partial T}\right)_V - P + P\right]\left(\frac{\partial V}{\partial P}\right)_T + V = T\left(\frac{\partial P}{\partial T}\right)_V\left(\frac{\partial V}{\partial P}\right)_T + V$$

$$= -T\left(\frac{\partial V}{\partial T}\right)_P + V = -TV\beta + V = V(1-\beta T)$$

In this equation, β is the volumetric thermal expansion coefficient defined in Equation (3.9).

EXAMPLE PROBLEM 3.10

Using Equation (3.43),

$$\left(\frac{\partial H}{\partial T}\right)_T = \left[\left(\frac{\partial U}{\partial V}\right)_T + P\right]\left(\frac{\partial V}{\partial P}\right)_T + V$$

show that $\mu_{J-T} = 0$ for an ideal gas.

Solution

$$\mu_{J-T} = -\frac{1}{C_P}\left(\frac{\partial H}{\partial P}\right)_T = -\frac{1}{C_P}\left[\left(\frac{\partial U}{\partial V}\right)_T\left(\frac{\partial V}{\partial P}\right)_T + P\left(\frac{\partial V}{\partial P}\right)_T + V\right]$$

$$= -\frac{1}{C_P}\left[0 + P\left(\frac{\partial V}{\partial P}\right)_T + V\right]$$

$$= -\frac{1}{C_P}\left[P\left(\frac{\partial[nRT/P]}{\partial P}\right)_T + V\right] = -\frac{1}{C_P}\left[-\frac{nRT}{P} + V\right] = 0$$

In this calculation, we have used the result that $(\partial U/\partial V)_T = 0$ for an ideal gas.

Example Problem 3.10 shows that for an ideal gas, μ_{J-T} is zero. It can be shown that for a van der Waals gas in the limit of zero pressure

$$\mu_{J-T} = \frac{1}{C_{P,m}}\left(\frac{2a}{RT} - b\right) \tag{3.54}$$

Vocabulary

cyclic rule
exact differential
heat capacity at constant
   pressure

heat capacity at constant
   volume
internal pressure
isenthalpic

isothermal compressibility
Joule–Thomson coefficient
Joule–Thomson experiment
partial derivatives

volumetric thermal
   expansion coefficient

Questions on Concepts

Q3.1 Why is $C_{P,m}$ a function of temperature for ethane, but not for argon?

Q3.2 Why is $q_v = \Delta U$ only for a constant volume process? Is this formula valid if work other than $P-V$ work is possible?

Q3.3 Refer to Figure 1.7 and explain why $(\partial U/\partial V)_T$ is generally small for a real gas.

Q3.4 Explain without using equations why $(\partial H/\partial P)_T$ is generally small for a real gas.

Q3.5 Why is it reasonable to write $dH \approx C_p dT + V dP$ for a liquid or solid sample?

Q3.6 Why is the equation $\Delta H = \int_{T_i}^{T_f} C_P(T)\,dT = n\int_{T_i}^{T_f} C_{P,m}(T)\,dT$ valid for an ideal gas even if P is not constant in the process? Is this equation also valid for a real gas? Why or why not?

Q3.7 The heat capacity $C_{P,m}$ is less than $C_{V,m}$ for $H_2O(l)$ near 4°C. Explain this result.

Q3.8 What is the physical basis for the experimental result that U is a function of V at constant T for a real gas? Under what conditions will U decrease as V increases?

Q3.9 Why does the relation $C_P > C_V$ always hold for a gas? Can $C_P > C_V$ be valid for a liquid?

Problems

Problem numbers in **RED** indicate that the solution to the problem is given in the *Student Solutions Manual.*

P3.1 A differential $dz = f(x,y)dx + g(x,y)dy$ is exact if the integral $\int f(x,y)dx + \int g(x,y)dy$ is independent of the path. Demonstrate that the differential $dz = 2xydx + x^2dy$ is exact by integrating dz along the paths $(1,1) \rightarrow (5,1) \rightarrow (5,5)$ and $(1,1) \rightarrow (3,1) \rightarrow (3,3) \rightarrow (5,3) \rightarrow (5,5)$. The first number in each set of parentheses is the x coordinate, and the second number is the y coordinate.

P3.2 The function $f(x,y)$ is given by $f(x,y) = xy\sin 5x + x^2\sqrt{y}\ln y + 3e^{-2x^2}\cos y$. Determine

$$\left(\frac{\partial f}{\partial x}\right)_y, \left(\frac{\partial f}{\partial y}\right)_x, \left(\frac{\partial^2 f}{\partial x^2}\right)_y, \left(\frac{\partial^2 f}{\partial y^2}\right)_x, \left(\frac{\partial}{\partial y}\left(\frac{\partial f}{\partial x}\right)_y\right)_x$$

and $\left(\dfrac{\partial}{\partial x}\left(\dfrac{\partial f}{\partial y}\right)_x\right)_y$

a. Is $\left(\dfrac{\partial}{\partial y}\left(\dfrac{\partial f}{\partial x}\right)_y\right)_x = \left(\dfrac{\partial}{\partial x}\left(\dfrac{\partial f}{\partial y}\right)_x\right)_y$?

b. Obtain an expression for the total differential df.

P3.3 This problem will give you practice in using the cyclic rule. Use the ideal gas law to obtain the three functions $P = f(V,T)$, $V = g(P,T)$, and $T = h(P,V)$. Show that the cyclic rule $(\partial P/\partial V)_T\,(\partial V/\partial T)_P(\partial T/\partial P)_V = -1$ is obeyed.

P3.4 Using the chain rule for differentiation, show that the isobaric expansion coefficient expressed in terms of density is given by $\beta = -1/\rho(\partial\rho/\partial T)_P$.

P3.5 A vessel is filled completely with liquid water and sealed at 25.0°C and a pressure of 1.00 bar. What is the pressure if the temperature of the system is raised to 60.0°C?

Under these conditions, $\beta_{water} = 2.04 \times 10^{-4}$ K^{-1}, $\beta_{vessel} = 1.02 \times 10^{-4}$ K^{-1}, and $\kappa_{water} = 4.59 \times 10^{-5}$ bar^{-1}.

P3.6 Because U is a state function, $(\partial/\partial V(\partial U/\partial T)_V)_T = (\partial/\partial T(\partial U/\partial V)_T)_V$. Using this relationship, show that $(\partial C_V/\partial V)_T = 0$ for an ideal gas.

P3.7 Because V is a state function, $(\partial/\partial P(\partial V/\partial T)_P)_T = (\partial/\partial T(\partial V/\partial P)_T)_P$. Using this relationship, show that the isothermal compressibility and isobaric expansion coefficient are related by $(\partial\beta/\partial P)_T = -(\partial\kappa/\partial T)_P$.

P3.8 Integrate the expression $\beta = 1/V(\partial V/\partial T)_P$ assuming that β is independent of pressure. By doing so, obtain an expression for V as a function of T and β at fixed P.

P3.9 The molar heat capacity $C_{P,m}$ of $SO_2(g)$ is described by the following equation over the range 300 K < 1700 K:

$$\frac{C_{P,m}}{R} = 3.093 + 6.967 \times 10^{-3}\frac{T}{K} - 45.81 \times 10^{-7}\frac{T^2}{K^2} + 1.035 \times 10^{-9}\frac{T^3}{K^3}$$

In this equation, T is the absolute temperature in kelvin. The ratios T^n/K^n ensure that $C_{P,m}$ has the correct dimension. Assuming ideal gas behavior, calculate q, w, ΔU, and ΔH if 1 mol of $SO_2(g)$ is heated from 75° to 1350°C at a constant pressure of 1 bar. Explain the sign of w.

P3.10 Starting with the van der Waals equation of state, find an expression for the total differential dP in terms of dV and dT. By calculating the mixed partial derivatives $(\partial/\partial T(\partial P/\partial V)_T)_V$ and $(\partial/\partial V(\partial P/\partial T)_V)_T$, determine if dP is an exact differential.

P3.11   Obtain an expression for the isothermal compressibility $\kappa = -1/V(\partial V/\partial P)_T$ for a van der Waals gas.

P3.12   Regard the enthalpy as a function of T and P. Use the cyclic rule to obtain the expression

$$C_P = -\frac{(\partial H/\partial P)_T}{(\partial T/\partial P)_H}$$

P3.13   Equation (3.38), $C_P = C_V + TV(\beta^2/\kappa)$, links C_P and C_V with β and κ. Use this equation to evaluate $C_P - C_V$ for an ideal gas.

P3.14   Use $(\partial U/\partial V)_T = (\beta T - \kappa P)/\kappa$ to evaluate $(\partial U/\partial V)_T$ for an ideal gas.

P3.15   An 80.0-g piece of gold at 650. K is dropped into 100.0 g of $H_2O(l)$ at 298 K in an insulated container at 1 bar pressure. Calculate the temperature of the system once equilibrium has been reached. Assume that $C_{P,m}$ for Au and H_2O is constant at their values for 298 K throughout the temperature range of interest.

P3.16   A mass of 35.0 g of $H_2O(s)$ at 273 K is dropped into 180.0 g of $H_2O(l)$ at 325 K in an insulated container at 1 bar of pressure. Calculate the temperature of the system once equilibrium has been reached. Assume that $C_{P,m}$ for H_2O is constant at its values for 298 K throughout the temperature range of interest.

P3.17   A mass of 20.0 g of $H_2O(g)$ at 373 K flows into 250. g of $H_2O(l)$ at 300. K and 1 atm. Calculate the final temperature of the system once equilibrium has been reached. Assume that $C_{P,m}$ for H_2O is constant at its values for 298 K throughout the temperature range of interest.

P3.18   Calculate w, q, ΔH, and ΔU for the process in which 1 mol of water undergoes the transition $H_2O(l, 373\ K) \rightarrow H_2O(g, 460.\ K)$ at 1 bar of pressure. The volume of liquid water at 373 K is $1.89 \times 10^{-5}\ m^3\ mol^{-1}$ and the volume of steam at 373 and 460. K is 3.03 and $3.74 \times 10^{-2}\ m^3\ mol^{-1}$, respectively. For steam, $C_{P,m}$ can be considered constant over the temperature interval of interest at 33.58 J mol^{-1} K^{-1}.

P3.19   Because $(\partial H/\partial P)_T = -C_P\mu_{J-T}$, the change in enthalpy of a gas expanded at constant temperature can be calculated. To do so, the functional dependence of μ_{J-T} on P must be known. Treating Ar as a van der Waals gas, calculate ΔH when 1 mol of Ar is expanded from 400. to 1.00 bar at 300. K. Assume that μ_{J-T} is independent of pressure and is given by $\mu_{J-T} = [(2a/RT) - b]/C_{P,m}$, and $C_{P,m} = 5/2R$ for Ar. What value would ΔH have if the gas exhibited ideal gas behavior?

P3.20   Using the result of Equation (3.8), $(\partial P/\partial T)_V = \beta/\kappa$, express β as a function of κ, and V_m for an ideal gas, and β as a function of b, κ, and V_m for a van der Waals gas.

P3.21   The Joule coefficient is defined by $(\partial T/\partial V)_U = 1/C_V[P - T(\partial P/\partial T)_V]$. Calculate the Joule coefficient for an ideal gas and for a van der Waals gas.

P3.22   Use the relation $(\partial U/\partial V)_T = T(\partial P/\partial T)_V - P$ and the cyclic rule to obtain an expression for the internal pressure, $(\partial U/\partial V)_T$, in terms of P, β, T, and κ.

P3.23   Derive the following relation,

$$\left(\frac{\partial U}{\partial V_m}\right)_T = \frac{3a}{2\sqrt{T}\ V_m\ (V_m + b)}$$

for the internal pressure of a gas that obeys the Redlich–Kwong equation of state,

$$P = \frac{RT}{V_m - b} - \frac{a}{\sqrt{T}}\frac{1}{V_m\ (V_m + b)}$$

P3.24   Derive an expression for the internal pressure of a gas that obeys the Bethelot equation of state,

$$P = \frac{RT}{V_m - b} - \frac{a}{TV_m^2}$$

P3.25   For a gas that obeys the equation of state

$$V_m = \frac{RT}{P} + B(T)$$

derive the result

$$\left(\frac{\partial H}{\partial P}\right)_T = B(T) - T\frac{dB(T)}{dT}$$

P3.26   Derive the following expression for calculating the isothermal change in the constant volume heat capacity: $(\partial C_V/\partial V)_T = T(\partial^2 P/\partial T^2)_V$.

P3.27   Use the result of Problem P3.26 to show that $(\partial C_V/\partial V)_T$ for the van der Waals gas is zero.

P3.28   Use the result of Problem P3.26 to derive a formula for $(\partial C_V/\partial V)_T$ for a gas that obeys the Redlich–Kwong equation of state,

$$P = \frac{RT}{V_m - b} - \frac{a}{\sqrt{T}}\frac{1}{V_m\ (V_m + b)}$$

P3.29   For the equation of state $V_m = RT/P + B(T)$, show that

$$\left(\frac{\partial C_{P,m}}{\partial P}\right)_T = -T\frac{d^2\ B(T)}{dT^2}$$

(*Hint:* Use Equation 3.44 and the property of state functions with respect to the order of differentiation in mixed second derivatives.)

P3.30   Use the relation

$$C_{P,m} - C_{V,m} = T\left(\frac{\partial V_m}{\partial T}\right)_P\left(\frac{\partial P}{\partial T}\right)_V$$

the cyclic rule, and the van der Waals equation of state to derive an equation for $C_{P,m} - C_{V,m}$ in terms of V_m, T, and the gas constants R, a, and b.

P3.31   Show that the expression $(\partial U/\partial V)_T = T(\partial P/\partial T)_V - P$ can be written in the form

$$\left(\frac{\partial U}{\partial V}\right)_T = T^2\left(\partial\left[\frac{P}{T}\right]\Big/\partial T\right)_V = -\left(\partial\left[\frac{P}{T}\right]\Big/\partial\left[\frac{1}{T}\right]\right)_V$$

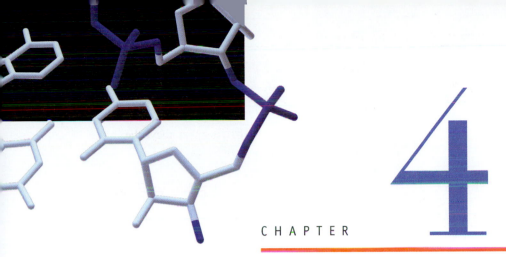

CHAPTER

4

Thermochemistry

Thermochemistry is the branch of thermodynamics that investigates the heat flow into or out of a reaction system and deduces the energy stored in chemical bonds. As reactants are converted into products, energy can either be taken up by the system or released to the surroundings. For a reaction that takes place at constant temperature and volume, the heat that flows to or out of the system is equal to ΔU for the reaction. For a reaction that takes place at constant temperature and pressure, the heat that flows to or out of the system is equal to ΔH for the reaction. The enthalpy of formation is defined as the heat flow into or out of the system in a reaction between pure elements that leads to the formation of 1 mol of product. Because H is a state function, the reaction enthalpy can be written as the enthalpies of formation of the products minus those of the reactants. This property allows ΔH and ΔU for a reaction to be calculated for many reactions without carrying out an experiment.

4.1 Energy Stored in Chemical Bonds Is Released or Taken Up in Chemical Reactions

A significant amount of the internal energy or enthalpy of a molecule is stored in the form of chemical bonds. As reactants are transformed to products in a chemical reaction, energy can be released or taken up as bonds are made or broken. For example, consider a reaction in which $N_2(g)$ and $H_2(g)$ dissociate into atoms, and the atoms recombine to form $NH_3(g)$. The enthalpy changes associated with individual steps and with the overall reaction $1/2\ N_2(g) + 3/2\ H_2(g) \rightarrow NH_3(g)$ are shown in Figure 4.1. Note that large enthalpy changes are associated with the individual steps, but that the enthalpy change in the overall reaction is much smaller.

The change in enthalpy or internal energy resulting from chemical reactions appears in the surroundings in the form of a temperature increase or decrease resulting from heat flow and/or in the form of expansion or nonexpansion work. For example, the combustion of gasoline in an automobile engine can be used to do expansion work on the surroundings. Nonexpansion electrical work is possible if the chemical reaction is carried out in an electrochemical cell. In Chapters 6 and 9 the extraction of nonexpansion work from chemical reactions will be discussed. In this chapter, the focus is on using measurements of heat flow to determine changes in U and H due to chemical reactions.

FIGURE 4.1
Enthalpy changes are shown for individual steps in the overall reaction $1/2\ N_2 + 3/2\ H_2 \rightarrow NH_3$.

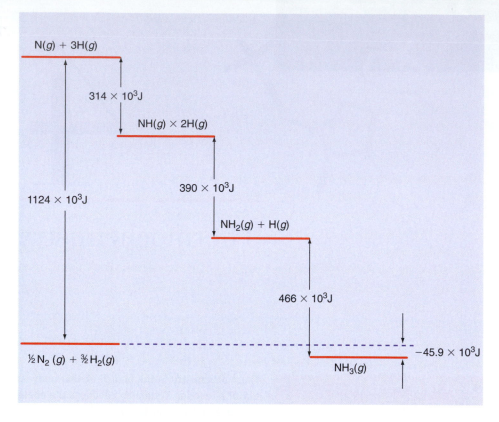

4.2 Internal Energy and Enthalpy Changes Associated with Chemical Reactions

In the previous chapters, we discussed how ΔU and ΔH are calculated from work and heat flow between the system and the surroundings for processes that do not involve phase changes or chemical reactions. In this section, this discussion is extended to reaction systems. Because most reactions of interest to chemists are carried out at constant pressure, chemists generally use ΔH rather than ΔU as a measure of the change in energy of the system. However, because ΔH and ΔU are related, one can be calculated from the other.

Imagine that a reaction involving a stoichiometric mixture of reactants is carried out in a constant pressure reaction vessel with diathermal walls immersed in a water bath. It is our experience that the temperature of the water bath will either increase or decrease as the reaction proceeds, depending on whether energy is taken up or released in the reaction. If the temperature of the water bath increases, heat flows from the system (the contents of the reaction vessel) to the surroundings (the water bath and the vessel). In this case, we say that the reaction is **exothermic.** If the temperature of the water bath decreases, the heat flows instead from the surroundings to the system, and we say that the reaction is **endothermic.**

The temperature and pressure of the system also increase or decrease as the reaction proceeds. The total measured values for ΔU (or ΔH) include contributions from two separate components: ΔU for the reaction at constant T and P and ΔU from the change in T and P. To determine ΔU for the reaction, it is necessary to separate these two components. Consider the combustion reaction for sucrose, $C_{12}H_{22}O_{11}(s)$, in Equation (4.1):

$$C_{12}H_{22}O_{11}(s) + 12O_2(g) \rightarrow 12CO_2(g) + 11H_2O(l) \qquad (4.1)$$

It is useful to tabulate all values for ΔU and ΔH for reactions at a standard temperature and pressure, generally 298.15 K and 1 bar. The standard state for gases is actually a hypothetical state in which the gas behaves ideally at a pressure of 1 bar. For most reactions, deviations from ideal behavior are very small. Note that the phase (solid, liquid, or gas) for each reactant and product has been specified, because U and H are different for each phase. Because U and H are also functions of T and P, their values must also be specified. Changes in enthalpy and internal energy at the standard pressure of 1 bar are indicated by $\Delta H°$ and $\Delta U°$. The **enthalpy of reaction,** $\Delta H_{reaction}$, at specific values of T and P is defined as the heat withdrawn from the surroundings as the reactants are transformed into products at conditions of constant T and P. It is, therefore, a negative quantity for an exothermic reaction and a positive quantity for an endothermic reaction. How can the reaction enthalpy and internal energy be determined? Because H is a state function, any path can be chosen that proceeds from reactants to products at the standard values of T and P. It is useful to visualize the following imaginary two-step process:

1. The reaction is carried out in an isolated system. The **standard state** value of $P = 1$ bar is chosen for the initial system conditions, and we choose $T = 298.15$ K. The walls of the constant pressure reaction vessel are switched to an adiabatic condition and the reaction is initiated. At 298.15 K, $C_{12}H_{22}O_{11}$ is a solid, O_2 and CO_2 are gases, and H_2O is a liquid. The temperature changes from T to T' in the course of the reaction. The reaction is described by

$$(\{C_{12}H_{22}O_{11}(s) + 12O_2(g)\}, 298.15 \text{ K}, P = 1 \text{ bar}) \rightarrow (\{12CO_2(g) \qquad (4.2)$$
$$+ 11H_2O(l)\}, T', P = 1 \text{ bar})$$

Because the process in this step is adiabatic, $q_{P,a} = \Delta H_a° = 0$.

2. In the second step, the system is brought to thermal equilibrium with the surroundings. The walls of the reaction vessel are switched to a diathermal condition, and heat flow occurs to a very large water bath in the surroundings. The temperature changes from T' back to its initial value of 298.15 K in this process. This process is described by

$$(\{12CO_2(g) + 11H_2O(l)\}, T', P = 1 \text{ bar}) \rightarrow (\{12CO_2(g) + 11H_2O(l)\}, \quad (4.3)$$
$$298.15 \text{ K}, P = 1 \text{ bar})$$

The heat flow in this step is $\Delta H_b° = q_{P,b}$.

The sum of these two steps is the overall reaction in which the initial and final states are at the same values of P and T, namely, 298.15 K and 1 bar:

$$(\{C_{12}H_{22}O_{11}(s) + 12O_2(g)\}, 298.15 \text{ K}, P = 1 \text{ bar}) \rightarrow (\{12CO_2(g) \qquad (4.4)$$
$$+ 11H_2O(l)\}, 298.15 \text{ K}, P = 1 \text{ bar})$$

The sum of the ΔH values for the individual steps at 298.15 K, $\Delta H_{reaction}$, is

$$\Delta H°_{reaction} = \Delta H_a° + \Delta H_b° = q_{P,a} + q_{P,b} = 0 + q_{P,b} = q_P \qquad (4.5)$$

Note that because H is a state function, it is unimportant if the temperature during the course of the reaction differs from 298.15 K. It is also unimportant if T' is higher than 373 K so that H_2O is present as a gas rather than a liquid. If the initial and final temperatures of the system are 298.15 K, then the measured value of q_P is $\Delta H°_{reaction}$ at 298.15 K.

The previous discussion suggests an operational way to determine $\Delta H°_{reaction}$ for a reaction at constant pressure. Because researchers do not have access to an infinitely large water bath, the temperature of the system and surroundings will change in the course of the reaction. The procedure just described, therefore, is modified in the following way. The reaction is allowed to proceed, and the ΔT that occurs in a finite-size water bath in the surroundings that is initially at 298.15 K is measured. If the temperature of the water bath decreases as a result of the reaction, the bath is heated to return it to 298.15 K using an electrical heater. By doing so, we ensure that the initial and final P and T are the same and, therefore, $\Delta H = \Delta H°_{reaction}$. How is $\Delta H°_{reaction}$ determined for this case? The electrical work done on the heater that restores the temperature to 298.15 K is equal to $\Delta H°_{298.15 \text{ K}}$, which is equal to $\Delta H°_{reaction}$. If the temperature of the water bath increases as a result of the reaction by the amount ΔT, the electrical work done on a heater in the water bath at 298.15 K that increases its temperature by ΔT in a separate experiment is measured. In this case, $\Delta H°_{reaction}$ is equal to the negative of the electrical work done on the heater.

Although an experimental method for determining $\Delta H^\circ_{reaction}$ has been described, to tabulate the reaction enthalpy for all possible chemical reactions would be a monumental undertaking. Fortunately, $\Delta H^\circ_{reaction}$ can be calculated from tabulated enthalpy values for individual reactants and products. This is advantageous because there are far fewer reactants and products than the number of possible among them. Consider $\Delta H^\circ_{298.15 \text{ K}}$ for the reaction of Equation (4.1) at $T = 298.15$ K and $P = 1$ bar. These value for P and T are chosen because tabulated values are available. However, standard enthalpies for reaction at other values of P and T can be calculated as discussed in Chapters 2 and 3.

$$\Delta H^\circ_{reaction, 298.15 \text{ K}} = H^\circ_{products} - H^\circ_{reactants} \tag{4.6}$$

$$= 12H^\circ_m (CO_2, g) + 11H^\circ_m (H_2O, l) - H^\circ_m (C_{12}H_{22}O_{11}, s) - 12H^\circ_m (O_2, g)$$

where the m subscripts refer to molar quantities. Although Equation (4.6) is correct, it does not provide a useful way to calculate $\Delta H^\circ_{298.15 \text{ K}}$ because there is no experimental way to determine the absolute enthalpy for any element or compound. This is the case because there is no unique reference zero against which individual enthalpies can be measured. For this reason, only ΔH and ΔU, as opposed to H and U, can be determined in an experiment.

Equation (4.6) can be transformed into a more useful form by introducing the enthalpy of formation. The standard **enthalpy of formation,** ΔH°_f, is the enthalpy associated with the reaction in which the only reaction product is 1 mol of the species of interest, and only pure elements in their most stable state of aggregation under the standard state appear as reactants. Note that with this definition, $\Delta H^\circ_f = 0$ for an element in its standard state because the reactants and products are identical. The compounds that are produced or consumed in the reaction $C_{12}H_{22}O_{11}(s) + 12O_2(g) \rightarrow 12CO_2(g) + 11H_2O(l)$ are $C_{12}H_{22}O_{11}(s)$, $CO_2(g)$, and $H_2O(l)$. The formation reactions for these compounds at 298.15 K and 1 bar are

$$H_2(g) + 1/2 \, O_2(g) \rightarrow H_2O(l)$$

$$\Delta H^\circ_{reaction} = \Delta H^\circ_f (H_2O, l) = H^\circ_m (H_2O, l) - H^\circ_m (H_2, g) - \frac{1}{2} H^\circ_m (O_2, g) \tag{4.7}$$

$$C(s, graphite) + O_2(g) \rightarrow CO_2(g)$$

$$\Delta H^\circ_{reaction} = \Delta H^\circ_f (CO_2, g) = H^\circ_m (CO_2, g) - H^\circ_m (O_2, g) - \Delta H^\circ_m(C, graphite) \tag{4.8}$$

$$12C(s, graphite) + 11H_2(g) + 11/2O_2(g) \rightarrow C_{12}H_{22}O_{11}(s)$$

$$\Delta H^\circ_{reaction} = \Delta H^\circ_f (C_{12}H_{22}O_{11}, s) = H^\circ_m (C_{12}H_{22}O_{11}, s) - 12H^\circ_m (C, graphite) \tag{4.9}$$

$$- 11H^\circ_m (H_2, g) - 11/2 \, H^\circ_m (O_2, g)$$

Note that carbon is a solid at $T = 298.15$ K and 1 bar. But carbon has two solid forms under these conditions: graphite (gr) and diamond. Graphite is selected as the standard state of carbon at 298.15 K and 1 bar. If Equation (4.6) is rewritten in terms of the enthalpies of formation, a simple equation for the reaction enthalpy is obtained:

$$\Delta H^\circ_{reaction} = 11\Delta H^\circ_f (H_2O, l) + 12\Delta H^\circ_f (CO_2, g) - \Delta H^\circ_f (C_{12}H_{22}O_{11}, s) \tag{4.10}$$

Note that elements in their standard state do not appear in this equation. This result can be generalized to any chemical transformation

$$\nu_A A + \nu_B B + \ldots \rightarrow \nu_X X + \nu_Y Y + \ldots \tag{4.11}$$

which we write in the form

$$\sum_i \nu_i X_i \tag{4.12}$$

The X_i refer to all species other than elements in their standard state that appear in the overall equation. The unitless stoichiometric coefficients ν_i are positive for products and negative for reactants. The enthalpy change associated with this reaction is

$$\Delta H^\circ_{reaction} = \sum_i \nu_i \Delta H^\circ_{f,i} \tag{4.13}$$

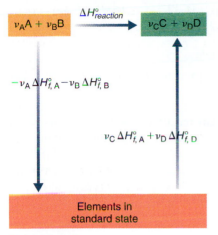

FIGURE 4.2
Equation (4.13) follows from the fact that the enthalpy of both paths is the same because they connect the same initial and final states.

The rationale behind Equation (4.13) can also be depicted as shown in Figure
are considered between the reactants A and B and the products C and D
$v_A A + v_B B \rightarrow v_C C + v_D D$. The first of these is a direct path for which ΔH
In the second path, A and B are first broken down into their elements, each in
state. Subsequently, the elements are combined to form C and D. The enthalpy change for
the second route is $\Delta H°_{reaction} = \sum_i v_i \Delta H°_{f, products} - \sum_i |v_i| \Delta H°_{f, reactants}$. Because H is a
state function, the enthalpy change is the same for both paths. This is stated in mathe-
matical form in Equation (4.13).

A specific application of this general equation is given in Equation (4.10). Writing
$\Delta H°_{reaction}$ in this form is a great simplification over compiling measured values of reaction enthalpies, because the reaction enthalpies can be calculated using only tabulated
values of the formation enthalpies of species other than elements in their standard state.
Formation enthalpies for inorganic atoms, ions, and molecules are listed in Table 4.1
(Appendix B, Data Tables) and formation enthalpies for selected organic compounds are
listed in Table 4.2 (Appendix B, Data Tables).

EXAMPLE PROBLEM 4.1

Using the data:

$$\Delta H°_f (C_{12}H_{22}O_{11}, s) = -2226.1 \text{ kJ mol}^{-1}$$

$$\Delta H°_f (CO_2, aq) = -412.9 \text{ kJ mol}^{-1} \text{ and}$$

$$\Delta H°_f (H_2O, l) = -285.8 \text{ kJ mol}^{-1}$$

calculate the heat evolved in the oxidation of 1 mol of sucrose ($C_{12}H_{22}O_{11}, s$) to
$CO_2(aq)$ and water (l) at 1 bar:

$$C_{12}H_{22}O_{11}(s) + 12O_2(g) \rightarrow 12CO_2(aq) + 11H_2O(l)$$

Solution

$$\Delta H°_{reaction} = 12\Delta H°_f (CO_2, aq) + 11\Delta H°_f (H_2O, l) - \Delta H°_f (C_{12}H_{22}O_{11}, s) - 12\Delta H°_f (O_2, g)$$

$$= 12 \times (-412.9 \text{ kJ}) + 11 \times (-285.8 \text{ kJ}) - (-2226.1 \text{ kJ}) - 12 \times (0 \text{ kJ})$$

$$= -4955 \text{ kJ} - 3144 \text{ kJ} + 2226.1 \text{ kJ} = -5873 \text{ kJ}$$

Another thermochemical convention is introduced at this point in order to be able to
calculate enthalpy changes involving electrolyte solutions. The reaction that occurs
when a salt such as NaCl is dissolved in water is

$$NaCl(s) \rightarrow Na^+(aq) + Cl^-(aq)$$

Because it is not possible to form only positive or negative ions in solution, the measured
enthalpy of solution of an electrolyte is the sum of the enthalpies of all anions and cations
formed. To be able to tabulate values for enthalpies of formation of individual ions, the enthalpy for the following reaction is set equal to zero at $P = 1$ bar for all temperatures:

$$1/2 \text{ H}_2(g) \rightarrow H^+(aq) + e^-(\text{metal electrode})$$

In other words, solution enthalpies of ions are measured relative to that for $H^+(aq)$. The
thermodynamics of electrolyte solutions will be discussed in detail in Chapter 9.

As the previous discussion shows, to calculate $\Delta H°_{reaction}$, only the $\Delta H°_f$ of the reactants and products are needed. The $\Delta H°_f$ are again a *difference* in enthalpy between the
compound and its constituent elements, rather than an absolute enthalpy. However,
there is a convention that allows absolute enthalpies to be specified using the experimentally determined values of the $\Delta H°_f$ of compounds. In this convention, the absolute
enthalpy of each pure element in its standard state is set equal to zero. With this convention, the absolute molar enthalpy of any chemical species in its standard state, $H°_m$,
is equal to $\Delta H°_f$ for that species. To demonstrate this convention, the reaction in Equation (4.9) is considered:

$$\Delta H^\circ_{reaction} = \Delta H^\circ_f (C_{12}H_{22}O_{11},s) = H^\circ_m (C_{12}H_{22}O_{11},s) - 12H^\circ_m (C, gr) \quad (4.9)$$
$$- 11H^\circ_m (H_2, g) - 11/2 \, H^\circ_m (O_2, g)$$

Setting $H^\circ = 0$ for each element in its standard state,

$$\Delta H^\circ_f (C_{12}H_{22}O_{11}, s) = H^\circ_m (C_{12}H_{22}O_{11}, s) - 12 \times 0 - 11 \times 0 - 11/2 \times 0 \quad (4.14)$$
$$= H^\circ_m (C_{12}H_{22}O_{11}, s)$$

Convince yourself that the value of $\Delta H^\circ_{reaction}$ for any reaction involving compounds and elements is unchanged by this convention. In fact, one could choose a different number for the absolute enthalpy of each pure element in its standard state, and it would still not change the value of $\Delta H^\circ_{reaction}$. However, it is much more convenient (and easier to remember) if one sets $H^\circ_m = 0$ for all elements in their standard state. This convention will be used again in Chapter 6 when the chemical potential is discussed.

4.3 Hess's Law Is Based on Enthalpy Being a State Function

As discussed in the previous section, it is extremely useful to have tabulated values of ΔH°_f for chemical compounds at one fixed combination of P and T. With access to these values of ΔH°_f, $\Delta H^\circ_{reaction}$ can be calculated for all reactions among these compounds at the tabulated values of P and T.

But how is ΔH°_f determined? Consider the formation reaction for urea, $H_2N(CO)NH_2(s)$:

$$C(s) + N_2(g) + 2H_2(g) + 1/2O_2(g) \rightarrow H_2N(CO)NH_2(s) \quad (4.15)$$

However, it is unlikely that one would obtain only urea if the reaction were carried out as written. Given this experimental hindrance, how can ΔH°_f for urea be determined? To determine ΔH°_f for urea, we take advantage of the fact that ΔH°_f is path independent. In this context, path independence means that the enthalpy change for any sequence of reactions that sum to the same overall reaction is identical. This statement is known as **Hess's law.** Therefore, one is free to choose any sequence of reactions that leads to the desired outcome. Combustion reactions are well suited for these purposes, because in general they proceed rapidly, go to completion, and produce only a few products. To determine ΔH°_f for urea, one can carry out the following combustion reactions:

$$H_2N(CO)NH_2(s) + 7/2 \, O_2(g) \rightarrow CO_2(g) + 2NO_2(g) + 2H_2O(l) \qquad \Delta H^\circ_I \quad (4.16)$$

$$C(s) + O_2(g) \rightarrow CO_2(g) \qquad \Delta H^\circ_{II} \quad (4.17)$$

$$H_2(g) + \tfrac{1}{2}O_2(g) \rightarrow H_2O(l) \qquad \Delta H^\circ_{III} \quad (4.18)$$

$$1/2 \, N_2(g) + O_2(g) \rightarrow NO_2(g) \qquad \Delta H^\circ_{IV} \quad (4.19)$$

These reactions are combined in the following way to obtain the desired reaction:

$$C(s) + O_2(g) \rightarrow CO_2(g) \qquad \Delta H^\circ_{II} \quad (4.20)$$

$$CO_2(g) + 2NO_2(g) + 2H_2O(l) \rightarrow H_2N(CO)NH_2(s) + 7/2 \, O_2(g) \qquad -\Delta H^\circ_I \quad (4.21)$$

$$2 \times [1/2 \, N_2(g) + O_2(g) \rightarrow NO_2(g)] \qquad 2\Delta H^\circ_{IV} \quad (4.22)$$

$$2 \times [H_2(g) + \tfrac{1}{2}O_2(g) \rightarrow H_2O(l)] \qquad 2\Delta H^\circ_{III} \quad (4.23)$$

$$C(s) + N_2(g) + 3H_2(g) + O_2(g) \rightarrow H_2N(CO)NH_2(s) + H_2O(l)$$

$$\Delta H^\circ_f = -\Delta H^\circ_I + \Delta H^\circ_{II} + 2\Delta H^\circ_{III} + 2\Delta H^\circ_{IV} \quad (4.24)$$

We emphasize again that it is not necessary for these reactions to be carried out at 298.15 K. The reaction vessel is immersed in a water bath at 298.15 K and the combustion reaction is initiated. If the temperature in the vessel rises during the course of the reaction, the heat flow that restores the system and surroundings to 298.15 K after completion of the reaction is measured, allowing $\Delta H^\circ_{reaction}$ to be determined at 298.15 K.

Several points should be made about enthalpy changes in relation to balanced overall equations describing chemical reactions. First, because H is an extensive function, multiplying all stoichiometric coefficients with any number changes $\Delta H^\circ_{reaction}$ by the same factor. Therefore, it is important to know which set of stoichiometric coefficients has been assumed if a numerical value of $\Delta H^\circ_{reaction}$ is given. Second, because the units of ΔH°_f for all compounds in the reaction are kJ mol^{-1}, the units of the reaction enthalpy $\Delta H^\circ_{reaction}$ are also kJ mol^{-1}. One might pose the question "per mole of what?" given that all the stoichiometric coefficients may differ from each other and from one. The answer to this question is per mole of the reaction *as written*. Doubling all the stoichiometric coefficients doubles $\Delta H^\circ_{reaction}$.

EXAMPLE PROBLEM 4.2

The average **bond enthalpy** of the O—H bond in water is defined as one-half of the enthalpy change for the reaction $H_2O(g) \rightarrow 2H(g) + O(g)$. The formation enthalpies, ΔH°_f, for H(g) and O(g) are 218.0 and 249.2 kJ mol^{-1}, respectively, at 298.15 K, and ΔH°_f for $H_2O(g)$ is -241.8 kJ mol^{-1} at the same temperature.

a. Use this information to determine the average bond enthalpy of the O—H bond in water at 298.15 K.

b. Determine the average **bond energy, ΔU**, of the O—H bond in water at 298.15 K. Assume ideal gas behavior.

Solution

a. We consider the sequence

$H_2O(g) \rightarrow H_2(g) + 1/2\, O_2(g)$ $\Delta H^\circ = 241.8$ kJ mol^{-1}

$H_2(g) \rightarrow 2H(g)$ $\Delta H^\circ = 2 \times 218.0$ kJ mol^{-1}

$1/2\, O_2(g) \rightarrow O(g)$ $\Delta H^\circ = 249.2$ kJ mol^{-1}

$H_2O(g) \rightarrow 2H(g) + O(g)$ $\Delta H^\circ = 927.0$ kJ mol^{-1}

This is the enthalpy change associated with breaking both O—H bonds under standard conditions. We conclude that the average bond enthalpy of the O—H bond in water is $1/2 \times 927.0$ kJ mol^{-1} = 463.5 kJ mol^{-1}. We emphasize that this is the average value because the values of ΔH for the transformations $H_2O(g) \rightarrow H(g) + OH(g)$ and $OH(g) \rightarrow O(g) + H(g)$ differ.

b. $\Delta U^\circ = \Delta H^\circ - \Delta(PV) = \Delta H^\circ - \Delta nRT$

$= 927.0$ kJ mol$^{-1} - 2 \times 8.314$ J mol^{-1}K$^{-1} \times 298.15$ K $= 922.0$ kJ mol^{-1}

The average value for ΔU° for the O—H bond in water is $\frac{1}{2} \times 922.0$ kJ mol^{-1} = 461.0 kJ mol^{-1}. The bond energy and the bond enthalpy are nearly identical.

Example Problem 4.2 shows how bond energies can be calculated from reaction enthalpies. The value of a bond energy is of particular importance for chemists in estimating the thermal stability of a compound as well as its stability with respect to reactions with other molecules. Values of bond energies tabulated in the format of the periodic table together with the electronegativities are shown in Table 4.3 [N. K. Kildahl, *J. Chemical Education*, 72 (1995), 423]. The value of the single bond energy, D_{A-B}, for a combination A–B not listed in the table can be estimated using the empirical relationship due to Linus Pauling in Equation (4.25):

$$D_{A-B} = \sqrt{D_{A-A}\, D_{B-B}} + 96.5(x_A - x_B)^2 \tag{4.25}$$

where x_A and x_B are the electronegativities of atoms A and B, respectively.

TABLE 4.3 **Mean Bond Energies**

Selected Bond Energies (kJ/mol)

Element	EN	Single bond with self	Double, triple with self	Bond with H	Single, double bond with O	Bond with F
H	2.20	432	---	432	459	565
He						
Li	0.98	105	---	243	---	573
Be	1.57	208	---	---	---,444	632
B	2.04	293	---	389	536,636	613
C	2.55	346	602,835	411	358,799	485
N	3.04	167	418,942	386	201,607	283
O	3.44	142	494	459	142,494	190
F	3.98	155	---	565	---	155
Ne						
Na	0.93	72	---	197	---	477
Mg	1.31	129	---	---	---,377	513
Al	1.61	--	---	272	---	583
Si	1.90	222	318	318	452,640	565
P	2.19	≈220	---,481	322	335,544	490
S	2.58	240	425	363	---,523	284
Cl	3.16	240	---	428	218	249
Ar						
K	0.82	49	---	180	---	490
Sr	1.00	105	---	---	---,460	550
Ga	1.81	113	---	---	---	≈469
Ge	2.01	188	272	---	---	≈470
As	2.18	146	---,380	247	301,389	≈440
Se	2.55	172	272	276	---	≈351
Br	2.96	190	---	362	201	250
Kr		50				
Rb	0.82	45	---	163	---	490
Sr	0.95	84	---	---	---,347	553
In	1.78	100	---	---	---	≈523
Sn	1.80	146	---	---	---	≈450
Sb	2.05	121	---,295	---	---	≈420
Te	2.10	126	218	238	----	≈393
I	2.66	149	--	295	201	278
Xe		84	≈131			
Cs	0.79	44	---	176	---	502
Ba	0.89	44	---	---	467,561	578
Tl	2.04	---	---	---	---	439
Pb	2.33	---	---	---	---	≈360
Bi	2.02	---	---,192	---	≈350	
Po	2.00	116				
At	2.20					
Rn						

KEY

Element symbol	**C**,2.55	Electronegativity	
	C—C	346	Single bond with self
C=C,	C≡C	602,835	Double, triple bond with self
	H—C	411	Bond with H
O—C,	O=C	358,799	Single, double bond with O
	C—F	485	Bond with F

4.4 The Temperature Dependence of Reaction Enthalpies

Consider a reaction that is mildly exothermic at 298.15 K. Suppose you want to carry out that same reaction at a higher temperature. Is the reaction endothermic or exothermic at the higher temperature? To answer this question, it is necessary to determine $\Delta H^{\circ}_{reaction}$ at the elevated temperature. It is assumed that no phase changes occur in the temperature interval of interest. The enthalpy of each reactant and product at temperature T is related to the value at 298.15 K by Equation (4.26), which accounts for the energy supplied in order to heat the substance to the new temperature at constant pressure:

$$H^{\circ}_T = H^{\circ}_{298.15K} + \int_{298.15K}^{T} C_P(T')dT' \qquad (4.26)$$

The prime in the integral indicates a "dummy variable" that is otherw[...]
temperature. This notation is needed because T appears in the upper li[...]
In Equation (4.26), $H^\circ_{298.15\,K}$ is the absolute enthalpy at 1 bar and 298.[...]
cause there are no unique values for absolute enthalpies, it is useful[...]
equations for all reactants and products to obtain the following equation for the rea[...]
enthalpy at temperature T:

$$\Delta H^\circ_{reaction,\,T} = \Delta H^\circ_{reaction,\,298.15\,K} + \int_{298.15\,K}^{T} \Delta C_P\,(T')dT' \tag{4.27}$$

where

$$\Delta C_p(T') = \sum_i v_i C_{p,i}(T') \tag{4.28}$$

In this equation, the sum is over all reactants and products, *including both elements and compounds*. Elements must be included in the sum over heat capacities. A calculation of $\Delta H^\circ_{reaction}$ at an elevated temperature is shown in Example Problem 4.3.

EXAMPLE PROBLEM 4.3

The standard enthalpies of formation of glucose and lactic acid are -1274.5 and -694.0 kJ/mol, respectively. The molar heat capacities of the two substances are 218.2 and 127.6 J $K^{-1}mol^{-1}$ at 298 K. Calculate the enthalpy change for the formation of lactic acid from glucose at $T = 298$ K and $T = 310$. K. Assume that all heat capacities are temperature independent between 298 K and 310. K.

Solution

The reaction under consideration is

$$C_6H_{12}O_6(s) \rightarrow 2CH_3CHOHCOOH(s)$$

Using Equation (4.27),

$$\Delta H^\circ_{reaction,\,310K} = \Delta H^\circ_{reaction,298K} + \int_{298}^{310} \Delta C^\circ_P\,(T)dT$$

$$\approx \Delta H^\circ_{reaction,298K} + \Delta C^\circ_P \int_{298}^{310} dT = \Delta H^\circ_{298K} + \Delta C^\circ_P \Delta T$$

$$\Delta H^\circ_{reaction,298K} = 2\Delta H^\circ_f\,(CH_3CHOHCOOH,\,s) - \Delta H^\circ_f\,(C_6H_{12}O_6,\,s)$$

$$= 2 \times (-694.0\text{ kJ mol}^{-1}) - (-1273.1\text{ kJ mol}^{-1}) = -114.9\text{ kJ mol}^{-1}$$

$$\Delta C^\circ_P = 2C^\circ_P(CH_3CHOHCOOH,\,s) - C^\circ_P(C_6H_{12}O_6,\,s)$$

$$= 2 \times 127.6\text{ J mol}^{-1}\text{ K}^{-1} - 219.2\text{ J mol}^{-1}\text{ K}^{-1} = 36.0\text{ J mol}^{-1}\text{ K}^{-1}$$

Therefore, $\Delta H^\circ_{reaction,\,310K} = \Delta H^\circ_{reaction,\,298K} + \Delta C^\circ_p \Delta T$

$$= -114,900\text{ J mol}^{-1} + 36.0\text{ J mol}^{-1}\text{ K}^{-1} \times (310.\text{ K} - 298\text{K})$$

$$= -114500\text{ J mol}^{-1} = -114.5\text{ kJ mol}^{-1}$$

In this particular case, the change in the reaction enthalpy with T is not large. This is the case because $\Delta C^\circ_P\,(T)$ is small, and not because the individual $C^\circ_{P,i}\,(T)$ are small, and because the temperature change is not large.

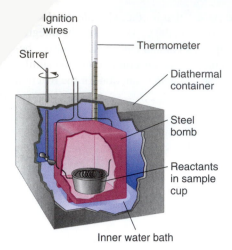

Ignition wires
Thermometer
Stirrer
Diathermal container
Steel bomb
Reactants in sample cup
Inner water bath

FIGURE 4.3
Schematic diagram of a bomb calorimeter. The liquid or solid sample is placed in a cup suspended in the thick-walled steel bomb, which is filled with O_2 gas. The vessel is immersed in an inner water bath and its temperature monitored. The diathermal container is immersed in an outer water bath (not shown) whose temperature is maintained at the same value as the inner bath through a heating coil. By doing so, there is no heat exchange between the inner water bath and the rest of the universe.

4.5 The Experimental Determination of ΔU and ΔH for Chemical Reactions

For chemical reactions, ΔU and ΔH are generally determined experimentally. In this section, we discuss how these experiments are carried out. If some or all of the reactants or products are volatile, it is necessary to contain the reaction mixture for which ΔU and ΔH are being measured. Such an experiment can be carried out in a **bomb calorimeter,** shown schematically in Figure 4.3. In a bomb calorimeter, the reaction is carried out at constant volume. The motivation for doing so is that if $dV = 0$, $\Delta U = q_V$. Therefore, a measurement of the heat flow provides a direct measurement of $\Delta U°_{reaction}$. Bomb calorimetry is restricted to reaction mixtures containing gases, because it is impractical to carry out chemical reactions at constant volume for systems consisting solely of liquids and solids, as shown in Example Problem 3.1. How are $\Delta U°_{reaction}$ and $\Delta H°_{reaction}$ determined in such an experiment?

The bomb calorimeter is a good illustration of how one can define the system and surroundings to simplify the analysis of an experiment. The system is defined as the contents of a stainless steel thick-walled pressure vessel, the pressure vessel itself, and the inner water bath. Given this definition of the system, the surroundings consist of the container holding the inner water bath, the outer water bath, and the rest of the universe. The outer water bath encloses the inner bath and, through a heating coil, its temperature is always held at the temperature of the inner bath. Therefore, no heat flow will occur between the system and the surroundings, and $q = 0$. Because the experiment takes place at constant volume, $w = 0$. Therefore, $\Delta U = 0$. These conditions describe an isolated system of finite size that is not coupled to the rest of the universe. We are only interested in one part of this system, namely, the reaction mixture.

What are the individual components that make up ΔU? Consider the system as consisting of three subsystems: the reactants in the calorimeter, the calorimeter vessel, and the inner water bath. These three subsystems are separated by rigid diathermal walls and are in thermal equilibrium. Energy is redistributed among the subsystems through the following processes. Reactants are converted to products, the energy of the inner water bath changes because of the change in temperature, ΔT, and the same is true of the calorimeter:

$$\Delta U° = \frac{m_s}{M_s} \Delta U°_{reaction,m} + \frac{m_{H_2O}}{M_{H_2O}} C_{H_2O,m} \Delta T + C_{calorimeter} \Delta T = 0 \quad (4.29)$$

In Equation (4.29), ΔT is the change in the temperature of the inner water bath. The mass of water in the inner bath, m_{H_2O}, its molecular weight, M_{H_2O}, its heat capacity, $C_{H_2O,m}$, and the mass of the sample, m_s, and its molecular weight, M_s, are known. Our interest is in determining $\Delta U°_{combustion,m}$. However, to determine $\Delta U°_{reaction,m}$, the heat capacity of the calorimeter, $C_{calorimeter}$, must first be determined by carrying out a reaction for which $\Delta U°_{reaction,m}$ is already known, as illustrated in Example Problem 4.4. To be more specific, we consider a combustion reaction between a compound and an excess of O_2.

EXAMPLE PROBLEM 4.4

Although ammonia plays a vital role in amino acid synthesis and degradation, its accumulation has toxic effects. Therefore, most terrestrial animals convert ammonia to uric acid, and subsequently excrete nitrogen waste as urea, NH_2CONH_2. Urea is a soluble, electrically neutral compound that does not affect pH when it accumulates in tissue fluids. Suppose a sample of crystalline urea weighing 1.372 g is burned in a bomb calorimeter for which the heat capacity is 5.00 kJ K^{-1}. The mass of water in the inner bath is 1.812×10^3 g, and $C_{P,m}$ of water is $75.3 \text{ J K}^{-1} \text{ mol}^{-1}$. The temperature is observed to rise by 1.92 K. Calculate $\Delta U°_{reaction}$.

Solution

$$\Delta U^{\circ}_{reaction} = -\frac{M_s}{m_s}\left(\frac{m_{H_2O}}{M_{H_2O}}C_{H_2O,m}\Delta T + C_{calorimeter}\Delta T\right)$$

$$= -\frac{60.05 \text{ g mol}^{-1}}{1.372 \text{ g}} \times \left(\begin{array}{l}\dfrac{1.812 \times 10^3 \text{ g}}{18.02 \text{ g mol}^{-1}} \times 75.3 \text{ J mol}^{-1}\text{K}^{-1} \times 1.92 \text{ K} \\ + 5.00 \text{ kJ K}^{-1} \times 1.92 \text{ K}\end{array}\right)$$

$$= -1.06 \times 10^3 \text{ kJ mol}^{-1}$$

Once ΔU has been determined, ΔH can be determined using the following equation:

$$\Delta H^{\circ}_{reaction} = \Delta U^{\circ}_{reaction} + \Delta(PV) \qquad (4.30)$$

For reactions involving only solids and liquids, $\Delta U \gg \Delta(PV)$ and $\Delta H \approx \Delta U$. If some of the reactants or products are gases, the small change in the temperature that is measured in a calorimetric experiment can be ignored and $\Delta(PV) = \Delta(nRT) = \Delta nRT$ and

$$\Delta H^{\circ}_{reaction} = \Delta U^{\circ}_{reaction} + \Delta nRT \qquad (4.31)$$

where Δn is the change in the number of moles of gas in the overall reaction. For the first reaction of Example Problem 4.4,

$$NH_2CONH_2(s) + \tfrac{3}{2}O_2(g) \rightarrow CO_2(g) + 2H_2O(l) + N_2(g)$$

and $\Delta n = \tfrac{1}{2}$. Then

$$\Delta H^{\circ}_{reaction} = \Delta U^{\circ}_{reaction} + \Delta(PV) = \Delta U^{\circ}_{reaction} + RT\Delta n$$

$$= -6.37 \times 10^5 \text{ J mol}^{-1} + \frac{8.314 \text{ J mol}^{-1}\text{ K}^{-1} \times 298.15 \text{ K}}{2}$$

$$= -635 \text{ kJ mol}^{-1} \qquad (4.33)$$

For this reaction, $\Delta H^{\circ}_{reaction}$ and $\Delta U^{\circ}_{reaction}$ differ by only 0.3%.

EXAMPLE PROBLEM 4.5

The amino acid glycine NH_2CH_2COOH has a standard enthalpy of combustion of 981.0 kJ mol^{-1}. Calculate q, w, ΔU, and ΔH when 10.0 g of glycine are burned at a constant pressure of 1.00 atm and $T = 298$ K. Assume the combustion products are carbon dioxide gas, nitrogen gas, and liquid water.

Solution

At constant pressure the heat of combustion is equal to the enthalpy change. The molecular weight of glycine is 75.0 g mol^{-1}. Then

$$q_P = \Delta H = -981 \text{ kJ mol}^{-1} \times \frac{10.0 \text{ g}}{75.0 \text{ g mol}^{-1}} = -132 \text{ kJ}. \text{ To determine the work,}$$

we use $W = -P_{ext}\Delta V$, where ΔV is the change in the volume of the system. This is dominated by the change in the moles of gaseous species, which can be determined from the balanced equation for the combustion of glycine:

$$NH_2CH_2COOH(s) + \tfrac{9}{4}O_2(g) \rightarrow 2CO_2(g) + \tfrac{5}{2}H_2O(l) + \tfrac{1}{2}N_2(g)$$

Salt — Thermometer

Stopper

H₂O

Solution

FIGURE 4.4
Schematic diagram of a constant pressure calorimeter suitable for measuring the enthalpy of solution of a salt in a solution.

The change in moles of gas is $\Delta n = 5/2$ moles $- 9/4$ moles $= 1/4$ moles. Then

$$\Delta V = \frac{RT}{P_{ext}} \Delta n = \frac{0.08206 \text{ L atm mol}^{-1} \text{ K}^{-1} \times 300. \text{ K}}{1.00 \text{ atm}} \times 0.250 \text{ mol} = 6.16 \text{ L}$$

$$w = -P_{ext}\Delta V = -6.16 \text{ L atm} \times \frac{101 \text{ J}}{\text{L atm}} \times 0.133 \text{ mol} = -83.0 \text{ J}$$

$$\Delta U = q_P + w = -132 \text{kJ} - 83.0 \text{ J} \approx -132 \text{ kJ}$$

Because $q_P = \Delta H$, $\Delta H = -132 \text{ kJ}$

If the reaction under study does not involve gases or highly volatile liquids, there is no need to operate under constant volume conditions. It is preferable to carry out the reaction at constant P, and $\Delta H^{\circ}_{reaction}$ is directly determined because $\Delta H^{\circ}_{reaction} = q_P$. A vacuum-insulated vessel with a loosely fitting stopper as shown in Figure 4.4 is adequate for many purposes and can be treated as an isolated composite system. Equation (4.27) takes the following form for constant pressure calorimetry involving the solution of a salt in water:

$$\Delta H^{\circ}_{reaction} = \frac{m_s}{M_s} \Delta H^{\circ}_{solution,m} + \frac{m_{H_2O}}{M_{H_2O}} C_{H_2O,m}\Delta T + C_{calorimeter} \Delta T = 0 \qquad (4.34)$$

Note that in this case the water in the round-bottomed reaction vessel is part of the reaction mixture. Because $\Delta(PV)$ is negligibly small for the solution of a salt in a solvent, $\Delta U_{solution} = \Delta H_{solution}$. The solution must be stirred to ensure that equilibrium is attained before ΔT is measured.

EXAMPLE PROBLEM 4.6

The enthalpy of solution for the reaction

$$Na_2SO_4(s) + H_2O(l) \rightarrow 2Na^+(aq) + SO_4{}^{2-}(aq)$$

is determined in a **constant pressure calorimeter.** The calorimeter constant was determined to be 342.5 J K^{-1}. When 1.423 g of Na$_2$SO$_4$ is dissolved in 100.34 g of H$_2$O(l), $\Delta T = 0.037$ K. Calculate ΔH_m for Na$_2$SO$_4$ from these data. Compare your result with that calculated using the standard enthalpies of formation in Table 4.1 (Appendix B, Data Tables) and in Chapter 9 in Table 9.1.

Solution

$$\Delta H^{\circ}_{solution,m} = -\frac{M_s}{m_s}\left(\frac{m_{H_2O}}{M_{H_2O}} C_{H_2O,m} \Delta T + C_{calorimeter}\Delta T\right)$$

$$= -\frac{142.04 \text{ g mol}^{-1}}{1.423 \text{ g}} \times \left(\begin{array}{c}\frac{100.34 \text{ g}}{18.02 \text{ g mol}^{-1}} \times 75.3 \text{ J K}^{-1}\text{mol}^{-1} \times 0.037 \text{ K}\\ + 342.5 \text{ J K}^{-1} \times 0.037 \text{ K}\end{array}\right)$$

$$= -2.8 \times 10^3 \text{ J mol}^{-1}$$

We next calculate $\Delta H^{\circ}_{solution,m}$ using the data tables:

$$\Delta H^{\circ}_{solution,m} = 2\Delta H^{\circ}_f(Na^+, aq) + \Delta H^{\circ}_f(SO_4^{2-}, aq) - \Delta H^{\circ}_f(Na_2SO_4, s)$$

$$= 2 \times (-240.1 \text{ kJ mol}^{-1}) - 909.3 \text{ kJ mol}^{-1} + 1387.1 \text{ kJ mol}^{-1}$$

$$= -2.4 \text{kJ mol}^{-1}$$

The agreement between the calculated and experimental results is satisfactory.

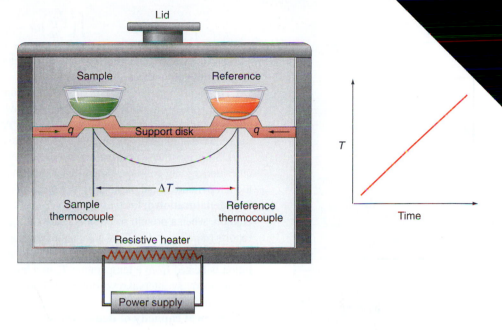

FIGURE 4.5

A differential scanning calorimeter consists of a massive enclosure and lid that are heated to the temperature T using a resistive heater. A support disk in good thermal contact with the enclosure supports multiple samples and a reference material. The temperatures of each of the samples and the reference are measured with a thermocouple.

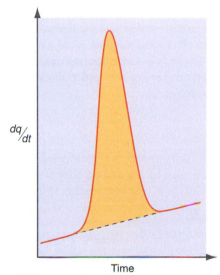

FIGURE 4.6

At temperatures different from the melting temperature, the rate of heat uptake of the sample and the reference are nearly identical. However, at the time corresponding to $T = T_{melting}$, the uptake rate of heat for the sample differs significantly from that of the reference. Kinetic delays due to finite rates of heat conduction broaden the sharp melting transition to finite range.

4.6 Differential Scanning Calorimetry

The constant volume and constant pressure calorimeters described in the preceding section are well suited for measurements on individual samples. Suppose that it is necessary to determine the **enthalpy of fusion** of a dozen related solid materials, each of which is only available in a small quantity. Because experiments are carried out individually, constant volume and constant pressure calorimeters are not well suited to the rapid determination of thermodynamic properties of a series of materials. The technique of choice for these measurements is **differential scanning calorimetry (DSC).** The experimental apparatus for such an experiment is shown schematically in Figure 4.5. The word *differential* appears in the name of the technique because the uptake of heat is measured relative to that for a reference material, and *scanning* refers to the fact that the temperature of the sample is varied linearly with time.

The temperature of the cylindrical enclosure is increased linearly with time using a power supply. Heat flows from the enclosure through the disk to the sample because of the temperature gradient generated by the heater. Because all samples and the reference are equidistant from the enclosure, the heat flow to each sample is the same. The reference material is chosen such that its melting point is not in the range of that of the samples.

The measured temperature difference between one sample and the reference, ΔT, is proportional to the difference in the rate of heat uptake of the sample and the reference:

$$\Delta T = \alpha \frac{dq_P}{dt} = \alpha \left(\frac{dq_{reference}}{dt} - \frac{dq_{sample}}{dt} \right) \qquad (4.35)$$

where α is a calorimeter constant that is determined experimentally. In the temperature range over which no phase transition occurs, $\Delta T \approx 0$. However, as a sample melts, it takes up more heat than the reference, and its temperature remains constant while that of the reference increases. The temperature difference between the sample and the reference has the form shown in Figure 4.6 as a function of time. Because the relationship between the time and temperature is known, the time scale can be converted to a temperature scale. Given that the sample has a well-defined melting point, why is the peak in Figure 4.6 broad? The system experiences kinetic delays because the heat flow through the disk is not instantaneous. Therefore, the experimentally determined function dq_P/dt has a finite width and the form of a Gaussian curve. Because the vertical axis is dq_P/dt and the horizontal axis is time, the integral of the area under the curve is

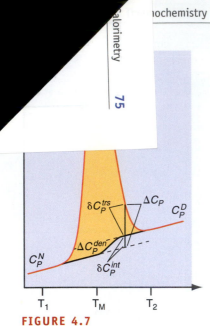

FIGURE 4.7

The heat capacity of a protein undergoing a reversible, thermal denaturation within the temperature interval $T_1 - T_2$. The temperature at the heat capacity maximum is the melting temperature T_m, defined as the temperature at which half the total protein has been denatured. The excess heat capacity $\Delta C_P = C_P(T) - C_P^N(T)$ is composed of two parts: the intrinsic excess heat capacity δC_P^{int} and the transition excess heat capacity δC_P^{trs}. See text.

$$\int \left(\frac{dq_P}{dt} \right) dt = q_P = \Delta H \tag{4.36}$$

Therefore, ΔH_{fusion} can be ascertained by determining the area under the curve representing the output data.

The significant advantage of differential scanning calorimetry over constant pressure or constant volume calorimetry is the ability to investigate several samples in parallel and the ability to work with small amounts of sample. As is apparent from Figure 4.5, the apparatus shown is limited to working with solids or to solutions in which phase transitions occur.

DSC is the most direct method for determining the energetics of biological macromolecules undergoing conformational transitions. In particular, DSC has been used to determine the temperature range over which proteins undergo the conformational changes associated with **denaturation** defined as the unfolding of the protein structure. From the preceding discussion, when a protein solution is heated at a constant rate and at constant pressure, DSC reports $(dq/dT)_P = C_p(T)$, which is the heat capacity of the protein solution. Protein structures are stabilized by the cooperation of numerous weak forces. Assume that a protein solution is heated from a temperature T_1 to a temperature T_2. If the protein denatures within this temperature interval reversibly and cooperatively, the $C_p(T)$ vs. T curve is peak shaped, as shown in Figure 4.7.

The **heat capacity of denaturation** is defined as

$$\Delta C_P^{den} = C_P^D(T) - C_P^N(T) \tag{4.37}$$

which is the difference between the heat capacity of the denatured protein, $C_P^D(T)$, and the heat capacity of the native (i.e., structured) protein, $C_P^N(T)$. The value of ΔC_P^{den} can be determined as shown in Figure 4.7. The value of ΔC_P^{den} is a positive quantity and for many globular proteins varies from 0.3 to 0.7 J K^{-1}g^{-1}. The higher heat capacity of the protein in the denatured state can be understood in the following way. In the structured state, internal motions of the protein are characterized by coupled bending and rotations of several bonds that occur at frequencies on the order of $k_B T/h$, where k_B is Boltzmann's constant and h is Planck's constant. When the protein denatures, these "soft" vibrations are shifted to lower frequencies, higher frequency bond vibrations are excited, and the heat capacity increases. In addition to increased contributions from vibrational motions, when a protein structure is disrupted, nonpolar amino acid residues formerly isolated from solvent within the interior of the protein are exposed to water. Water molecules now order about these nonpolar amino acids, further increasing the heat capacity.

The **excess heat capacity** is defined as

$$\Delta C_P(T) = C_P(T) - C_P^N(T) \tag{4.38}$$

where ΔC_p is obtained at a given temperature from the difference between the point on the heat capacity curve $C_p(T)$ and the linearly extrapolated value for the heat capacity of the native state $C_P^N(T)$; see Figure 4.7. The excess heat capacity is in turn composed of two parts:

$$\Delta C_P(T) = \delta C_P^{int} + \delta C_P^{trs} \tag{4.39}$$

Within the interval $T_1 - T_2$ the protein structure does not unfold suddenly. There occurs instead a gradual unfolding of the protein such that at any given temperature a fraction of the structured protein remains, and an ensemble of partly unfolded species is produced. The observed heat capacity and enthalpy change arise from an ensemble of physical forms of the partially unfolded protein, and as such these properties are averages. The component of the heat capacity that accounts for heat absorbed by the molecular species produced in the course of making a transition from the folded to the unfolded state is the **intrinsic excess heat capacity** or δC_P^{int} and results in the upward drift of the δC baseline. The second component of the heat capacity is called the **transition excess heat capacity** δC_P^{trs}. The transition excess heat capacity results from heat absorbed by transitions between different structural states. At the melting temperature T_m, the folded and unfolded states are equally populated, the conversion between folded and unfolded states is nearly reversible and the heat absorbed reaches a maximum. This is observed to be the peak in the heat capacity curve at T_m.

The intrinsic and transition excess heat capacities are determined by first extrapolating the functions $C_P^N(T)$ and $C_P^D(T)$ into the transition zone between T_1 and T_2. This is indicated in Figure 4.7 by the solid line connecting the heat capacity baselines above and below T_m. Once this baseline extrapolation is accomplished, δC_P^{int} is the difference at a given temperature between the extrapolated baseline curve and $C_P^N(T)$. The value of δC_P^{trs} is obtained from the difference between $C_p(T)$ and the extrapolated baseline curve. The excess enthalpy of thermal denaturation is finally given by $\Delta H_{den} = \int_{T_1}^{T_2} \delta C_P^{trs} \, dT$. In other words the excess enthalpy of thermal denaturation is the area under the peak in Figure 4.7 above the extrapolated baseline curve.

Vocabulary

bomb calorimeter
bond energy
bond enthalpy
constant pressure calorimeter

denaturation
differential scanning
   calorimetry (DSC)
endothermic
enthalpy of formation

enthalpy of fusion
enthalpy of reaction
excess heat capacity
exothermic

Hess's law
intrinsic excess heat capacity
standard state
transition excess heat
   capacity

Questions on Concepts

Q4.1 Under what conditions are ΔH and ΔU for a reaction involving gases and/or liquids or solids identical?

Q4.2 If the ΔH_f° for the chemical compounds involved in a reaction are available at a given temperature, how can the reaction enthalpy be calculated at another temperature?

Q4.3 Does the enthalpy of formation of compounds containing a certain element change if the enthalpy of formation of the element under standard state conditions is set equal to 100 kJ mol^{-1} rather than to zero? If it changes, how will it change for the compound A_nB_m if the formation enthalpy of element A is set equal to 100 kJ mol^{-1}?

Q4.4 Is the enthalpy for breaking the first C—H bond in methane equal to the average C—H bond enthalpy in this molecule? Explain your answer.

Q4.5 Why is it valid to add the enthalpies of any sequence of reactions to obtain the enthalpy of the reaction that is the sum of the individual reactions?

Q4.6 The reactants in the reaction $2NO(g) + O_2(g) \rightarrow 2NO_2(g)$ are initially at 298 K. Why is the reaction enthalpy the same if the reaction is (a) constantly kept at 298 K or (b) if the reaction temperature is not controlled and the heat flow to the surroundings is measured only after the temperature of the products is returned to 298 K?

Q4.7 In calculating $\Delta H_{reaction}^\circ$ at 298.15 K, only the ΔH_f° of the compounds that take part in the reactions listed in Tables 4.1 and 4.2 (Appendix B, Data Tables) are needed. Is this statement also true if you want to calculate $\Delta H_{reaction}^\circ$ at 500 K?

Q4.8 What is the point of having an outer water bath in a bomb calorimeter (see Figure 4.3), especially if its temperature is always equal to that of the inner water bath?

Q4.9 What is the advantage of a differential scanning calorimeter over a bomb calorimeter in determining the enthalpy of fusion of a series of samples?

Q4.10 You wish to measure the heat of solution of NaCl in water. Would the calorimetric technique of choice be at constant pressure or constant volume? Why?

Problems

Problem numbers in **RED** indicate that the solution to the problem is given in the *Student Solutions Manual*.

P4.1 Calculate $\Delta H_{reaction}^\circ$ and $\Delta U_{reaction}^\circ$ at 298.15 K for the following reactions:

  a. $4NH_3(g) + 6NO(g) \rightarrow 5N_2(g) + 6H_2O(g)$

  b. $2NO(g) + O_2(g) \rightarrow 2NO_2(g)$

  c. $TiCl_4(l) + 2H_2O(l) \rightarrow TiO_2(s) + 4HCl(g)$

  d. $2NaOH(aq) + H_2SO_4(aq) \rightarrow Na_2SO_4(aq) + 2H_2O(l)$. Assume complete dissociation of NaOH, H_2SO_4, and Na_2SO_4.

  e. $CH_4(g) + H_2O(g) \rightarrow CO(g) + 3H_2(g)$

  f. $CH_3OH(g) + CO(g) \rightarrow CH_3COOH(l)$

P4.2 Calculate $\Delta H_{reaction}^\circ$ and $\Delta U_{reaction}^\circ$ for the oxidation of benzene. Also calculate

$$\frac{\Delta H_{reaction}^\circ - \Delta U_{reaction}^\circ}{\Delta H_{reaction}^\circ}$$

P4.3 Use the tabulated values of the enthalpy of combustion of benzene and the enthalpies of formation of $CO_2(g)$ and $H_2O(l)$ to determine ΔH_f° for benzene.

P4.4   Calculate ΔH for the process $N_2(g, 298 \text{ K}) \rightarrow N_2(g, 650 \text{ K})$ using the temperature dependence of the heat capacities from the data tables. How large is the relative error if the molar heat capacity is assumed to be constant at its value of 298.15 K over the temperature interval?

P4.5   Several reactions and their standard reaction enthalpies at 25°C are given here:

$$\Delta H°_{reaction} \text{ (kJ mol}^{-1})$$

$CaC_2(s) + 2H_2O(l) \rightarrow Ca(OH)_2(s) + C_2H_2(g)$	-127.9
$Ca(s) + 1/2\ O_2(g) \rightarrow CaO(s)$	-635.1
$CaO(s) + H_2O(l) \rightarrow Ca(OH)_2(s)$	-65.2

The standard enthalpies of combustion of graphite and $C_2H_2(g)$ are -393.51 and -1299.58 kJ mol^{-1}, respectively. Calculate the standard enthalpy of formation of $CaC_2(s)$ at 25°C.

P4.6   From the following data at 25°C, calculate the standard enthalpy of formation of FeO(s) and of $Fe_2O_3(s)$:

$$\Delta H°_{reaction} \text{ (kJ mol}^{-1})$$

$Fe_2O_3(s) + 3C(graphite) \rightarrow 2Fe(s) + 3CO(g)$	492.6
$FeO(s) + C(graphite) \rightarrow Fe(s) + CO(g)$	155.8
$C(graphite) + O_2(g) \rightarrow CO_2(g)$	-393.51
$CO(g) + 1/2\ O_2(g) \rightarrow CO_2(g)$	-282.98

P4.7   Calculate $\Delta H°_f$ for NO(g) at 840 K assuming that the heat capacities of reactants and products are constant over the temperature interval at their values at 298.15 K.

P4.8   Calculate $\Delta H°_{reaction}$ at 650 K for the reaction $4NH_3(g) + 6NO(g) \rightarrow 5N_2(g) + 6H_2O(g)$ using the temperature dependence of the heat capacities from the data tables.

P4.9   From the following data at 298.15 K as well as data in Table 4.1 (Appendix B, Data Tables), calculate the standard enthalpy of formation of $H_2S(g)$ and of $FeS_2(s)$:

$$\Delta H°_{reaction} \text{ (kJ mol}^{-1})$$

$Fe(s) + 2H_2S(g) \rightarrow FeS_2(s) + 2H_2(g)$	-137.0
$H_2S(g) + 3/2\ O_2(g) \rightarrow H_2O(l) + SO_2(g)$	-562.0

P4.10   Calculate the average C—H bond enthalpy in methane using the data tables. Calculate the percent error in equating the average C—H bond energy in Table 4.3 with the bond enthalpy.

P4.11   Use the average bond energies in Table 4.3 to estimate ΔU for the reaction $C_2H_4(g) + H_2(g) \rightarrow C_2H_6(g)$. Also calculate $\Delta U°_{reaction}$ from the tabulated values of $\Delta H°_f$ for reactant and products (Appendix B, Data Tables). Calculate the percent error in estimating $\Delta U°_{reaction}$ from the average bond energies for this reaction.

P4.12   Calculate the standard enthalpy of formation of $FeS_2(s)$ at 300°C from the following data at 25°C. Assume that the heat capacities are independent of temperature.

Substance	Fe(s)	$FeS_2(s)$	$Fe_2O_3(s)$	S(rhombic)	$SO_2(g)$
$\Delta H°_f$ (kJ mol^{-1})			-824.2		-296.81
$C_{P,m}/R$	3.02	7.48		2.72	

You are also given that for the reaction $2FeS_2(s) + 11/2O_2(g) \rightarrow Fe_2O_3(s) + 4\ SO_2(g)$, $\Delta H°_{reaction} = -1655$ kJ mol^{-1}.

P4.13   At 1000 K, $\Delta H°_{reaction} = -123.77$ kJ mol^{-1} for the reaction $N_2(g) + 3H_2(g) \rightarrow 2NH_3(g)$, with $C_{P,m} = 3.502R$, 3.466R, and 4.217R for $N_2(g)$, $H_2(g)$, and $NH_3(g)$, respectively. Calculate $\Delta H°_f$ of $NH_3(g)$ at 300 K from this information. Assume that the heat capacities are independent of temperature.

P4.14   At 298 K, $\Delta H°_{reaction} = 131.28$ kJ mol^{-1} for the reaction $C(graphite) + H_2O(g) \rightarrow CO(g) + H_2(g)$, with $C_{P,m} = 8.53, 33.58, 29.12,$ and 28.82 J K^{-1} mol^{-1} for graphite, $H_2O(g)$, CO(g), and $H_2(g)$, respectively. Calculate $\Delta H°_{reaction}$ at 125°C from this information. Assume that the heat capacities are independent of temperature.

P4.15   From the following data, calculate $\Delta H°_{reaction,\ 391.4\ K}$ for the reaction $CH_3COOH(g) + 2O_2(g) \rightarrow 2H_2O(g) + 2CO_2(g)$:

$$\Delta H°_{reaction} \text{ (kJ mol}^{-1})$$

$CH_3COOH(l) + 2O_2(g) \rightarrow 2H_2O(l) + 2CO_2(g)$	-871.5
$H_2O(l) \rightarrow H_2O(g)$	40.656
$CH_3COOH(l) \rightarrow CH_3COOH(g)$	24.4

Values for $\Delta H°_{reaction}$ for the first two reactions are at 298.15 K, and for the third reaction at 391.4 K.

Substance	$CH_3COOH(l)$	$O_2(g)$	$CO_2(g)$	$H_2O(l)$	$H_2O(g)$
$C_{P,m}/R$	14.9	3.53	4.46	9.055	4.038

P4.16   Consider the reaction $TiO_2(s) + 2\ C(graphite) + 2\ Cl_2(g) \rightarrow 2\ CO(g) + TiCl_4(l)$ for which $\Delta H°_{reaction,\ 298\ K} = -80$ kJ mol^{-1}. Given the following data at 25°C, (a) calculate $\Delta H°_{reaction}$ at 135.8°C, the boiling point of $TiCl_4$, and (b) calculate $\Delta H°_f$ for $TiCl_4$ (l) at 25°C:

Substance	$TiO_2(s)$	$Cl_2(g)$	C(graphite)	CO(g)	$TiCl_4(l)$
$\Delta H°_f$ (kJ mol^{-1})	-945			-110.5	
$C_{P,m}$ (J K^{-1} mol^{-1})	55.06	33.91	8.53	29.12	145.2

Assume that the heat capacities are independent of temperature.

P4.17   Use the following data at 25°C to complete this problem:

$$\Delta H°_{reaction} \text{ (kJ mol}^{-1})$$

$1/2\ H_2(g) + 1/2\ O_2(g) \rightarrow OH(g)$	38.95
$H_2(g) + 1/2\ O_2(g) \rightarrow H_2O(g)$	-241.814
$H_2(g) \rightarrow 2H(g)$	435.994
$O_2(g) \rightarrow 2O(g)$	498.34

Calculate $\Delta H°_{reaction}$ for

a. $OH(g) \rightarrow H(g) + O(g)$

b. $H_2O(g) \rightarrow 2H(g) + O(g)$

c. $H_2O(g) \rightarrow H(g) + OH(g)$

Assuming ideal gas behavior, calculate $\Delta H°_{reaction}$ and $\Delta U°_{reaction}$ for all three reactions.

P4.18 Given the data in Table 4.1 (Appendix B, Data Tables) and the following information, calculate the single bond enthalpies and energies for Si—F, Si—Cl, C—F, N—F, O—F, H—F:

Substance	$SiF_4(g)$	$SiCl_4(g)$	$CF_4(g)$	$NF_3(g)$	$OF_2(g)$	$HF(g)$
$\Delta H°_f$ (kJ mol^{-1})	−1614.9	−657.0	−925	−125	−22	−271

P4.19 Given the data in Table 4.3 and the data tables, calculate the bond enthalpy and energy of the following:

a. The C—H bond in CH_4

b. The C—C single bond in C_2H_6

c. The C—C double bond in C_2H_4

Use your result from part (a) to solve parts (b) and (c).

P4.20 A sample of K(s) of mass 2.140 g undergoes combustion in a constant volume calorimeter. The calorimeter constant is 1849 J K^{-1}, and the measured temperature rise in the inner water bath containing 1450 g of water is 2.62 K. Calculate $\Delta U°_f$ and $\Delta H°_f$ for K_2O.

P4.21 Benzoic acid, 1.35 g, is reacted with oxygen in a constant volume calorimeter to form $H_2O(l)$ and $CO_2(g)$. The mass of the water in the inner bath is 1.240×10^3 g. The temperature of the calorimeter and its contents rise 3.45 K as a result of this reaction. Calculate the calorimeter constant.

P4.22 A sample of $Na_2SO_4(s)$ is dissolved in 225 g of water at 298 K such that the solution is 0.200 molar in Na_2SO_4. A temperature rise of 0.101°C is observed. The calorimeter constant is 330 J K^{-1}. Calculate the enthalpy of solution of Na_2SO_4 in water at this concentration. Compare your result with that calculated using the data in Table 4.1 (Appendix B, Data Tables).

P4.23 Oxygen gas reacts with solid glycylglycine $(C_4H_8N_2O_3)$ to form solid urea (CH_4N_2O), carbon dioxide gas, and liquid water. At T= 298 K and P = 1 bar, the standard enthalpy change is $\Delta H°_{298}$ = −1340.1 kJ mol^{-1}.

$$3O_2(g) + C_4H_8N_2O_3(s) \rightarrow CH_4N_2O(s) + 3CO_2(g) + 2H_2O(l)$$

The heat capacities $C°_{P,m}$ for the reactants and products are:

Substance	Glycylglycine $(C_4H_8N_2O_3)$	Urea (CH_4N_2O)	O_2	CO_2	H_2O
$C°_{P,m}$ (J mol^{-1} K^{-1})	163.6	93.1	29.4	37.1	75.3

Calculate the standard enthalpy change $\Delta H°_{reaction}$ at T = 330. K and P = 1.00 bar. Assume all heat capacities are constant between T = 298 K and T = 330. K.

P4.24 Consider the formation of glucose from carbon dioxide and water, that is, the reaction of the photosynthetic process: $6CO_2(g) + 6H_2O(l) \rightarrow C_6H_{12}O_6(s) + 6O_2(g)$.

The following table of information will be useful in working this problem:

T = 298 K	$CO_2(g)$	H_2O	$C_6H_{12}O_6$	O_2
$\Delta H°_f$ (kJ mol^{-1})	−393.5	−285.8	−1273.1	0.0
$C°_{P,m}$ (J mol^{-1} K^{-1})	37.1	75.3	218.2	29.4

Calculate the enthalpy change for this chemical system at T = 298 K and at T = 330. K.

P4.25 The total surface area of the earth consisting of forest, cultivated land, grass land, and desert is 1.49×10^8 km^2. Every year, the mass of carbon fixed by photosynthesis by vegetation covering this land surface is about 450. metric tons km^2. Using your result from Problem P 4.24, calculate the annual enthalpy change resulting from photosynthetic carbon fixation over the land surface given above. Assume standard conditions and T = 298 K, on average.

P4.26 The total surface area of the earth covered by ocean is 3.61×10^8 km^2. Carbon is fixed in the oceans via photosynthesis performed by marine plants. A lower range estimate of the mass of carbon fixed in the oceans is 46.0 metric tons km^2. Repeat the enthalpy calculation in Problem P 4.25 for carbon fixation in the oceans. Assume standard conditions and T=298K, on average.

P4.27 The total solar flux over the earth's surface is about 4×10^{24} J yr^{-1}. Using your results from Problems P 4.25 and P 4.26, calculate the fraction of solar flux that is eventually used to fix carbon via photosynthesis.

P4.28 Certain yeast can degrade glucose into ethanol and carbon dioxide in a process called alcoholic fermentation according to the equation:

$$C_6H_{12}O_6(s) \rightarrow 2C_2H_5OH(l) + 2CO_2(g)$$

a. Calculate the enthalpy of reaction. You can find appropriate heats of formation in the appendixes.

b. Calculate the work done at constant pressure of 1 bar and T = 298 K per mole of glucose fermented. Assume carbon dioxide behaves ideally.

c. Calculate the energy change ΔU when 1 mol of glucose ferments at T = 298 K and 1 bar.

P4.29 A good yield of photosynthesis for agricultural crops in bright sunlight is 2000. kg of carbohydrate (e.g., sucrose) per km^2 per hour. The net reaction for sucrose formation in photosynthesis is:

$$12CO_2(g) + 11H_2O(l) \xrightarrow{light} C_{12}H_{22}O_{11}(s) + 12O_2(g)$$

a. Use standard enthalpies of formation to calculate $\Delta H°_{reaction}$ for the production of 1 mol of sucrose at 298 K by the preceding reaction.

b. Calculate the rate at which energy is stored in carbohydrates (e.g., sucrose) per km^2 as a result of photosynthesis. (*Note:* 1 watt = 1 joule/s.)

c. Bright sunlight corresponds to radiation flux at the surface of the earth of about 1 kW/m^2. What percentage of this energy can be stored in the form of carbohydrates as a result of photosynthesis?

P4.30 In anaerobic cells, glucose, $C_6H_{12}O_6$, is converted to lactic acid in the reaction $C_6H_{12}O_6 \rightarrow 2CH_3CH(OH)COOH$. The ΔH_f° of glucose and lactic acid are -1273.1 and -694.0 kJ mol^{-1}, respectively. The $C_{P,m}^\circ$ for glucose and lactic acid are 219.2 and 127.6 J mol^{-1} K^{-1}, respectively.

  a. Calculate the molar enthalpy associated with the formation of lactic acid from glucose at 298 K.

  b. What would the quantity in part (a) be if the reaction proceeded at a physiological temperature of 310. K?

  c. Based on your answers in parts (a) and (b), how sensitive is the enthalpy of this reaction to moderate temperature changes?

P4.31 The figure below shows a DSC scan of a solution of a T4 lysozyme mutant. From the DSC data, determine T_m, the excess heat capacity δC_P and the intrinsic and transition excess heat capacities at $T = 308$ K. In your calculations, use the extrapolated curves, shown as dashed lines in the DSC scan.

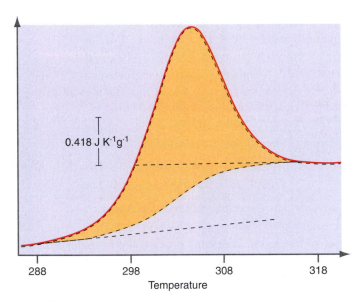

0.418 J $K^{-1}g^{-1}$

288   298   308   318
Temperature

P4.32 Using the protein DSC data from Problem P 4.31, calculate the enthalpy change between the $T=288$ K and $T=318$ K. Give your answer in units of kilojoules per mole. Assume the molecular weight of the protein is 14000. grams. Hint: You can perform the integration of the heat capacity by estimating the area under the DSC curve and above the dotted baseline in Problem P 4.31. This can be done by dividing the area up into small rectangles and summing the areas of the rectangles. Comment on the accuracy of this method.

P4.33 Suppose a 1.0-ton automobile has a mileage of 20. miles gal^{-1}. Compare the energy consumed by this vehicle per pound per mile with the energy consumed by a walking adult, who expends about 2.0 kJ per mile per pound of body weight. Use the enthalpy of combustion of hexane (C_6H_{14}), -4163 kJ mol^{-1}, as typical of automobile fuel. Assume a fuel density of 0.68 g mL^{-1}.

P4.34 Compare the heat evolved at constant pressure per mole of oxygen in the combustion of sucrose ($C_{12}H_{22}O_{11}$) and palmitic acid ($C_{16}H_{32}O_2$) with the combustion of a typical

protein, for which the empirical formula is $C_{4.3}H_{6.6}NO$. Assume for the protein that the combustion yields $N_2(g)$, $CO_2(g)$, and $H_2O(l)$. Assume that the enthalpies for combustion of sucrose, palmitic acid, and a typical protein are 5647 kJ mol^{-1}, 10035 kJ mol^{-1}, and 22.0 kJ g^{-1}, respectively. Based on these calculations, determine the average heat evolved per mole of oxygen consumed, assuming combustion of equal moles of sucrose, palmitic acid, and protein.

P4.35 A camper stranded in snowy weather loses heat by wind convection. The camper is packing emergency rations consisting of 60.% sucrose, 30.% fat, and 10.% protein by weight. Using the data provided in Problem P4.34 and assuming the fat content of the rations can be treated with palmitic acid data and the protein content similarly by the protein data in Problem P4.34, how much emergency rations must the camper consume in order to compensate for a reduction in body temperature of 4.0 K? Assume the heat capacity of the body equals that of water. Assume the camper weighs 70. kg. State any additional assumptions.

P4.36 The dipeptide glycylglycine ($C_4H_8O_3N_2$) has a standard enthalpy of combustion of 1969 kJ mol^{-1}. Calculate q, w, ΔU, and ΔH when 10.0 g of glycine are burned at $T = 298$ K and a constant pressure of 1.00 bar. Assume the combustion products are carbon dioxide gas, nitrogen gas, and liquid water.

P4.37 Nitrogen is a vital component of proteins and nucleic acids, and is therefore necessary for life. The atmosphere is composed of roughly 80% N_2, but to most organisms, N_2 is chemically inert and cannot be directly utilized for biosynthesis. Bacteria capable of "fixing" nitrogen, i.e. converting N_2 to a chemical form, e.g. NH_3, that can be utilized in the biosynthesis of proteins and nucleic acids are called diazotrophs. The ability of some plants like legumes to fix nitrogen is due to a symbiotic relationship between the plant and nitrogen fixing diazotrophs that live in the plant's roots. Nitrogen fixation by the Haber process illustrated in Figure 4.1 does not occur in bacteria. Assume the hypothetical reaction for fixing nitrogen biologically:

$$N_2(g) + 3H_2O(l) \rightarrow 2NH_3(aq) + \tfrac{3}{2}O_2(g)$$

  a. Calculate the standard enthalpy change for the biosynthetic fixation of nitrogen at $T = 298$ K. For $NH_3(aq)$, ammonia dissolved in aqueous solution, $\Delta H_f^\circ = -80.3$ kJ mol^{-1}.

  b. In some bacteria, glycine is produced from ammonia by the reaction

$$NH_3(g) + 2CH_4(g) + \tfrac{5}{2}O_2\,(g) \rightarrow NH_2CH_2COOH(s) + H_2O(l)$$

  Calculate the standard enthalpy change for the synthesis of glycine from ammonia. For glycine, $\Delta H_f^\circ = -537.2$ kJ mol^{-1}. Assume $T = 298$ K.

  c. Calculate the standard enthalpy change for the synthesis of glycine from nitrogen, water, oxygen, and methane. Assume $T = 298$ K.

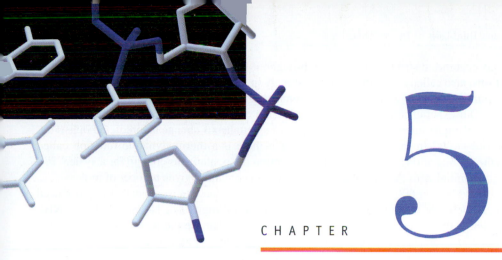

CHAPTER

5

CHAPTER

Entropy and the Second and Third Laws of Thermodynamics

Real-world processes have a natural direction of change. Heat flows from hotter bodies to colder bodies, and gases mix rather than separate. Entropy, designated by S, is the state function that predicts the direction of natural, or spontaneous change, and entropy increases for a spontaneous change in an isolated system. For a spontaneous change in a system interacting with its environment, the sum of the entropy of the system and that of the surroundings increases. In this chapter, we introduce entropy, derive the conditions for spontaneity, and show how S varies with the macroscopic variables P, V, and T.

5.1 The Universe Has a Natural Direction of Change

To this point, we have discussed q and w, as well as U and H. The first law of thermodynamics states that in any process, the total energy of the universe remains constant. Although the first law requires that a thermodynamic process conserve energy, it does not predict which of several possible processes will occur. Consider the following two examples. A metal rod initially at a uniform temperature could, in principle, undergo a spontaneous transformation in which one end becomes hot and the other end becomes cold without being in conflict with the first law, as long as the total energy of the rod remains constant. However, experience demonstrates that this does not occur. Similarly, an ideal gas that is uniformly distributed in a rigid adiabatic container could undergo a spontaneous transformation such that all of the gas moves to one-half of the container, leaving a vacuum in the other half. Because for an ideal gas, $(\partial U/\partial V)_T = 0$, the energy of the initial and final states is the same. Neither of these transformations violates the first law of thermodynamics—and yet neither occurs.

Experience tells us that there is a natural direction of change in these two processes. A metal rod with a temperature gradient always reaches a uniform temperature at some time after it has been isolated from a heat source. A gas confined to one-half of a container with a vacuum in the other half distributes itself uniformly throughout the container if a valve separating the two parts is opened. The transformations described in the previous paragraph are **unnatural transformations.** The word *unnatural* is used to indicate that in many human lifetimes, these transformations do not occur. By contrast, the reverse processes, in which the temperature gradient along the rod disappears and the gas becomes distributed uniformly throughout the container, are **natural transformations,** also called **spontaneous processes.** A spontaneous process is one that if repeated again and again over the course of a human lifetime leads to the same outcome.

Our experience is sufficient to predict the direction of spontaneous change for the two examples cited, but can the direction of spontaneous change be predicted in less obvious cases? In this chapter, we show that there is a thermodynamic function called *entropy* that allows one to predict the direction of spontaneous change for a system in a given initial state. Assume that a reaction vessel contains a given number of moles of N_2, H_2, and NH_3 at 600 K and at a total pressure of 280 bar. An iron catalyst is introduced that allows the mixture of gases to equilibrate according to $1/2\ N_2 + 3/2\ H_2 \rightleftarrows NH_3$. What is the direction of spontaneous change and what are the partial pressures of the three gases at equilibrium? The answer to this question is obtained by calculating the entropy change in the system and the surroundings.

Most students are initially uncomfortable when working with entropy, and there are good reasons for this unease. Entropy is further removed from direct experience than energy, work, or heat. At this stage, it is more important to learn how to calculate changes in entropy than it is to "understand" entropy. In a practical sense, understanding a concept is equivalent to being able to work with that concept. We all know how to work with gravity and can predict the direction in which a mass will move if it is released from our grasp. We can calculate the thrust that a rocket must develop in order to escape from Earth's gravitational field. However, how many of us could say something sensible if asked what the origin of gravity is? Why do two masses attract one another with an inverse square dependence on distance? The fact that these questions are not easily answered is not a hindrance to living in comfort with gravity. Similarly, if confidence is developed in being able to calculate the change in entropy for processes, the first steps are taken toward "understanding" entropy.

A deeper explanation of entropy will be presented in Chapter 23 on the basis of a microscopic model of matter. Historically, the importance of entropy as a state function that could predict the direction of natural change was known by Carnot in his analysis of steam engines in 1824, nearly 50 years before entropy was understood at a microscopic level by Boltzmann. Even though thermodynamics has no need for a microscopic basis, the underlying microscopic basis of entropy is discussed briefly here to make it less mysterious in the rest of this chapter.

At the microscopic level, matter consists of atoms or molecules that have energetic degrees of freedom (i.e., translational, rotational, vibrational, and electronic), each of which is associated with discrete energy levels that can be calculated using quantum mechanics. Quantum mechanics also characterizes a molecule by a state associated with a set of quantum numbers and a molecular energy. Entropy serves as a measure of the number of quantum states accessible to a macroscopic system at a given energy. Quantitatively, $S = k \ln W$, where W provides a measure of the number of states accessible to the system, and $k = R/N_A$, where R is the ideal gas constant and N_A is Avogadro's number. As demonstrated later in this chapter, the entropy of an isolated system is maximized at equilibrium. Therefore, the approach to equilibrium can be envisioned as a process in which the system achieves the distribution of energy among molecules that corresponds to a maximum value of W and, correspondingly, to a maximum in S.

5.2 Heat Engines and the Second Law of Thermodynamics

The development of entropy used here follows the historical route by which this state function was first introduced. The concept of entropy arose as 19th-century scientists attempted to maximize the work output of engines. In steam engines, heat produced by a wood or coal fire is partially converted to work done on the surroundings by the engine (the system). An automobile engine operates in a cyclical process of fuel intake, compression, ignition and expansion, and exhaust, which occurs several thousand times per minute and is used to perform work on the surroundings. Because the work produced by

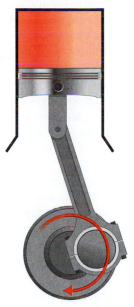

FIGURE 5.1

A schematic depiction of a heat engine is shown. Changes in temperature of the working substance brought about by contacting the cylinder with hot or cold reservoirs generate a linear motion that is mechanically converted to a rotary motion, which is used to do work.

such engines is a result of the heat released in a combustion process, they are referred to as **heat engines.** A heat engine is depicted in Figure 5.1. A working substance, which is the system (in this case an ideal gas), is confined in a piston and cylinder assembly with diathermal walls. This assembly can be brought into contact with a hot reservoir at T_{hot} or a cold reservoir at T_{cold}. The expansion or contraction of the gas caused by changes in its temperature drives the piston in or out of the cylinder. This linear motion is converted to circular motion using an eccentric, and the rotary motion is used to do work in the surroundings.

The efficiency of a heat engine is of particular interest. Experience shows that work can be converted to heat with 100% efficiency. Consider an example from calorimetry, discussed in Chapter 4. Electrical work can be done on a resistive heater immersed in a water bath. Observations reveal that the internal energy of the heater is not increased significantly in this process and that the temperature of the water bath increases. One can conclude that all of the electrical work done on the heater has been converted to heat, resulting in an increase in the temperature of the water. What is the maximum theoretical efficiency of the reverse process, the conversion of heat to work? As shown later, it is less than 100%, which limits the efficiency of an automobile engine. *There is a natural asymmetry between converting work to heat and converting heat to work. Thermodynamics provides an explanation for this asymmetry.*

As discussed in Section 2.7, the maximum work output in an isothermal expansion occurs in a reversible process. For this reason, the efficiency of a reversible heat engine is calculated, even though any real engine operates irreversibly, because the efficiency of a reversible engine is an upper bound to the efficiency of a real engine. The purpose of this engine is to convert heat into work by exploiting the spontaneous tendency of heat to flow from a hot reservoir to a cold reservoir. This engine operates in a cycle of reversible expansions and compressions of an ideal gas in a piston and cylinder assembly, and it does work on the surroundings.

This reversible cycle is shown in Figure 5.2 in a *P–V* diagram. The expansion and compression steps are designed so that the engine returns to its initial state after four steps. Recall that the area within the cycle equals the work done by the engine. Beginning at point *a*, the first segment is a reversible isothermal expansion in which the gas absorbs heat from the reservoir at T_{hot}, and also does work on the surroundings. In the second segment, the gas expands further, this time adiabatically. Work is also done on the surroundings in this step. At the end of the second segment, the gas has cooled to the temperature T_{cold}. The third segment is an isothermal compression in which the surroundings do work on the system and heat is absorbed by the cold reservoir. In the final segment, the gas is compressed to its initial volume, this time adiabatically. Work is done

FIGURE 5.2

A reversible Carnot cycle for a sample of an ideal gas working substance is shown on an indicator diagram. The cycle consists of two adiabatic and two isothermal segments. The arrows indicate the direction in which the cycle is traversed. The insets show the volume of gas and the coupling to the reservoirs at the beginning of each successive segment of the cycle. The coloring of the contents of the cylinder indicates the presence of the gas and not its temperature. The volume of the cylinder shown is that at the beginning of the appropriate segment.

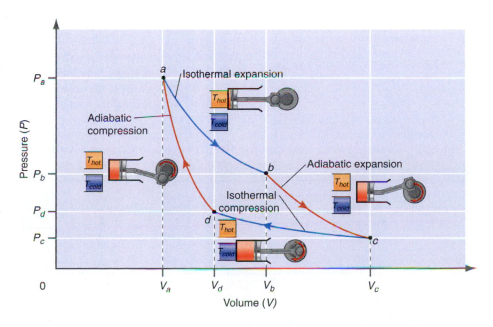

on the system in this segment, and the temperature rises to its initial value, T_{hot}. In summary, heat is taken up by the engine in the first segment at T_{hot}, and released to the surroundings in the third segment at T_{cold}. Work is done on the surroundings in the first two segments and on the system in the last two segments. An engine is only useful if net work is done on the surroundings, that is, if the magnitude of the work done in the first two steps is greater than the magnitude of the work done in the last two steps. The efficiency of the engine can be calculated by comparing the net work and the heat taken up by the engine from the hot reservoir.

Before carrying out this calculation, the rationale for the design of this reversible cycle should be discussed in more detail. To avoid losing heat to the surroundings at temperatures between T_{hot} and T_{cold}, the adiabatic segments 2 ($b \rightarrow c$) and 4 ($d \rightarrow a$) are used to move the gas between these temperatures. To absorb heat only at T_{hot} and release heat only at T_{cold}, segments 1 ($a \rightarrow b$) and 3 ($c \rightarrow d$) must be isothermal. The reason for using alternating isothermal and adiabatic segments is that no two isotherms at different temperatures intersect, and no two adiabats starting from two different temperatures intersect. Therefore, it is impossible to create a closed cycle of nonzero area in an indicator diagram out of isothermal or adiabatic segments alone. However, net work can be done using alternating adiabatic and isothermal segments. The reversible cycle depicted in Figure 5.2 is called a **Carnot cycle,** after the French engineer who first studied such cycles.

The efficiency of the Carnot cycle can be determined by calculating q, w, and ΔU for each segment of the cycle, assuming that the working substance is an ideal gas. The results are shown first in a qualitative fashion in Table 5.1. The appropriate signs for q and w are indicated. If $\Delta V > 0$, $w < 0$ for the segment, and the corresponding entry for work has a negative sign. For the isothermal segments, q and w have opposite signs from the first law because $\Delta U = 0$ for an isothermal process involving an ideal gas. From Table 5.1, it is seen that

$$w_{cycle} = w_{cd} + w_{da} + w_{ab} + w_{bc} \text{ and } q_{cycle} = q_{ab} + q_{cd} \qquad (5.1)$$

Because $\Delta U_{cycle} = 0$,

$$w_{cycle} = -(q_{cd} + q_{ab}) \qquad (5.2)$$

By comparing the areas under the two expansion segments with those under the two compression segments in the indicator diagram in Figure 5.2, you can see that the total work as seen from the system is negative, meaning that work is done on the surroundings in each cycle. Using this result, we arrive at an important conclusion that relates the heat flow in the two isothermal segments:

$$w_{cycle} < 0, \text{ so that } |q_{ab}| > |q_{cd}| \qquad (5.3)$$

In this engine, more heat is withdrawn from the hot reservoir than is deposited in the cold reservoir. It is useful to make a model of this heat engine that indicates the relative magnitude and direction of the heat and work flow, as shown in Figure 5.3a. The figure makes it clear that not all of the heat withdrawn from the higher temperature reservoir is converted to work done by the system (the green disk) on the surroundings.

the
Chemistry
place

5.1 The Reversible Carnot Cycle

TABLE 5.1 **Heat, Work, and ΔU for the Reversible Carnot Cycle**

Segment	Initial State	Final State	q	w	ΔU
$a \rightarrow b$	P_a, V_a, T_{hot}	P_b, V_b, T_{hot}	$q_{ab} (+)$	$w_{ab} (-)$	$\Delta U_{ab} = 0$
$b \rightarrow c$	P_b, V_b, T_{hot}	P_c, V_c, T_{cold}	0	$w_{bc} (-)$	$\Delta U_{bc} = w_{bc} (-)$
$c \rightarrow d$	P_c, V_c, T_{cold}	P_d, V_d, T_{cold}	$q_{cd} (-)$	$w_{cd} (+)$	$\Delta U_{cd} = 0$
$d \rightarrow a$	P_d, V_d, T_{cold}	P_a, V_a, T_{hot}	0	$w_{da} (+)$	$\Delta U_{da} = w_{da} (+)$
Cycle	P_a, V_a, T_{hot}	P_a, V_a, T_{hot}	$q_{ab} + q_{cd} (+)$	$w_{ab} + w_{bc} + w_{cd} + w_{da} (-)$	$\Delta U_{cycle} = 0$

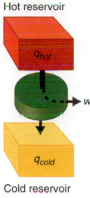

Hot reservoir

Cold reservoir

(a)

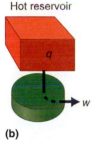

Hot reservoir

(b)

FIGURE 5.3
(a) A schematic model of the heat engine operating in a reversible Carnot cycle.
(b) The second law of thermodynamics asserts that it is impossible to construct a heat engine that operates using a single heat reservoir and converts the heat withdrawn from the reservoir into work with 100% efficiency as shown. The green disk represents the heat engine.

The efficiency, ε, of the reversible Carnot engine is defined as the ratio of the work output to the heat withdrawn from the hot reservoir. Referring to Table 5.1,

$$\varepsilon = \frac{q_{ab} + q_{cd}}{q_{ab}} = 1 - \frac{|q_{cd}|}{|q_{ab}|} < 1 \text{ because } |q_{ab}| > |q_{cd}|, q_{ab} > 0, \text{ and } q_{cd} < 0 \quad (5.4)$$

Equation (5.4) shows that the efficiency of a heat engine operating in a reversible Carnot cycle is always less than one. Equivalently, not all of the heat withdrawn from the hot reservoir can be converted to work.

These considerations on the efficiency of heat engines led to the Kelvin–Planck formulation of the **second law of thermodynamics:**

> It is impossible for a system to undergo a cyclic process whose sole effects are the flow of heat into the system from a heat reservoir and the performance of an equivalent amount of work by the system on the surroundings.

An alternative, but equivalent, statement of the second law by Clausius is as follows:

> It is impossible for a system to undergo a cyclic process whose sole effects are the flow of heat into the system from a cold reservoir and the flow of an equivalent amount of heat out of the system into a hot reservoir.

The second law asserts that the heat engine depicted in Figure 5.3b cannot be constructed. The second law has been put to the test many times by inventors who have claimed that they have invented an engine that has an efficiency of 100%. No such claim has ever been validated. To test the assertion made in this statement of the second law, imagine that such an engine has been invented. We mount it on a boat in Seattle and set off on a journey across the Pacific Ocean. Heat is extracted from the ocean, which is the single heat reservoir, and converted entirely to work in the form of a rapidly rotating propeller. Because the ocean is huge, the decrease in its temperature as a result of withdrawing heat is negligible. By the time we arrive in Japan, not a gram of diesel fuel has been used, because all the heat needed to power the boat has been extracted from the ocean. The money that was saved on fuel is used to set up an office and begin marketing this wonder engine. Does this scenario sound too good to be true? It is. Such an impossible engine is called a **perpetual motion machine of the second kind,** because it violates the second law of thermodynamics. A **perpetual motion machine of the first kind** violates the first law.

The first statement of the second law can be understood using an indicator diagram. For an engine to produce work, the area of the cycle in a P–V diagram must be greater than zero. However, this is impossible in a simple cycle using a single heat reservoir. If $T_{hot} = T_{cold}$ in Figure 5.2, the cycle $a \rightarrow b \rightarrow c \rightarrow d \rightarrow a$ collapses to a line, and the area of the cycle is zero. An arbitrary reversible cycle can be constructed that does not consist of individual adiabatic and isothermal segments. However, as shown in Figure 5.4, any reversible cycle can be approximated by a succession of adiabatic and isothermal segments, an approximation that becomes exact as the length of each segment approaches zero. It can be shown that the efficiency of such a cycle is also given by Equation (5.9) so that the efficiency of all heat engines operating in any reversible cycle between the same two temperatures, T_{hot} and T_{cold}, is identical.

A more useful form than Equation (5.4) for the efficiency of a reversible heat engine can be derived by using the fact that the working substance in the engine is an ideal gas. Calculating the work flow in each of the four segments of the Carnot cycle using the results of Sections 2.7 and 2.9,

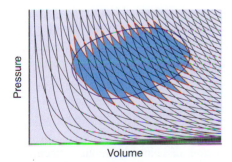

Pressure

Volume

FIGURE 5.4
An arbitrary reversible cycle, indicated by the ellipse, can be approximated to any desired accuracy by a sequence of alternating adiabatic and isothermal segments.

$$w_{ab} = -nRT_{hot} \ln \frac{V_b}{V_a} \qquad w_{ab} < 0 \text{ because } V_b > V_a$$

$$w_{bc} = nC_{V,m}(T_{cold} - T_{hot}) \qquad w_{bc} < 0 \text{ because } T_{cold} < T_{hot} \qquad (5.5)$$

$$w_{cd} = -nRT_{cold} \ln \frac{V_d}{V_c} \qquad w_{cd} > 0 \text{ because } V_d < V_c$$

$$w_{da} = nC_{V,m}(T_{hot} - T_{cold}) \qquad w_{da} > 0 \text{ because } T_{hot} > T_{cold}$$

As derived in Section 2.10, the volume and temperature in the reversible adiabatic segments are related by

$$T_{hot}V_b^{\gamma-1} = T_{cold}V_c^{\gamma-1} \text{ and } T_{cold}V_d^{\gamma-1} = T_{hot}V_a^{\gamma-1} \qquad (5.6)$$

You will show in the end-of-chapter problems that V_c and V_d can be eliminated from the set of Equations (5.5) to yield

$$w_{cycle} = -nR(T_{hot} - T_{cold}) \ln \frac{V_b}{V_a} < 0 \qquad (5.7)$$

Because $\Delta U_{a \rightarrow b} = 0$, the heat withdrawn from the hot reservoir is

$$q_{ab} = -w = nRT_{hot} \ln \frac{V_b}{V_a} \qquad (5.8)$$

and the efficiency of the reversible Carnot heat engine with an ideal gas as the working substance is

$$\varepsilon = \frac{|w_{cycle}|}{q_{ab}} = \frac{T_{hot} - T_{cold}}{T_{hot}} = 1 - \frac{T_{cold}}{T_{hot}} < 1 \qquad (5.9)$$

The efficiency of this reversible heat engine can approach one only as $T_{hot} \rightarrow \infty$ or $T_{cold} \rightarrow 0$, neither of which can be accomplished in practice. Therefore, heat can never be totally converted to work in a reversible cyclic process. Because w_{cycle} for an engine operating in an irreversible cycle is less than the work attainable in a reversible cycle, $\varepsilon_{irreversible} < \varepsilon_{reversible} < 1$.

EXAMPLE PROBLEM 5.1

Calculate the maximum work that can be done by a reversible heat engine operating between 500. and 200. K if 1000. J is absorbed at 500. K.

Solution

The fraction of the heat that can be converted to work is the same as the fractional fall in the absolute temperature. This is a convenient way to link the efficiency of an engine with the properties of the absolute temperature.

$$\varepsilon = 1 - \frac{T_{cold}}{T_{hot}} = 1 - \frac{200. \text{ K}}{500. \text{ K}} = 0.600$$

$$w = \varepsilon q_{ab} = 0.600 \times 1000. \text{ J} = 600. \text{ J}$$

In this section, only the most important features of heat engines have been discussed. It can also be shown that the efficiency of a reversible heat engine is independent of the working substance. For a more in-depth discussion of heat engines, the interested reader is referred to *Heat and Thermodynamics,* seventh edition, by M. W. Zemansky and R. H. Dittman (McGraw-Hill, 1997). We will return to a discussion of the efficiency of engines when we discuss refrigerators, heat pumps, and real engines in Section 5.11.

5.3 Introducing Entropy

Equating the two formulas for the efficiency of the reversible heat engine given in Equations (5.4) and (5.9),

$$\frac{T_{hot} - T_{cold}}{T_{hot}} = \frac{q_{ab} + q_{cd}}{q_{ab}} \quad \text{or} \quad \frac{q_{ab}}{T_{hot}} + \frac{q_{cd}}{T_{cold}} = 0 \tag{5.10}$$

The last expression in Equation (5.10) is the sum of the quantity $q_{reversible}/T$ around the Carnot cycle. This result can be generalized to any reversible cycle made up of any number of segments to give the important result stated in Equation (5.11):

$$\oint \frac{dq_{reversible}}{T} \tag{5.11}$$

This equation can be regarded as the mathematical statement of the second law. What conclusions can be drawn from Equation (5.11)? Because the cyclic integral of $dq_{reversible}/T$ is zero, this quantity must be the exact differential of a state function. This state function is called the **entropy,** and given the symbol S:

$$dS \equiv \frac{dq_{reversible}}{T} \tag{5.12}$$

For a macroscopic change,

$$\Delta S = \int \frac{dq_{reversible}}{T} \tag{5.13}$$

Note that whereas $dq_{reversible}$ is not an exact differential, multiplying this quantity by $1/T$ makes the differential exact.

EXAMPLE PROBLEM 5.2

a. Show that the following differential expression is not an exact differential:

$$\frac{RT}{P} dP + RdT$$

b. Show that $RTdP + PRdT$, obtained by multiplying the function in part (a) by P, is an exact differential.

Solution

a. For the expression $f(P,T)dP + g(P,T)dT$ to be an exact differential, the condition $\partial f(P,T)/\partial T = \partial g(P,T)/\partial P$ must be satisfied as discussed in Section 3.1. Because

$$\frac{\partial f\left(\dfrac{RT}{P}\right)}{\partial T} = \frac{R}{P} \quad \text{and} \quad \frac{\partial R}{\partial P} = 0$$

the condition is not fulfilled.

b. Because $\partial(RT)/\partial T = R$ and $\partial(RP)/\partial P = R$, $RTdP + RPdT$ is an exact differential.

Keep in mind that it has only been shown that S is a state function. It has not yet been demonstrated that S is a suitable function for measuring the natural direction of change in a process that the system may undergo. We will do so in Section 5.5.

5.4 Calculating Changes in Entropy

The most important thing to remember in doing entropy calculations using Equation (5.13) is that ΔS *must be calculated along a reversible path.* In considering an **irreversible process**, ΔS must be calculated for an equivalent reversible process that proceeds between the same initial and final states corresponding to the irreversible process. Because S is a state function, ΔS is necessarily path independent, *provided that the transformation is between the same initial and final states in both processes.*

Next consider two cases that require no calculation. For any reversible adiabatic process, $q_{reversible} = 0$, so that $\Delta S = \int (dq_{reversible}/T) = 0$. For any cyclic process,

$$\Delta S = \oint \frac{dq_{reversible}}{T} = 0,$$ because the change in any state function for a cyclic process

is zero.

Next consider ΔS for the reversible isothermal expansion or compression of an ideal gas, described by $V_i, T_i \rightarrow V_f, T_i$. Because $\Delta U = 0$ for this case,

$$q_{reversible} = -w_{reversible} = nRT \ln \frac{V_f}{V_i} \tag{5.14}$$

and

$$\Delta S = \int \frac{dq_{reversible}}{T} = \frac{1}{T} q_{reversible} = nR \ln \frac{V_f}{V_i} \tag{5.15}$$

Note that $\Delta S > 0$ for an expansion ($V_f > V_i$) and $\Delta S < 0$ for a compression ($V_f < V_i$). Although the preceding calculation is for a reversible process, ΔS has exactly the same value for any reversible or irreversible isothermal path that goes between the same initial and final volumes and satisfies the condition $T_f = T_i$. This is the case because S is a state function.

Consider next ΔS for an ideal gas that undergoes a reversible change in T at constant V or P. For a reversible process described by $V_i, T_i \rightarrow V_i, T_f$, $dq_{reversible} = C_V dT$, and

$$\Delta S = \int \frac{dq_{reversible}}{T} = \int \frac{nC_{V,m} dT}{T} \approx nC_{V,m} \ln \frac{T_f}{T_i} \tag{5.16}$$

For a constant pressure process described by $P_i, T_i \rightarrow P_i, T_f$, $dq_{reversible} = C_P dT$ and

$$\Delta S = \int \frac{dq_{reversible}}{T} = \int \frac{nC_{P,m} dT}{T} \approx nC_{P,m} \ln \frac{T_f}{T_i} \tag{5.17}$$

The last expressions in Equations (5.16) and (5.17) are valid if the temperature interval is small enough that the temperature dependence of $C_{V,m}$ and $C_{P,m}$ can be neglected. Again, although ΔS has been calculated for a reversible process, Equations (5.16) and (5.17) hold for any reversible or irreversible process between the same initial and final states for an ideal gas.

The results of the last two calculations can be combined in the following way. Because the macroscopic variables V, T or P, T completely define the state of an ideal gas, any change $V_i, T_i \rightarrow V_f, T_f$ can be separated into two segments, $V_i, T_i \rightarrow V_f, T_i$ and $V_f, T_i \rightarrow V_f, T_f$. A similar statement can be made about P and T. Because S is a state function, ΔS is independent of the path. Therefore, any reversible or irreversible process for an ideal gas described by $V_i, T_i \rightarrow V_f, T_f$ can be treated as consisting of two segments, one of which occurs at constant volume and the other of which occurs at constant temperature. For this two-step process, ΔS is given by

$$\Delta S = nR \ln \frac{V_f}{V_i} + nC_{V,m} \ln \frac{T_f}{T_i} \tag{5.18}$$

Similarly, for any reversible or irreversible process for an ideal gas described by $P_i, T_i \rightarrow P_f, T_f$,

$$\Delta S = -nR \ln \frac{P_f}{P_i} + nC_{P,m} \ln \frac{T_f}{T_i} \tag{5.19}$$

In writing Equations (5.18) and (5.19), it has been assumed that the temperature dependence of $C_{V,m}$ and $C_{P,m}$ can be neglected over the temperature range of interest.

EXAMPLE PROBLEM 5.3

One mole of a monatomic ideal gas initially at $T = 500.$ K and $P = 100.$ atm expands adiabatically against a constant external pressure of 1.00 atm until the gas pressure also equals 1.00 atm.

a. Calculate the initial and final volume of the gas and the final temperature of the gas.

b. Calculate the entropy change ΔS for the gas.

Solution

a. The initial volume is:

$$V_1 = \frac{nRT_1}{P_1} = \frac{1.00 \text{ mol} \times 0.08206 \text{ L atm mol}^{-1}\text{K}^{-1} \times 500. \text{ K}}{100. \text{ atm}} = 0.411 \text{ L}$$

The final volume is

$$V_2 = \frac{nRT_2}{P_2} = \frac{1.00 \text{ mol} \times RT_2}{1.00 \text{ atm}}$$

Now, $nC_{V,m}\Delta T = -P_{ext}\Delta V$ because the gas is ideal and the process is irreversible and adiabatic.

Substituting numerical values:

$$nC_{V,m}(T_2 - 5.00 \times 10^2 \text{K}) = -P_{ext}(V_2 - 0.411 \text{ L})$$

which after substituting $C_{V,m} = \frac{3R}{2}$, $V_2 = \frac{1.00 \text{ mol} \times RT_2}{1.00 \text{ atm}}$, and $P_{ext} = 1.00$ atm becomes:

$$1.00 \text{ mol} \times \left(\frac{3RT_2}{2} - \frac{1.50 \times 10^3 \text{ K} \times R}{2} \right) = -(1.00 \text{ mol} \times RT_2 - 0.411 \text{ L atm})$$

Rearranging terms:

$$\frac{5T_2}{2} \times 1 \text{ mol} = \frac{0.144 \text{ L atm}}{R} + 7.50 \times 10^2 \text{ K} \times 1 \text{ mol}$$

and solve for T_2:

$$T_2 = \frac{2}{5} \times \left(\frac{0.411 \text{ L atm mol}^{-1}}{0.08206 \text{ L atm mol}^{-1}\text{K}^{-1}} + 7.50 \times 10^2 \text{ K} \right) = 302 \text{ K}$$

Substituting the value for T_2 back into the ideal gas law yields V_2:

$$V_2 = \frac{nRT_2}{P_2} = \frac{1.00 \text{ mol} \times 0.08206 \text{ L atm mol}^{-1}\text{K}^{-1}}{1.00 \text{ atm}} \times 3.02 \times 10^2 \text{ K} = 24.8 \text{ L}$$

b. The entropy change can now be obtained using Equation (5.18):

$$\Delta S = nC_{V,m} \ln\left(\frac{T_2}{T_1}\right) + nR \ln\left(\frac{V_2}{V_1}\right) = 1.00 \text{ mol} \times \frac{3R}{2} \ln\left(\frac{5.00}{3.02}\right) + 1.00 \text{ mol} \times R \ln\left(\frac{24.8}{0.411}\right)$$

$$= 6.29 \text{ J K}^{-1} + 34.1 \text{ J K}^{-1} = 40.4 \text{ J K}^{-1}$$

Note that ΔS is nonzero even though the process is adiabatic. This occurs because even though the heat flow is zero, the process occurs irreversibly, and therefore $q_{irrev} = 0$. The entropy, however, is defined in terms of the reversible heat flow for which we have shown $q_{rev} > q_{irrev}$. For this irreversible process, we find accordingly that $\Delta S > 0$.

Suppose the gas expanded adiabatically and *reversibly*. According to Equation (2.43), for a monatomic ideal gas

$$\frac{T_2}{T_1} = \left(\frac{V_1}{V_2}\right)^{\frac{C_P}{C_V} - 1} = \left(\frac{V_1}{V_2}\right)^{\frac{5}{3} - 1} = \left(\frac{V_1}{V_2}\right)^{2/3}$$

Using this fact in Equation (5.18),

$$\Delta S = nC_{V,m} \ln\left(\frac{T_2}{T_1}\right) + nR \ln\left(\frac{V_2}{V_1}\right) = n\left(\frac{3R}{2}\right) \ln\left(\frac{V_1}{V_2}\right)^{2/3} + nR \ln\left(\frac{V_2}{V_1}\right)$$

$$= nR \ln\left(\frac{V_1}{V_2}\right) + nR \ln\left(\frac{V_2}{V_1}\right) = nR \ln\left(\frac{V_1}{V_2}\right) - nR \ln\left(\frac{V_1}{V_2}\right) = 0$$

For an adiabatic, reversible process $q_{rev} = 0$ and $\Delta S = 0$ as expected.

Next consider ΔS for phase changes. Experience shows that a liquid is converted to a gas at a constant boiling temperature through heat input if the process is carried out at constant pressure. Because $q_p = \Delta H$, ΔS, for this reversible process is given by

$$\Delta S_{vaporization} = \int \frac{dq_{reversible}}{T} = \frac{q_{reversible}}{T_{vaporization}} = \frac{\Delta H_{vaporization}}{T_{vaporization}} \qquad (5.20)$$

Similarly, for the phase change solid $\rightarrow$ liquid,

$$\Delta S_{fusion} = \int \frac{dq_{reversible}}{T} = \frac{q_{reversible}}{T_{fusion}} = \frac{\Delta H_{fusion}}{T_{fusion}} \qquad (5.21)$$

EXAMPLE PROBLEM 5.4

Supercooled water is liquid water that has been cooled below its normal freezing point. This state is thermodynamically unstable and converts to ice irreversibly. If 1 mol of supercooled water are transformed into ice at $T = 263.0$. K and at a constant pressure of 1.00 atm, calculate the enthalpy change and entropy change of the water. For water at $T = 273.0$ K, $\Delta H_{fusion} = -6.010 \text{ kJ mol}^{-1}$. Assume all heat capacities are constant between 263.0 and 273.0 K.

Solution

To address this problem, we exploit the fact that the entropy is a state function and set up a three-step pathway. Because S is a state function, ΔS is path independent. Therefore, ΔS for the transformation H_2O (*l*, 263 K) $\rightarrow$ H_2O (*s*, 263 K) is equal to ΔS for the transformation H_2O (*l*, 263 K) $\rightarrow$ H_2O (*l*, 273 K) $\rightarrow$ H_2O (*s*, 273 K) $\rightarrow$ H_2O (*s*, 263 K) because the initial and final states are identical for the two pathways. We use the second pathway to calculate ΔS for the process. See Table 4.1 (see Appendix B, Data Tables) for heat capacities for liquid water and ice.

$$H_2O(l, 263\ K) \quad \rightarrow \quad H_2O(s, 263\ K)$$
$$\downarrow \qquad\qquad\qquad \uparrow$$
$$H_2O(l, 273\ K) \quad \rightarrow \quad H_2O(s, 273\ K)$$

$$\Delta H = \Delta H_{sys} = nC_{P,m}(H_2O,l)\Delta T_{273-263} + \Delta H_{fusion} + nC_{P,m}(H_2O,s)\Delta T_{263-273}$$

$$= \Delta H_{fusion} + n\Delta C_{P,m}\Delta T = -6.010\ kJ + (75.3\ J\ K^{-1} - 36.2\ J\ K^{-1}) \times (273.0\ K - 263.0\ K)$$

$$= -6.010 \times 10^3\ J + 39.1\ J\ K^{-1} \times 10.0\ K = -6.010 \times 10^3\ J + 3.91 \times 10^2\ J = -5.619 \times 10^3\ J$$

$$\Delta S_{sys} = nC_{P,m}(H_2O,l)\ln\left(\frac{273.0}{263.0}\right) + \frac{\Delta H_{fusion}}{273\ K} + nC_{P,m}(H_2O,s)\ln\left(\frac{263.0}{273.0}\right)$$

$$= 75.3\ J\ K^{-1} \times \ln(1.038) - \frac{6.010 \times 10^3\ J}{2.730 \times 10^2\ K} + 36.2\ J\ K^{-1} \times \ln(0.9633)$$

$$= 2.81\ J\ K^{-1} - 22.02\ J\ K^{-1} - 1.35\ J\ K^{-1} = -20.6\ J\ K^{-1}$$

Finally, consider ΔS for an arbitrary process involving real gases, solids, and liquids for which β and κ, but not the equation of state, are known. The calculation of ΔS for such substances is described in Engel and Reid, *Physical Chemistry,* Supplemental Sections 5.12 and 5.13, in which the properties of S as a state function are fully exploited. The results are stated here. For the system undergoing the change $V_i, T_i \rightarrow V_f, T_f$,

$$\Delta S = \int_{T_i}^{T_f} \frac{C_V}{T}\,dT + \int_{V_i}^{V_f} \frac{\beta}{\kappa}\,dV = C_V\ln\frac{T_f}{T_i} + \frac{\beta}{\kappa}(V_f - V_i) \tag{5.22}$$

In deriving the last result, it has been assumed that κ and β are constant over the temperature and volume intervals of interest. Is this the case for solids, liquids, and gases? For the system undergoing a change $P_i, T_i \rightarrow P_f, T_f$,

$$\Delta S = \int_{T_i}^{T_f} \frac{C_P}{T}\,dT - \int_{P_i}^{P_f} V\beta\,dP \tag{5.23}$$

For a solid or liquid, the last equation can be simplified to

$$\Delta S = C_P\ln\frac{T_f}{T_i} - V\beta(P_f - P_i) \tag{5.24}$$

if V and β are assumed constant over the temperature and pressure intervals of interest. The integral forms of Equations (5.22) and (5.23) are valid for real gases, liquids, and solids. Examples of calculations using these equations are given in Example Problems 5.5 through 5.7. As Example Problem 5.5 shows, Equations (5.22) and (5.23) are also applicable to ideal gases.

EXAMPLE PROBLEM 5.5

One mole of CO gas is transformed from an initial state characterized by $T_i = 320.\ K$ and $V_i = 80.0\ L$ to a final state characterized by $T_f = 650.\ K$ and $V_f = 120.0\ L$. Using Equation (5.22), calculate ΔS for this process. Use the ideal gas values for β and κ. For CO,

$$\frac{C_{V,m}}{J\ mol^{-1}K^{-1}} = 31.08 - 0.01452\frac{T}{K} + 3.1415 \times 10^{-5}\frac{T^2}{K^2} - 1.4973 \times 10^{-8}\frac{T^3}{K^3}$$

Solution

For an ideal gas,

$$\beta = \frac{1}{V}\left(\frac{\partial V}{\partial T}\right)_P = \frac{1}{V}\left(\frac{\partial[nRT/P]}{\partial T}\right)_P = \frac{1}{T} \quad \text{and}$$

$$\kappa = -\frac{1}{V}\left(\frac{\partial V}{\partial P}\right)_T = -\frac{1}{V}\left(\frac{\partial[nRT/P]}{\partial P}\right)_T = \frac{1}{P}$$

Consider the following reversible process in order to calculate ΔS. The gas is first heated reversibly from 320. to 650. K at a constant volume of 80.0 L. Subsequently, the gas is reversibly expanded at a constant temperature of 650. K from a volume of 80.0 L to a volume of 120.0 L. The entropy change for this process is obtained using the integral form of Equation (5.22) with the values of β and κ cited earlier. The result is

$$\Delta S = \int_{T_i}^{T_f} \frac{C_V}{T}\, dT + nR \ln \frac{V_f}{V_i}$$

$$\Delta S(\text{J K}^{-1}\text{mol}^{-1}) = \int_{320.}^{650.} \frac{\left(31.08 - 0.01452\frac{T}{K} + 3.1415 \times 10^{-5}\frac{T^2}{K^2} - 1.4973 \times 10^{-8}\frac{T^3}{K^3}\right)}{\frac{T}{K}}\, d\frac{T}{K}$$

$$+ \; 8.314 \text{ J K}^{-1} \times \ln\frac{120.0 \text{ L}}{80.0 \text{ L}}$$

$$= 22.024 \text{ J K}^{-1}\text{mol}^{-1} - 4.792 \text{ J K}^{-1}\text{mol}^{-1} + 5.027 \text{ J K}^{-1}\text{mol}^{-1}$$
$$- 1.207 \text{ J K}^{-1}\text{mol}^{-1} + 3.371 \text{ J K}^{-1}\text{mol}^{-1}$$

$$= 24.4 \text{ J K}^{-1}\text{mol}^{-1}$$

EXAMPLE PROBLEM 5.6

In this problem, 2.5 mol of CO_2 gas is transformed from an initial state characterized by $T_i = 450.$ K and $P_i = 1.35$ bar to a final state characterized by $T_f = 800.$ K and $P_f = 3.45$ bar. Using Equation (5.23), calculate ΔS for this process. Assume ideal gas behavior and use the ideal gas value for β. For CO_2,

$$\frac{C_{P,m}}{\text{J mol}^{-1}\text{K}^{-1}} = 18.86 + 7.937 \times 10^{-2}\frac{T}{K} - 6.7834 \times 10^{-5}\frac{T^2}{K^2} + 2.4426 \times 10^{-8}\frac{T^3}{K^3}$$

Solution

Consider the following reversible process in order to calculate ΔS. The gas is first heated reversibly from 450. to 800. K at a constant pressure of 1.35 bar. Subsequently, the gas is reversibly compressed at a constant temperature of 800. K from a pressure of 1.35 bar to a pressure of 3.45 bar. The entropy change for this process is obtained using Equation (5.23) with the value of $\beta = 1/T$ from Example Problem 5.5.

$$\Delta S = \int_{T_i}^{T_f} \frac{C_P}{T}\, dT - \int_{P_i}^{P_f} V\beta\, dP = \int_{T_i}^{T_f} \frac{C_P}{T}\, dT - nR\int_{P_i}^{P_f} \frac{dP}{P} = \int_{T_i}^{T_f} \frac{C_P}{T}\, dT - nR \ln\frac{P_f}{P_i}$$

$$= 2.5 \times \int_{450.}^{800.} \frac{\left(18.86 + 7.937 \times 10^{-2}\frac{T}{K} - 6.7834 \times 10^{-5}\frac{T^2}{K^2} + 2.4426 \times 10^{-8}\frac{T^3}{K^3}\right)}{\frac{T}{K}}\, d\frac{T}{K}$$

$$- \; 2.5 \text{ mol} \times 8.314 \times \ln\frac{3.45 \text{ bar}}{1.35 \text{ bar}} \text{ J K}^{-1}$$

$$= 27.13 \text{ J K}^{-1} + 69.45 \text{ J K}^{-1} - 37.10 \text{ J K}^{-1} + 8.57 \text{ J K}^{-1}$$
$$- 19.50 \text{ J K}^{-1}$$
$$= 48.6 \text{ J K}^{-1}$$

EXAMPLE PROBLEM 5.7

In this problem, 3.00 mol of liquid mercury is transformed from an initial state characterized by $T_i = 300.$ K and $P_i = 1.00$ bar to a final state characterized by $T_f = 600.$ K and $P_f = 3.00$ bar.

a. Calculate ΔS for this process; $\beta = 1.81 \times 10^{-4} \text{ K}^{-1}$, $\rho = 13.54 \text{ g cm}^{-3}$, and $C_{P,m}$ for Hg(l) = 27.98 J mol^{-1} K^{-1}.

b. What is the ratio of the pressure-dependent term to the temperature-dependent term in ΔS? Explain your result.

Solution

a. Because the volume changes only slightly with temperature and pressure over the range indicated,

$$\Delta S = \int_{T_i}^{T_f} \frac{C_P}{T} dT - \int_{P_i}^{P_f} V \beta \, dp \approx n C_{P,m} \ln \frac{T_f}{T_i} - n V_{m,i} \beta (P_f - P_i)$$

$$= 3.00 \text{ mol} \times 27.98 \text{ J mol}^{-1} \text{ K}^{-1} \times \ln \frac{600. \text{ K}}{300. \text{ K}}$$

$$- 3.00 \text{ mol} \times \frac{200.59 \text{ g mol}^{-1}}{13.54 \text{ g cm}^{-3} \times \frac{10^6 \text{ cm}^3}{\text{m}^3}} \times 1.81 \times 10^{-4} \text{ K}^{-1}$$

$$\times 2.00 \text{ bar} \times 10^5 \text{ Pa bar}^{-1}$$

$$= 58.2 \text{ J K}^{-1} - 1.61 \times 10^{-3} \text{ J K}^{-1} = 58.2 \text{ J K}^{-1}$$

b. The ratio of the pressure-dependent to the temperature-dependent term is -3×10^{-4}. Because the volume change with pressure is very small, the contribution of the pressure-dependent term is negligible in comparison with the temperature-dependent term.

As Example Problem 5.7 shows, ΔS for a liquid or solid as both P and T change is dominated by the temperature dependence of S. *Unless the change in pressure is very large, ΔS for liquids and solids can be considered to be a function of temperature only.*

5.5 Using Entropy to Calculate the Natural Direction of a Process in an Isolated System

Is entropy useful in predicting the direction of spontaneous change? We now return to the two processes introduced in Section 5.1. The first process concerns the natural direction of change in a metal rod subject to a temperature gradient. Will the gradient become larger or smaller as the system approaches its equilibrium state? To model this process, consider the *isolated* composite system shown in Figure 5.5. Two systems, in the form of metal rods with uniform, but different temperatures $T_1 > T_2$, are brought into thermal contact.

In the following discussion, heat is withdrawn from the left rod; the same reasoning holds if the direction of heat flow is reversed. To calculate ΔS for this irreversible process using the heat flow, one must imagine a reversible process in which the initial and final

FIGURE 5.5
Two systems at constant P, each consisting of a metal rod, are placed in thermal contact. The temperatures of the two rods differ by ΔT. The composite system is contained in a rigid adiabatic enclosure (not shown) and is, therefore, an isolated system.

states are the same as for the irreversible process. In the imaginary reversible process, the rod is coupled to a reservoir whose temperature is lowered very slowly. The temperatures of the rod and the reservoir differ only infinitesimally throughout the process in which an amount of heat, q_P, is withdrawn from the rod. The total change in temperature of the rod, ΔT, is related to q_P by

$$dq_p = C_P dT \quad \text{or} \quad \Delta T = \frac{1}{C_P}\int dq_P = \frac{q_P}{C_P} \tag{5.25}$$

It has been assumed that $\Delta T = T_2 - T_1$ is small enough that C_P is constant over the interval.

Because the path is defined (constant pressure), $\int dq_P$ is independent of how rapidly the heat is withdrawn (the path); it depends only on C_P and ΔT. More formally, because $q_P = \Delta H$ and because H is a state function, q_P is independent of the path between the initial and final states. Therefore, $q_P = q_{reversible}$ if the temperature increment ΔT is identical for the reversible and irreversible processes.

Using this result, the entropy change for this irreversible process in which heat flows from one rod to the other is calculated. Because the composite system is isolated, $q_1 + q_2 = 0$, and $q_1 = -q_2 = q_P$. The entropy change of the composite system is the sum of the entropy changes in each rod:

$$\Delta S = \frac{q_{reversible,1}}{T_1} + \frac{q_{reversible,2}}{T_2} = \frac{q_1}{T_1} + \frac{q_2}{T_2} = q_P\left(\frac{1}{T_1} - \frac{1}{T_2}\right) \tag{5.26}$$

Because $T_1 > T_2$, the quantity in the parentheses is negative. This process has two possible directions:

- If heat flows from the hotter to the colder rod, the temperature gradient will become smaller. In this case, $q_p < 0$ and $dS > 0$.
- If heat flows from the colder to the hotter rod, the temperature gradient will become larger. In this case, $q_p > 0$ and $dS < 0$.

Note that ΔS has the same magnitude, but a different sign, for the two directions of change. Therefore, S appears to be a useful function for measuring the direction of natural change in an isolated system. Experience tells us that the temperature gradient will become less with time. *It can be concluded that the process in which S increases is the direction of natural change in an isolated system.*

Next, consider the second process introduced in Section 5.1 in which an ideal gas spontaneously collapses to half its initial volume without a force acting on it. This process and its reversible analog are shown in Figure 5.6. Recall that U is independent of V for an ideal gas. Because U does not change as V increases, and U is a function of T only for an ideal gas, the temperature remains constant in the irreversible process. Therefore, the spontaneous irreversible process shown in Figure 5.6a is both adiabatic and isothermal and is described by $V_i, T_i \rightarrow 1/2V_i, T_i$. The imaginary reversible process that we use to carry out the calculation of ΔS is shown in Figure 5.6b. In this process, which must have the same initial and final states as the irreversible process, sand is slowly and continuously added to the beaker on the piston to ensure that $P = P_{external}$. The ideal gas undergoes a reversible isothermal transformation described by $V_i, T_i \rightarrow 1/2V_i, T_i$. Because $\Delta U = 0$, $q = -w$. We calculate ΔS for this process:

$$\Delta S = \int \frac{dq_{reversible}}{T} = \frac{q_{reversible}}{T_i} = -\frac{w_{reversible}}{T_i} = nR \ln \frac{\frac{1}{2}V_i}{V_i} = -nR \ln 2 < 0 \tag{5.27}$$

For the reverse process, in which the gas spontaneously expands so that it occupies twice the volume, the reversible model process is an isothermal expansion for which

$$\Delta S = nR \ln \frac{2V_i}{V_i} = nR \ln 2 > 0 \tag{5.28}$$

Again, the process with $\Delta S > 0$ is the direction of natural change in this isolated system. The reverse process for which $\Delta S > 0$ is the unnatural direction of change.

The results obtained for isolated systems are generalized in the following statement:

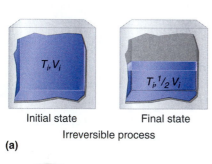

Initial state   Final state

Irreversible process

(a)

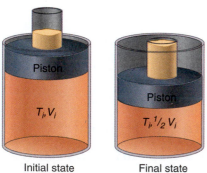

Initial state   Final state

Reversible process

(b)

FIGURE 5.6
(a) An irreversible process is shown in which an ideal gas confined in a container with rigid adiabatic walls is spontaneously reduced to half its initial volume. (b) A reversible isothermal compression is shown between the same initial and final states as for the irreversible process.

For any irreversible process in an isolated system, there is a unique direction of spontaneous change: $\Delta S > 0$ for the spontaneous process, $\Delta S < 0$ for the opposite or nonspontaneous direction of change, and $\Delta S = 0$ only for a reversible process. In a quasi-static reversible process, there is no direction of spontaneous change because the system is proceeding along a path, each step of which corresponds to an equilibrium state.

We cannot emphasize too strongly that $\Delta S > 0$ is a criterion for spontaneous change *only* if the system does not exchange energy in the form of heat or work with its surroundings. Note that if any process occurs in the isolated system, it is by definition spontaneous and the entropy increases. Whereas U can neither be created nor destroyed, S for an isolated system can be created ($\Delta S > 0$), but not destroyed.

5.6 The Clausius Inequality

In the previous section, it was shown using two examples that $\Delta S > 0$ provides a criterion to predict the natural direction of change in an isolated system. This result can also be obtained without considering a specific process. Consider the differential form of the first law for a process in which only $P-V$ work is possible:

$$dU = dq - P_{external}\,dV \tag{5.29}$$

Equation (5.29) is valid for both reversible and irreversible processes. If the process is reversible, we can write Equation (5.29) in the following form:

$$dU = dq_{reversible} - P\,dV = T\,dS - P\,dV \tag{5.30}$$

Because U is a state function, dU is independent of the path, and Equation (5.30) holds for both reversible and irreversible processes, as long as there are no phase transitions or chemical reactions, and only $P-V$ work occurs.

To derive the Clausius inequality, we equate the expressions for dU in Equations (5.29) and (5.30):

$$dq_{reversible} - dq = (P - P_{external})dV \tag{5.31}$$

If $P - P_{external} > 0$, the system will spontaneously expand, and $dV > 0$. If $P - P_{external} < 0$, the system will spontaneously contract, and $dV < 0$. In both possible cases, $(P - P_{external})dV > 0$. Therefore, we conclude that

$$dq_{reversible} - dq = T\,dS - dq \geq 0 \text{ or } T\,dS \geq dq \tag{5.32}$$

The equality holds only for a reversible process. We rewrite the **Clausius inequality** in Equation (5.32) for an irreversible process in the form

$$dS > \frac{dq}{T} \tag{5.33}$$

However, for an irreversible process in an isolated system, $dq = 0$. *Therefore, we have proved that for any irreversible process in an isolated system, $\Delta S > 0$.*

How can the result from Equations (5.29) and (5.30) that $dU = dq - P_{external}\,dV = T\,dS - PdV$ be reconciled with the fact that work and heat are path functions? The answer is that $dw \geq -PdV$ and $dq \leq TdS$, where the equalities hold only for a reversible process. The result $dq + dw = TdS - PdV$ states that the amount by which the work is greater than $-PdV$ and the amount by which the heat is less than TdS in an irreversible process involving only PV work are exactly equal. *Therefore, the differential expression for dU in Equation (5.30) is obeyed for both reversible and irreversible processes.* In Chapter 6, the Clausius inequality is used to generate two new state functions, the Gibbs energy and the Helmholtz energy. These functions allow predictions to be made about the direction of change in processes for which the system interacts with its environment.

The Clausius inequality is next used to evaluate the cyclic integral $\oint \dfrac{dq}{T}$ for an arbitrary process. Because $dS = dq_{reversible}/T$, the value of the cyclic integral is zero for a reversible process. Consider a process in which the transformation from state 1 to state 2 is reversible, but the transition from state 2 back to state 1 is irreversible:

$$\oint \frac{dq}{T} = \int_1^2 \frac{dq_{reversible}}{T} + \int_2^1 \frac{dq_{irreversible}}{T} \tag{5.34}$$

The limits of integration on the first integral can be interchanged to obtain

$$\oint \frac{dq}{T} = -\int_2^1 \frac{dq_{reversible}}{T} + \int_2^1 \frac{dq_{irreversible}}{T} \tag{5.35}$$

Exchanging the limits as written is only valid for a state function. Because $dq_{reversible} > dq_{irreversible}$

$$\oint \frac{dq}{T} \leq 0 \tag{5.36}$$

where the equality only holds for a reversible process. Note that the cyclic integral of an exact differential is always zero, but the integrand in Equation (5.36) is only an exact differential for a reversible cycle.

5.7 The Change of Entropy in the Surroundings and $\Delta S_{total} = \Delta S + \Delta S_{surroundings}$

As shown in Section 5.6, the entropy of an isolated system increases in a spontaneous process. Is it true that a process is spontaneous if ΔS for the system is positive? As shown later, this statement is only true for an isolated system. In this section, a criterion for spontaneity is developed that takes into account the entropy change of both the system and the surroundings.

In general, a system interacts only with the part of the universe that is *very close*. Therefore, one can think of the system and the interacting part of the surroundings as forming an interacting composite system that is isolated from the rest of the universe. The part of the surroundings that is relevant for entropy calculations is a thermal reservoir at a fixed temperature, T. The mass of the reservoir is sufficiently large that its temperature is only changed by an infinitesimal amount dT when heat is transferred between the system and the surroundings. Therefore, the surroundings always remain in internal equilibrium during heat transfer.

Next consider the entropy change of the surroundings, whereby the surroundings are at either constant V or constant P. The amount of heat absorbed by the surroundings, $q_{surroundings}$, depends on the process occurring in the system. If the surroundings are at constant V, $q_{surroundings} = \Delta U_{surroundings}$, and if the surroundings are at constant P, $q_{surroundings} = \Delta H_{surroundings}$. Because H and U are state functions, the amount of heat entering the surroundings is independent of the path. In particular, the system and surroundings need not be at the same temperature, and q is the same whether the transfer occurs reversibly or irreversibly. Therefore,

$$dS_{surroundings} = \frac{dq_{surroundings}}{T_{surroundings}} \text{ or for a macroscopic change, } \Delta S_{surroundings} = \frac{q_{surroundings}}{T_{surroundings}} \tag{5.37}$$

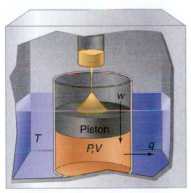

FIGURE 5.7
A sample of an ideal gas (the system) is confined in a piston and cylinder assembly with diathermal walls. The assembly is in contact with a thermal reservoir that holds the temperature at a value of 300 K. Sand falling on the outer surface of the piston increases the external pressure slowly enough to ensure a reversible compression. The directions of work and heat flow are indicated.

Note that the heat that appears in Equation (5.37) is the *actual* heat transferred. By contrast, in calculating ΔS for the system using the heat flow, $dq_{reversible}$ for a *reversible* process that connects the initial and final states of the system must be used, *not the actual dq for the process. It is essential to understand this reasoning in order to carry out calculations for ΔS and $\Delta S_{surroundings}$.*

This important difference is discussed in calculating the entropy change of the system as opposed to the surroundings with the aid of Figure 5.7. A gas (the system) is enclosed in a piston and cylinder assembly with diathermal walls. The gas is reversibly compressed by an external pressure generated by a stream of sand slowly falling on the external surface of the piston. The piston and cylinder assembly is in contact with a water bath thermal reservoir that keeps the temperature of the gas fixed at the value T. In Example Problem 5.8, ΔS and $\Delta S_{surroundings}$ are calculated for this reversible compression.

EXAMPLE PROBLEM 5.8

One mole of an ideal gas at 300. K is reversibly and isothermally compressed from a volume of 25.0 L to a volume of 10.0 L. Because it is very large, the temperature of the water bath thermal reservoir in the surroundings remains essentially constant at 300. K during the process. Calculate ΔS, $\Delta S_{surroundings}$, and ΔS_{total}.

Solution

Because this is an isothermal process, $\Delta U = 0$, and $q_{reversible} = -w$. From Section 2.7,

$$q_{reversible} = -w = nRT \int_{V_i}^{V_f} \frac{dV}{V} = nRT \ln \frac{V_f}{V_i}$$

$$= 1.00 \text{ mol} \times 8.314 \text{ J mol}^{-1} \text{ K}^{-1} \times 300. \text{ K} \times \ln \frac{10.0 \text{ L}}{25.0 \text{ L}} = -2.29 \times 10^3 \text{ J}$$

The entropy change of the system is given by

$$\Delta S = \int \frac{dq_{reversible}}{T} = \frac{q_{reversible}}{T} = \frac{-2.29 \times 10^3 \text{ J}}{300. \text{ K}} = -7.62 \text{ J K}^{-1}$$

The entropy change of the surroundings is given by

$$\Delta S_{surroundings} = \frac{q_{surroundings}}{T} = -\frac{q_{system}}{T} = \frac{2.29 \times 10^3 \text{ J}}{300. \text{ K}} = 7.62 \text{ J K}^{-1}$$

The total change in the entropy is given by

$$\Delta S_{total} = \Delta S + \Delta S_{surroundings} = -7.62 \text{ J K}^{-1} + 7.62 \text{ J K}^{-1} = 0$$

Because the process in Example Problem 5.8 is reversible, there is no direction of spontaneous change and, therefore, $\Delta S_{total} = 0$. In Example Problem 5.9, this calculation is repeated for an irreversible process that goes between the same initial states of the system.

EXAMPLE PROBLEM 5.9

One mole of an ideal gas at 300. K is isothermally compressed by a constant external pressure equal to the final pressure in Example Problem 5.8. At the end of the process, $P = P_{external}$. Because $P \neq P_{external}$ at all but the final state, this process is irreversible. The initial volume is 25.0 L and the final volume is 10.0 L. The temperature of the surroundings is 300. K. Calculate ΔS, $\Delta S_{surroundings}$, and ΔS_{total}.

Solution

We first calculate the external pressure and the initial pressure in the system:

$$P_{external} = \frac{nRT}{V} = \frac{1 \text{ mol} \times 8.314 \text{ J mol}^{-1}\text{K}^{-1} \times 300.\text{ K}}{10.0 \text{ L} \times \dfrac{1 \text{ m}^3}{10^3 \text{ L}}} = 2.49 \times 10^5 \text{ Pa}$$

$$P_i = \frac{nRT}{V} = \frac{1 \text{ mol} \times 8.314 \text{ J mol}^{-1}\text{K}^{-1} \times 300.\text{ K}}{25.0 \text{ L} \times \dfrac{1 \text{ m}^3}{10^3 \text{ L}}} = 9.98 \times 10^4 \text{ Pa}$$

Because $P_{external} > P_i$, we expect that the direction of spontaneous change will be the compression of the gas to a smaller volume. Because $\Delta U = 0$,

$$q = -w = P_{external}(V_f - V_i) = 2.49 \times 10^5 \text{ Pa} \times (10.0 \times 10^{-3}\text{ m}^3 - 25.0 \times 10^{-3}\text{ m}^3)$$

$$= -3.74 \times 10^3 \text{ J}$$

The entropy change of the surroundings is given by

$$\Delta S_{surroundings} = \frac{q_{surroundings}}{T} = -\frac{q}{T} = \frac{3.74 \times 10^3 \text{ J}}{300.\text{ K}} = 12.45 \text{ J K}^{-1}$$

The entropy change of the system must be calculated on a reversible path and has the value obtained in Example Problem 5.8:

$$\Delta S = \int \frac{dq_{reversible}}{T} = \frac{q_{reversible}}{T} = \frac{-2.29 \times 10^3 \text{ J}}{300.\text{ K}} = -7.62 \text{ J K}^{-1}$$

It is seen that $\Delta S < 0$, and $\Delta S_{surroundings} > 0$. The total change in the entropy is given by

$$\Delta S_{total} = \Delta S + \Delta S_{surroundings} = -7.62 \text{ J K}^{-1} + 12.45 \text{ J K}^{-1} = 4.83 \text{ J K}^{-1}$$

The previous calculations lead to the following conclusion: *If the system and the part of the surroundings with which it interacts are viewed as an isolated composite system, the criterion for spontaneous change is* $\Delta S_{total} = \Delta S + \Delta S_{surroundings} > 0$. This statement defines a unique direction of time. A decrease in the entropy of the universe will never be observed, because $\Delta S_{total} \geq 0$. The equality only holds for the universe at equilibrium. However, *any process* that occurs anywhere in the universe is by definition spontaneous and leads to an increase of S_{total}. Because such processes always occur, $\Delta S_{total} > 0$ as time increases. Consider the following consequence of this conclusion: If you view a movie in which two gases in contact with their surroundings are mixed, and the same movie is run backward, you cannot decide which direction corresponds to real time on the basis of the first law. However, using the criterion $\Delta S_{total} \geq 0$, the direction of real time can be established. The English astrophysicist Arthur Eddington coined the phrase "entropy is time's arrow" to emphasize this relationship between entropy and time.

Note that a spontaneous process in a system that interacts with its surroundings is not characterized by $\Delta S > 0$, but by $\Delta S_{total} > 0$. The entropy of the system can decrease in a spontaneous process, as long as the entropy of the surroundings increases by a greater amount. In Chapter 6, the spontaneity criterion $\Delta S_{total} = \Delta S + \Delta S_{surroundings} > 0$ will be used to generate two state functions, the Gibbs energy and the Helmholtz energy. These functions allow one to predict the direction of change in processes that interact with their environment using only the changes in state functions of the system.

5.8 Absolute Entropies and the Third Law of Thermodynamics

All elements and many compounds exist in three different states of aggregation. One or more solid phases are the most stable forms at low temperature, and when the temperature is increased to the melting point, a constant temperature transition to the liquid

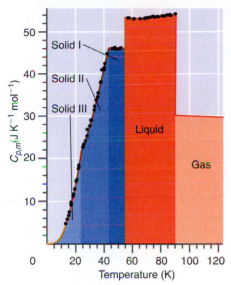

FIGURE 5.8
The experimentally determined heat capacity for O_2 is shown as a function of temperature below 125 K. The dots are data from Giauque and Johnston [*J. American Chemical Society* 51 (1929), 2300]. The red solid lines below 90 K are polynomial fits to these data. The red line above 90 K is a fit to data from the *NIST Chemistry Webbook*. The yellow line is an extrapolation from 12.97 to 0 K as described in the text. The vertical dashed lines indicate constant temperature-phase transitions, and the most stable phase at a given temperature is indicated in the figure.

phase is observed. After the temperature is increased further, a constant temperature phase transition to a gas is observed at the boiling point. At temperatures higher than the boiling point, the gas is the stable form.

The entropy of an element or a compound is experimentally determined using heat capacity data through the relationship $dq_{reversible} = C_p dT$. Just as for the thermochemical data discussed in Chapter 4, entropy values are generally tabulated for a standard temperature of 298.15 K and a standard pressure of 1 bar. We describe such a determination for the entropy of O_2 at 298.15 K, first in a qualitative fashion, and then quantitatively in Example Problem 5.10 later in this section.

The experimentally determined heat capacity of O_2 is shown in Figure 5.8 as a function of temperature for a pressure of 1 bar. O_2 has three solid phases, and transitions between them are observed at 23.66 and 43.76 K. The solid form that is stable above 43.76 K melts to form a liquid at 54.39 K. The liquid vaporizes to form a gas at 90.20 K. These phase transitions are indicated in Figure 5.8. Experimental measurements of $C_{P,m}$ are available above 12.97 K. Below this temperature, the data are extrapolated to 0 K by assuming that in this very low temperature range $C_{P,m}$ varies with temperature as T^3. This extrapolation is based on a model of the vibrational spectrum of a crystalline solid that will be discussed in Chapter 23. The explanation for the dependence of $C_{P,m}$ on T is the same as that presented for Cl_2 in Section 2.4.

Under constant pressure conditions, the molar entropy of the gas can be expressed in terms of the molar heat capacities of the solid, liquid, and gaseous forms and the enthalpies of fusion and vaporization as

$$S_m(T) = S_m(0\ \text{K}) + \int_0^{T_f} \frac{C_{P,m}^{solid}\, dT'}{T'} + \frac{\Delta H_{fusion}}{T_f} + \int_{T_f}^{T_b} \frac{C_{P,m}^{liquid}\, dT'}{T'} + \frac{\Delta H_{vaporization}}{T_b} \quad (5.38)$$

$$+ \int_{T_b}^{T} \frac{C_{P,m}^{gas}\, dT'}{T'}$$

If the substance has more than one solid phase, each will give rise to a separate integral. Note that the entropy change associated with the phase transitions solid → liquid and liquid → gas discussed in Section 5.4 must be included in the calculation. To obtain a numerical value for $S_m(T)$, the heat capacity must be known down to 0 K, and $S_m(0\ \text{K})$ must also be known.

We first address the issue of the entropy of a solid at 0 K. The **third law of thermodynamics** can be stated in the following form, due to Max Planck:

The entropy of a pure, perfectly crystalline substance (element or compound) is zero at 0 K.

A more detailed discussion of the third law will be presented in Chapter 23. Recall that in a perfectly crystalline atomic (or molecular) solid, the position of each atom is known. Because the individual atoms are indistinguishable, exchanging the positions of two atoms does not lead to a new state. Therefore, a perfect crystalline solid has only one state and $S = k \ln W = k \ln 1 = 0$. The purpose of introducing the third law at this point is that it allows calculations of the absolute entropies of elements and compounds to be carried out for any value of T. To calculate the entropy at a temperature T using Equation (5.38), the $C_{P,m}$ data of Figure 5.8 are graphed in the form $C_{P,m}/T$ as shown in Figure 5.9.

The entropy can be obtained as a function of temperature by numerically integrating the area under the curve in Figure 5.9 and adding the entropy changes associated with phase changes at the transition temperatures. The results for O_2 are shown in Figure 5.10. One can also make the following general remarks about the relative magnitudes of the entropy of different substances. Because these remarks will be justified on the basis of a microscopic model, a more detailed discussion will be deferred until Chapter 23.

FIGURE 5.9

$C_{P,m}/T$ as a function of temperature for O_2. The vertical dashed lines indicate constant temperature-phase transitions, and the most stable phase at a given temperature is indicated in the figure.

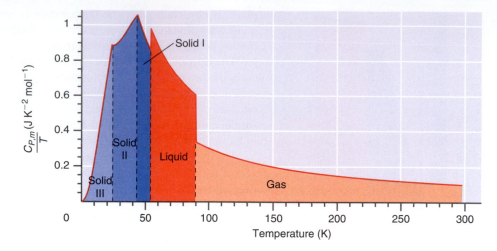

FIGURE 5.10

The molar entropy for O_2 is shown as a function of temperature. The vertical dashed lines indicate constant temperature-phase transitions, and the most stable phase at a given temperature is indicated in the figure.

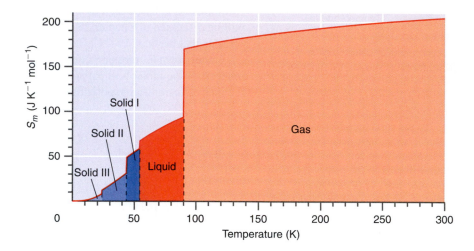

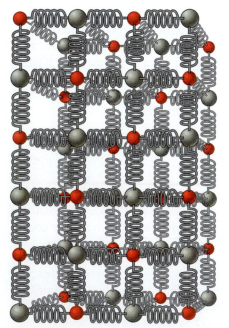

FIGURE 5.11

A useful model of a solid is a three-dimensional array of coupled harmonic oscillators. In solids with a high binding energy, the atoms are coupled by stiff springs.

- Because C_P/T in a single phase region and ΔS for melting and vaporization are always positive, S_m for a given substance is greatest for the gas-phase species. The molar entropies follow the order $S_m^{gas} > S_m^{liquid} > S_m^{solid}$.
- The molar entropy increases with the size of a molecule, because the number of degrees of freedom increases with the number of atoms. A nonlinear gas-phase molecule has three translational degrees of freedom, three rotational degrees of freedom, and $3n - 6$ vibrational degrees of freedom. A linear molecule has three translational, two rotational, and $3n - 5$ vibrational degrees of freedom. For a molecule in a liquid, the three translational degrees of freedom are converted to local vibrational modes.
- A solid has only vibrational modes. It can be modeled as a three-dimensional array of coupled harmonic oscillators as shown in Figure 5.11. This solid has a wide spectrum of vibrational frequencies, and solids with a large binding energy have higher frequencies than more weakly bound solids. Because modes with high frequencies are not activated at low temperatures, weakly bound solids have a larger molar entropy than strongly bound solids at low and moderate temperatures.
- The entropy of all substances is a monotonically increasing function of temperature.

EXAMPLE PROBLEM 5.10

The heat capacity of O_2 has been measured at 1 atm pressure over the interval 12.97 K $< T <$ 298.15 K. The data have been fit to the following polynomial series in T/K, in order to have a unitless variable:

$0 < T < 12.97$ K:

$$\frac{C_{P,m}(T)}{\text{J mol}^{-1}\,\text{K}^{-1}} = 2.11 \times 10^{-3}\,\frac{T^3}{\text{K}^3}$$

12.97 K $< T < 23.66$ K:

$$\frac{C_{P,m}(T)}{\text{J mol}^{-1}\text{K}^{-1}} = -5.666 + 0.6927\,\frac{T}{\text{K}} - 5.191 \times 10^{-3}\,\frac{T^2}{\text{K}^2} + 9.943 \times 10^{-4}\,\frac{T^3}{\text{K}^3}$$

23.66 K $< T < 43.76$ K:

$$\frac{C_{P,m}(T)}{\text{J mol}^{-1}\text{K}^{-1}} = 31.70 - 2.038\,\frac{T}{\text{K}} + 0.08384\,\frac{T^2}{\text{K}^2} - 6.685 \times 10^{-4}\,\frac{T^3}{\text{K}^3}$$

43.76 K $< T < 54.39$ K:

$$\frac{C_{P,m}(T)}{\text{J mol}^{-1}\,\text{K}^{-1}} = 46.094$$

54.39 K $< T < 90.20$ K:

$$\frac{C_{P,m}(T)}{\text{J mol}^{-1}\text{K}^{-1}} = 81.268 - 1.1467\,\frac{T}{\text{K}} + 0.01516\,\frac{T^2}{\text{K}^2} - 6.407 \times 10^{-5}\,\frac{T^3}{\text{K}^3}$$

90.20 K $< T < 298.15$ K:

$$\frac{C_{P,m}(T)}{\text{J mol}^{-1}\,\text{K}^{-1}} = 32.71 - 0.04093\,\frac{T}{\text{K}} + 1.545 \times 10^{-4}\,\frac{T^2}{\text{K}^2} - 1.819 \times 10^{-7}\,\frac{T^3}{\text{K}^3}$$

The transition temperatures and the enthalpies for the transitions indicated in Figure 5.8 are as follows:

Solid III → solid II   23.66 K   93.8 J mol^{-1}
Solid II → solid I   43.76 K   743 J mol^{-1}
Solid I → liquid   54.39 K   445 J mol^{-1}
Liquid → gas   90.20 K   6815 J mol^{-1}

a. Using these data, calculate S_m° for O_2 at 298.15 K.
b. What are the three largest contributions to S_m°?

Solution

a. $S_m^\circ(298.15\text{ K}) = \displaystyle\int_0^{23.66} \frac{C_{P,m}^{solid,III}\,dT}{T} + \frac{93.8\text{ J}}{23.66\text{ K}} + \int_{23.66}^{43.76} \frac{C_{P,m}^{solid,II}\,dT}{T} + \frac{743\text{ J}}{43.76\text{ K}}$

$\displaystyle + \int_{43.76}^{54.39} \frac{C_{P,m}^{solid,I}\,dT}{T} + \frac{445\text{ J}}{54.39\text{ K}} + \int_{54.39}^{90.20} \frac{C_{P,m}^{liquid}\,dT}{T} + \frac{6815\text{ J}}{90.20\text{ K}}$

$\displaystyle + \int_{90.20}^{298.15} \frac{C_{P,m}^{gas}\,dT}{T}$

$= 8.182\text{ J K}^{-1} + 3.964\text{ J K}^{-1} + 19.61\text{ J K}^{-1} + 16.98\text{ J K}^{-1}$
$+ 10.13\text{ J K}^{-1} + 8.181\text{ J K}^{-1} + 27.06\text{ J K}^{-1} + 75.59\text{ J K}^{-1}$
$+ 35.27\text{ J K}^{-1}$
$= 204.9\text{ J mol}^{-1}\text{K}^{-1}$

There is an additional small correction for nonideality of the gas at 1 bar. The currently accepted value is $S_m^\circ(298.15\text{ K}) = 205.152\text{ J mol}^{-1}\text{K}^{-1}$

(P. J. Linstrom and W. G. Mallard, Eds., *NIST Chemistry Webbook: NIST Standard Reference Database Number 69,* National Institute of Standards and Technology, Gaithersburg, MD; retrieved from *http://webbook.nist.gov.*)

b. The three largest contributions to S_m° are ΔS for the vaporization transition, ΔS for the heating of the gas from the boiling temperature to 298.15 K, and ΔS for heating of the liquid from the melting temperature to the boiling point.

5.9 Standard States in Entropy Calculations

As discussed in Chapter 4, changes in U and H are calculated using the result that ΔH_f values for pure elements in their standard state at a pressure of 1 bar and a temperature of 298.15 K are zero. For S, the third law provides a natural definition of zero, namely, the crystalline state at 0 K. Therefore, the absolute entropy of a compound can be experimentally determined from heat capacity measurements as described earlier. Because S is a function of pressure, tabulated values of entropies refer to a standard pressure of 1 bar. The value of S varies most strongly with P for a gas. From Equation (5.19), for an ideal gas at constant T,

$$\Delta S_m = R \ln \frac{V_f}{V_i} = -R \ln \frac{P_f}{P_i} \qquad (5.39)$$

Choosing $P_i = P^{\circ} = 1$ bar,

$$S_m(P) = S_m^{\circ} - R \ln \frac{P \,(bar)}{P^{\circ}} \qquad (5.40)$$

Figure 5.12 shows a plot of the molar entropy of an ideal gas as a function of pressure. It is seen that as $P \rightarrow 0$, $S_m \rightarrow \infty$. This is a consequence of the fact that as $P \rightarrow 0$, $V \rightarrow \infty$. As Equation (5.18) shows, the entropy becomes infinite in this limit.

Equation (5.40) provides a way to calculate the entropy of a gas at any pressure. For solids and liquids, S varies so slowly with P, as shown in Section 5.4 and Example Problem 5.7 that the pressure dependence of S can usually be neglected. The value of S for 1 bar of pressure can be used in entropy calculations for pressures that do not differ greatly from 1 bar.

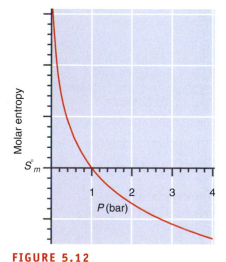

FIGURE 5.12
The molar entropy of an ideal gas is shown as a function of the gas pressure. By definition, at 1 bar, $S_m = S_m^{\circ}$, the standard state molar entropy.

5.10 Entropy Changes in Chemical Reactions

The entropy change in a chemical reaction is a major factor in determining the equilibrium concentration in a reaction mixture. In an analogous fashion to calculating ΔH_R° and ΔU_R° for chemical reactions, ΔS_R° is equal to the difference in the entropies of products and reactions, which can be written as

$$\Delta S_R^{\circ} = \sum_i v_i S_i^{\circ} \qquad (5.41)$$

In Equation (5.41), the stoichiometric coefficients v_i are positive for products and negative for reactants. For example, consider the formation of glucose from carbon dioxide and water, the basic reaction of photosynthesis. We propose to investigate whether this chemical reaction proceeds spontaneously at 298.15 K by determining the sign of the entropy change of the universe $\Delta S_{universe} = \Delta S_{system} + \Delta S_{surroundings}$. The system is defined as the reactants and the products:

$$6CO_2(g) + 6H_2O(l) \rightarrow C_6H_{12}O_6(s) + 6O_2\,(g) \qquad (5.42)$$

The entropy change under standard state conditions is given by

$$\Delta S_{system} = \Delta S^{\circ}_{298.15} = S^{\circ}_{298.15}(glucose, s) + 6S^{\circ}_{298.15K}(O_2, g) - 6S^{\circ}_{298.15}(CO_2, g)$$
$$- 6S^{\circ}_{298.15}(H_2O, l)$$
$$= 209.2 \text{ J K}^{-1}\text{mol}^{-1} + 6 \times 205.2 \text{ J K}^{-1}\text{mol}^{-1} - 6 \times 213.8 \text{ J K}^{-1}\text{mol}^{-1}$$
$$- 6 \times 70.0 \text{ J K}^{-1}\text{mol}^{-1} = -262 \text{ J K}^{-1}$$

To calculate the entropy of the surroundings, we first calculate the heat transfer between the system and the surroundings:

$$\Delta H^{\circ}_{system} = \Delta H^{\circ}_{f}(glucose, s) + 6\Delta H^{\circ}_{f}(O_2, g) - 6\Delta H^{\circ}_{f}(CO_2, g) - 6\Delta H^{\circ}_{f}(H_2O, l)$$
$$= -1 \times 1273.1 \text{ kJ} + 6 \times 0 \text{ kJ} + 6 \times 393.5 \text{ kJ} + 6 \times 285.8 \text{ kJ} = 2802 \text{ kJ}$$

Heat is transferred from the surroundings to the system reversibly so:

$$\Delta S_{surroundings} = \frac{\Delta H_{surroundings}}{T_{surroundings}} = -\frac{\Delta H^{\circ}_{system}}{T_{surroundings}} = -\frac{2802 \text{ kJ}}{298.15 \text{ K}} = -9398 \text{ J K}^{-1}$$

$$\Delta S_{universe} = \Delta S_{system} + \Delta S_{surrounding} = -262 \text{ J K}^{-1} - 9398 \text{ J K}^{-1} = -9.66 \text{ kJ K}^{-1}$$

For this reaction, ΔS for the universe is large and negative, primarily because gaseous species and liquid species are consumed in the reaction, and replaced by solid glucose and oxygen gas. We conclude that the formation of glucose from $CO_2(g)$ and $H_2O(l)$ is not a spontaneous process at 298 K.

Tabulated values of S° are generally available at the standard temperature of 298.15 K, and values for selected elements and compounds are listed in Tables 4.1 and 4.2 (see Appendix B, Data Tables). However, it is often necessary to calculate ΔS° at other temperatures. Such calculations are carried out using the temperature dependence of S discussed in Section 5.4:

$$\Delta S^{\circ}_T = \Delta S^{\circ}_{298.15} + \int_{298.15}^{T} \frac{\Delta C^{\circ}_p}{T'} dT' \qquad (5.43)$$

This equation is only valid if no phase changes occur in the temperature interval between 298.15 K and T. If phase changes occur, the associated entropy changes must be included as they were in Equation (5.38).

EXAMPLE PROBLEM 5.11

The standard entropies of CO, CO_2, and O_2 at 298.15 K are

$$S^{\circ}_{298.15}(CO, g) = 197.67 \text{ J K}^{-1}\text{mol}^{-1}$$
$$S^{\circ}_{298.15}(CO_2, g) = 213.74 \text{ J K}^{-1}\text{mol}^{-1}$$
$$S^{\circ}_{298.15}(O_2, g) = 205.138 \text{ J K}^{-1}\text{mol}^{-1}$$

The temperature dependence of constant pressure heat capacity for of CO, CO_2, and O_2 is given by

$$\frac{C^{\circ}_{P,m}(CO, g)}{\text{J K}^{-1}\text{mol}^{-1}} = 31.08 - 1.452 \times 10^{-2}\frac{T}{K} + 3.1415 \times 10^{-5}\frac{T^2}{K^2} - 1.4973 \times 10^{-8}\frac{T^3}{K^3}$$

$$\frac{C^{\circ}_{P,m}(CO_2, g)}{\text{J K}^{-1}\text{mol}^{-1}} = 18.86 + 7.937 \times 10^{-2}\frac{T}{K} - 6.7834 \times 10^{-5}\frac{T^2}{K^2} + 2.4426 \times 10^{-8}\frac{T^3}{K^3}$$

$$\frac{C^{\circ}_{P,m}(O_2, g)}{\text{J K}^{-1}\text{mol}^{-1}} = 30.81 - 1.187 \times 10^{-2}\frac{T}{K} + 2.3968 \times 10^{-5}\frac{T^2}{K^2}$$

Calculate ΔS° for the reaction $CO(g) + 1/2 \, O_2(g) \rightarrow CO_2(g)$ at 475 K.

$$\frac{\Delta C_{P,m}^{\circ}}{\text{J K}^{-1}\text{mol}^{-1}} = \left(18.86 - 31.08 - \frac{1}{2} \times 30.81\right)$$

$$+ \left(7.937 + 1.452 + \frac{1}{2} \times 1.187\right) \times 10^{-2}\frac{T}{\text{K}}$$

$$- \left(6.7834 + 3.1415 + \frac{1}{2} \times 2.3968\right) \times 10^{-5}\frac{T^2}{\text{K}^2}$$

$$+ (2.4426 + 1.4973) \times 10^{-8}\frac{T^3}{\text{K}^3}$$

$$= -27.63 + 9.983 \times 10^{-2}\frac{T}{\text{K}} - 1.112 \times 10^{-4}\frac{T^2}{\text{K}^2}$$

$$+ 3.940 \times 10^{-8}\frac{T^3}{\text{K}^3}$$

$$\Delta S^{\circ} = S_{298.15}^{\circ}(\text{CO}_2,g) - S_{298.15}^{\circ}(\text{CO},g) - \frac{1}{2} \times S_{298.15}^{\circ}(\text{O}_2,g)$$

$$= 213.74\text{ J K}^{-1}\text{mol}^{-1} - 197.67\text{ J K}^{-1}\text{mol}^{-1} - \frac{1}{2} \times 205.138\text{ J K}^{-1}\text{mol}^{-1}$$

$$= -86.50\text{ J K}^{-1}\text{mol}^{-1}$$

$$\Delta S_T^{\circ} = \Delta S_{298.15}^{\circ} + \int_{298.15}^{T/\text{K}} \frac{\Delta C_p^{\circ}}{T'}dT'$$

$$= -86.50\text{ J K}^{-1}\text{mol}^{-1}$$

$$+ \int_{298.15}^{475} \frac{\left(-27.63 + 9.983 \times 10^{-2}\frac{T}{\text{K}} - 1.112 \times 10^{-4}\frac{T^2}{\text{K}^2} + 3.940 \times 10^{-8}\frac{T^3}{\text{K}^3}\right)}{\frac{T}{\text{K}}} d\frac{T}{\text{K}}\text{ J K}^{-1}\text{mol}^{-1}$$

$$= -86.50\text{ J K}^{-1}\text{mol}^{-1} + (-12.866 + 17.654 - 7.605 + 1.0594)\text{ J K}^{-1}\text{mol}^{-1}$$

$$= -86.50\text{ J K}^{-1}\text{mol}^{-1} - 1.757\text{ J K}^{-1}\text{mol}^{-1} = -88.3\text{ J K}^{-1}\text{mol}^{-1}$$

The value of ΔS° is negative at both temperatures because the number of moles of gas is reduced in the reaction.

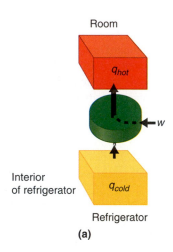

Room

q_{hot}

w

Interior of refrigerator

q_{cold}

Refrigerator

(a)

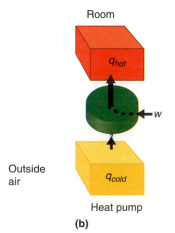

Room

q_{hot}

w

Outside air

q_{cold}

Heat pump

(b)

FIGURE 5.13
Reverse Carnot heat engines can be used to induce heat flow from a cold reservoir to a hot reservoir with the input of work. The hot reservoir in both parts (a) and (b) is a room. The cold reservoir can be configured as the inside of a refrigerator, or outside ambient air.

5.11 Refrigerators, Heat Pumps, and Real Engines

What happens if the Carnot cycle in Figure 5.2 is traversed in the opposite direction? The signs of w and q in the individual segments and the signs of the overall w and q are changed. Heat is now withdrawn from the cold reservoir and deposited in the hot reservoir. Because this is not a spontaneous process, work must be done on the system to effect this direction of heat flow. The heat and work flow in such a reverse engine is shown in Figure 5.13. The usefulness of such reverse engines is that they form the basis for refrigerators and heat pumps.

First consider the refrigerator (Figure 5.13a). The interior of a **refrigerator** acts as the system, and the room in which the refrigerator is situated is the hot reservoir. Because more heat is deposited in the room than is withdrawn from the interior of the refrigerator, the overall effect of such a device is to increase the temperature of the room. However, the usefulness of the device is that it provides a cold volume for food storage. The coefficient of performance, η_r, of a reversible Carnot refrigerator is defined as the ratio of the heat withdrawn from the cold reservoir to the work supplied to the device:

$$\eta_r = \frac{q_{cold}}{w} = \frac{q_{cold}}{q_{hot} + q_{cold}} = \frac{T_{cold}}{T_{hot} - T_{cold}} \tag{5.44}$$

This formula shows that as T_{cold} decreases from 0.9 T_{hot} to 0.1 T_{hot}, η_r decreases from 9 to 0.1. Equation (5.44) states that if the refrigerator is required to provide a lower temperature, more work is required to extract a given amount of heat.

A household refrigerator typically operates at 255 K in the freezing compartment, and 277 K in the refrigerator section. Using the lower of these temperatures for the cold reservoir and 294 K as the temperature of the hot reservoir (the room), the maximum η_r value is 6.5. This means that for every joule of work done on the system, 6.5 J of heat can be extracted from the contents of the refrigerator. This is the maximum coefficient of performance, and it is only applicable to a refrigerator operating in a reversible Carnot cycle with no dissipative losses. Real-world refrigerators operate in an irreversible cycle, and have dissipative losses associated with friction and turbulence. Therefore, it is difficult to achieve η_r values greater than ~1.5 in household refrigerators. This shows the significant loss of efficiency in an irreversible cycle.

Next consider the **heat pump** (Figure 5.13b). A heat pump is used to heat a building by extracting heat from a colder thermal reservoir such as a lake or the ambient air. The maximum coefficient of performance of a heat pump, η_{hp}, is defined as the ratio of the heat pumped into the hot reservoir to the work input to the heat pump:

$$\eta_{hp} = \frac{q_{hot}}{w} = \frac{q_{hot}}{q_{hot} + q_{cold}} = \frac{T_{hot}}{T_{hot} - T_{cold}} \tag{5.45}$$

Assume that $T_{hot} = 294$ K and $T_{cold} = 278$ K, typical for a mild winter day. The maximum η_{hp} value is calculated to be 18. Just as for the refrigerator, such high values cannot be attained for real heat pumps operating in an irreversible cycle with dissipative losses. Typical values for commercially available heat pumps lie in the range of 2 to 3.

Heat pumps become less effective as T_{cold} decreases, as shown by Equation (5.45). This decrease is more pronounced in practice, because in order to extract a given amount of heat from air, more air has to be brought into contact with the heat exchanger as T_{cold} decreases. Limitations imposed by the rate of heat transfer come into play. Therefore, heat pumps using ambient air as the cold reservoir are impractical in cold climates. Note that a coefficient of performance of 2 for a heat pump means that a house can be heated with a heat pump using half the electrical power consumption that would be required to heat the same house using electrical baseboard heaters. This is a significant argument for using heat pumps for residential heating.

It is also instructive to consider refrigerators and heat pumps from an entropic point of view. Transferring an amount of heat, q, from a cold reservoir to a hot reservoir is not a spontaneous process in an isolated system, because

$$\Delta S = q \left(\frac{1}{T_{hot}} - \frac{1}{T_{cold}} \right) < 0 \tag{5.46}$$

However, work can be converted to heat with 100% efficiency. Therefore, the coefficient of performance, η_r, can be calculated by determining the minimum amount of work input required to make ΔS for the withdrawal of q from the cold reservoir, together with the deposition of $q + w$, a spontaneous process.

We next discuss real engines, using the Otto engine, typically used in automobiles, and the diesel engine as examples. The **Otto engine** is the most widely used engine in automobiles. The engine cycle consists of four strokes as shown in Figure 5.14. The intake valve opens as the piston is moving downward, drawing a fuel–air mixture into the cylinder. The intake valve is closed, and the mixture is compressed as the piston moves upward. Just after the piston has reached its highest point, the fuel–air mixture is ignited by a spark plug, and the rapid heating resulting from the combustion process causes the gas to expand and the pressure to increase. This drives the piston down in a power stroke. Finally, the combustion products are forced out of the cylinder by the upward-moving piston as the exhaust valve is opened. To arrive at a maximum theoretical efficiency for the Otto engine, the reversible Otto cycle shown in Figure 5.15a is analyzed, assuming reversibility.

FIGURE 5.14
Illustration of the four-stroke cycle of an Otto engine, as explained in the text. The left valve is the intake valve, and the right valve is the exhaust valve.

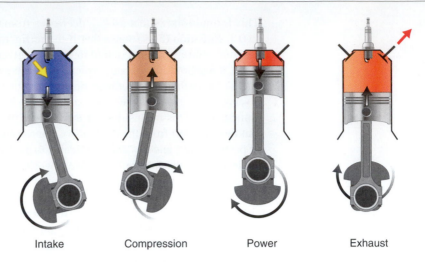

Intake   Compression   Power   Exhaust

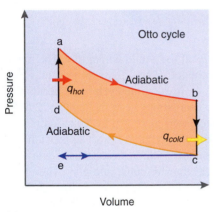

(a)

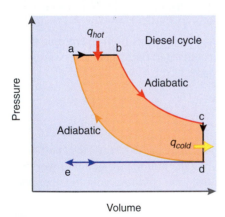

(b)

FIGURE 5.15
Idealized reversible cycles of the (a) Otto and (b) diesel engines.

The reversible Otto cycle begins with the intake stroke $e \rightarrow c$, which is assumed to take place at constant pressure. At this point, the intake valve is closed, and the piston compresses the fuel–air mixture along the adiabatic path $c \rightarrow d$ in the second step. This path can be assumed to be adiabatic because the compression occurs too rapidly to allow much heat to be transferred out of the cylinder. Ignition of the fuel–air mixture takes place at d. The rapid increase in pressure takes place at essentially a constant volume. In this reversible cycle, the combustion is modeled as a quasi-static heat transfer from a series of reservoirs at temperatures ranging from T_d to T_a. The power stroke is modeled as the adiabatic expansion $a \rightarrow b$. At this point, the exhaust valve opens and the gas is expelled. This step is modeled as the constant volume pressure decrease $b \rightarrow c$. The upward movement of the piston expels the remainder of the gas along the line $c \rightarrow e$, after which the cycle begins again.

The efficiency of this reversible cyclic engine can be calculated as follows. Assuming C_V to be constant along the segments $d \rightarrow a$ and $b \rightarrow c$, we write

$$q_{hot} = C_V(T_a - T_d) \text{ and } q_{cold} = C_V(T_b - T_c) \tag{5.47}$$

The efficiency is given by

$$\varepsilon = \frac{q_{hot} + q_{cold}}{q_{hot}} = 1 - \frac{q_{cold}}{q_{hot}} = 1 - \left(\frac{T_b - T_c}{T_a - T_d}\right) \tag{5.48}$$

The temperatures and volumes along the reversible adiabatic segments are related by

$$T_c V_c^{\gamma-1} = T_d V_d^{\gamma-1} \text{ and } T_b V_c^{\gamma-1} = T_a V_d^{\gamma-1} \tag{5.49}$$

because $V_b = V_C$ and $V_d = V_e$. Recall that $\gamma = C_P/C_V$. The T_a and T_b terms can be eliminated from Equation (5.48) to give

$$\varepsilon = 1 - \frac{T_c}{T_d} \tag{5.50}$$

where T_c and T_d are the temperatures at the beginning and end of the compression stroke $c \rightarrow d$. Temperature T_c is fixed at ~300 K, and the efficiency can be increased only by increasing T_d. This is done by increasing the compression ratio, V_c/V_d. However, if the compression ratio is too high, T_d will be sufficiently high that the fuel ignites before the end of the compression stroke. A reasonable upper limit for T_d is 600 K, and for $T_c = 300$ K, $\varepsilon = 0.50$. This value is an upper limit, because a real engine does not operate in a reversible cycle, and because heat is lost along the $c \rightarrow d$ segment. Achievable efficiencies in Otto engines used in passenger cars lie in the range of 0.20 to 0.30.

In the **diesel engine,** higher compression ratios are possible because only air is let into the cylinder in the intake stroke. The fuel is injected into the cylinder at the end of the compression stroke, thus avoiding spontaneous ignition of the fuel–air mixture during the compression stroke $d \rightarrow a$. Because the compressed air has a high temperature of ~950 K, combustion occurs spontaneously without a spark plug along the constant pressure segment $a \rightarrow b$ after fuel injection. Because the fuel is injected over a time period in which the piston is moving out of the cylinder, this step can be modeled as a constant pressure heat intake. In this segment, it is assumed that heat is absorbed from a series of reservoirs at temperatures between T_a and T_b in a quasi-static process. In the other segments, the same processes occur as described for the reversible Otto cycle.

The higher efficiency achievable with the diesel cycle in comparison with the Otto cycle is a result of the higher temperature attained in the compression cycle. Real diesel engines used in trucks and passenger cars have efficiencies in the range of 0.30 to 0.35.

The efficiency of widely used engines, the efficiency of the production of electrical power, and the effect of these sectors of human activity on the global climate are matters of current concern. In the 20th century, the world population increased by a factor of 4, and the total electrical work and heat output due to human activity increased by a factor of 16. Because more than 85% of the electrical work and heat output has been generated by burning fossil fuels, the CO_2 concentration in the atmosphere has risen dramatically. At the current rate of increase, it will reach 550 ppm by the end of the 21st century. Modeling studies indicate that this could trigger a global warming equal in magnitude, but opposite in sign, to the last Ice Age.

Two avenues are being pursued to counter this trend: new technologies and conservation. Efforts are under way to identify and optimize technologies that produce heat or electrical work without the emission of greenhouse gases. Technologies for electricity production such as photovoltaic cells, fuel cells based on H_2 combustion, nuclear power, capturing solar power in space and relaying it to the Earth using microwave transmitters, and wind turbines do not produce CO_2 as a by-product. Fuel cells are of particular interest because the work arising from a chemical reaction can be accessed directly, rather than first harvesting the heat from the reaction and subsequently (partially) converting the heat to work. Because a fuel cell converts electrical work to mechanical work, it is not subject to the efficiency limitations imposed on heat engines by the second law. These technologies are in varying stages of development along the path to widespread deployment. Lighting is very inefficient, with incandescent lights converting only 2% of the electrical work required to heat the tungsten filament to visible light. The remaining 98% is converted to heat. Fluorescent lighting is a factor of 5 to 6 more efficient.

Until technologies are available that do not affect global warming, large reductions in energy consumption and the emission of greenhouse gases can be realized by conservation. Significant reductions in heat loss can be achieved with more effective building insulation and with window coatings that reduce infrared transmission. Unnecessary heat generation can be reduced by using compact fluorescent lights in place of incandescent lighting. The increased use of commercially available ultraefficient cars that use a factor of 3 less gasoline for a given distance of travel than the current fleet would substantially reduce CO_2 emissions.

Vocabulary

Carnot cycle	heat pump	perpetual motion machine of	second law of
Clausius inequality	irreversible process	the first kind	thermodynamics
diesel engine	natural transformations	perpetual motion machine of	spontaneous process
entropy	Otto engine	the second kind	third law of thermodynamics
heat engine		refrigerator	unnatural transformations

Questions on Concepts

Q5.1 Which of the following processes is spontaneous?
  a. The reversible isothermal expansion of an ideal gas
  b. The vaporization of superheated water at 102°C and 1 bar
  c. The constant pressure melting of ice at its normal freezing point by the addition of an infinitesimal quantity of heat
  d. The adiabatic expansion of a gas into a vacuum

Q5.2 Why are ΔS_{fusion} and $\Delta S_{vaporization}$ always positive?

Q5.3 Why is the efficiency of a Carnot heat engine the upper bound to the efficiency of an internal combustion engine?

Q5.4 The amplitude of a pendulum consisting of a mass on a long wire is initially adjusted to have a very small value. The amplitude is found to decrease slowly with time. Is this process reversible? Would the process be reversible if the amplitude did not decrease with time?

Q5.5 A process involving an ideal gas is carried out in which the temperature changes at constant volume. For a fixed value of ΔT, the mass of the gas is doubled. The process is repeated with the same initial mass and ΔT is doubled. For which of these processes is ΔS greater? Why?

Q5.6 Under what conditions does the equality $\Delta S = \Delta H/T$ hold?

Q5.7 Under what conditions is $\Delta S < 0$ for a spontaneous process?

Q5.8 Is the equation

$$\Delta S = \int_{T_i}^{T_f} \frac{C_V}{T} dT + \int_{V_i}^{V_f} \frac{\beta}{\kappa} dV = C_V \ln \frac{T_f}{T_i} + \frac{\beta}{\kappa} (V_f - V_i)$$

valid for an ideal gas?

Q5.9 Without using equations, explain why ΔS for a liquid or solid is dominated by the temperature dependence of S as both P and T change.

Q5.10 You are told that $\Delta S = 0$ for a process in which the system is coupled to its surroundings. Can you conclude that the process is reversible? Justify your answer.

Problems

Problem numbers in **RED** indicate that the solution to the problem is given in the *Student Solutions Manual*.

P5.1 Beginning with Equation (5.5), use Equation (5.6) to eliminate V_c and V_d to arrive at the result $w_{cycle} = nR(T_{hot} - T_{cold})\ln V_b/V_a$.

P5.2 Consider the reversible Carnot cycle shown in Figure 5.2 with 1 mol of an ideal gas with $C_V = 3/2R$ as the working substance. The initial isothermal expansion occurs at the hot reservoir temperature of $T_{hot} = 600°C$ from an initial volume of 3.50 L (V_a) to a volume of 10.0 L (V_b). The system then undergoes an adiabatic expansion until the temperature falls to $T_{cold} = 150.°C$. The system then undergoes an isothermal compression and a subsequent adiabatic compression until the initial state described by $T_a = 600.°C$ and $V_a = 3.50$ L is reached.
  a. Calculate V_c and V_d.
  b. Calculate w for each step in the cycle and for the total cycle.
  c. Calculate ε and the amount of heat that is extracted from the hot reservoir to do 1.00 kJ of work in the surroundings.

P5.3 Using your results from Problem P5.2, calculate q, ΔU, and ΔH for each step in the cycle and for the total cycle described in Problem P5.2.

P5.4 Using your results from Problems P5.2 and P5.3, calculate ΔS, $\Delta S_{surroundings}$, and ΔS_{total} for each step in the cycle and for the total cycle described in Problem P5.2.

P5.5 Calculate ΔS if the temperature of 1 mol of an ideal gas with $C_V = 3/2 R$ is increased from 150. to 350. K under conditions of (a) constant pressure and (b) constant volume.

P5.6 One mole of N_2 at 20.5°C and 6.00 bar undergoes a transformation to the state described by 145°C and 2.75 bar. Calculate ΔS if

$$\frac{C_{P,m}}{J\,mol^{-1}\,K^{-1}} = 30.81 - 11.87 \times 10^{-3}\frac{T}{K} + 2.3968 \times 10^{-5}\frac{T^2}{K^2}$$
$$- 1.0176 \times 10^{-8}\frac{T^3}{K^3}$$

P5.7 One mole of an ideal gas with $C_V = 3/2 R$ undergoes the transformations described in the following list from an initial state described by $T = 300.$ K and $P = 1.00$ bar. Calculate q, w, ΔU, ΔH, and ΔS for each process.
  a. The gas is heated to 450. K at a constant external pressure of 1.00 bar.
  b. The gas is heated to 450. K at a constant volume corresponding to the initial volume.
  c. The gas undergoes a reversible isothermal expansion at 300 K until the pressure is half of its initial value.

P5.8 Calculate $\Delta S_{surroundings}$ and ΔS_{total} for part (c) of Problem P5.7. Is the process spontaneous? The state of the surroundings for each part is $T = 300.$ K, $P = 0.500$ bar.

P5.9 The average heat evolved by the oxidation of foodstuffs in an average adult per hour per kilogram of body weight is 7.20 kJ kg^{-1} h^{-1}. Suppose the heat evolved by this oxidation is transferred into the surroundings over a period lasting 1 day. Calculate the entropy change of the surroundings associated with this heat transfer. Assume the weight of an average adult is 70.0 kg. Assume also the surroundings are at $T = 293.0$ K.

P5.10   Calculate $\Delta S_{surroundings}$ and $\Delta S_{universe}$ for the process described in Example Problem 5.4.

P5.11   At the transition temperature of 95.4°C, the enthalpy of transition from rhombic to monoclinic sulfur is 0.38 kJ mol^{-1}.

   a. Calculate the entropy of transition under these conditions.

   b. At its melting point, 119°C, the enthalpy of fusion of monoclinic sulfur is 1.23 kJ mol^{-1}. Calculate the entropy of fusion.

   c. The values given in parts (a) and (b) are for 1 mol of sulfur; however, in crystalline and liquid sulfur, the molecule is present as S_8. Convert the values of the enthalpy and entropy of fusion in parts (a) and (b) to those appropriate for S_8.

P5.12

   a. Calculate ΔS if 1 mol of liquid water is heated from 0° to 100.°C under constant pressure if $C_{P,m} = 75.291$ J K^{-1} mol^{-1}.

   b. The melting point of water at the pressure of interest is 0°C and the enthalpy of fusion is 6.0095 kJ mol^{-1}. The boiling point is 100.°C and the enthalpy of vaporization is 40.6563 kJ mol^{-1}. Calculate ΔS for the transformation $H_2O(s, 0°C) \rightarrow H_2O(g, 100°C)$.

P5.13   One mole of an ideal gas with $C_{V,m} = 5/2\, R$ undergoes the transformations described in the following list from an initial state described by $T = 250.$ K and $P = 1.00$ bar. Calculate q, w, ΔU, ΔH, and ΔS for each process.

   a. The gas undergoes a reversible adiabatic expansion until the final pressure is half its initial value.

   b. The gas undergoes an adiabatic expansion against a constant external pressure of 0.500 bar until the final pressure is half its initial value.

   c. The gas undergoes an expansion against a constant external pressure of zero bar until the final pressure is equal to half of its initial value.

P5.14   The standard entropy of Pb(s) at 298.15 K is 64.80 J K^{-1} mol^{-1}. Assume that the heat capacity of Pb(s) is given by

$$\frac{C_{P,m}(\text{Pb},s)}{\text{J mol}^{-1}\text{K}^{-1}} = 22.13 + 0.01172\frac{T}{K} + 1.00 \times 10^{-5}\frac{T^2}{K^2}$$

The melting point is 327.4°C and the heat of fusion under these conditions is 4770. J mol^{-1}. Assume that the heat capacity of Pb(l) is given by

$$\frac{C_{P,m}(\text{Pb},l)}{\text{J K}^{-1}\text{mol}^{-1}} = 32.51 - 0.00301\frac{T}{K}$$

   a. Calculate the standard entropy of Pb(l) at 500°C.

   b. Calculate ΔH for the transformation Pb(s, 25°C) → Pb(l, 500°C).

P5.15   Between 0.00°C and 100.°C, the heat capacity of Hg(l) is given by

$$\frac{C_{P,m}(\text{Hg},l)}{\text{J K}^{-1}\text{mol}^{-1}} = 30.093 - 4.944 \times 10^{-3}\frac{T}{K}$$

Calculate ΔH and ΔS if 1 mol of Hg(l) is raised in temperature from 0.00° to 100.°C at constant P.

P5.16   One mole of a van der Waals gas at 27°C is expanded isothermally and reversibly from an initial volume of 0.020 m^3 to a final volume of 0.060 m^3. For the van der Waals gas, $(\partial U/\partial V)_T = a/V_m^2$. Assume that $a = 0.556$ Pa m^6 mol^{-2}, and that $b = 64.0 \times 10^{-6}$ m^3 mol^{-1}. Calculate q, w, ΔU, ΔH, and ΔS for the process.

P5.17   The heat capacity of α-quartz is given by

$$\frac{C_{P,m}(\text{α-quartz}, s)}{\text{J K}^{-1}\text{mol}^{-1}} = 46.94 + 34.31 \times 10^{-3}\frac{T}{K} - 11.30 \times 10^5\frac{T^2}{K^2}$$

The coefficient of thermal expansion is given by $\beta = 0.3530 \times 10^{-4}$ K^{-1} and $V_m = 22.6$ cm^3 mol^{-1}. Calculate ΔS_m for the transformation α-quartz (25.0°C, 1 atm) → α-quartz (225°C, 1000. atm).

P5.18   The amino acid glycine dimerizes to form the dipeptide glycylglycine according to the reaction

$$2\text{Glycine}(s) \rightarrow \text{Glycylglycine}(s) + H_2O(l)$$

Calculate ΔS, $\Delta S_{surroundings}$, and $\Delta S_{universe}$ at $T = 298$ K. Useful thermodynamic data are:

	Glycine	Glycylglycine	Water
ΔH_f° (kJ mol^{-1})	−537.2	−746.0	−285.8
S°(J K^{-1} mol^{-1})	103.5	190.0	70.0

P5.19   Under anaerobic conditions, glucose is broken down in muscle tissue to form lactic acid according to the reaction $C_6H_{12}O_6 \rightarrow 2CH_3CHOHCOOH$. Thermodynamic data at $T = 298$ K for glucose and lactic acid are given here:

	ΔH_f° (kJ mol^{-1})	C_P° (J K^{-1} mol^{-1})	S°(J K^{-1} mol^{-1})
Glucose	−1273.1	219.2	209.2
Lactic acid	−673.6	127.6	192.1

Calculate the entropy of the system, the surroundings, and the universe at $T = 310.$ K. Assume the heat capacities are constant between $T = 298$ K and $T = 330.$ K.

P5.20   Consider the formation of glucose from carbon dioxide and water, that is, the reaction of the following photosynthetic process: $6CO_2(g) + 6H_2O(l) \rightarrow C_6H_{12}O_6(s) + 6O_2(g)$. The following table of information will be useful in working this problem:

$T = 298$ K	CO_2 (g)	H_2O (l)	$C_6H_{12}O_6(s)$	$O_2(g)$
ΔH_f° kJ mol^{-1}	−393.5	−285.8	−1273.1	0.0
S° J mol^{-1}K^{-1}	213.8	70.0	209.2	205.2
$C_{P,m}$ J mol^{-1}K^{-1}	37.1	75.3	219.2	29.4

Calculate the entropy and enthalpy changes for this chemical system at $T = 298$ K and $T = 330.$ K. Calculate also the entropy of the surrounding and the universe at both temperatures.

P5.21   Pyruvic acid is produced as the final product in the metabolism of glucose via the glycolytic pathway. Prior to entry into the Krebs (i.e. the citric acid) cycle, pyruvate is oxidized to acetate. Consider the hypothetical oxidation of pyruvic acid.

$$CH_3COCOO^- (aq) + \tfrac{1}{2}O_2(g) \rightarrow CH_3COO^- (aq) + CO_2(g)$$

Calculate the entropy change involved in the oxidation of pyruvic acid to acetic acid at $T = 298$ K. For pyruvate ion $S° = 179.9$ J K^{-1} mol^{-1}. For acetate ion $S° = 113.0$ J K^{-1} mol^{-1}.

P5.22   One mole of an ideal gas with $C_{V,m} = 3/2\,R$ is transformed from an initial state $T = 600.$ K and $P = 1.00$ bar to a final state $T = 250.$ K and $P = 4.50$ bar. Calculate ΔU, ΔH, and ΔS for this process.

P5.23   15.0 g of steam at 373 K is added to 250.0 g of H$_2$O(l) at 298 K at constant pressure of 1 bar. Is the final state of the system steam or liquid water? Calculate ΔS for the process.

P5.24   An ideal gas sample containing 2.50 mol for which $C_{V,m} = 3/2R$ undergoes the following reversible cyclical process from an initial state characterized by $T = 450.$ K and $P = 1.00$ bar:

a. It is expanded reversibly and adiabatically until the volume doubles.

b. It is reversibly heated at constant volume until T increases to 450. K.

c. The pressure is increased in an isothermal reversible compression until $P = 1.00$ bar.

Calculate q, w, ΔU, ΔH and ΔS for each step in the cycle, and for the total cycle. The temperature of the surroundings is 300. K.

P5.25   One mole of H$_2$O(l) is compressed from a state described by $P = 1.00$ bar and $T = 298$ K to a state described by $P = 800.$ bar and $T = 450.$ K. In addition, $\beta = 2.07 \times 10^{-4}$ K^{-1} and the density can be assumed to be constant at the value 997 kg m^{-3}. Calculate ΔS for this transformation, assuming that $\kappa = 0$.

P5.26   A 25.0-g mass of ice at 273 K is added to 150.0 g of H$_2$O(l) at 360. K at constant pressure. Is the final state of the system ice or liquid water? Calculate ΔS for the process. Is the process spontaneous?

P5.27   In the Krebs cycle, degradation of the carbon skeleton of glucose is completed when acetate ion is oxidized to carbon dioxide and water. The net reaction for the Krebs cycle is:

$$CH_3COO^- (aq) + H^+ (aq) + 2O_2(g) \rightarrow 2CO_2(g) + 2H_2O(l)$$

Calculate the entropy change for this reaction at $T = 298$ K.

P5.28   The maximum theoretical efficiency of an internal combustion engine is achieved in a reversible Carnot cycle. Assume that the engine is operating in the Otto cycle and that $C_{V,m} = 5/2\,R$ for the fuel–air mixture initially at 298 K (the temperature of the cold reservoir). The mixture is compressed by a factor of 8.0 in the adiabatic compression step. What is the maximum theoretical efficiency of this engine? How much would the efficiency increase if the compression ratio could be increased to 30? Do you see a problem in doing so?

P5.29   One mole of H$_2$O(l) is supercooled to $-2.25°$C at 1 bar pressure. The freezing temperature of water at this pressure is 0.00°C. The transformation H$_2$O(l) → H$_2$O(s) is suddenly observed to occur. By calculating ΔS, $\Delta S_{surroundings}$, and ΔS_{total}, verify that this transformation is spontaneous at $-2.25°$C. The heat capacities are given by $C_P(H_2O(l)) = 75.3$ J K^{-1} mol^{-1} and $C_P(H_2O(s)) = 37.7$ J K^{-1} mol^{-1}, and $\Delta H_{fusion} = 6.008$ kJ mol^{-1} at 0.00°C. Assume that the surroundings are at $-2.25°$C. [*Hint:* Consider the two pathways at 1 bar: (a) H$_2$O(l, $-2.25°$C) → H$_2$O(s, $-2.25°$C) and (b) H$_2$O(l, $-2.25°$C) → H$_2$O(l, 0.00°C) → H$_2$O(s, 0.00°C) → H$_2$O(s, $-2.25°$C). Because S is a state function, ΔS must be the same for both pathways.]

P5.30   Calculate $\Delta S_{surroundings}$ and ΔS_{total} for the processes described in parts (a) and (b) of Problem P5.13. Which of the processes is a spontaneous process? The state of the surroundings for each part is as follows:

a. 250. K, 0.500 bar

b. 300. K, 0.500 bar

P5.31   The mean solar flux at the Earth's surface is ~4.00 J cm^{-2} min^{-1}. In a nonfocusing solar collector, the temperature can reach a value of 90.0°C. A heat engine is operated using the collector as the hot reservoir and a cold reservoir at 298 K. Calculate the area of the collector needed to produce one horsepower (1 hp = 746 watts). Assume that the engine operates at the maximum Carnot efficiency.

P5.32   A refrigerator is operated by a 0.25-hp (1 hp = 746 watts) motor. If the interior is to be maintained at $-20.°$C and the room temperature is 35°C, what is the maximum heat leak (in watts) that can be tolerated? Assume that the coefficient of performance is 75% of the maximum theoretical value.

P5.33   An electrical motor is used to operate a Carnot refrigerator with an interior temperature of 0.00°C. Liquid water at 0.00°C is placed into the refrigerator and transformed to ice at 0.00°C. If the room temperature is 20.°C, what mass of ice can be produced in 1 min by a 0.25-hp motor that is running continuously? Assume that the refrigerator is perfectly insulated and operates at the maximum theoretical efficiency.

P5.34   Calculate ΔS, ΔS_{total}, and $\Delta S_{surroundings}$ when the volume of 85.0 g of CO initially at 298 K and 1.00 bar increases by a factor of three in (a) an adiabatic reversible expansion, (b) an expansion against $P_{external} = 0$, and (c) an isothermal reversible expansion. Take $C_{P,m}$ to be constant at the value 29.14 J mol^{-1}K^{-1} and assume ideal gas behavior. State whether each process is spontaneous. The temperature of the surroundings is 298 K.

P5.35   Suppose someone claims: "I have developed a heat engine that will operate by using the temperature difference between the top and bottom water of a lake. The engine is ultimately solar-powered because the sun sustains this temperature difference by warming the top layer of lake water. Assume the lake has a surface area of 1.0×10^5 m². Assume the water at the surface of the lake has a temperature of 298 K and the water at the bottom of the lake has a temperature of 288 K. Assume the average solar flux at the lake surface is 500. Watts/m². Based on these numbers the engine can generate work at an average rate of 1.0×10^7 Watts."

Does this person's claims for the engine's performance violate the First Law of Thermodynamics? Does this person's claims for engine performance violate the Second Law of Thermodynamics? Explain your answer and show relevant calculations.

P5.36   From the following data, derive the absolute entropy of crystalline glycine at $T = 300$ K. [*Hint:* You can perform the integration numerically using either a spread sheet program or a curve-fitting routine and a graphing calculator (see Example Problem 5.10).]

Temperature (K)	Heat Capacity $C_{P,m}^{\circ}$ (J K^{-1} mol^{-1})
10	0.3
20	2.4
30	7.0
40	13.0
60	25.1
80	35.2
100	43.2
120	50.0
140	56.0
160	61.6
180	67.0
200	72.2
220	77.4
240	82.8
260	88.4
280	94.0
300	99.7

P5.37   An air conditioner is a refrigerator with the inside of the house acting as the cold reservoir and the outside atmosphere acting as the hot reservoir. Assume that an air conditioner consumes 1.50×10^3 W of electrical power, and that it can be idealized as a reversible Carnot refrigerator. If the coefficient of performance of this device is 2.50, how much heat can be extracted from the house in a 24-hour period?

P5.38   The interior of a refrigerator is typically held at 277 K and the interior of a freezer is typically held at 255 K. If the room temperature is 294 K, by what factor is it more expensive to extract the same amount of heat from the freezer than from the refrigerator? Assume that the theoretical limit for the performance of a reversible refrigerator is valid.

P5.39   The Chalk Point, Maryland, generating station supplies electrical power to the Washington, D.C., area. Units 1 and 2 have a gross generating capacity of 710 megawatts (MW). The steam pressure is 25×10^6 Pa, and the superheater outlet temperature (T_h) is 540°C. The condensate temperature (T_c) is 30.0°C.

a. What is the efficiency of a reversible Carnot engine operating under these conditions?

b. If the efficiency of the boiler is 91.2%, the overall efficiency of the turbine, which includes the Carnot efficiency and its mechanical efficiency, is 46.7%, and the efficiency of the generator is 98.4%, what is the efficiency of the total generating unit? (Another 5% needs to be subtracted for other plant losses.)

c. One of the coal burning units produces 355 MW. How many metric tons (1 metric ton = 1×10^6 g) of coal per hour are required to operate this unit at its peak output if the enthalpy of combustion of coal is 29.0×10^3 kJ kg^{-1}?

P5.40   Calculate ΔS° for the reaction $H_2(g) + Cl_2(g) \rightarrow 2HCl(g)$ at 650. K. Omit terms in the temperature-dependent heat capacities higher than T^2/K^2.

P5.41   For protein denaturations, the excess entropy of denaturation is defined as $\Delta S_{den} = \int_{T_1}^{T_2} \dfrac{\delta C_P^{trs}}{T} \, dT$, where δC_P^{trs} is the transition excess heat capacity (see Section 4.10). Using the equation for ΔS_{den} given here, calculate the excess entropy of denaturation from the differential scanning calorimetry data given in Problem P4.31. Assume the molecular weight of the protein is 14000. grams. [*Hint:* You can perform the integration numerically using either a spread sheet program or a curve-fitting routine and a graphing calculator (see Example Problem 5.10).]

P5.42   Using the data in Problem P5.42 and the equation

$$\Delta H_{den} = \int_{T_1}^{T_2} \delta C_P^{trs} dT,$$ calculate δS_{den}. Assume the denaturation

occurs reversibly. (*Hint:* Determine ΔH_{den} graphically then determine T_m.) [*Hint:* You can perform the integration numerically using either a spread sheet program or a curve-fitting routine and a graphing calculator (see Example Problem 5.10).]

P5.43   The following heat capacity data have been reported for L-alanine:

T (K)	10	20	40	60	80	100	140	180	220	260	300
$C_{P,m}$ (J K^{-1} mol^{-1})	0.49	3.85	17.45	30.99	42.59	52.50	68.93	83.14	96.14	109.6	122.7

By a graphical treatment, obtain the molar entropy of L-alanine at $T = 300$ K. [*Hint:* You can perform the integration numerically using either a spread sheet program or a curve-fitting routine and a graphing calculator (see Example Problem 5.10).]

P5.44   Using your result from Problem P5.43, extrapolate the absolute entropy of L-alanine to physiological conditions, $T = 310$. K. Assume the heat capacity is constant between $T = 300$. K and $T = 310$. K.

P5.45   From a DSC experiment it is determined that for the denaturation of a protein $\Delta H = 640.1$ kJ mol^{-1} at $T_m = 340$ K and $P = 1$ bar. Calculate ΔS for this denaturation at $T = 340$ K and $P = 1$ atm. Determine the entropy and enthalpy changes at $T = 310$ K assuming $\Delta C_P = 8.37$ kJ $K^{-1}mol^{-1}$ is independent of temperature between $T = 310.$ K and $T = 340.$ K.

P5.46   The metabolism of foodstuffs in an adult human releases about 1.00×10^4 kJ day^{-1}. If an adult human performs work at the rate of 24.8 Watts over a period of 8 hours day^{-1}, with what efficiency is the energy released by foodstuffs converted to work?

Web-Based Simulations, Animations, and Problems

W5.1   The reversible Carnot cycle is simulated with adjustable values of T_{hot} and T_{cold}, and ΔU, q, and w are determined for each segment and for the cycle. The efficiency is also determined for the cycle.

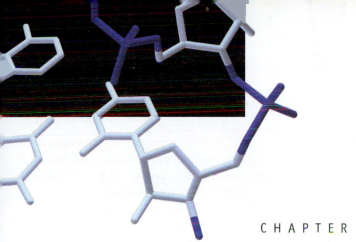

CHAPTER

6

The Gibbs Energy and Chemical Equilibrium

In the previous chapter, criteria for the spontaneity of arbitrary processes were developed. In this chapter, spontaneity is discussed in the context of a reactive mixture of gases. Two new state functions are introduced that express spontaneity in terms of the properties of the system only. The Helmholtz energy is the criterion for spontaneity for a process in a system whose initial and final states are at the same values of V and T. The Gibbs energy is the criterion for spontaneity for a process in a system whose initial and final states are at the same values of P and T. Using the Gibbs energy, a thermodynamic equilibrium constant, K_P, is derived that predicts the equilibrium concentrations of reactants and products in a mixture of reactive ideal gases.

6.1 The Gibbs Energy and the Helmholtz Energy

In Chapter 5, it was shown that the direction of spontaneous change for a process is predicted by $\Delta S + \Delta S_{surroundings} > 0$. In this section, this spontaneity criterion is used to derive two new state functions, the Gibbs and Helmholtz energies. These new state functions provide the basis for all further discussions of equilibrium. The fundamental expression governing spontaneity is the Clausius inequality [Equation (5.33)], written in the form

$$T\, dS \geq đq \tag{6.1}$$

The equality is satisfied only for a reversible process. Because $đq = dU - đw$,

$$T\, dS \geq dU - đw \text{ or, equivalently,} -dU + đw + T\, dS \geq 0 \tag{6.2}$$

As discussed in Section 2.2, a system can do different types of work on the surroundings. It is particularly useful to distinguish between expansion work, in which the work arises from a volume change in the system, and nonexpansion work. We rewrite Equation (6.2) in the form

$$-dU - P_{external}\, dV + đw_{nonexpansion} + T\, dS \geq 0 \tag{6.3}$$

This equation expresses the condition of spontaneity for an arbitrary process in terms of the changes in the state functions U, V, S, and T as well as the path-dependent functions $P_{external}\, dV$ and $w_{nonexpansion}$.

To make a connection with the discussion of spontaneity in Chapter 5, consider a special case of Equation (6.3). For an isolated system, $w = 0$ and $dU = 0$. Therefore, Equation (6.3) reduces to the familiar result derived in Section 5.5:

$$dS \geq 0 \tag{6.4}$$

Chemists are generally more interested in systems that interact with their environment than in isolated systems. Therefore, the next criteria that are derived define equilibrium and spontaneity for such systems. As was done in Chapters 1 through 5, it is useful to consider transformations at constant temperature and either constant volume or constant pressure. *Note that constant T and P (or V) does not imply that these variables are constant throughout the process, but rather that they are the same for the initial and final states of the process.*

For isothermal processes, $T\,dS = d(TS)$, and Equation (6.3) can be written in the following form:

$$-dU + TdS \geq -đw_{expansion} - đw_{nonexpansion} \text{ or, equivalently,} \tag{6.5}$$
$$d\,(U - TS) \leq đw_{expansion} + đw_{nonexpansion}$$

The combination of state functions $U - TS$, which has the units of energy, defines a new state function that we call the **Helmholtz energy,** abbreviated A. Using this definition, the general condition of spontaneity for isothermal processes becomes

$$dA \leq đw_{expansion} + đw_{nonexpansion} \tag{6.6}$$

Because the equality applies for a reversible transformation, Equation (6.6) provides a way to calculate the maximum work that a system can do on the surroundings in an isothermal process. Example Problem 6.1 illustrates the usefulness of A for calculating the maximum work available using the example of stretching a collagen fiber.

EXAMPLE PROBLEM 6.1

Collagen is the most abundant protein in the mammalian body. It is a fibrous protein that serves to strengthen and support tissues. Suppose a collagen fiber can be stretched reversibly with a force constant of $\kappa = 10.0 \text{ N m}^{-1}$ and that the tension, γ, (see Table 2.1) is given by $\gamma = \kappa l$. When a collagen fiber is contracted reversibly, it absorbs heat $q_{rev} = 5.0 \times 10^{-2}$ J. Calculate the change in the Helmholtz energy, ΔA, as the fiber contracts isothermally from $l = 0.20$ to 0.10 m. Calculate also the reversible work performed w_{rev}, ΔS, and ΔU. Assume that the temperature is constant at $T = 310.$ K.

Solution

The equation for the work performed by stretching a fiber reversibly is obtained from Table 2.1:

$$-w_{rev} = \int_{0.20m}^{0.10m} \gamma\,d\ell = -\kappa \int_{0.20m}^{0.10m} \ell\,d\ell = (10.0 \text{ N m}^{-1}) \times \frac{1.0 \times 10^{-2}\text{ m}^2}{2} = 0.050 \text{ J}$$

$$\Delta A = w_{rev} = -0.050 \text{ J}$$

By definition the entropy is

$$\Delta S = \frac{q_{rev}}{T} = \frac{0.050 \text{ J}}{310.\text{ K}} = 1.61 \times 10^{-4} \text{ J K}^{-1}.$$

ΔU is given by

$$\Delta U = \Delta A + T\Delta S = -0.050 \text{ J} + 310.\text{K} \times 1.61 \times 10^{-4} \text{ J K}^{-1} = 0.$$

In discussing the Helmholtz energy, $dT = 0$ was the only constraint applied. We now apply the additional constraint for a constant volume process, $dV = 0$. If only expansion work is possible in the transformation, then $dw_{expansion} = 0$, because $dV = 0$. In this case, the condition that defines spontaneity and equilibrium becomes

$$dA \leq 0 \tag{6.7}$$

The condition for spontaneity at constant T and V takes on a simple form using the Helmholtz energy rather than the entropy, if nonexpansion work is not possible.

Chemical reactions are more commonly studied under constant pressure than constant volume conditions. Therefore, the condition for spontaneity is considered next for an isothermal transformation that takes place at constant pressure, $P = P_{external}$. At constant pressure and temperature, $P\,dV = d(PV)$ and $T\,dS = d(TS)$. In this case, Equation (6.3) can be written in the form

$$d\,(U + PV - TS) = d\,(H - TS) \leq dw_{nonexpansion} \tag{6.8}$$

The combination of state functions $H - TS$, which has the units of energy, defines a new state function called the **Gibbs energy,** abbreviated G. Using the Gibbs energy, the condition for spontaneity and equilibrium for an isothermal process at constant pressure becomes

$$dG \leq dw_{nonexpansion} \tag{6.9}$$

For a reversible process, the equality holds, and the change in the Gibbs energy is a measure of the **maximum nonexpansion work** that can be produced in the transformation.

EXAMPLE PROBLEM 6.2

Assume the internal energy of an elastic fiber (see Example Problem 6.1) is given by $dU = T\,dS - P\,dV - \gamma\,d\ell$. Obtain an expression for $(\partial G/\partial \ell)_{P,T}$ and calculate the maximum nonexpansion work obtainable when a collagen fiber contracts from $\ell = 20.0$ to 10.0 cm at constant P and T. Assume other properties as described in Example Problem 6.1.

Solution

Determining the maximum nonexpansion work requires a calculation of the Gibbs energy change that accompanies fiber contraction at constant P and T.

$$
\begin{aligned}
dG &= dH - d(TS) = d(U + PV) - d(TS) \\
&= T\,dS - P\,dV - \gamma\,d\ell + P\,dV + V\,dP - TdS - SdT \\
&= -S\,dT + V\,dP - \gamma\,d\ell
\end{aligned}
$$

At constant P and T, $dT = dP = 0$ and we finally obtain

$$\left(\frac{\partial G}{\partial \ell}\right)_{P,T} = -\gamma = \kappa\ell \text{ or } dG = \kappa\ell\,d\ell.$$

Integrating the previous equation with respect to the length, we obtain the maximum nonexpansion work.

$$\Delta G = \kappa \int_{0.200m}^{0.100m} \ell\,d\ell = -10.0 \text{ N m}^{-1} \times \frac{1.00 \times 10^{-2}\,\text{m}^2}{2} = -0.0500 \text{ J}$$

Compare this result with that of Example Problem 6.1.

By integrating Equation 6.9, we obtain $\Delta G \leq w_{nonexpansion}$, which shows that the change in the Gibbs energy for a process can be obtained if the reversible work associated with that

(a)

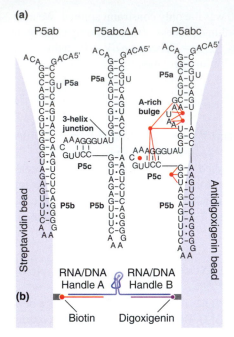

(b)

FIGURE 6.1

(a) The three types of RNA studied by Liphardt *et al.* [*Science* 292, (2001), 733] are shown. The letters G, C, A, and U refer to guanine, cytosine, adenine, and uracil, respectively. The strands are held together by hydrogen bonding between the complementary base pairs A and U, and G and C. The pronounced bulge at the center of the P5abcΔA form is a hairpin loop. Such loops are believed to play a special role in the folding of RNA. (b) An individual RNA molecule is linked by handles attached to the end of each strand to functionalized polystyrene spheres that are moved relative to one another to induce unfolding of the RNA mechanically. (Source: Jan Liphardt, Bibiana Onoa, Steven B. Smith, Ignacio Tinoco, Jr., Carlos Bustamante, "Reversible Unfolding of Single RNA Molecules by Mechanical Force, *Science* Vol. 292 27 April 2001)

process can be measured as shown in Example Problem 6.2. An interesting application of this result is the measurement of ΔG for the unfolding of single RNA molecules. RNA molecules must fold into very specific shapes in order to carry out their function as catalysts for biochemical reactions. It is not well understood how the molecule achieves its equilibrium shape, and the energetics of the folding is a key parameter in modeling the folding process. Liphardt *et al.* [*Science* 292, (2001), 733] have obtained ΔG for this process by unzipping a single RNA molecule. Figure 6.1 shows how the force associated with the unfolding of RNA can be measured. Each end of the RNA is attached to a chemical "handle" that allows the individual strands to be firmly linked to 2-μm-diameter polystyrene beads. Using techniques that will not be described here, the distance between the beads is increased, and the force needed to do so is measured.

The force versus distance plot for the P5abcΔA RNA is shown in Figure 6.2 for two different rates at which the force was increased with time. For a force of 10 pN s^{-1} (bottom panel), the folding and unfolding curves have different forms, showing that the process is not reversible. However, for a rate of 1 pN s^{-1} (top panel), the red and black curves are superimposable, showing that the process is essentially reversible. (No process that occurs at a finite rate is truly reversible.) We next discuss the form of the force versus distance curve.

The initial increase in length between ~210 and ~225 nm is due to the stretching of the handles. However, a marked change in the shape of the curve is observed for $F = 12.7$ pN, for which the length increases abruptly. This behavior is the signature of the unfolding of the RNA. The negative slope of the curve between 230 and 260 nm is determined by the experimental technique used to stretch the RNA. In a "perfect" experiment, this portion of the curve would be a horizontal line, corresponding to an increase in length at constant force. The observed length increase is identical to the length of the molecule, showing that the molecule unfolds all at once, in a process in which all bonds between the complementary base pairs are broken. The experimentally obtained value for ΔG is 174 kJ/mol, showing that a change in the thermodynamic state function G can be determined by a measurement of work. The unfolding can also be carried out by increasing the temperature of the solution containing the RNA to ~80°C, in which case the process is referred to as melting. In the biochemical environment, the unfolding is achieved through molecular forces exerted by enzymes called DNA polymerases, which act as molecular motors.

We next consider a transformation at constant P and T for which nonexpansion work is not possible, for example the burning of fuel in an internal combustion engine. In this case, Equation 6.9 becomes

$$dG \leq 0 \qquad (6.10)$$

so that the condition of spontaneity is that the Gibbs energy decreases in the process.

What is the advantage of using the state functions G and A as criteria for spontaneity rather than entropy? The Clausius inequality can be written in the form

$$dS - \frac{\bar{d}q}{T} \geq 0 \qquad (6.11)$$

Because, as was shown in Section 5.8, $dS_{surroundings} = -\bar{d}q/T$, the Clausius inequality is equivalent to the spontaneity condition:

$$dS + dS_{surroundings} \geq 0 \qquad (6.12)$$

By introducing G and A, the fundamental conditions for spontaneity have not been changed. However, G and A are expressed only in terms of the macroscopic state variables of the system. *By introducing G and A, it is no longer necessary to consider the surroundings explicitly. Knowledge of ΔG and ΔA for the system alone is sufficient to predict the direction of natural change.*

Our focus now turns to the use of G to determine the direction of spontaneous change. For macroscopic changes at constant P and T in which no nonexpansion work is possible, the condition for spontaneity is $\Delta G < 0$ where

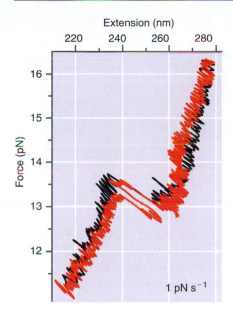

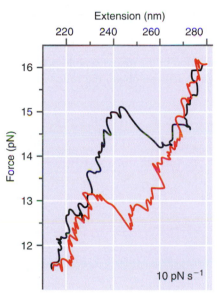

FIGURE 6.2
The measured force versus distance curve is shown for the P5abcΔA form of RNA. The black trace represents the stretching of the molecule, and the red curve represents the refolding of the molecule as the beads are brought back to their original separation. The top panel shows data for an increase in force of 1 pN s^{-1}, and the bottom panel shows the corresponding results for 10 pN s^{-1}. (Source: Jan Liphardt, Bibiana Onoa, Steven B. Smith, Ignacio Tinoco, Jr., Carlos Bustamante, "Reversible Unfolding of Single RNA Molecules by Mechanical Force, *Science* Vol. 292 27 April 2001, pgs. 733–737)

$$\Delta G = \Delta H - T\Delta S$$

Note that there are two contributions to ΔG that determine if an isothermal chemical transformation is spontaneous. They are the energetic contribution, ΔH, and the entropic contribution, $T\Delta S$. The following conclusions can be drawn based on Equation (6.13):

- The entropic contribution to ΔG is greater for higher temperatures.
- A chemical transformation is always spontaneous if $\Delta H < 0$ (an exothermic reaction) and $\Delta S > 0$.
- A chemical transformation is never spontaneous if $\Delta H > 0$ (an endothermic reaction) and $\Delta S < 0$.
- For all other cases, the relative magnitudes of ΔH and $T\Delta S$ determine if the chemical transformation is spontaneous.

Similarly, for macroscopic changes at constant V and T in which no nonexpansion work is possible, the condition for spontaneity is $\Delta A < 0$, where

$$\Delta A = \Delta U - T\Delta S \tag{6.14}$$

Again, two contributions determine if an isothermal chemical transformation is spontaneous: ΔU is an energetic contribution, and $T\Delta S$ is an entropic contribution to ΔA. The same conclusions can be drawn from this equation as for those listed earlier, with U substituted for H.

6.2 The Differential Forms of *U, H, A,* and *G*

To this point, the state functions $U, H, A,$ and G have been defined. In this section, we discuss how these state functions depend on the macroscopic system variables. To do so, the differential forms $dU, dH, dA,$ and dG are developed. Starting from the definitions

$$H = U + PV$$
$$A = U - TS$$
$$G = H - TS = U + PV - TS \tag{6.15}$$

the following total differentials can be formed:

$$dU = T\,dS - P\,dV \tag{6.16}$$
$$dH = T\,dS - P\,dV + P\,dV + V\,dP = T\,dS + V\,dP \tag{6.17}$$
$$dA = T\,dS - P\,dV - T\,dS - S\,dT = -S\,dT - P\,dV \tag{6.18}$$
$$dG = T\,dS + V\,dP - T\,dS - S\,dT = -S\,dT + V\,dP \tag{6.19}$$

These differential forms express the internal energy as $U(S,V)$, the enthalpy as $H(S,P)$, the Helmholtz energy as $A(T,V)$, and the Gibbs energy as $G(T,P)$. Although other combinations of variables can be used, these **natural variables** are used because the differential expressions are compact.

What information can be obtained from the differential expressions in Equations (6.16) through (6.19)? Because $U, H, A,$ and G are state functions, two different equivalent expressions such as those written for dU can be formulated:

$$dU = T\,dS - P\,dV = \left(\frac{\partial U}{\partial S}\right)_V dS + \left(\frac{\partial U}{\partial V}\right)_S dV \tag{6.20}$$

For Equation (6.20) to be valid, the coefficients of dS and dV on both sides of the equation must be equal. Applying this reasoning to Equations (6.16) through (6.19), the following expressions are obtained:

$$\left(\frac{\partial U}{\partial S}\right)_V = T \text{ and } \left(\frac{\partial U}{\partial V}\right)_S = -P \tag{6.21}$$

$$\left(\frac{\partial H}{\partial S}\right)_P = T \text{ and } \left(\frac{\partial H}{\partial P}\right)_S = V \tag{6.22}$$

$$\left(\frac{\partial A}{\partial T}\right)_V = -S \text{ and } \left(\frac{\partial A}{\partial V}\right)_T = -P \tag{6.23}$$

$$\left(\frac{\partial G}{\partial T}\right)_P = -S \text{ and } \left(\frac{\partial G}{\partial P}\right)_T = V \tag{6.24}$$

These expressions state how U, H, A, and G vary with their natural variables. For example, because T and V always have positive values, Equation (6.22) states that H increases as either the entropy or the pressure of the system increases. We discuss how to use these relations for macroscopic changes in the system variables in Section 6.4.

There is also a second way in which the differential expressions in Equations (6.16) through (6.19) can be used. From Section 3.1, we know that because dU is an exact differential:

$$\left(\frac{\partial}{\partial V}\left(\frac{\partial U(S,V)}{\partial S}\right)_V\right)_S = \left(\frac{\partial}{\partial S}\left(\frac{\partial U(S,V)}{\partial V}\right)_S\right)_V$$

Applying the condition that the order of differentiation in the mixed second partial derivative is immaterial to Equations (6.16) through (6.19), we obtain the following four **Maxwell relations:**

$$\left(\frac{\partial T}{\partial V}\right)_S = -\left(\frac{\partial P}{\partial S}\right)_V \tag{6.25}$$

$$\left(\frac{\partial T}{\partial P}\right)_S = \left(\frac{\partial V}{\partial S}\right)_P \tag{6.26}$$

$$\left(\frac{\partial S}{\partial V}\right)_T = \left(\frac{\partial P}{\partial T}\right)_V = \frac{\beta}{\kappa} \tag{6.27}$$

$$-\left(\frac{\partial S}{\partial P}\right)_T = \left(\frac{\partial V}{\partial T}\right)_P = V\beta \tag{6.28}$$

Equations (6.25) and (6.26) refer to a partial derivative at constant S. What conditions must a transformation at constant entropy satisfy? Because $dS = đq_{reversible}/T$, a transformation at constant entropy refers to a reversible adiabatic process.

The Maxwell relations have been derived using only the property that U, H, A, and G are state functions. These four relations are extremely useful in transforming seemingly obscure partial derivatives in other partial derivatives that can be directly measured. For example, the Maxwell relations can be used to express U, H, and heat capacities solely in terms of measurable quantities such as κ, β and the state variables P, V, and T.

6.3 The Dependence of the Gibbs and Helmholtz Energies on *P, V,* and *T*

State functions A and G are particularly important for chemists because of their roles in determining the direction of spontaneous change in a reaction mixture. Therefore, it is important to know how A changes with T and V, and how G changes with T and P. Because most reactions of interest to chemists are carried out under constant pressure rather than constant volume conditions, we focus more on the properties of G than on those of A.

We begin by asking how A changes with T and V. From Section 6.2,

$$\left(\frac{\partial A}{\partial T}\right)_V = -S \quad \text{and} \quad \left(\frac{\partial A}{\partial V}\right)_T = -P \tag{6.29}$$

where S and P always take on positive values. Therefore, the general statement can be made that the Helmholtz energy of a pure substance decreases as either the temperature or the volume increases.

We proceed in an analogous way for the Gibbs energy. From Section 6.2, Equation (6.24),

$$\left(\frac{\partial G}{\partial T}\right)_P = -S \quad \text{and} \quad \left(\frac{\partial G}{\partial P}\right)_T = V$$

Whereas the Gibbs energy decreases with increasing temperature, it increases with increasing pressure.

How can Equation (6.24) be used to calculate changes in G with macroscopic changes in the variables T and P? In doing such a calculation, each of the variables is considered separately. The total change in G as both T and P are varied is the sum of the separate contributions. This follows because G is a state function. We first discuss the change in G with P.

For a macroscopic change in P at constant T, the second expression in Equation (6.24) is integrated at constant T:

$$\int_{P^\circ}^{P} dG = G(T,P) - G^\circ(T,P^\circ) = \int_{P^\circ}^{P} V\, dP' \tag{6.30}$$

where we have chosen the initial pressure to be the standard state pressure $P^\circ = 1$ bar. This equation takes on different form for liquids and solids, and for gases. For liquids and solids, the volume is to a good approximation independent of P over a limited range in P and

$$G(T,P) = G^\circ(T,P^\circ) + \int_{P^\circ}^{P} V\, dp' \approx G^\circ(T,P^\circ) + V(P - P^\circ) \tag{6.31}$$

By contrast, the volume of a gaseous system changes appreciably with pressure. In calculating the change of G_m with P at constant T, any path connecting the same initial and final states gives the same result. Choosing the reversible path and assuming ideal gas behavior,

$$G(T,P) = G^\circ(T) + \int_{P^\circ}^{P} V\, dP' = G^\circ(T) + \int_{P^\circ}^{P} \frac{nRT}{P'}\, dP' = G^\circ(T) + nRT \ln \frac{P}{P^\circ} \tag{6.32}$$

The functional dependence of the molar Gibbs energy, G_m, on P for an ideal gas is shown in Figure 6.3, where G_m approaches minus infinity as the pressure approaches zero. This is a result of the volume dependence of S that was discussed in Section 5.5, $\Delta S = nR \ln(V_f/V_i)$ at constant T. As $P \to 0$, $V \to \infty$. Because the volume available to a gas molecule is maximized as $V \to \infty$, we find that $S \to \infty$ as $P \to 0$. Therefore, $G = H - TS \to -\infty$ in this limit. A convention for assigning values to the standard state molar Gibbs energies is discussed next.

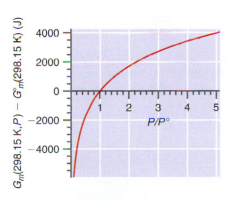

FIGURE 6.3
The molar Gibbs energy of an ideal gas relative to its standard state value is shown as a function of the pressure at 298.15 K.

It is generally not necessary to assign values to G_m°, because only differences in the Gibbs energy rather than absolute values can be obtained from experiments. However, to discuss the chemical potential in Section 6.4, it is useful to introduce a convention that allows an assignment of numerical values to G_m°. With the convention from Section 4.2 that $H_m^\circ = 0$ for an element in its standard state,

$$G_m^\circ = H_m^\circ - TS_m^\circ = -TS_m^\circ \tag{6.33}$$

for a pure *element* in its standard state at 1 bar. To obtain an expression for G_m° for a pure *compound* in its standard state, consider ΔG° for the formation reaction. Recall that 1 mol of the product appears on the right side in the formation reaction, and only elements in their standard states appear on the left side:

$$\Delta G_f^\circ = G_{m,product}^\circ + \sum_i v_i G_{m,reactant\ i}^\circ \tag{6.34}$$

$$G_{m,product}^\circ = \Delta G_f^\circ + \overset{elements}{\underset{i}{\overset{only}{\sum}}} v_i TS_{m,i}^\circ$$

where the v_i are the stoichiometric coefficients of the elemental reactants in the balanced formation reaction, all of which have negative values. Note that this result differs from the corresponding value for the enthalpy, $H_m^\circ = \Delta H_f^\circ$. The values for G_m° obtained in this way are called the **conventional molar Gibbs energies.**[1] This formalism is illustrated in Example Problem 6.3.

EXAMPLE PROBLEM 6.3

Calculate the conventional molar Gibbs energy of (a) Ar(g) and (b) $H_2O(l)$ at 1 bar and 298.15 K.

Solution

a. Because Ar(g) is an element in its standard state for these conditions, $H_m^\circ = 0$ and $G_m^\circ = H_m^\circ - TS_m^\circ = -TS_m^\circ = -298.15\ \text{K} \times 154.8\ \text{J K}^{-1}\ \text{mol}^{-1} = -4.617\ \text{kJ mol}^{-1}$.

b. The formation reaction for $H_2O(l)$ is $H_2(g) + 1/2\ O_2(g) \rightarrow H_2O(l)$:

$$G_{m,product}^\circ = \Delta G_f^\circ\ (product) + \overset{elements}{\underset{i}{\overset{only}{\sum}}} v_i TS_{m,i}^\circ$$

$$G_m^\circ(H_2O, l) = \Delta G_f^\circ(H_2O, l) - \left[TS_m^\circ(H_2, g) + \frac{1}{2} TS_m^\circ(O_2, g) \right]$$

$$= -237.1\ \text{kJ mol}^{-1} - 298.15\ \text{K} \times \left[\begin{array}{c} 130.7\ \text{J K}^{-1}\ \text{mol}^{-1} + \frac{1}{2} \\ \times\ 205.2\ \text{J K}^{-1}\ \text{mol}^{-1} \end{array} \right]$$

$$= -306.6\ \text{kJ mol}^{-1}$$

These calculations give G_m° at 298.15 K. To calculate the conventional molar Gibbs energy at another temperature, use the Gibbs–Helmholtz equation as discussed next.

We next investigate the dependence of G on T. As will be seen in the next section, the thermodynamic equilibrium constant, K, is related to G/T. Therefore, it is more useful to obtain an expression for the temperature dependence of G/T, than for the temperature dependence of G. The dependence of G/T on temperature is given by

---

[1]We note that this convention is not unique. Some authors set $G_m^\circ = 0$ for all elements in their standard state at 298.15 K.

$$\left(\frac{\partial \dfrac{G}{T}}{\partial T}\right)_P = \frac{1}{T}\left(\frac{\partial G}{\partial T}\right)_P - \frac{G}{T^2} = -\frac{S}{T} - \frac{G}{T^2} = -\frac{G + TS}{T^2} = -\frac{H}{T^2} \qquad (6.35)$$

This result is known as the **Gibbs–Helmholtz equation.** Because

$$\frac{d(1/T)}{dT} = -\frac{1}{T^2}$$

the Gibbs–Helmholtz equation can also be written in the form

$$\left(\frac{\partial \dfrac{G}{T}}{\partial \dfrac{1}{T}}\right) = \left(\frac{\partial \dfrac{G}{T}}{\partial T}\frac{dT}{d\left(\dfrac{1}{T}\right)}\right)_P = -\frac{H}{T^2}(-T^2) = H \qquad (6.36)$$

The preceding equation also applies to the change in *G* and *H* associated with a process such as a chemical reaction, in which *G* becomes ΔG. Integrating Equation (6.36) at constant *P*,

$$\int_{T_1}^{T_2} d\left(\frac{\Delta G}{T}\right) = \int_{T_1}^{T_2} \Delta H\, d\left(\frac{1}{T}\right) \qquad (6.37)$$

$$\frac{\Delta G(T_2)}{T_2} = \frac{\Delta G(T_1)}{T_1} + \Delta H\left(\frac{1}{T_2} - \frac{1}{T_1}\right)$$

It has been assumed in the second equation that ΔH is independent of *T* over the temperature interval of interest. If this is not the case, the integral must be evaluated numerically, using tabulated values of ΔH_f° and temperature-dependent expressions of $C_{P,m}$ for reactants and products.

EXAMPLE PROBLEM 6.4

The value of ΔG_f° for glycine(*s*), $H_2N(CH_2)COOH$, at 298.15 K is -377.69 kJ mol^{-1}, and ΔH_f° for glycine is -537.23 kJ mol^{-1} at the same temperature. Assuming that ΔH_f° is constant in the interval from 298.15 to 350. K, calculate ΔG_f° for the formation of glycine at 350. K.

Solution

$$\Delta G_f^\circ(T_2) = T_2\left[\frac{\Delta G_f^\circ(T_1)}{T_1} + \Delta H_f^\circ(T_1) \times \left(\frac{1}{T_2} - \frac{1}{T_1}\right)\right]$$

$$= 350\ \text{K} \times \left[\begin{array}{c}\dfrac{-377.69 \times 10^3\,\text{J mol}^{-1}}{298.15\ \text{K}} - 537.23 \times 10^3\,\text{J mol}^{-1} \\[2mm] \times \left(\dfrac{1}{350.\ \text{K}} - \dfrac{1}{298.15\ \text{K}}\right)\end{array}\right]$$

$$\Delta G_f^\circ(350\ \text{K}) = -350.\ \text{kJ mol}^{-1}$$

Similar calculations are carried out in Section 6.10, in relating the temperature dependence of the equilibrium constant and the reaction enthalpy.

6.4 The Gibbs Energy of a Mixture

To this point, the discussion has been limited to systems at a fixed composition. The fact that reactants are consumed and products are generated in chemical reactions requires the expressions derived for state functions such as U, H, S, A, and G to be revised to include these changes in composition. We focus on G in the following discussion. For a reaction mixture containing species 1, 2, 3, ..., G is no longer a function of the variables T and P only. Because it depends on the number of moles of each species, G is written in the form $G = G(T, P, n_1, n_2, n_3, \ldots)$. The total differential dG is

$$dG = \left(\frac{\partial G}{\partial T}\right)_{P,n_1,n_2\ldots} dT + \left(\frac{\partial G}{\partial P}\right)_{T,n_1,n_2\ldots} dP + \left(\frac{\partial G}{\partial n_1}\right)_{T,P,n_2\ldots} dn_1$$
$$+ \left(\frac{\partial G}{\partial n_2}\right)_{T,P,n_1\ldots} dn_2 + \ldots \qquad (6.38)$$

Note that if the concentration of all species is constant, all of the $dn_i = 0$, and Equation (6.38) reduces to Equation (6.19).

Equation (6.38) can be simplified by defining the **chemical potential, μ_i,** as

$$\mu_i = \left(\frac{\partial G}{\partial n_i}\right)_{P,T,n_j \neq n_i} \qquad (6.39)$$

It is important to realize that although μ_i is defined mathematically in terms of an infinitesimal change in the amount dn_i of species i, the chemical potential μ_i is the change in the Gibbs energy per mole of substance i added *at constant concentration*. These two requirements are not contradictory. To keep the concentration constant, one adds a mole of substance i to a huge vat containing many moles of the various species. In this case, the slope of a plot of G versus n_i is the same if the differential $(\partial G/\partial n_i)_{P,T,n_j \neq n_i}$ is formed, where $dn_i \rightarrow 0$, or the ratio $(\Delta G/\Delta n_i)_{P,T,n_j \neq n_i}$ is formed, where Δn_i is 1 mol. Using the notation of Equation (6.39), Equation (6.38) can be written as follows:

$$dG = \left(\frac{\partial G}{\partial T}\right)_{P,n_1,n_2\ldots} dT + \left(\frac{\partial G}{\partial P}\right)_{T,n_1,n_2\ldots} dP + \sum_i \mu_i \, dn_i \qquad (6.40)$$

Now imagine integrating Equation (6.40) at constant composition and at constant T and P from an infinitesimal size of the system where $n_i \rightarrow 0$ and therefore $G \rightarrow 0$ to a macroscopic size where the Gibbs energy has the value G. Because T and P are constant, the first two terms in Equation (6.40) do not contribute to the integral. Because the composition is constant, μ_i is constant,

$$\int_0^G dG' = \sum_i \mu_i \int_0^{n_i} dn'_i \qquad (6.41)$$

$$G = \sum_i n_i \mu_i$$

Note that because μ_i depends on the number of moles of each species present, it is a function of concentration. If the system consists of a single pure substance A, $G = n_A G_{m,A}$ because G is an extensive quantity. Applying Equation (6.39),

$$\mu_A = \left(\frac{\partial G}{\partial n_A}\right)_{P,T} = \left(\frac{\partial [n_A G_{m,A}]}{\partial n_A}\right)_{P,T} = G_{m,A}$$

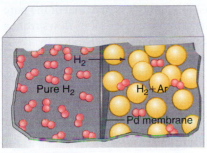

FIGURE 6.4
An isolated system consists of two subsystems. Pure H_2 gas is present on the left of a palladium membrane that is permeable to H_2, but not to argon. The H_2 is contained in a mixture with Ar in the subsystem to the right of the membrane.

showing that μ_A is equal to the molar Gibbs energy of A *for a pure substance.* As seen later, however, this statement is not true for mixtures.

Why is μ_i called the chemical potential of species i? This can be understood by assuming that the chemical potential for species i has the values μ_i^I in region I, and μ_i^{II} in region II of a given mixture with $\mu_i^I > \mu_i^{II}$. If dn_i moles of species i are transported from region I to region II, at constant T and p, the change in G is given by

$$dG = -\mu_i^I \, dn_i + \mu_i^{II} \, dn_i = (\mu_i^{II} - \mu_i^I)dn_i < 0 \qquad (6.42)$$

Because $dG < 0$, this process is spontaneous. *For a given species, transport will occur spontaneously from a region of high chemical potential to one of low chemical potential. The flow of material will continue until the chemical potential has the same value in all regions of the mixture.* Note the analogy between this process and the flow of mass in a gravitational potential or the flow of charge in an electrostatic potential. Therefore, the term *chemical potential* is appropriate.

In this discussion, we have defined an additional criterion for equilibrium in a multicomponent mixture: *At equilibrium, the chemical potential of each individual species is the same throughout a mixture.*

6.5 The Gibbs Energy of a Gas in a Mixture

In this and the next section, the conditions for equilibrium in a mixture of ideal gases are derived in terms of the μ_i of the chemical constituents. We will show that the partial pressures of all constituents of the mixture are related by the thermodynamic equilibrium constant, K_P. Consider first the simple system consisting of two volumes separated by a semipermeable membrane, as shown in Figure 6.4. On the left side, the gas consists solely of pure H_2. On the right side, H_2 is present as one constituent of a mixture. The membrane allows only H_2 to pass in both directions.

Once equilibrium has been reached with respect to the concentration of H_2 throughout the system, the hydrogen pressure is the same on both sides of the membrane and

$$\mu_{H_2}^{pure} = \mu_{H_2}^{mixture} \qquad (6.43)$$

Recall from Section 6.3 that the molar Gibbs energy of a pure ideal gas depends on its pressure as $G(T,P) = G°(T) + nRT \ln(P/P°)$. For a pure substance, i, $\mu_i = G_{m,i}$. Therefore, Equation (6.43) can be written in the form

$$\mu_{H_2}^{pure}(T,P_{H_2}) = \mu_{H_2}°(T) + RT \ln \frac{P_{H_2}}{P°} = \mu_{H_2}^{mixture}(T,P_{H_2}) \qquad (6.44)$$

The chemical potential of a gas in a mixture depends logarithmically on its partial pressure. Equation (6.44) applies to any mixture, not just to those for which an appropriate semipermeable membrane exists. We, therefore, generalize the discussion by referring to a component of the mixture as A.

The partial pressure of species A in the gas mixture, P_A, can be expressed in terms of x_A, its mole fraction in the mixture, and the total pressure, P:

$$P_A = x_A P \qquad (6.45)$$

Using this relationship, Equation (6.44) becomes

$$\mu_A^{mixture}(T,P) = \mu_A°(T) + RT \ln \frac{P}{P°} + RT \ln x_A \qquad (6.46)$$

$$= \left(\mu_A°(T) + RT \ln \frac{P}{P°} \right) + RT \ln x_A \text{ or}$$

$$\mu_A^{mixture}(T,P) = \mu_A^{pure}(T,P) + RT \ln x_A$$

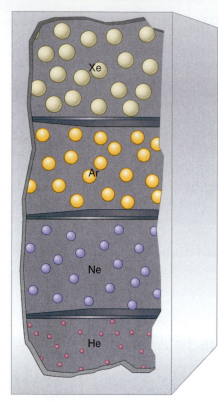

FIGURE 6.5
An isolated system consists of four separate subsystems containing He, Ne, Ar, and Xe, each at a pressure of 1 bar. The barriers separating these subsystems can be removed, leading to mixing.

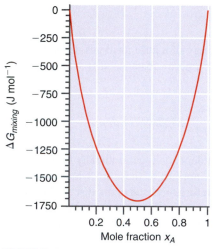

FIGURE 6.6
The Gibbs energy of mixing of the ideal gases A and B as a function of x_A, with $n_A + n_B = 1$ and $T = 298.15$ K.

Because $x_A < 1$, the chemical potential of a gas in a mixture is less than that of the pure gas if P is the same for the pure sample and the mixture. Because matter flows from a region of high to one of low chemical potential, a pure gas initially separated by a barrier from a mixture at the same pressure flows into the mixture as the barrier is removed. This shows that the mixing of gases at constant pressure is a spontaneous process.

6.6 Calculating the Gibbs Energy of Mixing for Ideal Gases

The previous section demonstrated that the mixing of gases is a spontaneous process. We next obtain a quantitative relationship between ΔG_{mixing} and the mole fractions of the individual constituents of the mixture. Consider the system shown in Figure 6.5. The four compartments contain the gases He, Ne, Ar, and Xe at the same temperature and pressure. The volumes of the four compartments differ. To calculate ΔG_{mixing}, we must compare G for the initial state shown in Figure 6.5 and the final state. For the initial state, in which we have four pure separated substances,

$$G_i = G_{He} + G_{Ne} + G_{Ar} + G_{Xe} = n_{He}G_{He}^{\circ} + n_{Ne}G_{Ne}^{\circ} + n_{Ar}G_{Ar}^{\circ} + n_{Xe}G_{Xe}^{\circ} \quad (6.47)$$

For the final state in which all four components are dispersed in the mixture, from Equation (6.46),

$$G_f = n_{He}(G_{He}^{\circ} + RT \ln x_{He}) + n_{Ne}(G_{Ne}^{\circ} + RT \ln x_{Ne}) \quad (6.48)$$
$$+ n_{Ar}(G_{Ar}^{\circ} + RT \ln x_{Ar}) + n_{Xe}(G_{Xe}^{\circ} + RT \ln x_{Xe})$$

The Gibbs energy of mixing is $G_f - G_i$ or

$$\Delta G_{mixing} = RTn_{He} \ln x_{He} + RTn_{Ne} \ln x_{Ne} + RTn_{Ar} \ln x_{Ar} + RTn_{Xe} \ln x_{Xe} \quad (6.49)$$
$$= RT \sum_i n_i \ln x_i = nRT \sum_i x_i \ln x_i$$

Note that because each term in the last expression of Equation (6.49) is negative, $\Delta G_{mixing} < 0$. Therefore, mixing is a spontaneous process at constant T and P.

Equation (6.49) allows one to calculate the value of ΔG_{mixing} for any given set of the mole fractions x_i. It is easiest to visualize graphically the results for a binary mixture of species A and B. To simplify the notation, we set $x_A = x$, so that $x_B = 1 - x$. It follows that

$$\Delta G_{mixing} = nRT \left[x \ln x + (1 - x) \ln(1 - x) \right] \quad (6.50)$$

A plot of ΔG_{mixing} versus x is shown in Figure 6.6 for a binary mixture. Note that ΔG_{mixing} is zero for $x_A = 0$ and $x_A = 1$ because only pure substances are present in these limits. Also, ΔG_{mixing} has a minimum for $x_A = 0.5$, because there is a maximal decrease in G that arises from the dilution of both A and B when A and B are present in equal amounts.

The entropy of mixing can be calculated from Equation (6.49):

$$\Delta S_{mixing} = -\left(\frac{\partial \Delta G_{mixing}}{\partial T} \right)_P = -nR \sum_i x_i \ln x_i \quad (6.51)$$

As shown in Figure 6.7, the entropy of mixing is greatest for $x_A = 0.5$. What lies behind the increase in entropy? Each of the two components of the mixture expands from its ini-

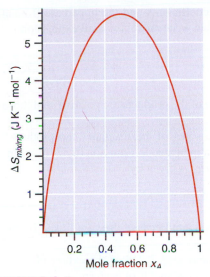

FIGURE 6.7
The entropy of mixing of the ideal gases A and B is shown as a function of the mole fraction of component A at 298.15 K, with $n_A + n_B = 1$.

tial volume to the same final volume. Therefore, ΔS_{mixing} arises purely from the dependence of S on V at constant T,

$$\Delta S = R\left(n_A \ln \frac{V_f}{V_{iA}} + n_B \ln \frac{V_f}{V_{iB}}\right) = R\left(nx_A \ln \frac{1}{x_A} + nx_B \ln \frac{1}{x_B}\right) = -nR(x_A \ln x_A + x_B \ln x_B)$$

if both components and the mixture are at the same pressure.

EXAMPLE PROBLEM 6.5

Consider the system shown in Figure 6.5. Assume that the separate compartments contain 1.0 mol of He, 3.0 mol of Ne, 2.0 mol of Ar, and 2.5 mol of Xe at 298.15 K. The pressure in each compartment is 1 bar.

a. Calculate ΔG_{mixing}.

b. Calculate ΔS_{mixing}.

Solution

a. $\Delta G_{mixing} = RTn_{He} \ln x_{He} + RTn_{Ne} \ln x_{Ne} + RTn_{Ar} \ln x_{Ar} + RTn_{Xe} \ln x_{Xe}$

$$= RT \sum_i n_i \ln x_i = nRT \sum_i x_i \ln x_i$$

$$= 8.5 \text{ mol} \times 8.314 \text{ J k}^{-1} \text{ mol}^{-1} \times 298.15 \text{ K}$$

$$\times \left(\frac{1.0}{8.5} \ln \frac{1.0}{8.5} + \frac{3.0}{8.5} \ln \frac{3.0}{8.5} + \frac{2.0}{8.5} \ln \frac{2.0}{8.5} + \frac{2.5}{8.5} \ln \frac{2.5}{8.5}\right)$$

$$= -2.8 \times 10^4 \text{ J}$$

b. $\Delta S_{mixing} = -nR \sum_i x_i \ln x_i$

$$= -8.5 \text{ mol} \times 8.314 \text{ J K}^{-1} \text{ mol}^{-1}$$

$$\times \left(\frac{1.0}{8.5} \ln \frac{1.0}{8.5} + \frac{3.0}{8.5} \ln \frac{3.0}{8.5} + \frac{2.0}{8.5} \ln \frac{2.0}{8.5} + \frac{2.5}{8.5} \ln \frac{2.5}{8.5}\right)$$

$$= 93 \text{ J K}^{-1}$$

What is the driving force for the mixing of gases? We saw in Section 6.1 that there are two contributions to ΔG, an enthalpic contribution ΔH and an entropic contribution $T\Delta S$. By calculating ΔH_{mixing} from $\Delta H_{mixing} = \Delta G_{mixing} + T\Delta S_{mixing}$ using Equations (6.50) and (6.51), you will see that for the mixing of ideal gases, $\Delta H_{mixing} = 0$. Because the molecules in an ideal gas do not interact, there is no enthalpy change associated with mixing. We conclude that the mixing of ideal gases is driven entirely by ΔS_{mixing} as shown earlier in Figure 6.7.

Although the mixing of gases is always spontaneous, the same is not true of liquids. Liquids can be either miscible or immiscible. How can this observation be explained? For gases or liquids, $\Delta S_{mixing} > 0$. Therefore, if two liquids are immiscible, $\Delta G_{mixing} > 0$ because $\Delta H_{mixing} > 0$ and $\Delta H_{mixing} > T\Delta S_{mixing}$. If two liquids mix, it is generally energetically favorable for one species to be surrounded by the other species. In this case, $\Delta H_{mixing} < T\Delta S_{mixing}$, and $\Delta G_{mixing} < 0$.

6.7 Expressing Chemical Equilibrium in Terms of the μ_i

Consider the balanced chemical reaction

$$\alpha A + \beta B + \chi C + \ldots \rightarrow \delta M + \varepsilon N + \gamma O + \ldots \qquad (6.52)$$

in which Greek letters represent stoichiometric coefficients and the uppercase Roman letters represent the reactants and products. We can write an abbreviated expression for this reaction in the form

$$\sum_i \nu_i X_i = 0 \tag{6.53}$$

In Equation (6.53), the stoichiometric coefficients of the products are positive, and those of the reactants are negative.

What determines the equilibrium partial pressures of the reactants and products? Imagine that the reaction proceeds in the direction indicated in Equation (6.52) by an infinitesimal amount. The change in the Gibbs energy is given by

$$dG = \sum_i \mu_i \, dn_i \tag{6.54}$$

In this equation, the individual dn_i are not independent because they are linked by the stoichiometric equation.

It is convenient at this point to introduce a parameter, ξ, called the **extent of reaction.** If the reaction advances by ξ moles, the number of moles of each species i changes according to

$$n_i = n_i^{initial} + \nu_i \xi \tag{6.55}$$

Differentiating this equation leads to $dn_i = \nu_i d\xi$. By inserting this result in Equation (6.54), we can write dG in terms of ξ. At constant T and P,

$$dG = \left(\sum_i \nu_i \mu_i\right) d\xi = \Delta G_{reaction} \, d\xi \tag{6.56}$$

An advancing reaction can be described in terms of the partial derivative of G with ξ:

$$\left(\frac{\partial G}{\partial \xi}\right)_{T,P} = \sum_i \nu_i \mu_i = \Delta G_{reaction} \tag{6.57}$$

The direction of spontaneous change is that in which $\Delta G_{reaction}$ is negative. This direction corresponds to $(\partial G/\partial \xi)_{T,P} < 0$. At a given composition of the reaction mixture, the partial pressures of the reactants and products can be determined, and the μ_i can be calculated. Because $\mu_i = \mu_i(T,P,n_A,n_B,\dots)$, $\sum_i \nu_i \mu_i = \Delta G_{reaction}$ must be evaluated at specific values of $T,P,n_A,n_B,\dots$. Based on the value of $\Delta G_{reaction}$, the following conclusions can be drawn:

- If $(\partial G/\partial \xi)_{T,P} < 0$, the reaction proceeds spontaneously as written.
- If $(\partial G/\partial \xi)_{T,P} > 0$, the reaction proceeds spontaneously in the opposite direction.
- If $(\partial G/\partial \xi)_{T,P} = 0$, the reaction system is at equilibrium, and there is no direction of spontaneous change.

The most important value of ξ is ξ_{eq}, corresponding to equilibrium. How can this value be found? To make the following discussion more concrete, we consider the reaction system $2NO_2(g) \rightleftharpoons N_2O_4(g)$, with $(2 - 2\xi)$ moles of $NO_2(g)$ and ξ moles of $N_2O_4(g)$ present in a vessel at a constant pressure of 1 bar and 298 K. The parameter ξ could, in principle, take on all values between zero and one, corresponding to pure $NO_2(g)$ and pure $N_2O_4(g)$, respectively. The Gibbs energy of the pure unmixed reagents and products, G_{pure}, is given by

$$G_{pure} = (2 - 2\xi) G_m^\circ(NO_2,g) + \xi G_m^\circ(N_2O_4,g) \tag{6.58}$$

where ΔG_m° is the conventional molar Gibbs energy defined by Equation 6.34. Note that G_{pure} varies linearly with ξ, as shown in Figure 6.8. Because the reactants and products are mixed throughout the range accessible to ξ, G_{pure} is not equal to $G_{mixture}$, which is given by

$$G_{mixture} = G_{pure} + \Delta G_{mixing} \tag{6.59}$$

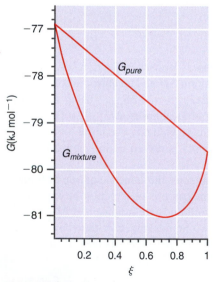

FIGURE 6.8

$G_{mixture}$ and G_{pure} are depicted for the $2NO_2(g) \rightleftharpoons N_2O_4(g)$ equilibrium. For this reaction system, the equilibrium position would correspond to $\xi = 1$ in the absence of mixing. The mixing contribution to $G_{mixture}$ shifts ξ_{eq} to values smaller than one.

Equation (6.59) shows that if G_{pure} alone determined ξ_{eq}, $\xi_{eq} = 0$ or $\xi_{eq} = 1$, depending on whether G_{pure} is lower for reagents or products. For the reaction under consideration, $\Delta G_f^{\circ}(N_2O_4, g) < \Delta G_f^{\circ}(NO_2, g)$ so that $\xi_{eq} = 1$ in this limit. In other words, if G_{pure} alone determined ξ_{eq}, every chemical reaction would go to completion, or the reverse reaction would go to completion. How does mixing influence the equilibrium position for the reaction system? The value of ΔG_{mixing} can be calculated using Equation (6.49). You will show in the end-of-chapter problems that $|\Delta G_{mixing}|$ for this reaction system has a maximum at $\xi_{eq} = 0.55$.

If G_{pure} alone determines ξ_{eq}, then $\xi_{eq} = 1$, and if ΔG_{mixing} alone determined ξ_{eq}, then $\xi_{eq} = 0.55$. However, the minimum in $\Delta G_{reaction}$ is determined by the minimum in $G_{pure} + \Delta G_{mixing}$ rather than in the minimum of the individual components. For our specific case, $\xi_{eq} = 0.72$ at 298 K. The decrease in G on mixing plays a critical role in determining the position of equilibrium in a chemical reaction in that ξ_{eq} is shifted from 1.0 to 0.72.

We summarize the roles of G_{pure} and ΔG_{mixing} in determining ξ_{eq}. For reactions in which $G_{pure}^{reactants}$ and $G_{pure}^{products}$ are very similar, ξ_{eq} will be largely determined by ΔG_{mixing}. For reactions in which $G_{pure}^{reactants}$ and $G_{pure}^{products}$ are very different, ξ_{eq} will not be greatly influenced by ΔG_{mixing} and the equilibrium mixture will essentially consist of pure reactants or pure products.

The direction of spontaneous change is determined by $\Delta G_{reaction}$. How can $\Delta G_{reaction}$ be calculated? We distinguish between calculating $\Delta G_{reaction}$ for the standard state in which $P_i = P^{\circ} = 1$ bar and including the pressure dependence of $\Delta G_{reaction}$. Tabulated values of ΔG_f° for compounds are used to calculate $\Delta G_{reaction}^{\circ}$ at $P^{\circ} = 1$ bar and $T = 298.15$ K as shown in Example Problem 6.6. Just as for enthalpy, ΔG_f° for a pure element in its standard state at the reference temperature is equal to zero because the reactants and products in the formation reaction are identical. Just as for $\Delta H_{reaction}^{\circ}$ discussed in Chapter 4,

$$\Delta G_{reaction}^{\circ} = \sum_i v_i \Delta G_{f,i}^{\circ} \tag{6.60}$$

EXAMPLE PROBLEM 6.6

Calculate the standard Gibbs energy of formation ΔG_f° for urea at 298.15 K using data tabulated in Tables 4.1 and 4.2 (see Appendix B, Data Tables).

Solution

$$\Delta G_f^{\circ} = \Delta H_f^{\circ}(CH_4N_2O, s) - T\Delta S^{\circ} = -3.331 \times 10^5 \text{ J mol}^{-1} + 298.15 \text{ K}$$
$$\times 457.0 \text{ J K}^{-1} \text{mol}^{-1} = -1.969 \times 10^5 \text{ J mol}^{-1}$$

EXAMPLE PROBLEM 6.7

Calculate $\Delta G_{reaction}^{\circ}$ for the reaction $6CO_2(g) + 6H_2O(l) \rightarrow C_6H_{12}O_6(s) + 6O_2(g)$ at 298.15 K.

Solution

$$\Delta G_{reaction}^{\circ} = \Delta G_f^{\circ}(C_6H_{12}O_6, s) + 6\Delta G_f^{\circ}(O_2, g) - 6\Delta G_f^{\circ}(H_2O, l) - 6\Delta G_f^{\circ}(CO_2, g)$$
$$= -1 \times 910.6 \text{ kJ mol}^{-1} + 6 \times 0 \text{ kJ mol}^{-1} + 6 \times 237.1 \text{ kJ mol}^{-1}$$
$$+ 6 \times 394.4 \text{ kJ mol}^{-1}$$
$$= 2879 \text{ kJ mol}^{-1}$$

At other temperatures, $\Delta G_{reaction}^{\circ}$ is calculated as shown in Example Problem 6.8.

EXAMPLE PROBLEM 6.8

Calculate $\Delta G^{\circ}_{reaction}$ for the reaction in Example Problem 6.7 at $T = 310.$ K.

Solution

To calculate $\Delta G^{\circ}_{reaction}$ at the elevated temperature, we use Equation (6.37) assuming that $\Delta H^{\circ}_{reaction}$ is independent of T:

$$\Delta G^{\circ}_{reaction}(T_2) = T_2\left[\frac{\Delta G^{\circ}_{reaction}(298.15 \text{ K})}{298.15 \text{ K}} + \Delta H^{\circ}_{reaction}(298.15 \text{ K})\left(\frac{1}{T_2} - \frac{1}{298.15 \text{ K}}\right)\right]$$

$$\Delta H^{\circ}_{reaction}(298.15 \text{ K}) = \Delta H^{\circ}_f(C_6H_{12}O_6, s) + 6\Delta H^{\circ}_f(O_2, g) - 6\Delta H^{\circ}_f(H_2O, l)$$
$$\qquad\qquad - 6\Delta H^{\circ}_f(CO_2, g)$$

$$= -1 \times 1273.1 \text{ kJ mol}^{-1} + 6 \times 0 \text{ kJ mol}^{-1} + 6 \times 285.8 \text{ kJ mol}^{-1}$$
$$\qquad + 6 \times 393.5 \text{ kJ mol}^{-1}$$

$$= 2803 \text{ kJ mol}^{-1}$$

$$\Delta G^{\circ}_{reaction}(310. \text{ K}) = 310. \text{ K} \times \left[\frac{\dfrac{2879 \times 10^3 \text{ J mol}^{-1}}{298.15 \text{ K}} + 2803}{\times 10^3 \text{J mol}^{-1}\left(\dfrac{1}{310. \text{ K}} - \dfrac{1}{298.15 \text{ K}}\right)}\right]$$

$$= 2993 \text{ kJ mol}^{-1} - 111 \text{ kJ mol}^{-1} = 2.882 \times 10^3 \text{ kJ mol}^{-1}$$

In the next section, the pressure dependence of G is used to calculate $\Delta G_{reaction}$ for arbitrary values of the pressure.

6.8 Calculating $\Delta G_{reaction}$ and Introducing the Equilibrium Constant for a Mixture of Ideal Gases

In this section, we introduce the pressure dependence of μ_i in order to calculate $\Delta G_{reaction}$ for reaction mixtures in which P_i is different from 1 bar. Consider the following reaction that takes place between ideal gas species A, B, C, and D:

$$\alpha A + \beta B \rightleftharpoons \gamma C + \delta D \tag{6.61}$$

Because all four species are present in the reaction mixture, the reaction Gibbs energy is given by

$$\Delta G_{reaction} = \sum_i \nu_i \Delta G_{f,i} = \gamma\mu^{\circ}_C + \gamma RT \ln\frac{P_C}{P^{\circ}} + \delta\mu^{\circ}_D + \delta RT \ln\frac{P_D}{P^{\circ}} \tag{6.62}$$

$$- \alpha\mu^{\circ}_A - \alpha RT \ln\frac{P_A}{P^{\circ}} - \beta\mu^{\circ}_B - \beta RT \ln\frac{P_B}{P^{\circ}}$$

The terms in the previous equation can be separated into those at the standard condition of $P^{\circ} = 1$ bar and the remaining terms:

$$\Delta G_{reaction} = \Delta G^{\circ}_{reaction} + \gamma RT \ln\frac{P_C}{P^{\circ}} + \delta RT \ln\frac{P_D}{P^{\circ}} - \alpha RT \ln\frac{P_A}{P^{\circ}} - \beta RT \ln\frac{P_B}{P^{\circ}} \tag{6.63}$$

$$= \Delta G^{\circ}_{reaction} + RT \ln\frac{\left(\dfrac{P_C}{P^{\circ}}\right)^{\gamma}\left(\dfrac{P_D}{P^{\circ}}\right)^{\delta}}{\left(\dfrac{P_A}{P^{\circ}}\right)^{\alpha}\left(\dfrac{P_B}{P^{\circ}}\right)^{\beta}}$$

where

$$\Delta G_{reaction}^{\circ} = \gamma\mu_C^{\circ}(T) + \delta\mu_D^{\circ}(T) - \alpha\mu_A^{\circ}(T) - \beta\mu_B^{\circ}(T) = \sum_i \nu_i \Delta G_{f,i}^{\circ} \quad (6.64)$$

Recall from Example Problem 6.7 that standard Gibbs energies of formation rather than chemical potentials are used to calculate $\Delta G_{reaction}^{\circ}$.

The combination of the partial pressures of reactants and products is called the **reaction quotient of pressures,** which is abbreviated Q_P and defined as follows:

$$Q_P = \frac{\left(\dfrac{P_C}{P^{\circ}}\right)^{\gamma}\left(\dfrac{P_D}{P^{\circ}}\right)^{\delta}}{\left(\dfrac{P_A}{P^{\circ}}\right)^{\alpha}\left(\dfrac{P_B}{P^{\circ}}\right)^{\beta}} \quad (6.65)$$

With these definitions of $\Delta G_{reaction}^{\circ}$ and Q_P, Equation (6.62) becomes

$$\Delta G_{reaction} = \Delta G_{reaction}^{\circ} + RT \ln Q_P \quad (6.66)$$

Note that $\Delta G_{reaction}$ can be separated into two terms, only one of which depends on the partial pressures of reactants and products.

We next show how Equation (6.66) can be used to predict the direction of spontaneous change for given partial pressures of the reactants and products. If the partial pressures of the products C and D are large, and those of the reactants A and B are small compared to their values at equilibrium, Q_P will be large. As a result, $RT \ln Q_P$ will be large and positive, and $\Delta G_{reaction} = \Delta G_{reaction}^{\circ} + RT \ln Q_P > 0$. In this case, the reaction as written in Equation (6.61) from left to right is not spontaneous, but the reverse of the reaction is spontaneous. Next, consider the opposite extreme. If the partial pressures of the reactants A and B are large, and those of the products C and D are small compared to their values at equilibrium, Q_P will be small. As a result, $RT \ln Q_P$ will be large and negative. In this case, $\Delta G_{reaction} = \Delta G_{reaction}^{\circ} + RT \ln Q_P < 0$ and the reaction will be spontaneous as written from left to right, because a portion of the reactants combines to form products.

Whereas the two cases that we have considered lead to a change in the partial pressures of the reactants and products, the most interesting case is equilibrium for which $\Delta G_{reaction} = 0$. At equilibrium, $\Delta G_{reaction}^{\circ} = -RT \ln Q_P$. We denote this special system configuration by adding a superscript eq to each partial pressure, and renaming Q_P as K_P. The quantity K_P is called the **thermodynamic equilibrium constant:**

$$0 = \Delta G_{reaction}^{\circ} + RT \ln \frac{\left(\dfrac{P_C^{eq}}{P^{\circ}}\right)^c\left(\dfrac{P_D^{eq}}{P^{\circ}}\right)^d}{\left(\dfrac{P_A^{eq}}{P^{\circ}}\right)^a\left(\dfrac{P_B^{eq}}{P^{\circ}}\right)^b} \quad \text{or, equivalently,} \quad \Delta G_{reaction}^{\circ} = -RT \ln K_P \quad (6.67)$$

or

$$\ln K_P = -\frac{\Delta G_{reaction}^{\circ}}{RT}$$

Because $\Delta G_{reaction}^{\circ}$ is a function of T only, K_P is also a function of T only. *The thermodynamic equilibrium constant K_P does not depend on the pressure.* Note also that K_P is a dimensionless number.

EXAMPLE PROBLEM 6.9

a. Using data from Table 4.1 (see Appendix B, Data Tables), calculate K_P at 298.15 K for the reaction $CO(g) + H_2O(l) \rightarrow CO_2(g) + H_2(g)$.

b. Based on the value that you obtained for part (a), do you expect the mixture to consist mainly of $CO_2(g)$ and $H_2(g)$ or mainly of $CO(g) + H_2O(l)$ at equilibrium?

Solution

a. $\ln K_P = -\dfrac{1}{RT}\Delta G_{reaction}^{\circ} = -\dfrac{1}{RT}\left(\begin{array}{l}\Delta G_f^{\circ}(CO_2,g) + \Delta G_f^{\circ}(H_2,g) - \Delta G_f^{\circ}(H_2O,l) \\ -\Delta G_f^{\circ}(CO,g)\end{array}\right)$

$= -\dfrac{1}{8.314\ \text{J mol}^{-1}\text{K}^{-1} \times 298.15\ \text{K}} \times \left(\begin{array}{l}-394.4 \times 10^3\ \text{J mol}^{-1} + 0 + 237.1 \\ \times 10^3\ \text{J mol}^{-1} + 137.2 \times 10^3\ \text{J mol}^{-1}\end{array}\right)$

$= 8.1087$

$K_P = 3.32 \times 10^3$

b. Because $K_P \gg 1$, the mixture will consist mainly of the products $CO_2(g) + H_2(g)$ at equilibrium.

6.9 Calculating the Equilibrium Partial Pressures in a Mixture of Ideal Gases

As shown in the previous section, the partial pressures of the reactants and products in a mixture of gases at equilibrium cannot take on arbitrary values because they are related through K_P. In this section, we show how the equilibrium partial pressures can be calculated for ideal gases. Similar calculations for real gases will be discussed in the next chapter. In Example Problem 6.10, we consider the dissociation of chlorine:

$$Cl_2(g) \rightarrow 2Cl(g) \qquad (6.68)$$

It is useful to set up the calculation in a tabular form as shown in the following example problem.

EXAMPLE PROBLEM 6.10

In this example, n_0 moles of chlorine gas are placed in a reaction vessel, whose temperature can be varied over a wide range, so that molecular chlorine can partially dissociate to atomic chlorine.

a. Derive an expression for K_P in terms of n_0, ξ, and P.

b. Define the degree of dissociation as $\alpha = \xi_{eq}/n_0$, where $2\xi_{eq}$ is the number of moles of $Cl(g)$ present at equilibrium, and n_0 represents the number of moles of $Cl_2(g)$ that would be present in the system if no dissociation occurred. Derive an expression for α as a function of K_P and P.

Solution

a. We set up the following table:

	$Cl_2(g) \rightarrow$	$2Cl(g)$
Initial number of moles	n_0	0
Moles present at equilibrium	$n_0 - \xi_{eq}$	$2\xi_{eq}$
Mole fraction present at equilibrium, x_i	$\dfrac{n_0 - \xi_{eq}}{n_0 + \xi_{eq}}$	$\dfrac{2\xi_{eq}}{n_0 + \xi_{eq}}$
Partial pressure at equilibrium, $P_i = x_i P$	$\left(\dfrac{n_0 - \xi_{eq}}{n_0 + \xi_{eq}}\right)P$	$\left(\dfrac{2\xi_{eq}}{n_0 + \xi_{eq}}\right)P$

We next express K_P in terms of n_0, ξ_{eq}, and P:

$$K_P(T) = \frac{\left(\dfrac{P_{Cl}^{eq}}{P°}\right)^2}{\left(\dfrac{P_{Cl_2}^{eq}}{P°}\right)} = \frac{\left[\left(\dfrac{2\xi_{eq}}{n_0+\xi_{eq}}\right)\dfrac{P}{P°}\right]^2}{\left(\dfrac{n_0-\xi_{eq}}{n_0+\xi_{eq}}\right)\dfrac{P}{P°}} = \frac{4\xi_{eq}^2}{(n_0+\xi_{eq})(n_0-\xi_{eq})}\frac{P}{P°} = \frac{4\xi_{eq}^2}{(n_0)^2-\xi_{eq}^2}\frac{P}{P°}$$

This expression is converted into one in terms of α:

b. $K_P(T) = \dfrac{4\xi_{eq}^2}{(n_0)^2 - \xi_{eq}^2}\dfrac{P}{P°} = \dfrac{4\alpha^2}{1-\alpha^2}\dfrac{P}{P°}$

$$\left(K_P(T) + 4\frac{P}{P°}\right)\alpha^2 = K_P(T)$$

$$\alpha = \sqrt{\frac{K_P(T)}{K_P(T) + 4\dfrac{P}{P°}}}$$

Because $K_P(T)$ depends strongly on temperature, α will also be a strong function of temperature. Note that α depends on both K_P and P.

Note that whereas $K_P(T)$ is independent of P, α as calculated in Example Problem 6.10 does depend on P. In the particular case considered, α decreases as P increases for constant T. As will be shown in Section 6.12, α depends on P if $\Delta\nu \neq 0$ for a reaction.

6.10 The Variation of K_P with Temperature

As shown in Example Problem 6.10, the degree of dissociation of Cl_2 is dependent on the temperature. Based on our chemical intuition, we expect the degree of dissociation to be high at high temperatures and low at room temperature. For this to be true, $K_P(T)$ must increase with temperature for this reaction. How can we understand the temperature dependence for this reaction and for reactions in general?

Starting with Equation (6.67), we write

$$\frac{d \ln K_P}{dT} = -\frac{d(\Delta G_{reaction}°/RT)}{dT} = -\frac{1}{R}\frac{d(\Delta G_{reaction}°/T)}{dT} \tag{6.69}$$

Using the Gibbs–Helmholtz equation [Equation (6.35)], the preceding equation reduces to

$$\frac{d \ln K_P}{dT} = -\frac{1}{R}\frac{d(\Delta G_{reaction}°/T)}{dT} = \frac{\Delta H_{reaction}°}{RT^2} \tag{6.70}$$

Equation (6.70) is called the van't Hoff equilibrium equation. Because tabulated values of $\Delta G_f°$ are available at 298.15 K, we can calculate $\Delta G_{reaction}°$ and K_P at this temperature; K_P can be calculated at the temperature T_f by integrating Equation (6.70) between the appropriate limits:

$$\int_{K_P(298.15K)}^{K_P(T_f)} d \ln K_P = \frac{1}{R}\int_{298.15K}^{T_f} \frac{\Delta H_{reaction}°}{T^2}\,dT \tag{6.71}$$

If the temperature T_f is not much different from 298.15 K, it can be assumed that $\Delta H_{reaction}°$ is constant over the temperature interval. This assumption is better than it might appear at first glance. Although H is strongly dependent on temperature, the temperature dependence of $\Delta H^{\circ}_{reaction}$ is governed by the difference in heat capacities ΔC_p between reactants and products (see Section 4.4). If the heat capacities of reactants and products are nearly the same, $\Delta H^{\circ}_{reaction}$ is nearly independent of temperature. With the assumption that $\Delta H_{reaction}$ is independent of temperature, Equation (6.71) becomes

$$\ln K_P(T_f) = \ln K_P(298.15\ \text{K}) - \frac{\Delta H^{\circ}_{reaction}}{R}\left(\frac{1}{T_f} - \frac{1}{298.15\ \text{K}}\right) \tag{6.72}$$

EXAMPLE PROBLEM 6.11

Using the result of Example Problem 6.10 and the data tables, consider the dissociation reaction $Cl_2(g) \rightarrow 2Cl(g)$.

a. Calculate K_p at 800, 1500, and 2000 K for $P = 0.010$ bar.

b. Calculate the degree of dissociation, α, at 300, 1500, and 2000 K.

Solution

a. $\Delta G^{\circ}_{reaction} = 2\Delta G^{\circ}_f(Cl,g) - \Delta G^{\circ}_f(Cl_2,g) = 2 \times 105.7 \times 10^3\ \text{J mol}^{-1} - 0$

$\qquad = 211.4\text{kJ mol}^{-1}$

$\Delta H^{\circ}_{reaction} = 2\Delta H^{\circ}_f(Cl,g) - \Delta H^{\circ}_f(Cl_2,g) = 2 \times 121.3 \times 10^3\ \text{J mol}^{-1} - 0$

$\qquad = 242.6\ \text{kJ mol}^{-1}$

$\ln K_P(T_f) = -\dfrac{\Delta G^{\circ}_{reaction}}{RT} - \dfrac{\Delta H^{\circ}_{reaction}}{R}\left(\dfrac{1}{T_f} - \dfrac{1}{298.15\ \text{K}}\right)$

$\qquad = -\dfrac{211.4 \times 10^3\ \text{J mol}^{-1}}{8.314\ \text{J K}^{-1}\text{mol}^{-1} \times 298.15\ \text{K}} - \dfrac{242.6 \times 10^3\ \text{J mol}^{-1}}{8.314\ \text{J K}^{-1}\text{mol}^{-1}}$

$\qquad \times \left(\dfrac{1}{T_f} - \dfrac{1}{298.15\ \text{K}}\right)$

$\ln K_P(800\ \text{K})$

$= -\dfrac{210.6 \times 10^3\ \text{J mol}^{-1}}{8.314\ \text{J K}^{-1}\text{mol}^{-1} \times 298.15\ \text{K}} - \dfrac{242.6 \times 10^3\ \text{J mol}^{-1}}{8.314\ \text{J K}^{-1}\text{mol}^{-1}}$

$\qquad \times \left(\dfrac{1}{800\ \text{K}} - \dfrac{1}{298.15\ \text{K}}\right) = -23.898$

$K_P(300\ \text{K}) = 4.18 \times 10^{-11}$

The values for K_p at 1500 and 2000 K are 1.03×10^{-3} and 0.134, respectively.

b. The value of α at 2000 K is given by

$$\alpha = \sqrt{\frac{K_P(T)}{K_P(T) + 4\dfrac{P}{P^{\circ}}}} = \sqrt{\frac{0.134}{0.134 + 4 \times 0.01}} = 0.878$$

The values of α at 1500 and 800 K are 0.159 and 3.23×10^{-5}, respectively.

The degree of dissociation of Cl_2 increases with temperature as shown in Example Problem 6.11. This is always the case for an endothermic reaction ($\Delta H_{reaction} > 0$), as will be proved in Section 6.13 when Le Chatelier's principle is discussed.

6.11 Equilibria Involving Ideal Gases and Solid or Liquid Phases

In the preceding sections, we discussed chemical equilibrium in a homogeneous system of ideal gases. However, many chemical reactions involve a gas phase in equilibrium with a solid or liquid phase. An example is the thermal decomposition of $CaCO_3(s)$:

$$CaCO_3(s) \rightarrow CaO(s) + CO_2(g) \tag{6.73}$$

In this case, a pure gas is in equilibrium with two solid phases. At equilibrium,

$$\Delta G_{reaction} = \sum_i n_i \mu_i = 0 \tag{6.74}$$

$$0 = \mu_{eq}(CaO,s,P) + \mu_{eq}(CO_2,g,P) - \mu_{eq}(CaCO_3,s,P)$$

Because the equilibrium pressure is $P \neq P°$, the pressure dependence of each species must be taken into account. From Section 6.3, we know that the pressure dependence of G for a solid or liquid is very small:

$$\mu_{eq}(CaO,s,P) \approx \mu°(CaO,s) \text{ and } \mu_{eq}(CaCO_3,s,P) \approx \mu°(CaCO_3,s) \tag{6.75}$$

You will verify the validity of Equation (6.75) in the end-of-chapter problems. Using the dependence of μ on P for an ideal gas, Equation (6.74) becomes

$$0 = \mu°(CaO,s) + \mu°(CO_2,g) - \mu°(CaCO_3, s) + RT \ln \frac{P_{CO_2}}{P°} \text{ or} \tag{6.76}$$

$$\Delta G°_{reaction} = \mu°(CaO, s) + \mu°(CO_2, g) - \mu°(CaCo_3, s) = -RT \ln \frac{P_{CO_2}}{P°}$$

Rewriting this equation in terms of K_P, we obtain

$$\ln K_P = \ln \frac{P_{CO_2}}{P°} = -\frac{\Delta G°_{reaction}}{RT} \tag{6.77}$$

EXAMPLE PROBLEM 6.12

Using the preceding discussion and the tabulated values of $\Delta G°_f$ and $\Delta H°_f$ in Appendix B, calculate the $CO_2(g)$ pressure in equilibrium with a mixture of $CaCO_3(s)$ and $CaO(s)$ at 1000, 1100, and 1200 K.

Solution

For the reaction $CaCO_3(s) \rightarrow CaO(s) + O_2(g)$, $\Delta G°_{reaction, 298.15 K} = 131.1 \text{ kJ mol}^{-1}$ and $\Delta H°_{reaction, 298.15 K} = 178.5 \text{ kJ mol}^{-1}$.

We use Equation (6.77) to calculate K_P (298.15 K) and Equation (6.72) to calculate K_P at elevated temperatures.

$$\ln \frac{P_{CO_2}}{P°} (1000 \text{ K}) = \ln K_P (1000 \text{ K})$$

$$= \ln K_P (298.15 \text{ K}) - \frac{\Delta H°_{R, 298.15 K}}{R} \left(\frac{1}{1000 \text{ K}} - \frac{1}{298.15 \text{ K}} \right)$$

$$= \frac{-131.1 \times 10^3 \text{ J mol}^{-1}}{8.314 \text{ J K}^{-1} \text{ mol}^{-1} \times 298.15 \text{ K}} - \frac{178.5 \times 10^3 \text{ J mol}^{-1}}{8.314 \text{ J K}^{-1} \text{ mol}^{-1}}$$

$$\times \left(\frac{1}{1000 \text{ K}} - \frac{1}{298.15 \text{ K}} \right)$$

$$= -2.348; P_{CO_2} (1000 \text{ K}) = 0.0956 \text{ bar}$$

The values for P_{CO_2} at 1100 and 1200 K are 0.673 and 1.23 bar, respectively.

If the reaction involves only liquids or solids, the pressure dependence of the chemical potential is generally small and can be neglected. However, it cannot be neglected if $P > 1$ bar as shown in Example Problem 6.13.

EXAMPLE PROBLEM 6.13

At 298.15 K, $\Delta G_f^\circ(\text{C,}graphite) = 0$, and $\Delta G_f^\circ(\text{C,}diamond) = 2.90 \text{ kJ mol}^{-1}$. Therefore, graphite is the more stable solid phase at this temperature at $P = P^\circ = 1$ bar. Given that the densities of graphite and diamond are 2.25 and 3.52 kg/L, respectively, at what pressure will graphite and diamond be in equilibrium at 298.15 K?

Solution

At equilibrium $\Delta G = G(\text{C, }graphite) - G(\text{C, }diamond) = 0$. Using the pressure dependence of G, $(\partial G_m/\partial P)_T = V_m$, we establish the condition for equilibrium:

$$\Delta G = \Delta G_f^\circ(\text{C, }graphite) - \Delta G_f^\circ(\text{C, }diamond) + (V_m^{graphite} - V_m^{diamond})(\Delta P) = 0$$

$$0 = 0 - 2900 \text{ J} + (V_m^{graphite} - V_m^{diamond})(P - 1 \text{ bar})$$

$$P = 1 \text{ bar} + \cfrac{2900 \text{ J}}{M_C\left(\cfrac{1}{\rho_{graphite}} - \cfrac{1}{\rho_{diamond}}\right)}$$

$$= 1 \text{ bar} + \cfrac{2900 \text{ J}}{12.00 \times 10^{-3} \text{ kg mol}^{-1} \times \left(\cfrac{1}{2.25 \times 10^3 \text{ kg m}^{-3}} - \cfrac{1}{3.52 \times 10^3 \text{ kg m}^{-3}}\right)}$$

$$= 10^5 \text{Pa} + 1.51 \times 10^9 \text{ Pa} = 1.51 \times 10^4 \text{ bar}$$

Fortunately for all those with diamond rings, although diamond is unstable with respect to graphite at 1 bar and 298 K, the rate of conversion is vanishingly small.

6.12 Expressing the Equilibrium Constant in Terms of Mole Fraction or Molarity

Chemists often find it useful to express the concentrations of reactants and products in units other than partial pressures. Two examples of other units that we will consider in this section are mole fraction and molarity. Note, however, that this discussion is still limited to a mixture of ideal gases. The extension of chemical equilibrium to include neutral and ionic species in aqueous solutions will be made in Chapter 9 after the concept of activity has been introduced.

We first express the equilibrium constant in terms of mole fractions. The mole fraction, x_i, and the partial pressure, P_i, are related by $P_i = x_i P$. Therefore,

$$K_P = \cfrac{\left(\cfrac{P_C^{eq}}{P^\circ}\right)^c \left(\cfrac{P_D^{eq}}{P^\circ}\right)^d}{\left(\cfrac{P_A^{eq}}{P^\circ}\right)^a \left(\cfrac{P_B^{eq}}{P^\circ}\right)^b} = \cfrac{\left(\cfrac{x_C^{eq} P}{P^\circ}\right)^c \left(\cfrac{x_D^{eq} P}{P^\circ}\right)^d}{\left(\cfrac{x_A^{eq} P}{P^\circ}\right)^a \left(\cfrac{x_B^{eq} P}{P^\circ}\right)^b} = \cfrac{(x_C^{eq})^c (x_D^{eq})^d}{(x_A^{eq})^a (x_B^{eq})^b}\left(\cfrac{P}{P^\circ}\right)^{d+c-a-b} \tag{6.78}$$

$$= K_x\left(\cfrac{P}{P^\circ}\right)^{\Delta v}$$

$$K_x = K_P\left(\cfrac{P}{P^\circ}\right)^{-\Delta v}$$

Note that just as for K_P, K_x is a dimensionless number.

Because the molarity, c_i, is defined as $c_i = n_i/V = P_i/RT$, we can write $P_i/P° = (RT/P°)c_i$. To work with dimensionless quantities, we introduce the ratio $c_i/c°$, which is related to $P_i/P°$ by

$$\frac{P_i}{P°} = \frac{c°RT}{P°}\frac{c_i}{c°} \tag{6.79}$$

Using this notation, we can express K_c in terms of K_P:

$$K_P = \frac{\left(\frac{P_C^{eq}}{P°}\right)^c \left(\frac{P_D^{eq}}{P°}\right)^d}{\left(\frac{P_A^{eq}}{P°}\right)^a \left(\frac{P_B^{eq}}{P°}\right)^b} = \frac{\left(\frac{c_C^{eq}}{c°}\right)^c \left(\frac{c_D^{eq}}{c°}\right)^d}{\left(\frac{c_A^{eq}}{c°}\right)^a \left(\frac{c_B^{eq}}{c°}\right)^b}\left(\frac{c°RT}{P°}\right)^{d+c-a-b} = K_c\left(\frac{c°RT}{P°}\right)^{\Delta\nu} \tag{6.80}$$

$$K_c = K_P\left(\frac{c°RT}{P°}\right)^{-\Delta\nu}$$

Recall that $\Delta\nu$ is the difference in the stoichiometric coefficients of products and reactants. Equations (6.78) and (6.80) show that K_x and K_c are in general different from K_P. They are only equal in the special case that $\Delta\nu = 0$.

6.13 The Dependence of ξ_{eq} on T and P

Suppose that we have a mixture of reactive gases at equilibrium. Does the equilibrium shift toward reactants or products as T or P is changed? To answer this question, we first consider the dependence of K_P on T. Because

$$\frac{d \ln K_P}{dT} = \frac{\Delta H°_{reaction}}{RT^2} \tag{6.81}$$

Note that K_P will change differently with temperature for an exothermic reaction than for an endothermic reaction. For an exothermic reaction, $d \ln K_P/dT < 0$, and ξ_{eq} will shift toward the reactants as T increases. For an endothermic reaction, $d \ln K_P/dT > 0$, and ξ_{eq} will shift toward the products as T increases.

The dependence of ξ_{eq} on pressure can be ascertained from the relationship between K_P and K_x:

$$K_x = K_P\left(\frac{P}{P°}\right)^{-\Delta\nu} \tag{6.82}$$

Because K_P is independent of pressure, the pressure dependence of K_x arises solely from $(P/P°)^{-\Delta\nu}$. If the number of moles of gaseous products increases as the reaction proceeds, K_x decreases as P increases, and ξ_{eq} shifts back toward the reactants. If the number of moles of gaseous products decreases as the reaction proceeds, K_x increases as P increases, and ξ_{eq} shifts forward toward the products. If $\Delta\nu = 0$, ξ_{eq} is independent of pressure.

A combined change in T and P leads to a superposition of the effects just discussed. According to the French chemist Le Chatelier, reaction systems at chemical equilibrium respond to an outside stress, such as a change in T or P, by countering the stress. Consider the $Cl_2(g) \rightarrow 2Cl(g)$ reaction discussed in Example Problems 6.10 and 6.11 for which $\Delta H_{reaction} > 0$. There, ξ_{eq} responds to an increase in T in such a way that heat is taken up by the system; in other words, more $Cl_2(g)$ dissociates. This counters the stress imposed on the system by an increase in T. Similarly, ξ_{eq} responds to an increase in P in such a way that the volume of the system decreases. Specifically, ξ_{eq} changes in the direction that $\Delta\nu < 0$, and for the reaction under consideration, $Cl(g)$ is converted to $Cl_2(g)$. This shift in the position of equilibrium counters the stress brought about by an increase in P.

Vocabulary

chemical potential
conventional molar Gibbs
   energies
extent of reaction

Gibbs energy
Gibbs–Helmholtz equation
Helmholtz energy

maximum nonexpansion
   work
Maxwell relations
natural variables

reaction quotient of pressures
thermodynamic equilibrium
   constant

Questions on Concepts

Q6.1   Under what conditions is $dA \leq 0$ a condition that defines the spontaneity of a process?

Q6.2   Under what conditions is $dG \leq 0$ a condition that defines the spontaneity of a process?

Q6.3   Which thermodynamic state function gives a measure of the maximum electric work that can be carried out in a fuel cell?

Q6.4   By invoking the pressure dependence of the chemical potential, show that if a valve separating a vessel of pure A from a vessel containing a mixture of A and B is opened, mixing will occur. Both A and B are ideal gases, and the initial pressure in both vessels is 1 bar.

Q6.5   Under what condition is $K_P = K_x$?

Q6.6   It is found that K_P is independent of T for a particular chemical reaction. What does this tell you about the reaction?

Q6.7   The reaction A + B → C + D is at equilibrium for $\xi = 0.1$. What does this tell you about the variation of G_{pure} with ξ?

Q6.8   The reaction A + B → C + D is at equilibrium for $\xi = 0.5$. What does this tell you about the variation of G_{pure} with ξ?

Q6.9   Why is it reasonable to set the chemical potential of a pure liquid or solid substance equal to its standard state chemical potential at that temperature independent of the pressure in considering chemical equilibrium?

Q6.10   Is the equation $(\partial U / \partial V)_T = (\beta T - \kappa P)/\kappa$ valid for liquids, solids, and gases?

Q6.11   What is the relationship between the K_P for the two reactions $3/2 H_2 + 1/2 N_2 \rightarrow NH_3$ and $3H_2 + N_2 \rightarrow 2NH_3$?

Problems

Problem numbers in **RED** indicate that the solution to the problem is given in the *Student Solutions Manual*.

P6.1   Calculate the maximum nonexpansion work that can be gained from the combustion of benzene(*l*) and of $H_2(g)$ on a per gram and a per mole basis under standard conditions. Is it apparent from this calculation why fuel cells based on H_2 oxidation are under development for mobile applications?

P6.2   Calculate ΔA for the isothermal compression of 2.00 mol of an ideal gas at 298 K from an initial volume of 35.0 L to a final volume of 12.0 L. Does it matter whether the path is reversible or irreversible?

P6.3   Calculate ΔG for the isothermal expansion of 2.50 mol of an ideal gas at 350 K from an initial pressure of 10.5 bar to a final pressure of 0.500 bar.

P6.4   A sample containing 2.50 mol of an ideal gas at 298 K is expanded from an initial volume of 10.0 L to a final volume of 50.0 L. Calculate ΔG and ΔA for this process for (a) an isothermal reversible path and (b) an isothermal expansion against a constant external pressure of 0.750 bar. Explain why ΔG and ΔA do or do not differ from one another.

P6.5   The pressure dependence of G is quite different for gases and condensed phases. Calculate G_m(C, *solid, graphite,* 100 bar, 298.15 K) and G_m(He, *g,* 100 bar, 298.15 K) relative to their standard state values. By what factor is the change in G_m greater for He than for graphite?

P6.6   Assuming that ΔH_f° is constant in the interval from 275 to 600 K, calculate ΔG for the process (H_2O, *g*, 298 K) → (H_2O, *g*, 525 K).

P6.7   Calculate $\Delta G_{reaction}^{\circ}$ for the reaction $CO(g) + 1/2\ O_2(g)$ → $CO_2(g)$ at 298.15 K. Calculate $\Delta G_{reaction}^{\circ}$ at 650 K assuming that $\Delta H_{reaction}^{\circ}$ is constant in the temperature interval of interest.

P6.8   Calculate $\Delta A_{reaction}^{\circ}$ and $\Delta G_{reaction}^{\circ}$ for the reaction $CH_4(g) + 2O_2(g) \rightarrow CO_2(g) + 2H_2O(l)$ at 298 K from the combustion enthalpy of methane and the entropies of the reactants and products.

P6.9   James Watt once observed that a hard-working horse can lift a 330.-lb weight 100. ft in 1 min. Assuming the horse generates energy to accomplish this work by metabolizing glucose:

$$C_6H_{12}O_6(s) + 6O_2\ (g) \rightarrow 6CO_2\ (g) + 6H_2O(l)$$

calculate how much glucose a horse must metabolize to sustain this rate of work for 1 hour at $T = 298.15$ K?

P6.10   The standard Gibbs energy of formation ΔG_f° for carbon dioxide gas is -394.4 kJ mol^{-1}.
Calculate the Gibbs energy of formation of carbon dioxide at its normal sea level partial pressure of 0.00031 atm.

P6.11   Consider the equilibrium $C_2H_6(g) \rightleftharpoons C_2H_4(g) + H_2(g)$. At 1000 K and a constant total pressure of 1 bar,

$C_2H_6(g)$ is introduced into a reaction vessel. The total pressure is held constant at 1 bar and at equilibrium the composition of the mixture in mole percent is as follows: $H_2(g)$, 26%; $C_2H_4(g)$, 26%; and $C_2H_6(g)$, 48%.

    a. Calculate K_P at 1000 K.

    b. If $\Delta H^\circ_{reaction} = 137.0$ kJ mol^{-1}, calculate the value of K_P at 298.15 K.

    c. Calculate $\Delta G^\circ_{reaction}$ for this reaction at 298.15 K.

P6.12 The industrial synthesis of ammonia has the form: $\frac{1}{2}N_2(g) + \frac{3}{2}H_2(g) \rightarrow NH_3(g)$. Calculate the standard Gibbs energy change and equilibrium constant for the industrial synthesis of ammonia. Using the text as a source for the enthalpy of formation of ammonia, and assuming the enthalpy is constant with temperature, determine the equilibrium constant for the industrial synthesis of ammonia at 400. K.

P6.13 Nitrogen is a vital element for all living systems; except for a few types of bacteria, blue-green algae, and some soil fungi, organisms cannot utilize N_2 from the atmosphere. The formation of "fixed" nitrogen is therefore necessary to sustain life and the simplest form of fixed nitrogen is ammonia NH_3. Living systems cannot fix nitrogen using the gas-phase components listed in Problem P6.12. A hypothetical ammonia synthesis by a living system might be:

$$\frac{1}{2}N_2(g) + \frac{3}{2}H_2O(l) \rightarrow NH_3(aq) + \frac{3}{4}O_2(g)$$

where (aq) means the ammonia is dissolved in water. Calculate the standard free energy change for the biological synthesis of ammonia and calculate the equilibrium constant as well. Based on your answer, would the biological synthesis of ammonia occur spontaneously? Note that $\Delta G^\circ_f(NH_3, aq) = -80.3$ kJ mol^{-1}.

P6.14 Consider the equilibrium $NO_2(g) \rightleftharpoons NO(g) + 1/2O_2(g)$. One mole of $NO_2(g)$ is placed in a vessel and allowed to come to equilibrium at a total pressure of 1 bar. An analysis of the contents of the vessel gives the following results:

T	700 K	800 K
P_{NO}/P_{NO_2}	0.872	2.50

    a. Calculate K_P at 700 and 800 K.

    b. Calculate $\Delta G^\circ_{reaction}$ for this reaction at 298.15 K, assuming that $\Delta H^\circ_{reaction}$ is independent of temperature.

P6.15 Consider the equilibrium $CO(g) + H_2O(g) \rightleftharpoons CO_2(g) + H_2(g)$. At 1000 K, the composition of the reaction mixture is

CO_2	$H_2(g)$	$CO(g)$	$H_2O(g)$	Substance (g)
27.1	27.1	22.9	22.9	Mole %

    a. Calculate K_P and $\Delta G^\circ_{reaction}$ at 1000 K.

    b. Given the answer to part (a), use the ΔH°_f of the reaction species to calculate $\Delta G^\circ_{reaction}$ at 298.15 K. Assume that $\Delta H^\circ_{reaction}$ is independent of temperature.

P6.16 Consider the reaction $FeO(s) + CO(g) \rightleftharpoons Fe(s) + CO_2(g)$ for which K_P is found to have the following values:

T	600°C	1000°C
K_P	0.900	0.396

    a. Calculate $\Delta G^\circ_{reaction}$, $\Delta S^\circ_{reaction}$, and $\Delta H^\circ_{reaction}$ for this reaction at 600°C. Assume that $\Delta H^\circ_{reaction}$ is independent of temperature.

    b. Calculate the mole fraction of $CO_2(g)$ present in the gas phase at 600°C.

P6.17 If the reaction $Fe_2N(s) + 3/2H_2(g) \rightleftharpoons 2Fe(s) + NH_3(g)$ comes to equilibrium at a total pressure of 1 bar, analysis of the gas shows that at 700 and 800 K, $P_{NH_3}/P_{H_2} = 2.165$ and 1.083, respectively, if only $H_2(g)$ was initially present in the gas phase and $Fe_2N(s)$ was in excess.

    a. Calculate K_P at 700 and 800 K.

    b. Calculate $\Delta S^\circ_{reaction}$ at 700 and 800 K and $\Delta H^\circ_{reaction}$ assuming that it is independent of temperature.

    c. Calculate $\Delta G^\circ_{reaction}$ for this reaction at 298.15 K.

P6.18 Many biological macromolecules undergo a transition called *denaturation*. Denaturation is a process whereby a structured, biological active molecule, called the native form, unfolds or becomes unstructured and biologically inactive. The equilibrium is

$$\text{native(folded)} \rightleftharpoons \text{denatured(unfolded)}$$

For a protein at pH = 2, the enthalpy change associated with denaturation is $\Delta H^\circ = 418.0$ kJ mol^{-1} and the entropy change is $\Delta S^\circ = 1.3$ kJ K^{-1} mol^{-1}.

    a. Calculate the Gibbs energy change for the denaturation of the protein at pH = 2 and $T = 303$ K. Assume the enthalpy and entropy are temperature independent between 298.15 and 303 K.

    b. Calculate the equilibrium constant for the denaturation of protein at pH = 2 and $T = 303$ K.

    c. Based on your answers for parts (a) and (b), is protein structurally stable at pH = 2 and $T = 303$ K?

P6.19 The melting temperature of a protein is defined as the temperature at which the equilibrium constant for denaturation has the value $K = 1$. Assuming that the enthalpy of denaturation is temperature independent, use the information in Problem P6.18 to calculate the melting temperature of the protein at pH = 2. where (aq) means the ammonia is dissolved in water. Calculate the standard free energy change for the biological synthesis of ammonia and calculate the equilibrium constant as well. Based on your answer, would the biological synthesis of ammonia occur spontaneously? Note that $\Delta G^\circ_f(NH_3, aq) = -80.3$ kJ mol^{-1}.

P6.20 Calculate the Gibbs energy change for the protein denaturation described in Problem 5.45 at $T = 310.$K and $T = 340.$K.

P6.21 For a protein denaturation at $T = 310.$ K and $P = 1.00$ atm, the enthalpy change is 911 kJ mol^{-1} and the entropy change is 3.12 J K^{-1} mol^{-1}. Calculate the Gibbs energy change at $T = 310.$ K and $P = 1.00$ atm. Calculate the Gibbs energy change at $T = 310.$ K and $P = 1.00 \times 10^3$ bar. Assume for the denaturation $\Delta V = 3.00$ mL mol^{-1}. State any assumptions you make in the calculation.

P6.22 For a protein denaturation the entropy change is 2.31 J K^{-1} mol^{-1} at $P = 1.00$ bar and at the melting temperature $T = 338$ K. Calculate the melting temperature at a pressure of $P = 1.00 \times 10^3$ bar if the heat capacity change $\Delta C_{P,m} = 7.98$ J K^{-1}mol^{-1} and if $\Delta V = 3.10$ mL mol^{-1}. State any assumptions you make in the calculation.

P6.23 Under anaerobic conditions, glucose is broken down in muscle tissue to form lactic acid according to the reaction: $C_6H_{12}O_6 \rightarrow 2CH_3CHOHCOOH$. Thermodynamic data at $T = 298$ K for glucose and lactic acid are given below.

	ΔH_f° (kJ mol^{-1})	$C_{P,m}^\circ$ (J K^{-1}mol^{-1})	S°(J K^{-1}mol^{-1})
Glucose	−1273.1	219.2	209.2
Lactic Acid	−673.6	127.6	192.1

Calculate ΔG° at $T = 298$ K and $T = 310.$ K. Assume all heat capacities are constant from $T = 298$ K to $T = 310.$K.

P6.24 At $T = 298$ K and pH=3 chymotrypsinogen denatures with $\Delta G^\circ = 30.5$ kJ mol^{-1}, $\Delta H^\circ = 163$ kJ mol^{-1}, and $\Delta C_{P,m} = 8.36$ kJ K^{-1} mol^{-1}. Determine ΔG° for the denaturation of chymotrypsinogen at $T = 320.$ K and pH=3. Assume $\Delta C_{P,m}$ is constant between $T = 298$ K and $T = 320.$ K.

P6.25 At 25°C, values for the formation enthalpy and Gibbs energy and $\log_{10} K_P$ for the formation reactions of the various isomers of C_5H_{10} in the gas phase are given by the following table:

Substance	ΔH_f° (kJ mol^{-1})	ΔG_f° (kJ mol^{-1})	$\log_{10}K_P$
A = 1-pentene	−20.920	78.605	−13.7704
B = cis-2-pentene	−28.075	71.852	−12.5874
C = trans-2-pentene	−31.757	69.350	−12.1495
D = 2-methyl-1-butene	−36.317	64.890	−11.3680
E = 3-methyl-1-butene	−28.953	74.785	−13.1017
F = 2-methyl-2-butene	−42.551	59.693	−10.4572
G = cyclopentane	−77.24	38.62	−6.7643

Consider the equilibrium A $\rightleftharpoons$ B $\rightleftharpoons$ C $\rightleftharpoons$ D $\rightleftharpoons$ E $\rightleftharpoons$ F $\rightleftharpoons$ G, which might be established using a suitable catalyst.

   a. Calculate the mole ratios A/G, B/G, C/G, D/G, E/G, and F/G present at equilibrium at 25°C.

   b. Do the ratios of part (a) depend on the total pressure?

   c. Calculate the mole percentages of the various species in the equilibrium mixture.

P6.26 In this problem, you calculate the error in assuming that $\Delta H_{reaction}^\circ$ is independent of T for a specific reaction. The following data are given at 25°C:

	CuO(s)	Cu(s)	O$_2$(g)
ΔH_f°(kJ mol^{-1})	−157		
ΔG_f°(kJ mol^{-1})	−130		
$C_{P,m}^\circ$(J K^{-1} mol^{-1})	42.3	24.4	29.4

   a. From Equation (6.71),

$$\int_{K_P(T_0)}^{K_P(T_f)} d \ln K_P = \frac{1}{R} \int_{T_0}^{T_f} \frac{\Delta H_{reaction}^\circ}{T^2} dT$$

To a good approximation, we can assume that the heat capacities are independent of temperature over a limited range in temperature, giving $\Delta H_{reaction}^\circ (T) = \Delta H_{reaction}^\circ (T_0) + \Delta C_P (T - T_0)$ where $\Delta C_P = \sum_i v_i C_{P,m} (i)$. By integrating Equation (6.71), show that

$$\ln K_P (T) = \ln K_P (T_0) - \frac{\Delta H_{reaction}^\circ(T_0)}{R} \left(\frac{1}{T} - \frac{1}{T_0} \right)$$
$$+ \frac{T_0 \times \Delta C_P}{R} \left(\frac{1}{T} - \frac{1}{T_0} \right) + \frac{\Delta C_P}{R} \ln \frac{T}{T_0}$$

   b. Using the result from part (a), calculate the equilibrium pressure of oxygen over copper and CuO(s) at 1200 K. How is this value related to K_P for the reaction $2CuO(s) \rightleftharpoons 2Cu(s) + O_2(g)$?

   c. What value would you obtain if you assumed that $\Delta H_{reaction}^\circ$ were constant at its value for 298.15 K up to 1200 K?

P6.27 Show that

$$\left[\frac{\partial (A/T)}{\partial (1/T)} \right]_V = U$$

Write an expression analogous to Equation (6.37) that would allow you to relate ΔA at two temperatures.

P6.28 Calculate $\mu_{O_2}^{mixture}$ (298.15 K, 1 bar) for oxygen in air, assuming that the mole fraction of O_2 in air is 0.200.

P6.29 A sample containing 2.25 mol of He (1 bar, 298 K) is mixed with 3.00 mol of Ne (1 bar, 298 K) and 1.75 mol of Ar (1 bar, 298 K). Calculate ΔG_{mixing} and ΔS_{mixing}.

P6.30 You have containers of pure H_2 and He at 298 K and 1 atm pressure. Calculate ΔG_{mixing} relative to the unmixed gases of

   a. a mixture of 10 mol of H_2 and 10 mol of He.

   b. a mixture of 10 mol of H_2 and 20 mol of He.

   c. Calculate ΔG_{mixing} if 10 mol of pure He are added to the mixture of 10 mol of H_2 and 10 mol of He.

P6.31 A gas mixture with 4 mol of Ar, x moles of Ne, and y moles of Xe is prepared at a pressure of 1 bar and a temperature of 298 K. The total number of moles in the mixture is three times that of Ar. Write an expression for ΔG_{mixing} in terms of x. At what value of x does ΔG_{mixing} have its minimum value? Calculate ΔG_{mixing} for this value of x.

P6.32 In Example Problem 6.9, K_P for the reaction $CO(g) + H_2O(l) \rightarrow CO_2(g) + H_2(g)$ was calculated to be 3.32×10^3 at 298.15 K. At what temperature does $K_P = 5.00 \times 10^3$? What is the highest value that K_P can have by changing the temperature? Assume that $\Delta H^\circ_{reaction}$ is independent of temperature.

P6.33 Calculate K_P at 550. K for the reaction $N_2O_4(l) \rightarrow 2NO_2(g)$ assuming that $\Delta H^\circ_{reaction}$ is constant over the interval from 298 to 600 K.

P6.34 Calculate K_P at 475 K for the reaction $NO(g) + 1/2\ O_2(g) \rightarrow NO_2(g)$ assuming that $\Delta H^\circ_{reaction}$ is constant over the interval from 298 to 600 K. Do you expect K_P to increase or decrease as the temperature is increased to 550 K?

P6.35 Calculate the degree of dissociation of N_2O_4 in the reaction $N_2O_4(g) \rightarrow 2NO_2(g)$ at 250 K and a total pressure of 0.500 bar. Do you expect the degree of dissociation to increase or decrease as the temperature is increased to 550 K? Assume that $\Delta H^\circ_{reaction}$ is independent of temperature.

P6.36 You wish to design an effusion source for Br atoms from $Br_2(g)$. If the source is to operate at a total pressure of 20 Torr, what temperature is required to produce a degree of dissociation of 0.50? What value of the pressure would increase the degree of dissociation to 0.65 at this temperature?

P6.37 A sample containing 2.00 mol of N_2 and 6.00 mol of H_2 is placed in a reaction vessel and brought to equilibrium at 20.0 bar and 750 K in the reaction $1/2 N_2(g) + 3/2 H_2(g) \rightarrow NH_3(g)$.

  a. Calculate K_P at this temperature.

  b. Set up an equation relating K_P and the extent of reaction as in Example Problem 6.10.

  c. Using a numerical equation solver, calculate the number of moles of each species present at equilibrium.

P6.38 Consider the equilibrium in the reaction $3O_2(g) \rightleftharpoons 2O_3(g)$, with $\Delta H^\circ_{reaction} = 285.4 \times 10^3$ J mol^{-1} at 298 K. Assume that $\Delta H^\circ_{reaction}$ is independent of temperature.

  a. Without doing a calculation, predict whether the equilibrium position will shift toward reactants or products as the pressure is increased.

  b. Without doing a calculation, predict whether the equilibrium position will shift toward reactants or products as the temperature is increased.

  c. Calculate K_P at 550 K.

  d. Calculate K_x at 550 K and 0.500 bar.

P6.39 You place 2.00 mol of $NOCl(g)$ in a reaction vessel. Equilibrium is established with respect to the decomposition reaction $NOCl(g) \rightleftharpoons NO(g) + 1/2\ Cl_2(g)$.

  a. Derive an expression for K_P in terms of the extent of reaction ξ.

  b. Simplify your expression for part (a) in the limit that ξ is very small.

  c. Calculate ξ and the degree of dissociation of NOCl in the limit that ξ is very small at 375 K and a pressure of 0.500 bar.

  d. Solve the expression derived in part (a) using a numerical equation solver for the conditions stated in part (c). What is the relative error in ξ made using the approximation of part (b)?

P6.40 $Ca(HCO_3)_2(s)$ decomposes at elevated temperatures according to the stoichiometric equation $Ca(HCO_3)_2(s) \rightarrow CaCO_3(s) + H_2O(g) + CO_2(g)$.

  a. If pure $Ca(HCO_3)_2(s)$ is put into a sealed vessel, the air is pumped out, and the vessel and its contents are heated, the total pressure is 0.115 bar. Determine K_P under these conditions.

  b. If the vessel initially also contains 0.225 bar $H_2O(g)$, what is the partial pressure of $CO_2(g)$ at equilibrium?

P6.41 Assume that a sealed vessel at constant pressure of 1 bar initially contains 2.00 mol of $NO_2(g)$. The system is allowed to equilibrate with respect to the reaction $2\ NO_2(g) \rightleftharpoons N_2O_4(g)$. The number of moles of $NO_2(g)$ and $N_2O_4(g)$ at equilibrium is $2.00 - 2\xi$ and ξ, respectively, where ξ is the extent of reaction.

  a. Derive an expression for the entropy of mixing as a function of ξ.

  b. Graphically determine the value of ξ for which ΔS_{mixing} has its maximum value.

  c. Write an expression for G_{pure} as a function of ξ. Use Equation (6.34) to obtain values of G°_m for NO_2 and N_2O_4.

  d. Plot $G_{mixture} = G_{pure} + \Delta G_{mixing}$ as a function of ξ for $T = 298$ K and graphically determine the value of ξ for which $G_{mixture}$ has its minimum value. Is this value the same as that for part (b)?

P6.42 Oxygen reacts with solid glycylglycine, $C_4H_8N_2O_3$, to form urea, CH_4N_2O, carbon dioxide, and water:

$$3O_2(g) + C_4H_8N_2O_3(s) \rightarrow CH_4N_2O(s) + 3CO_2\ (g) + 2H_2O(l).$$

At $T = 298$ K and 1.00 atm, solid glycylglycine has the following thermodynamic properties:

$$\Delta G^\circ_f = -491.5 \text{ kJ mol}^{-1},\ \Delta H^\circ_f = -746.0 \text{ kJ mol}^{-1},\ S^\circ_f$$
$$= 190.0 \text{ J K}^{-1} \text{ mol}^{-1}$$

Calculate $\Delta G^\circ_{reaction}$ at $T = 298$ K and at $T = 310$. K. State any assumptions.

P6.43 The shells of marine organisms contain calcium carbonate, $CaCO_3$, largely in a crystalline form known as calcite. A second crystalline form of calcium carbonate is known as aragonite. Some physical and thermodynamic properties of calcite and aragonite are given here:

Properties (T=298K, P=1.00atm)	Calcite	Aragonite
ΔH°_f (kJ mol^{-1})	−1206.9	−1207.0
ΔG°_f (kJ mol^{-1})	−1128.8	−1127.7
S° (J K^{-1} mol^{-1})	92.9	88.7
$C^\circ_{P,m}$ (J K^{-1} mol^{-1})	81.9	81.3
Density (g mL^{-1})	2.710	2.930

a. Based on the thermodynamic data given, would you expect an isolated sample of calcite at $T = 298$ K and $P = 1.00$ bar to convert to aragonite, given sufficient time? Explain.

b. Suppose the pressure applied to an isolated sample of calcite is increased. Can the pressure be increased to the point that isolated calcite will be converted to aragonite? Explain.

c. What pressure must be achieved to induce the conversion of calcite to aragonite at $T = 298$ K? Assume both calcite and aragonite are incompressible at $T = 298$ K.

d. Can isolated calcite be converted to aragonite at $P = 1.00$ atm if the temperature is increased? Explain.

P6.44 The standard Gibbs energy for the formation of double-stranded DNA from single strands follows the equation $\Delta G° = \Delta G°_{initiation} + \sum \Delta G°_{neighbors}$. The Gibbs energy of initiation for two terminal A—T pairs is $\Delta G°_{initiation} \approx 8.10$ kJ mol^{-1} when two separated strands come together to form a duplex. $\Delta G°_{neighbors}$ is the contribution to the Gibbs energy from the hydrogen bonding and stacking as successive nucleotide pairs are added. As such, this contribution is characteristic of the nearest neighbor pairs. A table of nearest-neighbor Gibbs energies for DNA duplex formation at $T = 310.$K is given as follows:

	5'-A-A-3' 3'-T-T-5'	5'-A-T-3' 3'-T-A-5'	5'-A-G-3' 3'-T-C-5'	5'-G-C-3' 3'-C-G-5'	5'-C-G-3' 3'-G-C-5'	5'-C-A-3' 3'-G-T-5'	5'-T-A-3' 3'-A-T-5'
$\Delta G°_{neighbors}$ (kJmol^{-1},310K)	−4.2	−3.7	−5.4	−9.3	−9.1	−6.0	−2.4
$\Delta H°_{neighbors}$ (kJmol^{-1})	−33.1	−30.2	−32.7	−41.0	−44.4	−35.6	−30.2
$\Delta S°_{neighbors}$ (JK^{-1} mol^{-1})	−93.0	−85.4	−87.9	− 102	−114	−95.0	−89.2

Source: H. T. Allawai & J. SantaLucia, Jr. 1997, Biochemistry, 36, 10581–10594.

Consider the duplex formation reaction, where the dot indicates a Watson-Crick base pair:

$5' - A - G - C - G - C - A - 3'$
$+$
$3' - T - C - G - C - G - T - 5'$

$\rightleftharpoons$

$5' - A - G - C - G - C - A - 3'$
$\quad \cdot \quad \cdot \quad \cdot \quad \cdot \quad \cdot \quad \cdot$
$3' - T - C - G - C - G - T - 5'$

Calculate the Gibbs energy change and equilibrium constant for this reaction at $T = 310.$K.

P6.45 Repeat the calculation in Problem 6.44 for $T = 330.$K. Assume the enthalpy and entropy changes are constant between $T = 310.$K and $T = 330.$K. Assume the enthalpy and entropy changes for the double-stranded DNA formation follow equations analogous to the Gibbs energy equation given in P.6.44. Assume for two terminal A-T pairs the enthalpy and entropy of initiation are: $\Delta H°_{initiation} = 19.3$ kJ mol^{-1} and $\Delta S°_{initiation} = 34.3$ JK^{-1}mol^{-1}. State any assumptions you make.

Web-Based Simulations, Animations, and Problems

W6.1 The equilibrium system $A \rightleftharpoons B$ is simulated. The variables $\Delta G°_{reactants}$ and $\Delta G°_{products}$ and T can be varied independently using sliders. The position of the equilibrium as well as the time required to reach equilibrium is investigated as a function of these variables.

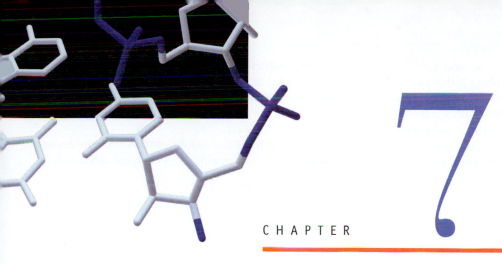

CHAPTER

7

Phase Equilibria

It is our experience that the solid form of matter is favored at low temperatures, and that most substances can exist in liquid and gaseous phases at higher temperatures. In this chapter, criteria are developed that allow one to determine which of these phases is favored at a given temperature and pressure. The conditions under which two or three phases of a pure substance can coexist in equilibrium are also discussed, as well as the unusual properties of supercritical fluids. A *P–T* phase diagram summarizes all of this information in a form that is very useful to chemists. In this chapter we also explore the properties of amphiphilic molecules and their aggregates. Amphiphiles are chimeric molecules composed of a polar segment that is attracted to water and a nonpolar segment that is not attracted to water. We explore an important property of systems of amphiphilic molecules called the hydrophobic effect. The hydrophobic effect results in the formation of micelles, bilayers, and surface monolayers when amphiphiles are placed in water. Finally, we explore the thermodynamics of other biologically relevant structural transitions including protein and nucleic acid conformational transitions.

7.1 What Determines the Relative Stability of the Solid, Liquid, and Gas Phases?

In general, substances are found in the solid, liquid, or gaseous phases at a given *P* and *T*. **Phase** refers to a form of matter that is uniform with respect to chemical composition and the state of aggregation on both microscopic and macroscopic length scales. For example, liquid water in a beaker is a single-phase system. A mixture of ice in liquid water consists of two distinct phases, each of which is uniform on microscopic and macroscopic length scales. Although a substance may exist in several different solid phases, it can only exist in a single gaseous state. In this section, the conditions under which a pure substance spontaneously forms a solid, liquid, or gas are discussed.

Experience demonstrates that as *T* is lowered from 300 to 250 K at atmospheric pressure, liquid water is converted to a solid phase. Similarly, as liquid water is heated to 400 K at atmospheric pressure, it vaporizes to form a gas. Experience also shows that if a solid block of carbon dioxide is placed in an open container at 1 bar, it sublimes over time without passing through a liquid phase. Because of this property, solid CO_2 is known as dry ice. These observations can be generalized to state that the solid phase is the most stable state of a substance at sufficiently low temperatures, and that the gas phase is the most stable state of a substance at sufficiently high temperatures. The liquid state is stable at intermediate temperatures if it exists at the pressure of interest. What determines which of the solid, liquid, or gas phases is most stable at a given temperature and pressure?

As discussed in Chapter 6, the criterion for phase stability at constant temperature and pressure is that the Gibbs energy, $G(T,P, n)$, is minimized. For a pure substance,

$$\mu = \left(\frac{\partial G}{\partial n}\right)_{T,P} = \left(\frac{\partial [nG_m]}{\partial n}\right)_{T,P} = G_m$$

where n designates the number of moles of substance in the system, $d\mu = dG_m$, and we can express $d\mu$ as

$$d\mu = -S_m\, dT + V_m\, dP \tag{7.1}$$

which is identical in content to Equation (6.19). From this equation, the variation of μ with P and T can be determined:

$$\left(\frac{\partial \mu}{\partial T}\right)_P = -S_m \quad \text{and} \quad \left(\frac{\partial \mu}{\partial P}\right)_T = V_m \tag{7.2}$$

Because S_m and V_m are always positive, μ decreases as the temperature increases, and increases as the pressure increases. In Section 5.4 we found that S varies slowly with T (as $\ln T$). Therefore, over a limited range in T, a plot of μ versus T at constant P is a straight line with a negative slope.

From experience it is known that a solid melts to form a liquid, and a liquid vaporizes to form a gas. Both processes are endothermic. This observation shows that $\Delta S = \Delta H/T$ is positive for both of these reversible constant temperature phase changes. Because the heat capacity is always positive for a solid, liquid, or gas, the entropy of the three phases follows this order:

$$S_m^{gas} > S_m^{liquid} > S_m^{solid} \tag{7.3}$$

The functional relation between μ and T for the solid, liquid, and gas phases is graphed at a given value of P in Figure 7.1. The entropy of a phase is the magnitude of the slope of the μ versus T line, and the relative entropies of the three phases are given by Equation (7.3). The stable state of the system at any given temperature is that phase which has the lowest μ.

Assume that the initial state of the system is described by the dot in Figure 7.1. It can be seen that the most stable phase is the solid phase, because μ for the liquid and gas phases is smaller than that for the solid. As the temperature is increased, the chemical potential falls as μ remains on the solid line. However, because the slopes of the liquid and gas lines are greater than that for the solid, each of these μ versus T lines will intersect the solid line for some value of T. In Figure 7.1, the liquid line intersects the solid line at T_m, which is called the melting temperature. At this temperature, the solid and liquid phases coexist and are in thermodynamic equilibrium. However, if the temperature is raised by an infinitesimal amount dT, the solid will melt completely because the liquid phase has the lower chemical potential at $T_m + dT$. Similarly, the liquid and gas phases are in thermodynamic equilibrium at T_b. For $T > T_b$, the system is entirely in the gas phase. Note that the progression of solid → liquid → gas as T increases at this value of P can be explained with no other information than that $(\partial\mu/\partial T)_P = -S_m$ and that $S_m^{gas} > S_m^{liquid} > S_m^{solid}$.

If the temperature is changed too quickly, the equilibrium state of the system may not be reached. For example, it is possible to form a superheated liquid, in which the liquid phase is metastable above T_b. Superheated liquids are dangerous, because of the large volume increase that occurs if the system suddenly converts to the stable vapor

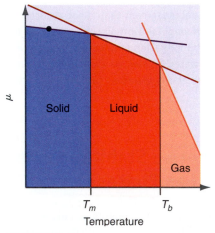

μ / Solid / Liquid / Gas

T_m T_b

Temperature

FIGURE 7.1
The chemical potential of a substance in the solid (blue line), liquid (red line), and gaseous (orange line) phases is plotted as a function of the temperature for a given value of pressure. The substance melts at the temperature T_m, corresponding to the intersection of the solid and liquid lines. The substance boils at the temperature T_b, corresponding to the intersection of the liquid and gas lines. The temperature ranges in which the different phases are the most stable are indicated by shaded areas.

phase. Boiling chips are often used in chemical laboratories to avoid the formation of superheated liquids. Similarly, it is possible to form a supercooled liquid, in which case the liquid is metastable below T_m. **Glasses** are made by cooling a viscous liquid fast enough to avoid crystallization. These disordered materials lack the periodicity of crystals but behave mechanically like solids. Seed crystals can be used to avoid supercooling if the viscosity of the liquid is not too high and the cooling rate is sufficiently slow.

Figure 7.1 shows how the chemical potential changes with T at constant P. How is the relative stability of the three phases affected if P is changed at constant T? From Equation (7.2), $(\partial \mu / \partial P)_T = V_m$ and $V_m^{gas} \gg V_m^{liquid}$. For most substances, $V_m^{liquid} > V_m^{solid}$. Therefore, the μ versus T line for the gas changes much more rapidly (by a factor of ~1000) with pressure than the liquid and solid lines. This behavior is illustrated in Figure 7.2, where it can be seen that the point at which the solid and the liquid lines intersect shifts as the pressure is increased. Because $V_m^{gas} \gg V_m^{liquid} > 0$, an increase in the pressure always leads to a **boiling point elevation.** An increase in the pressure leads to a **freezing point elevation** if $V_m^{liquid} > V_m^{solid}$ and to a **freezing point depression** if $V_m^{liquid} < V_m^{solid}$, as is the case for water. Few substances obey the relation $V_m^{liquid} < V_m^{solid}$ and the consequences of this unusual behavior for water will be discussed in Section 7.3.

The μ versus T line for a gas changes much more rapidly with P than the liquid and solid lines. As a consequence, changes in P can change the way in which a system progresses through the phases with increasing T from the "normal" order of solid $\rightarrow$ liquid $\rightarrow$ gas shown in Figure 7.1. For example, the sublimation of dry ice at 298 K and 1 bar can be explained using Figure 7.3a. For CO_2 at the given pressure, the μ versus T line for the liquid intersects the corresponding line for the solid at a higher temperature than the gaseous line. Therefore, the solid $\rightarrow$ liquid transition is energetically unfavorable with respect to the solid $\rightarrow$ gas transition at this pressure. Under these conditions, the solid sublimes and the transition temperature T_s is called the **sublimation temperature.** There is also a pressure at which the μ versus T lines for all three phases intersect. The P, V_m, and T values for this point specify the **triple point,** so named because all three phases coexist in equilibrium at this point. This case is shown in Figure 7.3b. Triple point temperatures for a number of substances are listed in Table 7.1 (see Appendix B, Data Tables).

7.2 The Pressure–Temperature Phase Diagram

As shown in the previous section, at a given value of P and T, a system containing a pure substance may consist of a single phase, two phases in equilibrium, or three phases in equilibrium. A **phase diagram** displays this information graphically. Although any two of the macroscopic system variables P, V, and T can be used to construct a phase diagram, the P–T diagram is particularly useful. In this section, the features of a P–T phase

FIGURE 7.2

The solid lines show μ as a function of T for all three phases at $P = P_1$. The dashed lines show the same information for $P = P_2$, where $P_2 > P_1$. The unprimed temperatures refer to $P = P_1$ and the primed temperatures refer to $P = P_2$. The left diagram applies if $V_m^{liquid} > V_m^{solid}$. The right diagram applies if $V_m^{liquid} < V_m^{solid}$. The shifts in the solid and liquid lines are greatly exaggerated. The colored areas correspond to the temperature range in which the phases are most stable. The shaded area between T_m and T_m' is either solid or liquid, depending on P. The shaded area between T_b and T_b' is either liquid or gas, depending on P.

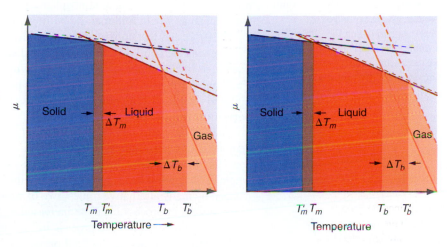

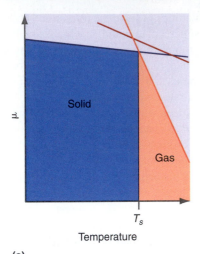

(a)

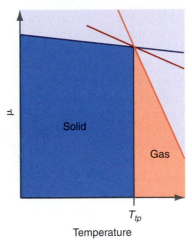

(b)

FIGURE 7.3
The chemical potential of a substance in the solid (blue line), liquid (red line), and gaseous (orange line) states is plotted as a function of the temperature for a fixed value of pressure. (a) The pressure lies below the triple point pressure, and the solid sublimes. (b) The pressure corresponds to the triple point pressure. At T_{tp}, all three phases coexist in equilibrium. The colored areas correspond to the temperature range in which the phases are the most stable. The liquid phase is not stable in part (a), and is only stable at the single temperature T_{tp} in part (b).

diagram that are common to pure substances are discussed. Phase diagrams are generally determined experimentally. Increasingly, computational techniques have become sufficiently accurate that major features of phase diagrams can be obtained using microscopic theoretical models. However, as shown in Section 7.4, thermodynamics can say a great deal about the phase diagram without considering the microscopic properties of the system.

A sample *P–T* **phase diagram** is shown in Figure 7.4. The diagram displays stability regions for a pure substance as a function of pressure and temperature. Most *P,T* points correspond to a single solid, liquid, or gas phase. At the triple point, all three phases coexist. For example, the triple point of water is 273.16 K and 611 Pa. All *P,T* points for which the same two phases coexist at equilibrium fall on a curve. Such a curve is called a **coexistence curve.** Three separate coexistence curves are shown in Figure 7.4, corresponding to solid–gas, solid–liquid, and gas–solid coexistence. As discussed in Section 7.4, the slopes of the solid–gas and liquid–gas curves are always positive. The slope of the solid–liquid curve can be either positive or negative.

The boiling point of a substance is defined as the temperature at which the vapor pressure of the substance is equal to the external pressure. The **standard boiling temperature** is the temperature at which the vapor pressure of the substance is 1 bar. The **normal boiling temperature** is the temperature at which the vapor pressure of the substance is 1 atm. Values of the normal boiling and freezing temperatures for a number of substances are shown in Table 7.2 (see Appendix B, Data Tables). Because 1 bar is slightly less than 1 atm, the standard boiling temperature is slightly less than the normal boiling temperature. Along two-phase curves in which one of the coexisting phases is the gas, *P* refers to the **vapor pressure** of the substance. In all regions, *P* refers to the hydrostatic pressure that would be exerted on the pure substance if it were confined in a piston and cylinder assembly.

The solid–liquid coexistence curve traces out the melting point as a function of *P*. The magnitude of the slope of this curve is large, as described in Section 7.4. Therefore, T_m is a weak function of *P*. The slope of this curve is positive, and the melting temperature increases with *P* if the solid is more dense than the liquid. This is the case for most substances. The slope is negative and the melting temperature decreases with pressure if the solid is less dense than the liquid. Water is one of the few substances that exhibits this behavior. The biological impact of this and other thermal properties of water will be discussed in Section 7.3.

As described in Section 7.4, the slope of the liquid–gas coexistence curve is much smaller than that of the solid–liquid coexistence curve. Therefore, the boiling point is a much stronger function of *P* than the freezing point. The boiling point always increases with *P*. This property is utilized in a pressure cooker. Increasing the pressure in a pressure cooker by 1 bar increases the boiling temperature of water by approximately 20°C. Because the rate of the chemical processes involved in cooking increases exponentially with *T*, a pressure cooker operating at *P* = 2 bar can cook food in 20% to 40% of the time required for cooking at atmospheric pressure. By contrast, a mountain climber in the Himalayas would find that the boiling temperature of water is reduced by approximately 10°C, and that cooking takes significantly longer under these conditions.

Whereas the solid–gas and liquid–solid coexistence curves extend indefinitely, the liquid–gas line terminates at the **critical point,** characterized by $T = T_c$ and $P = P_c$. For $T > T_c$ and $P > P_c$, the liquid and gas phases have the same density, so that it is not meaningful to refer to distinct phases. Substances for which $T > T_c$ and $P > P_c$ are called **supercritical fluids.**

Each of the paths labeled *a, b, c,* and *d* in Figure 7.4 corresponds to a process that demonstrates the usefulness of the *P–T* phase diagram. In the following, each process is considered individually. Process *a* follows a constant pressure (isobaric) path. An example of this path is heating ice. Note that the initial and final states in process *b* can be reached by many alternative routes, two of which are shown in Figure 7.4. The pressure of the system in process *b* can be increased to the value of the initial state of process *a* at constant temperature. The process follows that described for process *a,* after which the pressure is returned to the final pressure of process *b.* Invoking Hess's law, the enthalpy changes for this pathway and for pathway *b* are equal. Now imagine that the con-

FIGURE 7.4

A *P–T* phase diagram displays single-phase regions, coexistence curves for which two phases coexist at equilibrium, and a triple point. The processes corresponding to paths *a, b, c,* and *d* are described in the text. Two solid–liquid coexistence curves are shown. For most substances, the solid line, which has a positive slope, is observed. For water, the red dashed line corresponding to a negative slope is observed.

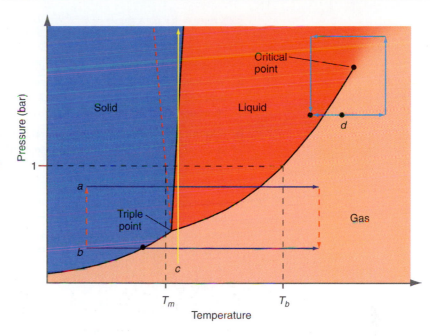

stant pressure for processes *a* and *b* differs only by an infinitesimal amount, although that for *a* is higher than the triple point pressure, and that for *b* is lower than the triple point pressure. We can express this mathematically by setting the pressure for process *a* equal to $P_{tp} + dP$, and that for process *b* equal to $P_{tp} - dP$. In the limit $dP \rightarrow 0$, $\Delta H \rightarrow 0$ for the two steps in the process indicated by the dashed arrows because $dP \rightarrow 0$. Therefore, ΔH for the transformation solid → liquid → gas in process *a* and for the transformation solid → gas in process *b* must be identical. We conclude that

$$\Delta H = \Delta H_{sublimation} = \Delta H_{fusion} + \Delta H_{vaporization} \qquad (7.4)$$

Path *c* indicates an isothermal process in which *P* increases. The initial state of the system is the single-phase gas region. As the gas is compressed, it is liquefied as it crosses the solid–liquid coexistence curve. As the pressure is increased further, the sample freezes as it crosses the liquid–solid coexistence curve. Crystallization is exothermic, therefore heat must flow to the surroundings as the liquid solidifies. If the process is reversed, heat must flow into the system to keep *T* constant as the solid melts.

If *T* is below the triple point temperature, the liquid exists at equilibrium only if the slope of the liquid–solid coexistence curve is negative, as is the case for water. Liquid water below the triple point temperature can freeze at constant *T* if the pressure is lowered sufficiently to cross the liquid–solid coexistence curve. In an example of such a process, a thin wire to which a heavy weight is attached on each end is stretched over a block of ice. With time, it is observed that the wire lies within the ice block and eventually passes through the block. There is no visible evidence of the passage of the wire in the form of a narrow trench. What happens in this process? Because the wire is thin, the force on the wire results in a high pressure in the area of the ice block immediately below the wire. This high pressure causes local melting of the ice below the wire. Melting allows the wire to displace the liquid water, which flows to occupy the volume immediately above the wire. Because in this region water no longer experiences a high pressure, it freezes again and hides the passage of the wire.

The consequences of having a critical point in the gas–liquid coexistence curve are illustrated in Figure 7.4. Process *d*, indicated by the double-headed arrow, is a constant pressure heating or cooling such that the gas–liquid coexistence curve is crossed. In a reversible process, a clearly visible interface will be observed along the two-phase gas–liquid coexistence curve. However, the same overall process can be carried out in four steps indicated by the single-headed arrows. In this case, two-phase coexistence is not observed, because the gas–liquid coexistence curve is not crossed. The overall transition is the same along both paths; namely, gas is transformed into liquid. However, no interface will be observed in this process.

EXAMPLE PROBLEM 7.1

Draw a generic $P-T$ phase diagram like that shown in Figure 7.4. Draw pathways in the diagram that correspond to the processes described here:

a. You hang wash out to dry at a temperature below the triple point. Initially, the water in the wet clothing has frozen. However, after a few hours in the sun, the clothing is warmer, dry, and soft.

b. A small amount of ethanol is contained in a thermos bottle. A test tube is inserted into the neck of the thermos bottle through a rubber stopper. A few minutes after filling the test tube with liquid nitrogen, the ethanol is no longer visible at the bottom of the bottle.

c. A transparent cylinder and piston assembly contains only a pure liquid in equilibrium with its vapor pressure. An interface is clearly visible between the two phases. When you increase the temperature by a small amount, the interface disappears.

Solution

The phase diagram with the paths is shown here:

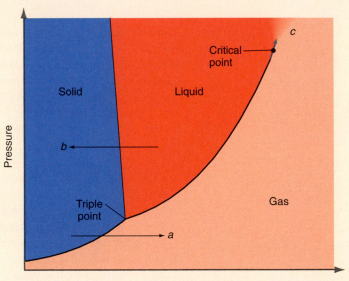

Paths a and b are not unique. Path a must occur at a pressure lower than the triple point pressure and process b must occur at a pressure greater than the triple point pressure. Path c will lie on the liquid–gas coexistence line up to the critical point, but can deviate once $T > T_c$ and $P > P_c$.

A $P-T$ phase diagram for water at high P values is shown in Figure 7.5. Water has several solid phases that are stable in different pressure ranges because they have different densities. Eleven different crystalline forms of ice have been identified up to a pressure of 10^{12} atm. For a comprehensive collection of material on the phase diagram of water, see *http://www.lsbu.ac.uk/water/phase.html*. Note in Figure 7.5 that ice VI does not melt until the temperature is raised to ~ 100°C for $P \approx 2000$ MPa.

Hexagonal ice (ice I) is the normal form of ice and snow. The structure shown in Figure 7.6 may be thought of as consisting of a set of parallel sheets, connected to one another through hydrogen bonds. Hexagonal ice has a fairly open structure with a density of 0.931 g cm^{-3} near the triple point. Figure 7.6 also shows the crystal structure of ice VI. All water molecules in this structure are hydrogen bonded to four other molecules. Ice VI is much more closely packed than hexagonal ice, and has a density of 1.31 g cm^{-3} at 1.6 GPa, at which pressure the density of liquid water is 1.18 g cm^{-3}. Note that ice VI will not float on liquid water.

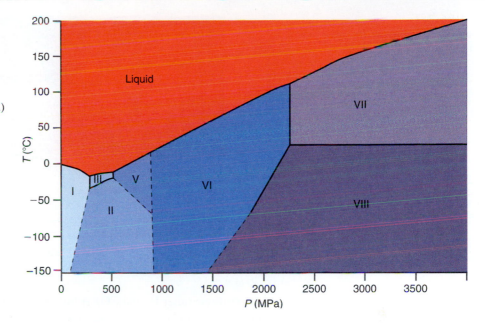

FIGURE 7.5

The *P–T* phase diagram for H₂O is shown for pressures up to 3.5×10^{10} bar. (Reprinted with permission from D. R. Lide, Ed., *CRC Handbook of Chemistry and Physics,* 83rd ed., Figure 3, page 12–202, CRC Press, Boca Raton, FL, 2002.)

As shown in Figure 7.5, phase diagrams can be quite complex for simple substances because a number of solid phases can exist as *P* and *T* are varied. A further example is sulfur, which can also exist in several different solid phases. A portion of the phase diagram for sulfur is shown in Figure 7.7, and the solid phases are described by the symmetry of their unit cells. Note that several points correspond to three-phase equilibria. By contrast, the CO_2 phase diagram shown in Figure 7.8 is simpler. It is similar in structure to that of water, but the solid–liquid coexistence curve has a positive slope. Several of the end-of-chapter problems and questions refer to the phase diagrams in Figures 7.6, 7.7, and 7.8.

7.3 Biological Impact of the Thermal Properties of Water

Having considered thus far the phases of water as a function of *P*, *V*, and *T*, it is interesting to consider how these phase properties make water the solvent absolutely crucial for life as we know it. Water is thought to be the third most abundant molecule in

FIGURE 7.6

Two different crystal structures are shown. Hexagonal ice (left) is the stable form of ice under atmospheric conditions. Ice VI (right) is only stable at elevated pressures as shown in the phase diagram of Figure 7.7. The dashed lines indicate hydrogen bonds.

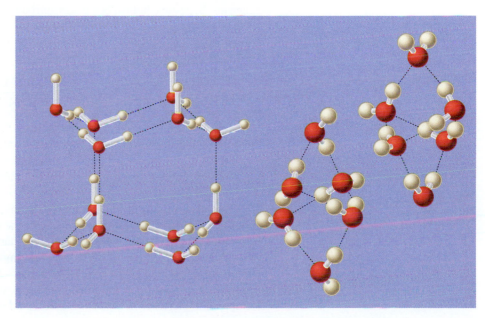

FIGURE 7.7
The *P–T* phase diagram for sulfur (not to scale).

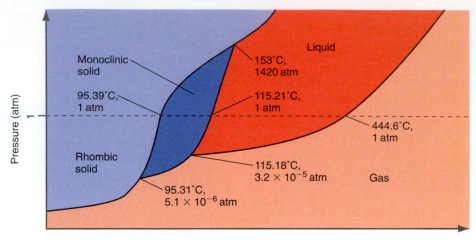

FIGURE 7.8
The *P–T* phase diagram for CO_2 (not to scale).

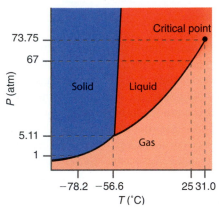

the universe (after H_2 and CO). A billion cubic kilometers of water is contained in the oceans of this planet. Fifty tons of water pass through a human's body in a lifetime. In view of its unique role in living systems, it is useful to consider the thermal properties of water and the biological impact of these properties.

Biologically relevant thermal properties of water include ΔH at phase transitions, the temperature dependence of water's molar volume, the heat capacity of liquid water, and the high melting and boiling points of water. As common as water is on this planet and as necessary as it is for life, many of these thermodynamic properties are considered anomalous for a small molecular liquid. The properties of water that distinguish it from other liquids originate from the fact that water molecules are polar. As shown earlier in Figure 7.6, ice is characterized by a long-range network of intermolecular hydrogen bonds, and even in the liquid state, water molecules form extensive hydrogen bond networks. The enthalpy of vaporization for water $\Delta H_{vap} = 40.65$ kJ mol^{-1} reflects the fact that vaporization disrupts virtually all intermolecular interactions that occur in liquid water. Because of the extensive hydrogen bonding that must be overcome to vaporize water, on a per mole basis, water has a much larger ΔH_{vap} than molecules of similar size that display less or no hydrogen bonding. In contrast, melting of ice disrupts the long-range translational order induced by this hydrogen bonding network, but because liquid water still displays considerable hydrogen bonding and local translational order, the enthalpy of fusion of water is only about $\Delta H_{fusion} = 60.010$ kJ mol^{-1}.

The high heat of vapoization of water provides a mechanism for heat dissipation in organisms. Suppose a human being was surrounded by an adiabatic barrier. Assuming heat is evolved by a human at the rate of 5.0 kJ per hour per kilogram of body mass, and approximating the heat capacity of the body with water's heat capacity of 4.18 kJ K^{-1} kg^{-1}, in 1 h the body temperature would rise by

$$\Delta T = \frac{q_P}{C_P} = \frac{5.0 \text{ kJ kg}^{-1}}{4.18 \text{ kJ K}^{-1}\text{kg}^{-1}} \approx 1.2 \text{ K} \tag{7.5}$$

Now assume the body dissipates this heat by vaporizing water in the form of perspiration. In this case, the amount of water that must be vaporized to absorb this heat is

$$\frac{q_P}{\Delta H_{vap}} \approx \frac{5.0 \text{ kJ per kg body weight}}{2.26 \text{ kJ per kg } H_2O} = 2.2 \text{ g } H_2O \text{ per kg body mass} \tag{7.6}$$

This means that a 100.-kg adult would have to lose more than 200. g of water per hour in the form of perspiration to dissipate the heat produced. In actuality, most adults do not sweat that much unless working very hard in very hot weather. Heat can also be dissipated by evaporation of water from the lungs, and the total mass of water evaporated in this form and in the form of perspiration is ~2.2 g H_2O per kilogram body mass. The dissipation of heat by evaporative cooling is an important mechanism for preventing temperature increases in organisms that would otherwise be harmful.

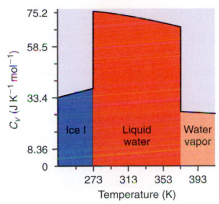

FIGURE 7.9
The heat capacity of water as a function of temperature and phase. Below $T = 273$ K, $C_{P,m}$ is actually plotted. (Source: D. Eisenberg and W. Kauzman *The Structure and Properties of Water,* Oxford University Press, New York and Oxford, 1969)

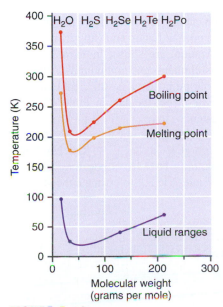

FIGURE 7.10
Melting points, boiling points, and liquid ranges for the Group VI hydrides.

Another interesting and biologically significant thermal property of water is the contraction of the molar volume V_m when water melts. There is an approximately 8% decrease in the molar volume $\Delta V_m / V_m \approx -0.08$ when ice melts. The V_m of water has its minimum value near 277 K, with the density decreasing at lower and higher temperatures. Water shares this unusual property with only a few other substances such as silicon, germanium, and diamond.

The volume contraction of water has profound biological consequences. Count Rumford did an experiment that illustrates the consequences if ice were to contract in volume upon freezing from the liquid. Rumford filled a vessel with liquid water and confined ice to the bottom of the vessel. Heating the water at the surface of the vessel to boiling did not melt much of the ice at the bottom of the vessel. Therefore, if ice had a higher density than liquid water, instead of floating on the surface of liquid water, ice would settle to the bottom of lakes, rivers, and oceans. Because the heating of the water surface by the sun would not be sufficient to thaw the ice at the bottoms of bodies of water, ice would accumulate from season to season in oceans and lakes, making the existence of marine life virtually impossible.

Water also has a high molar heat capacity relative to liquids composed of molecules of similar size. Water's heat capacity is of great biological importance because temperature fluctuations on Earth's surface are moderated by large bodies of water that can absorb very large quantities of heat, thus making the environment of the planet's surface more amenable to plants and animals, which would otherwise have to tolerate much larger temperature variations than occur on most of the planet's surface.

As shown in Figure 7.9, liquid water has a higher heat capacity (75.3 J K^{-1} mol^{-1}) than either ice (36.2 J K^{-1} mol^{-1} at $T = 273$ K) or water vapor (33.6 J K^{-1} mol^{-1} at $T = 298$ K). This trend is observed in many other substances also; see, for example, Figure 2.4. This general trend in substances was explained in Section 2.4 as being a result of the fact that the magnitude of the heat capacity is determined in part by the degree to which translational, rotational, and vibrational motions are thermally excited. In solids the dominant motions are vibrations, which are not thermally excited to a great extent at low temperatures, so the heat capacities of solids are small at low T. As the temperature increases, the heat capacity increases as vibrations are thermally excited to a greater and greater extent. When a solid melts, the resulting liquid is structurally more labile than the solid, and local structural deformations are now manifested as low-frequency hindered motions that are thermally excited to a much greater extent than the high-frequency vibrational motions that occur in the solid state. Therefore, the heat capacity of the liquid phase is higher than the heat capacity of the solid. When the liquid is vaporized, further temperature increases excite unhindered translational and rotational motions that do not absorb as much heat as the hindered motions that occur in the liquid or the high-frequency vibrations that occur in the solid. Therefore, the heat capacity of the gas phase is lower than the solid or liquid phases.

A final demonstration of the consequences of the anomalous properties of water is afforded by Figure 7.10, which shows the melting points, boiling points, and liquid ranges for the Group VI hydrides. As we progress downward in molecular weight from H$_2$Po to H$_2$S, the boiling points and melting points decrease, and the trend in liquid range is downward as well. Clearly, if water's melting and boiling points followed the trend extrapolated from the plot of the other Group VI hydrides, we would expect water to boil around 200 K instead of at 373 K; to melt around 175 K instead of at 273 K; and the liquid range of water, the difference between the boiling temperature and the melting temperature, would be only about 25 K instead of its real value which is 373 K $-$ 273 K $= 100$ K.

Water shows a dramatic deviation from the trends of the thermal properties of other Group VI elements due to its stronger and far more extensive hydrogen bonding. For example, in liquid water, strong hydrogen bonding interactions between molecules lower the tendency for molecules to escape from the liquid into the vapor phases, thus lowering the vapor pressure. Because boiling will not occur until the vapor pressure equals the atmospheric pressure, water boils at a much higher temperature than would be expected based on the behavior of the other Group VI hydrides. Clearly, if not for the extensive network hydrogen bonds found in solid and liquid water, and the anomalous physical properties that result from the presence of these bonds, liquid water would not exist to any great extent on this planet, and neither would life.

7.4 Providing a Theoretical Basis for the *P–T* Phase Diagram

In this section, a theoretical basis is provided for the coexistence curves that separate different single-phase regions in the *P–T* phase diagram. Along the coexistence curves, two phases are in equilibrium. From Section 6.7, we know that if two phases, α and β, are in equilibrium at a pressure P and temperature T, their chemical potentials must be equal:

$$\mu_\alpha(P,T) = \mu_\beta(P,T) \tag{7.7}$$

If the macroscopic variables are changed by a small amount, $P,T \rightarrow P + dP, T + dT$ such that the system pressure and temperature still lie on the coexistence curve, then

$$\mu_\alpha(P,T) + d\mu_\alpha = \mu_\beta(P,T) + d\mu_\beta \tag{7.8}$$

In order for the two phases to remain in equilibrium,

$$d\mu_\alpha = d\mu_\beta \tag{7.9}$$

Because $d\mu$ can be expressed in terms of dT and dP,

$$d\mu_\alpha = -S_{m\alpha} dT + V_{m\alpha} dP \text{ and } d\mu_\beta = -S_{m\beta} dT + V_{m\beta} dP \tag{7.10}$$

The expressions for $d\mu$ can be equated, giving

$$-S_{m\alpha} dT + V_{m\alpha} dP = -S_{m\beta} dT + V_{m\beta} dP \text{ or } (S_{m\beta} - S_{m\alpha}) dT = (V_{m\beta} - V_{m\alpha}) dP \tag{7.11}$$

Assume that as $P,T \rightarrow P + dP, T + dT$, an incremental amount of phase α is transformed to phase β. In this case, $\Delta S_m = S_{m\beta} - S_{m\alpha}$ and $\Delta V_m = V_{m\beta} - V_{m\alpha}$. Rearranging Equation (7.11) gives the **Clapeyron equation:**

$$\frac{dP}{dT} = \frac{\Delta S_m}{\Delta V_m} \tag{7.12}$$

The importance of the Clapeyron equation is that it allows one to calculate the slope of the coexistence curves in a *P–T* phase diagram if ΔS_m and ΔV_m for the transition are known. One can use the Clapeyron equation to estimate the slope of the solid–liquid coexistence curve. At the melting temperature,

$$\Delta G_m^{fusion} = \Delta H_m^{fusion} - T\Delta S_m^{fusion} = 0 \tag{7.13}$$

Therefore, the ΔS_m values for the fusion transition can be calculated from the enthalpy of fusion and the fusion temperature. Values of the normal fusion and vaporization temperatures, as well as ΔH_m for fusion and vaporization, are shown in Table 7.2 (see Appendix B, Data Tables) for a number of different elements and compounds. Although there is a significant variation in these values, for our purposes, it is sufficient to use the average value of $\Delta S_m^{fusion} = 22 \text{ J mol}^{-1} \text{ K}^{-1}$ calculated from the data in Table 7.2 in order to estimate the slope of the solid–liquid coexistence curve.

For the fusion transition, ΔV is small because the densities of the solid and liquid states are quite similar. The average ΔV_m^{fusion} for Ag, AgCl, Ca, $CaCl_2$, K, KCl, Na, NaCl, and H_2O is $+ 4 \times 10^{-6} \text{ m}^3$. Of these substances, only H_2O has a negative value for ΔV_m^{fusion}. We next use the average values of ΔS_m^{fusion} and ΔV_m^{fusion} to estimate the slope of the solid–liquid coexistence curve:

$$\left(\frac{dP}{dT}\right)_{fusion} = \frac{\Delta S_m^{fusion}}{\Delta V_m^{fusion}} \approx \frac{22 \text{ J mol}^{-1} \text{ K}^{-1}}{\pm 4 \times 10^{-6} \text{ m}^3 \text{ mol}^{-1}} \tag{7.14}$$
$$= \pm 5.5 \times 10^6 \text{ Pa K}^{-1} = \pm 55 \text{ bar K}^{-1}$$

Inverting this result, $(dT/dP)_{fusion} \approx \pm 0.02 \text{ K bar}^{-1}$. An increase of P by ~50 bar is required to change the melting temperature by one degree. This result explains the very steep solid–liquid coexistence curve shown in Figure 7.4.

The same analysis applies to the liquid–gas coexistence curve. Because $\Delta H_m^{vaporization}$ and $\Delta V_m^{vaporization} = V_m^{gas} - V_m^{liquid}$ are always positive, $(dP/dT)_{vaporization}$ is always positive. The average of $\Delta S_m^{vaporization}$ for the substances shown in Table 7.3 is 95 J mol^{-1} K^{-1}. This value is in accord with **Trouton's rule,** which states that $\Delta S_m^{vaporization} \approx$ 90 J mol^{-1} K^{-1} for liquids. The rule fails for liquids in which there are strong interactions between molecules such as —OH or —NH$_2$ groups capable of forming hydrogen bonds.

The molar volume of an ideal gas is approximately 20 L mol^{-1} in the temperature range in which many liquids boil. Because, $V_m^{gas} \gg V_m^{liquid}$, $\Delta V_m^{vaporization} \approx 20 \times 10^{-3}$ m^3 mol^{-1}. The slope of the liquid–gas coexistence line is given by

$$\left(\frac{dP}{dT}\right)_{vaporizaion} = \frac{\Delta S_m^{vaporization}}{\Delta V_m^{vaporization}} \approx \frac{95 \text{ J mol}^{-1}\text{K}^{-1}}{2 \times 10^{-2} \text{ m}^3\text{ mol}^{-1}} \approx 5 \times 10^3 \text{ Pa K}^{-1} \quad (7.15)$$

$$= 5 \times 10^{-2} \text{ bar K}^{-1}$$

This slope is a factor of 10^3 smaller than the slope for the liquid–solid coexistence curve. Inverting this result, $(dP/dT)_{vaporization} \approx 20$ K bar^{-1}. This result shows that it takes only a modest increase in the pressure to increase the boiling point of a liquid. For this reason, a pressure cooker does not need to be able to withstand high pressures. Note that the slope of the liquid–gas coexistence curve in Figure 7.4 is much less than that of the solid–liquid coexistence curve. The slope of both curves increases with T because ΔS increases with T.

The solid–gas coexistence curve can also be analyzed using the Clapeyron equation. Because entropy is a state function, the entropy change for the processes solid $(P,T) \rightarrow$ gas (P,T) and solid $(P,T) \rightarrow$ liquid $(P,T) \rightarrow$ gas (P,T) must be the same. Therefore, $\Delta S_m^{sublimation} = \Delta S_m^{fusion} + \Delta S_m^{vaporization} > \Delta S_m^{vaporization}$. Because the molar volume of the gas is much larger than that of the solid or liquid, $\Delta V_m^{sublimation} \approx \Delta V_m^{vaporization}$. We conclude that $(dP/dT)_{sublimation} > (dP/dT)_{vaporization}$. Therefore, the slope of the solid–gas coexistence curve will be greater than that of the liquid–gas coexistence curve. Because this comparison applies to a common value of the temperature, it is best made for temperatures just above and just below the triple point temperature. This difference in slope of these two coexistence curves is exaggerated in Figure 7.4.

7.5 Using the Clapeyron Equation to Calculate Vapor Pressure as a Function of T

From watching a pot of water as it is heated on a stove, it is clear that the vapor pressure of a liquid increases rapidly with increasing temperature. The same conclusion holds for a solid below the triple point. To calculate the vapor pressure at different temperatures, the Clapeyron equation must be integrated. Consider the solid–liquid coexistence curve:

$$\int_{P_i}^{P_f} dP = \int_{T_i}^{T_f} \frac{\Delta S_m^{fusion}}{\Delta V_m^{fusion}} dT = \int_{T_i}^{T_f} \frac{\Delta H_m^{fusion}}{\Delta V_m^{fusion}} \frac{dT}{T} \approx \frac{\Delta H_m^{fusion}}{\Delta V_m^{fusion}} \int_{T_i}^{T_f} \frac{dT}{T} \quad (7.16)$$

where the integration is along the solid–liquid coexistence curve. In the last step, it has been assumed that ΔH_m^{fusion} and ΔV_m^{fusion} are independent of T over the temperature range of interest. Assuming that $(T_f - T_i)/T_i$ is small, the previous equation can be simplified to give

$$P_f - P_i = \frac{\Delta H_m^{fusion}}{\Delta V_m^{fusion}} \ln\frac{T_f}{T_i} = \frac{\Delta H_m^{fusion}}{\Delta V_m^{fusion}} \ln\frac{T_i + \Delta T}{T_i} \approx \frac{\Delta H_m^{fusion}}{\Delta V_m^{fusion}} \frac{\Delta T}{T_i} \quad (7.17)$$

The last step uses the result $\ln(1 + x) = x$ for $x \ll 1$, obtained by expanding $\ln(1 + x)$ in a Taylor series about $x = 0$. We see that ΔP varies linearly with ΔT in this limit. The value of the slope dP/dT was discussed in the previous section.

Because $\Delta V \approx V^{gas}$, for the liquid–gas coexistence curve, we have a different result. Assuming that the ideal gas law holds, we obtain the **Clausius–Clapeyron equation**.

$$\frac{dP}{dT} = \frac{\Delta S_m^{vaporization}}{\Delta V_m^{vaporization}} \approx \frac{\Delta H_m^{vaporization}}{TV^{gas}} = \frac{P\Delta H_m^{vaporization}}{RT^2} \quad (7.18)$$

$$\frac{dP}{P} = \frac{\Delta H_m^{vaporization}}{R}\frac{dT}{T^2}$$

Assuming that $\Delta H_m^{vaporization}$ remains constant over the range of temperature of interest,

$$\int_{P_i}^{P_f} \frac{dP}{P} = \frac{\Delta H_m^{vaporization}}{R} \times \int_{T_i}^{T_f} \frac{dT}{T^2} \quad (7.19)$$

$$\ln\frac{P_f}{P_i} = -\frac{\Delta H_m^{vaporization}}{R} \times \left(\frac{1}{T_f} - \frac{1}{T_i}\right)$$

The same procedure is followed for the solid–gas coexistence curve. The result is the same as Equation (7.19) with $\Delta H_m^{sublimation}$ substituted for $\Delta H_m^{vaporization}$. Equation (7.19) provides a way to determine the enthalpy of vaporization for a liquid by measuring its vapor pressure as a function of temperature, as shown in Example Problem 7.2. In this discussion, it has been assumed that $\Delta H_m^{vaporization}$ is independent of temperature. More accurate values of the vapor pressure as a function of temperature can be obtained by fitting experimental data. This leads to an expression for the vapor pressure as a function of temperature. These functions for selected liquids and solids are listed in Tables 7.3 and 7.4 (see Appendix B, Data Tables).

EXAMPLE PROBLEM 7.2

The normal boiling temperature of benzene is 353.24 K, and the vapor pressure of liquid benzene is 1.00×10^4 Pa at 20.0°C. The enthalpy of fusion is 9.95 kJ mol^{-1}, and the vapor pressure of solid benzene is 88.0 Pa at −44.3°C. Calculate the following:

a. $\Delta H_m^{vaporization}$

b. $\Delta S_m^{vaporization}$

c. Triple point temperature and pressure

Solution

a. We can calculate $\Delta H_m^{vaporization}$ using the Clapeyron equation because we know the vapor pressure at two different temperatures:

$$\ln\frac{P_f}{P_i} = -\frac{\Delta H_m^{vaporization}}{R}\left(\frac{1}{T_f} - \frac{1}{T_i}\right)$$

$$\Delta H_m^{vaporization} = -\frac{R\ln\dfrac{P_f}{P_i}}{\left(\dfrac{1}{T_f} - \dfrac{1}{T_i}\right)} = -\frac{8.314\ \text{J mol}^{-1}\text{K}^{-1} \times \ln\dfrac{101{,}325\ \text{Pa}}{1.00 \times 10^4\ \text{Pa}}}{\left(\dfrac{1}{353.24\ \text{K}} - \dfrac{1}{273.15 + 20.0\ \text{K}}\right)}$$

$$= 33.2\ \text{kJ mol}^{-1}$$

b. $\Delta S_m^{vaporization} = \dfrac{\Delta H_m^{vaporization}}{T_b} = \dfrac{33.2 \times 10^3\ \text{J mol}^{-1}}{353.24\ \text{K}} = 93.9\ \text{J mol}^{-1}\text{K}^{-1}$

c. At the triple point, the vapor pressures of the solid and liquid are equal:

$$\ln\frac{P_{tp}^{liquid}}{P^\circ} = \ln\frac{P_i^{liquid}}{P^\circ} - \frac{\Delta H_m^{vaporization}}{R}\left(\frac{1}{T_{tp}} - \frac{1}{T_i^{liquid}}\right)$$

$$\ln\frac{P_{tp}^{solid}}{P^\circ} = \ln\frac{P_i^{solid}}{P^\circ} - \frac{\Delta H_m^{sublimation}}{R}\left(\frac{1}{T_{tp}} - \frac{1}{T_i^{solid}}\right)$$

$$\ln\frac{P_i^{liquid}}{P^\circ} - \ln\frac{P_i^{solid}}{P^\circ} - \frac{\Delta H_m^{sublimation}}{RT_i^{solid}} + \frac{\Delta H_m^{vaporization}}{RT_i^{liquid}} = \frac{(\Delta H_m^{vaporization} - \Delta H_m^{sublimation})}{RT_{tp}}$$

$$T_{tp} = \frac{(\Delta H_m^{vaporization} - \Delta H_m^{sublimation})}{R\left(\ln\dfrac{P_i^{liquid}}{P^\circ} - \ln\dfrac{P_i^{solid}}{P^\circ} - \dfrac{\Delta H_m^{sublimation}}{RT_i^{solid}} + \dfrac{\Delta H_m^{vaporization}}{RT_i^{liquid}}\right)}$$

$$= \frac{9.95 \times 10^3\ \text{J mol}^{-1}}{8.314\ \text{J K}^{-1}\,\text{mol}^{-1} \times \left(\begin{array}{c}\ln\dfrac{10{,}000\ \text{Pa}}{1\ \text{Pa}} - \ln\dfrac{88.0\ \text{Pa}}{1\ \text{Pa}} - \dfrac{(33.2\times10^3 + 9.95\times10^3)\ \text{J mol}^{-1}}{8.314\ \text{J K}^{-1}\,\text{mol}^{-1} \times 228.9\ \text{K}} \\ + \dfrac{33.2\times10^3\ \text{J mol}^{-1}}{8.314\ \text{J K}^{-1}\,\text{mol}^{-1} \times 293.15\ \text{K}}\end{array}\right)}$$

$$= 277\ \text{K}$$

We calculate the triple point pressure using the Clapeyron equation:

$$\ln\frac{P_f}{P_i} = -\frac{\Delta H_m^{vaporization}}{R}\left(\frac{1}{T_f} - \frac{1}{T_i}\right)$$

$$\ln\frac{P_{tp}}{101{,}325} = -\frac{33.2\times10^3\ \text{J mol}^{-1}}{8.314\ \text{J mol}^{-1}\,\text{K}^{-1}} \times \left(\frac{1}{277\ \text{K}} - \frac{1}{353.24\ \text{K}}\right)$$

$$\ln\frac{P_{tp}}{P^\circ} = 8.41465$$

$$P_{tp} = 4.51 \times 10^3\ \text{Pa}$$

7.6 Surface Tension

In discussing the liquid phase, the effect of the boundary surface on the properties of the liquid has been neglected. In the absence of a gravitational field, a liquid droplet will assume a spherical shape, because in this geometry the maximum number of molecules is surrounded by neighboring molecules, and the surface area is minimized. Because the interaction between molecules in a liquid is attractive, minimizing the surface-to-volume ratio minimizes the energy. How does the energy of the droplet depend on its surface area? Starting with the equilibrium spherical shape, assume that the droplet is distorted to create more area while keeping the volume constant. The work associated with the creation of additional surface area at constant V and T is

$$dA = \gamma\, d\sigma \tag{7.20}$$

where A is the Helmholtz energy, γ is the surface tension, and σ is the unit element of area. The **surface tension** has the units of energy/area or J m^{-2}, which is equivalent to N m^{-1} (Newtons per meter). Because $dA < 0$ for a spontaneous process at constant V and T, Equation (7.20) predicts that a liquid, or a bubble, or a liquid film suspended in a wire frame will tend to minimize its surface area.

Consider the spherical droplet depicted in Figure 7.11. There must be a force acting on the droplet in the radially inward direction for the liquid to assume a spherical shape.

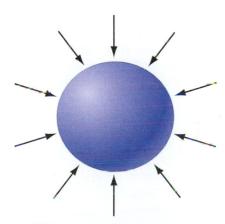

FIGURE 7.11
The forces acting on a spherical droplet that arise from surface tension.

An expression for the force can be generated as follows. If the radius of the droplet is increased from r to $r + dr$, the area increases by

$$d\sigma = 4\pi(r + dr)^2 - 4\pi r^2 = 4\pi(r^2 + 2r\,dr + (dr)^2) - 4\pi r^2 \approx 8\pi r\,dr \tag{7.21}$$

From Equation (7.20), the work done in the expansion of the droplet is $8\pi\gamma r\,dr$. The force, which is normal to the surface of the droplet, is the work divided by the distance or

$$F = 8\pi\gamma r \tag{7.22}$$

The net effect of this force is to generate a pressure differential across the droplet surface. At equilibrium, there is a balance between the inward and outward acting forces. The inward acting force is the sum of the force exerted by the external pressure and the force arising from the surface tension, whereas the outward acting force arises solely from the pressure in the liquid:

$$4\pi r^2 P_{outer} + 8\pi\gamma r = 4\pi r^2 P_{inner} \quad \text{or} \tag{7.23}$$

$$P_{inner} = P_{outer} + \frac{2\gamma}{r}$$

Note that $P_{inner} - P_{outer} \to 0$ as $r \to \infty$. Therefore, the pressure differential exists only for a curved surface. From the geometry in Figure 7.11 it is apparent that the higher pressure is always on the concave side of the interface. Values for the surface tension for a number of liquids are listed in Table 7.5.

Equation (7.23) has interesting implications for the relative stabilities of bubbles with different curvatures $1/r$. Consider two air bubbles of liquid with the same surface tension γ, one with a large radius R_1, and one with a smaller radius R_2. Assume there is a uniform pressure outside both bubbles. From Equation (7.23) we obtain the difference between the pressures P_1 and P_2 within each bubble:

$$P_1 - P_2 = \frac{2\gamma}{R_1} - \frac{2\gamma}{R_2} = 2\gamma\left(\frac{1}{R_1} - \frac{1}{R_2}\right) \tag{7.24}$$

Now suppose the two bubbles come into contact as shown at the top of Figure 7.12. Because $R_1 > R_2$, from Equation (7.24) the pressure P_2 inside bubble 2 (with respect to some common exterior pressure) is greater than the pressure P_1 in bubble 1, so air will flow from bubble 2 to bubble 1 until the smaller bubble disappears entirely as shown at the bottom of Figure 7.12. In foams this is called "coarsening" or in crystals "Ostwald ripening."

A biologically relevant issue that follows from Equations (7.23) and (7.24) is the stability of lung tissue. Lung tissue is composed of small water-lined, air-filled chambers called alveoli. According to Equation (7.23), the pressure difference across the surface of a spherical air-filled cavity is proportional to the surface tension and inversely proportional to the radius of the cavity. Therefore, if two air-filled alveoli, approximated as spheres with radii r and R and for which $r < R$, are interconnected, the smaller alveolus would collapse and the larger alveolus would expand because the pressure within the smaller alveolus is greater than the pressure within the larger alveolus. In theory, due to the surface tension of water that lines the alveoli, these cavities would "coarsen" as smaller alveoli collapse, leaving an ever diminishing number of alveoli of increasing size. Coarsening of the alveoli according to Equation 7.23 would result eventually in the collapse of the lungs.

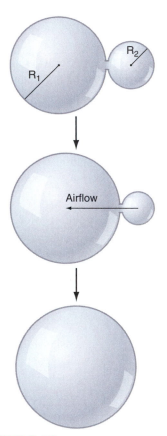

FIGURE 7.12
Two bubbles with unequal radii R_1 and R_2 make contact (top). Because the pressure within a bubble varies inversely with the bubble radius, air flows from the smaller into the larger bubble until a single large bubble remains (bottom series).

TABLE 7.5   Surface Tension of Selected Liquids at 298 K

Formula	Name	γ (mN m^{-1})	Formula	Name	(mN m^{-1})
Br_2	Bromine	40.95	CS_2	Carbon disulfide	31.58
H_2O	Water	71.99	C_2H_5OH	Ethanol	21.97
Hg	Mercury	485.5	C_6H_5N	Pyridine	36.56
CCl_4	Carbon tetrachloride	26.43	C_6H_6	Benzene	28.22
CH_3OH	Methanol	22.07	C_8H_{18}	Octane	21.14

Source: Data from D. R. Lide, Ed., *Handbook of Chemistry and Physics,* 83rd ed. CRC Press, Boca Raton, FL, 2002.

As will be explained in more detail in Section 7.9, molecules called surfactants lower the surface tension of water. In the alveoli, this is accomplished by a lipid called phosphatidylcholine, which lowers the surface tension of the water lining the alveoli, thus stabilizing the lungs. But in persons lacking sufficient surfactant in the lining of their lungs, such as prematurely born infants and heavy smokers, the surface tension of the water in the alveoli is not lowered, and as a consequence lung tissue is destabilized by coarsening.

Another consequence of Equation (7.23) is that the vapor pressure of a droplet depends on its radius. The condition for phase equilibrium between the vapor on the outer side of the surface and the liquid on the inner side of the surface is

$$\mu_{outer} = \mu_{inner} \text{ or } \mu_{vapor} = \mu_{liquid} \tag{7.25}$$

We can take the differential of both sides of Equation (7.25) and at constant temperature obtain

$$d\,\mu_{vapor} = V_m^{vapor}\,dP_{vapor} = d\,\mu_{liquid} = V_m^{liquid}\,dP_{liquid} \tag{7.26}$$

where V_m^{vapor} and V_m^{liquid} are the molar volumes of water vapor and liquid water at constant T, respectively. We can also take the differential of both sides of Equation (7.23):

$$dP_{inner} - dP_{outer} = dP_{liquid} - dP_{vapor} = d\left(\frac{2\gamma}{r}\right) = 2\gamma d\left(\frac{1}{r}\right) \tag{7.27}$$

Equations (7.26) and (7.27) can be combined to eliminate dP_{liquid}:

$$\left(\frac{V_m^{vapor} - V_m^{liquid}}{V_m^{liquid}}\right) dP_{vapor} = 2\gamma d\left(\frac{1}{r}\right) \tag{7.28}$$

The volume of the vapor greatly exceeds the volume of the liquid so

$$\frac{V_m^{vapor} - V_m^{liquid}}{V_m^{liquid}} \approx \frac{V_m^{vapor}}{V_m^{liquid}}$$

If the vapor also behaves ideally, then $V_m^{vapor} = RT/P_{vapor}$ and Equation (7.28) becomes

$$d\left(\frac{1}{r}\right) = \frac{RT}{2\gamma V_m^{liquid}}\frac{dP_{vapor}}{P_{vapor}} = \frac{\rho RT}{2\gamma M}\frac{dP_{vapor}}{P_{vapor}} \tag{7.29}$$

where M is the molar mass of the substance and $\rho = M/V_m^{liquid}$ is the density of the liquid.

Now we integrate both sides of Equation (7.29). Specifically we integrate the left-hand side from $1/r = 0$, to a surface with an inverse curvature $1/r$. The right-hand side is correspondingly integrated from the vapor pressure P^0 over a flat surface to the vapor pressure P over a surface with an inverse curvature $1/r$:

$$\int_0^{\frac{1}{r}} d\left(\frac{1}{r}\right) = \frac{\rho RT}{2\gamma M}\int_{P^0}^{P} \frac{dP_{vapor}}{P_{vapor}} \tag{7.30}$$

Completing the integrations in Equation (7.30):

$$\ln\left(\frac{P}{P^0}\right) = \frac{2\gamma M}{r\rho RT} \tag{7.31}$$

By substituting numbers into Equation (7.31) to calculate the vapor pressure P, we find that the vapor pressure of 10^{-7}-m-radius water droplet is increased by 1%, that of a 10^{-8}-m-radius droplet is increased by 11%, and that of a 10^{-9}-m-radius droplet is increased by 270%. However, at such a small diameter, the application of Equation (7.31) is questionable because the size of an individual water molecule is comparable to the droplet diameter. Equation (7.31) plays a role in the formation of liquid droplets in a condensing gas such as fog. Small droplets evaporate more rapidly than large droplets, and the vapor condenses on the larger droplets, allowing them to grow at the expense of small droplets.

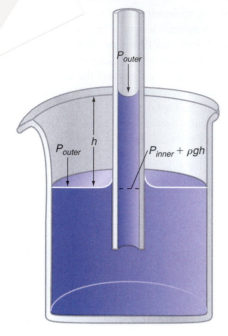

(a)

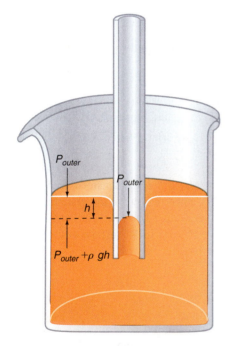

(b)

FIGURE 7.13
(a) If the liquid wets the interior wall of the capillary, a capillary rise is observed. The combination Pyrex–water exhibits this behavior. (b) If the liquid does not wet the capillary surface, a capillary depression is observed. The combination Pyrex–mercury exhibits this behavior.

Capillary rise and **capillary depression** are other consequences of the pressure differential across a curved surface. Assume that a capillary of radius r is partially immersed in a liquid. When the liquid comes in contact with a solid surface, there is a natural tendency to minimize the energy of the system. If the surface tension of the liquid is lower than that of the solid, the liquid will wet the surface, as shown in Figure 7.13a. However, if the surface tension of the liquid is higher than that of the solid, the liquid will avoid the surface, as shown in Figure 7.13b. In either case, there is a pressure differential in the capillary across the gas–liquid interface, because the interface is curved. If we assume that the liquid–gas interface is tangent to the interior wall of the capillary at the solid–liquid interface, the radius of curvature of the interface is equal to the capillary radius.

The difference in the pressure across the curved interface, $2\gamma/r$, is balanced by the weight of the column in the gravitational field, $\rho g h$. Therefore, the capillary rise or depression is given by

$$h = \frac{2\gamma}{\rho g r} \tag{7.32}$$

In the preceding discussion, it was assumed that either (1) the liquid completely wets the interior surface of the capillary, in which case the liquid coats the capillary walls, but does not fill the core, or (2) the liquid is completely nonwetting, in which case the liquid does not coat the capillary walls, but fills the core. In a more realistic model, the interaction is intermediate between these two extremes. In this case, the liquid–surface is characterized by the **contact angle** θ, as shown in Figure 7.14.

Complete **wetting** corresponds to $\theta = 0°$ and complete **nonwetting** corresponds to $\theta = 180°$. For intermediate cases,

$$P_{inner} = P_{outer} + \frac{2\gamma \cos \theta}{r} \quad \text{and} \quad h = \frac{2\gamma \cos \theta}{\rho g r} \tag{7.33}$$

The measurement of the contact angle is one of the main experimental methods used to measure the difference in surface tension at the solid–liquid interface.

EXAMPLE PROBLEM 7.3

The six-legged water strider supports itself on the surface of a pond on four of its legs.

(Source: Bernard Photo Productions/Animals Animals/Earth Scenes)

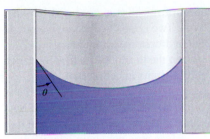

FIGURE 7.14
For cases intermediate between wetting and nonwetting, the contact angle θ lies in the range $0° < \theta < 180°$.

Each of these legs causes a depression to be formed in the pond surface. Assume that each depression can be approximated as a hemisphere of radius 1.2×10^{-4} m and that θ (as in Figure 7.14) is $0°$. Calculate the force that one of the insect's legs exerts on the pond.

Solution

$$\Delta P = \frac{2\gamma \cos \theta}{r} = \frac{2 \times 71.99 \times 10^{-3}\,\text{N m}^{-1} \times 1}{1.2 \times 10^{-4}\,\text{m}} = 1.20 \times 10^{3}\,\text{Pa}$$

$$F = PA = P \times \pi r^2 = 1.20 \times 10^{3}\,\text{Pa} \times \pi(1.2 \times 10^{-4}\,\text{m})^2 = 5.4 \times 10^{-5}\,\text{N}$$

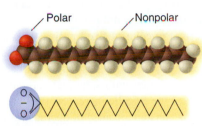

(a) Stearate ion

(b) Oleate ion

FIGURE 7.15
The fatty acids (a) stearic acid and (b) oleic acid display the characteristics of a typical amphiphilic molecule, a polar "hydrophilic" head group and a nonpolar "hydrophobic" tail. (Source: C.K. Matthews, K.E. van Holde, K.G. Ahern *Biochemistry,* 3rd ed., Benjamin Cummings, 2000)

7.7 Amphiphilic Molecules and the Hydrophobic Effect

Amphiphilic molecules, or **amphiphiles,** are composed of a polar segment and a nonpolar segment. Polar derivatives of aliphatic hydrocarbons are amphiphiles. The nonpolar hydrocarbon "tail" of an amphiphile is sparingly soluable in water. The polar "head group" is water soluable. Amphiphilic molecules include fatty acids (see Figure 7.15) and **lipids** (see Figure 7.18 in the next section).

An important property displayed by aqueous solutions of amphiphiles is the hydrophobic effect. The hydrophobic effect, manifested in the tendency of amphiphilic molecules to arrange themselves in such a way as to remove their hydrophobic tail from the aqueous phase and at the same time expose their hydrophilic head groups to water, has its origins primarily in the strong attractive forces between water molecules, which must be disrupted or distorted when a solute molecule is introduced. As a result of the hydrophobic effect, amphiphilic molecules form a number of structures in aqueous environments including monolayers, micelles, and bilayers. Phospholipid bilayers are a fundamental structural component of biological membranes. Before discussing these various structures, let us describe the origins of the hydrophobic effect in more detail.

When a nonpolar solute such as benzene is removed from a hydrocarbon (HC) solvent and transferred into water (W) the chemical potential change is observed to be positive:

$$\Delta\mu^0 = \mu_W^0 - \mu_{HC}^0 > 0 \qquad (7.34)$$

However, the enthalpy change that accompanies the transfer of a nonpolar solute from a nonpolar solvent into water is negative. As discussed in Sections 7.2 and 7.3, intermolecular hydrogen bonding makes water a highly structured liquid. Suppose the hydrophobic effect is due to disruption of the network of hydrogen bonds by the introduction of nonpolar molecules into water. It follows that the disruption of the hydrogen bonding network can only be accomplished at the cost of an increase in enthalpy, and a positive change in the chemical potential would thus result from a positive enthalpy change. But experiments show that the enthalpy change is negative for this process and, therefore, the hydrophobic effect must have its origins in a negative entropy change.

Why do both the enthalpy and entropy decrease when a nonpolar solute is transferred into an aqueous environment? When nonpolar molecules are introduced into water, water molecules form cages around the nonpolar solute molecules. In doing so, water molecules structure themselves in order to reestablish hydrogen bonding, and locally this bonding may be stronger than in pure water, leading to a decrease in enthalpy. The entropy also decreases because of increased local structuring of water molecules around the solute cavities.

When amphiphilic molecules such as fatty acids are dissolved in water, the hydrophobic effect can lead to the formation of aggregates that enable the amphiphiles to remove their nonpolar segments from the aqueous phase. These aggregates are called **micelles.** As shown in Figure 7.16a, amphiphiles in the micelle have their hydrophobic "tails" directed inward and away from the aqueous environment, whereas the hydrophilic "heads" point

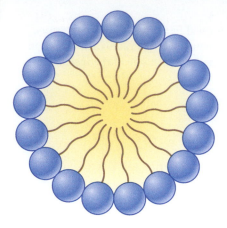

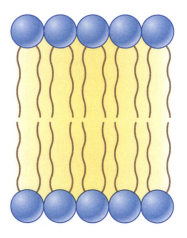

(a)   Micelle

(b)   Bilayer

FIGURE 7.16
Single-chain amphiphiles such as fatty acids form spherical aggregates in aqueous solution called (a) micelles, whereas double-chain amphiphiles such as phospholipids form (b) bilayers. (Source: C.K. Matthews, K.E. van Holde, K.G. Ahern *Biochemistry*, 3rd ed., Benjamin Cummings, 2000)

outward into the aqueous environment. The formation of micelles is governed by what Charles Tanford called the principle of opposing forces. Two opposing forces give rise to micellar formation: an attractive force arising from the hydrophobic effect, and a repulsive force that originates from electrostatic repulsion between the head groups. The attractive force favors micellar formation, while the repulsive force limits micellar size.

What determines the shape of a micelle? Suppose the surface area of the micelle is σ and the total number of chains in the micelle is m. Micellar shape is determined by a number of parameters including m, the maximum length of the hydrophobic chain that is embedded in the micellar core, ℓ_{max}, the micellar surface area per chain σ/m, and the number of chains attached to each head group: one for fatty acids and two for phospholipids. Small micelles composed of amphiphiles with a single chain attached to each head group are spherical (that is, globular), and the radius of a spherical micelle may not exceed ℓ_{max}. If the micelle is to increase in size it must assume a nonspherical shape because if it remained spherical and the radius exceeded ℓ_{max}, a hole would be created in the center of the micelle, a situation that would be energetically unfavorable.

As the number of amphiphile chains per spherical micelle of fixed radius ℓ_{max} increases, σ/m decreases, and the repulsive interaction between the head groups increases. Above a limiting value of σ/m, the spherical micelle is destabilized through the head group repulsion, and the micelle shape changes to a cylinder with spherical end caps. The cylindrical shape is favored because as the size of the micelle increases, the cylinder can grow in length whereby the ratio σ/m remains constant, instead of decreasing as is the case for a spherical micelle of fixed radius.

EXAMPLE PROBLEM 7.4

Experimental measurements show that the volume of a micelle may be written in terms of the number of chains in the micelle m and the number of carbon atoms per chain n_C. $V = (27.4 + 26.9n_C)m$ where V is in units of cubic angstroms. An expression for the maximum radius is similarly given in units of angstroms as $\ell_{max} = 1.5 + 1.26n_C$. Assuming n_C is a very large number, calculate the ratio σ/m for a spherical micelle and a cylindrical micelle of length L. Show that for a cylinder σ/m is independent of L.

Solution

For a spherical micelle the surface area is

$$\sigma = 4\pi\ell_{max}^2 = 4\pi \times (1.5 + 1.26n_C)^2 \approx 4\pi \times (1.26n_C)^2$$

Similarly, the volume of a spherical micelle is $V \approx 26.9n_C m = \dfrac{4\pi}{3}\ell_{max}^3$ $\approx \dfrac{4\pi}{3}(1.26n_C)^3$.

We can use the volume equation to solve for m: $m = (1.26)^3 \dfrac{4\pi}{3} \dfrac{n_C^2}{29.6}$.

Using the equations for σ and m, we can obtain the surface area to chain ratio:

$$\frac{\sigma}{m} = \frac{3 \times 29.6}{1.26} = 70.5$$

For a micelle shaped like a long cylinder of length L the surface area is

$$\sigma = 2\pi\,\ell_{max}L \approx 2\pi \times 1.26n_C \times L$$

The volume is $V = \pi\ell_{max}^2 L = \pi L(1.26n_C)^2 \approx 29.6n_C m$.

We can again use the volume equation to solve for m: $m = \dfrac{\pi(1.26)^2\, n_C L}{29.6}$.

We finally obtain the surface area to chain ratio for a cylindrical micelle:

$$\frac{\sigma}{m} = \frac{2 \times 29.6}{1.26} = 47.0$$

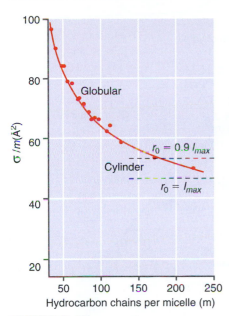

FIGURE 7.17
Surface area per amphiphile chain σ/m plotted as a function of hydrocarbon chains per micelle m for single-chain amphiphiles. The surface area per amphiphile chain σ/m decreases as the volume and m increase for spherical (i.e., globular) micelles. Eventually cylindrical micelles form because the volume of a cylinder can change without markedly affecting σ/m. (Reprinted with permission from C. Tanford, The *Hydrophobic Effect: Formation of Micelles and Biological Membranes,* John Wiley and Sons, New York, 1973.)

Note that for a cylindrical micelle with maximum cross-sectional radius ℓ_{max} the volume can grow without limit by increasing L, but the ratio σ/m will remain constant.

Figure 7.17 shows calculations of σ/m, as the number of amphiphiles per micelle increases. Note that as the spherical micellar radius r_0 approaches its maximum value ℓ_{max}, cylindrical micelles appear. Like the radius of a spherical micelle, the radius of the cylindrical micelle cannot exceed ℓ_{max}, but the length L can increase without limit. Thus, the maximum surface area of a cylindrical micelle $\sigma_{cylinder} = 2\pi\ell_{max}L$ can become much larger than that of a spherical micelle whose radius cannot exceed ℓ_{max}. Once cylindrical micelles formation occurs, σ/m becomes insensitive to further increases in the number of amphiphiles because additional amphiphiles can be accommodated simply by increasing the length of the cylinder.

7.8 Lipid Bilayers and Biological Membranes

The membranes of cells are composed of **phospholipids,** which are amphiphilic molecules with two hydrocarbon chains per head group. Although amphiphiles with one chain per head group do not form **bilayer** structures, phospholipids membranes cell membranes have a bilayer structure where the maximum thickness of the bilayers is $t_{max} = 2\ell_{max}$. See Figure 7.16b. The reason bilayers composed of phospholipids are energetically favorable under physiological conditions and yet fatty acids or other single-chain amphiphiles do not form bilayers under similar conditions is again due to the opposing forces of Coulombic repulsion between head groups and the attractive interaction between the hydrocarbon chains of the amphiphiles. For single-chain amphiphiles σ/m decreases by a factor of 2 as bilayers form compared to cylinders, so repulsive forces make the bilayer energetically unfavorable compared to cylinders in those systems.

EXAMPLE PROBLEM 7.5

Using the experimental micellar volume and ℓ_{max} expressions from Example Problem 7.4, calculate the ratio σ/m for a wide planar bilayer. Compare this ratio to the value calculated for a cylindrical micelle with cross-sectional radius ℓ_{max}.

Solution

For a wide planar bilayer, ignoring edge effects the surface area is

$$\sigma = 2\frac{V}{t_{max}} = \frac{V}{\ell_{max}} \approx \frac{29.6 n_C m}{1.26 n_C} = 23.9m$$

Then $\sigma/m = 23.9$. From Example Problem 7.4 the same ratio for a cylindrical micelle is 47.0. Therefore, a bilayer composed of single-chain amphilphiles will be less stable due to increased head group repulsion than a cylindrical micelle.

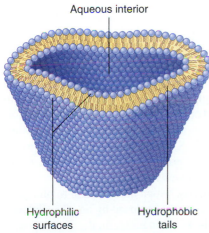

FIGURE 7.18
Cross-section through a lipid bilayer vesicle. (Source: M.K. Campbell, Biochemistry, 2nd ed. Saunder Colleege Publishing, Philadelphia, 1995)

For phospholipids and other amphiphiles with two hydrocarbon tails, the surface area per chain is twice that for bilayers composed of single-chain amphiphiles. For example, for a single-chain amphiphile the number of chains in a micelle m equals the number of head groups n_{hg} so that $\dfrac{\sigma}{m} = \dfrac{\sigma}{n_{hg}}$. For a double-chain amphiphile the number of chains is twice the number of head groups: $2m = n_{hg}$. The surface area per chain σ/m for a phospholipids in a bilayer is $\dfrac{\sigma}{m} = \dfrac{\sigma}{n_{hg}/2} = \dfrac{2\sigma}{n_{hg}}$. Phospholipids and other double-chain amphiphiles can withstand smaller σ/m values because for equivalent chain lengths, the chemical potential change that accompanies transfer of a double-chain amphiphile from water to a hydrocarbon environment is 60% more negative than for a single-chain amphiphile, reflecting a larger attractive force. For this reason, phospholipids form bilayers under physiological conditions. Bilayers can form spherical vesicles (Figure 7.18) with σ/m values that differ little from planar bilayers.

FIGURE 7.19
Schematic of the fluid mosaic model, which attributes to the cell membrane a dynamic aspect wherein lipid molecules undergo both reorientational and translational motions amid a complex mixture of transmembrane proteins, cholesterol, glycolipids, glycoproteins, and so on. (Source: Adapted from W.M. Becker, L.J. Kleinsmith, and J. Hardin, *The World of the Cell,* 6th ed. (San Francisco, CA: Pearson Benjamin Cummings), p. 161. Copyright 2006. Reprinted by permission of Pearson Education, Inc.)

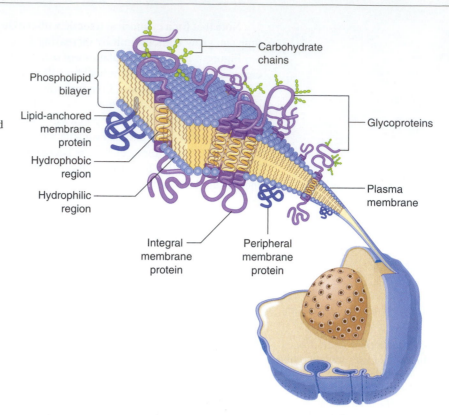

Lipid bilayers are critical components of cell structure. For example, cells are surrounded by bilayer membranes that separate the working architecture of the cell from the surroundings and from other cells. In eukaryotic cells, membranes also surround certain cellular organelles such as mitochondria and nuclei. In addition to being a boundary layer that protects the cell and prevents its contents from dissipating away, cell membranes act to transport material between the cell interior and the exterior. In fact, it is estimated that almost half the energy consumed by the human body is used by nerve cells in the brain to pump ions across nerve cell membranes. In addition, cell membranes act as platforms for proteins, which in turn have numerous functions including providing the mechanism for pumping ions in and out of the cell.

The prevailing view of the structure of the cellular membrane is called the **fluid mosaic model,** schematized in Figure 7.19. The fluid mosaic model attributes a complex, dynamic structure to the membrane. The mosaic aspect of the model refers to the numerous component molecules found in a biological membrane. The amphiphilic component is structurally arranged into a bilayer, which has long-range curvature because it has to enclose the cytosol, which is the fluid within the cell together with the organelles. Rather than being composed of a single type of amphiphile, the cell membrane is a mixture of several types of amphiphilic species. The dominant membrane component is the phosphoglyceride (see Figure 7.20a).

A phosphoglyceride is composed of four basic components: two fatty acid with acyl chains R_1 and R_2 are connected to a three-carbon glyceride framework. The carbon atom not connected to a fatty acid is phosphorylated and connected via the phosphate to a moiety X. The glyceride, phosphate, and X groups are collectively called the "head group." The head group is negatively charged at pH 7 if the X group is neutral or zwitterionic, or the head group may be zwitterionic at pH7 if the X group is positively charged. Common X moieties and appropriate nomenclatures are given in Table 7.6.

Other membrane components include plasmogens, which are identical to phosphoglycerides except that one or two of the ester linkages are replaced by an ether linkage, sphingomyelin (see Figure 7.20b), and glycolipids. In sphingomyelin the X group is

$$\overset{\displaystyle O^-}{\underset{\displaystyle O}{\overset{\displaystyle |}{\underset{\displaystyle |}{P}}}} - O{-}CH_2{-}CH_2{-}N(CH_3)_3^+ .$$

(a)

(b)

(c)

FIGURE 7.20
General formulas for a
(a) phosphoglyceride, (b) sphingolipid, and
(c) cholesterol.

TABLE 7.6 **Some Common Phosphoglycerides Found in Biological Membranes**

Head Group Negative at pH 7	
Lipid Name	X
Phosphatidic acid	$-H$
Phosphatidylserine	$-CH_2 CH (NH_3^+) COO^-$
Phosphatidylinositol	$-C_6H_6(OH)_5$
Phosphatidylglycerol	$-CH_2CH(OH)CH_2OH$
Head Group Zwitterionic at pH 7	
Lipid Name	X
Phosphatidylethanolamine	$-CH_2 CH_2 NH_3^+$
Phosphatidylcholine	$-CH_2 CH_2 N (CH_3)_3^+$

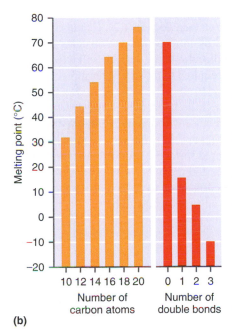

Gel — Liquid crystal

Heat → ← Cool

Head groups tightly packed; tails regular; membrane thicker

Head groups loosely packed; tails disordered; membrane thinner

(a)

Melting point (°C)

Number of carbon atoms   Number of double bonds

(b)

FIGURE 7.21

(a) Schematic of the change in ordering of lipid chains at the melting temperature T_m. Below T_m, the hydrocarbon chains are highly ordered and lipids are rigidly packed into the gel phase. Above T_m conformational ordering of the chains decreases in the liquid crystalline phase. (b) Dependence of the melting temperature T_m on the number of carbon atoms n_c and the number of double bonds. (Source b: (Source: Adapted from W.M. Becker, L.J. Kleinsmith, and J. Hardin, *The World of the Cell*, 6th ed. (San Francisco, CA: Pearson Benjamin Cummings), p. 169. Copyright 2006. Reprinted by permission of Pearson Education, Inc.)

Glycolipids have a similar structure to sphingomyelin except that X is a carbohydrate moiety, which is extruded beyond the membrane surface. Finally, an important component of membranes is cholesterol (see Figure 7.20c). Cholesterol is not a lipid at all but is amphiphilic by virtue of having a hydroxide group connected to a steroid ring.

The fluid aspect of the biological membrane refers to the fact that individual component phospholipids are free to undergo a variety of motions within the membrane including translational diffusion, rotation around the long axis of the lipid molecule, and reorientational motions of the molecule as a whole as well as conformational flexibility of the acyl chains. The fluidity of the membrane is sensitive to temperature. At low temperature the lipid components of the membrane are translationally, rotationally, and orientationally ordered into a gel phase (see Figure 7.21a). Ordering breaks down cooperatively at the melting temperature T_m. Above the melting temperature the membrane is fluid and individual phospholipid molecules translate freely but retain some degree of orientational order. A phase that is translationally disordered yet displays a partial degree of orientational ordering is called liquid crystalline.

The fluidity of the membrane is not only determined by temperature but also by the structural details of the acyl chains and the membrane composition. As shown in Figure 7.21b, membrane fluidity is higher for lipids with short acyl chains or with double bonds in their acyl chains, presumably because the "kink" induced in the chain by the unsaturated bond lowers chain packing. Such lipids have low melting temperatures. Lipids with long, saturated chains have higher melting temperatures.

7.9 Surface Films of Amphiphiles

In 1774, Benjamin Franklin read a paper to the Royal Society of London describing the result of pouring olive oil onto the surface of a pond at Clapham Common in London: "The oil, though not more than a teaspoonful, produced an instant calm over a space of several yards square, which spread amazingly, and extended itself gradually till it reached the leeside, making all that quarter of the pond, perhaps half an acre, as smooth as a looking glass." By estimating the area of the film and knowing the original volume of the oil, Franklin calculated the thickness of the oil layer to be about 2.5 nm.

Franklin was observing another manifestation of the hydrophobic effect: the tendency of amphiphilic molecules to form **monolayers** at the air–water interface when an amphiphilic is added to the water surface. As shown in Figure 7.22, an amphiphile in a monolayer will point its polar head group into the aqueous environment and the hydrophobic tail will be directed away from the aqueous environment.

As discussed in Section 7.6, an important effect of monolayers composed of amphiphiles called surfactants is to lower the surface tension of water. The way in which monolayers composed of amphiphiles lowers surface tension may be appreciated from

FIGURE 7.22

A monolayer of stearic and isostearic acid
formed at an air–water interface.

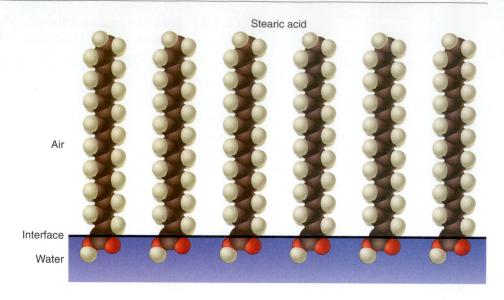

FIGURE 7.22

A monolayer of stearic and isostearic acid
formed at an air–water interface.

Figure 7.22. In Section 7.6, the work required to expand the surface area of a liquid was
shown to be $dA = \gamma\,d\sigma$ where A is the Helmholtz free energy, γ is the surface tension,
and σ is the surface area. In liquid water each molecule is hydrogen bonded to four
neighbors arranged approximately as a tetrahedron (see Section 7.2). Expansion of the
surface area of water can only be accomplished by moving water molecules from the
bulk environment to the surface. But because molecules at the surface will have in general fewer hydrogen-bonded neighbors, work is required to move a water molecule from
the bulk solvent environment to the surface.

However, if a monolayer composed of amphiphilic molecules forms at the surface,
with the polar head groups directed into the aqueous environment and the hydrocarbon
tails pointed away from the aqueous phase (see Figure 7.22), the work required to move
a molecule to the vicinity of the monolayer-covered surface is reduced as a result of attractive interactions between the water molecules and the polar head groups of the amphiphiles. Therefore, monolayer formation is accompanied by a decrease in the surface
tension γ of the solution according to the Gibbs adsorption equation:

$$\Gamma = -\frac{C_2}{RT}\frac{\partial\gamma}{\partial C_2} \qquad (7.35)$$

where C_2 is the total concentration of amphiphile, and Γ is the **surface adsorption**,
which has units of moles per square meter. Amphiphilic substances that lower surface
tension by monolayer formation are called surfactants.

EXAMPLE PROBLEM 7.6

Suppose the total concentration of surfactant is c_2 and the change in surface
tension of the solvent is $\gamma_0 - \gamma$ where γ_0 is the surface tension of the pure solvent
and γ is the surface tension of the solution. For aliphatic acids $\gamma_0 - \gamma$ and c_2 are
empirically related by the equation $\gamma_0 - \gamma = a \log (1 + bc_2)$. The constants a and
b are determined for caproic acid: $a = 0.0298 \text{ N m}^{-1}$ and $b = 232.7 \text{ L mol}^{-1}$.
What is the surface tension of a 0.1 M solution of caproic acid?

Solution

$$\gamma_0 - \gamma = a \log (1 + bc_2)$$
$$\gamma = \gamma_0 - a \log (1 + bc_2) = 0.07199 \text{ N m}^{-1} - 0.0298 \text{ N m}^{-1}$$
$$\times \log (1 + 232.70 \cdot 0.1)$$
$$= 0.07199 \text{ N m}^{-1} - 0.0298 \text{ N m}^{-1} \times 1.385 = 0.03072 \text{ N m}^{-1}$$

EXAMPLE PROBLEM 7.7

Calculate the surface adsorption Γ for a $0.10\ M$ solution of caproic acid.

Solution

$$\gamma = \gamma_0 - a\log(1 + bc_2) = \gamma_0 - \frac{a}{2.303}\ln(1 + bc_2)$$

$$\Gamma = -\frac{c_2}{RT}\left(\frac{d\gamma}{dc_2}\right)_{P,T} = -\frac{c_2}{RT}\frac{d}{dc_2}\left(\gamma_0 - \frac{a}{2.303}\ln(1 + bc_2)\right)$$

$$= \frac{c_2}{RT}\frac{ab}{2.303(1 + bc_2)}$$

$$= \frac{0.10\,\mathrm{M} \times 0.0298\ \mathrm{J\ m^{-2}} \times 232.7\ M^{-1}}{8.314\ \mathrm{J\ mol^{-1}\ K^{-1}} \times 298\ \mathrm{K} \times 2.303 \times (1 + 232.7\ M^{-1} \times 0.10\ M)}$$

$$= \frac{0.69\ \mathrm{J\ m^{-2}}}{1.4 \times 10^5\ \mathrm{J\ mol^{-1}}} = 4.9 \times 10^{-6}\ \mathrm{mol\ m^{-2}}$$

The physical state of the monolayer is a function of the surface adsorption Γ. A lateral pressure (Π) can be applied to the monolayer using a device called a Langmuir trough (see Figure 7.23). A solution of a surfactant dissolved in a volatile organic liquid is dripped onto the water surface contained in a Teflon trough. After the organic solvent evaporates, the surfactant is dispersed over the water surface.

A movable barrier spans the width of the trough and can be swept over the water surface. This exerts on the monolayer a lateral pressure Π, which changes Γ. If as a result of applying a lateral pressure Π the surface adsorption increases, the surface tension decreases as a result of the presence of more amphiphilic molecules at the surface; see Equation (7.35). The surface pressure Π is equal to the difference between the surface tension of pure water γ_0 and the surface tension of the water containing the surfactant, that is,

$$\Pi = \gamma_0 - \gamma \tag{7.36}$$

The surface pressure and thus the surface tension of a solution can be measured using a number of methods. One of the oldest methods, developed in the 1950s and still widely used to measure the surface tension of lung surfactants is the Wilhelmy balance method. Basically the Wilhelmy method uses a platinum blade connected to a balance and suspended over the liquid surface in a Langmuir trough. As the blade is brought into contact with the surface, a meniscus forms. The force F resulting from the surface tension is determined from the weight change registered by the balance. The force F divided by the blade perimeter p is related to the surface tension γ by

$$\frac{F}{p} = \frac{\Delta mg}{p} = \gamma\cos\theta \tag{7.37}$$

where θ is the contact angle (see Section 7.6).

FIGURE 7.23
Schematic of a Langmuir trough. The trough is shown equipped with a Wilhelmy blade for detecting the surface tension of the monolayer-covered water. The change in surface tension is directly related to the lateral pressure exerted by the moving barrier on the monolayer. See text.

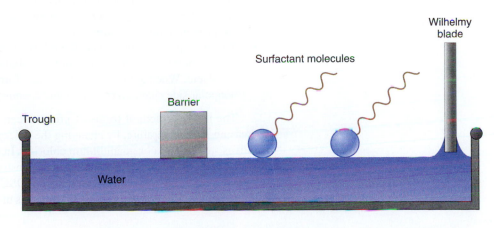

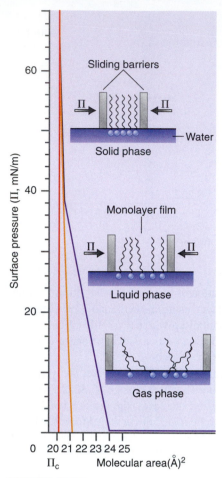

FIGURE 7.24
As the lateral pressure Π on a monolayer increases, the area per molecule decreases suddenly. The effect is reminiscent of phase transitions in bulk materials.

The surface adsorption Γ and lateral pressure Π are related by equations analogous to equations of state in bulk materials. For example, at low surface adsorption amphiphiles within the monolayer do not interact, and experiment shows that the lateral pressure Π is related to Γ by

$$\Pi = \Gamma RT \tag{7.38}$$

which is the two-dimensional analogue of the ideal gas equation. As the surface concentration increases, interactions between amphiphiles become appreciable and nonidealities appear in the equation of state. For example, at higher surface concentrations the equation of state is

$$\Pi(1 - b\Gamma) = \Gamma RT \tag{7.39}$$

which is analogous to the van der Waals equation corrected only for excluded volume b.

As surface concentration increases, the area per molecule in the monolayer decreases, and trends in the surface pressure resembling phase transitions occur. In the "gas" phase amphiphiles interact weakly, are translationally mobile, and the hydrocarbon chains are not closely packed. As the surface adsorption increases, a compressible "liquid" phase appears where amphiphiles change the orientation of their hydrocarbon chains as lateral pressure increases, lying flat on the surface at low Π and moving the tails perpendicular to the surface at higher Π. The "solid" phase observed at high surface concentrations is characterized by loss of both translational and orientational mobility. Figure 7.24 plots surface pressure versus the surface area occupied per molecule. As the lateral pressure Π on a monolayer increases, the area per molecule decreases suddenly. The effect is reminiscent of phase transitions in bulk materials.

7.10 Conformational Transitions of Biological Polymers

It is well known that under physiological conditions, many biological polymers, including proteins and nucleic acids, assume well-defined structural forms. For example, under physiological conditions, DNA has a double helix structure as described by Watson and Crick, and the polypeptide chains of many functional proteins are folded into tight, globular structures. Experimental and theoretical investigations of protein folding are active fields of research with the ultimate goal of understanding how biological polymers assume their native structures. In this section we describe structural transitions of proteins and nucleic acids from a thermodynamic point of view.

The stability of a protein structure depends on the interplay of a number of factors including these:

- The unfolded state is entropically favored (that is, $\Delta S > 0$) over the folded state in the sense that the unfolded state is less conformationally restricted.
- Folding of a protein is enthalpically favored ($\Delta H < 0$) as a result of the formation of hydrogen bonds between various side chain and backbone functional groups of the protein in the structured state.
- The hydrophobic effect favors the folded state of the protein where nonpolar side chains are buried in the interior of the globular protein structure and thus inaccessible to solvent. When a protein unfolds, water forms highly structured cages around nonpolar side chains, a condition that is entropically unfavorable ($\Delta S < 0$).

The native structural forms of proteins can be unfolded, that is, denatured, by increasing the temperature, by changing the pH, or by the addition of organic solvent additives such as urea or guanidinium chloride. In Section 4.6 we discussed the transition from a native, structured protein (N) to the unstructured, denatured form (D) in the context of a differential scanning calorimetry (DSC) experiment. At low temperatures and in the absence of chemical perturbants such as urea, the folded state is favored. At higher

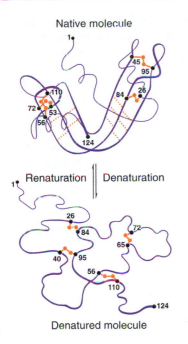

Native molecule

Renaturation ‖ Denaturation

Denatured molecule

(a)

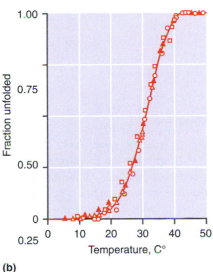

(b)

FIGURE 7.25

(a) Thermal denaturation of the protein ribonuclease is schematized. Note that ribonuclease has four disulfide bonds between sulfhydryl groups in the side chains of eight cysteine amino acids (pairs numbered 26/84, 65/72, 40/95, 58/110, shown in orange). The disulfide bonds remain intact after the protein has melted. (b) The fraction of unfolded ribonuclease species monitored by several physical methods upon heating: increase in solution viscosity (□), change in optical rotation (○), change in UV absorbance (△). Closed triangles are the same measurements made upon cooling, indicating the transition is reversible. (Reprinted with Permission from: A. Ginsburg and W.R. Carroll, Some Specific Ion Effects on the Conformation and Thermal Stability of Ribonuclease, *Biochemistry* 4, 2159–2174 (1965).

temperatures the unfolded state dominates. As shown in Figure 7.25b for the protein ribonuclease, as the temperature approaches the melting temperature T_m, the fraction of native ribonuclease f_N decreases abruptly. At the melting temperature T_m, the fraction of native ribonuclease f_N equals the fraction of denatured ribonuclease f_D. Above the melting temperature the unfolded state dominates and f_D approaches 1. Changes in pH perturb protein structural stability by affecting the charges on protein side chains, thus perturbing the hydrogen bonding interactions that enthalpically favor the folded state. The effect of solvent additives such as urea is more complicated but appears to involve preferential solvation of the unfolded state.

In Section 4.6 we discussed protein denaturation from the point of view of thermochemistry and the measurement by DSC of the enthalpy of denaturation ΔH_{den}. We now extend discussion of protein denaturation to the Gibbs energy associated with protein conformations, and the changes in Gibbs energy associated with protein denaturation. As the schematic in Figure 7.26 shows, the native protein conformation exists in a global Gibbs energy minimum relative to the various unfolded species.

At a given temperature $\Delta G_{den} = \Delta H_{den} - T\Delta S_{den}$. At temperatures below the melting temperature and in the absence of denaturants such as urea, the Gibbs energy change accompanying protein denaturation is positive, as the data in Table 7.7 show. The enthalpy of denaturation ΔH_{den}, also called ΔH_{cal} (that is, the calorimetrically measured enthalpy change), is also positive, as explained in Section 4.6. The entropy change associated with denaturation is likewise positive.

An idealized expression for the denaturation equilibrium of a protein is

$$N \rightleftharpoons D \qquad (7.40)$$

where, once again, N designates the native or folded protein and D denotes the denatured or unfolded protein. This equilibrium expression implies that the denaturation transition is completely cooperative in the sense that only two states exist: the folded state N and the unfolded state D. In this simplified two-state limit, the equilibrium constant has the form

$$K = \frac{C_D}{C_N} = \frac{f_D C}{f_N C} = \frac{f_D}{1 - f_D} \qquad (7.41)$$

where the total concentration of protein is C, the concentration of denatured protein is C_D, and the concentration of native protein is C_N. Note that in Chapter 6 the equilibrium constant was quantified in terms of the equilibrium composition of a gas-phase reaction mixture, where the convenient measurement in such circumstances is the set of partial pressures. The equilibrium constant of Equation (7.40) however does not describe a gas-phase reaction and Equation (7.41) expresses the equilibrium constant for denaturation in terms of the equilibrium concentrations of the native and denatured species. More specific formulations of the equilibrium constant for solution-phase reactions in terms of functions of equilibrium concentrations (that is, activities) of reactants and products will be given in Chapter 8. Note that the melting temperature T_m is defined as the point where the folded and unfolded species are equal in number; hence, $f_D = f_N = 0.5$ and so at $T = T_m$, $K = 1$.

A fundamental question is whether the denaturation of a protein can be described by a two-state model or whether additional equilibria involving partially unfolded species are also involved. There are a number of qualitative indications of cooperative "two-state" behavior. When a physical property that monitors the populations of the folded

TABLE 7.7 Thermodynamic Parameters for Protein Denaturation

Protein (Temperature, pH)	$\Delta G°$ (kJ mol^{-1})	$\Delta H°$ (kJ mol^{-1})	$\Delta S°$ (J K^{-1} mol^{-1})	$\Delta C_P°$ (kJ K^{-1} mol^{-1})
Ribonuclease (303 K, 2.5)	3.8	240	774	8
Chymotrypsinogen (298 K, 3)	31	160	439	11
Myoglobin (298 K, 9)	56.9	180	400	5.9
β-Lactoglobulin (298 K, 3, 5 M urea)	3	−88	−300	9.2

Source: C. Tanford, *Adv. Protein Chem.* 23 (1968), 121.

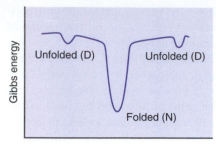

FIGURE 7.26

Gibbs energy profile of the folded and unfolded forms of a protein. Although various partially folded species may occupy local minima in the Gibbs energy, the folded (i.e., native, N) form occupies a global Gibbs energy minimum. There may be many unfolded (i.e., denatured, D) forms, but the Gibbs energies of the unfolded species differ slightly relative to the differences with the Gibbs energy of the folded protein.

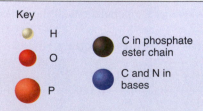

FIGURE 7.27

A space-filling model of DNA showing the double helix structure. (Source: C.K Matthews, K.E. van Holde, K.G. Ahern, *Biochemistry,* 3rd ed., Benjamin Cummings, San Francisco, 2000)

and unfolded states changes abruptly near T_m, cooperative behavior is indicated. In Figure 7.25, several physical measurements that monitor ribonuclease folding including UV absorbance, solution viscosity, and optical rotation change abruptly near T_m, all indicating that the folding of ribonuclease can be described by a two-state model. Similarly in a DSC experiment when the heat capacity peak is very narrow around T_m, the denaturation of the protein likely involves only two states. However, many protein denaturations involve formation of partially folded intermediate species. In such cases the denaturation is not fully cooperative and cannot be described accurately by a two-state model.

Can the deviation of protein denaturation from fully cooperative or two-state behavior be quantified? To answer this question, we consider how thermodynamic properties can be determined experimentally. As shown in Figure 7.25, many physical measurements can be used to determine f_N—and hence the equilibrium constant—using Equation (7.41). We should keep in mind, however, that this is an apparent equilibrium constant because we have assumed that no forms of the protein other than N and D exist, and hence no other equilibria need be considered. Given these assumptions we obtain the following from the apparent K:

$$\Delta G_{den}^{\circ} = -RT \ln K \tag{7.42}$$

$$\Delta H_{den}^{\circ} = RT^2 \left(\frac{\partial \ln K}{\partial T} \right)_P$$

$$\Delta S_{den}^{\circ} = \frac{\Delta H_{den}^{\circ} - \Delta G_{den}^{\circ}}{T}$$

$$\Delta C_{den}^{\circ} = \left(\frac{\partial \Delta H_{den}^{\circ}}{\partial T} \right)_P$$

Of these four equations the enthalpy equation $\Delta H_{den}^{\circ} = RT^2 \left(\frac{\partial \ln K}{\partial T} \right)_P$, also called the van't Hoff equation, which was derived in Chapter 6, is key to quantifying deviations from two-state behavior. The van't Hoff equation indicates that the enthalpy of denaturation can be determined indirectly from the temperature dependence of the equilibrium constant. Assuming ΔH_{den}° is independent of temperature within a range T_1–T_2, which includes T_m, a plot of $\ln K$ versus $1/T$ will be a straight line with a negative slope equal to $-\Delta H_{den}^{\circ}/R$.

EXAMPLE PROBLEM 7.8

A protein has a melting temperature of $T_m = 330.$ K. At $T = 310.$ K, UV absorbance determines that the fraction of native protein is $f_N = 0.985$. At $T = 340.$ K, $f_N = 0.012$. Assuming a two-state model and assuming also that the enthalpy is constant between $T = 320.$ and $340.$ K, determine the enthalpy of denaturation.

Solution

Using a two-state model at $T = 310.$ and $340.$ K,

$$K_{310} = \frac{f_D}{f_N} = \frac{1 - f_N}{f_N} = \frac{1.000 - 0.985}{0.985} = 0.0152$$

$$K_{340} = \frac{1 - 0.012}{0.012} = 82.3$$

Integrating the van't Hoff equation and assuming the enthalpy is constant between $T_1 = 310.$ K and $T_2 = 340.$ K:

$$\ln \left(\frac{K_2}{K_1} \right) = -\frac{\Delta H^{\circ}}{R} \left(\frac{1}{T_2} - \frac{1}{T_1} \right)$$

Using this equation,

$$\ln\left(\frac{K_{340.}}{K_{310.}}\right) = \ln\frac{82.3}{0.0152} = 8.60 = -\frac{\Delta H^\circ}{8.314 \text{ J K}^{-1} \text{mol}^{-1}}\left(\frac{1}{340. \text{ K}} - \frac{1}{310. \text{ K}}\right)$$

$$\Delta H^\circ = \Delta H^\circ_{den} = \frac{8.60 \times 8.314 \text{ J K}^{-1} \text{mol}^{-1}}{2.85 \times 10^{-4} \text{ K}^{-1}} = 251 \text{ kJ mol}^{-1}$$

The van't Hoff equation and the two-state model is used whenever a physical method measures the enthalpy indirectly. However, as explained in Section 4.6, differential scanning calorimetry measures the enthalpy of denaturation directly from the area under the heat capacity versus temperature curve. If we call the enthalpy measured by indirect methods using the van't Hoff equation ΔH_{vH} and the enthalpy measured directly using DSC, ΔH_{cal}, under what circumstances will these two measurements agree or disagree?

Differential scanning calorimetry measures the enthalpy contributions from all equilibria in the solution, whereas the enthalpy deduced using the van't Hoff equation assumes only a single equilibrium between N and D. If the denaturation process indeed involves only two species N and D, then we would expect that the DSC enthalpy measurement ΔH_{cal} should equal the enthalpy calculated using a two-state model ΔH_{vH}. Suppose denaturation of a monomeric protein involves formation of intermediate, partially folded forms and thus departs from full cooperativity. Because the van't Hoff equilibrium equation neglects the equilibria involving intermediate, partially folded forms, $\Delta H_{vH} < \Delta H_{cal}$.

We have thus far concentrated on conformational transitions in proteins, but nucleic acids likewise undergo changes in conformation that can be described thermodynamically. DNA in particular exists in the cell as a Watson–Crick double helix (see Figure 7.27).

The double helix structure is stabilized by a number of interactions. First there is the base stacking interaction, which refers to the fact that planar purine and pyrimidine bases stack vertically in aqueous solution. As shown in Figure 7.28, bases pairs in DNA are stacked vertically with a base-base distance of about 0.34 nm. Several factors drive base stacking in solution. For example, aggregation of purine and pyrimidine bases is entropically favored as a result of the hydrophobic effect. If purine and pyrimidine bases stack vertically, nonpolar parts of these molecules are isolated from water and the formation of entropically unfavorable water cages is avoided. The second factor favoring the formation of double-stranded DNA is the formation of hydrogen bonds between purine and pyrimidine bases. Two hydrogen bonds form between adenine (A) and thymine (T) bases in a Watson–Crick base pair, whereas three hydrogen bonds are formed between guanine (G) and cytosine (C) (see Figure 7.28a).

As with proteins, the structures of DNA and other polynucleotides can be denatured thermally. This means that as the temperature is increased, each double-stranded DNA dissociates or "melts" into two single strands. As with proteins, the decrease in the duplex DNA component can be monitored as a function of temperature using a number of physical techniques. The most common way to monitor the progress of structural change in polynucleotides is a melting curve, which is the UV absorbance of an aqueous solution of a polynucleotide plotted as a function of temperature. As the temperature of an oligonucleotide or polynucleotide solution increases, the UV absorbance increases. Although the UV absorbance by all aqueous solutions decreases slightly with heating due to dilution of the solution when the solvent expands, the large increases in UV absorbance by solutions of polynucleotides are indicative of conformational changes.

Figure 7.29 shows melting curves for several polynucleotides and oligonucleotides. Note the similarity between these melting curves and the plot of f_N versus T for ribonuclease shown in Figure 7.25b. Like the ribonuclease data, polynucleotide melting curves monitor the progressive dissociation or denaturation of the structures of polynucleotides. In solutions of single-stranded nucleic acids such as polyadenylic acid (poly(A)), there occurs a gradual increase in UV absorption with heating as a result of a decrease in base

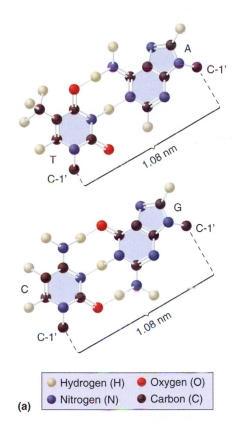

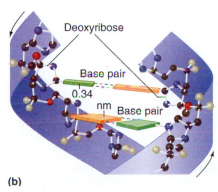

(a)

Hydrogen (H)   Oxygen (O)
Nitrogen (N)   Carbon (C)

(b)

FIGURE 7.28

(a) AT and GC Watson–Crick base pairs in DNA. Three hydrogen bonds form in a GC base pair and two hydrogen bonds form in a AT base pair. (b) Vertical base stacking in double helical DNA. (Source: C.K Matthews, K.E. van Holde, K.G. Ahern, *Biochemistry,* 3rd ed., Benjamin Cummings, San Francisco, 2000)

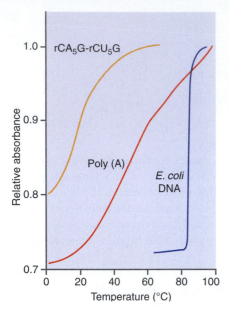

FIGURE 7.29

UV absorbance melting curves for various oligonucleotides and polynucleotides. The single-stranded polynucleotide polyadenylic acid (poly(A)) displays a broad melting transition as bases unstack with heating. The dissociation of polynucleotide duplexes display sharp, cooperative transitions. (Source: Modified from Figure 6–5 in V.A. Bloomfield, D.M. Crothers, I. Tinoco, *Nucleic Acids: Structure, Properties, and Functions,* University Science Books, Sausilito, Calif., 2000)

stacking. Double-stranded oligonucleotides and polynucleotides show much sharper transitions in UV absorption as a result of a transition from double helix structures at temperatures below the melting temperature T_m, to single strands at temperatures above T_m. Figure 7.29 also shows that the position of T_m, which is obtained from the inflection point of the melting curve, and the sharpness of the UV absorbance curve near T_m vary with the size of the polynucleotide. Small oligonucleotides such as rCA$_5$G-rCU$_5$G, a heptaribonucleotide with a duplex structure in which one-strand rCAAAAAG is base paired to rGUUUUUC, have low melting temperatures and broad UV transitions near T_m, indicating a low degree of cooperativity. High molecular weight polynucleotides such as DNA derived from the bacterium *E. coli* melt at higher temperatures and display highly cooperative conformational transitions. The melting temperature is also influenced by the composition of the polynucleotide. Due to the fact that there are a greater number of hydrogen bonds in a GC base pair (three) versus an AT base pair (two), the melting temperature increases linearly as a function of the relative number of GC base pairs in a polynucleotide duplex. Therefore, given two duplex polynucleotides of identical length, the polynucleotide with the greater number of GC base pairs will have a higher T_m.

Thermodynamic properties can be obtained from UV absorbance data if a model is used to describe the dissociation of a polynucleotide from a duplex structure to single strands. As is the case with proteins such as ribonuclease, which show highly cooperative denaturation transitions, the highly cooperative dissociations of polynucleotide duplexes can be described similarly by a two-state model. As an example, consider the fully cooperative transition of a self-complementary DNA from a duplex to single strands. An example of a self-complementary DNA is the dodecamer d(CGCGAATTCGCG)$_2$ where the deoxynucleotide strand dCGCGAATTCGCG is paired with the strand dGCGCTTAAGCGC. Note that these two strands are identical. Now if we designate a self-complementary duplex A_2, and the single strands individually as A, the two-state model gives the dissociation equilibrium

$$A_2 \rightleftharpoons 2A \qquad (7.43)$$

The equilibrium constant for this dissociation is

$$K = \frac{C_A^2}{C_{A_2}} = \frac{2}{fC} \times (1-f)^2 C^2 = \frac{2(1-f)^2 C}{f} \qquad (7.44)$$

where f is the fraction of the total strand concentration C that is in duplex form, C_{A_2} is the concentration of duplex, and C_A is the concentration of single strands.

At the melting temperature, $f = 0.5$ by definition. Substituting this expression into Equation (7.44), we obtain at $T = T_m$, $K = C$. Substituting this expression for K into

$$\Delta G° = -RT \ln K = \Delta H° - T\Delta S° \qquad (7.45)$$

we obtain

$$-RT_m \ln C = \Delta H° - T_m \Delta S° \qquad (7.46)$$

Rearranging Equation (7.46) we find:

$$\frac{1}{T_m} = -\frac{R \ln C}{\Delta H°} + \frac{\Delta S°}{\Delta H°} \qquad (7.47)$$

Equation (7.47) means that for a self-complementary polynucleotide, a plot of $1/T_m$ versus $\ln C$ has a slope of $-R/\Delta H°$ and a y intercept of $\Delta S°/\Delta H°$. As in the case of protein denaturations, an enthalpy for the melting of a polynucleotide duplex structure that is determined indirectly using a two-state model may deviate from calorimetrically determined enthalpy if the structural transition deviates from full cooperativity.

Vocabulary

amphiphile	contact angle	nonwetting	supercritical fluids
bilayer	critical point	normal boiling temperature	surface adsorption
boiling point elevation	fluid mosaic model	phase	surface tension
capillary depression	freezing point depression	phase diagram	triple point
capillary rise	freezing point elevation	phospholipids	Trouton's rule
Clapeyron equation	glass	P–T phase diagram	vapor pressure
Clausius-Clapeyron	lipids	standard boiling temperature	wetting
equation	micelle	sublimation temperature	
coexistence curve	monolayer		

Questions on Concepts

Q7.1 At a given temperature, a liquid can coexist with its gas at a single value of the pressure. However, you can sense the presence of $H_2O(g)$ above the surface of a lake by the humidity, and it is still there if the barometric pressure rises or falls at constant temperature. How is this possible?

Q7.2 Why is it reasonable to show the μ versus T segments for the three phases as straight lines as is done in Figure 7.1? More realistic curves would have some curvature. Is the curvature upward or downward on a μ versus T plot?

Q7.3 Why is $\Delta H_{sublimatation} = \Delta H_{fusion} + \Delta H_{vaporization}$?

Q7.4 A triple point refers to a point in a P–T phase diagram for which three phases are in equilibrium. Do all triple points correspond to a gas–liquid–solid equilibrium?

Q7.5 Why are the triple point temperature and the normal freezing point very close in temperature for most substances?

Q7.6 As the pressure is increased at $-45°C$, ice I is converted to ice II. Which of these phases has the lower density?

Q7.7 What is the physical origin of the pressure difference across a curved liquid–gas interface?

Q7.8 Give a molecular level explanation as to why the surface tension of $Hg(l)$ is not zero.

Problems

Problem numbers in **RED** indicate that the solution to the problem is given in the *Student Solutions Manual*.

P7.1 In this problem, you will calculate the differences in the chemical potentials of ice and supercooled water, and of steam and superheated water all at 1 bar pressure shown schematically in Figure 7.1. For this problem, $S°_{H_2O,s} = 48.0 \text{ J mol}^{-1} \text{ K}^{-1}$, $S°_{H_2O,l} = 70.0 \text{ J mol}^{-1} \text{ K}^{-1}$ and $S°_{H_2O,g} = 188.8 \text{ J mol}^{-1} \text{ K}^{-1}$.

  a. By what amount does the chemical potential of water exceed that of ice at $-5.00°C$?

  b. By what amount does the chemical potential of water exceed that of steam at $105.00°C$?

P7.2 The phase diagram of NH_3 can be characterized by the following information. The normal melting and boiling temperatures are 195.2 and 239.82 K, respectively, and the triple point pressure and temperature are 6077 Pa and 195.41 K, respectively. The critical point parameters are 112.8×10^5 Pa and 405.5 K. Make a sketch of the P–T phase diagram (not necessarily to scale) for NH_3. Place a point in the phase diagram for the following conditions. State which and how many phases are present.

  a. 195.41 K, 1050 Pa

  b. 195.41 K, 6077 Pa

  c. 237.51 K, 101,325 Pa

  d. 420 K, 130×10^5 Pa

  e. 190 K, 6077 Pa

P7.3 Within what range can you restrict the values of P and T if the following information is known about CO_2? Use Figure 7.8 to answer this problem.

  a. As the temperature is increased, the solid is first converted to the liquid and subsequently to the gaseous state.

  b. As the pressure on a cylinder containing pure CO_2 is increased from 65 to 80 atm, no interface delineating liquid and gaseous phases is observed.

  c. Solid, liquid, and gas phases coexist at equilibrium.

  d. An increase in pressure from 10 to 50 atm converts the liquid to the solid.

  e. An increase in temperature from $-80°$ to $20°C$ converts a solid to a gas with no intermediate liquid phase.

P7.4 Within what range can you restrict the values of P and/or T if the following information is known about sulfur? Use Figure 7.7 to answer this problem.

  a. Only the monoclinic solid phase is observed for $P = 1$ atm.

  b. When the pressure on the vapor is increased, the liquid phase is formed.

c. Solid, liquid, and gas phases coexist at equilibrium.

d. As the temperature is increased, the rhombic solid phase is converted to the liquid directly.

e. As the temperature is increased at 1 atm, the monoclinic solid phase is converted to the liquid directly.

P7.5   The vapor pressure of liquid SO_2 is 2232 Pa at 201 K, and $\Delta H_{vaporization} = 24.94$ kJ mol^{-1}. Calculate the normal and standard boiling points. Does your result for the normal boiling point agree with that in Table 7.1? If not, suggest a possible cause.

P7.6   For water, $\Delta H_{vaporization}$ is 40.65 kJ mol^{-1}, and the normal boiling point is 373.15 K. Calculate the boiling point for water on the top of a mountain of height 5500 m, where the normal barometric pressure is 380 Torr.

P7.7   Use the values for ΔG_f° (ethanol, l) and ΔG_f° (ethanol, g) from Appendix B to calculate the vapor pressure of ethanol at 298.15 K.

P7.8   Use the vapor pressures of ClF_3 given in the following table to calculate the enthalpy of vaporization using a graphical method or a least-squares fitting routine.

$T(°C)$	P (Torr)	T (°C)	P (Torr)
−246.97	29.06	−233.14	74.31
−41.51	42.81	−30.75	86.43
−35.59	63.59	−27.17	107.66

P7.9   Use the following vapor pressures of 1-butene given here to calculate the enthalpy of vaporization using a graphical method or a least-squares fitting routine.

T (K)	P (atm)
273.15	1.268
275.21	1.367
277.60	1.490
280.11	1.628
283.15	1.810

P7.10   Use the vapor pressures of Cl_2 given in the following table to calculate the enthalpy of vaporization using a graphical method or a least-squares fitting routine.

T (K)	P (atm)	T (K)	P (atm)
227.6	0.585	283.15	4.934
238.7	0.982	294.3	6.807
249.8	1.566	305.4	9.173
260.9	2.388	316.5	12.105
272.0	3.483	327.6	15.676

P7.11   Use the vapor pressures of n-butane given in the following table to calculate the enthalpy of vaporization using a graphical method or a least-squares fitting routine.

T (K)	P (Torr)	T (K)	P (Torr)
187.45	5.00	220.35	60.00
195.35	10.00	228.95	100.00
204.25	20.00	241.95	200.0
214.05	40.00	256.85	400.0

P7.12   Use the vapor pressures of ice given here to calculate the enthalpy of sublimation using a graphical method or a least-squares fitting routine.

T (°C)	P (Torr)
−28.00	0.3510
−29.00	0.3169
−30.00	0.2859
−31.00	0.2575
−32.00	0.2318

P7.13   Carbon tetrachloride melts at 250 K. The vapor pressure of the liquid is 10,539 Pa at 290 K and 74,518 Pa at 340 K. The vapor pressure of the solid is 270 Pa at 232 K and 1092 Pa at 250 K.

a. Calculate $\Delta H_{vaporization}$ and $\Delta H_{sublimation}$.

b. Calculate ΔH_{fusion}.

c. Calculate the normal boiling point and $\Delta S_{vaporization}$ at the boiling point.

d. Calculate the triple point pressure and temperature.

P7.14   It has been suggested that the surface melting of ice plays a role in enabling speed skaters to achieve peak performance. Carry out the following calculation to test this hypothesis. At 1 atm pressure, ice melts at 273.15 K, ΔH_{fusion} = 6010 J mol^{-1}, the density of ice is 920 kg m^{-3}, and the density of liquid water is 997 kg m^{-3}.

a. What pressure is required to lower the melting temperature by 5.0°C?

b. Assume that the width of the skate in contact with the ice has been reduced by sharpening to 25×10^{-3} cm, and that the length of the contact area is 15 cm. If a skater of mass 85 kg is balanced on one skate, what pressure is exerted at the interface of the skate and the ice?

c. What is the melting point of ice under this pressure?

d. If the temperature of the ice is −5.0°C, do you expect melting of the ice at the ice−skate interface to occur?

P7.15   Solid iodine, $I_2(s)$, at 25°C has an enthalpy of sublimation of 56.30 kJ mol^{-1}. The $C_{P,m}$ of the vapor and solid phases at that temperature are 36.9 and 54.4 J K^{-1} mol^{-1}, respectively. The sublimation pressure at 25°C is 0.30844 Torr. Calculate the sublimation pressure of the solid at the melting point (113.6°C) assuming

a. that the enthalpy of sublimation is constant.

b. that the enthalpy of sublimation at temperature T can be calculated from the equation
$$\Delta H_{sublimation}^\circ (T) = \Delta H_{sublimation}^\circ (T_0) + \Delta C_P(T - T_0).$$

P7.16   Carbon disulfide, $CS_2(l)$, at 25°C has a vapor pressure of 0.4741 bar and an enthalpy of vaporization of 27.66 kJ mol^{-1}. The $C_{P,m}$ of the vapor and liquid phases at that temperature are 45.4 and 75.7 J K^{-1} mol^{-1}, respectively. Calculate the vapor pressure of $CS_2(l)$ at 100.0°C assuming

a. that the enthalpy of sublimation and the heat capacities do not change with temperature.

b. that the enthalpy of sublimation at temperature T can be calculated from the equation

$$\Delta H^{\circ}_{sublimation}(T) = \Delta H^{\circ}_{sublimation}(T_0) + \Delta C_P(T - T_0).$$

P7.17   Consider the transition between two forms of solid tin, $Sn(s, gray) \rightleftharpoons Sn(s, white)$. The two phases are in equilibrium at 1 bar and 18°C. The densities for gray and white tin are 5750 and 7280 kg m^{-3}, respectively, and $\Delta H_{transition} = 8.8$ J K^{-1} mol^{-1}. Calculate the temperature at which the two phases are in equilibrium at 200 bar.

P7.18   The UV absorbance of a solution of a double-stranded DNA is monitored at 260 nm as a function of temperature. Data appear in the following table. From the data determine the melting temperature.

Temperature (K)	343	348	353	355	357	359	361	365	370
Absorbance (260 nm)	0.30	0.35	0.50	0.75	1.22	1.40	1.43	1.45	1.47

P7.19   A protein has a melting temperature of $T_m = 335$ K. At $T = 315$ K, UV absorbance determines that the fraction of native protein is $f_N = 0.965$. At $T = 345$. K, $f_N = 0.015$. Assuming a two-state model and assuming also that the enthalpy is constant between $T = 315$ and 345 K, determine the enthalpy of denaturation. Also, determine the entropy of denaturation at $T = 335$ K. By DSC, the enthalpy of denaturation was determined to be 251 kJ mol^{-1}. Is this denaturation accurately described by the two-state model?

P7.20   You have collected a tissue specimen that you would like to preserve by freeze drying. To ensure the integrity of the specimen, the temperature should not exceed -10.5°C. The vapor pressure of ice at 273.16 K is 611 Pa; $\Delta H^{\circ}_{fusion} = 6.01$ kJ mol^{-1} and $\Delta H^{\circ}_{vaporization} = 40.65$ kJ mol^{-1}. What is the maximum pressure at which the freeze drying can be carried out?

P7.21   The vapor pressure of methanol(l) is given by

$$\ln\left(\frac{P}{Pa}\right) = 23.593 - \frac{3.6791 \times 10^3}{\dfrac{T}{K} - 31.317}$$

a. Calculate the standard boiling temperature.

b. Calculate $\Delta H_{vaporization}$ at 298 K and at the standard boiling temperature.

P7.22   Suppose a DNA duplex is not self-complementary in the sense that the two polynucleotide strands composing the double helix are not identical. Call these strands A and B. Call the duplex AB. Consider the association equilibrium of A and B to form duplex AB;

$$A + B \rightleftharpoons AB$$

Assume the total strand concentration is C and, initially, A and B have equal concentrations; that is, $C_{A,0} = C_{B,0} = C/2$. Obtain an expression for the equilibrium constant at a point where the fraction of the total strand concentration C that is duplex is defined as f. If the strand concentration is $1.00 \times 10^{-4} M$, calculate the equilibrium constant at the melting temperature.

P7.23   For a self-complementary DNA oligonucleotide, the melting temperature is monitored as a function of total strand concentration:

T_m (K)	315	318	321	324	327	329
Strand Concentration (M)	5.00×10^{-5}	9.00×10^{-5}	1.50×10^{-4}	2.70×10^{-4}	4.00×10^{-4}	8.20×10^{-3}

Determine the enthalpy and entropy changes associated with the melting of this DNA duplex. Assume the enthalpy and entropy are both constants in this temperature interval. State any assumptions.

P7.24   The vapor pressure of a liquid can be written in the empirical form known as the Antoine equation, where $A(1)$, $A(2)$, and $A(3)$ are constants determined from measurements:

$$\ln\frac{P(T)}{Pa} = A(1) - \frac{A(2)}{\dfrac{T}{K} + A(3)}$$

Starting with this equation, derive an equation giving $\Delta H_{vaporization}$ as a function of temperature.

P7.25   The vapor pressure of an unknown solid is approximately given by $\ln(P/Torr) = 22.413 - 2035(K/T)$, and the vapor pressure of the liquid phase of the same substance is approximately given by $\ln(P/Torr) = 18.352 - 1736(K/T)$.

a. Calculate $\Delta H_{vaporization}$ and $\Delta H_{sublimation}$.

b. Calculate ΔH_{fusion}.

c. Calculate the triple point temperature and pressure.

P7.26   For a self-complementary DNA, the melting temperature T_m is related to the total strand concentration C by

$$\frac{1}{T_m} = \frac{R \ln C}{\Delta H^{\circ}} + \frac{\Delta S^{\circ}}{\Delta H^{\circ}}.$$ Using your result from Problem P7.22, obtain the relationship between T_m and C for a nonself-complementary DNA. Using this expression and the data in Problem P7.23, obtain the enthalpy and entropy of melting of a nonself-complementary DNA oligonucleotide.

P7.27   Butanoic acid, $CH_3CH_2CH_2COOH$, is a surfactant. The surface tension γ (N m^{-1}) of aqueous solutions of butanoic acid is given in the following table as a function of concentration (mol L^{-1}).

Concentration (mol L^{-1})	Surface Tension (N m^{-1})
0.01	0.000704
0.05	0.000639
0.10	0.000587
0.20	0.000521
0.30	0.000477
0.40	0.000445
0.50	0.000419
0.60	0.000397
0.70	0.000379
0.80	0.000363
0.90	0.000348

From a graph of the surface tension as a function of concentration determine the surface adsorption of butanoic acid at a concentration of 0.40 M. Assume $T = 291$ K.

P7.28   For the formation of a self-complementary duplex DNA from single strands $\Delta H^{\circ} = -177.2$ kJ mol^{-1}, and $T_m = 311$ K for strand concentrations of $1.00 \times 10^{-4} M$. Calculate the equilibrium constant and Gibbs energy change for duplex formation at $T = 335$ K. Assume the enthalpy change for duplex formation is constant between $T = 311$ K and $T = 335$ K.

P7.29 The densities of a given solid and liquid of molecular weight 122.5 at its normal melting temperature of 427.15 K are 1075 and 1012 kg m^{-3}, respectively. If the pressure is increased to 120 bar, the melting temperature increases to 429.35 K. Calculate ΔH°_{fusion} and ΔS°_{fusion} for this substance.

P7.30 In Equation (7.15), $(dP/dT)_{vaporization}$ was calculated by assuming that $V_m^{gas} \gg V_m^{liquid}$. In this problem, you will test the validity of this approximation. For water at its normal boiling point of 373.13 K, $\Delta H_{vaporization} = 40.65 \times 10^3$ J mol^{-1} ρ_{liquid} = 958.66 kg m^{-3}, and $\rho_{gas} = 0.58958$ kg m^{-3}. Compare the calculated values for $(dP/dT)_{vaporization}$ with and without the approximation of Equation (7.15). What is the relative error in making the approximation?

P7.31 Suppose the denaturation temperature of a protein is $T = 325$ K and $\Delta H^\circ = 245$ kJ mol^{-1}. Assuming the denaturation of a protein can be described by a two state model, calculate the equilibrium constant and standard Gibbs energy change for the denaturation of the protein at $T = 310$. K. Assume the enthalpy change is constant between $T = 310$. K and $T = 325$ K.

P7.32 The variation of the vapor pressure of the liquid and solid forms of a pure substance near the triple point are given by $\ln(P_{solid}/\text{Pa}) = -8750(\text{K}/T) + 31.143$ and $\ln(P_{liquid}/\text{Pa}) = -4053(\text{K}/T) + 21.10$. Calculate the temperature and pressure at the triple point.

P7.33 Calculate the vapor pressure of CS_2 at 298 K if He is added to the gas phase at a partial pressure of 200 bar. The vapor pressure of CS_2 is given by the empirical equation

$$\ln \frac{P(T)}{\text{Pa}} = 20.801 - \frac{2.6524 \times 10^3}{\dfrac{T}{K} - 33.402}$$

The density of CS_2 at this temperature is 1255.5 kg m^{-3}. By what factor does the vapor pressure change?

P7.34 Use the vapor pressures for PbS given in the following table to estimate the temperature and pressure of the triple point and also the enthalpies of fusion, vaporization, and sublimation.

Phase	T (°C)	P (Torr)
Solid	1048	40.0
Solid	1108	100.
Liquid	1221	400.
Liquid	1281	760.

P7.35 Use the vapor pressures for C_2N_2 given in the following table to estimate the temperature and pressure of the triple point and also the enthalpies of fusion, vaporization, and sublimation.

Phase	T (°C)	P (Torr)
Solid	−62.7	40.0
Solid	−51.8	100.
Liquid	−33.0	400.
Liquid	−21.0	760.

P7.36 A reasonable approximation to the vapor pressure of krypton is given by

$$\log_{10}(P/\text{Torr}) = b - 0.05223(a/T)$$

For solid krypton, $a = 10065$ and $b = 7.1770$. For liquid krypton, $a = 9377.0$ and $b = 6.92387$. Use these formulas to estimate the triple point temperature and pressure and also the enthalpies of vaporization, fusion, and sublimation of krypton.

P7.37 The vapor pressure of $H_2O(l)$ is 23.766 Torr at 298.15 K. Use this value to calculate $\Delta G^\circ_f (H_2O, g) - \Delta G^\circ_f (H_2O, l)$. Compare your result with those in Table 4.1.

P7.38 The normal melting point of H_2O is 273.15 K, and $\Delta H_{fusion} = 6010$ J mol^{-1}. Calculate the decrease in the normal freezing point at 100 and 500 bar assuming that the density of the liquid and solid phases remains constant at 997 and 917 kg m^{-3}, respectively.

P7.39 Autoclaves that are used to sterilize surgical tools require a temperature of 120°C to kill bacteria. If water is used for this purpose, at what pressure must the autoclave operate? The normal boiling point of water is 373.15 K, and $\Delta H_{vaporication} = 40.656 \times 10^3$ J mol^{-1} at the normal boiling point.

P7.40 Calculate the difference in pressure across the liquid–air interface for a water droplet of radius 150 nm.

P7.41 Calculate the factor by which the vapor pressure of a droplet of methanol of radius 1.00×10^{-4} m at 45.0°C in equilibrium with its vapor is increased with respect to a very large droplet. Use the tabulated value of the density and the surface tension at 298 K from Appendix B for this problem. (*Hint:* You need to calculate the vapor pressure of methanol at this temperature.)

P7.42 Calculate the vapor pressure of water droplets of radius 1.25×10^{-8} m at 360 K in equilibrium with water vapor. Use the tabulated value of the density and the surface tension at 298 K from Appendix B for this problem. (*Hint:* You need to calculate the vapor pressure of water at this temperature.)

P7.43 In Section 7.7, it is stated that the maximum height of a water column in which cavitation does not occur is ~9.7 m. Show that this is the case at 298 K.

P7.44 Calculate the vapor pressure for a mist of water droplets, where the droplets are spherical with radius 1.00×10^{-8} m. Assume $T = 293$ K. The vapor pressure of water with a flat interface is 23.75 Torr.

P7.45 In Example Problem 7.8 we found that the surface tension of solutions of aliphatic acids is given by $\gamma_0 - \gamma = a \log(1 + bc_2)$ where c_2 is the solute concentration and γ_0 is the surface tension of pure water. The constants a and b are determined at $T = 291$ K for n-butanoic acid: $a = 0.0298$ N m^{-1} and $b = 19.6$ L mol^{-1}. Calculate the surface adsorption for $c_2 = 0.01$M, 0.10M, 0.20M, 0.40M, 0.8M, and 1.0M. What is the maximum surface adsorption that can be achieved in solutions of *n*-butanoic acid?

P7.46 Because of the high surface tension of water, macromolecules like proteins in water solutions exhibit the phenomenon of hydrophobic interactions whereby hydrophobic groups in the protein contact each other in order to avoid contact with water. Assume a spherical protein has a molecular weight of $M_2 = 60,000.$ g mol^{-1} and a specific volume of $\overline{V}_2 = 0.73$ mL g^{-1} Assuming 25% of the total surface area of the protein is hydrophobic, estimate ΔG per mole for the hydrophobic interactions. Assume the interfacial tension of water is 0.0720 N m^{-1}. State any simplifications that you make in obtaining this estimate.

P7.47 For films of egg albumin, the surface pressure Π was obtained as a function of surface area (per unit mass of surface protein) at $T = 298$ K

Π(mNm^{-1})	0.07	0.11	0.18	0.20	0.26	0.33	0.38
A(m^{2}mg^{-1})	2.00	1.64	1.50	1.45	1.38	1.36	1.32

Calculate the molecular weight of egg albumin.

P7.48 From the data provided in Problem 7.47, determine the cross sectional area of egg albumin on the surface. Assuming the specific volume of egg albumin is $V_2 = 0.72$ mL g^{-1} calculate the volume of a single protein. Egg albumin is a globular protein, so approximating the shape of egg albumin as a sphere, calculate the radius and the surface area of egg albumin. Compare the surface area you have calculated to the cross sectional area obtained from surface pressure measurements. Account for any differences.

P7.49 The table lists the chemical potential difference $\mu^\circ_{water} - \mu^\circ_{ethanol}$ for transferring various amino acids from water into ethanol. $T = 298$ K.

Amino Acid	Glycine	Alanine	Valine	Phenylalanine	Proline
$\mu^\circ_{water} - \mu^\circ_{ethanol}$ (kJ mol^{-1})	−19.4	−16.3	−12.3	−8.28	−8.61

Source: Klotz, I. M. "Energy Changes in Biochemical Reactions." New York, Academinc Press, 1967.

All of the quantities in the table are negative because all amino acids consist of amino and carboxyl groups, both of which are polar, in addition to side chains, which may be polar or nonpolar. Assume the contributions to $\mu^\circ_{water} - \mu^\circ_{ethanol}$ from the polar and nonpolar groups are simply additive. Assume further that for glycine only polar groups contribute to $\mu^\circ_{water} - \mu^\circ_{ethanol}$ because a side chain is essentially absent. For each amino acid, calculate the contribution to $\mu^\circ_{water} - \mu^\circ_{ethanol}$ from the side chain alone. Assuming ethanol is representative of the interior of a protein, explain how amino acids with nonpolar side chains contribute to folding. If in a protein a phenylalanine were replaced by a glycine, would the stability of the folded state be increased or decreased? Explain.

P7.50 A cell is roughly spherical with a radius of 20.0×10^{-6} m. If the cell grows to double its volume, calculate the work required to expand the cell surface. Assume the cell is surrounded by pure water and that $T = 298$ K.

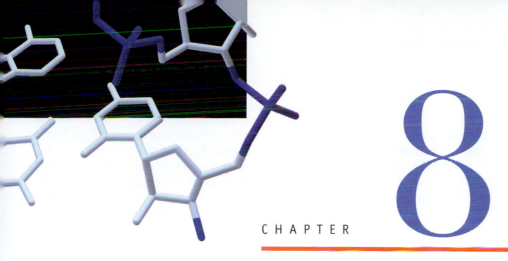

CHAPTER

8

Ideal and Real Solutions

In an ideal solution of A and B, the A–B interactions are the same as the A–A and B–B interactions. In this case, the equilibrium between the solution concentration and gas-phase partial pressure is described by Raoult's law for each component. In an ideal solution, the vapor over a solution is enriched in the most volatile component, allowing a separation into its components through fractional distillation. Nonvolatile solutes lead to a decrease of the vapor pressure above a solution. Such solutions exhibit a freezing point depression and a boiling point elevation. These properties depend only on the concentration and not on the identity of the nonvolatile solutes for an ideal solution. Real solutions are described by a modification of the ideal dilute solution model. In an ideal dilute solution, the solvent obeys Raoult's law, and the solute obeys Henry's law. This model is limited in its applicability. To quantify deviations from the ideal dilute solution model, we introduce the concept of the *activity*. The activity of a component of the solution is defined with respect to a standard state. Knowledge of the activities of the various components of a reactive mixture is essential in modeling chemical equilibrium. The thermodynamic equilibrium constant for real solutions is calculated by expressing the reaction quotient Q in terms of activities rather than concentrations.

8.1 Defining the Ideal Solution

In an ideal gas, the atoms or molecules do not interact with one another. Clearly, this is not a good model for a liquid, because without attractive interactions, a gas will not condense. The attractive interaction in liquids varies greatly, as shown by the large variation in boiling points among the elements; for instance, helium has a normal boiling point of 4.2 K, whereas hafnium has a boiling point of 5400 K.

In developing a model for solutions, the vapor phase that is in equilibrium with the solution must be taken into account. Consider pure liquid benzene in a beaker placed in a closed room. Because the liquid is in equilibrium with the vapor phase above it, there is a nonzero partial pressure of benzene in the air surrounding the beaker. This pressure is called the vapor pressure of benzene at the temperature of the liquid. What happens when toluene is added to the beaker? The partial pressure of benzene is reduced, and the vapor phase now contains both benzene and toluene. For this particular mixture, the partial pressure of each component (*i*) above the liquid is given by

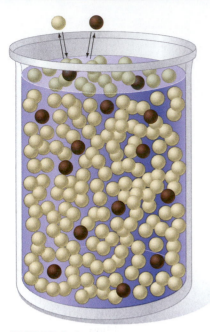

FIGURE 8.1
Schematic model of a solution. The white and black spheres represent solvent and solute molecules, respectively.

$$P_i = x_i P_i^* \text{ for } i = 1, 2 \tag{8.1}$$

where x_i is the mole fraction of that component in the liquid. This equation states that the partial pressure of each of the two components is directly proportional to the vapor pressure of the corresponding pure substance, P_i^*, and that the proportionality constant is x_i. Equation (8.1) is known as **Raoult's law** and is the definition of an ideal solution. Raoult's law holds for each substance in an **ideal solution** over the range from $0 \le x_i \le 1$. In binary solutions, one refers to the component that has the higher value of x_i as the **solvent** and the component that has the lower value of x_i as the **solute.**

Few solutions satisfy Raoult's law. However, it is useful to study the thermodynamics of ideal solutions and to introduce departures from ideal behavior later. Why is Raoult's law not generally obeyed over the whole concentration range of a binary solution consisting of molecules A and B? Equation (8.1) is only obeyed if the A–A, B–B, and A–B interactions are all equally strong. This criterion is satisfied for a mixture of benzene and toluene because the two molecules are very similar in size, shape, and chemical properties. However, it is not satisfied for arbitrary molecules. Raoult's law is an example of a **limiting law;** the solvent in a real solution obeys Raoult's law as the solution becomes highly dilute.

Raoult's law is derived in Example Problem 8.1 and can be rationalized using the model depicted in Figure 8.1. In the solution, molecules of solute are distributed in the solvent. The solution is in equilibrium with the gas phase, and the gas-phase composition is determined by a dynamic balance between evaporation from the solution and condensation from the gas phase, as indicated for one solvent and one solute molecule in Figure 8.1.

EXAMPLE PROBLEM 8.1

Assume that the rates of evaporation, R_{evap}, and condensation, R_{cond}, of the solvent from the surface of pure liquid solvent are given by the expressions

$$R_{evap} = Ak_{evap}$$

$$R_{cond} = Ak_{cond}P_{solvent}^*$$

where A is the surface area of the liquid and k_{evap} and k_{cond} are the rate constants for evaporation and condensation, respectively. Derive a relationship between the vapor pressure of the solvent above the solution and above the pure solvent.

Solution

For the pure solvent, the equilibrium vapor pressure is found by setting the rates of evaporation and condensation equal:

$$R_{evap} = R_{cond}$$

$$Ak_{evap} = Ak_{cond}P_{solvent}^*$$

$$P_{solvent}^* = \frac{k_{evap}}{k_{cond}}$$

Next, consider the ideal solution. In this case, the rate of evaporation is reduced by the factor $x_{solvent}$.

$$R_{evap} = Ak_{evap}x_{solvent}$$

$$R_{cond} = Ak_{cond}P_{solvent}$$

and at equilibrium

$$R_{evap} = R_{cond}$$

$$Ak_{evap}x_{solvent} = Ak_{cond}P_{solvent}$$

$$P_{solvent} = \frac{k_{evap}}{k_{cond}}x_{solvent} = P_{solvent}^* x_{solvent}$$

The derived relationship is Raoult's law.

8.2 The Chemical Potential of a Component in the Gas and Solution Phases

If the liquid and vapor phases are in equilibrium, the following equation holds for each component of the solution, where μ_i is the chemical potential of species i:

$$\mu_i^{solution} = \mu_i^{vapor} \tag{8.2}$$

Recall from Section 6.3 that the chemical potential of a substance in the gas phase is related to its partial pressure, P_i, by

$$\mu_i^{vapor} = \mu_i^{\circ} + RT \ln \frac{P_i}{P^{\circ}} \tag{8.3}$$

where μ_i° is the chemical potential of pure component i in the gas phase at the standard state pressure $P^{\circ} = 1$ bar. Because at equilibrium $\mu_i^{solution} = \mu_i^{vapor}$, Equation (8.3) can be written in the form

$$\mu_i^{solution} = \mu_i^{\circ} + RT \ln \frac{P_i}{P^{\circ}} \tag{8.4}$$

For pure liquid i in equilibrium with its vapor, $\mu_i^*(\text{liquid}) = \mu_i^*(\text{vapor}) = \mu_i^*$. Therefore, the chemical potential of the pure liquid is given by

$$\mu_i^* = \mu_i^{\circ} + RT \ln \frac{P_i^*}{P^{\circ}} \tag{8.5}$$

Subtracting Equation (8.5) from (8.4) gives

$$\mu_i^{solution} = \mu_i^* + RT \ln \frac{P_i}{P_i^*} \tag{8.6}$$

For an ideal solution, $P_i = x_i P_i^*$. Combining Equations (8.6) and (8.1), the central equation describing ideal solutions is obtained:

$$\mu_i^{solution} = \mu_i^* + RT \ln x_i \tag{8.7}$$

This equation relates the chemical potential of a component in an ideal solution to the chemical potential of the pure liquid form of component i and the mole fraction of that component in the solution. This equation is most useful in describing the thermodynamics of solutions in which all components are volatile and miscible in all proportions.

Keeping in mind that $\mu_i = G_{i,m}$, the form of Equation (8.7) is identical to that derived for the Gibbs energy of a mixture of gases in Section 6.5. Therefore, one can derive relations for the thermodynamics of mixing to form ideal solutions that are identical to those developed in Section 6.6. We repeat them here as Equation (8.8). Note in particular that ΔH_{mixing} and ΔV_{mixing} are zero for an ideal solution:

$$\Delta G_{mixing} = nRT \sum_i x_i \ln x_i$$

$$\Delta S_{mixing} = -\left(\frac{\partial \Delta G_{mixing}}{\partial T}\right)_P = -nR \sum_i x_i \ln x_i$$

$$\Delta V_{mixing} = \left(\frac{\partial \Delta G_{mixing}}{\partial P}\right)_{T,n_1,n_2} = 0 \text{ and} \tag{8.8}$$

$$\Delta H_{mixing} = \Delta G_{mixing} + T\Delta S_{mixing} = nRT \sum_i x_i \ln x_i - T\left(nR \sum_i x_i \ln x_i\right) = 0$$

EXAMPLE PROBLEM 8.2

An ideal solution is made from 5.00 mol of benzene and 3.25 mol of toluene. Calculate ΔG_{mixing} and ΔS_{mixing} at 298 K and 1 bar pressure. Is mixing a spontaneous process?

Solution

The mole fractions of the components in the solutions are $x_{benzene}$ = 0.606 and $x_{toluene}$ = 0.394.

$$\Delta G_{mixing} = nRT \sum_i x_i \ln x_i$$

$$= 8.25 \text{ moles} \times 8.314 \text{ J mol}^{-1}\text{K}^{-1} \times 298 \text{ K}$$

$$\times (0.606 \ln 0.606 + 0.394 \ln 0.394)$$

$$= -13.7 \times 10^3 \text{ J}$$

$$\Delta S_{mixing} = -nR \sum_i x_i \ln x_i$$

$$= -8.25 \text{ moles} \times 8.314 \text{ J mol}^{-1}\text{K}^{-1} \times (0.606 \ln 0.606 + 0.394 \ln 0.394)$$

$$= 46.0 \text{ J K}^{-1}$$

Mixing is spontaneous because $\Delta G_{mixing} < 0$. If two liquids are miscible, it is always true that $\Delta G_{mixing} < 0$.

8.3 Applying the Ideal Solution Model to Binary Solutions

Although the ideal solution model can be applied to any number of components, the focus in this chapter is on simplifying the mathematics, so binary solutions, which consist of only two components, will be used. Because Raoult's law holds for both components of the mixture, $P_1 = x_1 P_1^*$ and $P_2 = x_2 P_2^* = (1 - x_1)P_2^*$. The total pressure in the gas phase varies linearly with the mole fraction of each of its components in the liquid:

$$P_{total} = P_1 + P_2 = x_1 P_1^* + (1 - x_1)P_2^* = P_2^* + (P_1^* - P_2^*)x_1 \qquad (8.9)$$

The individual partial pressures as well as P_{total} above a benzene–ethene chloride solution are shown in Figure 8.2. Small deviations from Raoult's law are seen. Such deviations are

FIGURE 8.2
The vapor pressure of benzene (yellow), ethene chloride (red), and the total vapor pressure (blue) above the solution are shown as a function of $x_{ethene\ chloride}$. The symbols are data points [J. v. Zawidski, *Zeitschrift für Physikalische Chemie*, 35 (1900) 129]. The solid lines are polynomial fits to the data. The dashed lines are calculated using Raoult's law.

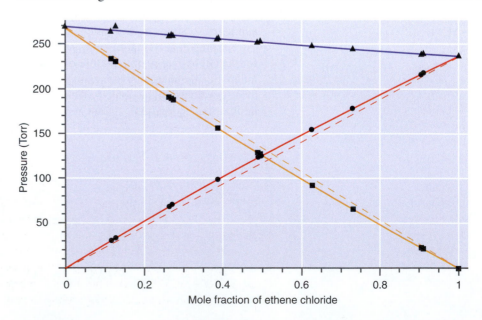

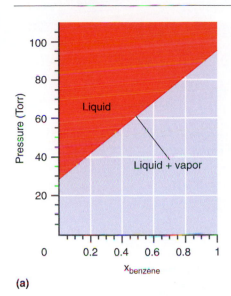

(a)

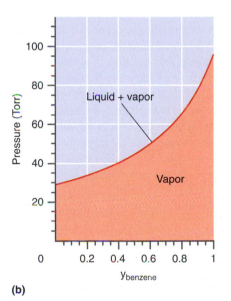

(b)

FIGURE 8.3
The total pressure above a benzene–toluene ideal solution is shown for different values of (a) the mole fraction of benzene in the solution and (b) as a function of the mole fraction of benzene in the vapor. Points on the curves correspond to vapor–liquid coexistence. Only the curves and the shaded areas are of physical significance as explained in text.

typical, because few solutions obey Raoult's law exactly. Nonideal solutions, which generally exhibit large deviations from Raoult's law, are discussed in Section 8.9.

The concentration unit used in Equation (8.9) is the mole fraction of each component in the liquid phase. The mole fraction of each component in the gas phase can also be calculated. Using the symbols y_1 and y_2 to denote the gas-phase mole fractions and the definition of the partial pressure, we can write

$$y_1 = \frac{P_1}{P_{total}} = \frac{x_1 P_1^*}{P_2^* + (P_1^* - P_2^*)x_1} \qquad (8.10)$$

To obtain the pressure in the vapor phase as a function of y_1, we first solve Equation (8.10) for x_1:

$$x_1 = \frac{y_1 P_2^*}{P_1^* + (P_2^* - P_1^*)y_1} \qquad (8.11)$$

and obtain P_{total} from $P_{total} = P_2^* + (P_1^* - P_2^*)x_1$:

$$P_{total} = \frac{P_1^* P_2^*}{P_1^* + (P_2^* - P_1^*)y_1} \qquad (8.12)$$

Equation (8.12) can be rearranged to give an equation for y_1 in terms of the vapor pressures of the pure components and the total pressure:

$$y_1 = \frac{P_1^* P_{total} - P_1^* P_2^*}{P_{total}(P_1^* - P_2^*)} \qquad (8.13)$$

The variation of the total pressure with x_1 and y_1 is not the same, as is seen in Figure 8.3. In Figure 8.3a, the system consists of a single-phase liquid for pressures above the curve and of a two-phase vapor–liquid mixture for $P–x_1$ points lying on the curve. Only points lying above the curve are meaningful, because points lying below the curve do not correspond to equilibrium states at which the liquid is present. If the system were placed in such an unstable state, liquid would evaporate to bring the pressure up to the value on the liquid–vapor coexistence curve. If there is not enough liquid to bring the pressure up to the curve, all the liquid will evaporate and x_1 is no longer a meaningful variable. In Figure 8.3b, the system consists of a single-phase vapor for pressures below the curve and of a two-phase vapor–liquid mixture for $P–y_1$ points lying on the curve. For the reason discussed earlier, points lying above the curve do not correspond to equilibrium states. The excess vapor would condense to form liquid.

Note that the pressure is plotted as a function of different variables in the two parts of Figure 8.3. To compare the gas phase and liquid composition at a given total pressure, both are graphed as a function of $Z_{benzene}$, which is called the average composition of benzene in the whole system in Figure 8.4. The **average composition** Z is defined by

$$Z_{benzene} = \frac{n_{benzene}^{liquid} + n_{benzene}^{vapor}}{n_{toluene}^{liquid} + n_{toluene}^{vapor} + n_{benzene}^{liquid} + n_{benzene}^{vapor}} = \frac{n_{benzene}}{n_{total}}$$

In the region labeled "Liquid" in Figure 8.4, the system consists entirely of a liquid phase, and $Z_{benzene} = x_{benzene}$. In the region labeled "Vapor," the system consists entirely of a gaseous phase and $Z_{benzene} = y_{benzene}$. The area separating the single-phase liquid and vapor regions corresponds to the two-phase liquid–vapor coexistence region.

To demonstrate the usefulness of this **pressure–average composition (P–Z) diagram,** consider a constant temperature process in which the pressure of the system is decreased from the value corresponding to point a in Figure 8.4. Because $P > P_{total}$, the

FIGURE 8.4

A P–Z phase diagram is shown for a benzene–toluene ideal solution. The upper curve shows the vapor pressure as a function of $x_{benzene}$. The lower curve shows the vapor pressure as a function of $y_{benzene}$. Above the two curves, the system is totally in the liquid phase, and below the two curves, the system is totally in the vapor phase. The area intermediate between the two curves shows the liquid–vapor coexistence region. The horizontal lines connecting the curves are called tie lines.

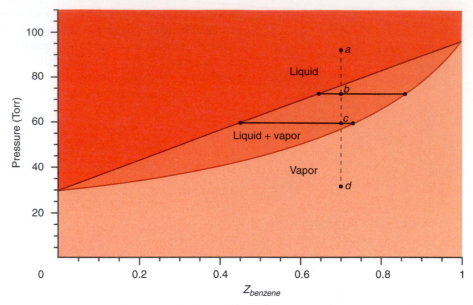

system is initially entirely in the liquid phase. As the pressure is decreased at constant composition, the system remains entirely in the liquid phase until the constant composition line intersects the P versus $x_{benzene}$ curve. At this point, the system enters the two-phase vapor–liquid coexistence region. As point b is reached, what are the values for $x_{benzene}$ and $y_{benzene}$, the mole fractions of the vapor and liquid phases, respectively? These values can be determined by constructing a tie line in the two-phase coexistence region.

A **tie line** (see Figure 8.4) is a horizontal line at the pressure of interest that connects the P versus $x_{benzene}$ and P versus $y_{benzene}$ curves. Note that for all values of the pressure, $y_{benzene}$ is greater than $x_{benzene}$, showing that the vapor phase is always enriched in the more volatile or higher vapor pressure component in comparison with the liquid phase.

EXAMPLE PROBLEM 8.3

An ideal solution is made from 5.00 mol of benzene and 3.25 mol of toluene. At 298 K, the vapor pressure of the pure substances are $P^*_{benzene} = 96.4$ Torr and $P^*_{toluene} = 28.9$ Torr.

a. The pressure above this solution is reduced from 760 Torr. At what pressure does the vapor phase first appear?

b. What is the composition of the vapor under these conditions?

Solution

a. The mole fractions of the components in the solution are $x_{benzene} = 0.606$ and $x_{toluene} = 0.394$. The vapor pressure above this solution is

$$P_{total} = x_{benzene}P^*_{benzene} + x_{toluene}P^*_{toluene} = 0.606 \times 96.4 \text{ Torr} + 0.394 \times 28.9 \text{ Torr}$$

$$= 69.8 \text{ Torr}$$

No vapor will be formed until the pressure has been reduced to this value.

b. The composition of the vapor at a total pressure of 69.8 Torr is given by

$$y_{benzene} = \frac{P^*_{benzene}P_{total} - P^*_{benzene}P^*_{toluene}}{P_{total}(P^*_{benzene} - P^*_{toluene})}$$

$$= \frac{96.4 \text{ Torr} \times 69.8 \text{ Torr} - 96.4 \text{ Torr} \times 28.9 \text{ Torr}}{69.8 \text{ Torr} \times (96.4 \text{ Torr} - 28.9 \text{ Torr})} = 0.837$$

$$y_{toluene} = 1 - y_{benzene} = 0.163$$

Note that the vapor is enriched relative to the liquid in the more volatile component, which has the lower boiling temperature.

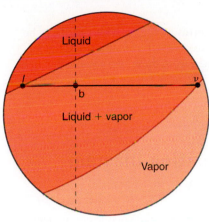

FIGURE 8.5

An enlarged region of the two-phase coexistence region of Figure 8.4 is shown. The vertical line through point b indicates a transition at constant system composition. The lever rule (see text) is used to determine what fraction of the system is in the liquid and vapor phases.

To calculate the relative amount of material in each of the two phases in a coexistence region, we derive the lever rule for a binary solution of components A and B. Figure 8.5 shows a magnified portion of Figure 8.4 centered at the tie line that passes through point b. We derive the lever rule using the following geometrical argument. The lengths of the line segments lb and bv are given by

$$lb = Z_B - x_B = \frac{n_B^{tot}}{n^{tot}} - \frac{n_B^{liq}}{n_{liq}^{tot}} \tag{8.14}$$

$$bv = y_B - Z_B = \frac{n_B^{vapor}}{n_{vapor}^{tot}} - \frac{n_B^{tot}}{n^{tot}} \tag{8.15}$$

The subscripts and superscripts on n indicate the number of moles of component B in the vapor and liquid phases and the total number of moles of B in the system. If Equation (8.14) is multiplied by n_{liq}^{tot}, Equation (8.15) is multiplied by n_{vapor}^{tot}, and the two equations are subtracted, we find that

$$lb\, n_{liq}^{tot} - bv\, n_{vapor}^{tot} = \frac{n_B^{tot}}{n^{tot}}(n_{liq}^{tot} + n_{vapor}^{tot}) - (n_B^{liq} + n_B^{vapor}) = n_B^{tot} - n_B^{tot} = 0$$

we conclude that $\qquad \dfrac{n_{liq}^{tot}}{n_{vap}^{tot}} = \dfrac{bv}{lb} \tag{8.16}$

It is convenient to restate this result as

$$n_{liq}^{tot}(Z_B - x_B) = n_{vapor}^{tot}(y_B - Z_B) \tag{8.17}$$

Equation (8.17) is called the **lever rule** in analogy with the torques acting on a lever of length $lb + bv$ with the fulcrum positioned at point b. For the specific case shown in Figure 8.5, $n_{liq}^{tot}/n_{vapor}^{tot} = 2.34$. Therefore, 70% of the total number of moles in the system is in the liquid phase, and 30% is in the vapor phase.

EXAMPLE PROBLEM 8.4

For the benzene–toluene solution of Example Problem 8.3, calculate

a. the total pressure

b. the liquid composition

c. the vapor composition

when 1.50 mol of the solution has been converted to vapor.

Solution

The lever rule relates the average composition, $Z_{benzene} = 0.606$, and the liquid and vapor compositions:

$$n_{vapor}(y_{benzene} - Z_{benzene}) = n_{liq}(Z_{benzene} - x_{benzene})$$

Entering the parameters of the problem, this equation simplifies to

$$6.75 x_{benzene} + 1.50 y_{benzene} = 5.00$$

The total pressure is given by

$$P_{total} = x_{benzene}P_{benzene}^* + (1 - x_{benzene})P_{toluene}^* = [96.4 x_{benzene} + 28.9(1 - x_{benzene})]\ \text{Torr}$$

and the vapor composition is given by

$$y_{benzene} = \frac{P_{benzene}^* P_{total} - P_{benzene}^* P_{toluene}^*}{P_{total}(P_{benzene}^* - P_{toluene}^*)} = \left[\frac{96.4\dfrac{P_{total}}{\text{Torr}} - 2786}{67.5\dfrac{P_{total}}{\text{Torr}}}\right]$$

These three equations in three unknowns can be solved by using an equation solver or by eliminating the variables by combining equations. For example, the

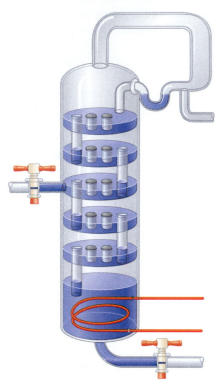

first equation can be used to express $y_{benzene}$ in terms of $x_{benzene}$. This result can be substituted into the second and third equations to give two equations in terms of $x_{benzene}$ and p_{total}. The solution for $x_{benzene}$ obtained from these two equations can be substituted in the first equation to give $y_{benzene}$. The answers are $x_{benzene} = 0.561$, $y_{benzene} = 0.810$, and $P_{total} = 66.8$ Torr.

8.4 The Temperature–Composition Diagram and Fractional Distillation

The enrichment of the vapor phase above a solution in the more volatile component is the basis for fractional distillation, an important separation technique in chemistry and in the chemical industry. It is more convenient to discuss fractional distillation using a **temperature–composition diagram** than using the pressure–average composition diagram discussed in the previous section. The temperature–composition diagram gives the temperature of the solution as a function of the average system composition for a predetermined total vapor pressure, P_{total}. It is convenient to use the value $P_{total} = 1$ atm, so that the vertical axis is the normal boiling point of the solution. Figure 8.6 shows a boiling temperature–composition diagram for a benzene–toluene solution. Neither the $T_b - x_{benzene}$ nor the $T_b - y_{benzene}$ curves are linear in a temperature–composition diagram. Note that because the more volatile component has the lower boiling point, the vapor and liquid regions are inverted when compared with the pressure–average composition diagram.

The principle of **fractional distillation** can be illustrated using the sequence of lines labeled a through k in Figure 8.6. The vapor above the solution at point a is enriched in benzene. If this vapor is separated from the original solution and condensed by lowering the temperature, the resulting liquid will have a higher mole fraction of benzene than the original solution. As for the original solution, the vapor above this separately collected liquid is enriched in benzene. As this process is repeated, the successively collected vapor samples become more enriched in benzene. In the limit of a very large number of steps, the last condensed samples are essentially pure benzene. The multistep procedure described earlier is very cumbersome because it requires the collection and evaporation of many different samples. In practice, the separation into pure benzene and toluene is accomplished using a distillation column, shown schematically in Figure 8.7.

FIGURE 8.7

Schematic of a fractional distillation column. The solution to be separated into its components is introduced at the bottom of the column. The resistive heater provides the energy needed to vaporize the liquid. It can be assumed that the liquid and vapor are at equilibrium at each level of the column. The equilibrium temperature decreases from the bottom to the top of the column.

FIGURE 8.6

The boiling temperature of an ideal solution of the components benzene and toluene is plotted versus the average system composition, $Z_{benzene}$. The upper red curve shows the boiling temperature as a function of $y_{benzene}$, and the lower red curve shows the boiling temperature as a function of $x_{benzene}$. The area intermediate between the two curves shows the vapor–liquid coexistence region.

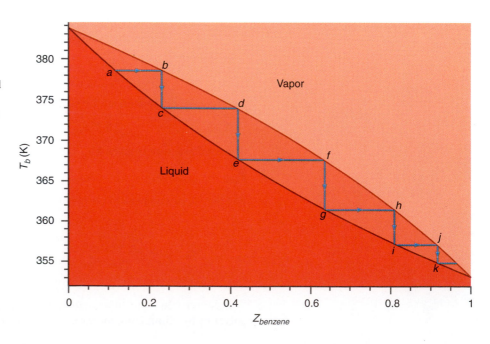

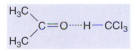

FIGURE 8.8
Hydrogen bond formation between acetone and chloroform leads to the formation of a maximum boiling azeotrope.

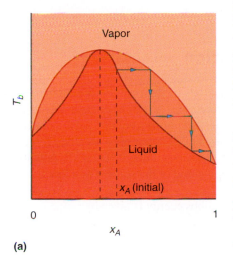

(a)

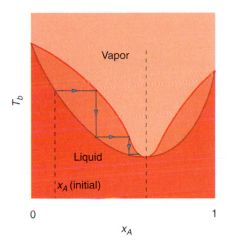

FIGURE 8.9
A boiling point diagram is shown for (a) maximum and (b) minimum boiling point azeotropes. The dashed lines indicate the initial composition of the solution and the composition of the azeotrope. The sequence of horizontal segments with arrows corresponds to successively higher (cooler) portions of the column.

Rather than have the whole system at a uniform temperature, a distillation column operates with a temperature gradient so that the top part of the column is at a lower temperature than the solution being boiled off. The horizontal segments *ab*, *cd*, *ef*, and so on in Figure 8.6 correspond to successively higher (cooler) portions of the column. Each of these segments corresponds to one of the distillation stages in Figure 8.7. Because the vapor is moving upward through the condensing liquid, heat and mass exchange facilitates equilibration between the two phases.

Distillation plays a major role in the separation of crude oil into its various components. The lowest boiling liquid is gasoline, followed by kerosene, diesel fuel and heating oil, and gas oil, which is further broken down into lower molecular weight components by catalytic cracking. It is important in distillation to keep the temperature of the boiling liquid as low as possible to avoid the occurrence of thermally induced reactions. A useful way to reduce the boiling temperature is to distill the liquid under a partial vacuum by pumping on the gas phase. The boiling point of a typical liquid mixture can be reduced by approximately 100°C if the pressure is reduced from 760 to 20 Torr.

Although the principle of fractional distillation is the same for real solutions, it is not possible to separate a binary solution into its pure components if the nonideality is strong enough. If the A–B interactions are more attractive than the A–A and B–B interactions, the boiling point of the solution will go through a maximum at a concentration intermediate between $x_A = 0$ and $x_A = 1$. An example of such a case is an acetone–chloroform mixture. Chloroform forms a hydrogen bond with the oxygen in acetone as shown in Figure 8.8, leading to stronger A–B than A–A and B–B interactions.

A boiling point diagram for this case is shown in Figure 8.9a. At the maximum boiling temperature, the liquid and vapor composition lines are tangent to one another. Fractional distillation of such a solution beginning with an initial value of x_A greater than that corresponding to the maximum boiling point is shown schematically in the figure. The component with the lowest boiling point will initially emerge at the top of the distillation column. In this case, it will be pure component A. However, the liquid left in the heated flask will not be pure component B, but rather the solution corresponding to the concentration at which the maximum boiling point is reached. Continued boiling of the solution at this composition will lead to evaporation at constant composition. Such a mixture is called an azeotrope, and because $T_{b,azeotrope} > T_{b,A}$, $T_{b,B}$, it is called a maximum boiling azeotrope. An example for a maximum boiling azeotrope is a mixture of H_2O ($T_b = 100°C$) and HCOOH ($T_b = 100°C$) at $x_{H_2O} = 0.427$, which boils at 107.2°C. Other commonly occurring azeotropic mixtures are listed in Table 8.1.

If the A–B interactions are less attractive than the A–A and B–B interactions, a minimum boiling azeotrope can be formed. A schematic boiling point diagram for such an azeotrope is also shown in Figure 8.9b. Fractional distillation of such a solution beginning with an initial value of x_A less than that corresponding to the minimum boiling point

TABLE 8.1 **Composition and Boiling Temperatures of Selected Azeotropes**

Azeotropic Mixture	Boiling Temperature of Components (°C)	Mole Fraction of First Component	Azeotrope Boiling Point (°C)
Water–ethanol	100/78.5	0.096	78.2
Water–trichloromethane	100/61.2	0.160	56.1
Water–benzene	100/80.2	0.295	69.3
Water–toluene	100/111	0.444	84.1
Ethanol–hexane	78.5/68.8	0.332	58.7
Ethanol–benzene	78.5/80.2	0.440	67.9
Ethyl acetate–hexane	78.5/68.8	0.394	65.2
Carbon disulfide–acetone	46.1/56.2	0.608	39.3
Toluene–acetic acid	111/118	0.625	100.7

Source: D. R. Lide, Ed., *Handbook of Chemistry and Physics*, 83rd ed. CRC Press, Boca Raton, FL, 2002.

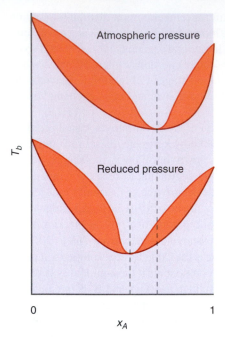

FIGURE 8.10
Because the azeotropic composition depends on the total pressure, pure B can be recovered from the A–B mixture by first distilling the mixture under atmospheric pressure and, subsequently, under a reduced pressure. Note that the boiling temperature is lowered as the pressure is reduced.

leads to a liquid with the azeotropic composition initially emerging at the top of the distillation column. An example for a minimum boiling azeotrope is a mixture of CS_2 ($T_b = 46.1°C$) and acetone ($T_b = 56.2°C$) at $x_{CS_2} = 0.608$, which boils at $39.3°C$.

It is still possible to collect one component of an azeotropic mixture using the property that the azeotropic composition is a function of the total pressure as shown in Figure 8.10. The mixture is first distilled at atmospheric pressure and the volatile distillate is collected. This vapor is condensed and subsequently distilled at a reduced pressure for which the azeotrope contains less A (more B). If this mixture is distilled, the azeotrope evaporates, leaving some pure B behind.

8.5 The Gibbs–Duhem Equation

In this section, we show that the chemical potentials of the two components in a binary solution are not independent. This is an important result, because it allows the chemical potential of a nonvolatile solute such as sucrose in a volatile solvent such as water to be determined. Such a solute has no measurable vapor pressure; therefore, its chemical potential cannot be measured directly. As shown later, its chemical potential can be determined knowing only the chemical potential of the solvent as a function of concentration.

From Chapter 6, the differential form of the Gibbs energy is given by

$$dG = -S\,dT + V\,dP + \sum_i \mu_i\,dn_i \tag{8.18}$$

For a binary solution at constant T and P, this equation reduces to

$$dG = \mu_1\,dn_1 + \mu_2\,dn_2 \tag{8.19}$$

Imagine starting with an infinitesimally small amount of a solution at constant T and P. The amount is gradually increased at constant composition. Because the composition remains constant, the chemical potentials are unchanged as the size of the system is changed. Therefore, the μ_i can be taken out of the integral:

$$\int_0^G dG' = \mu_1 \int_0^{n_1} dn_1' + \mu_2 \int_0^{n_2} dn_2' \text{ or} \tag{8.20}$$
$$G = \mu_1 n_1 + \mu_2 n_2$$

The primes have been introduced to avoid using the same symbol for the integration variable and the upper limit. The total differential of the last equation is

$$dG = \mu_1\,dn_1 + \mu_2\,dn_2 + n_1\,d\mu_1 + n_2\,d\mu_2 \tag{8.21}$$

The previous equation differs from Equation (8.19) because, in general, we have to take changes of the composition of the solution into account. Therefore, μ_1 and μ_2 must be regarded as variables. Setting Equations (8.19) and (8.21) equal, one obtains the **Gibbs–Duhem equation** for a binary solution, which can be written in either of two forms:

$$n_1\,d\mu_1 + n_2\,d\mu_2 = 0 \quad \text{or} \quad x_1\,d\mu_1 + x_2\,d\mu_2 = 0 \tag{8.22}$$

This equation states that the chemical potentials of the components in a binary solution are not independent. If the change in the chemical potential of the first component is $d\mu_1$, the change of the chemical potential of the second component is given by

$$d\mu_2 = -\frac{n_1\,d\mu_1}{n_2} \tag{8.23}$$

The use of the Gibbs–Duhem equation is illustrated in Example Problem 8.5.

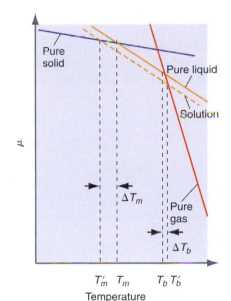

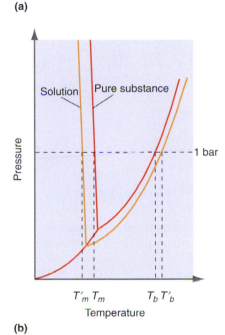

(a)

(b)

FIGURE 8.11
Two different illustrations of the boiling point elevation and freezing point depression: (a) These effects arise because the chemical potential of the solution is lowered through addition of the nonvolatile solute, whereas the chemical potential of the vapor and liquid is unaffected at constant pressure. (b) The same information from part (a) is shown using a P–T phase diagram.

EXAMPLE PROBLEM 8.5

One component in a solution follows Raoult's law, $\mu_1^{solution} = \mu_1^* + RT \ln x_1$ *over the entire range* $0 \le x_1 \le 1$. Using the Gibbs–Duhem equation, show that the second component must also follow Raoult's law.

Solution

From Equation (8.20),

$$d\mu_2 = -\frac{x_1 \, d\mu_1}{x_2} = -\frac{x_1}{x_2}d(\mu_1^* + RT \ln x_1) = -RT\frac{x_1}{x_2}\frac{dx_1}{x_1}$$

Because $x_1 + x_2 = 1$, then $dx_2 = -dx_1$ and $d\mu_2 = RT \, dx_2/x_2$. Integrating this equation, one obtains $\mu_2 = RT \ln x_2 + C$, where C is a constant of integration. This constant can be evaluated by examining the limit $x_2 \to 1$. This limit corresponds to the pure substance 2 for which $\mu_2 = \mu_2^* = RT \ln 1 + C$. We conclude that $C = \mu_2^*$ and, therefore, $\mu_2^{solution} = \mu_2^* + RT \ln x_2$.

8.6 Colligative Properties

Many solutions consist of nonvolatile solutes that have limited solubility in a volatile solvent. Examples are solutions of sucrose or sodium chloride in water. Important properties of these solutions, including boiling point elevation, freezing point depression, and osmotic pressure are found to depend only on the solute concentration—not on the nature of the solute. These properties are called **colligative properties.** In this section, colligative properties are discussed using the model of the ideal solution. Corrections for nonideality require that activities be used in place of concentrations, as shown in Section 8.13.

As discussed in connection with Figure 8.1, the vapor pressure above a solution containing a solute is lowered with respect to the pure solvent. As will be shown in this and in the next three sections, the solution exhibits a freezing point depression, a boiling point elevation, and an osmotic pressure. Generally, the solute does not crystallize out with the solvent during freezing because the solute cannot be easily integrated into the solvent crystal structure. For this case, the change in chemical potential on dissolution of a nonvolatile solute can be understood using Figure 8.11a.

Only the liquid chemical potential is affected through formation of the solution. Although the gas pressure is lowered by the addition of the solute, the chemical potential of the gas is unaffected because it remains pure and because the comparison in Figure 8.11a is made at constant pressure. The chemical potential of the solid is unaffected because of the assumption that the solute does not crystallize out with the solvent. As shown in the figure, the melting temperature T_m, defined as the intersection of the solid and liquid μ versus T curves, is lowered by dissolution of a nonvolatile solute in the solvent. Similarly, the boiling temperature T_b is raised by dissolution of a nonvolatile solute in the solvent.

The same information is shown in a P–T phase diagram in Figure 8.11b. Because the vapor pressure above the solution is lowered by the addition of the solute, the liquid–gas coexistence curve intersects the solid–gas coexistence curve at a lower temperature than for the pure solvent. This intersection defines the triple point, and it must also be the origin of the solid–liquid coexistence curve. Therefore, the solid–liquid coexistence curve is shifted to lower temperatures through the dissolution of the nonvolatile solute. The overall effects of the shifts in the solid–gas and solid–liquid coexistence curves are a **freezing point depression** and a boiling point elevation, both of which depend only on the concentration, and not on the identity, of the solute. The preceding discussion is qualitative in nature. In the next two sections, quantitative relationships are developed between the colligative properties and the concentration of the nonvolatile solute.

8.7 The Freezing Point Depression and Boiling Point Elevation

If the solution is in equilibrium with the pure solid solvent, the following relation must be satisfied:

$$\mu_{solution} = \mu^*_{solid} \tag{8.24}$$

Recall that $\mu_{solution}$ refers to the chemical potential of the solvent in the solution, and μ^*_{solid} refers to the chemical potential of the pure solvent in solid form. From Equation (8.7), we can express $\mu_{solution}$ in terms of the chemical potential of the pure solvent and its concentration and rewrite Equation (8.24) in the form

$$\mu^*_{solvent} + RT \ln x_{solvent} = \mu^*_{solid} \tag{8.25}$$

This equation can be solved for $\ln x_{solvent}$:

$$\ln x_{solvent} = \frac{\mu^*_{solid} - \mu^*_{solvent}}{RT} \tag{8.26}$$

The difference in chemical potentials $\mu^*_{solid} - \mu^*_{solvent} = -\Delta G_{fusion,m}$, so that

$$\ln x_{solvent} = \frac{-\Delta G_{fusion,m}}{RT} \tag{8.27}$$

Because we are interested in how the freezing temperature is related to $x_{solvent}$ at constant pressure, the partial derivative $(\partial T/\partial x_{solvent})_P$ is needed. This quantity can be obtained by differentiating Equation (8.27) with respect to $x_{solvent}$:

$$\left(\frac{\partial \ln x_{solvent}}{\partial x_{solvent}} \right)_P = \frac{1}{x_{solvent}} = -\frac{1}{R} \left(\frac{\partial \dfrac{\Delta G_{fusion,m}}{T}}{\partial T} \right)_P \left(\frac{\partial T}{\partial x_{solvent}} \right)_P \tag{8.28}$$

The first partial derivative on the right side of Equation (8.28) can be simplified using the Gibbs–Helmholtz equation (see Section 6.3), giving

$$\frac{1}{x_{solvent}} = \frac{\Delta H_{fusion,m}}{RT^2} \left(\frac{\partial T}{\partial x_{solvent}} \right)_P \text{ or} \tag{8.29}$$

$$\frac{dx_{solvent}}{x_{solvent}} = d \ln x_{solvent} = \frac{\Delta H_{fusion,m}}{R} \frac{dT}{T^2} \text{(constant } P)$$

This equation can be integrated between the limits given by the pure solvent ($x_{solvent} = 1$) for which the fusion temperature is T_{fusion}, and an arbitrarily small concentration of solute for which the fusion temperature is T:

$$\int_1^{x_{solvent}} \frac{dx}{x} = \int_{T_{fusion}}^T \frac{\Delta H_{fusion,m}}{R} \frac{dT'}{T'^2} \tag{8.30}$$

For $x_{solvent}$ not very different from 1, $\Delta H_{fusion.m}$ is assumed to be independent of T, and Equation (8.30) simplifies to

$$\frac{1}{T} = \frac{1}{T_{fusion}} - \frac{R \ln x_{solvent}}{\Delta H_{fusion,m}} \tag{8.31}$$

It is more convenient in discussing solutions to use the molality of the solute rather than the mole fraction of the solvent as the concentration unit. For dilute solutions, $\ln x_{solvent} = -\ln (n_{solvent}/n_{solvent} + m_{solute} M_{solvent} n_{solvent}) = -\ln (1 + M_{solvent} n_{solute})$, and Equation (8.31) can be rewritten in terms of the molality (m) rather than the mole fraction. The result is

$$\Delta T_f = -\frac{R M_{solvent} T_{fusion}^2}{\Delta H_{fusion,m}} m_{solute} = -K_f m_{solute} \tag{8.32}$$

where $M_{solvent}$ is the molecular mass of the solvent. In going from Equation (8.31) to Equation (8.32) we have made the approximation $\ln(1 + M_{solvent} m_{solute}) \approx M_{solvent} m_{solute}$ and $1/T - 1/T_{fusion} \approx -\Delta T_f / T_{fusion}^2$. Note that K_f depends only on the properties of the solvent and is primarily determined by the molecular mass and the enthalpy of fusion. For most solvents, the magnitude of K_f lies between 1.5 and 10 as shown in Table 8.2. However, K_f can reach unusually high values, for example, 30 K kg mol^{-1} for carbon tetrachloride and 40 K kg mol^{-1} for camphor.

The **boiling point elevation** can be calculated using the same argument with $\Delta G_{vaporization}$ and $\Delta H_{vaporization}$ substituted for ΔG_{fusion} and ΔH_{fusion}. The resulting equations are

$$\left(\frac{\partial T}{\partial m_{solute}} \right)_{p,m \to 0} = \frac{R M_{solvent} T_{vaporization}^2}{\Delta H_{vaporization,m}} \tag{8.33}$$

and

$$\Delta T_b = \frac{R M_{solvent} T^2}{\Delta H_{vaporization,m}} m_{solute} = K_b m_{solute} \tag{8.34}$$

Because $\Delta H_{vaporization.m} > \Delta H_{fusion.m}$, it follows that $K_f > K_b$. Typically, K_b values range between 0.5 and 5.0 K kg mole^{-1} as shown in Table 8.2. Note also that, by convention, both K_f and K_b are positive; hence, the negative sign in Equation (8.32) is replaced by a positive sign in Equation (8.34).

EXAMPLE PROBLEM 8.6

In this example, 4.50 g of a substance dissolved in 125 g of CCl_4 leads to an elevation of the boiling point of 0.65 K. Calculate the freezing point depression, the molecular mass of the substance, and the factor by which the vapor pressure of CCl_4 is lowered.

Solution

$$\Delta T_m = \left(\frac{K_f}{K_b} \right) \Delta T_b = \frac{30.\,K \text{ kg mole}^{-1}}{4.95\,K \text{ kg mole}^{-1}} \times 0.65\,K = 3.9\,K$$

TABLE 8.2 **Freezing Point Depression and Boiling Point Elevation Constants**

Substance	Standard Freezing Point (K)	K_f (K kg mol^{-1})	Standard Boiling Point (K)	K_b (K kg mol^{-1})
Acetic acid	289.6	3.59	391.2	3.08
Benzene	278.6	5.12	353.3	2.53
Camphor	449	40	482.3	5.95
Carbon disulfide	161	3.8	319.2	2.40
Carbon tetrachloride	250.3	30	349.8	4.95
Cyclohexane	279.6	20.0	353.9	2.79
Ethanol	158.8	2.0	351.5	1.07
Phenol	314	7.27	455.0	3.04
Water	273.15	1.86	373.15	0.51

Source: D. R. Lide, Ed., *Handbook of Chemistry and Physics,* 83rd ed. CRC Press, Boca Raton, FL, 2002.

To avoid confusion, we use the symbol m for molality and m for mass. We solve for the molecular mass m_{solute} using Equation (8.34):

$$\Delta T_b = K_b m_{solute} = K_b \times \left(\frac{m_{solute}/M_{solute}}{m_{solvent}(kg)} \right)$$

$$M_{solute} = \frac{K_b m_{solute}}{m_{solvent} \Delta T_b}$$

$$M_{solute} = \frac{4.95 \text{ K kg mol}^{-1} \times 4.50 \text{ g}}{0.125 \text{ kg} \times 0.65 \text{ K}} = 274 \text{ g mol}^{-1}$$

We solve for the factor by which the vapor pressure of the solvent is reduced by using Raoult's law:

$$\frac{P_{solvent}}{P_{solvent}^*} = x_{solvent} = 1 - x_{solute} = 1 - \frac{n_{solute}}{n_{solute} + n_{solvent}}$$

$$= 1 - \frac{\dfrac{4.50 \text{ g}}{274 \text{ g mol}^{-1}}}{\left(\dfrac{4.50 \text{ g}}{274 \text{ g mol}^{-1}} \right) + \left(\dfrac{125 \text{ g}}{153.8 \text{ g mol}^{-1}} \right)} = 0.98$$

8.8 The Osmotic Pressure

Some membranes allow the passage of small molecules such as water through them, yet do not allow larger molecules such as sucrose to pass. Such a **semipermeable membrane** is an essential component in medical technologies such as kidney dialysis, which is described later. If a sac composed of such a membrane containing a solute that cannot pass through the membrane is immersed in a beaker containing the pure solvent, then initially the solvent diffuses into the sac. Diffusion ceases when equilibrium is attained, and at equilibrium, the pressure is higher in the sac than in the surrounding solvent. This result is shown schematically in Figure 8.12. The process in which the solvent diffuses through a membrane and dilutes a solution is known as **osmosis.** The amount by which the pressure in the solution is raised is known as the **osmotic pressure.**

To understand the origin of the osmotic pressure, denoted by π, the equilibrium condition is applied to the contents of the sac and the surrounding solvent:

$$\mu_{solvent}^{solution}(T, P + \pi, x_{solvent}) = \mu_{solvent}^*(T, P) \tag{8.35}$$

Using Raoult's law to express the concentration dependence of $\mu_{solution}$,

$$\mu_{solvent}^{solution}(T, P + \pi, x_{solvent}) = \mu_{solvent}^*(T, P + \pi) + RT \ln x_{solvent} \tag{8.36}$$

Because μ for the solvent is lower in the solution than in the pure solvent, only an increased pressure in the solution can raise its μ sufficiently to achieve equilibrium with the pure solvent. The dependence of μ on pressure and temperature is given by $d\mu = dG_m = V_m dP - S_m dT$. At constant T we can write

$$\mu_{solvent}^*(T, P + \pi, x_{solvent}) - \mu_{solvent}^*(T, P) = \int_P^{p+\pi} V_m^* dP' \tag{8.37}$$

where V_m^* is the molar volume of the pure solvent and P is the pressure in the solvent outside the sac. Because a liquid is nearly incompressible, it is reasonable to assume that V_m^* is independent of P to evaluate the integral in the previous equation. Therefore, $\mu_{solvent}^*(T, P + \pi, x_{solvent}) - \mu_{solvent}^*(T, P) = V_m^* \pi$, and Equation (8.36) reduces to

$$\pi V_m^* + RT \ln x_{solvent} = 0 \tag{8.38}$$

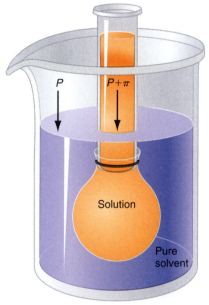

FIGURE 8.12

An osmotic pressure arises if a solution containing a solute that cannot pass through the membrane boundary is immersed in the pure solvent.

For a dilute solution, $n_{solvent} \gg n_{solute}$, and

$$\ln x_{solvent} = \ln(1 - x_{solute}) \approx -x_{solute} = -\frac{n_{solute}}{n_{solute} + n_{solvent}} \approx -\frac{n_{solute}}{n_{solvent}} \quad (8.39)$$

Equation (8.39) can be simplified further by recognizing that for a dilute solution, $V \approx n_{solvent}V_m^*$. With this substitution, Equation (8.38) becomes

$$\pi = \frac{n_{solute}RT}{V} \quad (8.40)$$

which is known as the **van't Hoff osmotic equation.** Note the similarity in form between this equation and the ideal gas law.

An important application of the selective diffusion of the components of a solution through a membrane is dialysis. In healthy individuals, the kidneys remove waste products from the bloodstream, whereas individuals with damaged kidneys use a dialysis machine for this purpose. Blood from the patient is shunted through tubes made of a selectively porous membrane surrounded by a flowing sterile solution made up of water, sugars, and other components. Blood cells and other vital components of blood are too large to fit through the pores in the membranes, but urea and salt flow out of the bloodstream through membranes into the sterile solution and are removed as waste.

EXAMPLE PROBLEM 8.7

Calculate the osmotic pressure generated at 298 K if a cell with a total solute concentration of 0.500 mol L^{-1} is immersed in pure water. The cell wall is permeable to water molecules, but not to the solute molecules.

Solution

$$\pi = \frac{n_{solute}RT}{V} = 0.500 \text{ mol L}^{-1} \times 8.206 \times 10^{-2} \text{ L atm K}^{-1} \text{ mol}^{-1} \times 298 \text{ K} = 12.2 \text{ atm}$$

As this calculation shows, the osmotic pressure generated for moderate solute concentrations can be quite high. Hospital patients have died after pure water has accidentally been injected into their blood vessels, because the osmotic pressure is sufficient to burst the walls of blood cells.

The stress on the cell generated by osmotic pressure is one of the important physical factors that determines the degree to which cells can proliferate in their environment. This is because the osmolarity, the total concentration of solutes in the environment of bacteria, for example, fluctuates constantly. Therefore, the cell must respond to stresses that are constantly changing. The process whereby organisms respond to changes in stresses generated by changes in external osmolarity is called osmotic regulation or osmoregulation.

To understand the stress exerted by osmotic forces on the cell, suppose a plant cell or a bacterial cell is placed in pure water. Because the cell membrane is permeable to the passage of water, water will move into the cell as explained earlier. However, the cell membrane cannot expand indefinitely as a result of the uptake of water because the cytoplasmic membranes of plant and bacterial cells are surrounded by rigid cell walls, which limit the expansion of the membrane. In reality, cells rarely exist in pure water but rather are in contact with a medium consisting of water containing varying level of solutes. In response to immersion into media with solute levels unequal to cytoplasmic solute levels, there builds up across the plant or bacterial cell wall a turgor pressure, defined as

$$P_{turgor} = \pi_{cell} - \pi_{medium} = (c_{cell} - c_{medium})RT \quad (8.41)$$

where $c_{cell} = n_{cell}/V_{cell}$ is the total molar concentration of solutes within the cell and $c_{medium} = n_{medium}/V_{medium}$ is the total molar concentration of solutes in the surrounding medium. Turgor pressures as high as 20 atm can be generated across bacterial cell walls as a result of fluctuations in external osmolarity. Because the bacterial cell wall is not perfectly rigid, it has been hypothesized that turgor pressure provides the mechanical force for expansion of the cell wall during cell growth.

The influx of water into the cell in response to high levels of solutes in the cytoplasm relative to the medium is called hypoosmotic shock, and the medium is said to be hypotonic. In reality, hypoosmotic shock results in only modest changes in bacterial or plant cell volume because the cell wall can withstand pressures of up to 100 atm. So in response to hypoosmotic shock, the plant or bacterial cell membrane moves against the cell wall, resulting in an increase in turgor pressure, but unlike animal cells that lack cell walls, bacterial and plant cells will not rupture as a result of hypoosmotic shock.

EXAMPLE PROBLEM 8.8

Calculate the turgor pressure generated across a bacterial cell wall at 298 K if the bacterial cytoplasm with a total solute concentration of 0.600 mol L^{-1} is immersed in water with a total solute concentration 0.100 mol L^{-1}. The cell wall is permeable to water molecules, but not to the solute molecules. Will this turgor pressure cause rupture of the bacterial cell wall?

Solution

$$P = (c_{cell} - c_{medium})RT = 0.500 \text{ mol L}^{-1} \times 8.206 \times 10^{-2} \text{ L atm K}^{-1} \text{ mol}^{-1}$$
$$\times 298 \text{ K} = 12.2 \text{ atm}$$

As this calculation shows, the turgor pressure generated for moderate solute concentrations can be quite high but not sufficient to burst a bacterial cell wall, which can withstand pressures of up to 100 atm.

A more serious challenge to cell viability is hyperosmotic shock, the efflux of water from the cell when solute levels within the cell are lower than solute levels in the external medium. Such a medium is said to be hypertonic. In this case the cell membrane will move away from the cell wall, a condition called plasmolysis. Mechanical stability in plants is influenced by hyper- and hypoosmotic conditions. Plant cells are in their healthiest state when experiencing hypoosmosis, where the cell membrane expands against the cell wall in response to a hypotonic medium, generating a high turgor pressure. The wilting of a plant during a drought is due to plasmolysis in response to hyperosmotic conditions. If isotonic conditions prevail, the levels of solutes within the cell are equivalent to levels in the external medium, net water transport into the cell ceases, and the cell becomes flaccid. See Figure 8.13. Unlike plant

FIGURE 8.13

(a) Animal cells fare best under isotonic conditions, where net transport of water into or out of the cell ceases. (b) Plant cells fare best under hypoosmotic conditions, where net transport of water into the cell generates a turgor pressure as a result of the presence of a rigid cell wall. (Source: Adpated from N.A. Campbell and J.B. Reece, *Biology,* 7th ed. (San Francisco, CA: Pearson Benjamin Cummings), p. 133X. Copyright 2005. Reprinted by permission of Pearson Education, Inc.)

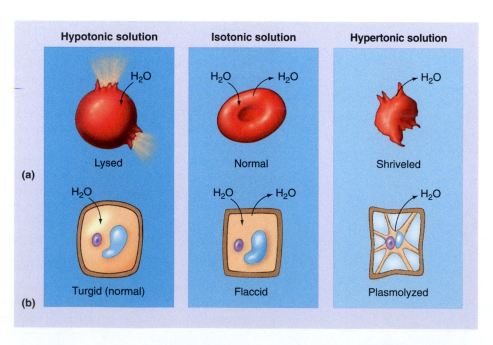

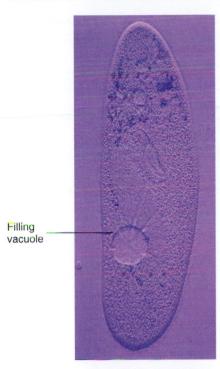

Filling vacuole

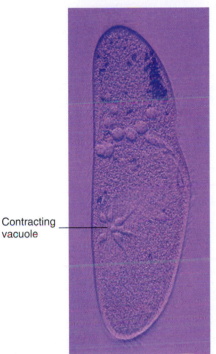

Contracting vacuole

FIGURE 8.14

Top: A contractile vacuole of a *Paramecium* filling with water as a result of hyperosmotic conditions. *Bottom*: The vacuole contracts, expelling water that has entered the cell. (Source: Adapted from N.A. Campbell and J.B. Reece, *Biology*, 7th ed. (San Francisco, CA: Pearson Benjamin Cummings), p. 133X. Copyright 2005. Reprinted by permission of Pearson Education, Inc.)

cells that thrive under turgid conditions, animal cells fare best under isotonic conditions. Prolonged plasmolysis followed by lowered levels of water within the cell is a serious condition because many cellular processes have evolved to function only at certain levels of water. Processes ranging from nutrient uptake to DNA replication are retarded during plasmolysis.

In response to plasmolysis, cells have evolved osmotic regulatory systems that generate increased levels of selected solutes including potassium ion, glutamic acid, glutamine, proline, alanine, and sucrose until total solute concentrations within the cytoplasm are restored to levels sufficient to relieve hyperosmosis. Some cells living in hypertonic environments have evolved organelles to effect water balance. For example, the protist *Paramecium* lives in hypertonic pond water and has evolved a cell membrane that is much less permeable to water than the membranes of other cells. In addition, the *Paramecium* has contractile vacuoles, organelles that act like tiny bilge pumps, that expel water from the cell as fast as it enters during hyperosmotic episodes. See Figure 8.14.

Osmotic pressure measurements have also been used to determine the molecular weight of macromolecular solutes. Beginning with the van't Hoff osmotic equation, we make the substitution $n_{solute} = w_2/M_2$ where M_2 is the molecular mass of the solute and w_2 is the mass of the solute in the solution. This converts the van't Hoff equation to

$$\pi = \frac{w_2}{M_2}\frac{RT}{V} \qquad (8.42)$$

If we define a mass concentration as $C_2 = \frac{w_2}{V}$, the molecular mass is determined by the expression:

$$M_2 = \frac{RT}{\pi}C_2 \qquad (8.43)$$

EXAMPLE PROBLEM 8.9

1.553 grams of chymotrypsinogen are placed into 100.0 mL of water. When this solution was separated from pure water by a semipermeable membrane, the resulting osmotic pressure supported a column of water 157 mm in height. What is the molecular mass of chymotrypsinogen? Assume $T = 298$ K.

Solution

First we have to convert the height of the water column to mmHg, which then can be converted to atmospheres. The density of Hg is 13.596 g mL^{-1}. The height of the water column can be converted to mm Hg by dividing by the ratio of the densities, which converts the osmotic pressure to mm Hg:

$$\pi = \frac{157 \text{ mm}}{13.595 \text{ g mL}^{-1}/1.00 \text{ g mL}^{-1}} = 11.6 \text{ mm Hg} \times \frac{1 \text{ atm}}{760 \text{ mm Hg}} = 0.0152 \text{ atm}$$

Then

$$M_2 = \frac{RT}{\pi}C_2 = \frac{0.0821 \text{ L atm K}^{-1}\text{ mol}^{-1} \times 298 \text{ K}}{0.0152 \text{ atm}} \times \frac{1.533 \text{ g}}{0.100 \text{ L}} = 2.47 \times 10^4 \text{ g}$$

The simple expression given in Equation (8.43) must be used with caution. For example, many macromolecules have net charge at physiological pH, and may exist in solution as polycations or polyanions. DNA is a polyanion in solution at physiological pH, and in solid form it may exist as a sodium or lithium salt. Salts of biological polymers may dissociate in solution, which will drastically affect an osmotic pressure measurement of molecular weight. This topic will be treated in the next chapter.

Another limitation of Equation (8.43) is that it assumes ideal solution behavior. Equation (8.43) should be properly written:

$$M_2 = \lim_{C_2 \to 0} \frac{RT}{\pi/C_2} \qquad (8.44)$$

because the ideal solution assumption is only valid in the infinitely dilute limit. An application of Equation (8.44) is given in Example Problem 8.10.

EXAMPLE PROBLEM 8.10

The following table gives the osmotic pressure as a function of concentration for solutions of polyisobutylene dissolved in benzene. Using the data and Equation (8.44), determine the molecular weight of polyisobutylene. Assume $T = 298$ K.

C_2 (g mL^{-1})	0.0200	0.0150	0.0100	0.0050
Osmotic Pressure (atm)	0.00208	0.00152	0.00099	0.00049

Solution

Calculate the ratios π/C_2:

C_2 (g mL^{-1})	0.0200	0.0150	0.0100	0.0050
π/C_2 (atm mL g^{-1})	0.104	0.101	0.099	0.098

Now plot the data and extrapolate to $C_2 = 0$.

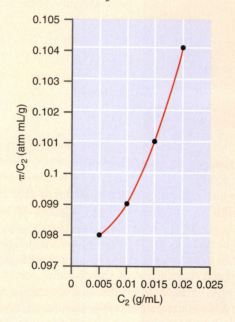

The plot extrapolates to $\pi/C_2 \approx 0.097$ atm mL g^{-1} at $C_2 = 0$. Now, using Equation (8.44):

$$M_2 = \lim_{C_2 \to 0} \frac{RT}{\pi/C_2} = \frac{0.0821 \text{ L atm K}^{-1}\text{mol}^{-1} \times 298 \text{ K}}{0.097 \text{ atm mL g}^{-1} \times 0.001 \text{ L mL}^{-1}} = 2.5 \times 10^5 \text{ g mol}^{-1}$$

In many cases macromolecular solutions are polydisperse with respect to the masses of the macromolecular solutes. This means that there exists a distribution of solute sizes within the solution. How will this affect the osmotic pressure measurement of molecular weight? The osmotic pressures arising from multiple solutes are additive, so

$$\pi_{total} = \sum_i \pi_i = \sum_i \frac{n_i}{V} RT = \sum_i c_i RT = \sum_i \frac{C_i}{M_i} RT \qquad (8.45)$$

where M_i is the molecular weight of the ith solute species and C_i is the mass concentration of the ith solute species.

Now divide both sides of Equation (8.45) by $C_T = \sum_i C_i$:

$$\frac{\pi_{total}}{C_T} = \left(\sum_i \frac{C_i}{M_i} RT \right) \times \frac{1}{\sum_i C_i} \qquad (8.46)$$

We assume that the osmotic pressure obtained from a solution with a distribution of solute masses will detect a generalized mass average so:

$$\pi_{total} = \frac{C_T}{\overline{M}} RT \tag{8.47}$$

Comparing Equations (8.46) and (8.47), we find:

$$\overline{M} = \frac{\sum_i C_i}{\sum_i \frac{C_i}{M_i}} = \frac{\sum_i \frac{C_i}{M_i} M_i}{\sum_i \frac{C_i}{M_i}} \tag{8.48}$$

Recall that $C_i/M_i = n_i/V$. Hence, Equation (8.48) is clearly a number-weighted average:

$$\overline{M} = \frac{\sum_i C_i}{\sum_i \frac{C_i}{M_i}} = \frac{\sum_i \frac{n_i}{V} M_i}{\sum_i \frac{n_i}{V}} = \frac{1}{n_{total}} \sum_i n_i M_i \tag{8.49}$$

where $n_{total} = \sum_i n_i$.

Another important application involving osmotic pressure is the desalination of seawater using reverse osmosis. Seawater is typically 1.1 molar in NaCl. If such a solution is separated from pure water with a semipermeable membrane that allows the passage of water, but not of the solvated Na^+ and Cl^- ions, Equation (8.40) shows that an osmotic pressure of 27 bar is built up. If the side of the membrane on which the seawater is found is subjected to a pressure greater than 27 bar, H_2O from the seawater will flow through the membrane so that separation of pure water from the seawater occurs. This process is called reverse osmosis. The challenge in carrying out reverse osmosis on the industrial scale needed to provide a coastal city with potable water is to produce robust membranes that (1) accommodate the necessary flow rates without getting fouled by algae and (2) also effectively separate the ions from the water. The mechanism that leads to rejection of the ions is not fully understood. However, it is not based on the size of pores within the membrane alone. The major mechanism involves polymeric membranes that carry a surface charge within the pores in their hydrated state. These membrane-anchored ions repel the mobile Na^+ and Cl^- ions, while allowing the passage of the neutral water molecule.

8.9 Real Solutions Exhibit Deviations from Raoult's Law

In Sections 8.1 through 8.8, the discussion has been limited to ideal solutions. However, in general, if two volatile and miscible liquids are combined to form a solution, Raoult's law is not obeyed. This is the case because the A–A, B–B, and A–B interactions in a binary solution of A and B are unequal. If the A–B interactions are less (more) attractive than the A–A and B–B interactions, positive (negative) deviations from Raoult's law will be observed. An example of a binary solution with positive deviations from Raoult's law is CS_2–acetone. Experimental data for this system are shown in Table 8.3 and plotted in Figure 8.15. How can a thermodynamic framework analogous to that presented for the ideal solution in Section 8.2 be developed for real solutions? This issue is addressed throughout the rest of this chapter.

Figure 8.15 shows that the partial and total pressures above a real solution can differ substantially from the behavior predicted by Raoult's law. Another way in which ideal and real solutions differ is that the set of equations denoted Equation (8.8), which describes the

FIGURE 8.15

The data in Table 8.3 are plotted versus x_{CS_2}. The dashed lines show the expected behavior if Raoult's law were obeyed.

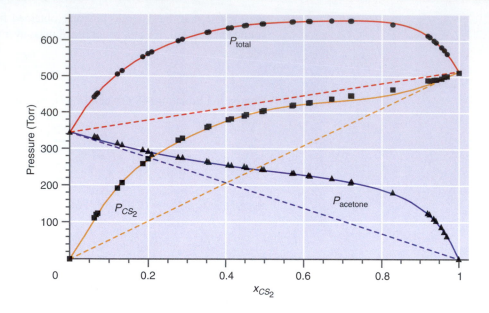

change in volume, entropy, enthalpy, and Gibbs energy that results from mixing, are not applicable to real solutions. For real solutions, these equations can only be written in a much less explicit form. Assuming that A and B are miscible,

$$
\begin{aligned}
\Delta G_{mixing} &< 0 \\
\Delta S_{mixing} &> 0 \\
\Delta V_{mixing} &\neq 0 \\
\Delta H_{mixing} &\neq 0
\end{aligned}
\tag{8.50}
$$

TABLE 8.3   **Partial and Total Pressures above a CS_2–Acetone Solution**

x_{CS_2}	P_{CS_2} (Torr)	$P_{acetone}$ (Torr)	P_{total} (Torr)	x_{CS_2}	P_{CS_2} (Torr)	$P_{acetone}$ (Torr)	P_{total} (Torr)
0	0	343.8	343.8	0.4974	404.1	242.1	646.2
0.0624	110.7	331.0	441.7	0.5702	419.4	232.6	652.0
0.0670	119.7	327.8	447.5	0.5730	420.3	232.2	652.5
0.0711	123.1	328.8	451.9	0.6124	426.9	227.0	653.9
0.1212	191.7	313.5	505.2	0.6146	427.7	225.9	653.6
0.1330	206.5	308.3	514.8	0.6161	428.1	225.5	653.6
0.1857	258.4	295.4	553.8	0.6713	438.0	217.0	655.0
0.1991	271.9	290.6	562.5	0.6713	437.3	217.6	654.9
0.2085	283.9	283.4	567.3	0.7220	446.9	207.7	654.6
0.2761	323.3	275.2	598.5	0.7197	447.5	207.1	654.6
0.2869	328.7	274.2	602.9	0.8280	464.9	180.2	645.1
0.3502	358.3	263.9	622.2	0.9191	490.7	123.4	614.1
0.3551	361.3	262.1	623.4	0.9242	490.0	120.3	610.3
0.4058	379.6	254.5	634.1	0.9350	491.9	109.4	601.3
0.4141	382.1	253.0	635.1	0.9407	492.0	103.5	595.5
0.4474	390.4	250.2	640.6	0.9549	496.2	85.9	582.1
0.4530	394.2	247.6	641.8	0.9620	500.8	73.4	574.2
0.4933	403.2	242.8	646.0	0.9692	502.0	62.0	564.0
				1	512.3	0	512.3

Source: J. v. Zawidski, *Zeitschrift für Physikalische Chemie* 35 (1900) 129.

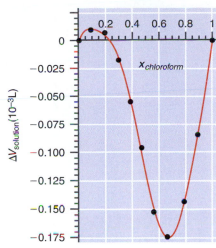

FIGURE 8.16

Deviations in the volume from the behavior expected for 1 mol of an ideal solution [Equation (8.51)] are shown for the acetone–chloroform system as a function of the mole fraction of chloroform.

Whereas $\Delta G_{mixing} < 0$ and $\Delta S_{mixing} > 0$ always hold for miscible liquids, ΔV_{mixing} and ΔH_{mixing} can be positive or negative, depending on the nature of the A–B interaction in the solution.

As indicated in Equation (8.50), the volume change on mixing is not generally zero. Therefore, the volume of a solution will not be given by

$$V_m^{ideal} = x_A V_{m,A}^* + (1 - x_A)V_{m,B}^* \tag{8.51}$$

as expected for 1 mol of an ideal solution, where $V_{m,A}^*$ and $V_{m,B}^*$ are the molar volumes of the pure substances A and B. Figure 8.16 shows $\Delta V_m = V_m^{real} - V_m^{ideal}$ for an acetone–chloroform solution as a function of $x_{chloroform}$. Note that ΔV_m can be positive or negative for this solution, depending on the value of $x_{chloroform}$. The deviations from ideality are small, but are clearly evident.

The deviation of the volume from ideal behavior can best be understood by defining the concept of **partial molar quantities.** This concept is illustrated by discussing the **partial molar volume.** The volume of 1 mol of pure water at 25°C is 18.1 cm³. However, if 1 mol of water is added to a large volume of an ethanol–water solution with $x_{H_2O} = 0.75$, the volume of the solution increases by only 16 cm³. This is the case because the local structure around a water molecule in the solution is more compact than in pure water. The partial molar volume of a component in a solution is defined as the volume by which the solution changes if 1 mol of the component is added to such a large volume that the solution composition can be assumed constant. This statement is expressed mathematically in the following form:

$$\overline{V}_1(P,T,n_1,n_2) = \left(\frac{\partial V}{\partial n_1}\right)_{P,T,n_2} \tag{8.52}$$

With this definition, the volume of a binary solution is given by

$$V = n_1\overline{V}_1(P,T,n_1,n_2) + n_2\overline{V}_2(P,T,n_1,n_2) \tag{8.53}$$

Note that because the partial molar volumes depend on the concentration of all components, the same is true of the total volume.

One can form partial molar quantities for any extensive property of a system (for example U, H, G, A, and S). Partial molar quantities (other than the chemical potential, which is the partial molar Gibbs energy) are usually denoted by the appropriate symbol topped by a horizontal bar. The partial molar volume is a function of P, T, n_1, and n_2, and $\overline{V}_i$ can be greater than or less than the molar volume of the pure component. Therefore, the volume of a solution of two miscible liquids can be greater than or less than the sum of the volumes of the pure components of the solution. Figure 8.17 shows data for the partial volumes of acetone and chloroform in an acetone–chloroform binary solution at 298 K. Note that the changes in the partial molar volumes with concentration are small, but not negligible.

In Figure 8.17, we can see that $\overline{V}_1$ increases if $\overline{V}_2$ decreases and vice versa. This is the case because partial molar volumes are related in the same way as the chemical potentials are related in the Gibbs–Duhem equation [Equation (8.22)]. In terms of the partial molar volumes, the Gibbs–Duhem equation takes the form

$$x_1\,d\overline{V}_1 + x_2\,d\overline{V}_2 = 0 \text{ or } d\overline{V}_1 = -\frac{x_2}{x_1}\,d\overline{V}_2 \tag{8.54}$$

Therefore, as seen in Figure 8.17, if $\overline{V}_2$ changes by an amount $d\overline{V}_2$ over a small concentration interval, $\overline{V}_1$ will change in the opposite direction. The Gibbs–Duhem equation is applicable to both ideal and real solutions.

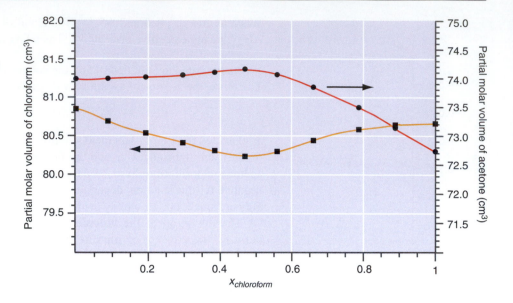

8.10 The Ideal Dilute Solution

Although no simple model exists that describes all real solutions, there is a limiting law that merits further discussion. In this section, we describe the model of the ideal dilute solution, which provides a useful basis for describing the properties of real solutions if they are dilute. Just as for an ideal solution, at equilibrium the chemical potentials of a component in the gas and solution phases of a real solution are equal. As for an ideal solution, the chemical potential of a component in a real solution can be separated into two terms, a standard state chemical potential and a term that depends on the partial pressure:

$$\mu_i^{solution} = \mu_i^* + RT \ln \frac{P_i}{P_i^*} \tag{8.55}$$

Recall that for a pure substance μ_i^* (vapor) $= \mu_i^*$ (liquid) $= \mu_i^*$. Because the solution is not ideal, $P_i \neq x_i P_i^*$.

First consider only the solvent in a dilute binary solution. To arrive at an equation for $\mu_i^{solution}$ that is similar to Equation (8.7) for an ideal solution, we define the dimensionless **activity, $a_{solvent}$,** of the solvent by

$$a_{solvent} = \frac{P_{solvent}}{P_{solvent}^*} \tag{8.56}$$

Note that $a_{solvent} = x_{solvent}$ for an ideal solution. For a nonideal solution, the activity and the mole fraction are related through the **activity coefficient, $\gamma_{solvent}$,** defined by

$$\gamma_{solvent} = \frac{a_{solvent}}{x_{solvent}} \tag{8.57}$$

The activity coefficient quantifies the degree to which the solution is nonideal. The activity plays the same role for a component of a solution that the fugacity plays for a real gas in expressing deviations from ideal behavior. In both cases, ideal behavior is observed in the appropriate limit, namely, $P \rightarrow 0$ for the gas, and $x_{solute} \rightarrow 0$ for the solution. To the extent that there is no atomic-scale model that tells us how to calculate γ, it should be regarded as a correction factor that exposes the inadequacy of our model, rather than as a fundamental quantity. As will be discussed in Chapter 9, there is such an atomic-scale model for dilute electrolyte solutions.

How is the chemical potential of a component related to its activity? Combining Equations (8.55) and (8.56), one obtains a relation that holds for all components of a real solution:

$$\mu_i^{solution} = \mu_i^* + RT \ln a_i \tag{8.58}$$

Equation (8.58) is a central equation in describing real solutions. It is the starting point for the discussion in the rest of this chapter.

The preceding discussion focused on the solvent in a dilute solution. However, the ideal dilute solution is defined by the conditions $x_{solute} \to 0$ and $x_{solvent} \to 1$. Because the solvent and solute are considered in different limits, we use different expressions to relate the mole fraction of a component and the partial pressure of the component above the solution.

Consider the partial pressure of acetone as a function of x_{CS_2} shown in Figure 8.18. Although Raoult's law is not obeyed over the whole concentration range, it is obeyed in the limit that $x_{acetone} \to 1$ and $x_{CS_2} \to 0$. In this limit, the average acetone molecule at the surface of the solution is surrounded by acetone molecules. Therefore, to a good approximation, $P_{acetone} = x_{acetone} P_{acetone}^*$ as $x_{acetone} \to 1$. Because the majority species is defined to be the solvent, we see that Raoult's law is obeyed for the solvent in a dilute solution. This limiting behavior is also observed for CS$_2$ *in the limit in which it is the solvent*, as seen earlier in Figure 8.15.

Consider the opposite limit in which $x_{acetone} \to 0$. In this case, the average acetone molecule at the surface of the solution is surrounded by CS$_2$ molecules. Therefore, the molecule experiences very different interactions with its neighbors than if it were surrounded by acetone molecules. For this reason, $P_{acetone} \neq x_{acetone} P_{acetone}^*$ as $x_{acetone} \to 0$. However, it is apparent from Figure 8.18 that $P_{acetone}$ also varies linearly with $x_{acetone}$ in this limit. This behavior is described by the following equation:

$$P_{acetone} = x_{acetone} k_H^{acetone} \text{ as } x_{acetone} \to 0 \tag{8.59}$$

This relationship is known as **Henry's law**, and the constant k_H is known as the **Henry's law constant.** The value of the constant depends on the nature of the solute and solvent and quantifies the degree to which deviations from Raoult's law occur. As the solution approaches ideal behavior, $k_H^i \to P_i^*$. For the data shown in Figure 8.15, $k_H^{CS_2} = 2010$ Torr and $k_H^{acetone} = 1950$ Torr. Note that these values are substantially greater than the vapor pressures of the pure substances, which are 512.3 and 343.8 Torr, respectively. The Henry's law constants are less than the vapor pressures of the pure substances if the system exhibits negative deviations from Raoult's law. Henry's law constants for aqueous solutions are listed for a number of solutes in Table 8.4.

Based on these results, the **ideal dilute solution** can be defined: *an ideal dilute solution is a solution in which the solvent is described using Raoult's law and the solute is*

FIGURE 8.18
The partial pressure of acetone from Table 8.3 is plotted as a function of x_{CS_2}. Note that the $P_{acetone}$ plot follows Raoult's law as $x_{CS_2} \to 0$, and Henry's law as $x_{CS_2} \to 1$.

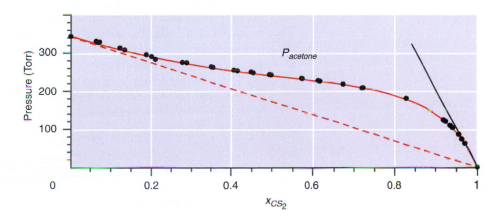

TABLE 8.4   Henry's Law Constants for Aqueous Solutions Near 298 K

Substance	k_H (Torr)	k_H (bar)
Ar	2.80×10^7	3.72×10^4
C_2H_6	2.30×10^7	3.06×10^4
CH_4	3.07×10^7	4.08×10^4
CO	4.40×10^6	5.84×10^3
CO_2	1.24×10^6	1.65×10^3
H_2S	4.27×10^5	5.68×10^2
He	1.12×10^8	1.49×10^6
N_2	6.80×10^7	9.04×10^4
O_2	3.27×10^7	4.95×10^4

Source: R. A. Alberty and R. S. Silbey, *Physical Chemistry*, John Wiley & Sons, New York, 1992.

described using Henry's law. As shown by the data in Figure 8.15, the partial pressures above the CS_2–acetone mixture are consistent with this model in either of two limits, $x_{acetone} \rightarrow 1$ or $x_{CS_2} \rightarrow 1$. In the first of these limits, we consider acetone to be the solvent and CS_2 to be the solute. Acetone is the solute and CS_2 is the solvent in the second limit.

8.11 Activities Are Defined with Respect to Standard States

The ideal dilute solution model's predictions that Raoult's law is obeyed for the solvent and Henry's law is obeyed for the solute are not obeyed over a wide range of concentration. The concept of the activity coefficient introduced in Section 8.10 is used to quantify these deviations. In doing so, it is useful to define the activities in such a way that the solution approaches ideal behavior in the limit of interest, which is generally $x_A \rightarrow 0$ or $x_A \rightarrow 1$. With this choice, the activity approaches the concentration, and it is reasonable to set the activity coefficient equal to one. Specifically, $a_i \rightarrow x_i$ as $x_i \rightarrow 1$ for the solvent, and $a_i \rightarrow x_i$ as $x_i \rightarrow 0$ for the solute. The reason for this choice is that numerical values for activity coefficients are generally not known. Choosing the standard state as described earlier ensures that the concentration (divided by the unit concentration to make it dimensionless), which is easily measured, is a good approximation to the activity.

In Section 8.10 the activity and activity coefficient for the solvent in a dilute solution ($x_{solvent} \rightarrow 1$) were defined by the relations

$$a_i = \frac{P_i}{P_i^*} \text{ and } \gamma_i = \frac{a_i}{x_i} \tag{8.60}$$

As shown in Figure 8.15, the activity approaches unity as $x_{solvent} \rightarrow 1$. We refer to an activity calculated using Equation (8.60) as being based on a **Raoult's law standard state.** The standard state chemical potential based on Raoult's law is $\mu_{solvent}^*$, which is the chemical potential of the pure solvent.

However, this definition of the activity and choice of a standard state are not optimal for the solute at a low concentration, because the solute obeys Henry's law rather than Raoult's law and, therefore, the activity coefficient will differ appreciably from one. In this case,

$$\mu_{solute}^{solution} = \mu_{solute}^* + RT \ln \frac{k_H^{solute} x_{solute}}{P_{solute}^*} = \mu_{solute}^{*H} + RT \ln x_{solute} \text{ as } x_{solute} \rightarrow 0 \tag{8.61}$$

The standard state chemical potential is the value of the chemical potential when $x_i = 1$. We see that the Henry's law standard state chemical potential is given by

$$\mu_{solute}^{*H} = \mu_{solute}^* + RT \ln \frac{k_H^{solute}}{P_{solute}^*} \qquad (8.62)$$

The activity and activity coefficient based on Henry's law are defined, respectively, by

$$a_i = \frac{P_i}{k_i^H} \quad \text{and} \quad \gamma_i = \frac{a_i}{x_i} \qquad (8.63)$$

Note that Henry's law is still obeyed for a solute that has such a small vapor pressure. Such a solute is referred to as nonvolatile.

The **Henry's law standard state** is a state in which the pure solute has a vapor pressure $k_{H,solute}$ rather than its actual value P_{solute}^*. It is a hypothetical state that does not exist. Recall that the value $k_{H,solute}$ is obtained by extrapolation from the low coverage range in which Henry's law is obeyed. Although this definition may seem peculiar, this is the only way we can ensure that $a_{solute} \rightarrow x_{solute}$ and $\gamma_{solute} \rightarrow 1$ as $x_{solute} \rightarrow 0$. We reiterate the reason for this choice. If the preceding conditions are satisfied, the concentration (divided by the unit concentration to make it dimensionless) is, to a good approximation, equal to the activity. Therefore, the equilibrium constant can be calculated without having numerical values for activity coefficients. We emphasize the difference in the Raoult's law and Henry's law standard states because it is the standard state chemical potentials μ_{solute}^* and μ_{solute}^{*H} that are used to calculate ΔG° and the thermodynamic equilibrium constant, K. Different choices for standard states will result in different numerical values for K. The standard chemical potential μ_{solute}^{*H} refers to the hypothetical standard state in which $x_{solute} = 1$, and each solute species is in an environment characteristic of the infinitely dilute solution.

We now consider a less well-defined situation. For solutions in which the components are miscible in all proportions, such as the CS_2–acetone system discussed earlier, either a Raoult's law or a Henry's law standard state can be defined, as we show with sample calculations in Example Problems 8.11 and 8.12. This is the case because there is no unique choice for the standard state over the entire concentration range in such a system. Numerical values for the activities and activity coefficients will differ, depending on whether the Raoult's law or the Henry's law standard state is used.

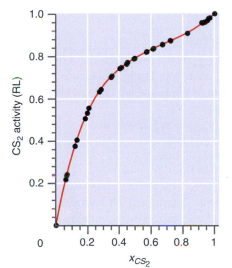

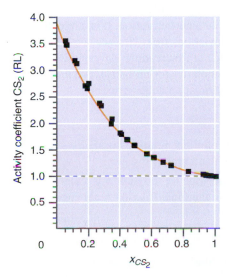

FIGURE 8.19

The activity and activity coefficient for CS_2 in a CS_2–acetone solution based on a Raoult's law standard state are shown as a function of x_{CS_2}.

EXAMPLE PROBLEM 8.11

Calculate the activity and activity coefficient for CS_2 at $x_{CS_2} = 0.3502$ using data from Table 8.3. Assume a Raoult's law standard state.

Solution

$$a_{CS_2}^R = \frac{P_{CS_2}}{P_{CS_2}^*} = \frac{358.3 \text{ Torr}}{512.3 \text{ Torr}} = 0.6994$$

$$\gamma_{CS_2}^R = \frac{a_{CS_2}^R}{x_{CS_2}} = \frac{0.6994}{0.3502} = 1.997$$

The activity and activity coefficients for CS_2 are concentration dependent. Results calculated as in Example Problem 8.11 using a Raoult's law standard state are shown in Figure 8.19 as a function of x_{CS_2}. For this solution, $\gamma_{CS_2}^R > 1$ for all values of the concentration for which $x_{CS_2} < 1$. Note that $\gamma_{CS_2}^R \rightarrow 1$ as $x_{CS_2} \rightarrow 1$ as the model requires. The activity and activity coefficients for CS_2 using a Henry's law standard state are shown in Figure 8.20 as a function of x_{CS_2}. For this solution, $\gamma_{CS_2}^H < 1$ for all values of the concentration for which $x_{CS_2} > 0$. Note that $\gamma_{CS_2}^H \rightarrow 1$ as $x_{CS_2} \rightarrow 0$ as the model requires. Which of these two possible standard states should be chosen? There is a good answer to this question only in the limits $x_{CS_2} \rightarrow 0$ or $x_{CS_2} \rightarrow 1$. For intermediate concentrations, either standard state can be used.

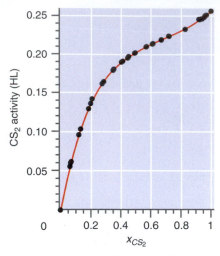

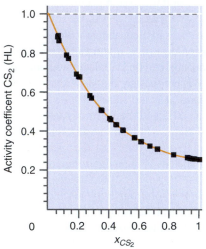

FIGURE 8.20

The activity and activity coefficient for CS_2 in a CS_2–acetone solution based on a Henry's law standard state are shown as a function of x_{CS_2}.

EXAMPLE PROBLEM 8.12

Calculate the activity and activity coefficient for CS_2 at $x_{CS_2} = 0.3502$ using data from Table 8.3. Assume a Henry's law standard state.

Solution

$$a_{CS_2}^H = \frac{P_{CS_2}}{k_{H,CS_2}} = \frac{358.3 \text{ Torr}}{2010 \text{ Torr}} = 0.1783$$

$$\gamma_{CS_2}^H = \frac{a_{CS_2}^H}{x_{CS_2}} = \frac{0.1783}{0.3502} = 0.5090$$

The Henry's law standard state just discussed is defined with respect to concentration measured in units of the mole fraction. This is not a particularly convenient scale, and either the molarity or molality concentration scales are generally used in laboratory experiments. The mole fraction of the solute can be converted to the molality scale by dividing the first expression in Equation (8.64) by $n_{solvent} M_{solvent}$:

$$x_{solute} = \frac{n_{solute}}{n_{solvent} + n_{solute}} = \frac{m_{solute}}{\dfrac{1}{M_{solvent}} + m_{solute}} \tag{8.64}$$

where m_{solute} is the molality of the solute, and $M_{solvent}$ is the molecular mass of the solvent in kg mol^{-1}. We see that $m_{solute} \rightarrow x/M_{solvent}$ as $x \rightarrow 0$. Using molality as the concentration unit, the activity and activity coefficient of the solute are defined by

$$a_{solute}^{molality} = \frac{P_{solute}}{k_H^{molality}} \text{ with } a_{solute}^{molality} \rightarrow m_{solute} \text{ as } m_{solute} \rightarrow 0 \text{ and} \tag{8.65}$$

$$\gamma_{solute}^{molality} = \frac{a_{solute}^{molality}}{m_{solute}} \text{ with } \gamma_{solute}^{molality} \rightarrow 1 \text{ as } m_{solute} \rightarrow 0 \tag{8.66}$$

The Henry's law constants and activity coefficients determined on the mole fraction scale must be recalculated to describe a solution characterized by its molality or molarity, as shown in Example Problem 8.13. Similar conversions can be made if the concentration of the solution is expressed in terms of molality.

What is the standard state if concentrations rather than mole fractions are used? The standard state in this case is the hypothetical state in which Henry's law is obeyed by a solution that is 1.0 molar (or 1.0 molal) in the solute concentration. It is a hypothetical state because at this concentration, substantial deviations from Henry's law will be observed. Although this definition may seem peculiar at first reading, only in this way can we ensure that the activity becomes equal to the molarity (or molality), and the activity coefficient approaches 1, as the solute concentration approaches zero.

EXAMPLE PROBLEM 8.13

a. Derive a general equation, valid for dilute solutions, relating the Henry's law constants for the solute on the mole fraction and molarity scales.

b. Determine the Henry's law constants for acetone on the molarity scale. Use the results for the Henry's law constants on the mole fraction scale cited in Section 8.10. The density of acetone is 789.9 g L^{-1}.

Solution

a. We use the symbol c_{solute} to designate the solute molarity, and $c°$ to indicate a 1 molar concentration.

$$\frac{dP}{d\left(\dfrac{c_{solute}}{c°}\right)} = \frac{dP}{dx_{solute}} \frac{dx_{solute}}{d\left(\dfrac{c_{solute}}{c°}\right)}$$

To evaluate this equation, we must determine $\dfrac{dx_{solute}}{d\left(\dfrac{c_{solute}}{c^{\circ}}\right)}$:

$$x_{solute} = \frac{n_{solute}}{n_{solute} + n_{solvent}} \approx \frac{n_{solute}}{n_{solvent}} = \frac{n_{solute}M_{solvent}}{V_{solution}\rho_{solution}} = c_{solute}\frac{M_{solvent}}{\rho_{solution}}$$

Therefore,

$$\frac{dP}{d\left(\dfrac{c_{solute}}{c^{\circ}}\right)} = \frac{c^{\circ}M_{solvent}}{\rho_{solution}}\frac{dP}{dx_{solute}}$$

b. $\dfrac{dP}{d\left(\dfrac{c_{solute}}{c^{\circ}}\right)} = \dfrac{1\ \text{mol L}^{-1} \times 58.08\ \text{g mol}^{-1}}{789.9\ \text{g L}^{-1}} \times 1950\ \text{Torr} = 143.4\ \text{Torr}$

The colligative properties discussed for an ideal solution in Sections 8.6 through 8.8 refer to the properties of the solvent in a dilute solution. The Raoult's law standard state applies to this case, and in an ideal dilute solution, Equations (8.32), (8.34), and (8.40) can be used with activities replacing concentrations:

$$\Delta T_f = -K_f \gamma m_{solute}$$
$$\Delta T_b = K_b \gamma m_{solute} \tag{8.67}$$
$$\pi = \gamma c_{solute} RT$$

The activity coefficients are defined with respect to molality for the boiling point elevation ΔT_b and freezing point depression ΔT_f, and with respect to molarity for the osmotic pressure π. Equations (8.67) provide a useful way to determine activity coefficients as shown in Example Problem 8.14.

EXAMPLE PROBLEM 8.14

In 500 g of water, 24.0 g of a nonvolatile solute of molecular weight 241 g mol^{-1} is dissolved. The observed freezing point depression is 0.359°C. Calculate the activity coefficient of the solute.

Solution

$$\Delta T_f = -K_f \gamma m_{solute};\ \gamma = -\frac{\Delta T_f}{K_f m_{solute}}$$

$$\gamma = \frac{0.359\ \text{K}}{1.86\ \text{K kg mol}^{-1} \times \dfrac{24.0}{241 \times 0.500}\ \text{mol kg}^{-1}} = 0.969$$

8.12 Henry's Law and the Solubility of Gases in a Solvent

The ideal dilute solution model can be applied to the solubility of gases in liquid solvents. An example for this type of solution equilibrium is the amount of N_2 absorbed by water at sea level, which is considered in Example Problem 8.17. In this case, one of the components of the solution is a liquid and the other is a gas. The equilibrium of interest between the solution and the vapor phase is

$$N_2(aqueous) \underset{\longleftarrow}{\longrightarrow} N_2(vapor) \tag{8.68}$$

The chemical potential of the dissolved N_2 is given by

$$\mu_{N_2}^{solution} = \mu_{N_2}^{*H}(\text{vapor}) + RT \ln a_{solute} \tag{8.69}$$

In this case, a Henry's law standard state is the appropriate choice, because the nitrogen is sparingly soluble in water. The mole fraction of N_2 in solution, x_{N_2}, is given by

$$x_{N_2} = \frac{n_{N_2}}{n_{N_2} + n_{H_2O}} \approx \frac{n_{N_2}}{n_{H_2O}} \tag{8.70}$$

The amount of dissolved gas is given by

$$n_{N_2} = n_{H_2O} x_{N_2} = n_{H_2O} \frac{P_{N_2}}{k_H^{N_2}} \tag{8.71}$$

Henry's law predicts that as the pressure increases the amount of a gas dissolved in a solution will also increase. This fact can be applied to describing how gases dissolve in tissue fluids. Example Problem 8.15 illustrates this point.

EXAMPLE PROBLEM 8.15

Calculate the concentration of oxygen in water at a pressure of 1 atmosphere. Assume air is 21% oxygen and that $T = 298$ K. Assume Henry's law behavior. Express your answer in milligrams of oxygen dissolved per liter of water.

Solution

If the atmosphere is assumed to be 21% oxygen, then the partial pressure of oxygen is:

$$P_{O_2} = 1 \text{ atm} \times 0.21 \times 1.01 \times 10^5 \text{ Pa atm}^{-1} = 2.1 \times 10^4 \text{ Pa}$$

Using Henry's law and the data in Table 8.4, $P_{O_2} = k_H^{O_2} x_{O_2}$, we can solve for the mole fraction of oxygen dissolved in water:

$$x_{O_2} = \frac{P_{O_2}}{k_H^{O_2}} = \frac{2.1 \times 10^4 \text{ Pa}}{4.95 \times 10^4 \text{ bar} \times 10^5 \text{ Pa bar}^{-1}} = 4.25 \times 10^{-6}$$

We can easily convert this to a molar concentration because the solution is so dilute. Assuming only water and oxygen are in the solution:

$$x_{O_2} = \frac{n_{O_2}}{n_{O_2} + n_{H_2O}} \approx \frac{n_{O_2}}{n_{H_2O}} \times \frac{V_{m,H_2O}}{V_{m,H_2O}} = \frac{n_{O_2}}{V_{H_2O}} \times V_{m,H_2O}$$

where $V_{m,H_2O} = 0.018 \text{ L mol}^{-1}$ is the molar volume of water.

Now rearrange and assume the volume of the water is about equal to the volume of the solution:

$$\frac{n_{O_2}}{V_{H_2O}} \approx \frac{n_{O_2}}{V_{solution}} = c_{O_2} = \frac{x_{O_2}}{V_{m,H_2O}} = \frac{4.5 \times 10^{-6}}{0.018 \text{ L mol}^{-1}} = 2.5 \times 10^{-4} \text{ mol L}^{-1}$$

This corresponds to $2.4 \times 10^{-4} \text{ mol L}^{-1} \times 32 \text{ g mol}^{-1} = 0.0077 \text{ g L}^{-1}$ or about 7.7 milligrams of oxygen per liter of water.

A useful application of Henry's law is determining the concentrations of gases in tissue fluids and blood during deep sea diving. A simple way to breathe under water is to use a snorkel. But to be useful snorkels cannot exceed lengths of more than a meter or so because as a diver descends to depths lower than 1 m, inhaling air through a snorkel becomes difficult. The reason for this is that as the diver descends to ever greater depths the weight of water on the diver and the hydrostatic pressure that it exerts increase. As a result of the greater density of water versus air, a diver need only descend to about 10 m for the sum of the atmospheric pressure and the hydrostatic pressure to equal about 2 atm. For every 10 additional meters of descent, the applied pressure increases by about 1 atm.

The problem with breathing at great depths becomes clear. To inhale air through a snorkel at a depth of 50 m, the muscles that allow our lungs to inhale and are accustomed to working at 1 atm would have to work against a pressure of 4 atm exerted by the surrounding water and the atmosphere. This is a difficult if not impossible job for our lungs to perform. There are two solutions to this problem: atmospheric pressure diving and ambient pressure diving. Atmospheric pressure diving uses a rigid diving chamber to protect the diver from the surrounding pressure, which is exerted against the rigid side walls of the chamber. The hollow interior of the chamber is maintained at 1 atm so breathing can proceed normally. In ambient pressure diving, a mixture of oxygen and other gases is inhaled at the pressure being exerted by the atmosphere and the surrounding water. If the diver descends to 30 m and thus encounters an ambient pressure of 4 atm, the pressure of the gas mixture that the diver inhales is 4 atm. According to Henry's law, ambient pressure diving increases the amount of a gas dissolved in the tissue fluids of a diver.

EXAMPLE PROBLEM 8.16

A person performing ambient pressure diving descends to a depth of 30.0 m. Calculate the total pressure exerted on the diver by the atmosphere and the surrounding water. Assuming the pressure of the gas mixture inhaled by the diver is adjusted to equal the ambient pressure at that depth, calculate the concentration of oxygen dissolved in the diver's tissues. Assume the gas mixture is 21% oxygen. Assume the tissue fluid can be treated as water and the solubility of oxygen in tissue fluid is governed by Henry's law. Also assume that $T = 298$ K.

Solution

The pressure is given by

$$P = 1 \text{ atm} + \rho G_n h = 1 \text{ atm} + 1.00 \times 10^3 \text{ kg m}^{-3} \times 9.81 \text{ ms}^{-2} \times 30.0 \text{ m}$$

$$= 1 \text{ atm} \times \frac{1.01 \times 10^5 \text{ Pa}}{1 \text{ atm}} + 2.94 \times 10^5 \text{ Pa} = 3.95 \times 10^5 \text{ Pa}$$

If the inhaled gas mixture is assumed to be 21% oxygen, then the partial pressure of oxygen is:

$$P_{O_2} = 0.21 \times 3.95 \times 10^5 \text{ Pa atm}^{-1} = 8.3 \times 10^4 \text{ Pa}$$

Using Henry's law and the data in Table 8.4, $P_{O_2} = k_H^{O_2} x_{O_2}$, we can solve for the mole fraction of oxygen dissolved in water:

$$x_{O_2} = \frac{P_{O_2}}{k_H^{O_2}} = \frac{8.3 \times 10^4 \text{ Pa}}{4.95 \times 10^4 \text{ bar} \times 10^5 \text{ Pa bar}^{-1}} = 17.0 \times 10^{-6}$$

This is four times the concentration of oxygen versus Example Problem 8.15, as expected. So there are now about 32 mg of oxygen per liter of tissue fluid in our diver.

Although elevated pressures enable divers to inhale oxygen at great depths, elevated levels of oxygen in tissue fluids can have adverse effects on the central nervous system. Prolonged breathing of oxygen at $P_{O_2} \gg 1.2 - 1.4$ atm can induce central nervous system or acute oxygen toxicity. Symptoms include tunnel vision, muscular twitching, and uncontrolled convulsions.

Other dangers in deep sea diving involve the inhalation of nitrogen gas at elevated pressure. Prolonged breathing of nitrogen at elevated levels induces nitrogen narcosis, sometimes called "rapture of the deep" because its effects are likened to alcoholic inebriation. Some divers have reported symptoms of nitrogen narcosis at depths as shallow as 90 m.

A far more serious malady accompanying breathing of air at great depth is decompression sickness, also called the "bends." Decompression sickness occurs when people who have breathed air at elevated pressure quickly return to normal sea level pressure. Symptoms include mild to severe pain in the joints, dizziness, nausea, and disorientation. The origin of the bends is again explained by Henry's law. As a person who has breathed air for prolonged periods returns to the surface, the external pressure drops, and the levels of nitrogen dissolved in the blood and tissue fluids must drop according to Henry's law. The result is the formation of nitrogen bubbles in the blood vessels that block circulation to various organs and induce the symptoms of the bends. Example Problem 8.17 shows how Henry's law is used to model the dissolution of nitrogen gas in a liquid.

EXAMPLE PROBLEM 8.17

The average human with a body weight of 70. kg has a blood volume of 5.00 L. The Henry's law constant for the solubility of N_2 in H_2O is 9.04×10^4 bar at 298 K. Assume that this is also the value of the Henry's law constant for blood and that the density of blood is 1.00 kg L^{-1}.

a. Calculate the number of moles of nitrogen absorbed in this amount of blood in air of composition 80% N_2 at sea level, where the pressure is 1 bar, and at a pressure of 50.0 bars.

b. Assume that a diver accustomed to breathing compressed air at a pressure of 50.0 bars is suddenly brought to sea level. What volume of N_2 gas is released as bubbles in the diver's bloodstream?

Solution

a. $n_{N_2} = n_{H_2O} \dfrac{P_{N_2}}{k_H^{N_2}}$

$$= \frac{5.0 \times 10^3 \text{ g}}{18.02 \text{ g mol}^{-1}} \times \frac{0.80 \text{ bar}}{9.04 \times 10^4 \text{ bar}}$$

$$= 2.5 \times 10^{-3} \text{ mol at 1 bar total pressure}$$

At 50.0 bars, $n_{N_2} = 50 \times 2.5 \times 10^{-3}$ mol $= 0.125$ mol.

b. $V = \dfrac{nRT}{P}$

$$= \frac{(0.125 \text{ mol} - 2.5 \times 10^{-3} \text{ mol}) \times 8.314 \times 10^{-2} \text{ L bar mol}^{-1} \text{ K}^{-1} \times 300. \text{ K}}{1.00 \text{ bar}}$$

$$= 3.1 \text{ L}$$

The volume of N_2 just calculated is far more than is needed to cause the formation of arterial blocks due to gas-bubble embolisms.

8.13 Chemical Equilibrium in Solutions

The concept of activity can be used to express the thermodynamic equilibrium constant in terms of activities for real solutions. Consider a reaction between solutes in a solution. At equilibrium, the following relation must hold:

$$\left(\sum_j v_j \mu_j(\text{solution}) \right)_{equilibrium} = 0 \tag{8.72}$$

where the subscript states that the individual chemical potentials must be evaluated under equilibrium conditions. Each of the chemical potentials in Equation (8.72) can be expressed in terms of a standard state chemical potential and a concentration-dependent term. Assume a Henry's law standard state for each solute. Equation (8.72) then takes the form

$$\sum_j v_j \mu_j^{*H}(\text{solution}) + RT \sum_j \ln (a_i^{eq})^{v_i} = 0 \qquad (8.73)$$

Using the relation between the Gibbs energy and the chemical potential, the previous equation can be written in the form

$$\Delta G_{reaction}^{\circ} = -RT \sum_j \ln (a_i^{eq})^{v_j} = -RT \ln K \qquad (8.74)$$

The equilibrium constant in terms of activities is given by

$$K = \prod_i (a_i^{eq})^{v_j} = \prod_i (\gamma_i^{eq})^{v_j} \left(\frac{c_i^{eq}}{c^{\circ}} \right)^{v_j} \qquad (8.75)$$

where the symbol $\prod$ indicates that the terms following the symbol are multiplied with one another. This equilibrium constant is the fundamental thermodynamic equilibrium for all systems. It can be viewed as a generalization of K_P, defined in Equation (6.67). The equilibrium constant K defined by Equation (8.75) can be applied to equilibria involving gases, liquids, dissolved species, and solids.

To obtain a numerical value for K, the standard state Gibbs reaction energy $\Delta G_{reaction}^{\circ}$ must be known. As for gas-phase reactions, the $\Delta G_{reaction}^{\circ}$ must be determined experimentally. This can be done by measuring the individual activities of the species in solution and calculating K from these results. After a series of $\Delta G_{reaction}^{\circ}$ values for different reactions has been determined, they can be combined to calculate the $\Delta G_{reaction}^{\circ}$ for other reactions, as discussed for reaction enthalpies in Chapter 4. Because of the significant interactions between the solutes and the solvent, K values depend on the nature of the solvent, and for the electrolyte solutions to be discussed in Chapter 9, they additionally depend on the ionic strength.

An equilibrium constant in terms of molarities or molalities can also be defined starting from Equation (8.75) and setting all activity coefficients equal to 1. This is most appropriate for a dilute solution of a nonelectrolyte, using a Henry's law standard state:

$$K = \prod_i (\gamma_i^{eq})^{v_j} \left(\frac{c_i^{eq}}{c^{\circ}} \right)^{v_j} \approx \prod_i \left(\frac{c_i^{eq}}{c^{\circ}} \right)^{v_j} \qquad (8.76)$$

EXAMPLE PROBLEM 8.18

a. Write the equilibrium constant for the reaction $N_2(aq, m) \rightleftharpoons N_2(g, P)$ in terms of activities at 25°C, where m is the molarity of $N_2(aq)$.

b. By making suitable approximations, convert the equilibrium constant of part (a) into one in terms of pressure and molarity only.

Solution

a. $K = \prod_i (a_i^{eq})^{v_j} = \prod_i (\gamma_i^{eq})^{v_j} \left(\frac{c_i^{eq}}{c^{\circ}} \right)^{v_j} = \dfrac{\left(\dfrac{\gamma_{N_2,g} P}{P^{\circ}} \right)}{\left(\dfrac{\gamma_{N_2,aq} m}{m^{\circ}} \right)}$

$= \dfrac{\gamma_{N_2,g}}{\gamma_{N_2,aq}} \dfrac{\left(\dfrac{P}{P^{\circ}} \right)}{\left(\dfrac{m}{m^{\circ}} \right)}$

b. Using a Henry's law standard state for dissolved N_2, $\gamma_{N_2,aq} \approx 1$, because the concentration is very low. Similarly, because N_2 behaves like an ideal gas up to quite high pressures at 25°C, $\gamma_{N_2,g} \approx 1$. Therefore,

$$K \approx \frac{\left(\dfrac{P}{P°}\right)}{\left(\dfrac{m}{m°}\right)}$$

Note that in this case, the equilibrium constant is simply the Henry's law constant in terms of molarity.

The numerical values for the dimensionless thermodynamic equilibrium constant depend on the choice of the standard states for the components involved in the reaction. The same is true for $\Delta G°_{reaction}$. Therefore, it is essential to know which standard state has been assumed before an equilibrium constant is used. The activity coefficients of most neutral solutes are close to 1 with the appropriate choice of standard state. Therefore, example calculations of chemical equilibrium using activities will be deferred until electrolyte solutions are discussed in Chapter 9. For such solutions, γ_{solute} differs substantially from 1, even for dilute solutions.

Vocabulary

activity	Henry's law	osmotic pressure	semipermeable membrane
activity coefficient	Henry's law constant	partial molar quantities	solute
average composition	Henry's law standard state	partial molar volume	solvent
boiling point elevation	ideal dilute solution	pressure–average	temperature–composition
colligative properties	ideal solution	composition diagram	diagram
fractional distillation	lever rule	Raoult's law	tie line
freezing point depression	limiting law	Raoult's law standard state	van't Hoff osmotic equation
Gibbs–Duhem equation	osmosis		

Questions on Concepts

Q8.1   Using the differential form of G, $dG = VdP - SdT$, show that if $\Delta G_{mixing} = nRT\sum_i x_i \ln x_i$, then $\Delta H_{mixing} = \Delta V_{mixing} = 0$.

Q8.2   For a pure substance, the liquid and gaseous phases can only coexist for a single value of the pressure at a given temperature. Is this also the case for an ideal solution of two volatile liquids?

Q8.3   Fractional distillation of a particular binary liquid mixture leaves behind a liquid consisting of both components in which the composition does not change as the liquid is boiled off. Is this behavior characteristic of a maximum or a minimum boiling point azeotrope?

Q8.4   Why is the magnitude of the boiling point elevation less than that of the freezing point depression?

Q8.5   Why is the preferred standard state for the solvent in an ideal dilute solution the Raoult's law standard state? Why is the preferred standard state for the solute in an ideal dilute solution the Henry's law standard state? Is there a preferred standard state for the solution in which $x_{solvent} = x_{solute} = 0.5$?

Q8.6   Is a whale likely to get the bends when it dives deep into the ocean and resurfaces? Answer this question by considering the likelihood of a diver getting the bends if he or she dives and resurfaces on one lung full of air as opposed to breathing air for a long time at the deepest point of the dive.

Q8.7   The statement "The boiling point of a typical liquid mixture can be reduced by approximately 100°C if the pressure is reduced from 760 to 20 Torr" is found in Section 8.4. What figure(s) in Chapter 8 can you identify to support this statement in a qualitative sense?

Q8.8   Explain why chemists doing quantitative work using liquid solutions prefer to express concentration in terms of molality rather than molarity.

Q8.9   Explain the usefulness of a tie line on a P–Z phase diagram such as that of Figure 8.4.

Q8.10   Explain why colligative properties depend only on the concentration, and not on the identity of the molecule.

Problems

Problem numbers in **RED** indicate that the solution to the problem is given in the *Student Solutions Manual*.

P8.1 At 303 K, the vapor pressure of benzene is 118 Torr and that of cyclohexane is 122 Torr. Calculate the vapor pressure of a solution for which $x_{benzene} = 0.25$ assuming ideal behavior.

P8.2 A volume of 5.50 L of air is bubbled through liquid toluene at 298 K, thus reducing the mass of toluene in the beaker by 2.38 g. Assuming that the air emerging from the beaker is saturated with toluene, determine the vapor pressure of toluene at this temperature.

P8.3 An ideal solution is formed by mixing liquids A and B at 298 K. The vapor pressure of pure A is 180. Torr and that of pure B is 82.1 Torr. If the mole fraction of A in the vapor is 0.450, what is the mole fraction of A in the solution?

P8.4 A and B form an ideal solution. At a total pressure of 0.900 bar, $y_A = 0.450$ and $x_A = 0.650$. Using this information, calculate the vapor pressure of pure A and of pure B.

P8.5 A and B form an ideal solution at 298 K, with $x_A = 0.600$, $P_A^* = 105$ Torr, and $P_B^* = 63.5$ Torr.

  a. Calculate the partial pressures of A and B in the gas phase.

  b. A portion of the gas phase is removed and condensed in a separate container. Calculate the partial pressures of A and B in equilibrium with this liquid sample at 298 K.

P8.6 The vapor pressures of 1-bromobutane and 1-chlorobutane can be expressed in the form

$$\ln \frac{P_{bromo}}{Pa} = 17.076 - \frac{1584.8}{\frac{T}{K} - 111.88}$$

and

$$\ln \frac{P_{chloro}}{Pa} = 20.612 - \frac{2688.1}{\frac{T}{K} - 55.725}$$

Assuming ideal solution behavior, calculate x_{bromo} and y_{bromo} at 300.0 K and a total pressure of 8741 Pa.

P8.7 Assume that 1-bromobutane and 1-chlorobutane form an ideal solution. At 273 K, $P_{chloro}^* = 3790$ Pa and $P_{bromo}^* = 1394$ Pa. When only a trace of liquid is present at 273 K, $y_{chloro} = 0.75$.

  a. Calculate the total pressure above the solution.

  b. Calculate the mole fraction of 1-chlorobutane in the solution.

  c. What value would Z_{chloro} have in order for there to be 4.86 mol of liquid and 3.21 mol of gas at a total pressure equal to that in part (a)? [*Note*: This composition is different from that of part (a).]

P8.8 An ideal solution at 298 K is made up of the volatile liquids A and B, for which $P_A^* = 125$ Torr and $P_B^* = 46.3$ Torr.

As the pressure is reduced from 450 Torr, the first vapor is observed at a pressure of 70.0 Torr. Calculate x_A.

P8.9 At $-47°C$, the vapor pressure of ethyl bromide is 10.0 Torr and that of ethyl chloride is 40.0 Torr. Assume that the solution is ideal. Assume there is only a trace of liquid present and the mole fraction of ethyl chloride in the vapor is 0.80 and then answer these questions:

  a. What is the total pressure and the mole fraction of ethyl chloride in the liquid?

  b. If there are 5.00 mol of liquid and 3.00 mol of vapor present at the same pressure as in part (a), what is the overall composition of the system?

P8.10 At $-31.2°C$, pure propane and *n*-butane have vapor pressures of 1200 and 200 Torr, respectively.

  a. Calculate the mole fraction of propane in the liquid mixture that boils at $-31.2°C$ at a pressure of 760 Torr.

  b. Calculate the mole fraction of propane in the vapor that is in equilibrium with the liquid of part (a).

P8.11 In an ideal solution of A and B, 3.50 mol are in the liquid phase and 4.75 mol are in the gaseous phase. The overall composition of the system is $Z_A = 0.300$ and $x_A = 0.250$. Calculate y_A.

P8.12 Given the vapor pressures of the pure liquids and the overall composition of the system, what are the upper and lower limits of pressure between which liquid and vapor coexist in an ideal solution?

P8.13 At 39.9°C, a solution of ethanol ($x_1 = 0.9006$, $P_1^* = 130.4$ Torr) and isooctane ($P_2^* = 43.9$ Torr) forms a vapor phase with $y_1 = 0.6667$ at a total pressure of 185.9 Torr.

  a. Calculate the activity and activity coefficient of each component.

  b. Calculate the total pressure that the solution would have if it were ideal.

P8.14 Ratcliffe and Chao [*Canadian Journal of Chemical Engineering* 47 (1969), 148] obtained the following tabulated results for the variation of the total pressure above a solution of isopropanol ($P_1^* = 1008$ Torr) and *n*-decane ($P_2^* = 48.3$ Torr) as a function of the mole fraction of the *n*-decane in the solution and vapor phases. Using these data, calculate the activity coefficients for both components using a Raoult's law standard state.

P (Torr)	x_2	y_2
942.6	0.1312	0.0243
909.6	0.2040	0.0300
883.3	0.2714	0.0342
868.4	0.3360	0.0362
830.2	0.4425	0.0411
786.8	0.5578	0.0451
758.7	0.6036	0.0489

P8.15 At 39.9°C, the vapor pressure of water is 55.03 Torr (component A) and that of methanol (component B) is 255.6 Torr. Using data from the following table, calculate the activity coefficients for both components using a Raoult's law standard state.

x_A	y_A	P (Torr)
0.0490	0.0175	257.9
0.3120	0.1090	211.3
0.4750	0.1710	184.4
0.6535	0.2550	156.0
0.7905	0.3565	125.7

P8.16   The partial pressures of Br_2 above a solution containing CCl_4 as the solvent at 25°C are found to have the values listed in the following table as a function of the mole fraction of Br_2 in the solution [G. N. Lewis and H. Storch, *J. American Chemical Society 39* (1917), 2544]. Use these data and a graphical method to determine the Henry's law constant for Br_2 in CCl_4 at 25°C.

x_{Br_2}	P (Torr)	x_{Br_2}	P (Torr)
0.00394	1.52	0.0130	5.43
0.00420	1.60	0.0236	9.57
0.00599	2.39	0.0238	9.83
0.0102	4.27	0.0250	10.27

P8.17   The data from Problem P8.16 can be expressed in terms of the molality rather than the mole fraction of Br_2. Use the data from the following table and a graphical method to determine the Henry's law constant for Br_2 in CCl_4 at 25°C in terms of molality.

m_{Br_2}	P (Torr)	m_{Br_2}	P (Torr)
0.026	1.52	0.086	5.43
0.028	1.60	0.157	9.57
0.039	2.39	0.158	9.83
0.067	4.27	0.167	10.27

P8.18   The partial molar volumes of ethanol in a solution with $x_{H_2O} = 0.60$ at 25°C are 17.0 and 57.0 cm^3 mol^{-1}, respectively. Calculate the volume change on mixing sufficient ethanol with 2 mol of water to give this concentration. The densities of water and ethanol are 0.997 and 0.7893 g cm^{-3}, respectively, at this temperature.

P8.19   A solution is prepared by dissolving 32.5 g of a nonvolatile solute in 200. g of water. The vapor pressure above the solution is 21.85 Torr and the vapor pressure of pure water is 23.76 Torr at this temperature. What is the molecular weight of the solute?

P8.20   The heat of fusion of water is 6.008×10^3 J mol^{-1} at its normal melting point of 273.15 K. Calculate the freezing point depression constant K_f.

P8.21   The dissolution of 5.25 g of a substance in 565 g of benzene at 298 K raises the boiling point by 0.625°C. Note that $K_f = 5.12$ K kg mol^{-1}, $K_b = 2.53$ K kg mol^{-1}, and the density of benzene is 876.6 kg m^{-3}. Calculate the freezing point depression, the ratio of the vapor pressure above the solution to that of the pure solvent, the osmotic pressure, and the molecular weight of the solute. Note that $P^*_{benzene} = 103$ Torr at 298 K.

P8.22   A sample of glucose ($C_6H_{12}O_6$) of mass 1.25 g is placed in a test tube of radius 1.00 cm. The bottom of the test tube is a membrane that is semipermeable to water. The tube is partially immersed in a beaker of water at 298 K so that the bottom of the test tube is only slightly below the level of the water in the beaker. The density of water at this temperature is 997 kg m^{-3}. After equilibrium is reached, how high is the water level of the water in the tube above that in the beaker? What is the value of the osmotic pressure? You may find the approximation In $(1/1 + x) \approx -x$ useful.

P8.23   The osmotic pressure of an unknown substance is measured at 298 K. Determine the molecular weight if the concentration of this substance is 25.5 kg m^{-3} and the osmotic pressure is 4.50×10^4 Pa. The density of the solution is 997 kg m^{-3}.

P8.24   To extend the safe diving limit, both oxygen and nitrogen must be reduced in the breathing mixture. One way to do this is to mix oxygen with helium. Assume a mixture of 10.% oxygen and 90.% helium. Assuming Henry's law behavior, calculate the levels of oxygen and helium in the blood of a diver at 100 m. Assume $T = 298$ K.

P8.25   1.053 g of beef heart myoglobin dissolved in 50.0 ml of water at T = 298 K generates sufficient osmotic pressure to support a column of solution of height d. If the molar mass of myoglobin is 16.9 kg per mole, calculate d.

P8.26   The average osmotic pressure of human blood versus water is 7.60 atm. at a temperature of $T = 310$. K. Calculate the total solute concentration in human blood.

P8.27   An ideal dilute solution is formed by dissolving the solute A in the solvent B. Write expressions equivalent to Equations (8.9) through (8.13) for this case.

P8.28   A solution is made up of 184.2 g of ethanol and 108.1 g of H_2O. If the volume of the solution is 333.4 cm^3 and the partial molar volume of H_2O is 17.0 cm^3, what is the partial molar volume of ethanol under these conditions?

P8.29   Calculate the solubility of H_2S in 1 L of water if its pressure above the solution is 3.25 bar. The density of water at this temperature is 997 kg m^{-3}.

P8.30   The densities of pure water and ethanol are 997 and 789 kg m^{-3}, respectively. The partial molar volumes of ethanol and water in a solution with $x_{ethanol} = 0.20$ are 55.2 and 17.8×10^{-3} L mol^{-1}, respectively. Calculate the change in volume relative to the pure components when 1.00 L of a solution with $x_{ethanol} = 0.20$ is prepared.

P8.31   At a given temperature, a nonideal solution of the volatile components A and B has a vapor pressure of 832 Torr. For this solution, $y_A = 0.404$. In addition, $x_A = 0.285$, $P^*_A = 591$ Torr, and $P^*_B = 503$ Torr. Calculate the activity and activity coefficient of A and B.

P8.32   Calculate the activity and activity coefficient for CS_2 at $x_{CS_2} = 0.7220$ using the data in Table 8.3 for both a Raoult's law and a Henry's law standard state.

P8.33   When a solute shows very small activity coefficients at relatively low concentrations, this is evidence for strong interactions between the solute molecules. For example, the activity coefficient for 6-methylpurine is 0.329 at 0.20 m, whereas the activity coefficient of cytidine is more than twice that value at a comparable concentration. This is an indication that 6-methylpurine may be aggregating.

The simplest form of aggregation is dimer formation. If M forms a dimer D in solution we have the equilibrium: $2M \rightleftarrows D$. Assume the equilibrium constant is $K = \dfrac{C_D}{C_M^2}$ where C_D is the concentration of dimer, C_M is the concentration of monomer and $C_T = 2\,C_D + C_M$. The chemical potential for the monomer is

$$\mu_M = \mu_M^0 + RT \ln C_M$$

Dimerization can be treated as a type of non-ideality where $\mu_M = \mu_M^0 + RT \ln \gamma C_T$ where the activity coefficient $\gamma = 1$ if no dimer forms and $\gamma < 1$ if a dimer forms.

a. Using the two equations for the chemical potential of M, given above, obtain an expression for the activity coefficient γ in terms of C_M and C_T.

b. Using the expression for the equilibrium constant K and your result from part a show that

$$\gamma = \frac{(1 + 8KC_T)^{1/2}}{4KC_T} - \frac{1}{4KC_T}$$

c. Using the data given above, calculate the equilibrium constant for the dimerization of 6-methylpurine. Estimate the equilibrium constant for the dimerization of cytidine.

P8.34   Suppose the drug actinomycin at a concentration of 0.00122 M and $T = 278$ K forms a dimer with an equilibrium constant of $K = 3600$. Using your result from Problem P8.33b, calculate the activity coefficient of 0.00122 M actinomycin at $T = 278$ K.

P8.35   Calculate the change in the freezing point of water if 0.0053 g of a protein with molecular weight 10083 g mol^{-1} is dissolved in 100. mL of water.

P8.36   Assume sucrose and water form an ideal solution. What is the equilibrium vapor pressure of a solution of 2.0 grams of sucrose (molecular weight 342g mol^{-1}) at $T = 293$ K if the vapor pressure of pure water at 293 K is 17.54 Torr. What is the osmotic pressure of the sucrose solution versus pure water? Assume 100.0 g of water.

P8.37   Red blood cells do not swell or contract in a solution composed of 103 g of sucrose dissolved in a kilogram of water. Assuming the red blood cell membrane is impermeable to sucrose, calculate the osmotic pressure exerted by the cell cytoplasm at $T = 310$ K relative to pure water. Assume sucrose forms an ideal solution in water.

P8.38   When the cells of the skeletal vacuole of a frog were placed in a series of NaCl solutions of different concentrations at 298 K, it was found that the cells remained unchanged in a 0.70 % (by weight) NaCl solution. For a 0.70 % NaCl solution the freezing point is depressed by -0.406 K. What is the osmotic pressure of the cell cytoplasm at $T = 298$ K?

P8.39   At high altitudes, mountain climbers are unable to absorb a sufficient amount of O_2 into their bloodstreams to maintain a high activity level. At a pressure of 1 bar, blood is typically 95% saturated with O_2, but near 18,000 feet where the pressure is 0.50 bar, the corresponding degree of saturation is 71%. Assuming that the Henry's law constant for blood is the same as for water, calculate the amount of O_2 dissolved in 1.00 L of blood for pressures of 1 bar and 0.500 bar. Air contains 20.99% O_2 by volume. Assume that the density of blood is 998 kg m^{-3}.

P8.40   The osmotic pressure developed by a protein solution containing 10. mg mL^{-1}, is 0.0244 atm. Calculate the molecular weight of the protein. Assume $T = 298$ K.

P8.41   The following data on the osmotic pressure of solutions of bovine serum albumin were obtained from the work of Scatchard et al., *J. American Chemical Society* **68** (1946), 2320:

Concentration (g L^{-1})	27.28	56.20	8.95	17.69
Osmotic Pressure (mmHg)	8.35	19.33	2.51	5.07

Calculate the molecular weight of bovine serum albumin. Assume $T = 298$ K.

P8.42   The following data on the osmotic pressure of solutions of polyisobutylene in cyclohexane have been reported:

Concentration (g L^{-1})	0.0200	0.0150	0.0100	0.0075	0.0050	0.0025
Osmotic Pressure (atm)	0.0117	0.0066	0.0030	0.00173	0.00090	0.00035

Calculate the molecular weight of polyisobutylene.

P8.43   Assume an aqueous solution contains equal numbers of polynucleotides with molecular weights of roughly 20,000, 50,000, and 100,000, units of g mol^{-1}. Calculate the osmotic pressure of this solution at $T = 298$ K. Assume 1.00 g of polynucleotide is dissolved in 1.00 L of water.

P8.44   Because of problems associated with oxygen toxicity, nitrogen narcosis, and decompression sickness, the maximum safe limit depth for divers breathing air is 61 m. Calculate the pressure on a diver at this depth. Assuming air is about 20% oxygen and 80% nitrogen, calculate the levels of oxygen and nitrogen in the blood at 61 m. Assume Henry's law behavior. Assume $T = 298$ K.

P8.45   The concentrations in moles per kilogram of water for the dominant salts in sea water are:

Ion	Cl$^-$	Na$^+$	Mg^{2+}	SO$_4^{2-}$	Ca^{2+}
Mol kg^{-1}	0.546	0.456	0.053	0.028	0.010

Calculate the osmotic pressure exerted by sea water at $T = 298$ K. Suppose sea water is separated from pure water by a membrane that is permeable to water but impermeable to the ions in sea water. Assuming the density of sea water is about equal to pure water at $T = 298$ K, calculate the column of sea water that would be supported by osmotic pressure.

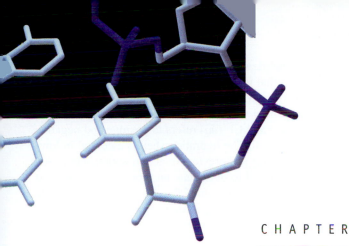

CHAPTER

9

Electrolyte Solutions, Electrochemical Cells, and Redox Reactions

Electrolyte solutions are quite different from the ideal and real solutions of neutral solutes discussed in Chapter 8. The fundamental reason for this difference is that solutes in electrolyte solutions exist as solvated positive and negative ions. Why are electrolyte and nonelectrolyte solutions so different? The Coulomb interactions between ions in an electrolyte solution are of much longer range than the interactions between neutral solutes. For this reason, electrolyte solutions deviate from ideal behavior at much lower concentrations than do solutions of neutral solutes. Although a formula unit of an electrolyte dissociates into positive and negative ions, only the mean activity and activity coefficient of these ions are accessible through experiments. The Debye–Hückel limiting law provides a useful way to calculate activity coefficients for dilute electrolyte solutions. Redox reactions that take place in electrolyte solutions can be used to construct electrochemical cells. Aside from their practical application as batteries, electrochemical cells can be used to determine equilibrium constants for redox reactions and to determine activity coefficients for the species in an electrolyte solution.

Redox reactions are also vital components of an organism's metabolic system and comprise a large part of the machinery of the respiratory systems of aerobic cells, for example. Biological redox chemistry will be reviewed in Chapters 10 and 11, but the foundation for understanding the energetics of cellular redox reactions is laid down in this chapter. In addition, biological polymers exist in solutions that contain high levels of electrolytes including Na^+, K^+, Ca^{2+}, and Cl^-. At physiological pH, many proteins and nucleic acids exist in charged form. Understanding the properties of solutions that contain charged biological polymers and counter ions is another subject treated in this chapter.

9.1 The Enthalpy, Entropy, and Gibbs Energy of Ion Formation in Solutions

In this chapter, materials, called **electrolytes,** are discussed that dissociate into positively and negatively charged mobile solvated ions when dissolved in an appropriate solvent. Consider the following overall reaction in water:

$$1/2\ H_2(g) + 1/2\ Cl_2(g) \rightarrow H^+(aq) + Cl^-(aq) \tag{9.1}$$

in which $H^+(aq)$ and $Cl^-(aq)$ represent mobile solvated ions. Although similar in structure, Equation (9.1) represents a reaction that is quite different than the gas-phase dissociation of an HCl molecule to give $H^+(g)$ and $Cl^-(g)$. For the aqueous phase reaction, $\Delta H_{reaction}$ is -167.2 kJ mol^{-1} when measured with constant pressure calorimetry (see Chapter 4). The shorthand notation $H^+(aq)$ and $Cl^-(aq)$ refers to positive and negative ions as well as their associated hydration shell. The hydration shell is essential in lowering the energy of the ions, thereby making the previous reaction spontaneous. Although energy flow into the system is required to dissociate and ionize hydrogen and chlorine, even more energy is gained in the orientation of the dipolar water molecule around the ions in the **solvation shell.** Therefore, the reaction is exothermic.

The standard state enthalpy for this reaction can be written in terms of formation enthalpies:

$$\Delta H_{reaction}^\circ = \Delta H_f^\circ(H^+,aq) + \Delta H_f^\circ(Cl^-,aq) \tag{9.2}$$

There is no contribution of $H_2(g)$ and $Cl_2(g)$ to $\Delta H_{reaction}^\circ$ in Equation (9.2) because ΔH_f° for a pure element in its standard state is zero.

Unfortunately, no direct calorimetric experiment can measure only the heat of formation of the solvated anion or cation. This is the case because the solution must remain electrically neutral; therefore, any dissociation reaction of a neutral solute must produce both anions and cations. As we saw in Chapter 4, tabulated values of formation enthalpies, entropies, and Gibbs energies for various chemical species are very useful. How can this information be obtained for individual solvated cations and anions?

The discussion in the rest of this chapter is restricted to aqueous solutions, for which water is the solvent. Values of thermodynamic functions for anions and cations in aqueous solutions can be obtained by making an appropriate choice for the zero of ΔH_f°, ΔG_f°, and S°. By convention, the formation Gibbs energy for $H^+(aq)$ at unit activity is set equal to zero at all temperatures:

$$\Delta G_f^\circ(H^+,aq) = 0 \text{ for all } T \tag{9.3}$$

With this choice,

$$S^\circ(H^+,aq) = -\left(\frac{\partial \Delta G_f^\circ(H^+,aq)}{\partial T}\right)_P = 0 \text{ and} \tag{9.4}$$

$$\Delta H_f^\circ(H^+,aq) = \Delta G_f^\circ(H^+,aq) + TS^\circ(H^+,aq) = 0$$

Using the convention of Equation (9.3), which has the consequences shown in Equation (9.4), the values of $\Delta H_f^\circ, \Delta G_f^\circ$, and S° for an individual ion can be assigned numerical values, as shown next.

As discussed earlier, $\Delta H_{reaction}^\circ$ for the reaction $1/2 \, H_2(g) + 1/2 \, Cl_2(g) \rightarrow H^+(aq) + Cl^-(aq)$ can be directly measured. The value of $\Delta G_{reaction}^\circ$ can be determined from $\Delta G_{reaction}^\circ = -RT \ln K$ by measuring the degree of dissociation in the reaction, using the solution conductivity, and $\Delta S_{reaction}^\circ$ can be determined from the relation

$$\Delta S_{reaction}^\circ = \frac{\Delta H_{reaction}^\circ - \Delta G_{reaction}^\circ}{T}$$

Using the conventions stated in Equations (9.3) and (9.4) together with the conventions regarding ΔH_f° and ΔG_f° for pure elements, $\Delta H_{reaction}^\circ = \Delta H_f^\circ(Cl^-,aq)$, $\Delta G_{reaction}^\circ = \Delta G_f^\circ(Cl^-,aq)$ and

$$\Delta S_{reaction}^\circ = S^\circ(Cl^-,aq) - \frac{1}{2}S^\circ(H_2,g) - \frac{1}{2}S^\circ(Cl_2,g)$$

for the reaction under discussion. In this way, the numerical values $\Delta H_f(Cl^-,aq) = -167.2$ kJ mol^{-1}, $S^\circ(Cl^-,aq) = 56.5$ J K^{-1} mol^{-1}, and $\Delta G_f^\circ(Cl^-,aq) = -131.2$ kJ mol^{-1} can be obtained.

These values can be used to determine the formation data for other ions. To illustrate how this is done, consider the following reaction:

$$NaCl(s) \rightarrow Na^+(aq) + Cl^-(aq) \tag{9.5}$$

for which the standard reaction enthalpy is found to be $+3.90 \text{ kJ mol}^{-1}$. For this reaction,

$$\Delta H^\circ_{reaction} = \Delta H^\circ_f(\text{Cl}^-, aq) + \Delta H^\circ_f(\text{Na}^+, aq) - \Delta H^\circ_f(\text{NaCl}, s) \quad (9.6)$$

We use the tabulated value of $\Delta H^\circ_f(\text{NaCl}, s) = -411.2 \text{ kJ mol}^{-1}$ and the value for $\Delta H^\circ_f(\text{Cl}^-, aq)$ just determined to obtain a value for $\Delta H^\circ_f(\text{Na}^+, aq) = -240.1 \text{ kJ mol}^{-1}$. Proceeding to other reactions that involve either $\text{Na}^+(aq)$ or $\text{Cl}^-(aq)$, the enthalpies of formation of the counter ions can be determined. This procedure can be extended to include other ions. Values for ΔG°_f and S° can be determined in a similar fashion. Values for ΔH°_f, ΔG°_f, and S° for aqueous ionic species are tabulated in Table 9.1. These thermodynamic quantities are called **conventional formation enthalpies, conventional Gibbs energies of formation,** and **conventional formation entropies** because of the convention described earlier.

Note that ΔH°_f, ΔG°_f, and S°_f for ions are defined relative to $\text{H}^+(aq)$. Negative values for ΔH°_f indicate that the formation of the solvated ion is more exothermic than the formation of $\text{H}^+(aq)$. A similar statement can be made for ΔG°_f. Generally speaking, ΔH°_f for multiply charged ions is more negative than that of singly charged ions, and ΔH°_f for a given charge is more negative for smaller ions because of the stronger electrostatic attraction between the multiply charged or smaller ion and the water in the solvation shell.

Recall from Section 5.8 that the entropy of an atom or molecule was shown to be always positive. This is not the case for solvated ions because the entropy is measured relative to $\text{H}^+(aq)$. The entropy decreases as the hydration shell is formed because liquid water molecules are converted to relatively immobile molecules. Ions with a negative value for the conventional standard entropy such as $\text{Mg}^{2+}(aq)$, $\text{Zn}^{2+}(aq)$, and $\text{PO}_4^{3-}(aq)$ have a larger charge-to-size ratio than $\text{H}^+(aq)$. For this reason, the solvation shell is more tightly bound. Conversely, ions with a positive value for the standard entropy such as $\text{Na}^+(aq)$, $\text{Cs}^+(aq)$, and $\text{NO}_3^-(aq)$ have a smaller charge-to-size ratio than $\text{H}^+(aq)$.

TABLE 9.1 **Conventional Formation Enthalpies, Gibbs Energies, and Entropies of Selected Aqueous Anions and Cations**

Ion	ΔH°_f (kJ mol^{-1})	ΔG°_f (kJ mol^{-1})	S° (J K^{-1} mol^{-1})
$\text{Ag}^+(aq)$	105.6	77.1	72.7
$\text{Br}^-(aq)$	−121.6	−104.0	82.4
$\text{Ca}^{2+}(aq)$	−542.8	−553.6	−53.1
$\text{Cl}^-(aq)$	−167.2	−131.2	56.5
$\text{Cs}^+(aq)$	−258.3	−292.0	133.1
$\text{Cu}^+(aq)$	71.7	50.0	40.6
$\text{Cu}^{2+}(aq)$	64.8	65.5	−99.6
$\text{F}^-(aq)$	−332.6	−278.8	−13.8
$\text{H}^+(aq)$	0	0	0
$\text{I}^-(aq)$	−55.2	−51.6	111.3
$\text{K}^+(aq)$	−252.4	−283.3	102.5
$\text{Li}^+(aq)$	−278.5	−293.3	13.4
$\text{Mg}^{2+}(aq)$	−466.9	−454.8	−138.1
$\text{NO}_3^-(aq)$	−207.4	−111.3	146.4
$\text{Na}^+(aq)$	−240.1	−261.9	59.0
$\text{OH}^-(aq)$	−230.0	−157.2	−10.9
$\text{PO}_4^{3-}(aq)$	−1277.4	−1018.7	−220.5
$\text{SO}_4^{2-}(aq)$	−909.3	−744.5	20.1
$\text{Zn}^{2+}(aq)$	−153.9	−147.1	−112.1

Source: D. R. Lide, Ed., *Handbook of Chemistry and Physics,* 83rd ed., CRC Press, Boca Raton, FL, 2002.

9.2 Understanding the Thermodynamics of Ion Formation and Solvation

As discussed in the preceding section, ΔH_f°, ΔG_f°, and S° cannot be determined for an individual ion in a calorimetric experiment. However, as seen next, values for thermodynamic functions associated with individual ions can be calculated with a reasonable level of confidence using a thermodynamic model. This result allows the conventional values of ΔH_f°, ΔG_f°, and S° to be converted to absolute values for individual ions. In the following discussion, the focus is on ΔG_f°.

We first discuss the individual contributions to ΔG_f°, and do so by analyzing the following sequence of steps that describe the formation of $H^+(aq)$ and $Cl^-(aq)$:

$1/2\ H_2(g) \rightarrow H(g)$	$\Delta G_{reaction}^\circ = 203.3\ \text{kJ mol}^{-1}$
$1/2\ Cl_2(g) \rightarrow Cl(g)$	$\Delta G_{reaction}^\circ = 105.7\ \text{kJ mol}^{-1}$
$H(g) \rightarrow H^+(g) + e^-$	$\Delta G_{reaction}^\circ = 1312\ \text{kJ mol}^{-1}$
$Cl(g) + e^- \rightarrow Cl^-(g)$	$\Delta G_{reaction}^\circ = -349\ \text{kJ mol}^{-1}$
$Cl^-(g) \rightarrow Cl^-(aq)$	$\Delta G_{reaction}^\circ = \Delta G_{solvation}^\circ(Cl^-,aq)$
$H^+(g) \rightarrow H^+(aq)$	$\Delta G_{reaction}^\circ = \Delta G_{solvation}^\circ(H^+,aq)$

$$1/2\ H_2(g) + 1/2\ Cl_2(g) \rightarrow H^+(aq) + Cl^-(aq) \qquad \Delta G_{reaction}^\circ = -131.2\ \text{kJ mol}^{-1}$$

This information is shown pictorially in Figure 9.1. Because G is a state function, both the green and tan shaded paths must have the same ΔG value. The first two reactions in this sequence are the dissociation of the molecules in the gas phase. The second two reactions are the formation of ions, and ΔG° is determined from the known ionization energy and electron affinity.

The change in the Gibbs energy for the overall process is

$$\Delta G_{reaction}^\circ = \Delta G_{solvation}^\circ(Cl^-,aq) + \Delta G_{solvation}^\circ(H^+,aq) + 1272\ \text{kJ mol}^{-1} \qquad (9.7)$$

Equation (9.7) allows us to relate the $\Delta G_{solution}$ of the solvated H^+ and Cl^- ions with ΔG for the overall reaction.

As Equation (9.7) shows, $\Delta G_{solvation}^\circ$ plays a critical role in the determination of the Gibbs energies of ion formation. Although $\Delta G_{solvation}^\circ$ of an individual cation or anion cannot be determined experimentally, it can be estimated using a model developed by Max Born. In this model, the solvent is treated as a uniform fluid with the appropriate dielectric constant, and the ion is treated as a charged sphere. How can $\Delta G_{solvation}^\circ$ be calculated with these assumptions? At constant T and P, the nonexpansion work for a reversible process equals ΔG for the process. Therefore, if the reversible work associated with solvation can be calculated, ΔG for the process is known. Imagine a process in which a neutral atom A gains the charge q, first in a vacuum and secondly in a uniform dielectric medium. The value of $\Delta G_{solvation}^\circ$ of an ion with a charge q is the reversible work for the process $(A \rightarrow A^q)_{solution}$ minus that for the reversible process $(A \rightarrow A^q)_{vacuum}$.

FIGURE 9.1
The change in the Gibbs energy, ΔG°, is shown pictorially for two different paths starting with $1/2\ H_2(g)$ and $1/2\ Cl_2(g)$ and ending with $H^+(aq) + Cl^-(aq)$. The units for the numbers are kJ mol^{-1}. Because ΔG° is the same for both paths, $\Delta G_{solvation}^\circ(H^+,aq)$ can be expressed in terms of gas-phase dissociation and ionization energies.

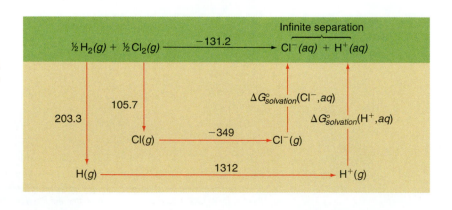

The electrical potential around a sphere of radius r with the charge q' is given by $\phi = q'/4\pi\varepsilon r$. From electrostatics, the work in charging the sphere by the additional amount dq is $\phi\,dq$. Therefore, the work in charging a neutral sphere in vacuum to the charge q is

$$w = \int_0^q \frac{q'\,dq'}{4\pi\varepsilon_0 r} = \frac{1}{4\pi\varepsilon_0 r}\int_0^q q'dq' = \frac{q^2}{8\pi\varepsilon_0 r} \tag{9.8}$$

where ε_0 is the permittivity of free space. The work of the same process in a solvent is $q^2/8\pi\varepsilon_0\varepsilon_r r$, where ε_r is the relative permittivity (dielectric constant) of the solvent. Consequently, the electrostatic component to the molar solvation Gibbs energy for an ion of charge $q = ze$ is given by

$$\Delta G^\circ_{solvation} = \frac{z^2 e^2 N_A}{8\pi\varepsilon_0 r}\left(\frac{1}{\varepsilon_r} - 1\right) \tag{9.9}$$

Because $\varepsilon_r > 1$, $\Delta G^\circ_{solvation} < 0$, showing that solvation is a spontaneous process. Values for ε_r for a number of solvents are listed in Table 9.2 (see Appendix B, Data Tables).

To test this model, one compares measured values of the absolute values of $\Delta G^\circ_{solvation}$ with the functional form proposed in Equation (9.9). This requires knowledge of $\Delta G^\circ_{solvation}(H^+, aq)$ and experimentally determined values of $\Delta G^\circ_{solvation}$ referenced to $H^+(aq)$. It turns out that $\Delta G^\circ_{solvation}(H^+, aq)$ can be calculated by comparing values of ΔG_f for gaseous positive and negative ions of the negatively charged halogen ions and the positively charged alkali metal ions. A similar analysis can be used to obtain $\Delta H^\circ_{solvation}(H^+, aq)$ and $S^\circ_{solvation}(H^+, aq)$. Because the justification is involved, the results are simply stated here. Because of the models used, these results exhibit greater uncertainty than do the conventional values referenced to $H^+(aq)$:

$$\Delta H^\circ_{solvation}(H^+, aq) \approx -1090 \text{ kJ mol}^{-1}$$

$$\Delta G^\circ_{solvation}(H^+, aq) \approx -1050 \text{ kJ mol}^{-1} \tag{9.10}$$

$$S^\circ_{solvation}(H^+, aq) \approx -130 \text{ J mol}^{-1}\text{ K}^{-1}$$

The values listed in Equation (9.10) can be used to calculate absolute values of $\Delta H^\circ_{solvation}$, $\Delta G^\circ_{solvation}$, and $S^\circ_{solvation}$ for other ions from the conventional values referenced to $H^+(aq)$. These calculated absolute values can be used to test the validity of the Born model. If the model is valid, a plot of $\Delta G^\circ_{solvation}$ versus z^2/r should give a straight line as shown by Equation (9.9) and the data points for individual ions should lie on the line. The results of such a test of validity are shown in Figure 9.2, where r is the ionic radius obtained from crystal structure determinations.

The first and second clusters of data points in Figure 9.2 are for singly and doubly charged ions, respectively. The data are compared with the result predicted by Equation (9.9) in Figure 9.2a. As can be seen from the figure, the trends are reproduced, but there is no quantitative agreement. The agreement can be considerably improved by using an effective radius for the ion rather than the ionic radius from crystal structure determinations. The effective radius is defined as the distance from the center of the ion to the center of charge in the dipolar water molecule. Latimer, Pitzer, and Slansky [*J. Chemical Physics* 7 (1939), 109] found the best agreement with the Born equation by adding 0.085 nm to the crystal radius of positive ions, and 0.100 nm to the crystal radius for negative ions to account for the fact that the H_2O molecule is not a point dipole. This difference is explained by the fact that the center of charge in the water molecule is closer to positive ions than to negative ions. Figure 9.2b shows that the agreement obtained between the predictions of Equation (9.9) and experimental values is very good if this correction to the ionic radii is made.

Figure 9.2 shows good agreement between the predictions of the Born model and calculated values for $\Delta G^\circ_{solvation}$. However, because of uncertainties about the numerical values of the ionic radii and for the dielectric constant of the immobilized water in the solvation shell of an ion, values obtained from different models differ by about ± 50 kJ mol^{-1} for the solvation enthalpy and Gibbs energy and ± 10 J K^{-1} mol^{-1} for

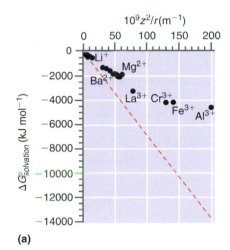

(a)

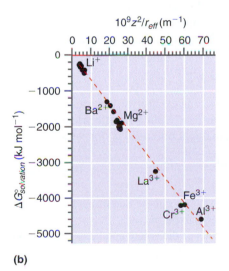

(b)

FIGURE 9.2

(a) The solvation energy calculated using the Born model is shown as a function of z^2/r. (b) The same results are shown as a function of z^2/r_{eff}. (See text.) The dashed line shows the behavior predicted by Equation (9.9).

the solvation entropy. Because this absolute uncertainty is large compared to the relative uncertainty of the thermodynamic functions using the convention described in Equations (9.3) and (9.4), absolute values of ΔH_f°, ΔG_f°, and S° are not generally used by chemists for solvated ions in aqueous solutions.

9.3 Activities and Activity Coefficients for Electrolyte Solutions

The ideal dilute solution model presented for the activity and activity coefficient of components of real solutions in Chapter 8 is not valid for electrolyte solutions. This is the case because solute–solute interactions are dominated by the long-range electrostatic forces present between ions in electrolyte solutions. For example, the reaction that occurs when NaCl is dissolved in water is

$$\text{NaCl}(s) + \text{H}_2\text{O}(l) \rightarrow \text{Na}^+(aq) + \text{Cl}^-(aq) \tag{9.11}$$

In dilute solutions, NaCl is completely dissociated. Therefore, the solute–solute interactions are electrostatic in nature.

Although the concepts of activity and activity coefficients introduced in Chapter 8 are applicable to electrolytes, these concepts must be formulated differently for electrolytes to include the Coulomb interactions among ions. We next discuss a model for electrolyte solutions that is stated in terms of the chemical potentials, activities, and activity coefficients of the individual ionic species. However, only mean activities and activity coefficients (to be defined later) can be determined through experiments.

Consider the Gibbs energy of the solution, which can be written as

$$G = n_{solvent}\mu_{solvent} + n_{solute}\mu_{solute} \tag{9.12}$$

For the general electrolyte $A_{v_+}B_{v_-}$ that dissociates completely, one can also write an equivalent expression for G:

$$G = n_{solvent}\mu_{solvent} + n_+\mu_+ + n_-\mu_- = n_{solvent}\mu_{solvent} + n_{solute}(v_+\mu_+ + v_-\mu_-) \tag{9.13}$$

where v_+ and v_- are the stoichiometric coefficients of the cations and anions, respectively, produced on dissociation of the electrolyte. In shorthand notation, an electrolyte is called a 1–1 electrolyte if $z_+ = 1$ and $z_- = 1$. Similarly, for a 2–3 electrolyte, $z_+ = 2$ and $z_- = 3$. Because Equations (9.12) and (9.13) describe the same solution, they are equivalent. Therefore,

$$\mu_{solute} = v_+\mu_+ + v_-\mu_- \tag{9.14}$$

Although this equation is formally correct for a strong electrolyte, one can never make a solution of either cations or anions alone, because any solution is electrically neutral. Therefore, it is useful to define a **mean ionic chemical potential** $\mu_\pm$ for the solute

$$\mu_\pm = \frac{\mu_{solute}}{v} = \frac{v_+\mu_+ + v_-\mu_-}{v} \tag{9.15}$$

where $v = v_+ + v_-$. The reason for doing so is that $\mu_\pm$ can be determined experimentally, whereas μ_+ and μ_- are not accessible through experiments.

The next task is to relate the chemical potentials of the solute and its individual ions to the activities of these species. For the individual ions,

$$\mu_+ = \mu_+^\circ + RT \ln a_+ \text{ and } \mu_- = \mu_-^\circ + RT \ln a_- \tag{9.16}$$

where the standard chemical potentials of the ions, μ_+° and μ_-°, are based on a Henry's law standard state. Substituting Equation (9.16) into Equation (9.15), an equation for the mean ionic chemical potential is obtained that is similar in structure to the expressions that we derived for the ideal dilute solution:

$$\mu_\pm = \mu_\pm^\circ + RT \ln a_\pm \tag{9.17}$$

The **mean ionic activity** $a_\pm$ is related to the individual ion activities by

$$a_\pm^\nu = a_+^{\nu_+} a_-^{\nu_-} \quad \text{or} \quad a_\pm = (a_+^{\nu_+} a_-^{\nu_-})^{1/\nu} \tag{9.18}$$

EXAMPLE PROBLEM 9.1

Write the mean ionic activities of $NaCl$, K_2SO_4, and H_3PO_4 in terms of the ionic activities of the individual anions and cations. Assume complete dissociation.

Solution

$$a_{NaCl}^2 = a_{Na^+} a_{Cl^-} \quad \text{or} \quad a_{NaCl} = \sqrt{a_{Na^+} a_{Cl^-}}$$

$$a_{K_2SO_4}^3 = a_{K^+}^2 a_{SO_4^{2-}} \quad \text{or} \quad a_{K_2SO_4} = (a_{K^+}^2 a_{SO_4^{2-}})^{1/3}$$

$$a_{H_3PO_4}^4 = a_{H^+}^3 a_{PO_4^{3-}} \quad \text{or} \quad a_{H_3PO_4} = (a_{H^+}^3 a_{PO_4^{3-}})^{1/4}$$

If the ionic activities are referenced to the concentration units of molality, then

$$a_+ = \frac{m_+}{m^\circ} \gamma_+ \quad \text{and} \quad a_- = \frac{m_-}{m^\circ} \gamma_- \tag{9.19}$$

where $m_+ = \nu_+ m$ and $m_- = \nu_- m$. Because the activity is unitless, the molality must be referenced to a standard state concentration chosen to be $m^\circ = 1$ mol kg^{-1}. As in Chapter 8, a hypothetical standard state based on molality is defined. In this standard state, Henry's law, which is valid in the limit $m \to 0$, is obeyed up to a concentration of $m = 1$ molal. Substitution of Equation (9.19) in Equation (9.18) shows that

$$a_\pm^\nu = \left(\frac{m_+}{m^\circ}\right)^{\nu_+} \left(\frac{m_-}{m^\circ}\right)^{\nu_-} \gamma_+^{\nu_+} \gamma_-^{\nu_-} \tag{9.20}$$

To simplify this notation, we define the **mean ionic molality** $m_\pm$ and **mean ionic activity coefficient** $\gamma_\pm$ by

$$m_\pm^\nu = m_+^{\nu_+} m_-^{\nu_-}$$

$$m_\pm = (\nu_+^{\nu_+} \nu_-^{\nu_-})^{1/\nu} m \quad \text{and} \tag{9.21}$$

$$\gamma_\pm^\nu = \gamma_+^{\nu_+} \gamma_-^{\nu_-}$$

$$\gamma_\pm = (\gamma_+^{\nu_+} \gamma_-^{\nu_-})^{1/\nu}$$

With these definitions, the mean ionic activity is related to the mean ionic activity coefficient and mean ionic molality as follows:

$$a_\pm^\nu = \left(\frac{m_\pm}{m^\circ}\right)^\nu \gamma_\pm^\nu \quad \text{or} \quad a_\pm = \left(\frac{m_\pm}{m^\circ}\right) \gamma_\pm \tag{9.22}$$

Equations (9.19) through (9.22) relate the activities, activity coefficients, and molalities of the individual ionic species to mean ionic quantities and measurable properties of the system such as the molality and activity of the solute. With these definitions, Equation (9.17) defines the chemical potential of the electrolyte solute in terms of its activity:

$$\mu_{solute} = \nu\mu_\pm^\circ + RT \ln a_\pm^\nu \tag{9.23}$$

Equations (9.20) and (9.21) can be used to express the chemical potential of the solute in terms of measurable or easily accessible quantities:

$$\mu_{solute} = [\nu\mu_\pm^\circ + RT \ln(\nu_+^{\nu_+} \nu_-^{\nu_-})] + \nu RT \ln\left(\frac{m}{m^\circ}\right) + \nu RT \ln \gamma_\pm \tag{9.24}$$

The first term in the square bracket is defined by the "normal" standard state, which is usually taken to be a Henry's law standard state. The second term is obtained from the chemical formula for the solute. These two terms can be combined to create a new standard state $\mu_\pm^{\circ\circ}$ defined by the terms in the square brackets in Equation (9.24):

$$\mu_{solute} = \mu_{solute}^{\circ\circ} + vRT \ln\left(\frac{m}{m^{\circ}}\right) + vRT \ln \gamma_{\pm} \qquad (9.25)$$

The first two terms in Equation (9.25) correspond to the "ideal" ionic solution, which is associated with $\gamma_{\pm} = 1$.

The last term in Equation (9.25), which is the most important term in this discussion, contains the deviations from ideal behavior. The mean activity coefficient $\gamma_{\pm}$ can be obtained through experiment. For example, the activity coefficient of the solvent can be determined by measuring the boiling point elevation, the freezing point depression, or the lowering of the vapor pressure above the solution upon solution formation. The activity of the solute is obtained from that of the solvent using the Gibbs–Duhem equation. As shown in Section 9.8, $\gamma_{\pm}$ can also be determined through measurements on electrochemical cells. In addition, a very useful theoretical model allows $\gamma_{\pm}$ to be calculated for dilute electrolytic solutions. This model is discussed in the next section.

9.4 Calculating $\gamma_{\pm}$ Using the Debye–Hückel Theory

None of the available models adequately explains the deviations from ideality for the solutions of neutral solutes discussed in Sections 8.1 through 8.4. This is the case because the deviations arise through A–A, B–B, and A–B interactions that are specific to components A and B. This precludes a general model that holds for arbitrary A and B. However, the situation for solutions of electrolytes is different.

Deviations from ideal solution behavior occur at a much lower concentration for electrolytes than for nonelectrolytes, because the dominant interaction between the ions in an electrolyte is a long-range electrostatic Coulomb interaction rather than a short-range van der Waals or chemical interaction. Because of its long range, the Coulomb interaction among the ions cannot be neglected even for very dilute solutions of electrolytes. The Coulomb interaction allows a model of electrolyte solutions to be formulated for the following reason: the attractive or repulsive interaction of two ions depends only on their charge and separation, and not on their chemical identity. Therefore, the solute–solute interactions can be modeled knowing only the charge on the ions, and the model becomes independent of the identity of the solute species.

Measurements of activity coefficients in electrolyte solutions show that $\gamma_{\pm} < 1$ for dilute solutions in the limit $m \to 0$. Because $\gamma_{\pm} < 1$, the chemical potential of the solute in a dilute solution is lower than that for a solution of uncharged solute species. Why is this the case? The lowering of μ_{solute} arises because the net electrostatic interaction among the ions surrounding an arbitrarily chosen central ion is attractive rather than repulsive. The model that describes the lowering of the energy of electrolytic solutions is due to Peter Debye and Erich Hückel. Rather than derive their results, the essential features of their model are described next.

The solute ions in the solvent give rise to a spatially dependent electrostatic potential, ϕ, which can be calculated if the spatial distribution of ions is known. In dilute electrolyte solutions, the energy increase or decrease experienced by an ion of charge $\pm ze$ if the potential ϕ could be turned on suddenly is small compared to the thermal energy. This condition can be expressed in the form

$$|\pm ze\phi| \ll kT \qquad (9.26)$$

In Equation (9.26), e is the charge on a proton, and k is Boltzmann's constant. In this limit, the dependence of ϕ on the spatial coordinates and the spatial distribution of the ions around an arbitrary central ion can be calculated. In contrast to the potential around an isolated ion in a dielectric medium, which is described by

$$\phi_{isolated\ ion}(r) = \frac{\pm ze}{4\pi\varepsilon_r\varepsilon_0 r} \qquad (9.27)$$

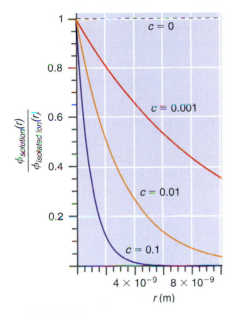

FIGURE 9.3
The ratio of the falloff in the electrostatic potential in the electrolyte solution to that for an isolated ion is shown as a function of the radial distance for three different molarities, c, of a 1–1 electrolyte such as NaCl.

the potential in the dilute electrolyte solution has the form

$$\phi_{solution}(r) = \frac{\pm ze}{4\pi\varepsilon_r\varepsilon_0 r}\exp(-\kappa r) \tag{9.28}$$

In Equations (9.27) and (9.28), ε_0 and ε_r are the permittivity of free space and the relative permittivity (dielectric constant) of the dielectric medium or solvent, respectively.

The Debye–Hückel theory shows that κ is related to the individual charges on the ions and to the solute molality m by

$$\kappa^2 = e^2 N_A (1000 \text{ L m}^{-3})\left(\frac{\nu_+ z_+^2 + \nu_- z_-^2}{\varepsilon_0\varepsilon_r kT}\right)\rho_{solvent} \tag{9.29}$$

From this formula, it can be seen that the screening becomes more effective as the concentration of the ionic species increases. Screening is also more effective for multiply charged ions and for larger values of ν_+ and ν_-.

The ratio

$$\frac{\phi_{solution}(r)}{\phi_{isolated\ ion}(r)} = e^{-\kappa r}$$

is plotted in Figure 9.3 for different values of c for an aqueous solution of a 1–1 electrolyte. Note that the potential falls off much more rapidly with the radial distance, r, in the electrolyte solution than in the uniform dielectric medium. Note also that the potential falls off more rapidly with increasing concentration of the electrolyte. The origin of this effect is that ions of sign opposite to the central ion are more likely to be found close to the central ion. These surrounding ions form a diffuse ion cloud around the central ion, as shown pictorially in Figure 9.4. If a spherical surface is drawn centered at the central ion, the net charge within the surface can be calculated. Calculations show that the net charge has the same sign as the central charge, falls off rapidly with distance, and is close to zero for $\kappa r \sim 8$. For larger values of κr, the central ion is completely screened by the diffuse ion cloud, meaning that the net charge is zero. Because of the lowering of energy that arises from the electrostatic interactions within the diffuse charge cloud, $\gamma_{\pm} < 1$ in dilute electrolyte solutions. The net effect of the diffuse ion cloud is to screen the central ion from the rest of the solution, and the quantity $1/\kappa$ is known as the **Debye–Hückel screening length.** Larger values of κ correspond to a smaller diffuse cloud, and a more effective screening.

It is convenient to combine the concentration-dependent terms that contribute to the screening length in the **ionic strength I,** which is defined by

$$I = \frac{m}{2}\sum_i (\nu_{i+}z_{i+}^2 + \nu_{i-}z_{i-}^2) = \frac{1}{2}\sum_i (m_{i+}z_{i+}^2 + m_{i-}z_{i-}^2) \tag{9.30}$$

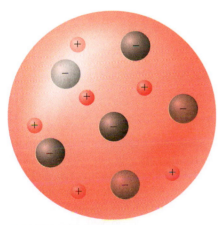

FIGURE 9.4
Pictorial rendering of the arrangement of ions about an arbitrary ion in an electrolyte solution. The central ion is more likely to have oppositely charged ions as neighbors. The large circle represents a sphere of radius $r \sim 8/\kappa$. From a point outside of this sphere, the charge on the central ion is essentially totally screened.

EXAMPLE PROBLEM 9.2

Calculate I for (a) a 0.050 molal solution of NaCl and for (b) a Na_2SO_4 solution of the same molality.

Solution

a. $I_{NaCl} = \dfrac{m}{2}(\nu_+ z_+^2 + \nu_- z_-^2) = \dfrac{0.050 \text{ mol kg}^{-1}}{2} \times (1 + 1) = 0.050 \text{ mol kg}^{-1}$

b. $I_{Na_2SO_4} = \dfrac{m}{2}(\nu_+ z_+^2 + \nu_- z_-^2) = \dfrac{0.050 \text{ mol kg}^{-1}}{2} \times (2 + 4) = 0.15 \text{ mol kg}^{-1}$

EXAMPLE PROBLEM 9.3

The molal concentrations of the principal ions in a sample of intracellular fluid are:

K^+, $m_{K^+} = 0.15\ m$; Na^+, $m_{Na^+} = 0.01\ m$; Mg^{2+}, $m_{Mg^{2+}} = 0.04\ m$; HPO_4^{2-}, $m_{HPO_4^{2-}} = 0.12\ m$

Calculate the ionic strength of the intracellular fluid.

Solution

$$I = \tfrac{1}{2}\sum_i m_i z_i^2 = \tfrac{1}{2}(0.15\ m + 0.01\ m + 0.04\ m \times 2^2 + 0.12\ m \times (-2)^2) = 0.40$$

Using the definition of the ionic strength, Equation (9.29) can be written in the form

$$\kappa = \sqrt{\frac{2e^2 N_A}{\varepsilon_0 kT}(1000\ \text{L m}^{-3})}\sqrt{\left(\frac{I}{\varepsilon_r}\right)\rho_{solvent}} \qquad (9.31)$$

$$= 2.91 \times 10^8 \sqrt{\frac{I/\text{mol kg}^{-1}}{\varepsilon_r}\frac{\rho_{solvent}}{\text{kg L}^{-1}}}\ \text{m}^{-1}\ \text{at 298 K}$$

The first term in this equation contains only fundamental constants that are independent of the solvent and solute as well as the temperature. The second term contains the ionic strength of the solution and the unitless relative permittivity of the solvent. For the more conventional units of mol L^{-1}, and for water, for which $\varepsilon_r = 78.5$, $\kappa = 3.29 \times 10^9\ \sqrt{I}\ \text{m}^{-1}$ at 298 K.

By calculating the charge distribution of the ions around the central ion and the work needed to charge these ions up to their charges z_+ and z_- from an initially neutral state, Debye and Hückel were able to obtain an expression for the mean ionic activity coefficient. It is given by

$$\ln \gamma_{\pm} = -|z_+ z_-| \frac{e^2 \kappa}{8\pi \varepsilon_0 \varepsilon_r kT} \qquad (9.32)$$

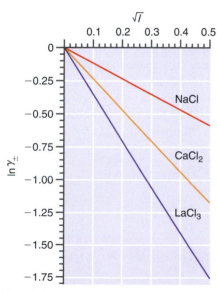

FIGURE 9.5
The decrease in the Debye–Hückel mean activity coefficient with the square root of the ionic strength is shown for a 1–1, a 1–2, and a 1–3 electrolyte, all of the same molality in the solute.

This equation is known as the **Debye–Hückel limiting law**. It is called a limiting law because Equation (9.32) is only obeyed for small values of the ionic strength. Note that because of the negative sign in Equation (9.32), $\gamma_{\pm} < 1$. From the concentration dependence of κ shown in Equation (9.31), the model predicts that $\ln \gamma_{\pm}$ decreases with the ionic strength as $\sqrt{I}$. This dependence is shown in Figure 9.5. Although all three solutions have the same solute concentration, they have different values for z^+ and z^-. For this reason, the three lines have a different slope.

Equation (9.32) can be simplified for a particular choice of solvent and temperature. For aqueous solutions at 298.15 K, the result is

$$\log \gamma_{\pm} = -0.5092 |z_+ z_-| \sqrt{I}\ \text{or}\ \ln \gamma_{\pm} = -1.173 |z_+ z_-| \sqrt{I} \qquad (9.33)$$

How well does the Debye–Hückel limiting law agree with experimental data? Figure 9.6 shows a comparison of the model with data on the aqueous solutions of $AgNO_3$ and $CaCl_2$. In both cases, $\gamma_{\pm}$ is plotted versus $\sqrt{I}$. The Debye–Hückel limiting law predicts that the data will fall on the line indicated in each figure. The data points deviate from the predicted behavior above $\sqrt{I} = 0.1$ for $AgNO_3$ ($m = 0.01$), and above $\sqrt{I} = 0.06$ for $CaCl_2$ ($m = 0.001$). In the limit that $I \to 0$, the limiting law is obeyed. However, the deviations are noticeable at a concentration for which a neutral solute would exhibit ideal behavior.

The deviations continue to increase with increasing ionic strength. Figure 9.7 shows experimental data for $ZnBr_2$ out to $\sqrt{I} = 5.5$, corresponding to $m = 10$. Note that, although the Debye–Hückel limiting law is obeyed as $I \to 0$, $\ln \gamma_{\pm}$ goes through a minimum and begins to increase with increasing ionic strength. At the highest value of the

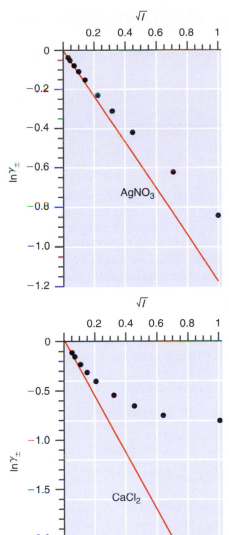

FIGURE 9.6
Activity coefficients for $AgNO_3$ and $CaCl_2$ are shown as a function of $\sqrt{I}$. The solid lines are the prediction of the Debye–Hückel theory.

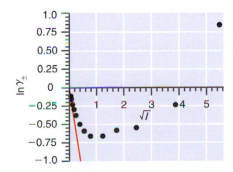

FIGURE 9.7
Mean activity coefficient for $ZnBr_2$ is shown as a function of $\sqrt{I}$. The solid line is the prediction of the Debye–Hückel theory.

ionic strength plotted, $\gamma_\pm = 2.32$, which is greater than 1. Although the deviations from ideal behavior are less pronounced in Figure 9.6, the trend is the same for all the solutes. The mean ionic activity coefficient $\gamma_\pm$ falls off more slowly with the ionic strength than is predicted by the Debye–Hückel limiting law. The behavior shown in Figure 9.7 is typical for most electrolytes: after passing through a minimum, $\gamma_\pm$ rises with increasing ionic strength, and for high values of I, $\gamma_\pm > 1$. Experimental values for $\gamma_\pm$ for a number of solutes at different concentrations in aqueous solution are listed in Table 9.3 (see Appendix B, Data Tables).

There are a number of reasons why the experimental values of $\gamma_\pm$ differ at high ionic strength from those calculated from the Debye–Hückel limiting law. They mainly involve the assumptions made in the model. The model assumes that the ions can be treated as point charges with zero volume, whereas ions and their associated solvation shells occupy a finite volume. As a result, there is an increase in the repulsive interaction among ions in an electrolyte over that predicted for point charges; this increase becomes more important as the concentration increases. Repulsive interactions raise the energy of the solution and, therefore, increase $\gamma_\pm$. The Debye–Hückel model also assumes that the solvent can be treated as a structureless dielectric medium. However, the ion is surrounded by a relatively ordered primary solvation shell, as well as by more loosely bound water molecules. The atomic-level structure of the solvation shell is not adequately represented by using the dielectric strength of bulk solvent. Another factor that has not been taken into account is that as the concentration increases, some ion pairing occurs such that the concentration of ionic species is less than would be calculated assuming complete dissociation.

Additionally, consider the fact that the water molecules in the solvation shell have effectively been removed from the solvent. For example, in an aqueous solution of H_2SO_4, approximately nine H_2O molecules are tightly bound per dissolved H_2SO_4 formula unit. Therefore, the number of moles of H_2O as solvent in 1 L of a 1 molar H_2SO_4 solution is reduced from 55 for pure H_2O to 46 in the solution. Consequently, the actual solute molarity is larger than that calculated by assuming that all the H_2O is in the form of solvent. Because the activity increases linearly with the actual molarity, $\gamma_\pm$ increases as the solute concentration increases. If there were no change in the enthalpy of solvation with concentration, all of the H_2O molecules would be removed from the solvent at a concentration of six molar H_2SO_4. Clearly, this assumption is unreasonable. What actually happens is that solvation becomes energetically less favorable as the H_2SO_4 concentration increases. This corresponds to a less negative value of $\ln \gamma_\pm$, or equivalently to an increase in $\gamma_\pm$. Summing up, many factors explain why the Debye–Hückel limiting law is only valid for small concentrations. Because of the complexity of these different factors, there is no simple formula based on theory that can replace the Debye–Hückel limiting law. However, the main trends exhibited in Figures 9.6 and 9.7 are reproduced in more sophisticated theories of electrolyte solutions.

Because none of the usual models are valid at high concentrations, empirical models that "improve" on the Debye–Hückel model by predicting an increase in $\gamma_\pm$ for high concentrations are in widespread use. An empirical modification of the Debye–Hückel limiting law that has the form

$$\log_{10} \gamma_\pm = -0.51 |z_+ z_-| \left[\frac{\left(\dfrac{I}{m^\circ}\right)^{1/2}}{1 + \left(\dfrac{I}{m^\circ}\right)^{1/2}} - 0.20\left(\dfrac{I}{m^\circ}\right) \right] \tag{9.34}$$

is known as the **Davies equation.** As seen in Figure 9.8, this equation for $\gamma_\pm$ shows the correct limiting behavior for low I values, and the trend at higher values of I is in better agreement with the experimental results shown in Figures 9.6 and 9.7.

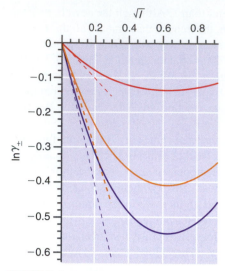

FIGURE 9.8

Comparison between the predictions of the Debye–Hückel limiting law (dashed lines) and the Davies equation (solid curves) for 1–1 (red), 1–2 (yellow), and 1–3 (blue) electrolytes.

9.5 Chemical Equilibrium in Electrolyte Solutions

As discussed in Section 8.13, the equilibrium constant in terms of activities is given by Equation (8.75):

$$K = \prod_i (a_i^{eq})^{\nu_j} \tag{9.35}$$

It is convenient to define the activity of a species relative to its molarity. In this case,

$$a_i = \gamma_i \frac{c_i}{c^\circ} \tag{9.36}$$

where γ_i is the activity coefficient of species i. We next specifically consider chemical equilibrium in electrolyte solutions, illustrating that activities rather than concentrations must be taken into account to accurately model equilibrium concentrations. We restrict our considerations to the range of ionic strengths for which the Debye–Hückel limiting law is valid. As an example, we calculate the degree of dissociation of MgF_2 in water. The equilibrium constant in terms of molarities for ionic salts is usually given the symbol K_{sp}, where the subscript refers to the solubility product. The equilibrium constant K_{sp} is unitless and has the value of 6.4×10^{-9} for the reaction shown in Equation (9.37). Values for K_{sp} are generally tabulated for reduced concentration units of molarity (c/c°) rather than molality (m/m°), and values for selected substances are listed in Table 9.4. For highly dilute solutions, the numerical value of the concentration is the same on both scales.

We next consider dissociation of MgF_2 in an aqueous solution:

$$MgF_2(s) \rightarrow Mg^{2+}(aq) + 2F^-(aq) \tag{9.37}$$

Because the activity of the pure solid can be set equal to 1,

$$K_{sp} = a_{Mg^{2+}} a_{F^-}^2 = \left(\frac{c_{Mg^{2+}}}{c^\circ}\right)\left(\frac{c_{F^-}}{c^\circ}\right)^2 \gamma_\pm^3 = 6.4 \times 10^{-9} \tag{9.38}$$

From the stoichiometry of the overall equation, we know that $c_{F^-} = 2c_{Mg^{2+}}$, but Equation (9.38) still contains two unknowns, $\gamma_\pm$ and c_{F^-}, that we solve for iteratively.

First assume that $\gamma_\pm = 1$ and solve Equation (9.38) for c_{F^-}, giving $c_{Mg^{2+}} = 1.17 \times 10^{-3}$ mol L^{-1}. Next calculate the ionic strength from

$$I = \frac{1}{2}(z_+^2 m_+ + z_-^2 m_-) = \frac{1}{2}(4 \times 1.17 \times 10^{-3} + 2.34 \times 10^{-3}) \tag{9.39}$$

$$= 3.51 \times 10^{-3} \text{ mol } L^{-1}$$

and recalculate $\gamma_\pm$ from the Debye–Hückel limiting law of Equation (9.29), giving $\gamma_\pm = 0.870$. Substitute this improved value of $\gamma_\pm$ in Equation (9.38), giving $c_{Mg^{2+}} = 1.34 \times 10^{-3}$ mol L^{-1}. A second iteration gives $\gamma_\pm = 0.862$ and $c_{Mg^{2+}} = 1.36 \times 10^{-3}$ mol L^{-1},

TABLE 9.4 **Solubility Product Constants (Molarity Based) for Selected Salts**

Salt	K_{sp}	Salt	K_{sp}
AgBr	4.9×10^{-13}	$CaSO_4$	4.9×10^{-6}
AgCl	1.8×10^{-10}	$Mg(OH)_2$	5.6×10^{-11}
AgI	8.5×10^{-17}	$Mn(OH)_2$	1.9×10^{-13}
$Ba(OH)_2$	5.0×10^{-3}	$PbCl_2$	1.6×10^{-5}
$BaSO_4$	1.1×10^{-10}	$PbSO_4$	1.8×10^{-8}
$CaCO_3$	3.4×10^{-9}	ZnS	1.6×10^{-23}

Source: D. R. Lide, Ed., *Handbook of Chemistry and Physics,* 83rd ed., CRC Press, Boca Raton, FL, 2002.

and a third iteration gives $\gamma_{\pm} = 0.861$ and $c_{Mg^{2+}} = 1.36 \times 10^{-3}$ mol L^{-1}. This result is sufficiently accurate. These results show that assuming that $\gamma_{\pm} = 1$ leads to an unacceptably large error of 14% in $c_{Mg^{2+}}$.

Another effect of the ionic strength on solubility is described by the terms *salting in* and *salting out*. The behavior shown in Figure 9.7, in which the activity coefficient first decreases and subsequently increases with concentration, affects the solubility of a salt in the following way. For a salt such as MgF$_2$, the product $[Mg^{2+}][F^-]^2 \gamma_{\pm}^3 = K$ is constant as the concentration varies for constant T, because the thermodynamic equilibrium constant K depends only on T. Therefore, the concentrations $[Mg^{2+}]$ and $[F^-]$ change in an opposite way to $\gamma_{\pm}$. At small values of the ionic strength, $\gamma_{\pm} < 1$, and the solubility increases as $\gamma_{\pm}$ decreases with concentration until the minimum in a plot of $\gamma_{\pm}$ versus I is reached. This effect is known as **salting in.** For high values of the ionic strength, $\gamma_{\pm} > 1$, and the solubility is less than at low values of I. This effect is known as **salting out.**

Salting in and salting out are frequently encountered in studies of the solubility of proteins in aqueous electrolyte solutions. Assume a macromolecule is a 1:1 electrolyte. By definition,

$$K_{sp} = \gamma_{\pm}^2 s^2 \qquad (9.40)$$

where s is the solubility expressed in units of molarity. Rearranging Equation (9.40) taking the logarithm, and combining the result with Equation (9.33), we obtain:

$$\log\left(\frac{s}{K_{sp}^{1/2}}\right) = -\log \gamma_{\pm} = 0.5092|z_+ z_-|\sqrt{I} \qquad (9.41)$$

Therefore Debye–Hückel theory predicts an increase in protein solubility or salting in as the ionic strength of the solution increases. However, as the ionic strength of the solution increases, much of the water that would hydrate the protein is bound up in the hydration spheres of the salt ions. As a result, at high ionic strength the solubility of the protein decreases and follows the empirical equation:

$$\log\left(\frac{s}{K_{sp}^{1/2}}\right) = B - K'I \qquad (9.42)$$

It is common practice to salt in and salt out proteins using ammonium sulfate, (NH$_4$)$_2$SO$_4$, mainly because of its high solubility in water. Figure 9.9 shows the solubility of horse hemoglobin as a function of ionic strength in the presence of (NH$_4$)$_2$SO$_4$. Because of differences in the binding of water to different proteins, proteins salt out at difference ionic strengths and thus the behavior displayed in Figure 9.9 can be used to purify proteins.

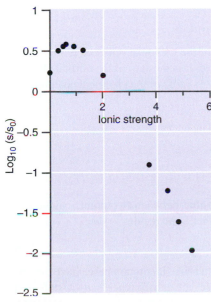

FIGURE 9.9

A graph of log(s/s_0) versus I where $s_0 = K_{sp}^{1/2}$ for horse hemoglobin in the presence of (NH$_4$)$_2$SO$_4$. (Source: Data derived from E. Cohn & J. Edsall "Proteins, Amino Acids and Peptides" Litton Educational Publishing 1943)

9.6 The Electrochemical Potential

In the previous sections, we outlined a model of electrolyte solutions. Because solvated ions bear an electrical charge, their energy is affected by an electrical field. As discussed in this section, the electrical potential associated with the field affects the chemical potential of ions in solution. Because the equilibrium in a chemical reaction is determined by the chemical potentials of reactants and products, we will see that it is possible to change the direction in which a reaction proceeds simply by applying an electrical potential between the electrodes in an electrochemical cell. We first discuss what happens if we partially immerse a metal electrode in an aqueous solution.

If a Zn electrode is partially immersed in an aqueous solution of ZnSO$_4$, it is observed that a slight negative charge is built up on the Zn electrode, and that an equally large positive charge is built up in the surrounding solution. As a result of this charging,

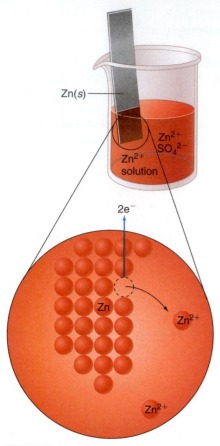

FIGURE 9.10
When a Zn electrode is immersed in water, it is observed that a very small amount of the Zn goes into solution as $Zn^{2+}(aq)$, leaving two electrons behind on the Zn electrode per ion formed. Although the charge buildup on the electrode and in the solution is very small, it leads to a difference in the electrical potential of the solution and the electrode on the order of a volt.

there is a difference in the electrical potential of the electrode and the solution, as depicted in Figure 9.10. The charge separation in the system arises through the dissociation equilibrium

$$Zn(s) \rightleftarrows Zn^{2+}(aq) + 2e^- \qquad (9.43)$$

Whereas the Zn^{2+} ions go into the solution, the electrons remain on the Zn electrode. The equilibrium position in this reaction lies far toward $Zn(s)$. At equilibrium, fewer than 10^{-14} mol of the $Zn(s)$ electrode dissolve. However, this miniscule amount of charge transfer between the electrode and the solution is sufficient to create a difference of approximately 1 V in the electrical potential between the Zn electrode and the electrolyte solution. Because the value of the equilibrium constant depends on the identity of the metal electrode, the difference in the electrical potential between the electrode and the solution also depends on the identity of the metal. It is important to realize that this electrical potential affects the energy of all charged particles in the solution.

The chemical potential of a neutral atom or molecule is not affected if a small electrical potential is applied to the environment containing the species. However, this is not the case for a charged species such as a Na^+ ion in an electrolyte solution. We calculate the work needed to transfer dn moles of charge reversibly from a chemically uniform phase at an electrical potential ϕ_1 to a second otherwise identical phase at an electrical potential ϕ_2. From electrostatics, the work is equal to the product of the charge and the difference in the electrical potential between the two locations.

$$dw_{rev} = (\phi_2 - \phi_1)\, dQ \qquad (9.44)$$

In this equation, $dQ = -zF\, dn$ is the charge transferred through the potential, where z is the charge in units of the electron charge (that is, $+1, -1, +2, -2, \ldots$), and the **Faraday constant** F is the absolute magnitude of the charge associated with 1 mol of a singly charged species. The constant F has the numerical value $F = 96{,}485$ coulombs $mol^{-1}(C\, mol^{-1})$.

Because the work being carried out in this reversible process is nonexpansion work, $dw_{rev} = dG$, which is the difference in the **electrochemical potential,** $\tilde{\mu}$, of the charged particle in the two phases.

$$dG = \tilde{\mu}_2 z\, dn - \tilde{\mu}_1 z\, dn \qquad (9.45)$$

The electrochemical potential is a generalization of the chemical potential to include the effect of an electrical potential on a charged particle. It is the sum of the normal chemical potential, μ, and a term that results from the nonzero value of the electrical potential:

$$\tilde{\mu} = \mu + z\phi \qquad (9.46)$$

Note that with this definition $\tilde{\mu} \to \mu$ as $\phi \to 0$.

Combining Equations (9.44) and (9.46) gives

$$\tilde{\mu}_2 - \tilde{\mu}_1 = +z(\phi_2 - \phi_1)F \quad \text{or} \quad \tilde{\mu}_2 = \tilde{\mu}_1 + z(\phi_2 - \phi_1)F \qquad (9.47)$$

Because only the difference in the electrical potential between two points can be measured, one can set $\phi_1 = 0$ in Equation (9.47) to obtain the result

$$\tilde{\mu}_2 = \tilde{\mu}_1 + z\phi_2 F \qquad (9.48)$$

This result shows that charged particles in two otherwise identical phases have different values for the electrochemical potential if the phases are at different electrical potentials. Because the particles will flow in a direction that decreases their electrochemical potential, the flow of negatively charged particles in a conducting phase is toward a region of more positive electric potential. The opposite is true for positively charged particles.

Equation (9.48) is the basis for understanding all electrochemical reactions. In an electrochemical environment, the equilibrium condition is

$$\Delta G_{reaction} = \sum_i v_i \tilde{\mu}_i = 0 \quad \text{rather than} \quad \Delta G_{reaction} = \sum_i v_i \mu_i = 0 \qquad (9.49)$$

Chemists have a limited ability to change the chemical potential μ_i of a neutral species in solution by varying P, T, and concentration. However, because the electrochemical potential can be varied through the application of an electrical potential, it can be changed easily. It is possible to vary $\tilde{\mu}_i$ to a far greater extent than μ_i, because $z_i\phi F$ can be larger in magnitude than μ_i, and can have either the same or the opposite sign. As Equation (9.46) shows, this also applies to $\Delta G_{reaction}$. Because a change in the electrical potential can lead to a change in the sign of $\Delta G_{reaction}$, the direction of spontaneous change in a reaction system can be changed simply by applying suitable electrical potentials within the system.

9.7 Electrochemical Cells and Half-Cells

We next consider an **electrochemical cell** such as the one shown in Figure 9.11. This particular cell is known as the **Daniell cell,** after its inventor. On the left, a Zn electrode is immersed in a solution of $ZnSO_4$. The solute is completely dissociated to form $Zn^{2+}(aq)$ and $SO_4^{2-}(aq)$. On the right, a Cu electrode is immersed in a solution of $CuSO_4$, which is completely dissociated to form $Cu^{2+}(aq)$ and $SO_4^{2-}(aq)$. Each of the two parts of the cell are referred to as **half-cells.** The two half-cells are connected by an ionic conductor known as a salt bridge. The **salt bridge** consists of an electrolyte such as KCl suspended in a gel. A salt bridge allows current to flow between the half-cells while preventing the mixing of the solutions. A metal wire fastened to each electrode allows the electron current to flow through the external part of the circuit.

The measurement of electrical potentials is discussed using this cell. Because of the equilibrium established between $Zn(s)$ and $Zn^{2+}(aq)$ and between $Cu(s)$ and $Cu^{2+}(aq)$ [see Equations (9.63) and (9.64) in Section 9.8], there is a difference in the electrical potential between the metal electrode and the solution in each of the two half-cells. Can this electrical potential be measured directly? Assume that the probes are two chemically inert Pt wires. We place one Pt wire on the Zn electrode, and the second Pt wire in the $ZnSO_4$ solution. In this measurement, the measured voltage is the difference in electrical potential between a Pt wire connected to a Zn electrode in a $ZnSO_4$ solution, and a Pt electrode in a $ZnSO_4$ solution. A difference in electrical potential can only be measured between one phase and a second phase *of identical composition*. For example, the difference in electrical potential across a resistor is measured by contacting the metal wire at each end of the resistor with metal probes connected to the terminals of a voltmeter.

On the basis of this discussion, it is clear that the electrical potential in the solution cannot be directly measured. Therefore, the convenient standard state $\phi = 0$ is chosen for the electrical potential of all ions in the solution. With this choice,

$$\tilde{\mu}_i = \mu_i \quad \text{(ions in solution)} \qquad (9.50)$$

Adopting this standard state simplifies calculations, because μ_i can be calculated from the solute concentration at low concentrations using the Debye–Hückel limiting law discussed in Section 9.4 or at higher concentrations if the activity coefficients are known.

Next consider appropriate standard states for electrons and ions in a metal electrode. As shown in Equation (9.43), the electrochemical potential consists of two parts, a chemical component and a component that depends on the electrical potential. For an electron in a metal, there is no way to determine the relative magnitude of the two components. Therefore, the convention is established in which the standard state chemical potential of an electron in a metal electrode is zero so that

$$\tilde{\mu}_{e^-} = -\phi F \qquad \text{(electrons in metal electrode)} \qquad (9.51)$$

For a metal electrode at equilibrium,

$$M^{z+} + ze^- \rightleftarrows M \qquad (9.52)$$

FIGURE 9.11
Schematic diagram of the Daniell cell. Zn^{2+}/Zn and Cu^{2+}/Cu half-cells are connected through a salt bridge in the internal circuit. A voltmeter is shown in the external circuit. The insets show the atomic-level processes that occur at each electrode.

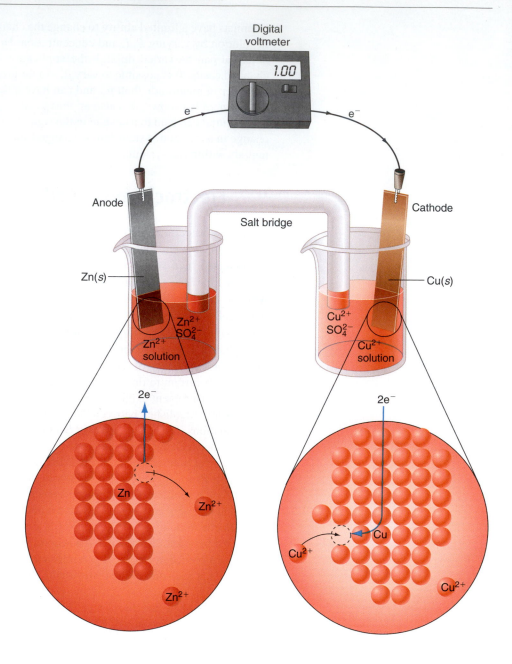

Because $\tilde{\mu}_M = \mu_M$, at equilibrium, the electrochemical potentials of the metal, the metal cation, and the electron are related by

$$\mu_M = \tilde{\mu}_{M^{z+}} + zF\tilde{\mu}_{e^-} \tag{9.53}$$

Using the relations $\tilde{\mu}_{e^-} = -\phi F$, $\tilde{\mu}_M = \mu_M$, and $\tilde{\mu}_{M^{z+}} = \mu_{M^{z+}} + zF\phi$ in Equation (9.53),

$$\mu_M = \mu_{M^{z+}} + zF\phi - zF\phi = \mu_{M^{z+}} \tag{9.54}$$

and for a pure metal at 1 bar, $\mu_M^\circ = \mu_{M^{z+}}^\circ = 0$

Recall that $G_{M,m}^\circ = \mu_M^\circ = 0$ because the electrode is a pure element in its standard state. Note that this convention refers to the chemical potential rather than the electrochemical potential. Combining Equations (9.53) and (9.54),

$$\tilde{\mu}_{M^{z+}} = \mu_{M^{z+}}^\circ + zF\phi = zF\phi \tag{9.55}$$

Equations (9.51), (9.54), and (9.55) define the standard states for a neutral metal atom in the electrode, the metal cation, and the electron in the electrode.

Even with these conventions, there is a further problem in obtaining numerical values for electrical potentials of different half-cells. Only the electrical potential difference between two half-cells can be measured, as opposed to the absolute electrical potential of each half-cell. Therefore, it is convenient to choose one half-cell as a reference and arbitrarily assign a fixed electrical potential to this half-cell. Once this is done, the electrical potential associated with any other half-cell can be determined by combining it with the reference half-cell. The measured potential difference across the cell is associated with the half-cell of interest. It is next shown that the standard hydrogen electrode (SHE) fulfills the role of a reference half-cell of zero potential without further assumptions.

The equilibrium reaction in this half-cell is

$$H^+(aq) + e^- \rightleftarrows \frac{1}{2} H_2(g) \tag{9.56}$$

and the equilibrium in the half-cell is described by

$$\mu_{H^+}(aq) + \tilde{\mu}_{e^-} \rightleftarrows \frac{1}{2} \mu_{H_2}(g) \tag{9.57}$$

It is useful to separate $\mu_{H_2}(g)$ and $\mu_{H^+}(aq)$ into a standard state portion and one that depends on the activity and use Equation (9.48) for the electrochemical potential of the electron. The preceding equation then takes the form

$$\mu_{H^+}^{\circ} + RT \ln a_{H^+} - F\phi_{H^+/H_2} = \frac{1}{2}\mu_{H_2}^{\circ} + \frac{1}{2}RT \ln f_{H_2} \tag{9.58}$$

where f is the fugacity of the hydrogen gas. Solving Equation (9.58) for ϕ_{H^+/H_2},

$$\phi_{H^+/H_2} = \frac{\mu_{H^+}^{\circ} - \frac{1}{2}\mu_{H_2}^{\circ}}{F} - \frac{RT}{F} \ln \frac{f_{H_2}^{1/2}}{a_{H^+}} \tag{9.59}$$

For unit activities of all species, the cell has its standard state potential, designated ϕ_{H^+/H_2}°. Because $\mu_{H_2}^{\circ} = 0$,

$$\phi_{H^+/H_2}^{\circ} = \frac{\mu_{H^+}^{\circ} - \frac{1}{2}\mu_{H_2}^{\circ}}{F} = \frac{\mu_{H^+}^{\circ}}{F} \tag{9.60}$$

Recall that in Section 9.1, the convention that $\Delta G_f^{\circ}(H^+, aq) = \mu_{H^+}^{\circ} = 0$ was established. As a result of this convention,

$$\phi_{H^+/H_2}^{\circ} = 0 \tag{9.61}$$

The standard hydrogen electrode is a convenient **reference electrode** against which the potentials of all other half-cells can be measured. A schematic drawing of the SHE electrode is shown in Figure 9.12. To achieve equilibrium on a short timescale, the reaction $H^+(aq) + e^- \rightleftarrows 1/2\, H_2(g)$ is carried out over a Pt catalyst. It is also necessary to establish a standard state for the activity of $H^+(aq)$. It is customary to use a Henry's law standard state based on molarity. Therefore, $a_i \rightarrow c_i$ and $\gamma_i \rightarrow 1$ as $c_i \rightarrow 0$. The standard state is a (hypothetical) aqueous solution of $H^+(aq)$ that shows ideal solution behavior at a concentration of

$$c^{\circ} = 1 \text{ mol L}^{-1} \tag{9.62}$$

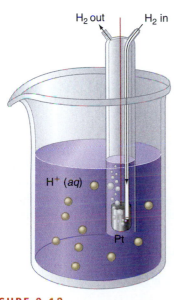

H₂ out   H₂ in

H⁺ (aq)

Pt

FIGURE 9.12
The standard hydrogen electrode (SHE) consists of a solution of an acid such as HCl, H_2 gas, and a catalyst that allows the equilibrium in the half-cell reaction to be established rapidly. The activities of H_2 and H^+ are equal to one.

FIGURE 9.13

In a cell consisting of an arbitrary half-cell and the standard hydrogen electrode, the entire cell voltage is assigned to the arbitrary half-cell.

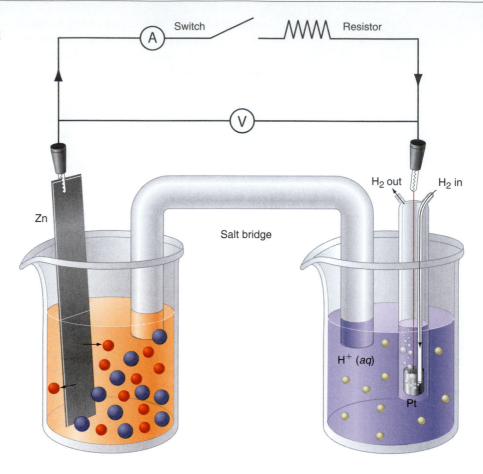

The usefulness of the result $\phi_{H^+/H_2}^{\circ} = 0$ is that values for the electrical potential can be assigned to individual half-cells by measuring their potential relative to the H_2/H^+ half-cell. For example, the cell potential of the electrochemical cell in Figure 9.13 is assigned to the Zn/Zn^{2+} half-cell if the $H^+(aq)$ and $H_2(g)$ activities both have the value 1.

9.8 Redox Reactions in Electrochemical Cells and the Nernst Equation

What reactions occur in the Daniell cell shown in Figure 9.11? If the half-cells are connected through the external circuit, Zn atoms leave the Zn electrode to form Zn^{2+} in solution, and Cu^{2+} ions are deposited as Cu atoms on the Cu electrode. In the external circuit, electrons are observed to flow through the wires and the resistor in the direction from the Zn electrode to the Cu electrode. These observations are consistent with the following electrochemical reactions:

Left half-cell: $\quad Zn(s) \rightleftarrows Zn^{2+}(aq) + 2e^-$ (9.63)

Right half-cell: $\quad Cu^{2+}(aq) + 2e^- \rightleftarrows Cu(s)$ (9.64)

Overall: $\quad Zn(s) + Cu^{2+}(aq) \rightleftarrows Zn^{2+}(aq) + Cu(s)$ (9.65)

In the left half-cell, Zn is being oxidized to Zn^{2+}, and in the right half-cell, Cu^{2+} is being reduced to Cu. By convention, the electrode at which oxidation occurs is called the **anode,** and the electrode at which reduction occurs is called the **cathode.** Each half-cell in an electrochemical cell must contain a species that can exist in an oxidized and a reduced form. For a general redox reaction, the reactions at the anode, the cathode, and the overall reaction can be written as follows:

Anode:	$Red_1 \rightarrow Ox_1 + v_1e^-$	(9.66)
Cathode:	$Ox_2 + v_2e^- \rightarrow Red_2$	(9.67)
Overall:	$v_2Red_1 + v_1Ox_2 \rightarrow v_2Ox_1 + v_1Red_2$	(9.68)

Note that electrons do not appear in the overall reaction, because the electrons produced at the anode are consumed at the cathode.

How are the cell voltage and the ΔG for the overall reaction related? This important relationship can be determined from the electrochemical potentials of the species involved in the overall reaction of the Daniell cell:

$$\Delta G_{reaction} = \tilde{\mu}_{Zn^{2+}} + \tilde{\mu}_{Cu} - \tilde{\mu}_{Cu^{2+}} - \tilde{\mu}_{Zn} = \tilde{\mu}^\circ_{Zn^{2+}} - \tilde{\mu}^\circ_{Cu^{2+}} \qquad (9.69)$$

$$+ RT \ln \frac{a_{Zn^{2+}}}{a_{Cu^{2+}}} = \Delta G^\circ_{reaction} + RT \ln \frac{a_{Zn^{2+}}}{a_{Cu^{2+}}}$$

In this equation, $\tilde{\mu}_{Cu} = \tilde{\mu}^\circ_{Cu} = 0$, and the same is true for Zn because the cell is at a pressure of 1 bar. If this reaction is carried out reversibly, the electrical work done is equal to the product of the charge and the potential difference through which the charge is moved. However, the reversible work at constant pressure is also equal to ΔG. Therefore, we can write the following equation:

$$\Delta G_{reaction} = -nF\Delta\phi \qquad (9.70)$$

In Equation (9.70), $\Delta\phi$ is the measured potential difference generated by the spontaneous chemical reaction for particular values of $a_{Zn^{2+}}$ and $a_{Cu^{2+}}$ and n is the number of moles of electrons involved in the redox reaction. It is seen that the measured cell voltage is directly proportional to ΔG. For a reversible reaction, the symbol E is used in place of $\Delta\phi$, and E is referred to as the **electromotive force (emf)**. Using this definition, we rewrite Equation (9.70) as

$$-2FE = \Delta G^\circ_{reaction} + RT \ln \frac{a_{Zn^{2+}}}{a_{Cu^{2+}}} \qquad (9.71)$$

For standard state conditions, $a_{Zn^{2+}} = a_{Cu^{2+}} = 1$, and Equation (9.70) takes the form $\Delta G^\circ = -2FE^\circ$. This definition of E° allows Equation (9.71) to be rewritten as

$$E = E^\circ - \frac{RT}{2F} \ln \frac{a_{Zn^{2+}}}{a_{Cu^{2+}}} \qquad (9.72)$$

For a general overall electrochemical reaction involving the transfer of n moles of electrons,

$$E = E^\circ - \frac{RT}{nF} \ln Q \qquad (9.73)$$

where Q is the familiar reaction quotient. The preceding equation is known as the **Nernst equation.** At 298.15 K, the Nernst equation can be written in the form

$$E = E^\circ - \frac{0.05916}{n} \log_{10} Q \qquad (9.74)$$

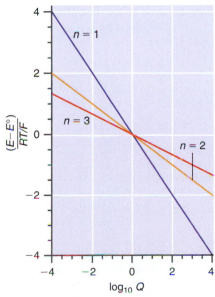

FIGURE 9.14

The cell potential E varies linearly with log Q. The slope of a plot of $(E - E^\circ)/(RT/F)$ is inversely proportional to the number of electrons transferred in the redox reaction.

This function is graphed in Figure 9.14. The Nernst equation allows the emf for an electrochemical cell to be calculated if the activity is known for each species and E° is known.

The Nernst equation has been derived on the basis of the overall cell reaction. For a half-cell, an equation of a similar form can be derived. The equilibrium condition for the half-cell reaction

$$Ox + ve^- \rightarrow Red \tag{9.75}$$

is given by

$$\mu_{Ox}^{n+} + n\tilde{\mu}_{e^-} \rightleftarrows \mu_{Red} \tag{9.76}$$

Using the convention for the electrochemical potential of an electron in a metal electrode [Equation (9.51)], Equation (9.76) can be written in the form

$$\mu_{Ox^{n+}}^{\circ} + RT \ln a_{Ox^{n+}} - nF\phi_{Ox/Red} = \mu_{Red}^{\circ} + RT \ln a_{Red}$$

$$\phi_{Ox/Red} = \frac{\mu_{Red}^{\circ} - \mu_{Ox^{n+}}^{\circ}}{nRT} - \frac{RT}{nF} \ln \frac{a_{Red}}{a_{Ox^{n+}}} \tag{9.77}$$

$$E_{Ox/Red} = E_{Ox/Red}^{\circ} - \frac{RT}{nF} \ln \frac{a_{Red}}{a_{Ox^{n+}}}$$

The last line in Equation (9.77) has the same form as the Nernst equation, but the activity of the electrons does not appear in Q.

9.9 Combining Standard Electrode Potentials to Determine the Cell Potential

A representative set of standard potentials is listed in Tables 9.5 and 9.6 (see Appendix B, Data Tables). By convention, half-cell emfs are always tabulated as reduction potentials. However, whether the reduction or the oxidation reaction is spontaneous in a half-cell is determined by the relative emfs of the two half-cells that make up the electrochemical cell. Because $\Delta G^{\circ} = -nFE^{\circ}$, and values of ΔG for the oxidation and reduction reactions are equal in magnitude and opposite in sign,

$$E_{reduction}^{\circ} = -E_{oxidation}^{\circ} \tag{9.78}$$

How is the cell potential related to the potentials of the half-cells? The standard potential of a half-cell is given by $E^{\circ} = -\Delta G^{\circ}/nF$ and is an intensive property, because although both ΔG° and n are extensive quantities, the ratio $\Delta G^{\circ}/n$ is an intensive quantity. In particular, E° is not changed if all of the stoichiometric coefficients are multiplied by any integer, because both ΔG° and n are changed by the same factor. Therefore,

$$E_{cell}^{\circ} = E_{reduction}^{\circ} + E_{oxidation}^{\circ} \tag{9.79}$$

regardless of how the balanced half-cell is written. In this equation, the standard potentials on the right refer to the half-cells.

EXAMPLE PROBLEM 9.4

Nicotine adenine dinucleotide (NAD) is a cellular redox reagent that is involved in redox chemistry throughout the respiratory system (see Chapter 11). The reduced form of NAD is designated NADH and the oxidized form is designated NAD$^+$.

An electrochemical cell is constructed using a half-cell for which the reduction reaction is given by

$$NAD^+ + H^+ + 2e^- \rightarrow NADH \quad E^{\circ} = -0.105 \text{ V}$$

which is combined with the half-cells for which the reduction reaction is given by

a. $CO_2 + H^+ + 2e^- \rightarrow HCOO^- \quad E^{\circ} = -0.20 \text{ V}$

b. $O_2 + 2H^+ + 2e^- \rightarrow H_2O_2 \quad E^{\circ} = 0.69 \text{ V}$

Write the overall reaction for the cells in the direction of spontaneous change. Is the NAD reduced or oxidized in the spontaneous reactions?

Solution

The emf for the cell is the sum of the emfs for the half-cells, one of which is written as an oxidation reaction, and the other of which is written as a reduction reaction. For the direction of spontaneous change, $E° = E°_{reduction} + E°_{oxidation} > 0$. Note that the two half-cell reactions are combined with the appropriate stoichiometry so that electrons do not appear in the overall equation. Note also that the half-cell emfs are not changed in multiplying the overall reactions by the integers necessary to eliminate the electrons in the overall equation, because $E°$ is an intensive rather than an extensive quantity.

a. $NAD^+ + HCOO^- \rightarrow NADH + CO_2$

$$E° = -0.105 \text{ V} - (-0.200 \text{ V}) = 0.095 \text{ V}$$

NAD is reduced.

b. $O_2 + H^+ + NADH \rightarrow H_2O_2 + NAD^+$

$$E° = +0.69V - (-0.105V) = 0.80V$$

NAD is oxidized.

In Section 9.8 it was shown that $\Delta G_{reaction} = -nF\Delta\phi$. Therefore, if the cell potential is measured under standard conditions,

$$\Delta G°_{reaction} = -nFE° \tag{9.80}$$

If $E°$ is known, $\Delta G°_R$ can be determined using Equation (9.80). For example, $E°$ for the Daniell cell is $+1.10$ V. Therefore, $\Delta G°_{reaction}$ for the reaction $Zn(s) + Cu^{2+}(aq) \rightleftarrows Zn^{2+}(aq) + Cu(s)$ is

$$\Delta G°_{reaction} = -nFE° = -2 \times 96,485 \text{ C mol} \times 1.10 \text{ V} = -212 \text{ kJ mol}^{-1} \tag{9.81}$$

The reaction entropy is related to $\Delta G°_{reaction}$ by

$$\Delta S°_{reaction} = -\left(\frac{\partial \Delta G°_{reaction}}{\partial T}\right)_P = nF\left(\frac{\partial E°}{\partial T}\right)_P \tag{9.82}$$

Therefore, a measurement of the temperature dependence of $E°$ can be used to determine $\Delta S°_{reaction}$, as shown in Example Problem 9.5.

EXAMPLE PROBLEM 9.5

The standard potential of the cell formed by combining the $Cl_2/Cl^-(aq)$ half-cell with the standard hydrogen electrode is $+1.36$ V, and $(\partial E°/\partial T)_P = -1.20 \times 10^{-3}$ V K^{-1}. Calculate $\Delta S°_{reaction}$ for the reaction $H_2(g) + Cl_2(g) \rightarrow 2H^+(aq) + 2Cl^-(aq)$. Compare your result with the value for that you obtain by using the data tables.

Solution

From Equation (9.79),

$$\Delta S°_{reaction} = -\left(\frac{\partial \Delta G°_{reaction}}{\partial T}\right)_P = nF\left(\frac{\partial E°}{\partial T}\right)_P$$

$$= 2 \times 96485 \text{ C mol}^{-1} \times (-1.2 \times 10^{-3} \text{ V K}^{-1}) = -2.3 \times 10^2 \text{ J K}^{-1}$$

From Table 9.1,

$$\Delta S_{reaction}^{\circ} = 2S^{\circ}(H^{+}, aq) + 2S^{\circ}(Cl^{-}, aq) - S^{\circ}(H_2, g) - S^{\circ}(Cl_2, g)$$

$$= 2 \times 0 + 2 \times 56.5 \text{ J K}^{-1} \text{ mol}^{-1} - 130.7 \text{ J K}^{-1} \text{ mol}^{-1} - 223.1 \text{ J K}^{-1} \text{ mol}^{-1}$$

$$= -240.8 \text{ J K}^{-1} \text{ mol}^{-1}$$

The uncertainty in the temperature dependence of E° limits the precision in the determination of $\Delta S_{reaction}^{\circ}$.

9.10 The Relationship Between the Cell emf and the Equilibrium Constant

If the redox reaction is allowed to proceed until equilibrium is reached, $\Delta G = 0$, so that $E = 0$. For the equilibrium state, the reaction quotient $Q = K$. Therefore,

$$E^{\circ} = \frac{RT}{nF} \ln K \tag{9.83}$$

Equation (9.83) shows that a measurement of the standard state cell potential, for which $a_i = 1$ for all species in the redox reaction, allows K for the overall reaction to be determined. Although this statement is true, it is not practical to adjust all activities to the value 1. The determination of E° is discussed in the next section.

Electrochemical measurements provide a powerful way to determine the equilibrium constant in an electrochemical cell. To determine K, the overall reaction must be separated into the oxidation and reduction half-reactions, and the number of electrons transferred must be determined. For example, consider the reaction

$$NAD^{+}(aq) + HCOO^{-}(aq) \rightleftarrows CO_2(g) + NADH(aq) \tag{9.84}$$

The oxidation and reduction half-reactions are

$$NAD^{+}(aq) + H^{+}(aq) + 2e^{-} \rightarrow NADH(aq) \quad E^{\circ} = -0.105 \text{ V} \tag{9.85}$$

$$HCOO^{-}(aq) \rightarrow CO_2(g) + H^{+}(aq) + 2e^{-} \quad E^{\circ} = 0.200 \text{ V} \tag{9.86}$$

$$\Delta G^{\circ} = -nFE^{\circ} = -2 \times 96,485 \text{ C mol}^{-1} \times (-0.105 \text{ V} + 0.200 \text{ V}) = -18.3 \text{ kJ mol}^{-1}$$

$$\ln K = -\frac{\Delta G^{\circ}}{RT} = \frac{18.3 \text{ kJ mol}^{-1}}{8.314 \text{ J K}^{-1} \text{ mol}^{-1} \times 298.15 \text{ K}} = 7.40 \tag{9.87}$$

$$K = e^{7.40} \approx 1640$$

As this result shows, the equilibrium lies to the right, that is, toward conversion of reactants to products. Example Problem 9.6 shows an even more extreme case.

EXAMPLE PROBLEM 9.6

For the oxidation of acetaldehyde to acetate by Fe^{3+} present in the heme group of cytochrome c, the cell potential is $E^{\circ \prime} = 0.835 \text{ V}$ at pH = 7. Calculate the equilibrium constant.

Solution

$$\ln K' = \frac{nF}{RT} E^{\circ \prime} = \frac{2 \times 96,485 \text{ C mol}^{-1} \times 0.835 \text{ V}}{8.314 \text{ J K}^{-1} \text{ mol}^{-1} \times 298.15 \text{ K}}$$

$$= 65.0$$

$$K' = 1.70 \times 10^{28}$$

Note that the equilibrium constant calculated in Example Problem 9.6 is so large that it could not have been measured by determining the activities of NADH, NAD$^+$, acetaldehyde, and acetate by spectroscopic methods. This would require a measurement technique that is accurate to almost 30 orders of magnitude in the activity. By contrast, the equilibrium constant in an electrochemical cell can be determined with high accuracy using only a voltmeter.

A further example of the use of electrochemical measurements to determine equilibrium constants is the solubility constant for a weakly soluble salt. If the overall reaction corresponding to dissolution can be generated by combining half-cell potentials, then the solubility constant can be calculated from the potentials. For example, the following half-cell reactions can be combined to calculate the solubility product of AgBr:

$$AgBr(s) + e^- \rightarrow Ag(s) + Br^-(aq) \qquad E° = 0.07133 \text{ V and}$$

$$\underline{Ag(s) \rightarrow Ag^+(aq) + e^- \qquad\qquad\qquad E° = -0.7996}$$

$$AgBr(s) \rightarrow Ag^+(aq) + Br^-(aq) \qquad E° = -0.7283 \text{ V}$$

$$\ln K_{sp} = \frac{nF}{RT} E° = \frac{1 \times 96{,}485 \text{ C mol}^{-1} \times (-0.7283 \text{ V})}{8.314 \text{ J K}^{-1} \text{ mol}^{-1} \times 298.15 \text{ K}} = -28.35$$

The value of the solubility constant is $K_{sp} = 4.88 \times 10^{-13}$.

9.11 The Determination of $E°$ and Activity Coefficients Using an Electrochemical Cell

The main problem in determining standard potentials lies in knowing the value of the activity coefficient $\gamma_\pm$ for a given solute concentration. The best strategy is to carry out measurements of the cell potential at low concentrations, where $\gamma_\pm \rightarrow 1$, rather than near unit activity, where $\gamma_\pm$ differs appreciably from 1. Consider an electrochemical cell consisting of the Ag$^+$/Ag and SHE half-cells at 298 K. Assume that the Ag$^+$ arises from the dissociation of AgNO$_3$. Recall that the activity of an individual ion cannot be measured directly. It must be calculated from the measured activity $a_\pm$ and the definition $a_\pm^\nu = a_+^{\nu_+} a_-^{\nu_-}$. In this case, $a_\pm^2 = a_{Ag^+} a_{Cl^-}$ and $a_\pm = a_{Ag^+} = a_{Cl^-}$. Similarly, $\gamma_\pm = \gamma_{Ag^+} = \gamma_{Cl^-}$ and $m_{Ag^+} = m_{Cl^-} = m_\pm$. The cell potential E is given by

$$E = E°_{Ag^+/Ag} + \frac{RT}{F} \ln a_{Ag^+} = E°_{Ag^+/Ag} + \frac{RT}{F} \ln m_\pm + \frac{RT}{F} \ln \gamma_\pm \qquad (9.88)$$

At low enough concentrations, the Debye–Hückel limiting law is valid and $\log \gamma_\pm = -0.50926\sqrt{m_\pm}$ at 298 K. Using this relation, Equation (9.85) can be rewritten in the form

$$E - 0.05916 \log_{10} m_\pm = E°_{Ag^+/Ag} - 0.03013 \log_{10} \sqrt{m_\pm} \qquad (9.89)$$

The quantities on the left-hand side of this equation can be calculated from measurements and plotted as a function of $\sqrt{m}$. The results will resemble the graph in Figure 9.15. An extrapolation of the line that best fits the data to $m = 0$ gives $E°$ as the intercept with the vertical axis. Once $E°$ has been determined, Equation (9.88) can be used to calculate $\gamma_\pm$.

Electrochemical cells provide a powerful method of determining activity coefficients, because cell potentials can be measured more accurately and more easily than colligative properties such as the freezing point depression or boiling point elevation. Note that although the Debye–Hückel limiting law was used to determine $E°$, it is not necessary to use the limiting law to determine activity coefficients once $E°$ is known.

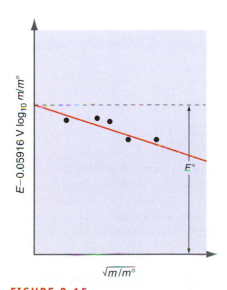

FIGURE 9.15
The value of $E°$ and the activity coefficient can be measured by plotting the left-hand side of Equation (9.89) against the square root of the molality.

9.12 The Biochemical Standard State

From the discussion in Sections 9.7 through 9.10 the numerical values of many properties of redox reactions including the equilibrium constant, the standard Gibbs energy change, and the standard electrochemical potentials depend on the definition of the standard reference states defined for the species involved in these reactions. It has been shown that for chemical species involved in redox reactions, the reference states are set to unity, that is, $c^o = 1$ mol L^{-1}. Of fundamental importance to the evaluation of electrochemical cell potentials is the reference state of $H^+(aq)$, which according to Equation (9.61) is similarly set to 1 mol L^{-1}.

However, in most biological applications, equilibrium constants, standard Gibbs energies, and standard electrochemical potentials are calculated assuming a different reference state for $H^+(aq)$ than 1 M. In a convention adopted in the biochemical literature, the reference state for $H^+(aq)$ is 1.00×10^{-7} mol L^{-1}. This difference in standard state definition affects the numerical values of several properties of chemical reactions that involve $H^+(aq)$.

Consider the general chemical equation

$$\nu_A A(aq) + \nu_B B(aq) \rightleftharpoons \nu_D D(aq) + \nu_E E(aq) + x H^+(aq) \qquad (9.90)$$

Assume for simplicity that all species in Equation (9.90) are in a dilute enough state so that all activity coefficients approach unity; that is, $\gamma_i \approx 1$ for i = A, B, C, D, and H^+. This simplification will not affect the generality of the following discussion. To calculate the equilibrium constant, we use the following expression:

$$K = \frac{\left(\dfrac{c_D}{c_D^\circ}\right)^{\nu_D} \left(\dfrac{c_E}{c_E^\circ}\right)^{\nu_E} \left(\dfrac{c_{H^+}}{c_{H^+}^\circ}\right)^x}{\left(\dfrac{c_A}{c_A^\circ}\right)^{\nu_A} \left(\dfrac{c_B}{c_B^\circ}\right)^{\nu_B}} \qquad (9.91)$$

where activities have been replaced by concentrations because of the assumption $\gamma_i \approx 1$. Now if we calculate the equilibrium constant in Equation (9.91) using the biochemical standard state conventions, all reference concentrations are set to unity except for H^+ where

$$c_{H^+}^\circ = 1.00 \times 10^{-7} \text{ mol } L^{-1} \qquad (9.92)$$

Using the convention given in Equation (9.92), the equilibrium constant is evaluated as follows:

$$K' = \frac{\left(\dfrac{c_D}{1.00\ M}\right)^{\nu_D} \left(\dfrac{c_E}{1.00\ M}\right)^{\nu_E} \left(\dfrac{c_{H^+}}{1.00 \times 10^{-7}\ M}\right)^x}{\left(\dfrac{c_A}{1.00\ M}\right)^{\nu_A} \left(\dfrac{c_B}{1.00\ M}\right)^{\nu_B}} \qquad (9.93)$$

The prime accompanying the equilibrium constant in Equation (9.93) indicates that the biochemical convention definition of the standard state of $H^+(aq)$ from Equation (9.92) is being used. If the value of the equilibrium constant using the chemical convention for the standard state of $H^+(aq)$ given by Equation (9.62) is indicated as K (that is, without a prime), the relationship between the equilibrium constants is

$$K' = K \times 10^{7x} \qquad (9.94)$$

where x is a positive number if $H^+(aq)$ is a product and x is a negative number if $H^+(aq)$ is a reactant.

The standard Gibbs energy change is directly dependent on the logarithm of the equilibrium constant so the convention used to define the concentration of $H^+(aq)$ in the reference state must also be noted in the standard Gibbs energy. The notation ΔG° is reserved for the standard Gibbs energy defined using the "chemical" convention given by Equation (9.62). If the "biochemical" convention given in Equati standard Gibbs energy is designated by $\Delta G^{\circ\prime}$:

$$\Delta G^{\circ\prime} = -RT \ln K'$$

The standard Gibbs energies for the two conventions given in (9.92) are easily related:

$$\Delta G^{\circ\prime} = -RT \ln K' = -RT \ln (K \times 10^{7x}) = -RT \ln K - 7xRT \ln 10 \quad (9.96)$$

$$= \Delta G^{\circ} - 7xRT \ln 10$$

Equation (9.83) relates the equilibrium constant to the standard electrochemical cell potential. This allows us to relate the standard cell potentials evaluated according to the chemical and biochemical reference state conventions:

$$E^{\circ\prime} = -\frac{\Delta G^{\circ\prime}}{nF} = \frac{RT}{nF} \ln K' = \frac{RT}{nF} \ln (K \times 10^{7x}) \quad (9.97)$$

$$= \frac{RT}{nF} \left[\ln K + 7x \ln 10 \right] = E^{\circ} + \frac{RT}{nF} \times 7x \ln 10$$

Standard reduction potentials for some biochemically relevant half reactions evaluated at pH = 7 appear in Table 9.7.

EXAMPLE PROBLEM 9.7

You are given the following reduction reactions and $E^{\circ\prime}$ values at pH = 7.

$$CH_3COO^-(aq) + 3H^+(aq) + 2e^- \rightarrow CH_3CHO(aq) + H_2O \quad E^{\circ\prime} = -0.581 \text{ V}$$

$$CH_3CHO(aq) + 2H^+(aq) + 2e^- \rightarrow CH_3CH_2OH(aq) \quad E^{\circ\prime} = -0.197 \text{ V}$$

where $E^{\circ\prime}$ indicates the biological standard state. Calculate $E^{\circ\prime}$ for the half-cell reaction

$$CH_3COO^-(aq) + 5H^+(aq) + 4e^- \rightarrow CH_3CH_2OH(aq) + H_2O(l)$$

Solution

We calculate the desired value of $E^{\circ\prime}$ by converting the given $E^{\circ\prime}$ values to ΔG° values, and combining these reduction reactions to obtain the desired equation:

$$CH_3COO^-(aq) + 3H^+(aq) + 2e^- \rightarrow CH_3CHO(aq) + H_2O(l)$$

$$\Delta G^{\circ\prime} = -nFE^{\circ\prime} = -2 \times 96{,}485 \text{ C mol}^{-1} \times (-0.581 \text{ V}) = 112 \text{ kJ mol}^{-1}$$

TABLE 9.7   **Standard Reduction Potentials at pH 7 for Some Reactions of Biological Importance** ($T = 298$ K)

	Reaction	$E^{\circ\prime}(V)$
	$1/2\ O_2 + 2H^+ + 2e^- \rightarrow H_2O$	0.815
	$O_2 + 2H^+ + 2e^- \rightarrow H_2O_2$	0.295
Fumarate/malate	$^-OOCCH = CHCOO^- + 2H^+ + 2e^- \rightarrow\ ^-OOCCH_2CH_2COO^-$	0.031
Oxaloacetate/malate	$^-OOCCOCH_2COO^- + 2H^+ + 2e^- \rightarrow\ ^-OOCCHOHCH_2COO^-$	−0.166
Pyruvate/lactate	$CH_3(CO)COO^- + 2H^+ + 2e^- \rightarrow CH_3CH(OH)COO^-$	−0.185
Acetaldehyde/ethanol	$CH_3CHO + 2H^+ + 2e^- \rightarrow CH_3CH_2OH$	−0.197
Nicotine adenine dinucleotide	$NAD^+ + H^+ + 2e^- \rightarrow NADH$	−0.320
	$2H^+ + 2e^- \rightarrow H_2$	−0.421
CO_2/formate	$CO_2 + H^+ + 2e^- \rightarrow HCOO^-$	−0.414
Acetic acid/acetaldehyde	$CH_3COOH + 2H^+ + 2e^- \rightarrow CH_3CHO + H_2O$	−0.581
Acetic acid/pyruvic acid	$CH_3COOH + CO_2 + 2H^+ + 2e^- \rightarrow CH_3(CO)COOH + H_2O$	−0.700

$$CH_3CHO(aq) + 2H^+(aq) + 2e^- \rightarrow CH_3CH_2OH(aq)$$

$$\Delta G^{\circ\prime} = -nFE^{\circ\prime} = -2 \times 96,485 \text{ C mol}^{-1} \times (-0.197 \text{ V}) = 38.0 \text{ kJ mol}^{-1}$$

$$CH_3COO^-(aq) + 5H^+(aq) + 4e^- \rightarrow CH_3CH_2OH(aq) + H_2O(l)$$

$$\Delta G^{\circ\prime} = 112 \text{ kJ mol}^{-1} + 38.0 \text{ kJ mol}^{-1} = 150. \text{ kJ mol}^{-1}$$

$$E^{\circ\prime}_{CH_3COO^-/CH_3CH_2OH} = -\frac{\Delta G^{\circ\prime}}{nF} = \frac{-150 \times 10^3 \text{ J mol}^{-1}}{4 \times 96,485 \text{ C mol}^{-1}} = -0.388 \text{ V}$$

The E° values cannot be combined directly, because they are intensive rather than extensive quantities.

The preceding calculation can be generalized as follows. Assume that n_1 electrons are transferred in the reaction with the potential $E^\circ_{A/B}$, and n_2 electrons are transferred in the reaction with the potential $E^\circ_{B/C}$. If n_3 electrons are transferred in the reaction with the potential $E^\circ_{A/C}$, then $n_3 E^\circ_{A/C} = n_1 E^\circ_{A/B} + n_2 E^\circ_{B/C}$.

EXAMPLE PROBLEM 9.8

Calculate the standard cell potential, the standard Gibbs energy change, and the equilibrium constant for the oxidation of the formate ion to carbon dioxide using the data in Table 9.7. Repeat the calculation using the chemical convention for the reference state concentration of $H^+(aq)$. Assume $T = 298$ K.

Solution

The overall chemical reaction is:

$$1/2 \, O_2(g) + HCOO^-(aq) + H^+(aq) \rightarrow CO_2(g) + H_2O(l)$$

The half reactions are:

$$1/2 \, O_2(g) + 2H^+(aq) + 2e^- \rightarrow H_2O(l) \quad E^{\circ\prime} = 0.815 \text{ V}$$

$$HCOO^-(aq) \longrightarrow CO_2(g) + H^+(aq) + 2e^- \quad E^{\circ\prime} = -(-0.414 \text{ V})$$

The standard cell potential is $E^\circ_{cell}{}' = E^\circ_{red}{}' + E^\circ_{ox}{}' = 0.815 \text{ V} + 0.414 \text{ V} = 1.229 \text{ V}$

The standard Gibbs energy is
$$\Delta G^{\circ\prime} = -nFE^{\circ\prime} = -2 \times 96,485 \text{ C mol}^{-1} \times 1.229 \text{ V} = -237.2 \text{ kJ mol}^{-1}$$

The equilibrium constant is

$$K' = e^{-\Delta G^{\circ\prime}/RT} = \exp\left[-\frac{237.2 \times 10^3 \text{ J mol}^{-1}}{8.314 \text{ J K}^{-1} \text{ mol}^{-1} \times 298 \text{ K}}\right] = e^{95.7} = 3.79 \times 10^{41}$$

Converting to the chemical convention:

$$K = K' \times 10^{-7x} = 3.79 \times 10^{41} \times 10^{-7x(-1)} = 3.79 \times 10^{48}$$

$$\Delta G^\circ = -RT \ln K = -8.314 \text{ J K}^{-1} \text{ mol}^{-1} \times 2.98 \text{ K} \times \ln(3.79 \times 10^{48})$$

$$= -2478 \text{ J mol}^{-1} \times (\ln 3.79 + 48 \times \ln 10) = -2478 \text{ kJ mol}^{-1}$$

$$\times (1.33 + 110.52) = 277.2 \text{ kJ mol}^{-1}$$

$$E^\circ_{cell} = -\frac{\Delta G^\circ}{nF} = \frac{-277.2 \text{ kJ}}{2 \times 96,485 \text{ C}} = 1.435 \text{ V}$$

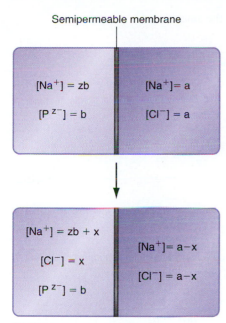

Semipermeable membrane

$[Na^+] = zb$ $[Na^+] = a$

$[P^{z-}] = b$ $[Cl^-] = a$

$[Na^+] = zb + x$ $[Na^+] = a - x$

$[Cl^-] = x$ $[Cl^-] = a - x$

$[P^{z-}] = b$

FIGURE 9.16
Schematic of two compartments separated by a membrane that is impermeable to an anionic macromolecule P^{z-} but permeable to Na^+ and Cl^-. Concentrations of all three species are shown before (left) and after (right) equilibrium is reached. The unequal distribution of permeant species on opposite sides of the membrane that results when an impermeant species occurs on only one side of the membrane is called the Donnan effect.

9.13 The Donnan Potential

The Donnan potential is a consequence of the alteration of the concentrations of permeant ionic species on both sides of a semipermeable membrane as a result of the presence of an impermeant species on one side of the membrane. Consider a volume divided into two compartments by a semipermeable membrane, as shown in Figure 9.16. Assume the membrane is completely permeable to the ionic species arising from the dissociation of a salt, which for the present we will assume is NaCl. Therefore, the membrane is freely permeable to Na^+ and Cl^-. Into the solution in the left-hand compartment, the sodium salt of a biopolymer, which we designate NaP, is introduced. We assume that this salt dissociates completely into P^- and Na^+, where the membrane is impermeable to P^- but freely permeable to Na^+.

Because the membrane is permeable to Na^+ and Cl^-, the system can come to equilibrium with respect to these species. The condition for equilibrium is

$$\mu_{\pm}^L = \mu_{\pm}^R \tag{9.98}$$

Using Equation (9.98) and the fact that the standard mean chemical potentials for the left- and right-hand compartments are equal (that is, $\mu_{\pm}^{\circ L} = \mu_{\pm}^{\circ R}$), the condition for equilibrium reduces to

$$a_{\pm}^L = a_{\pm}^R \tag{9.99}$$

or

$$a_{+}^L a_{-}^L = a_{+}^R a_{-}^R \tag{9.100}$$

To simplify calculations, we assume the activity coefficients of the ionic species Na^+, Cl^-, and P^- approach 1. Then we can approximate the activities in Equation (9.100) with concentrations and the condition for equilibrium becomes

$$c_{+}^L c_{-}^L = c_{+}^R c_{-}^R \tag{9.101}$$

If the biopolymer were absent from both compartments, and thus the only species present were freely permeable to the membrane, the equilibrium concentrations would be related by

$$\frac{c_{+,eq}^L}{c_{+,eq}^R} = \frac{c_{-,eq}^R}{c_{-,eq}^L} = 1 \tag{9.102}$$

However, the presence of a charged biopolymer that is impermeant to the membrane will cause the ratio in Equation (9.102) to deviate from 1. The deviation of the ratios in Equation (9.102) from 1 as a result of the presence of an impermeant species is called the **Donnan effect:**

$$\frac{c_{+,eq}^L}{c_{+,eq}^R} = \frac{c_{-,eq}^R}{c_{-,eq}^L} = r_D \neq 1 \tag{9.103}$$

where r_D is the Donnan ratio.

To quantify r_D, assume as shown in Figure 9.16 that the initial concentrations of Na^+ and Cl^- in the right-hand compartment are $c_{Na^+}^R = c_{Cl^-}^R = a$. Assume the polymer P is an anion with a charge of -1 (that is, $z = 1$), the concentrations of P^- and Na^+ in the left-hand compartment are both initially equal to b, and that Cl^- is initially absent from the left-hand compartment. Then we have $c_{Na^+}^L = c_{P^-}^L = b$ and $c_{Cl^-}^L = 0$.

To reach equilibrium, the system must transport chloride ion from the right-hand compartment to the left-hand compartment. Call x the amount of Cl^- transported to reach equilibrium. To maintain charge neutrality, an identical amount of Na^+ must be transported from the left- to the right-hand compartment. The condition for equilibrium given by Equation (9.101) now becomes

$$(b + x)x = (a - x)(a - x) \tag{9.104}$$

Equation (9.104) can be easily solved to obtain the amount of Cl^- transported:

$$x = \frac{a^2}{b + 2a} \tag{9.105}$$

Equation (9.105) indicates that at equilibrium the concentrations of Na^+ and Cl^- in the left- and right-hand compartments are:

$$c_{Na^+,eq}^L = b + x = b + \frac{a^2}{b + 2a} = \frac{(a + b)^2}{b + 2a}$$

$$c_{Cl^-,eq}^L = x = \frac{a^2}{b + 2a} \tag{9.106}$$

$$c_{Na^+,eq}^R = c_{Cl^-,eq}^R = a - x = a - \frac{a^2}{b + 2a} = \frac{a^2 + ab}{b + 2a}$$

Because the membrane is impermeable to P^- the concentration of that species remains at b. We can use the expressions in Equation (9.106) to calculate r_D:

$$r_D = \frac{c_{Na^+,eq}^L}{c_{Na^+,eq}^R} = \frac{(a + b)^2}{b + 2a} \frac{b + 2a}{a^2 + ab} = \frac{(a + b)^2}{a(a + b)} = \frac{a + b}{a}$$

$$\frac{c_{Cl^-,eq}^R}{c_{Cl^-,eq}^L} = \frac{a^2 + ab}{b + 2a} \frac{b + 2a}{a^2} = \frac{a(a + b)}{a^2} = \frac{a + b}{a} = r_D \tag{9.107}$$

If electrodes are placed in the left and right compartments in the system shown at the right in Figure 9.16, the net potential difference is $\Delta\phi = \phi_L - \phi_R = 0$, because the system is at equilibrium. The net potential difference can be expressed as a Nernst equation:

$$\Delta\phi = 0 = \phi_D + \frac{RT}{F} \ln\left(\frac{c_{Na^+,eq}^L}{c_{Na^+,eq}^R}\right) = \phi_D + \frac{RT}{F} \ln r_D \tag{9.108}$$

The second term on the right side of Equation (9.108) is the potential difference arising from solution with different ion concentrations. Equation (9.108) can be rearranged to obtain:

$$\phi_D = -\frac{RT}{F} \ln r_D = -\frac{RT}{F} \ln\left(\frac{a + b}{a}\right) \tag{9.109}$$

where ϕ_D is the Donnan potential. The Donnan potential arises as a result of the polarization of the membrane by the ion concentration gradients.

Because the equilibrium concentrations of the solute species in the left- and right-hand compartments are unequal, a net osmotic pressure arises, which according to the van't Hoff osmotic equation will be proportional to the difference between the equilibrium concentrations in the left- and right-hand chambers:

$$\pi_{Net} = [b + b + x + x - 2(a - x)]RT = [2(b - a) + 4x]RT$$

$$= \left[2(b - a) + \frac{4a^2}{b + 2a}\right]RT$$

$$= \left[\frac{2b^2 + 2ab - 4a^2}{b + 2a} + \frac{4a^2}{b + 2a}\right]RT = \left[\frac{2b^2 + 2ab}{b + 2a}\right]RT \tag{9.110}$$

Equation (9.110) applies only to the case where the impermeant species is monovalent, that is, P^-. Equation (9.110) can be easily corrected for the case where P has a charge of $-z$. In this case the net osmotic pressure has the form

$$\pi_{Net} = \left[\frac{zb^2 + 2ab + z^2b^2}{zb + 2a}\right]RT \tag{9.111}$$

The biological relevance of the Donnan effect is as follows. Cells contain a large amount of anionic macromolecules, mostly proteins and organic phosphates that cannot cross the cell membrane. At the same time, the extracellular fluid contains very low levels of anionic macromolecules. The result is an alteration of the concentrations of mobile ionic species on both sides of the cell membrane as a result of the Donnan effect, and the appearance of a net osmotic pressure according to Equation (9.110). This osmotic pressure would lead to rupture of the cell membrane if left unopposed. What compensates for the macromolecular Donnan effect is the fact that levels of sodium ion are kept high outside the cell membrane and low within the cell by the action of the sodium pump, as discussed in Chapter 10, resulting in a membrane that is essentially impermeable to sodium ion.

EXAMPLE PROBLEM 9.9

Suppose a volume is divided into two compartments by a semipermeable membrane, as shown in Figure 9.16. In the left-hand "protein compartment" is a solution containing the sodium salt of a protein P, NaP, with initial concentration $1.00 \times 10^{-2} M$. In the right-hand "salt compartment" is a NaCl solution initially at $1.00 \times 10^{-1} M$. Calculate the equilibrium concentrations of all ionic species in both compartments. Calculate also the Donnan ratio and the junction potential resulting from the Donnan effect. Assume $T = 298$ K.

Solution

Using the notation of Figure 9.16, $b = 1.00 \times 10^{-2} M$ and $a = 1.00 \times 10^{-1} M$. Because the membrane is impermeable to passage by the protein, the protein concentration remains $b = 1.00 \times 10^{-2} M$ in the left-hand compartment and zero in the right-hand compartment. All other concentrations can be obtained using Equations (9.106):

$$c_-^L = x = \frac{a^2}{b + 2a} = \frac{(1.00 \times 10^{-1} M)^2}{1.00 \times 10^{-2} M + 2 \times 1.00 \times 10^{-1} M} = 4.76 \times 10^{-2} M$$

$$c_+^L = b + x = 1.00 \times 10^{-2} M + 4.76 \times 10^{-2} M = 5.76 \times 10^{-2} M$$

$$c_+^R = c_-^R = a - x = 1.00 \times 10^{-1} M - 0.476 \times 10^{-1} M = 0.052 M$$

The Donnan ratio is $r_D = \dfrac{c_+^L}{c_+^R} = \dfrac{0.0576}{0.052} = 1.1$.

The junction potential is

$$\phi_D = -\frac{RT}{F} \ln r_D = -\frac{8.314 \text{ J K}^{-1} \text{ mol}^{-1} \times 298 \text{ K}}{96{,}485 \text{ C mol}^{-1}} \ln (1.1) = -2.4 \text{ mV.}$$

Vocabulary

anode	Daniell cell	electrolyte	mean ionic chemical potential
cathode	Davies equation	electromotive force (emf)	mean ionic molality
conventional formation enthalpies	Debye-Hückel limiting law	Faraday constant	Nernst equation
conventional formation entropies	Debye-Hückel screening length	half-cell	reference electrode
conventional Gibbs energies of formation	Donnan effect	ionic strength	salt bridge
	electrochemical cell	mean ionic activity	salting in
	electrochemical potential	mean ionic activity coefficient	salting out
			solvation shell

Questions on Concepts

Q9.1   Tabulated values of standard entropies of some aqueous ionic species are negative. Why is this statement consistent with the third law of thermodynamics?

Q9.2   Why is the value for the dielectric constant for water in the solvation shell around ions less than that for bulk water?

Q9.3   Why is it possible to formulate a general theory for the activity coefficient for electrolyte solutions, but not for nonelectrolyte solutions?

Q9.4   Why are activity coefficients calculated using the Debye-Hückel limiting law always less than 1?

Q9.5   Why does an increase in the ionic strength in the range where the Debye-Hückel law is valid lead to an increase in the solubility of a weakly soluble salt?

Q9.6   Discuss how the Debye-Hückel screening length changes as the (a) temperature, (b) dielectric constant, and (c) ionic strength of an electrolyte solution are increased.

Q9.7   What is the correct order of the following inert electrolytes in their ability to increase the degree of dissociation of acetic acid?

  a. 0.001 m NaCl

  b. 0.001 m KBr

  c. 0.10 m CuCl$_2$

Q9.8   Why is it not appropriate to use ionic radii from crystal structures to calculate $\Delta G^\circ_{solvation}$ of ions using the Born model?

Q9.9   Why is it not possible to measure the activity coefficient of Na$^+(aq)$?

Q9.10   What can you conclude about the interaction between ions in an electrolyte solution if the mean ionic activity coefficient is greater than 1?

Q9.11   What is the difference in the chemical potential and the electrochemical potential for an ion and for a neutral species in solution? Under what conditions is the electrochemical potential equal to the chemical potential for an ion?

Q9.12   Show that if $\Delta G^\circ_f(H^+, aq) = 0$ for all T, the potential of the standard hydrogen electrode is zero.

Q9.13   To determine standard cell potentials, measurements are carried out in very dilute solutions rather than at unit activity. Why is this the case?

Q9.14   The temperature dependence of the potential of a cell is vanishingly small. What does this tell you about the thermodynamics of the cell reaction?

Problems

Problem numbers in **RED** indicate that the solution to the problem is given in the *Student Solutions Manual*.

P9.1   Calculate $\Delta H^\circ_{reaction}$ and $\Delta G^\circ_{reaction}$ for the reaction AgNO$_3(aq)$ + KCl(aq) → AgCl(s) + KNO$_3(aq)$.

P9.2   Calculate $\Delta H^\circ_{reaction}$ and $\Delta G^\circ_{reaction}$ for the reaction Ba(NO$_3$)$_2(aq)$ + 2KCl(aq) → BaCl$_2(s)$ + 2KNO$_3(aq)$.

P9.3   Calculate $\Delta S^\circ_{reaction}$ for the reaction AgNO$_3(aq)$ + KCl(aq) → AgCl(s) + KNO$_3(aq)$.

P9.4   Calculate $\Delta S^\circ_{reaction}$ for the reaction Ba(NO$_3$)$_2(aq)$ + 2KCl(aq) → BaCl$_2(s)$ + 2KNO$_3(aq)$.

P9.5   Calculate $\Delta G^\circ_{solvation}$ in an aqueous solution for Cl$^-(aq)$ using the Born model. The radius of the Cl$^-$ ion is 1.81×10^{-10} m.

P9.6   Calculate the value of $m_\pm$ in 5.0×10^{-4} molal solutions of (a) KCl, (b) Ca(NO$_3$)$_2$, and (c) ZnSO$_4$. Assume complete dissociation.

P9.7   Express $\mu_\pm$ in terms of μ_+ and μ_- for (a) NaCl, (b) MgBr$_2$, (c) Li$_3$PO$_4$, and (d) Ca(NO$_3$)$_2$. Assume complete dissociation.

P9.8   Express $a_\pm$ in terms of a_+ and a_- for (a) Li$_2$CO$_3$, (b) CaCl$_2$, (c) Na$_3$PO$_4$, and (d) K$_4$Fe(CN)$_6$. Assume complete dissociation.

P9.9   Express $\gamma_\pm$ in terms of γ_+ and γ_- for (a) SrSO$_4$, (b) MgBr$_2$, (c) K$_3$PO$_4$, and (d) Ca(NO$_3$)$_2$. Assume complete dissociation.

P9.10   Calculate the ionic strength in a solution that is 0.0050 m in K$_2$SO$_4$, 0.0010 m in Na$_3$PO$_4$, and 0.0025 m in MgCl$_2$.

P9.11   Calculate the mean ionic activity of a 0.0150 m K$_2$SO$_4$ solution for which the mean activity coefficient is 0.465.

P9.12   Calculate the mean ionic molality and mean ionic activity of a 0.150 m Ca(NO$_3$)$_2$ solution for which the mean ionic activity coefficient is 0.165.

P9.13   In the Debye-Hückel theory, the counter charge in a spherical shell of radius r and thickness dr around the central

ion of charge $+q$ is given by $-q\kappa^2 re^{-\kappa r}\, dr$. Calculate the most probable value of r, r_{mp}, from this expression. Evaluate r_{mp} for a 0.050 m solution of NaCl at 298 K.

P9.14   Calculate the Debye–Hückel screening length $1/\kappa$ at 298 K in a 0.00100 m solution of NaCl.

P9.15   Calculate the probability of finding an ion at a distance greater than $1/\kappa$ from the central ion.

P9.16   Using the Debye–Hückel limiting law, calculate the value of $\gamma_\pm$ in 5.0×10^{-3} m solutions of (a) KCl, (b) Ca(NO$_3$)$_2$, and (c) ZnSO$_4$. Assume complete dissociation.

P9.17   Calculate I, $\gamma_\pm$, and $a_\pm$ for a 0.0250 m solution of AlCl$_3$ at 298 K. Assume complete dissociation.

P9.18   Calculate I, $\gamma_\pm$, and $a_\pm$ for a 0.0250 m solution of K$_2$SO$_4$ at 298 K. Assume complete dissociation. How confident are you that your calculated results will agree with experimental results?

P9.19   Calculate I, $\gamma_\pm$, and $a_\pm$ for a 0.0325 m solution of K$_4$Fe(CN)$_6$ at 298 K.

P9.20   Calculate the solubility of BaSO$_4$ ($K_{sp} = 1.08 \times 10^{-10}$) (a) in pure H$_2$O and (b) in an aqueous solution with $I = 0.0010$ mol kg^{-1}.

P9.21   Dichloroacetic acid has a dissociation constant of $K_a = 3.32 \times 10^{-2}$. Calculate the degree of dissociation for a 0.125 m solution of this acid (a) using the Debye–Hückel limiting law and (b) assuming that the mean ionic activity coefficient is one.

P9.22   Chloroacetic acid has a dissociation constant of $K_a = 1.38 \times 10^{-3}$. (a) Calculate the degree of dissociation for a 0.0825 m solution of this acid using the Debye–Hückel limiting law. (b) Calculate the degree of dissociation for a 0.0825 m solution of this acid that is also 0.022 m in KCl using the Debye–Hückel limiting law.

P9.23   The equilibrium constant for the hydrolysis of dimethylamine,

$$(CH_3)_2NH(aq) + H_2O(aq) \rightarrow CH_3NH_3^+(aq) + OH^-(aq)$$

is 5.12×10^{-4}. Calculate the extent of hydrolysis for (a) a 0.125 m solution of $(CH_3)_2NH$ in water and (b) a solution that is also 0.045 m in $NaNO_3$.

P9.24 From the data in Table 9.3 (see Appendix B, Data Tables), calculate the activity of the electrolyte in 0.100 m solutions of

 a. KCl

 b. H_2SO_4

 c. $MgCl_2$

P9.25 Calculate the mean ionic molality, $m_{\pm}$, in 0.0500 m solutions of (a) $Ca(NO_3)_2$, (b) NaOH, (c) $MgSO_4$, and (d) $AlCl_3$.

P9.26 Calculate the ionic strength of each of the solutions in Problem P9.25.

P9.27 At $25°C$, the equilibrium constant for the dissociation of acetic acid, K_a, is 1.75×10^{-5}. Using the Debye–Hückel limiting law, calculate the degree of dissociation in 0.100 and 1.00 m solutions. Compare these values with what you would obtain if the ionic interactions had been ignored.

P9.28 The principal ions of human blood plasma and their molal concentrations are

$$m_{Na^+} = 0.14\ m, \quad m_{Cl^-} = 0.10\ m, \quad m_{HCO_3^-} = 0.025\ m.$$

Calculate the ionic strength of blood plasma.

P9.29 Estimate the degree of dissociation of a 0.100 m solution of acetic acid ($K_a = 1.75 \times 10^{-5}$) that is also 0.500 m in the strong electrolyte given in parts (a)–(c). Use the data tables in Appendix B to obtain $\gamma_{\pm}$, because the electrolyte concentration is too high to use the Debye–Hückel limiting law.

 a. $Ca(Cl)_2$

 b. KCl

 c. $MgSO_4$

P9.30 Using Equations (9.41) and (9.42), explain the trends in the solubility of hemoglobin observed in Figure 9.9.

P9.31 Calculate ΔG_r° and the equilibrium constant at 298.15 K for the reaction

$$Hg_2Cl_2(s) \rightarrow 2Hg(l) + Cl_2(g).$$

P9.32 Calculate ΔG_r° and the equilibrium constant at 298.15 K for the reaction

$$Cr_2O_7^{2-}(aq) + 3H_2(g) + 8H^+(aq) \rightarrow 2Cr^{3+}(aq) + 7H_2O(l).$$

P9.33 Using half-cell potentials, calculate the equilibrium constant at 298.15 K for the reaction $2H_2O(l) \rightarrow 2H_2(g) + O_2(g)$. Compare your answer with that calculated using Table 9.1. What is the value of E° for the overall reaction that makes the two methods agree exactly?

P9.34 For the half-cell reaction $AgBr(s) + e^- \rightarrow Ag(s) + Br^-(aq)$, $E^{\circ} = +0.0713$ V. Using this result and the data tables in Appendix B, determine ΔG_f° (Br^-, aq).

P9.35 For the half-cell reaction $Hg_2Cl_2(s) + 2e^- \rightarrow 2Hg(l) + 2Cl^-(aq)$, $E^{\circ} = +0.27$ V. Using this result and the data tables in Appendix B, determine ΔG_f° (Cl^-, aq).

P9.36 By finding appropriate half-cell reactions, calculate the equilibrium constant at 298.15 K for the following reactions:

 a. $4NiOOH + 2H_2O \rightarrow 4Ni(OH)_2 + O_2$

 b. $4NO_3^- + 4H^+ \rightarrow 4NO + 2H_2O + 3O_2$

P9.37 The oxidation of NADH by molecular oxygen occurs in the cellular respiratory system:

$$O_2(g) + 2NADH(aq) + 2H^+(aq) \rightarrow 2H_2O(l) + 2NAD^+(aq)$$

Using the information in Table 9.7, calculate the standard Gibbs energy change that results from the oxidation of NADH by molecular oxygen. The standard reduction potential for nicotine adenine dinucleotide is

$$NAD^+(aq) + H^+(aq) + 2e^- \rightarrow NADH(aq)\quad E^{\circ\prime} = -0.320V$$

P9.38 Although the numerical values of the equilibrium constant, the standard Gibbs energy change ΔG°, and the standard cell potential E_{cell}° depend on the definition of the concentrations of the reference states c°, prove that the Gibbs energy change ΔG and the emf E are independent of the concentrations defined for the reference states.

P9.39 Using the data in Table 9.7, calculate the standard cell potential and the equilibrium constant for the oxidation of ethanol to acetaldehyde. Calculate the equilibrium constant. Assume $T = 298$ K. At pH $= 7$ the equilibrium concentrations of NADH and NAD^+ are in the ratio $c_{NAD^+,eq}/c_{NADH,eq} = 255$, calculate the ratio of the equilibrium concentrations of ethanol and acetaldehyde.

P9.40 By finding appropriate half-cell reactions, calculate the equilibrium constant at 298.15 K for the following reactions:

 a. $2Cd(OH)_2 \rightarrow 2Cd + O_2 + 2H_2O$

 b. $2MnO_4^{2-} + 2H_2O \rightarrow 2MnO_2 + 4OH^- + O_2$

P9.41 The half-cell potential for the reaction $O_2(g) + 4H^+(aq) + 4e^- \rightarrow 2H_2O$ is $+1.03$ V at 298.15 K when $a_{O_2} = 1$. Determine a_{H^+}.

P9.42 You are given the following half-cell reactions:

$$Pd^{2+}(aq) + 2e^- \rightleftarrows Pd(s) \qquad\qquad E^{\circ} = 0.83\ V$$
$$PdCl_4^{2-}(aq) + 2e^- \rightleftarrows Pd(s) + 4Cl^-(aq) \quad E^{\circ} = 0.64$$

 a. Calculate the equilibrium constant for the reaction
  $Pd^{2+}(aq) + 4Cl^-(aq) \rightleftarrows PdCl_4^{2-}(aq)$

 b. Calculate ΔG° for this reaction.

P9.43 Determine E° for the reaction $Cr^{2+} + 2e^- \rightarrow Cr$ from the one-electron reduction potential for Cr^{3+} and the three-electron reduction potential for Cr^{3+} given in Table 9.5.

P9.44 Determine K_{sp} for AgBr at 298.15 K using the electrochemical cell described by

$$Ag(s)|AgBr(s)|Br^-(aq, a_{Br^-})\|Ag^+(aq, a_{Ag^+})|Ag(s).$$

P9.45 The standard potential E° for a given cell is 1.100 V at 298.15 K and

$$\left(\frac{\partial E^{\circ}}{\partial T}\right)_P = -6.50 \times 10^{-5} \text{ V K}^{-1}$$

Calculate ΔG_R°, ΔS_R°, and ΔH_R°.

P9.46 For a given overall cell reaction,
$\Delta S_R^{\circ} = 17.5$ J mol^{-1} K^{-1} and $\Delta H_R^{\circ} = -225.0$ kJ mol^{-1}.

Calculate E° and $\left(\dfrac{\partial E^{\circ}}{\partial T}\right)_P$.

P9.47 The standard half-cell potential for the reaction $O_2(g)$ $+ 4H^+(aq) + 4e^- \rightarrow 2H_2O$ is $+1.03$ V at 298.15 K. Calculate E for a 0.50 molal solution of HCl for $a_{O_2} = 1$ assuming (a) that the a_{H^+} is equal to the molality and (b) using the measured mean ionic activity coefficient for this concentration of $\gamma_{\pm} = 0.757$. How large is the relative error if the concentrations, rather than the activities, are used?

P9.48 Consider the half-cell reaction $O_2(g) + 4H^+(aq) +$ $4e^- \rightarrow 2H_2O$. By what factor are n, Q, E, and E° changed if all of the stoichiometric coefficients are multiplied by a factor of 2? Justify your answers.

P9.49 Consider the Daniell cell, for which the overall cell reaction is

$Zn(s) + Cu^{2+}(aq) \rightarrow Zn^{2+}(aq) + Cu(s)$. The concentrations of $CuSO_4$ and $ZnSO_4$ are 2.5×10^{-3} m and 1.1×10^{-3} m, respectively.

a. Calculate E setting the activities of the ionic species equal to their molalities.

b. Calculate $\gamma_{\pm}$ for each of the half-cell solutions using the Debye–Hückel limiting law.

c. Calculate E using the mean ionic activity coefficients determined in part (b).

P9.50 Consider the couple $Ox + e^- \rightleftarrows Red$ with the oxidized and reduced species at unit activity. What must be the value of E° for this half-cell if the reductant R is to liberate hydrogen at 1 atm from

a. an acid solution with $a_{H^+} = 1$

b. water at pH = 7?

c. Is hydrogen a better reducing agent in acid or basic solution?

P9.51 Consider the half-cell reaction
$AgCl(s) + e^- \rightleftarrows Ag(s) + Cl^-(aq)$. If $\mu^{\circ}(AgCl(s)) =$ -109.71 kJ mol^{-1}, and if $E^{\circ} = +0.222$ V for this half-cell, calculate the standard Gibbs energy of formation of $Cl^-(aq)$.

P9.52 The following data have been obtained for the potential of the cell
$Pt(s)|H_2(g, f = 1 \text{ atm})|HCl(aq, m)|AgCl(s)|Ag(s)$ as a function of m at 25°C.

m(mol kg^{-1})	E(V)	m(mol kg^{-1})	E(V)	m(mol kg^{-1})	E(V)
0.00100	0.597915	0.0200	0.43024	0.500	0.27231
0.00200	0.54425	0.0500	0.38588	1.000	0.23328
0.00500	0.49846	0.100	0.35241	1.500	0.20719
0.0100	0.46417	0.200	0.31874	2.000	0.18631

Calculate E° and $\gamma_{\pm}$ for HCl at $m = 0.00100$, 0.0100, 0.100, and 1.000.

P9.53 Consider the reaction $Sn(s) + Sn^{4+}(aq) \rightleftarrows 2Sn^{2+}(aq)$. If metallic tin is in equilibrium with a solution of Sn^{2+} in which $a_{Sn^{2+}} = 0.100$, what is the activity of Sn^{4+} at equilibrium?

P9.54 Consider the reaction of pyruvate with NADH form to lactate:

$$CH_3(CO)COO^-(aq) + NADH(aq) + H^+(aq)$$
$$\rightleftarrows CH_3(CHOH)COO^-(aq) + NAD^+(aq)$$

Suppose at equilibrium [pyruvate] = 0.0010 M, [lactate] = 0.1000 M, [NADH] = 0.0010 M, and [NAD$^+$] = 0.25210 M. Calculate K, K', ΔG°, and $\Delta G^{\circ\prime}$ at pH = 7.

P9.55 For the reaction in Problem P9.54, calculate the standard cell potentials E_{cell}° and $E_{cell}^{\circ}{}'$ at pH = 7.

P9.56 Using the data in Table 9.7, calculate the standard Gibbs energy change and the equilibrium constant for the oxidation of acetaldehyde to acetic acid by NAD$^+$ at pH = 7. For concentrations

$$c_{NAD^+} = 1.00 \times 10^{-2} M, c_{NADH} = 2.50 \times 10^{-4} M,$$
$$c_{acetaldehyde} = 3.30 \times 10^{-1} M, c_{acetic} = 3.60 \times 10^{-3} M$$

Calculate the reaction quotient Q and the Gibbs energy change.

P9.57 For the oxidation of NADH by hydrogen
$NADH(aq) + H^+(aq) \rightarrow NAD^+(aq) + H_2(g)$
assume $C_{NADH} = 3.2 \times 10^{-2} M$, $c_{NAD^+} = 1.5 \times 10^{-3} M$, and $p_{H_2} = 0.015$ bar and pH = 4.5. Using the data in Table 9.7, calculate the emf and the Gibbs energy change.

P9.58 Referring to Figure 9.16, suppose a volume is divided into two compartments by a semipermeable membrane. In the left-hand "protein compartment" is a solution containing the sodium salt of a protein P, NaP, with initial concentration 1.00 $\times 10^{-2} M$. In the right-hand "salt compartment" is a NaCl solution initially at $3.00 \times 10^{-1} M$. Calculate the equilibrium concentrations of all ionic species in both compartments. Calculate also the net osmotic pressure across the membrane. Assume T = 298 K.

P9.59 Referring to Figure 9.16, obtain expressions for $c_{+,eq}^L$, $c_{+,eq}^R$, $c_{-,eq}^L$, and $c_{-,eq}^R$ for the case in which the net charge on the polymer is $-z$. Using these expressions, derive Equation (9.111).

P9.60 Suppose a volume is divided into two compartments by a semipermeable membrane, as shown in Figure 9.16. In the left-hand "protein compartment" is a solution containing the sodium salt of a protein P, Na$_2$P, with initial concentration $1.00 \times 10^{-2} M$. In the right-hand "salt compartment" is a NaCl solution initially at $1.00 \times 10^{-1} M$. Calculate the equilibrium concentrations of all ionic species in both compartments. Calculate also the Donnan ratio and the junction potential resulting from the Donnan effect. Assume $T = 298$ K.

P9.61 Calculate the net osmotic pressure across the membrane for the system described in Problem P9.58.

P9.62   Referring to Figure 9.16, for the case in which the mobile ion concentrations exceed the concentration of the impermeant macromolecule, show that Equation (9.111) reduces to $\pi_{Net} \approx bRT$. If on the other hand the impermeant macromolecule exists at high levels compared to the mobile ions, show that $\pi_{Net} \approx (z + 1)bRT$. What are the physical consequences of these two limiting equations?

P9.63   In our discussion of the Donnan potential, we neglected H^+ and OH^- ions because they are usually in such low concentrations. However, as a result of the Donnan effect, the concentrations of H^+ and OH^- on each side of the membrane are fixed. Given a Donnan ratio r_D, and referring to Figure 9.16, show that

$$pH_R - pH_L = \log_{10} r_D$$

P9.64   Calculate the pH gradient across a semipermeable membrane if the concentrations initially are:

$$c_{Na^+}^L = c_{P^-}^L = 3.00 \times 10^{-3} \, M; \, c_{Cl^-}^L = 0$$

$$c_{Na^+}^R = c_{Cl^-}^R = 1.05 \times 10^{-2} \, M$$

Calculate the concentrations of these species when the system reaches equilibrium and calculate the pH gradient across the membrane.

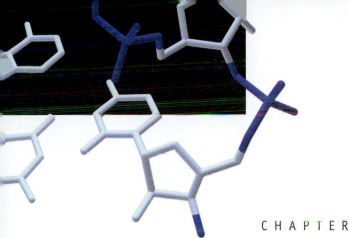

CHAPTER

10

Principles of Biochemical Thermodynamics

Applying thermodynamics to living cells requires care because it is rare that a particular chemical or physical process occurring in a living cell is isolated energetically or physically from other processes. The more common circumstance is that chemical processes within a cell are energetically coupled to other processes and chemical components within a cell may be shared by several processes.

Chemical processes in living cells are maintained far from equilibrium in a condition called dynamic steady state. Under steady-state conditions, cellular processes that release Gibbs energy (e.g., biological oxidations) are coupled energetically to those processes that utilize Gibbs energy (e.g., biosynthesis, transport, mechanical work). In this chapter we present basic principles relevant to understanding the nature of coupled biological reactions. We mention several types of transfers involved in metabolic processes including phosphate group transfer, proton transfer (i.e., acid–base equilibria), and electron transfer (oxidation–reduction reactions). The formalism of the electrochemical potential will be applied to quantifying the energetics of biological oxidation–reduction reactions and ion transport across biological membranes. Finally, the detailed energetics of ATP hydrolysis will be reviewed.

10.1 Thermodynamics and Living Systems

Living cells are open systems that undergo irreversible changes and function far from equilibrium in a condition in which the concentrations of reagents do not change with time. This condition is called **dynamic steady state.** For open systems not at equilibrium, transport processes—where matter, heat, charge, and/or momentum "flow" into and out of the system in response to driving forces that arise from gradients in concentration, temperature, electrical field, and/or fluid velocity—determine the properties of the system, including the concentration of individual species in the system, temperatures, and so forth. To maintain dynamic steady state, the concentration of reagents and products in the cell must remain constant, and this means the rates of cellular reactions must be maintained within particular ranges or regulated, a task performed by protein catalysts called enzymes.

Living systems perform mechanical work associated with muscle movement, osmotic work associated with the transport of metabolites against concentration gradients, and electrical work associated with the movement of charged species across cell membranes.

Although it is clear that cells must produce work, exactly how this is done at levels sufficient to sustain life may not be clear to the reader. Recall in Section 5.2 that the limitation imposed by the second law on the efficiency ε of work production by cyclic heat engines was stated to be

$$\varepsilon = \frac{|w_{cycle}|}{q_{ab}} = \frac{T_{hot} - T_{cold}}{T_{hot}} \tag{10.1}$$

Living cells do not function as heat engines. They do not produce work by exploiting internal thermal gradients. Departing from this knowledge for a moment, suppose that solely for the purpose of work production, a cell exploited the thermal gradient between physiological temperatures (ca. 310 K) and typical ambient temperatures (ca. 298 K). From Equation (5.1) the efficiency is $\varepsilon \approx (310 \text{ K} - 298 \text{ K})/310 \text{ K} \approx 0.04$, a level of work production too low to sustain life. Clearly, the rules imposed on the efficiency of energy transformations as applied to heat engines are not immediately applicable to energy transformations occurring within living cells. But although cells function essentially isothermally, we will show that the efficiencies of some cellular energy transformations exceed 70%.

Essential to understanding energy transformations in the biological world is the transport of energy and matter, schematized in Figure 10.1. The flow of energy and the movement of matter in the biological world may be viewed as a series of energy transformations and exchanges of water, oxygen, and nutrient molecules starting with the absorption of radiant energy, water, and carbon dioxide by plants. Plants contain **autotrophic cells,** or cells that have the ability to produce complex molecules such as glucose from simple starting materials such as carbon dioxide and water. In photoautotrophic cells, radiant energy is used to generate complex molecules such as carbohydrates from carbon dioxide and water via a process called photosynthesis. The net reaction for the production of glucose $C_6H_{12}O_6$ from water and carbon dioxide via photosynthesis is

$$6CO_2(g) + 6H_2O(l) \xrightarrow{\text{light}} C_6H_{12}O_6(s) + 6O_2(g) \tag{10.2}$$

In biological thermodynamics, processes for which $\Delta G > 0$ are called **endergonic.** Photosynthesis is an endergonic process for which $\Delta G^{\circ\prime}_{298.15} = 2868 \text{ kJ mol}^{-1}$, and is therefore not spontaneous at $T = 298$ K. The radiant energy source for photosynthesis is the sun, For the net reaction of photosynthesis [Equation (10.2)] it can also be shown that $\Delta H^{\circ\prime}_{298.15} = 2813 \text{ kJ mol}^{-1}$ and so the net entropy change in photosynthesis is negative:

$$\Delta S^{\circ\prime}_{298.15} = \frac{\Delta H^{\circ\prime} - \Delta G^{\circ\prime}}{T} = -182.3 \text{ J K}^{-1} \text{ mol}^{-1}$$

The entropy of the system (i.e., the plant) decreases because simple, "low Gibbs energy" molecules such as carbon dioxide and water are being assembled via the photosynthetic process into more complex "high Gibbs energy" molecules such as glucose. By high Gibbs energy, we mean that the magnitude of $\Delta G^{\circ\prime}_f$ is large.

FIGURE 10.1

Transport of energy and matter in the biological world.

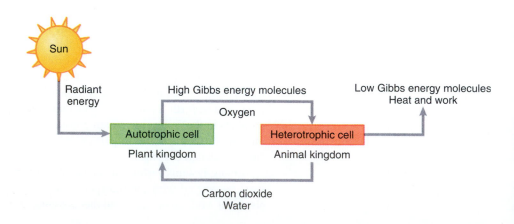

Another important transformation of energy is performed by heterotrophic cells in the animal kingdom. **Heterotrophic cells** are incapable of performing photosynthesis and thus cannot produce complex molecules such as glucose from simple starting materials alone. Fundamental to heterotrophic cells is a process or series of processes in which complex "high Gibbs energy" fuel molecules such as carbohydrates, proteins, and fats are broken down or "metabolized" into simpler "low Gibbs energy" waste molecules, such as carbon dioxide and water.

For example, in some heterotrophic cells each glucose molecule is metabolized into two molecules of pyruvic acid by a process called anaerobic glycolysis (see Figure 11.3). For anaerobic glycolysis the standard Gibbs energy change is $\Delta G^{\circ\prime}_{298.15} = -217 \text{ kJ mol}^{-1}$. In biological thermodynamics, processes for which $\Delta G < 0$ are called **exergonic.** The metabolism of glucose to pyruvic acid under standard conditions is an example of an exergonic reaction.

In aerobic heterotrophic cells, a series of oxidation–reduction reactions is used to convert pyruvic acid to carbon dioxide and reduce O_2 to H_2O. This sequence of oxidation–reduction reactions is called a respiratory system. As opposed to anaerobic cells, which metabolize glucose to pyruvic acid in the absense of oxygen, aerobic cells use oxygen to accomplish the complete metabolism of glucose to carbon dioxide and water:

$$C_6H_{12}O_6(s) + 6O_2(g) \rightarrow 6CO_2(g) + 6H_2O(l) \qquad (10.3)$$

Because aerobic glucose metabolism is the reverse of the net photosynthesis reaction, the Gibbs energy change is $\Delta G^{\circ\prime}_{298.15} = -2868 \text{ kJ mol}^{-1}$, so under standard conditions and at $T = 298$ K the complete metabolism of glucose to water and carbon dioxide is spontaneous. In aerobic cells the efficiency with which the combination of glycolysis and respiration converts the Gibbs energy of glucose to work exceeds 60%.

Figure 10.2 shows an important property of heterotrophic cells: the Gibbs energy released by the degradation of glucose in the respiratory system is used to drive endergonic processes such as the transport of molecules and ions across cell membranes, muscle contraction, and the biosynthesis of complex molecules such as proteins, nucleic acids, and carbohydrates. Exergonic and endergonic reactions are energetically coupled in cells using phosphorylated molecules such as **adenosine triphosphate (ATP)** and **adenosine diphosphate (ADP).** See Figure 10.3.

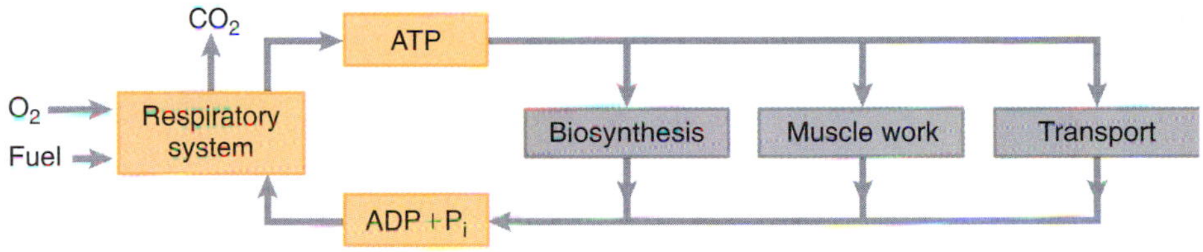

FIGURE 10.2

The coupling of exergonic processes (e.g., respiration) to endergonic processes (e.g., biosynthesis, mechanical work, transport) via ATP hydrolysis and ADP phosphorylation.

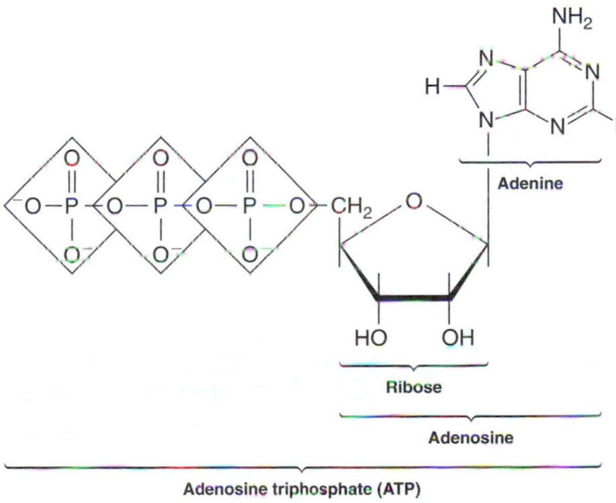

FIGURE 10.3

Molecular structure of adenosine triphosphate, ATP. When ATP is hydrolyzed, one of the P–O bonds is broken, which removes the terminal phosphate group to form adenosine diphosphate, ADP. ADP can be similarly hydrolyzed to adenosine monophosphate, AMP.

There is a resemblance between the energy-coupling scheme in Figure 10.2 and heat engines. Heat engines exploit a thermal gradient to produce work. In a heat engine, heat energy absorbed by a working fluid is used to produce mechanical work. The respiratory system consists of a chain of reactions where stepwise metablism of glucose produces products with smaller Gibbs energies. The energy liberated by this system could be evolved as heat but biological processes do not function as heat engines. Rather the Gibbs energy liberated by the respiratory system is "captured" in the form of ATP, which interacts with the various biological process to produce mechanical work, osmotic work, and so forth.

The phosphorylation of ADP to form ATP and the hydrolysis of ATP to ADP are intermediate processes that are used to couple the Gibbs energy released by respiration to endergonic processes such as muscle contraction, ion transport, and biosynthesis. The living cell then exploits a Gibbs energy gradient to produce work, with ATP acting as the carrier of Gibbs energy between exergonic and endergonic processes.

10.2 The Principle of Common Intermediates

In earlier chapters we learned that if $\Delta G > 0$, an isolated process will not proceed. In biological systems chemical processes rarely occur in isolation. For example, it may occur that $\Delta G > 0$ for transport of potassium ions across a cell membrane, but this does not imply that this transport cannot occur in the cell because the transport process does not occur in isolation. Processes that are endergonic and thus would not proceed in isolation are energetically coupled within the cell to exegonic processes.

If the Gibbs energy released in the cell by an exergonic reaction is to be used to drive an endergonic reaction, some means must exist by which the two reactions are energetically coupled. In the absence of such a coupling, the energy released in the cell by glucose metabolism, for example, would eventually be lost as heat. To be energetically coupled, endergonic and exergonic processes must share a common chemical intermediate. To illustrate the thermodynamic aspects of the **common intermediate** principle, consider the synthesis of the disaccharide sucrose from the monosaccharides glucose and fructose:

$$(10.4)$$

Because $\Delta G^{\circ\prime}_{reaction} = 23.0 \text{ kJ mol}^{-1}$ the reaction in Equation (10.4) is highly endergonic. The **hydrolysis** of ATP to ADP and inorganic phosphate (P_i) may be written

$$ATP(aq) + H_2O(l) \rightarrow ADP(aq) + P_i(aq) \tag{10.5}$$

In most of the discussion that follows, biochemical reactions are assumed to occur in aqueous solution, so we will omit the (aq) designation. Exceptions will be noted. We will also frequently refer to the hydronium ion H_3O^+ simply as H^+. At pH 7 and $T = 310$ K, this hydrolysis is exergonic: $\Delta G^{\circ\prime} = -30.5$ kJ mol^{-1}.

EXAMPLE PROBLEM 10.1

For the hydrolysis of ATP, shown in Equation (10.5), standard conditions do not prevail in the cellular environment. In the cell, typical concentrations of ATP, ADP, and inorganic phosphate are $c_{ATP} = 1850$ µM, $c_{ADP} = 138$ µM, and $c_{P_i} = 1.00$ mM . Calculate the Gibbs energy of hydrolysis in the cellular environment, assuming pH $= 7$ and $T = 310$ K.

Solution

Start by calculating the reaction quotient

$$Q' = \frac{c_{ADP}\, c_{P_i}}{c_{ATP}} \times \frac{c^\circ_{ATP}}{c^\circ_{ADP}\, c^\circ_{P_i}}$$

$$= \frac{1.38 \times 10^{-4}\, M \times 1.00 \times 10^{-3}\, M}{1.85 \times 10^{-3}\, M} \times \frac{1.00\, M}{1.00\, M \times 1.00\, M}$$

$$= 7.46 \times 10^{-5}$$

Then

$$\Delta G_{reaction} = \Delta G^{\circ\prime}_{reaction} + RT \ln Q'$$

$$= -30.5\ \text{kJ mol}^{-1} \times 1\ \text{mol} + 8.314\ \text{J K}^{-1}\text{mol}^{-1} \times 310.\ \text{K} \times \ln(7.46 \times 10^{-5}) \times 1\ \text{mol}$$

$$= -55.0\ \text{kJ}$$

Therefore, under cellular conditions, the hydrolysis of ATP is even more exergonic than it is under standard conditions.

If the hydrolysis of ATP were coupled energetically to the formation of sucrose from fructose and glucose, the net reaction would be

$$\text{fructose} + \text{glucose} + \text{ATP} \rightarrow \text{sucrose} + \text{ADP} + \text{P}_i \qquad (10.6)$$

and the standard Gibbs energy change would be

$$\Delta G^{\circ\prime}_{reaction} = 23.0\ \text{kJ mol}^{-1} - 30.5\ \text{kJ mol}^{-1} = -7.5\ \text{kJ mol}^{-1} \qquad (10.7)$$

In fact, the coupling of ATP hydrolysis to sucrose synthesis occurs in two steps using a common intermediate glucose-1-phosphate (G-1-P):

$$(1)\ \text{ATP} + \text{glucose} \rightarrow \text{ADP} + \text{G-1-P}$$

$$(2)\ \text{G-1-P} + \text{fructose} \rightarrow \text{sucrose} + \text{P}_i \qquad (10.8)$$

For the first reaction where glucose is phosphorylated by ATP, $\Delta G^\circ_1{}' = -9.6$ kJ mol^{-1}, and for the reaction of G-1-P with fructose to form sucrose, $\Delta G^\circ_2{}' = 2.1$ kJ mol^{-1}.

Again, because standard conditions do not in general prevail in the cellular environment, Equation (10.6) may be more or less exergonic depending on the actual concentrations of the reactants and products that occur in the cellular environment. As long as the reaction quotient deviates markedly from the equilibrium constant such that the concentrations of ATP, fructose, and glucose exceed those of ADP, inorganic phosphate, and sucrose, the Gibbs energy will be more negative than the standard Gibbs energy and the reaction under cellular conditions will be more exergonic than under standard conditions.

EXAMPLE PROBLEM 10.2

Equation (10.5) does not attain equilibrium in the living cell where concentrations of ATP, ADP, and P_i are approximately $c_{ATP} = 1850 \ \mu M$, $c_{ADP} = 138 \ \mu M$, and $c_{P_i} = 1.00 \ mM$. In addition, glucose has a cellular concentration of about 5.00 mM. Assuming the cellular concentrations of fructose and sucrose are 1.00 mM and 100. μM, respectively, calculate the Gibbs energy change ΔG for the reaction in Equation (10.6) with these cellular concentrations, at pH = 7 and at $T = 310.$ K.

Solution

$$Q' = \frac{c_{ADP}c_{P_i}c_{sucrose}}{c_{ATP}c_{fructose}c_{glucose}} \times \frac{c^{\circ}_{ATP}c^{\circ}_{fructose}c^{\circ}_{glucose}}{c^{\circ}_{ADP}c^{\circ}_{P_i}c^{\circ}_{sucrose}}$$

$$= \frac{1.38 \times 10^{-4} \ M \times 1.00 \times 10^{-3} \ M \times 1.00 \times 10^{-4} \ M}{1.00 \times 10^{-3} \ M \times 5.00 \times 10^{-3} \ M \times 1.850 \times 10^{-3} \ M}$$

$$\times \frac{1 \ M \times 1 \ M \times 1 \ M}{1 \ M \times 1 \ M \times 1 \ M} = 1.49 \times 10^{-3}$$

$$\therefore \Delta G_{reaction} = \Delta G^{\circ \prime}_{reaction} + RT \ln Q'$$

$$= -7.50 \times kJ \ mol^{-1} \times 1 \ mol + 8.314 \ J \ K^{-1} mol^{-1}$$
$$\times 310. \ K \ ln(0.00149) \times 1 \ mol = -2.43 \ kJ$$

Note that because the common intermediate G-1-P is a reactant in the first reaction and a product in the second, the reaction quotient for two coupled reactions will not be dependent on the concentration of the common intermediate G-1-P. Note also that under the cellular conditions specified, the reaction given by Equation (10.6) is even more exergonic than that obtained under standard conditions. We will find in Chapter 11 that cellular levels of metabolites in the glycolytic pathway are maintained at levels far from standard conditions, with the result that the Gibbs energy for the glycolytic pathway under cellular conditions differs markedly from the Gibbs energy that would result if the same pathway occurred under standard conditions. This has important implications for how the glycolytic pathway is regulated and how much energy is eventually liberated by the metabolism of glucose.

10.3 Phosphate Transfer Potentials

In the preceding example we found that ATP acts as a phosphate group donor to produce a phosphorylated common intermediate G-1-P, which couples the exergonic phosphate transfer reaction to the endergonic reaction of G-1-P with fructose to produce sucrose. This reaction is schematized in Figure 10.2 as the biosynthesis of a carbohydrate being driven by hydrolysis of ATP, with ADP emerging as a product of the synthesis. But ADP may also act as a phosphate group acceptor from a common intermediate whose hydrolysis is more exergonic than that of ATP.

Figure 10.2 also shows ATP emerging as a product from the respiratory system, where the highly endergonic phosphorylation of ADP is accomplished by coupling to various exergonic reactions. As an example of how the phosphorylation of ADP can be accomplished by coupling to an exergonic reaction via a common intermediate, consider the oxidation of glyceraldehyde-3-phosphate (RCHO) to 3-phosphoglycerate (RCOO⁻) by nicotine adenine dinucleotide (NAD). NAD is a ubiquitous oxidizing agent involved in numerous redox reactions in respiring cells. The structure of NAD is shown in Figure 10.4.

As will be discussed in more detail in Chapter 11, the oxidation of glyceraldehyde-3-phosphate to 3-phosphoglycerate by NAD^+ is an important component of the glycolytic pathway. We will also discuss the structure and chemistry of NAD in Chapter 11, but for the present it is sufficient to recognize that the oxidized form of NAD is NAD^+ and the reduced form is NADH. The oxidation of glyceraldehyde-3-phosphate by NAD^+ is

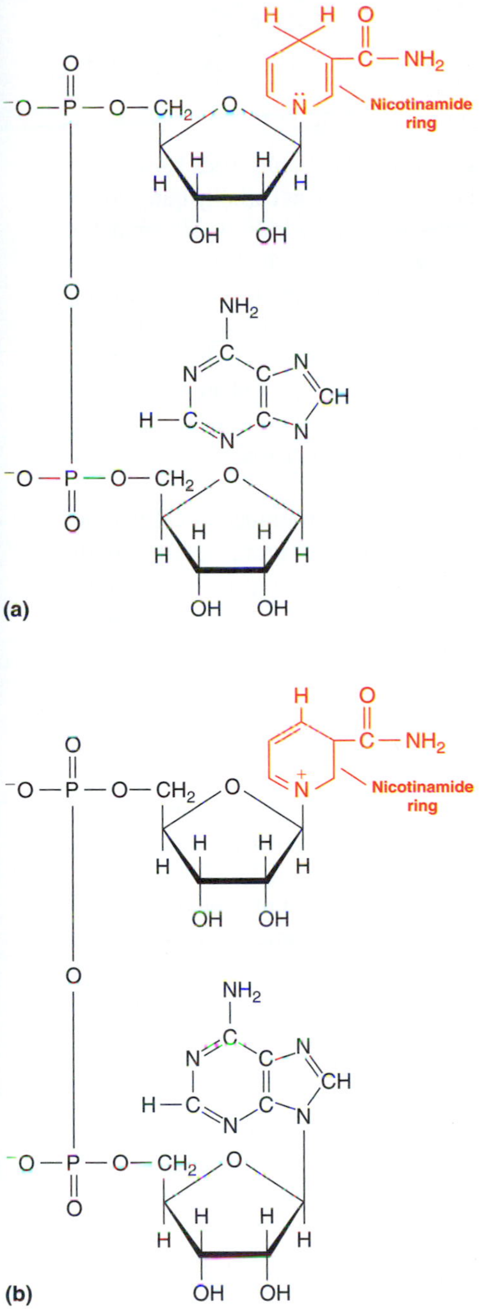

(a)

(b)

FIGURE 10.4
(a) The reduced form of nicotine adenine dinucleotide (NADH); (b) the oxidized form NAD$^+$. NAD is composed of an adenine ring, a nicotinamide ring, and two sugar-phosphate groups directly linked. In its oxidized form, NAD is designated NAD$^+$, to indicate a positive charge on the nicotinamide ring at pH 7. The nicotinamide ring is the site of redox chemistry, the reduced form being designated NADH.

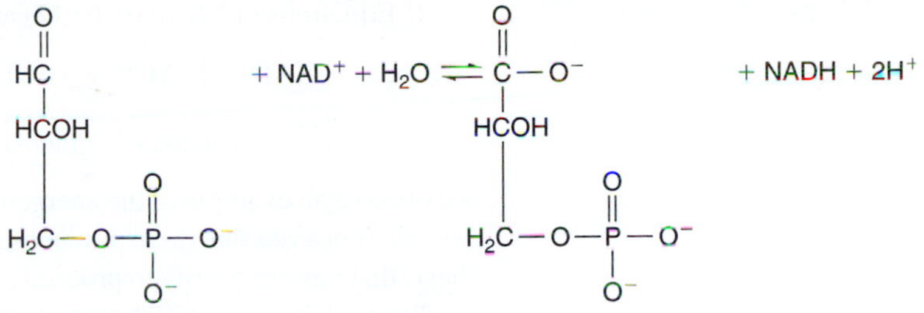

Equation (10.9) can be written in shorthand as

$$RCHO + NAD^+ + H_2O \rightarrow RCOO^- + NADH + 2H^+ \qquad (10.9')$$

where R = $-CH(OH)CH_2OPO_3^{2-}$. For this oxidation–reduction reaction, $\Delta G^{\circ\prime}_{reaction} = -43.1$ kJ mol^{-1}.

The oxidation of glyceraldehyde-3-phosphate is coupled to the phosphorylation of ADP by the common intermediate 1,3-bisphosphoglycerate ($RCOPO_3^{2-}$), where production of this species is accomplished by a direct reaction between 3-phosphoglycerate and inorganic phosphate, also called a substrate-level phosphorylation. The reaction is

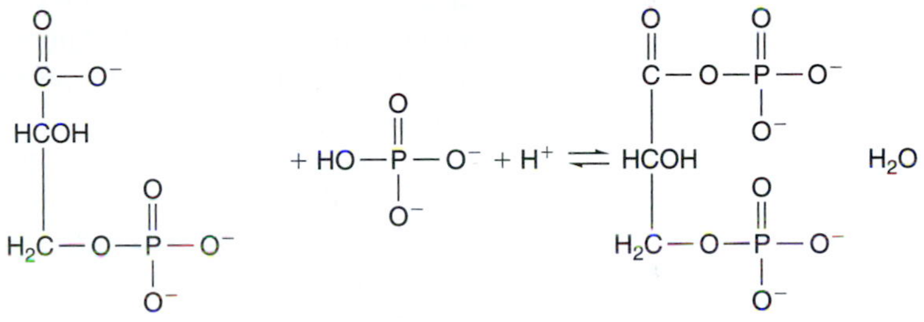

$$RCOO^- + H^+ + HPO_4^{2-} \rightleftharpoons RCOPO_3^{2-} + H_2O \qquad (10.10)$$

for which $\Delta G^{\circ\prime}_{reaction} = 49.3$ kJ mol^{-1}. In turn, 1,3-bisphosphoglycerate transfers a phosphate group to ADP:

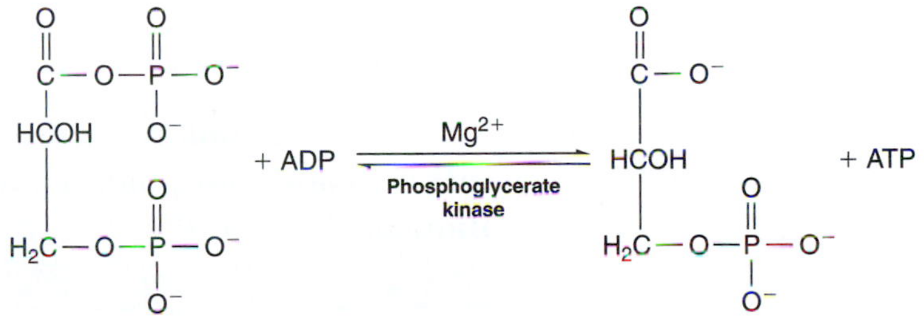

$$RCOPO_3^{2-} + ADP \rightarrow RCOO^- + ATP \qquad (10.11)$$

For this reaction $\Delta G^{\circ\prime}_{reaction} = -18.8$ kJ mol^{-1}. Equations (10.10) and (10.11) are coupled by the common intermediate 1,3-bisphosphoglycerate. The net reaction is exergonic:

$$RCHO + NAD^+ + ADP + HPO_4^{2-} \rightarrow RCOO^- + NADH + H^+ + ATP \qquad (10.12)$$

for which $\Delta G^{\circ\prime}_{reaction} = -43.1$ kJ mol^{-1} + 49.3 kJ mol^{-1} − 18.8 kJ mol^{-1} = −12.6 kJ mol^{-1}.

If Equations (10.6) and (10.12) are juxtaposed

$$RCHO + NAD^+ + ADP + HPO_4^{2-} \rightarrow RCOO^- + NADH + H^+ + ATP \quad (10.12)$$

$$\text{fructose} + \text{glucose} + ATP \rightarrow \text{sucrose} + ADP + P_i \quad (10.6)$$

we observe an example of the energetic coupling, schematized in Figure 10.2, between exergonic processes that drive production of ATP and processes such as carbohydrate biosynthesis that require energy to proceed and must be driven by ATP hydrolysis. The oxidation of glyceraldehyde-3-phosphate to 3-phosphoglycerate was accomplished in two steps: first by the production of the phosphorylated intermediate 1,6-bisphosphoglycerate, which in the second step phosphorylates ADP. The endergonic biosynthesis of sucrose from fructose and glucose was similarly accomplished in two steps: first through the phosphorylation of glucose by ATP to produce the intermediate glucose-1-phosphate, and second through the reaction of glucose-1-phosphate with fructose to yield sucrose. ATP is the common intermediate between processes that require energy to proceed, and processes that release energy.

The direction of phosphate group transfer is of pivotal importance in cellular energy coupling. In the preceding examples, ADP was a phosphate group acceptor in the reaction with 1,6-Bisphosphoglycerate, but ATP was a phosphate group donor in the reaction with glucose. The physical property that determines the direction of phosphate group transfer in a given reaction, and that defines the key role of ATP in cellular energetics, is the **phosphate transfer potential.** The phosphate transfer potential is simply the negative of the Gibbs energy of hydrolysis in units of kilojoules. Table 10.1 shows phosphate transfer potentials for a number of biologically important compounds.

In a given reaction, the phosphate donor is the molecule with the higher phosphate transfer potential while the phosphate acceptor is the molecule whose phosphorylated form has the lower phosphate transfer potential. In Equation (10.11), 1,3-bisphosphoglycerate is the phosphate donor and ADP is the phosphate acceptor because 1,3-bisphosphoglycerate has a phosphate transfer potential of 49.4, whereas ATP has a transfer potential of 30.5. Similarly, ATP is the phosphate donor in Equation (10.10) and glucose is the phosphate acceptor because ATP has a phosphate transfer potential of 30.5, whereas glucose-1-phosphate has a transfer potential of only 20.9. Clearly, the key role played by ATP in cellular energetics (schematized in Figure 10.3) is due to the intermediate value of its phosphate transfer potential. The net production of ATP by the respiratory system is due to the coupling of ADP phosphorylation with exergonic reactions involving compounds such as phosphoenolpyruvate and 1,3-bisphosphoglycerate, both of which have phosphate transfer potentials higher than ATP and thus will phosphorylate ADP exergonically.

TABLE 10.1 **Standard Gibbs Energies of Phosphate Hydrolysis (pH = 7) and Phosphate Transfer Potentials of Compounds of Biological Interest***

Compound	$\Delta G^{\circ\prime}_{hydrolysis}$ (kJ mol^{-1})	Phosphate Transfer Potential
Phosphoenolpyruvate	−61.9	61.9
1,3-bisphosphoglycerate	−49.4	49.4
Phosphocreatine	−43.1	43.1
Acetyl phosphate	−43.1	43.1
ATP	−30.5	30.5
Glucose-1-phosphate	−20.9	20.9
2-Phosphoglycerate	−17.6	17.6
Fructose-6-phosphate	−13.8	13.8
Glucose-6-phosphate	−13.8	13.8
Glycerol-3-phosphate	−9.2	9.2

*The phosphate transfer potential equals $-\Delta G'$ for phosphate hydrolysis in kJoules per mole.

EXAMPLE PROBLEM 10.3

The phosphate transfer potential of the phosphorylated amino acid phosphoarginine is 33.2. Calculate the Gibbs energy change for the reaction

$$arginine + ATP \rightarrow phosphoarginine + ADP$$

Solution

Because the phosphate transfer potential of phosphoarginine is 33.2, the hydrolysis of phosphoarginine has a standard Gibbs energy change of $\Delta G°'_{hydrolysis} = -33.2 \text{ kJ mol}^{-1}$. The reaction of arginine with ATP can be written as the sum of a hydrolysis and a phosphorylation:

$$(1)\ ATP + H_2O \rightarrow ADP + P_i \quad \Delta G°'_1 = -31.5 \text{ kJ mol}^{-1}$$

$$(2)\ arginine + P_i \rightarrow phosphoarginine + H_2O \quad \Delta G°'_2 = 33.2 \text{ kJ mol}^{-1}$$

Note that the second reaction is the reverse of phosphoarginine hydrolysis so the Gibbs energy change reverses sign. Then the overall change in the Gibbs energy under standard conditions is

$$\Delta G°'_{reaction} = \Delta G°'_1 + \Delta G°'_2 = -31.5 \text{ kJ mol}^{-1} + 33.2 \text{ kJ mol}^{-1} = 1.7 \text{ kJ mol}^{-1}$$

This means that under standard conditions the phosphorylation of arginine by ATP is endergonic and must be coupled to an exergonic process by a common intermediate in order to occur. The coupling of ATP hydrolysis to endergonic reactions such as sucrose biosynthesis occurs via phosphorylated common intermediates, such as glucose-1-phosphate, that have phosphate transfer potentials that are lower than ATP.

10.4 Biological Membranes and the Energetics of Ion Transport

In this section we discuss an important class of endergonic processes that is coupled to ATP hydrolysis: ion transport across biological membranes. For our present purpose the structure of a biological membrane may be schematized as shown in Figure 10.5. As described in Chapter 7, biological membranes are composed of phospholipids arranged in a bilayer structure, with the polar head groups of the lipids directed into the aqueous phases that exist inside and outside the biological membrane and with the hydrophobic "tails" of the lipids directed into the interior of the membrane bilayer. Integrated into the membrane bilayer structure are assemblies of proteins called channels. Protein channels are involved in the transport of molecules into and out of the cell.

Transport of charged particles, that is, ions, across a membrane produces a charge current. Suppose a number of ionic species are distributed with different concentrations inside and outside a cell membrane. If only one of these species can cross the membrane, and the

FIGURE 10.5
General schematic of the bilayer structure of a membrane with an embedded channel protein.

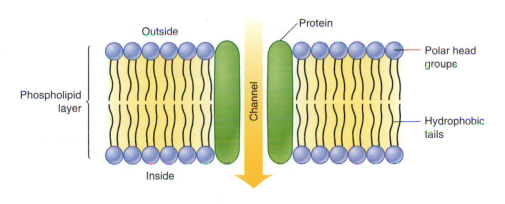

membrane is impermeable to all other species, the permeable species will eventually reach equilibrium. Once equilibrium is reached transport ceases by definition, so the transmembrane current must go to zero.

However, with regard to its electrostatic properties, the cell membrane is not at equilibrium. Multiple ionic species, including sodium (Na^+), potassium (K^+), chloride (Cl^-), and so on, can cross the membrane with varying permeabilities, and distributed inside and outside the cell membrane are relatively impermeable species including charged proteins. Most models of membrane transport assume ionic concentrations are maintained at steady state, so that there is no net ion current across the membrane.

To calculate the energetics of membrane transport at steady state, all that is required is a method for computing the Gibbs energy required to achieve such transfers. From the thermodynamic point of view, membrane transport is characterized by whether the transport is an endergonic process requiring energy to occur or an exergonic process that releases energy. Key to determining the energetics of membrane transport is the chemical potential, or in the case of ion transport, the electrochemical potential.

Let us start with a simple model of the chemical potential of an uncharged molecule that is to be transported across a membrane. The chemical potential for the species of interest outside the membrane is

$$\mu_{out} = \mu_{out}^\circ + RT \ln c_{out} \tag{10.13}$$

where c_{out} is the concentration of the species outside the membrane.

The chemical potential of the same molecule inside the membrane is similarly

$$\mu_{in} = \mu_{in}^\circ + RT \ln c_{in} \tag{10.14}$$

The reversible work involved in transporting dn moles of the molecule from inside the membrane to outside the membrane is

$$dw_{rev} = dG = \mu_{out} dn - \mu_{in} dn = (\mu_{out} - \mu_{in})dn \tag{10.15}$$

The work of reversible transport is equal to the change in the Gibbs energy, which we write in terms of the chemical potential change:

$$w_{rev} = \Delta\mu = \mu_{out} - \mu_{in} = \mu_{out}^\circ - \mu_{in}^\circ + RT \ln\left(\frac{c_{out}}{c_{in}}\right) \tag{10.16}$$

We assume the standard chemical potential of the molecule is independent of its specific location relative to the membrane so Equation (10.16) simplifies to

$$\Delta\mu = \mu_{out} - \mu_{in} = RT \ln\left(\frac{c_{out}}{c_{in}}\right) \tag{10.17}$$

Equation (10.17) states that the energetics of the transport of uncharged molecules across a membrane depends on the direction of transport relative to the concentration gradient of the molecule being transported. In passive transport, a molecule across a membrane moves from a region of high concentration to a region of low concentration. Using Equation (10.17), the transport of a molecule from inside a membrane to outside a membrane is passive if $c_{in} > c_{out}$, in which case

$$\Delta\mu = RT \ln\left(\frac{c_{out}}{c_{in}}\right) < 0 \tag{10.18}$$

Equation (10.18) illustrates that passive transport is exergonic. Passive transport may occur as a result of the molecule diffusing directly through the membrane or through a protein channel. The molecule may also undergo facilitated diffusive transport where it binds to a carrier protein that assists with or facilitates transport.

If, in contrast, the molecule is transported against a concentration gradient (i.e., $c_{in} < c_{out}$), the chemical potential change is positive:

$$\Delta\mu = RT \ln\left(\frac{c_{out}}{c_{in}}\right) > 0 \tag{10.19}$$

This type of transport is called active transport, and it is clearly endergonic. In order for this process to occur, it must be coupled to an exergonic process, as discussed in Section 10.2. For uncharged species the criterion for equilibrium is $\Delta\mu = 0$, which will occur according to Equation (10.19) when $c_{out} = c_{in}$.

Reversible work calculations can be extended to include the transport of charged species if we use the electrochemical potential. In Section 9.6 the electrochemical potential $\tilde{\mu}$, was defined as

$$\tilde{\mu} = \mu + zF\phi \tag{10.20}$$

where μ is the chemical potential, F is Faraday's constant, ϕ is the electrical potential at the position of the charged particle, and the ionic charge z is expressed in units of the electron charge ($+1, -1, +2, -2, \dots$). The work involved in moving reversibly 1 mol of ions with charge z from inside the membrane where the potential is ϕ_{in} to outside the membrane where the potential is ϕ_{out} is

$$w_{elec} = zF(\phi_{out} - \phi_{in}) \tag{10.21}$$

Using Equations (10.16) and (10.21), the relevant quantity in calculating the reversible work of transporting a mole of a charged species from the interior to the exterior of a cell is the change in the electrochemical potential $\tilde{\mu}$:

$$\Delta\tilde{\mu} = \tilde{\mu}_{out} - \tilde{\mu}_{in} = RT \ln\left(\frac{c_{out}}{c_{in}}\right) + zF(\phi_{out} - \phi_{in}) = RT \ln\left(\frac{c_{out}}{c_{in}}\right) + zF\Delta\phi \tag{10.22}$$

The criterion for equilibrium with respect to the distribution of a single charged species is $\Delta\tilde{\mu} = 0$, which leads to

$$0 = RT \ln\left(\frac{c_{out}}{c_{in}}\right) + zF\Delta\phi \tag{10.23}$$

Equation (10.23) can be rearranged and solved for the potential difference $\Delta\phi$:

$$\Delta\phi = -\frac{RT}{zF} \ln\left(\frac{c_{out}}{c_{in}}\right) \tag{10.24}$$

Equation (10.24) is a special case of a more general relationship called the Nernst equation.

The most extensively studied case of active ionic transport in cells is the transport of sodium and potassium ions across the cell membrane, a process that is called the **sodium-potassium pump**. In general, the concentration of extracellular potassium is less than the concentration of intracellular potassium: $c_{in}^{K^+} > c_{out}^{K^+}$. Conversely the concentration of sodium outside the cell exceeds the concentration inside the cell: $c_{in}^{Na^+} < c_{out}^{Na^+}$. Therefore, the transport of potassium from outside to inside the membrane and the simultaneous transport of sodium from inside to the outside are both endergonic. The active transport of potassium and sodium becomes exergonic through coupling to the hydrolysis of ATP.

A single membrane-associated enzyme complex that hydrolyzes ATP (i.e., ATPase) also serves as a transport protein. The functional details of the sodium-potassium pump are under current investigation, but Figure 10.6 schematizes the general way in which the sodium-potassium pump is believed to operate. The molecular machinery of the sodium-potassium pump consists of a large subunit called α and a small subunit called β. The α subunit has a high-affinity sodium ion binding site that faces toward the cell interior. Call the α subunit in the initial unphosphorylated condition E_1. ATP will phosphorylate the α subunit only in the presence of sodium ion at this binding site, forming E_1-P. When phosphorylated the α subunit undergoes a structural change that allows passage of three sodium ions from the cell interior to the exterior, and causes E_1-P to be converted to a lower energy complex called E_2-P. E_2-P has a high-affinity potassium ion binding site that faces the cell exterior. When two potassium ions bind to E_2-P, the complex is hydrolyzed, causing a conformational change that allows passage of the two potassium ions into the cell. Once the potassium ions have passed into the cell and sodium ion is restored to the E_1 binding site, the cycle repeats itself. The action of the

FIGURE 10.6
Schematic of a single cycle of the sodium-potassium pump. (Source: Adapted from C. K. Mathews, K. E. van Holde, and K. G. Ahern, *Biochemistry,* 3rd ed., San Francisco CA: Pearson Benjamin Cummings) p. 343. Copyright 2000. Reprinted by Permission of Pearson Education.)

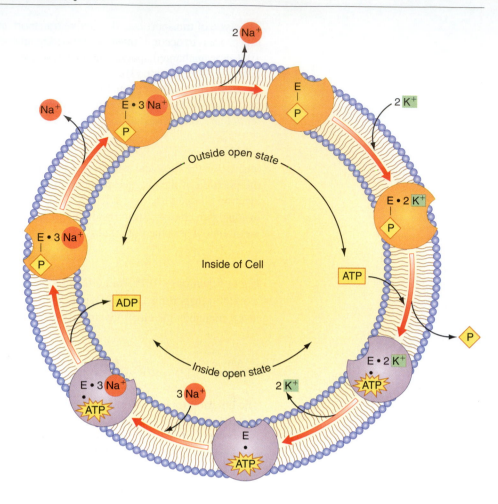

pump can be reversed if potassium ion is depleted from the extracellular medium and if sodium ion builds up. Subsequently, ATP is produced by phosphorylation of ADP.

From the preceding discussion and Equation (10.22), we can calculate the reversible work required to operate a cycle of the sodium-potassium pump. The net equation for a cycle of the ion pump is

$$\left.\begin{array}{c} 3Na^+ \text{ (inside)} \\ + \\ 2K^+ \text{ (outside)} \end{array}\right\} + ATP \rightarrow ADP + \text{phosphate} + \left\{\begin{array}{c} 3Na^+ \text{ (outside)} \\ + \\ 2K^+ \text{ (inside)} \end{array}\right. \quad (10.25)$$

From Equation (10.25), the reversible work required to transport 3 mol of sodium ions and 2 mol of potassium ions is

$$\Delta G = 3\Delta\tilde{\mu}_{in \rightarrow out}^{Na^+} + 2\Delta\tilde{\mu}_{out \rightarrow in}^{K^+} \quad (10.26)$$

Using Equation (10.22), Equation (10.26) becomes

$$\Delta G = 3\left(RT \ln\left(\frac{c_{out}^{Na^+}}{c_{in}^{Na^+}}\right) + F(\phi_{out} - \phi_{in})\right) + 2\left(RT \ln\left(\frac{c_{in}^{K^+}}{c_{out}^{K^+}}\right) + F(\phi_{in} - \phi_{out})\right)$$

$$= RT\left(3 \ln\left(\frac{c_{out}^{Na^+}}{c_{in}^{Na^+}}\right) + 2 \ln\left(\frac{c_{in}^{K^+}}{c_{out}^{K^+}}\right)\right) + F\Delta\phi \quad (10.27)$$

EXAMPLE PROBLEM 10.4

The distributions of sodium and potassium ions inside and outside the cell membrane are $c_{out}^{Na^+} = 1.40 \times 10^{-1}\,M$, $c_{out}^{K^+} = 5.00 \times 10^{-3}\,M$, $c_{in}^{Na^+} = 1.00 \times 10^{-2}\,M$, and $c_{in}^{K^+} = 1.00 \times 10^{-1}\,M$.

a. Calculate the total free energy change involved in transporting 3 mol of sodium ion out of the cell and 2 mol of potassium into the cell at $T = 310$ K. Assume a potential difference of $\Delta\phi = \phi_{out} - \phi_{in} = 0.070$ V.

Solution

Referring to Equation (10.22), we calculate the individual changes in chemical potential that accompany transfers of sodium and potassium ions:

$$\Delta\tilde{\mu}_{in \to out}^{Na^+} = RT \ln\left(\frac{c_{out}^{Na^+}}{c_{in}^{Na^+}}\right) + F(\phi_{out} - \phi_{in})$$

$$= (8.314 \text{ J mol}^{-1}\text{K}^{-1})(310.\text{ K}) \ln\left(\frac{140.}{10.0}\right) + (96{,}485 \text{ C mol}^{-1})(0.070 \text{ V}) = 13.6 \text{ kJ mol}^{-1}$$

$$\Delta\tilde{\mu}_{out \to in}^{K^+} = RT \ln\left(\frac{c_{in}^{K^+}}{c_{out}^{K^+}}\right) + F(\phi_{in} - \phi_{out})$$

$$= (8.314 \text{ J mol}^{-1}\text{K}^{-1})(310.\text{ K})\ln\left(\frac{100.}{5.00}\right) + (96485 \text{ C mol}^{-1})(-0.070 \text{ V}) = 1.00 \text{ kJ mol}^{-1}$$

Now using Equation (10.25), we calculate the reversible work involved in performing the transfer of 3 mol of sodium ion and 2 mol of potassium ion:

$$\Delta G_{transport} = 3\Delta\tilde{\mu}_{in \to out}^{Na^+} + 2\Delta\tilde{\mu}_{out \to in}^{K^+} = (3)(13.6 \text{ kJ}) + 2(1.00 \text{ kJ}) = 42.8 \text{ kJ}$$

b. The standard free energy change for the hydrolysis of ATP at 310 K is $\Delta G^0 = -30.5$ kJ mol^{-1}. If the total concentration of inorganic phosphate is 0.010 M, calculate the ratio of ATP to ADP (i.e., in the reaction quotient Q) that will furnish the work required to accomplish the transport described in part (a).

Solution:

For ATP hydrolysis $Q' = \dfrac{c_{ADP}c_{P_i}}{c_{ATP}} \times \dfrac{c_{ATP}^\circ}{c_{ADP}^\circ c_{P_i}^\circ} = \dfrac{c_{ADP}c_{P_i}}{c_{ATP}} \times \dfrac{1.00\ M}{1.00\ M \times 1.00\ M} = \dfrac{c_{ADP}c_{P_i}}{c_{ATP}}$

where the final ratio is dimensionless.

$$-\Delta G_{transport} = \Delta G_{ATP \to ADP} = \Delta G^{\circ\prime}_{ATP \to ADP} + RT \ln Q'$$

$$\therefore -42.8 \text{ kJ} = -30.5 \text{ kJ mol}^{-1} + (8.314 \text{ J K}^{-1}\text{ mol}^{-1})(310.\text{ K}) \ln\left(\frac{c_{P_i}c_{ADP}}{c_{ATP}}\right)$$

$$-42.8 \text{ kJ} + 30.5 \text{ kJ} = -12.3 \text{ kJ} = (8.314 \text{ J K}^{-1})(310 \text{ K}) \ln\left(\frac{(0.010)c_{ADP}}{c_{ATP}}\right)$$

$$\ln\left(\frac{0.010 \times c_{ADP}}{c_{ATP}}\right) = -4.77 \quad \text{or} \quad \frac{(0.010)c_{ADP}}{c_{ATP}} = e^{-4.77} = 8.5 \times 10^{-3}$$

Finally, $\dfrac{c_{ADP}}{c_{ATP}} = 0.85.$

10.5 Thermodynamics of Adenosine Triphosphate Hydrolysis

In Section 10.3 we showed how ATP's central role in cellular energetics is due to the intermediate value of its phosphate transfer potential, and that the transfer of phosphate groups between the ADP/ATP system and other metabolites is of central importance in cellular energetics. However, in addition to being involved in phosphate transfer reactions, ATP is also involved in proton exchanges and is also in equilibria with divalent metal ions, notably Mg^{2+}. How these equilibria impact the hydrolysis thermodynamics of ATP is considered in this section.

The hydrolysis of ATP is commonly written as:

$$ATP + H_2O \rightleftarrows ADP + P_i \qquad (10.28)$$

But because ATP undergoes simultaneous equilibria involving exchanges of phospate groups, protons, and metal ions, the hydrolysis of Equation (10.28) is shorthand in the sense that ATP, ADP, and P_i each represents several species. For example, in Equation (10.28) H^+ is not accounted for explicitly, and in fact there is no attempt made in Equation (10.28) to balance hydrogen atoms or charge. To describe the species present in the hydrolysis of ATP in more detail, and to illustrate the impact of other equilibria on the phosphate transfer potential of ATP, we start with a simple experimental circumstance: ATP undergoes hydrolysis near pH 7 in the absence of metal ions. Figure 10.7 lists the important species involved in Equation (10.28) near pH 7 and in the absence of metal ions.

For example, in aqueous solution, ATP is an ion with a -4 charge. An ATP ion can bind five protons at low pH, four on the phosphate groups and one on the adenine ring. But near pH 7 and neglecting binding to metals, proton binding to the terminal phosphate group is the dominant proton exchange:

$HATP^{3-} \rightleftarrows ATP^{4-} + H^+$ for which $pK \approx 6.95$ (see Table 10.2). For ADP, the dominant proton exchange equilibrium near pH 7 is $HADP^{2-} \rightleftarrows ADP^{3-} + H^+$ for which $pK \approx 6.88$. The dominant equilibrium for inorganic phosphate is $H_2PO_4^- \rightleftarrows HPO_4^{2-} + H^+$ for which $pK \approx 6.78$. With the information in Table 10.2 and Figure 10.7, the "observed" hydrolysis equilibrium constant at pH 7 is

$$K_{obs} = \frac{\left(\dfrac{c_{ADP}}{c_{ADP}^{\circ}}\right)\left(\dfrac{c_{P_i}}{c_{P_i}^{\circ}}\right)}{\left(\dfrac{c_{ATP}}{c_{ATP}^{\circ}}\right)} \approx \frac{\left(\dfrac{c_{ADP}}{1\,M}\right)\left(\dfrac{c_{P_i}}{1\,M}\right)}{\left(\dfrac{c_{ATP}}{1\,M}\right)} \approx \frac{(c_{ADP^{3-}} + c_{HADP^{2-}})(c_{HPO_4^{2-}} + c_{H_2PO_4^-})}{c_{ATP^{4-}} + c_{HATP^{3-}}}$$

$$(10.29)$$

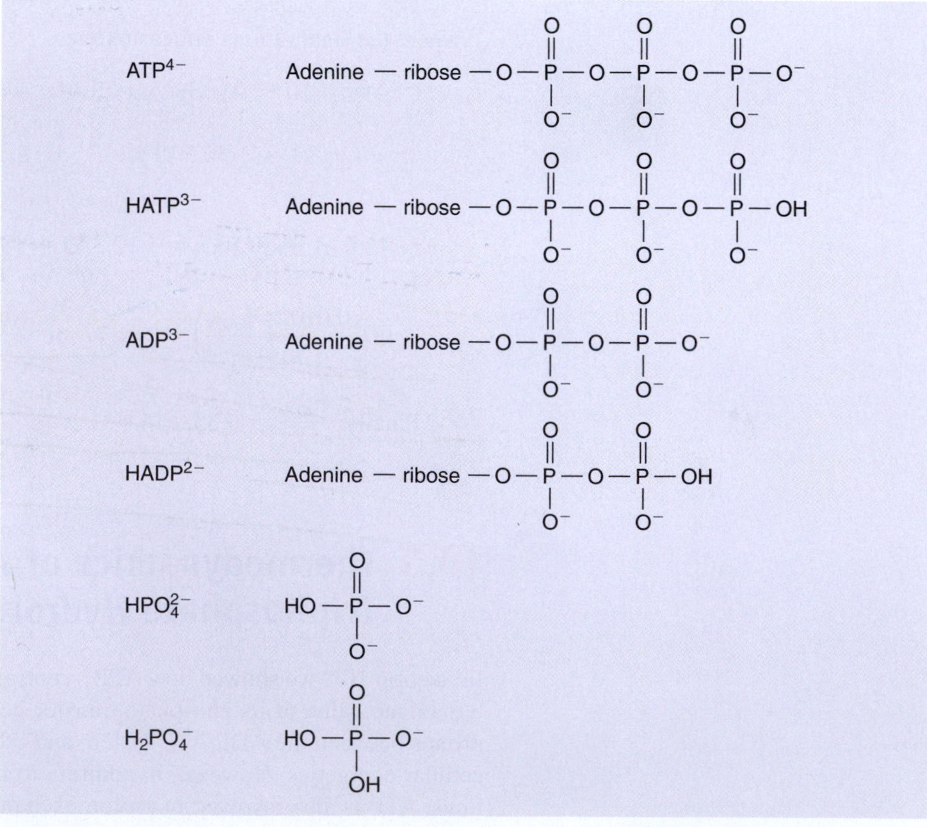

FIGURE 10.7
Dominant species of ATP, ADP, and P_i near pH 7.

where the reference state for each species is assumed to be 1 M. Either the biochemical or the chemical conventions could be used in this section for the proton reference state $C_{H^+}^\circ$. Here we find the chemical convention $C_{H^+}^\circ = 1.00\ M$ more convenient because this is the convention used in the definition of the dissociation constants in Table 10.2. Equation (10.29) can be reduced to a form containing simply C_{H^+} and the four relevant equilibrium constants K_1, K_{1ATP}, K_{1ADP}, and K_{1P}, obtained from Table 10.2,

Using the expressions for the equilibrium constants $K_{1,ADP} = \dfrac{\left(\frac{c_{H^+}}{1\ M}\right)\left(\frac{c_{ADP^{3-}}}{1\ M}\right)}{\left(\frac{c_{HADP^{2-}}}{1\ M}\right)}$,

$$K_{1,ATP} = \frac{\left(\frac{c_{H^+}}{1\ M}\right)\left(\frac{c_{ATP^{4-}}}{1\ M}\right)}{\left(\frac{c_{HATP^{3-}}}{1\ M}\right)}, \text{ and } K_{1,P_i} = \frac{\left(\frac{c_{H^+}}{1\ M}\right)\left(\frac{c_{HPO_4^{2-}}}{1M}\right)}{\left(\frac{c_{HPO_4^{2-}}}{1\ M}\right)}: \quad \text{← use for P/O. 17} \quad (10.30)$$

$$K_{obs} = \frac{(c_{ADP^{3-}} + c_{HADP^{2-}})(c_{HPO_4^{2-}} + c_{H_2PO_4^-})}{c_{ATP^{4-}} + c_{HATP^{3-}}} = \frac{c_{ADP^{3-}}\left(1 + \frac{c_{H^+}}{K_{1,ADP}}\right)c_{HPO_4^{2-}}\left(1 + \frac{c_{H^+}}{K_{1,P_i}}\right)}{c_{ATP^{4-}}\left(1 + \frac{c_{H^+}}{K_{1,ATP}}\right)}$$

$$= \frac{c_{ADP^{3-}}\,c_{HPO_4^{2-}}}{c_{ATP^{4-}}}\frac{c_{H^+}}{c_{H^+}}\frac{\left(1 + \frac{c_{H^+}}{K_{1,ADP}}\right)\left(1 + \frac{c_{H^+}}{K_{2,P}}\right)}{\left(1 + \frac{c_{H^+}}{K_{1,ATP}}\right)} = \frac{K_1}{c_{H^+}}\frac{(1 + c_{H^+}K_{1,ADP}^{-1})(1 + c_{H^+}K_{2P}^{-1})}{(1 + c_{H^+}K_{1,ATP}^{-1})}$$

where we used the expression from Table 10.3 $K_1 = \dfrac{c_{ADP^{3-}}\,c_{HPO_4^{2-}}\,c_{H^+}}{c_{ATP^{4-}}}$.

Each of the terms in parentheses in Equation (10.30) has a particular physical meaning. For each the inverse is the fraction of a particular reactant or production chosen:

$$x_{ATP} = \frac{c_{ATP^{4-}}}{c_{ATP}} = (1 + c_{H^+}K_{1ATP}^{-1})^{-1} \quad (10.31a)$$

$$x_{ADP} = \frac{c_{ADP^{3-}}}{c_{ADP}} = (1 + c_{H^+}K_{1ADP}^{-1})^{-1} \quad (10.31b)$$

TABLE 10.2   Equilibrium Constants, Standard Gibbs Energy, Enthalpy, and Entropy for the ATP Series*

Reaction	Name	pK($I = 0.2$)	ΔG°	ΔH°	ΔS°
$HADP^{2-} \rightleftarrows ADP^{3-} + H^+$	$K_{1,ADP}$	6.88	39.3	−5.7	−151.
$H_2ADP^- \rightleftarrows HADP^{2-} + H^+$	$K_{2,ADP}$	3.93	22.4	4.2	−61.0
$HATP^{3-} \rightleftarrows ATP^{4-} + H^+$	$K_{1,ATP}$	6.95	39.6	−7.0	−156.
$H_2ATP^{2-} \rightleftarrows HATP^{3-} + H^+$	$K_{2,ATP}$	4.06	23.2	0.0	−77.8
$MgATP^{2-} \rightleftarrows ATP^{4-} + Mg^{2+}$	K_{MgATP}	4.00	22.8	13.8	−123.
$MgHATP^{1-} \rightleftarrows HATP^{3-} + Mg^{2+}$	K_{MgHATP}	1.49	8.5	−7.1	−55.2
$MgADP^- \rightleftarrows ADP^{3-} + Mg^{2+}$	K_{MgADP}	3.01	17.2	−15.0	−108.
$H_2PO_4^- \rightleftarrows HPO_4^{2-} + H^+$	$K_{2,P}$	6.78	38.7	3.3	−119.
$MgHPO_4 \rightleftarrows HPO_4^{2-} + Mg^{2+}$	K_{MgP}	1.88	10.7	−12.1	−76.5
$ATP^{4-} + H_2O \rightleftarrows ADP^{3-} + HPO_4^{2-} + H^+$	K_1	0.202	1.15	−19.7	−69.8

*At $T = 298.15$ K and for ionic strength (I) 0.20 M. $pK = -\log K$. Units for ΔG° and ΔH° are kJ mol^{-1}. Units for ΔS° are J K^{-1} mol^{-1}.

Source: R. A. Alberty and R. N. Goldberg, *Biochemistry* 31 (1992), 10,612.

$$x_P = \frac{c_{HPO_4^{2-}}}{c_{P_i}} = (1 + c_{H^+}K_{2P}^{-1})^{-1} \tag{10.31c}$$

EXAMPLE PROBLEM 10.5

Calculate x_{ATP}, x_{ADP}, and x_P at pH 7.

Solution

Using the pK values in Table 10.2, we obtain:

$$x_{ATP} = (1 + c_{H^+}K_{1ATP}^{-1})^{-1} = (1 + 10^{-7.00+6.95})^{-1} = 0.529$$

Similarly, $x_{ADP} = 0.568$ and $x_P = 0.624$.

Near pH 7 ATP and ADP species are almost equally divided between the anionic species ATP^{4-} and ADP^{3-} and the singly protonated species $HATP^{3-}$ and $HADP^{2-}$. Inorganic phosphate is predominantly in the form HPO_4^{2-}.

Note that Equations (10.31a–c) are dimensionless because each C_{H^+} term is divided by the $1\,M$ reference state concentration and each equilibrium constant is dimensionless. With Equations (10.31a–c), Equation (10.30) can be written in a more compact form:

$$K_{obs} = \frac{K_1}{c_{H^+}} \frac{x_{ATP}}{x_{ADP}x_P} \tag{10.32}$$

To obtain the standard Gibbs energy we use

$$\Delta G_{obs}^{\circ} = -RT \ln K_{obs} = -RT \ln\left(\frac{K_1}{c_{H^+}} \frac{x_{ATP}}{x_{ADP}x_P}\right) \tag{10.33}$$

The heat evolved from ATP hydrolysis at constant pressure at a given pH is the enthalpy change ΔH_{obs}°, and can be obtained using Equation (6.36):

$$\Delta H_{obs}^{\circ} = \left(\frac{\partial(\Delta G_{obs}^{\circ}/T)}{\partial(1/T)}\right)_{c_{H^+}} = RT^2\left(\frac{\partial \ln K_{obs}}{\partial T}\right)_{c_{H^+}} \tag{10.34}$$

Using Equations (10.31) and (10.32), Equation (10.34) becomes

$$\Delta H_{obs}^{\circ} = RT^2\left(\frac{\partial \ln K_{obs}}{\partial T}\right)_{c_{H^+}} = RT^2 \frac{\partial}{\partial T} \ln\left(\frac{K_1}{c_{H^+}} \frac{(1 + c_{H^+}K_{1,ADP}^{-1})(1 + c_{H^+}K_{2P}^{-1})}{(1 + c_{H^+}K_{1,ATP}^{-1})}\right)_{c_{H^+}}$$

$$= RT^2\left(\frac{\partial \ln K_1}{\partial T}\right)_{c_{H^+}} - x_{ADP}\frac{c_{H^+}}{K_{1ADP}} RT^2\left(\frac{\partial \ln K_{1ADP}}{\partial T}\right)_{c_{H^+}}$$

$$- x_P \frac{c_{H^+}}{K_{2P}} RT^2\left(\frac{\partial \ln K_{2P}}{\partial T}\right)_{c_{H^+}}$$

$$+ x_{ATP}\frac{c_{H^+}}{K_{1ATP}} RT^2\left(\frac{\partial \ln K_{1ATP}}{\partial T}\right)_{c_{H^+}} =$$

$$\Delta H_1^{\circ} - x_{ADP}\frac{c_{H^+}}{K_{1ADP}} \Delta H_{1ADP}^{\circ}$$

$$- x_P \frac{c_{H^+}}{K_{2P}} \Delta H_{2P}^{\circ} + x_{ATP}\frac{c_{H^+}}{K_{1ATP}} \Delta H_{1ATP}^{\circ} \tag{10.35}$$

Equation (10.35) has the form of a weighted average of the enthalpy changes associated with the phosphate exchange equilibrium between ADP^{3-} and ATP^{4-} and the three proton exchange equilibria involving $ADP^{3-}/HADP^{2-}$, $H_2PO_4^-/HPO_4^{2-}$, and $ADP^{4-}/HADP^{3-}$. The enthalpies ΔH_{1ADP}°, ΔH_{2P}°, ΔH_{1ATP}°, and ΔH_1° are provided in Table 10.2.

EXAMPLE PROBLEM 10.6

Calculate ΔH°_{obs} for ATP hydrolysis at $T = 298.15$ K and pH 7.

Solution

From Table 10.2, Equation (10.35), and Example Problem 10.5:

$$\Delta H^{\circ}_{obs} = \Delta H^{\circ}_1 - x_{ADP}\frac{c_{H^+}}{K_{1,ADP}}\Delta H^{\circ}_{1,ADP} - x_P\frac{c_{H^+}}{K_{2P}}\Delta H^{\circ}_{1P_i} + x_{ATP}\frac{c_{H^+}}{K_{1ATP}}\Delta H^{\circ}_{1ADP}$$

$$= -19.7 \text{ kJ mol}^{-1} - (0.568)(10^{-7+6.88})(-5.7 \text{ kJ mol}^{-1})$$
$$- (0.624)(10^{-7+6.78})(3.30 \text{ kJ mol}^{-1})$$

$$+ (0.529)(10^{-7+6.95})(-7.0 \text{ kJ mol}^{-1}) = -19.7 \text{ kJ mol}^{-1}$$
$$+ 2.46 \text{ kJ mol}^{-1} - 1.24 \text{ kJ mol}^{-1} - 3.30 \text{ kJ mol}^{-1}$$

$$= -21.8 \text{ kJ mol}^{-1}$$

The ratios preceding ΔH°_{1ADP}, ΔH°_{2P}, and ΔH°_{1ATP} are the mole fractions of $HADP^{2-}$, $H_2PO_4^-$, and $HATP^{3-}$, respectively.

In general, the entropy change observed for ATP hydrolysis can be obtained using Equation (6.29):

$$\Delta S^{\circ}_{obs} = -\left(\frac{\partial \Delta G^{\circ}_{obs}}{\partial T}\right)_{c_{H^+}} = R\left(\frac{\partial(T \ln K_{obs})}{\partial T}\right) \qquad (10.36)$$

At constant temperature we can obtain ΔS°_{obs} from Equations (10.33) and (10.35).

The approach outlined in Equations (10.29) through (10.36) can be extended easily to include other equilibria. Accordingly, we can extend the pH range for calculating the Gibbs energy, enthalpy, and entropy of ATP hydrolysis by incorporating additional ADP and ATP proton exchange equilibria. Suppose we want to extend these calculations to the range pH 4–10. Note the additional proton exchange equilibria of phosphoric acid have pK's less than 4 or greater than 10 so H_3PO_4 or PO_4^{3-} will be negligible in the pH range 4–10. For pH 4–10 the additional equilibria that we need are:

$$H_2ATP^{2-} \rightleftarrows HATP^{3-} + H^+ \quad pK_{2,ATP} = 4.06 \qquad (10.37)$$

$$H_2ADP^- \rightleftarrows HADP^{2-} + H^+ \quad pK_{2,ADP} = 3.93 \qquad (10.37a)$$

Using Equations (10.37a,b), Equations (10.31a–c) become

$$x_{ATP} = \frac{c_{ATP^{4-}}}{c_{ATP}} = \frac{c_{ATP^{4-}}}{c_{ATP^{4-}} + c_{HATP^{3-}} + c_{H_2ATP^{2-}}} \qquad (10.31a')$$

$$= (1 + c_{H^+}K_{1ATP}^{-1} + c_{H^+}^2 K_{1ATP}^{-1}K_{2ATP}^{-1})^{-1}$$

$$x_{ADP} = \frac{c_{ADP^{3-}}}{c_{ADP}} = \frac{c_{ADP^{3-}}}{c_{ADP^{3-}} + c_{HADP^{2-}} + c_{H_2ADP^{1-}}} \qquad (10.31b')$$

$$= (1 + c_{H^+}K_{1ADP}^{-1} + c_{H^+}^2 K_{1ADP}^{-1}K_{2ADP}^{-1})^{-1}$$

$$x_P = \frac{c_{HPO_4^{2-}}}{c_{P_i}} = (1 + c_{H^+}K_{2P}^{-1})^{-1} \qquad (10.31c')$$

Equations (10.31a',b',c') can be used in combination with Equations (10.32) through (10.36) to generate the Gibbs energy, enthalpy, and entropy of ATP hydrolysis at $T = 298.15$ K. The variation of these thermodynamic quantities with pH is shown in Figure 10.8.

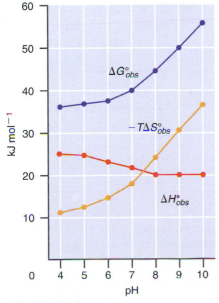

FIGURE 10.8
The standard Gibbs energy, enthalpy, and entropy at $T = 298.15$ K observed for ATP hydrolysis as a function of pH obtained using Equations 10.31a–c and data in Table 10.2.

ATP, ADP, and P_i are involved not only in phosphate and proton exchange equilibria, in the cellular environment they also form complexes with metal cations, notably Mg^{2+} and Ca^{2+}. In particular, if we include equilibria between Mg^{2+} ADP, ATP, and P_i the dependence of the Gibbs energy, enthalpy, and entropy of hydrolysis on pH and metal ion concentration can be determined using a set of equations analogous to Equations (10.31a–c). For example, if we incorporate the additional ATP equilibria

$$MgATP^{2-} \rightleftarrows ATP^{4-} + Mg^{2+} \qquad pK_{MgATP} = 4.00$$

$$MgHATP^{1-} \rightleftarrows HATP^{3-} + Mg^{2+} \quad pK_{MgHATP} = 1.49 \tag{10.38}$$

the mole fraction expression for ATP becomes

$$x_{ATP} = \frac{c_{ATP^{4-}}}{c_{ATP}} = \frac{c_{ATP^{4-}}}{c_{ATP^{4-}} + c_{HATP^{3-}} + c_{H_2ATP^{2-}} + c_{MgATP^{2-}} + c_{MgHATP^{1-}}} \tag{10.39}$$

$$= \left(1 + c_{H^+}K_{1ATP}^{-1} + c_{Mg^{2+}}K_{MgATP}^{-1} + c_{H^+}^2 K_{1ATP}^{-1} K_{2ATP}^{-1} + c_{H^+}c_{Mg^{2+}}K_{1ATP}^{-1}K_{MgATP}^{-1}\right)^{-1}$$

[handwritten: K^{-1}_{MgHATP}]

Analogous equations incorporating equilibria phosphate and ADP species and Mg^{2+} can be obtained:

$$x_{ADP} = \frac{c_{ADP^{3-}}}{c_{ADP}} = \frac{c_{ADP^{3-}}}{c_{ADP^{3-}} + c_{HADP^{2-}} + c_{H_2ADP^{1-}} + c_{MgADP^{1-}} + c_{MgHADP}} \tag{10.40}$$

[handwritten: ADP^{3-}]

$$= \left(1 + c_{H^+}K_{1ADP}^{-1} + c_{Mg^{2+}}K_{MgADP}^{-1} + c_{H^+}^2 K_{1ADP}^{-1}K_{2ADP}^{-1} + c_{H^+}c_{Mg^{2+}}K_{1ADP}^{-1}K_{MgADP}^{-1}\right)^{-1}$$

[handwritten: K^{-1}_{MgHADP}]

$$x_P = \frac{c_{HPO_4^{2-}}}{c_{P_i}} = \left(1 + c_{H^+}K_{1P_i}^{-1} + c_{Mg^{2+}}K_{MgP}^{-1}\right)^{-1} \tag{10.41}$$

[handwritten: $K^{-1}_{MgHADP} \to 0.0909$]

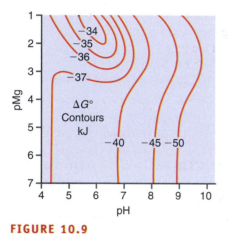

FIGURE 10.9
Gibbs energy observed for the hydrolysis of ATP as a function of pH and pMg. (Source: R. A. Alberty, *J. Biol. Chem.* 244 (1969), 3290.)

Using Equations (10.39), (10.40), and (10.41) in combination with Equation (10.33), the Gibbs energy of hydrolysis can be obtained as a function of pH and pMg $= -\log C_{Mg^{2+}}$. Results are shown in Figure 10.9. Although still vastly simplified versus actual cellular conditions, the data in Figures 10.8 and 10.9 illustrate how multiple exchange equilibria influence the thermodynamics of ATP hydrolysis and in particular give specific physical meaning to the components of the ATP hydrolysis equilibrium in Equation (10.28).

Vocabulary

adenosine diphosphate (ADP)
adenosine triphosphate (ATP)
autotrophic cell

common intermediate
coupled biological reactions
dynamic steady state

endergonic
exergonic
heterotrophic cell

hydrolysis
phosphate transfer potential
sodium-potassium pump

Questions and Concepts

Q10.1 What is the ultimate source of all biological energy? Explain.

Q10.2 In the energetic sense is it more efficient for a person to eat grain or to feed the grain to an animal then eat the animal? Explain.

Q10.3 ATP's pivotal role in cellular energetics has sometimes been attributed to its "high energy phosphate bonds." Comment on this putative property of ATP.

Q10.4 The equation for the efficiency of a heat engine is $\varepsilon = \frac{T_H - T_L}{T_H}$. As explained in Section 10.1, the difference between the temperature within a living cell and ambient

temperature is so small that the cell could not function efficiently as a heat engine. Nevertheless the efficiency with which heterotrophic cells convert chemical energy into work can exceed 50%. Explain how a cell might do this.

Q10.5 Why is aerobic respiration more efficient than anaerobic respiration? What advantage, if any, does anaerobic respiration have over aerobic respiration? Give an example of this advantage at work.

Q10.6 Biological evolution has been invoked to explain the origin of species. However, in the time course of biological evolution, simple organisms eventually evolve into more complex organisms, seemingly contradicting the second law of thermodynamics. Discuss whether or not biological evolution violates the second law.

Q10.7 According to Section 10.3, many compounds like phosphocreatine and phosphoenolpyruvate have higher Gibbs energies of hydrolysis than ATP. This being the case, why would these "high-energy" compounds not be more useful than ATP for coupling exergonic and endergonic processes?

Q10.8 How is the transport of sodium and potassium across the cell membrane coupled to ATP hydrolysis?

Q10.9 Discuss the validity of the following statement: "Transport of ions across a cell membrane against concentration gradients cannot occur. It would be like water flowing uphill."

Q10.10 Explain how the photosynthetic process in plants can approach 85% efficiency although the plant performs this process at ambient temperatures.

Problems

Problem numbers in **RED** indicate that the solution to the problem is given in the *Student Solutions Manual*.

P10.1 In anaerobic cells glucose $C_6H_{12}O_6$ is converted to lactic acid:

$C_6H_{12}O_6 \rightarrow 2CH_3CH(OH)COOH$. The standard enthalpies of formation of glucose and lactic acid are -1274.45 and -694.04 kJ/mol, respectively. The molar heat capacities (at constant pressure) for glucose and lactic acid are 218.86 and 127.6 J/mol K, respectively.

a. Calculate the molar enthalpy associated with forming lactic acid from glucose at 298 K.

b. What would the quantity in part (a) be if the reaction proceeded at a physiological temperature of 310 K?

c. Based on your answers in parts (a) and (b) how sensitive is the enthalpy of this reaction to moderate temperature changes?

P10.2 Using the data in Table 10.1, calculate the equilibrium constant for the phosphorylation of fructose by ATP to form fructose-6-phosphate.

P10.3 The cellular concentrations of glucose, glucose-6-phosphate, ADP, and ATP are $5.00 \times 10^{-3}\,M$, $8.3 \times 10^{-5}\,M$, $1.38 \times 10^{-4}\,M$, and $1.85 \times 10^{-3}\,M$, respectively. Using the data in Table 10.1, calculate the Gibbs energy change for the phosphorylation of glucose by ATP.

P10.4 Glutamine (G-ine) is an amino acid formed from the reaction of ammonium ion NH_4^+ with glutamate (G-ate$^-$):

$$NH_4^+ + \text{G-ate}^- \rightleftarrows \text{G-ine} + H_2O$$

a. At pH 7 and $T = 298$ K, the equilibrium constant is 0.003. Calculate the standard Gibbs energy change for the formation of glutamine.

b. Glutamine synthesis can be coupled to ATP hydrolysis as follows:

$$NH_4^+ + \text{G-ate}^- + ATP \rightleftarrows \text{G-ine} + H_2O + ADP + P_i$$

Assuming the equilibrium constant for this reaction is 9600, calculate the standard Gibbs energy change.

c. From the data in parts (a) and (b), calculate the equilibrium constant for ATP hydrolysis and the Gibbs energy of hydrolysis of ATP.

P10.5 Calculate the reversible work required to transport 1 mol of sucrose from a region where $C_{sucrose} = 0.140\,M$ to a region where $C_{sucrose} = 0.340\,M$. Assume $T = 298$ K.

P10.6 Calculate the reversible work required to transport 1 mol of K$^+$ from a region where $C_{K^+} = 5.25$ mM to a region where $C_{K^+} = 35.5$ mM if the potential change accompanying this movement is $\Delta\phi = 0.055$ V. Assume $T = 298$ K.

P10.7 Suppose the concentration of protons inside a cell membrane is $C_{in}^{H^+}$ and similarly the concentration outside is $C_{out}^{H^+}$. Assume the electrostatic potential difference is $\Delta\phi = \phi_{out} - \phi_{in}$.

a. Obtain an expression for the electrochemical potential difference $\Delta\tilde{\mu} = \tilde{\mu}_{out} - \tilde{\mu}_{in}$ in terms of the pH difference and the electrostatic potential difference $\Delta\phi$.

b. Suppose the pH inside the inner mitochondrial membrane is 0.75 units higher than outside the membrane and $\Delta\phi = \phi_{out} - \phi_{in} = -0.168V$. Calculate the electrochemical potential difference between protons inside versus outside the membrane.

P10.8 Suppose the equilibrium distributions of chloride ions inside and outside a membrane are $C_{in}^{eq} = 0.005\,M$ and $C_{out}^{eq} = 0.120\,M$. Calculate the equilibrium potential difference $\Delta\phi = \phi_{out} - \phi_{in}$. Assume $T = 298$ K.

P10.9 Suppose a membrane is passively permeable to water and to Cl$^-$ ion, but not to H$^+$. The electrostatic potential ϕ and the equilibrium concentrations of Cl$^-$ and H$^+$ inside and outside the membrane are given here. Assume that the ionic strength inside the membrane is sufficiently small so that the activity coefficients of univalent ions may be taken to be 1 inside the membrane. Assume $T = 298$ K.

Outside Membrane	Inside Membrane
$\phi = 0.150$ V	$\phi = 0.000$ V
$C_{H^+} = 5 \times 10^{-6}\,M$	$C_{H^+} = 10^{-7}\,M$
$C_{Cl^-} = 5 \times 10^{-2}\,M$	

a. What is the equilibrium concentration of Cl$^-$ inside the membrane ?

b. Calculate the difference between the chemical potentials of [H$^+$] inside and outside the membrane.

c. The reaction ATP $\rightleftharpoons$ ADP + P$_i$, $\Delta G° = -31$ kJ/mol is coupled to the transport of H$^+$ with an efficiency of 50% (i.e., half the free energy from ATP hydrolysis is recovered and available to drive the active transport of H^+). Assuming the concentration of inorganic phosphate is 0.01 M, what ratio of ATP to ADP is required to drive the transport of H^+?

P10.10   The net reaction for active transport of sodium and potassium ions is thought to be:

$$\left.\begin{array}{c} 3Na^+ \text{ (inside)} \\ + \\ 2K^+ \text{ (outside)} \end{array}\right\} + ATP \rightarrow ADP + phosphate + \left\{\begin{array}{c} 3Na^+ \text{ (outside)} \\ + \\ 2K^+ \text{ (inside)} \end{array}\right.$$

The concentrations of sodium and potassium ions inside and outside a cell, and the electrical potential E inside and outside the cell are as follows:

	C_{Na^+} (mol L^{-1})	C_{K^+} (mol L^{-1})	ϕ (V)
Outside	1.40×10^{-1}	5.00×10^{-3}	0.00
Inside	1.00×10^{-2}	1.00×10^{-1}	-7.00×10^{-2}

a. Calculate the free energy change involved in transporting 1 mol of sodium ion out of the cell. Assume the activity coefficients of sodium ion inside and outside the cell are unity. Assume the temperature is 310 K.

b. Calculate the free energy change involved in transporting 1 mol of potassium ion into the cell. Assume the activity coefficients of potassium ion inside and outside the cell are unity. Assume the temperature is 310 K.

c. Calculate the total free energy change involved in transporting 3 mol of sodium ion out of the cell and 2 mol of potassium into the cell at $T = 310$ K. Assume, as in parts (a) and (b), that all activity coefficients are unity.

P10.11   The standard free energy change for the hydrolysis of ATP, i.e., (ATP $\rightleftharpoons$ ADP + phosphate) at 310 K is $\Delta G^{\circ\prime} = -30.5$ kJ mol^{-1}. If the total concentration of inorganic phosphate is 0.01 M, calculate the ratio of ATP to ADP (i.e., in the reaction quotient Q), which will furnish the work required to accomplish the transport described in Problem P10.10.

P10.12   Suppose the cellular concentrations of ATP, ADP, and inorganic phosphate are $1.85 \times 10^{-3} M$, $0.138 \times 10^{-3} M$, and $10^{-3} M$, respectively.

a. Assuming the standard Gibbs energy of ATP hydrolysis is -30.5 kJ mol^{-1}, calculate the cellular Gibbs energy change for the hydrolysis of a mole of ATP. Assume $T = 312$ K.

b. Suppose 1 mol of ATP is hydrolyzed to run the sodium-potassium pump described in Problem P10.10. Calculate the efficiency of energy coupling between ATP hydrolysis and this ion transfer.

P10.13   For the hydrolysis $ATP^{4-} + H_2O \rightarrow ADP^{3-} + HPO_4^{2-} + H^+$, $\Delta G^{\circ\prime} = -30.5$ kJ mol^{-1} at pH $= 7$ and $T = 310$ K. Calculate K', K, and ΔG° for the hydrolysis of ATP^{4-}.

P10.14   Prove that $x_{ADP} = \frac{C_{H^+}}{K_{1ADP}}$, $x_P = \frac{C_{H^+}}{K_{2P}}$, and $x_{ATP} = \frac{C_{H^+}}{K_{1ATP}}$ are the mole fractions of $HADP^{2-}$, $H_2PO_4^-$, and $HATP^{3-}$, respectively. Use these facts to calculate ΔH°_{obs} for ATP hydrolysis at $T = 298.15$ K and pH 7.

P10.15   Calculate K_{obs}, ΔG°_{obs}, and ΔS°_{obs} for ATP hydrolysis at pH 7 and at $T = 298.15$ K.

P10.16   Derive Equations (10.31a′), (10.31b′), and (10.31c′). Calculate the mole fractions χ_{ATP}, χ_{ADP}, and χ_{P_i} at pH $= 4.5$.

P10.17   Derive Equations (10.39) through (10.41). Using the data in Table 10.2, calculate the mole fractions χ_{ATP}, χ_{ADP}, and χ_{P_i} at pH $= 7$ and for $C_{Mg^+} = 1.00 \times 10^{-3} M$.

P10.18   Show that near pH $= 7$ the enthalpy of hydrolysis of ATP is given by

$$\Delta H^{\circ}_{obs} = RT^2 \frac{\partial}{\partial T} \ln \left(\frac{K_1}{C_{H^+}} \frac{(1 + C_{H^+}K_{1,ADP}^{-1})(1 + C_{H^+}K_{2P}^{-1})}{(1 + C_{H^+}K_{1,ATP}^{-1})} \right)_{C_{H^+}}$$

P10.19   The net reaction for the aerobic metabolism of glucose is:

$$C_6H_{12}O_6(s) + 6O_2(g) \rightarrow 6CO_2(g) + 6H_2O(l)$$

The Gibbs energy associated with this reaction is $\Delta G^{\circ\prime} = -2878$ kJ mol^{-1} at pH $= 7$ and $T = 310$ K. Assuming the oxidation of glucose is coupled with 100% efficiency to the phosphorylation of ADP, calculate the number of moles of ADP that can be phosphorylated under standard conditions by the oxidation of a single mole of glucose.

P10.20   In cells functioning anaerobically glucose is converted to lactic acid:

$$C_6H_{12}O_6 \rightarrow 2CH_3CH(OH)COOH$$

for which $\Delta G^{\circ\prime} = -217$ kJ mol^{-1} at $T = 310$ K and pH $= 7$. Calculate the number of moles of ADP that can be phosphorylated under standard conditions if the coupling of anaerobic glucose metabolism to ADP phosphorylation is 100% efficient.

CHAPTER **11**

Biochemical Equilibria

The central topic of this chapter is the nature of several important classes of biochemical equilibria. This chapter begins with a discussion of the energy transformations associated with the glycolytic pathway and the respiratory system of the living cell, a topic called *bioenergetics*. The respiratory system is the cell's energy mainspring. It is an assembly of chemical reactions that is exergonic in its net effect. As we learned in Chapter 10, this assembly of exergonic reactions is coupled to endergonic processes such as active transport of ions across cell membranes and mechanical motions of muscles. Critical to this exergonic–endergonic coupling is ATP, whose synthesis in the respiratory system we describe in this chapter. In this chapter we also cover a broad class of equilibria involving the binding of biological molecules to one another. Intermolecular associations are critical to living cells and in this chapter we also discuss in detail the thermodynamics of molecular binding equilibria. In particular, we discuss the thermodynamics of allosteric interactions, that is, long-range interactions between spatially remote ligand binding sites that are mediated by changes in protein structure.

11.1 Bioenergetics Overview

Chapter 10 describes how in living cells, exergonic processes that release energy are coupled through common intermediate compounds such as ATP to endergonic processes that require energy in order to proceed. Here we describe the cell's apparatus for producing energy. In anaerobic cells, the degradation of glucose to pyruvate in the cellular fluids constitutes the primary energy-producing process. In aerobic cells glycolysis is augmented by a system of membrane-associated oxidation–reduction reactions that further degrade pyruvate to carbon dioxide and reduce oxygen to water, thus releasing far more energy than that can be released anaerobically. These additional aerobic reactions constitute the Krebs (i.e., citric acid) cycle and the electron transport chain. See Figure 11.1. The energetics of the three primary cellular energy-producing pathways—glycolysis, the Krebs cycle, and the electron transport chain—are a central topic of Chapter 11.

In summary, heterotrophic cells acquire complex "high Gibbs energy" molecules as "fuel." Complex molecules such as proteins, carbohydrates, and fats are first degraded to simpler molecules such as amino acids, simple sugars, and fatty acids. When excess carbohydrates are ingested, for example, the body stores the surplus in the form of a polymer called glycogen, a highly branched polymer composed of repeating units of α-D-glucose (see Figure 11.2), connected by 1,4-glycosidic bonds between adjacent monomers. At branch points, 1,6-glycosidic bonds connect adjacent monomers. A single

FIGURE 11.1

The metabolic pathways involved in the degradation of glucose to carbon dioxide and water. Degradation of glucose to pyruvate is accomplished in the absence of oxygen by anaerobic glycolysis. Aerobic cells further degrade pyruvic acid to carbon dioxide in the citric acid or Krebs cycle. The citric acid cycle also generates reducing agents (e.g., NADH), which subsequently reduce oxygen, O_2, to water in a series of oxidation–reduction reactions called the electron transport chain. As a result of the Gibbs energy made available by electron transport, ATP is synthesized from ADP by a process called oxidative phosphorylation. (Source: Adapted from N.A. Campbell and J.B. Reece, *Biology,* 7th ed. (San Francisco, CA: Pearson Benjamin Cummings), p. 177. Copyright 2005. Reprinted by permission of Pearson Education, Inc.)

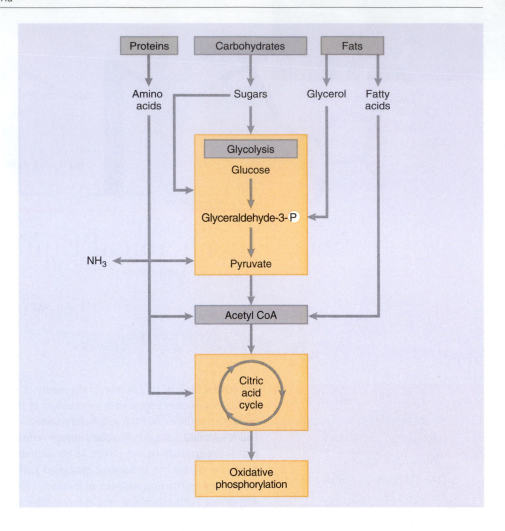

glycogen polymer can be as large as 40,000 monomer units, and have a molecular weight of more than $10,000 \text{ kg mol}^{-1}$.

When an organism needs energy quickly, it will first access stored glycogen, and the branched structure of glycogen is especially advantageous for the purpose of rapid degradation by enzymes. Only when glycogen stores have been depleted will an organism begin to break down fat. Glycogen is broken down first to glucose-1-phospate through phosphorylation by the enzyme glycogen phosphorylase. Glucose-1-phosphate is then converted to glucose-6-phosphate by the enzyme phosphoglucomutase.

FIGURE 11.2

(a) α-D-glucose is the monomeric unit of glycogen. (b) Coupling of glucose monomers in glycogen via 1,4-glycosidic bonds. Glycogen forms a branched structure by formation of 1,6-glycosidic bonds.

FIGURE 11.3

Schematic of the reactions of the glycolytic pathway. At left, abbreviations for each reagent are given (e.g., glucose-6-phosphate = G6P). Endergonic steps are shown coupled to ATP hydrolysis and exergonic steps are shown coupled to ADP phosphorylation. ΔG values are obtained from the approximate intracellular concentrations of glycolytic intermediates in rabbit skeletal muscle. (Source: Adapted from C.K. Mathews, K.E. von Holde, and K.G. Ahern, *Biochemistry*, 3rd ed. (San Francisco, CA: Pearson Benjamin Cummings) p. 458. Copyright 2000. Reprinted by Permission of Pearson Education.)

Reaction	Enzyme	ATP Yield	$\Delta G^{\circ\prime}$ (kJ/mol)	ΔG° (kJ/mol)
ENERGY INVESTMENT PHASE				
① (ATP → ADP)	Hexokinase (HK)	−1	−16.7	−33.5
Glucose-6-phosphate (G6P)				
②	Phosphoglucoisomerase (PGI)		+1.7	−2.5
Fructose-6-phosphate (F6P)				
③ (ATP → ADP)	Phosphofructokinase (PFK)	−1	−14.2	−22.2
Fructose-1,6-bisphosphate (FBP)				
④	Aldolase (ALD)		+23.9	−1.3
Glyceraldehyde-3-phosphate + dihydroxyacetone phosphate (DHAP)				
⑤	Triose phosphate isomerase (TPI)		+7.6	+2.5
Two glyceraldehyde-3-phosphate				
ENERGY GENERATION PHASE				
⑥ ($2NAD^+ + 2P_i$ → $2NADH + 2H^+$)	Glyceraldehyde-3-phosphate dehydrogenase (G3PDH)		+12.6	−3.4
Two 1,3-biphosphoglycerate (BPG)				
⑦ (2ADP → 2 ATP)	Phosphoglycerate kinase (PGK)	+2	−37.6	+2.6
Two 3-phosphoglycerate (3PG)				
⑧	Phosphoglycerate mutase (PGM)		+8.8	−1.6
Two 2-phosphoglycerate (2PG)				
⑨ (→ $2H_2O$)	Enolase (ENO)		+3.4	−6.6
Two 2-phosphoenolpyruvate (PEP)				
⑩ (2ADP → 2 ATP)	Pyruvate kinase (PK)	+2	−62.8	−33.4
Two pyruvate (Pyr)				

Net: Glucose + 2ADP + 2P$_i$ + 2NAD$^+$ $\longrightarrow$ 2 pyruvate + 2ATP + 2NADH + 2H$^+$ + 2H$_2$O

		+2	−73.3	−96.2

Glucose is broken down in the cell into simpler molecules by a process called **glycolysis**. Glycolysis, or the glycolytic pathway, is a series of 10 chemical reactions in which the six-carbon skeleton of glucose is broken down into two three-carbon molecules of pyruvate (see Figure 11.3). The process is exergonic and as such it can be coupled through common intermediates to endergonic processes such as the phosphorylation of ADP.

11.2 Glycolysis

The detailed chemical reactions and energetics of glycolysis are shown in Figure 11.3. Each step is enzymatically catalyzed and, except for glucose and pyruvate, all compounds are phosphates. Because the cell membrane lacks a mechanism for transporting phosphorylated sugars, reagents such as glucose-6-phosphate cannot diffuse out of the cell. If it were otherwise, reagents for glycolysis would be lost and the output of the pathway would be diminished.

Although standard free energy changes $\Delta G^{\circ\prime}$ for the reactions of the glycolytic pathway for glycolysis are shown in Figure 11.3, steady-state concentrations of glycolytic pathway intermediates are not at standard levels. Therefore Figure 11.3 also shows Gibbs energy changes for the reaction steps of glycolysis under cellular conditions in rabbit skeletal muscle, where ΔG is computed using Equation (6.66):

$$\Delta G_{reaction} = \Delta G^{\circ\prime}_{reaction} + RT \ln Q' \qquad (6.66)$$

where the primes indicate use of the biochemical standard states. Table 11.1 shows cellular concentration levels for glycolytic intermediates in human red blood cells. In Example Problem 11.1 we use the data in Figure 11.3 with Equation (6.66) to obtain a Gibbs energy change under particular cellular conditions.

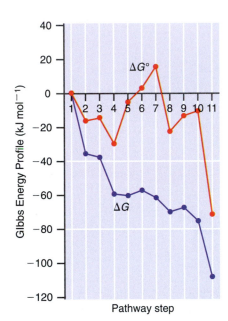

FIGURE 11.4
Gibbs energy profile for the steps of the glycolytic pathway (see Figure 11.3 under standard and cellular conditions). (Source: Reprinted with permission from C.K. Mathews, K.E. van Holde, and K.G. Ahern *Biochemistry*, 3rd ed., p. 459, Benjamin Cummings, Redwood City, CA, 2000.)

EXAMPLE PROBLEM 11.1

Using the data in Table 11.1 and Figure 11.3, calculate the Gibbs energy change for the phosphorylation of glucose by ATP to produce glucose-6-phosphate. Assume physiological conditions of $T = 310$ K and pH $= 7$.

Solution

$$\Delta G = \Delta G^{\circ\prime} + RT \ln\left(\frac{C_{G6P}C_{ADP}}{C_{Glucose}C_{ATP}}\right) = -16.7 \text{ kJ mol}^{-1}$$

$$+ (8.314 \text{ J mol}^{-1} \text{ K}^{-1})(310. \text{ K}) \ln\left(\frac{83.0 \times 138}{5.00 \times 10^3 \times 1850}\right)$$

$$= -16.7 \text{ kJ mol}^{-1} + (2580 \text{ kJ mol}^{-1})(-6.70) = -16.7 \text{ kJ mol}^{-1}$$
$$-17.3 \text{ kJ mol}^{-1} = -33.9 \text{ kJ mol}^{-1}$$

Note that because steady-state conditions vary somewhat between cells types, this Gibbs energy change for human red blood cell conditions differs slightly from that shown in Figure 11.3 for rabbit skeletal muscle cells.

Figure 11.4 compares the Gibbs energy profile for the ten steps of glycolysis under standard conditions and at pH 7 ($\Delta G^{\circ\prime}$) and the Gibbs energy profile for cellular conditions found in rabbit skeletal muscle.

According to Figures 11.3 and 11.4, the majority of steps in glycolysis (steps 2, 4–9) are accompanied by relatively small Gibbs energy changes under both standard and cellular conditions. Exceptions are steps 1, 3, and 10, which are highly exergonic under standard and cellular conditions. Step 1 is crucial because, as mentioned earlier, phos-

TABLE 11.1 **Steady-State Concentrations of Glycolysis Intermediates in the Human Red Blood Cell**

Reagent*	Glu	G6P	F6P	FBP	DHAP	GAP	3PG	2PG	PEP	Pyr	ATP	ADP	P$_i$	Lact
Steady-State Concentration (μM)	5.00×10^3	83.0	14.0	31.0	138	18.5	118	29.5	23.0	51.0	1850	138	1.00×10^3	2.90×10^3

*Glu = glucose, Pyr = pyruvate, Lact = lactate. All other abbreviations are as found in Figure 11.3.

Source: A. L. Lehninger, *Biochemistry*, Worth Publishers, New York, 1970.

phorylation of glucose to G6P ensures that this reagent remains confined to the cell. But if accomplished by a substrate-level phosphorylation of glucose, G6P production would be highly endergonic under standard conditions. Due to the high steady-state concentrations of glucose and inorganic phosphate maintained in the cell (see Table 11.1), the reaction would be even more endergonic under cellular conditions.

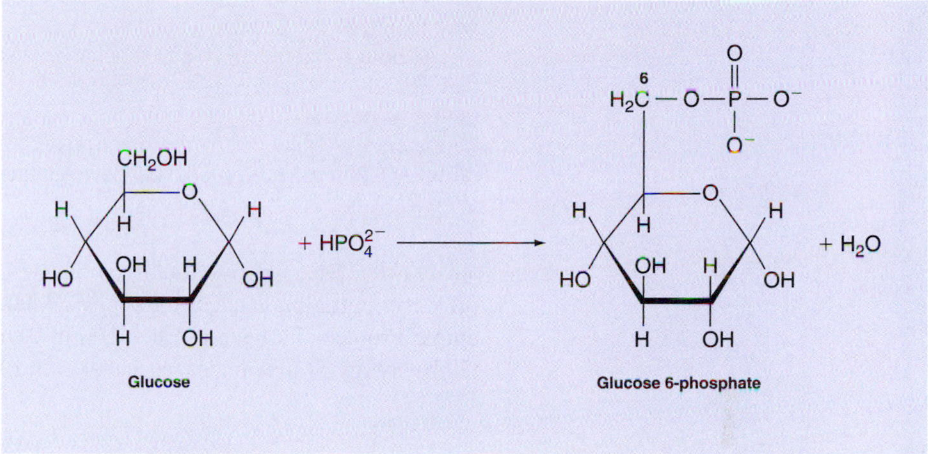

Glucose

Glucose 6-phosphate

$$\text{Glucose} + P_i \rightarrow \text{G6P} + H_2O \quad \Delta G^{\circ\prime} = +13.8 \text{ kJ mol}^{-1} \quad (11.1)$$

Many of the steps in the glycolytic pathways consist of coupled reactions with common intermediates. Phosphorylation of glucose must be coupled to ATP hydrolysis in order to proceed:.

$$\text{Glucose} + P_i \rightarrow \text{G6P} + H_2O \quad \Delta G^{\circ\prime} = +13.8 \text{ kJ mol}^{-1}$$

$$\text{ATP} + H_2O \rightarrow \text{ADP} + P_i \quad \Delta G^{\circ\prime} = -30.5 \text{ kJ mol}^{-1} \quad (11.2)$$

For this coupled pair of reactions $\Delta G^{\circ\prime} = 13.8 \text{ kJ mol}^{-1} - 30.5 \text{ kJ mol}^{-1} = -16.7 \text{ kJ mol}^{-1}$. Under cellular conditions the net reaction is even more exergonic: $\Delta G = -33.5 \text{ kJ mol}^{-1}$. In the reaction pair shown in Equation (11.2) the common intermediate is inorganic phosphate, and the coupling is accomplished by the enzyme hexokinase. A kinase is an enzyme that transfers phosphoryl groups between ATP and a metabolite. As the name implies, hexokinase is a protein that is not specific for glucose, but will catalyze phosphorylation of other six-carbon sugars. The net cellular reaction is

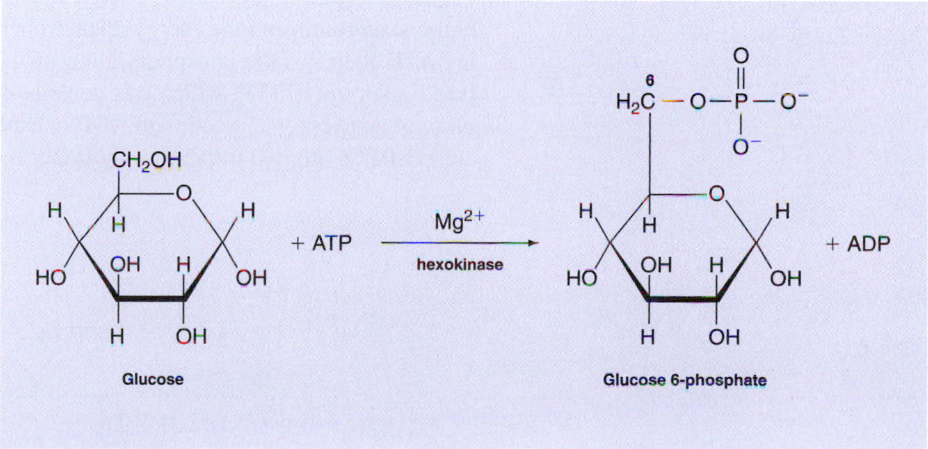

Glucose

Glucose 6-phosphate

(11.3)

The Mg^{2+} shown over the reaction arrow indicates the involvement of magnesium ion in a complex with ATP^{4-}.

The other two highly exergonic steps of the glycolytic pathway, steps 3 and 10, serve the purposes of regulating the pathway and generating ATP, respectively. Step 3 involves the phosphorylation of fructose-6-phosphate (F6P) to produce fructose-1,6-bisphosphate (FBP):

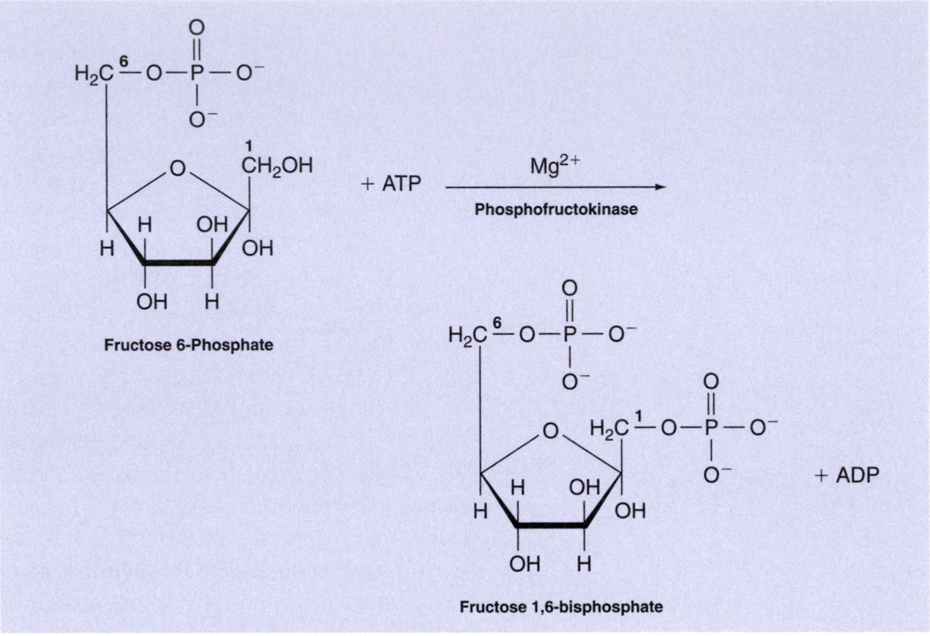

Fructose 6-Phosphate

Fructose 1,6-bisphosphate

(11.4)

Like step 1, the phosphorylation of F6P is highly endergonic and must also be coupled to ATP hydrolysis, so that the net reaction proceeds with a $\Delta G^{\circ\prime} = -14.2 \text{ kJ mol}^{-1}$. Under cellular conditions, $\Delta G = -22.2 \text{ kJ mol}^{-1}$.

The glycolytic pathway is regulated at a single step. Although pathway regulation is a kinetic property, the choice for the regulation step can be partly understood in terms of the Gibbs energy landscape. Step 3 is the regulated step of glycolysis for two reasons. First, it is highly exergonic and thus for all practical purposes irreversible. Second, unlike glucose, G6P, and F6P, all of which are involved in other cellular functions, FBP has the unique role of being the glycolytic precursor of pyruvate. At this point the chemical fate of the six-carbon species is determined and thus this irreversible step is an ideal control point. Accordingly, phosphofructokinase (PFK), the enzyme that catalyzes phosphorylation of F6P, is the dominant regulatory enzyme of the glycolytic pathway.

Steps 7 and 10 are called "energy-conserving" steps in the sense that each step is highly exergonic, and the energy released in each step is conserved for use in synthesizing ATP. Step 7 is the phosphorylation of ADP by 1,3-bisphosphoglycerate (1,3-BPG) [see Equation (10.11)], which was described in Section 10.3 as an example of the coupling of an exergonic reaction to ADP phosphorylation through the common intermediate 1,3-BPG. Step 10 is the phosphorylation of ADP by phosphoenolpyruvate (PEP):

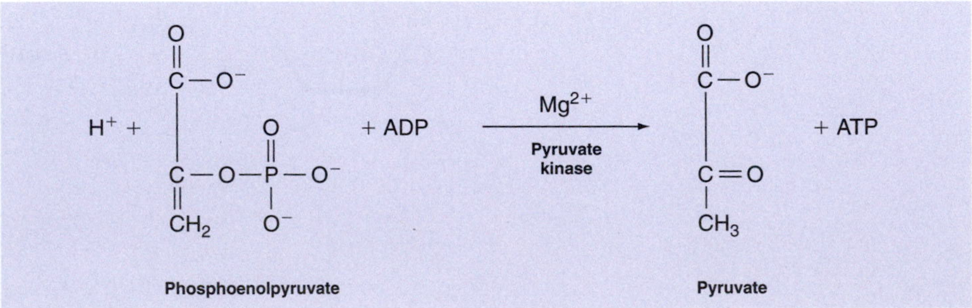

Phosphoenolpyruvate

Pyruvate

$$PEP + ADP \rightarrow \text{pyruvate} + ATP \qquad (11.5)$$

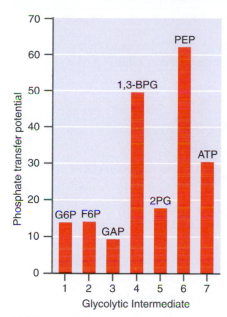

FIGURE 11.5

Phosphate transfer potentials for various glycolytic intermediates versus ATP. See also Figure 11.3

For the hydrolysis of PEP, $\Delta G°' = -61.9$ kJ mol^{-1}. Therefore, the overall standard Gibbs energy change for step 10 is $\Delta G°' = -61.9$ kJ mol^{-1} + 30.5 kJ mol^{-1} = -31.4 kJ mol^{-1}. Because two molecules of PEP are produced per molecule of glucose that enters the glycolytic pathway, the total Gibbs energy change for step 10 is $\Delta G°' = 2 \times -31.4$ kJ mol^{-1} = -62.8 kJ mol^{-1}, which is the number that appears in Figure 11.3.

To phosphorylate ADP, the glycolytic pathway must produce intermediates with higher phosphate transfer potentials than ATP. Given that steps 1 and 3 together hydrolyze two ATP molecules, it is necessary to have at least two steps where in each step a three-carbon phosphoester is produced that has a higher phosphate transfer potential than ATP. Figure 11.5 shows phosphate transfer potentials for the sequence of intermediates in the glycolytic pathway compared to the transfer potential of ATP. Only 1,3-BPG and PEP have transfer potentials higher than ATP, 61.9 for PEP and 49.4 for 1,3-BPG versus 30.5 for ATP, so only these intermediates can phosphorylate ADP. Given that steps 7 and 10 each deal with two three-carbon compounds that arise from the degradation of one glucose molecule, four ATP molecules are produced by steps 7 and 10 per glucose molecule input to the pathway. Because steps 1 and 3 each hydrolyze a ATP molecule, the net ATP production by glycolysis is two ATP molecules per glucose molecule.

Anaerobic reactions that further metabolize pyruvate include the production of lactate in stressed muscle tissue and the degradation of pyruvate to ethanol via fermentation. Physical exercise requires rapid production of ATP via glycolysis in order to effect muscle contraction. In aerobic cells, more ATP can be produced by further oxidizing pyruvate to carbon dioxide and water in the respiratory system (see later discussion). But ATP production by anaerobic glycolysis can occur almost 100 times as rapidly as oxidative phosphorylation, so without time for oxygen to diffuse through the blood and into the mitochondria for oxidative phosphorylation to occur, cells in muscles stressed by a burst of exercise will often undergo anaerobic glycolysis. Reduction of pyruvate to lactate by NADH is catalyzed by the enzyme lactate dehydrogenase:

$$CH_3(CO)COO^- + H^+ + NADH \xrightarrow{\text{lactate dehydrogenase}} CH_3(CHOH)COO^- + NAD^+ \quad (11.6)$$

This is an exergonic reaction for which $\Delta G°' = -25.1$ kJ mol^{-1}. Because two pyruvates are produced per glucose molecule that enters the glycolytic pathway, we double the Gibbs energy change for the reduction of pyruvate to lactate, and add the reduction to the net glycolysis equation in Figure 11.3

$$C_6H_{12}O_6 + 2ADP + 2P_i + NAD^+ \rightarrow 2CH_3(CO)COO^- + 2ATP + NADH + H^+$$
$$\Delta G°' = -73.3 \text{ kJ mol}^{-1}$$

$$2CH_3(CO)COO^- + H^+ + NADH \rightarrow 2CH_3(CHOH)COO^- + NAD^+$$
$$\Delta G°' = 2 \times -25.1 \text{ kJ mol}^{-1} = -50.2 \text{ kJ mol}^{-1} \quad (11.7)$$

to obtain the net reaction:

$$C_6H_{12}O_6 + 2ADP + 2P_i \rightarrow 2CH_3(CHOH) COO^- + 2ATP \quad (11.8)$$

For Equation (11.8), the Gibbs energy change is

$$\Delta G°' = -73.3 \text{ kJ mol}^{-1} + -50.2 \text{ kJ mol}^{-1} = -123.5 \text{ kJ mol}^{-1} \quad (11.9)$$

Although lactate can be recycled into pyruvate and used in the synthesis of glucose (gluconeogenesis), lactate production is a metabolic dead end and a waste of Gibbs energy in the sense that lactate cannot be further degraded for ATP production and will eventually accumulate in working muscle tissue.

We can now calculate the efficiency with which the Gibbs energy released by anaerobic glucose metabolism is recovered and applied toward ADP phosphorylation. Previously we gave the standard Gibbs energy change for anaerobic metabolism of glucose to lactate as $\Delta G°' = -123.5$ kJ mol^{-1}. The net production of ATP is 2 mol per mole of glucose metabolized anaerobically so that $\Delta G°' = 2 \times 30.5$ kJ mol^{-1} = 61 kJ mol^{-1}. Under standard conditions the efficiency with which the Gibbs energy released by anaerobic glucose metabolism is recovered as Gibbs energy in ATP is $\dfrac{61 \text{ kJ}}{123.5 \text{ kJ} + 61 \text{ kJ}} \times 100\% \approx 33\%$.

EXAMPLE PROBLEM 11.2

Calculate the Gibbs energy change for the conversion of glucose to lactate [Equation (11.8)] under the cellular conditions described in Table 11.1. Assume $T = 310.$ K and pH $= 7$.

Solution

$$\Delta G = \Delta G^{\circ\prime} + RT \ln Q = \Delta G^{\circ\prime} + RT \ln\left(\frac{C_{Lact}^2 C_{ATP}^2}{C_{glucose} C_{ADP}^2 C_{P_i}^2}\right)$$

$$= -123.5 \text{ kJ mol}^{-1} + (8.31 \text{ J mol}^{-1} \text{ K}^{-1})(310 \text{ K}) \ln\left(\frac{(2900.)^2(1850.)^2}{(5.00 \times 10^{-3})(138)^2(1000.)^2}\right)$$

$$= -123.5 \text{ kJ mol}^{-1} + (2.58 \text{ kJ mol}^{-1})\ln(3.02 \times 10^5) = -91.0 \text{ kJ mol}^{-1}$$

EXAMPLE PROBLEM 11.3

Calculate the efficiency with which the Gibbs energy released by the conversion of glucose to lactate is coupled to the phosphorylation of ADP. Assume the cellular conditions described in Table 11.1.

Solution

For the cellular conditions described in Table 11.1, the Gibbs energy change for Equation (11.8) is $\Delta G = -111.5 \text{ kJ mol}^{-1}$, as shown in Example Problem 11.2. We now need the Gibbs energy change for ADP phosphorylation under celular conditions. This is

$$\Delta G = \Delta G^{\circ\prime} + RT \ln Q = \Delta G^{\circ\prime} + RT \ln\left(\frac{C_{ATP}^2}{C_{ADP}^2 C_{P_i}^2}\right)$$

$$= 61. \text{ kJ mol}^{-1} + (8.31 \text{ J mol}^{-1} \text{ K}^{-1})(310 \text{ K}) \ln\left(\frac{(1850.)^2}{(138)^2(1000. \times 10^{-6})^2}\right)$$

$$= 61. \text{ kJ mol}^{-1} + (2.58 \text{ kJ mol}^{-1})(19.0) = 110 \text{ kJ mol}^{-1}$$

Therefore, the efficiency under cellular conditions is $\frac{110.}{111.5 + 110.} \times 100\% \approx 50\%$.

11.3 The Krebs Cycle

The primary means of ATP production in aerobic cells is the respiratory system. As shown in Figure 11.1, the respiratory system is composed of two parts: the **Krebs cycle** or **citric acid cycle,** and the **electron transport chain,** also called the terminal transport chain. Unlike glycolysis, which occurs in the cytosol, respiration occurs in the plasma or cellular membrane in prokaryotic cells (i.e., cells whose genetic material is not confined to an organized nucleus), whereas in eukaryotic cells (i.e., cells whose genetic material is organized into a nucleus) respiration occurs in an organelle called the **mitochondrion.** See Figure 11.6.

The inner membrane of the mitochondrion encompasses the **matrix.** Between the inner and outer membranes is the intermembrane space. Most of the chemical reactions comprising the Krebs cycle occur in the matrix. The proteins and other molecules comprising the electron transport chain are largely associated with the inner membrane. As shown in Figure 11.7, the purpose of the Krebs cycle is to complete the degradation of the three-carbon skeleton of pyruvate into carbon dioxide and, in doing so, to produce cellular reducing agents, which will be used in the electron transport chain to reduce oxygen to water. The process by which ADP is phosphorylated as a result of the transfer of elec-

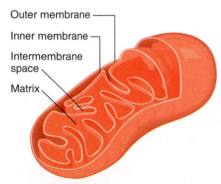

Outer membrane
Inner membrane
Intermembrane space
Matrix

FIGURE 11.6

The mitochondrion has two membranes. The inner membrane encompasses the matrix where the Krebs cycle occurs. The electron transport chain occurs within the inner membrane.

FIGURE 11.7

The Krebs or citric acid cycle oxidizes the three-carbon pyruvate to carbon dioxide. Oxidizing agents NADH and FADH$_2$ produced by the Krebs cycle are input to the electron transport or respiratory chain. (Source Adapted from C.K. Mathews, K.E. von Holde, and K.G. Ahern, *Biochemistry*, 3rd ed. (San Francisco, CA: Pearson Benjamin Cummings) p. 485. Copyright 2000. Reprinted by Permission of Pearson Education.)

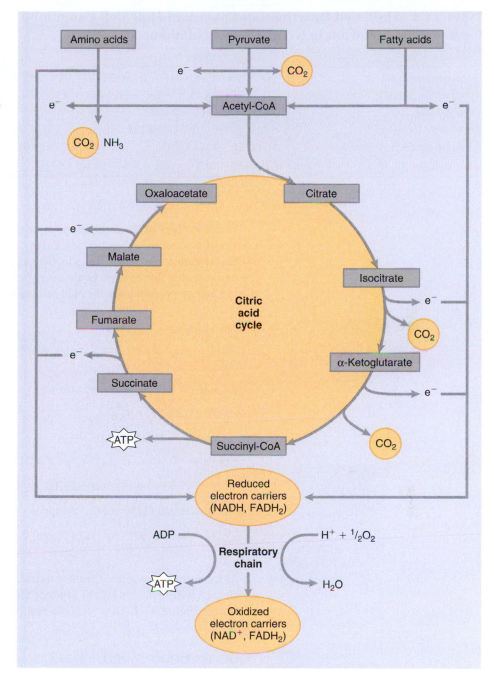

trons from the reducing agents produced by the Krebs cycle to oxygen is called oxidative phosphorylation. Oxidative phosphorylation is the primary source of ATP in aerobic cells.

The operating principle of the Krebs cycle is simple. By a process called oxidative decarboxylation, the Krebs cycle degrades the three-carbon pyruvate molecule into three carbon dioxide molecules. In doing so, the Krebs cycle transfers electrons to molecules such as NAD$^+$ to produce NADH. These molecules later perform as reducing agents in the electron transport chain. However, the Krebs cycle accepts only two-carbon oxidation products (or the equivalent) from the degradation pathways for three major food stuffs: carbohydrates, proteins, and fats. So prior to entry into the cycle, pyruvate is oxidized to acetate CH$_3$COO$^-$ by NAD$^+$ with the production of NADH and CO$_2$. This redox reaction can be written as:

$$CH_3COCOO^- + H_2O + NAD^+ \rightarrow CH_3COO^- + H^+ + CO_2 + NADH \quad (11.10)$$

TABLE 11.2   Half-Cell Reactions and Standard Half-Cell Potentials for Reactions Relevant to Krebs Cycle Oxidations

System	Half-Cell Reaction	$E^{\circ\prime}$ (V)
Fumarate/succinate	$^-OOCCH=CHCOO^- + 2e^- + 2H^+ \rightarrow {}^-OOCCH_2CH_2COO^-$	+0.310
Oxaloacetate/malate	$^-OOCCOCH_2COO^- + 2e^- + 2H^+ \rightarrow {}^-OOCCHOHCH_2COO^-$	−0.166
FAD/FADH$_2$	$FAD + 2e^- + 2H^+ \rightarrow FADH_2$	−0.219
NAD$^+$/NADH	$NAD^+ + 2e^- + H^+ \rightarrow NADH$	−0.320
α-Ketoglutarate/isocitrate	$^-OOCCH_2CH_2COCOO^- + 2e^- + 2H^+ \rightarrow {}^-OOCCH_2CH_2CHOHCOO^-$	−0.380
Acetate/pyruvate	$CH_3COO^- + CO_2 + 2e^- + 2H^+ \rightarrow CH_3COCOO^- + H_2O$	−0.700

Because many of the reactions of the Krebs cycle and the electron transport chain consist of oxidations and reductions, energy calculations for these pathways frequently use the concept of the oxidation–reduction potential, which was introduced in Chapter 9. We can calculate the standard Gibbs energy change for Equation (11.10) using the standard reduction potentials for the half-cell reactions given in Table 11.2.

EXAMPLE PROBLEM 11.4

Calculate the standard Gibbs energy change for Equation (11.10) using the data in Table 11.2.

Solution

From Table 11.2:

$$NAD^+ + 2e^- + H^+ \rightarrow NADH \qquad E^{\circ\prime} = -0.32 \text{ V}$$

$$CH_3COO^- + CO_2 + 2H^+ + 2e^- \rightarrow CH_3COCOO^- + H_2O \qquad E^{\circ\prime} = -0.70 \text{ V}$$

$$\Delta E^{\circ\prime} = (-0.320\text{V} + (0.700 \text{ V})) = 0.380 \text{ V}$$

and the standard Gibbs energy change is

$$\Delta G^{\circ\prime} = -nF\,\Delta E^{\circ\prime} = -(2)(96{,}485 \text{ C mol}^{-1})(0.380 \text{ V}) = -73.3 \text{ kJ mol}^{-1}$$

Once acetate is available from the oxidation of pyruvate, it must in principle be condensed with oxaloacetate to produce citrate, which is the first step of the Krebs cycle, see Figure 11.1. However, the reaction of oxaloacetate with acetate to form citrate,

$$^-OOCCH_2COCOO^- + H_2O + CH_3COO^- \rightarrow {}^-OOCCH_2C(COO^-)OHCOO^- \qquad (11.11)$$

is actually endergonic with $\Delta G^{\circ\prime} \approx 8 \text{ kJ mol}^{-1}$. Therefore, if acetate itself were fed to the Krebs cycle, the first step would be energetically unfavorable. Therefore, prior to entering the Krebs cycle, acetate is reacted with a nucleotide called coenzyme A (CoA); see Figure 11.8, top. CoA has a structure reminiscent of NAD, where the nucleotide adenosine is connected via a pyrophosphate linkage to a moiety that performs a specific chemical task. In the case of NAD the moiety is nicotinamide, which we have seen acts as a carrier of electrons in redox reactions. In CoA, the moiety is a vitamin called pantothenic acid. Pantothenic acid has a thiol group at its terminus that forms a thioester bond with acetate. The result is called acetyl-CoA. So just as the nicotinamide group of NAD$^+$ receives electrons in redox reactions, pantothenic acid in CoA is a carrier of acetate groups. The net reaction in which pyruvate is oxidized to acetate and acetate forms a thioester with CoA is

$$CH_3COCOO^- + NAD^+ + CoA\text{-}SH \rightarrow CH_3COS\text{-}CoA + CO_2 + NADH \qquad (11.12)$$

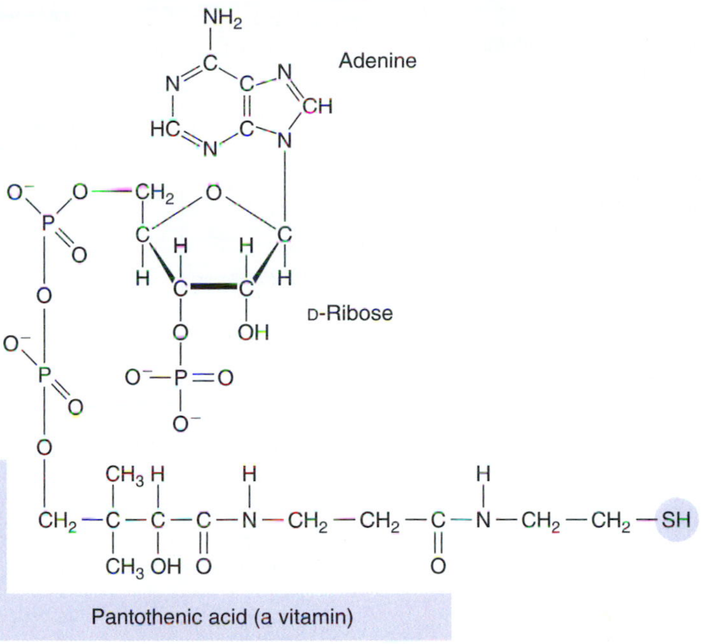

FIGURE 11.8
Coenzyme A has a structure reminiscent of NAD (see Figure 10.4). It is composed of adenine, a D-ribose sugar, and a pyrophosphate bridge to a second group. For CoA this second group is pantothenic acid. The acetate group forms a thioester bond at the thiol (−SH) group to form acetyl-CoA.

where the abbreviation CoA-SH highlights the thiol group that reacts with acetate to form CH_3COS-CoA, acetyl-CoA. The Gibbs energy change for Equation (11.12) is $\Delta G°' = -33.4$ kJ mol^{-1}, and the difference in the Gibbs energy between Equations (11.11) and (11.12) is due to the formation of the thioester acetyl-CoA.

Once acetyl-CoA is formed, acetate can enter the first step of the Krebs cycle and react with the four-carbon molecule oxaloacetate to form citrate. Because this reaction involves the hydrolysis of the thioester bond that separates CoA from acetate, the overall reaction

$$^-OOCCH_2COCOO^- + H_2O + CH_3COS\text{-CoA} \rightarrow {}^-OOCCH_2C(COO^-)OHCOO^- + CoA\text{-SH} + H^+$$

$$(11.13)$$

is exergonic: $\Delta G°' = -32.2$ kJ mol^{-1}. Gibbs energy changes for the Krebs cycle are given in Table 11.3.

Although standard Gibbs energy changes are known for the Krebs cycle, because components of the Krebs cycle are partitioned between the mitochondrion and the cytosol, concentrations for Krebs cycle intermediates within the matrix are not known accurately.

TABLE 11.3   **Standard Gibbs Energy Changes for the Krebs Cycle and Ranges for Gibbs Energy Changes under Cellular Conditions**

Step	Reaction	$\Delta G°'$ (kJ mol^{-1})	ΔG
	Pyruvate + CoA-SH + NAD$^+$ → acetyl-CoA + NADH + CO$_2$ + H$^+$	−33.4	
1	Acetyl-CoA + oxaloacetate + H$_2$O → citrate + CoA-SH + H$^+$	−32.2	<0
2	Citrate → isocitrate	6.3	~0
3	Isocitrate + NAD$^+$ → α-ketoglutarate + NADH + CO$_2$	−7.1	<0
4	α-Ketoglutarate + NAD$^+$ + CoA-SH → succinyl-CoA + NADH + CO$_2$ + H$^+$	−33.4	<0
5	Succinyl-CoA + GDP + P$_i$ → succinate + GTP + CoA-SH	−3.3	~0
6	Succinate + FAD → fumarate + FADH$_2$	~0	~0
7	Fumarate + H$_2$O → L-malate	−3.8	~0
8	L-Malate + NAD$^+$ → oxaloacetate + NADH + H$^+$	+29.2	~0
Net	Pyruvate + NAD$^+$ + FAD + GDP + P$_i$ + 2H$_2$O → 3CO$_2$ + 4NADH + FADH$_2$ + GTP + 4H$^+$	−77.7	<0

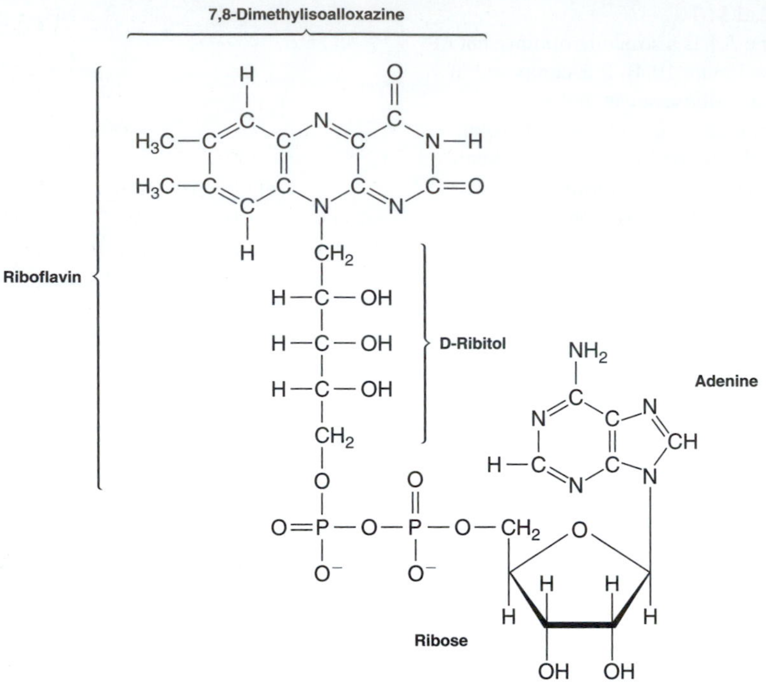

FIGURE 11.9
The oxidized form of flavin adenine dinucleotide (FAD).

The net reaction for the Krebs cycle, including the initial oxidation of pyruvate, yields three molecules of carbon dioxide per molecule of pyruvate that enters the cycle, four molecules of NADH, a molecule of flavin adenine dinucleotide ($FADH_2$), and a molecule of guanosine triphosphate (GTP). Like NAD^+, flavin adenine dinucleotide (FAD) is a cellular oxidizing agent. It consists of adenine and ribose connected via a pyrophosphate linkage to riboflavin. See Figure 11.9.

The site of electron transfer is the 7,8-dimethylisoalloxazine ring. Guanosine triphosphate is a nucleotide similar to ATP. The enzyme nucleoside diphosphotase kinase catalyzes the transfer of a phosphoryl group from GTP to ADP to form ATP:

$$GTP + ADP \rightarrow GDP + ATP \qquad (11.14)$$

The net reaction in Table 11.3 shows that the Krebs cycle is an efficient supplier of reducing agents to the electron transport chain. In the next section we find that each NADH molecule fed to the electron transport chain effects the phosphorylation of three ADP molecules. Each $FADH_2$ molecule causes the phosphorylation of two ADP molecules. Therefore, the products of the Krebs cycle eventually yield the following ATP harvest per pyruvate molecule fed to the cycle:

$$4NADH \Rightarrow 12ATP$$

$$1FADH_2 \Rightarrow 2ATP$$

$$1GTP \Rightarrow 1ATP$$

Total = 15 ATP molecules per pyruvate

Because two pyruvate molecules originate from each glucose molecule that enters the glycolytic pathway, 30 ATP molecules are produced for each pyruvate pair that enters the Krebs cycle. Moreover, the glycolytic pathway itself produces 2 ATP molecules and 2 NADH molecules or 8 ATP molecules total because the 2 NADH molecules will eventually enter the electron transport chain and effect the phosphorylation of 6 ADP molecules. Therefore, in a respiring cell, the Gibbs energy released by the metabolism of 1 glucose molecule is coupled to the phosphorylation of 38 molecules of ADP. We can calculate the efficiency of this system for standard conditions once we know the Gibbs energy change associated with electron transport.

11.4 The Electron Transport Chain

Of the 38 ATP molecules produced by the metabolism of a single glucose molecule into carbon dioxide and water, 30 ATP molecules are produced by the combined action of the Krebs cycle and the electron transport chain. Of the 15 molecules of ATP produced per pyruvate molecule that enters the Krebs cycle, 14 are produced by reducing agents NADH or $FADH_2$ that enter the electron transport chain. Because the majority of ATP is produced by the electron transport chain, this assembly of redox reactions is the cell's energy mainspring.

The purpose of the electron tranport chain is to release energy by moving electrons obtained by the degradation of the carbon skeleton of pyruvate and carried by cellular reducing agents such as NADH through a series of membrane-assocated redox reactions. Associated with each redox step of the electron transport chain is a negative change in the Gibbs energy, making the entire chain of redox reactions highly exergonic.

The physical apparatus that effects electron transport and the consequent reduction of oxygen to water is associated with the inner membrane of the mitochondrion. If mitochondria are extracted intact from a cell, the oxidation–reduction reactions that comprise the transport chain can be observed to occur, and the components of the transport chain can be resolved. The electron transport chain is composed of four multienzyme complexes associated with the inner mitochondrial membrane. The arrangement of the oxidation–reduction reactions among these complexes is schematized in Figure 11.10. Relevant half-cell reduction potentials for the transport chain are given in Table 11.4.

Complex I of the electron transport chain is called NADH-CoQ oxidoreductase, and as the name implies, this complex of enzymes catalyzes the oxidation of NADH by coenzyme Q (CoQ). Coenzyme Q is also called ubiquinone and has the oxidized and reduced forms shown in Figure 11.11a. Like NADH, coenzyme Q can carry an electron pair in a fully reduced state, called ubiquinol or hydroquinone, which is designated $CoQH_2$. But coenzyme Q also has a stable radical intermediate state called semiquinone that carriers a single unpaired electron. This intermediate state is of fundamental importance when CoQ in its fully reduced ubiquinol state must reduce proteins that contain a single Fe^{3+} and that can only carry a single electron. The net reaction in complex I is

$$NADH + CoQ + H^+ \rightarrow NAD^+ + CoQH_2 \tag{11.15}$$

FIGURE 11.10

Schematic of the arrangement of oxidation–reduction reactions among the complexes of the electron transport chain. The yellow arrow traces the path of electrons between the complexes of the electron transport chain. Redox reactions are black arrows. Electron carriers CoQ and cytochrome c are shown as "Q" and "Cyt c," respectively. (Source: Adapted from N.A. Campbell, and J.B. Reece, *Biology,* 7th ed. (San Francisco, CA: Pearson Benjamin Cummings), p. 172. Copyright 2005. Reprinted by permission of Pearson Education, Inc.)

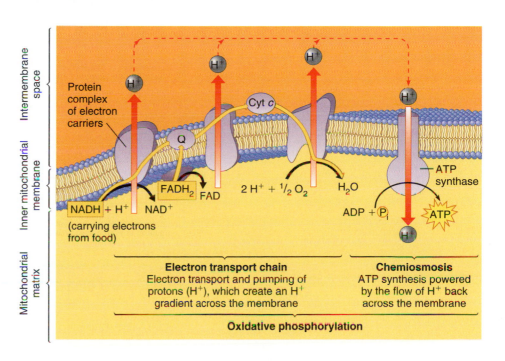

Equation (11.15) actually occurs in three steps involving oxidation and reducing agents that are bound to proteins within complex I. Two important proteins are called flavoprotein, which binds a redox reagent called flavin mononucleotide (FMN; see Figure 11.11b), and ferredoxin, which bind an iron-sulfur moiety. FMN has a structure reminiscent of the flavin ring in FAD and, like CoQ, FMN carries an electron pair in its fully reduced state but it can form a stable radical intermediate that carries a single electron.

TABLE 11.4 Standard Half-Cell Reduction Potentials for Components of the Electron Transport Chain

System	Reaction	$E^{\circ\prime}$ (V)
O_2/H_2O	$\frac{1}{2}O_2 + 2H^+ + 2e^- \rightarrow H_2O$	0.815
Fe^{3+}/Fe^{2+}(Cyt a_3)	Fe^{3+}(Cyt a_3) + $e^- \rightarrow Fe^{2+}$(Cyt a_3)	0.35
Fe^{3+}/Fe^{2+}(Cyt a)	Fe^{3+}(Cyt a) + $e^- \rightarrow Fe^{2+}$(Cyt a)	0.290
Fe^{3+}/Fe^{2+}(Cyt c)	Fe^{3+}(Cyt c) + $e^- \rightarrow Fe^{2+}$(Cyt c)	0.254
Fe^{3+}/Fe^{2+}(Cyt c_1)	Fe^{3+}(Cyt c_1) + $e^- \rightarrow Fe^{2+}$(Cyt c_1)	0.220
$CoQ/CoQH_2$	$CoQ + 2H^+ + 2e^- \rightarrow CoQH_2$	0.10
Fe^{3+}/Fe^{2+}(Cyt b)	Fe^{3+}(Cyt b) + $e^- \rightarrow Fe^{2+}$ (Cyt b)	0.08
$FAD/FADH_2$	$FAD + 2e^- + 2H^+ \rightarrow FADH_2$	−0.219
$NAD^+/NADH$	$NAD^+ + 2e^- + H^+ \rightarrow NADH$	−0.32
Fe^{3+}/Fe^{2+}(ferredoxin)	Fe^{3+} (ferredoxin) + $e^- \rightarrow Fe^{2+}$ (ferredoxin)	−0.432

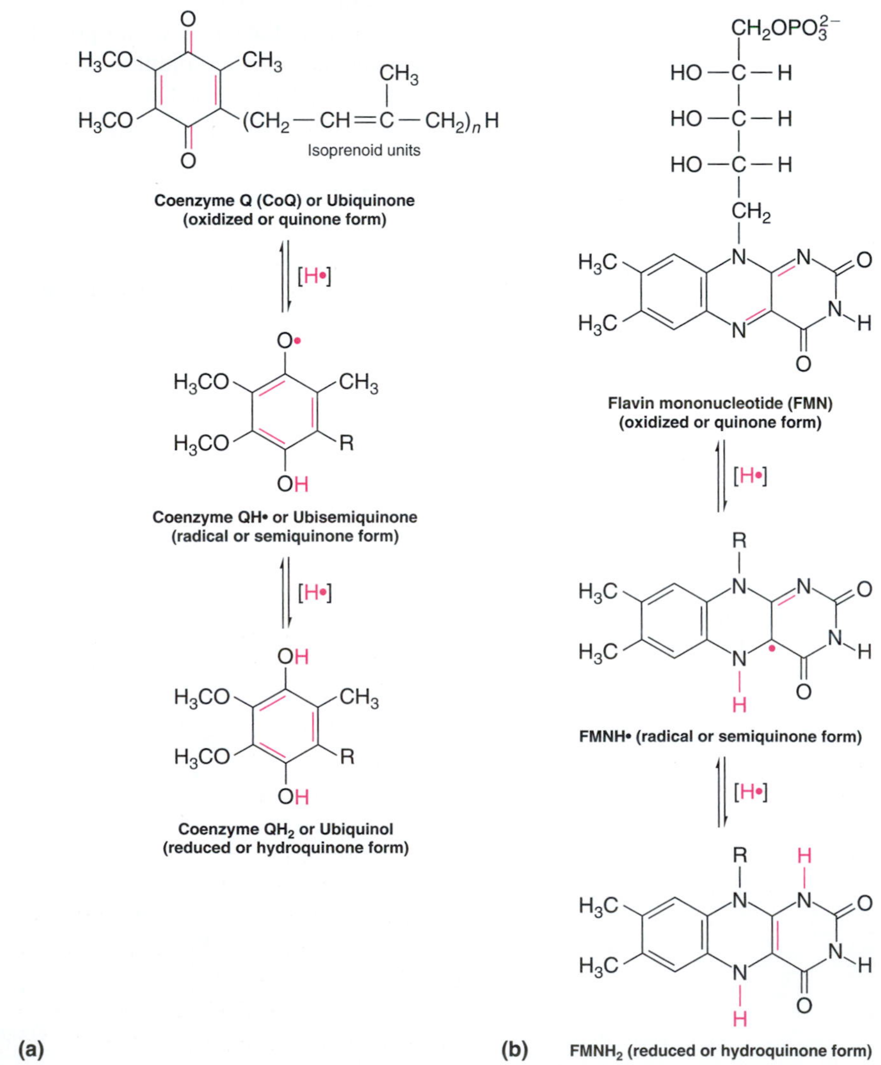
H3CO
CH3
(CH2—CH=C—CH2)nH
Isoprenoid units
Coenzyme Q (CoQ) or Ubiquinone
(oxidized or quinone form)
[H•]
R
OH
Coenzyme QH• or Ubisemiquinone
(radical or semiquinone form)
Coenzyme QH2 or Ubiquinol
(reduced or hydroquinone form)
CH2OPO3 2−
HO—C—H
CH2
H3C
Flavin mononucleotide (FMN)
(oxidized or quinone form)
FMNH• (radical or semiquinone form)
(a)
(b)
FMNH2 (reduced or hydroquinone form)

FIGURE 11.11
Redox reagents of the electron transport chain: the three oxidation states of (a) coenzyme Q (CoQ) and (b) flavin mononucleotide (FMN).

The redox active moiety in ferredoxin is iron. Iron appears throughout the electron transport chain as a redox reagent where it alternates between the $+3$ and $+2$ oxidation states. However, as Table 11.4 shows, the Fe^{3+}/Fe^{2+} half-cell potentials vary from protein to protein in the electron transport chain due to differences in the local structural environments of the iron. The constituent steps of Equation (11.15) that occur within complex 1 are:

Step 1: $NADH + H^+ + FMN \rightarrow NAD^+ + FMNH_2$

Step 2: $FMNH_2 + 2Fe^{3+}$ (ferredoxin) $\rightarrow FMN + 2Fe^{2+}$ (ferredoxin) $+ 2H^+$ (11.16)

Step 3: $2Fe^{2+}$ (ferredoxin) $+ CoQ + 2H^+ \rightarrow 2Fe^{3+}$ (ferredoxin) $+ CoQH_2$

EXAMPLE PROBLEM 11.5

Calculate the standard Gibbs energy change for the net reaction of complex I [Equation (11.15)] using the half-cell potentials in Table 11.4. Assuming the Gibbs energy released by this reaction is coupled with 40% efficiency to the phosphorylation of ADP, how much ATP can be produced under standard conditions by the oxidation of 1 mol of NADH in complex I?

Solution

Using the reduction half-cell potentials for NAD and CoQ given in Table 11.4, we obtain for the cell potential $E^{\circ\prime} = 0.10\ V - (-0.32\ V) = 0.42\ V$, and the Gibbs energy is $\Delta G^{\circ\prime} = -nFE^{\circ\prime} = (2)(96{,}485\ C\ mol^{-1})(0.42\ V) = -81.1\ kJ\ mol^{-1}$ for the oxidation of NADH by CoQ. Now assuming 40% of this Gibbs energy is coupled to ADP phosphorylation the energy available is $\Delta G^{\circ\prime} = (-81.1\ kJ\ mol^{-1})(0.40) = -32.4\ kJ\ mol^{-1}$. Dividing this result by the energy required to phosphorylate 1 mol of ADP under standard conditions: moles ATP $= 32.4\ kJ/30.5\ kJ\ mol^{-1} = 1.1$. The oxidation of NADH by CoQ is one of the most exergonic redox reactions of the electron transport chain.

The second complex of the electron transport chain is called succinate-CoQ oxidoreductase. This complex essentially uses components of the Krebs cycle, specifically succinate and the enzyme that catalyzes oxidation of succinate by FAD to fumarate, to reduce CoQ in the following sequence of reactions:

$$\text{Step 1: succinate} + FAD \rightarrow \text{fumarate} + FADH_2$$
$$\text{Step 2: } FADH_2 + CoQ \rightarrow FAD + CoQH_2$$
(11.17)

The net reaction for complex II, therefore, is

$$\text{succinate} + CoQ \rightarrow \text{fumarate} + CoQH_2 \qquad (11.18)$$

Although $FADH_2$ is oxidized by CoQ in complex II to produce $CoQH_2$, which in turn continues through the electron transport chain just as does $CoQH_2$, which originates from oxidation of NADH in complex I, there is a significant difference. The redox reactions of complex II are not coupled to ADP phosphorylation. So, although for every NADH that enters the electron transport chain via complex I three ADPs are phosphorylated, in the case of $FADH_2$, which enters via complex II, only two ADPs are phosphorylated.

We now consider the unique role played by CoQ in passing electrons to the proteins of complex III of the electron transport chain. With its aromatic ring and long isoprenoid tail, it is not surprising that $CoQH_2$ is hydrophobic and soluble in the lipid bilayer that constitutes the inner mitochondrial membrane. Therefore, after emerging from complex I near the inner (i.e., matrix) surface of the inner mitochondrial membrane, $CoQH_2$ diffuses through the membrane to a complex of enzymes called complex III or $CoQH_2$-cytochrome c oxidoreductase.

In complex III, $CoQH_2$ encounters proteins called cytochromes that contain iron as the redox reagent, that is, Fe^{3+}/Fe^{2+}. But unlike ferredoxin in complex I and the other iron-sulfur proteins in complex II, in the cytochrome proteins the iron resides in a porphyrin ring. The complex of iron and the porphyrin is called a heme group. Cytochrome proteins are labeled by letters (e.g., cytochrome c) that are meant to indicate chemical substituents on the porphyrin ring of the heme group; see Figure 11.12a and b.

The net reaction for complex III is the transfer of two electrons from $CoQH_2$ to the the iron in two cytochrome c_1 (Cyt c_1) heme groups contained in complex III:

$$CoQH_2 + 2Fe^{3+}(Cyt\ c_1) \rightarrow CoQ + 2Fe^{2+}\ (Cyt\ c_1) + 2H^+ \quad (11.19)$$

The Gibbs energy change resulting from the reduction of Fe^{3+} to Fe^{2+} in cytochrome c_1 by $CoQH_2$ yields about -31 kJ mol^{-1}, which is sufficient energy for a second phosphorylation of ADP.

Cytochome c_1 is bound to complex III and thus cannot directly transfer an electron to the next component of the electron transfer chain, which is called complex IV. So cytochrome c_1 undergoes a redox reacton with a second heme protein called cytochrome c (Cyt c). Cytochrome c is loosely associated with the P side of the inner mitochondrial membrane, which is the side adjacent to the intermembrane space. Cytochrome c diffuses to complex IV once it has been reduced by cytochrome c_1; see Figure 11.10. The net reaction of cytochrome c_1 with cytochrome c is

$$2Fe^{3+}\ (Cyt\ c) + 2Fe^{2+}\ (Cyt\ c_1) \rightarrow 2Fe^{2+}\ (Cyt\ c) + 2Fe^{3+}(Cyt\ c_1) \quad (11.20)$$

where the stoichiometric coefficients of 2 in Equation (11.20) account for the transfer of an electron pair that originated with the $CoQH_2$ molecule that entered complex III. Using the half-cell potentials given in Table 11.5, the Gibbs energy change for Equation (11.20) is about -6 kJ mol^{-1}, which is too small to phosphorylate ADP.

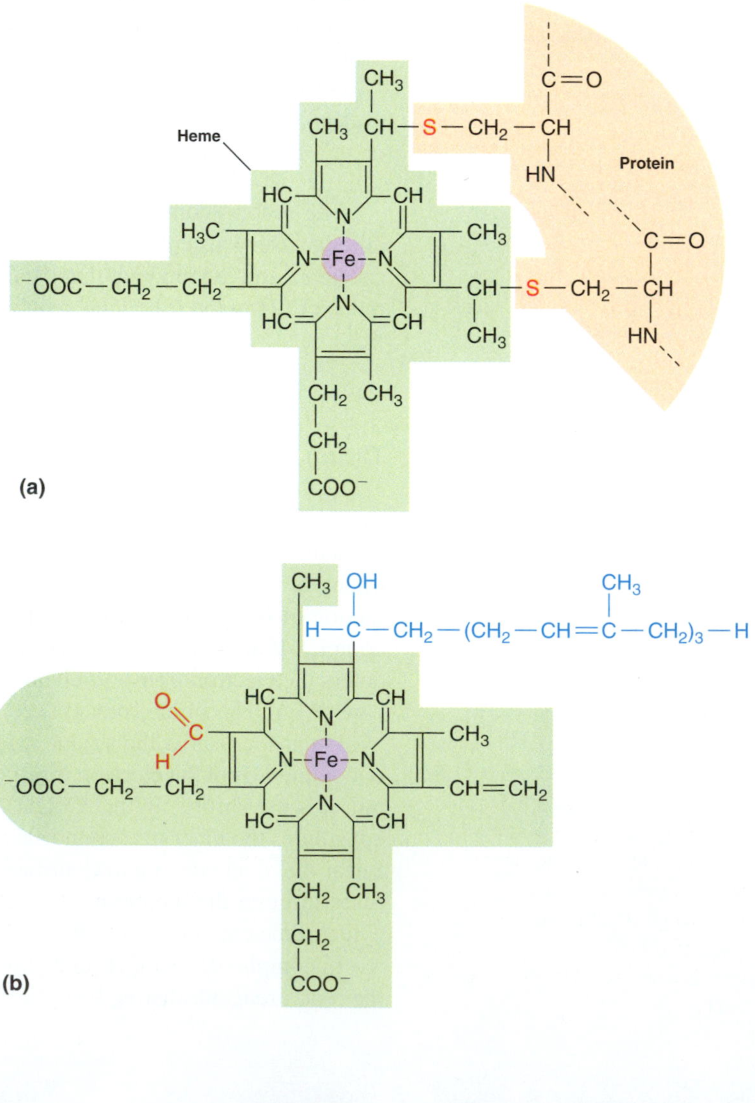

FIGURE 11.12

Heme groups found in cytochrome proteins of the electron transport chain. (Source Adapted from C.K. Mathews, K.E. von Holde, and K.G. Ahern, *Biochemistry,* 3rd ed. (San Francisco, CA: Pearson Benjamin Cummings) p. 532. Copyright 2000. Reprinted by Permission of Pearson Education.)

TABLE 11.5   **Summary of Gibbs Energy Changes for the Successive Stages of Electron Transport**

Reaction	EMF ($\Delta E°'$)	$\Delta G°'$ (kJ mol^{-1})
$CoQ + H^+ + NADH \rightarrow CoQH_2 + NAD^+$	0.42	-81.0
$CoQH_2 + 2Fe^{3+}$ (Cyt c_1) $\rightarrow CoQ + 2H^+ + 2Fe^{2+}$(Cyt c_1)	0.12	-23.0
$2Fe^{2+}$ (Cyt c_1) $+ 2Fe^{3+}$ (Cyt c) $\rightarrow 2Fe^{3+}$ (Cyt c_1) $+ 2Fe^{2+}$ (Cyt c)	0.03	-6.0
$2Fe^{2+}$ (Cyt c) $+ 2Fe^{3+}$ (Cyt $a + a_3$) $\rightarrow 2Fe^{3+}$ (Cyt c) $+ 2Fe^{2+}$(Cyt $a + a_3$)	0.04	-8.0
$2Fe^{2+}$ (Cyt $a + a_3$) $+ \frac{1}{2}O_2 + 2H^+ \rightarrow 2Fe^{3+}$ (Cyt $a + a_3$) $+ H_2O$	0.53	-102
Net Reaction		
$NADH + \frac{1}{2}O_2 + H^+ \rightarrow H_2O + NAD^+$	1.14	$-220.$

Upon arriving at complex IV, cytochrome c with its Fe^{2+}-containing heme group encounters a complex of two cytochrome proteins a and a_3, which together are called cytochrome oxidase. The net reaction for the oxidation of cytochrome c by cytochrome oxidase is

$$2Fe^{2+} \text{ (Cyt c)} + 2Fe^{3+} \text{ (Cyt } a + a_3) \rightarrow 2Fe^{3+} \text{ (Cyt c)} + 2Fe^{2+} \text{ (Cyt } a + a_3) \quad (11.21)$$

The redox reaction in Equation (11.21) is weakly exergonic and proceeds with a standard free energy change of -8 kJ mol^{-1}. Once reduced, cytochrome oxidase is oxidized by oxygen with the net reaction

$$2Fe^{2+} \text{ (Cyt } a + a_3) + \frac{1}{2}O_2 + 2H^+ \rightarrow H_2O + 2Fe^{+3} \text{ (Cyt } a + a_3) \quad (11.22)$$

The reaction in Equation (11.22) is highly exergonic and proceeds under standard conditions with a Gibbs energy change of -102 kJ mol^{-1}. The phosphorylation of ADP is associated with this step.

11.5 Oxidative Phosphorylation

In Section 11.4, we determined standard Gibbs energy changes for the reactions of electron transport based on standard reduction potentials for the components of the chain. We made the qualitative observation that three of the net oxidation reactions associated with electron chain complexes—the oxidation of NADH by CoQ in complex I, the oxidation of $CoQH_2$ and reduction of cytochrome c_1 in complex III, and the reduction of oxygen in complex IV by cytochrome $a + a_3$—each release sufficient Gibbs energy under standard conditions to phosphorylate ADP. In fact, for each electron pair initially introduced into the transport chain via a single NADH, three ATP molecules are formed. We have not, however, explained just how electron transport through complexes I, III, and IV is coupled to ATP synthesis.

Several theories have been proposed to explain how the redox reactions of the electron transport chain are coupled to phosphorylation of ADP. The theory of oxidative phosphorylation most widely accepted today is called chemiosmotic theory. Proposed by Peter Mitchell in 1961, chemiosmotic theory states that electron transport through complexes I, III, and IV is coupled to the production of a proton gradient across the inner mitochondrial membrane, and that the discharge of this proton gradient is coupled to ATP synthesis (see Figure 11.13). Chemiosmotic theory is supported by a number of experimental observations:

- The chemiosmotic hypothesis requires an intact vesicular membrane for oxidative phosphorylation. In fact, the inner mitochondrial membrane must be intact for oxidative phosphorylation to occur. Compartmentalization appears to be critical for oxidative phosphorylation to proceed.

- The pH outside the inner mitochondrial membrane is observed to decrease during electron transport. This observation is explained by chemiosmotic theory, which

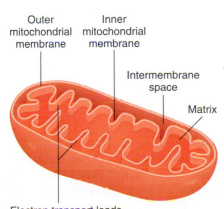

Outer mitochondrial membrane   Inner mitochondrial membrane

Intermembrane space

Matrix

Electron transport leads to proton pumping across the inner mitochondrial membrane

FIGURE 11.13

The chemiosmotic hypothesis: the Gibbs energy released by the redox reactions of the electron transport chain is coupled to the production of a proton gradient across the inner mitochondrial membrane. The Gibbs energy released by the discharge of this gradient is used by ATP synthetase to produce ATP.

hypothesizes the pumping of protons from the matrix across the inner membrane into the intermembrane space during electron transport

- Chemicals that increase the permeability of the inner mitochondrial membrane to protons do not suppress electron transport, but ATP production is suppressed. This observation is explained by the chemiosmotic hypothesis because increased proton permeability will discharge the gradient across the inner mitochondrial membrane that is supposed to generate ATP production.

- Artificially decreasing the pH of the intermembrane space increases ATP production. This observation is explained by the chemiosmotic hypothesis, which hypothesizes that ATP production is coupled to the discharge of the pH gradient across the inner mitochondrial membrane, which in turn is created when protons are pumped into the inner mitochondrial space during electron transport.

Fundamental to chemiosmotic theory is the hypothesis that complexes I, III, and IV pump protons from the matrix into the intermembrane space during the course of performing the chain of redox reactions summarized in Table 11.5. In fact the pH of the intermembrane space in the mitochondrion, that is, outside the inner membrane, is about 0.75 unit lower than the pH inside the inner membrane, that is, in the matrix: $\Delta pH = pH_{out} - pH_{in} = -0.75$. The measured difference between electrostatic potential outside the inner mitochondrial membrane, and the potential inside the inner membrane is $\Delta\phi = \phi_{out} - \phi_{in} = 0.168$ V.

EXAMPLE PROBLEM 11.6

Calculate the Gibbs energy change for transport of 1 mol of protons from the matrix of the mitochondrion to the intermembrane space. Assume the pH of the intermembrane space is 0.750 unit lower than the pH of the matrix. Assume also that the potential differs outside versus inside the inner membrane by $\Delta\phi = \phi_{out} - \phi_{in} = 0.168$ V. Assume $T = 310.$ K.

Solution

$$\Delta G_{H^+} = \mu_{H^+}^{out} - \mu_{H^+}^{in} = \mu_{H^+}^{out\circ} - \mu_{H^+}^{in\circ} + RT \ln\left(\frac{c_{H^+}^{out}}{c_{H^+}^{in}}\right) + zF\Delta\phi$$

$$= 2.30\, RT \log\left(\frac{c_{H^+}^{out}}{c_{H^+}^{in}}\right) + zF\Delta\phi = RT\,[pH^{in} - pH^{out}] + F\Delta\phi$$

$$= (2.30)(8.314 \text{ J mol}^{-1}\text{ K}^{-1})(310.\text{ K})(0.750) + (96,485 \text{ C mol}^{-1})(0.168 \text{ V})$$

$$= 4440 \text{ J mol}^{-1} + 16,200 \text{ J mol}^{-1} = 20.6 \text{ kJ mol}^{-1}$$

From Example Problem 11.6, the transport of protons from the matrix across the inner mitochondrial membrane and into the inner membrane space is highly endergonic and must be coupled to the redox reaction chain in complexes I, III, and IV. To understand how this occurs, consider Figure 11.10, which juxtaposes the path of electron transport and the path of proton transport through and across the inner mitochondrial membrane.

Chemiosmotic theory proposes a "redox loop" mechanism to explain the coupling of electron transport to proton pumping. As Figure 11.10 shows, the protein components of complexes I, III, and IV are oriented in the membrane such that an electron is transferred into each complex from a reducing agent (e.g., NADH, CoQ) at the matrix surface of the membrane. At the same time, a proton is required for this redox reaction. For example, at the matrix side of complex I we have

$$NADH + H^+ + FMN \rightarrow NAD^+ + FMNH_2 \qquad (11.23)$$

where the proton on the left-hand side originates from the matrix as shown in Figure 11.10. However, the oxidation of $FMNH_2$ by Fe^{3+} in the iron-sulfur center of ferredoxin,

$$FMNH_2 + 2Fe^{3+} \text{ (ferredoxin)} \rightarrow FMN + 2Fe^{2+} \text{ (ferredoxin)} + 2H^+ \quad (11.24)$$

occurs on the opposite surface of the membrane, and the two protons released by the oxidation of $FMNH_2$ enter the intermembrane space also, as shown in Figure 11.10. The net reaction for complex I, including proton transport is:

$$NADH + 2Fe^{3+} \text{ (ferredoxin)} \rightarrow NAD^+ + 2Fe^{2+} \text{ (ferredoxin)} + H^+ \quad (11.25)$$

where the proton appearing on the right of Equation (11.25) is released into the intermembrane space.

The redox loop mechanism also accounts for how $CoQH_2$, a reducing agent that yields two electrons on oxidation, actually transfers its two electrons to two separate cytochrome c_1 proteins, each of which can only receive a single electron when its heme Fe^{3+} is reduced to Fe^{2+}. A brief explanation is that when $CoQH_2$ emerges from complex I it is near the surface of the inner mitochondrial membrane adjacent to the matrix, also called the N side. $CoQH_2$ subsequently diffuses through the membrane to complex III at the opposite bilayer surface adjacent to the intermembrane space, called the P side. As shown in Figure 11.11, coenzyme Q can actually exist in three forms: an oxidized form CoQ (i.e., quinone); a fully reduced form bearing two electrons, which we call $CoQH_2$ (i.e., ubiquinol); and a stable anionic intermediate bearing a single electron called CoQ^- (i.e., semiquinone). When $CoQH_2$ arrives at complex III at the P side of the inner mitochondrial membrane, it donates a single electron through an intermediate iron-sulfur (FeS) protein to a single cytochrome c_1, reducing the Fe^{3+} in the cytochrome c_1 heme group to Fe^{2+} and forming CoQ^- as a stable intermediate. The net reaction for this step is

$$\text{Step 1: } CoQH_2 + Fe^{3+} \text{ (Cyt } c_1) \rightarrow CoQ^- + Fe^{2+} \text{ (Cyt } c_1) + 2H_P^+ \quad (11.26)$$

where the subscript P on the proton in Equation (11.26) indicates that because this redox reaction occurs near the bilayer surface adjacent to the intermembrane space of the mitochondrion, that is, on the P side, two protons are ejected into the intermembrane space. CoQ^- itself actually undergoes a sequence of redox reactions with a set of cytochome b proteins, but the net reaction of Equation (11.26) suffices. In a second step involving semiquinone CoQ^- and a second molecule of $CoQH_2$, an electron is again transferred to a second cytochrome c_1 molecule, forming CoQ and giving off two more protons into the intermembrane space. But now two protons from the matrix, designated H_N^+, enter the membrane and regenerate $CoQH_2$ from CoQ^-. The net reaction for step 2 is

Step 2:
$$CoQH_2 + CoQ^- + 2H_N^+ + Fe^{3+} \text{ (Cyt } c_1) \rightarrow CoQH_2 + CoQ + 2H_P^+ + Fe^{2+} \text{ (cyt } c_1) + 2H_P^+$$
$$(11.27)$$

where H_P^+ designates a proton emerging on the P side of the inner mitochnodrial membrane. Equations (11.26) and (11.27) constitute a model proposed by Peter Mitchell for the redox reactions of complex III called the **Q cycle.** Combining Equations (11.26) and (11.27), we obtain a net equation in which we keep track of the locations of the protons that enter the inner mitochondrial membrane from the matrix H_N^+, and emerge from the membrane into the intermembrane space H_P^+:

$$CoQH_2 + 2H_N^+ + 2Fe^{3+} \text{ (Cyt } c_1) \rightarrow CoQ + 2Fe^{2+} \text{ (Cyt } c_1) + 4H_P^+ \quad (11.28)$$

Equation (11.28) means that, coupled to the transfer of an electron pair from $CoQH_2$ to two cytochrome c_1 molecules via the redox reactions of the Q cycle, is a net transfer of two protons from the matrix of the mitochondrion to the intermembrane space.

A mechanism must exist in the mitochondrion that couples the pH gradient that electron transport creates across the inner membrane with the phosphorylation of ADP. As shown in Figure 11.10, covering the inner mitochondrial membrane are lollipop-shaped protein complexes called **ATP synthase,** which are the cell's machinery for coupling the discharge of the pH gradient produced by electron transport with ATP production. The stalk of the lollipop is a multiprotein complex called F_0. F_0 consists of at least 10 to 14 protein subunits, which span the inner mitochondrial membrane and constitute the channel through which protons pass from the intermembrane space back into the matrix. The

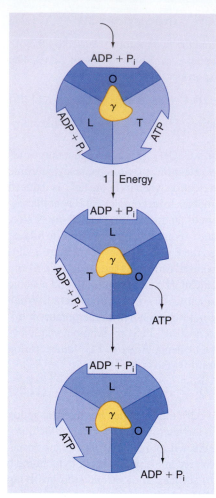

FIGURE 11.14

Conformational change mechanism for ATP synthesis by F_1. The γ subunit contacts the three β subunits of F_1 at the intersection of the three lines. A 120-degree counterclockwise rotation of the γ subunit interconverts the conformations of the three $\alpha\beta$ subunits. The $\alpha\beta$ subunit in the T (tight) conformations tightly binds newly synthesized ATP, but will not release it. The $\alpha\beta$ subunit in the L (loose) conformation contains ADP and inorganic phosphate (P_i). When the γ subunit rotates, the T conformation is converted into the O (open) conformation. In this form ATP is released. The $\alpha\beta$ subunit in the L conformation is also converted to T where ADP is now phosphorylated and the cycle repeats itself. (Source Adapted from C.K. Mathews, K.E. von Holde, and K.G. Ahern, *Biochemistry,* 3rd ed. (San Francisco, CA: Pearson Benjamin Cummings) p. 546. Copyright 2000. Reprinted by Permission of Pearson Education.)

head of the lollipop is called F_1 and consists of three protein dimers called $\alpha\beta$ subunits, which are symmetrically arranged around a fourth protein assembly called the γ subunit (see schematic in Figure 11.14). The $\alpha\beta$ subunits of F_1 contain binding sites for ADP and ATP, whereas the γ subunit contains a long helix that passes through the $\alpha\beta$ subunits of F_1.

The structure of F_1 is reminiscent of a wheel connected to a camshaft, and all the more fascinating is the fact that the "wheel," composed of the three $\alpha\beta$ subunits, actually rotates with the γ protein "camshaft." As shown in Figure 11.14, each $\alpha\beta$ subunit can exist in one of three possible conformations. In the T or tight conformation the subunit is catalytically active and will phosphorylate ADP, but the ATP product will not be released. The L or loose conformation binds ADP and P_i tightly but is not catalytically active. Finally, the O or open conformation releases bound ATP.

At any given time, one $\alpha\beta$ subunit is the T conformation, one is O, and one is L. According to the binding change mechanism proposed by Peter Boyer, in response to the translocation of protons through the channel of F_0, the γ subunit rotates. It is believed that rotation of the γ subunit induces conformational interconversion of the three $\alpha\beta$ subunits. Therefore, ATP bound to an $\alpha\beta$ subunit in the T conformation will be released when a 120-degree rotation of the γ subunit induces a conformational change from the T state to the O state, as shown in Figure 11.14.

The response of ATP synthase to proton flux therefore amounts to a biomolecular motor, where F_0, acting as a turbine powered by the transmembrane proton flux, drives the rotation of the F_1 complex. The transmembrane proton gradient that gives rise to the flux originates, as we have shown, from electron transport, thus enabling us to relate electron transport to ATP synthesis. For every four electrons that enter the electron transport chain, a single oxygen molecule is reduced to two water molecules. In the process eight protons are abstracted from the matrix of the mitochondrion, of which four are used to produce water and four are pumped into the intermembrane space of the mitochondrion, thus producing the observed pH gradient. According to Example Problem 11.6, the translocation of 4 mol of protons back into the matrix would release 4×20, 640 J = 82,560 J which is more than twice the energy required to phosphorylate a mole of ADP under standard conditions.

11.6 Overview of Binding Equilibria

Molecules often physically bind to one another or to the surfaces of metals or inorganic crystals. In biological systems, molecular binding serves a vast array of purposes: substrate molecules bind to enzymes prior to conversion to product, inhibitory molecules bind to enzymes to suppress catalytic action, small ligands such as oxygen bind to transport proteins, and proteins bind to nucleic acids in the course of gene expression and replication. Although in this chapter we cover the thermodynamics of binding equilibria with an emphasis on biological systems, the formulations in this have nonbiological applications, such as chemisorption and physisorption of molecules from the gas phase onto solid surfaces. We begin with a basic description of binding equilibria, followed by descriptions of several models for binding of ligands to biomolecular targets.

We assume that a small molecule, which we call a **ligand** and designate L, exists in solution in free form and is bound to a macromolecular target that we assume is a biological polymer and designate P. If the biological polymer contains N binding sites, at equilibrium we will find in solution a mixture of free ligand at concentration c_L, free polymer at concentration c_P, and in general all forms of partially occupied polymers ranging from complexes PL composed of polymers with one site occupied at concentration c_{PL}, to complexes composed of polymers with all N sites occupied by ligands, that is, PL_N at concentration c_{PL_N}. If the binding equilibrium were simply $P + NL \underset{}{\overset{K}{\rightleftharpoons}} PL_N$, we could quantify the equilibrium with the equilibrium constant $K = \frac{a_{PL_N}}{a_P a_L^N} \approx \frac{c_{PL_N}}{c_P c_L^N}$, where we assume dilute enough conditions so that all activity coefficients equal unity. In general, however, this expression is not applicable to the equilibrium we described earlier because more than three species exist at equilibrium. How then do we quantify this binding equilibrium?

In an equilibrium mixture of biopolymers P and ligands L, on average the number of sites occupied per biopolymer $\bar{\nu}$ is given by

$$\bar{\nu} = \frac{\text{number of bound ligands}}{\text{number of polymers}} = \frac{c_{PL} + 2c_{PL_2} + 3c_{PL_3} + \cdots + Nc_{PL_N}}{c_P + c_{PL} + c_{PL_2} + c_{PL_3} + \cdots + c_{PL_N}} \quad (11.29)$$

Note in the numerator of Equation (11.29) that each concentration is weighted by the number of ligands involved in the complex, whereas in the denominator all terms are equally weighted because each type of complex contains a single polymer.

The simplest case is a single ligand binding site where the equilibrium can be written as

$$P + L \underset{}{\overset{K}{\rightleftharpoons}} PL \quad (11.30)$$

and where the equilibrium constant $K = c_{PL}/c_P c_L$. For a single binding site Equation (11.29) simplifies to

$$\bar{\nu} = \frac{c_{PL}}{c_P + c_{PL}} \quad (11.31)$$

When Equation (11.31) is combined with $K = c_{PL}/c_P c_L$, we obtain

$$\bar{\nu} = \frac{c_{PL}}{c_P + c_{PL}} = \frac{Kc_P c_L}{c_P + Kc_P c_L} = \frac{Kc_L}{1 + Kc_L} \quad (11.32)$$

In practical terms, the three concentrations in $K = c_{PL}/c_P c_L$ can be measured and $\bar{\nu}$ quantified by performing a dialysis experiment. Consider Figure 11.15. Into a large volume of solution is introduced a bag composed of a semipermeable membrane. A biopolymer P and ligand L are in solution in the bag and a binding equilibrium of the type shown in Equation (11.30) occurs. Although the membrane that composes the bag is permeable to the small molecular ligand L, it is impermeable to the polymer P and to the complex PL. When the system reaches equilibrium, the chemical potential of free L outside the bag equals the chemical potential of free L inside the bag, and therefore the chemical potentials are equal:

$$\mu_L^{out} = \mu_L^{in} \quad (11.33)$$

Using the assumptions stated in deriving Equation (11.32), it is simple to show from Equation (11.33) that at equilibrium the concentrations of free ligand inside and outside the dialysis bag are equal:

$$\mu_L^{\circ,out} + RT \ln c_L^{out}(free) = \mu_L^{\circ,in} + RT \ln c_L^{in}(free) \quad (11.34)$$

Because $\mu_L^{\circ,out} = \mu_L^{\circ,in}$ it follows from Equation (11.34) that at equilibrium

$$c_L^{out}(free) = c_L^{in}(free) \quad (11.35)$$

The amount of bound ligand in the dialysis bag is easily calculated

$$c_L(bound) = c_L^{in}(total) - c_L^{in}(free) = c_L^{in}(total) - c_L^{out}(free) \quad (11.36)$$

where c_t is the amount of ligand introduced initially into the bag. Because the complex is 1:1 stoichiometry the amount of bound ligand equals the amount of bound polymer:

$$c_L(bound) = c_P(bound) \quad (11.37)$$

Finally, the amount of free polymer is calculated as the difference between the amount of polymer introduced initially into the bag $c_P(total)$ and the amount of bound polymer:

$$c_P(free) = c_P(total) - c_P(bound) = c_P(total) - c_L(bound) \quad (11.38)$$

Equations (11.35) through (11.38) mean that the three concentrations required to determine the binding constant K can be determined by measuring the concentration of free ligand from a dialysis experiment. Then the binding equilibrium constant is

$$K = \frac{c_{PL}}{c_P(free)\,c_L(free)} = \frac{c_L(bound)}{(c_P(total) - c_L(bound))\,c_L(free)} \quad (11.39)$$

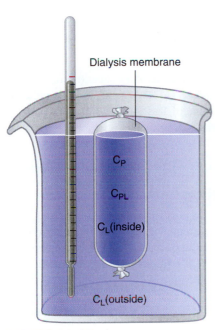

Dialysis membrane

C_P

C_{PL}

$C_L(inside)$

$C_L(outside)$

FIGURE 11.15
Schematic of an equilibrium dialysis experiment.

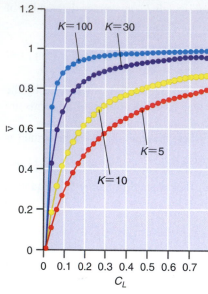

FIGURE 11.16
The saturation parameter $\bar{\nu}$ plotted as a function of free ligand concentration C_L using Equation (11.33) for binding constant values $K = 5$, 10, 30, and 100.

The quantities measured in a dialysis experiment can also be related to $\bar{\nu}$, the average number of bound ligands per polymer, also called the saturation parameter:

$$\bar{\nu} = \frac{c_L \, (bound)}{c_P \, (total)} \tag{11.40}$$

Combining Equations (11.39) and (11.40), we obtain:

$$K = \frac{c_L \, (bound)}{(c_P \, (total) - c_L \, (bound))c_L} = \frac{\bar{\nu}c_P \, (total)}{(c_P \, (total) - \bar{\nu}c_P \, (total))c_L} = \frac{\bar{\nu}}{(1 - \bar{\nu})c_L} \tag{11.41}$$

which can be rearranged to recover Equation (11.32):

$$\bar{\nu} = \frac{Kc_L}{1 + Kc_L} \tag{11.42}$$

where from now on we drop the "free" designation on the ligand concentration.

Figure 11.16 shows a plot of $\bar{\nu}$ as given by Equation (11.33) versus free ligand concentration c_L. Note that $\bar{\nu}$ approaches a limiting value of 1 as the free ligand concentration approaches infinity. This limiting value corresponds to a state in which all polymers are involved in 1:1 complexes with ligand.

11.7 Independent Site Binding

The term **independent binding** refers to binding to a given site or sites that has no impact on simultaneous or subsequent binding to other unoccupied sites. We can therefore focus on binding of a ligand to each individual site. If a biopolymer has N ligand binding sites, the average number of ligands bound to site j per biopolymer is:

$$\bar{\nu}_j = \frac{K_j c_L}{1 + K_j c_L} \tag{11.43}$$

where K_j is the equilibrium constant governing ligand binding to site j. The average number of sites bound per biopolymer is then

$$\bar{\nu} = \sum_{j=1}^{N} \bar{\nu}_j = \sum_{j=1}^{N} \frac{K_j c_L}{1 + K_j c_L} \tag{11.44}$$

If all N sites are equivalent in the sense that all equilibrium constants are equal, Equation (11.44) reduces to

$$\bar{\nu} = \sum_{j=1}^{N} \frac{Kc_L}{1 + Kc_L} = \frac{NKc_L}{1 + Kc_L} \tag{11.45}$$

A plot of Equation (11.45) will resemble the data shown in Figure 11.16 except at large c_L the binding curves will approach N instead of 1. If a polymer is thought to possess N equivalent ligand binding sites, the affinity constants may be obtained by simulation of a plot of experimental data according to Equation (11.45). It is convenient, however, to obtain N and K from a linear plot. To obtain this result, start with Equation (11.45) and solve for $\bar{\nu}/c_L$. In doing so, we obtain

$$\frac{\bar{\nu}}{c_L} = KN - K\bar{\nu} \tag{11.46}$$

Equation (11.46) is called the **Scatchard equation.** A **Scatchard plot** of the ligand binding data for $K = 100$ and $N = 4$ is shown in Figure 11.17. A Scatchard plot permits extraction of K and N from the slope and the y or x intercept. For example, from the Scatchard equation [Equation (11.46)] the slope of a plot of $\bar{\nu}/c_L$ versus $\bar{\nu}$ is $-K$, whereas the y intercept is the product of K and N. On the other hand, if data are available for high ligand concentrations, N may be extracted directly from the x intercept as shown in Figure 11.17.

Scatchard plots performed according to Equation (11.46) will be linear if binding is independent and if the binding affinities to all N sites are equivalent. Scatchard plots can

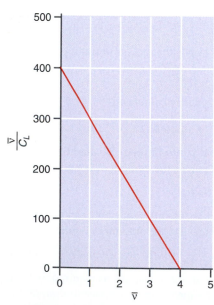

FIGURE 11.17
Scatchard plot of independent binding for $K = 100$ and $N = 4$. Note that the y intercept is $K \times N = 400$, the slope is $-K = -100$, and the x intercept is 4.

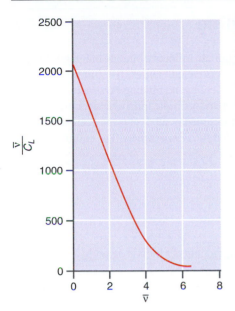

FIGURE 11.18

A biphasic Scatchard plot based on Equation (11.47) for which $N_1 = N_2 = 4$, $K_1 = 10$, and $K_2 = 500$.

be nonlinear if either condition is not fulfilled, that is, if independent binding occurs to nonequivalent sites or if multiple site binding displays cooperativity. Cooperative binding will be treated in the next section. Here we consider the case in which independent binding occurs to in-equivalent sites. Assume a ligand binds to two types of sites. The sites differ by way of their affinity constants K_1 and K_2, and they may also differ in their numbers N_1 and N_2. The total average number of sites bound is the sum of the average number of bound sites for each type of site:

$$\bar{\nu} = \bar{\nu}_1 + \bar{\nu}_2 = \frac{N_1 K_1 c_L}{1 + K_1 c_L} + \frac{N_2 K_2 c_L}{1 + K_2 c_L} \tag{11.47}$$

The appearance of the Scatchard plot will now depend on the relative values of K_1 versus K_2 and N_1 versus N_2. In general the y intercept will be $N_1 K_1 + N_2 K_2$ and the x intercept is $N_1 + N_2$. Figure 11.18 shows a biphasic Scatchard plot that results from $N_1 = N_2 = 4$ and $K_1 = 10$ and $K_2 = 500$.

As shown in Figure 11.18, if $K_2 \gg K_1$ it may be difficult to discern the smaller equilibrium constant at low ligand concentrations (i.e., small values of $\bar{\nu}$) where binding to sites with the larger equilibrium constant dominates. At larger values of $\bar{\nu}$ after the sites with large binding affinities saturate, the slope of the Scatchard plot diminishes and the curve intersects the x axis at $N_1 + N_2$.

EXAMPLE PROBLEM 11.7

The following data show the binding of NADH to beef heart lactate dehydrogenase, taken from the data of S. Anderson and G. Weber, *Biochemistry* 4 (1965), 1948. The NADH binding sites on beef heart LDH act independently and bind NADH with equal affinities.

$\bar{\nu}$	3.80	3.50	3.00	2.50	2.00	1.50	1.00	0.50
$C_{NADH}(10^{-7}M)$	77.0	17.5	9.09	5.81	3.51	2.21	1.21	0.50

From a Scatchard plot determine the number of NADH binding sites on beef heart LDH and the binding constant K.

Solution

The Scatchard equation is $\frac{\bar{\nu}}{c_{NADH}} = K \cdot (N - \bar{\nu})$ where K is the binding constant and N is the number of binding sites. Therefore, the slope of a plot of $\bar{\nu}/c_{NADH}$ as a function of $\bar{\nu}$ is $-K$, the y intercept is KN, and the x intercept is N.

The slope is $-K = -0.3 \times 10^7$, approximately (best fit by linear regression is -2.7×10^6). Therefore, $K = 2.7 \times 10^6$. The y intercept is about 1.2×10^7. Therefore, $N = 4$. The latter fact can also be discerned from the x intercept.

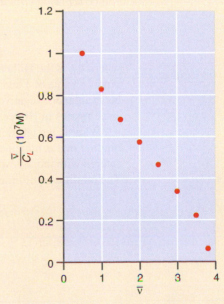

11.8 Cooperative Site Binding

The binding of a ligand to a site on a target molecule can influence binding of ligands to other unoccupied sites on the same target. This situation is called **cooperative binding.** Cooperative binding plays an important role in biology, where ligand binding exerts a regulatory effect on protein function, a property called **allosterism.** A simple form of cooperative binding occurs when either all N sites on a protein are unoccupied or all N sites are occupied, and no other situations are possible. This situation is called *full cooperativity.* The equilibrium is

$$P + NL \rightleftharpoons PL_N \tag{11.48}$$

and the equilibrium binding constant is

$$K = \frac{c_{PL_N}}{c_P c_L^N} \tag{11.49}$$

Now c_{PL_N}/c_P is the ratio of occupied to unoccupied sites, which is given by

$$\frac{c_{PL_N}}{c_P} = \frac{\bar{\nu}}{N - \bar{\nu}} \tag{11.50}$$

Combining Equations (11.49) and (11.50), we obtain

$$K = \frac{\bar{\nu}}{c_L^N (N - \bar{\nu})} \tag{11.51}$$

which can be rearranged to solve for the saturation parameter:

$$\theta = \frac{\bar{\nu}}{N} = \frac{c_L^N K}{1 + c_L^N K} \tag{11.52}$$

where θ is the fraction of sites bound per polymer. Alternatively, take the logarithm of both sides of Equation (11.51) and rearrange to obtain

$$\ln\left(\frac{\theta}{1 - \theta}\right) = \ln\left(\frac{\bar{\nu}}{N - \bar{\nu}}\right) = \ln K + N \ln c_L \tag{11.53}$$

Equation (11.53) is called the **Hill equation,** and assuming full binding cooperativity the Hill equation states that a plot of $\ln(\theta/1 - \theta)$ as a function of $\ln(c_L)$ yields a straight line with a slope of N.

In practice, **Hill plots** are not always linear and do not display a slope equal to the number of binding sites, reflecting a degree of cooperativity that is less extreme than the fully cooperative binding scenario discussed earlier. More realistically, Equation (11.52) can be replaced with the empirical equation

$$\theta = \frac{c_L^{\alpha_H} K}{1 + c_L^{\alpha_H} K} \tag{11.54}$$

where α_H is the Hill coefficient. The Hill coefficient can vary between the independent binding value of 1 and the full cooperative limit of N. The Hill coefficient need not even be an integer. Negative cooperativity, that is, reduction of binding affinity to remaining sites, results in Hill coefficients of less than 1.

What is the physical origin of a nonlinear Hill plot and how can we quantify ligand binding that is not independent but less than fully cooperative? This question can be answered with the following simple model. Assume a protein has four ligand binding sites. The binding equilibria are

$$L + P \xrightarrow{K_1} PL$$

$$L + PL \xrightarrow{K_2} PL_2$$

$$L + PL_2 \xrightarrow{K_3} PL_3$$

$$L + PL_3 \xrightarrow{K_3} PL_4 \tag{11.55}$$

For this series of equilibria, we define as usual the binding equilibrium constants:

$$K_j = \frac{c_{PL_j}}{c_{PL_{j-1}} c_L} \qquad (11.56)$$

The binding equilibrium constants given in Equation (11.56) will be unequal for a number of reasons. First, even if binding to all four sites is fully independent and the sites equivalent, the equilibrium constants K_j will be unequal because ligands have more choices of binding sites on unoccupied polymers than on partly occupied polymers. Second, occupation of binding sites by ligands may influence subsequent binding to unoccupied sites through some structural interaction.

To distinguish the first effect, which is statistical in nature, from actual interactions between binding sites of the type that result in cooperative binding, we define microscopic equilibrium constants. For example, we define

$$k_1 = \frac{c_{PL'}}{c_P c_L} \qquad (11.57)$$

a microscopic constant that refers to a binding equilibrium involving binding of a ligand to a particular site on a previously unoccupied polymer. Designate as PL' the ligand–polymer complex formed when a ligand binds to a particular site on a previously unoccupied polymer. The concentration of PL' is $c_{PL'}$. If there are four equivalent binding sites then $N = 4$ and $c_{PL} = Nc_{PL'} = 4c_{PL'}$, where we assume that concentrations of all singly occupied polymers are equal. Putting this fact into Equation (11.57), we obtain:

$$k_1 = \frac{c_{PL'}}{c_P c_L} = \frac{c_{PL}}{4 c_P c_L} = \frac{K_1}{4} \qquad (11.58)$$

Equation (11.58) gives the relationship between the microscopic binding equilibrium constant k_1 and the macroscopic or apparent binding equilibrium constant K_1. Applying similar arguments to the other equilibria in Equation (11.55), we obtain:

$$k_1 = \frac{K_1}{4}; \quad k_2 = \frac{2}{3} K_2; \quad k_3 = \frac{3}{2} K_3; \quad k_4 = 4K_4 \qquad (11.59)$$

The usefulness of Equation (11.59) in describing cooperative binding becomes clear when we realize that if binding to any given site is independent of events at the other three sites, then $k_1 = k_2 = k_3 = k_4 = k$. If we take this fact and substitute it into Equation (11.59) and then combine the resulting macroscopic constants with the concentrations given in Equation (11.29), we obtain for the average number of sites occupied per polymer

$$\bar{\nu} = \frac{c_{PL} + 2c_{PL_2} + 3c_{PL_3} + 4c_{PL_4}}{c_P + c_{PL} + c_{PL_2} + c_{PL_3} + c_{PL_4}} = \frac{4kc_L}{1 + kc_L} \qquad (11.60)$$

which is identical to Equation (11.45) for independent binding to four equivalent sites. Note that even in the independent binding limit, where the microscopic binding constants are equal, for purely statistical reasons the macroscopic or apparent binding constants are unequal: $K_1 > K_2 > K_3 > K_4$.

Suppose we now assume that binding to the four sites is not independent so that the microscopic binding constants are unequal. We assume the simple model:

$$k_1 = \alpha^3 k; \quad k_2 = \alpha^2 k; \quad k_3 = \alpha k; \quad k_4 = k \qquad (11.61)$$

If $\alpha < 1$, the site binding affinity increases as a function of the number of filled sites, a situation called *positive cooperativity*. The situation $\alpha > 1$ is called *negative cooperativity*. Combining the relationships in Equation (11.61) with Equations (11.59), (11.56), and (11.29) as before, we obtain

$$\bar{\nu} = 4c_L \alpha^3 k \frac{1 + 3\alpha^2 kc_L + 3\alpha^3 k^2 c_L^2 + \alpha^3 k^3 c_L^3}{1 + 4\alpha^3 kc_L + 6\alpha^5 k^2 c_L^2 + 4\alpha^6 k^3 c_L^3 + \alpha^6 k^4 c_L^4} \qquad (11.62)$$

A plot of $\bar{\nu}$ versus the ligand concentration c_L using Equation (11.62) with $k = 100$ and $\alpha = 0.25$ gives the lower plot in Figure 11.19. By way of comparison a plot of $\bar{\nu}$ for $K = 100$ with independent binding (i.e., $\alpha = 1$) to four sites is also shown by the upper plot.

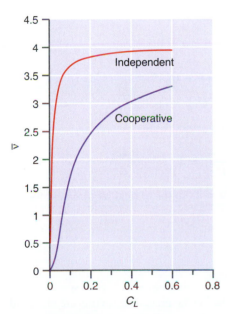

FIGURE 11.19

Comparison of independent and cooperative binding to four sites. For independent binding $k = 100$. Cooperative binding is characterized by four microscopic binding constants of the form given in Equation (11.62) for which $\alpha = 0.25$ and $K = 100$.

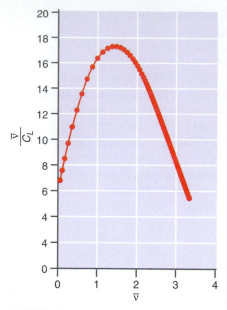

FIGURE 11.20

Scatchard plot of cooperative binding to four sites using the saturation parameter in Equation (11.62). Cooperative binding is characterized by four microscopic binding constants of the form given in Equation (11.61) for which $\alpha = 0.25$ and $k = 100$.

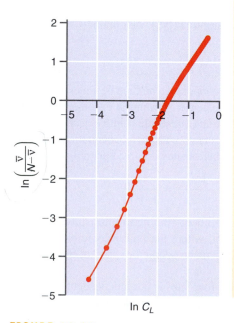

FIGURE 11.21

Hill plot of cooperative ligand binding to four sites. This Hill plot is based on a model described by Equation (11.62) and assuming $\alpha = 0.25$ and $K = 100$.

Comparing the plots in Figure 11.19, positive cooperativity produces a characteristic "sigmoidal" pattern in the isotherm plot that is particularly pronounced in the low ligand concentration region of the plot. An even more dramatic demonstration of cooperativity is shown in the Scatchard plot of Figure 11.20 that uses the expression for $\bar{\nu}$ given in Equation (11.62) and four microscopic binding constants of the form given in Equation (11.61) with $\alpha = 0.25$ and $k = 100$. The Scatchard plot deviates most strongly from independent behavior for values of $\bar{\nu} < 2$ where binding is dominated by the weaker binding constants. For $\bar{\nu} > 2$ the Scatchard plot becomes linear and extrapolates to an x intercept of 4, which correctly identifies the number of binding sites.

As a result of positive cooperativity the Hill coefficient in Equation (11.54) varies as c_L increases. Figure 11.21 illustrates the effect of using the model of cooperative four-site binding given by Equation (11.62) with $\alpha = 0.25$ and $k = 100$ in the Hill equation [Equation (11.53)]. In the low c_L limit the slope of the Hill plot is 1, reflecting near-independent binding behavior at low ligand concentrations where occupation of binding sites is at low levels and cooperative effects are minimal because ligands predominantly bind to different, unoccupied molecules. At very high ligand concentrations, the slope of the Hill plot reflects binding of the last ligand, which similarly displays minimal cooperative effects, again because ligands are binding to different molecules. The intermediate ligand concentration region displays a Hill plot with nonlinear behavior reflecting the complicated behavior in Equation (11.62). By convention, the maximum slope of this intermediate region is the Hill coefficient and is taken as a qualitative measure of cooperativity.

EXAMPLE PROBLEM 11.8

Show that Equation (11.62) reduces to independent binding if $\alpha = 1$.

Solution

Apply $\alpha = 1$ to Equation (11.62):

$$\bar{\nu} = 4c_L\alpha^3k\frac{1 + 3\alpha^2kc_L + 3\alpha^3k^2c_L^2 + \alpha^3k^3c_L^3}{1 + 4\alpha^3kc_L + 6\alpha^5k^2c_L^2 + 4\alpha^6k^3c_L^3 + \alpha^6k^4c_L^4}$$

$$= 4c_Lk\frac{1 + 3kc_L + 3k^2c_L^2 + k^3c_L^3}{1 + 4kc_L + 6k^2c_L^2 + 4k^3c_L^3 + k^4c_L^4}$$

Factor the numerator and denominator:

$$1 + 3kc_L + 3k^2c_L^2 + k^3c_L^3 = (1 + kc_L)^3$$

$$1 + 4kc_L + 6k^2c_L^2 + 4k^3c_L^3 + k^4c_L^4 = (1 + kc_L)^4$$

Substitute these results into Equation (11.62):

$$\bar{\nu} = 4c_Lk\frac{1 + 3kc_L + 3k^2c_L^2 + k^3c_L^3}{1 + 4kc_L + 6k^2c_L^2 + 4k^3c_L^3 + k^4c_L^4} = 4c_Lk\frac{(1 + kc_L)^3}{(1 + kc_L)^4} = \frac{4c_Lk}{1 + kc_L}$$

which is equivalent to Equation (11.45) for $N = 4$. So the critical requirement for cooperativity is that $\alpha \neq 1$.

11.9 Protein Allosterism

Cooperative binding equilibria like those described in the preceding section are observed in many biological systems. Allosterism, the regulatory effect on protein function exerted by cooperative ligand binding, is vital to many biological functions. Cooperative binding in proteins is believed to originate with changes in protein conformation associated with partial occupation of ligand binding sites. Proteins in which allosteric effects originate from ligand binding sites are commonly composed of a number of structurally similar polypeptide subunits. In such proteins, ligand binding is accompanied by structural rearrangements of these subunits.

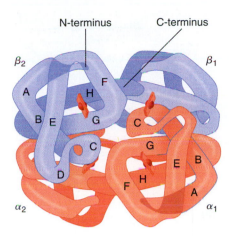

N-terminus   C-terminus

β_2 β_1

A   F   H

B E   G   C

C   G

C   E B

D   H   A

F

α_2 α_1

(a)   Hemoglobin

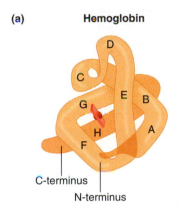

D

C

G   E B

H   A

F

C-terminus

N-terminus

(b)   Myoglobin

FIGURE 11.22

Comparison of the tertiary structures of (a) hemoglobin and (b) myoglobin. Hb is composed of two protein dimers called $\alpha_1\beta_1$ and $\alpha_2\beta_2$. Each protein subunit of the dimer contains a heme group that binds oxygen. Mb is a monomeric oxygen binding protein that has only a single heme group (see Figure 11.25). (Source Adapted from C.K. Mathews, K.E. von Holde, and K.G. Ahern, *Biochemistry,* 3rd ed. (San Francisco, CA: Pearson Benjamin Cummings) p. 214. Copyright 2000. Reprinted by Permission of Pearson Education.)

The oxygen transport protein **hemoglobin (Hb)** is perhaps the best studied of the allosteric proteins. In animals smaller than earthworms, oxygen is supplied by simple diffusion. In larger organisms, a complex oxygen transport system is required. Because the solubility of oxygen in blood plasma is too low to sustain life, the blood of virtually all large organisms (certain types of fish adapted to living in cold arctic waters are notable exceptions) must contain a transport protein. Hb is a tetrameric protein, or more exactly, a dimer of two polypeptide types of subunits called α and β. One dimer is designated $\alpha_1\beta_1$ and the second is designated $\alpha_2\beta_2$. Each monomeric unit of Hb has a single oxygen binding site for a total of four oxygen binding sites in tetrameric Hb. The function of Hb is to transport oxygen from the lungs or gills to the tissues. In the tissues, oxygen is stored in protein called **myoglobin (Mb)**. Mb is a monomeric protein with a single oxygen binding site. See Figure 11.22.

The chemical moiety in both Mb and Hb responsible for binding oxygen is called a heme group (see Figure 11.23). The heme group consists of a porphyrin ring, which itself consists of four pyrrole rings (A–D) and a centrally bound iron atom, Fe(II), which remains in the +2 oxidation state whether or not oxygen is bound. Heme is responsible for the red color of blood.

Oxygen binding to Hb displays cooperative effects. Figure 11.24 shows binding curves and Hill plots for Mb and Hb. In Figure 11.24, note that the plots for oxygen binding to Hb and Mb have the fractional binding on the y axis defined in terms of $\theta = \bar{v}/N$ and the x axis graduated in terms of the partial pressure of O_2, p_{O_2}, which according to Henry's law, is proportional to the concentration of oxygen in solution. Because Mb has only a single oxygen binding site, oxygen binding is independent so the binding curve follows Equation (11.45) with $N = 1$. This results in the hyperbolic curve shown in Figure 11.24a. Accordingly the Hill plot for Mb is linear with a slope of 1, as shown in Figure 11.24b. In contrast, the binding curve for Hb is sigmoidal and the Hill plot displays a pattern reminiscent of Figure 11.21. The Hill plot for Hb in Figure 11.24b is linear with a slope of 1 for low p_{O_2} values where oxygen molecules bind predominantly to deoxyhemoglobin molecules. Therefore, at low p_{O_2}, binding is independent because it occurs predominantly on different molecules. In the high p_{O_2} limit, oxygen molecules are binding predominantly to Hb molecules with three sites occupied. Therefore, oxygen molecules are binding to sites on different Hb molecules and the binding is again independent. In the intermediate p_{O_2} region, the curve is nonlinear with a maximum slope that yields a Hill coefficient of about 2.9 to 3.0. This is less than the $N = 4$ values, indicating less than fully cooperative binding.

Oxygen binding data to Hb shown in Figure 11.24 can be fitted to an empirical equation that models the binding of oxygen to Hb as occurring sequentially with unequal microscopic binding constants, that is,

$$k_1 \neq k_2 \neq k_3 \neq k_4 \tag{11.63}$$

Combining the microscopic binding constants in Equation (11.63) with Equations (11.59), (11.56), and (11.29) as before, we obtain

$$\theta = \frac{\bar{v}}{4} = \frac{k_1 p_{O_2} + 3k_1 k_2 (p_{O_2})^2 + 3k_1 k_2 k_3 (p_{O_2})^3 + k_1 k_2 k_3 k_4 (p_{O_2})^4}{1 + 4k_1 p_{O_2} + 6k_1 k_2 (p_{O_2})^2 + 4k_1 k_2 k_3 (p_{O_2})^3 + k_1 k_2 k_3 k_4 (p_{O_2})^4} \tag{11.64}$$

Equation (11.64) is similar in form to Equation (11.62), where the partial pressure of oxygen, p_{O_2}, has been substituted for the ligand solution concentration, c_L. Equation (11.64) is called the **Adair equation** after G. S. Adair who first used the expression in 1925. Measurements of θ versus p_{O_2} have been used in conjunction with the Adair equation to obtain values for the microscopic binding constants k_1, k_2, k_3, and k_4. Table 11.6 gives values for the k_1 through k_4 constants in the absence of NaCl and in 0.10 M NaCl.

In the absence of NaCl, there is a more than 30-fold difference in the oxygen binding affinity from k_1 to k_4. The addition of NaCl diminishes all binding affinities relative

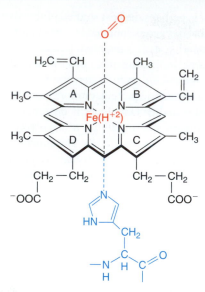

FIGURE 11.23
The heme group, which binds oxygen in Mb and Hb, is shown with a bound oxygen molecule above the plane of the porphyrin ring. The Fe(II) is additionally coordinated to five nitrogen atoms in a square pyramid arrangement. The square base of the pyramid has each corner occupied by a nitrogen from each of the four pyrrole rings (A–D). The fifth nitrogen located at the apex of the pyramid is contributed by the imidazole ring of a histidine side chain in Mb or Hb. Below the porphyrin ring is shown the side chain of a histidine amino acid.

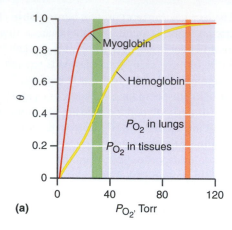

(a)

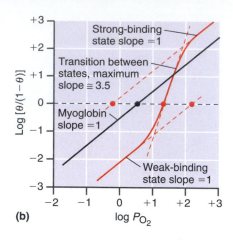

(b)

FIGURE 11.24
(a) Oxygen binding curves for myoglobin (Mb) and tetrameric hemoglobin (Hb). For Mb, p_{50} = 2.8 Torr = 3.68×10^{-3} atm at T = 273 K and N = 1. As expected the oxygen binding curve for Mb is hyperbolic, indicating independent binding. For tetrameric Hb, p_{50} = 26 Torr = 0.034 atm and the binding curve is sigmoidal, indicating cooperative binding.. The vertical green line on the left marks the pressure typically found in tissues (ca. 30 Torr), whereas the vertical orange line on the right marks the pressure in the lungs (ca. 100 Torr). (b) Hill plots for Mb and Hb. The linear Hill plot for Mb has a slope of unity, which is the independent binding limit. The Hill plot for Hb has the general form shown in Figure 11.24 for cooperative binding. The maximum slope is about 3.0–3.5. (Source Adapted from C.K. Mathews, K.E. von Holde, and K.G. Ahern, *Biochemistry,* 3rd ed. (San Francisco, CA: Pearson Benjamin Cummings) pp. 22 and 223. Copyright 2000. Reprinted by Permission of Pearson Education.)

TABLE 11.6 **Values for the Microscopic Binding Constants k_j (in atm^{-1})***

	k_1	k_2	k_3	k_4
No NaCl	86.3	125	894	3040
0.1 M NaCl	18.1	58.5	63.3	5429

*Microscopic binding constant for human Hb estimated using the Adair equation. Binding constants are obtained for solution buffered at T = 298 K and pH = 7.4.

Source: I. Tyuma, K. Imai, and K. Shimizu, *Biochemistry* 12 (1973), 1491.

to the unsalted solution except for k_4. There is a >300-fold enhancement of oxygen binding affinity in this case. The diminishing of k_1 through k_3 with the addition of salt is thought to occur from the stabilization of the structures of deoxyhemoglobin and partially oxygenated Hb by NaCl.

Although the binding of oxygen to Hb can be fitted to the Adair equation, empirical equations do not provide insight into the structural origins of positive cooperative binding in Hb. It is now realized that there is a fundamental link between protein structural flexibility and cooperativity. Because the heme groups in Hb are separated by at least 25 angstroms, direct interactions between oxygen binding sites are impossible and thus cannot serve as the basis for cooperativity. Long-range structural changes induced by oxygen binding appear to induce cooperativity, and the structural association of the αβ dimers of Hb appears to be fundamental to the cooperative nature of oxygen binding.

What is the long-range structural rearrangement that oxygen binding induces in tetrameric Hb? X-ray diffraction studies of deoxy-Hb and oxy-Hb show that the two αβ dimers that compose the Hb tetramer structurally assemble in one of two ways, as shown

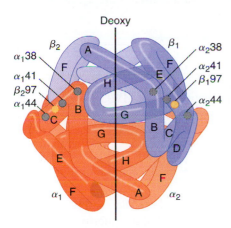

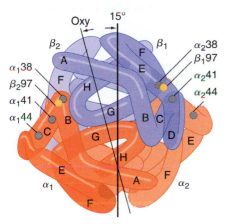

FIGURE 11.25

A low-resolution tertiary structural model of Hb showing the $\alpha_1\beta_1/\alpha_2\beta_2$ dimer interface and the locations of the heme groups. The heme groups are separated by >25Å. (b) Hb tetramer view obtained by rotating the structure by 90 degrees. The top figure shows the orientation of the dimers in deoxy-Hb; the bottom figure shows the orientation of the dimers in oxy-Hb. (Source Adapted from C.K. Mathews, K.E. von Holde, and K.G. Ahern, *Biochemistry,* 3rd ed. (San Francisco, CA: Pearson Benjamin Cummings) p. 224. Copyright 2000. Reprinted by Permission of Pearson Education.)

in Figure 11.25. In the T (tense, also called deoxy) form, the $\alpha_1\beta_1$ dimer and the $\alpha_2\beta_2$ dimer are oriented relative to the plane of dimer contact as shown in the left-hand portion of Figure 11.25. In the R (relaxed or oxy) structural form shown in the right-hand side of Figure 11.25, the $\alpha_1\beta_1$ dimer is rotated 15 degrees through the dimer–dimer plane of contact. X-ray diffraction data have also shown that as oxygen binds to the heme groups located in each subunit, there is a change of local structure from the oxygen-free or deoxy "t" to oxy or "r." The local t form has low oxygen affinity, whereas the "r" form has higher oxygen affinity. These local structural changes drive the global T→R structural rearrangement shown in Figure 11.25.

Two basic lines of thought seek to account for how structural changes that accompany oxygen binding induce allostery. The **concerted** or symmetry **model** of allosterism was introduced in 1965 by J. Monod, J. Wyman, and J.-P. Changeux. This model is also called the **MWC model.** The second approach is called the sequential or induced fit model. This approach was introduced in 1966 by D. E. Koshland, G. Nemethy, and D. Filmer and is called the **KNF model.** The relationship between these models is shown in Figure 11.26

In both the MWC and the KNF models, the α and β subunits exist in either low oxygen affinity "t" forms or high oxygen affinity "r" forms, as explained earlier. In the MWC model, oxygen binds to sites that are either in t or r states. Regardless of whether a site is occupied by oxygen or is unoccupied, the structure of the site is preserved. Moreover, the MWC model invokes a symmetry constraint which states that all sites in a tetramer, regardless of the degree of oxygen saturation, must be t or r. Therefore, the MWC model identifies two types of tetramers: the T state in which all four binding sites are in the low oxygen affinity t state, and the R state in which all four binding sites are in the high affinity r state. It is possible for a site to make a transition from a t state to an r state, but this can only occur in concert, that is, in an all-or-none fashion. For every degree of oxygen occupation, including the deoxy state, the T state of the Hb tetramer is in equilibrium with the R state. In the absence of oxygen, we represent the T state as T_0 and the R state as R_0. These two forms are in equilibrium, defined by the relationship

$$K_0 = \frac{c_{T_0}}{c_{R_0}} \tag{11.65}$$

This equilibrium favors the T form, so $K_0 \gg 1$. As oxygen fills one, two, three, and finally four sites, the equilibrium shifts toward the R form.

In contrast, the KNF model hypothesizes that oxygen binding induces a t → r subunit structural transition at specific sites. No concerted transition is required, so tetramers change from a low-affinity T_0 state through a series of partly t and partly r intermediates to a high-affinity RL_4 state, as shown in Figure 11.26. The KNF model includes pairwise interactions between occupied and unoccupied sites. These interactions increase the affinity of the unoccupied site for oxygen binding.

A simple expression for the average number of oxygen binding sites occupied per Hb tetramer can be obtained using the MWC model. According to Figure 11.27, the basic parameters of the MWC model are the equilibrium constant K_0, given by Equation (11.65) and the microscopic equilibrium constants k_T and k_R, which define the equilibria between the components of the T species and the R species, respectively. The ligand is indicated by L and the ligand concentration is therefore c_L. A protein in the T form complexed with j ligands bound is indicated by TL_j with concentration c_{TL_j}, and similarly for complexes composed of ligands bound to a protein in the R form. The simplest way to proceed is to adapt Equation (11.29) for the presence of the two families of Hb tetramers:

$$\bar{\nu} = \frac{\sum\limits_{j=1}^{4} j c_{TL_j} + \sum\limits_{j=1}^{4} j c_{RL_j}}{\sum\limits_{j=0}^{4} c_{TL_j} + \sum\limits_{j=0}^{4} c_{RL_j}} \tag{11.66}$$

To simplify Equation (11.66) we use the definition of the equilibrium constant K_0 given in Equation (11.65) and the fact that all microscopic equilibrium constants for T equilibria

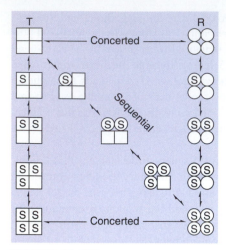

FIGURE 11.26

Juxtaposition of the concerted and sequential cooperative binding models. The sequential (KNF) model asserts that ligand binding induces a conformational change at a binding site from a low binding-affinity "t" conformation (squares) to a high affinity "r" conformation (circles). The concerted (MWC) model imposes a symmetry condition where all sites must be in low affinity or high affinity conformations. In the MWC model a protein with four "t" binding sites is said to be in the T conformation and similarly the R conformation corresponds to four "r" binding sites. In the MWC model, the T and R conformations are in equilibrium. Successive occupation of binding sites by ligands, indicated by labeling t and/or r sites with S, shifts the equilibrium from the T to the R form.

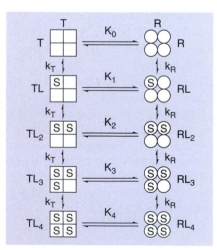

FIGURE 11.27

The concerted binding model for hemoglobin allosterism proposed by Monod, Wyman, and Changeux. Occupied t and/or r binding sites are labeled with "S."

are assumed equal to k_T and similarly all R equilibrium constants are k_R, as shown in Figure 11.27. After some algebra the result is

$$\bar{\nu} = 4c_L k_R \frac{cL(1 + ck_R c_L)^3 + (1 + k_R c_L)^3}{L(1 + ck_R c_L)^4 + (1 + k_R c_L)^4} \tag{11.67}$$

where $c = k_T/k_R$.

Using Equation (11.67), we obtain the Hill equation for the MWC model:

$$\frac{\bar{\nu}}{4 - \bar{\nu}} = \frac{\theta}{1 - \theta} = k_R c_L \left(\frac{cK_0 \left(\frac{1 + ck_R c_L}{1 + k_R c_L} \right)^3 + 1}{K_0 \left(\frac{1 + ck_R c_L}{1 + k_R c_L} \right)^3 + 1} \right) \tag{11.68}$$

Although Equation (11.68) seems complicated, we can check the validity of this expression by considering the expression for $\frac{\theta}{1 - \theta}$ in the limit of low substrate concentration and in the limit of high substrate concentration. In the limit that $c_L \to 0$, Equation (11.69) simplifies to

$$\frac{\theta}{1 - \theta} \approx k_R c_L \left(\frac{cK_0 + 1}{K_0 + 1} \right) \tag{11.69}$$

If the T state predominates over the R state, $K_0 >> 1$ and so

$$\frac{\theta}{1 - \theta} \approx ck_R c_L = k_T c_L \quad \text{or} \quad \ln\left(\frac{\theta}{1 - \theta} \right) = \ln c_L + \ln k_T \tag{11.70}$$

Equation (11.70) means that in the low c_L limit the Hill plot has a slope of 1 and a y intercept of $\ln k_T$.

Because the R state has a higher binding affinity than the T state we assume $k_R >> k_T$ and therefore $c = k_T/k_R << 1$. In the limit that $c_L \to \infty$, Equation (11.68) becomes

$$\frac{\theta}{1 - \theta} \approx k_R c_S \left(\frac{cK_0 \left(\frac{ck_R c_L}{k_R c_L} \right)^3 + 1}{K_0 \left(\frac{ck_R c_L}{k_R c_L} \right)^3 + 1} \right) = k_R c_L \left(\frac{c^4 K_0 + 1}{c^3 K_0 + 1} \right) \tag{11.71}$$

Assuming $c << 1$ Equation (11.71) simplifies to

$$\frac{\theta}{1 - \theta} \approx k_R c_L \quad \text{or} \quad \ln\left(\frac{\theta}{1 - \theta} \right) = \ln c_L + \ln k_R \tag{11.72}$$

Equation (11.72) means that in the high c_L limit, the Hill plot is also linear with a slope of 1 and a y intercept of $\ln k_R$. The combination of Equations (11.68), (11.70), and (11.72) accounts for the form of the Hill plot for Hb in Figure 11.24b.

EXAMPLE PROBLEM 11.9

The MWC model outlined in Figure 11.29 has five equilibria between T and R species:

$$T \xrightleftharpoons{K_0} R$$

$$TL \xrightleftharpoons{K_1} RL$$

$$TL_2 \xrightleftharpoons{K_2} RL_2$$

$$TL_3 \xrightleftharpoons{K_3} RL_3$$

$$TL_4 \xrightleftharpoons{K_4} RL_4$$

Obtain expressions for the macroscopic K_1-K_4 in terms of $K_0 = c_{T0}/c_{R0}$ and $c = k_T/k_R$

Solution

Begin with the definition $K_1 = c_{TL}/c_{RL}$. Use the expressions for the *T/TL* and *R/RL* equilibria: $k_T = c_{TL}/c_T$ and $k_R = c_{RL}/c_R$ to eliminate c_{TL} and c_{RL} from the expression for K_1:

$$K_1 = \frac{c_{TL}}{c_{RL}} = \frac{c_T k_T}{c_R k_R} = K_0 c$$

We use the same approach with $K_2 = \dfrac{c_{TL_2}}{c_{RL_2}} = \dfrac{c_{TL}}{c_{RL}} \dfrac{k_T}{k_R} = \dfrac{c_T}{c_R}\left(\dfrac{k_T}{k_R}\right)^2 = K_0 c^2$.

In general, $K_j = K_0 c^j$ for $j = 1, 2, 3, 4$. Note that because we assume the T species binds oxygen weakly compared to the R species, $c \ll 1$ and so $K_4 < K_3 < K_2 < K_1 < K_0$. In other words, the equilibria between the TL_j and RL_j species ($j = 0-4$) are shifted to the right as j increases, thus favoring the RL_j species. This accounts for the positive cooperativity of Hb.

11.10 Isothermal Titration Calorimetry

Enthalpies of binding between two molecules in solution may be measured by **isothermal titration calorimetry (ITC)**. A schematic of ITC instrumentation is shown in Figure 11.28. Two vessels are thermally isolated from their surroundings. The control or reference vessel contains a buffered solution. The reaction vessel contains a buffered solution of a macromolecule or some other binding target. Initially the temperatures of the control and reaction vessels are identical ($\Delta T = 0$). The injection system delivers a ligand solution into the reaction vessel at a measured rate. Depending on whether the binding is exothermic or endothermic, the temperature of the reaction vessel will increase ($\Delta T > 0$) or decrease ($\Delta T < 0$). The heat of binding q_b can be obtained given a knowledge of the solution heat capacity c:

$$q_b = C\Delta T \tag{11.73}$$

The standard enthalpy of binding is related to the heat of binding by

$$q_b = V\,\Delta H_b^\circ c_L \,(bound) \tag{11.74}$$

where V is the volume of the solution in the reaction vessel and $c_L(bound)$ is the concentration of bound ligand. To calculate the enthalpy of binding from Equation (11.74), we require the concentration of bound ligand. The total ligand delivered into the reaction vessel will partition into free ligand with concentration $c_L(free)$ and bound ligand with concentration $c_L(bound)$:

$$c_L(total) = c_L(bound) + c_L(free) = \bar{\nu}c_P + c_L \tag{11.75}$$

where in the last step c_P is the total concentration of polymer and c_L is the free ligand concentration. Equation (11.75) can be solved for c_L if we assume an expression for $\bar{\nu}$. The simplest approach is to assume a model of independent ligand binding. Using Equation (11.45):

$$c_L(total) = \bar{\nu}c_P + c_L = \left(\frac{NK_b c_L}{1 + K_b c_L}\right)c_P + c_L \tag{11.76}$$

where N is the number of ligand binding sites and K_b is the binding equilibrium constant. Equation (11.76) can be rearranged into standard quadratic form:

$$K_b c_L^2 + (NK_b c_P - K_b c_L(total) + 1)\,c_L - c_L(total) = 0 \tag{11.77}$$

and solved using the quadratic formula

$$c_L = \frac{-(NK_b c_P - K_b c_L(total) + 1) \pm \left[NK_b c_P - K_b c_L(total) + 1)^2 + 4K_b c_L(total)\right]^{1/2}}{2K_b} \tag{11.78}$$

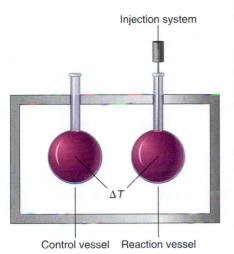

FIGURE 11.28
Schematic of an isothermal titration calorimeter.

Injection system

ΔT

Control vessel   Reaction vessel

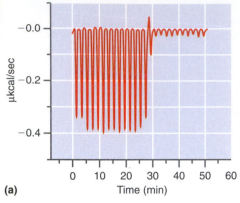

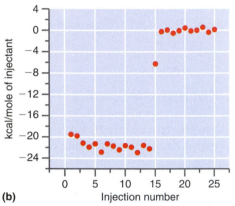

(a)

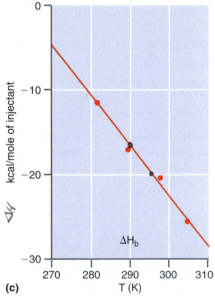

(b)

(c)

FIGURE 11.29
Isothermal titration calorimetry applied to the study of binding between monoclonal antibodies MAb 5F8 and MAb 2B5 to horse heart cytochrome c. (a) Calorimetric heat obtained by adding cytochrome c to a solution of MAb 5F8 ($T = 298$ K, pH = 7). (b) A plot of q_b versus time, obtained by integration of each peak in part (a). (c) Binding enthalpy for MAb 2B5 measured by ITC as a function of temperature. (Source: C.S. Raman, M.J. Allen, and B.T. Nall, *Biochemistry* 34 (1995), 5831–5838.)

Combining Equation (11.78) with Equation (11.75) we obtain a final expression for the bound ligand concentration:

$$c_L (bound) = c_L (total) - c_L \tag{11.79}$$

$$= c_L (total)$$

$$- \left\{ \frac{-(NK_b c_P - K_b c_L (total) + 1) + \left[(NK_b c_P - K_b c_L (total) + 1)^2 + 4K c_L (total)\right]^{1/2}}{2K_b} \right\}$$

Near the saturation point where virtually all N sites on the target protein are filled, the bound ligand concentration approaches Nc_P in which case Equation (11.75) can be approximated by

$$q_b \approx VNc_P \Delta H_b^\circ \tag{11.80}$$

and the standard enthalpy of binding is

$$\Delta H_b^\circ \approx \frac{q_b}{VNc_P} \tag{11.81}$$

Pierce, Raman, and Nall have used ITC to study the binding of two monoclonal antibodies, 2B5 and 5F8, to horse heart cytochrome c. Because the volume of the solution in the reaction vessel changes on addition of the titrant, calorimetric heat has to be corrected for the heat of dilution. Figure 11.29 shows an experimental calorimetric titration of a solution of MAb 5F8 with cytochrome c corrected for the heat of dilution. The quantities N, ΔH_b°, and K_b are obtained by fitting the calorimetric data in Figure 11.29a using Equations (11.74) and (11.79).

EXAMPLE PROBLEM 11.10

An ITC study of binding between horse heart cytochrome c and MAb 2B5 yields $K_b = 2.00 \times 10^9$ and $\Delta H_b^\circ = -87.8$ kJ mol^{-1}. From these data calculate the standard Gibbs energy of binding ΔG_b° and the standard entropy of binding ΔS_b°. Assume $T = 298.15$ K.

Solution

The Gibbs energy is obtained from the relation

$$\Delta G_b^\circ = -RT \ln K_b = -(8.315 \text{ JK}^{-1} \text{ mol}^{-1}) (298.15 \text{ K}) \ln (2.00 \times 10^9) = -52.5 \text{ kJ mol}^{-1}$$

At constant temperature the entropy is obtained from

$$\Delta S_b^\circ = \frac{\Delta H_b^\circ - \Delta G_b^\circ}{T} = \frac{-87.8 \text{ kJ mol}^{-1} - (-52.5 \text{ kJ mol}^{-1})}{298.15 \text{ K}} =$$

$$-1.18 \times 10^2 \text{ J K}^{-1} \text{ mol}^{-1}$$

Vocabulary

allosterism	glycolysis	KNF model (of protein	myoglobin (Mb)
Adair equation	hemoglobin (Hb)	allosterism)	oxidative phosphorylation
ATP synthase	Hill equation	Krebs cycle	Q cycle
citric acid cycle	Hill plot	ligand	Scatchard equation
concerted model for	independent binding	matrix	Scatchard plot
cooperative binding	isothermal titration	mitochondrion	sequential model for
cooperative binding	calorimetry (ITC)	MWC model (of protein	cooperative binding
electron transport chain		allosterism)	tertiary structure

Questions and Concepts

Q11.1 Explain the advantages of aerobic respiration over anaerobic glycolysis.

Q11.2 Under cellular conditions, is the biological oxidation of glucose reversible, in the thermodynamic sense of the word? Explain.

Q11.3 Discuss the validity of the following statement: Biological cells are nearly in a state of thermodynamic equilibrium so they can derive maximum work from the metabolism of high Gibbs energy compounds.

Q11.4 Explain why the glycolytic pathway is regulated at step 3.

Q11.5 Explain how synthesis of ATP is coupled to discharge of the pH gradient across the inner mitochondrial membrane.

Q11.6 Explain the significance of the Q cycle in the functioning of the electron transport chain.

Q11.7 How does the functioning of the ATP synthase complex resemble a molecular motor?

Q11.8 What main lines of evidence support the chemiosmotic hypothesis?

Q11.9 Explain how electron transfer is coupled to the creation of a proton gradient across the inner mitochondrial membrane.

Q11.10 Explain the purpose of the Krebs cycle.

Q11.11 Explain the difference between microscopic binding constants and macroscopic binding constants.

Q11.12 If a Scatchard plot is linear, what assumptions can be made about the binding of a ligand to a protein?

Q11.13 Explain the relationship between cooperative binding and protein allosterism.

Q11.14 Explain the difference between the MWC and KNF models for oxygen binding to hemoglobin.

Q11.15 What is the purpose of isothermal titration calorimetry?

Q11.16 Explain why Hb, which transports oxygen from the lungs to the tissues, binds oxygen cooperatively, whereas Mb, which carries oxygen within the tissues, is noncooperative.

Problems

Problem numbers in **RED** indicate that the solution to the problem is given in the *Student Solutions Manual*.

P11.1 The standard Gibbs energy change for step 1 in Equation (11.16) is $\Delta G^{\circ\prime} = -38.6 \text{ kJ mol}^{-1}$. Using this fact and the data in Table 11.4 calculate the half-cell potential for $FMN + 2e^- + 2H^+ \rightarrow FMNH_2$.

P11.2 Derive the relationship between the standard cell potential E° and the equilibrium constant K.

P11.3 For the general reaction $A + B \rightleftarrows C + D + xH^+$, obtain the relationship between E° and $E^{\circ\prime}$.

P11.4 Prove that the cell potential E is independent of standard state conventions.

P11.5 Consider the reduction of pyruvate to lactate by NADH:

$CH_3 (CO) COO^- + NADH + H^+ \rightleftarrows CH_3 (CHOH) COO^- + NAD^+$

Suppose at equilibrium the concentrations of reactants and products are $C_{lactate} = 0.00200 \ M$, $C_{pyruvate} = 0.05 \ M$, $C_{NADH} = 0.00200 \ M$, $c_{NAD^-} = 0.504 \ M$. Calculate $K, K', \Delta G^\circ, \Delta G^{\circ\prime}$ at pH = 7. Assume $T = 298$ K.

P11.6 For the reaction and conditions given in Problem P11.5 calculate E° and $E^{\circ\prime}$.

P11.7 Calculate the Gibbs energy change that occurs on phosphorylation of fructose-6-phosphate (F6P) by ATP to form fructose-1,6-bisphosphate (FBP). Assume concentrations that prevail in the human red blood cell. Assume also $T = 310$ K.

P11.8 Consider the oxidation of ethanol by NAD^+ to form acetaldehyde:

$$NAD^+ + C_2H_5OH \rightarrow NADH + CH_3CHO + H^+$$

Assuming standard half-cell reduction potentials $E^{\circ\prime}_{NAD^+/NADH} = -0.32$ V and $E^{\circ\prime}_{CH_3CHO/EtOH} = -0.18$ V, calculate the standard cell potential $E^{\circ\prime}$, the standard Gibbs energy change $\Delta G^{\circ\prime}$, and the equilibrium constant K'. Based on your answer, does the system favor the formation of acetaldehyde or ethanol at equilibrium? Assume $T = 310$ K.

P11.9 Repeat the calculation in Problem 11.8 only using the standard state convention $C^0_{H^+} = 1 \ M$.

P11.10 Suppose the oxidation of lactate to pyruvate,

$$CH_3CH(OH)COO^- \rightarrow CH_3(CO)COO^- + 2H^+ + 2e^-$$

is coupled energetically to the reduction of Fe^{3+}-cytochrome c. Assume the half-cell standard reduction potential for pyruvate/lactate is $E^{\circ\prime} = -0.19$ V. When the system reaches equilibrium, the concentrations of Fe^{3+}-cytochrome c and Fe^{2+}-cytochrome c are equal. Calculate the ratio of the concentration of pyruvate to the concentration of lactate. Assume $T = 298$ K and pH = 7.

P11.11 Studies of the binding of ATP to the enzyme tetrahydrofolate synthetase were conducted at $T = 293$ K and appear in the following table, where C_{ATP} is in molar units.

$\bar{v}$	0.25	0.50	1.0	1.5	2.0	2.5	3
C_{ATP} (M)	6.67×10^{-6}	1.43×10^{-5}	3.33×10^{-5}	6.00×10^{-5}	1.00×10^{-4}	1.67×10^{-4}	3×10^{-4}

a. From a Scatchard plot, determine K_{293} and N.

b. Assume $K_{293}/K_{310} = 2$. Using the information from the Scatchard plot of part (a), calculate the standard enthalpy $\Delta H°$ for the binding of ATP to tetrahydrofolate synthetase. Assume $\Delta H°$ is constant between $T = 293$ K and $T = 310$ K.

c. Calculate $\Delta G°$, the standard Gibbs energy change for the binding of ATP to tetrahydrofolate synthetase, at $T = 293$ K.

d. Calculate $\Delta S°$, the standard entropy change for the binding of ATP to tetrahydrofolate synthetase.

P11.12 The cooperative binding of oxygen to tetrameric Hb can be best viewed by rearranging Equation (11.54) to obtain the expression for the fractional saturation in terms of the partial pressure of oxygen p_{O_2}:

$$\theta = \frac{Kp_{O_2}^{\alpha_H}}{1 + Kp_{O_2}^{\alpha_H}}$$

a. Define p_{50} as the oxygen pressure at which the fractional saturation $\theta = 0.5$. Show that the Hill equation becomes

$$\ln\left(\frac{\theta}{1-\theta}\right) = \alpha_H \ln p_{O_2} - \alpha_H \ln p_{50}$$

b. The partial pressure of oxygen is 0.13 atm in the lungs and 0.026 atm in the tissues. Using the data in Figure 11.24, calculate the fractional saturation of Hb in the lungs. Assume the Hill coefficient α_H is 2.9. Repeat the calculation assuming independent binding.

P11.13 Using the microscopic binding constants in Table 11.6, calculate the average number of bound oxygen sites per Hb tetramer at a partial oxygen pressure of 60 Torr. Assume the solution is free of salt.

P11.14 At pH = 7.40 and in 2 mM bisphosphoglycerate (BPG), hemoglobin binds oxygen with the following microscopic binding constants: $k_1 = 10.3$ atm^{-1}, $k_2 = 6.79$ atm^{-1}, $k_3 = 33.0$ atm^{-1}, and $k_4 = 3167$ atm^{-1}. Determine the degree to which BPG inhibits oxygen binding to Hb. Assume $p_{O_2} = 60$ Torr.

P11.15 Prove that in the absence of the T structural form or when $k_T << k_R$, the MWC model predicts that ligand binding is independent.

P11.16 The change in heat capacity ΔC_p that accompanies molecular binding can be obtained from isothermal titration calorimetry if the binding enthalpy is measured as a function of temperature. From Figure 11.29c, where ΔH_b for the monoclonal antibody MAb 2B5 binding to cytochrome c is plotted as a function of temperature, determine the heat capacity of binding.

P11.17 The following data are for the binding of oxygen to squid hemocyanin. The percent saturation is the parameter $\frac{\bar{v}}{N} \times 100\% = \theta \times 100\%$.

P_{O_2}	Percent Saturation
1.13	0.30
7.72	1.92
31.71	8.37
100.5	32.9
136.7	55.7
203.2	73.4
327.0	83.4
566.9	89.4
736.7	91.3

Construct a Hill plot of the data shown and determine whether the binding is independent.

P11.18 It is possible to obtain a binding equation analogous to Equation (11.42) for gases onto solid surfaces. We designate molecules in the gas phase as A and occupied surface sites SA. The unoccupied surface is S. At equilibrium the rates of adsorption and desorption are equal and the equilibrium equation is written $A + S \rightleftarrows SA$ with $K = \frac{c_{SA}}{c_S c_A}$ were c_{SA} quantifies the filled surface, c_S is the unoccupied surface, and c_A quantifies the amount of free gas.

a. Assume c_{SA} is proportional to θ and c_A is proportional to the pressure P. Show that $\theta = \frac{bP}{1 + bP}$ where $b = \frac{K}{c}$ and where c is a constant of proportionality. This equation is called the Langmuir equation.

b. If P_0 is the pressure at which $\theta = 0.5$, and if $\theta = \frac{V}{V_m}$ where V is the volume of gas adsorbed onto the surface and V_m is the volume that is adsorbed onto the surface to form a monolayer, show that the Langmuir equation is

$$V = \frac{V_m P}{P_0 + P}.$$

c. Starting with the form for the Langmuir equation in part (b), derive an equation analogous to the Scatchard equation.

P11.19 Confirm that the data in the following table for the adsorption of CO onto a surface obey the Langmuir equation (see Problem P11.18). Determine P_0 and V_m.

P (atm)	0.132	0.264	0.396	0.528	0.660	0.792
V (mL)	0.130	0.150	0.162	0.166	0.175	0.180

P11.20 The sequential binding of ligands L to four binding sites on a polymer P is quantified by the macroscopic or apparent equilibrium binding constants:

$$P + L \underset{}{\overset{K_1}{\rightleftarrows}} PL$$
$$PL + L \underset{}{\overset{K_2}{\rightleftarrows}} PL_2$$
$$PL_2 + L \underset{}{\overset{K_3}{\rightleftarrows}} PL_3$$
$$PL_3 + L \underset{}{\overset{K_4}{\rightleftarrows}} PL_4$$

For these equilibria, derive the relationships between the microscopic equilibrium constants and the macroscopic equilibrium constants:

$$k_1 = \frac{K_1}{4}; \quad k_2 = \frac{2}{3}K_2; \quad k_3 = \frac{3}{2}K_3; \quad k_4 = 4K_4$$

12

From Classical to Quantum Mechanics

As scientists became able to investigate the atomic realm, quantum mechanics was developed to explain the inconsistencies found between the results of measurements and the predictions of classical physics. Classical physics incorrectly predicts that all bodies at a temperature other than zero kelvin radiate an infinite amount of energy. It incorrectly predicts that the kinetic energy of electrons produced on illuminating a metal surface in vacuum with light is proportional to the light intensity, and it cannot explain the diffraction of electrons by a crystalline solid. Researchers know from experiments that atoms consist of a small, positively charged nucleus surrounded by a diffuse cloud of electrons. Classical physics, however, predicts that the atom is unstable and that the electrons will spiral into the nucleus while radiating energy to the environment. These inconsistencies between classical theory and experimental observations provided the stimulus for the development of quantum mechanics.

12.1 Why Study Quantum Mechanics?

Chemistry is a molecular science; the goal of chemists is to understand macroscopic behavior in terms of the properties of individual molecules. For example, H_2 is a good fuel because the energy released in forming H_2O is much greater than that needed to break the bonds in the reactants O_2 and H_2. As you will learn in this chapter, in the first decade of the 20th century, it became clear that classical physics was unable to explain why the atomic structure of a positively charged nucleus surrounded by electrons was stable. Classical physics was also unable to explain why the light emitted by a hydrogen discharge lamp appears at only a small number of wavelengths, and why the bond angle in H_2O is different from that in H_2S.

These deficiencies in classical physics made it clear that another physical model was needed to describe atoms and molecules. Over a period of about 20 years, quantum mechanics was developed and scientists have found that all of the phenomena just cited that cannot be understood within classical physics can be explained using quantum mechanics. As you will learn, the central feature that distinguishes quantum and classical mechanics is wave-particle duality. At the atomic level, electrons, protons, and light all behave as waves or particles. It is the experiment that determines whether wave or particle behavior will be observed.

Although you may not know it, you are already a user of quantum mechanics. You take for granted the stability of the atom with its central positively charged nucleus and surrounding electron cloud, the laser in your CD player, the integrated circuit in your computer, and the chemical bond. You know that infrared spectroscopy provides a useful way to identify chemical compounds and that nuclear magnetic resonance spectroscopy provides a powerful tool to image the internal organs of humans. However, spectroscopic techniques would not be possible if atoms and molecules could have *any* value of energy as predicted by classical physics. Quantum mechanics predicts that atoms and molecules can only have discrete energies and thereby provides a basis for understanding all spectroscopies.

Technology is increasingly based on quantum mechanics. For example, quantum computing, in which a state can be described by zero *and* one rather than zero *or* one, is a very active area of research. If quantum computers can ultimately be realized, they will be much more powerful than current computers. Quantum mechanical calculations of chemical properties of biologically important molecules are now sufficiently accurate that molecules can be designed for a specific application before they are tested at the laboratory bench. As many sciences such as biology become increasingly focused on the molecular level, more scientists will need to be able to think in terms of quantum mechanical models. Therefore, quantum mechanics is an essential part of the chemist's knowledge base.

12.2 Quantum Mechanics Arose Out of the Interplay of Experiments and Theory

Scientific theories gain acceptance if they can make the world around us understandable. A key feature of validating theories as useful models is to compare the result of a new experiment with the prediction of currently accepted theories. If the experiment and the theory agree, we gain confidence in the model underlying the theory; if they do not, the model needs to be modified. At the end of the 19th century, Maxwell's electromagnetic theory unified existing knowledge in the areas of electricity, magnetism, and waves. This theory, combined with the well-established field of classical mechanics, ushered in a new era of maturity for the physical sciences. Many scientists of that era believed that there was little left in the natural sciences that was unknown. However, the growing ability of scientists to probe natural phenomena at an atomic level soon showed that this was incorrect. The field of quantum mechanics arose in the early 1900s as scientists became able to investigate natural phenomena at this newly accessible atomic level. A number of key experiments showed that the predictions of classical physics were inconsistent with experimental outcomes. Several of these experiments are described in more detail in this chapter in order to show the important role that experiments have had—and continue to have—in stimulating the development of theories to describe the natural world. These experiments stimulated the leading scientists of the era to formulate quantum mechanics.

In the rest of this chapter, experimental evidence is presented for two key properties that have come to distinguish classical and quantum physics. The first of these is **quantization.** Energy at the atomic level is not a continuous variable, but comes in discrete packets called *quanta.* The second key property is **wave-particle duality.** At the atomic level, light waves have particle-like properties, and atoms as well as subatomic particles such as electrons have wave-like properties. Neither quantization nor wave-particle duality were known concepts until the experiments described in Sections 12.3 through 12.7 were conducted.

FIGURE 12.1

An idealized blackbody. A cubical solid at a high temperature emits photons from an interior spherical surface. The photons reflect several times before emerging through a narrow channel. The reflections ensure that the radiation is in thermal equilibrium with the solid.

12.3 Blackbody Radiation

Think of the heat that you feel from the embers of a campfire on a cold night. The energy that your body absorbs is radiated from the glowing coals. An idealization of this system that is more amenable to theoretical study is a red-hot block of metal with a spherical cavity in its interior that can be observed through a hole small enough that the conditions inside the block are not perturbed. An **ideal blackbody** is shown in Figure 12.1.

Web-Based Problems 12.1 and 12.2
Blackbody Radiation

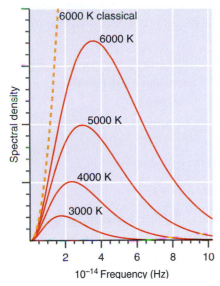

FIGURE 12.2
The red curves show the light intensity emitted from an ideal blackbody as a function of the frequency for different temperatures. The dashed curve shows the predictions of classical theory for $T = 6000.$ K.

It is possible to measure the **spectral density,** which is the energy at frequency v per unit volume and unit frequency stored in the electromagnetic field of the blackbody radiator, and the results are shown in Figure 12.2 for several temperatures together with the result predicted by classical theory. The experimental curves have a common behavior. The spectral density is peaked in a broad maximum and falls off to both lower and higher frequencies. The shift of the maxima to higher frequencies with increasing temperatures is consistent with our experience that if more power is put into an electrical heater, its color will change from dull red to yellow (increasing frequency).

The comparison of the spectral density distribution, ρ, predicted by classical theory with that observed experimentally for $T = 6000$ K is particularly instructive. The two curves show similar behavior at low frequencies, but the theoretical curve keeps on increasing with frequency. Because the area under the $\rho(v, T)$ versus v curves gives the total energy per unit volume of the field of the blackbody, classical theory predicts that a blackbody will emit an infinite amount of energy at all temperatures above absolute zero! It is clear that this prediction is incorrect, but at the beginning of the 20th century scientists were greatly puzzled about where the theory went wrong.

In looking at data such as that shown in Figure 12.2, the German physicist Max Planck was able to develop some important insights that ultimately led to an understanding of **blackbody radiation.** It was understood at the time that the origin of blackbody radiation was the vibration of electric dipoles that emit radiation at the frequency at which they oscillate. Planck saw that the discrepancy between experiment and classical theory occurred at high, not low, frequencies. The absence of high-frequency radiation at low temperatures showed that the high-frequency dipole oscillators were active only at high temperatures. This led him to postulate that the energy of the radiation was proportional to the frequency. This was a radical departure from classical theory, in which the energy stored in a radiation field is proportional to the square of the amplitude, but independent of the frequency. We can follow Planck's line of reasoning by realizing that unless a large amount of energy is put into the blackbody (high temperature), it will not be possible to excite the high-energy (high-frequency) oscillators.

Planck found that he could obtain agreement between theory and experiment only if he assumed that the energy radiated by the dipoles was given by the relation

$$E = nh\nu \tag{12.1}$$

Planck's constant, h, was initially an unknown proportionality constant and n is a positive integer ($n = 0, 1, 2, \ldots$). The frequency v is continuous, but for a given v, the energy is *quantized* according to Equation (12.1). It was the introduction of this relationship between energy and frequency that ushered in a new era of physics. Energy in classical theory is a *continuous* quantity, which means that it can take on all values. Equation (12.1) states that the energies radiated by a blackbody are not continuous, but can take on only a set of *discrete* values for each frequency. This assumption did not arise out of classical theory, and its main justification was that agreement between theory and experiment could be obtained. Using Equation (12.1) and some classical physics, Planck obtained the following result for the spectral density, $\rho(v, T)$:

$$\rho(\nu, T)d\nu = \frac{8\pi h \nu^3}{c^3} \frac{1}{e^{h\nu/kT} - 1} d\nu \tag{12.2}$$

The value of the constant h was not known and Planck used it as a parameter to fit the data. He was able to reproduce the experimental data at all temperatures with the single adjustable parameter h, which through more accurate measurements has the value $h = 6.626 \times 10^{-34}$ J s. Although obtaining this degree of agreement was a remarkable achievement, Planck's explanation was not accepted initially. However, soon afterward, Einstein's explanation of the photoelectric effect gave support to Planck's hypothesis.

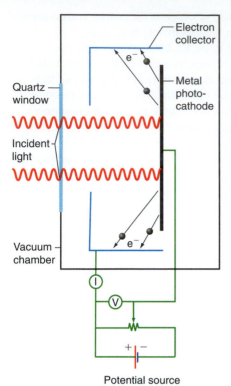

FIGURE 12.3

The electrons emitted by the surface on illumination are incident on the collector, which is at an appropriate electrical potential to attract them. The experiment is carried out in a vacuum chamber to avoid collisions and capture of electrons by gas molecules.

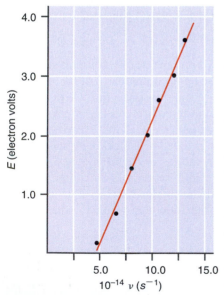

FIGURE 12.4

The energy of photoejected electrons is shown as a function of the light frequency. The individual data points are well fit by a straight line as shown.

12.4 The Photoelectric Effect

Imagine a copper plate in a vacuum. Light incident on the plate can be absorbed, leading to the excitation of electrons to unoccupied energy levels. Sufficient energy can be transferred to the electrons such that some of them leave the metal and are ejected into the vacuum. If another electrode, called the *collector* in the vacuum system, is added, the electrons that have been emitted from the copper on illumination can be collected. This is called the **photoelectric effect.** A schematic apparatus is shown in Figure 12.3. Invoking conservation of energy, we conclude that the absorbed light energy must be balanced by the kinetic energy of the emitted electrons and the energy required to eject an electron at equilibrium, because the energy of the system is constant. Classical theory makes the following predictions:

- Light is incident as a plane wave over the whole copper plate. Therefore, the light is absorbed by many electrons in the solid. Any one electron can absorb only a small fraction of the incident light.

- Electrons are emitted to the collector for all light frequencies, provided that the light is sufficiently intense.

- The kinetic energy per electron increases with the light intensity.

The results of the experiment can be summarized as follows:

- The number of emitted electrons is proportional to the light intensity, but their kinetic energy is independent of the light intensity.

- No electrons are emitted unless the frequency ν is above a threshold frequency ν_0 even for high light intensities.

- The kinetic energy of the emitted electrons depends on the frequency in the manner depicted in Figure 12.4.

- Electrons are emitted even at such low light intensities that all of the light absorbed by the entire copper plate is barely enough to eject a single electron, based on energy conservation considerations.

As in the case of blackbody radiation, the inability of classical theory to correctly predict experimental results stimulated a new theory. In 1905, Albert Einstein hypothesized that the energy of the light was proportional to its frequency:

$$E = \beta\nu \tag{12.3}$$

where β is a constant to be determined. This is a marked departure from classical electrodynamics, in which there is no relation between the energy of a light wave and its frequency. From energy conservation considerations, the energy of the electron, E_e, is related to that of the light by

$$E_e = \beta\nu - \phi \tag{12.4}$$

The binding energy of the electron in the solid, which is analogous to the ionization energy of an atom, is designated by ϕ in this equation and is called the **work function.** In words, this equation says that the kinetic energy of the photoelectron that has escaped from the solid is smaller than the photon energy by the amount with which the electron is bound to the solid. Einstein's theory gives a prediction of the dependence of the kinetic energy of the photoelectrons as a function of the light frequency for a given solid that can be compared directly with experiment. Because ϕ can be determined independently, only β is unknown. It can be obtained by fitting the data points in Figure 12.4 to Equation (12.4). The results shown by the red line in Figure 12.4 not only reproduce the data very well, but they yield the result that β, the slope of the line, is identical to Planck's constant, h. This result, which relates the energy of light to its frequency,

$$E = h\nu \qquad (12.5)$$

is one of the most widely used equations in quantum mechanics and earned Albert Einstein a Nobel Prize in physics.

The agreement between the theoretical prediction and the experimental data validates the fundamental assumption that Einstein made, namely, that the energy of light is proportional to its frequency. This result also suggested that h is a "universal constant" that appears in seemingly unrelated phenomena. Its appearance in this context made the assumptions Planck used to explain blackbody radiation more widely accepted.

EXAMPLE PROBLEM 12.1

Light with a wavelength of 300. nm is incident on a potassium surface for which the work function, ϕ, is 2.26 eV. Calculate the kinetic energy and speed of the ejected electrons.

Solution

Using Equation (12.4), we write $E_e = h\nu - \phi = (hc/\lambda) - \phi$ and convert the units of ϕ from electron-volts to joules: $\phi = (2.26 \text{ eV})(1.602 \times 10^{-19} \text{ J/eV}) = 3.62 \times 10^{-19}$ J. Electrons will only be ejected if the photon energy, $h\nu$, is greater than ϕ. The photon energy is calculated to be

$$\frac{hc}{\lambda} = \frac{(6.626 \times 10^{-34} \text{ J s})(2.998 \times 10^8 \text{ m s}^{-1})}{300. \times 10^{-9} \text{ m}} = 6.62 \times 10^{-19} \text{ J}$$

which is sufficient to eject electrons.

Using Equation (12.4), we obtain $E_e = (hc/\lambda) - \phi = 2.99 \times 10^{-19}$ J. Using $E_e = \frac{1}{2}(mv^2)$, we calculate that

$$v = \sqrt{\frac{2E_e}{m}} = \sqrt{\frac{2(2.99 \times 10^{-19} \text{ J})}{9.109 \times 10^{-31} \text{ kg}}} = 8.10 \times 10^5 \text{ m s}^{-1}$$

Another important conclusion can be drawn from the observation that even at very low light intensities, photoelectrons are emitted from the solid. More precisely, photoelectrons are detected even at intensities so low that the total integrated radiation intensity incident on the solid surface is only slightly more than the threshold energy required to yield a single photoelectron. This result is not consistent with the concept that the light that liberates the photoelectron is uniformly distributed over the surface. If this were true, no individual electron could receive enough energy to escape into the vacuum. The surprising conclusion that can be drawn from this result is that all of the incident light energy can be concentrated in a single electron excitation. This led to the coining of the term **photon** to describe a spatially localized packet of light. Because this is how particles are conceptualized, the conclusion is that light can exhibit particle-like behavior under some circumstances.

Many experiments have shown that light exhibits wave-like behavior. We have long known that light can be diffracted by an aperture or slit. However, the photon in the photoelectric effect that shows particle-like properties and the photon in a diffraction experiment that shows wave-like properties are one and the same. This forces us to conclude that light exhibits a wave-particle duality, which means that, depending on the experiment, it can manifest itself as a wave or as a particle. This important recognition leads us to the third fundamental experiment to be described: the diffraction of electrons by a crystalline solid. Because diffraction is clearly wave-like behavior, if particles can be diffracted, they exhibit a particle-wave duality just as light does.

12.5 Particles Exhibit Wave-Like Behavior

In 1924, Louis de Broglie suggested that a relationship that had been derived to relate momentum and wavelength for light should also apply to particles. The **de Broglie relation** states that

$$\lambda = \frac{h}{p} \tag{12.6}$$

in which h is the now-familiar Planck constant and p is the particle momentum given by $p = mv$, in which the momentum is expressed in terms of the particle mass and velocity. This proposed relation was confirmed in 1927 by Davisson and Germer, who carried out a diffraction experiment. Recall from your physics course that diffraction is observed under conditions for which the characteristic dimension of the diffraction grating is on the order of one wavelength. Putting numbers in Equation (12.6) should convince you that it is difficult to obtain wavelengths much longer than 1 nm even with particles as light as the electron, as shown next in Example Problem 12.2. Therefore, diffraction requires a grating with atomic dimensions, and an ideal candidate is a crystalline solid. Davisson and Germer diffracted electrons from crystalline NiO in their classic experiment to verify the de Broglie relation. Diffraction of He and H_2 from crystalline surfaces has also been observed in the intervening years.

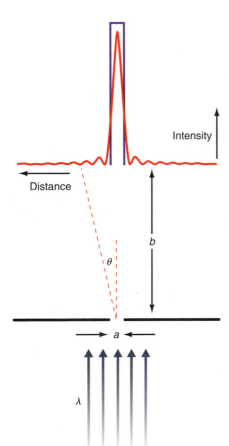

FIGURE 12.5
Diffraction of light of wavelength λ from a slit whose long axis is perpendicular to the page. The arrows from the bottom indicate parallel rays of light incident on an opaque plate containing the slit. Instead of seeing a sharp image of the slit on the screen, a diffraction pattern will be seen. This is schematically indicated in a plot of intensity versus distance. In the absence of diffraction, the intensity versus distance indicated by the blue lines would be observed.

Web-Based Problem 12.3 Diffraction of Light

EXAMPLE PROBLEM 12.2

Electrons are used to determine the structure of crystal surfaces. To have diffraction, the wavelength, λ, of the electrons should be on the order of the lattice constant, which is typically 0.30 nm. What energy do such electrons have, expressed in electron-volts and joules?

Solution

Using Equation (12.6) and the expression $E = p^2/2m$ for the kinetic energy, we obtain

$$E = \frac{p^2}{2m} = \frac{h^2}{2m\lambda^2} = \frac{(6.626 \times 10^{-34}\ \text{J s})^2}{2(9.109 \times 10^{-31}\ \text{kg})(3.0 \times 10^{-10}\ \text{m})^2} = 2.7 \times 10^{-18}\ \text{J or 17 eV}$$

The Davisson–Germer experiment was critical in the development of quantum mechanics, in that it showed that particles exhibit wave behavior. If this is the case, there must be a wave equation that relates the spatial and time dependencies of the wave amplitude for the (wave-like) particle. It was Erwin Schrödinger who formulated this wave equation, which will be discussed in Chapter 13.

12.6 Diffraction by a Double Slit

There is probably no single experiment that exhibits the surprising nature of measurement in quantum mechanical experiments as well as the diffraction of particles by a double slit. An idealized version of this experiment is described, but everything in the following explanation has been confirmed by experiments carried out with particles such as neutrons. We first briefly review classical diffraction of waves.

Diffraction is a phenomenon that is widely exploited in science. For example, the atomic-level structure of DNA was in large part determined by analyzing the diffraction of X-rays from crystalline DNA samples. Diffraction is a phenomenon that can occur with any waves, including sound waves, water waves, and electromagnetic (light) waves. Figure 12.5 illustrates diffraction of light from a thin slit in an opaque wall.

It turns out that the analysis of this problem is much simpler if the screen on which the image is projected is far away from the slit. Mathematically, this requires that $b \gg a$. In ray optics, which is used to determine the focusing effect of a lens on light, the light incident on the slit from the left in Figure 12.5 would give a sharp image of the slit on the screen. In this case parallel light is assumed to be incident on the slit and, therefore, the image and slit dimensions are identical. This would give rise to an intensity pattern on the screen like that shown by the blue lines in the figure. Instead, for the condition that the light wavelength is comparable in magnitude to the slit width, an intensity distribution like that shown by the red curve will be observed.

The origin of this pattern of alternating maxima and minima (which lies well outside the profile expected from ray optics) is wave interference. Its origin can be understood by treating each point in the plane of the slit as a source of cylindrical waves (Huygens' construction). Maxima and minima arise as a result of a path difference between the sources of the cylindrical waves and the screen, as shown in Figure 12.6. The condition that the minima satisfy is

$$\sin \theta = \frac{n\lambda}{a}, n = \pm 1, \pm 2, \pm 3, \ldots \qquad (12.7)$$

This equation helps us to understand under what conditions we might observe diffraction. The wavelength of light in the middle of the visible spectrum is about 600 nm or 6.00×10^{-4} mm. If this light is allowed to pass through a 1.00-mm-wide slit and the angle calculated at which the first minimum will appear, the result is $\theta = 0.03°$ for $n = 1$. This minimum is not easily observable because it lies so close to the maximum, and we expect to see a sharp image of the slit on our screen, just as in ray optics. However, if the slit width is decreased to 1.00×10^{-2} mm, then $\theta = 3.4°$. This minimum is easily observable and successive bands of light and darkness will be observed instead of a sharp image. Note that there is no clear demarcation between ray optics and diffraction. The crossover between the two is continuous and depends on the resolution of our experimental techniques. The exact same behavior is observed in wave-particle duality in quantum mechanics. If the slit is much larger than the wavelength, then diffraction will not be observed and ray optics holds.

Consider the experimental setup designed to detect the diffraction of particles shown in Figure 12.7. The essential feature of the apparatus is a metal plate in which two rectangular slits of width a have been cut. The long axis of the rectangles is perpendicular to the plane of the page. Why two slits? Rather than detecting the diffraction from the individual slits, the apparatus is designed to detect diffraction from the *combination of the two slits*. Diffraction will only be observed for case 2 if the particle passes through both slits simultaneously. Let's take a closer look at diffraction to determine what it is about the previous sentence that is difficult to accept from the vantage point of classical physics.

Web-Based Problem 12.4 Diffraction from Double Slit

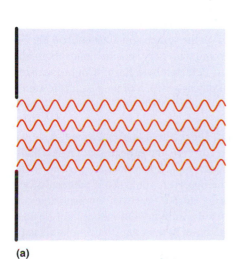

(a)

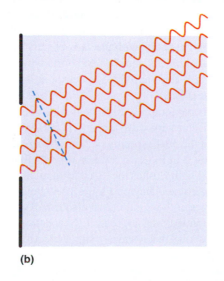

(b)

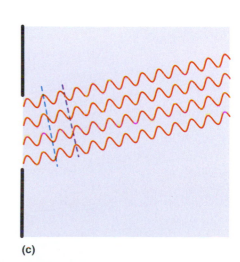

(c)

FIGURE 12.6

Each segment of a slit through which light is diffracted can be viewed as a source of waves that interfere with one another. (a) The waves that emerge perpendicular to the slit are all in phase and give rise to the principal maximum in the diffraction pattern. (b) Successive waves that emerge at the angle shown are exactly out of phase. They will interfere destructively and a minimum intensity will be observed. (c) Every other wave is out of phase and destructive interference with a minimum intensity will be observed. The wavelength and slit width are not drawn to scale.

The double-slit diffraction experiment.
Case 1 describes the outcome of the
diffraction when one of the slits is blocked.
Case 2 describes the outcome when both
slits are open.

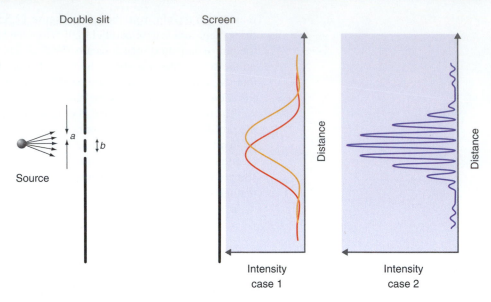

First, we need a source of particles, for instance, an electron gun. By controlling the energy of the electron, the wavelength is varied. Each electron has a **random phase angle** with respect to every other electron. Consequently, two electrons can never interfere with one another to produce a diffraction pattern. One electron gives rise to the diffraction pattern, but many electrons are needed to amplify the signal so that we can easily see the diffraction pattern. A more exact way to say this is that the intensities of the electron waves add together rather than the amplitudes.

A phosphorescent screen that gives rise to light when an incident wave or particle transfers energy to it (as in a television picture tube) is mounted behind the plate with the slits. The electron energy is adjusted so that diffraction by the single slits of width a results in broad maxima with the first intensity minimum at a large diffraction angle. The distance between the two slits, $b,$ has been chosen such that we will observe a number of intensity oscillations for small diffraction angles. The diffraction patterns in case 2 of Figure 12.7 were calculated for the ratio $b/a = 5$.

Now let's talk about the results. If one slit is closed off and an observer looks at the screen, he will see the broad intensity versus distance pattern shown as case 1 in Figure 12.7. Depending on which slit has been closed, the observer will see one of the two curves shown. The observer sees diffraction from the slit, which demonstrates that the electron can act like a wave. If the device that was used to close the slit also measures the electron current, we will find that, for a large number of electrons that pass through the plate with the slits, exactly 50% land on the screen and 50% land on the device that blocks one slit. Additionally, when working with very sensitive phosphors, the arrival of each individual electron at the screen is detected by a flash of light localized to a small area of a screen. Viewed from classical physics, which of the results presented in this paragraph is consistent with both wave *and* particle behavior, and which is only consistent with wave *or* particle behavior?

In the next experiment, both slits are left open. The result of this experiment is shown as case 2 in Figure 12.7. For very small electron currents, the observer again sees individual light flashes localized to small areas of the screen in a random pattern. This looks like particle behavior. Figure 12.8 shows what the observer would see if the results of a number of these individual electron experiments are stored. However, unmistakable diffraction features are seen if we accumulate the results of many individual light flashes. This shows that a wave (a single electron) is incident on both of the slits simultaneously.

How can the results of this experiment be interpreted? The fact that diffraction is seen from a single slit as well as from the double slit shows the wave-like behavior of the electron. Yet, individual light flashes are observed on the screen, which is what we expect from particle trajectories rather than the diffraction pattern shown in Figure 12.8d. To add to the complexity, the spatial distribution of the individual flashes on the screen is what

FIGURE 12.8

Simulation of the diffraction pattern observed in the double-slit experiment for (a) 60, (b) 250, (c) 1000, and (d) 3000 particles. The bottom panel shows what would be expected for a wave incident on the apparatus. Bright red corresponds to high intensity and blue corresponds to low intensity. Note that the diffraction pattern only becomes obvious after a large number of particles have passed through the apparatus, although intensity minima are evident even for 60 incident particles.

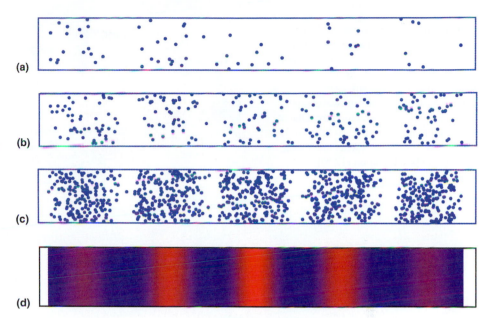

we expect from waves rather than from particles. The measurement of the electron current to the slit blocker seems to indicate that the electron *either* went through one slit *or* through the other. However, this conclusion is inconsistent with the appearance of a diffraction pattern because a diffraction pattern can only arise if one and the same electron goes through both slits!

Regardless of how you turn these results around, you will find that all the results are inconsistent with the logic chemists are accustomed to from classical mechanics, namely, *either* the electron goes through one slit and not the other, *or* it goes through both slits. This logic cannot explain the results! In quantum mechanical terms, the electron wave function is a superposition of wave functions for going through the top slit and the bottom slit, which is equivalent to saying that the electron goes through both slits. We will have much more to say about wave functions in the next few chapters. The act of measurement, such as blocking one slit, changes the wave function such that it goes through either the top slit or the bottom one. The results represent a mixture of particle and wave behavior. Individual electrons move through the slits and generate points of light on the screen. This behavior is particle-like. However, the location of the points of light on the screen is not what is expected from classical trajectories; it is governed by the diffraction pattern. This behavior is wave-like. Whereas in classical mechanics, the operative word concerning several possible modes of behavior is *or,* in quantum mechanics it is *and.* If all of this seems strange to you at first sight, welcome to the crowd! In 1997, this classic double-slit experiment was carried out using a collimated beam of He atoms. The diffraction pattern observed was exactly as described earlier, suggesting that a He atom also goes through both slits.

12.7 Atomic Spectra

The most direct evidence of energy quantization comes from the analysis of the light emitted from highly excited atoms in a plasma. The structure of the atom was not known until fundamental studies using the scattering of alpha particles were carried out in Ernest Rutherford's laboratory beginning in 1910. These experiments showed that the positive and negative charges in an atom were separated. The positive charge is contained in a small volume called the *nucleus,* whereas the negative charge of the electrons occupies a much greater volume that is centered at the nucleus. In analogy to our solar system, the first picture that emerged of the atom was of electrons orbiting the nucleus.

FIGURE 12.9
Light emitted from a hydrogen discharge lamp is passed through a narrow slit and separated into its component wavelengths by a dispersing element. As a result, multiple images of the slit, each corresponding to a different wavelength, are seen on the photographic film. One of the different series of spectral lines for H is shown. The term $\tilde{\nu}$ represents the inverse wavelength [see Equation (12.8)].

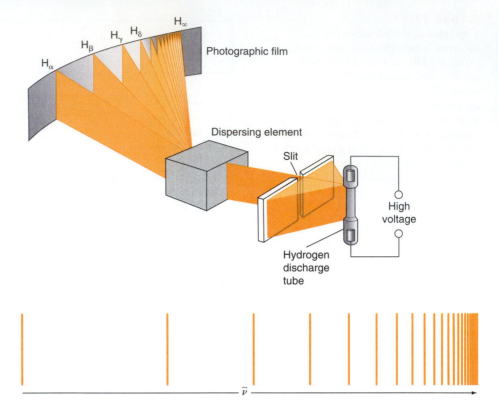

However, this picture of the atom is inconsistent with electrodynamic theory. An electron orbiting the nucleus is constantly accelerating and must therefore radiate energy. In a classical picture, the electron would continually radiate away its kinetic energy and eventually fall into the nucleus. This clearly was not happening, but why? We will answer this question when we discuss the hydrogen atom in Chapter 15. Even before Rutherford's experiments, it was known that if an electrical arc was placed across a vacuum tube with a small partial pressure of a gas like hydrogen, light is emitted. Our present picture of this phenomenon is that the atom takes up energy from the electromagnetic field and makes a transition to an excited state. The excited state has a limited lifetime, and when the transition to a state of lower energy occurs, light is emitted. An apparatus used to obtain atomic spectra and a typical spectrum are shown schematically in Figure 12.9.

How did scientists working in the 1890s explain these spectra? The most important experimental observation that these scientists made is that over a wide range of wavelengths, light is only observed at certain discrete wavelengths, that is, it is *quantized*. This was not understandable on the basis of classical theory because in classical physics, energy is a continuous variable. Even more baffling to these first spectroscopists was that they could derive a simple relationship to explain all of the frequencies that appeared in the emission spectrum. For the emission spectra observed, the inverse of the wavelength, $1/\lambda = \tilde{\nu}$ of all lines in an atomic hydrogen spectrum is given by equations of the type

$$\tilde{\nu} \ (\text{cm}^{-1}) = R_H \ (\text{cm}^{-1}) \left(\frac{1}{n_1^2} - \frac{1}{n^2} \right), \quad \text{for } n > n_1 \qquad (12.8)$$

in which only a single parameter, n_1, appears. In this equation, n is an integer that takes on the values $n_1 + 1, n_1 + 2, n_1 + 3, \dots$, and R_H is called the **Rydberg constant.** For hydrogen, R_H has the value 109,677.581 cm^{-1}. What gives rise to such a simple relationship and why does n take on only integral values? We will see that it is the wave character of the electron that gives rise to this equation.

Vocabulary

blackbody radiation	photoelectric effect	quantization	spectral density
de Broglie relation	photon	random phase angle	wave-particle duality
ideal blackbody	Planck's constant	Rydberg constant	work function

Questions on Concepts

Q12.1 How did Planck conclude that the discrepancy between experiments and classical theory for blackbody radiation occurred at high and not low frequencies?

Q12.2 The inability of classical theory to explain the spectral density distribution of a blackbody was called the *ultraviolet catastrophe*. Why is this name appropriate?

Q12.3 Why does the analysis of the photoelectric effect based on classical physics predict that the kinetic energy of electrons will increase with increasing light intensity?

Q12.4 What did Einstein postulate to explain that the kinetic energy of the emitted electrons in the photoelectric effect depends on the frequency? How does this postulate differ from the predictions of classical physics?

Q12.5 Which of the experimental results for the photoelectric effect suggests that light can display particle-like behavior?

Q12.6 In the diffraction of electrons by crystals, the volume sampled by the diffracting electrons is on the order of 3 to 10 atomic layers. If He atoms are incident on the surface, only the topmost atomic layer is sampled. Can you explain this difference?

Q12.7 In the double-slit experiment, researchers found that an equal amount of energy passes through each slit. Does this result allow you to distinguish between purely particle-like and purely wave-like behavior?

Q12.8 Is the intensity observed from the diffraction experiment depicted in Figure 12.6 the same for the angles shown in parts (b) and (c)?

Q12.9 What feature of the distribution depicted as case 1 in Figure 12.7 tells you that the broad distribution arises from diffraction?

Q12.10 Why were investigations at the atomic and subatomic levels required to detect the wave nature of particles?

Problems

Problem numbers in **RED** indicate that the solution to the problem is given in the *Student Solutions Manual*.

P12.1 The distribution in wavelengths of the light emitted from a radiating blackbody is a sensitive function of the temperature. This dependence is used to measure the temperature of hot objects, without making physical contact with those objects, in a technique called *optical pyrometry*. In the limit $(hc/\lambda kT) >> 1$, the maximum in a plot of $\rho(\lambda,T)$ versus λ is given by $\lambda_{max} = hc/5kT$. At what wavelength does the maximum in $\rho(\lambda,T)$ occur for $T = 450.$, $1500.$, and $4500.$ K?

P12.2 For a monatomic gas, one measure of the "average speed" of the atoms is the root mean square speed, $v_{rms} = |v^2|^{1/2} = \sqrt{3kT/m}$, in which m is the molecular mass and k is the Boltzmann constant. Using this formula, calculate the de Broglie wavelength for He and Ar atoms at 100. and at 500. K.

P12.3 Using the root mean square speed, $v_{rms} = |v^2|^{1/2} = \sqrt{3kT/m}$, calculate the gas temperatures of He and Ar for which $\lambda = 0.20$ nm, a typical value needed to resolve diffraction from the surface of a metal crystal. On the basis of your result, explain why Ar atomic beams are not suitable for atomic diffraction experiments.

P12.4 Electrons have been used to determine molecular structure by diffraction. Calculate the speed of an electron for which the wavelength is equal to a typical bond length, namely, 0.150 nm.

P12.5 Calculate the speed that a gas-phase oxygen molecule would have if it had the same energy as an infrared photon ($\lambda = 10^4$ nm), a visible photon ($\lambda = 500.$ nm), an ultraviolet photon ($\lambda = 100.$ nm), and an X-ray photon ($\lambda = 0.1$ nm). What temperature would the gas have if it had the same energy as each of these photons? Use the root mean square speed, $v_{rms} = |v^2|^{1/2} = \sqrt{3kT/m}$, for this calculation.

P12.6 Pulsed lasers are powerful sources of nearly monochromatic radiation. Lasers that emit photons in a pulse of 10.-ns duration with a total energy in the pulse of 0.10 J at 1000. nm are commercially available.

   a. What is the average power (energy per unit time) in units of watts (1 W = 1 J/s) associated with such a pulse?

   b. How many 1000.-nm photons are emitted in such a pulse?

P12.7 Assume that water absorbs light of wavelength 3.00×10^{-6} m with 100% efficiency. How many photons are required to heat 1.00 g of water by 1.00 K? The heat capacity of water is 75.3 J mol^{-1} K^{-1}.

P12.8 A 1000.-W gas discharge lamp emits 3.00 W of ultraviolet radiation in a narrow range centered near 280. nm. How many photons of this wavelength are emitted per second?

P12.9 A newly developed substance that emits 225 W of photons with a wavelength of 225 nm is mounted in a small rocket such that all of the radiation is released in the same direction. Because momentum is conserved, the rocket will be accelerated in the opposite direction. If the total mass of the rocket is 5.25 kg, how fast will it be traveling at the end of 365 days in the absence of frictional forces?

P12.10 What speed does a H_2 molecule have if it has the same momentum as a photon of wavelength 280. nm?

P12.11   The following data were observed in an experiment on the photoelectric effect from potassium:

10^{19} Kinetic Energy (J)	4.49	3.09	1.89	1.34	0.700	0.311
Wavelength (nm)	250.	300.	350.	400.	450.	500.

Graphically evaluate these data to obtain values for the work function and Planck's constant.

P12.12   Show that the energy density radiated by a blackbody

$$\frac{E_{total}(T)}{V} = \int_0^\infty \rho(\nu,T)\, d\nu = \int_0^\infty \frac{8\pi h\nu^3}{c^3}\frac{1}{e^{h\nu/kT}-1}\, d\nu$$

depends on the temperature as T^4. (*Hint:* Make the substitution of variables $x = h\nu/kT$.) The definite integral

$$\int_0^\infty [x^3/(e^x - 1)]\, dx = \pi^4/15.$$ Using your result, calculate the energy density radiated by a blackbody at 800. and 4000. K.

P12.13   The power per unit area emitted by a blackbody is given by $P = \sigma T^4$ with $\sigma = 5.67 \times 10^{-8}$ W m^{-2} K^{-4}. Calculate the energy radiated per second by a spherical blackbody of radius 0.500 m at 1000. K. What would the radius of a blackbody at 2500. K be if it emitted the same energy as the spherical blackbody of radius 0.500 m at 1000. K?

P12.14   The power per unit area radiated by blackbody per unit area of surface expressed in units of W m^{-2} is given by $P = \sigma T^4$ with $\sigma = 5.67 \times 10^{-8}$ W m^{-2} K^{-4}. The radius of the sun is 7.00×10^5 km and the surface temperature is 6000. K. Calculate the total energy radiated per second by the sun. Assume ideal blackbody behavior.

P12.15   The observed lines in the emission spectrum of atomic hydrogen are given by

$$\tilde{\nu}\,(\text{cm}^{-1}) = R_H\,(\text{cm}^{-1})\left(\frac{1}{n_1^2} - \frac{1}{n^2}\right)\text{cm}^{-1}, \quad \text{for } n > n_1$$

In the notation favored by spectroscopists, $\tilde{\nu} = 1/\lambda = E/hc$ and $R_H = 109,677$ cm^{-1}. The Lyman, Balmer, and Paschen series refers to $n_1 = 1$, 2, and 3, respectively, for emission from atomic hydrogen. What is the highest value of $\tilde{\nu}$ and E in each of these series?

P12.16   A beam of electrons with a speed of 3.50×10^4 m/s is incident on a slit of width 200. nm. The distance to the detector plane is chosen such that the distance between the central maximum of the diffraction pattern and the first diffraction minimum is 0.500 cm. How far is the detector plane from the slit?

P12.17   If an electron passes through an electrical potential difference of 1.00 V, it has an energy of 1.00 electron-volt. What potential difference must it pass through in order to have a wavelength of 0.100 nm?

P12.18   What is the maximum number of electrons that can be emitted if a potassium surface of work function 2.40 eV absorbs 3.25×10^{-3} J of radiation at a wavelength of 300. nm? What is the kinetic energy and velocity of the electrons emitted?

P12.19   The work function of platinum is 5.65 eV. What is the minimum frequency of light required to observe the photoelectric effect on Pt? If light with a 150.-nm wavelength is absorbed by the surface, what is the velocity of the emitted electrons?

P12.20   X-rays can be generated by accelerating electrons in a vacuum and letting them impact on atoms in a metal surface. If the 1000.-eV kinetic energy of the electrons is completely converted to photon energy, what is the wavelength of the X-rays produced? If the electron current is 1.50×10^{-5} A, how many photons are produced per second?

P12.21   When a molecule absorbs a photon, both the energy and momentum are conserved. If a H_2 molecule at 300. K absorbs an ultraviolet photon of wavelength 100. nm, what is the change in its velocity Δv? Given that its average speed is $v_{rms} = \sqrt{3kT/m}$, what is $\Delta v/v_{rms}$?

Web-Based Simulations, Animations, and Problems

W12.1   The maximum in a plot of the spectral density of blackbody radiation versus T is determined for a number of values of T using numerical methods. Using these results, the validity of the approximation $\lambda_{max} = hc/5kT$ is tested graphically.

W12.2   The total radiated energy of blackbody radiation is calculated numerically for the temperatures of W12.1. Using these results, the exponent in the relation $E = CT^a$ is determined graphically.

W12.3   Diffraction of visible light from a single slit is simulated. The slit width and light wavelength are varied using sliders. The student is asked to draw conclusions about how the diffraction pattern depends on these parameters.

W12.4   Diffraction of a particle from single and double slits is simulated. The intensity distribution on the detector plane is updated as each particle passes through the slits. The slit width and light wavelength are varied using sliders. The student is asked to draw conclusions about how the diffraction pattern depends on these parameters.

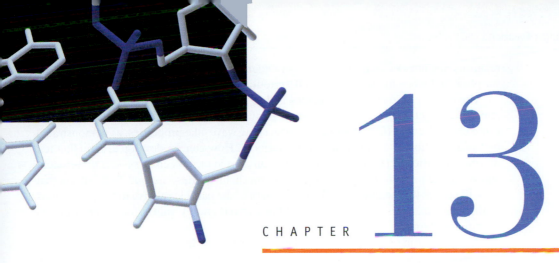

CHAPTER

13

The Schrödinger Equation

The key to understanding why classical mechanics does not provide an appropriate framework for understanding phenomena at the atomic level is the recognition that wave-particle duality needs to be integrated into the physics. Rather than solving Newton's equations of motion for a particle, an appropriate wave equation needs to be solved for the wave-particle. Erwin Schrödinger was the first to formulate such an equation successfully. A quantum mechanical system is characterized by a wave function, and the wave function is a complete description of the system in that any measurable (observable) property can be obtained if the wave function is known.

13.1 What Determines If a System Needs to Be Described Using Quantum Mechanics?

Quantum mechanics was viewed as a radically different way of looking at matter at the molecular, atomic, and subatomic level in the 1920s. However, the historical distance we have traveled from what was a revolution at the time makes the quantum view much more familiar today. It is important to realize that classical and quantum mechanics are not two competing ways to describe the world around us. Each has its usefulness in a different regime of physical properties that describe reality. Quantum mechanics merges seamlessly into classical mechanics and one can show that classical mechanics can be derived from quantum mechanics in the limit in which allowed energy values are continuous rather than discrete. Some complexities, however, will require hard thinking from you as you gain an understanding of quantum mechanics. For instance, it is not correct to say that whenever one talks about atoms, a quantum mechanical description must be used.

To illustrate this point, consider a container filled with argon gas at a low pressure. At the molecular level, the origin of pressure is the collision of rapidly and randomly moving argon atoms with the container walls. Even at the level of considering the force exerted by a single argon atom against the wall, classical mechanics gives a perfectly good description of the origin of pressure, although we are talking about atoms. However, if we pass light of a very short wavelength through this same gas and ask how much energy can be taken up by an argon atom, we must use a quantum mechanical description. At first, this seems puzzling; why do we need quantum mechanics in one case but not the other? On further consideration, we discover that a very few important relationships govern whether a classical description suffices in a given case. We discuss these relationships next in order to develop an understanding of when to use a classical description and when to use a quantum description for a given system.

The essence of quantum mechanics is that particles and waves are not really separate and distinct entities. Waves can show particle-like behavior. An illustration for this is the photoelectric effect. Particles can also show wave-like properties, as demonstrated by the diffraction of atomic beams from surfaces. How can we develop criteria that tell us when a particle (classical) description of an atomic or molecular system is sufficient and when we need to use a wave (quantum mechanical) description? Two criteria that we will use are the magnitude of the wavelength of the particle relative to the dimensions of the problem, and the degree to which the allowed energy values form a **continuous energy spectrum.**

A good starting point is to think about diffraction of light of wavelength λ passing through a slit of width a. Ray optics is a good description as long as $\lambda << a$. Diffraction is only observed when the wavelength is comparable to the slit width. How big is the wavelength of a molecule? Of a macroscopic mass like a baseball? By putting numbers into Equation (12.6), you will find that the wavelength for a room temperature H_2 molecule is about 100 pm and that for a baseball is about 2×10^{-26} m. Keep in mind that because p rather than v appears in the denominator of Equation (12.6), the wavelength of a toluene molecule with the same velocity as a H_2 molecule is about a factor of 50 smaller. When interacting with matter, an H_2 molecule will only be diffracted if it encounters an opening that has a size similar to its wavelength. As we learned in discussing the Davisson–Germer experiment in Chapter 12, crystalline solids have atomic spacings that are appropriate for the diffraction of electrons as well as light atoms and molecules. Particle diffraction is a demonstration of wave-particle duality. To see the wave character of a baseball, we will need to come up with a diffraction experiment. We cannot see diffraction of a baseball because we cannot construct an opening whose size is $\sim 2 \times 10^{-26}$ m. This does not mean that wave-particle duality breaks down for macroscopic masses; it simply says that we have no experimental way to demonstrate the wave character of a baseball. There is no sharp boundary so that above a certain value for the momentum we are dealing with a particle and below it we are dealing with a wave. The degree to which each of these properties is exhibited flows smoothly from one extreme to the other. Does this mean that hydrogen molecules in a 1- $\times$ 1- $\times$ 1-m box behave totally classically with respect to all kinds of ways in which we add energy to them? No, it doesn't because the localization of an electron to a small volume around the nuclei brings out the wave-like behavior of the particles (protons and electrons) that make up the molecule.

We next discuss the second criterion for when we need a quantum mechanical description of a system. It is based on the energy spectrum of the system. Because all values of the energy are allowed for a classical system, we say that it has a continuous energy spectrum. In a quantum mechanical system, only certain values of the energy are allowed, and we say that such a system has a **discrete energy spectrum.** To make this criterion quantitative, we need to discuss the Boltzmann distribution.

You will learn more about Boltzmann's work in Chapter 23. At this point, we will try to make his most important result plausible so that we can apply it in our studies of quantum mechanics. Consider a 1-L container filled with an ideal atomic gas at the standard conditions of 1 bar pressure (10^5 Pascal) and a temperature of 298.15 K. Because the atoms have no rotational or vibrational degrees of freedom, all of their energy is in the form of translational kinetic energy. At equilibrium, not all of the atoms have the same kinetic energy. In fact, the atoms exhibit a broad range of energies. To define the distribution of atoms having a given energy, we use descriptors such as the *mean* or the *median* or the *most probable energy per atom*. For the atoms we are considering, the root mean square energy is simply related to the absolute temperature T by

$$\bar{E} = \frac{3}{2} kT \tag{13.1}$$

The Boltzmann constant, k, is the familiar ideal gas law constant, R, divided by Avogadro's number.

We said that there is a broad distribution of kinetic energy in the gas for the individual atoms. What governs the probability of observing one value of the energy as opposed to another? This question led Ludwig Boltzmann to one of the most important equations in physics and chemistry. Looking specifically at our case, it relates the number of atoms n_i that have energy ε_i to the number of atoms n_j that have energy ε_j by the equation

$$\frac{n_i}{n_j} = \frac{g_i}{g_j} e^{-\left(\frac{\varepsilon_i - \varepsilon_j}{kT}\right)}$$

(13.2)

This formula is called the **Boltzmann distribution.** An important concept to keep in mind is that a formula is just a shorthand way of describing phenomena that occur in the real world. It is critical that you understand what lies behind the formula. To emphasize this approach, take a closer look at this equation. It says that the ratio of the number of atoms having the energy ε_i to the number having the energy ε_j depends on three things. It depends on the difference in the energies and the temperature with an exponential dependence. This means that this ratio varies rapidly with temperature and $\varepsilon_i - \varepsilon_j$. The equation also states that it is the ratio of the energy difference to kT that is important. What is kT? It has the units of energy as it must, and it is approximately the average energy that an atom has at temperature T. We can understand this exponential term as telling us that the larger the temperature, the closer the ratio n_i/n_j will be to unity. This means that the probability of an atom having a given energy will fall off exponentially with increasing energy at high temperatures.

The third factor that influences the ratio n_i/n_j is the ratio g_i/g_j. The quantities g_i and g_j are the degeneracies of the energy values i and j. The **degeneracy** of an energy value counts the number of ways in which an atom can have an energy ε_i within the interval $\varepsilon - \Delta\varepsilon < \varepsilon_i < \varepsilon + \Delta\varepsilon$. The degeneracy can depend on the energy. In our example, degeneracy can be illustrated as follows. The energy of an atom $\varepsilon_i = 1/2\, mv_i^2$ is determined by $v_i^2 = v_{xi}^2 + v_{yi}^2 + v_{zi}^2$. We have explicitly written that the energy depends only on the speed of the atom and not on any of its individual velocity components. For a fixed value of $\Delta\varepsilon$, there are many more ways of combining different individual velocity components to give the same speed at large speeds than there are for low speeds. Therefore, the degeneracy corresponding to a particular energy ε_i can be large and will increase with the speed.

The importance of these considerations will become clearer as we begin to apply quantum mechanics to atoms and molecules. As we have already stated, a quantum mechanical system has a discrete rather than continuous energy spectrum. If kT is small compared to the spacing between allowed energies, the distribution of states in energy will be very different from classical mechanics for which we have a continuous energy spectrum. On the other hand, if kT is much larger than the energy spacing, classical and quantum mechanics will give the same result for the relative numbers of atoms or molecules of different energy. This can occur in either of two limits: large T or small $\varepsilon_i - \varepsilon_j$. This illustrates how we can have a continuous transition between classical and quantum mechanics. A large increase in T could cause a system that exhibited quantum behavior at low temperatures to exhibit classical behavior at high temperatures. A calculation using the Boltzmann distribution for a two-level system is carried out in Example Problem 13.1.

EXAMPLE PROBLEM 13.1

Consider a system of 1000. particles that can only have two energies, ε_1 and ε_2, with $\varepsilon_2 > \varepsilon_1$. The difference in the energy between these two values is $\Delta\varepsilon = \varepsilon_2 - \varepsilon_1$. Assume that $g_1 = g_2 = 1$.

a. Graph the number of particles, n_1 and n_2, in states ε_1 and ε_2 as a function of $kT/\Delta\varepsilon$. Explain your result.

b. At what value of $kT/\Delta\varepsilon$ do 750. of the particles have the energy ε_1?

Solution

Using information from the problem and Equation (13.2), we can write down the following two equations: $n_2/n_1 = e^{-\Delta\varepsilon/kT}$ and $n_1 + n_2 = 1000$. We can solve these two equations for n_2 and n_1. We obtain

$$n_2 = \frac{1000. \, e^{-\Delta\varepsilon/kT}}{1 + e^{-\Delta\varepsilon/kT}} \quad \text{and} \quad n_1 = \frac{1000.}{1 + e^{-\Delta\varepsilon/kT}}$$

Plotting these functions as a function of $kT/\Delta\varepsilon$, we obtain the graphs shown here.

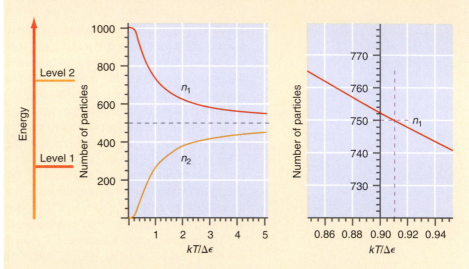

We see from the left graph that as long as $kT/\Delta\varepsilon$ is small, all the particles will have the lower energy. How do we interpret this result? As long as the thermal energy of the particle, which is about kT, is much less than the difference in energy between the two allowed values, the particles with the lower energy are unable to gain energy through collisions with other particles and increase their energy. However, as $kT/\Delta\varepsilon$ increases (which is equivalent to a temperature increase for a fixed energy difference between the two values), the random thermal energy available to the particles enables some of them to jump up to the higher energy value. Therefore, n_1 decreases, and n_2 increases. However, we see that for all finite temperatures, $n_1 > n_2$. As T approaches infinity, n_1 becomes equal to n_2.

We solve part (b) graphically. The n_1 term is shown as a function of $kT/\Delta\varepsilon$ on an expanded scale on the right side of the preceding figure. We see that $n_1 = 750$ for $kT/\Delta\varepsilon = 0.91$.

As we have seen, the population of states that we label by their adjacent allowed energy values ε_i and ε_j are very different if

$$\frac{(\varepsilon_i - \varepsilon_j)}{kT} \geq 1 \tag{13.3}$$

and very similar if the left side of Equation (13.3) is much smaller than one. What are the consequences of this result?

Consider a quantum mechanical system that, unlike a classical system, has a discrete energy spectrum. We refer to the allowed values of energy as the energy levels. Anticipating a system that we will deal with in Chapter 14, we refer specifically to the vibrational energy levels of a molecule. The allowed levels are equally spaced with an interval ΔE. We number these discrete energy levels with integers, beginning with one. Under what conditions will this quantum mechanical system *appear* to follow classical behavior? It will do so if the discrete energy spectrum *appears* to be continuous. How can this occur? In a gas at equilibrium, the total energy of an individual molecule will fluctuate within a range of $\Delta E \approx kT$ through collisions of molecules with one another. Therefore,

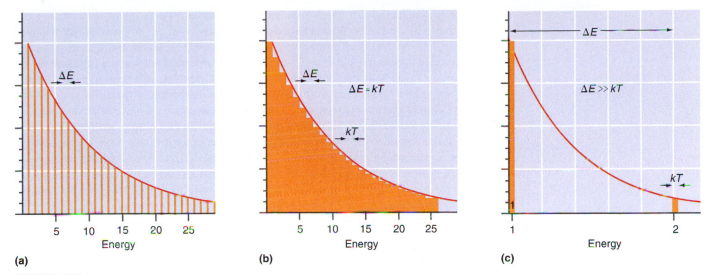

FIGURE 13.1
The relative population in the different energy levels designated by the integers 1–25 is plotted at constant T as a function of energy (a) for sharp energy levels, (b) for $\Delta E \approx kT$, and (c) for $\Delta E >> kT$. In both (b) and (c), the Boltzmann distribution describes the relative populations. However, the system behaves as if it has a continuous energy spectrum only if $\Delta E \approx kT$ or $\Delta E < kT$.

the energy of a molecule with a particular vibrational quantum number will fluctuate within a range of width kT. A plot of the relative number of molecules having a vibrational energy E as a function of E is shown in Figure 13.1 for sharp energy levels and for the two indicated limits, $\Delta E \approx kT$, and $\Delta E >> kT$. The plot is generated using the Boltzmann distribution.

For $\Delta E \approx kT$, each discrete energy level is sufficiently broadened through energy fluctuations that adjacent energy levels can no longer be distinguished. This is indicated by the overlap of the orange bars representing individual states shown in Figure 13.1b. In this limit, any energy that we choose in the range shown lies in the orange area. It corresponds to an allowed value and therefore the discrete energy spectrum *appears* to be continuous. Classical behavior will be observed under these conditions. However, if $\Delta E >> kT$, an arbitrarily chosen energy in the range will lie in the blue area with high probability, because the orange bars of width kT are widely separated. The blue area corresponds to forbidden energies and therefore the discontinuous nature of the energy spectrum will be observable. Quantum mechanical behavior will be observed under these conditions.

If the allowed energies form a continuum, classical mechanics is sufficient to describe *that feature* of the system. If they are discrete, a quantum mechanical description is needed. The words *that feature* require emphasis. The pressure exerted by the H_2 molecules in the box arises from momentum transfer governed by their translational energy spectrum, which *appears* to be continuous, as we will learn in Chapter 14. Therefore we do not need quantum mechanics to discuss the pressure in the box. However, if we discuss light absorption by the same H_2 molecules, a quantum mechanical description of light absorption is required. This is so because light absorption involves an electronic excitation within the molecule, and the spacing between electronic energy levels is much larger than kT, so that these levels remain discrete at all reasonable temperatures.

13.2 Classical Waves and the Nondispersive Wave Equation

In Chapter 12, we learned that particles exhibit wave character under certain conditions. This suggests that there is a wave equation that should be used to describe particles. This equation is called the Schrödinger equation, and it is the fundamental equation used to describe atoms and molecules. However, before discussing the quantum mechanical wave equation, we briefly review classical waves and the classical wave equation.

What characteristics capture the essence of waves? Think about the collision between two billiard balls. We can treat the balls as point masses (any pool player will recognize this as an idealization) and apply Newton's laws of motion to calculate trajectories, momenta, and energies as a function of time if we know all the forces acting on the balls. Now think of yourself shouting. Often you will hear an echo. What is happening here? Your vocal cords create a local compression of the air in your larynx. This compression zone propagates away from its source as a wave with the speed of sound. The louder the sound, the larger the pressure is in the compressed zone. The pressure is the amplitude of the wave and the energy contained in this wave is proportional to the square of the maximum pressure. The sound reflects from a surface and comes back to you as a weakened local compression of the air. When the wave is incident on your eardrum, a signal is generated that you recognize as sound. Note that this energy transfer is fundamentally different from the direct transfer that occurs in the collision of the two billiard balls. The energy stored in a wave is not localized to a small volume except at the point of origin. A further important characteristic of a wave is that it has a characteristic velocity and frequency with which it propagates. The velocity and frequency govern the variation of the amplitude with time for a fixed position of the observer with respect to the source and the variation of the amplitude with distance between the source and the observer at a fixed time:

A wave can be represented pictorially by a succession of **wave fronts,** corresponding to surfaces over which the amplitude of the wave has a maximum or minimum value. A point source emits spherical waves as shown in Figure 13.2b, and the light passing through a rectangular slit can be represented by cylindrical waves as shown in Figure 13.2c. The waves sent out from a faraway source such as the sun when viewed from the Earth are spherical waves with such little curvature that they can be represented as plane waves as shown in Figure 13.2a.

Mathematically, the amplitude of a wave can be described by a **wave function.** The wave function describes how the amplitude of the disturbance depends on the variables x and t. The variable x is measured along the direction of propagation. For convenience, only sinusoidal waves of wavelength λ and the single **frequency** $\nu = 1/T$, where T is the **period,** are considered. The velocity, v, frequency, ν, and **wavelength,** λ, are related by v $= \lambda\nu$. The peak-to-peak amplitude of the wave is $2A$:

$$\Psi(x,t) = A \sin 2\pi \left(\frac{x}{\lambda} - \frac{t}{T} \right) \tag{13.4}$$

In this equation, we have arbitrarily chosen our zero of time and distance such that $\Psi(0,0) = 0$. This equation represents a wave that is moving in the direction of positive x. You can convince yourself of this by considering how a specific feature of this wave changes with time. The wave amplitude is zero for

$$2\pi \left(\frac{x}{\lambda} - \frac{t}{T} \right) = n\pi \tag{13.5}$$

where n is an integer. Solving for x, the location of the nodes is obtained:

$$x = \lambda \left(\frac{n}{2} + \frac{t}{T} \right) \tag{13.6}$$

Note that x increases as t increases, showing that the wave is moving in the direction of positive x. Figure 13.3 shows a graph of the wave functions. To graph this function in two dimensions, one of the variables is kept constant.

The functional form in Equation (13.4) appears so often that it is convenient to combine some of the constants and variables to write the wave amplitude as

$$\Psi(x,t) = A \sin (kx - \omega t) \tag{13.7}$$

the Chemistry place

Web-Based Problem 13.1 Transverse, Longitudinal, and Surface Waves

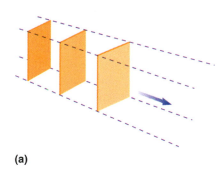

(a)

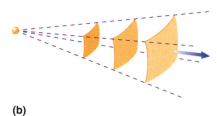

(b)

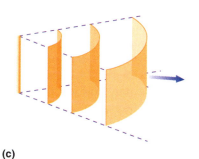

(c)

FIGURE 13.2
Waves can be represented by a succession of surfaces over which the amplitude of the wave has its maximum or minimum value. The distance between successive surfaces is the wavelength. Representative surfaces are shown for (a) plane waves, (b) spherical waves, and (c) cylindrical waves. The direction of propagation of the waves is perpendicular to the surfaces as indicated by the blue arrows.

FIGURE 13.3
The upper panel shows the wave amplitude as a function of time at a fixed point. The wave is completely defined by the period, the maximum amplitude, and the amplitude at $t = 0$. The lower panel shows the analogous information when the wave amplitude is plotted as a function of distance for a given time; $\lambda = 1.46$ m and $T = 1.00 \times 10^{-3}$ s.

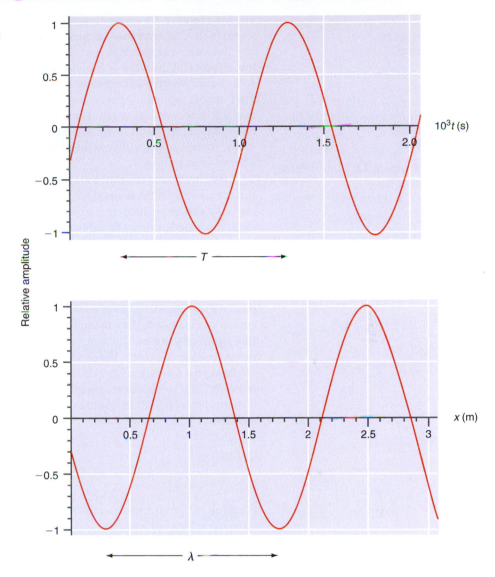

The quantity k is called the **wave vector** and is defined by $k = 2\pi/\lambda$. The quantity $\omega = 2\pi\nu$ is called the **angular frequency.**

Because the wave amplitude is a simple sine function in our case, it has the same value as the argument changes by 2π. The choice of a zero in position or time is arbitrary and is chosen at our convenience. To illustrate this, consider Equation (13.7) rewritten in the form

$$\Psi(x,t) = A \sin(kx - \omega t + \phi) \tag{13.8}$$

in which the quantity ϕ has been added to the argument of the sine function. This is appropriate when $\Psi(0,0) \neq 0$. The argument of the wave function is called the **phase** and a change in the initial phase ϕ shifts the wave function to the right or left relative to the horizontal axes in Figure 13.3 depending on the sign of ϕ.

Figure 13.3 shows important information about a traveling wave because the axes give the values of the amplitude and either x or t. It is incomplete because, for example, the amplitude versus distance curve only shows us one direction in space. A wave propagating in three-dimensional space can have quite different amplitudes in different directions. The two common types of waves encountered are **plane waves** and **spherical waves.**

When two or more waves are present in the same region of space, their time-dependent amplitudes add together, and the waves are said to interfere with one another.

Web-Based Problem 13.2 Interference of Two Traveling Waves

Web-Based Problem 13.3 Interference of Two Standing Waves

The **interference** between two waves gives rise to an enhancement in a region of space (**constructive interference**) if the wave amplitudes are both positive or both negative. It can also lead to a cancellation of the wave amplitude in a region of space (**destructive interference**) if the wave amplitudes are opposite in sign and equal in amplitude. At the constructive interference condition, waves originating from two sources are in phase. They are out of phase at the destructive interference condition. This is true because the maxima of the wave functions from the two sources will line up (constructive interference) only if the phases of the two functions are the same to within an integral multiple of 2π. At the positions of destructive interference, the phases must differ by $(2n + 1)\pi$, where n is an integer.

Standing waves are a second category of waves that differ from **traveling waves** in that the position of the nodes does not change with time. The wave function for a standing wave can be wrtitten as the product of two factors, one of which depends only on time, and the other depends only on distance as shown in Equation (13.9):

$$\Psi(x,t) = A\psi(x) \cos \omega t \tag{13.9}$$

The form that the standing-wave amplitudes take is shown in Figure 13.4. Standing waves arise if the space in which the waves can propagate is bounded. For instance, plucking a guitar string gives rise to a standing wave because the string is fixed at both ends. Standing waves play an important role in quantum mechanics because they represent **stationary states,** which are states of the system in which the measurable properties of the system do not change with time.

As we have seen, the functional dependencies of the wave amplitude on time and distance are not independent, except for a standing wave. For wave propagation in a medium for which all frequencies propagate with the same velocity (a nondispersive medium), we can write

$$\frac{\partial^2 \Psi(x,t)}{\partial x^2} = \frac{1}{v^2} \frac{\partial^2 \Psi(x,t)}{\partial t^2} \tag{13.10}$$

Equation (13.10) is known as the **classical nondispersive wave equation** and v designates the velocity at which the wave propagates. This equation provides a starting point in justifying the Schrödinger equation, which is the fundamental quantum mechanical wave equation. (See the Math Supplement, Appendix A, for a discussion of partial differentiation.) It turns out that the mathematics of dealing with wave functions is much simpler if these functions are represented in the complex number plane as is also discussed in the Math Supplement.

FIGURE 13.4

Time evolution of a standing wave at a fixed point. The time intervals are shown as a function of the period T. The vertical lines indicate the nodal positions x_0. Note that the wave function has temporal nodes for $t = T/4$ and $3T/4$.

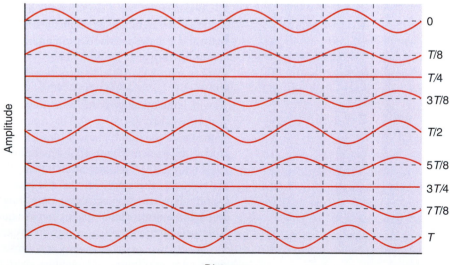

Distance

13.3 Quantum Mechanical Waves and the Schrödinger Equation

In this section, we justify the time-independent Schrödinger equation by combining the classical nondispersive wave equation and the de Broglie relation. If we substitute the wave function for a standing wave [Equation (13.9)] in Equation (13.10), we obtain

$$\frac{d^2 \psi(x)}{dx^2} + \frac{\omega^2}{v^2} \psi(x) = 0 \tag{13.11}$$

The time-dependent part $\cos \omega t$ cancels because it appears on both sides of the equation after the derivative with respect to time is taken. Using the relations $\omega = 2\pi\nu$ and $\nu\lambda = v$, Equation (13.11) becomes

$$\frac{d^2\psi(x)}{dx^2} + \frac{4\pi^2}{\lambda^2} \psi(x) = 0 \tag{13.12}$$

To this point, everything that we have written is for a classical wave. We introduce quantum mechanics by using the de Broglie relation, $\lambda = h/p$, for the wavelength. The momentum is related to the total energy, E, and the potential energy, $V(x)$, by

$$\frac{p^2}{2m} = E - V(x) \quad \text{or} \quad p = \sqrt{2m(E - V(x))} \tag{13.13}$$

Introducing this expression for the momentum into the de Broglie relation, and substituting the expression obtained for λ into Equation (13.12), we obtain

$$\frac{d^2\psi(x)}{dx^2} + \frac{8\pi^2 m}{h^2} [E - V(x)]\psi(x) = 0 \tag{13.14}$$

Using the abbreviation $h = h/2\pi$ and rewriting Equation (13.14), we obtain the **time-independent Schrödinger equation** in one dimension:

$$-\frac{h^2}{2m} \frac{d^2\psi(x)}{dx^2} + V(x)\psi(x) = E\psi(x) \tag{13.15}$$

This is the fundamental equation that we will use to study quantum mechanical systems.

13.4 Quantum Mechanics and Experimental Measurements

You are familiar with the use of Newton's second law to calculate the trajectory of a classical particle. In that case, you solve a differential equation for the position as a function of time. The time-independent Schrödinger equation is a wave equation that, when solved for the problem of interest, gives the amplitude of the wave as a function of position. We use this equation to describe systems such as a particle moving without any forces acting on it that is confined to a certain region of space or the vibrational motion of two atoms connected by a chemical bond, which we model as a spring. If we solve the time-independent Schrödinger equation for $\psi(x)$ and calculate $\Psi(x,t)$, using the relation $\Psi(x,t) = \psi(x)e^{-i\frac{E}{\hbar}t}$, how do we relate $\Psi(x,t)$ to reality? In other words, is $\Psi(x,t)$ something that we can measure in the laboratory with a scientific instrument as we can measure the position of a classical particle as a function of time?

To answer this question, we consider a specific case. For a sound wave, the wave function $\Psi(x,t)$, which is the solution of the nondispersive classical wave equation, is associated with the pressure at a time t and position x. For a water wave, $\Psi(x,t)$ is the height of the wave. What meaning does $\Psi(x,t)$ have as a solution of the Schrödinger equation? The fundamental link between the theory and reality is associated with the position of the particle. In classical physics, the position of a particle can be determined exactly, but it makes no sense to ask for the location of a wave. In quantum mechanics, particles also have wave character, and so the certainty of being able to say where a particle is located is replaced with the probability of finding the particle at a specific location. Quantum mechanics says that the probability $P(x_0,t_0)$ of finding a wave-particle at position x_0 at time t_0 within an interval dx is

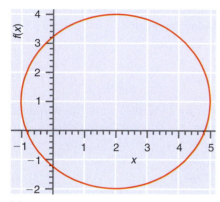

$$P(x_0,t_0) = \Psi^*(x_0,t_0)\Psi(x_0,t_0)\,dx = |\Psi(x_0,t_0)|^2\,dx \tag{13.16}$$

where $\Psi^*(x,t)$ is the **complex conjugate** of $\Psi(x,t)$. As discussed in the Math Supplement (see Appendix A), the complex conjugate of a wave function is obtained by changing the sign of $i \equiv \sqrt{-1}$, wherever it appears in the wave function, and the product of a complex function with its complex conjugate is a real function. Unlike the classical mechanics of wave motion, the wave amplitude $\Psi(x,t)$ itself has no physical meaning in quantum mechanics. Because the probability is related to the *square of the magnitude* of $\Psi(x,t)$, given by $\Psi^*(x,t)\Psi(x,t)$, the wave function can be complex or negative and still be associated with a probability that lies between 0 and 1.

This association of the wave function with the probability has an important consequence, called **normalization.** The probability that the particle is found in an interval of width dx centered at the position x must lie between 0 and 1. The sum of the probabilities over all intervals accessible to the particle is 1, because the particle is somewhere in its range. Consider a particle that is confined to a one-dimensional space of infinite extent. This leads to the following normalization condition:

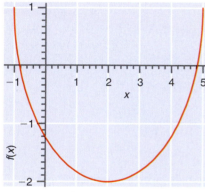

(a)

$$\int_{-\infty}^{\infty} \Psi^*(x,t)\Psi(x,t)\,dx = 1 \tag{13.17}$$

Such a definition is obviously meaningless if the integral does not exist. Therefore, $\Psi(x,t)$ must satisfy several mathematical conditions to ensure that it represents a possible physical state. These conditions are as follows:

- The wave function must be a **single-valued function** of the spatial coordinates. If this were not the case, a particle would have more than one probability of being found in the same interval. For example, for the ellipse depicted in Figure 13.5a, $f(x)$ is a double-valued function of x. If only the part of the ellipse is considered for which $f(x) < 1$, as in Figure 13.5b, $f(x)$ is a single-valued function of x.

- The first derivative of the wave function must be continuous so that the second derivative exists and is well behaved. If this were not the case, we could not set up the Schrödinger equation. As shown in Figure 13.6, $\sin x$ is a **continuous function** of x, but $\tan x$ has discontinuities at values of x for which the function becomes infinite.

- The wave function cannot have an infinite amplitude over a finite interval. If this were the case, the wave function could not be normalized. For example, the function $\Psi(x,t) = e^{-i(E/\hbar)t}\tan\dfrac{2\pi x}{a}$, for $0 \leq x \leq a$, cannot be normalized.

(b)

FIGURE 13.5

(a) A double-valued function of x and (b) a single-valued function of x.

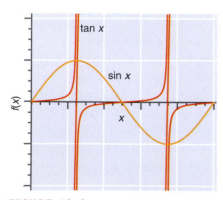

FIGURE 13.6

Examples of continuous and discontinuous functions.

Solving the time-independent Schrödinger equation for a particular case gives a set of wave functions that describe the possible states of the system and the total energy for each state. What about other observables of the system such as linear or angular momentum, kinetic energy, or potential energy? The set of wave functions that describes a quantum mechanical system is of great interest to us because all observable properties of the system can be determined if the set is known. How the values of these observables are determined from the wave function is a central topic in quantum mechanics that we will not discuss here, but the interested reader can find such a discussion in Engel and Reid, *Physical Chemistry*, Chapter 14. We focus instead on the calculation of allowed values for three observables of the system in this text. They are the total (kinetic plus potential) energy of the system, the position, and the value of the potential energy. As we have seen, the allowed values of the total energy of a quantum mechanical system are obtained by solving the Schrödinger equation. We next give a rule that allows us to determine average values of certain observables. The average value of any observable, $f(x)$, that depends only on the spatial coordinates such as the position, x, is given by

$$\langle f(x) \rangle = \frac{\int \psi^*(x) f(x) \psi(x)\, dx}{\int \psi^*(x) \psi(x)\, dx} \qquad (13.18)$$

Taking a specific example, the average value of the potential energy of a diatomic molecule modeled as two atoms connected by a spring of force constant k following Hooke's law $V(x) = 1/2\, kx^2$ is given by

$$\langle V(x) \rangle = \frac{\int \psi^*(x) \frac{1}{2} kx^2 \psi(x)\, dx}{\int \psi^*(x) \psi(x)\, dx} \qquad (13.19)$$

The formulas in Equations (13.18) and (13.19) can be readily understood using the association of $\Psi^*(x)\Psi(x)dx$ with probability. The average value of $f(x)$ is simply the value of $f(x)$ multiplied by the probability that the particle will be found at the position x, integrated over all possible values of $f(x)$. A similar argument can be made for the average value of the potential energy. We use $\psi(x)$ rather than $\Psi(x,t)$ in Equations (13.18) and (13.19) because the factors $e^{-i(E/h)t}$ and $e^{+i(E/h)t}$ cancel because $f(x)$ and $V(x)$ are not functions of time.

13.5 The Solutions of the Schrödinger Equation Are Orthogonal

A further important property of the solutions of the Schrödinger equation for a particular system is that they are orthogonal to one another. Orthogonality is a concept that is familiar to you in a vector space. For example, orthogonality in three-dimensional Cartesian coordinate space is defined by

$$\mathbf{x} \cdot \mathbf{y} = \mathbf{x} \cdot \mathbf{z} = \mathbf{y} \cdot \mathbf{z} = 0 \qquad (13.20)$$

in which the scalar product between the unit vectors along the x, y, and z axes is 0. In function space, the analogous expression that defines **orthogonality** between the eigenfunctions $\psi_i(x)$ and $\psi_j(x)$ of a quantum mechanical operator is

$$\int_{-\infty}^{\infty} \psi_i^*(x)\psi_j(x)\, dx = \begin{cases} 0, i \neq j \\ 1, i = j \end{cases} \qquad (13.21)$$

Example Problem 13.2 shows that graphical methods can be used to determine if two functions are orthogonal.

EXAMPLE PROBLEM 13.2

Show graphically that $\sin x$ and $\cos 3x$ are orthogonal functions. Also show graphically that $\displaystyle\int_{-\infty}^{\infty} (\sin mx)(\sin nx)\, dx \neq 0$ for $n = m = 1$.

Solution

The functions are shown in the following graphs. The vertical axes have been offset to avoid overlap and the horizontal line indicates the zero for each plot. Because the functions are periodic, we can draw conclusions about their behavior in an infinite interval by considering their behavior in any interval that is an integral multiple of the period.

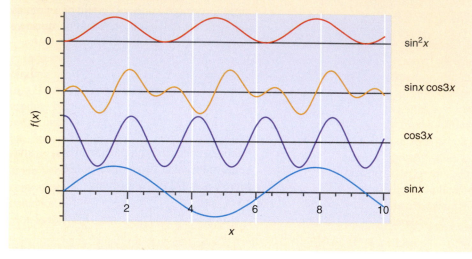

The integral of these functions equals the sum of the areas between the curves and the zero line. Areas above and below the line contribute with positive and negative signs, respectively, and indicate that $\int_{-\infty}^{\infty} \sin x \cos 3x \, dx = 0$ and $\int_{-\infty}^{\infty} \sin x \sin x \, dx > 0$. By similar means, we could show that any two functions of the type $\sin mx$ and $\sin nx$ or $\cos mx$ and $\cos nx$ are orthogonal unless $n = m$. Are the functions $\cos mx$ and $\sin mx(m = n)$ orthogonal? If, in addition to satisfying Equation (13.21), the integral has the value 1 for $i = j$, we say that the functions are normalized and form an **orthonormal** set. Wave functions must be normalized so that they can be used to calculate probabilities. We show how to normalize wave functions in Example Problems 13.3 and 13.4.

EXAMPLE PROBLEM 13.3

Normalize the function $a(a - x)$ over the interval $0 \leq x \leq a$.

Solution

To normalize a function $\psi(x)$ over the given interval, we multiply it by a constant N, and then calculate N from the equation $N^2 \int_0^a \psi^*(x) \, \psi(x) \, dx = 1$. In this particular case,

$$N^2 \int_0^a [a(a - x)]^2 \, dx = 1$$

$$N^2 a^2 \int_0^a [a^2 - 2ax + x^2] \, dx = 1$$

$$N^2 \left(a^4 x - a^3 x^2 + a^2 \frac{x^3}{3} \right)_0^a = 1$$

$$N^2 \frac{a^5}{3} = 1 \text{ so that } N = \sqrt{\frac{3}{a^5}}$$

The normalized wave function is $\sqrt{\dfrac{3}{a^5}} \, a(a - x)$.

FIGURE 13.7

Defining variables and the volume element in spherical coordinates.

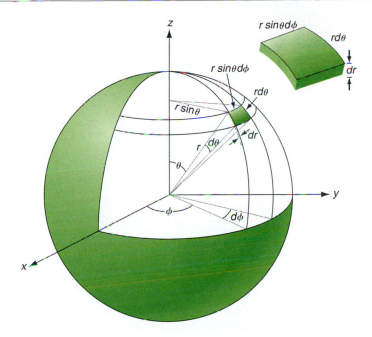

Up until now, we have considered functions of a single variable. This restricts us to dealing with a single spatial dimension. As we will see in Chapter 14, important problems such as the harmonic oscillator can be solved in a one-dimensional framework. The extension to three independent variables becomes important in describing three-dimensional systems. The three-dimensional system of most importance to us is the atom. Closed-shell atoms are spherically symmetric, so that we might expect atomic wave functions to be best described by spherical coordinates, as shown in Figure 13.7. Therefore, you should become familiar with integrations in these coordinates. The Math Supplement (see Appendix A) provides a more detailed discussion of working with spherical coordinates. Note in particular that the volume element in spherical coordinates is $r^2 \sin \theta \, dr \, d\theta \, d\phi$ and not $dr \, d\theta \, d\phi$. A function is normalized in spherical coordinates in Example Problem 13.4.

EXAMPLE PROBLEM 13.4

Normalize the function e^{-r} over the interval $0 \le r \le \infty$; $0 \le \theta \le \pi$; $0 \le \phi \le 2\pi$.

Solution

We proceed as in Example Problem 13.3, remembering that the volume element in spherical coordinates is $r^2 \sin \theta \, dr \, d\theta \, d\phi$:

$$N^2 \int_0^{2\pi} d\phi \int_0^{\pi} \sin \theta \, d\theta \int_0^{\infty} r^2 e^{-2r} \, dr = 1$$

Evaluating the integral over the angles

$$\int_0^{2\pi} d\phi \int_0^{\pi} \sin \theta \, d\theta = 2\pi \int_0^{\pi} \sin \theta \, d\theta = 4\pi$$

$$4\pi N^2 \int_0^{\infty} r^2 e^{-2r} \, dr = 1$$

Using the standard integral $\int_0^{\infty} x^n e^{-ax} \, dx = n!/a^{n+1}$ ($a > 0$, n is a positive integer), we obtain $4\pi N^2 \, (2!/2^3) = 1$ so that $N = \sqrt{1/\pi}$. The normalized wave function is $\sqrt{1/\pi} \, e^{-r}$. Note that the integration of any function involving r, even if it does not explicitly involve θ or ϕ, requires integration over all three variables.

Vocabulary

angular frequency	degeneracy	period	time-independent
Boltzmann distribution	destructive interference	phase	Schrödinger equation
classical nondispersive wave equation	discrete energy spectrum	plane wave	traveling wave
	frequency	single-valued function	wave front
complex conjugate	interference	spherical wave	wave function
constructive interference	normalization	standing wave	wave vector
continuous energy spectrum	orthogonality	stationary state	wavelength
continuous function	orthonormal		

Questions on Concepts

Q13.1 By discussing the diffraction of a beam of particles by a single slit, justify the statement that there is no sharp boundary between particle-like and wave-like behavior.

Q13.2 Why does a quantum mechanical system with discrete energy levels behave as if it has a continuous energy spectrum if the energy difference between energy levels ΔE satisfies the relationship $\Delta E << kT$?

Q13.3 Why can we conclude that the wave function $\Psi(x,t) = \psi(x)e^{-i(E/\hbar)t}$ represents a standing wave?

Q13.4 Is it correct to say that because the de Broglie wavelength of a H_2 molecule at 300 K is on the order of atomic dimensions that all properties of H_2 are quantized?

Q13.5 If $\Psi(x,t) = A \sin(kx - \omega t)$ describes a wave traveling in the plus x direction, how would you describe a wave moving in the minus x direction?

Q13.6 A traveling wave with arbitrary phase ϕ can be written as $\Psi(x,t) = A \sin(kx - \omega t + \phi)$. What are the units of ϕ? Show that ϕ could be used to represent a shift in the origin of time or distance.

Q13.7 One source emits spherical waves and another emits plane waves. For which source does the intensity measured by a detector of fixed size fall off more rapidly with distance? Why?

Problems

Problem numbers in **RED** Sindicate that the solution to the problem is given in the *Student Solutions Manual*.

P13.1 Assume that a system has a very large number of energy levels given by the formula $\varepsilon = \varepsilon_0 l^2$ with $\varepsilon_0 = 2.34 \times 10^{-22}$ J, where l takes on the integral values 1, 2, 3, ... Assume further that the degeneracy of a level is given by $g_l = 2l$. Calculate the ratios n_5/n_1 and n_{10}/n_1 for $T = 100$. K and $T = 650$. K.

P13.2 Consider a two-level system with $\varepsilon_1 = 3.10 \times 10^{-21}$ J and $\varepsilon_2 = 6.10 \times 10^{-21}$ J. If $g_2 = g_1$, what value of T is required to obtain $n_2/n_1 = 0.150$? What value of T is required to obtain $n_2/n_1 = 0.999$?

P13.3 To plot $\Psi(x,t) = A \sin(kx - vt)$ as a function of one of the variables x and t, the other variable needs to be set at a fixed value, x_0 or t_0. If $\Psi(x_0,0)/\Psi_{max} = -0.280$, what is the constant value of x_0 in the upper panel of Figure 13.3? If $\Psi(0,t_0)/\Psi_{max} = -0.309$, what is the constant value of t_0 in the lower panel of Figure 13.3? (*Hint:* The inverse sine function has two solutions within an interval of 2π. Make sure that you choose the correct one.)

P13.4 A wave traveling in the z direction is described by the wave function $\Psi(z,t) = A_1 \mathbf{x} \sin(kz - \omega t + \phi_1) + A_2 \mathbf{y} \sin(kz - \omega t + \phi_2)$, where $\mathbf{x}$ and $\mathbf{y}$ are vectors of unit length along the x and y axes, respectively. Because the amplitude is perpendicular to the propagation direction, $\Psi(z,t)$ represents a transverse wave.

   a. What requirements must A_1 and A_2 satisfy for a plane polarized wave in the x-z plane?

   b. What requirements must A_1 and A_2 satisfy for a plane polarized wave in the y-z plane?

   c. What requirements must A_1 and A_2 and ϕ_1 and ϕ_2 satisfy for a plane polarized wave in a plane oriented at 45° to the xz plane?

   d. What requirements must A_1 and A_2 and ϕ_1 and ϕ_2 satisfy for a circularly polarized wave?

P13.5 Show that

$$\frac{a + ib}{c + id} = \frac{ac + bd + i(bc - ad)}{c^2 + d^2}$$

P13.6 Does the superposition $\Psi(x,t) = A \sin(kx - \omega t) + 2A \sin(kx + \omega t)$ generate a standing wave? Answer this question by using trigonometric identities to combine the two terms.

P13.7 Express the following complex numbers in the form $re^{i\theta}$.

   a. $2 - 4i$ **c.** $\dfrac{3 + i}{4i}$

   b. 6 **d.** $\dfrac{8 + i}{2 - 4i}$

P13.8 Express the following complex numbers in the form $a + ib$.

   a. $2e^{i\pi/2}$ **c.** $e^{i\pi}$

   b. $2\sqrt{5}\,e^{-i\pi/2}$ **d.** $\dfrac{3\sqrt{2}}{5 + \sqrt{3}}\,e^{i\pi/4}$

P13.9   Using the exponential representation of the sine and cosine functions,

$$\cos\theta = \frac{1}{2}(e^{i\theta} + e^{-i\theta}) \quad \text{and} \quad \sin\theta = \frac{1}{2i}(e^{i\theta} - e^{-i\theta})$$

show that

a. $\cos^2\theta + \sin^2\theta = 1$.

b. $d(\cos\theta)/d\theta = -\sin\theta$.

c. $\sin(\theta + \frac{\pi}{2}) = \cos\theta$

P13.10   Show that the set of functions $\phi_n(\theta) = c^{in\theta}, 0 \le \theta \le 2\pi$, is orthogonal if n and m are integers. To do so, you need to show that the integral $\int_0^{2\pi}\phi_m^*(\theta)\phi_n(\theta)\,d\theta = 0$ for $m \ne n$ if n and m are integers.

P13.11   Show by carrying out the integration that $\sin(m\pi x/a)$ and $\cos(m\pi x/a)$, where m is an integer, are orthogonal over the interval $0 \le x \le a$. Would you get the same result if you used the interval $0 \le x \le 3a/4$? Explain your result.

P13.12   Normalize the set of functions $\phi_n(\theta) = e^{in\theta}, 0 \le \theta \le 2\pi$. To do so, you need to multiply the functions by a so-called normalization constant N so that the integral

$$N N^*\int_0^{2\pi}\phi_m^*(\theta)\phi_n(\theta)\,d\theta = 1 \quad \text{for } m = n$$

P13.13   In normalizing wave functions, the integration is over all space in which the wave function is defined. The following examples allow you to practice your skills in two- and three-dimensional integration.

a. Normalize the wave function $\sin(n\pi x/a)\sin(m\pi y/a)$ over the range $0 \le x \le a, 0 \le y \le b$. The element of area in two-dimensional Cartesian coordinates is $dxdy$, n and m are integers, and a and b are constants.

b. Normalize the wave function $e^{-(r/a)}\cos\theta\sin\phi$ over the interval $0 \le r < \infty, 0 \le \theta \le \pi, 0 \le \phi \le 2\pi$. The volume element in three-dimensional spherical coordinates is $r^2\sin\theta\,dr\,d\theta\,d\phi$.

P13.14   Show that the following pairs of wave functions are orthogonal over the indicated range:

a. $e^{(-1/2)\alpha x^2}$ and $(2\alpha x^2 - 1)e^{(-1/2)\alpha x^2}, -\infty \le x < \infty$ where α is a constant that is greater than zero.

b. $(2 - r/\alpha_0)e^{-r/2\alpha_0}$ and $(r/\alpha_0)e^{-r/2\alpha_0}\cos\theta$ over the interval $0 \le r < \infty, 0 \le \theta \le \pi, 0 \le \phi \le 2\pi$.

P13.15   Carry out the following coordinate transformations:

a. Express the point $x = 3, y = 2$, and $z = 1$ in spherical coordinates.

b. Express the point $r = 5, \theta = \frac{\pi}{4}$, and $\phi = \frac{3\pi}{4}$ in Cartesian coordinates.

Web-Based Simulations, Animations, and Problems

W13.1   The motion of transverse, longitudinal, and surface traveling waves is analyzed by varying the frequency and amplitude.

W13.2   Two waves of the same frequency traveling in opposite directions are combined. The relative amplitude is changed with sliders and the relative phase of the waves is varied. The effect of these changes on the superposition wave is investigated.

W13.3   Two waves, both of which are standing waves, are combined. The effect of varying the wavelength, period, and phase of the waves on the resulting wave using sliders is investigated.

CHAPTER

14

Using Quantum Mechanics on Simple Systems: The Free Particle, the Particle in a Box, and the Harmonic Oscillator

In this chapter, we use quantum mechanics to solve four problems in a quantum mechanical framework that are familiar from classical mechanics. The first two involve the motion of a particle on which no forces are acting. In the first case, the particle is not constrained. In the second, it is constrained to move within the confines of a box, but has no other forces acting on it. We find that unlike in classical mechanics, where the energy spectrum is continuous and the particle is equally likely to be found anywhere in the box, the quantum mechanical particle in the box has a discrete energy spectrum and has preferred positions that depend on the quantum mechanical state. The third problem is the motion of a particle in a box that is not infinitely deep, which allows the particle to escape if it has sufficient energy. The particle in a box is a simple model that can be used to explore concepts such as why core electrons are not involved in chemical bonds, the stabilizing effect of delocalized π electrons in aromatic molecules, and the ability of metals to conduct electrons. It also provides a framework for understanding the tunneling of quantum mechanical particles through (not over!) barriers and size quantization, both of which play important roles in chemical phenomena. The fourth system that we discuss is the quantum mechanical harmonic oscillator, which is a useful model for describing the vibration of molecules. Like the particle in the box, the quantum harmonic oscillator has a discrete energy spectrum.

14.1 The Free Particle

The simplest classical system imaginable is the free particle, a particle in a one-dimensional space on which no forces are acting. We begin with

$$F = ma = m\frac{d^2x}{dt^2} = 0 \tag{14.1}$$

This differential equation can be solved to obtain

$$x = x_0 + v_0 t \tag{14.2}$$

Verify that this is a solution by substitution in Equation (14.1). The initial position x_0 and initial velocity v_0 arise from the constants of integration. To give them explicit values, the boundary conditions of the problem, namely, the initial position and velocity, must be known.

How is this problem solved using quantum mechanics? The condition that no forces can be acting on the particle means that the potential energy is constant and independent of t. Therefore, we use the time-independent Schrödinger equation in one dimension,

$$-\frac{\hbar^2}{2m}\frac{d^2\psi(x)}{dx^2} + V(x)\psi(x) = E\psi(x) \tag{14.3}$$

to solve for the dependence of the wave function $\psi(x)$ on x. Whenever the potential energy, $V(x)$, is constant, we can choose to make it zero because there is no fixed reference point for the zero of potential energy, and only changes in this quantity are measurable. The Schrödinger equation for this problem reduces to

$$\frac{d^2\psi(x)}{dx^2} = -\frac{2m}{\hbar^2}E\psi(x) \tag{14.4}$$

In words, $\psi(x)$ is a function that can be differentiated twice to return the same function multiplied by a constant. The most appropriate particular solutions for these purposes are

$$\psi_+(x) = A_+ e^{+i\sqrt{(2mE/\hbar^2)}x} = A_+ e^{+ikx}$$
$$\psi_-(x) = A_- e^{-i\sqrt{(2mE/\hbar^2)}x} = A_- e^{-ikx} \tag{14.5}$$

in which the constants in the exponent have been combined using $k = 2\pi/\lambda = \sqrt{2mE/\hbar^2}$. We have been working with $\psi(x)$ rather than $\Psi(x,t)$. To obtain $\Psi(x,t)$, these two solutions are multiplied by $e^{-i(E/\hbar)t}$ or equivalently $e^{-i\omega t}$, where the relation $E = \hbar\omega$ has been used.

These solutions are plane waves, one moving to the right (positive x direction), the other moving to the left (negative x direction). The allowed values for the total energy can be found by substituting the wave functions of Equation (14.5) into Equation (14.4). For both particular solutions, $E = \hbar^2 k^2/2m$. Using the de Broglie relation to relate k with v, $k = mv/\hbar = \sqrt{2mE/\hbar^2}$. Because k is a constant, these functions represent waves moving at a constant velocity that is determined by their initial velocity. Therefore, the quantum mechanical solution of this problem contains the same information as the classical particle problem, namely, motion with a constant velocity. One other important similarity between the classical and quantum mechanical free particle is that both can take on all values of energy, because k is a continuous variable. The quantum mechanical free particle has a **continuous energy spectrum.** Why is this the case? We will learn the answer to this question in Section 14.3.

Of course, because a plane wave is not localized in space, we cannot speak of its position as is done for a particle. However, the **probability** of finding the particle in an interval of length dx can be calculated. The free-particle wave functions cannot be normalized over the interval $-\infty < x < \infty$, but if x is restricted to the interval $-L \le x \le L$, then

$$P(x)\, dx = \frac{\psi_\pm^*(x)\psi_\pm(x)\, dx}{\int_{-L}^{L}\psi_\pm^*(x)\psi_\pm(x)\, dx} = \frac{A_\pm A_\pm e^{\mp ikx}e^{\pm ikx}\, dx}{A_\pm A_\pm \int_{-L}^{L}e^{\mp ikx}e^{\pm ikx}\, dx} = \frac{dx}{2L} \tag{14.6}$$

The coefficients A_+ or A_- cancel because they appear in both the numerator and the denominator. Surprisingly, $P(x)\, dx$ is independent of x. This result states that the particle is equally likely to be anywhere in the interval, which is equivalent to saying that nothing is known about the position of the particle.

14.2 The Heisenberg Uncertainty Principle

As shown in the previous section, it is equally probable that a free particle for which the wave vector k is known exactly will be found anywhere. Recall that the probability of finding the particle in the length dx is $dx/2L$, and let the interval length L become arbitrarily large. The probability of finding the particle within the interval dx centered at $x = x_0$ approaches zero! *We conclude that if a particle is prepared in a state in which the momentum is exactly known, that is,* ($\Delta p = 0$), *then its position is completely unknown,* ($\Delta x \rightarrow \infty$). It turns out that if a particle is prepared such that its position is exactly known (the wave function is an eigenfunction of the position operator and $\Delta x = 0$), then its momentum is completely unknown ($\Delta p \rightarrow \infty$).

This result is completely at variance with expectations based on classical mechanics, because a simultaneous knowledge of position and momentum is essential to calculate the trajectories of particles subject to forces. How can this counterintuitive result be understood? Since the uncertainty in position arises because the momentum is precisely known, is it possible to construct a wave function for which the momentum is not precisely known?

It turns out that this can be done by making a linear superposition of individual plane waves, each of which has an exactly defined momentum. Such a superposition is called a **wave packet.** *For a wave packet, the position of the particle is no longer completely unknown, and the momentum of the particle is no longer exactly known.* The take-home lesson is that both position and momentum cannot be known *exactly and simultaneously* in quantum mechanics. We must accept a trade-off between the uncertainty in p, Δp and that of x, Δx. This result was quantified by Heisenberg in his famous uncertainty principle:

$$\Delta p \, \Delta x \geq \frac{\hbar}{2} \tag{14.7}$$

If the right-hand side of the inequality were equal to zero instead of $\hbar/2$, then it would be possible to know both the position and momentum exactly. This is not the case; therefore, it is not possible to calculate the trajectory of a quantum mechanical particle exactly. The trajectory of a particle for which the momentum and energy are exactly known is not a well-defined concept in quantum mechanics. However, you can get a good approximation for a "trajectory" in quantum mechanical systems by using wave packets, because $\hbar$ is a very small number.

Web-Based Problem 14.1 The Heisenberg Uncertainty Principle

Web-Based Problem 14.2 Wave Packets and the Uncertainty Principle

---

EXAMPLE PROBLEM 14.1

Assume that the double-slit experiment could be carried out with electrons using a slit spacing of $b = 10.0$ nm. To be able to observe diffraction, we choose $\lambda = b$, and because diffraction requires reasonably monochromatic radiation, we choose $\Delta p/p = 0.010$. Show that with these parameters, the uncertainty in the position of the electron is greater than the slit spacing b.

Solution

Using the de Broglie relation, the mean momentum is given by

$$\langle p \rangle = \frac{h}{\lambda} = \frac{6.626 \times 10^{-34} \text{ J s}}{100 \times 10^{-10} \text{ m}} = 6.626 \times 10^{-26} \text{ kg m s}^{-1}$$

and $\Delta p = 6.626 \times 10^{-28} \text{ kg m s}^{-1}$. The minimum uncertainty in position is given by

$$\Delta x = \frac{\hbar}{2\Delta p} = \frac{1.055 \times 10^{-34} \text{ J s}}{2(6.626 \times 10^{-28} \text{ kg m s}^{-1})} = 7.9 \times 10^{-8} \text{ m}$$

which is greater than the slit spacing. Note that the concept of an electron trajectory is not well defined under these conditions. This offers an explanation for the observation that the electron appears to go through both slits simultaneously!

What is the practical effect of the uncertainty principle? Does this mean that you have no idea what trajectories the electrons in your TV picture tube will follow or where a baseball thrown by a pitcher will pass a waiting batter? As mentioned earlier, it comes down to what is meant by *exact*. An exact trajectory could be calculated if $\hbar$ were equal to zero, rather than being a small number. However, because $\hbar$ is a very small number, the uncertainty principle does not affect the calculation of the trajectories of baseballs, rockets, or other macroscopic objects. Although the uncertainty principle holds for both electrons and baseballs, the effect is so small that it is not detectable for large masses.

EXAMPLE PROBLEM 14.2

The electrons in a TV picture tube have an energy of about 1.0×10^4 eV. If $\Delta p/p = 0.010$ for this case, calculate the minimum uncertainty in the position that defines where the electrons land on the phosphor in the picture tube.

Solution

Using the relation $\langle p \rangle = \sqrt{2mE}$, the momentum is calculated as follows:

$$\langle p \rangle = \sqrt{2(9.11 \times 10^{-31} \text{ kg})(1.00 \times 10^4 \text{ eV})(1.602 \times 10^{-19} \text{ J/eV})}$$

$$= 5.41 \times 10^{-23} \text{ kg m s}^{-1}$$

Proceeding as in Example Problem 14.1,

$$\Delta x = \frac{\hbar}{2\Delta p} = \frac{1.055 \times 10^{-34} \text{ J s}}{2(5.41 \times 10^{-25} \text{ kg m s}^{-1})} = 9.8 \times 10^{-11} \text{ m}$$

This distance is much smaller than the size of the electron beam and, therefore, the uncertainty principle has no effect in this instance.

EXAMPLE PROBLEM 14.3

An (over)educated baseball player tries to convince his manager that he cannot hit a 100 mile per hour (26.8 m s^{-1}) baseball that has a mass of 140. g and relative momentum uncertainty of 1% because the uncertainty principle does not allow him to estimate its position within 0.100 mm. Is his argument valid?

Solution

The momentum is calculated using the following equation:

$$p = mv = 0.140 \text{ kg} \times 26.8 \text{ m s}^{-1} = 3.75 \text{ kg m s}^{-1} \text{ and } \Delta p = 0.0375 \text{ kg m s}^{-1}$$

Substituting in the uncertainty principle,

$$\Delta x = \frac{\hbar}{2\Delta p} = \frac{1.055 \times 10^{-34} \text{ J s}}{2 \times 0.0375 \text{ kg m s}^{-1}} = 1.41 \times 10^{-33} \text{ m}$$

The uncertainty is not zero, but it is well below the experimental sensitivity. Sorry, back to the minor leagues.

This result—that it is not possible to know the exact values of two observables simultaneously—is not restricted to position and momentum. Energy and time are another example of two observables that are linked by an uncertainty principle. The energy of the H atom with the electron in the $1s$ state can only be known to high accuracy because it has a very long lifetime. This is the case because there is no lower state to which it can decay. Excited states that rapidly decay to the **ground state** can have an appreciable uncertainty in their energy.

14.3 The Particle in a One-Dimensional Box

The next case to be considered is the particle confined to a box. To keep the mathematics simple, the box is one dimensional; that is, it is the one-dimensional analogue of a single atom moving freely in a cube that has impenetrable walls. The impenetrable walls are modeled by making the potential energy infinite outside of a region of width a. The potential is depicted in Figure 14.1.

$$V(x) = 0, \text{ for } a > x > 0$$
$$V(x) = \infty, \text{ for } x \geq a, x \leq 0 \tag{14.8}$$

How does this change in the potential affect the solutions that were obtained for the free particle? To answer this question, the Schrödinger equation is written in the following form:

$$\frac{d^2\psi(x)}{dx^2} = \frac{2m}{\hbar^2}[V(x) - E]\psi(x) \tag{14.9}$$

Outside of the box, where the potential energy is infinite, the second derivative of the wave function would be infinite if $\psi(x)$ were not zero for all x values outside the box. Because $d^2\psi(x)/dx^2$ must exist and be well behaved, $\psi(x)$ must be zero everywhere outside of the box. Moreover, because the wave function must be continuous, for the potential energy function given in Equation (14.8),

$$\psi(0) = \psi(a) = 0 \tag{14.10}$$

Equation (14.10) lists **boundary conditions** that any well-behaved wave function for the one-dimensional box must satisfy.

Inside the box, where $V(x) = 0$, the Schrödinger equation is identical to that for a free particle [Equation (14.4)], so the solutions must be the same. For ease in applying the boundary conditions, the solution is written in a form equivalent to that of Equation (14.5):

$$\psi(x) = A \sin kx + B \cos kx \tag{14.11}$$

Now the boundary conditions given by Equation (14.10) are applied. Putting the values $x = 0$ and $x = a$ in Equation (14.11), we obtain

$$\psi(0) = 0 + B = 0$$

$$\psi(a) = A \sin ka = 0 \tag{14.12}$$

The first condition can only be satisfied by the condition that $B = 0$. The second condition can be satisfied if either $A = 0$ or if $ka = n\pi$ with n being an integer. Setting A equal to zero would mean that the wave function is always zero, which is unacceptable because then there is no particle in the interval. Therefore, we conclude that

$$\psi_n(x) = A \sin\left(\frac{n\pi x}{a}\right) \tag{14.13}$$

The requirement that $ka = n\pi$ will turn out to have important consequences for the energy spectrum of the particle in the box. The preceding discussion shows that acceptable wave functions for this problem must have the form

$$\psi_n(x) = A \sin\left(\frac{n\pi x}{a}\right), \text{ for } n = 1, 2, 3, \ldots \tag{14.14}$$

Each different value of n corresponds to a different solution of the Schrödinger equation.

Note the undefined constant A in these equations. This constant can be determined by normalization, that is, by realizing that $\psi^*(x)\psi(x)\,dx$ represents the probability of finding the particle in the interval of width dx centered at x. Because the probability of finding the particle somewhere in the entire interval is one,

the **Chemistry place**

Web-Based Problem 14.3 The Classical Particle in a Box

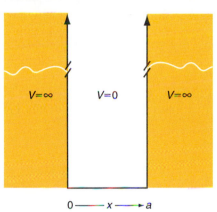

$V = \infty$　　$V = 0$　　$V = \infty$

$0 \longrightarrow x \longrightarrow a$

FIGURE 14.1
The potential described by Equation (14.8) is depicted. Because the particle is confined to the range $0 < x < a$, we say that it is confined to a one-dimensional box.

$$\int_0^a \psi^*(x)\psi(x)dx = A^*A\int_0^a \sin^2\left(\frac{n\pi x}{a}\right)dx = 1 \tag{14.15}$$

This integral is evaluated using the standard integral

$$\int \sin^2(by)\,dy = \frac{y}{2} - \frac{\sin 2by}{4b}$$

resulting in $A = \sqrt{2/a}$, so the normalized solutions are

$$\psi_n(x) = \sqrt{\frac{2}{a}}\sin\left(\frac{n\pi x}{a}\right) \tag{14.16}$$

What are the allowed values of the total energy that go with these solutions? Solving the Schrödinger equation, we find that

$$-\frac{\hbar^2}{2m}\frac{d^2\psi_n(x)}{dx^2} = \frac{\hbar^2}{2m}\left(\frac{n\pi}{a}\right)^2\sqrt{\frac{2}{a}}\sin\left(\frac{n\pi x}{a}\right) \tag{14.17}$$

Because

$$-\frac{\hbar^2}{2m}\frac{d^2\psi_n(x)}{dx^2} = E_n\psi_n(x)$$

the following result is obtained:

the Chemistry place

Web-Based Problem 14.4 Energy Levels for the Particle in a Box

$$E_n = \frac{\hbar^2}{2m}\left(\frac{n\pi}{a}\right)^2 = \frac{h^2n^2}{8ma^2}, \quad \text{for } n = 1, 2, 3, \ldots \tag{14.18}$$

An important difference is seen when this result is compared to that obtained for the free particle. The energy for the particle in the box can only take on discrete values, and we say that the energy of the particle in the box is **quantized** and the integer n is a **quantum number.** Another important result of this calculation is that the lowest allowed energy is greater than zero. The particle has a nonzero minimum energy, known as a **zero point energy.**

Why are quite different results obtained for the free and the confined particle? A comparison of these two problems reveals that quantization entered through the confinement of the particle. Because the particle is confined to the box, the amplitude of all allowed wave functions must be zero everywhere outside the box. By considering the limit $a \to \infty$, the confinement condition is removed.

The lowest four energy levels for the particle in the box are shown in Figure 14.2 superimposed on an energy versus distance diagram. The solutions are also shown in this figure. Keep in mind that the time-independent part of the wave function is graphed. The full wave function is obtained by multiplying the wave functions shown in Figure 14.2 by $e^{-i(E/\hbar)t}$. If this is done, you will see that the variation of the total wave function with time is exactly what was shown in Figure 13.4 for a **standing wave,** if the real and imaginary parts of $\Psi(x,t)$ are considered separately. This result turns out to be general: the wave function for a stationary state is a standing wave and not a **traveling wave.** This result becomes clear by considering the boundary conditions for the particle in the box. These conditions state that the amplitude of the wave function is zero at the ends of the box for all times, whereas the nodes move in time for a traveling wave. For this reason, the boundary conditions cannot be satisfied for a traveling wave.

The quantization of the energy ultimately has its origin in the coupling of wave properties and boundary conditions. In moving from $\psi_n(x)$ to $\psi_{n+1}(x)$, the number of half-wavelengths, and therefore the number of **nodes,** has been increased by one. There is no way to add anything other than an integral number of half-wavelengths and still have

FIGURE 14.2
The first few solutions for the particle in a box are shown together with the corresponding allowed values of the total energy. The energy scale is shown on the left. The wave function amplitude is shown on the right with the zero for each level indicated by the dashed line.

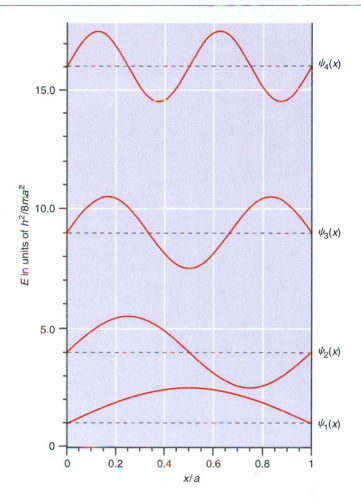

nodes at the ends of the box. Therefore, the **wave vector** k will increase in discrete increments rather than continuously in going from one stationary state to another. Because

$$k = \frac{2\pi}{\lambda} = \frac{p}{\hbar} = \frac{\sqrt{2mE}}{\hbar}$$

the allowed energies, E, also increase in jumps rather than in a continuous fashion as in classical mechanics. Thinking in this way also helps in understanding the origin of the zero point energy. Because $E = h^2/2m\lambda^2$, zero energy corresponds to an infinite wavelength. But the longest wavelength that has standing-wave nodes at the ends of the box is $\lambda = 2a$. Substituting this value in the equation for E gives exactly the zero point energy.

The total energy is one example of an observable that can be calculated once the solutions of the time-independent Schrödinger equation are known. Another observable that comes directly from solving this equation is the quantum mechanical analogue of position. Recall that the probability of finding the particle in any interval of width dx in the one-dimensional box is given by $\psi^*(x)\psi(x)dx$. The **probability density** $\psi^*(x)\psi(x)$ at a given point is shown in Figure 14.3 for the first few solutions.

How can these results be understood? Looking back at the discussion of waves in Chapter 13, recall that to ask for the position of a wave is not meaningful because the wave is not localized at a point. Wave-particle duality modifies the classical picture of being able to specify the location of a particle. Figure 14.3 shows the probability density of finding the particle in the vicinity of a given value of x rather than the position of that particle. We see that the probability of finding the particle outside of the box is zero, but that the probability of finding the particle within an interval dx in the box depends on the position and the wave function of the particle. Although $|\psi(x)|^2$ can be zero at nodal positions, $|\psi(x)|^2 \, dx$ is never zero for a finite interval dx inside the box. This means that

Web-Based Problem 14.5 Probability of Finding the Particle in a Given Interval

FIGURE 14.3
The square of the magnitude of the wave function, or probability density, is shown as a function of distance together with the corresponding allowed values of the total energy. The energy scale is shown on the left. The square of the wave function amplitude is shown on the right with the zero for each level indicated by the dashed line.

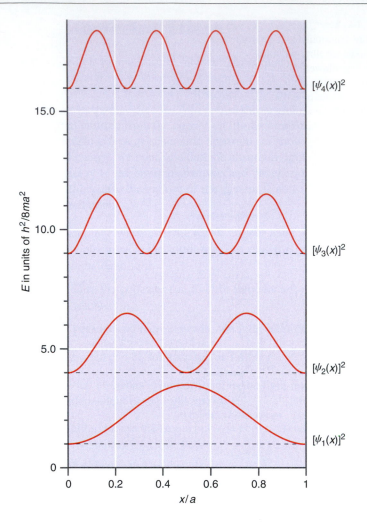

there is no finite length interval inside the box in which the particle is not found. For the ground state, it is much more likely that the particle is found near the center of the box than at the edges. A classical particle would be found with the same probability everywhere. Does this mean that quantum mechanics and classical mechanics are in conflict? No, because we need to consider large values of n as well as the resolution of the experimental method to compare with the **classical limit.** In this limit, the classical and quantum mechanical results are identical.

This first attempt to apply quantum rather than classical mechanics to two familiar problems has led to several useful insights. By representing a wave-particle as a wave, familiar questions that can be asked in classical mechanics become inappropriate. An example is "Where is the particle at time t_0?" The appropriate question in quantum mechanics is "What is the probability of finding the particle at time t_0 in an interval of length dx centered at the position x_0?" For the free particle, we found that the relationship between momentum and energy is the same as in classical mechanics and that there are no restrictions on the allowed energy. Restricting the motion of a particle to a finite region on the order of its wavelength has a significant effect on many observables associated with the particle. We saw that the origin of the effect is the requirement that the amplitude of the wave function be zero at the ends of the box for all times. This requirement changes the solutions of the Schrödinger equation from the traveling waves of the free particle to standing waves. Only discrete values of the particle momentum are allowed because of the condition $ka = n\pi, n = 1, 2, 3, \ldots$. Because $E = \hbar^2 k^2 / 2m$, the particle can only have certain values for the energy, and these values are determined by the dimensions of the box. Wave-particle duality also leads to a nonuniform probability of finding the particle in the box.

EXAMPLE PROBLEM 14.4

What is the probability, P, of finding the particle in the central third of the box if it is in its lowest energy state?

Solution

For the lowest energy state, $\psi_1(x) = \sqrt{2/a}\,\sin(\pi x/a)$. From the postulate, P is the sum of all the probabilities of finding the particle in intervals of width dx within the central third of the box. This probability is given by the integral

$$P = \frac{2}{a} \int_{a/3}^{2a/3} \sin^2\left(\frac{\pi x}{a}\right) dx$$

Solving this integral as in Section 14.2,

$$P = \frac{2}{a}\left[\frac{a}{6} - \frac{a}{4\pi}\left(\sin\frac{4\pi}{3} - \sin\frac{2\pi}{3}\right)\right] = 0.609$$

Although we cannot predict the outcome of a single measurement, we can predict that for 60.9% of a large number of individual measurements, the particle is found in the central third of the box. What is the probability of finding a classical particle in this interval?

14.4 Two- and Three-Dimensional Boxes

the Chemistry place

Web-Based Problem 14.6

Eigenfunctions for the Two-Dimensional Box

The one-dimensional box is a useful model system because the conceptual simplicity allows us to focus on the quantum mechanics rather than on the mathematics. The extension of the formalism that we have developed for the one-dimensional problem to two and three dimensions has several aspects that will help us to understand topics, such as the electronic structure of atoms, that cannot be reduced to one-dimensional problems.

We focus on the three-dimensional box because, as you will see, the reduction in dimensionality from three to two is straightforward. The potential energy is given by

$$V(x, y, z) = 0 \text{ for } 0 < x < a; 0 < y < b; 0 < z < c$$
$$= \infty \text{ otherwise} \tag{14.19}$$

As before, the amplitude of the eigenfunctions of the total energy operator will be identically zero outside the box. Inside the box, the Schrödinger equation can be written as

$$-\frac{\hbar^2}{2m}\left(\frac{\partial^2}{\partial x^2} + \frac{\partial^2}{\partial y^2} + \frac{\partial^2}{\partial z^2}\right)\psi(x, y, z) = E\psi(x, y, z) \tag{14.20}$$

We solve this differential equation assuming that $\psi(x, y, z)$ has the form

$$\psi(x, y, z) = X(x)Y(y)Z(z) \tag{14.21}$$

in which $\psi(x, y, z)$ is the product of three functions, each of which depends on only one of the variables. The assumption is valid in our case because $V(x, y, z)$ is independent of x, y, and z inside the box. It would also be valid for a potential of the form $V(x, y, z) = V_x(x) + V_y(y) + V_z(z)$. Substituting Equation (14.21) in Equation (14.20), we obtain

$$-\frac{\hbar^2}{2m}\left(Y(y)Z(z)\frac{d^2X(x)}{dx^2} + X(x)Z(z)\frac{d^2Y(y)}{dy^2} + X(x)Y(y)\frac{d^2Z(z)}{dz^2}\right)$$
$$= E\,X(x)Y(y)Z(z)$$

Note that we no longer have partial derivatives because each of the three functions X, Y, and Z depends on only one variable. Dividing by the product $X(x)Y(y)Z(z)$ results in

$$-\frac{\hbar^2}{2m}\left(\frac{1}{X(x)}\frac{d^2X(x)}{dx^2} + \frac{1}{Y(y)}\frac{d^2Y(y)}{dy^2} + \frac{1}{Z(z)}\frac{d^2Z(z)}{dz^2}\right) = E \quad (14.22)$$

The form of this equation shows that we can express E as having independent contributions from the three coordinates, $E = E_x + E_y + E_z$, and our original differential equation in three variables reduces to three differential equations, each in one variable:

$$-\frac{\hbar^2}{2m}\frac{d^2X(x)}{dx^2} = E_x\,X(x); \quad -\frac{\hbar^2}{2m}\frac{d^2Y(y)}{dy^2} = E_y\,Y(y); \quad -\frac{\hbar^2}{2m}\frac{d^2Z(z)}{dz^2} = E_z\,Z(z)$$

$$(14.23)$$

Each of these equations has the same form as the equation that we solved for the one-dimensional problem. Therefore, the total energy eigenfunctions have the form

$$\psi_{n_x n_y n_z}(x, y, z) = N\sin\frac{n_x\pi x}{a}\sin\frac{n_y\pi y}{b}\sin\frac{n_z\pi z}{c} \quad (14.24)$$

and the total energy has the form

$$E = \frac{h^2}{8m}\left(\frac{n_x^2}{a^2} + \frac{n_y^2}{b^2} + \frac{n_z^2}{c^2}\right) \quad (14.25)$$

This is a general result. *If the total energy can be written as a sum of independent terms corresponding to different degrees of freedom, then the wave function will be a product of individual terms, each corresponding to one of the degrees of freedom.*

Because this is a three-dimensional problem, the eigenfunctions will depend on three quantum numbers. Because more than one set of the three quantum numbers may have the same energy [for example, (1,2,1), (2,1,1), and (1,1,2) if $a = b = c$], several distinct eigenfunctions of the total energy operator may have the same energy. In this case, we say that the energy level is *degenerate,* and the number of states that have the same energy is the **degeneracy** of the level.

What form do you expect ψ and E to take for the two-dimensional box? How many quantum numbers are needed to characterize ψ and E for the two-dimensional problem? Additional issues related to the functional form, degeneracy, and normalization of the total energy eigenfunctions are covered in the end-of-chapter problems.

14.5 The Particle in the Finite Depth Box

Before applying the particle in a box model to the "real world," the box must be modified to make it more realistic. This is done by letting the box have a finite depth. The potential is defined by

$$V(x) = 0, \text{ for } a/2 > x > -a/2$$
$$= V_0, \text{ for } x \geq a/2, x \leq -a/2 \quad (14.26)$$

The origin of the x coordinate has been changed from one end of the box to the center of the box to simplify the mathematics of solving the Schrödinger equation. The shift of the origin changes the functional form of the total energy solutions. However, it has no physical consequences in that graphs of the solutions superimposed on the potential are identical for both choices of the point $x = 0$.

How does the finite depth of the potential affect the solutions and allowed values of the total energy for the Schrödinger equation that were obtained for the infinitely deep potential? We distinguish between two cases. For $E > V(x)$ (inside the box), the solutions have the **oscillatory behavior** that was exhibited for the infinitely deep box. However, because $V_0 < \infty$, the amplitude of the solutions is not zero at the ends of the box. For $E < V(x)$ (outside of the box), the solutions decay exponentially with distance from the box. These two regions are considered separately. Inside the box, $V(x) = 0$, and

$$\frac{d^2\psi(x)}{dx^2} = -\frac{2mE}{\hbar^2}\psi(x) \tag{14.27}$$

Outside of the box, the Schrödinger equation has the form

$$\frac{d^2\psi(x)}{dx^2} = \frac{2m(V_0 - E)}{\hbar^2}\psi(x) \tag{14.28}$$

The difference in sign on the right-hand side makes a big difference in the solutions! Inside the box, the solutions have the same general form as discussed in Section 14.3, but outside the box, they have the form

$$\psi(x) = A\,e^{-\kappa x} + Be^{+\kappa x} \text{ for } \infty \geq x \geq a/2 \text{ and}$$

$$= A'e^{-\kappa x} + B'e^{+\kappa x} \text{ for } -\infty \leq x \leq -a/2 \text{ where } \kappa = \sqrt{\frac{2m(V_0 - E)}{\hbar^2}}$$

$$\tag{14.29}$$

Convince yourself that the functions of Equation (14.29) are the correct general solutions to Equation (14.28). The coefficients (A, B and A', B') are different on each side of the box. Because $\psi(x)$ must remain finite for very large positive and negative values of x, $B = A' = 0$. By matching the wave functions and their derivatives at the boundaries of the three regions of the potential and imposing a normalization condition, the Schrödinger equation can be solved for the solutions and allowed values of the total energy in the potential for given values of m, a, and V_0. (See Problem P5.1 in Engel and Reid's *Physical Chemistry.*) The allowed energy levels and the corresponding solutions for a finite depth potential are shown in Figure 14.4. The yellow areas correspond to the classically forbidden region, for which $E_{potential} > E_{total}$. It is called the classically forbidden region because $E_{kinetic} < 0$, implying that the velocity is an imaginary quantity.

Two major differences in the solutions between the finite and the infinite depth box are immediately apparent. First, the potential has only a finite number of bound states, which are the allowed energies. The number depends on m, a, and V_0 as can be seen in carrying out the simulation in Web-Based Problem 14.7. Second, the amplitude of the wave function does not go to zero at the edge of the box. We explore the consequences of this second difference further in Sections 14.6, 14.11, and 14.12.

14.6 Differences in Overlap between Core and Valence Electrons

Figure 14.4 shows that weakly bound states have wave functions that leak quite strongly into the region outside of the box. What are the consequences of this behavior? Take this potential as a crude model for electrons in an atom. Strongly bound levels correspond to **core electrons** and weakly bound levels correspond to **valence electrons**. What happens when a second atom is placed close enough to the first atom that a chemical bond is formed? The result is shown in Figure 14.5. In this figure, the falloff of the wave functions for the weakly bound states in the box is gradual enough that both wave functions have a nonzero amplitude in the region between the wells. *These wave functions have a significant overlap.* Note that this is not the case for the strongly bound levels; these energy solutions have a small overlap.

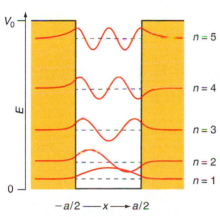

FIGURE 14.4

Solutions and allowed energy levels are shown for an electron in a well of depth $V_0 = 1.20 \times 10^{-18}$ J and width 1.00×10^{-9} m.

Web-Based Problem 14.7 Energy Solutions and Allowed Values of the Total Energy for a Finite Depth Box

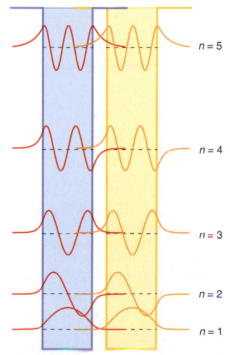

FIGURE 14.5

Overlap of wave functions from two closely spaced finite depth wells. The vertical scale has been expanded relative to Figure 14.4 to better display the overlap.

We conclude that a correlation exists between the nonzero overlap required for chemical bond formation and the position of the energy level in the potential. This is our first application of the particle-in-the-box model. It provides an understanding of why chemical bonds involve the least strongly bound, or valence, electrons and not the more strongly bound, or core, electrons. We will have more to say on this topic later when the chemical bond is discussed.

14.7 Pi Electrons in Conjugated Molecules Can Be Treated as Moving Freely in a Box

The absorption of light in the visible and ultraviolet (UV) part of the electromagnetic spectrum in molecules is a result of the excitation of electrons from occupied to unoccupied energy levels. If the electrons are delocalized as in an organic molecule with a **π-bonded network,** the maximum in the absorption spectrum shifts from the UV into the visible range. The greater the degree of **delocalization,** the more the absorption maximum shifts toward the red end of the visible spectrum. The energy levels for such a conjugated system can be described quite well with a one-dimensional particle-in-a-box model. The series of dyes, 1,4-diphenyl–1,3-butadiene, 1,6-diphenyl–1,3,5-hexatriene, and 1,8-diphenyl–1,3,5,7-octatetraene, can be used to demonstrate that the maximum wavelength at which these molecules absorb can be predicted with reasonable accuracy. These molecules consist of a planar backbone of alternating C—C and C=C bonds and have phenyl groups attached to the ends. The phenyl groups serve the purpose of decreasing the volatility of the compound. The π-bonded network does not include the phenyl groups, but does include the terminal carbon–phenyl group bond length.

The longest wavelength at which light is absorbed occurs when one of the electrons in the highest occupied energy level is promoted to the lowest lying unoccupied level. As Equation (14.18) shows, the energy level spacing depends on the length of the π-bonded network. For 1,4-diphenyl–1,3-butadiene, 1,6-diphenyl–1,3,5-hexatriene, and 1,8-diphenyl–1,3,5,7-octatetraene, the maximum wavelengths at which absorption occurs are 345, 375, and 390 nm, respectively. From these data, and taking into account the quantum numbers corresponding to the highest occupied and lowest unoccupied levels, the apparent network length can be calculated. We demonstrate the calculation for 1,6-diphenyl–1,3,5-hexatriene, for which the highest occupied level corresponds to $n = 3$:

$$a = \sqrt{\frac{(n_f^2 - n_i^2)h^2}{8m\Delta E}} = \sqrt{\frac{(n_f^2 - n_i^2)h\lambda_{max}}{8mc}}$$

$$= \sqrt{\frac{(4^2 - 3^2)(6.626 \times 10^{-34}\,\text{J s})(375 \times 10^{-9}\,\text{m})}{8(9.11 \times 10^{-31}\,\text{kg})(2.998 \times 10^8\,\text{m s}^{-1})}}$$

$$= 892\,\text{pm} \tag{14.30}$$

The apparent and calculated network length has been compared for each of the three molecules by B. D. Anderson [*J. Chemical Education* 74 (1997), 985]. Values are shown in Table 14.1. The agreement is reasonably good, given the simplicity of the model. Most importantly, the model correctly predicts that because λ_{max} is proportional to a^2, shorter π-bonded networks show absorption at smaller wavelengths. This trend is confirmed by experiment.

TABLE 14.1   Calculated Network Length for Conjugated Molecules

Compound	Apparent Network Length (pm)	Calculated Network Length (pm)
1,4-Diphenyl–1,3-butadiene	723	695
1,6-Diphenyl–1,3,5-hexatriene	892	973
1,8-Diphenyl–1,3,5,7-octatetraene	1030	1251

FIGURE 14.6
At large distances, the valence level on each Na atom is localized on that atom. When they are brought close enough together to form the dimer, the barrier between them is lowered, and the level is delocalized over both atoms. The term x_e represents the equilibrium distance of the dimer.

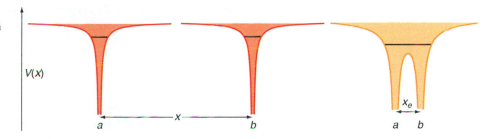

14.8 Why Does Sodium Conduct Electricity and Why Is Diamond an Insulator?

We have just learned that valence electrons on adjacent atoms in a molecule or a solid can have an appreciable overlap. This means that the electrons can "hop" from one atom to the next. Consider sodium (Na), which has one valence electron per atom. If we bond two Na atoms together to form a dimer, the valence level that was localized on each atom will now be delocalized over both atoms. This is illustrated in Figure 14.6. Now we add additional Na atoms to form a one-dimensional Na crystal. A crystalline metal can be thought of as a box with a periodic corrugated potential at the bottom. To illustrate this, we show in Figure 14.7 the potential of a one-dimensional periodic array of Na$^+$ potentials arising from the atomic cores at lattice sites. Because the Na $3s$ valence electrons can hop from one atom to another, we can view one electron per atom as being delocalized over the whole metal sample. This is exactly the model of the particle in the box.

We can idealize the potential of Figure 14.7 to a box as shown in Figure 14.8. This box differs from the simple boxes that we have discussed in two essential ways. There are many atoms in the atomic chain that we are currently considering (large a), so that the energy levels are much more closely spaced. Secondly, we have included the localized core electrons in Figure 14.8. What is the energy level spacing for the delocalized electrons in a 1.00-cm-long box? About 2×10^7 atoms will fit into the box. If each donates one electron to the band, you can easily show that at the highest filled level,

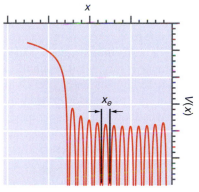

FIGURE 14.7
The potential energy resulting from a one-dimensional periodic array of Na$^+$ ions. One valence electron per Na is delocalized over this box. The term x_e represents the lattice spacing.

$$E_{n+1} - E_n = \frac{(n+1)^2 h^2}{8ma^2} - \frac{n^2 h^2}{8ma^2} = \frac{h^2}{8ma^2}(2n+1) = (2n+1)(6.03 \times 10^{-34}\,\text{J})$$

$$(14.31)$$

This spacing is only $3 \times 10^{-6}\,kT$ for $T = 300$ K. This tells us that the energy spectrum is essentially continuous. All energies within the range bounded by the dark black

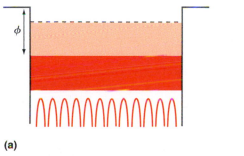

(a)

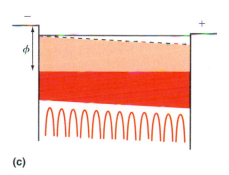

(c)

FIGURE 14.8
Idealization of a metal in the particle-in-the-box model. The horizontal scale is greatly expanded to show the periodic potential. Actually, more than 10^7 atoms will fit into a 1-cm-long box. The dark shaded band shows the range of energies filled by the valence electrons of the individual atoms. The highest energy that can be occupied in this band is indicated by the dashed line. The energy required to remove an electron from the highest occupied state is the *work function*, ϕ. Part (a) shows the metal without an applied potential. Part (b) shows the effect on the energy levels of applying an electric field. Part (c) shows the response of the metal to the change in the energy levels induced by the electric field. The thin solid line at the top of the band in parts (b) and (c) indicates where the energy of the highest level would lie in the absence of an electric field.

shaded area to low energies and the dashed line at high energies are accessible. We refer to this set of continuous energy levels as an **energy band.** The band shown in Figure 14.8a extends from the bottom of the dark shaded area up to the dashed line, beyond which there are no allowed energy levels. An energy range in which there are no allowed states is called a **band gap.** For Na, not all available states in the band are filled, as shown in Figure 14.8. The range of energies between the top of the dark black shaded area and the dashed line corresponds to unfilled **valence band** states. The fact that the band is only partially filled is critical in making Na a **conductor** as will be explained later in this section.

What happens when we apply an electrical potential between the two ends of our box? The field gives rise to a gradient of potential energy along the box superimposed on the original potential as shown in Figure 14.8b. On the side of the metal with the more positive electrical potential, we see that the unoccupied states have a lower energy than the occupied states with the more negative electrical potential. This makes it energetically favorable for the electrons to move toward the end of the box with the more positive voltage as shown in Figure 14.8c. This flow of electrons through the metal is the current that flows through the "wire." It occurs because of the **overlap of wave functions** on adjacent atoms, which leads to hopping, and also because the energy levels are so close together that they form a continuous energy spectrum.

What makes diamond an **insulator** in this picture? Bands are separated from one another by band gaps, in which there are no allowed eigenfunctions of the total energy operator. In diamond, all quantum states in the band accessible to the delocalized valence electrons are filled. In Figure 14.8, this would correspond to extending the dark shaded area up to the dashed line. As we go higher in energy, a range is reached at which there are no allowed states of the system until the next band of allowed energy levels is reached. This means that although we could draw diagrams just like the lower two panels of Figure 14.8 for diamond, the system cannot respond as shown in the upper panel. There are no unoccupied states in the valence band that can be used to transport electrons through the crystal. Therefore, diamond is an insulator. **Semiconductors** also have a band gap separating the fully occupied valence and the empty valence band. However, in semiconductors, the band gap is smaller than for insulators, allowing them to become conductors at elevated temperatures.

14.9 Tunneling through a Barrier

Consider a particle with energy E that is confined to a very large box. Within this box, a barrier of height V_0 separates two regions in which $E < V_0$. Classically, the particle will not pass the barrier region because it has insufficient energy to get over the barrier. This situation can look quite different in quantum mechanics. As long as the particle that leaks into the barrier sees a potential energy such that $V_0 > E$, the wave function will decay exponentially with distance. However, something surprising happens if the barrier is thin, meaning that $V_0 > E$ only over a distance comparable to the particle wavelength. The particle can escape through the barrier even though it does not have sufficient energy to go over the barrier. This effect, which is depicted in Figure 14.9, is known as **tunneling.**

To investigate tunneling, we focus on the region near the edge of the box. The **finite depth box** is modified by making it wider and letting the barrier have a finite thickness on the right-hand side. The potential is now described by

$$V(x) = 0, \text{ for } x < 0$$
$$= V_0, \text{ for } 0 \leq x \leq a$$
$$= 0, \text{ for } x > a \tag{14.32}$$

The incident wave functions are shown to the left of the barrier in Figure 14.9. If the **barrier width,** a, is small enough that $\psi(x)$ has not decayed to a negligibly small value at $x = a$, the wave function in the region $x > a$ has a finite amplitude. Because $V(x) = 0$

FIGURE 14.9
Waves corresponding to the indicated energy are incident from the left on a barrier of height $V_0 = 1.60 \times 10^{-19}$ J and width 9.00×10^{-10} m.

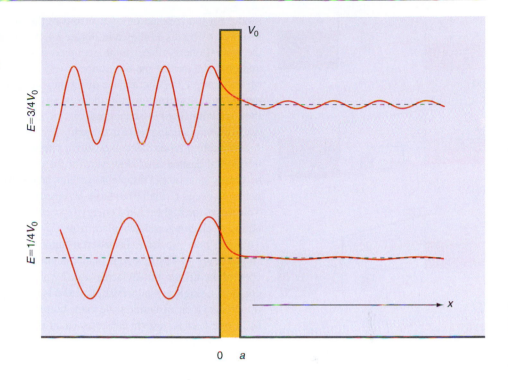

Web-Based Problem 14.8 Tunneling through a Barrier

for $x > a$, the wave function is a traveling wave. This means that the particle has a finite probability of escaping from the well, although its energy is less than the height of the barrier. An exponentially decaying wave function is shown inside the barrier and the incident and transmitted wave functions are shown in the other two regions.

We see that tunneling is much more likely for particles with energies near the top of the barrier. This is a direct consequence of the degree to which the wave function in the barrier falls off with distance as $e^{-\kappa x}$, in which $1/\kappa$, called the **decay length,** is given by $\sqrt{\hbar^2/2m(V_0 - E)}$.

14.10 The Scanning Tunneling Microscope

Researchers have known how the phenomenon of tunneling applies to light under special conditions. However, they did not know it also applied to particles until the advent of quantum mechanics. In the early 1980s, the tunneling of electrons between two solids was used to develop an atomic resolution microscope. For this development, Gerd Binnig and Heinrich Rohrer received the Nobel Prize in physics for the invention of the **scanning tunneling microscope (STM)** in 1986.

The STM allows the imaging of solid surfaces with atomic resolution with a surprisingly minimal mechanical complexity. The STM and a closely related device called the atomic force microscope have been successfully used to study phenomena at atomic and near atomic resolution in a wide variety of areas including chemistry, physics, biology, and engineering. The essential elements of an STM are a sharp metallic tip and a conducting sample over which the tip is scanned to create an image of the sample surface. In an STM, the barrier between these two conductors is usually a vacuum or air, and electrons tunnel across this barrier as discussed later. As might be expected, the barrier width needs to be on the order of atomic dimensions to observe tunneling. Electrons with a typical energy of 5 eV can tunnel from the metal tip to the surface. This energy corresponds to the **work function,** as well as to the barrier height $V_0 - E$ in Figure 14.9. The decay length $\sqrt{\hbar^2/2m(V_0 - E)}$ for such an electron in the barrier is about 0.1 nm. Therefore, if the tip and sample are brought to within a nanometer of one another, electron tunneling will be observed between them.

How does a scanning tunneling microscope work? We address this question first in principle and then from a practical point of view. Because the particle in a box is a good

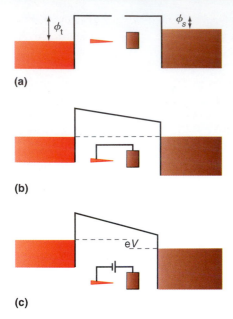

FIGURE 14.10

(a) If the conducting tip and surface are electrically isolated from one another, their energy diagrams line up. (b) If they are connected by a wire in an external circuit, charge flows from the lower work function material into the higher work function material until the highest occupied states have the same energy in both materials. (c) By applying a voltage V between the two materials, the highest occupied levels have an offset of energy eV. This allows tunneling to occur from left to right. The subscripts t and s refer to tip and surface.

model for the conduction of electrons in the metal solid, the tip and surface can be represented by boxes as shown in Figure 14.10. For convenience, the part of the box below the lowest energy that can be occupied by the conduction electrons has been omitted, and only the part of the box immediately adjacent to the tip–sample gap is shown. The tip and sample in general have different work functions as indicated. If they are not connected in an external circuit, their energy diagrams line up as in Figure 14.10a. When they are connected in an external circuit, charge flows between the tip and sample until the highest occupied level is the same everywhere as shown in Figure 14.10b.

Tunneling takes place at constant energy, which in Figure 14.10 corresponds to the horizontal dashed line. However, for the configuration shown in Figure 14.10b, there is no empty state on the sample into which an electron from the tip can tunnel. To allow tunneling to occur, a small (0.01- to 1-V) electrical potential is placed between the two metals. This raises the highest filled energy level of the tip relative to that of the sample. Now tunneling of electrons can take place from tip to sample, resulting in a net current flow.

Until now, we have discussed a tunneling junction, not a microscope. Figure 14.11 shows how an STM functions in an imaging mode. A radius of curvature of 100 nm at the apex of the tip is routinely achievable by electrolytically etching a metal wire. The sample could be a single crystal whose structure is to be investigated at an atomic scale. This junction is shown on an atomic scale in the bottom part of Figure 14.11. No matter how blunt the tip is, one atom is closer to the surface than all the others. At a tunneling gap distance of about

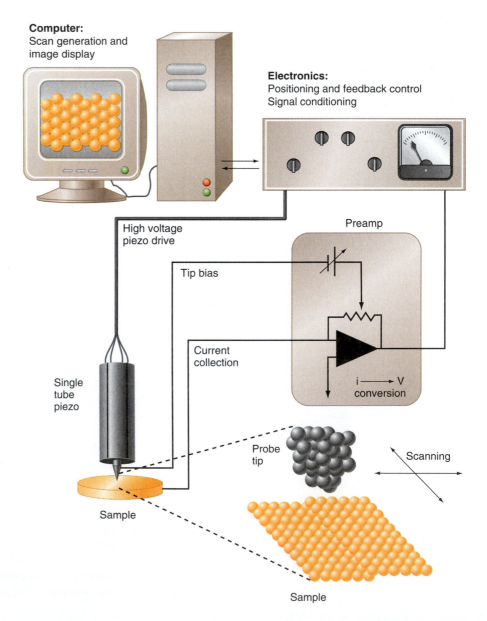

FIGURE 14.11

Schematic representation of a scanning tunneling microscope. (Source: Courtesy of Kevin Johnson, University of Washington thesis, 1991.)

0.5 nm, the tunneling current decreases by an order of magnitude for every 0.1 nm that the gap is increased. Therefore, the next atoms back from the apex of the tip make a negligible contribution to the tunneling current and the whole tip acts like a single atom for tunneling.

The tip is mounted on a segmented tubular scanner made of a piezoelectric material that changes its length in response to an applied voltage. In this way the tip can be brought close to the surface by applying a voltage to the piezoelectric tube. Assume that we have managed to bring the tip within tunneling range of the surface. On the magnified scale shown in Figure 14.11, the individual atoms in the tip and surface are seen at a tip–surface spacing of about 0.5 nm. Keep in mind that the wave functions for the electrons in the tip that tunnel into the sample decay rapidly in the region between tip and sample, as shown earlier in Figure 14.9. If the end of the tip is directly over a surface atom, the amplitude of the wave function will be large at the surface atom, and the tunneling current will be high. If the end of the tip is between surface atoms, the wave functions will have decayed and the tunneling current will be lower. To scan over the surface, different voltages are applied to the four segmented electrodes on the piezo tube. This allows a topographical image of the surface to be obtained. Because the tunneling current varies exponentially with the tip–surface distance, the microscope provides a very high sensitivity to changes in the height of the surface that occur on an atomic scale.

In this abbreviated description, some details have been glossed over. The current is usually kept constant as the tip is scanned over the surface using a feedback circuit to keep the tip–surface distance constant. This is done by changing the voltage to the piezo tube electrodes as the tip scans over the surface. Additionally, a vibrational isolation system is required to prevent the tip from crashing into the surface. Figure 14.12 provides an example of the detail that can be seen with a scanning tunneling microscope. The individual planes, which are stacked together to make the silicon crystal, and the 0.3-nm height change between planes are clearly seen. Defects in the crystal structure are also clearly resolved. Researchers are using this microscope in many new applications aimed at understanding the structure of solid surfaces and modifying surfaces atom by atom.

EXAMPLE PROBLEM 14.5

As was found for the finite depth well, the wave function amplitude decays in the barrier according to $\psi(x) = A \exp[-\sqrt{2m(V_0 - E)/\hbar^2}\, x]$. This result will be used to calculate the sensitivity of the scanning tunneling microscope. Assume that the tunneling current through a barrier of width a is proportional to $|A|^2 \exp[-2\sqrt{2m(V_0 - E)/\hbar^2}\, a]$.

a. If $V_0 - E$ is 4.50 eV, how much larger would the current be for a barrier width of 0.20 nm than for 0.30 nm?

b. A friend suggests to you that a proton tunneling microscope would be equally effective as an electron tunneling microscope. For a 0.20-nm barrier width, by what factor is the tunneling current changed if protons are used instead of electrons?

Solution

a. Substituting the numbers into the formula given, we obtain

$$\frac{I(a = 2.0 \times 10^{-10}\ \text{m})}{I(a = 3.0 \times 10^{-10}\ \text{m})} = \exp\left[-2\sqrt{\frac{2m(V_0 - E)}{\hbar^2}}\,(2.0 \times 10^{-10}\ \text{m} - 3.0 \times 10^{-10}\ \text{m})\right]$$

$$= \exp\left[-2\sqrt{\frac{2(9.11 \times 10^{-31}\ \text{kg})(4.50\ \text{eV})(1.602 \times 10^{-19}\ \text{J/eV})}{(1.055 \times 10^{-34}\ \text{J})^2}} \times (-1.0 \times 10^{-10}\ \text{m})\right]$$

$$= 8.8$$

Even a small distance change results in a substantial change in the tunneling current.

FIGURE 14.12
STM images of the (111) surface of Si. The upper image shows a 200- × 200-nm region with a high density of atomic steps, and the light dots correspond to individual Si atoms. The lower image shows how the image is related to the structure of parallel crystal planes separated by steps of one atom in height. The step edges are shown as dark ribbons. (Source: Courtesy of Kevin Johnson, University of Washington thesis, 1991.)

b. We find that the tunneling current for protons is appreciably smaller than that for electrons.

$$\frac{I(proton)}{I(electron)} = \frac{\exp\left[-2\sqrt{\dfrac{2m_{proton}(V_0 - E)}{\hbar^2}}\,a\right]}{\exp\left[\left[-2\sqrt{\dfrac{2m_{electron}(V_0 - E)}{\hbar^2}}\,a\right]\right]}$$

$$= \exp\left[-2\sqrt{\dfrac{2(V_0 - E)}{\hbar^2}}\left(\sqrt{m_{proton}} - \sqrt{m_{electron}}\right)a\right]$$

$$= \exp\left[-2\sqrt{\dfrac{2(4.50\text{ eV})(1.602 \times 10^{-19}\text{ J/eV})}{(1.055 \times 10^{-34}\text{ J s})^2}}\left(\begin{array}{c}\sqrt{1.67 \times 10^{-27}\text{ kg}}\\[4pt] -\sqrt{9.11 \times 10^{-31}\text{ kg}}\end{array}\right)\right.$$
$$\left.\times (2.0 \times 10^{-10}\text{ m})\right]$$

$$= 1.2 \times 10^{-79}$$

This result does not make the proton tunneling microscope look very promising.

Scanning tunneling microscopy can also be used to image individual molecules if they are adsorbed on a very flat substrate so that the topography seen in the STM image arises primarily form the molecule rather than the substrate. Particular care must be taken when imaging molecules. Although the tunneling current is very small (20 to 100×10^{-12} A), the current density can be as high as 1×10^6 A cm^{-2} because the current is localized in a cylinder of a few atoms in diameter. Such high current densities can lead to decomposition of the molecule. Figure 14.13 shows an image of palladium phthalocyanine adsorbed on an atomically flat graphite surface together with a molecular model consistent with the image. Note that some regions of the molecule are lighter in color, corresponding to an apparent increase in the height of the molecule. This can be explained with the aid of the images of the highest occupied molecular orbital (HOMO) and lowest unoccupied molecular orbital (LUMO) obtained through quantum mechanical calculations. At this point, rely on what you have learned in general and organic chemistry about molecular orbitals; the topic will be discussed in detail in Chapter 16. A comparison of Figures 14.13 and 14.14 shows that the "structure" obtained under the imaging conditions corresponds to a map of the LUMO rather than to a map of the atomic positions. STM can also be used to obtain images of biomolecules. Figure 14.15 shows an image of a 2739-base-pair double-helix plasmid DNA on a copper single crystal. The helix repeat unit is clearly imaged.

14.11 Tunneling in Chemical Reactions

Most chemical reactions are thermally activated; they proceed faster as the temperature of the reaction mixture is increased. This behavior is typical of reactions for which an energy barrier must be overcome in order to transform reactants into products. This barrier is referred to as the **activation energy** for the reaction. By increasing the temperature of the reactants, the fraction that has an energy that exceeds the activation energy is increased, allowing the reaction to proceed.

Tunneling provides another mechanism to convert reactants to products that does not require an increase in energy of the reactants for the reaction to proceed. It is well known that hydrogen transfer reactions can involve tunneling. An example is the reaction $R_1OH + R_2O^- \rightarrow R_2OH + R_1O^-$, where R_1 and R_2 are two different organic groups. The test for tunneling in this case is to substitute deuterium for hydrogen. If the reaction is thermally activated, the change in reaction rate is small and can be attributed to the different ground-state vibrational frequency of —OH and —OD bonds (see Chapter 18). However, if tunneling

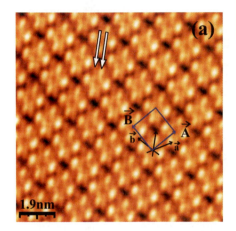
(a)

1.9nm

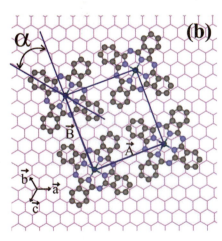
(b)

FIGURE 14.13
The top panel shows the STM image of palladium phthalocyanine and the unit cell of the periodic structure formed. The bottom panel shows a molecular model consistent with the STM image. (Source: Courtesy of T.G. Gopalumar, M. Lackinger, M. Hackert, F. Mueller and M. Hietschold, "Adsorption of Palladium Phthalocyanine on Graphite: STM and LEED Study", *J. Phys. Chem. B*, 2004, 108, 7839–7843, American Chemical Society)

FIGURE 14.14
Molecular orbitals calculated using quantum mechanical methods are shown for palladium phthalocyanine: (a) molecular structure, (b) HOMO, and (c, d) two LUMOs, both of which have the same energy. (Source: T. G. Gopakumar, M. Lackinger, M. Hackert, F. Mueller, and M. Hietschold, "Adsorption of Palladium Phthalocyanine on Graphite: STM and LEED Study," *J. Phys. Chem. B* 2004, 108, 7839–7843)

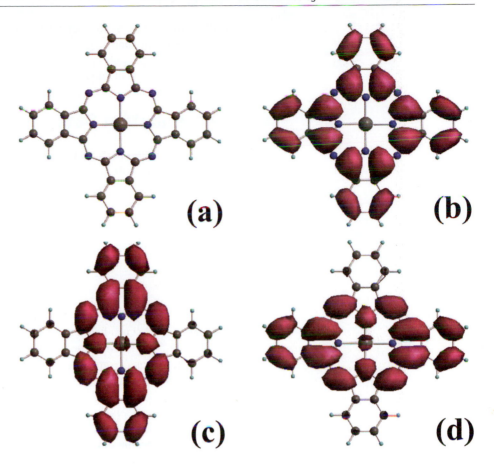

(a) (b) (c) (d)

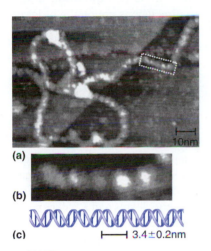

FIGURE 14.15
(a) STM image of DNA deposited on a copper surface. The diagonal line in the image is a single atomic height step separating two atomically flat regions. The white dots are components of the buffer solution from which the DNA was deposited on the surface. (b) A magnified image of the 6.4- × 19.3-nm rectangular region outlined in part (a), showing a repeat periodicity of the double helix. (c) The Watson-Crick double helix model of DNA. (Source: Vol. 38 (1999) L606-L607, Part 2, NO. 6A, B, 15 June 1999, Real Space Observation of Double-Helix DNA Structure Using a Low Temperature Scanning Tunneling Microscopy by Takashi Kanno, Hiroyuki Tanaka,Tomohiko Nakamura, Hitoshi Tabata and Tomoji Kawai, *Japanese Journal of Applied Physics,* reprinted with permission of The Institute of Pure and Applied Physics)

FIGURE 14.16
The structures of four species along the reaction path from reactant to product are shown together with a schematic energy diagram. The reaction occurs not by surmounting the barrier, but by tunneling through the barrier at the energy indicated by the wavy line. (Source: Adapted with permission from *Science* and Wes Borden, University of Washington. Copyright 2003 American Association for the Advancement of Science.)

occurs, the rate decreases greatly because the tunneling rate depends exponentially on the decay length $\sqrt{\hbar^2/2m(V_0 - E)}$. It is not widely appreciated that tunneling can be important for heavier atoms such as C and O. A report by Zuev *et al.* [*Science* 299 (2003), 867] shows that the ring expansion reaction depicted in Figure 14.16 is faster than the predicted

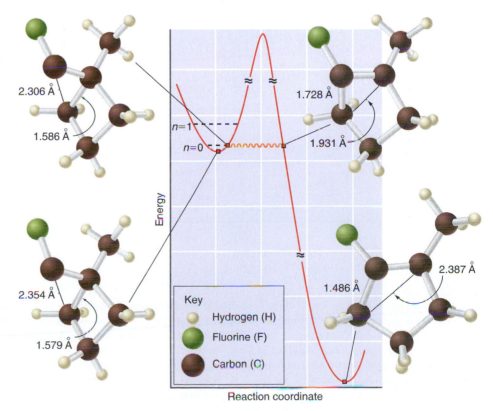

Energy

$n=1$

$n=0$

2.306 Å
1.586 Å
2.354 Å
1.579 Å

1.728 Å
1.931 Å
1.486 Å
2.387 Å

Key
Hydrogen (H)
Fluorine (F)
Carbon (C)

Reaction coordinate

thermally activated rate by the factor 10^{152} at 10 K! This increase is due to the tunneling pathway. Because the tunneling rate depends exponentially on the product $-2m(V_0 - E)$, heavy atom tunneling is only appreciable if $(V_0 - E)$ is very small. In a number of reactions, particularly in the fields of chemical catalysis and enzymology, this condition is met.

14.12 Quantum Dots

We have seen that the particle in the box is a useful model for understanding electrical conduction in solids, which are macroscopic in size. Particles that are large compared to a single atom, but small on the macroscopic scale are called **mesoscopic particles.** An example of a mesoscopic particle is a **quantum dot,** a three-dimensional piece of a crystalline solid whose dimensions are on the order of nanometers that can be modeled as a 3D box. If we assume that the solid has the form of a cube with edge length b, the energy levels for the electron in this 3D box are given by

$$E_{n_x n_y n_z} = \frac{h^2}{8mb^2}(n_x^2 + n_y^2 + n_z^2) \tag{14.33}$$

Assume that all states below the band gap are filled and all states above the band gap of width E_{bg} are filled in the ground state of the quantum dot, making it a semiconductor. Transitions from states below the band gap to those above it can occur through absorption of visible light. Subsequently, the electron in the excited state can drop to an empty state below the band gap, emitting a photon in a process called fluorescence with a wavelength $\lambda = hc/E_{bg}$. Because the energy levels and E_{bg} depend on the length b, λ also depends on b. This property is illustrated in Figure 14.17a. For CdSe quantum dots, the emission wavelength increases from 450 nm (blue light) to 650 nm (red light) as the dot diameter increases from 2 to 8 nm. Figure 14.17b shows another important property of quantum dots. Although they absorb light over a wide range of wavelengths, they emit light in a much smaller range of wavelengths. This occurs because electrons excited from occupied states just below the band gap to states

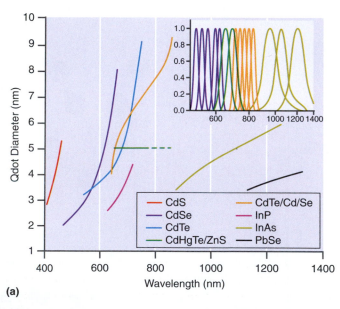

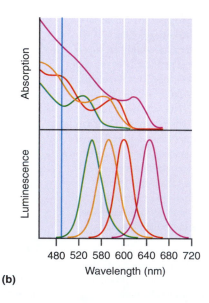

(a)

(b)

FIGURE 14.17

(a) The dependence of the wavelength of the light emitted in a transition from just above to just below the band gap is shown as a function of the quantum dot diameter for a number of materials. (b) The top panel shows the absorption spectrum of four CdSe/ZnS quantum dots of different diameters, and the bottom panel shows the corresponding emission spectrum. Note that absorption occurs over a much larger range of wavelengths than emission. The vertical bar indicates the wavelength of a 488-nm argon ion laser, which can be used to excite electrons from below to above the band gap for all four diameters. Using this laser ensures that absorption and emission occur at distinctly different wavelengths. (Source: X. Michalet, F.F. Pinaud, L.A. Bentolila, J.M. Tsay, S. Doose, J.J. Li, G. Sundaresan, A.M. Wu, S. Gambhir, S. Weiss, "Quantum Dots for Live Cells, in Vivo Imaging, and Diagnostics," *Science,* 28 January 2005, Vol. 307, No. 5709, pp. 538–544, Fig. 1)

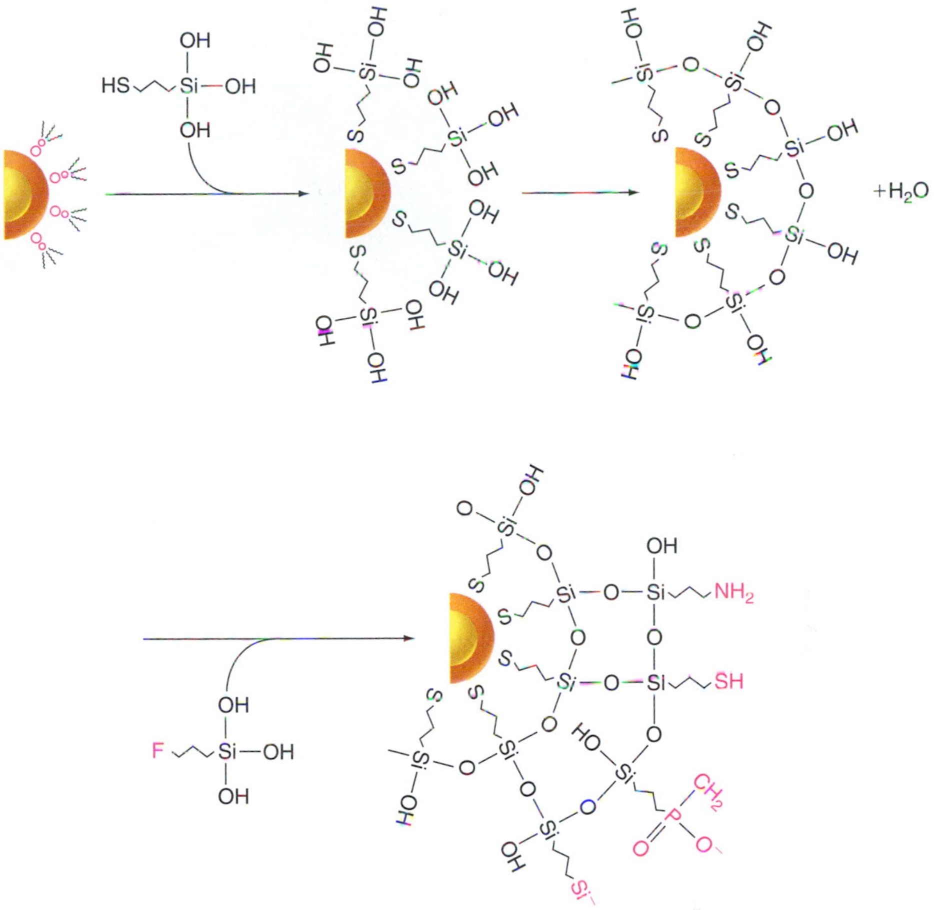

FIGURE 14.18
Quantum dots consisting of a CdSe core and a ZnS shell were capped with a trioctyl phosphine oxide (TOPO) layer (left image) and dissolved in mercaptopropyltris(methoxy)silane (MPS). By making the solution basic, the MPS replaces the TOPO layer. The methoxysilane ($Si-OCH_3$) groups hydrolyze into silanol ($Si-OH$) groups. Additional silane precursors containing desired functional groups such as $-SH$ or $-NH_2$ are then incorporated into the outer layer. (Source: Reprinted with permission from Gerion et al., *Journal of Physical Chemistry* B105 (2001), 8861.)

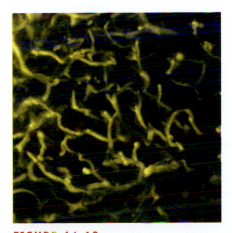

FIGURE 14.19
This 160- × 160-μm image was obtained by projecting the capillary structure in a 250-μm-thick specimen of adipose tissue in the skin of a living mouse using CdSe quantum dots that fluoresce at 550 nm. (Source: D. R. Larson et al., *Science* 300 (May 2003) 30. Copyright 2003 American Association for the Advancement of Science.)

well above the band gap in absorption lose energy and relax to states just above the band gap before emitting light in fluorescence. The absorbed light is re-emitted in a narrow frequency range determined by the band-gap energy of the semiconducting quantum dot.

By functionalizing such quantum dots with an appropriate molecular layer as shown in Figure 14.18, they can be made soluble in aqueous solutions and sufficiently chemically inert that they are compatible with the biological environment.

Because a quantum dot absorbs strongly over a wide range of wavelengths, but fluoresces in a narrow range of wavelengths, it can be used as an internal light source for imaging the interior of semitransparent specimens. Figure 14.19 shows an image obtained by projecting the capillary structure in a 250-μm-thick specimen of adipose tissue in the skin of a living mouse on a plane [D. R. Larson et al., *Science* 300 (2003), 1434]. Additionally, the rate of blood flow and the differences in systolic and diastolic pressure can be directly observed in these experiments. It is not possible to obtain such images with X-ray-based techniques, because of the absence of a contrast mechanism.

The usefulness of quantum dots in imaging tissues *in vivo* has significant potential applications in surgery provided that toxicity issues currently associated with quantum dots can be resolved. Figure 14.20 shows images obtained with a near-infrared camera resulting from the injection of quantum dots emitting in the near-infrared region (840 to 860 nm) into the paw of a mouse. The quantum dots had a diameter of ~15 nm including the functionalization layer, which is a suitable diameter for trapping in lymph nodes. Fluorescence from the quantum dots allows the lymph node to be imaged through the overlying tissue layers. Because near-infrared light is invisible to the human eye and visible light is not registered by the camera, the mouse can be simultaneously illuminated with both types of light. A comparison of images obtained with near-infrared and visible light can be used to guide the surgeon in the removal of tumors and to verify that all affected tissues have been removed. Near-infrared light is

FIGURE 14.20

Quantum dots were injected into the paw of a mouse. (a) The middle panel shows a video taken 5 minutes after the quantum dots were introduced. The right panel is a fluorescence image that shows localization of the quantum dots in the lymph node. The left panel shows that without the quantum dots, no fluorescence is observed. (b) Surgery after injection of a chemical mapping agent that is known to localize in lymph nodes confirms that the quantum dots are localized in the lymph nodes. (Source: Sungee Kim, Yong Taik Lim, Edward G. Soltexz, Alec M. De Grand, Jaihyoung Lee, Akira Nakayama, J. Anthony Parker, Tomislav Mihaljevic, Rita G. Laurence, Delphine M. Dor, Lawrence H. Cohn, Moungi G. Bawendi & John V. Frangioni, "Near-infrared fluorescent type II quantum dots for sentinel lymph node mapping," *Nature Biotechnology*, Volume 22, Number 1, January 2004, Fig. 2a-b)

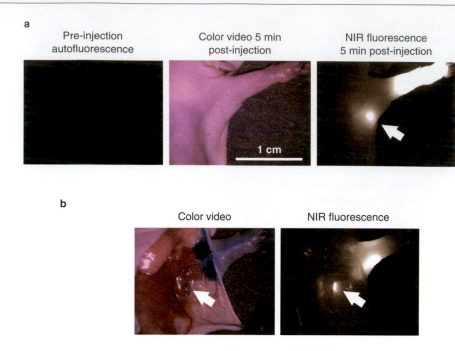

a
Pre-injection autofluorescence | Color video 5 min post-injection | NIR fluorescence 5 min post-injection

1 cm

b
Color video | NIR fluorescence

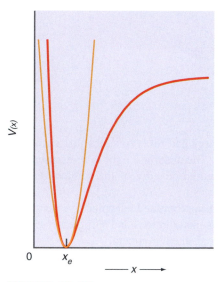

FIGURE 14.21

Potential energy, $V(x)$, as a function of the bond length, x, for a diatomic molecule. The zero of energy is chosen to be the bottom of the potential. The red curve depicts a realistic potential in which the molecule dissociates for large values of x. The yellow curve shows a harmonic potential, $V(x) = (1/2)kx^2$, which is a good approximation to the realistic potential near the bottom of the well. The term x_e represents the equilibrium bond length.

the Chemistry place

Web-Based Problem 14.9 The Classical Harmonic Oscillator

also useful because this wavelength minimizes the absorption of light by overlying tissues, allowing tissues containing quantum dots to be imaged through an overlying tissue layer of 6 to 10 cm.

Quantum dots have other applications that are in developmental stages. It may be possible to use them to couple electrical signal amplification, currently based on charge carrier conduction in semiconductors with light amplification, in an application known as optoelectronics. Additionally, the reduced dimensions of quantum dots utilized as wavelength-tunable lasers allow them to be integrated into conventional silicon-based microelectronics.

14.13 The Quantum Mechanical Harmonic Oscillator

The discussion of the free particle and the particle in the box was useful for understanding how **translational motion** in various potentials is described in the context of wave-particle duality. In applying quantum mechanics to molecules, there are two other types of motion that molecules can undergo: **vibration** and **rotation.** A quantum mechanical model for vibration is formulated in this section based on the **harmonic oscillator.**

The simplest vibrational motion that can be imagined is that which occurs in a diatomic molecule. Vibration involves a displacement of the atoms from their equilibrium positions, which is dictated by the chemical bond length between the atoms. The energy needed to stretch the chemical bond can be described by a simple potential function such as that shown in Figure 14.21. The existence of a stable chemical bond implies that a minimum energy exists at some internuclear distance, called the **bond length.** The configuration of atoms in a molecule is dynamic rather than static. Think of the chemical bond as a spring rather than a rigid bar connecting the two atoms. Thermal energy increases the vibrational amplitude of the atoms about their equilibrium positions. The potential becomes steeply repulsive at short distances as the electron clouds of the atoms interpenetrate. It levels out at large distances because the overlap of electrons between the atoms required for chemical bond formation falls to zero.

The exact form of $V(x)$ as a function of x depends on the molecule under consideration. However, only the lowest one or two vibrational energy levels are occupied for most molecules for $T \sim 300$ K. Therefore, it is a good approximation to say that the functional form of the potential energy near the equilibrium bond length can be approximated by the harmonic potential

$$V(x) = \frac{1}{2} kx^2 \qquad (14.34)$$

In Equation (14.34), k is the **force constant** and is not to be confused with the wave vector k or the Boltzmann constant. Vibration is studied in the **center of mass coordinates** because the relative motion of the atoms is of interest. Transformation to these coordinates means that rather than having two atoms of mass m_1 and m_2 oscillating about their center of mass, we consider the mathematically equivalent problem of a single reduced mass $\mu = (m_1 m_2)/(m_1 + m_2)$ tethered to a wall of infinite mass with a spring of force constant k.

From solving the particle-in-the-box problem, we expect the wave-particle of mass μ vibrating around its equilibrium distance to be described by a set of wave functions $\psi_n(x)$. To find these wave functions and the corresponding allowed energies in the vibrational motion, the following form of the Schrödinger equation must be solved:

$$-\frac{\hbar^2}{2\mu} \frac{d^2\psi_n(x)}{dx^2} + \frac{kx^2}{2} \psi_n(x) = E_n\psi_n(x) \qquad (14.35)$$

The solution of this second-order differential equation was well known in the mathematical literature from other contexts well before the development of quantum mechanics. We simply state that the normalized wave functions are

$$\psi_n(x) = A_n H_n(\alpha^{1/2} x)e^{-\alpha x^2/2} \text{ for } n = 0, 1, 2, \dots \qquad (14.36)$$

EXAMPLE PROBLEM 14.6

Show that the function $e^{-\beta x^2}$ satisfies the Schrödinger equation for the quantum harmonic oscillator. What conditions does this place on β? What is E?

Solution

$$-\frac{\hbar^2}{2\mu} \frac{d^2\psi_n(x)}{dx^2} + V(x)\psi_n(x) = E_n\psi_n(x)$$

$$-\frac{\hbar^2}{2\mu} \frac{d^2(e^{-\beta x^2})}{dx^2} + V(x)(e^{-\beta x^2}) = -\frac{\hbar^2}{2\mu} \frac{d(-2\beta xe^{-\beta x^2})}{dx} + \frac{1}{2} kx^2(e^{-\beta x^2})$$

$$= -\frac{\hbar^2}{2\mu}(-2\beta e^{-\beta x^2}) + \frac{\hbar^2}{2\mu}(-4\beta^2 x^2 e^{-\beta x^2})$$

$$+ \frac{1}{2} kx^2(e^{-\beta x^2})$$

The function satisfies the Schrödinger equation only if the last two terms cancel

which requires that $\beta^2 = +\frac{1}{4} \frac{k\mu}{\hbar^2}$

Finally, $E = \frac{\hbar^2\beta}{\mu} = \frac{\hbar^2}{\mu} \sqrt{\frac{1}{4} \frac{k\mu}{\hbar^2}} = \frac{\hbar}{2}\sqrt{\frac{k}{\mu}}$.

In Equation (14.36), several constants have been combined to give $\alpha = \sqrt{k\mu/\hbar^2}$, and the normalization constant A_n is given by

$$A_n = \frac{1}{\sqrt{2^n n!}} \left(\frac{\alpha}{\pi}\right)^{1/4} \qquad (14.37)$$

The solution is written in this manner because the set of functions $H_n(\alpha^{1/2}x)$ is well known in mathematics as **Hermite polynomials.** The first few solutions of the Schrödinger equation $\psi_n(x)$ are given by

$$\psi_0(x) = \left(\frac{\alpha}{\pi}\right)^{1/4} e^{-(1/2)\alpha x^2}$$

$$\psi_1(x) = \left(\frac{4\alpha^3}{\pi}\right)^{1/4} x e^{-(1/2)\alpha x^2}$$

(14.38)

$$\psi_2(x) = \left(\frac{\alpha}{4\pi}\right)^{1/4} (2\alpha x^2 - 1) e^{-(1/2)\alpha x^2}$$

$$\psi_3(x) = \left(\frac{\alpha^3}{9\pi}\right)^{1/4} (2\alpha x^3 - 3x) e^{-(1/2)\alpha x^2}$$

where $\psi_0, \psi_2, \psi_4, \ldots$ are even functions of x, for which $\psi(x) = \psi(-x)$, whereas $\psi_1, \psi_3, \psi_5, \ldots$ are odd functions of x for which $\psi(x) = -\psi(-x)$.

Just as a boundary condition was applied to the particle in the box, here the amplitude of the wave functions is required to remain finite at large values of x. As for the particle in the box, this boundary condition gives rise to quantization. In this case, the quantization condition is not as easy to derive. However, it can be shown that the amplitude of the wave functions approaches zero for large x values only if the following condition is met:

Web-Based Problem 14.10 Energy Levels and Solutions of the Schrödinger Equation for the Harmonic Oscillator

$$E_n = \hbar \sqrt{\frac{k}{\mu}}\left(n + \frac{1}{2}\right) = h\nu\left(n + \frac{1}{2}\right) \quad \text{with } n = 0, 1, 2, 3, \ldots \quad (14.39)$$

Once again, we see that the imposition of boundary conditions has led to a discrete energy spectrum. Unlike the classical analogue, the energy stored in the quantum mechanical harmonic oscillator can only take on certain values. As for the particle in the box, the lowest state accessible to the system still has a nonzero energy, referred to as a **zero point energy.** The **frequency of oscillation** is given by

$$\nu = \frac{1}{2\pi}\sqrt{\frac{k}{\mu}}$$

(14.40)

EXAMPLE PROBLEM 14.7

What is the average value of the potential energy for a quantum mechanical oscillator in the state described by $\psi_1(x) = (4\alpha^3/\pi)^{1/4} x e^{-(1/2)\alpha x^2}$?

Solution

Because the potential energy depends only on the coordinate x, the average value of the potential energy is given by

$$\langle E_{potential} \rangle = \int \psi_1^*(x) V(x) \psi_1(x) dx$$

$$= \int_{-\infty}^{\infty} \left(\frac{4\alpha^3}{\pi}\right)^{1/4} x e^{-(1/2)\alpha x^2} \left(\frac{1}{2}kx^2\right) \left(\frac{4\alpha^3}{\pi}\right)^{1/4} x e^{-(1/2)\alpha x^2} dx$$

$$= \frac{1}{2}k\left(\frac{4\alpha^3}{\pi}\right)^{1/2} \int_{\infty}^{\infty} x^4 e^{-\alpha x^2} dx = k\left(\frac{4\alpha^3}{\pi}\right)^{1/2} \int_{0}^{\infty} x^4 e^{-\alpha x^2} dx$$

The limits can be changed as indicated in the last integral because the integrand is an even function of x. To obtain the solution, the following standard integral is used:

$$\int_0^\infty x^{2n} e^{-ax^2} \, dx = \frac{1 \cdot 3 \cdot 5 \cdots (2n-1)}{2^{n+1} a^n} \sqrt{\frac{\pi}{a}}$$

The calculated value for the average potential energy is

$$\langle E_{potential} \rangle = \frac{1}{2} k \left(\frac{4\alpha^3}{\pi} \right)^{1/2} \left(\sqrt{\frac{\pi}{\alpha}} \right) \frac{3}{4\alpha^2}$$

$$= \frac{3k}{4\alpha} = \frac{3}{4} \hbar \sqrt{\frac{k}{\mu}}$$

In general, we find that for the nth state,

$$\langle E_{potential,n} \rangle = \frac{\hbar}{2} \sqrt{\frac{k}{\mu}} \left(n + \frac{1}{2} \right)$$

As was done for the particle in the box, it is useful to plot $\psi(x)$ and $\psi^2(x)$ against x. They are shown superimposed on the potential energy function in Figures 14.22 and 14.23.

It is instructive to compare the quantum mechanical with the classical results. A good starting point is to create an analogy from what we know about the particle in the box. It is not possible to state what the vibrational amplitude is at a given time; only the probability of the vibrational amplitude having a particular value of x within an interval dx can be calculated. This probability is given by $\psi^2(x)dx$. For the classical harmonic oscillator, the probability of finding a particular value of x within the interval dx can also be calculated. Because the probability density varies inversely with the velocity, its maximum values are found at the turning points and its minimum value is found at $x = 0$. To visualize this behavior, imagine a frictionless ball rolling on a parabolic track under the influence of gravity. The ball moves fastest at the lowest point on the track, and stops momentarily as it reverses its direction at the highest points on either side of the track. Figure 14.24 shows a comparison of $\psi_{12}^2(x)$ and the probability density of finding a particular amplitude for a classical oscillator with the same total energy as a function of x. A large quantum number has been used for comparison because in the limit of high energies (very large quantum numbers), classical and quantum mechanics give the same result.

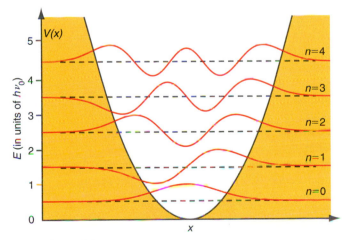

FIGURE 14.22
The first few solutions of the Schrödinger equation of the quantum mechanical harmonic oscillator are plotted together with the potential function. The energy scale is shown on the left. The amplitude is shown superimposed on the energy level, with the zero of amplitude for each solution of the Schrödinger equation indicated by the dashed lines. The yellow area indicates the classically forbidden region.

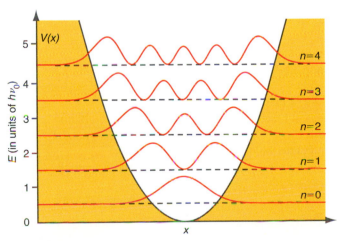

FIGURE 14.23
The square of the first few solutions of the Schrödinger equation for the quantum mechanical harmonic oscillator (the probability density) is superimposed on the energy spectrum and plotted together with the potential function. The yellow area indicates the classically forbidden region.

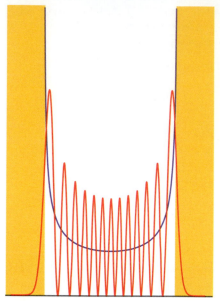

FIGURE 14.24
The calculated probability density for the vibrational amplitude is shown for the solution of the Schrödinger equation for $n = 12$ (red curve). The classical result is shown by the blue curve. The yellow area indicates the classically forbidden region.

Web-Based Problem 14.11 Probability of Finding the Oscillator in the Classically Forbidden Region

We have been working with the time-independent Schrödinger equation, whose solutions allow the probability density to be calculated. To describe the time dependence of the oscillation amplitude, the total wave function, $\psi_n(x,t) = e^{-i\omega t}\psi_n(x)$, is considered. The spatial amplitude shown in Figure 14.22 is modulated by the factor $e^{-i\omega t}$, which has a frequency given by $\omega = \sqrt{k/\mu}$. Because $\psi_n(x,t)$ is a standing wave, the nodal positions shown in Figures 14.22 and 14.23 do not move with time.

In looking at Figures 14.22 and 14.23, several similarities are seen with Figures 14.2 and 14.3, in which the equivalent results were shown for the particle in the box. The solutions of the Schrödinger equation are again standing waves, but they are now in a box with a more complicated shape. Successive solutions of the Schrödinger equation add one more oscillation within the "box," and the amplitude of the wave function is small at the edge of the "box." (The reason why it is small rather than zero follows the same lines as the discussion of the particle in the finite depth box.) The quantum mechanical harmonic oscillator also has a zero point energy, meaning that the lowest possible energy state still has vibrational energy. The origin of this zero point energy is similar to that for the particle in the box. By attaching a spring to the particle, its motion has been constrained. As the spring is made stiffer (larger k), the particle is more constrained and the zero point energy increases. This is the same trend observed for the particle in the box as the length is decreased.

Note, however, that important differences exist in the two problems that are a result of the more complicated shape of the harmonic oscillator "box." Although the wave functions show oscillatory behavior, they are no longer represented by simple sine functions. Because the classical probability density is not independent of x (see Figure 14.24), simple sine functions for the quantum mechanical wave functions do not lead to the correct classical limit. The energy spacing is the same between adjacent energy levels; that is, it does not increase with the quantum number as was the case for the particle in the box. These differences show the sensitivity of the solutions of the Schrödinger equation and allowed values of the total energy to the functional form of the potential. We will revisit the quantum mechanical oscillator in Chapter 18, when we discuss vibrational spectroscopy.

Vocabulary

activation energy	delocalization	overlap of wave functions	standing waves
band gap	energy band	particle in a box	translational motion
barrier width	finite depth box	π-bonded network	traveling waves
bond length	force constant	probability	tunneling
boundary condition	frequency of oscillation	probability density	valence band
center of mass coordinates	ground state	quantized	valence electrons
classical limit	harmonic oscillator	quantum dot	vibration
conductor	Hermite polynomial	quantum number	wave packet
continuous energy spectrum	insulator	rotation	wave vector
core electrons	mesoscopic particle	scanning tunneling	work function
decay length	node	microscope (STM)	zero point energy
degeneracy	oscillatory behavior	semiconductor	

Questions on Concepts

Q14.1   Why are standing-wave solutions for the free particle not compatible with the classical result $x = x_0 + v_0 t$?

Q14.2   Why is it not possible to normalize the free-particle wave functions over the whole range of motion of the particle?

Q14.3   Show that for the particle-in-the-box total energy solutions, $\psi_n(x) = \sqrt{2/a}\sin(n\pi x/a)$, $\psi(x)$ continuous

function at the edges of the box. Is $d\psi/dx$ a continuous function of x at the edges of the box?

Q14.4   The degeneracy of an energy level is defined as the number of distinct quantum states (different quantum numbers) all of which have the same energy. Can the particles in a one-dimensional box, for which the energy levels are given by Equation (14.18), a square two-dimensional box of length b, for which the energy levels are given by $E_{n_x n_y} = (h^2/8mb^{2/3})(n_x^2 + n_y^2)$, and a cubic three-dimensional box, for which the energy levels are given by Equation (14.25), all have degenerate energy levels?

Q14.5   We set the potential energy in the particle in the box equal to zero and justified it by saying that there is no absolute scale for potential energy. Is this also true for kinetic energy?

Q14.6   Why are traveling-wave solutions for the particle in the box not compatible with the boundary conditions?

Q14.7   Why is the zero point energy lower for a He atom in a box than for an electron?

Q14.8   Invoke wave-particle duality to address the following question: How does a particle get through a node in a wave function to get to the other side of the box?

Q14.9   What is the difference between probability and probability density?

Q14.10   What is the functional dependence of the total energy of the quantum harmonic oscillator on the position variable x?

Q14.11   The zero point energy of the particle in the box goes to zero as the length of the box approaches infinity. What is the appropriate analogue for the quantum harmonic oscillator?

Q14.12   Explain in words why the amplitude of the solutions of the Schrödinger equation for the quantum mechanical harmonic oscillator increases with $|x|$ as shown in Figure 14.24.

Problems

Problem numbers in **RED** indicate that the solution to the problem is given in the *Student Solutions Manual*.

P14.1   Show by examining the position of the nodes that $\text{Re}[A_+e^{i(kx - \omega t)}]$ and $\text{Re}[A_- e^{i(-kx - \omega t)}]$ represent plane waves moving in the positive and negative x directions, respectively. The notation Re[ ] refers to the real part of the function in the brackets.

P14.2   Show that the allowed values of the total energy for the free particle, $E = \hbar^2 k^2/2m$, are consistent with the classical result $E = (1/2)mv^2$.

P14.3   Consider a particle in a one-dimensional box defined by $V(x) = 0$, $a > x > 0$ and $V(x) = \infty$, $x \geq a$, $x \leq 0$. Explain why each of the following unnormalized functions is or is not an acceptable wave function based on criteria such as being consistent with the boundary conditions, and with the association of $\psi^*(x)\psi(x)\,dx$ with probability.

a. $A \cos \dfrac{n\pi x}{a}$

b. $B(x + x^2)$

c. $Cx^3(x - a)$

d. $\dfrac{D}{\sin(n\pi x/a)}$

P14.4   Evaluate the normalization integral for the solutions of the Schrödinger equation for the particle in the box $\psi_n(x) = A \sin(n\pi x/a)$ using the trigonometric identity $\sin^2 y = (1 - \cos 2y)/2$.

P14.5   Use the wave function $\psi(x) = A'e^{+ikx} + B'e^{-ikx}$ rather than $\psi(x) = A \sin kx + B \cos kx$ to apply the boundary conditions for the particle in the box.

a.   How do the boundary conditions restrict the acceptable choices for A' and B' and for k?

b.   Do these two functions give different probability densities if each is normalized?

P14.6   Calculate the probability that a particle in a one-dimensional box of length a is found between $0.31a$ and $0.35a$ when it is described by the following wave functions:

a. $\sqrt{\dfrac{2}{a}} \sin\left(\dfrac{\pi x}{a}\right)$

b. $\sqrt{\dfrac{2}{a}} \sin\left(\dfrac{3\pi x}{a}\right)$

What would you expect for a classical particle? Compare your results in the two cases with the classical result.

P14.7   It is useful to consider the result for the allowed values of the total energy for the one-dimensional box $E_n = h^2 n^2/8ma^2$, $n = 1, 2, 3, \ldots$ as a function of n, m, and a.

a.   By what factor do you need to change the box length to decrease the zero point energy by a factor of 400 for a fixed value of m?

b.   By what factor would you have to change n for fixed values of a and m to increase the energy by a factor of 400?

c.   By what factor would you have to increase a at constant n to have the zero point energies of an electron be equal to the zero point energy of a proton in the box?

P14.8   Is the superposition wave function $\psi(x) = \sqrt{2/a}[\sin(n\pi x/a) + \sin(m\pi x/a)]$ a solution of the Schrödinger equation for the particle in the box?

P14.9   The function $\psi(x) = Ax[1 - (x/a)]$ is an acceptable wave function for the particle in the one-dimensional infinite depth box of length a. Calculate the normalization constant A and the average values $\langle x \rangle$ and $\langle x^2 \rangle$.

P14.10   Derive an equation for the probability that a particle characterized by the quantum number n is in the first quarter $(0 \leq x \leq a/4)$ of an infinite depth box. Show that this probability approaches the classical limit as $n \to \infty$.

P14.11   What is the solution of the time-dependent Schrödinger equation $\psi(x,t)$ for the solution of the time-independent Schrödinger equation $\psi_4(x) = \sqrt{2/a} \sin(4\pi x/a)$ in the particle-in-the-box model? Write $\omega = E/\hbar$ explicitly in terms of the parameters of the problem.

P14.12   For a particle in a two-dimensional box, the total energy solutions are

$$\psi_{n_x n_y}(x, y) = N \sin \frac{n_x \pi x}{a} \sin \frac{n_y \pi y}{b}$$

a.   Obtain an expression for $E_{n_x n_y}$ in terms of n_x, n_y, a, and b by substituting this wave function into the two-dimensional analogue of Equation (14.9),

$$-\frac{\hbar^2}{2m}\left(\frac{\partial^2}{\partial x^2} + \frac{\partial^2}{\partial y^2}\right)\psi_n(x, y) = E_n \psi(x, y)$$

b. Contour plots of several solutions of the Schrödinger equation for the two-dimensional box are shown here. The x and y directions of the box lie along the horizontal and vertical directions, respectively. The amplitude has been displayed as a gradation in colors. Regions of positive and negative amplitude are indicated. Identify the values of the quantum numbers n_x and n_y for plots (a) through (f).

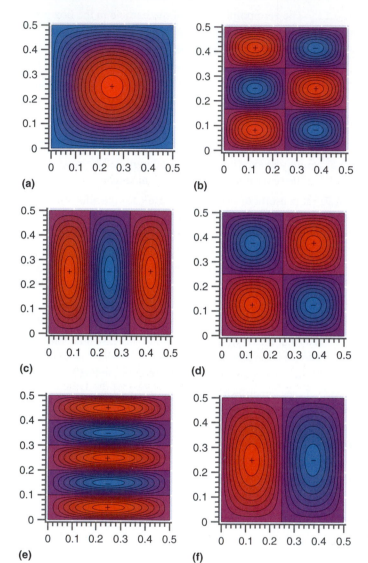

(a)   (b)

(c)   (d)

(e)   (f)

P14.13   Normalize the total energy eigenfunction for the rectangular two-dimensional box,

$$\psi_{n_x,n_y}(x, y) = N \sin\left(\frac{n_x\pi x}{a}\right) \sin\left(\frac{n_y\pi y}{b}\right)$$

in the interval $0 \le x \le a$, $0 \le y \le b$.

P14.14   Consider the contour plots of Problem P14.12.

a. What are the most likely area or areas $dx\,dy$ to find the particle for each of the solutions of the Schrödinger equation depicted in plots (a) through (f)?

b. For the one-dimensional box, the nodes are points. What form do the nodes take for the two-dimensional box? Where are the nodes located in plots (a) through (f)? How many nodes are there in each contour plot?

P14.15

a. Show by substitution into

$$-\frac{\hbar^2}{2m}\left(\frac{\partial^2}{\partial x^2} + \frac{\partial^2}{\partial y^2} + \frac{\partial^2}{\partial z^2}\right)\psi(x, y, z) = E\psi(x, y, z)$$

that the solutions of the Schrödinger equation for a box with lengths along the x, y, and z directions of a, b, and c, respectively, are

$$\psi_{n_x,n_y,n_z}(x, y, z) = N \sin\left(\frac{n_x\pi x}{a}\right) \sin\left(\frac{n_y\pi y}{b}\right) \sin\left(\frac{n_z\pi z}{c}\right).$$

b. Obtain an expression for E_{n_x,n_y,n_z} in terms of n_x, n_y, n_z and a, b, and c.

P14.16   Normalize the total energy solutions for the three-dimensional box in the interval $0 \le x \le a$, $0 \le y \le b$, $0 \le z \le c$.

P14.17   The degeneracy of an energy level is defined as the number of distinct quantum states (different quantum numbers) all of which have the same energy.

a. Using your answer to Problem P14.12a, what is the degeneracy of the energy level $5h^2/8ma^2$ for the square two-dimensional box of edge length a?

b. Using your answer to Problem P14.15b, what is the degeneracy of the energy level $9h^2/8ma^2$ for a three-dimensional cubic box of edge length a?

P14.18   This problem explores under what conditions the classical limit is reached for a macroscopic cubic box of edge length a. An argon atom of average translational energy $3/2\,kT$ is confined in a cubic box of volume $V = 0.500\ \text{m}^3$ at 298 K. Use the result obtained in Problem P14.15b for the dependence of the energy levels on a and on the quantum numbers n_x, n_y, and n_z.

a. What is the value of the "reduced quantum number" $\alpha = \sqrt{n_x^2 + n_y^2 + n_z^2}$ for $T = 298$ K?

b. What is the energy separation between the levels a and $a + 1$? (Hint: Subtract E_{a+1} from E_a before plugging in numbers.)

c. Calculate the ratio $(E_{a+1} - E_a)/kT$ and use your result to conclude whether a classical or quantum mechanical description is appropriate for the particle.

P14.19   Generally, the quantization of translational motion is not significant for atoms because of their mass. However, this conclusion depends on the dimensions of the space to which they are confined. Zeolites are structures with small pores that we describe by a cube with edge length 1 nm. Calculate the energy of a H_2 molecule with $n_x = n_y = n_z = 10$. Compare this energy to kT at $T = 300.$ K. Is a classical or a quantum description appropriate?

P14.20   Two wave functions are distinguishable if they lead to a different probability density. Which of the following wave functions are distinguishable from $\sin\theta + i\cos\theta$?

a. $(\sin\theta + i\cos\theta)\left(+\dfrac{\sqrt{2}}{2} + i\dfrac{\sqrt{2}}{2}\right)$

b. $i(\sin\theta + i\cos\theta)$

c. $-(\sin\theta + i\cos\theta)$

d. $(\sin\theta + i\cos\theta)\left(-\dfrac{\sqrt{2}}{2} + i\dfrac{\sqrt{2}}{2}\right)$

e. $\sin\theta - i\cos\theta$

f. $e^{i\theta}$

P14.21 For the π-network of β-carotene modeled using the particle in the box, the position-dependent probability density of finding 1 of the 22 electrons is given by

$$P_n(x) = |\psi_n(x)|^2 = \frac{2}{a}\sin^2\left(\frac{n\pi x}{a}\right)$$

The quantum number n in this equation is determined by the energy level of the electron under consideration. As we learned in this chapter, this function is strongly position dependent. The question addressed in this problem is as follows: Would you also expect the total probability density defined by $P_{total}(x) = \sum_n|\psi_n(x)|^2$ to be strongly position dependent? The sum is over all the electrons in the π-network.

a. Calculate the total probability density $P_{total}(x) = \sum_n|\psi_n(x)|^2$ using the box length $a = 29.0$ nm and plot your results as a function of x. Does $P_{total}(x)$ have the same value near the ends and at the middle of the molecule?

b. Determine $\Delta P_{total}(x)/\langle P_{total}(x)\rangle$, where $\Delta P_{total}(x)$ is the peak-to-peak amplitude of $P_{total}(x)$ in the interval between 12.0 and 16.0 nm.

c. Compare the result of part (b) with what you would obtain for an electron in the highest occupied energy level.

d. What value would you expect for $P_{total}(x)$ if the electrons were uniformly distributed over the molecule? How does this value compare with your result from part (a)?

P14.22 Calculate the energy levels of the π-network in butadiene, C_4H_6, using the particle in the box model. To calculate the box length, assume that the molecule is linear and use the values 135 and 154 pm for C=C and C—C bonds. What is the wavelength of light required to induce a transition from the ground state to the first excited state? How does this compare with the experimentally observed value of 290. nm? What does the comparison made suggest to you about estimating the length of the π-network by adding bond lengths for this molecule?

P14.23 Calculate the energy levels of the π-network in octatetraene, C_8H_{10}, using the particle-in-the-box model. To calculate the box length, assume that the molecule is linear and use the values 135 and 154 pm for C=C and C—C bonds. What is the wavelength of light required to induce a transition from the ground state to the first excited state?

P14.24 Semiconductors can become conductive if their temperature is raised sufficiently to populate the (empty) conduction band from the highest filled levels in the valence band. The ratio of the populations in the highest level of the conduction band to that of the lowest level in the conduction band is

$$\frac{n_{conduction}}{n_{valence}} = \frac{g_{conduction}}{g_{valence}}e^{-\Delta E/kT}$$

where ΔE is the band gap, which is 1.12 eV for Si and 5.5 eV for diamond. Assume for simplicity that the ratio of the

degeneracies is one and that the semiconductor becomes sufficiently conductive when

$$\frac{n_{conduction}}{n_{valence}} = 5.5 \times 10^{-7}$$

At what temperatures will silicon and diamond become sufficiently conductive? Given that the most stable form of carbon at normal pressures is graphite and that graphite sublimates near 3700 K, could you heat diamond enough to make it conductive and not sublimate it?

P14.25 The maximum safe current in a copper wire with a diameter of 3.0 mm is about 20. A. In an STM, a current of 1.0 $\times$ 10^{-9} A passes from the tip to the surface in a filament of diameter ~1.0 nm. Compare the current density in the copper wire with that in the STM.

P14.26 The force constant for a $H^{19}F$ molecule is 966 N m^{-1}.

a. Calculate the zero point vibrational energy for this molecule for a harmonic potential.

b. Calculate the light frequency needed to excite this molecule from the ground state to the first excited state.

P14.27 By substituting in the Schrödinger equation for the harmonic oscillator, show that the ground-state vibrational wave function is a solution of this equation. Determine the energy eigenvalue.

P14.28 Show by carrying out the appropriate integration that the solutions of the Schrödinger equation for the harmonic oscillator $\psi_0(x) = (\alpha/\pi)^{1/4}e^{-(1/2)\alpha x^2}$ and $\psi_2(x) = (\alpha/4\pi)^{1/4}(2\alpha x^2 - 1)e^{-(1/2)\alpha x^2}$ are orthogonal over the interval $-\infty < x < \infty$ and that $\psi_2(x)$ is normalized over the same interval. In evaluating integrals of this type, $\int_{-x}^{x}f(x)dx = 0$ if $f(x)$ is an odd function of x and $\int_{-x}^{x}f(x)dx = 2\int_{0}^{x}f(x)dx$ if $f(x)$ is an even function of x.

P14.29 Evaluate the average potential energy, $\langle E_{potential}\rangle$, for the ground state ($n = 0$) of the harmonic oscillator by carrying out the appropriate integration.

P14.30 Evaluate the average potential energy, $\langle E_{potential}\rangle$, for the second excited state ($n = 2$) of the harmonic oscillator by carrying out the appropriate integration.

P14.31 Evaluate the average vibrational amplitude of the quantum harmonic oscillator about its equilibrium value, $\langle x\rangle$, for the ground state ($n = 0$) and first two excited states ($n = 1$ and $n = 2$). Use the hint about evaluating integrals given in Problem P14.28.

P14.32 Evaluate the average of the square of the vibrational amplitude of the quantum harmonic oscillator about its equilibrium value, $\langle x^2\rangle$, for the ground state ($n = 0$) and first two excited states ($n = 1$ and $n = 2$). Use the hint about evaluating integrals in Problem P14.28.

P14.33 The vibrational frequency of $^1H^{35}Cl$ is 8.963 $\times$ 10^{13} s^{-1}. Calculate the force constant of the molecule. How large a mass would be required to stretch a classical spring with this force constant by 1.00 cm? Use the gravitational acceleration on Earth at sea level for this problem.

P14.34   Two 1.00-g masses are attached by a spring with a force constant of $k = 500.$ kg s^{-2}. Calculate the zero point energy of the system and compare it with the thermal energy kT. If the zero point energy were converted to translational energy, what would be the speed of the masses?

P14.35   Use $\sqrt{\langle x^2 \rangle}$ as calculated in Problem P14.32 as a measure of the vibrational amplitude for a molecule. What fraction is $\sqrt{\langle x^2 \rangle}$ of the 127-pm bond length of the HCl molecule for $n = 0, 1,$ and 2? The force constant for the ^{1}H^{35}Cl molecule is 516 N m^{-1}.

P14.36 ^{1}H^{35}Cl has a force constant of 516 N m^{-1} and a moment of inertia of 2.644×10^{-47} kg m^2. Calculate the frequency of the light corresponding to the lowest energy vibrational transition. In what region of the electromagnetic spectrum does the transition lie?

P14.37   The vibrational frequency for N$_2$ expressed in wave numbers is 2358 cm^{-1}. What is the force constant associated with the N$\equiv$N triple bond? How much would a classical spring with this force constant be elongated if a mass of 1.00 kg were attached to it? Use the gravitational acceleration on Earth at sea level for this problem.

P14.38   The force constant for the ^{1}H^{35}Cl molecule is 516 N m^{-1}. Calculate the vibrational zero point energy of this molecule. If this amount of energy were converted to translational energy, how fast would the molecule be moving? Compare this speed to the root mean square speed from the kinetic gas theory, $|\mathbf{v}|_{rms} = \sqrt{3kT/m}$ for $T = 300.$ K.

Web-Based Simulations, Animations, and Problems

W14.1   The simulation of particle diffraction from a single slit is used to illustrate the dependence between the uncertainty in the position and momentum. The slit width and particle velocity are varied using sliders.

W14.2   The Heisenberg uncertainty principle states that $\Delta p \Delta x > \hbar/2$. In an experiment, it is more likely that λ is varied rather than p, where λ is the de Broglie wavelength of the particle. The relationship between Δx and $\Delta \lambda$ will be determined using a simulation. The value of Δx will be measured as a function of $\Delta \lambda$ at a constant value of λ, and as a function of λ for a constant value of $\Delta \lambda$.

W14.3   The motion of a classical particle-in-a-box potential is simulated. The particle energy and the potential in the two halves of the box are varied using sliders. The kinetic energy is displayed as a function of the position x, and the result of measuring the probability of detecting the particle at x is displayed as a density plot. The student is asked to use the information gathered to explain the motion of the particle.

W14.4   Wave functions for $n = 1 - 5$ are shown for the particle in the infinite depth box and the energy levels are calculated. Sliders are used to vary the box length and the mass of the particle. The student is asked questions that clarify the relationship between the level energy, the mass, and the box length.

W14.5   The probability is calculated for finding a particle in the infinite depth box in the interval $0 \rightarrow 0.1a$, $0.1a \rightarrow 0.2a$, $\dots$, $0.9a \rightarrow 1.0a$ for $n = 1$, $n = 2$, and $n = 50$. The student is asked to explain these results.

W14.6   Contour plots are generated for the total energy solutions of the particle in the two-dimensional infinite depth box,

$$\psi_{n_x n_y}(x, y) = N \sin\frac{n_x \pi x}{a} \sin\frac{n_y \pi y}{b}$$

The student is asked questions about the nodal structure of these solutions and asked to assign quantum numbers n_x and n_y to each contour plot.

W14.7   The Schrödinger equation is solved numerically for the particle in the finite height box. Using the condition that the wave function must approach zero amplitude in the classically forbidden region, the energy levels are determined for a fixed particle mass, box depth, and box length. The particle mass and energy and the box depth and length are varied with sliders to demonstrate how the number of bound states varies with these parameters.

W14.8   The Schrödinger equation is solved numerically to calculate the tunneling probability for a particle through a thin finite barrier. Sliders are used to vary the barrier width and height and the particle energy and mass. The dependence of the tunneling probability on these variables is investigated.

W14.9   The motion of a particle in a harmonic potential is investigated, and the particle energy and force constant k are varied using sliders. The potential and kinetic energy are displayed as a function of the position x, and the result of measuring the probability of detecting the particle at x is displayed as a density plot. The student is asked to use the information gathered to explain the motion of the particle.

W14.10   The allowed energy levels for the harmonic oscillator are determined by numerical integration of the Schrödinger equation, starting in the classically forbidden region to the left of the potential. The criterion that the energy is an eigenvalue for the problem is that the wave function decays to zero in the classically forbidden region to the right of the potential. The zero point energy is determined for different values of k. The results are graphed to obtain a functional relationship between the zero point energy and k.

W14.11   The probability of finding the harmonic oscillator in the classically forbidden region, P_n, is calculated. The student generates a set of values for P_n for $n = 0, 1, 2, \dots, 20$ and graphs them.

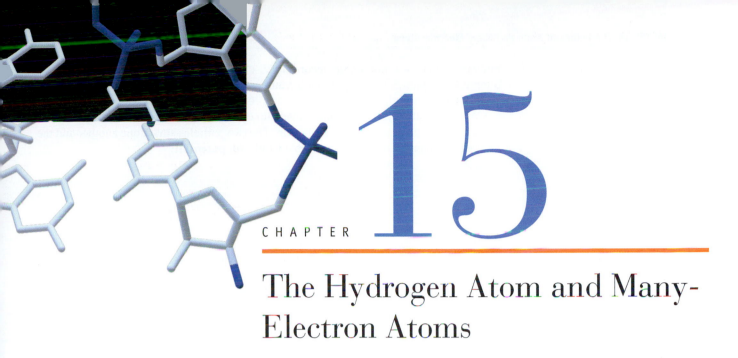

CHAPTER

15

The Hydrogen Atom and Many-Electron Atoms

Classical physics is unable to explain the stability of atoms. In a classical picture, the electrons radiate energy because they undergo accelerated motion as they orbit the positively charged nucleus. As a consequence, the electrons fall into the nucleus. This is clearly incorrect! To emphasize the similarities and differences between quantum mechanical and classical models, in this chapter a comparison is made between the quantum mechanical picture of the hydrogen atom and the popularly depicted shell picture of the atom. We next consider atoms containing more than one electron. The Schrödinger equation cannot be solved analytically for such atoms because of the electron–electron repulsion term in the potential energy. Instead, approximate numerical methods can be used to obtain the eigenfunctions and eigenvalues of the Schrödinger equation for many-electron atoms. Having more than one electron in an atom also raises new issues that we have not considered, including the indistinguishability of electrons and the electron spin.

15.1 Formulating the Schrödinger Equation for the Hydrogen Atom

With the background we have obtained by applying quantum mechanics to a number of simple problems, we turn to one of the triumphs of quantum mechanics: the understanding of atomic structure and spectroscopy. As discussed later, the complexities associated with the quantum mechanical calculations required to understand many-electron atoms force us to solve the Schrödinger equation numerically. However, for the hydrogen atom, the Schrödinger equation can be solved analytically, and the important results we obtain from that solution can be generalized to many-electron atoms.

To set the stage historically, experiments by Rutherford had established that the positive charge associated with an atom was localized at the center of the atom and that the electrons were spread out over a large volume (relative to nuclear dimensions) centered at the nucleus. The **shell model** in which the electrons are confined in spherical shells centered at the nucleus had a major flaw when viewed from the vantage point of classical physics. An electron orbiting around the nucleus undergoes accelerated motion and radiates energy. Therefore, it will eventually fall into the nucleus. Atoms are not stable according to classical mechanics. The challenge for quantum mechanics was to provide a framework within which the stability of atoms could be understood.

We picture the hydrogen atom as made up of an electron moving about a proton located at the origin of the coordinate system. The two particles attract one another and the interaction potential is given by a simple **Coulomb potential:**

$$V(\mathbf{r}) = -\frac{e^2}{4\pi\varepsilon_0|\mathbf{r}|} = -\frac{e^2}{4\pi\varepsilon_0 r} \tag{15.1}$$

We have abbreviated the magnitude of the vector $\mathbf{r}$ as r, e is the electron charge, and ε_0 is the permittivity of free space. Because the potential is spherically symmetrical, we choose spherical polar coordinates (r,θ,ϕ) to formulate the Schrödinger equation for this problem. In doing so, it takes on the formidable form

$$-\frac{\hbar^2}{2m_e}\left[\frac{1}{r^2}\frac{\partial}{\partial r}\left(r^2\frac{\partial\psi(r,\theta,\phi)}{\partial r}\right) + \frac{1}{r^2\sin\theta}\frac{\partial}{\partial\theta}\left(\sin\theta\,\frac{\partial\psi(r,\theta,\phi)}{\partial\theta}\right) \right.$$
$$\left. + \frac{1}{r^2\sin^2\theta}\frac{\partial^2\psi(r,\theta,\phi)}{\partial\phi^2}\right]$$

$$-\frac{e^2}{4\pi\varepsilon_0 r}\,\psi(r,\theta,\phi) = E\psi(r,\theta,\phi) \tag{15.2}$$

In this equation, m_e is the electron mass.

15.2 Eigenvalues and Eigenfunctions for the Total Energy

Equation (15.2) can be solved using standard mathematical methods, so we concern ourselves only with the results. Because the potential energy depends only on the variable r and not on θ or ϕ, it can be shown that the wave function $\psi(r,\theta,\phi)$ is the product of three functions, each of which depends on only one variable, or $\psi(r,\theta,\phi) = R(r)\Theta(\theta)\Phi(\phi)$. The functions $\Theta(\theta)\Phi(\phi)$ are known as the **spherical harmonic functions.** The quantization condition that results from the restriction that $R(r)$ be well behaved at large values of r $[R(r) \to 0$ as $r \to \infty]$ is

$$E_n = -\frac{m_e e^4}{8\varepsilon_0^2 h^2 n^2} \text{ for } n = 1, 2, 3, \ldots \tag{15.3}$$

This formula is usually simplified by combining a number of constants in the form $a_0 = \varepsilon_0 h^2/\pi m_e e^2$. The quantity a_0 has the value 0.529 Å and is called the **Bohr radius.** Use of this definition leads to the following formula:

$$E_n = -\frac{e^2}{8\pi\varepsilon_0 a_0 n^2} = -\frac{2.179 \times 10^{-18} \text{ J}}{n^2} = -\frac{13.60 \text{ eV}}{n^2} \text{ for } n = 1, 2, 3, \ldots \tag{15.4}$$

Note that E_n goes to zero as $n \to \infty$. As previously emphasized, the zero of energy is a matter of convention rather than being a quantity that can be determined. As n approaches infinity, the electron is on average farther and farther from the nucleus, and the zero of energy corresponds to the electron at infinite separation from the nucleus. All negative energies correspond to bound states of the electron in the Coulomb potential.

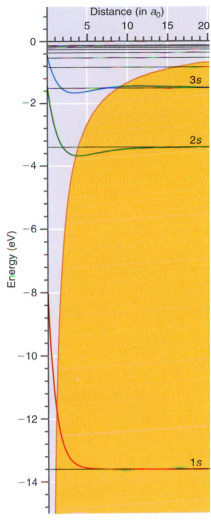

Distance (in a_0)

FIGURE 15.1

The border of the yellow classically forbidden region is the Coulomb potential, which is shown together with E_n for $n = 1$ through $n = 10$ and the 1s, 2s, and 3s wave functions.

As has been done previously for the particle in the box and the harmonic oscillator, the energy eigenvalues can be superimposed on a potential energy diagram, as shown in Figure 15.1. The potential forms a "box" that acts to confine the particle. This box has a peculiar form in that it is infinitely deep at the center of the atom, and the depth falls off inversely with distance from the proton. Figure 15.1 shows that the two lowest energy levels have an appreciable separation in energy and that the separation for adjacent energy levels becomes rapidly smaller as $n \to \infty$. All states for which $5 < n < \infty$ have an energy in the narrow range between $\sim -1 \times 10^{-19}$ J and zero. Although this seems strange at first, it is exactly what is expected based on the results for the particle in the box. Because of the shape of the potential, the H atom box is very narrow for the first few energy eigenstates, but becomes very wide for large n. The particle-in-the-box formula [Equation (14.20)] predicts that the energy spacing varies as the inverse of the square of box length. This is the trend seen in Figure 15.1. Note also that the wave functions penetrate into the classically forbidden region just as for the particle in the finite depth box and the harmonic oscillator.

Although the energy depends on a single quantum number, n, the eigenfunctions $\psi(r,\theta,\phi)$ are associated with three quantum numbers because three boundary conditions arise in a three-dimensional problem. The other two quantum numbers are l and m_l, which arise from the angular coordinates. Their relationship is given by

$$n = 1, 2, 3, \ldots$$
$$l = 0, 1, 2, 3, \ldots, n - 1$$
$$m_l = 0, \pm 1, \pm 2, \pm 3, \ldots \pm l \qquad (15.5)$$

The quantum number l refers to the angular momentum of the electron. If $l = 0$, the electron has no angular momentum, and the electron distribution will be spherical. For $l > 0$, the electron has a nonzero angular momentum and the electron distribution is non-spherical as will be discussed in Section 15.3. Although we do not present a justification of the relationship between n, l, and m_l here, all the conditions in Equation (15.5) emerge naturally out of the boundary conditions in the solution of the differential equations.

The radial functions, $R(r)$, are products of an exponential function with a polynomial in the dimensionless variable r/a_0. Their functional form depends on the quantum numbers n and l. The first few radial functions $R_{nl}(r)$ are as follows:

$$n = 1, l = 0 \quad R_{10}(r) = 2\left(\frac{1}{a_0}\right)^{3/2} e^{-r/a_0}$$

$$n = 2, l = 0 \quad R_{20}(r) = \frac{1}{\sqrt{8}}\left(\frac{1}{a_0}\right)^{3/2}\left(2 - \frac{r}{a_0}\right)e^{-r/2a_0}$$

$$n = 2, l = 1 \quad R_{21}(r) = \frac{1}{\sqrt{24}}\left(\frac{1}{a_0}\right)^{3/2}\frac{r}{a_0}e^{-r/2a_0}$$

$$n = 3, l = 0 \quad R_{30}(r) = \frac{2}{81\sqrt{3}}\left(\frac{1}{a_0}\right)^{3/2}\left(27 - 18\frac{r}{a_0} + 2\frac{r^2}{a_0^2}\right)e^{-r/3a_0}$$

$$n = 3, l = 1 \quad R_{31}(r) = \frac{4}{81\sqrt{6}}\left(\frac{1}{a_0}\right)^{3/2}\left(6\frac{r}{a_0} - \frac{r^2}{a_0^2}\right)e^{-r/3a_0}$$

$$n = 3, l = 2 \quad R_{32}(r) = \frac{4}{81\sqrt{30}}\left(\frac{1}{a_0}\right)^{3/2}\frac{r^2}{a_0^2}e^{-r/3a_0}$$

To form the hydrogen atom eigenfunctions, we combine $R_{nl}(r)$ with the spherical harmonic functions and list here the first few of the infinite set of normalized wave functions $\psi(r,\theta,\phi) = R(r)\Theta(\theta)\Phi(\phi)$ for the hydrogen atom. Note that, in general, the eigenfunctions depend on r, θ, and ϕ, but are not functions of θ and ϕ for

$l = 0$. The quantum numbers are associated with the wave functions using the notation ψ_{nlm_l}:

$$n = 1, l = 0, m_l = 0 \quad \psi_{100}(r) = \frac{1}{\sqrt{\pi}} \left(\frac{1}{a_0}\right)^{3/2} e^{-r/a_0}$$

$$n = 2, l = 0, m_l = 0 \quad \psi_{200}(r) = \frac{1}{4\sqrt{2\pi}} \left(\frac{1}{a_0}\right)^{3/2} \left(2 - \frac{r}{a_0}\right) e^{-r/2a_0}$$

$$n = 2, l = 1, m_l = 0 \quad \psi_{210}(r,\theta,\phi) = \frac{1}{4\sqrt{2\pi}} \left(\frac{1}{a_0}\right)^{3/2} \frac{r}{a_0} e^{-r/2a_0} \cos\theta$$

$$n = 2, l = 1, m_l = \pm 1 \quad \psi_{21\pm 1}(r,\theta,\phi) = \frac{1}{8\sqrt{\pi}} \left(\frac{1}{a_0}\right)^{3/2} \frac{r}{a_0} e^{-r/2a_0} \sin\theta\, e^{\pm i\phi}$$

$$n = 3, l = 0, m_l = 0 \quad \psi_{300}(r) = \frac{1}{81\sqrt{3\pi}} \left(\frac{1}{a_0}\right)^{3/2} \left(27 - 18\frac{r}{a_0} + 2\frac{r^2}{a_0^2}\right) e^{-r/3a_0}$$

$$n = 3, l = 1, m_l = 0 \quad \psi_{310}(r,\theta,\phi) = \frac{1}{81} \left(\frac{2}{\pi}\right)^{1/2} \left(\frac{1}{a_0}\right)^{3/2} \left(6\frac{r}{a_0} - \frac{r^2}{a_0^2}\right) e^{-r/3a_0} \cos\theta$$

$$n = 3, l = 1, m_l = \pm 1 \quad \psi_{31\pm 1}(r,\theta,\phi) = \frac{1}{81\sqrt{\pi}} \left(\frac{1}{a_0}\right)^{3/2} \left(6\frac{r}{a_0} - \frac{r^2}{a_0^2}\right) e^{-r/3a_0} \sin\theta\, e^{\pm i\phi}$$

$$n = 3, l = 2, m_l = 0 \quad \psi_{320}(r,\theta,\phi) = \frac{1}{81\sqrt{6\pi}} \left(\frac{1}{a_0}\right)^{3/2} \frac{r^2}{a_0^2} e^{-r/3a_0} (3\cos^2\theta - 1)$$

$$n = 3, l = 2, m_l = \pm 1 \quad \psi_{32\pm 1}(r,\theta,\phi) = \frac{1}{81\sqrt{\pi}} \left(\frac{1}{a_0}\right)^{3/2} \frac{r^2}{a_0^2} e^{-r/3a_0} \sin\theta \cos\theta\, e^{\pm i\phi}$$

$$n = 3, l = 2, m_l = \pm 2 \quad \psi_{32\pm 2}(r,\theta,\phi) = \frac{1}{162\sqrt{\pi}} \left(\frac{1}{a_0 1}\right)^{3/2} \frac{r^2}{a_0^2} e^{-r/3a_0} \sin^2\theta\, e^{\pm 2i\phi}$$

These functions are referred to both as the H atom eigenfunctions and the H atom **orbitals.** A shorthand notation for the quantum numbers is to give the numerical value of n followed by a symbol indicating the values of l and m_l. The letters s, p, d, and f are used to denote $l = 0, 1, 2,$ and 3, respectively, and $\psi_{100}(r)$ is referred to as the $1s$ orbital or wave function and all three wave functions with $n = 2$ and $l = 1$ are referred to as $2p$ orbitals. The wave functions are real functions when $m_l = 0$, and complex functions otherwise. The angular and radial portions of the wave functions have nodes that are discussed in more detail later in this chapter. These functions have been normalized in keeping with the association between probability density and $\psi(r,\theta,\phi)$.

EXAMPLE PROBLEM 15.1

Normalize the functions $e^{-r/2a_0}$ and $(r/a_0)e^{-r/2a_0} \sin\theta\, e^{+i\phi}$ in three-dimensional spherical coordinates.

Solution

In general, a wave function $\psi(\tau)$ is normalized by multiplying it by a constant N defined by $N^2 \int \psi^*(\tau)\psi(\tau)\, d\tau = 1$. In three-dimensional spherical coordinates, $d\tau = r^2 \sin\theta\, dr\, d\theta\, d\phi$, as discussed in Section 13.6. The normalization integral becomes $N^2 \int_0^\pi \sin\theta\, d\theta \int_0^{2\pi} d\phi \int_0^\infty \psi^*(r,\theta,\phi)\psi(r,\theta,\phi) r^2\, dr = 1$.

For the first function,

$$N^2 \int_0^\pi \sin\theta\, d\theta \int_0^{2\pi} d\phi \int_0^\infty e^{-r/2a_0} e^{-r/2a_0} r^2\, dr = 1$$

We use the standard integral

$$\int_0^\infty x^n e^{-ax}\,dx = \frac{n!}{a^{n+1}}$$

Integrating over the angles θ and ϕ, we obtain $4\pi N^2 \int_0^\infty e^{-r/2a_0} e^{-r/2a_0}\, r^2\, dr = 1$. Evaluating the integral over r,

$$4\pi N^2 \frac{2!}{1/a_0^3} = 1 \quad \text{or} \quad N = \frac{1}{2\sqrt{2\pi}}\left(\frac{1}{a_0}\right)^{3/2}$$

For the second function,

$$N^2 \int_0^\pi \sin\theta\, d\theta \int_0^{2\pi} d\phi \int_0^\infty \left(\frac{r}{a_0} e^{-r/2a_0} \sin\theta\, e^{-i\phi}\right)\left(\frac{r}{a_0} e^{-r/2a_0} \sin\theta\, e^{+i\phi}\right) r^2\, dr = 1$$

This simplifies to

$$N^2 \int_0^\pi \sin^3\theta\, d\theta \int_0^{2\pi} d\phi \int_0^\infty \left(\frac{r}{a_0}\right)^2 e^{-r/a_0}\, r^2\, dr = 1$$

Integrating over the angles θ and ϕ using the result $\int_0^\pi \sin^3\theta\, d\theta = 4/3$, we obtain

$$\frac{8\pi}{3} N^2 \int_0^\infty \left(\frac{r}{a_0}\right)^2 e^{-r/a_0}\, r^2\, dr = 1$$

Using the same standard integral as in the first part of the problem,

$$\frac{8\pi}{3} N^2 \frac{1}{a_0^2}\left(\frac{4!}{1/a_0^5}\right) = 1 \quad \text{or} \quad N = \frac{1}{8\sqrt{\pi}}\left(\frac{1}{a_0}\right)^{3/2}$$

Each eigenfunction listed before Example Problem 15.1 describes a separate state of the hydrogen atom. However, as we have seen, the energy depends only on the quantum number n. Therefore, all states with the same value for n, but different values for l and m_l, have the same energy and we say that the energy levels are degenerate. Using the formulas given in Equation (15.5), we can see that the degeneracy of a given level is n^2. Therefore, the $n = 2$ level has a fourfold degeneracy and the $n = 3$ level has a ninefold degeneracy.

The angular part of each hydrogen atom total energy eigenfunction is a spherical harmonic function. These functions are complex unless $m_l = 0$. To facilitate making graphs, it is useful to form combinations of those hydrogen orbitals $\psi_{nlm_l}(r,\theta,\phi)$ for which $m_l \neq 0$ are real functions of r, θ, and ϕ. This is done by forming linear combinations of $\psi_{nlm_l}(r,\theta,\phi)$ and $\psi_{nl-m_l}(r,\theta,\phi)$. The first few of these combinations, resulting in the $2p$, $3p$, and $3d$ orbitals, are shown here:

$$\psi_{2p_x}(r,\theta,\phi) = \frac{1}{4\sqrt{2\pi}}\left(\frac{1}{a_0}\right)^{3/2} \frac{r}{a_0} e^{-r/2a_0} \sin\theta \cos\phi$$

$$\psi_{2p_y}(r,\theta,\phi) = \frac{1}{4\sqrt{2\pi}}\left(\frac{1}{a_0}\right)^{3/2} \frac{r}{a_0} e^{-r/2a_0} \sin\theta \sin\phi$$

$$\psi_{2p_z}(r,\theta,\phi) = \frac{1}{4\sqrt{2\pi}}\left(\frac{1}{a_0}\right)^{3/2} \frac{r}{a_0} e^{-r/2a_0} \cos\theta$$

$$\psi_{3p_x}(r,\theta,\phi) = \frac{\sqrt{2}}{81\sqrt{\pi}}\left(\frac{1}{a_0}\right)^{3/2} \left(6\frac{r}{a_0} - \frac{r^2}{a_0^2}\right) e^{-r/3a_0} \sin\theta \cos\phi$$

$$\psi_{3p_y}(r,\theta,\phi) = \frac{\sqrt{2}}{81\sqrt{\pi}}\left(\frac{1}{a_0}\right)^{3/2} \left(6\frac{r}{a_0} - \frac{r^2}{a_0^2}\right) e^{-r/3a_0} \sin\theta \sin\phi$$

$$\psi_{3p_z}(r,\theta,\phi) = \frac{\sqrt{2}}{81\sqrt{\pi}}\left(\frac{1}{a_0}\right)^{3/2} \left(6\frac{r}{a_0} - \frac{r^2}{a_0^2}\right) e^{-r/3a_0} \cos\theta$$

$$\psi_{3d_{z^2}}(r,\theta,\phi) = \frac{1}{81\sqrt{6\pi}}\left(\frac{1}{a_0}\right)^{3/2}\frac{r^2}{a_0^2}e^{-r/3a_0}(3\cos^2\theta - 1)$$

$$\psi_{3d_{xz}}(r,\theta,\phi) = \frac{\sqrt{2}}{81\sqrt{\pi}}\left(\frac{1}{a_0}\right)^{3/2}\frac{r^2}{a_0^2}e^{-r/3a_0}\sin\theta\cos\theta\cos\phi$$

$$\psi_{3d_{yz}}(r,\theta,\phi) = \frac{\sqrt{2}}{81\sqrt{\pi}}\left(\frac{1}{a_0}\right)^{3/2}\frac{r^2}{a_0^2}e^{-r/3a_0}\sin\theta\cos\theta\sin\phi$$

$$\psi_{3d_{x^2-y^2}}(r,\theta,\phi) = \frac{1}{81\sqrt{2\pi}}\left(\frac{1}{a_0}\right)^{3/2}\frac{r^2}{a_0^2}e^{-r/3a_0}\sin^2\theta\cos 2\phi$$

$$\psi_{3d_{xy}}(r,\theta,\phi) = \frac{1}{81\sqrt{2\pi}}\left(\frac{1}{a_0}\right)^{3/2}\frac{r^2}{a_0^2}e^{-r/3a_0}\sin^2\theta\sin 2\phi$$

The challenge we posed for quantum mechanics at the beginning of this chapter was to provide an understanding for the stability of atoms. By verifying that there is a set of eigenfunctions and eigenvalues of the time-independent Schrödinger equation for a system consisting of a proton and an electron, we have demonstrated that there are states whose energy is independent of time. Because the energy eigenvalues are all negative numbers, all of these states are more stable than the reference state of zero energy that corresponds to the proton and electron separated by an infinite distance. Because $n \geq 1$, the energy cannot approach $-\infty$, corresponding to the electron falling into the nucleus. These results show that when the wave nature of the electron is taken into account, the H atom is stable.

As with any new theory, the true test is consistency with experimental data. Although the wave functions are not directly observable, we know that the spectral lines from a hydrogen arc lamp measured as early as 1885 must involve transitions between two stable states of the hydrogen atom. Therefore, the frequencies measured by the early experimentalists in emission spectra must be given by

$$\nu = \frac{1}{h}(E_{initial} - E_{final}) \tag{15.6}$$

In a more exact treatment, the origin of the coordinate system describing the H atom is placed at the center of mass of the proton and electron rather than at the position of the proton. Using Equation (15.3) with the reduced mass of the atom in place of m_e and Equation (15.6), quantum theory predicts that the frequencies of all spectral lines are given by

$$\nu = \frac{\mu e^4}{8\varepsilon_0^2 h^3}\left(\frac{1}{n_{initial}^2} - \frac{1}{n_{final}^2}\right) \tag{15.7}$$

where $\mu = \dfrac{m_e m_p}{m_e + m_p}$ is the reduced mass of the atom. Spectroscopists commonly refer to spectral lines in units of wave numbers. Rather than reporting values of ν, they use the units $\tilde{\nu} = \nu/c = 1/\lambda$. The combination of constants $m_e e^4/8\varepsilon_0^2 h^3 c$ is called the **Rydberg constant.** It has the units of energy and is equal to 2.180×10^{-18} J, which corresponds to 109,737 cm^{-1}. The reduced mass for H is 0.05% less than m_e.

Equation (15.7) quantitatively predicts all observed spectral lines for the hydrogen atom. It also correctly predicts the very small shifts in frequency observed for the isotopes of hydrogen, which have slightly different reduced masses. The agreement between theory and experiment verifies that the quantum mechanical model for the hydrogen atom is valid and accurate. Some of the possible transitions between states of the H atom induced by absorption or emission of a photon are shown superimposed on a set of energy levels in Figure 15.2.

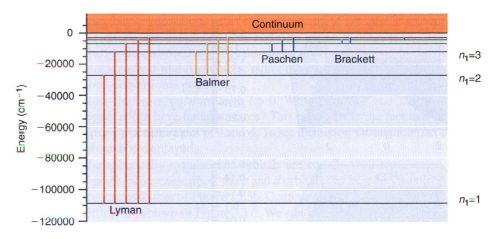

FIGURE 15.2

Energy-level diagram for the hydrogen atom showing the allowed transitions for n < 6. Because for E > 0, the energy levels are continuous, the absorption spectrum will be continuous above an energy that depends on the initial n value. The different sets of transitions are named after the scientists who first investigated them.

15.3 The Hydrogen Atom Orbitals

We now turn to the total energy eigenfunctions (or orbitals) of the hydrogen atom. What insight can be gained from them? Recall the early quantum mechanics shell model of atoms proposed by Niels Bohr. It depicted electrons as orbiting around the nucleus and associated orbits of small radius with more negative energies. Only certain orbits were allowed in order to give rise to a discrete energy spectrum. This model was discarded because defining orbits exactly is inconsistent with the Heisenberg uncertainty principle. The model postulated by Schrödinger and other pioneers of quantum theory replaced knowledge of the location of the electron in the hydrogen atom with knowledge of the probability of finding it in a small volume element at a specific location. As we have seen in considering the particle in the box and the harmonic oscillator, this probability is proportional to $\psi^*(r,\theta,\phi)\psi(r,\theta,\phi)\,d\tau$.

To what extent does the exact quantum mechanical solution resemble the shell model? To answer this question, information must be extracted from the H atom orbitals. A new concept, the radial distribution function, is introduced for this purpose. We begin our discussion by focusing on the wave functions $\psi_{nlm_l}(r,\theta,\phi)$. Next we discuss what can be learned about the probability of finding the electron in a particular region in space, $\psi_{nlm_l}^2(r,\theta,\phi)r^2\sin\theta\,dr\,d\theta\,d\phi$. Finally, we define the radial distribution function and look at the similarities and differences between quantum mechanical and shell models of the hydrogen atom.

The initial step is to look at the ground-state (lowest energy state) wave function for the hydrogen atom, and to find a good way to visualize this function. Because

$$\psi_{100}(r) = \frac{1}{\sqrt{\pi}}\left(\frac{1}{a_0}\right)^{3/2} e^{-r/a_0}$$

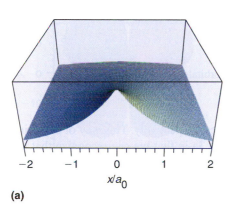

(a)

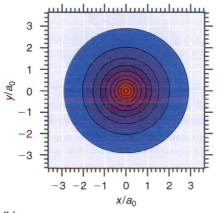

(b)

FIGURE 15.3

(a) 3-D perspective and (b) contour plot of $\psi_{100}(r)$. Red and blue contours correspond to the most positive and least positive values, respectively, of the wave function.

is a function of the three spatial coordinates x, y, and z, we need a four-dimensional space to plot ψ_{100} as a function of all of its variables. Because such a space is not readily available, the number of variables will be reduced. The dimensionality of the representation can be reduced by evaluating $r = \sqrt{x^2 + y^2 + z^2}$ in one of the x-y, x-z, or y-z planes by setting the third coordinate equal to zero. Three common ways of depicting $\psi_{100}(r)$ are shown in Figures 15.3 through 15.5. In Figure 15.3a, a three-dimensional plot of $\psi_{100}(r)$ evaluated in the x-y half-plane ($z = 0$, $y \geq 0$) is shown in perspective. Although it is difficult to extract quantitative information from such a plot directly, it allows a good visualization of the function. We clearly see that the wave function has its maximum value at $r = 0$ (the nuclear position) and that it falls off rapidly with increasing distance from the nucleus.

More quantitative information is available in the contour plot shown in Figure 15.3b in which $\psi_{100}(r)$ is evaluated in the x-y plane from a vantage point on the z axis. In this case, the outermost contour represents 10% of the maximum value, and successive contours are spaced at equal intervals. The shading also indicates the value of the function,

FIGURE 15.15

Radial probability distributions for the 3s, 3p, and 3d orbitals of the H atom as a function of distance in units of a_0.

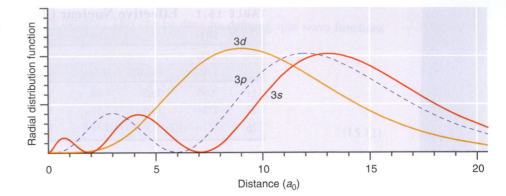

FIGURE 15.15

Radial probability distributions for the 3s, 3p, and 3d orbitals of the H atom as a function of distance in units of a_0.

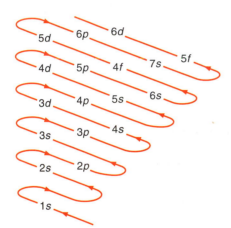

FIGURE 15.16

The order in which orbitals in many-electron atoms are filled for most atoms is described by the red line.

result predicts that the F$^-$ ion is less stable than the neutral F atom, contrary to experiment. A better estimate of the electron affinity of F is obtained by comparing the total energies of F and F$^-$. This gives a positive value for the electron affinity of 1.22 eV. However, this value still differs significantly from the experimental value of 3.34 eV. More accurate calculations, including electron correlation, are necessary to obtain accurate results for the ionization energy and electron affinity of atoms.

An important result of Hartree–Fock calculations is that the ε_i for many-electron atoms depend on both the principal quantum number n and on the angular momentum quantum number l. Within a shell of principal quantum number n, $\varepsilon_{ns} < \varepsilon_{np} < \varepsilon_{nd} < \ldots$. This was not the case for the H atom. This result can be understood by considering the radial distribution functions shown in Figure 15.15. This function gives the probability of finding an electron at a given distance from the nucleus. The potential energy associated with the attraction between the nucleus and the electron falls off as $1/r$, so that its magnitude increases substantially as the electron comes closer to the nucleus. The subsidiary maximum near $r = a_0$ in the 3s radial distribution function indicates that there is a higher probability of finding the 3s electron close to the nucleus than is the case for the 3p and 3d electrons. Therefore, the 3s electron does not experience the same degree of shielding from the nuclear charge as the 3p and 3d electrons do. This leads to the 3s electron being bound more tightly to the nucleus and, therefore, the orbital energy is less than that for the 3p and 3d electrons. The same argument can be used to understand why the 3p orbital energy is lower than that for the 3d orbital.

The **configuration** of an atom gives the order in which the atomic orbitals are filled, that is, $1s^2 2s^1$ for Li. The known configurations of most atoms can be obtained by using Figure 15.16, which shows the general order in which the atomic orbitals are filled. This sequence is known as the **Aufbau principle.** Note that a number of atoms show departures from this order.

15.9 Understanding Trends in the Periodic Table from Hartree–Fock Calculations

We briefly summarize the main results of Hartree–Fock calculations for atoms:

- The orbital energy depends on both n and l. Within a shell of principal quantum number n, $\varepsilon_{ns} < \varepsilon_{np} < \varepsilon_{nd} < \ldots$.
- Electrons in a many-electron atom are shielded from the full nuclear charge by other electrons. Shielding can be modeled in terms of an effective nuclear charge. Core electrons are more effective in shielding outer electrons than electrons in the same shell.

The single most important predictor of trends in the periodic table is the effective charge for a valence electron, ζ. Accurate calculations of atomic properties require including electron correlation; therefore, such calculations go beyond the Hartree–Fock model. However, the Hartree–Fock model is sufficient to explain observed trends in the periodic table. We focus on three properties of atoms, namely, the atomic radius, the first ionization energy, and the electronegativity. Good agreement with measured atomic radii is obtained by calculating the radius of the sphere that contains ~90% of the electron charge. This radius is essentially determined by the effective charge felt by the electrons in the valence shell; large values of ζ correspond to small values of the atomic radius and vice versa.

The **electronegativity,** χ, quantifies the degree to which atoms accept or donate electrons to other atoms in a reaction. It is closely related to the first ionization energy and the electron affinity. For example, how do we know whether the nearly ionic NaCl species is better described by Na^+Cl^- or Na^-Cl^+? Formation of Na^+ and Cl^- ions at infinite separation requires the energy

$$\Delta E = E^{Na}_{ionization} - E^{Cl}_{electron\ affinity} = 5.14\ eV - 3.61\ eV = 1.53\ eV \quad (15.15)$$

Formation of the oppositely charged ions requires

$$\Delta E = E^{Cl}_{ionization} - E^{Na}_{electron\ affinity} = 12.97\ eV - 0.55\ eV = 12.42\ eV \quad (15.16)$$

In each case, additional energy is gained by bringing the ions together. Because less energy is required, the formation of Na^+Cl^- is strongly favored over Na^-Cl^+. To be able to estimate the direction of electron transfer between two atoms, we need a quantitative definition of electronegativity.

Several definitions of electronegativity (which has no units) exist, and all lead to similar results when scaled to the same numerical range. For example, χ as defined by Robert Mulliken in 1934 is proportional to the average of the first ionization energy and the electron affinity. For an atom that has a large value of ζ for the outermost or valence electron, the first ionization energy and electron affinity are large because the highest occupied and lowest unoccupied energy levels lie well below the zero of energy. Such an atom has a large electronegativity. Similarly, an atom that has a small value of ζ for the outermost or valence electron has a small electronegativity. Small values of χ favor electron donation, and large values favor acceptance. Because of this correlation, in solid elements with small χ values, one or more valence electrons is easily detached from an atom and delocalized over the solid. These elements conduct electricity well and are called metals. By contrast, the electrons in solid elements with large χ values are strongly bound to individual atoms. These elements do not conduct electricity and are called nonmetals. Chemical bonds between atoms with large differences in χ have a strong ionic character because significant electron transfer occurs to the atom that has the larger χ value. Chemical bonds between atoms that have similar χ values are covalent, because there is no driving force for electron transfer.

Figure 15.17 compares values for the atomic radius, first ionization energy, and electronegativity as a function of atomic number up to $Z = 55$. This range spans one period in which only the $1s$ orbital is filled, two short periods in which only s and p orbitals are filled, and two longer periods in which d orbitals are also filled. These trends can be explained by examining how the value of ζ for the valence electrons varies in going across or down the periodic table. Because Hartree–Fock results are based on numerical calculations, it is not possible to write a simple formula relating ζ with the atomic radius. However, it is useful to fall back on the crude model of a hydrogen atom with an effective charge ζ to model the valence electrons in a many-electron atom. For such an H-like atom, the orbital energies are proportional to $-\zeta^2/n^2$ and the radius is approximately proportional to n^2/ζ. Similar behavior can be expected for many-electron atoms. The abbreviated set of ζ values shown earlier in Table 15.1 shows that ζ increases in going across a period and decreases in going down a group. Therefore, we understand why the radius decreases and the first ionization energy increases across a period, where n is constant. Similarly, going down a group, the radius increases because n^2 increases more rapidly than ζ.

The electronegativity follows the same pattern as the ionization energy. This is understandable, because an atom with a high ionization energy is an unlikely candidate to donate an electron to another atom. Because a high ionization energy corresponds to a low-lying highest occupied atomic orbital, it is likely that the lowest unoccupied atomic orbital (the next higher level) will also be relatively strongly binding. Therefore, an atom with a large ionization energy is also likely to have a large electron affinity. This qualitative picture can be illustrated with fluorine, which is an electron acceptor, and lithium,

FIGURE 15.17
The covalent atomic radius, first ionization energy, and electronegativity are plotted as a function of the atomic number for the first 55 elements. Dashed vertical lines mark the completion of each period.

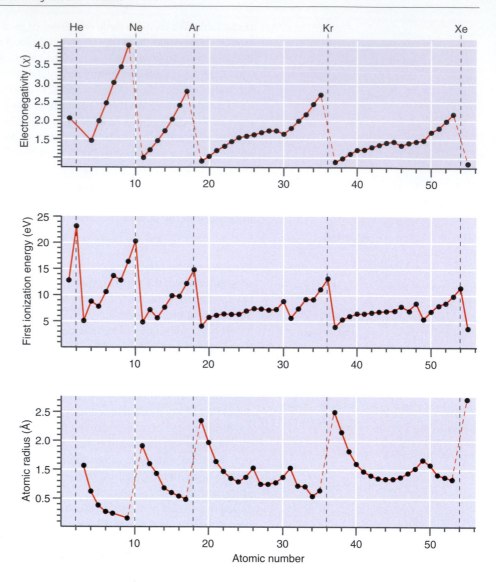

which is an electron donor. Fluorine has ionization energy and electron affinity values of 17.42 and 3.40 eV, respectively, and its χ is 4.10. Because lithium has a χ of 0.97, we expect it to have much smaller values for the ionization energy and electron affinity. In fact, these values are 0.98 and 0.62 eV, respectively. Note that these values are consistent with Li being a metal and fluorine a nonmetal.

Vocabulary

Aufbau principle	electron correlation	nodal surface	Rydberg constant
Bohr radius	electron–electron repulsion	orbital	shell
configuration	electronegativity	orbital approximation	shell model
Coulomb potential	electron spin	orbital energy	shielding
effective nuclear charge	Hartree–Fock self-consistent	Pauli exclusion principle	spherical harmonic functions
effective potential	field method	radial distribution function	subshell
electron affinity	ionization energy		

Questions on Concepts

Q15.1   What are the units of the H atom total energy eigen-functions? Why is $a_0^{3/2}R(r)$ graphed in Figure 15.5 rather than $R(r)$?

Q15.2   Use an analogy with the particle in the box to explain why the energy levels for the H atom are not equally spaced.

Q15.3   What transition gives rise to the highest frequency spectral line in the Lyman series shown in Figure 15.2?

Q15.4   Explain why the radial distribution function rather than the square of the magnitude of the wave function should be used to make a comparison with the shell model of the atom.

Q15.5   Is it always true that the probability of finding the electron in the H atom is greater in the interval $r - dr < r < r + dr$ than in the interval $r - dr < r < r + dr$ $\theta - d\theta < \theta < \theta + d\theta$ $\phi - d\phi < \phi < \phi + d\phi$?

Q15.6   What possible geometrical forms can the nodes in the angular function for p and d orbitals in the H atom have? What possible geometrical forms can the nodes in the radial function for s, p, and d orbitals in the H atom have?

Q15.7   If the probability density of finding the electron in the $1s$ orbital in the H atom has its maximum value for $r = 0$, does this mean that the proton and electron are located at the same point in space?

Q15.8   Why is the s, p, d, . . . nomenclature derived for the H atom also valid for many-electron atoms?

Q15.9   Explain why shielding is more effective by electrons in a shell of lower principal quantum number than by electrons having the same principal quantum number.

Q15.10   Why does the effective nuclear charge ζ for the $1s$ orbital increase by 0.99 in going from oxygen to fluorine, but ζ for the $2p$ orbital only increases by 0.65?

Q15.11   The angular functions, $\Theta(\theta)\Phi(\phi)$, for the one-electron Hartree–Fock orbitals are the same as for the hydrogen atom, and the radial functions and radial probability functions are similar to those for the hydrogen atom. For this question, assume that the latter two functions are identical to those for the hydrogen atom. The following figure shows (a) a contour plot in the x-y plane with the y axis being the vertical axis, (b) the radial function, and (c) the radial probability distribution for a one-electron orbital. Identify the orbital ($2s$, $4d_{xz}$, and so on). Apply the same conditions to Questions Q15.12 through Q15.19 and identify the orbitals as requested here.

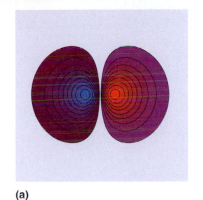

(a)

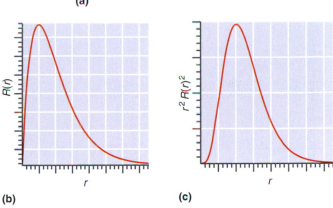

(b)   **(c)**

Q15.12   See Question Q15.11.

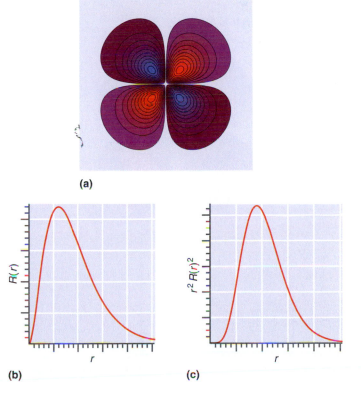

(a)

(b)   **(c)**

Q15.13   See Question Q15.11.

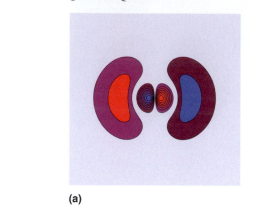

(a)

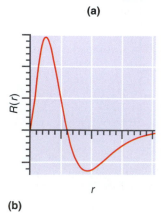

(b)

r

(c)

r

Q15.14   See Question Q15.11.

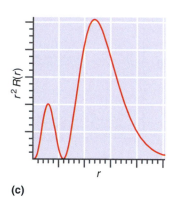

(a)

(b)

r

(c)

r

Q15.15   See Question Q15.11.

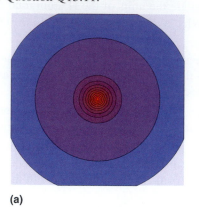

(a)

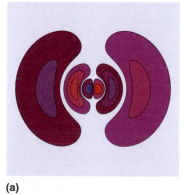

(b)

r

(c)

r

Q15.16   See Question Q15.11.

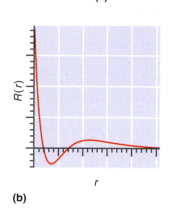

(a)

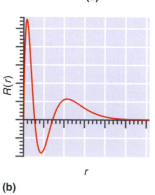

(b)

r

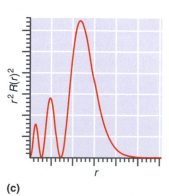

(c)

r

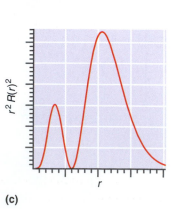

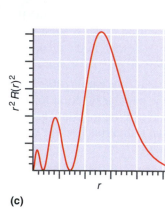

Q15.17   See Question Q15.11.

(a)

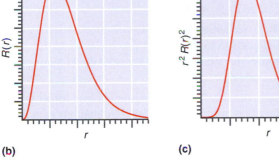

(b)   (c)

Q15.18   See Question Q15.11.

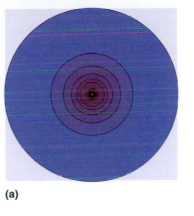

(a)

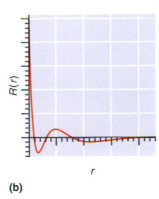   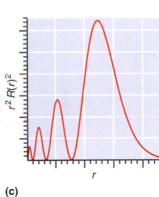

(b)   (c)

Problems

Problem numbers in **RED** indicate that the solution to the problem is given in the *Student Solutions Manual*.

P15.1   Assume that the radius of a hydrogen atom is 25 pm and that the nuclear radius is 1.2×10^{-15} m. (a) What fraction of the atomic volume is occupied by the nucleus? (b) What fraction of the atomic mass is due to the nucleus?

P15.2   Show by substitution that $\psi_{100}(r,\theta,\phi) = 1/\sqrt{\pi} \, (1/a_0)^{3/2} \, e^{-r/a_0}$ is a solution of Equation 15.2. What is the eigenvalue for total energy? Use the relation $a_0 = \varepsilon_0 h^2/(\pi m_e e^2)$.

P15.3   The total energy eigenvalues for the hydrogen atom are given by $E_n = -e^2/8\pi\varepsilon_0 a_0 n^2$, $n = 1, 2, 3, \ldots$, and the three quantum numbers associated with the total energy eigenfunctions are related by $n = 1, 2, 3, 4, \ldots$; $l = 0, 1, 2, 3, \ldots, n-1$; and $m_l = 0, \pm1, \pm2, \pm3, \ldots \pm l$. Using the nomenclature ψ_{nlm_l}, list all eigenfunctions that have the following total energy eigenvalues:

a. $E = -\dfrac{e^2}{32\pi\varepsilon_0 a_0}$

b. $E = -\dfrac{e^2}{72\pi\varepsilon_0 a_0}$

c. $E = -\dfrac{e^2}{128\pi\varepsilon_0 a_0}$

d. What is the degeneracy of each of these energy levels?

P15.4   Show that the total energy eigenfunctions $\psi_{100}(r)$ and $\psi_{210}(r)$ are orthogonal.

P15.5   Show that the total energy eigenfunctions $\psi_{100}(r)$ and $\psi_{200}(r)$ are orthogonal.

P15.6   Show that the total energy eigenfunctions $\psi_{210}(r,\theta,\phi)$ and $\psi_{211}(r,\theta,\phi)$ are orthogonal. Do you have to integrate over all three variables to show that the functions are orthogonal?

P15.7   Locate the radial and angular nodes in the H orbitals $\psi_{3p_x}(r,\theta,\phi)$ and $\psi_{3p_z}(r,\theta,\phi)$.

P15.8   How many radial and angular nodes are there in the following H orbitals?

a. $\psi_{2p_x}(r,\theta,\phi)$

b. $\psi_{2s}(r)$

c. $\psi_{3d_{xz}}(r,\theta,\phi)$

d. $\psi_{3d_{x^2-y^2}}(r,\theta,\phi)$

P15.9   In this problem, you will calculate the probability density of finding an electron within a sphere of radius r for the H atom in its ground state.

a. Show using integration by parts, $\int u \, dv = uv - \int v \, du$, that $\int r^2 e^{-r/\alpha} \, dr = e^{-r/\alpha}(-2\alpha^3 - 2\alpha^2 r - \alpha r^2)$.

b. Using this result, show that the probability density of finding the electron within a sphere of radius r for the hydrogen atom in its ground state is

$$1 - e^{-2r/a_0} - \frac{2r}{a_0}\left(1 + \frac{r}{a_0}\right)e^{-2r/a_0}$$

c. Evaluate this probability density for $r = 0.10a_0$, $r = 1.0a_0$, and $r = 4.0a_0$.

P15.10 Using the result of Problem P15.9, calculate the probability of finding the electron in the $1s$ state outside a sphere of radius $0.5a_0$, $3a_0$, and $5a_0$.

P15.11 Show that the function $2(1/a_0)^{3/2} e^{-r/a_0}$ is a solution of the following differential equation for $l = 0$:

$$R(r) - \frac{\hbar^2}{2\mu r^2}\frac{d}{dr}\left[r^2 \frac{dR(r)}{dr}\right] + \left[\frac{\hbar^2 l(l+1)}{2\mu r^2} - \frac{e^2}{4\pi\varepsilon_0 r}\right]R(r)$$
$$= ER(r)$$

What is the eigenvalue? Using this result, what is the value for the principal quantum number n for this function?

P15.12 Show that the function $(r/a_0)e^{-r/2a_0}$ is a solution of the differential in Problem P15.11 for $l = 1$. What is the eigenvalue? Using this result, what is the value for the principal quantum number n for this function?

P15.13 Calculate the mean value of the radius, $\langle r \rangle$, at which you would find the electron if the H atom wave function is $\psi_{100}(r)$.

P15.14 The wave function for the ground state of the He$^+$ atom is given by $\frac{1}{\sqrt{\pi}}\left(\frac{2}{a_0}\right)^{3/2} e^{-2r/a_0}$. Calculate the mean value of the radius $\langle r \rangle$ at which you would find the $1s$ electron in He.

P15.15 Calculate the quantity $(r - \langle r \rangle)^2$ if the H atom wave function is $\psi_{100}(r)$.

P15.16 Calculate the mean value of the radius $\langle r \rangle$ at which you would find the electron if the H atom wave function is $\psi_{210}(r,\theta,\phi)$.

P15.17 Ions with a single electron such as He$^+$, Li^{2+}, and Be^{3+} are described by the H atom wave functions with Z/a_0 substituted for $1/a_0$, where Z is the nuclear charge. The $1s$ wave function becomes $\psi(r) = 1/\sqrt{\pi}\,(Z/a_0)^{3/2}e^{-Zr/a_0}$. Using this result, compare the mean value of the radius $\langle r \rangle$ at which you would find the $1s$ electron in H, He$^+$, Li^{2+}, and Be^{3+}.

P15.18 The energy levels for ions with a single electron such as He$^+$, Li^{2+}, and Be^{3+} are given by $E_n = -Z^2 e^2/8\pi\varepsilon_0 a_0 n^2$, $n = 1, 2, 3, \ldots$. Calculate the ionization energies of H, He$^+$, Li^{2+}, and Be^{3+} in their ground states in units of electron-volts (eV).

P15.19 The d orbitals have the nomenclature $d_{z^2}, d_{xy}, d_{xz}, d_{yz}$, and $d_{x^2-y^2}$. Show how the d orbital

$$\psi_{3d_{yz}}(r,\theta,\phi) = \frac{\sqrt{2}}{81\sqrt{\pi}}\left(\frac{1}{a_0}\right)^{3/2}\frac{r^2}{a_0^2} e^{-r/3a_0} \sin\theta\cos\theta\sin\phi$$

can be written in the form $yzF(r)$.

P15.20 Core electrons shield valence electrons so that they experience an effective nuclear charge Z_{eff} rather than the full nuclear charge. Given that the first ionization energy of Li is 5.39 eV, use the formula in Problem P15.18 to estimate the effective nuclear charge experienced by the $2s$ electron in Li.

P15.21 The diameter of the Earth is 12,756.3 km and the radius of an iron atom is 156 pm. You make a one-dimensional chain of iron atoms equal in length to the Earth's diameter. (a) How many atoms do you need to make the chain? (b) What is the mass of the chain in grams?

P15.22 In spherical coordinates, $z = r\cos\theta$. Calculate $\langle z \rangle$ and $\langle z^2 \rangle$ for the H atom in its ground state. Without doing the calculation, what would you expect for $\langle x \rangle$ and $\langle y \rangle$, and $\langle x^2 \rangle$ and $\langle y^2 \rangle$? Why?

P15.23 Calculate the average value of the potential energy for the H atom in its ground state.

P15.24 The force acting between the electron and the proton in the H atom is given by $F = -e^2/4\pi\varepsilon_0 r^2$. Calculate the average value $\langle F \rangle$ for the $1s$ and $2p_z$ states of the H atom in terms of e, ε_0, and a_0.

P15.25 Calculate the distance from the nucleus for which the radial distribution function for the $2p$ orbital has its main and subsidiary maxima.

P15.26 Calculate $\langle r \rangle$ and the most probable value of r for the H atom in its ground state. Explain why they differ with a drawing.

P15.27 What is the shortest wavelength of light that Johannes Balmer expected to see emitted from the hydrogen discharge lamp based on his formula $\frac{1}{\lambda} = R_H\left(\frac{1}{2^2} - \frac{1}{n^2}\right)$, $n = 3, 4, 5, \ldots$?

P15.28 Calculate the wave number corresponding to the most and least energetic spectral lines in the Lyman, Balmer, and Paschen series shown in Figure 15.2 for the hydrogen atom.

P15.29 As the principal quantum number n increases, the electron is more likely to be found far from the nucleus. It can be shown that for H and for ions with only one electron such as He$^+$,

$$\langle r \rangle_{nl} = \frac{n^2 a_0}{Z}\left[1 + \frac{1}{2}\left(1 - \frac{l(l+1)}{n^2}\right)\right]$$

Calculate the value of n for an s state in the hydrogen atom such that $\langle r \rangle = 1000a_0$. Round up to the nearest integer. What is the ionization energy of the H atom in this state in electron-volts (eV)? Compare your answer with the ionization energy of the H atom in the ground state.

P15.30 You have commissioned a measurement of the second ionization energy from two independent research teams. You find that they do not agree and decide to plot the data together with known values of the first ionization energy. The results are shown here:

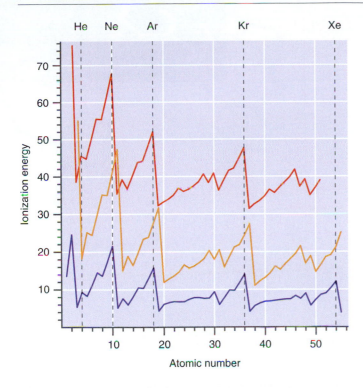

The lowest curve is for the first ionization energy and the upper two curves are the results for the second ionization energy from the two research teams. The uppermost curve has been shifted vertically to avoid an overlap with the other new data set. On the basis of your knowledge of the periodic table, you suddenly know which of the two sets of data is correct and the error that one of the teams of researchers made. Which data set is correct? Explain your reasoning.

P15.31   In this problem you will show that the charge density of the filled $n = 2$, $l = 1$ subshell is spherically symmetrical. The angular distribution of the electron charge is simply the sum of the squares of the magnitude of the angular part of the wave functions for $l = 1$ and $m_l = -1, 0,$ and 1.

a. Given that the angular part of these wave functions is

$$Y_1^0(\theta,\phi) = \left(\frac{3}{4\pi}\right)^{1/2} \cos\theta$$

$$Y_1^1(\theta,\phi) = \left(\frac{3}{8\pi}\right)^{1/2} \sin\theta\, e^{i\phi}$$

$$Y_1^{-1}(\theta,\phi) = \left(\frac{3}{8\pi}\right)^{1/2} \sin\theta\, e^{-i\phi}$$

write an expression for $|Y_1^0(\theta,\phi)|^2 + |Y_1^1(\theta,\phi)|^2 + |Y_1^{-1}(\theta,\phi)|^2$.

b. Show that $|Y_1^0(\theta,\phi)|^2 + |Y_1^1(\theta,\phi)|^2 + |Y_1^{-1}(\theta,\phi)|^2$ does not depend on θ and ϕ.

c. Why does this result show that the charge density for the filled $n = 2$, $l = 1$ subshell is spherically symmetrical?

P15.32   List the allowed quantum numbers m_l and m_s for the following subshells and determine the maximum occupancy of the subshells:

a. $2p$

b. $3d$

c. $4f$

d. $5g$

P15.33   An approximate formula for the energy levels in a multielectron atom is $E_n \approx -Z_{eff}^2 e^2/8\pi\varepsilon_0 a_0 n^2$, $n = 1, 2, 3, \dots$, where Z_{eff} is the effective nuclear charge felt by an electron in a given orbital. Calculate values for Z_{eff} from the first ionization energies for the elements Li through Ne (See www.webelements.com). Compare these values for Z_{eff} with those listed in Table 15.1. How well do they compare?

CHAPTER 16

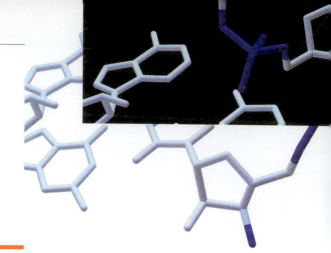

Chemical Bonding in Diatomic Molecules

The chemical bond is at the heart of chemistry. In this chapter we provide a qualitative model for chemical bonding, first using the H_2^+ molecule as an example. We show that H_2^+ is more stable than widely separated H and H^+ because of delocalization of the electron over the molecule and localization of the electron in the region between the two nuclei. The molecular orbital model developed for chemical bonding in H_2^+ is next used to discuss chemical bonding in other diatomic molecules. We discuss how to express molecular orbitals (MOs) in terms of atomic orbitals (AOs) and introduce the important concepts basis set and molecular orbital energy diagram. The MO model of the chemical bond is used to understand the bond order, bond energy, and bond length of homonuclear diatomic molecules. The formalism is extended to describe bonding in strongly polar molecules such as HF.

16.1 The Simplest One-Electron Molecule: H_2^+

A chemical bond is formed between two atoms if the energy of the molecule is lower than the energy of the separated atoms. What is it about the electron distribution around the nuclei that makes this happen? As discussed in Chapter 15, the introduction of a second electron vastly complicates the task of finding solutions to the Schrödinger equation for atoms. This is also true for molecules. For this reason, we initially focus our attention on the simplest molecule possible, the one-electron H_2^+ molecular ion. The Schrödinger equation can be solved essentially exactly for this molecule using numerical methods. However, just as for atoms, the Schrödinger equation cannot be solved exactly for any molecule containing more than one electron. Therefore, we approach H_2^+ using an approximate model. This model gives considerable insight into chemical bonding and, most importantly, can be extended easily to other molecules.

The motion of the nuclei and the electrons in Figure 16.1 can be separated using the **Born–Oppenheimer approximation.** Because the electron is lighter than the proton by a factor of nearly 2000, the electron charge quickly rearranges in response to the slower motion of the nuclei; this motion gives rise to molecular vibrations (that is, periodic changes in R). Because of the very different timescales for nuclear and electron motion, the two motions can be decoupled and we can solve the Schrödinger equation for a fixed nuclear separation and then calculate the energy of the molecule for that distance. If this procedure is repeated for many values of the internuclear separation, we can determine an energy function, $E(R)$.

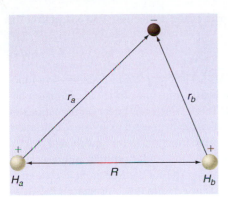

FIGURE 16.1
The two protons and the electron are shown at one instant in time. The quantities R, r_a, and r_b represent the distances between the charged particles.

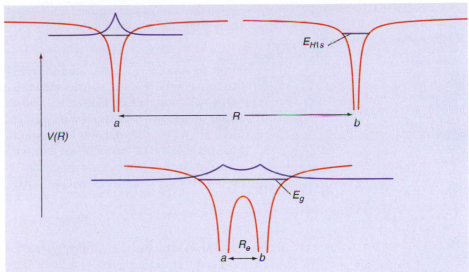

FIGURE 16.2
The potential energy of the H$_2^+$ molecule is shown for two different values of R (red curves). At large distances, the electron will be localized in a $1s$ orbital either on nucleus a or b. However, at the equilibrium bond length R_e, the two Coulomb potentials overlap, allowing the electron to be delocalized over the whole molecule. The blue curve represents the amplitude of the atomic (top panel) and molecular (bottom panel) wave functions, and the solid horizontal lines represent the corresponding energy eigenvalues.

From experimental results, we know that H$_2^+$ is a stable species, so that a solution of the Schrödinger equation for H$_2^+$ must give at least one bound state. We define the zero of energy as an H atom and an H$^+$ ion that are infinitely separated. Given this choice of the zero of energy, a stable molecule has a negative energy. Therefore, the molecule is described by an energy function $E(R)$, which has a minimum value for a distance R_e, which is the equilibrium bond length.

We next discuss how to construct approximate wave functions for the H$_2^+$ molecule. Imagine slowly bringing together an H atom and an H$^+$ ion. At infinite separation, the electron is in a $1s$ orbital on either one nucleus or the other. However, as the internuclear distance approaches R_e, the potential energy wells for the two species overlap, and the barrier between them is lowered. Consequently, the electron can move back and forth between the Coulomb wells on the two nuclei. It is equally likely to be on nucleus a as on nucleus b so that the molecular wave function looks like the superposition of a $1s$ orbital on each nucleus. This model is shown pictorially in Figure 16.2.

16.2 The Molecular Wave Function for Ground-State H$_2^+$

What does the electron distribution in H$_2^+$ look like? In answering this question, we consider the relative energies of two H atoms compared to the H$_2$ molecule. Two H atoms are more stable than the four infinitely separated charges by 2624 kJ mol^{-1}. The H$_2$ molecule is more stable than two infinitely separated H atoms by 436 kJ mol^{-1}. Therefore, the chemical bond lowers the total energy of the two protons and two electrons by only 17%. For chemical bonds in general, the **bond energy** is a small fraction of the total energy of the widely separated electrons and nuclei. This suggests that the charge distribution in a molecule is not very different from a superposition of the charge distribution of the individual atoms. It also suggests that the wave functions for molecules look like a linear superposition of the wave functions of the individual atoms. This is our starting point for the following discussion of the chemical bond in H$_2^+$.

On the basis of this reasoning, the spatial part of the **molecular wave function** for H_2^+ is given in the following equation:

$$\psi = c_a \phi_{H1s_a} + c_b \phi_{H1s_b} \tag{16.1}$$

In this equation, ϕ_{H1s} is a H1s **atomic orbital (AO)**. The complete molecular wave function includes the spin and satisfies the Pauli exclusion principle. To avoid confusion between AOs, one electron **molecular orbitals (MOs),** which are linear combinations of atomic orbitals and molecular wave functions, which are in general many electronic wave functions, we use the symbol ψ for a molecular wave function, ϕ for an AO, and σ for a MO. Because the H_2^+ has only one electron, both the molecular orbitals and molecular wave functions are one electron functions. To allow an AO for the free atom to change as the bond is formed, a **variational parameter,** ζ, is inserted in the hydrogen AO:

$$\phi_{H1s} = \frac{1}{\sqrt{\pi}} \left(\frac{\zeta}{a_0}\right)^{3/2} e^{-\zeta r/a_0}$$

This parameter looks like an effective nuclear charge. You will see in the end-of-chapter problems that varying ζ allows the size of the orbital to change.

Note that in contrast to the AOs, MOs are delocalized over the molecule. This idea is expressed mathematically in Equation (16.1) by making the molecular orbital a linear combination of the 1s AO on nucleus a and the 1s AO on nucleus b. What are the values of the coefficients c_a and c_b? Because identical quantum mechanical particles are indistinguishable, it is not meaningful to put a and b labels on the nuclei. Observables, and in particular the probability density $\psi^*\psi$, must not change when the two nuclei in a homonuclear diatomic molecule are interchanged. This requires that

$$|c_a| = |c_b| \quad \text{or} \quad c_a = \pm c_b \tag{16.2}$$

Although the signs of c_a and c_b can be the same or different, the magnitudes of the coefficients are equal.

Equation (16.2) enables us to generate two MOs from the two AOs:

$$\psi_g = c_g(\phi_{H1s_a} + \phi_{H1s_b})$$
$$\psi_u = c_u(\phi_{H1s_a} - \phi_{H1s_b}) \tag{16.3}$$

The wave functions for this homonuclear diatomic molecule are classified as g or u based on whether they change signs upon undergoing inversion through the center of the molecule. The subscripts g and u refer to the German words *gerade* and *ungerade,* respectively, for **symmetric** and **antisymmetric.** If the origin of the coordinate system is placed at the center of the molecule, inversion corresponds to $\psi(x,y,z) \rightarrow \psi(-x,-y,-z)$. If this operation leaves the wave function unchanged, that is, $\psi(x,y,z) = \psi(-x,-y,-z)$, it has g *symmetry.* If $\psi(x,y,z) = -\psi(-x,-y,-z)$, the wave function has u *symmetry.* See also Figure 16.13 later in this chapter for an illustration of g and u MOs. We will see later that only ψ_g describes a stable, chemically bonded H_2^+ molecule.

The values of c_g and c_u can be determined by normalizing ψ_u and ψ_g. Note that the integrals used in the normalization are over all three spatial coordinates. Normalization requires that

$$1 = \int c_g^* (\phi_{H1s_a}^* + \phi_{H1s_b}^*) c_g (\phi_{H1s_a} + \phi_{H1s_b}) d\tau$$
$$= c_g^2 \left(\int \phi_{H1s_a}^* \phi_{H1s_a} d\tau + \int \phi_{H1s_b}^* \phi_{H1s_b} d\tau + 2 \int \phi_{H1s_a}^* \phi_{H1s_b} d\tau \right) \tag{16.4}$$

The first two integrals have the value one because the H_{1s} orbitals are normalized. The third integral is called the **overlap integral** and is abbreviated $S_{ab} = \int \phi_{H1s_b}^* \phi_{H1s_a} d\tau$. The overlap is a new concept that was not encountered in atomic systems. The meaning of S_{ab} is indicated pictorially in Figure 16.3. Carrying out the integrations in Equation (16.4), we obtain $1 = c_g^2(2 + 2S_{ab})$. The result for c_g is

FIGURE 16.3
The amplitude of two H1s atomic orbitals is shown along an axis connecting the atoms. The overlap is appreciable only for regions in which the amplitude of both AOs is significantly different from zero. Such a region is shown in yellow.

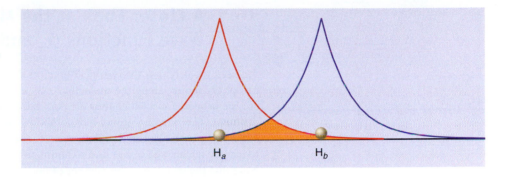

FIGURE 16.4
Schematic energy functions $E(R)$ are shown for the g and u states in the approximate solution discussed. The zero of energy corresponds to widely separated H and H$^+$ species.

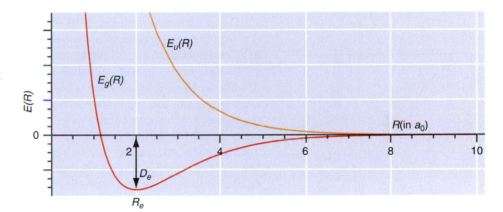

$$c_g = \frac{1}{\sqrt{2 + 2S_{ab}}} \tag{16.5}$$

The coefficient c_u has a similar form, as you will see in the end-of-chapter problems.

If the AOs under the integral in Equation (16.4) were both centered on the same nucleus, the value of S_{ab} would be one. Contributions to S_{ab} arise only when both of the orbital amplitudes are significantly different from zero, which makes it a sensitive function of the internuclear separation. The value of S_{ab} for two 1s AOs is a number that lies between one as $R \rightarrow 0$ and zero as $R \rightarrow \infty$. The electron can be thought of as delocalized over the molecule only if S_{ab} is significantly greater than zero.

A numerical solution of the Schrödinger equation at $R = R_e$ using the approximate wave functions of Equation (16.3), shows that $\Delta E_g < 0$ and $\Delta E_u > 0$, where the zero of energy corresponds to widely separated H and H$^+$. *Therefore, we conclude that only an H$_2^+$ molecule described by ψ_g is a stable molecule.* For a molecule described by ψ_u, the energy is greater than for the nonbonded state, and the molecule is unstable with respect to dissociation. It is also found that the u state is raised in energy relative to the nonbonded state more than the g state is lowered.

The $E(R)$ curves that result from a minimization of the energy with respect to ζ are shown schematically in Figure 16.4. The most important conclusion that can be drawn from this figure is that the ψ_g state describes a stable H$_2^+$ molecule, because the energy has a well-defined minimum at $R = R_e$. We also conclude that ψ_u does not describe a bound state of H and H$^+$ because $E_u(R) > 0$ for all R.

The distance corresponding to the minimum energy of $E_g(R)$ defines the equilibrium bond length, R_e. The calculated binding energy D_e in this simple model is 2.36 eV, which is reasonably close to the exact value of 2.70 eV. The exact and calculated R_e values are both 2.00 a_0. The fact that the calculated D_e and R_e values are quite close to the exact values validates the assumption that the exact molecular wave function is quite similar to ψ_g. The ψ_g and ψ_u wave functions are referred to as **bonding** and **antibonding orbitals,** respectively, to emphasize their relationship to the chemical bond.

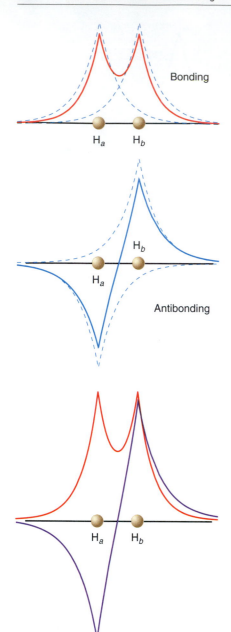

FIGURE 16.5

Molecular wave functions ψ_g and ψ_u (solid lines), evaluated along the internuclear axis, are shown in the top two panels. The unmodified ($\zeta = 1$) H1s orbitals from which they were generated are shown as dashed lines. The bottom panel shows a direct comparison of ψ_g and ψ_u.

16.3 A Closer Look at the Molecular Wave Functions ψ_g and ψ_u

We seek the origin of the chemical bond in the differences between ψ_g and ψ_u. The values of ψ_g and ψ_u along the molecular axis are shown in Figure 16.5 together with the atomic orbitals from which they are derived. Note that the two wave functions are quite different. The bonding orbital has no nodes, and the amplitude of ψ_g is quite high between the nuclei. The antibonding wave function has a node that is located midway between the nuclei, and ψ_u has its maximum positive and negative amplitudes at the nuclei. Note that the increase in the number of nodes in the wave function with energy is similar to the other quantum mechanical systems that have been studied to this point. Both wave functions are correctly normalized in three dimensions.

Figure 16.6 shows contour plots of ψ_g and ψ_u evaluated in the $z = 0$ plane. If we compare Figures 16.5 and 16.6, we can see that the node midway between the H atoms in ψ_u corresponds to a nodal plane.

The probability density of finding an electron at various points along the molecular axis is given by the square of the wave function, which is shown in Figure 16.7. For the antibonding and bonding orbitals, the probability density of finding the electron in H_2^+ is compared with the probability density of finding the electron in a hypothetical nonbonded case. For the nonbonded case, the electron is equally likely to be found on each nucleus in H1s, AOs, and $\zeta = 1$. Two important conclusions can be drawn from this figure. First, for both ψ_g and ψ_u, the volume in which the electron can be found is large compared with the volume accessible to an electron in a hydrogen atom. This tells us that the electron is delocalized over the whole molecule in both the bonding and antibonding orbitals. Second, we can see that the probability of finding the electron in the region between the nuclei is quite different for ψ_g and ψ_u. For the antibonding orbital, the probability is zero midway between the two nuclei, but for the bonding orbital, it is quite high. This difference is what makes the g state a bonding state and the u state an antibonding state.

This pronounced difference between ψ_g^2 and ψ_u^2 is explored further in Figure 16.8. The *difference* between the probability density for these orbitals and the hypothetical nonbonding state is shown in this figure. This difference tells us how the electron density would change if we could suddenly switch on the interaction at the equilibrium geometry. We see that for the antibonding state, electron density would move from the region between the two nuclei to the outer regions of the molecule. For the bonding state, electron density would move both to the region between the nuclei and closer to each nucleus. The origin of the density increase between the nuclei for the bonding orbital is the interference term $2\psi_{H1s_a}\psi_{H1s_b}$ in $(\psi_{H1s_a} + \psi_{H1s_b})^2$. The origin of the density increase near each nucleus is the increase in ζ from 1.00 to 1.24 in going from the free atom to the H_2^+ molecule.

In looking at Figure 16.8, note that for ψ_g the probability density is increased relative to the nonbonding case in the region between the nuclei, and decreased by the same amount outside of this region. The opposite is true for ψ_u. This is true because the integrated probability density over all space is one because the molecular wave functions are normalized. Although it may not be apparent in Figures 16.5 to 16.8, the wave functions satisfy this requirement. Only small changes in the probability density outside of the region between the nuclei are needed to balance larger changes in this region, because the integration volume outside of the region between the nuclei is much larger. The data shown as a line plot in Figure 16.8 are shown as a contour plot in Figure 16.9. Red and blue correspond to the most positive and least positive values for $\Delta\psi_g^2$ and $\Delta\psi_u^2$, respectively. The outermost contour for $\Delta\psi_g^2$ in Figure 16.9 corresponds to a negative value, and it is seen that the corresponding area is large. The product of the small negative charge in $\Delta\psi_g^2$ with the large volume corresponding to the contour area is equal in magnitude and opposite in sign to the increase in $\Delta\psi_g^2$ in the bonding region.

A comparison of the electron charge densities associated with ψ_g and ψ_u helps us to understand the important ingredients in chemical bond formation. For both states, the

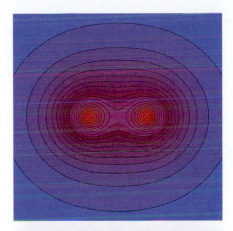

FIGURE 16.6
Contour plots of ψ_g (left) and ψ_u (right). The minimum amplitude is shown as blue, and the maximum amplitude is shown as red for each plot. The dashed line indicates the position of the nodal plane in ψ_u.

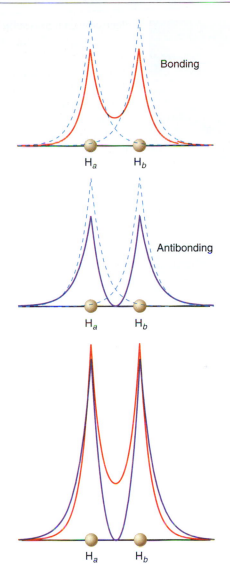

FIGURE 16.7
The upper two panels show the probability densities ψ_g^2 and ψ_u^2 along the internuclear axis for the bonding and antibonding wave functions. The dashed lines show $1/2\,\psi_{H1s_a}^2$ and $1/2\,\psi_{H1s_b}^2$, which are the probability densities for unmodified ($\zeta = 1$) H1s orbitals on each nucleus. The lowest panel shows a direct comparison of ψ_g^2 and ψ_u^2. Both molecular wave functions are correctly normalized in three dimensions.

FIGURE 16.8
The red curve shows ψ_g^2 (left panel) and the dark blue curve shows ψ_u^2 (right panel). The light blue curves show the differences $\Delta\psi_g^2 = \psi_g^2 - 1/2(\psi_{H1s_a})^2 - 1/2(\psi_{H1s_b})^2$ (left panel) and $\Delta\psi_u^2 = \psi_u^2 - 1/2(\psi_{H1s_a})^2 - 1/2(\psi_{H1s_b})^2$ (right panel). These differences are a measure of the change in electron density near the nuclei due to bond formation. A charge buildup occurs for the bonding orbital and a charge depletion occurs for the antibonding orbital in the region between the nuclei.

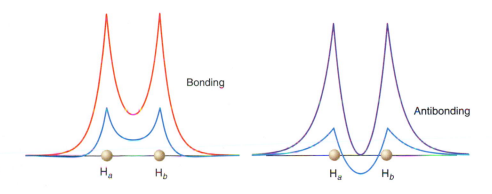

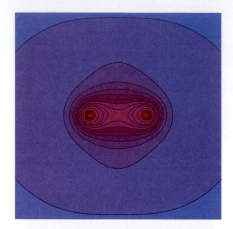

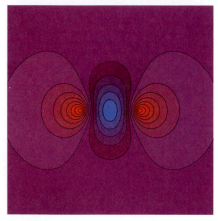

FIGURE 16.9
Contour plots of $\Delta\psi_g^2$ (top) and $\Delta\psi_u^2$ (bottom). The red areas in the top image correspond to positive values for $\Delta\psi_g^2$, and the gray area corresponds to negative values for $\Delta\psi_g^2$. The blue area in the bottom image corresponds to negative values for $\Delta\psi_u^2$, and the red areas just outside of the bonding region correspond to positive values for $\Delta\psi_u^2$. The color in the corners of each contour plot corresponds to $\Delta\psi^2 = 0$.

electronic charge is subject to **delocalization** over the whole molecule. However, charge is also localized in the molecular orbitals, and this **localization** is different in the bonding and antibonding states. In the bonding state, the electronic charge redistribution relative to the nonbonded state leads to a charge buildup both near the nuclei and between the nuclei. In the antibonding state, the electronic charge redistribution leads to a charge buildup outside of the region between the nuclei. We conclude that electronic charge buildup between the nuclei is an essential ingredient of a chemical bond.

At this point, we summarize what we have learned about the chemical bond. We have developed a way to generate molecular orbitals, starting with atomic orbitals. We conclude that both charge delocalization and localization play a role in chemical bond formation. Delocalization promotes bond formation because the kinetic energy is lowered as the electron occupies a larger region in the molecule than it would in the atom. However, localization through the contraction of atomic orbitals and the accumulation of electron density between the atoms in the state described by ψ_g lowers the total energy even further. Although electron delocalization is sufficient to stabilize the molecule, the contraction of the atomic orbitals leads to an additional increase in the bond energy. Both localization and delocalization play a role in bond formation, and it is this complex interplay between opposites that leads to a strong chemical bond. The preceding discussion is based on the one-electron molecule H_2^+. The simplest many-electron molecule, H_2, is considered next.

16.4 The H₂ Molecule: Molecular Orbital and Valence Bond Models

Adding a second electron to the H_2^+ molecule to give H_2 forces us to deal with the issues of electron spin. Just as for a many-electron atom, the Schrödinger equation can only be solved numerically for a many-electron molecule. Such calculations are extremely valuable because they give reliable values for a host of important chemical parameters. In this text, we focus on obtaining a qualitative understanding of chemical bonding in many-electron molecules. We begin this discussion with H_2.

Two distinctly different descriptions of the chemical bond are those of the **valence bond (VB) model** and the **molecular orbital (MO) model.** How do the VB and MO models differ? The valence bond model is a localized description of the chemical bond. As in a Lewis structure, each bond in a molecule is associated with an electron pair. The pair is made up of one electron from each of the two atoms involved in the bond and has a net spin of zero. By contrast, the molecular orbital model is a delocalized description of chemical bonding. MOs that extend over the whole molecule are constructed by making linear combinations of AOs (LCAOs). For this reason, one refers to the LCAO-MO model. Electrons are placed into these MOs just as they are placed in the AOs of many-electron atoms. To keep track of the electrons, the symbol for an AO, MO, or molecular wave function is followed by parentheses in which the electrons associated with the AO, MO, or molecular wave function are listed (that is, $\phi_{H1s_a}(1)$, $\sigma(1)$ or $\psi(1,2)$. To satisfy the Pauli exclusion principle, the molecular wave function for any many-electron molecule or atom must change sign if the indices of two electrons are interchanged. This requirement places limits on the form of the molecular wave function as shown in Example Problem 16.1.

EXAMPLE PROBLEM 16.1

Does the molecular wave function

$$\psi(1,2) = \frac{1}{\sqrt{2 + 2S_{ab}^2}} \left[\phi_{H1s_a}(1)\phi_{H1s_b}(2) \right]$$

$$+ \ \phi_{H1s_b}(1)\phi_{H1s_a}(2) \times \left[\frac{1}{\sqrt{2}}(\alpha(1)\beta(2) + \alpha(2)\beta(1)) \right]$$

satisfy the Pauli exclusion principle? Answer this question by exchanging the indices of the two electrons and determining if the wave function changes sign.

Solution

Exchanging indices 1 and 2 gives the wave function

$$\psi(2,1) = \frac{1}{\sqrt{2 + 2S_{ab}^2}} \left[\phi_{H1s_a}(2)\phi_{H1s_b}(1) \right.$$

$$\left. + \phi_{H1s_b}(2)\phi_{H1s_a}(1) \times \left[\frac{1}{\sqrt{2}}(\alpha(2)\beta(1) + \alpha(1)\beta(2)) \right] \right]$$

Because $\psi(2,1) = \psi(1,2)$ rather than $\psi(2,1) = -\psi(1,2)$, the wave function does not satisfy the Pauli exclusion principle.

We begin our discussion of bonding with the VB model. For H_2, the lowest energy VB molecular wave function including spin is

$$\psi_{bonding}^{VB}(1,2) = \frac{1}{\sqrt{2 + 2S_{ab}^2}} \left[\phi_{H1s_a}(1)\phi_{H1s_b}(2) + \phi_{H1s_b}(1)\phi_{H1s_a}(2) \right]$$

$$\times \left[\frac{1}{\sqrt{2}}(\alpha(1)\beta(2) - \alpha(2)\beta(1)) \right] \tag{16.6}$$

Convince yourself that this wave function satisfies the Pauli exclusion principle. You will verify that this molecular wave function is correctly normalized in the end-of-chapter problems. For a molecule with n bonds, the molecular wave function is a sum of terms, each of which is the product of $2n$ AOs.

We next discuss H_2 using the MO model. Following the same line of reasoning as for H_2^+, we construct a one-electron MO for H_2:

$$\sigma_g 1s(1) = \psi_g = \frac{1}{\sqrt{2 + 2S_{ab}}}(\phi_{H1s_a}(1) + \phi_{H1s_b}(1)) \tag{16.7}$$

in which a maximum of two electrons can be placed. Recall that ψ refers to a molecular wave function, ϕ to an AO, and σ to a MO. Note that in the VB model, the molecular wave functions are generated from the AOs, whereas in the MO model, the molecular wave functions are generated from the MOs, which are linear combinations of the AOs:

$$\psi_{MO}^{bonding}(1,2) = \sigma_g 1s(1)\sigma_g 1s(2) \left[\frac{1}{\sqrt{2}}(\alpha(1)\beta(2) - \alpha(2)\beta(1)) \right] \tag{16.8}$$

Just as for the VB model, the spin part of the MO molecular wave function for the ground state is a singlet. This result is in keeping with our chemical intuition that in the ground state, the two electrons in H_2 are paired. Because the Schrödinger equation does not contain terms that depend on spin, we need only consider the spatial part of the wave function in calculating the energy. Substituting Equation (16.7) into the spatial part of Equation (16.8), we write the spatial part of the MO wave function in terms of the AOs:

$$\sigma_g 1s(1)\sigma_g 1s(2) = \frac{1}{2 + 2S_{ab}} \left[1s_a(1) + 1s_b(1) \right]\left[1s_a(2) + 1s_b(2) \right]$$

$$= \frac{1}{2 + 2S_{ab}} \left[1s_a(1)1s_b(2) + 1s_b(1)1s_a(2) + 1s_a(1)1s_a(2) \right.$$

$$\left. + 1s_b(1)1s_b(2) \right] \tag{16.9}$$

$$= \frac{1}{2 + 2S_{ab}} \left[\begin{array}{l} \{1s_a(1)1s_b(2) + 1s_b(1)1s_a(2)\}_{VB} \\ + \{1s_a(1)1s_a(2) + 1s_b(1)1s_b(2)\}_{ionic} \end{array} \right]$$

How do we interpret this molecular wave function? It is a linear combination of four separate terms, in which the first two terms are identical with the two terms appearing in $\psi_{bonding}^{VB}(1,2)$. In these terms, one electron is associated with each of the H atoms, and each of these terms corresponds to a covalent H—H molecule. The second two terms have two electrons associated with a single H atom, corresponding to the ionic forms H^+H^- and H^-H^+. The weighting of the ionic terms is clearly unrealistic, because the molecular wave function of Equation (16.9) predicts that 50% of dissociation events for a H_2 molecule lead to H^+ and H^- products. Experimentally, only neutral H atoms are observed. However, there is a nonzero probability of finding both electrons in the vicinity of one nucleus in H_2, and this corresponds to an ionic term. Therefore, the VB wave function is unrealistic in that it does not include any "ionic" terms. Although the LCAO-MO wave function gives a good description of the molecule near the equilibrium position, it does not give a realistic description of the dissociation limit of large R. However, as we will see later, this is not a failure of the MO method, but rather a shortcoming of the approximate wave function used.

16.5 Comparing the Valence Bond and Molecular Orbital Models of the Chemical Bond

Can we conclude that the valence bond model is a good description of the chemical bond and that the molecular orbital model is not as good? No, we cannot, because the deficiencies in the MO calculation are due to the approximate wave function used in the calculation. Accurate results are obtained if the molecular wave function is expanded in a larger set of functions. The same can be said of the VB model. However, these more exact wave functions contain many terms that are not easily interpreted. Because our focus in this chapter is on approximate methods that give physical insight, rather than on obtaining accurate values, we next discuss how to improve the simple MO and VB wave functions.

Calculations show that the calculated total energy can be lowered by writing the MO wave function as

$$\psi^{MO}(1,2) = N(1s_a(1)1s_b(2) + 1s_b(1)1s_a(2) + \lambda\{1s_a(1)1s_a(2) + 1s_b(1)1s_b(2)\}) \quad (16.10)$$

where λ is a variational coefficient. Lambda turns out to be much less than one for the lowest energy. What is achieved by making λ less than one? By doing so, the ionic character of the MO wave function has been decreased. It can be shown that including ionic character in the MO wave function is equivalent to writing the wave function as a sum of ground-state and excited-state configurations. Therefore, a recipe for improving the MO wave function is to include a number of excited-state configurations when describing the molecule.

Similar calculations show that the VB wave function can be improved by adding a small amount of ionic character. In other words, the approximate MO and VB wave functions become quite similar as the simplest wave function in each model is improved. If the H_2 wave functions in the VB and MO models become similar as they approach the exact solution, are the models really different? As we will show in the next chapter, it is very useful to distinguish between the concept of the chemical bond implied by these models. The valence bond wave function is a mathematical description of the familiar Lewis picture of electron pair bonding, in that each of the terms in the wave function describes shared electron bonds. The model reinforces the idea that chemical bonds are localized between two adjacent atoms. For example, a localized picture is more useful than a delocalized model in visualizing C—Cl bond cleavage in ethyl chloride.

By contrast, MO theory initially assumes that electrons are delocalized over the entire molecule. However, calculations show that some MOs are largely localized between two adjacent atoms. Therefore, the MO model is capable of describing both localized and delocalized bonds. When is a delocalized picture of chemical bonding useful? Consider benzene or a metal. The π electrons in benzene and the conduction electrons in a metal are truly delocalized. Therefore, a localized picture of chemical bonding is unable to describe the energy lowering that arises in an aromatic system or in a metal. A delocalized model must be used instead. These examples show that both models are useful.

As discussed earlier, chemical bonding can be understood using either the VB model or the MO model. In this section, the MO model is used to discuss the electronic structure of diatomic molecules, because it is more cumbersome to use the valence bond model for this purpose. However, both the VB and MO models are used in discussing molecular structure in Chapter 17. We begin our discussion of many-electron molecules by applying the Hartree–Fock self-consistent field method in order to obtain one-electron MOs and the corresponding energy eigenvalues. Just as for atoms, this method allows a reduction of the n-electron Schrödinger equation to n one-electron Schrödinger equations. Two energies are of interest: the individual orbital energies and the total energy of the molecule. To avoid confusion, the symbol ε is used to designate an MO energy, and E is used to designate the total energy of the molecule. As for atoms, this set of equations is solved by iteration. The rest of this chapter gives you a qualitative understanding of how the molecular orbitals $\sigma_i(\mathbf{r}_i)$ and the corresponding orbital energies ε_i are used in understanding the electronic structure and properties of homonuclear and heteronuclear diatomic molecules.

16.6 Expressing Molecular Orbitals as a Linear Combination of Atomic Orbitals

In MO theory, an electron is associated with a wave function that is delocalized over the entire molecule. The MOs are constructed from linear combinations of AOs, leading to the term **LCAO-MO model.** As discussed for the H_2^+ molecule, the AOs must in general be modified (that is, expanded or contracted) to make them useful in constructing MOs. For many-electron molecules, MOs conceptually based on the MOs for H_2^+ are filled with two electrons, in order of increasing orbital energy, until all electrons are accommodated. Each of the MOs is expressed as a linear combination of AOs denoted by ϕ_i:

$$\sigma_j(1) = \sum_i c_{ij}\, \phi_i(1) \qquad (16.11)$$

The sum is over AOs on a given atom and on other atoms in the molecule, and the number in parentheses refers to the electron under consideration. For example, in H_2, both electrons are in the σ_1 molecular orbital and

$$\sigma_1(1) = \frac{1}{\sqrt{2 + 2S_{ab}}}\, [\phi_{H1s_a}(1) + \phi_{H1s_b}(1)] \qquad (16.12)$$

The set of the AOs ϕ_i used to construct the molecular orbital in Equation (16.11) is called the **basis set.** Because the infinite set of AOs on an atom forms a complete set, the expansion of MOs in AOs is approximate only because the infinite sum in Equation (16.11) must be truncated to carry out a calculation. Which and how many AOs should be included in the basis set used to construct a molecular orbital? No more functions are used than is required by the accuracy desired of the calculation, because the cost of a calculation increases rapidly with the size of the basis set.

Observables for a diatomic molecule such as the dipole moment, the bond dissociation energy, the bond length, and first and higher ionization energies can all be calculated using quantum mechanics. Ultimately, all of these quantities can be related to the MOs and to the expansion coefficients c_{ij} in Equation (16.11). Therefore, we need a method to determine the values for the c_{ij} in a given MO. How are the MOs and the corresponding energy levels ε_i determined? Assume that the basis set has been chosen. The best values for the c_{ij} are determined by minimizing the total energy of the molecule using the Hartree–Fock method. The accuracy of such a MO calculation increases as the size of the basis set increases. However, the cost of a calculation also increases as a

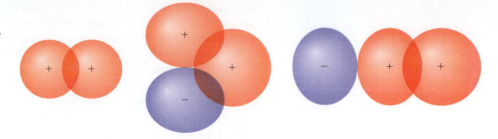

FIGURE 16.10
The overlap between two $1s$ orbitals ($\sigma + \sigma$), a $1s$ and a $2p_x$ or $2p_y$ ($\sigma + \pi$), and a $1s$ and a $2p_z$ ($\sigma + \sigma$) are depicted from left to right. Note that the two shaded areas in the middle panel have opposite signs, so the net overlap of these two atomic orbitals of different symmetry is zero.

power of the basis set size. Therefore, practical calculations require that the basis set be as small as possible. How big must this basis set be? A few empirical rules can be used to minimize the size of the basis set required to describe the electronic structure of the first and second row diatomic molecules at a quantitative level:

- The $1s$ core AOs and all valence shell AOs, both occupied and unoccupied, are included in the basis set. For H and He, this includes only $1s$ orbitals, whereas for Li $\rightarrow$ Ne, $2s$, $2p_x$, $2p_y$, and $2p_z$ atomic orbitals are also included.
- Only atomic orbitals of the same symmetry will combine with one another. Figure 16.10 provides a schematic explanation of this rule. For this example, we consider only s and p electrons. A net nonzero overlap between two atomic orbitals occurs only if both AOs are either cylindrically symmetric with respect to the molecular axis (σ AOs) or if both have a common nodal plane that coincides with the molecular axis (π AOs).

The basis set constructed using these rules is known as the **minimal basis set.** Much larger basis sets are used for quantum chemical calculations requiring a high accuracy.

16.7 The Molecular Orbital Energy Diagram

We present the MO energy results for H_2 and HF in the form of a **molecular orbital energy diagram** in Figure 16.11. By convention, the AOs for the bonded atoms are shown on the left and right sides of this diagram. The MOs generated by combining the AOs are shown in the middle. The MOs are depicted by drawing symbols at each atom whose size is proportional to the magnitude of the AO. The sign of the coefficient is indicated

FIGURE 16.11
Molecular orbital energy diagram for a qualitative description of bonding in H_2 and HF. The atomic orbitals are shown to the left and right, and the molecular orbitals are shown in the middle. Dashed lines connect the MO with the AOs from which it was constructed. Shaded circles have a diameter proportional to the coefficients c_{ij}. Red and blue shading signifies positive and negative signs of the AO coefficients, respectively.

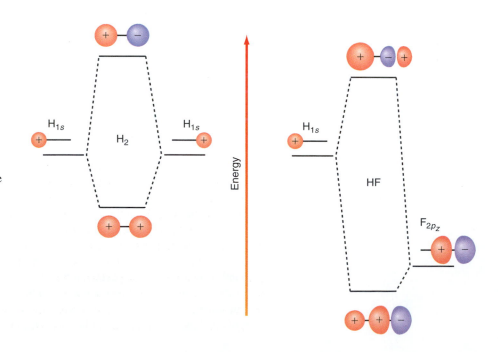

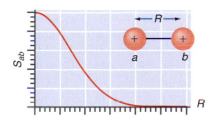

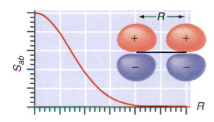

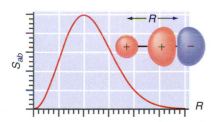

FIGURE 16.12
Variation of overlap with internuclear
distance for s and p orbitals.

by the color of the symbol. Note that the wave function with blue and red interchanged is not distinguishable from the original wave function because only the magnitude of the wave function is directly related to an observable.

What can be learned from the molecular orbital energy diagrams using HF as an example? Most importantly, we can determine the energies of the MOs relative to the energies of the AOs. We see that the bonding MO is lower in energy than the lower of the two AOs and that the antibonding MO is higher in energy than the higher of the two AOs. The results obtained for H_2 and HF from molecular orbital energy diagrams for diatomic molecules can be generalized as follows:

- Two interacting AOs give rise to two MOs. Each of these MOs has a different energy than the AO energies from which the MOs originated. A necessary condition for this splitting of energy levels to occur is that the AOs have a nonzero overlap in the molecule.
- The energy of the MO that has the form $\psi = |c_a|\phi_a + |c_b|\phi_b$, with **in-phase AOs,** is lower than that of the lower lying AO by $\Delta\varepsilon_+$. This MO is called the bonding orbital. The energy of the MO that has the form $\psi = |c_a'|\phi_a - |c_b'|\phi_b$, with **out-of-phase AOs,** is higher than the higher lying AO by $\Delta\varepsilon_-$. This MO is called the antibonding orbital.
- The energy splitting $\Delta\varepsilon_+ + \Delta\varepsilon_-$ increases with the overlap S_{ab}. The inequality $|\Delta\varepsilon_+| < |\Delta\varepsilon_-|$ is always satisfied.
- The relative contribution of two AOs in a MO is measured by the relative magnitude of their coefficients. These coefficients can be calculated as discussed in Engel and Reid's *Physical Chemistry,* Chapter 24. If two interacting AOs have the same energy, their coefficients in the MO have the same magnitude. They have the same sign in the bonding orbital and opposite signs in the antibonding orbital. This last statement is also true if the AOs do not have the same energy. If the AO energies are not equal, the magnitude of the coefficient of the lower lying AO is greater in the bonding orbital and smaller in the antibonding orbital.

These points usually suffice to draw qualitative molecular orbital energy diagrams for diatomic molecules such as that shown in Figure 16.11.

As discussed earlier, a net nonzero overlap of atomic orbitals is necessary for chemical bond formation. Figure 16.12 indicates how the overlap varies with distance for different orbitals. The molecular bond lies on the z axis. Whereas the overlap of two s or two p_x or p_y orbitals is maximized when the bond distance is zero, the overlap of an s orbital with a p_z orbital has its maximum value at a larger internuclear distance. Do not assume that an increase in overlap always leads to stronger chemical bonds. When the atoms are so close together that electron–electron repulsion among the **core electrons** or nuclear–nuclear repulsion becomes more important than electron delocalization of the **valence electrons,** a further decrease in internuclear distance will destabilize the bond.

16.8 Molecular Orbitals for Homonuclear Diatomic Molecules

It is useful to have a qualitative picture of the shape and spatial extent of molecular orbitals for diatomic molecules. Following the same path used in going from the H atom to many-electron atoms, we construct MOs for many-electron molecules on the basis of the excited states of the H_2^+ molecule. These MOs are useful in describing bonding in first and second row homonuclear diatomic molecules. Heteronuclear diatomic molecules are discussed in Section 16.11.

All MOs for homonuclear diatomics can be divided into two groups with regard to each of two **symmetry operations.** The first of these is rotation about the molecular axis. If this rotation leaves the MO unchanged, it has no nodes along this axis, and the MO has σ **symmetry.** Combining s AOs always gives rise to σ MOs for diatomic molecules. If the MO has one nodal plane containing the molecular axis, the MO has π **symmetry.**

Combining p_x or p_y AOs always gives rise to π MOs if they have a common nodal plane. The second operation is inversion through the center of the molecule. Placing the origin at the center of the molecule, inversion corresponds to $\sigma(x,y,z) \to \sigma(-x,-y,-z)$. If this operation leaves the MO unchanged, the MO has **g symmetry.** If $\sigma(x,y,z) = -\sigma(-x,-y,-z)$, the MO has **u symmetry.** All MOs constructed using $n = 1$ and $n = 2$ AOs have either σ or π and either g or u symmetry. Molecular orbitals for H_2^+ of g and u symmetry are shown in Figure 16.13. Note that $1\sigma_g$ and $1\pi_u$ are bonding MOs, whereas $1\sigma_u^*$ and $1\pi_g^*$ are antibonding MOs.

Two different notations are commonly used to describe MOs in homonuclear diatomic molecules. In the first, the MOs are classified according to symmetry and increasing energy. For instance, a $2\sigma_g$ orbital has the same symmetry, but a higher energy than the $1\sigma_g$ orbital. In the second notation, the integer indicating the relative energy is omitted, and the AOs from which the MOs are generated are listed instead. For instance, the $\sigma_g(2s)$ MO has a higher energy than the $\sigma_g(1s)$ MO. The superscript * is used to designate antibonding orbitals. Two types of MOs can be generated from $2p$ AOs. If the axis of the $2p$ orbital lies on the intermolecular axis (by convention the z axis), a σ MO is generated. This specific molecular orbital is called a $3\sigma_u$ or $\sigma_u(2p_z)$ orbital. Combining $2p_x$ (or $2p_y$) orbitals on each atom gives a π molecular orbital because the MO has a nodal plane containing the molecular axis. These MOs are degenerate in energy and are called $1\pi_u$ or $\pi_u(2p_x)$ and $\pi_u(2p_y)$ MOs.

In principle, we should take linear combinations of all the basis functions of the same symmetry (either σ or π) when constructing MOs. However, an additional criterion, based on orbital energies, reduces the number of AOs for which c_{ij} is significantly different from zero in this qualitative model. Little mixing occurs between AOs of the same symmetry if they have greatly different orbital energies. For example, the mixing between $1s$ and $2s$ AOs for the second row homonuclear diatomics can be neglected at our level of discussion. However, for these same molecules, the $2s$ and $2p_z$ AOs both have σ symmetry and will mix if their energies are not greatly different. Because the energy dif-

FIGURE 16.13
Contour plots of several bonding and antibonding orbitals of H_2^+. Red and blue contours correspond to the most positive and least positive amplitudes, respectively. The yellow arrows show the transformation $(x,y,z) \to (-x,-y,-z)$ for each orbital. If the amplitude of the wave function changes sign under this transformation, it has u symmetry. If it is unchanged, it has g symmetry.

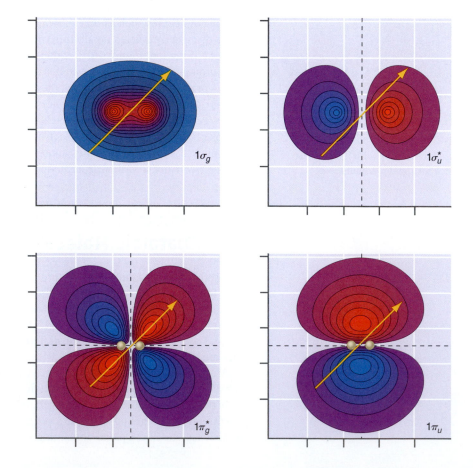

ference between the $2s$ and $2p_z$ atomic orbitals increases in the sequence Li $\rightarrow$ F, **s–p mixing** decreases for the second row diatomics in the order Li_2, B_2, ..., O_2, F_2. It is useful to think of MO formation in these molecules as a two-step process. We first create separate MOs from the $2s$ and $2p$ AOs, and subsequently combine the MOs of the same symmetry to create new MOs that include s–p mixing.

Are the contributions from the s and p AOs equally important in MOs that exhibit s–p mixing? The answer is no because the AO closest in energy to the resulting MO has the largest coefficient c_{ij} in $\sigma_j(1) = \Sigma_i c_{ij} \phi_i(1)$. Therefore, the $2s$ AO is the major contributor to the 2σ MO because the MO energy is closer to the $2s$ than to the $2p$ orbital energy. Applying the same reasoning, the $2p_z$ atomic orbital is the major contributor to the 3σ MO. The MOs used to describe chemical bonding in first and second row homonuclear diatomic molecules are shown in Table 16.1. The AO that is the major contributor to the MO is shown in the last column, and the minor contribution is shown in parentheses. For the sequence of molecules $H_2 \rightarrow N_2$, the orbital energy calculated using the Hartree–Fock method increases in the sequence $1\sigma_g < 1\sigma_u^* < 2\sigma_g < 2\sigma_u^* < 1\pi_u < 3\sigma_g < 1\pi_g^* < 3\sigma_u^*$. Moving across the periodic table to O_2 and F_2, the relative order of the $1\pi_u$ and $3\sigma_g$ MOs changes. Note that the first four MO energies follow the AO sequence, but that the σ and π MO energies generated from $2p$ AOs have different energies.

It is useful to have an understanding of the spatial extent of these MOs. Figure 16.14 shows contour plots of the first few H_2^+ MOs, including only the major AO in each case (no s–p mixing). The orbital exponent has not been optimized and $\zeta = 1$ for all AOs. Inclusion of the minor AO for the $2\sigma_g, 2\sigma_u^*, 3\sigma_g$, and $3\sigma_u^*$ MOs alters the plots in Figure 16.14 at a minor rather than a major level.

We next discuss the most important features of these plots. As might be expected, the $1\sigma_g$ orbital has no nodes, whereas the $2\sigma_g$ orbital has a nodal surface and the $3\sigma_g$ orbital has two nodal surfaces. All σ_u orbitals have a nodal plane perpendicular to the internuclear axis. The π orbitals have a nodal plane containing the internuclear axis. The amplitude for all antibonding σ MOs is zero midway between the atoms on the molecular axis. This means that the probability density for finding electrons in this region will be small. The antibonding $1\sigma_u^*$ and $3\sigma_u^*$ orbitals have a nodal plane, and the $2\sigma_u^*$ orbital has both a nodal plane and a nodal surface. The $1\pi_u$ orbital has no nodal plane other than on the intermolecular axis, whereas the $1\pi_g$ orbital has one nodal plane in the bonding region.

Note that the MOs made up of AOs with $n = 1$ do not extend as far away from the nuclei as the MOs made up of AOs with $n = 2$. In other words, electrons that occupy valence MOs are more likely to overlap with their counterparts on other molecules than are electrons in core MOs. This fact is important in understanding which electrons participate in making bonds in molecules, as well as in understanding reactions between molecules. The MOs shown in Figure 16.14 are specific to H_2^+ and have been calculated using $R_e = 2.00\, a_0$ and $\zeta = 1$. The detailed shape of these MOs varies from molecule to molecule and depends primarily on the effective nuclear charge, ζ, and the bond length. We can get a qualitative idea of what the MOs look like for other molecules by using the H_2^+ MOs with the effective nuclear charge obtained from Hartree–Fock calculations for the molecule of interest.

TABLE 16.1   Molecular Orbitals Used to Describe Chemical Bonding in Homonuclear Diatomic Molecules

MO Designation	Alternate	Character	Atomic Orbitals
$1\sigma_g$	$\sigma_g(1s)$	Bonding	$1s$
$1\sigma_u^*$	$\sigma_u^*(1s)$	Antibonding	$1s$
$2\sigma_g$	$\sigma_g(2s)$	Bonding	$2s\ (2p_z)$
$2\sigma_u^*$	$\sigma_u^*(2s)$	Antibonding	$2s\ (2p_z)$
$3\sigma_g$	$\sigma_g(2p_z)$	Bonding	$2p_z\ (2s)$
$3\sigma_u^*$	$\sigma_u^*(2p_z)$	Antibonding	$2p_z\ (2s)$
$1\pi_u$	$\pi_u(2p_x, 2p_y)$	Bonding	$2p_x, 2p_y$
$1\pi_g^*$	$\pi_g^*(2p_x, 2p_y)$	Antibonding	$2p_x, 2p_y$

FIGURE 16.14

MOs based on the ground and excited states for H_2^+ generated from $1s$, $2s$, and $2p$ atomic orbitals. Contour plots are shown on the left and line scans along the path indicated by the yellow arrow are shown on the right. Red and blue contours correspond to the most positive and least positive amplitudes, respectively. Dashed lines and curves indicate nodal surfaces. Lengths are in units of a_0, and $R_e = 2.00 \ a_0$.

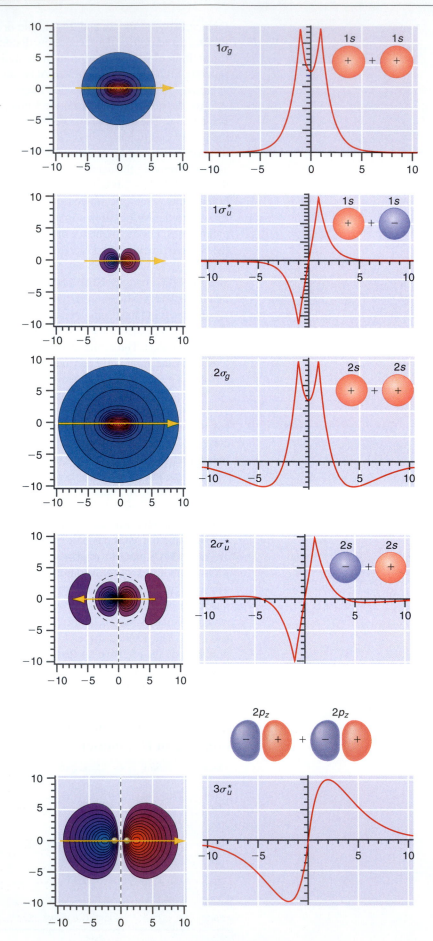

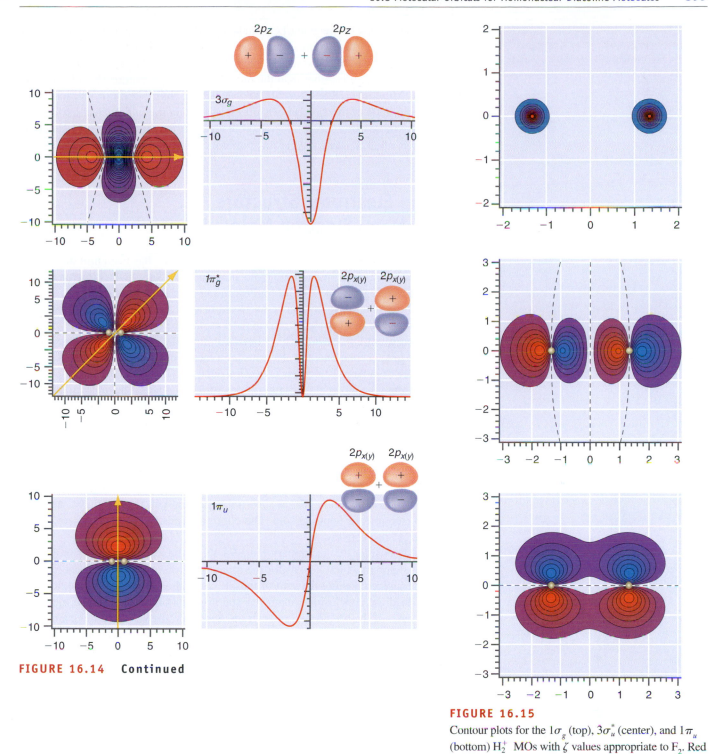

FIGURE 16.14   Continued

FIGURE 16.15

Contour plots for the $1\sigma_g$ (top), $3\sigma_u^*$ (center), and $1\pi_u$ (bottom) H_2^+ MOs with ζ values appropriate to F_2. Red and blue contours correspond to the most positive and least positive amplitudes, respectively. Dashed lines indicate nodal surfaces. Light circles indicate position of nuclei. Lengths are in units of a_0, and $R_e = 2.66\ a_0$.

For example, the bond length for F_2 is ~35% greater than that for H_2^+ and $\zeta = 8.65$ and 5.1 for the $1s$ and $2p$ orbitals, respectively. Because $\zeta > 1$, the amplitude of the fluorine AOs falls off much more rapidly with the distance from the nucleus than is the case for the H_2^+ molecule. Figure 16.15 shows $1\sigma_u\ 3\sigma_u^*$ and $1\pi_u$ MOs for these ζ values generated using the H_2^+ AOs. Note how much more compact the AOs and MOs are compared with $\zeta = 1$. The overlap between the $1s$ orbitals used to generate the lowest energy MO in F_2 is very small. For this reason, electrons in this MO do not make a major contribution to the chemical bond in F_2. Note also that the $3\sigma_u^*$ orbital for F_2 exhibits three nodal surfaces between the atoms rather than the one node shown in Figure 16.14 for $\zeta = 1$. Unlike the $1\pi_u$ MO for H_2^+, the $F_2\,1\pi_u$ MO shows distinct contributions from each atom, because the amplitude of the $2p$ AOs falls off rapidly along the internuclear axis. However, apart from these differences, the general features shown in Figure 16.14 are common to the MOs of all first and second row homonuclear diatomics.

16.9 The Electronic Structure of Many-Electron Molecules

A framework of molecular orbitals, based on the H_2^+ orbitals, has been introduced that can be used for many-electron diatomic molecules. In discussing many-electron atoms, the concept of a configuration is useful for describing their electronic structure, and the concept of a molecular configuration is discussed next. A **molecular configuration** is obtained by putting at most two electrons in each molecular orbital, in order of increasing orbital energy, until all electrons have been accommodated. If the degeneracy of an energy level is greater than one, the electrons are placed in the MOs in such a way that the total number of unpaired electrons is maximized.

We first discuss the molecular configurations for H_2 and He_2. The molecular orbital diagrams in Figure 16.16 show the number and spin of the electrons rather than the magnitude and sign of the AO coefficients as was the case in Figure 16.11. What can you say about the magnitude and sign of the AO coefficients for each of the four MOs of Figure 16.16?

In the minimal basis set, only $1s$ atomic orbitals need be considered in generating molecular orbitals for H_2 and He_2. The interaction of $1s$ orbitals on each atom gives rise to a bonding and an antibonding MO as shown schematically in Figure 16.16. Each MO can hold two electrons of opposite spin. The configurations for H_2 and He_2 are $(1\sigma_g)^2$ and $(1\sigma_g)^2(1\sigma_u^*)^2$, respectively.

You should consider two cautionary remarks about the interpretation of molecular orbital energy diagrams. First, just as for the many-electron atom, the total energy of a many-electron molecule is not the sum of the MO orbital energies. Therefore, it is not always valid to draw conclusions about the stability or bond strength of a molecule solely on the basis of the orbital energy diagram. Secondly, the words *bonding* and *antibonding* give information about the relative signs of the AO coefficients in the MO, but they do not convey whether the electron is bound to the molecule. The total energy for any stable molecule is lowered by adding electrons to any orbital for which the energy is less than zero. For example, O_2^- is a stable species compared to O_2 and an electron at infinity, even though the electron is placed in an antibonding MO on O_2^-.

For H_2, both electrons are in the $1\sigma_g$ MO, which is lower in energy than the $1s$ AOs. Calculations show that the $1\sigma_g$ MO energy is less than zero and that the $1\sigma_u^*$ MO energy is greater than zero. In this case, the total energy is lowered by putting electrons in the $1\sigma_g$ orbital and rises again if electrons are put into the $1\sigma_u^*$ orbital. In the MO model, He_2 has two electrons in each of the $1\sigma_g$ and $1\sigma_u^*$ orbitals. Because the energy of the $1\sigma_u^*$ orbital is greater than zero, He_2 is not a stable molecule in this model. In fact, He_2 is stable only below ~5 K as a result of a very weak van der Waal's interaction, rather than chemical bond formation.

The preceding examples used a single $1s$ orbital on each atom to form molecular orbitals. We now discuss the molecules F_2 and N_2, which have both s and p valence electrons. First, consider F_2, for which s–p mixing can be neglected, because the $2s$ AO lies 21.6 eV below the $2p$ AO. The configuration for F_2 is $(1\sigma_g)^2(1\sigma_u^*)^2(2\sigma_g)^2(2\sigma_u^*)^2(3\sigma_g)^2(1\pi_u)^2(1\pi_u)^2(1\pi_g^*)^2(1\pi_g^*)^2$. For this molecule, the 2σ MO is quite well described by a single $2s$ AO on each atom, and the 3σ MO is quite

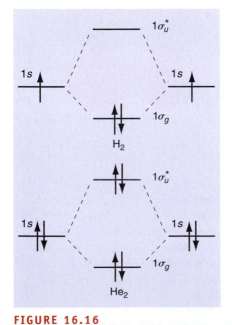

FIGURE 16.16
Atomic and molecular orbital energies and occupation for H_2 and He_2. Upward- and downward-pointing arrows indicate α and β spins. The energy splitting between the MO levels is not to scale.

FIGURE 16.17

Schematic MO energy diagram for the valence electrons in F_2. The degenerate p and π orbitals are shown slightly offset in energy. The dominant atomic orbital contributions to the MOs are shown as solid lines. Minor contributions due to s–p mixing have been neglected. The MOs are schematically depicted to the right of the figure. The $1\sigma_g$ and $1\sigma_u^*$ MOs are not shown.

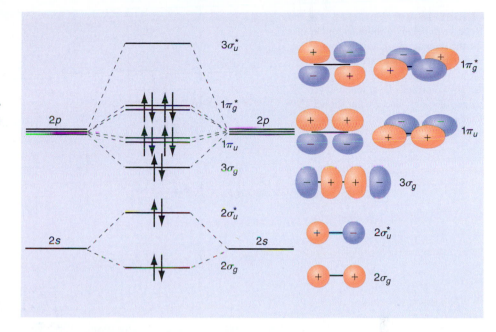

well described by a single $2p_z$ AO on each atom. Figure 16.17 shows a molecular orbital energy diagram for F_2. Because the $2p_x$ and $2p_y$ AOs have a net zero overlap with each other and with the s AOs, each of the doubly degenerate $1\pi_u$ and $1\pi_g^*$ molecular orbitals originates from a single AO on each atom.

For N_2, the configuration is $(1\sigma_g)^2(1\sigma_u^*)^2(2\sigma_g)^2(2\sigma_u^*)^2(1\pi_u)^2(1\pi_u)^2(3\sigma_g)^2$. The $2s$ AO lies below the $2p$ AO by only 12.4 eV, and in comparison to F_2, s–p mixing is not negligible. Therefore, the 2σ and 3σ MOs have significant contributions from both $2s$ and $2p_z$ AOs. A molecular orbital energy diagram for N_2 is shown in Figure 16.18. Note that mixing has changed the shape of the 2σ and 3σ N_2 MOs somewhat from that of the F_2 MOs. The $2\sigma_g$ MO has more bonding character, because the probability of finding the electron between the atoms is higher than it was without s–p mixing. Applying the same reasoning, the $2\sigma_u^*$ MO has become less antibonding and the $3\sigma_g$ MO has become less bonding for N_2 in comparison with F_2. We can see from the overlap in the AOs that the triple bond in N_2 arises from electron occupation of the $3\sigma_g$ and the pair of $1\pi_u$ MOs.

FIGURE 16.18

Schematic MO energy diagram for the valence electrons in N_2. The degenerate p and π orbitals are shown slightly offset in energy. The dominant AO contributions to the MOs are shown as solid lines. Lesser contributions arising from s–p mixing are shown as dashed lines. The MOs are schematically depicted to the right of the figure. The $1\sigma_g$ and $1\sigma_u^*$ MOs are not shown.

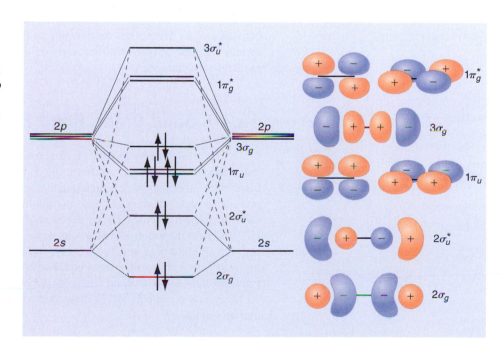

FIGURE 16.19
Relative molecular orbital energy levels for the second row diatomics (not to scale). Both notations are given for the molecular orbitals. The $1\sigma_g (\sigma_g 1s)$ and $1\sigma_u^* (\sigma_u^* 1s)$ orbitals lie at much lower values of energy and are not shown. (Not to scale.)

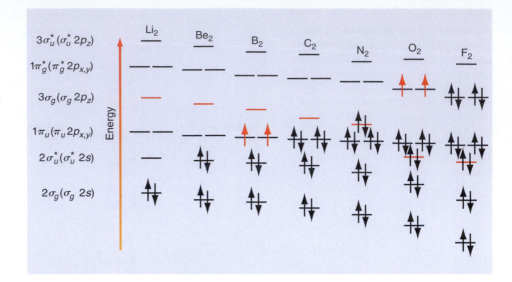

On the basis of this discussion of H_2, He_2, N_2, and F_2, the MO formalism is extended to all first and second row homonuclear diatomic molecules. After the relative energies of the molecular orbitals are established from numerical calculations, the MOs are filled in the sequence of increasing energy, and the number of unpaired electrons for each molecule can be predicted. The results for the second row are shown in Figure 16.19. We see that both B_2 and O_2 are predicted to have two unpaired electrons because the energy level is twofold degenerate. Therefore, these molecules should have a net magnetic moment (they are paramagnetic), whereas all other homonuclear diatomics should have a zero net magnetic moment (they are diamagnetic). These predictions are in good agreement with experimental measurements, which provides strong support for the validity of the MO model.

Figure 16.19 shows that the energy of the molecular orbitals tends to decrease with increasing atomic number in this series. This is a result of the increase in ζ in going across the periodic table. The larger effective nuclear charge and the smaller atomic size leads to a lower AO energy, which in turn leads to a lower MO energy. However, the $3\sigma_g$ orbital energy falls more rapidly across this series than the $1\pi_u$ orbital. This occurs because the degree of s–p mixing falls when going from Li_2 to F_2. Because the $2p_x$ and $2p_y$ AOs do not mix with the $2s$ AO, the $1\pi_u$ orbital energy remains essentially constant across this series. As a result, an inversion occurs in the order of molecular orbital energies between the $1\pi_u$ and $3\sigma_g$ orbitals for O_2 and F_2 relative to the other molecules in this series.

16.10 Bond Order, Bond Energy, and Bond Length

Molecular orbital theory has shown its predictive power by providing an explanation of the observed net magnetic moment in B_2 and O_2 and the absence of a net magnetic moment in the other second row diatomic molecules. We now show that the theory can also provide an understanding of trends in the binding energy and the vibrational force constant for these molecules. Figure 16.20 shows data for these observables for the series $H_2 \rightarrow Ne_2$. As the number of electrons in the diatomic molecule increases, the bond energy has a pronounced maximum for N_2 and a smaller maximum for H_2. The vibrational force constant k shows the same trend. The bond length increases as the bond energy and force constant decrease in the series $Be_2 \rightarrow N_2$, but it exhibits a more complicated trend for the lighter molecules. All of these data can be qualitatively understood using molecular orbital theory.

FIGURE 16.20
Bond energy, bond length, and vibrational force constant of the first 10 diatomic molecules as a function of the number of electrons in the molecule. The upper panel shows the calculated bond order for these molecules. The dashed line indicates the dependence of the bond length on the number of electrons if the He_2 data point is omitted.

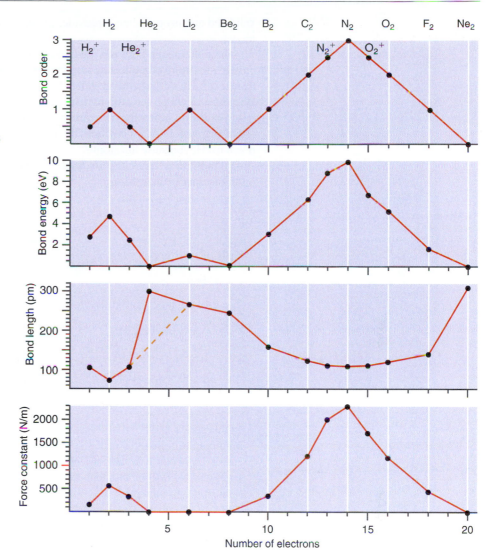

Consider combining two atomic orbitals to create two molecular orbitals in H_2 and He_2. For simplicity, we assume that the total energy of the molecules can be approximated by the sum of the orbital energies. Because the bonding orbital is lower in energy than the atomic orbitals from which it was created, placing electrons into a bonding orbital leads to an energy lowering with respect to the atoms. This makes the molecule more stable than the separated atoms, which is characteristic of a chemical bond. Similarly, placing two electrons into each of the bonding and antibonding orbitals leads to a total energy that is slightly greater than that of the separated molecules. Therefore, the molecule is unstable with respect to dissociation into two atoms. This result suggests that stable bond formation requires more electrons to be in bonding than in antibonding orbitals. We introduce the concept of **bond order,** which is defined as

Bond order = 1/2[(total bonding electrons) – (total antibonding electrons)]

We expect the bond energy to be very small for a bond order of zero, and to increase with increasing bond order. As shown in Figure 16.20, the bond order shows the same trend as the bond energies. With the exception of Li_2, the bond order also tracks the vibrational force constant very well. Again, we can explain the data by associating a stiffer bond with a higher bond order. This agreement is a good example of how a model becomes validated and useful when it provides an understanding for different sets of experimental data.

The relationship between the bond length and the number of electrons in the molecule is a bit more complicated. For a given atomic radius, the bond length is expected to vary inversely with the bond order. This trend is approximately followed for the series $Be_2 \rightarrow N_2$ in which the atomic radii are not constant, but decrease steadily. The trend is not followed in going from He_2 to Li_2 because the valence electron in Li is in the $2s$ AO rather than the $1s$ AO. The correlation between bond order and bond length also breaks down for He_2 because the atoms are not really chemically bonded. On balance, the trends shown in Figures 16.19 and 16.20 provide significant support for the concepts underlying molecular orbital theory.

EXAMPLE PROBLEM 16.2

Arrange the following in terms of increasing bond energy and bond length on the basis of their bond order: N_2^+, N_2, N_2^-, and N_2^{2-}.

Solution

The ground-state configurations for these species are

$$N_2^+ : (1\sigma_g)^2(1\sigma_u^*)^2(2\sigma_g)^2(2\sigma_u^*)^2(1\pi_u)^2(1\pi_u)^2(3\sigma_g)^1$$

$$N_2 : (1\sigma_g)^2(1\sigma_u^*)^2(2\sigma_g)^2(2\sigma_u^*)^2(1\pi_u)^2(1\pi_u)^2(3\sigma_g)^2$$

$$N_2^- : (1\sigma_g)^2(1\sigma_u^*)^2(2\sigma_g)^2(2\sigma_u^*)^2(3\sigma_g)^2(1\pi_u)^2(1\pi_u)^2(1\pi_g^*)^1$$

$$N_2^{2-} : (1\sigma_g)^2(1\sigma_u^*)^2(2\sigma_g)^2(2\sigma_u^*)^2(3\sigma_g)^2(1\pi_u)^2(1\pi_u)^2(1\pi_g^*)^1(1\pi_g^*)^1$$

In this series, the bond order is 2.5, 3, 2.5, and 2. Therefore, the bond energy is predicted to follow the order $N_2 > N_2^+$, $N_2^- > N_2^{2-}$ using the bond order alone. However, because of the extra electron in the antibonding $1\pi_g^*$ MO, the bond energy in N_2^- will be less than that in N_2^+. Because bond lengths decrease as the bond strength increases, the bond length will follow the opposite order.

Looking back at what we have learned about homonuclear diatomic molecules, several important concepts stand out. Combining atomic orbitals on each atom to form molecular orbitals provides a way to understand how electrons are distributed among the energy levels in molecules. Although including many AOs on each atom (that is, using a larger basis set) will improve the accuracy of the molecular orbitals and orbital energies, important trends can be predicted using the minimal basis set. The symmetry of atomic orbitals is important in predicting whether they contribute to a given molecular orbital. Molecular orbitals originating from s and p atomic orbitals are either of the σ or π type and, for a homonuclear diatomic molecule, have either u or g symmetry. The concept of bond order allows us to understand why He_2, Be_2, and Ne_2 are not stable and why the bond in N_2 is so strong.

16.11 Heteronuclear Diatomic Molecules

New issues arise when we consider chemical bond formation in heteronuclear diatomic molecules. For the case of two interacting orbitals of different energies, the coefficients of the atomic orbitals in the MO have different magnitudes. The coefficient of the lower energy AO is greater than that of the higher energy AO in the bonding MO. The opposite is true for the antibonding MO. The coefficients have the same sign in the bonding orbital and opposite signs in the antibonding orbital. Because the two atoms are dissimilar, the u and g symmetries do not apply (inversion interchanges the nuclei). Therefore, the MOs on a heteronuclear diatomic molecule are numbered differently as indicated here for the order in energy exhibited in the molecules $Li_2–N_2$:

FIGURE 16.21
Schematic energy diagram showing the relationship between the atomic and molecular orbital energy levels for the valence electrons in HF. The degenerate p and π orbitals are shown slightly offset in energy. The dominant atomic orbital contributions to the MOs are shown as solid lines. Lesser contributions are shown as dashed lines. The MOs are depicted to the right of the figure.

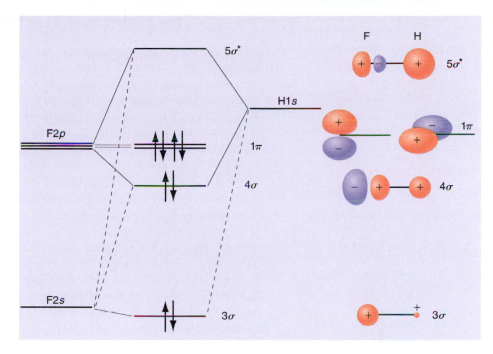

Homonuclear	$1\sigma_g$	$1\sigma_u^*$	$2\sigma_g$	$2\sigma_u^*$	$1\pi_u$	$3\sigma_g$	$1\pi_g^*$	$3\sigma_u^*$	$\ldots$
Heteronuclear	1σ	2σ	3σ	4σ	1π	5σ	2π	6σ	$\ldots$

However, the MOs will still have either σ or π symmetry. The symbol * is usually added to the MOs for the heteronuclear molecule to indicate an antibonding MO, as shown for HF in Figure 16.21.

To illustrate the differences between homonuclear and heteronuclear diatomic molecules, we consider HF. The basis set includes the $1s$ atomic orbital on H and the $2s$ and $2p$ orbitals on F. The molecular orbital energy diagram for HF is shown in Figure 16.21. The AOs on the two atoms that give rise to the MOs are shown on the right side of the diagram, with the size of the orbital proportional to its coefficient in the MO. The $2s$ electrons are almost completely localized on the F atom. The 1π electrons are completely localized on the F atom because the $2p_x$ and $2p_y$ orbitals on F have a zero net overlap with the $1s$ orbital on H. Electrons localized on a single atom are referred to as nonbonding electrons. The admixture of s and p in the 4σ and $5\sigma^*$ MOs changes the electron distribution in the HF molecule somewhat when compared with a homonuclear diatomic molecule. The 4σ MO takes on more antibonding character and the $5\sigma^*$ MO takes on more bonding character. Note that the total bond order is one because the 3σ MO is largely localized on the F atom, the 4σ MO is not totally bonding, and the 1π MOs are completely localized on the F atom. The MO energy diagram depicts the MOs in terms of their constituent AOs. MOs 2 through 4 obtained by forming the sum $\sigma_j(1) = \Sigma_i c_{ij}\phi_i(1)$ are shown in Figure 16.22.

FIGURE 16.22
The 3σ, 4σ, and 1π MOs for HF are shown from left to right.

16.12 The Molecular Electrostatic Potential

The charge on an atom in a molecule is not a quantum mechanical observable and, consequently, atomic charges cannot be assigned uniquely. However, we know that the electron charge is not uniformly distributed in a polar molecule. For example, the region around the oxygen atom in H_2O has a net negative charge, whereas the region around the hydrogen atoms has a net positive charge. How can this nonuniform charge distribution be discussed without assigning charges to the atoms in the molecule? To do so, we introduce the **molecular electrostatic potential,** which is the electrical potential felt by a test charge near a molecule.

The molecular electrostatic potential must be calculated numerically using the Hartree–Fock method or other numerical methods because the Schrödinger equation can only be solved exactly for a one-electron system. To visualize the polarity in a molecule, it is convenient to display a contour of constant electron density around the molecule and then display the values of the molecular electrostatic potential on the density contour using a color scale, as shown for HF in Figure 16.23. Negative values of the electrostatic potential, shown in red, are found near atoms to which electron charge transfer occurs. For HF, this is the region around the fluorine atom. Positive values of the molecular electrostatic potential, shown in blue, are found around atoms from which electron transfer occurs, as for the hydrogen atom in HF.

The calculated molecular electrostatic potential function identifies regions of a molecule that are either electron rich or depleted in electrons. We use this function to predict regions of a molecule that are susceptible to nucleophilic or electrophilic attack as in enzyme–substrate reactions.

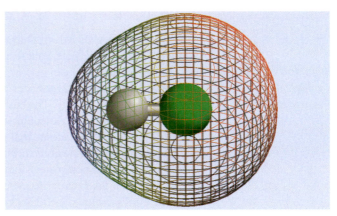

FIGURE 16.23
The grid shows a surface of constant electron density for the HF molecule. The fluorine atom is shown in green. The color shading on the grid indicates the value of the molecular electrostatic potential. Red and blue correspond to negative and positive values, respectively.

Vocabulary

antibonding orbital
antisymmetric (wave function)
atomic orbital (AO)
basis set
bond energy
bonding orbital
bond order
Born-Oppenheimer approximation

core electrons
delocalization
g symmetry
in-phase AO
LCAO-MO model
localization
minimal basis set
molecular configuration
molecular electrostatic potential

molecular orbital (MO)
molecular orbital energy diagram
molecular orbital (MO) model
molecular wave function
out-of-phase AO
overlap integral
π symmetry
σ symmetry

s–p mixing
symmetric (wave function)
symmetry operation
u symmetry
valence bond model
valence electrons
variational parameter

Questions on Concepts

Q16.1 Justify the Born–Oppenheimer approximation based on vibrational frequencies and electron relaxation times.

Q16.2 Why are the magnitudes of the coefficients c_a and c_b in the H_2^+ wave functions ψ_g and ψ_u equal?

Q16.3 If there is a node in ψ_u, is the electron in this wave function really delocalized? How does it get from one side of the node to the other?

Q16.4 For the case of two H1s AOs, the value of the overlap integral S_{ab} is never exactly zero. Explain this statement.

Q16.5 Why can you conclude that the energy of the antibonding MO in H_2^+ is raised more than the energy of the bonding MO is lowered?

Q16.6 Using Figures 16.8 and 16.9, explain why $\Delta\psi_g^2 < 0$ and $\Delta\psi_u^2 > 0$ outside of the bonding region of H_2^+.

Q16.7 Why is it necessary to include some ionic character in the wave function for the ground state of H_2 to better approximate the true wave function?

Q16.8 How can you conclude that the molecular wave function for H_2 in the molecular orbital model given in Equation (16.10) does not describe dissociation correctly?

Q16.9 What is the difference between a molecular orbital and a molecular wave function?

Q16.10 The following images show contours of constant electron density for H_2 calculated by accurate methods. The values of electron density are (a) 0.10, (b) 0.15, (c) 0.20, (d) 0.25, and (e) 0.30 electron/a_0^3.

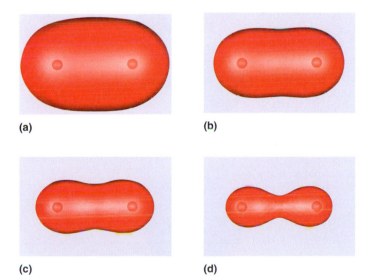

(a)   **(b)**

(c)   **(d)**

(e)

a. Explain why the apparent size of the H_2 molecule as approximated by the volume inside the contour varies in the sequence (a)–(e).

b. Notice the neck that forms between the two hydrogen atoms in contours (c) and (d). What does neck formation tell you about the relative density in the bonding region and in the region near the nuclei?

c. Explain the shape of the contours in image (e) by comparing this image with Figures 16.8 and 16.9.

d. Estimate the electron density in the bonding region midway between the H atoms by estimating the value of the electron density at which the neck disappears.

Q16.11 What is the justification for saying that, in expanding MOs in terms of AOs, the equality $\sigma_j(1) = \Sigma_i c_{ij} \phi_i(1)$ can in principle be satisfied?

Q16.12 Distinguish between the following concepts used to describe chemical bond formation: basis set, minimal basis set, atomic orbital, molecular orbital, and molecular wave function.

Q16.13 Give examples of AOs for which the overlap reaches its maximum value only as the internuclear separation approaches zero in a diatomic molecule. Also give examples of AOs for which the overlap goes through a maximum value and then decreases as the internuclear separation approaches zero.

Q16.14 Does the total energy of a molecule rise or fall when an electron is put in an antibonding orbital?

Q16.15 Are the coefficients of the AOs of different atoms in MOs responsible for bonding in an ionically bonded diatomic molecule similar or different in magnitude?

Q16.16 Explain why s–p mixing is more important in Li_2 than in F_2.

Q16.17 Why are MOs on heteronuclear diatomics not labeled with g and u subscripts?

Q16.18 Explain how the amplitude of the $1\sigma_g$ MOs midway between the atoms differs in H_2 and F_2.

Q16.19 Why do we neglect the bond length in He_2 when discussing the trends shown in Figure 16.20?

Q16.20 Consider the molecular electrostatic potential map for the LiH molecule shown here. Is the hydrogen atom (shown as a white sphere) an electron acceptor or an electron donor in this molecule?

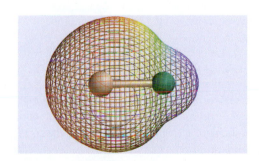

Q16.21   Consider the molecular electrostatic potential map for the BeH_2 molecule shown here. Is the hydrogen atom (shown as a white sphere) an electron acceptor or an electron donor in this molecule?

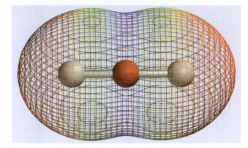

Q16.22   Consider the molecular electrostatic potential map for the BH_3 molecule shown here. Is the hydrogen atom (shown as a white sphere) an electron acceptor or an electron donor in this molecule?

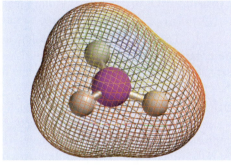

Q16.23   Consider the molecular electrostatic potential map for the NH_3 molecule shown here. Is the hydrogen atom (shown as a white sphere) an electron acceptor or an electron donor in this molecule?

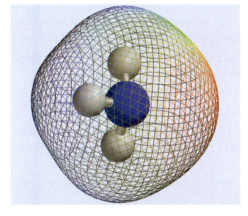

Q16.24   Consider the molecular electrostatic potential map for the H_2O molecule shown here. Is the hydrogen atom (shown as a white sphere) an electron acceptor or an electron donor in this molecule?

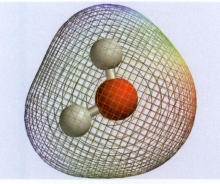

Q16.25   The molecular electrostatic potential maps for LiH and HF are shown here. Does the apparent size of the hydrogen atom (shown as a white sphere) tell you whether it is an electron acceptor or an electron donor in these molecules?

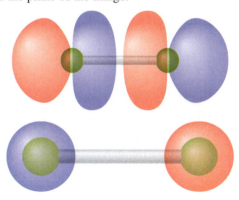

Q16.26   Identify the molecular orbitals for F_2 in the images shown here in terms of the two designations discussed in Section 16.9. The molecular axis is the z axis, and the y axis is tilted slightly out of the plane of the image.

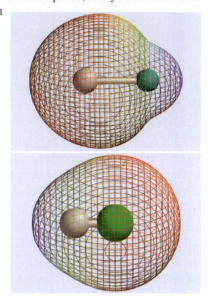

Q16.27   See Question Q16.26.

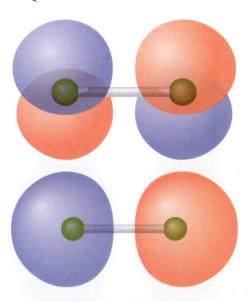

Q16.28 See Question Q16.26.

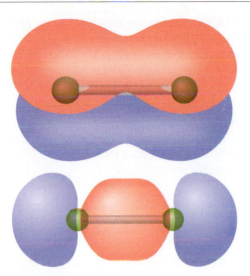

Q16.29 See Question Q16.26.

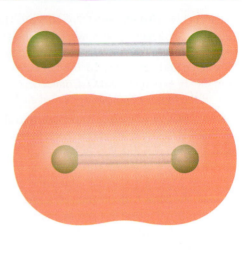

Problems

Problem numbers in **RED** indicate that the solution to the problem is given in the *Student Solutions Manual*.

P16.1 Follow the procedure outlined in Section 16.2 to determine c_u in Equation (16.3).

P16.2 Using ζ as a variational parameter in the normalized function $\psi_{H1s} = 1/\sqrt{\pi} \, (\zeta/a_0)^{3/2} e^{-\zeta r/a_0}$ allows one to vary the size of the orbital. Show this by calculating the probability of finding the electron inside a sphere of radius $2a_0$ for different values of ζ using the standard integral

$$\int x^2 e^{-ax} dx = -e^{-ax}\left(\frac{2}{a^3} + 2\frac{x}{a^2} + \frac{x^2}{a}\right)$$

   a. Obtain an expression for the probability as a function of ζ.

   b. Evaluate the probability for $\zeta = 1, 2,$ and 3.

P16.3 By evaluating the appropriate integral, show that the normalization constant for the H_2 VB wave function is $N = 1/\sqrt{2 + 2S_{ab}^2}$.

P16.4 Although He_2 is not a stable molecule, He_2^+ is stable. Explain this difference using the molecular configurations and bond order of the two species

P16.5 The bond dissociation energies of the species NO, CF^-, and CF^+ follow the trend $CF^+ > NO > CF^-$. Explain this trend using MO theory.

P16.6 What is the electron configuration corresponding to O_2, O_2^-, and O_2^+? What do you expect the relative order of bond strength to be for these species? Which, if any, have unpaired electrons?

P16.7 The ionization energy of CO is greater than that of NO. Explain this difference based on the electron configuration of these two molecules.

P16.8 Write the molecular configurations of the following homonuclear diatomic molecules.

   a. N_2^+ **b.** N_2^- **c.** O_2^+ **d.** O_2^-

P16.9 Calculate the bond order in each of the following species. Which of the species in parts (a) through (d) do you expect to have the shorter bond length?

   a. Li_2 or Li_2^+ **b.** C_2 or C_2^+

   c. O_2 or O_2^+ `**d.** F_2 or F_2^-

P16.10 Predict the bond order in the following species:

   a. N_2^+ **d.** H_2^-

   b. Li_2^+ **e.** C_2^+

   c. O_2^-

P16.11 An electron in the highest occupied MO in the following molecules is excited to the lowest unoccupied MO. For each molecule, specify whether the bond length increases or decreases. Explain your reasoning.

   a. Li_2 **b.** N_2 **c.** Be_2

P16.12 Make a sketch of the highest occupied molecular orbital (HOMO) for the following species:

   a. N_2^+ **b.** Li_2^+ **c.** O_2 **d.** H_2^- **e.** C_2^+

P16.13 Consider the molecules F_2 and F_2^+.

   a. Write the molecular configuration for each molecule

   b. Specify the bond order for each molecule.

   c. Are either of these molecules paramagnetic?

   d. Which of the molecules has the greater bond strength?

   e. Which of the molecules has the longer bond length?

P16.14 Calculate the bond order in each of the following species. Predict which of the two species in the following pairs has the higher vibrational frequency:

   a. Li_2 or Li_2^+ **b.** C_2 or C_2^+

   c. O_2 or O_2^+ **d.** F_2 or F_2^-

P16.15 Write the molecular configurations of the following heteronuclear diatomic molecules.

   a. CN   **b.** OH   **c.** BF   **d.** NH

P16.16 Arrange the following in terms of decreasing bond energy and bond length: O_2^+, O_2, O_2^-, and O_2^{2-}.

P16.17 Sketch a molecular orbital energy diagram for CO and place the electrons in the levels appropriate for the ground state. The AO ionization energies are $O2s$: 32.3 eV; $O2p$:

15.8 eV; C2s: 19.4 eV; and C2p: 10.9 eV. The MO energies follow the sequence (from lowest to highest) $1\sigma, 2\sigma, 3\sigma, 4\sigma$, $1\pi, 5\sigma, 2\pi, 6\sigma$. Assume that the $1s$ orbital need not be considered and define the 1σ orbital as originating primarily from the $2s$ AOs on C and O. Connect each MO level with the level of the major contributing AO on each atom.

P16.18 Images of molecular orbitals for LiH calculated using the minimal basis set are shown here. In these images, the smaller atom is H. The H1s AO has a lower energy than the Li2s AO. The energy of the MOs is (left to right) -63.9, -7.92, and $+2.14$ eV. Make a molecular orbital diagram for this molecule, associate the MOs with the images, and designate the MOs in the images shown here as filled or empty. Which MO is the HOMO? Which MO is the LUMO? Do you expect the dipole moment in this molecule to have the negative end on H or Li?

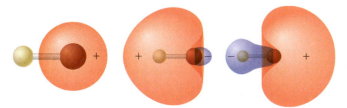

P16.19 Explain the difference in the appearance of the MOs in Problem P16.18 with those for HF. Based on the MO energies, do you expect LiH$^+$ to be stable? Do you expect LiH$^-$ to be stable?

P16.20 A surface displaying a contour of the total charge density in LiH is shown here. What is the relationship between this surface and the MOs displayed in Problem P16.18 Why does this surface closely resemble one of the MOs?

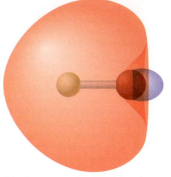

P16.21 Sketch the molecular orbital energy diagram for the radical OH based on what you know about the corresponding diagram for HF. How will the diagrams differ? Characterize the HOMO and LUMO as antibonding, bonding, or nonbonding.

P16.22 A Hartree–Fock calculation using the minimal basis set of the $1s$, $2s$, $2p_x$, $2p_y$, and $2p_z$ AOs on each of N and O generated the energy eigenvalues and AO coefficients listed in the following table:

MO	ε(eV)	c_{N1s}	c_{N2s}	c_{N2p_z}	c_{N2p_x}	c_{N2p_y}	c_{O1s}	c_{O2s}	c_{O2p_z}	c_{O2p_x}	c_{O2p_y}
3	-41.1	-0.13	$+0.39$	$+0.18$	0	0	-0.20	$+0.70$	$+0.18$	0	0
4	-24.2	-0.20	0.81	-0.06	0	0	0.16	-0.71	-0.30	0	0
5	-18.5	0	0	0	0	0.70	0	0	0	0	0.59
6	-15.2	$+0.09$	-0.46	$+0.60$	0	0	$+0.05$	-0.25	-0.60	0	0
7	-15.0	0	0	0	0.49	0	0	0	0	0.78	0
8	-9.25	0	0	0	0	0.83	0	0	0	0	-0.74

a. Designate the MOs in the table as σ or π symmetry and as bonding or antibonding. Assign the MOs to the following images, in which the O atom is red. The molecular axis is the z axis. In doing so, keep the following criteria in mind:

- If AOs of the same type on the two atoms have the same sign, they are in phase; if they have the opposite sign, they are out of phase.

- The energy increases with the number of nodes.

- If the coefficients of the s and p_z AOs are zero, the MO has π symmetry.

- If the coefficients of the p_x and p_y AOs are zero, the MO has σ symmetry.

b. Explain why MOs 6 and 7 do not have the same energy. Why is the energy of MO 6 lower than that of MO 7?

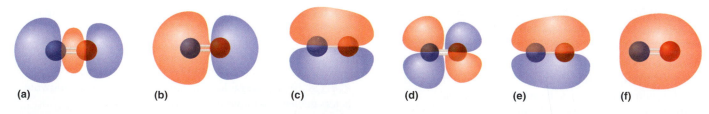

(a)   (b)   (c)   (d)   (e)   (f)

Web-Based Simulations, Animations, and Problems

W16.1 Two atomic orbitals are combined to form two molecular orbitals. The energy levels of the molecular orbitals and the coefficients of the atomic orbitals in each MO are cal- culated by varying the relative energy of the AOs and the over- lap, S_{12}, using sliders.

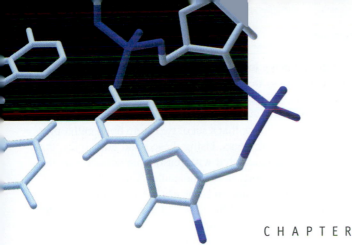

CHAPTER

17

Molecular Structure and Energy Levels for Polyatomic Molecules

For diatomic molecules, the only structural element is the bond length, whereas in polyatomic molecules, both bond lengths and bond angles determine the energy of the molecule. In this chapter, we discuss both localized and delocalized bonding models that enable the structures of small molecules to be predicted.

17.1 Lewis Structures and the VSEPR Model

In the previous chapter, we discussed chemical bonding and the electronic structure of diatomic molecules. Molecules with more than two atoms introduce a new aspect to our discussion of chemical bonding, namely, bond angles. In this chapter, the discussion of bonding is expanded to include the structure of small molecules. This will allow us to answer questions such as "Why is the bond angle 104.5° in H_2O and 92.2° in H_2S?" One way to answer this question is to say that the angles 104.5° and 92.2° in H_2O and in H_2S minimize the total energy of these molecules. This statement is correct, but it does not really answer the question. Numerical quantum mechanical calculations of bond angles are generally in very good agreement with experimentally determined values. This result confirms that the approximations made in the calculation are valid and gives confidence that bond angles in molecules for which there are no data can be calculated. However, these numerical calculations do not provide an understanding about *why* a bond angle of 104.5° minimizes the energy for H_2O, whereas a bond angle of 92.2° minimizes the energy for H_2S.

Qualitative theoretical models that address how the bond angle in a class of molecules—such as H_2X with X equal to O, S, Se, or Te—depends on X are useful for this purpose, because they can be applied to a class of molecules and, therefore, have predictive power. Gaining a qualitative understanding of why small molecules have an articular structure is the first goal of this chapter.

Molecular structure is addressed from two different vantage points. The significant divide between these points of view is in their description of the electrons in a molecule as being localized, as in the valence bond (VB) model, or delocalized, as in the molecular orbital (MO) model. As discussed in Chapter 16, MO theory is based on one-electron orbitals that are delocalized over the entire molecule. By contrast, a **Lewis structure** represents molecular fluorine as $:\ddot{F} - \ddot{F}:$, which is a description in terms of localized bonds and lone pairs. These two viewpoints seem to be irreconcilable at first glance.

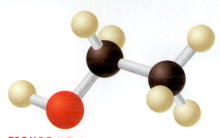

FIGURE 17.1

Ethanol depicted in the form of a ball-and-stick model.

We first discuss molecular structure using **localized bonding models.** We do so because there is a long tradition in chemistry of describing chemical bonds in terms of the interaction between neighboring atoms. A great deal was known about the thermochemical properties, stoichiometry, and structure of molecules before the advent of quantum mechanics. For instance, scientists knew that a set of two atom bond enthalpies could be extracted from experimental measurements. Using these bond enthalpies, the enthalpy of formation of molecules can be calculated with reasonable accuracy as the sum of the enthalpies of all the individual bonds in the molecule. Similarly, the bond length between two specific atoms, O—H, is found to be nearly the same in many different compounds, and the characteristic vibrational frequency of a group such as —OH is largely independent of the composition of the rest of the molecule. Results such as these give strong support for the idea that a molecule can be described by a set of coupled, but nearly independent, chemical bonds between adjacent molecules. The molecule can be assembled by linking these chemical bonds.

Figure 17.1 shows how a structural formula is used to describe ethanol. This structural formula provides a pictorial statement of a localized bonding model. However, a picture like this raises a number of questions. How can the line connecting two bonded atoms be described using the language of quantum mechanics? Localized bonding models imply that electrons are constrained within certain boundaries. However, we know from studying the particle in the box, the hydrogen atom, and many-electron atoms that the localizing of electrons results in high energy costs. We also know that it is not possible to distinguish one electron from another, so does it make sense to assign some electrons in F_2 to lone pairs and others to the bond? Quantum mechanics seems better suited to a delocalized picture, with orbitals extending over the whole molecule, rather than a localized picture. Yet, the preceding discussion provides ample evidence for a local model of chemical bonding. Which of these models is "correct"? It can be shown (see Engel and Reid's *Physical Chemistry*, Section 25.6) that both the localized and delocalized models of chemical bonding and molecular structure are equally valid and useful, but that no experiment can answer this question, because the model is only a useful construct, and is not an observable in quantum mechanics.

A useful place to start a discussion of localized bonding is with Lewis structures. Lewis structures emphasize the pairing of electrons as the basis for chemical bond formation. Bonds are shown as connecting lines, and electrons not involved in the bonds are indicated by dots. Lewis structures for a few representative small molecules are shown here:

$$H—H \qquad H—\overset{..}{\underset{..}{C}l}: \qquad \begin{matrix} H & & H \\ & \diagdown \!\! \diagup & \\ & C & \\ & \diagup \!\! \diagdown & \\ H & & H \end{matrix} \qquad H—\overset{..}{N}—H \qquad :\overset{..}{\underset{..}{O}}—H$$
$$\qquad\qquad\qquad\qquad\qquad\qquad\qquad\qquad\qquad\qquad\ \ | \qquad\qquad\qquad | $$
$$\qquad\qquad\qquad\qquad\qquad\qquad\qquad\qquad\qquad\qquad H \qquad\qquad\quad H$$

Lewis structures are useful in understanding the stoichiometry of a molecule and in emphasizing the importance of nonbonding electron pairs, also called **lone pairs.** Lewis structures are less useful in predicting the geometrical structure of molecules.

However, the **valence shell electron pair repulsion (VSEPR) model** provides a qualitative theoretical understanding of molecular structures using the Lewis concepts of localized bonds and lone pairs. The basic assumptions of the VSEPR model can be summarized in the following statements about a central atom bonded to several atomic ligands:

- The ligands and lone pairs around a central atom act as if they repel one another. They adopt an arrangement in three dimensions that maximizes their angular separation.
- A lone pair occupies more angular space than a ligand.
- The amount of angular space occupied by a ligand increases with its electronegativity and decreases as the electronegativity of the central atom increases.
- A multiply bonded ligand occupies more angular space than a singly bonded ligand.

As Figure 17.2 shows, the structure of a large number of molecules can be understood using the VSEPR model. For example, the decrease in the bond angle in the se-

FIGURE 17.2
Examples of correctly predicted molecular shapes using the VSEPR model.

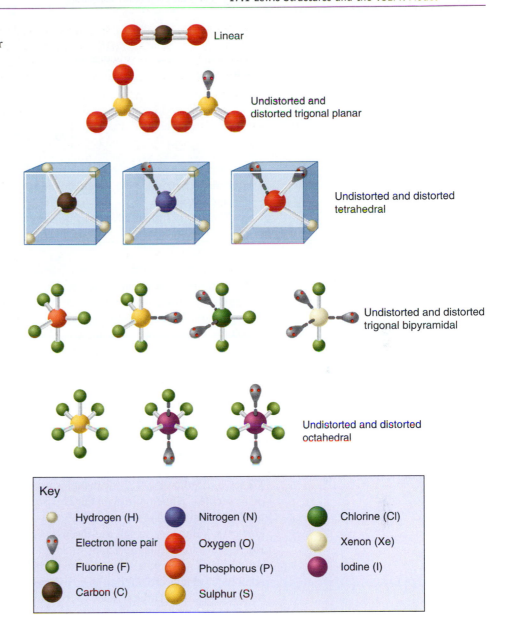

Linear

Undistorted and distorted trigonal planar

Undistorted and distorted tetrahedral

Undistorted and distorted trigonal bipyramidal

Undistorted and distorted octahedral

Key

Hydrogen (H)	Nitrogen (N)	Chlorine (Cl)
Electron lone pair	Oxygen (O)	Xenon (Xe)
Fluorine (F)	Phosphorus (P)	Iodine (I)
Carbon (C)	Sulphur (S)	

quence 109.5°, 107.3°, and 104.5° for the molecules CH_4, NH_3, and H_2O can be explained on the basis of the greater angular space occupied by lone pairs than by ligands. The tendency of lone pairs to maximize their angular separation also explains why XeF_2 is linear, SO_2 is bent, and IF_4^- is planar. However, in some cases the model is inapplicable or does not predict the correct structure. For instance, a radical, such as CH_3, that has an unpaired electron is planar and, therefore, does not fit into the VSEPR model. Alkaline earth dihalides such as CaF_2 and $SrCl_2$ are angular rather than linear as would be predicted by the model. SeF_6^{2-} and $TeCl_6^{2-}$ are octahedral even though they each have a lone pair in addition to the six ligands. This result indicates that lone pairs do not always exert an influence on molecular shape. In addition, lone pairs do not play as strong a role as the model suggests in transition metal complexes.

EXAMPLE PROBLEM 17.1

Using the VSEPR model, predict the shape of NO_3^- and OCl_2.

Solution

The following Lewis structure shows one of the three resonance structures of the nitrate ion:

$$\left[\begin{array}{c} :\overset{\displaystyle :O:}{\underset{\displaystyle :\underset{..}{O}.\quad .\underset{..}{O}:}{\overset{\|}{N}}} \end{array} \right]^{-}$$

Because the central nitrogen atom has no lone pairs and the three oxygens are equivalent, the nitrate ion should be planar with a 120° bond angle. This is the observed structure.

The Lewis structure for OCl_2 is

$$:\overset{..}{\underset{..}{Cl}}.\overset{\displaystyle :\overset{..}{O}.}{\diagdown}\quad.\overset{..}{\underset{..}{Cl}}:$$

The central oxygen atom is surrounded by two ligands and two lone pairs. The ligands and lone pairs are described by a distorted tetrahedral arrangement, leading to a bent molecule. The bond angle should be near the tetrahedral angle of 109.5°. The observed bond angle is 111°.

17.2 Describing Localized Bonds Using Hybridization for Methane, Ethene, and Ethyne

We have seen that the VSEPR model is useful in predicting the shape of a wide variety of molecules. Although the model is based on quantum mechanical results, the rules used in its application do not specifically use the vocabulary of quantum mechanics. However, VB theory, which was discussed in Chapter 16, does use the concept of localized orbitals to explain molecular structure. In the VB model, AOs on the same atom are combined to generate a set of directed orbitals in a process called **hybridization.** The combined orbitals are referred to as **hybrid orbitals.** In keeping with a local picture of bonding, the hybrid orbitals need to contribute independently to the electron density and to the energy of the molecule to the maximum extent possible, because this allows the assembly of the molecule out of separate and largely independent parts. This requires the set of hybrid orbitals to be orthogonal.

How is hybridization used to describe molecular structure? Consider the sequence of molecules methane, ethene, and ethyne. From previous chemistry courses, you know that carbon in these molecules is characterized by the *sp³, sp²,* **and** *sp* **hybridizations,** respectively. What is the functional form associated with these different hybridizations? We construct the hybrid orbitals for ethene to illustrate the procedure.

To model the three σ bonds in ethene, the carbon AOs are hybridized to the configuration $1s^2 2p_y^1 (\psi_a)^1 (\psi_b)^1 (\psi_c)^1$ rather than to the configuration $1s^2 2s^2 2p^2$, which is appropriate for an isolated carbon atom. The orbitals ψ_a, ψ_b, and ψ_c are the wave functions that are used in a valence bond model for the three σ bonds in ethene. We next formulate ψ_a, ψ_b, and ψ_c in terms of the $2s$, $2p_x$, and $2p_z$ AOs on carbon.

The three sp^2-hybrid orbitals ψ_a, ψ_b, and ψ_c must satisfy the geometry shown schematically in Figure 17.3. They lie in the x–z plane and are oriented at 120° to one another. The appropriate linear combination of carbon AOs is

$$\psi_a = c_1 \phi_{2p_z} + c_2 \phi_{2s} + c_3 \phi_{2p_x}$$

$$\psi_b = c_4 \phi_{2p_z} + c_5 \phi_{2s} + c_6 \phi_{2p_x} \qquad (17.1)$$

$$\psi_c = c_7 \phi_{2p_z} + c_8 \phi_{2s} + c_9 \phi_{2p_x}$$

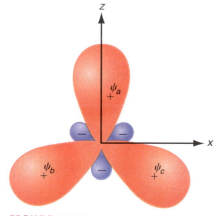

FIGURE 17.3

Geometry of the sp^2-hybrid orbitals used in Equation (17.1). In this and in most of the figures in this chapter, we use a "slimmed down" picture of hybrid orbitals to separate individual orbitals. A more correct form for s–p hybrid orbitals is shown in Figure 17.5.

How can c_1 through c_9 be determined? A few aspects of the chosen geometry simplify the task of determining the coefficients. Because the $2s$ orbital is spherically symmetrical, it will contribute equally to each of the hybrid orbitals. Therefore, $c_2 = c_5 = c_8$. These three coefficients must satisfy the equation $\sum_i (c_{2si})^2 = 1$, where the subscript $2s$ refers to the $2s$ AO. This equation states that all of the $2s$ contributions to the hybrid orbitals must be accounted for. We choose $c_2 < 0$ in the preceding equations to make the $2s$ orbital,

$$\psi_{200}(r) = \frac{1}{\sqrt{32\pi}} \left(\frac{1}{a_0}\right)^{3/2} \left(2 - \frac{r}{a_0}\right) e^{-r/2a_0}$$

have a positive amplitude in the bonding region. (For graphs of the $2s$ AO amplitude versus r, see Figures 15.4 and 15.5.) Therefore, we conclude that

$$c_2 = c_5 = c_8 = -\frac{1}{\sqrt{3}}$$

From the orientation of the orbitals seen in Figure 17.3, $c_3 = 0$ because ψ_a is oriented on the z axis. Because the hybrid orbital points along the positive z axis, $c_1 > 0$. We can also conclude that $c_4 = c_7$, that both are negative, and that $-c_6 = c_9$ with $c_9 > 0$. Based on these considerations, Equation (17.1) simplifies to

$$\psi_a = c_1 \phi_{2p_z} - \frac{1}{\sqrt{3}} \phi_{2s}$$

$$\psi_b = c_4 \phi_{2p_z} - \frac{1}{\sqrt{3}} \phi_{2s} - c_6 \phi_{2p_x} \qquad (17.2)$$

$$\psi_c = c_4 \phi_{2p_z} - \frac{1}{\sqrt{3}} \phi_{2s} + c_6 \phi_{2p_x}$$

As shown in Example Problem 17.2, the remaining unknown coefficients can be determined by normalizing and orthogonalizing ψ_a, ψ_b, and ψ_c.

EXAMPLE PROBLEM 17.2

Determine the three unknown coefficients in Equation (17.2) by normalizing and orthogonalizing the hybrid orbitals.

Solution

We first normalize ψ_a. Terms such as $\int \phi_{2p_y}^* \phi_{2p_z} d\tau$ and $\int \phi_{2s}^* \phi_{2p_y} d\tau$ do not appear in the following equations because all of the AOs are orthogonal to one another. Evaluation of the integrals is simplified because the individual AOs are normalized.

$$\int \psi_a^* \psi_a d\tau = (c_1)^2 \int \phi_{2p_z}^* \phi_{2p_z} d\tau + \left(-\frac{1}{\sqrt{3}}\right)^2 \int \phi_{2s}^* \phi_{2s} d\tau = 1$$

$$= (c_1)^2 + \frac{1}{3} = 1$$

which tells us that $c_1 = \sqrt{2/3}$. Orthogonalizing ψ_a and ψ_b, we obtain

$$\int \psi_a^* \psi_b d\tau = c_4 \sqrt{\frac{2}{3}} \int \phi_{2p_z}^* \phi_{2p_z} d\tau + \left(-\frac{1}{\sqrt{3}}\right)^2 \int \phi_{2s}^* \phi_{2s} d\tau = 0$$

$$= c_4 \sqrt{\frac{2}{3}} + \frac{1}{3} = 0 \quad \text{and} \quad c_4 = -\sqrt{\frac{1}{6}}$$

Normalizing ψ_b, we obtain

$$\int \psi_b^* \psi_b d\tau = \left(-\sqrt{\frac{1}{6}}\right)^2 \int \phi_{2p_z}^* \phi_{2p_z} d\tau + \left(-\frac{1}{\sqrt{3}}\right)^2 \int \phi_{2s}^* \phi_{2s} d\tau + (-c_6)^2 \int \phi_{2p_x}^* \phi_{2p_x} d\tau$$

$$= (c_6)^2 + \frac{1}{3} + \frac{1}{6} = 1 \quad \text{and} \quad c_6 = +\sqrt{\frac{1}{2}}$$

We have chosen the positive root so that the coefficient of ϕ_{2px} in ψ_b is negative. Using these results, the normalized and orthogonal set of hybrid orbitals is

$$\psi_a = \sqrt{\frac{2}{3}} \phi_{2p_z} - \frac{1}{\sqrt{3}} \phi_{2s}$$

$$\psi_b = -\sqrt{\frac{1}{6}} \phi_{2p_z} - \frac{1}{\sqrt{3}} \phi_{2s} - \sqrt{\frac{1}{2}} \phi_{2p_x}$$

$$\psi_c = -\sqrt{\frac{1}{6}} \phi_{2p_z} - \frac{1}{\sqrt{3}} \phi_{2s} + \sqrt{\frac{1}{2}} \phi_{2p_x}$$

Convince yourself that ψ_c is normalized and orthogonal to ψ_a and ψ_b.

How can the $2s$ and $2p$ character of the hybrids be quantified? Because the sum of the squares of the coefficients for each hybrid orbital equals one, the p and s character of the hybrid orbital can be calculated. The fraction of $2p$ character in ψ_b is $\left[\left(\sqrt{1/6}\right)^2 + \left(\sqrt{1/2}\right)^2\right]/1 = 1/6 + 1/2 = 2/3$. The fraction of $2s$ character is $1/3$. Because the ratio of the $2p$ to $2s$ character is 2:1, one refers to sp^2 hybridization.

By following the procedure outlined earlier, it can be shown that the set of orthonormal sp-hybrid orbitals that are oriented 180° apart is

$$\psi_a = \frac{1}{\sqrt{2}}(-\phi_{2s} + \phi_{2p_z})$$

$$\psi_b = \frac{1}{\sqrt{2}}(-\phi_{2s} - \phi_{2p_z})$$

(17.3)

and that the set of tetrahedrally oriented orthonormal hybrid orbitals for sp^3 hybridization that are oriented 109.4° apart is

$$\psi_a = \frac{1}{2}(-\phi_{2s} + \phi_{2p_x} + \phi_{2p_y} + \phi_{2p_z})$$

$$\psi_b = \frac{1}{2}(-\phi_{2s} - \phi_{2p_x} - \phi_{2p_y} + \phi_{2p_z})$$

$$\psi_c = \frac{1}{2}(-\phi_{2s} + \phi_{2p_x} - \phi_{2p_y} - \phi_{2p_z})$$

$$\psi_d = \frac{1}{2}(-\phi_{2s} - \phi_{2p_x} + \phi_{2p_y} - \phi_{2p_z})$$

(17.4)

By combining s and p orbitals, at most four hybrid orbitals can be generated. To describe bonding around a central atom with coordination numbers greater than four, d orbitals need to be included in forming the hybrids. Although hybrid orbitals with d character are not discussed here, the principles used in constructing them are the same as those outlined earlier.

TABLE 17.1   C—C Bond Types

Carbon—Carbon Bond Types	σ Bond Hybridization	s-to-p Ratio	Angle between Equivalent σ Bonds ($°$)	Carbon—Carbon Single Bond Length (pm)
$\ce{>C-C<}$	sp^3	1:3	109.4	154
$\ce{>C-C<}$	sp^2	1:2	120	146
$\ce{=C-C=}$	sp	1:1	180	138

The properties of C—C single bonds depend on the hybridization of the carbon atoms, as shown in Table 17.1. The most important conclusion that can be drawn from this table for the discussion in the next section is that increasing the *s* character in *s—p* hybrids increases the bond angle. Note also that the C—C single bond length becomes shorter as the *s* character of the hybridization increases, and that the C—C single bond energy increases as the *s* character of the hybridization increases.

17.3 Constructing Hybrid Orbitals for Nonequivalent Ligands

In the preceding section, the construction of hybrid orbitals for equivalent ligands was considered. However, in general, molecules contain nonequivalent ligands as well as nonbonding electron lone pairs. How can hybrid orbitals be constructed for such molecules if the bond angles are not known? By considering the experimentally determined structures of a wide variety of molecules, Henry Bent formulated the following guidelines:

- Central atoms that obey the octet rule can be classified into three structural types. Central atoms that are surrounded by a combination of four single bonds or electron pairs are to a first approximation described by a tetrahedral geometry and sp^3 hybridization. Central atoms that form one double bond and a combination of two single bonds or electron pairs are to a first approximation described by a trigonal geometry and sp^2 hybridization. Central atoms that form two double bonds or one triple bond and either a single bond or an electron pair are to a first approximation described by a linear geometry and sp hybridization.
- The presence of different ligands is taken into account by assigning a different hybridization to all nonequivalent ligands and lone pairs. The individual hybridization is determined by the electronegativity of each ligand. A nonbonding electron pair can be considered to be electropositive or, equivalently, to have a small electronegativity. Bent's rule states that atomic *s* character concentrates in hybrid orbitals directed toward electropositive ligands and that *p* character concentrates in hybrid orbitals directed toward electronegative ligands.

We now apply these guidelines to H_2O. The oxygen atom in H_2O is to a first approximation described by a tetrahedral geometry and sp^3 hybridization. However, because the H atoms are more electronegative than the electron pairs, the *p* character of the hybrid orbitals directed toward the hydrogen atoms will be greater than that of sp^3 hybridization. Because Table 17.1. shows that increasing the *p* character decreases the bond angle, Bent's rule says that the H—O—H bond angle will be less than 109.5°. Note that the effect of Bent's rule is the same as the effect of the VSEPR rules listed in Section 17.1. However, the hybridization model provides a basis for the rules.

Although useful in predicting bond angles, Bent's rule is not quantitative. To make it predictive, a method is needed to assign a hybridization to a specific combination of two atoms that is independent of the other atoms in the molecule. Several authors have developed methods that meet this need, for example, D. M. Root *et al.* [*J. American Chemical Society* 115 (1993), 4201–4209].

EXAMPLE PROBLEM 17.3

a. Use Bent's rule to estimate the change in the X—C—X bond angle in $\begin{smallmatrix} X \\ X \end{smallmatrix}C = O$ when going from H_2CO to F_2CO.

b. Use Bent's rule to estimate the deviation of the H—C—H bond angle in FCH_3 and $ClCH_3$ from 109.4°.

Solution

a. To first order, the carbon atom exhibits sp^2 hybridization. Because F is more electronegative than H, the hybridization of the C—F ligand will contain more p character than the C—H ligand. Therefore, the F—C—F bond angle will be smaller than the H—C—H bond angle.

b. For both FCH_3 and $ClCH_3$, H is more electropositive than the halogen atom so that the C—H bonds have greater s character than the C—halogen bond. This makes the H—C—H bond angle greater than 109.5° in both molecules.

17.4 Using Hybridization to Describe Chemical Bonding

By using the hybridization model to create local bonding orbitals, the concepts inherent in Lewis structures can be given a quantum mechanical basis. As an example, consider BeH_2, which is not observed as an isolated molecule because it forms a solid through polymerization of BeH_2 units stabilized by hydrogen bonds. We consider only a single BeH_2 unit. Be has the configuration $1s^2 2s^2 2p^0$, and because it has no unpaired electrons, it is not obvious how bonding to the H atoms can be explained in the Lewis model. Within the framework outlined earlier, the $2s$ and $2p$ orbitals are hybridized to create bonding hybrids on the Be atom. Because the bond angle is known to be 180°, two equivalent and orthogonal sp-hybrid orbitals are constructed as given by Equation (17.3). This allows Be to be described as $1s^2(\psi_a)^1(\psi_b)^1$. In this configuration, Be has two unpaired electrons and, therefore, the hybridized atom is divalent. The orbitals are depicted schematically in Figure 17.4. To make a connection to Lewis structures, the bonding electron pair is placed in the overlap region between the Be and H orbitals as indicated by the dots. In reality, the bonding electron density is distributed over the entire region in which the orbitals have a nonzero amplitude.

As a second example, Figure 17.5 is a valence bond hybridization picture that depicts bonding in ethene and ethane. For ethene, each carbon atom is promoted to the $1s^2 2p_y^1(\psi_a)^1(\psi_b)^1(\psi_c)^1$ configuration before forming four C—H σ bonds, a C—C σ bond, and a C—C π bond. The maximal overlap between the p orbitals to create a π_1 bond in ethene occurs when all atoms lie in the same plane. The double bond is made from one σ and one π bond. For ethane, each carbon atom is promoted to the $1s^2(\psi_a)^1(\psi_b)^1(\psi_c)^1(\psi_d)^1$ configuration before forming six C—H σ bonds and a C—C σ bond.

In closing this discussion of hybridization, it may be useful to emphasize the advantages of the model and to point out some of its shortcomings. The main advantage of hybridization is that it is an easily understandable model with considerable predictive power. It retains the main features of Lewis structures in describing local orthogonal bonds between adjacent atoms in terms of electron pairing, and it justifies Lewis structures in the language of quantum mechanics. Hybridization also provides a theoretical basis for the VSEPR rules through the construction of localized orbitals for bonding electrons and lone pairs.

Hybridization also offers more than a useful framework for understanding bond angles in molecules. Because the $2s$ AO is lower in energy than the $2p$ level in many-electron atoms, the electronegativity of a hybridized atom increases with increasing s

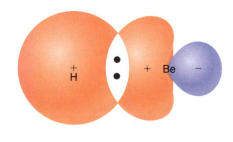

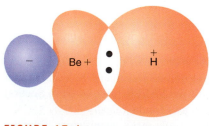

FIGURE 17.4

Bonding in BeH_2 using two sp-hybrid orbitals on Be. The two Be–H hybrid bonding orbitals are shown separately.

FIGURE 17.5
The top panel shows the arrangement of the hybrid orbitals for sp^2 and sp^3 carbon. The bottom panel shows a schematic depiction of bonding in ethene (left) and ethane (right) using hybrid bonding orbitals.

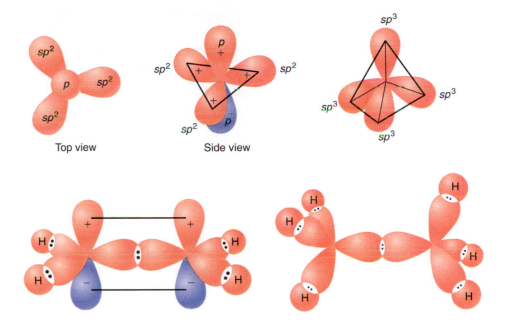

character. Therefore, the hybridization model predicts that an sp-hybridized carbon atom is more electronegative than an sp^3-hybridized carbon atom. Evidence for this effect can be seen in the observation that the positive end of the dipole moment in $N\equiv C-Cl$ is on the Cl atom. We conclude that the carbon atom in the cyanide group is more electronegative than a chlorine atom. Because increased s character leads to shorter bond lengths, and because shorter bonds generally have a greater bond strength, the hybridization model provides a correlation of s character and bond strength. These examples illustrate that the hybridization model has greater predictive power than the VSEPR model.

The hybridization model also has a few shortcomings. For known bond angles, the hybridization can be calculated as was done for ethane and H_2O. However, semiempirical prescriptions must be used to estimate the s and p character of a hybrid orbital for a molecule in the absence of structural information. It is also more straightforward to construct an appropriate hybridization for symmetric molecules such as methane than for molecules with electron lone pairs and several different ligands bonded to the central atom. Additionally, the depiction of bonding hybrids that is usually used (as in Figure 17.5) seems to imply that the electron density is highly concentrated along the bonding directions. This is not true as can be seen by looking at the realistic representation of hybrid orbitals for water in Figure 17.6.

Finally, the conceptual formalism used in creating hybrid orbitals assumes much more detail than can be verified by experiments. The individual steps of promoting electrons to unoccupied orbitals and then hybridizing them may be useful as a rationalization of the observed geometry, but these steps should not be taken literally.

17.5 Predicting Molecular Structure Using Molecular Orbital Theory

We now consider a **delocalized bonding model** of the chemical bond. MO theory approaches the structure of molecules quite differently than the VSEPR and hybridization local models of chemical bonding. The electrons involved in bonding are assumed to delocalize over the molecule. Each one-electron molecular orbital σ_j is expressed as a linear combination of atomic orbitals such as $\sigma_j(k) = \sum_i c_{ij} \phi_i(k)$, which refers to the jth molecular orbital for electron k. The many-electron wave function ψ is written as a Slater determinant in which the individual entries are the $\sigma_j(k)$.

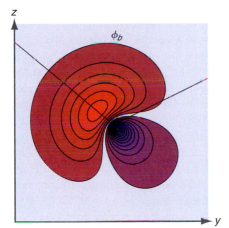

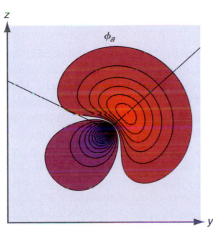

FIGURE 17.6
Directed hybrid bonding orbitals for H_2O. The black lines show the desired bond angle and orbital orientation. Red and blue contours correspond to the most positive and least positive values of the amplitude.

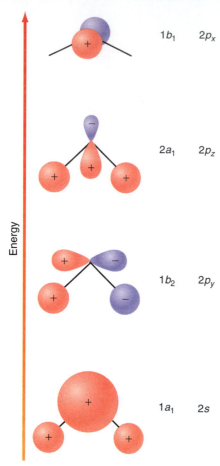

Energy →

$1b_1$	$2p_x$
$2a_1$	$2p_z$
$1b_2$	$2p_y$
$1a_1$	$2s$

FIGURE 17.7
The valence MOs occupied in the ground state of water are shown in order of increasing orbital energy. The MOs are depicted in terms of the AOs from which they are constructed. The second column gives the MO symmetry, and the third column lists the dominant AO orbital on the oxygen atom.

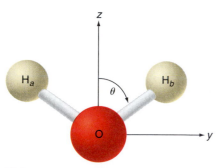

FIGURE 17.8
Coordinate system used to generate the hybrid orbitals on the oxygen atom that are suitable to describe the structure of H_2O.

In quantitative molecular orbital theory, structure emerges naturally as a result of solving the Schrödinger equation. The total energy of the molecule is minimized with respect to all parameters of the occupied MOs, such as the effective charge and the co-efficients c_{ij}, and with respect to the positions of all atoms in the molecule. The atomic positions for which the energy has its minimum value define the equilibrium structure. Although this concept can be formulated in a few sentences, carrying out this procedure is a complex exercise in numerical computing, hence our focus is on a more qualitative approach that conveys the spirit of molecular orbital theory, but which can be written down without extensive mathematics.

To illustrate this approach, we use qualitative MO theory to understand the bond angle in two triatomic molecules of the type H_2A, where A is one of the atoms in the sequence Be → O, and we also show that a qualitative picture of the optimal bond angle can be obtained by determining how the energy of the individual occupied molecular orbitals varies with the bond angle. In doing so, we assume that the total energy of the molecule is proportional to the sum of the orbital energies. This assumption can be justified, although we do not do so here. Experimentally, we know that BeH_2 is a linear molecule, and H_2O has a bond angle of 104.5°. How can this difference be explained using MO theory?

We begin by making educated guesses about the nature of the occupied MOs for these molecules. The basis set used here consists of the $1s$ orbitals on H and the $1s$, $2s$, $2p_x$, $2p_y$, and $2p_z$ orbitals on atom A. Seven MOs can be generated using these seven AOs. Water has 10 electrons, and four MOs accommodate the eight valence electrons. As in other textbooks, we assume that the $1s$ electrons on oxygen are localized and begin numbering the MOs with those generated from the $2s$ electrons on oxygen. Recall that the orbital energy increases with the number of nodes for the particle in the box, the harmonic oscillator, and the H atom. We also know that the lower the AO energies, the lower the MO energy will be. The MOs for water are shown in Figure 17.7, and each MO is depicted in terms of the AOs from which it is constructed. The relative MO orbital energies are discussed later. The MOs are labeled according to their symmetry with respect to a set of rotation and reflection operations that leaves the water molecule unchanged (see Engel and Reid's *Physical Chemistry,* Chapter 28). In the present context it is sufficient to think of these designations simply as labels. In the following discussion, we omit the two lowest energy MOs, which are linear combinations of the $1s$ oxygen AOs. Because the $2s$ AOs are lower in energy than the $2p$ AOs, the MO with no nodes designated $1a_1$ in Figure 17.7 is expected to have the lowest energy of all possible valence MOs. The next higher MOs involve $2p$ orbitals on the O atom.

The three $2p$ orbitals are differently oriented with respect to the plane containing the H atoms. As a result, the MOs that they generate are quite different in energy. Assume that the H_2A molecule lies in the y–z plane with the z axis bisecting the H—A—H angle as shown in Figure 17.8.

The $1b_2$ MO, generated using the $2p_y$ AO, and the $2a_1$ MO, generated using the $2p_z$ AO, each have no nodes in the O—H region and, therefore, have binding character. However, because each has one node, both MOs have a higher energy than the $1a_1$ MO. It turns out that the MO generated using the $2p_y$ AO has a lower orbital energy than that generated using the $2p_z$ AO. As depicted in Figures 17.9 and 17.10, some s–p mixing has been incorporated in the $2a_1$ and $3a_1$ MO generated from the $2s$ and $2p_z$ AO. Having discussed the MOs formed from the $2p_y$ and $2p_z$ AOs, we turn to the $2p_x$ AO. The $2p_x$ orbital has no net overlap with the H atoms and gives rise to the $1b_1$ nonbonding MO that is localized on the O atom. Because this MO is not stabilized through interaction with the H $1s$ AOs, it has the highest energy of all the MOs considered. Antibonding MOs that are higher in energy can be generated by combining out-of-phase AOs. Numerically calculated molecular orbitals are depicted in Figure 17.9.

The preceding discussion about the relative energy of the MOs is sufficient to allow us to draw the MO energy diagram shown in Figure 17.10. The MO energy levels in this figure are drawn for a particular bond angle near 105°, but the energy levels vary with 2θ, as shown in Figure 17.11, in what is known as a **Walsh correlation diagram.** You should make sure you understand the trends shown in this figure because the variation

FIGURE 17.9
The first five valence MOs for H_2O are depicted. The $1b_1$ and $3a_1$ MOs are the HOMO and LUMO, respectively. Note that the $1b_1$ MO is the AO corresponding to the nonbonding $2p_x$ electrons on oxygen.

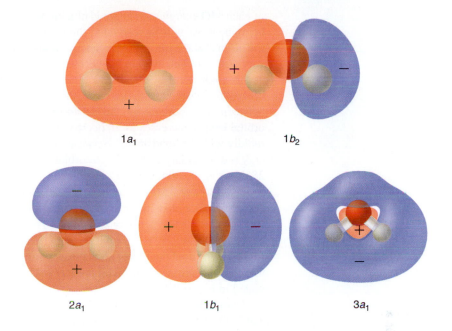

$1a_1$ $1b_2$

$2a_1$ $1b_1$ $3a_1$

FIGURE 17.10
Molecular orbital energy-level diagram for H_2O at its equilibrium geometry. To avoid clutter, minor AO contributions to the MOs and the $1a_1$ MO generated from the O1s AO are not shown.

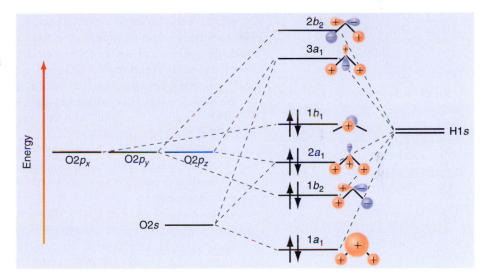

FIGURE 17.11
Schematic variation of the MO energies for water with bond angle. The symbols used on the left to describe the MOs are based on symmetry considerations and are valid for $2\theta < 180°$.

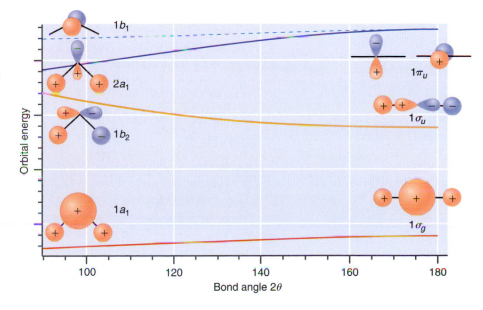

Bond angle 2θ

of each MO energy with angle is ultimately responsible for BeH_2 being linear and H_2O being bent. As will be seen later, the variation with 2θ of the overlap between the central atom A and hydrogen AOs determines the bond angle. How does the $1a_1$ MO energy vary with 2θ? The overlap between the s orbitals on A and H is independent of 2θ, but as this angle decreases from $180°$, the overlap between the H atoms increases. This stabilizes the molecule and, therefore, the $1a_1$ orbital energy falls. By contrast, the overlap between the $2p_y$ orbital and the H $1s$ orbitals is a maximum at $180°$ and therefore the $1b_2$ orbital energy increases as 2θ decreases. Which of these orbital energies changes more rapidly with the bond angle? Because the effect of the H—H overlap on the $1a_1$ MO energy is a secondary effect, the $1b_2$ orbital energy decreases more rapidly with increasing 2θ than the $1a_1$ orbital energy increases.

We now consider the $2a_1$ and $1b_1$ MO energies. The $2p_y$ and $2p_z$ orbitals are nonbonding and degenerate for a linear H_2A molecule. However, as 2θ decreases from $180°$, the $2p_z$ orbital has a net overlap with the H $1s$ AOs and has increasingly more bonding character. Therefore, the $2a_1$ MO energy decreases as 2θ decreases from $180°$. The $1b_1$ MO remains nonbonding as 2θ decreases from $180°$, but electron repulsion effects lead to a slight decrease of the orbital energy. These variations in the MO energies are depicted in Figure 17.11.

Using the MO energy diagram of Figure 17.10, let's consider the molecules BeH_2 and H_2O. BeH_2 has four valence electrons that are placed in the two lowest-lying MOs, $1a_1$ and $1b_2$. Because the $1b_2$ orbital energy decreases with increasing 2θ more than the $2a_1$ orbital energy increases, the total energy of the molecule is minimized if $2\theta = 180°$. This qualitative argument predicts that BeH_2, as well as any other four-valence electron molecule, is linear and has the valence electron configuration $(1\sigma_g)^2(1\sigma_u)^2$. Note that the description of H_2A in terms of σ MOs with g or u symmetry applies only to a linear molecule, whereas a description in terms of $1a_1$ and $1b_2$ applies to all bent molecules.

We now consider H_2O, which has eight valence electrons. In this case, the lowest four MOs are doubly occupied. At what angle is the total energy of the molecule minimized? For water, the decrease in the energy of the $1a_1$ and $2a_1$ MOs as 2θ decreases more than offsets the increase in energy for the $1b_2$ MO. Therefore, H_2O is bent rather than linear and has the valence electron configuration $(1a_1)^2(1b_2)^2(2a_1)^2(1b_1)^2$. The degree of bending depends on how rapidly the energy of the MOs changes with angle. Numerical calculations for water using this approach predict a bond angle that is very close to the experimental value of $104.5°$. These examples for BeH_2 and H_2O illustrate how qualitative MO theory can be used to predict bond angles.

These examples also illustrate the different approach that MO theory takes to predict the structure of molecules compared to the VSEPR or hybridization methods. The same procedure can be carried out to determine molecular shapes for large molecules. Although this procedure involves finding a minimum in an energy surface in a space whose dimension is equal to the number of bond distances and bond angles needed to define the molecular geometry, it is a process that can readily be implemented in numerical calculations. This method is a powerful tool for calculating the energy levels and molecular geometry for many molecules of interest to chemists.

EXAMPLE PROBLEM 17.4

Predict the equilibrium shape of H_3^+, LiH_2, and NH_2 using qualitative MO theory.

Solution

H_3^+ has two valence electrons and is bent as predicted by the variation of the $1a_1$ MO energy with angle shown in Figure 17.11. LiH_2 or any molecule of the type H_2A with four electrons is predicted to be linear. NH_2 has one electron fewer than H_2O, and using the same reasoning as for water, is bent.

Qualitative **molecular orbital theory** can be used to gain insight into chemical reactions. Recall from Chapter 16 that higher lying MOs extend farther out from the center of the molecule than do low-lying MOs. Therefore, the higher lying MOs will be the most involved MOs in chemical reactions between molecules. In MO theory, the **highest occupied molecular orbital (HOMO)** and **lowest unoccupied molecular orbital (LUMO)** play an important role in chemical reactions and are called **frontier orbitals.** Consider a reaction involving electron transfer between molecules A and B. The energies of the HOMO and LUMO orbitals are specific to each molecule and could be aligned in either of two ways as is indicated in Figure 17.12.

The degree to which MOs mix is proportional to $S^2/\Delta\varepsilon$, in which S is the overlap, and $\Delta\varepsilon$ is the difference between the MO energies on species A and B. Assuming that S is not very different for the orbitals being considered, the mixing is dominated by $\Delta\varepsilon$. For the relative energies of the orbitals on the left side of Figure 17.12, the energy separation between HOMO A and LUMO B is much smaller than that between LUMO A and HOMO B. This means that HOMO A and LUMO B mix more readily than LUMO A and HOMO B, which results in a finite probability of finding additional electron density on B. This is equivalent to a partial charge transfer from A to B. For the opposite relative energies of the orbitals, shown on the right side of Figure 17.12, the same reasoning predicts that electron transfer will occur from B to A rather than from A to B.

17.6 Dispersion Forces Act between Nonbonded Atoms and Molecules

Although no chemical bond forms between two He atoms, helium dimers exist at low enough temperatures, and liquid He is the most stable phase of He below its normal boiling point of 4.2 K. Similarly, there is no chemical bond between two water molecules, but the normal boiling point of 373.15 K for H_2O is much higher than the value 212.5 K for the similar molecule H_2S. This difference indicates that there is an appreciable attractive interaction between water molecules. What is the origin of these interactions that allows He and H_2O to form stable liquid phases at quite different temperatures? As we show in this section, **dispersion forces** are responsible for the attraction between He atoms. Although dispersion forces also act between H_2O molecules, the dominant interaction in this case is hydrogen bonding, which is discussed in Section 17.7.

Although the origin of dispersion forces is best understood using quantum mechanics, we describe these forces in terms of the classical electrostatic interaction between the positive and negative charges in neighboring neutral molecules. Figure 17.13 shows

FIGURE 17.12
Interaction between two species A and B. The difference in energies between the HOMO and LUMO orbitals on A and B will determine the direction of charge transfer.

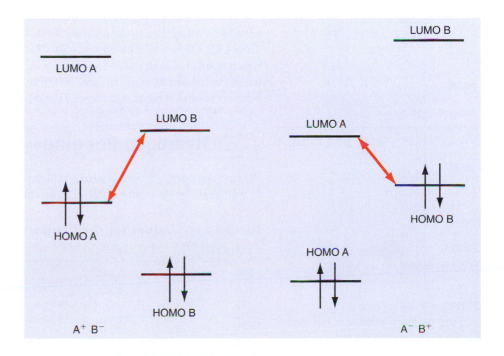

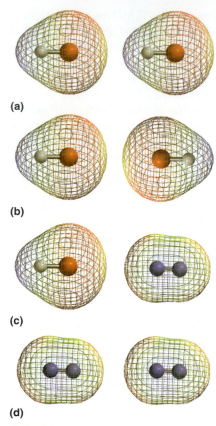

(a)

(b)

(c)

(d)

FIGURE 17.13

(a) Two dipolar HCl molecules with this relative orientation attract one another. (b) Two dipolar HCl molecules with this relative orientation repel one another. (c) The dipolar HCl molecule indices a dipole moment in N_2. (d) Two N_2 molecules experience a net attractive interaction through the time fluctuating dipole moments on each molecule.

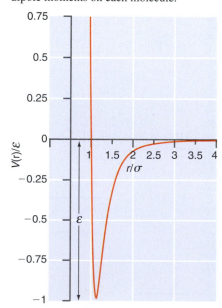

FIGURE 17.14

The Lennard–Jones potential of Equation (17.5) is plotted against the reduced length r/σ.

the possible interactions between two molecules that have a permanent dipole moment, a dipolar molecule (that is, HCl) and a molecule that does not have a permanent dipole moment (that is, N_2), and two molecules that do not have a permanent dipole moment (i.e., N_2). The molecules are represented by their electrostatic potentials in which red and blue shading corresponds to regions of negative and positive charge, respectively.

Although the interaction between two dipolar molecules can be attractive or repulsive as shown in Figures 17.13a and b, the net interaction when averaged over all possible interactions is attractive, and falls off as $1/r^6$, where r is the distance between the centers of mass of the molecules. We next consider the interaction between a dipolar molecule and one that has no permanent dipole moment. A molecule such as HCl induces a dipole moment in N_2 because the negatively charged Cl atom repels the electrons on the end of the N_2 molecule nearest it. This attractive interaction also falls off as $1/r^6$. As Figure 17.13d shows, a molecule with no permanent dipole moment has an instantaneous dipole moment that fluctuates rapidly in time. This transient dipole moment arises as the electrons circulate around the nuclei in the molecule. Although these dipoles on neighboring molecules may give rise to an attractive or repulsive interaction, the net effect when averaged over all orientations is attractive and also falls off as $1/r^6$. All of these interactions are called *dispersion forces*.

Even though all three types of interactions fall off as $1/r^6$, they are not equally strong. The dipole–dipole interaction is stronger than the dipole–nonpolar molecule interaction, which is stronger than the nonpolar–nonpolar interaction if the comparison is made between similar molecules. The strength of the interaction involving a nonpolar molecule is determined by the **polarizability** of a molecule, which is a measure of how much the electron distribution can be distorted by an external charge or electromagnetic field. The polarizability is approximately proportional to the number of electrons in the molecule. For all three types of interactions, the molecules will repel one another if they are brought close enough to one another that their electron distributions overlap. This repulsive interaction changes more rapidly with distance than the attractive interactions and is well approximated by the $1/r^{12}$ dependence. The dependence of the interaction energy between two molecules that arises through dispersion forces is often represented by a Lennard–Jones potential, which has the functional form

$$V(r) = 4\varepsilon\left[\left(\frac{\sigma}{r}\right)^{12} - \left(\frac{\sigma}{r}\right)^6\right] \tag{17.5}$$

This potential is shown in Figure 17.14. It has two adjustable parameters, namely, the well depth ε and the distance parameter σ, $V(r) = 0$ for $r/\sigma = 1$.

The values of the parameters are different from molecule to molecule and can be obtained from experimental data. Values for these parameters for several gases are shown in Table 17.2. Of the molecules listed, only CO and C_3H_8 have a permanent dipole moment, which is small in both cases. The increase in ε from He to C_3H_8 arises because the number of electrons in the molecule and, therefore, the polarizability increase along this series. Note the strong correlation of the well depth, ε, with the normal boiling temperature.

17.7 Hydrogen Bonding

We next consider hydrogen bonding, which plays an important role in aqueous solutions as well as in the chemistry of DNA replication. A **hydrogen bond** can form between a

TABLE 17.2   Values for the Lennard–Jones Parameters ε and σ

Atom/ Molecule	$10^{21}\varepsilon$(J)	Normal Boiling Temperature (K)	σ(pm)
He	0.141	4.2	256
N_2	1.31	77.4	370
CO	1.38	81.7	358
CH_4	2.05	90.8	378
C_3H_8	3.34	231	564

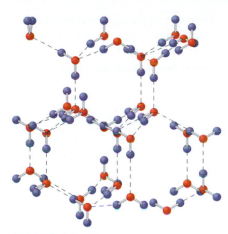

FIGURE 17.15

The crystal structure of hexagonal ice, which is the most stable solid phase at 1 bar, is shown. Note that hydrogen bonds connect each of the atoms in the water molecule to its neighbors.

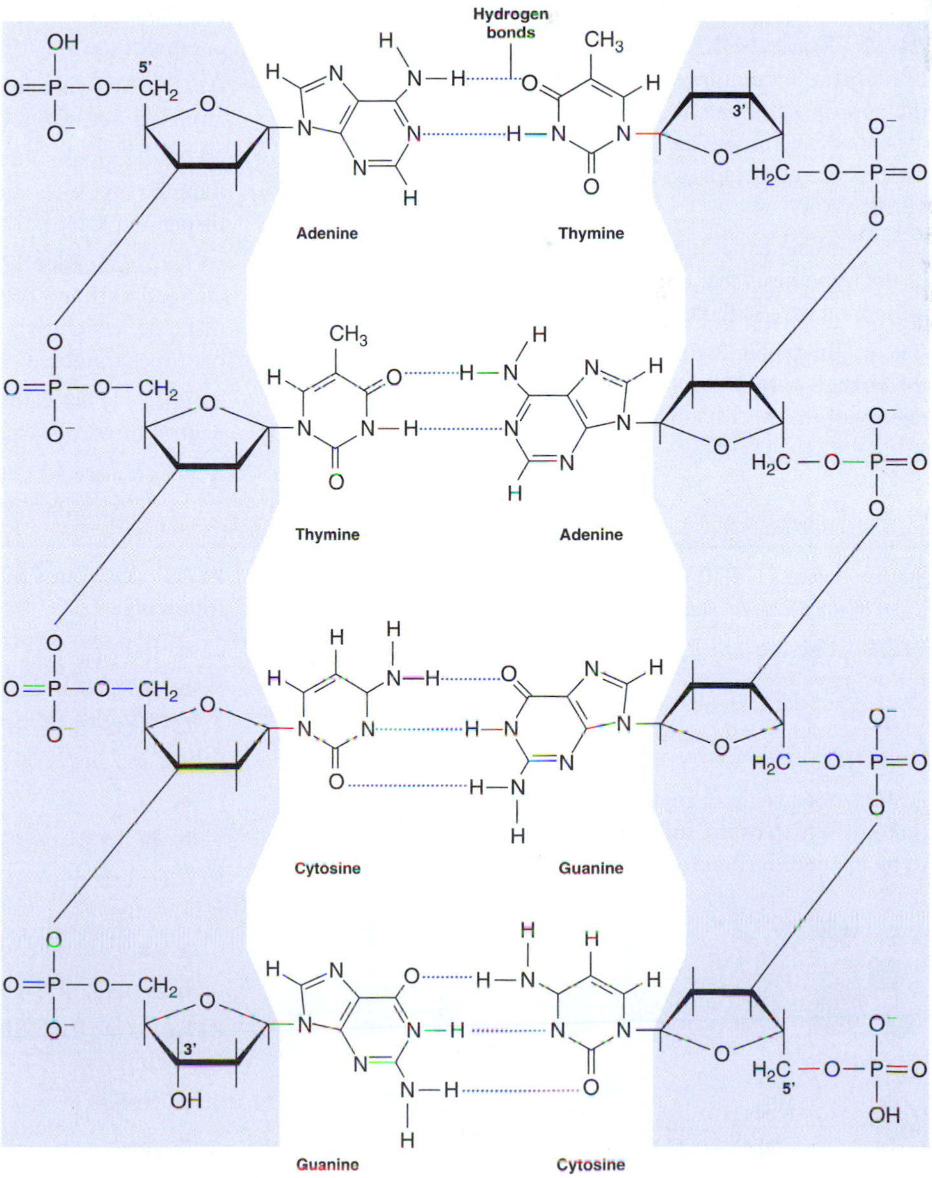

FIGURE 17.16

Hydrogen bonding between thymine and adenine and between guanine and cytosine are the dominant interactions that stabilize the double helix structure of DNA.

hydrogen atom that is chemically bound to the strongly electronegative atoms N, O, or F with a N, O, or F atom on a neighboring molecule. Unlike dispersion forces, there is no theoretical model that predicts the strength of the hydrogen bond or its dependence on the distance between the atoms involved in the hydrogen bond.

Liquid and solid water provide a good example of hydrogen bonding. If the intermolecular interactions were dominated by dispersion forces, one would expect H_2O and CH_4 to have nearly the same boiling temperature, because their molecular masses are nearly the same. However, the normal boiling temperature of methane is 111.65 K, whereas that for water is 373.15 K. This shows that hydrogen bonds are much stronger than dispersion forces for molecules that have approximately the same polarizability. Typical values for the hydrogen bond strength are ~2 to 7×10^{-20} J. Figure 17.15 shows the hydrogen bonding network in hexagonal ice. The relatively open structure of this solid phase results in it having a lower density than liquid water. This leads to the unusual property that the solid phase of water floats on the liquid in a two-phase mixture. Very few other substance show this behavior.

Hydrogen bonds also play a major role in stabilizing the double-helix structure of DNA and in the mechanism by which genetic material is copied. Each of the strands in the DNA molecule is connected to its partner strand through hydrogen bonds. The optimal geometry that maximizes the bonding between strands is achieved if adenine in one strand is paired with thymine and guanine is paired with cytosine as shown in Figure 17.16. Two hydrogen bonds are formed between adenine and thymine and three are formed between guanine and cytosine, which are referred to as complementary base pairs. As Watson and Crick noted in their seminal publication explaining the structure of DNA, it is the specificity of adenine–thymine pairing and guanine–cytosine pairing that provides a copying mechanism for the genetic material.

Vocabulary

delocalized bonding model	hybridization	lone pair	sp^3, sp^2, and sp hybridization
dispersion forces	hybrid orbital	lowest unoccupied molecular	valence shell electron pair
frontier orbitals	hydrogen bond	orbital (LUMO)	repulsion (VSEPR) model
highest occupied molecular	Lewis structure	molecular orbital theory	Walsh correlation diagram
orbital (HOMO)	localized bonding model	polarizability	

Questions on Concepts

Q17.1 On the basis of what you know about the indistinguishability of electrons and the difference between the wave functions for bonding electrons and lone pairs, discuss the validity and usefulness of the Lewis structure for the fluorine molecule, $\ddot{\underset{..}{F}} - \ddot{\underset{..}{F}}$.

Q17.2 The hybridization model assumes that atomic orbitals are recombined to prepare directed orbitals that have the appropriate bond angles for a given molecule. What aspects of the model can be tested by experiment, and what aspects are conjectures that are not amenable to experimental verification?

Q17.3 What experimental evidence can you cite in support of the hypothesis that the electronegativity of a hybridized atom increases with increasing s character?

Q17.4 In explaining molecular structure, the MO model uses the change in MO energy with bond angle. Explain how the energy of the $1a_1$ MO changes with the bond angle 2θ.

Q17.5 In explaining molecular structure, the MO model uses the change in MO energy with bond angle. Explain how the energy of the $2a_1$ MO changes with the bond angle 2θ.

Q17.6 In explaining molecular structure, the MO model uses the change in MO energy with bond angle. Explain how the energy of the $1b_2$ MO changes with the bond angle 2θ.

Q17.7 Why is a hydrogen bond formed with a N, O, or F atom but not with a C atom on a neighboring molecule?

Q17.8 Explain why the dipole–dipole interaction is stronger than the dipole–nonpolar molecule interaction in determining dispersion forces.

Q17.9 Consider the interaction between two oppositely charged ions and between two neutral molecules. At a given distance, which pair experiences the greater force? For which pair does the force change more rapidly with distance?

Q17.10 What is the origin of the attractive force between neutral molecules such as O_2 that have no dipole moment?

Problems

Problem numbers in **RED** indicate that the solution to the problem is given in the *Student Solutions Manual*.

P17.1 Use the VSEPR method to predict the structures of the following:

  a. PF_3

  b. CO_2

P17.2 SiF_4 has four ligands and one lone pair on the central S atom. Which of the following two structures do you expect to be the equilibrium form? Explain your reasoning.

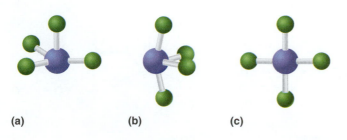

(a)   (b)   (c)

P17.3 Use the VSEPR method to predict the structures of the following:

  a. BrF_5

  b. SO_3^{2-}

P17.4 Use the VSEPR method to predict the structures of the following:

  a. SCl_2

  b. BCl_3

P17.5 Use the VSEPR method to predict the structures of the following:

  a. PCl_5

  b. SO_2

P17.6 Use the VSEPR method to predict the structures of the following:

  a. XeF_2

  b. XeF_5

P17.7 Show that the set of orthonormal *sp*-hybrid orbitals that are oriented 180° apart is

$$\psi_a = \frac{1}{\sqrt{2}}(-\phi_{2s} + \phi_{2p_z}) \text{ and } \psi_b = \frac{1}{\sqrt{2}}(-\phi_{2s} - \phi_{2p_z})$$

P17.8 Show that the water hybrid bonding orbitals given by

$$\psi_a = 0.55\phi_{2p_z} + 0.71\phi_{2p_y} - 0.45\phi_{2s} \text{ and}$$
$$\psi_b = 0.55\phi_{2p_z} - 0.71\phi_{2p_y} - 0.45\phi_{2s} \text{ are orthogonal.}$$

P17.9 Calculate the *s* and *p* character of the water lone pair hybrid orbitals

$$\psi_c = -0.45\phi_{2p_z} - 0.55\phi_{2s} + 0.71\phi_{2p_x} \text{ and}$$
$$\psi_d = -0.45\phi_{2p_z} - 0.55\phi_{2s} - 0.71\phi_{2p_x}.$$

P17.10 Show that two of the set of four equivalent orbitals appropriate for *sp³* hybridization,

$$\psi_a = \frac{1}{2}(-\phi_{2s} + \phi_{2p_x} + \phi_{2p_y} + \phi_{2p_z}) \text{ and}$$

$$\psi_b = \frac{1}{2}(-\phi_{2s} - \phi_{2p_x} - \phi_{2p_y} + \phi_{2p_z})$$

are orthogonal.

P17.11 Show that the set of four equivalent tetrahedrally oriented hybrid orbitals for *sp³* hybridization that are oriented 109.4° is

$$\psi_a = \frac{1}{2}(-\phi_{2s} + \phi_{2p_x} + \phi_{2p_y} + \phi_{2p_z})$$

$$\psi_b = \frac{1}{2}(-\phi_{2s} - \phi_{2p_x} - \phi_{2p_y} + \phi_{2p_z})$$

$$\psi_c = \frac{1}{2}(-\phi_{2s} + \phi_{2p_x} - \phi_{2p_y} - \phi_{2p_z})$$

$$\psi_d = \frac{1}{2}(-\phi_{2s} - \phi_{2p_x} + \phi_{2p_y} - \phi_{2p_z})$$

P17.12 Use the framework described in this chapter to construct normalized hybrid bonding orbitals on the central oxygen in O_3 that are derived from $2s$ and $2p$ atomic orbitals. The bond angle in ozone is 116.8°.

P17.13 Use the framework described in this chapter to derive the normalized hybrid lone pair orbital on the central oxygen in O_3 that is derived from $2s$ and $2p$ atomic orbitals. The bond angle in ozone is 116.8°.

P17.14 Use Bent's rule to estimate the change in the H—C—H bond angle when going from H_2CO to H_2CS.

P17.15 Use Bent's rule to estimate the change in the H—C bond length when going from H_2CO to F_2CO.

P17.16 Predict whether BH_2^+ is linear or bent using the Walsh correlation diagram in Figure 17.11 Explain your answers.

P17.17 Predict whether LiH_2^+ and NH_2^- should be linear or bent based on the Walsh correlation diagram in Figure 17.11 Explain your answers.

P17.18 Predict which of the bent molecules, BH_2 or NH_2, should have the larger bond angle on the basis of the Walsh correlation diagram in Figure 17.11. Explain your answer.

P17.19 Predict whether the ground state or the first excited state of CH_2 should have the larger bond angle on the basis of the Walsh correlation diagram shown in Figure 17.11 Explain your answer.

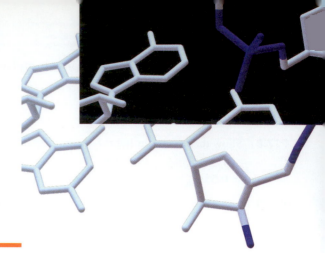

CHAPTER

18

Vibrational and Rotational Spectroscopy

the **Chemistry place**

Web-Based Problem 18.1 Energy Levels and Emission Spectra

Chemists have a wide range of spectroscopic techniques available to them. With these techniques, unknown molecules can be identified, bond lengths can be measured, and the force constants associated with chemical bonds can be determined. Spectroscopic techniques are based on transitions that occur between different energy states of molecules when they interact with electromagnetic radiation. In this chapter, we describe how light interacts with molecules to induce transitions between states. In particular, we discuss the absorption of electromagnetic radiation in the infrared and microwave regions of the spectrum. Light of these wavelengths induces transitions between eigenstates of vibrational and rotational energy.

18.1 An Introduction to Spectroscopy

The various forms of **spectroscopy** are among the most powerful tools that chemists have at their disposal to probe the world at an atomic and molecular level. The information that is accessible through molecular spectroscopy includes bond lengths (rotational spectroscopy) and the frequencies of the characteristic oscillatory motions associated with the stretching and bending of chemical bonds (vibrational spectroscopy). In addition, the allowed energy levels for electrons in molecules can be determined with electronic spectroscopy, which is crucial for a deeper understanding of the chemical bonding and the reactivity of molecules. In all of these spectroscopies, atoms or molecules absorb electromagnetic radiation and undergo transitions between allowed quantum states.

In the spectroscopies of interest to chemists, the atoms or molecules are confined in a container that is partly transparent to electromagnetic radiation. In most experiments, the attenuation or enhancement of the incident radiation resulting from excitation or de-excitation of the atoms or molecules of interest is measured as a function of the incident wavelength or frequency. Because quantum mechanical systems have a discrete energy spectrum, an absorption or emission spectrum consists of individual peaks, each of which is associated with a transition between two allowed energy levels of the system. The frequency at which energy is absorbed or emitted is related to the energy levels involved in the transitions by

$$h\nu = |E_2 - E_1| \tag{18.1}$$

FIGURE 18.1

The electromagnetic spectrum depicted on a logarithmic wavelength scale.

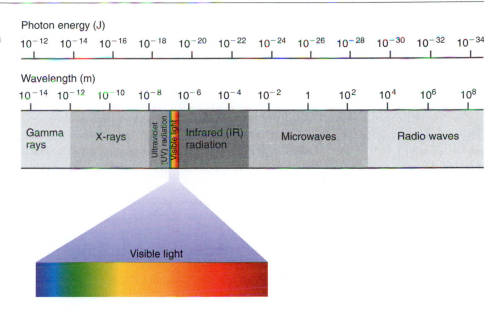

The photon energy that is used in the spectroscopies discussed in this chapter spans nine orders of magnitude in going from the microwave to the X-ray region. This is an indication of the very different energy-level spacing probed by these techniques. The energy-level spacing is smallest in nuclear magnetic resonance (NMR) spectroscopy, and largest for electronic spectroscopy in which electrons are promoted from lower to higher energy levels. Transitions between rotational and vibrational energy levels are intermediate between these two extremes, with rotational energy levels being more closely spaced than vibrational energy levels. The electromagnetic spectrum is depicted schematically in Figure 18.1. An examination of this figure shows that visible light is a very small part of this spectrum.

The spectral regions associated with various spectroscopies are shown in Table 18.1. Spectroscopists commonly use the quantity **wavenumber,** $\tilde{\nu} = 1/\lambda$, which has units of inverse centimeters, rather than the wavelength λ or frequency ν to designate spectral transitions for historical reasons. The relationship between ν and $\tilde{\nu}$ is given by $\nu = \tilde{\nu}c$, where c is the speed of light. It is important to use consistent units when calculating the energy difference between states associated with a frequency in units of inverse seconds and in units of inverse centimeters. Equation (18.1) expressed for both units is $|E_2 - E_1| = h\nu = hc\tilde{\nu}$.

The fact that atoms and molecules possess a set of discrete energy levels is an essential feature of all spectroscopies. If all molecules had a continuous energy spectrum, it would be very difficult to distinguish them on the basis of their absorption spectra. However, as discussed in Section 18.4, not all transitions between arbitrarily chosen states occur. **Selection rules** tell us which transitions will be experimentally observed. Because spectroscopies involve transitions between quantum states, we must first describe how electromagnetic radiation interacts with molecules.

We begin with a qualitative description of energy transfer from the electromagnetic field to a molecule leading to vibrational excitation. Light is an electromagnetic traveling wave

TABLE 18.1   Important Spectroscopies and Their Spectral Range

Spectral Range	λ (nm)	$\nu/10^{14}$ (Hz)	$\tilde{\nu}$ (cm)$^{-1}$	Energy (kJ/mol)	Spectroscopy
Radio	~1×10^9	~10^{-6}	~0.01	~10^{-8}	NMR
Microwave	~100,000	~10^{-2}	~100	~10^{-2}	Rotational
Infrared	~1000	~3.0	~10,000	~10^3	Vibrational
Visible (red)	~700	~4	~14,000	~10^5	Electronic
Visible (blue)	~450	~6	~22,000	~3×10^5	Electronic
Ultraviolet	<300	>10	>30,000	>5×10^5	Electronic

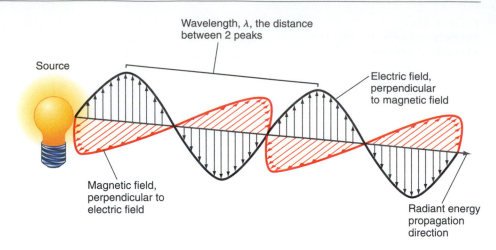

Source

Wavelength, λ, the distance between 2 peaks

Electric field, perpendicular to magnetic field

Magnetic field, perpendicular to electric field

Radiant energy propagation direction

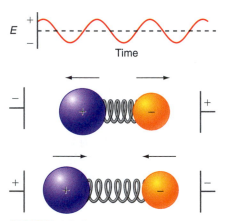

FIGURE 18.3
Schematic of the interaction of a classical harmonic oscillator constrained to move in one dimension under the influence of an electric field. The sinusoidally varying electric field, *E,* shown at the top of the figure is generated by a voltage between a pair of capacitor plates. The direction of *E* is parallel to the molecular axis. The arrows indicate the direction of force on each of the two charged masses. If the phases of the field and vibration are as shown, the oscillator will absorb energy in both the stretching and compression half-cycles.

that has perpendicular magnetic and electric field components as shown in Figure 18.2. Consider the effect of a time-dependent electric field on a classical dipolar diatomic "molecule" constrained to move in one dimension. Such a molecule is depicted in Figure 18.3. If the spring were replaced by a rigid rod, the molecule could not take up energy from the field. However, the spring allows the two masses to oscillate about their equilibrium distance, thereby generating a periodically varying dipole moment. If the electric field and oscillation of the dipole moment have the same frequency and if they are in phase, the molecule can absorb energy from the field. For a classical "molecule," any amount of energy can be taken up and the absorption spectrum is continuous.

For a real quantum mechanical molecule, the interaction with the electromagnetic field is similar. The electric field that acts on a dipole moment within the molecule can be of two types: permanent and dynamic. Polar molecules like HCl have a **permanent dipole moment.** As molecules vibrate, an additional induced **dynamic dipole moment** can be generated. How does the dynamic dipole arise? The magnitude of the dipole moment depends on the bond length and the degree to which charge is transferred from H to Cl. In turn, the charge transfer depends on the overlap of the electron densities of the two atoms and is, therefore, sensitive to the internuclear distance. As the molecule vibrates, its dipole moment may change because of these effects, generating a dynamic dipole moment. Because the vibrational amplitude is a small fraction of the bond distance, the dynamic dipole moment is generally small compared to the permanent dipole moment.

As will be seen in the next section, it is the dynamic rather than the permanent dipole that determines if a molecule will absorb energy in the infrared portion of the light spectrum. By contrast, it is the permanent dipole moment that determines if a molecule will undergo rotational transitions by absorbing energy in the microwave portion of the light spectrum. Homonuclear diatomic molecules have neither permanent nor dynamic dipole moments and cannot take up energy in the manner described here. However, vibrational spectroscopy on these molecules can be carried out using the Raman effect as discussed in Section 18.8.

18.2 Absorption, Spontaneous Emission, and Stimulated Emission

We now move from a classical picture to a quantum mechanical description involving discrete energy levels. The basic processes by which photon-assisted transitions between energy levels occur are **absorption, spontaneous emission,** and **stimulated emission.** For simplicity, only transitions in a two-level system are considered as shown in Figure 18.4.

In absorption, the incident photon induces a transition to a higher level, and in emission, a photon is emitted as an excited state relaxes to a state of lower energy. Absorption and stimulated emission are initiated by a photon incident on the molecule of interest. As the name implies, spontaneous emission is a random event and its rate is related to the life-

FIGURE 18.4
The three basic processes by which photon-assisted transitions occur. Orange- and red-filled circles indicate empty and occupied levels, respectively.

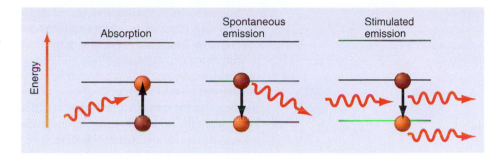

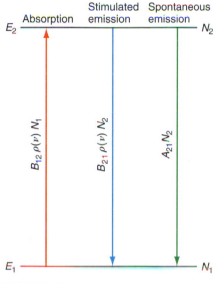

FIGURE 18.5
The rate at which transitions occur between two levels. It is in each case proportional to the product of the appropriate rate coefficient A_{21}, B_{12}, or B_{21} for the process and the population in the originating state, N_1 or N_2. For absorption and stimulated emission, the rate is additionally proportional to the radiation density, $\rho(\nu)$.

time of the excited state. These three processes are not independent in a system at equilibrium, as can be seen by considering Figure 18.5. At equilibrium, the overall transition rate from level 1 to 2 must be the same as that from 2 to 1. This means that

$$B_{12}\rho(\nu)N_1 = B_{21}\rho(\nu)N_2 + A_{21}N_2 \tag{18.2}$$

Whereas spontaneous emission is independent of the radiation density at a given frequency, $\rho(\nu)$, the rates of absorption and stimulated emission are directly proportional to $\rho(\nu)$. The proportionality constants for spontaneous emission, absorption, and stimulated emission are A_{21}, B_{12}, and B_{21}, respectively. Each of these rates is directly proportional to the number of molecules (N_1 or N_2) in the state from which the transition originates. This means that unless the lower state is populated, a signal will not be observed in an absorption experiment. Similarly, unless the upper state is populated, a signal will not be observed in an emission experiment.

The appropriate function to use for $\rho(\nu)$ in the equilibrium equation, Equation (18.2), is the blackbody spectral density function of Equation (12.2), because this is the distribution of frequencies at equilibrium for a given temperature. Following this reasoning, Einstein concluded that

$$B_{12} = B_{21} \quad \text{and} \quad \frac{A_{21}}{B_{21}} = \frac{16\pi^2\hbar\nu^3}{c^3} \tag{18.3}$$

This result is derived in Example Problem 18.1.

EXAMPLE PROBLEM 18.1

Derive the equations $B_{12} = B_{21}$ and $A_{21}/B_{21} = 16\pi^2\hbar\nu^3/c^3$ using these two pieces of information: (1) the overall rate of transition between levels 1 and 2 (see Figure 18.5) is zero at equilibrium, and (2) the ratio of N_2 to N_1 is governed by the Boltzmann distribution.

Solution

The rate of transitions from level 1 to level 2 is equal and opposite to the transitions from level 2 to level 1. This gives the equation $B_{12}\rho(\nu)N_1 = B_{21}\rho(\nu)N_2 + A_{21}N_2$. The Boltzmann distribution function states that

$$\frac{N_2}{N_1} = \frac{g_2}{g_1}e^{-h\nu/kT}$$

In this case, $g_2 = g_1$. These two equations can be solved for $\rho(\nu)$, giving $\rho(\nu) = A_{21}/(B_{12}e^{h\nu/kT} - B_{21})$. As Planck showed, $\rho(\nu)$ has the form shown in Equation (12.2) so that

$$\rho(\nu) = \frac{A_{21}}{B_{12}e^{h\nu/kT} - B_{21}} = \frac{8\pi h\nu^3}{c^3}\frac{1}{e^{h\nu/kT} - 1}$$

For these two expressions to be equal, $B_{12} = B_{21}$ and $A_{21}/B_{21} = 8\pi h\nu^3/c^3 = 16\pi^2\hbar\nu^3/c^3$.

Spontaneous emission and stimulated emission differ in an important respect. Spontaneous emission is a completely random process, and the emitted photons are incoherent, by which we mean that their phases and propagation direction are random. In stimulated emission, the phase and direction of propagation are the same as that of the incident photon. This is referred to as coherent photon emission. A lightbulb is an **incoherent photon source.** The phase relation between individual photons is random, and because the propagation direction of the photons is also random, the intensity of the source falls off as the square of the distance. A laser is a **coherent source** of radiation. All photons are in phase, and because they have the same propagation direction, the divergence of the beam is very small. This explains why a laser beam that is reflected from the moon still has a measurable intensity when it returns to Earth. We will have more to say about lasers in Section 19.4.

18.3 Basics of Vibrational Spectroscopy

We now have a framework with which we can discuss spectroscopy as a chemical tool. Two features have enabled vibrational spectroscopy to achieve the importance that it has as a tool in chemistry. The first is that the vibrational frequency depends much more on the identity of the two vibrating atoms on either end of the bond than on the other atoms farther away from the bond. This property generates characteristic frequencies for atoms joined by a bond known as **group frequencies.** We discuss group frequencies further in Section 18.6. The second feature is that a particular vibrational mode in a molecule has only one characteristic frequency of appreciable intensity. We discuss this feature next.

In any spectroscopy, transitions occur from one energy level to another. As discussed in Section 18.2, the energy level from which the transition originates must be occupied in order to generate a spectral signal. Which of the infinite set of vibrational levels has a substantial probability of being occupied? Table 18.2 shows the number of diatomic molecules in the first excited vibrational state (N_1) relative to those in the ground state (N_0) at 300 and 1000 K. The calculations have been carried out using the Boltzmann distribution. We can see that nearly all of the molecules in a macroscopic sample are in their ground vibrational state at room temperature because $N_1/N_0 << 1$. Even at 1000 K, N_1/N_0 is very small except for Br_2. This means that for these molecules, absorption of light at the characteristic frequency will occur predominantly from molecules in the $n = 0$ state. What final states are possible? As shown in the next section, for absorption by a quantum mechanical harmonic oscillator, $\Delta n = n_{final} - n_{initial} = +1$. Because only the $n = 0$ state has an appreciable population, in most cases only the $n = 0 \rightarrow n = 1$ transition is observed in vibrational spectroscopy.

TABLE 18.2   **Vibrational State Populations for Selected Diatomic Molecules**

Molecule	$\tilde{v}$ (cm^{-1})	v (s^{-1})	N_1/N_0 for 300 K	N_1/N_0 for 1000 K
H—H	4400	1.32×10^{14}	6.88×10^{-10}	1.78×10^{-3}
H—F	4138	1.24×10^{14}	2.42×10^{-9}	2.60×10^{-3}
H—Br	2649	7.94×10^{13}	3.05×10^{-6}	2.21×10^{-2}
N≡N	2358	7.07×10^{13}	1.23×10^{-5}	3.36×10^{-2}
C≡O	2170	6.51×10^{13}	3.03×10^{-5}	4.41×10^{-2}
Br—Br	323	9.68×10^{12}	0.213	0.628

EXAMPLE PROBLEM 18.2

A strong absorption of infrared radiation is observed for $^1H^{35}Cl$ at 2991 cm^{-1}.

a. Calculate the force constant, k, for this molecule.

b. By what factor do you expect this frequency to shift if deuterium is substituted for hydrogen in this molecule? The force constant is unaffected by this substitution.

Solution

a. Using Equation (14.40), we first write $\Delta E = h\nu = hc/\lambda = \hbar\sqrt{k/\mu}$. Solving for k,

$$\Delta E = h\nu = \frac{hc}{\lambda} = \frac{h}{2\pi}\sqrt{\frac{k}{\mu}} \qquad \text{and}$$

$$k = 4\pi^2\left(\frac{c}{\lambda}\right)^2\mu$$

$$= 4\pi^2(2.998\times10^8\ ms^{-1})^2\left(\frac{2991}{cm}\frac{100\ cm}{1\ m}\right)^2\frac{(1.008)(34.969)\ \text{amu}}{35.977}$$

$$\times\left(\frac{1.661\times10^{-27}\ kg}{1\ \text{amu}}\right)$$

$$= 516.3\ N\ m^{-1}$$

b. $\dfrac{\nu_{DCl}}{\nu_{HCl}} = \sqrt{\dfrac{\mu_{HCl}}{\mu_{DCl}}} = \sqrt{\dfrac{m_H m_{Cl}}{m_D m_{Cl}}\dfrac{(m_D + m_{Cl})}{(m_H + m_{Cl})}} = \sqrt{\left(\dfrac{1.0078}{2.0140}\right)\left(\dfrac{36.983}{35.977}\right)} = 0.717$

The vibrational frequency for DCl is lower by a substantial amount. Would the shift be as great if ^{37}Cl were substituted for ^{35}Cl? The fact that vibrational frequencies are so strongly shifted by isotopic substitution of deuterium for hydrogen makes infrared spectroscopy a valuable tool for determining the presence of hydrogen atoms in molecules.

the Chemistry place

Web-Based Problem 18.2 The Morse Potential

The high sensitivity available in modern instrumentation to carry out vibrational spectroscopy does make it possible in favorable cases to see vibrational transitions originating from the $n = 0$ state for which $\Delta n = +2, +3, \dots$. These **overtone** transitions are much weaker than the $\Delta n = +1$ absorption, but are possible because the selection rule $\Delta n = +1$ is not rigorously obeyed for an anharmonic potential, as discussed later.

The overtone transitions are useful because they allow us to determine the degree to which real molecular potentials differ from the simple **harmonic potential**, $V(x) = (1/2)kx^2$. To a good approximation, a realistic **anharmonic potential** can be described in analytical form by the **Morse potential:**

$$V(x) = D_e[1 - e^{-\alpha(x-x_e)}]^2 \tag{18.4}$$

in which D_e is the dissociation energy relative to the bottom of the potential and $\alpha = \sqrt{k/2D_e}$. The force constant, k, for the Morse potential is defined by $k = (d^2V/dx^2)_{x=x_e}$. Note that this definition for k is also valid for the harmonic potential. The **bond energy** D_0 is defined with respect to the lowest allowed level, rather than to the bottom of the potential, as shown in Figure 18.6.

The energy levels for this potential are given by

$$E_n = h\nu\left(n + \frac{1}{2}\right) - \frac{(h\nu)^2}{4D_e}\left(n + \frac{1}{2}\right)^2 \tag{18.5}$$

The second term gives the anharmonic correction to the energy levels. Measurements of the frequencies of the overtone vibrations allow the parameter D_e in the Morse potential

FIGURE 18.6
Morse potential, $V(x)$ (red curve), as a function of the bond length, x, for HCl, using the parameters from Example Problem 18.3. The zero of energy is chosen to be the bottom of the potential. The yellow curve shows a harmonic potential, which is a good approximation to the Morse potential near the bottom of the well. The horizontal lines indicate allowed energy levels in the Morse potential; D_e and D_0 represent the bond energies defined with respect to the bottom of the potential and the lowest state, respectively; and x_e is the equilibrium bond length.

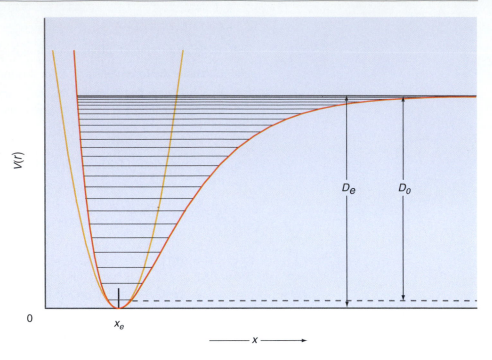

to be determined for a specific molecule. This provides a useful method for determining the details of the interaction potential in a molecule.

EXAMPLE PROBLEM 18.3

The Morse potential can be used to model dissociation as illustrated in this example. The $^1H^{35}Cl$ molecule can be described by a Morse potential with $D_e = 7.41 \times 10^{-19}$ J. The force constant k for this molecule is 516.3 N m^{-1} and $\nu = 8.97 \times 10^{13}$ s^{-1}. Calculate the number of allowed vibrational states in this potential and the bond energy for the $^1H^{35}Cl$ molecule.

Solution

We solve the equation

$$E_n = h\nu\left(n + \frac{1}{2}\right) - \frac{(h\nu)^2}{4D_e}\left(n + \frac{1}{2}\right)^2$$

to obtain the highest value of n consistent with the relation $E_{n_{max}} \leq D_0$. Using the parameters given earlier, we obtain the following equation for n:

$$-1.1918 \times 10^{-21} n^2 + 5.8243 \times 10^{-20} n + 2.942 \times 10^{-20} = 7.41 \times 10^{-19}$$

Both solutions to this quadratic equation give $n = 24.4$, so we conclude that the potential has 24 allowed levels. If the left side of the equation is graphed versus n, we obtain the results shown in the following figure:

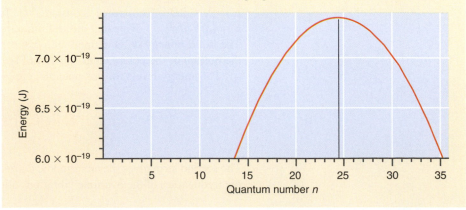

Note that E_n decreases for $n > 25$. This is mathematically correct, but unphysical because for $n > 24$, the molecule has a continuous energy spectrum, and Equation (18.5) is no longer valid.

The bond energy, D_0, is not D_e, but $D_e - E_0$ where

$$E_0 = \frac{h\nu}{2} - \frac{(h\nu)^2}{16D_e}$$

from Equation (18.5), because the molecule has a zero point vibrational energy. Using the parameters given earlier, the bond energy is 7.11×10^{-19} J. The Morse and harmonic potentials as well as the allowed energy levels for this molecule are shown in Figure 18.6.

The material-dependent parameters that determine the frequencies observed in vibrational spectroscopy for diatomic molecules are the force constant, k, and the reduced mass, μ. The corresponding parameters for rotational spectroscopy (see Section 18.5) are the rotational constant, B, and the bond length, r. These parameters, along with well depth D_e, are listed in Table 18.3 for selected molecules. The quantities B (defined in Section 18.5) and $\tilde{\nu}$ are expressed in units of inverse centimeters.

18.4 The Origin of Selection Rules

Every spectroscopy has selection rules that govern the transitions that can occur between eigenstates of a system. This is a great simplification in the interpretation of spectra, because far fewer transitions occur than would be the case if there were no selection rules. How do these selection rules arise? We can use the general model for transitions between states discussed earlier to derive the selection rules for vibrational spectroscopy based on the quantum mechanical harmonic oscillator.

According to quantum mechanics, the transition probability from state n to state m is only nonzero if the **transition dipole moment,** μ_x^{mn}, satisfies the following condition:

$$\mu_x^{mn} = \int \psi_m^*(x)\, \mu_x(x)\, \psi_n(x)\, d\tau \neq 0 \qquad (18.6)$$

In this equation, x is the time-dependent amplitude of the vibration and μ_x is the dipole moment along the electric field direction, which we take to be the x axis.

TABLE 18.3   Values of Molecular Constants for Selected Diatomic Molecules

	$\tilde{\nu}$ (cm)$^{-1}$	ν (s^{-1})	x_e (pm)	k (N m^{-1})	B (cm^{-1})	D_0 (kJ mol^{-1})	D_0 (J molecule^{-1})
H_2	4401	1.32×10^{14}	74.14	575	60.853	432	7.17×10^{-19}
D_2	3115	9.33×10^{13}	74.15	577	30.444	439	7.29×10^{-19}
$^1H^{81}Br$	2649	7.94×10^{13}	141.4	412	8.4649	366	6.08×10^{-19}
$^1H^{35}Cl$	2991	8.97×10^{13}	127.5	516	10.5934	432	7.17×10^{-19}
$^1H^{19}F$	4138	1.24×10^{14}	91.68	966	20.9557	570	9.46×10^{-19}
$^1H^{127}I$	2309	6.92×10^{13}	160.92	314	6.4264	298	4.95×10^{-19}
$^{35}Cl_2$	559.7	1.68×10^{13}	198.8	323	0.244	243	4.03×10^{-19}
$^{79}Br_2$	325.32	9.75×10^{12}	228.1	246	0.082107	193	3.2×10^{-19}
$^{19}F_2$	916.64	2.75×10^{13}	141.2	470	0.89019	159	2.64×10^{-19}
$^{127}I_2$	214.5	6.43×10^{12}	266.6	172	0.03737	151	2.51×10^{-19}
$^{14}N_2$	2358.6	7.07×10^{13}	109.8	2295	1.99824	945	1.57×10^{-18}
$^{16}O_2$	1580	4.74×10^{13}	120.8	1177	1.44563	498	8.27×10^{-19}
$^{12}C^{16}O$	2170	2.56×10^{13}	112.8	1902	1.9313	1080	1.79×10^{-18}

In the following discussion, we show how selection rules for vibrational excitation arise from Equation (18.6). As discussed in Section 18.1, the dipole moment μ_x may change slightly as the molecule vibrates. Because the amplitude of vibration x is an oscillatory function of t, the molecule will have a time-dependent dynamic dipole moment. We take this into account by expanding μ_x in a Taylor series about the equilibrium bond length. Because x is the amplitude of vibration, the equilibrium bond length corresponds to $x = 0$:

$$\mu_x(x(t)) = \mu_{0x} + x(t)\left(\frac{d\mu_x}{dx}\right)_{x=0} + \ldots \tag{18.7}$$

in which the values of μ_{0x}, the permanent dipole moment, and $(d\mu_x/dx)_{x=0}$ depend on the molecule under consideration. Note that because $x = x(t)$, $\mu_x(x(t))$ is a function of time. The first term on the right-hand side in Equation (18.7) is the permanent dipole moment, and the second term is the dynamic dipole moment. As we saw earlier, for absorption experiments, it is reasonable to assume that only the $n = 0$ state is populated. Using Equation (14.36), which gives explicit expressions for the eigenfunctions m:

$$\mu_x^{m0} = A_m A_0\, \mu_{0x} \int_{-\infty}^{\infty} H_m(\alpha^{1/2}x)\, H_0(\alpha^{1/2}x) e^{-\alpha x^2}\, dx$$

$$+ A_m A_0\left[\left(\frac{d\mu_x}{dx}\right)_{x=0}\right] \int_{-\infty}^{\infty} H_m(\alpha^{1/2}x)x\, H_0(\alpha^{1/2}x) e^{-\alpha x^2}\, dx \tag{18.8}$$

The first integral is zero because different eigenfunctions are orthogonal. To solve the second integral, we need to use the specific functional form of $H_m(\alpha^{1/2}x)$. However, because the integration is over the symmetric interval $-\infty < x < \infty$, this integral is zero if the integrand is an odd function of x. As Equation (14.38) shows, the Hermite polynomials $H_m(\alpha^{1/2}x)$ are odd functions of x if m is odd and even functions of x if m is even. The term $x H_0(\alpha^{1/2}x)e^{-\alpha x^2}$ in the integrand is an odd function of x and, therefore, μ_x^{m0} is zero if $H_m(\alpha^{1/2}x)$ is even. This simplifies the problem because only transitions of the type

$$n = 0 \rightarrow m = 2b + 1, \text{ for } b = 0, 1, 2, \ldots \tag{18.9}$$

can have nonzero values for μ_x^{m0}.

Do all the transitions indicated in Equation (18.9) lead to nonzero values of μ_x^{m0}? To answer this question, the integrand $H_m(\alpha^{1/2}x)x\, H_0(\alpha^{1/2}x)e^{-\alpha x^2}$ is graphed against x for the transitions $n = 0 \rightarrow m = 1$, $n = 0 \rightarrow m = 3$, and $n = 0 \rightarrow m = 5$ in Figure 18.7. Whereas the integrand is positive everywhere for the $n = 0 \rightarrow n = 1$ transition, the areas above the dashed line exactly cancel those below the line for the $n = 0 \rightarrow m = 3$ and $n = 0 \rightarrow m = 5$ transitions, showing that the value of $\mu_x^{mn} = 0$. Therefore, $\mu_x^{m0} \neq 0$ only for the first of the three transitions under consideration. It can be shown more generally that in the **dipole approximation,** the selection rule for absorption is $\Delta n = +1$, and for emission, it is $\Delta n = -1$. We have derived this selection rule for the particular case of vibrational spectroscopy, and selection rules are different for different spectroscopies. However, within the dipole approximation, the selection rules are calculated using Equation (18.6) and the appropriate total energy eigenfunctions for any spectroscopy.

Note that because we found that the first integral in Equation (18.8) was zero, the absence or presence of a permanent dipole moment μ_{0x} that remains constant as the molecule vibrates is not relevant for the absorption of infrared radiation. For vibrational excitation to occur, the dynamic dipole moment must satisfy the condition $d\mu_x/dx \neq 0$. Because of this condition, homonuclear diatomic molecules do not absorb light in the infrared. This has important consequences for our environment. The temperature of the Earth is determined primarily by an energy balance between visible and ultraviolet (UV) radiation absorbed from the sun and infrared radiation emitted by the planet. The molecules N_2, O_2, and H_2, which have a zero permanent and dynamic dipole moment, together with the rare gases make up 99.93% of the atmosphere. These gases do not absorb the infrared radiation emitted by the Earth. Therefore, almost all of the emitted infrared

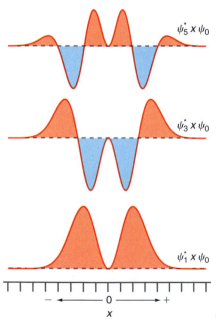

FIGURE 18.7
The integrand $H_m(\alpha^{1/2}x)x\, H_0(\alpha^{1/2}x)e^{-\alpha x^2}$ is graphed as a function of x for the transitions $n = 0 \rightarrow m = 1$, $n = 0 \rightarrow m = 3$, and $n = 0 \rightarrow m = 5$. The dashed line shows the zero level for each graph.

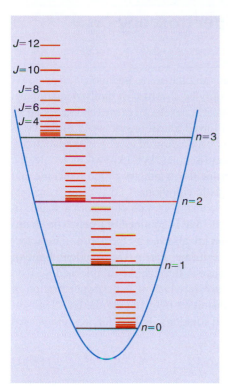

FIGURE 18.8
Schematic representation of rotational and vibrational levels. Each vibrational level has a set of rotational levels associated with it. Therefore, vibrational transitions usually also involve rotational transitions. The rotational levels are shown on an expanded energy scale and are much more closely spaced for real molecules.

radiation passes through the atmosphere and escapes into space. By contrast, greenhouse gases such as CO_2, NO_x, and hydrocarbons absorb infrared radiation emitted by the Earth and radiate it back to the Earth. This absorption increases the Earth's temperature and leads to global warming. However, as you will conclude in the end-of-chapter problems, not all of the vibrational modes of CO_2 are infrared active.

18.5 Rotational Energy Levels and Bond Lengths

The Schrödinger equation can be solved for the rotation of a molecule in 3-D space. Although we will not derive this result, the dependence of the rotational energy for a diatomic molecule on the rotational quantum number J is given by

$$E_J = \frac{\hbar^2}{2\mu r_0^2}J(J+1) = \frac{h^2}{8\pi^2\mu r_0^2}J(J+1) = hcBJ(J+1) \qquad (18.10)$$

where J can take on the integral values 0, 1, 2, 3, . . . , and each of the energy levels has the degeneracy $2J+1$. In Equation (18.10), the constants specific to a molecule are combined in the so-called **rotational constant** $B = h/8\pi^2 c\mu r_0^2$. The factor c is included in B so that it has the units of cm^{-1} rather than s^{-1}. Equation (18.10) can be compared with the energy of a classically rotating dumbbell, which is given by $E = \frac{|l|^2}{2\mu r_0^2}$ where l is the angular momentum. In the classical limit of very large J values, $|l|^2 = \hbar^2 J(J+1) \rightarrow \hbar^2 J^2$ and the rotational energy levels become continuous rather than discrete. Rotational energy levels are much more closely spaced than vibrational energy levels as shown in Figure 18.8. Consequently, whereas vibrational transitions are observed in the infrared region, rotational transitions are observed in the microwave region of the electromagnetic spectrum.

As for the harmonic oscillator, there is a selection rule that governs the absorption of electromagnetic energy for a molecule to change its rotational energy, namely,

$$\Delta J = J_{final} - J_{initial} = \pm 1$$

EXAMPLE PROBLEM 18.4

Because of the very high precision of frequency measurements, bond lengths can be determined with a correspondingly high precision, as illustrated in this example. From the rotational microwave spectrum of $^1H^{35}Cl$, it is found that $B = 10.59342\ cm^{-1}$. Given that the masses of ^1H and ^{35}Cl are 1.0078250 and 34.9688527 amu, respectively, determine the bond length of the $^1H^{35}Cl$ molecule.

Solution

$$B = \frac{h}{8\pi^2 c\mu r_0^2}$$

$$r_0 = \sqrt{\frac{h}{8\pi^2 c\mu B}}$$

$$= \sqrt{\frac{6.6260755 \times 10^{-34}\ J\ s}{8\pi^2 c\left(\dfrac{(1.0078250)(34.9688527)\ \text{amu}}{1.0078250 + 34.9688527}\right)(1.66054 \times 10^{-27}\ kg\ amu^{-1})(10.59342\ cm^{-1})}}$$

$$= 1.274553 \times 10^{-10}\ m$$

18.6 Infrared Absorption Spectroscopy in the Gas Phase

The two most basic results of quantum mechanics are that atoms and molecules possess a discrete energy spectrum and that energy can only be absorbed or emitted in amounts that correspond to the difference between two energy levels. Because the energy spectrum for each chemical species is unique, the allowed transitions between these levels provide a "fingerprint" for that species. Using such a fingerprint to identify and quantify the species is a primary role of all chemical spectroscopies. For a known molecule, the vibrational spectrum can also be used to determine the symmetry of the molecule and the force constants associated with the characteristic vibrations.

In absorption spectroscopies in general, electromagnetic radiation from a source of the appropriate wavelength is incident on a sample that is confined in a cell. The chemical species in the sample undergo transitions that are allowed by the appropriate selection rules among rotational, vibrational, or electronic states. The incident light of intensity $I_0(\lambda)$ is attenuated when passing a distance dl through the sample as described by the differential form of the **Beer–Lambert law** in which M is the concentration of the absorber, and $I(\lambda)$ is the intensity of the transmitted light leaving the cell. Units of moles per liter are commonly used for M in liquid solutions, and partial pressure is used for gas mixtures:

$$dI(\lambda) = \varepsilon(\lambda)M\,I(\lambda)dl \tag{18.11}$$

This equation can be integrated to give

$$\frac{I(\lambda)}{I_0(\lambda)} = e^{-\varepsilon(\lambda)M\,l} \tag{18.12}$$

The Beer–Lambert law is often stated in the form

$$\ln\frac{I(\lambda)}{I_0(\lambda)} = A(\lambda) = -\varepsilon(\lambda)M\,l \tag{18.13}$$

and A is called the **absorbance.** The ratio $I(\lambda)/I_0(\lambda)$ is called the **transmittance.** The information on the discrete energy spectrum of the chemical species in the cell is contained in the wavelength dependence of the **molar absorption coefficient, $\varepsilon(\lambda)$.** It is evident that the strength of the absorption is proportional to $I(\lambda)/I_0(\lambda)$, which increases with $\varepsilon(\lambda)$, M, and path length l. Because $\varepsilon(\lambda)$ is a function of the wavelength, absorption spectroscopy experiments typically consist of the elements shown in Figure 18.9. In the most basic form of this spectroscopy, a **monochromator** is used to separate the broadband radiation from the source into its constituent wavelengths. After passing through the sample, the transmitted light impinges on the detector. With this setup, only one wavelength can be measured at a time; therefore, this experiment is time consuming in comparison with Fourier transform techniques in which all wavelengths are detected simultaneously.

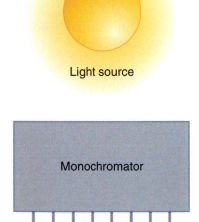

Light source

Monochromator

Sample cell

Detector

FIGURE 18.9
In an absorption experiment, the dependence of the sample absorption on wavelength is determined. A monochromator is used to filter out a particular wavelength from the broadband light source.

EXAMPLE PROBLEM 18.5

The molar absorption coefficient $\varepsilon(\lambda)$ for ethane is 40 $(\text{cm bar})^{-1}$ at a wavelength of 12 μm. Calculate $I(\lambda)/I_0(\lambda)$ in a 10-cm-long absorption cell if ethane is present at a contamination level of 2.0 ppm in 1 bar of air. What cell length is required to make $I(\lambda)/I_0(\lambda) = 0.90$?

Solution

Using $I(\lambda)/I_0(\lambda) = e^{-\varepsilon(\lambda)M\,l}$

$$\frac{I(\lambda)}{I_0(\lambda)} = \exp\{-[40\,(\text{cm bar})^{-1}(2.0\times10^{-6}\,\text{bar})(1.0\,\text{cm})]\} = 0.9992 \approx 1.0$$

This result shows that for this cell length, light absorption is difficult to detect.

Rearranging the Beer–Lambert equation, we have

$$l = -\frac{1}{M\varepsilon(\lambda)}\ln\left(\frac{I(\lambda)}{I_0(\lambda)}\right) = -\frac{1}{(40\ cm\ bar)^{-1}(2.0 \times 10^{-6}\ bar)}\ln 0.90 = 1.3 \times 10^3 cm$$

Path lengths of this order are possible in sample cells in which the light undergoes multiple reflections from mirrors outside of the cell. Even much longer path lengths are possible in cavity ringdown spectroscopy. In this method, the absorption cell is mounted between two highly focusing mirrors with a reflectivity greater than 99.99%. Because of the many reflections that take place between the mirrors without appreciable attenuation of the light, the effective length of the cell is very large. The detection sensitivity to molecules such as NO_2 is less than 10 parts per billion using this technique.

How does $\varepsilon(\lambda)$ depend on the wavelength or frequency? We know that for a harmonic oscillator, $\nu = (1/2\pi)\sqrt{k/\mu}$ so hetero that the masses of the atoms and the force constant of the bond determine the resonant frequency. Now consider a molecule such as

$$\begin{matrix} & O \\ & \| \\ R{-}&C{-}R' \end{matrix}$$

. The vibrational frequency of the C and O atoms in the carbonyl group is determined by the force constant for the C=O bond. This force constant is primarily determined by the chemical bond between these atoms and to a much lesser degree by the adjacent R and R' groups. For this reason, the carbonyl group has a characteristic frequency at which it absorbs infrared radiation that varies in a narrow range for different molecules. These group frequencies are very valuable in determining the structure of molecules, and an illustrative set is shown in Table 18.4.

We have shown that a diatomic molecule has a single vibrational peak of appreciable intensity, because the overtone frequencies have a low intensity. How many vibrational peaks are observed for larger molecules in an infrared absorption experiment? A molecule consisting of n atoms has three translational degrees of freedom, and two or three rotational degrees of freedom depending on whether it is a linear or nonlinear molecule. The remaining $3n - 6$ (nonlinear molecule) or $3n - 5$ (linear) degrees of freedom give rise to vibrational modes. For example, benzene has 30 vibrational modes. However, some of these modes have the same frequency (they are degenerate in energy), so that benzene has only 20 distinct vibrational frequencies.

We now examine some experimental data. Vibrational spectra for gas-phase CO and CH_4 are shown in Figure 18.10. Because CO and CH_4 are linear and nonlinear molecules, we expect one and nine vibrational modes, respectively. However, the spectrum for CH_4 shows two rather than nine peaks that we might associate with vibrational transitions. We also see several unexpected broad peaks in the CH_4 spectrum. The single peak in the CO spectrum is much broader than would be expected for a vibrational peak, and it has a deep minimum at the central frequency.

The spectra in Figure 18.10 look different than expected for two reasons. The broadening in the CO absorption peak and the broad envelopes of additional peaks for CH_4 result from transitions between different rotational energy states that occur simultaneously with the transition $n = 0 \rightarrow n = 1$ between vibrational energy levels. This occurs because many rotational levels lie between adjacent vibrational levels. The broad unresolved

TABLE 18.4   Selected Group Frequencies

Group	Frequency (cm^{-1})	Group	Frequency (cm^{-1})
O—H stretch	3600	C=O stretch	1700
N—H stretch	3350	C=C stretch	1650
C—H stretch	2900	C—C stretch	1200
C—H bend	1400	C—Cl stretch	700

FIGURE 18.10
Infrared absorption spectra of gaseous CO
and CH$_4$. The curves are offset vertically
for clarity.

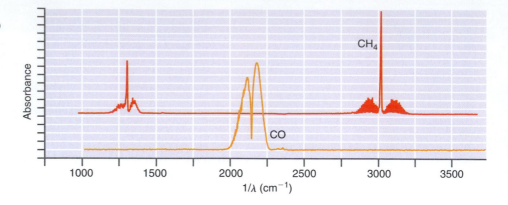

peaks seen for CO between 2000 and 2250 cm^{-1} correspond to transitions in which both the rotational and vibrational quantum numbers change. The minimum near 2200 cm^{-1} corresponds to the forbidden $\Delta J = 0$ transition. The broad and only partially resolved peaks for CH$_4$ seen around the sharp peaks centered near 1300 and 3000 cm^{-1} are again transitions in which both the rotational and vibrational quantum numbers change. The $\Delta J = 0$ transition is allowed for methane, and is the reason why the sharp central peaks are observed in the methane spectrum seen in Figure 18.10.

The second unexpected feature in Figure 18.10 is that two and not nine peaks are observed in the CH$_4$ spectrum. Why is this? To discuss the vibrational modes of polyatomic species in more detail, information about molecular symmetry and group theory is needed. At this point, we simply state the results and note that infrared absorption will not occur for a vibration that does not give rise to a change in the dipole moment of the molecule. In applying group theory to the CH$_4$ molecule, the 1306-cm^{-1} peak can be associated with three degenerate C—H bending modes, and the 3020-cm^{-1} peak can be associated with three degenerate C—H stretching modes. This still leaves three vibrational modes unaccounted for. Again applying group theory to the CH$_4$ molecule, one finds that these modes are symmetric and do not satisfy the condition $d\mu_x/dx \neq 0$. Therefore, they are infrared inactive.

However, all modes for CO and CH$_4$ are active in Raman spectroscopy. Of the 30 vibrational modes for benzene, four peaks (corresponding to 7 of the 30 modes) are observed in infrared spectroscopy, and seven peaks (corresponding to 12 of the 30 modes) are observed in Raman spectroscopy. None of the frequencies is observed in both Raman and infrared spectroscopy. Eleven vibrational modes are neither infrared nor Raman active.

Although the discussion to this point might lead us to believe that each bonded pair of atoms in a molecule vibrates independently of the others, this is not the case. For example, we might think that the linear CO$_2$ molecule has a single C=O stretching frequency, because the two C=O bonds are equivalent. However, experiments show that this molecule has two distinct C=O stretching frequencies. Why is this the case? Although the two bonds are equivalent, the vibrational motion in these bonds is not independent. When one C=O bond vibrates, the atomic positions and electron distribution throughout the molecule are changed, thereby influencing the other C=O bond. In other words, we can view the CO$_2$ molecule as consisting of two coupled harmonic oscillators. In the center of mass coordinates, each of the two C=O groups is modeled as a mass coupled to a wall by a spring with force constant k_1. We model the coupling as a second spring with force constant k_2 that connects the two oscillators. The model is depicted in Figure 18.11.

The coupled system shown in Figure 18.11 has two vibrational frequencies: the symmetric and antisymmetric modes. In the symmetric mode, the vibrational amplitude is equal in both magnitude and sign for the individual oscillators. In this case, the C atom does not move. This is equivalent to the coupling spring having the same length during the whole vibrational period. Therefore, the vibrational frequency is unaffected by the coupling and is given by

FIGURE 18.11
The coupled oscillator model of CO_2 is shown. The blue spheres represent C and O treated as a single reduced mass. The vertical dashed lines show the equilibrium positions. The symmetric vibrational mode is shown in the upper part of the figure, and the antisymmetric vibrational mode is shown in the lower part of the figure.

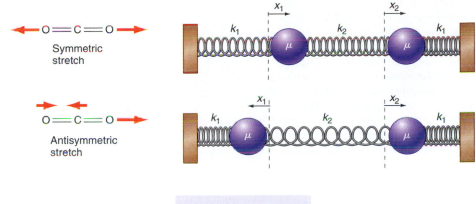

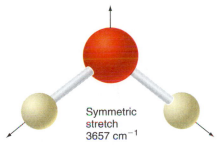

Symmetric
stretch
3657 cm^{-1}

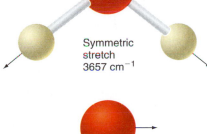

Antisymmetric
stretch
3756 cm^{-1}

FIGURE 18.12
The two O—H stretching modes for the H_2O molecule are shown. Note that neither of them corresponds to the stretching of a single localized O—H bond. This is the case because, although equivalent, the two bonds are coupled, rather than independent.

Web-Based Problem 18.3 Normal Modes for H_2O

Web-Based Problem 18.4 Normal Modes for CO_2

Web-Based Problem 18.5 Normal Modes for NH_3

$$\nu_{symmetric} = \frac{1}{2\pi}\sqrt{\frac{k_1}{\mu}} \qquad (18.14)$$

In the antisymmetric mode, the C atom does move. This is equivalent to the vibrational amplitude being equal in both magnitude and opposite sign for the individual oscillators. In this mode, the spring representing the coupling is doubly stretched, once by each of the oscillators. The resulting force on each mass is

$$F = -(k_1 + 2k_2)x \qquad (18.15)$$

and the resulting frequency of this antisymmetric mode is

$$\nu_{antisymmetric} = \frac{1}{2\pi}\sqrt{\frac{(k_1 + 2k_2)}{\mu}} \qquad (18.16)$$

We see that the C=O bond coupling gives rise to two different vibrational stretching frequencies and that the antisymmetric mode has the higher frequency. This effect is illustrated in Figure 18.12, where the symmetric and antisymmetric stretching modes of H_2O are shown.

Based on theoretical calculations, the intrinsic line width for vibrational spectra is less than ~10^{-3} cm^{-1}. This is very small compared with the resolution of conventional infrared spectrometers, which is typically no better than 0.1 cm^{-1}. Therefore, the width of peaks in a spectrum is generally determined by the instrumental function, which is the response of the instrument to a spectral line of very narrow line width. As shown in the top panel of Figure 18.13, a peak with the width of the instrument function gives no information about the intrinsic line width. However, peaks that are broader than the instrument function are obtained if a sample contains many different local environments for the entity generating the peak. For example, the O—H stretching region in an infrared spectrum in liquid water is very broad. This is the case because of the many different local geometries that arise from hydrogen bonding between H_2O molecules, and each of them gives rise to a slightly different O—H stretching frequency. This effect is referred to as **inhomogeneous broadening** and is illustrated in the bottom panel of Figure 18.13.

18.7 Vibrational Spectroscopy of Polypeptides in Solution

Vibrational spectra for molecules of biochemical interest in aqueous solution differ from those shown in Figure 18.10 in that broad overlapping peaks are generally observed, meaning that absorption of radiation occurs over a range in wavelength. There are two principal reasons for this peak broadening relative to gas-phase spectra. The first is that a nonlinear molecule made up of n atoms has $3n - 6$ modes of vibration, where n can

An experimental spectrum, shown on the right, arises from the convolution (indicated by the symbol *) of the instrument function and the intrinsic line width of the transition. For a homogeneous sample (top panel) the width of the spectrum is generally determined by the instrument function. For an inhomogeneous sample, the intrinsic line width is the sum of the line widths of the many different local environments. For inhomogeneous samples, the width of the spectrum can be determined by the intrinsic line width, rather than the instrument function.

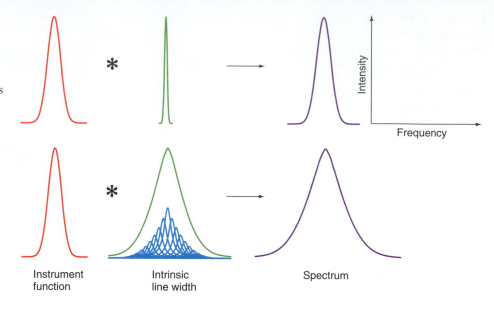

Instrument function   Intrinsic line width   Spectrum

Web-Based Problem 18.6 Normal Modes for Formaldehyde

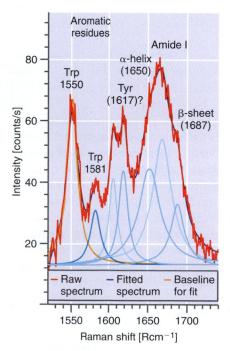

FIGURE 18.14

A vibrational spectrum of endothelin–1 obtained using Raman spectroscopy (see Section 18.8) is shown. A deconvolution of the spectrum into individual peaks is also shown. (Source: Reprinted from Analytical Biochemistry, Vol. 277, John T. Pelton and Larry R. McLean, *Spectroscopic Methods for Analysis of Protein Secondary Structure,* pages 167–176, copyright 2000, with permission from Elsevier)

be in the range of 100 to 1000. Therefore, the many vibrational transitions overlap, leading to absorption over a range of wavelengths. The second reason is that biomolecules dissolved in water interact strongly with the solvent and that there may be a variety of water–molecule complexes, each of which exhibits slightly different vibrational frequencies. Therefore, the absorption spectrum will exhibit inhomogeneous broadening as discussed in the previous section.

Figure 18.14 shows a vibrational spectrum in aqueous solution of the 21-amino-acid peptide endothelin-1, which is a drug that causes blood vessels to constrict. Whereas the observed widths of vibrational peaks in a gas-phase absorption spectrum are typically 2 cm^{-1} and instrument resolution limited, the peaks in this figure are typically ~40 to 100 cm^{-1} wide.

The spectrum in Figure 18.14 shows regions characteristic of the tryptophan and tyrosine residues, and a broad region labeled amide 1. The frequency corresponding to this spectral feature is very useful because it can be used to determine the secondary structure (that resulting from folding) of the molecule. The secondary structure of a protein plays a crucial role in its chemical function. Two common secondary structures of protein are the **α-helix** and the **β-sheet.** Each is stabilized by hydrogen bonds formed between the hydrogen on the —NH— amide group and the electronegative oxygen on a carbonyl group. In the α-helix, the hydrogen bonds are intramolecular because they are between groups in the same polypeptide chain. In the β-sheet, the hydrogen bonds are intermolecular because they are between groups in adjacent polypeptide chains. Each successive turn of the α-helix is stabilized through the formation of hydrogen bonds. These hydrogen bonds lead to a shift in the frequency of the amide group vibrational frequencies from those observed for an amide group without hydrogen bonding, and characteristic values for the maximum of this band in an α-helix structure lie between 1650 and 1655 cm^{-1}. A second common secondary structure is the β-sheet shown together with the α-helix in Figure 18.15. As for the α-helix, this structure is stabilized by hydrogen bonds, but the coupling of adjacent polypeptide chains leads to a sheet-like structure rather than a helical structure. A strong amide 1 band between 1612 and 1640 cm^{-1} is observed for the β-sheet structure.

18.8 Basics of Raman Spectroscopy

As discussed in the previous sections, absorption of light in the infrared portion of the spectrum can lead to transitions between eigenstates of the vibrational and rotational energy. Another interaction between a molecule and an electromagnetic field can also lead to vibrational and rotational excitation. It is called the **Raman effect** after its discoverer

FIGURE 18.15

The (a) α-helix and (b) β-sheet are two important forms in which proteins are found in aqueous solution. In both structures, hydrogen bonds form between imino (—NH—) groups and carbonyl groups. (Source: Reprinted with permission from Becker et al Fig. 3.8.)

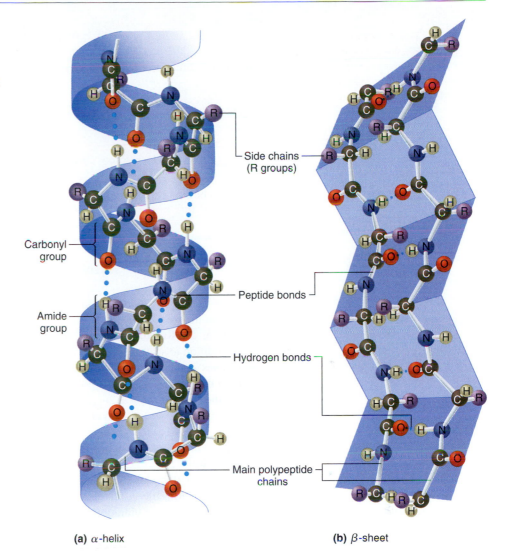

Side chains (R groups)

Carbonyl group

Amide group

Peptide bonds

Hydrogen bonds

Main polypeptide chains

(a) α-helix   **(b)** β-sheet

and involves scattering of a photon by the molecule. You can think of scattering as the collision between a molecule and a photon in which energy and momentum are transferred between the two collision partners. Raman spectroscopy complements infrared absorption spectroscopy because it obeys different selection rules. For instance, the stretching mode in a homonuclear diatomic molecule is Raman active, but infrared inactive. The reasons for this difference are discussed in Chapter 28 of Engel and Reid, *Physical Chemistry*.

Consider a molecule with a characteristic vibrational frequency ν_{vib} in an electromagnetic field that has a time-dependent electric field given by

$$E = E_0 \cos 2\pi\nu t \qquad (18.17)$$

The electric field distorts the molecule slightly because the negative valence electrons and the positive nuclei and their core electrons experience forces in opposite directions. This induces a time-dependent dipole moment of magnitude $\mu_{induced}(t)$ in the molecule of the same frequency as the field. The dipole moment is linearly proportional to the magnitude of the electric field, and the proportionality constant is the **polarizability,** α. The polarizability is an anisotropic quantity and its value depends on the direction of the electric field relative to the molecular axes:

$$\mu_{induced}(t) = \alpha E_0 \cos 2\pi\nu t \qquad (18.18)$$

The polarizability depends on the bond length $x_e + x(t)$, where x_e is the equilibrium value. The polarizability α can be expanded in a Taylor–Mclaurin series (see the Math Supplement, Appendix A) in which terms beyond the first order have been neglected:

$$\alpha(x) = \alpha(x_e) + x\left(\frac{d\alpha}{dx}\right)_{x=x_e} + \dots \tag{18.19}$$

Due to the vibration of the molecule, $x(t)$ is time dependent and is given by

$$x(t) = x_{max}\cos 2\pi\nu_{vib}t \tag{18.20}$$

Combining this result with Equation (18.19), we can rewrite Equation (18.18) in the form

$$\mu_{induced}(t) = \alpha E = E_0\cos 2\pi\nu t\left[\alpha(x_e) + \left[\left(\frac{d\alpha}{dx}\right)_{x=x_e}\right]x_{max}\cos 2\pi\nu_{vib}t\right] \tag{18.21}$$

which can be simplified, using the trigonometric identity $\cos x\cos y = \frac{1}{2}[\cos(x-y) + \cos(x+y)]$ to

$$\mu_{induced}(t) = \alpha E = \alpha(x_e)E_0\cos 2\pi\nu t$$

$$+ \left[\left(\frac{d\alpha}{dx}\right)_{x=x_e}\right]x_{max}E_0[\cos(2\pi\nu + 2\pi\nu_{vib})t + \cos(2\pi\nu - 2\pi\nu_{vib})t] \tag{18.22}$$

The time-varying dipole moment radiates light of the same frequency as the dipole moment, or at the frequencies ν, $(\nu - \nu_{vib})$, and $(\nu + \nu_{vib})$. These three frequencies are referred to as the **Rayleigh, Stokes,** and **anti-Stokes frequencies,** respectively. We see that in addition to scattered light at the incident frequency, light will also be scattered at frequencies corresponding to vibrational excitation and de-excitation. Higher order terms in the expansion for the polarizability [Equation (18.19)] also lead to scattered light at the frequencies $\nu \pm 2\nu_{vib}$, $\nu \pm 3\nu_{vib}$, ..., but the scattered intensity at these frequencies is much weaker than at the primary frequencies.

Equation (18.22) illustrates that the intensity of the Stokes and anti-Stokes peaks is zero unless $d\alpha/dx \neq 0$. We conclude that for vibrational modes to be Raman active, the polarizability of the molecule must change as it vibrates. This condition is satisfied for many vibrational modes and, in particular, it is satisfied for the stretching vibration of a homonuclear molecule. Although $d\mu_x/dx = 0$ for these molecules, making them infrared inactive, the stretching vibration of a homonuclear molecule is Raman active. Not all vibrational modes that are active for the absorption of infrared light are Raman active and vice versa. This is why infrared and Raman spectroscopies provide a valuable complement to one another.

A schematic picture of the scattering event in Raman spectroscopy on an energy scale is shown in Figure 18.16. This diagram is quite different from that considered earlier in depicting a transition between two states. Both the initial and final states are the $n = 0$ and $n = 1$ states at the bottom of the figure. To visualize the interaction of the molecule with the photon of energy $h\nu$, which is much greater than the vibrational energy spacing, we imagine the photon to be absorbed by the molecule, resulting in a much higher intermediate energy "state." This very short-lived "state" quickly decays to the final state. Whereas the initial and final states are eigenfunctions of the time-independent Schrödinger equation, the upper "state" in this energy diagram need not satisfy this condition. Therefore, it is referred to as a virtual state.

Are the intensities of the Stokes and anti-Stokes peaks equal? We know that their relative intensity is governed by the relative number of molecules in the originating states. For the Stokes line, the transition originates from the $n = 0$ state, whereas for the anti-Stokes line, the transition originates from the $n = 1$ state. Therefore, the relative intensity of the Stokes and anti-Stokes peaks can be calculated using the Boltzmann distribution:

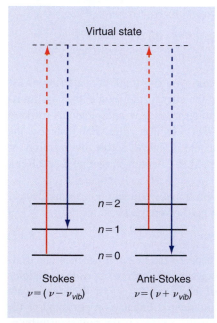

FIGURE 18.16
Schematic depiction of the Raman scattering event. The spectral peak resulting in vibrational excitation is called the Stokes peak, and the spectral peak originating from vibrational de-excitation is called the anti-Stokes peak.

$$\frac{I_{anti-Stokes}}{I_{Stokes}} = \frac{n_{excited}}{n_{ground}} = \frac{e^{-3h\nu/2kT}}{e^{-h\nu/2kT}} = e^{-h\nu/kT} \tag{18.23}$$

For vibrations for which $\tilde{\nu}$ is in the range of 1000 to 3000 cm^{-1}, this ratio ranges between 8×10^{-3} and 5×10^{-7} at 300 K. This calculation shows that the intensities of the Stokes and anti-Stokes peaks will be quite different. In this discussion of the Raman effect, we have only considered vibrational transitions. However, just as for infrared absorption spectra, Raman spectra show peaks originating from both vibrational and rotational transitions.

Raman and infrared absorption spectroscopy are complementary and both can be used to study the vibration of molecules. Both techniques can be used to determine the identities of molecules in a complex mixture by comparing the observed spectral peaks with characteristic group frequencies. The most significant difference between these two spectroscopies is the light source needed to implement the technique. For infrared absorption spectroscopy, the light source must have significant intensity in the infrared portion of the spectrum. Because Raman spectroscopy is a scattering technique, the frequency of the light used need not match the frequency of the transition being studied. Therefore, a source in the visible part of the spectrum is generally used to study rotational and vibrational modes. This has several advantages over infrared sources. By shifting the vibrational spectrum from the infrared into the visible part of the spectrum, intense and commonly available lasers can be used to obtain Raman spectra. Intense light sources are necessary because the probability for Raman scattering is generally on the order of 10^{-6} or less. Furthermore, shifting the frequency of the source from the infrared into the visible part of the spectrum can reduce interference with absorbing species that are not of primary interest. For instance, infrared spectra of aqueous solutions always contain strong water peaks that may mask other peaks of interest. By shifting the source frequency to the visible part of the spectrum, such interferences can be eliminated. If the photon energy used approaches the energy of an electronic excitation in the molecule, the virtual state in Figure 18.16 becomes a real state, and the intensity of the Raman peak can increase dramatically. Spectroscopy carried out under this condition is called *resonance Raman spectroscopy*. This type of spectroscopy has important analytical applications.

Another interesting application of the Raman effect is the Raman microscope or microprobe. Because Raman spectroscopy is done in the visible part of the light spectrum, it can be combined with optical microscopy to obtain spectroscopic information with a spatial resolution of better than 0.01 mm. An area in which this technique has proved particularly useful is as a nondestructive probe of the composition of gas inclusions such as CH_4, CO, H_2S, N_2, and O_2 in mineral samples. Raman microscopy has also been used in biopsy analyses to identify mineral particles in the lung tissues of silicosis victims and to analyze the composition of gallstones.

18.9 Using Raman Spectroscopy to Image Living Cells with Chemical Specificity

Imaging techniques such as optical and electron microscopy are essential tools of the scientist in probing the workings of a complex system such as a cell. It is particularly useful if the imaging technique can also provide a chemical differentiation of the sample under investigation. In optical microscopy, this can be done to a limited degree using stains, and X-rays emitted in electron microscopy have a photon energy that is element specific, also allowing chemical analysis. However, neither stains nor the vacuum environment needed to carry out electron microscopy are compatible with living cells. Fluorescence, which is light emission based on a spectroscopy involving transitions between electronic states (see Chapter 19), allows chemical imaging of biological systems such as cells, but has the limitation that most molecules found in cells do not undergo fluorescence. This issue can be addressed by adding fluorescent tags to molecules of interest, but in some cases, the fluorescent tag acts as a poison and is therefore not compatible with a living cell.

The development of **coherent anti-Stokes Raman spectroscopy (CARS)** microscopy has the potential to provide a chemically specific image technique compatible with living cells. Because it has a spatial resolution comparable to that of light microscopy and a time resolution of ~1 s, it can be used to probe dynamic processes with

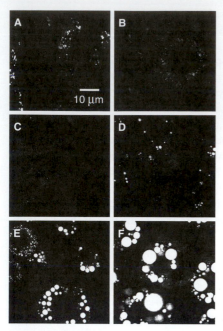

FIGURE 18.17

CARS images taken at 2845 cm^{-1} of the same cell culture at different times after adding induction media: (A) 0 h, (B) 24 h, (C) 48 h, (D) 60 h, (E) 96 h, and (F) 192 h. At each time point, several images at different areas were taken on the same cell sample. The image that best represents the average LD content and distribution at each time is shown. (Source: Xiaolin Nan, Ji-Xin Cheng and X. Sunney Xie, "Vibrational imaging of lipid droplets in live fibroblast cells with coherent anti-Stokes Raman scattering microscopy," *Journal of Lipid Research,* Vol. 44, 2002–2008, November 2003, Copyright 2003 by American Society for Biochemistry and Molecular Biology, Fig. 5))

moderate spatial resolution. In this technique, two laser beams of frequency ω_1 and ω_2 are focused on the sample. The electric field in the region in which the beams intersect has a frequency component $\omega_1 - \omega_2$. If this difference frequency is equal to a characteristic vibrational frequency of the sample, the vibrational modes of different molecules become synchronized, producing an absorption signal that can be up to a factor of 10^5 larger than signals normally encountered in Raman spectroscopy. We next discuss a specific application of this method to demonstrate its usefulness.

Lipids are water-insoluble biomolecules that are soluble in organic solvents. They store energy, have a role in signaling, and are a major structural component of biological membranes. They are typically found in fat and liver cells in the form of small droplets. CARS has been used to study the conversion of fibroblasts to fat cells [Nan et al., *J. Lipid Research* 44 (2003), 2202]. A fibroblast is a cell type found in connective tissues that migrates and multiplies readily in wound repair. Lipids are rich in —CH$_2$— groups and can be imaged with high sensitivity with CARS at a frequency of 2845 cm^{-1}, which corresponds to a C—H stretch.

Figure 18.17 shows images of representative cells taken at intervals over 192 h. The fibroblasts initially contained some lipid droplets with diameters of ~1 to 2 μm, scattered in the cytoplasm as is seen in part A. An initial reduction of lipid droplets was observed 24 h after adding the induction medium as seen in part B, and at 48 h most of the cells contained few or no lipid droplets as seen in part C. For longer times, cells were seen to grow lipid droplets again as seen in parts E and F. In the course of the conversion, the appearance of clear boundaries between cells was seen, and this was accompanied by the accumulation of lipid droplets. Although the details of this conversion are not well understood, this example illustrates the usefulness of CARS in obtaining chemically differentiated images using vibrational energy transitions.

Vocabulary

absorbance	coherent source	monochromator	selection rule
absorption	dipole approximation	Morse potential	spectroscopy
α-helix	dynamic dipole moment	overtone	spontaneous emission
anharmonic potential	group frequencies	permanent dipole moment	stimulated emission
anti-Stokes frequency	harmonic potential	polarizability	Stokes frequency
Beer–Lambert law	incoherent photon source	Raman effect	transmittance
β-sheet	inhomogeneous broadening	Rayleigh frequency	transition dipole moment
bond energy	molar absorption coefficient	rotational constant	wavenumber
coherent anti-Stokes Raman spectroscopy (CARS)			

Questions on Concepts

Q18.1 A molecule in an excited state can decay to the ground state either by stimulated emission or spontaneous emission. Use the Einstein coefficients to predict how the relative probability of these processes changes as the frequency of the transition doubles.

Q18.2 What feature of the Morse potential makes it suitable for modeling dissociation of a diatomic molecule?

Q18.3 What is the difference between a permanent and a dynamic dipole moment?

Q18.4 Solids generally expand as the temperature increases. Such an expansion results from an increase in the bond length between adjacent atoms as the vibrational amplitude increases. Will a harmonic potential between adjacent atoms lead to thermal expansion? Will a Morse potential lead to thermal expansion?

Q18.5 Does the initial excitation in Raman spectroscopy take place to a stationary state of the system? Explain your answer.

Q18.6 If a spectral peak is broadened, can you always conclude that the excited state has a short lifetime?

Q18.7 What is the intermolecular interaction that gives rise to α-helix or β-sheet formation?

Q18.8 What gives rise to the contrast observed in CARS images?

Problems

Problem numbers in **RED** indicate that the solution to the problem is given in the *Student Solutions Manual*.

P18.1 A strong absorption band in the infrared region of the electromagnetic spectrum is observed at $\tilde{\nu} = 2170 \text{ cm}^{-1}$ for $^{12}\text{C}^{16}\text{O}$. Assuming that the harmonic potential applies, calculate the fundamental frequency ν in units of inverse seconds, the vibrational period in seconds, and the zero point energy for the molecule in joules and electron-volts.

P18.2 Isotopic substitution is used to identify characteristic groups in an unknown compound using vibrational spectroscopy. Consider the C=C bond in ethene (C_2H_4). By what factor would the frequency change if deuterium were substituted for all the hydrogen atoms? Treat the H and D atoms as being rigidly attached to the carbon.

P18.3 The force constants for H_2 and Br_2 are 575 and 246 N m^{-1}, respectively. Calculate the ratio of the vibrational state populations n_1/n_0 and n_2/n_0 at $T = 300$ and at 1000 K.

P18.4 The $^1\text{H}^{35}\text{Cl}$ molecule can be described by a Morse potential with $D_e = 7.41 \times 10^{-19}$ J. The force constant k for this molecule is 516.3 N m^{-1} and $\nu = 8.97 \times 10^{13}$ s^{-1}.

  a. Calculate the lowest four energy levels for a Morse potential using the following formula:

  $$E_n = h\nu\left(n + \frac{1}{2}\right) - \frac{(h\nu)^2}{4D_e}\left(n + \frac{1}{2}\right)^2$$

  b. Calculate the fundamental frequency ν_0 corresponding to the transition $n = 0 \rightarrow n = 1$ and the frequencies of the first three overtone vibrations. How large would the relative error be if you assume that the first three overtone frequencies are $2\nu_0$, $3\nu_0$, and $4\nu_0$?

P18.5 Using the formula for the energy levels for the Morse potential,

$$E_n = h\nu\left(n + \frac{1}{2}\right) - \frac{(h\nu)^2}{4D_e}\left(n + \frac{1}{2}\right)^2$$

show that the energy spacing between adjacent levels is given by

$$E_{n+1} - E_n = h\nu - \frac{(h\nu)^2}{2D_e}(1 - n)$$

For $^1\text{H}^{35}\text{Cl}$, $D_e = 7.41 \times 10^{-19}$ J and $\nu = 8.97 \times 10^{13}$ s^{-1}. Calculate the smallest value of n for which $E_{n+1} - E_n < 0.5(E_1 - E_0)$.

P18.6 Show that the Morse potential approaches the harmonic potential for small values of the vibrational amplitude. (*Hint:* Expand the Morse potential in a Taylor–Mclaurin series.)

P18.7 A measurement of the vibrational energy levels of $^{12}\text{C}^{16}\text{O}$ gives the relationship

$$\tilde{\nu}(n) = 2170.21\left(n + \frac{1}{2}\right) \text{ cm}^{-1} - 13.461\left(n + \frac{1}{2}\right)^2 \text{ cm}^{-1}$$

where n is the vibrational quantum number. The fundamental vibrational frequency is $\tilde{\nu}_0 = 2170.21$ cm^{-1}. From these data, calculate the depth D_e of the Morse potential for $^{12}\text{C}^{16}\text{O}$. Calculate the bond energy of the molecule.

P18.8 The fundamental vibrational frequencies for $^1\text{H}^{19}\text{F}$ and $^2\text{D}^{19}\text{F}$ are 4138.52 and 2998.25 cm^{-1}, respectively, and D_e for both molecules is 5.86 eV. What is the difference in the bond energy of the two molecules?

P18.9 Derive a general relationship between the absorbance and the transmittance using the Beer–Lambert law.

P18.10 Greenhouse gases generated from human activity absorb infrared radiation from the Earth and keep it from being dispersed outside our atmosphere. This is a major cause of global warming. Compare the path length required to absorb 99% of the Earth's radiation near a wavelength of 7 μm for CH_3CCl_3 [$\varepsilon(\lambda) = 1.8$ (cm atm)$^{-1}$] and the chlorofluorocarbon CFC-14 [$\varepsilon(\lambda) = 4.1 \times 10^3$ (cm atm)$^{-1}$] assuming that each of these gases has a partial pressure of 2.0×10^{-6} bar.

P18.11 For a 10.0-mm-thick piece of fused silica quartz glass, 50.% of the light incident on it passes through the glass. What percentage of the light will pass through a 20.0-mm-thick piece of the same glass?

P18.12 An infrared absorption spectrum of an organic compound is shown in the following figure. Use the characteristic group frequencies listed in Section 18.6 to decide whether this compound is more likely to be ethyl amine, pentanol, or acetone.

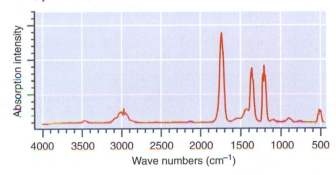

P18.13 An infrared absorption spectrum of an organic compound is shown in the following figure. Use the characteristic group frequencies listed in Section 18.6 to decide whether this compound is more likely to be hexene, hexane, or hexanol.

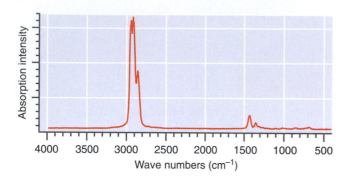

P18.14 Calculating the motion of individual atoms in the vibrational modes of molecules (called normal modes) is an advanced topic. Given the normal modes shown in the following figure, decide which of the normal modes of CO_2 and H_2O have a nonzero dynamic dipole moment and are therefore infrared active. The motion of the atoms in the second of the two doubly degenerate bend modes for CO_2 is identical to the first, but is perpendicular to the plane of the page.

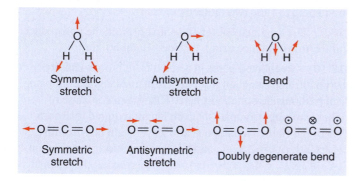

P18.15 Purification of water for drinking using UV light is a viable way to provide potable water in many areas of the world. Experimentally, the decrease in UV light of wavelength 250 nm follows the empirical relation $I/I_0 = e^{-\varepsilon' l}$ where l is the distance that the light passed through the water and ε' is an effective absorption coefficient. $\varepsilon' = 0.070$ cm^{-1} for pure water and 0.30 cm^{-1} for water exiting a waste water treatment plant. What distance corresponds to a decrease in I of 10% from its incident value for a) pure water, and b) waste water?

P18.16 The rotational constant for $^{127}I^{79}Br$ determined from microwave spectroscopy is 0.1141619 cm^{-1}. The atomic masses of ^{127}I and ^{79}Br are 126.904473 and 78.918336 amu, respectively. Calculate the bond length in $^{127}I^{79}Br$ to the maximum number of significant figures consistent with this information.

P18.17 The rotational constant for $^2D^{19}F$ determined from microwave spectroscopy is 11.007 cm^{-1}. The atomic masses of ^{19}F and ^2D are 18.9984032 and 2.0141018 amu, respectively. Calculate the bond length in $^2D^{19}F$ to the maximum number of significant figures consistent with this information.

P18.18 Calculate the moment of inertia and the energy in the $J = 1$ rotational state for $^1H_2$ in which the bond length of $^1H_2$ is 74.6 pm. The atomic mass of ^1H is 1.007825 amu.

P18.19 Calculate the moment of inertia and the energy in the $J = 1$ rotational state for $^{16}O_2$ in which the bond length is 120.8 pm. The atomic mass of ^{16}O is 15.99491 amu.

P18.20 The rigid rotor model can be improved by recognizing that in a realistic anharmonic potential, the bond length increases with the vibrational quantum number n. Therefore, the rotational constant depends on n, and it can be shown that $B_n = B - (n + 1/2)\alpha$, where B is the rigid rotor value. The constant α can be obtained from experimental spectra. For $^1H^{79}Br$, $B = 8.473$ cm^{-1} and $\alpha = 0.226$ cm^{-1}. Using this more accurate formula for B_n, calculate the bond length for HBr in the ground state and for $n = 3$.

P18.21 Overtone transitions in vibrational absorption spectra for which $\Delta n = +2, +3, \ldots$ are forbidden for the harmonic potential $V = (1/2)kx^2$ because $\mu_x^{mn} = 0$ for $|m - n| \neq 1$ as shown in Section 18.4. However, overtone transitions are allowed for the more realistic anharmonic potential. In this problem, you will explore how the selection rule is modified by including anharmonic terms in the potential. We do so in an indirect manner by including additional terms in the expansion of the dipole moment $\mu_x(x) = \mu_{0x} + x(d\mu_x/dx)_{r_e} + \ldots$, but assuming that the harmonic oscillator total energy eigenfunctions are still valid. This approximation is valid if the anharmonic correction to the harmonic potential is small. You will show that including the next term in the expansion of the dipole moment, which is proportional to x^2, allows for the transitions $\Delta n = \pm 2$.

a. Show that Equation (18.6) becomes

$$\mu_x^{m0} = A_m A_0 \mu_{0x} \int_{-\infty}^{\infty} H_m(\alpha^{1/2}x) H_0(\alpha^{1/2}x) e^{-\alpha x^2} dx$$

$$+ A_m A_0 \left(\frac{d\mu_x}{dx}\right)_{x=0} \int_{-\infty}^{\infty} H_m(\alpha^{1/2}x) x H_0(\alpha^{1/2}x) e^{-\alpha x^2} dx$$

$$+ \frac{A_m A_0}{2!} \left(\frac{d^2\mu_x}{dx^2}\right)_{x=0} \int_{-\infty}^{\infty} H_m(\alpha^{1/2}x) x^2 H_0(\alpha^{1/2}x) e^{-\alpha x^2} dx$$

b. Evaluate the effect of adding the additional term to μ_x^{mn}. You will need the recursion relationship
$\alpha^{1/2}x H_n(\alpha^{1/2}x) = n H_{n-1}(\alpha^{1/2}x) + \frac{1}{2}H_{n+1}(\alpha^{1/2}x)$.

Show that both the transitions $n = 0 \to n = 1$ and $n = 0 \to n = 2$ are allowed in this case.

P18.22 Selection rules in the dipole approximation are determined by the integral $\mu_x^{mn} = \int \psi_m^*(\tau)\mu_x(\tau)\psi_n(\tau)d\tau$. If this integral is nonzero, the transition will be observed in an absorption spectrum. If the integral is zero, the transition is "forbidden" in the dipole approximation. It actually occurs with low probability because the dipole approximation is not exact. Consider the particle in the one-dimensional box and set $\mu_x = -ex$.

a. Calculate μ_x^{12} and μ_x^{13} in the dipole approximation. Can you see a pattern and discern a selection rule? You may need to evaluate a few more integrals of the type μ_x^{1m}. The standard integral

$$\int x \sin\frac{\pi x}{a} \sin\frac{n\pi x}{a}\,dx = \frac{1}{2}\left(\frac{a^2\cos\frac{(n-1)\pi x}{a}}{(n-1)^2\pi^2} + \frac{a x \sin\frac{(n-1)\pi x}{a}}{(n-1)\pi}\right)$$

$$-\frac{1}{2}\left(\frac{a^2\cos\frac{(n+1)\pi x}{a}}{(n+1)^2\pi^2} + \frac{a x \sin\frac{(n+1)\pi x}{a}}{(n+1)\pi}\right)$$

is useful for solving this problem.

b. Determine the ratio μ_x^{12}/μ_x^{14}. On the basis of your result, would you modify the selection rule that you determined in part (a)?

P18.23 The fundamental vibration frequency of gaseous $^{14}N^{16}O$ is $\tilde{\nu} = 1904$ cm^{-1}.

a. Calculate the force constant using formula for a simple harmonic oscillator.

b. When $^{14}N^{16}O$ is bound to hemoglobin the oxygen is anchored to the protein so its effective mass is much greater (i.e., essentially infinite) compared to the mass of the nitrogen atom. Using the force constant from part a, calculate the fundamental vibration frequency of NO bound to hemoglobin.

P18.24 The fundamental vibrational frequency for $^{12}C^{16}O$ is $\tilde{\nu} = 2170$ cm^{-1}. Calculate the force constant using a simple harmonic oscillator model and calculate the vibrational frequency if $^{12}C^{16}O$ binds to the protein via the oxygen.

P18.25 Resonance Raman spectroscopy has been used to study the binding of O_2 to oxygen transport proteins like hemerythrin and hemocyanin (T.B. Freedman, J.S. Loehr, and T.M. Loehr, "A Resonance Raman Studies of the Copper Protein Hemcyanin" *J. Amer. Chem. Soc.* 98:10, 2809–2815). These transport proteins are found in brachiopods and in arthropods, respectively. Hemerythrin binds oxygen to two iron atoms while hemocyanin binds oxygen to two copper atoms. Although O_2 is IR inactive, the bond stretch of O_2 can be detected by Raman spectroscopy. Assume both iron atoms in deoxyhemerythrin are in the +2 valence state (i.e., Fe^{2+}). The oxygen stretch occurs at 845 cm^{-1} in oxyhemerythrin. In contrast free oxygen has a Raman band at 1555 cm^{-1} and peroxide (O_2^{2-}) occurs at 738 cm^{-1}. Based on the Raman data, write a scheme for the binding of O_2 to the two Fe^{2+} atoms of hemerythrin. Hint: The Raman data indicate that charge transfer accompanies oxygen binding. Include this fact in your reaction scheme.

P18.26 If the bond stretch for $^{16}O_2$ in hemacyanin occurs at 845 cm^{-1}, calculate the frequency of the Raman band expected for the bound stretch of $^{18}O_2$ in hemacyanin.

P18.27 In P18.26 the frequencies of the Raman bands for the bond stretches of $^{16}O_2$ and $^{18}O_2$ were considered. When $^{18}O-^{16}O$ is bond to hemacyanin, the single Raman band for

the O_2 bond stretch is split into a doublet. Based on this fact, is the binding geometry of O_2 to hemacyanin symmetric or asymmetric? Explain your answer. Examples of symmetric and asymmetric binding geometries are given here. Estimate the frequencies of the two bands and state any assumptions you make in your calculation.

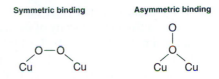

P18.28 The tautomeric forms of pyrimidine bases in a polynucleotide structure were determined by H. Todd Miles using IR absorption spectroscopy (H. T. Miles (1961) "Tautomeric Forms in a Polynucleotide Helix and their Bearing on DNA Structure" *Proc. Natl. Acad. Sci.* (USA) 47, 791–802). At issue is whether in the polynucleotide structure the labile proton is located on the amino nitrogen (i.e., structure I) or N3 in the cytosine pyrimidine ring (structure II). To determine this the IR spectra of A) cytidine, B) 1-(β-D-glucopyranosyl-4-dimethylamino-2-pyrimidone), i.e., structure III where R=CH$_3$, and C) 1,3-dimethylcytosine, i.e., structure IV where R=CH$_3$, were obtained. It is considered probable that the strong band at 1651 cm^{-1} is from the vibration of the C=O bond, while the weaker bands are from C=N stretches. Identify the tautomeric form of cytidine in DNA. Explain your answer.

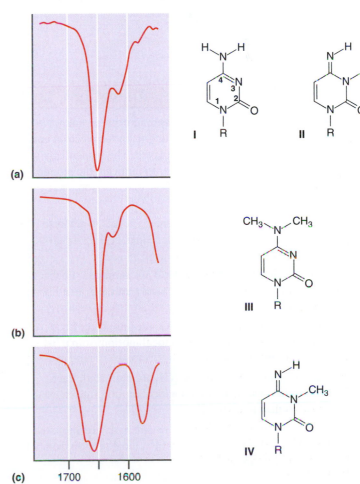

P18.29   The IR study by H. T. Miles of tautomeric forms of bases in polynucleotide helices described in P18.28 was also applied to purine bases. Inosine, for example, a labile proton may be attached to N1 (V) or to the oxygen bonded to C6 (VI). Compare the IR spectra of inosine (G), 1-methylinosine (VII), and 6-methyloxy-9-β-D-ribofuranosylpurine (VIII) and determine the tautomeric form of inosine.

P18.30   Using only your answers to P18.28 and P18.29, based on the data derived from the IR study of the tautomeric forms of purine and pyrimidine bases by H. T. Miles, determine the pattern of hydrogen bonding between the purine and pyrimine bases in a guanidine-cytidine (i.e., G-C) Watson-Crick base pair in DNA.

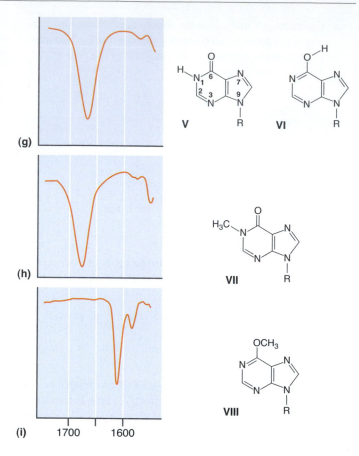

Web-Based Simulations, Animations, and Problems

W18.1   A pair of emission spectra, one from an unknown (hypothetical) atom and one resulting from the electron energy levels entered using sliders, is displayed. The student adjusts the displayed energy levels in order to replicate the atomic spectrum and, hence, determine the actual electron energy levels in the atom.

W18.2   The number of allowed energy levels in a Morse potential is determined for variable values of the vibrational frequency and the well depth.

W18.3   The normal modes for H_2O are animated. Each normal mode is associated with a local motion from a list displayed in the simulation.

W18.4   The normal modes for CO_2 are animated. Each normal mode is associated with a local motion from a list displayed in the simulation.

W18.5   The normal modes for NH_3 are animated. Each normal mode is associated with a local motion from a list displayed in the simulation.

W18.6   The normal modes for formaldehyde are animated. Each normal mode is associated with a local motion from a list displayed in the simulation.

W18.7   Simulated rotational (microwave) spectra are generated for one or more of the diatomic molecules $^{12}C^{16}O$, $^{1}H^{19}F$, $^{1}H^{35}Cl$, $^{1}H^{79}Br$, and $^{1}H^{127}I$. Using a slider, the temperature is varied. The J value corresponding to the maximum intensity peak is determined and compared with the prediction from the formula

$$J_{max} = \frac{1}{2}\left[\sqrt{\frac{4I\,kT}{\hbar^2}} - 1 \right]$$

The number of peaks that have an intensity greater than half of that for the largest peak is determined at different temperatures. The frequencies of the peaks are then used to generate the rotational constants B and α_e.

W18.8   Simulated rotational-vibrational (infrared) spectra are generated for one or more diatomic molecules including $^{12}C^{16}O$, $^{1}H^{19}F$, $^{1}H^{35}Cl$, $^{1}H^{79}Br$, or $^{1}H^{127}I$ for predetermined temperatures. The frequencies of the peaks are then used to generate the rotational constants B and α_e, and the force constant k.

Because the discrete energy levels for atoms differ, atomic spectroscopies give information on the identity and concentration of atoms in a sample. For this reason, atomic spectroscopies are widely used in analytical chemistry. The discrete energy spectra of atoms and the difference in the rates of transition between quantum states can be used to construct lasers that provide an intense and coherent source of monochromatic radiation. Absorption of visible or ultraviolet light by molecules can lead to transitions between the ground state and excited electronic states of atoms and molecules. Vibrational transitions that occur together with electronic transitions are governed by the Franck–Condon factors rather than the $\Delta n = \pm 1$ dipole selection rule. The excited state can relax to the ground state through a combination of fluorescence, internal conversion, intersystem crossing, and phosphorescence. Fluorescence is very useful in analytical chemistry and can detect as little as 2×10^{-13} mol/L of a strongly fluorescing species. Circular dichroism is a widely used technique to determine the conformation of biomolecules.

19.1 The Essentials of Atomic Spectroscopy

All spectroscopies involve the absorption or emission of energy that induces transitions between states of a quantum mechanical system. Whereas the energies involved in rotational and vibrational transitions are on the order of 1 and 10 kJ mol^{-1}, respectively, photon energies associated with electronic transitions are on the order of 200 to 1000 kJ mol^{-1}. Typically, such energies are associated with the visible to X-ray range for electromagnetic radiation.

The electronic states of many electron atoms arise through the coupling of the angular momenta of individual electrons. Because the description of these states goes beyond our discussion of quantum mechanics in Chapters 12 through 17, the discussion here is limited to the transitions between levels of the one electron atom hydrogen. Because the energy levels of H are given by

$$E_n = -\frac{m_e e^4}{8\varepsilon_0^2 h^2 n^2} \tag{19.1}$$

where n is the principal quantum number, the frequency for absorption lines in the hydrogen spectrum is given by

$$\tilde{\nu} = \frac{m_e e^4}{8\varepsilon_0^2 h^3 c}\left(\frac{1}{n_{initial}^2} - \frac{1}{n_{final}^2}\right) = R_H\left(\frac{1}{n_{initial}^2} - \frac{1}{n_{final}^2}\right) \tag{19.2}$$

where R_H is the Rydberg constant. The derivation of this formula was one of the early major triumphs of quantum mechanics. The Rydberg constant is one of the most precisely known fundamental constants, and it has the value 109677.581 cm^{-1}. (See Problem 19.2 for a more exact equation for $\tilde{\nu}$ that is based on having the coordinate system centered at the center of mass rather than the nucleus.) The series of spectral lines associated with $n_{initial} = 1$ is called the Lyman series, and the series associated with $n_{initial} = 2, 3, 4,$ and 5 are called the Balmer, Paschen, Brackett, and Pfund series, respectively, named after the spectroscopists who identified them.

EXAMPLE PROBLEM 19.1

The absorption spectrum of the hydrogen atom shows lines at 82,258; 97,491; 102,823; 105,290; and 106,631 cm^{-1}. There are no lower frequency lines in the spectrum. Use graphical methods to determine $n_{initial}$ and the ionization energy of the hydrogen atom in this state.

Solution

The knowledge that frequencies for transitions follow a formula like that of Equation (19.2) allows $n_{initial}$ and the ionization energy to be determined from a limited number of transitions between bound states. The plot of $\tilde{\nu}$ versus assumed values of $1/n_{final}^2$ has a slope of $-R_H$ and an intercept with the frequency axis of $R_H/n_{initial}^2$. However, both $n_{initial}$ and n_{final} are unknown, so that in plotting the data, n_{final} values have to be assigned to the observed frequencies. Because there are no lower frequency lines in the spectrum, the lowest value for n_{final} is $n_{initial} + 1$. We try different combinations of n_{final} and $n_{initial}$ values to see if the slope and intercept are consistent with the expected values of $-R_H$ and $R_H/n_{initial}^2$. In this case, the sequence of spectral lines is assumed to correspond to $n_{final} = 2, 3, 4, 5,$ and 6 for an assumed value of $n_{initial} = 1$; $n_{final} = 3, 4, 5, 6,$ and 7 for an assumed value of $n_{initial} = 2$; and $n_{final} = 4, 5, 6, 7$ and 8 for an assumed value of $n_{initial} = 3$. The plots are shown in the following figure:

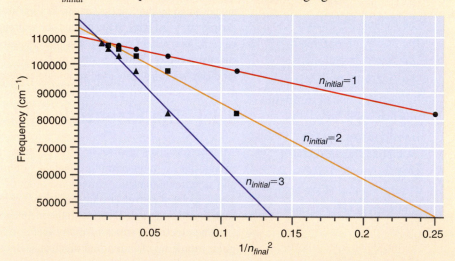

The slopes and intercepts calculated for these assumed values of $n_{initial}$ are:

Assumed $n_{initial}$	Slope (cm^{-1})	Intercept (cm^{-1})
1	-1.10×10^5	1.10×10^5
2	-2.71×10^5	1.13×10^5
3	-5.23×10^5	1.16×10^5

By examining the consistency of these values with the expected values, we can conclude that $n_{initial} = 1$. The ionization energy of the hydrogen atom in this state is hcR_H, corresponding to $n_{final} \to \infty$, or 2.18×10^{-18} J. The appropriate number of significant figures for the slope and intercept is approximate in this example and must be based on an error analysis of the data.

19.2 Analytical Techniques Based on Atomic Spectroscopy

The absorption and emission of light that occur during transitions between different atomic levels provide a powerful tool for qualitative and quantitative analysis of samples of chemical interest. For example, the concentration of lead in human blood and the presence of toxic metals in drinking water are routinely determined using **atomic emission** and **atomic absorption spectroscopy.** Figure 19.1 illustrates how these two spectroscopic techniques are implemented. A sample, ideally in the form of very small droplets (~1 − 10 μm in diameter) of a solution or suspension, is injected into the heated zone of the spectrometer. The heated zone may take the form of a flame, an electrically heated graphite furnace, or a plasma arc source. The main requirement of the heated zone is that it must convert a portion of the molecules in the sample of interest into atoms in their ground and excited states.

We first discuss atomic emission spectroscopy. In this technique, the light emitted by excited-state atoms as they undergo transitions back down to the ground state is dispersed into its component wavelengths by a monochromator, and the intensity of the radiation is measured as a function of wavelength. Because the emitted light intensity is proportional to the number of excited-state atoms and because the wavelengths at which emission occurs are characteristic for the atom, the technique can be used for both qualitative and quantitative analyses. Temperatures in the range of 1800 to 3500 K can be achieved in flames and carbon furnaces and up to 10,000 K can be reached in plasma arc sources. These high temperatures are required to produce sufficient excited-state atoms that emit light.

Atomic absorption spectroscopy differs from atomic emission spectroscopy in that light is passed through the heated zone, and the absorption associated with transitions from the lower to the upper state is detected. Because this technique relies on the population of low-lying rather than highly excited atomic states, it has some advantages in sensitivity over atomic emission spectroscopy. It became a very widely used technique when researchers realized that the sensitivity would be greatly enhanced if the light source were nearly monochromatic with a wavelength centered at $\lambda_{transition}$. The advantage of this arrangement can be seen from Figure 19.2.

Only a small fraction of the broadband light that passes through the heated zone is absorbed in the transition of interest. To detect the absorption, the light needs to

FIGURE 19.1
Schematic diagram of atomic emission and atomic absorption spectroscopies.

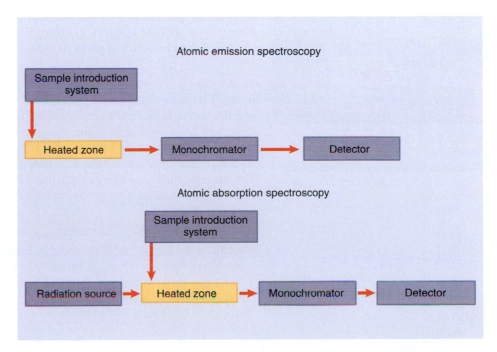

FIGURE 19.2
The intensity of light as a function of its frequency is shown at the entrance to the heated zone and at the detector for broadband and monochromatic sources. The absorption spectrum of the atom to be detected is shown in the middle column.

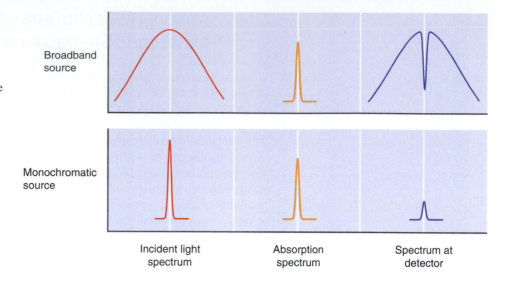

Broadband source

Monochromatic source

Incident light spectrum   Absorption spectrum   Spectrum at detector

FIGURE 19.2
The intensity of light as a function of its frequency is shown at the entrance to the heated zone and at the detector for broadband and monochromatic sources. The absorption spectrum of the atom to be detected is shown in the middle column.

be dispersed with a grating and the intensity of the light must be measured as a function of frequency. Because the monochromatic source matches the transition both in frequency and in line width, detection is much easier. Only a simple monochromator is needed to remove background light before the light is focused on the detector. The key to the implementation of this technique was the development of hollow cathode gas discharge lamps that emit light at the characteristic frequencies of the cathode materials. By using an array of these relatively inexpensive lamps on a single spectrometer, analyses for a number of different elements of interest can be carried out.

The sensitivity of atomic emission and absorption spectroscopy depends on the element and ranges from 10^{-4} µg/mL for Mg to 10^{-2} µg/mL for Pt. These techniques are used in a wide variety of applications, including drinking water analysis and engine wear, by detecting trace amounts of abraded metals in lubricating oil.

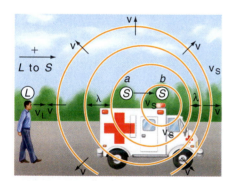

FIGURE 19.3
The frequency of light or sound at the position of the observer L depends on the relative velocity of the source S, V_S, and observer, V_L.

19.3 The Doppler Effect

A further application of atomic spectroscopy results from the **Doppler effect.** If a source is radiating light and moving relative to an observer, the observer sees a change in the frequency of the light as shown in Figure 19.3.

The shift in frequency is given by the formula

$$\omega = \omega_0 \sqrt{(1 \pm v_z/c)/(1 \mp v_z/c)} \qquad (19.3)$$

In this formula, v_z is the velocity component in the observation direction, c is the speed of light, and ω_0 is the light frequency in the frame in which the source is stationary. The upper and lower signs refer to the object approaching and receding from the observer, respectively. Note that the frequency shift is positive for objects that are approaching (a so-called "blue shift") and negative for objects that are receding (a so-called "red shift"). The **Doppler shift** is used to measure the speed at which stars and other radiating astronomical objects are moving relative to the Earth.

EXAMPLE PROBLEM 19.2

A line in the Lyman emission series for atomic hydrogen ($n_{final} = 1$), for which the wavelength is at 121.6 nm for an atom at rest, is seen for a particular quasar at 445.1 nm. Is the source approaching toward or receding from the observer? What is the magnitude of the velocity?

Solution

Because the frequency observed is less than that which would be observed for an atom at rest, the object is receding. The relative velocity is given by

$$\left(\frac{\omega}{\omega_0}\right)^2 = \left(1 - \frac{v_z}{c}\right)\bigg/\left(1 + \frac{v_z}{c}\right), \text{ or}$$

$$\frac{v_z}{c} = \frac{1 - (\omega/\omega_0)^2}{1 + (\omega/\omega_0)^2} = \frac{1 - (\lambda_0/\lambda)^2}{1 + (\lambda_0/\lambda)^2}$$

$$= \frac{1 - (121.6/445.1)^2}{1 + (121.6/445.1)^2} = 0.8611; \ v_z = 2.582 \times 10^8 \text{ ms}^{-1}$$

For source velocities much less than the speed of light, the nonrelativistic formula

$$\omega = \omega_0\left(1\bigg/1 \mp \frac{v_z}{c}\right) \tag{19.4}$$

applies. This formula is appropriate for a gas of atoms or molecules for which the distribution of speeds is given by the Maxwell–Boltzmann distribution. Because all velocity directions are equally represented for a particular speed, v_z has a large range for a gas at a given temperature, centered at $v_z = 0$. Therefore, the frequency is not shifted; instead, the spectral line is broadened. This is called **Doppler broadening.** Because atomic and molecular velocities are very small compared with the speed of light, the broadening of a line of frequency ω_0 is on the order of 1 part in 10^6. This effect is not as dramatic as the shift in frequency for the quasar, but still of importance in determining the line width of a laser, as we will see in the next section.

19.4 The Helium-Neon Laser

In this section, we demonstrate the relevance of the basic principles discussed earlier to the functioning of a laser. We focus on the common He-Ne laser. To understand the He-Ne laser, the concepts of absorption, spontaneous emission, and stimulated emission introduced in Chapter 18 are used.

Spontaneous and stimulated emission differ in an important respect. **Spontaneous emission** is a completely random process, and the photons that are emitted are incoherent, meaning that their phases and propagation directions are random. A lightbulb is an **incoherent photon source.** The phase relation between individual photons is random, and because the propagation direction of the photons is random, the intensity of the source falls off as the square of the distance. In **stimulated emission,** the phase and direction of propagation are the same as that of the incident photon. This is referred to as coherent photon emission. A **laser** is a coherent source of radiation. All photons are in phase, and because they have the same propagation direction, the divergence of the beam is very small. This explains why a laser beam that is reflected from the moon still has a measurable intensity when it returns to the Earth.

This discussion makes it clear that a **coherent photon source** must be based on stimulated rather than spontaneous emission. However, $B_{12} = B_{21}$, as was shown in Section 18.2. Therefore, the rates of absorption and stimulated emission are equal for $N_1 = N_2$. Stimulated emission with an amplification of the light will only dominate over absorption if $N_2 > N_1$. This is called a **population inversion** because, for equal level degeneracies, the lower energy state has the higher population at equilibrium. The key to making a practical laser is to create a stable population inversion. Although a population inversion is not possible under equilibrium conditions, it is possible to maintain such a distribution under steady-state conditions if the relative rates of the transitions between levels are appropriate. This is illustrated in Figure 19.4.

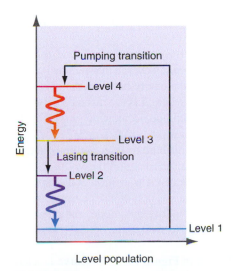

FIGURE 19.4

Schematic representation of a four-state laser. The energy is plotted vertically, and the level population is plotted horizontally.

FIGURE 19.5

Schematic representation of a He-Ne laser operated as an optical resonator. The parallel lines in the resonator represent coherent stimulated emission that is amplified by the resonator, and the waves represent incoherent spontaneous emission events.

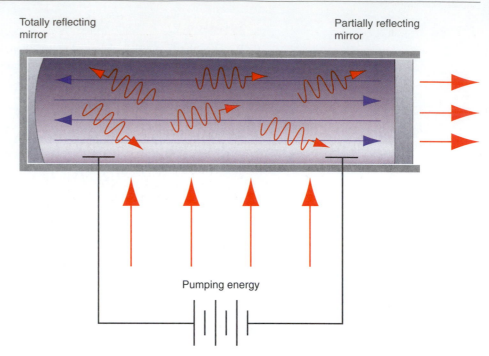

Totally reflecting mirror

Partially reflecting mirror

Pumping energy

Figure 19.4 can be used to understand how the population inversion between the levels involved in the lasing transition is established and maintained. The lengths of the horizontal lines representing the levels are proportional to the level populations N_1 to N_4. The initial step involves creating a significant population in level 4 by transitions from level 1. This is accomplished by an external source, which for the He-Ne laser is an electrical discharge in a tube containing the gas mixture. Relaxation to level 3 can occur through spontaneous emission of a photon as indicated by the curvy arrow. Similarly, relaxation from level 2 to level 1 can also occur through spontaneous emission of a photon. If this second relaxation process is fast compared to the first, N_3 will be maintained at a higher level than N_2. In this way, a population inversion is established. The advantage of having the lasing transition between levels 3 and 2 rather than 2 and 1 is that N_2 can be kept low if relaxation to level 1 from level 2 is fast. It is not possible to keep N_1 at a low level because atoms in the ground state cannot decay to a lower state.

This discussion shows how a population inversion can be established. How can a continuous lasing transition based on stimulated emission be maintained? This is made possible by carrying out the process indicated in Figure 19.4 in an **optical resonator** as shown in Figure 19.5.

The He-Ne mixture is put into a glass tube with carefully aligned parallel mirrors on each end. Electrodes are inserted to maintain the electrical discharge that pumps level 4 from level 1. Light reflected back and forth in the optical cavity between the two mirrors interferes constructively only if $n\lambda = n(c/v) = 2d$, where d is the distance between mirrors and n is an integer. The next constructive interference occurs when $n \rightarrow n + 1$. The difference in frequency between these two modes is $\Delta v = (c/2d)$, which defines the bandwidth of the cavity. The number of modes that contribute to laser action is determined by two factors: the frequencies of the **resonator modes** and the width in frequency of the stimulated emission transition. The width of the transition is determined by Doppler broadening, which arises through the thermal motion of gas-phase atoms or molecules. A schematic diagram of a He-Ne laser, including the anode, cathode, and power supply needed to maintain the electrical discharge as well as the optical resonator, is shown in Figure 19.6.

Six of the possible resonator modes are indicated in Figure 19.7. The curve labeled "Doppler line width" gives the relative number of atoms in the resonator as a function of the frequency at which they emit light. The product of these two functions gives the relative intensities of the stimulated emission at the different frequencies supported by the resonator. This is shown in Figure 19.7b. Because of losses in the cavity, a laser transition can only be sustained if enough atoms in the cavity are in the excited state at a sup-

FIGURE 19.6
Schematic diagram of a He-Ne laser.

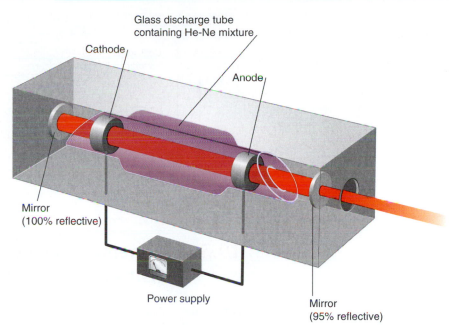

Web-Based Problem 19.1 Simulation of a Laser

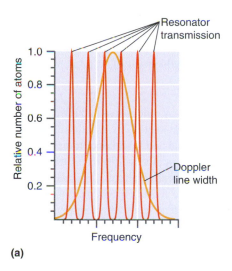

(a)

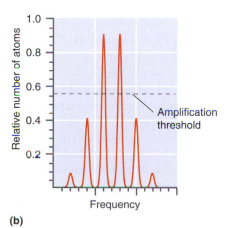

(b)

FIGURE 19.7
The line width of a transition in a He-Ne laser is Doppler broadened through the Maxwell–Boltzmann velocity distribution. (a) The resonator transmission decreases the line width of the lasing transition to less than the Doppler limit. (b) The amplification threshold further reduces the number of frequencies supported by the resonator.

ported resonance. In Figure 19.7, only two resonator modes lead to a sufficient intensity to sustain the laser. The main effect of the optical resonator is to decrease the $\Delta\nu$ associated with the frequency of the lasing transition to less than the Doppler limit.

By adding a further optical filter in the laser tube, it is possible to have only one mode enhanced by the resonator. This means that only the transition of interest is enhanced by multiple reflections in the optical resonator. In addition, all light of the correct frequency, but with a propagation direction that is not perpendicular to the mirrors, is not enhanced through multiple reflections. In this way, the resonator establishes a standing wave at the lasing frequency that has a propagation direction aligned along the laser tube axis. This causes further photons to be emitted from the lasing medium (He-Ne mixture) through stimulated emission. As discussed previously, these photons are exactly in phase with the photons that stimulate the emission and have the same propagation direction. These photons amplify the standing wave, which because of its greater intensity, causes even more stimulated emission. Allowing one of the end mirrors to be partially transmitting lets some of the light escape, and the result is a coherent, well-collimated laser beam.

To this point, the laser has been discussed at a schematic level. How does this discussion relate to the atomic energy levels of He and Ne? The relationship is shown in Figure 19.8. The electrical discharge in the laser tube produces electrons and positively charged ions. The electrons are accelerated in the electric field and can excite the He atoms from states in the $1s^2$ configuration to several states, all of which belong to the the $1s2s$ configuration. In the higher energy level, which corresponds to a single state, the electrons spins are paired. In the lower energy level, which consists of three separate states, they are unpaired. The lower and higher energy states correspond to triplet and singlet states, respectively, which will be discussed in Section 19.7. This is the **pumping transition** in the scheme shown in Figure 19.4. The excited He atoms efficiently transfer their energy through collisions to states in the $2p^55s$ and $2p^54s$ configurations of Ne. This creates a population inversion relative to Ne states in the $2p^54p$ and $2p^53p$ configurations. These levels are involved in the **lasing transition** through stimulated emission. Spontaneous emission to states in the $2p^53s$ configuration and collisional deactivation at the inner surface of the optical resonator depopulate the lower state of the lasing transitions and ensure that the population inversion is maintained. The manifold of these states is indicated in the figure by thicker lines, indicating a range of energies.

Note that a number of wavelengths can lead to lasing transitions. Coating the mirrors in the optical resonator ensures that they are reflective only in the range of interest. The resonator is usually configured to support the 632.8-nm transition in the visible part of the spectrum (see Figure 19.8). This corresponds to the red light usually seen from He-Ne lasers.

FIGURE 19.8
Transitions in the He-Ne laser. The slanted solid lines in the upper right side of the figure show three possible lasing transitions.

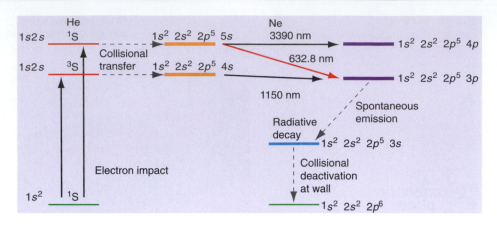

FIGURE 19.8
Transitions in the He-Ne laser. The slanted solid lines in the upper right side of the figure show three possible lasing transitions.

19.5 The Energy of Electronic Transitions in Molecules

For a molecule, the energy spacing between rotational levels is much less than the spacing between vibrational levels. Extending this comparison to electronic states, it is found that

$$\Delta E_{electronic} >> \Delta E_{vibrational} >> \Delta E_{rotational} \qquad (19.5)$$

Whereas rotational and vibrational transitions are induced by microwave and infrared radiation, electronic transitions are induced by visible and ultraviolet (UV) radiation. Just as an absorption spectrum in the infrared exhibits both rotational and vibrational transitions, an absorption spectrum in the visible and UV range exhibits a number of electronic transitions, and a specific electronic transition will contain vibrational and rotational fine structure.

Electronic excitations are responsible for giving color to the objects we observe, because the human eye is sensitive to light only in the limited range of wavelengths in which some electronic transitions occur. Either the reflected or the transmitted light is observed, depending on whether the object is transparent or opaque. Transmitted and reflected light complement the absorbed light. For example, a leaf is green because chlorophyll absorbs in the blue (450-nm) and red (650-nm) regions of the visible light spectrum. Electronic excitations can be detected (at a limited resolution) without the aid of a spectrometer because the human eye is a very sensitive detector of radiation. At a wavelength of 500 nm, the human eye can detect 1 part in 10^6 of the intensity of sunlight on a bright day. This corresponds to as few as 500 photons per second incident on an area of 1 mm^2.

Because the electronic spectroscopy of a molecule is directly linked to its energy levels, which are in turn determined by its structure and chemical composition, UV-visible spectroscopy provides a very useful qualitative tool for identifying molecules. In addition, for a given molecule, electronic spectroscopy can be used to determine energy levels in molecules. However, the UV and visible photons that initiate an electronic excitation perturb a molecule far more than rotational or vibrational excitation. For example, the bond length in electronically excited states of O_2 is as much as 30% longer than that in the ground state. Whereas in its ground state, formaldehyde is a planar molecule, it is pyramidal in its lowest two excited states. As might be expected from such changes in geometry, the chemical reactivity of excited-state species can be quite different from the reactivity of the ground-state molecule.

19.6 The Franck–Condon Principle

Each of the bound states of a molecule has well-defined vibrational and rotational energy levels, and changes in the vibrational state can occur together with a change in the rotational state. Similarly, the vibrational and rotational quantum numbers can change during electronic excitation. We next discuss the vibrational excitation and de-excitation associated with electronic transitions, but do not discuss the associated rotational transitions. We will see that the $\Delta n = \pm 1$ selection rule for vibrational transitions within a given electronic state does not hold for transitions between two electronic states.

What determines Δn in a vibrational transition between electronic states? This question can be answered by looking more closely at the **Born–Oppenheimer approximation.** The essence of this approximation is to decouple the motion of the electrons and the nuclei in a vibrating molecule. The rationale for this decoupling is that electrons adjust quickly to the much slower motion of the heavier nuclei, so that the electron density at any time in a vibrational cycle can be calculated assuming that the nuclei are fixed in space. This approximation can be expressed mathematically by stating that the total wave function for the molecule can be factored into two parts at a fixed position of all the nuclei. One part depends only on the position of the nuclei, given by $\mathbf{R}_1, ..., \mathbf{R}_m$, and the second part depends only on the position of the electrons, given by $\mathbf{r}_1, ..., \mathbf{r}_n$:

$$\psi(\mathbf{r}_1, ..., \mathbf{r}_n, \mathbf{R}_1, ..., \mathbf{R}_m) = \psi^{electronic}(\mathbf{r}_1, ..., \mathbf{r}_n, \mathbf{R}_1^{fixed}, ..., \mathbf{R}_m^{fixed})$$
$$\times \phi^{vibrational}(\mathbf{R}_1, ..., \mathbf{R}_m) \quad (19.6)$$

As discussed in Section 18.4, the spectral line corresponding to an electronic transition (initial $\rightarrow$ final) has a measurable intensity only if the value of the transition dipole moment is different from zero

$$\mu^{fi} = \int \psi_f^*(\mathbf{r}_1, ..., \mathbf{r}_n, \mathbf{R}_1, ..., \mathbf{R}_m) \times \left(-e \sum_{i=1}^{n} \mathbf{r}_i\right) \times \psi_i(\mathbf{r}_1, ..., \mathbf{r}_n, \mathbf{R}_1, ..., \mathbf{R}_m) d\tau \neq 0 \quad (19.7)$$

where the summation is over the positions of the electrons. The superscripts and subscripts f and i refer to the final and initial states, respectively, in the transition.

Because the total wave function can be written as a product of electronic and vibrational parts, Equation (19.7) becomes

$$\mu^{fi} = S \int \psi_f^*(\mathbf{r}_1, ..., \mathbf{r}_n, \mathbf{R}_1^{fixed}, ..., \mathbf{R}_m^{fixed}) \times \left(-e \sum_{i=1}^{n} \mathbf{r}_i\right)$$
$$\times \psi_i(\mathbf{r}_1, ..., \mathbf{r}_n, \mathbf{R}_1^{fixed}, ..., \mathbf{R}_m^{fixed}) d\tau$$
$$= \int (\phi_f^{vibrational}(\mathbf{R}_1, ..., \mathbf{R}_m))^* \phi_i^{vibrational}(\mathbf{R}_1, ..., \mathbf{R}_m) d\tau$$
$$\times \int (\psi_f^{electronic}(\mathbf{r}_1, ..., \mathbf{r}_n, \mathbf{R}_1^{fixed}, ..., \mathbf{R}_m^{fixed}))^* + \times \left(-e \sum_{i=1}^{n} \mathbf{r}_i\right)$$
$$\times \psi_i^{electronic}(\mathbf{r}_1, ..., \mathbf{r}_n, \mathbf{R}_1^{fixed}, ..., \mathbf{R}_m^{fixed}) d\tau \quad (19.8)$$

Note that the first of the two product integrals in Equation (19.8) represents the overlap between the vibrational wave functions in the ground and excited states. The magnitude of the square of this integral for a given transition is known as the **Franck–Condon factor** and is a measure of the expected intensity of an electronic transition. The Franck–Condon factor, S^2, replaces the selection rule $\Delta n = \pm 1$ obtained for pure vibrational transitions derived in Section 18.4 as a criterion for the intensity of a transition:

$$S^2 = \left| \int (\phi_f^{vibrational})^* \phi_i^{vibrational} d\tau \right|^2 \quad (19.9)$$

The **Franck–Condon principle** states that transitions between electronic states correspond to vertical lines on an energy versus internuclear distance diagram. The basis of this principle is that electronic transitions occur on a timescale that is very short compared to the vibrational period of a molecule. Therefore, the atomic positions are fixed during the transition. As Equation (19.9) shows, the probability of a vibrational–electronic transition is governed by the overlap between the final and initial vibrational wave functions at fixed values of the internuclear distances. Is it necessary to consider all vibrational levels in the ground state as an initial state for an electronic transition? As discussed in Chapter 18, nearly all of the molecules in the ground state have the vibrational quantum number $n = 0$, for which the maximum amplitude of the wave function is at the equilibrium bond length. As shown in Figure 19.9, vertical transitions predominantly occur from this ground vibrational state to several vibrational states in the upper electronic state.

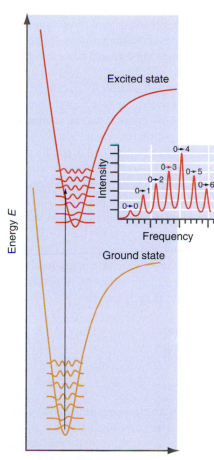

FIGURE 19.9

The relation between energy and bond length is shown for two electronic states. Only the lowest vibrational energy levels and the corresponding wave functions are shown. The vertical line shows the most probable transition predicted by the Franck–Condon principle. The inset shows the relative intensities of different vibrational lines in an absorption spectrum for the potential energy curves shown.

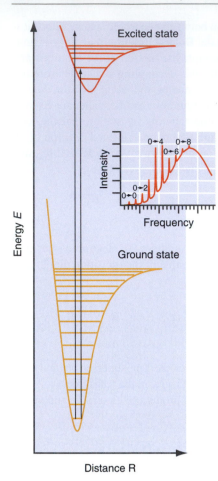

FIGURE 19.10

For absorption from the ground vibrational state of the ground electronic state to the excited electronic state, a continuous energy spectrum will be observed for sufficiently high photon energy. A discrete energy spectrum is observed for an incident light frequency $v < E/h$. A continuous spectrum is observed for higher frequencies.

How does the Franck–Condon principle determine the n values in the excited state that give the most intense spectral lines? The most intense spectral transitions occur to vibrational states in the upper electronic state that have the largest overlap with the ground vibrational state in the lower electronic state. As Figure 14.23 shows, the vibrational wave functions have their largest amplitude near the R value at which the energy level meets the potential curve, because this corresponds to the classical turning point. For the example shown in Figure 19.9, the overlap $|\int (\phi_f^{vibrational})^* \phi_i^{vibrational} d\tau|$ is greatest between the $n = 0$ vibrational state of the ground electronic state and the $n = 4$ vibrational state of the excited electronic state. Although this transition has the maximum overlap and generates the most intense spectral line, other states close in energy to the most probable state will also give rise to spectral lines that have a lower intensity. Their intensity is lower because the vibrational wave functions of the ground and excited states have a smaller overlap.

The fact that a number of vibrational transitions are observed in an electronic transition is very useful in obtaining detailed information about both the ground electronic state potential energy surface and that of the electronic state to which the transition occurs. For example, vibrational transitions are observed in the electronic spectra of O_2 and N_2, although neither of these molecules absorbs energy in the infrared. Because multiple vibrational peaks are often observed in electronic spectra, the bond strength of the molecule in the excited states can be determined by fitting the observed frequencies of the transitions to a model potential such as the Morse potential discussed in Section 18.3. Because the excited state can also correspond to a photodissociation product, electronic spectroscopy can be used to determine the vibrational force constant and bond energy of highly reactive species such as the CN radical.

For the example shown in Figure 19.9, the molecule will exhibit a discrete energy spectrum in the visible or UV region of the spectrum. However, for some conditions the electronic absorption spectrum for a diatomic molecule is continuous. A continuous spectrum is observed if the photon energy is sufficiently high that excitation occurs to an unbound region of an excited state. This is illustrated in Figure 19.10. In this case, a discrete energy spectrum is observed for low photon energy and a **continuous energy spectrum** is observed for incident light frequencies $v > E/h$, where E corresponds to the energy of the transition to the highest bound state in the excited state potential. A purely continuous energy spectrum at all energies is observed if the excited state is a nonbinding state, such as that corresponding to the first excited state for H_2^+.

The preceding discussion briefly summarizes the most important aspects of the electronic spectroscopy of diatomic molecules. In general, the vibrational energy levels for these molecules are sufficiently far apart that individual transitions can be resolved. We next consider polyatomic molecules, for which this is not usually the case.

19.7 UV-Visible Light Absorption in Polyatomic Molecules

Many rotational and vibrational transitions are possible if an electronic transition occurs in polyatomic molecules. A large molecule may have ~1000 rotational levels in an interval of 1 cm^{-1}. For this reason, individual spectral lines overlap so that broad bands are often observed in UV-visible absorption spectroscopy. This is schematically indicated in Figure 19.11. An electronic transition in an atom gives a sharp line. An electronic transition in a diatomic molecule has additional structure resulting from vibrational and rotational transitions that can often be resolved into individual peaks. However, the many rotational and vibrational transitions possible in a polyatomic molecule generally overlap, giving rise to a broad, nearly featureless band. This overlap makes it difficult to extract information on the initial and final states involved in an electronic transition in polyatomic molecules.

The main selection rule that applies to electronic transitions between states of a molecule is associated with the total spin angular momentum of the electrons in the molecule. For a molecule in which all spins are paired (for example N_2 in its ground state), we refer to the state as a **singlet state,** which has a value of zero for the angular momentum

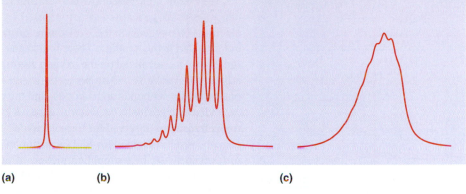

(a)   **(b)**   **(c)**

FIGURE 19.11

The intensity of absorption in a small part of the UV-visible range of the electromagnetic spectrum is shown schematically for (a) an atom, (b) a diatomic molecule, and (c) a polyatomic molecule.

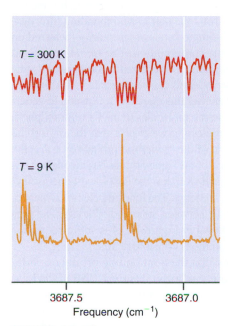

3687.5   3687.0
Frequency (cm^{-1})

FIGURE 19.12

A small portion of the electronic absorption spectrum of methanol is shown at 300 and 9 K using expansion of a dilute mixture of methanol in He through a nozzle into a vacuum. At 300 K, the molecule absorbs almost everywhere in the frequency range. At 9 K, very few rotational and vibrational states are populated, and individual spectral features corresponding to rotational fine structure are observed. (Source: Reproduced by permission of Elsevier Science from P. Carrick, R. F. Curl, M. Dawes, E. Koester, K. K. Murray, M. Petri, and M. L. Richnow, "OH Stretching Fundamental of Methanol," Figure 4, *J. Molecular Structure* 223 (1990), 171–184. Copyright 1990 by Elsevier Science Ltd.)

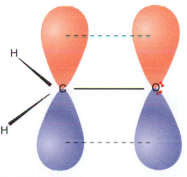

FIGURE 19.13

Valence bond picture of the formaldehyde molecule. The solid lines indicate σ bonds and the dashed lines indicate a π bond. The nonequivalent lone pairs on oxygen are also shown.

quantum number S. The molecule is *diamagnetic,* meaning that it is (weakly) repelled by a magnetic field. For a molecule that has a net number of unpaired spins, $S > 0$ and the molecule is *paramagnetic,* meaning that it is (weakly) attracted by a magnetic field. An example is NO in its ground state. For a paramagnetic molecule with two unpaired electrons (for example O_2 in its ground state), $S = 1$. As a result of the magnetic interactions between the unpaired electrons, three different quantum states are associated with $S = 1$ that have nearly the same energy. For this reason, we refer to a **triplet state.** The principal selection rule in electronic spectroscopy is that $\Delta S = 0$, so that singlet $\rightarrow$ singlet and triplet $\rightarrow$ triplet transitions are allowed, but singlet $\rightarrow$ triplet transitions are forbidden.

The number of transitions observed can be reduced dramatically by obtaining spectra at low temperatures, because only states of very low energy are populated. Therefore, the number of possible transitions is greatly reduced in comparison to a molecule at 300 K. Low-temperature spectra for individual molecules can be obtained either by imbedding the molecule of interest in a solid rare gas matrix at cryogenic temperatures or by expanding gaseous He containing the molecules of interest in dilute concentration through a nozzle into a vacuum. The He gas as well as the molecules of interest are cooled to very low temperatures in the expansion. An example of the elimination of spectral congestion through such a gas expansion is shown in Figure 19.12. The temperature of 9 K is reached by simply expanding the 300 K gas mixture into a vacuum using a molecular beam apparatus.

The concept of chromophores is particularly useful for discussing the electronic spectroscopy of polyatomic molecules. As discussed in Chapter 18, characteristic vibrational frequencies are associated with two neighboring atoms in larger molecules. Similarly, the absorption of UV and visible light in larger molecules can be understood by visualizing the molecule as a system of coupled atoms, such as —C=C— or —O—H, that are called **chromophores.** A chromophore is a chemical entity embedded within a molecule that absorbs radiation at nearly the same wavelength in different molecules. Common chromophores in electronic spectroscopy are C=C, C=O, C≡N, or C=S groups. Each chromophore has one or several characteristic absorption frequencies in the UV; and the UV absorption spectrum of the molecule, to a first approximation, can be thought of as arising from the sum of the absorption spectra of its chromophores. The wavelengths and absorption strength associated with specific chromophores are discussed in Section 19.9.

In viewing the transitions involved in electronic spectroscopy, it is in general useful to use a localized bonding model because transitions can be associated with absorption of light by chromophores. However, the electrons in radicals and those in delocalized π bonds in conjugated and aromatic bonds need to be described in a delocalized rather than a localized binding model.

What transitions are most likely to be observed in electronic spectroscopy? The fundamental excitation is the promotion of an electron from its HOMO to an excited-state MO. Consider the electronic ground-state configuration of formaldehyde, H_2CO, and those of its lowest lying electronically excited states. In a localized bonding model, the $2s$ and $2p$ electrons combine to form sp^2-hybrid orbitals on the carbon atom as shown in Figure 19.13.

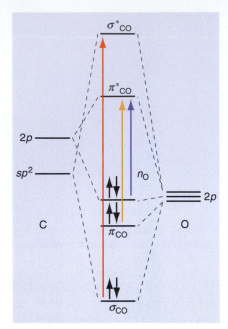

FIGURE 19.14

A simplified MO energy diagram is shown for the C—O bonding interaction in formaldehyde. The most important allowed transitions between these levels are shown. Only one of the sp^2 orbitals on carbon is shown because the other two hybrid orbitals form σ_{CH} bonds.

We write the ground-state configuration in the localized orbital notation $(1s_O)^2(1s_C)^2(2s_{CH})^2(\sigma)^2(\sigma'_{CH})^2(\sigma_{CO})^2(\pi_{CO})^2(n_O)^2(\pi^*_{CO})^0$ to emphasize that the $1s$ and $2s$ electrons on oxygen and the $1s$ electrons on carbon remain localized on the atoms and are not involved in the bonding. There is also an electron lone pair in a nonbonding MO, designated by n_O, localized on the oxygen atom. Bonding orbitals are primarily localized on adjacent C—H or C—O atoms as indicated in the configuration. The C—H bonds and one of the C—O bonds are σ bonds, and the remaining C—O bond is a π bond.

What changes in the occupation of the MO energy levels can be associated with the electronic transitions observed for formaldehyde? To answer this question, it is useful to generalize the results obtained for MO formation in diatomic molecules to the CO chromophore in formaldehyde. In a simplified picture of this molecule, we expect that the σ_{co} orbital formed primarily from the $2p_z$ orbital on O and one of the sp^2-hybrid orbitals on C has the lowest energy, and that the antibonding combination of the same orbitals has the highest energy. The π orbital formed from the $2p$ levels on each atom has the next lowest energy, and the antibonding π^* combination has the next highest energy. The lone pair electrons that occupy the $2p$ orbital on O have an energy intermediate between the π and π^* levels. The very approximate molecular orbital energy diagram shown in Figure 19.14 is sufficient to discuss the transitions that formaldehyde undergoes in the UV-visible region.

From the MO energy diagram, we conclude that the nonbonding orbital on O derived from the $2p$ AO is the HOMO, and the empty π^* orbital is the LUMO. The lowest excited state is reached by promoting an electron from the n_o to the π^*_{CO} orbital and is called a $n \rightarrow \pi^*$ **transition.** The resulting state is associated with the configuration $(1s_O)^2(1s_C)^2(2s_O)^2(\sigma_{CH})^2(\sigma'_{CH})^2(\sigma_{CO})^2(\pi_{CO})^2(n_O)^1(\pi^*_{CO})^1$. The next excited state is reached by promoting an electron from the π_{co} to the π^*_{CO} MO and is called a $\pi \rightarrow \pi^*$ **transition.** The resulting state is associated with the configuration $(1s_O)^2(1s_C)^2(2s_O)^2(\sigma_{CH})^2(\sigma'_{CH})^2(\sigma_{CO})^2(\pi_{CO})^1(n_O)^2(\pi^*_{CO})^1$.

These configurations do not completely describe the quantum states because the alignment of the spins (paired versus unpaired) in the unfilled orbitals is not specified by the configuration. Because each of the excited-state configurations just listed has two half-filled orbitals, both singlet and triplet states arise from each configuration. The relative energy of these states is indicated in Figure 19.15. It is a general rule that for the same configuration, triplet states lie lower in energy than singlet states. The difference in energy between the singlet and triplet states is specific to a molecule, but typically lies between 2 and 10 eV.

FIGURE 19.15

The ground state of formaldehyde is a singlet and is designated S_0. Successively higher energy singlet and triplet states are designated S_1, S_2, T_1, and T_2. The electron configurations and the alignment of the unpaired spins for the states involved in the most important transitions are also shown. The energy separation between the singlet and triplet states has been exaggerated in this figure.

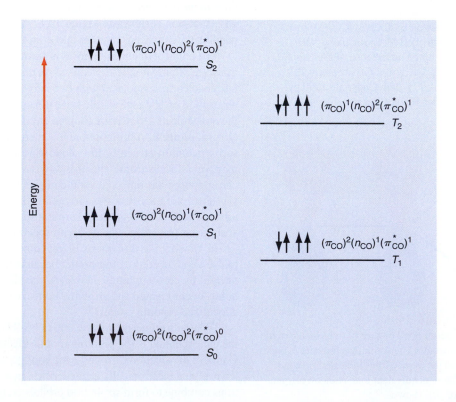

The energy difference between the initial and final states determines the frequency of the spectral line. Although large variations can occur among different molecules for a given type of transition, generally the energy increases in the sequence $n \rightarrow \pi^*$, $\pi \rightarrow \pi^*$, and $\sigma \rightarrow \sigma^*$. The $\pi \rightarrow \pi^*$ transitions require multiple bonds, and occur in alkenes, alkynes, and aromatic compounds. The $n \rightarrow \pi^*$ transitions require both a nonbonding electron pair and multiple bonds and occur in molecules containing carbonyls, thiocarbonyls, and nitro, azo, and imine groups and in unsaturated halocarbons. The $\boldsymbol{\sigma \rightarrow \sigma^*}$ **transitions** are observed in many molecules, particularly in alkanes, in which none of the other transitions is possible.

19.8 Transitions among the Ground and Excited States

We next generalize the preceding discussion for formaldehyde to an arbitrary molecule. What transitions can take place among ground and excited states? Consider the energy levels for such a molecule shown schematically in Figure 19.16. The ground state is in general a singlet state and the excited states can be either a singlet or triplet state. We include only one excited singlet and triplet state in addition to the ground state and consider the possible transitions among these states. The restriction is justified because an initial excitation to higher lying states will rapidly decay to the lowest lying state of the same multiplicity through a process called internal conversion, which is discussed later. The diagram also includes vibrational levels associated with each of the electronic levels. Rotational levels are omitted to simplify the diagram. The fundamental rule governing transitions is that all transitions must conserve energy and angular momentum. For transitions within a molecule, this condition can be satisfied by transferring energy between electronic, vibrational, and rotational states. Alternatively, energy can be conserved by transferring energy between a molecule and its surroundings.

Three types of transitions are indicated in Figure 19.16. **Radiative transitions,** in which a photon is absorbed or emitted, are indicated by solid vertical lines. **Nonradiative transitions,** in which energy is transferred between different degrees of freedom of a molecule or to the surroundings, are indicated by wavy vertical lines. The dashed line indicates nonradiative transitions between singlet and triplet states, which are forbidden by the dipole selection rule. The pathway by which a molecule in an excited state decays to the ground state depends on the rates of a number of competing processes. In the next two sections, these processes are discussed individually.

FIGURE 19.16
Possible transitions among the ground and excited electronic states are indicated. The spacing between vibrational levels is exaggerated in this diagram. Rotational levels have been omitted for reasons of clarity.

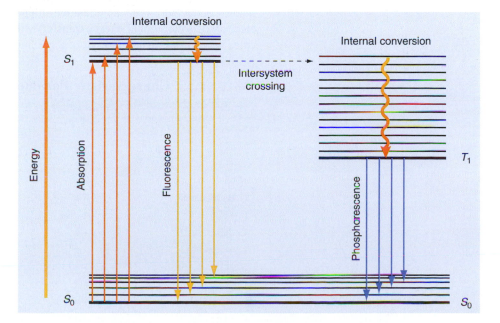

19.9 Singlet–Singlet Transitions: Absorption and Fluorescence

As discussed in Section 19.7, an absorption band in an electronic spectrum can be associated with a specific chromophore. Whereas in atomic spectroscopy the selection rule $\Delta S = 0$ is strictly obeyed, in molecular spectroscopy one finds instead that spectral lines for transitions corresponding to $\Delta S = 0$ are much stronger than those for which this condition is not fulfilled. It is useful to quantify what is meant by strong and weak absorption. If I_0 is the incident light intensity at the frequency of interest and I_t is the intensity of transmitted light, the dependence of I_t/I_0 on the concentration c and the path length l is described by the **Beer–Lambert law:**

$$\log\left(\frac{I_t}{I_0}\right) = -\varepsilon lc \tag{19.10}$$

The **molar extinction coefficient** ε is a measure of the strength of the transition. It is independent of the path length and concentration and is characteristic of the chromophore. The **integral absorption coefficient,** $A = \int \varepsilon(\nu)\, d\nu$, in which the integration over the spectral line includes associated vibrational and rotational transitions, is a measure of the probability that an incident photon will be absorbed in a specific electronic transition. The terms A and ε depend on the frequency, and ε measured at the maximum of the spectral line, ε_{max}, has been tabulated for many chromophores. Some characteristic values for spin-allowed transitions are given in Table 19.1.

In Table 19.1, note the large enhancement of ε_{max} that occurs for conjugated bonds. As a general rule, ε_{max} lies between 10 and 5×10^4 $dm^3\ mol^{-1}\ cm^{-1}$ for spin-allowed transitions ($\Delta S = 0$), and between 1×10^{-4} and 1 $dm^3\ mol^{-1}\ cm^{-1}$ for singlet–triplet transitions ($\Delta S = 1$). Therefore, the attenuation of light passing through the sample resulting from singlet–triplet transitions will be smaller by a factor of ~10^4 to 10^7 than the attenuation from singlet–singlet transitions. This illustrates that in an absorption experiment, transitions for which $\Delta S = 1$ are not totally forbidden if spin-orbit coupling is not negligible, but are typically too weak to be of much importance. However, as discussed in Section 19.10, singlet–triplet transitions are important when discussing phosphorescence.

The excited-state molecule can return to the ground state through radiative or nonradiative transitions involving collisions with other molecules. What determines which of these two pathways will be followed? An isolated excited-state molecule (for instance, in interstellar space) cannot exchange energy with other molecules through collisions and, therefore, nonradiative transitions (other than isoenergetic internal electronic-to-vibrational energy transfer) will not occur. However, excited-state molecules in a crystal, in solution, or in a gas undergo frequent collisions with other mole-

TABLE 19.1 **Characteristic Parameters for Common Chromophores**

Chromophore	Transition	λ_{max} (nm)	ε_{max} ($dm^3\ mol^{-1}\ cm^{-1}$)
N=O	$n \rightarrow \pi^*$	660	200
N=N	$n \rightarrow \pi^*$	350	100
C=O	$n \rightarrow \pi^*$	280	20
NO_2	$n \rightarrow \pi^*$	270	20
C_6H_6 (benzene)	$\pi \rightarrow \pi^*$	260	200
C=N	$\pi \rightarrow \pi^*$	240	150
C=C—C=O	$\pi \rightarrow \pi^*$	220	2×10^5
C=C—C=C	$\pi \rightarrow \pi^*$	220	2×10^5
S=O	$\pi \rightarrow \pi^*$	210	1.5×10^3
C=C	$\pi \rightarrow \pi^*$	180	1×10^3
C—C	$\sigma \rightarrow \sigma^*$	<170	1×10^3
C—H	$\sigma \rightarrow \sigma^*$	<170	1×10^3

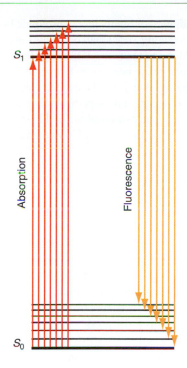

cules in which they lose energy and return to the lowest vibrational state of S_1. This process generally occurs much faster than a radiative transition directly from a vibrationally excited state in S_1 to a vibrational state in S_0. An example of a nonradiative transition induced by collisions is **internal conversion,** which is the decay from a higher vibrational state to the ground vibrational state of the same multiplicity indicated in Figure 19.16. Once in the lowest vibrational state of S_1, either of two events can occur. The molecule can undergo a radiative transition to a vibrational state in S_0 in a process called **fluorescence,** or it can make a nonradiative transition to an excited vibrational state of T_1 in a process called **intersystem crossing.** Intersystem crossing is forbidden in the Born–Oppenheimer approximation and, therefore, occurs at a very low rate in comparison with the other processes depicted in Figure 19.16.

Because internal conversion to the ground vibrational state of S_1 is generally fast in comparison with fluorescence to S_0, the vibrationally excited-state molecule will relax to the ground vibrational state of S_1 before undergoing fluorescence. As a result of the relaxation, the fluorescence spectrum is shifted to lower energies relative to the absorption spectrum, as shown in Figure 19.17. When comparing absorption and fluorescence spectra, it is often seen that for potentials that are symmetric about the minimum (for example, a harmonic potential), the bands of lines corresponding to absorption and fluorescence are mirror images of one another. This relationship is shown in Figure 19.17.

19.10 Intersystem Crossing and Phosphorescence

Although intersystem crossing between singlet and triplet electronic states is forbidden in the Born–Oppenheimer approximation, for many molecules the probability of this happening is high. The probability of intersystem crossing transitions is enhanced by two factors: a very similar molecular geometry in the excited singlet and triplet states, and a strong spin-orbit coupling, which allows the spin flip associated with a singlet–triplet transition to occur. The processes involved in phosphorescence are illustrated in a simplified fashion in Figure 19.18 for a diatomic molecule.

Imagine that a molecule is excited from S_0 to S_1. This is a dipole-allowed transition, so it has a high probability of occurring. Through collisions with other molecules, the excited-state molecule loses vibrational energy and decays to the lowest vibrational state of S_1. As shown in Figure 19.18, the potential energy curves can cross such that an excited vibrational state in S_1 intermediate in energy between the state to which the original excitation occurs and the lowest vibrational state in S_1 has approximately the same energy as an excited vibrational state in T_1. In this case, the molecule has the same geometry and energy in both singlet and triplet states. In Figure 19.18, this occurs for $n = 4$ in the state S_1. If the spin-orbit coupling is strong enough to initiate a spin flip, the molecule can cross over to the triplet state without a change in geometry or energy. Through internal conversion, it will rapidly relax to the lowest vibrational state of T_1. At this point, it can no longer make a transition back to S_1 because the ground vibrational state of T_1 is lower than any state in S_1.

However, from the ground vibrational state of T_1, the molecule can decay radiatively to the ground state in the dipole transition forbidden process called **phosphorescence.** Is this a high probability event? As discussed earlier, the lifetime of the ground vibrational state of T_1 can be very long compared to a vibrational period. Therefore, nonradiative processes involving collisions between molecules or with the walls of the reaction vessel can compete effectively with phosphorescence. Because it is a forbidden transition and because of the competition from nonradiative processes, the probability for a $T_1 \rightarrow S_0$ phosphorescence transition is generally much lower than for fluorescence. It usually lies in the range of 10^{-2} to 10^{-5}. The distinction between allowed and forbidden transitions is experimentally accessible through the lifetime of the excited state. Fluorescence is an allowed transition, and the excited-state lifetime is short, typically less than 10^{-7} s. By contrast, phosphorescence is a forbidden transition and, therefore, the excited-state lifetime is long, typically longer than 10^{-3} s.

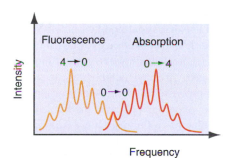

FIGURE 19.17
Illustration of the absorption and fluorescence bands expected if internal conversion is fast relative to fluorescence. The relative intensities of individual transitions within the absorption and fluorescence bands are determined by the Franck–Condon principle.

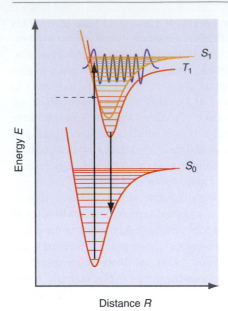

Energy E

Distance R

FIGURE 19.18

Process giving rise to phosphorescence illustrated for a diatomic molecule. Absorption from S_0 leads to population of excited vibrational states in S_1. The molecule has a finite probability of making a transition to an excited vibrational state of T_1 if it has the same geometry in both states and if there are vibrational levels of the same energy in both states. The dashed arrow indicates the coincidence of vibrational energy levels in T_1 and S_1. For reasons of clarity, only the lowest vibrational levels in T_1 are shown. The initial excitation to S_1 occurs to a vibrational state of maximum overlap with the ground state of S_0 as indicated by the blue vibrational wave function.

Fluorescence can be induced using broadband radiation or highly monochromatic laser light. Fluorescence spectroscopy is well suited for detecting very small concentrations of a chemical species if the wavelength of the emission lies in the visible-UV part of the electromagnetic spectrum where there is little background noise near room temperature. As shown in Figure 19.17, relaxation to lower vibrational levels within the excited electronic state has the consequence that the fluorescence signal occurs at a longer wavelength than the light used to create the excited state. Therefore, the contribution of the incident radiation to the background at the wavelength used to detect the fluorescence is very small.

19.11 Fluorescence Spectroscopy and Analytical Chemistry

We now describe a particularly powerful application of fluorescence spectroscopy, namely, the sequencing of the human genome. The goal of the human genome project is to determine the sequence of the four bases, A, C, T, and G, in DNA that encode all the genetic information necessary for propagating the human species. A sequencing technique based on laser-induced fluorescence spectroscopy that has been successfully used in this effort can be divided into three parts.

In the first part, a section of DNA is cut into small lengths of 1000 to 2000 base pairs using mechanical shearing. Each of these pieces is replicated to create many copies, and these replicated pieces are put into a solution with a mixture of the four nucleotide triphosphate corresponding to A, C, T, and G. A reaction is set in motion that leads to the strands growing in length through replication. A small fraction of each of the A, C, T, and G bases in solution that are incorporated into the pieces of DNA has been modified in two ways. The modified base terminates the replication process. It also contains a dye chosen to fluoresce strongly at a known wavelength. The initial segments continue to grow if they incorporate unmodified nucleotides, and no longer grow if they incorporate one of the modified nucleotides. As a result of these competing processes involving the incorporation of modified or unmodified nucleotides, a large number of partial replicas of the whole DNA are created, each of which is terminated in the base that has a fluorescent tag built into it. The ensemble of these partial replicas contains all possible lengths of the original DNA segment that terminate in the particular base chosen. If the lengths of these segments can be measured, then the positions of the particular base in the DNA segment can be determined.

The lengths of the partial replicas are measured using capillary electrophoresis coupled with detection using laser-induced fluorescence spectroscopy. In this method, a solution containing the partial replicas is passed through a glass capillary filled with a gel. An electrical field along the capillary causes the negatively charged DNA partial replicas to travel down the capillary with a speed that depends inversely on their length. Because of the different migration speeds, a separation in length occurs as the partial replicas pass through the capillary. At the end of the capillary, the partial replicas emerge from the capillary into a buffer solution that flows past the capillary, forming a sheath. The flow pattern of the buffer solution is carefully controlled to achieve a focusing of the emerging stream containing the partial replicas to a diameter somewhat smaller than the inner diameter of the capillary. A schematic diagram of such a sheath flow cuvette electrophoresis apparatus is shown in Figure 19.19. An array of capillaries is used rather than a single capillary in order to obtain the multiplexing advantage of carrying out several experiments in parallel.

The final part of the sequencing procedure is to measure the time that each of the partial replicas spent in transit through the capillary, which determines its length, and to identify the terminating base. The latter task is accomplished by means of laser-induced fluorescence spectroscopy. A narrow beam of visible laser light is passed through all the capillaries in series. Because of the very dilute solutions involved, the attenuation of the laser beam by each successive capillary is very small. The fluorescent light emitted from each of the capillaries is directed to light-sensing photodiodes by means of a microscope objective and individual focusing lenses. A rotating filter wheel between the microscope objective and the focusing lenses allows a discrimination to be made among the four different fluorescent dyes with which the bases were tagged. The sensitivity of the system shown in Figure 19.19 is 130±30

FIGURE 19.19
Schematic diagram of the application of fluorescence spectroscopy in the sequencing of the human genome. (Source: After J. Zhang et al., *Nucleic Acids Research* 27 (1999), 36e.)

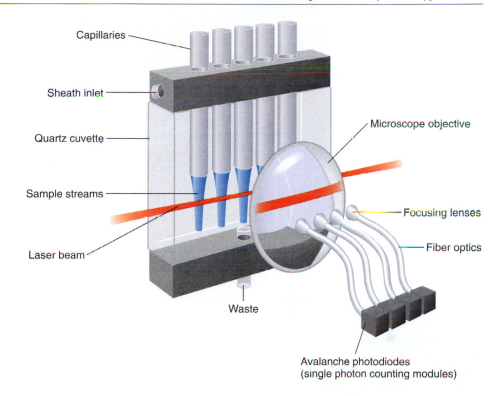

molecules in the volume illuminated by the laser. This corresponds to a concentration of 2×10^{-13} mol/L! This extremely high sensitivity is a result of coupling the sensitive fluorescence technique to a sample cell designed with a very small sampling volume. Matching the laser beam diameter to the sample size and reducing the size of the cuvette result in a significant reduction in background noise. Commercial versions of this approach utilizing 96 parallel capillaries played a major part in the first phase of the sequencing of the human genome.

19.12 Single-Molecule Spectroscopy

Spectroscopic measurements as described earlier and in the previous chapter are generally carried out in a sample cell in which a very large number of the molecules of interest, called an ensemble, are present. In general, the local environment of the molecules in an ensemble is not identical, which leads to inhomogeneous broadening of an absorption line as discussed in Section 18.6. Figure 19.20 shows how a broad absorption band arises if the corresponding narrow bands for individual molecules in the ensemble are slightly shifted in frequency because of variations in the immediate environment of a molecule. Clearly, more information is obtained from the spectra of the individual molecules than from the broadened band that arises from inhomogeneous broadening. The "true" absorption band for an individual molecule is observed only if the number of molecules in the volume being sampled is very small, for example, the bottom spectrum in Figure 19.20.

Single-molecule spectroscopy is particularly useful in understanding the structure–function relationship for biomolecules. The **conformation** of a biomolecule refers to the arrangement of its constituent atoms in space and can be discussed in terms of primary, secondary, and tertiary structures. The **primary structure** is determined by the backbone of the molecule, for example, the peptide bonds in a polypeptide. The term **secondary structure** refers to the local conformation of a part of the polypeptide. Two common secondary structures of polypeptides are the α-helix and the β-sheet (see Figure 18.15). **Tertiary structure** refers to the overall shape of the molecule; globular proteins are folded into a spherical shape, whereas fibrous proteins have polypeptide chains that arrange into parallel strands or sheets.

Keep in mind that the conformation of a biomolecule in solution is not static. Collisions with solvent and other solute molecules continuously change the energy and the confirmation of a dissolved biomolecule with time. What are the consequences of such

FIGURE 19.20
The absorption spectrum of an individual molecule is narrow, but the peak occurs over a range of frequencies for different molecules as shown in the lowest curve for 10 molecules in the sampling volume. As the number of molecules in the sampling volume is increased, the observed peak shows inhomogeneous broadening and is characteristic of the ensemble rather than of an individual molecule. (Source: After Ph. Tamarat et al, *J. Physical Chemistry A*, 104, 1, 2000.)

conformational changes for an enzyme? Because the activity is intimately linked to structure, conformational changes lead to fluctuations in activity, making an individual enzyme molecule alternately active and inactive as a function of time. Spectroscopic measurements carried out on an ensemble of enzyme molecules give an average over all possible conformations, and hence over all possible activities for the enzyme. Such measurements are of limited utility in understanding how structure and chemical activity are related. As we will show in the next section, single-molecule spectroscopy can go beyond the ensemble limit and gives information on the possible conformations of biomolecules and on the timescales on which transitions to different conformations take place.

To carry out single-molecule spectroscopy, the number of molecules in the sampling volume must be reduced to approximately one. How is a spectrum of individual biomolecules in solution obtained? Only molecules in the volume that is both illuminated by the light source and imaged by the detector contribute to the measured spectrum. For this number to be approximately one, a laser focused to a small diameter is used to excite the molecules of interest, and a confocal microscope is used to collect the photons emitted in fluorescence from a small portion of the much larger cylindrical volume of the solution illuminated by the laser. In a confocal microscope, the sampling volume is at one focal point and the detector is behind a pinhole aperture located at the other focal point of the microscope imaging optics. Because photons that originate outside of a volume of approximately $1 \times 1 \times 1$ μm centered at the focal point are not imaged on the pinhole aperture, they cannot reach the detector. Therefore, the confocal arrangement rather than the solution volume illuminated with the laser determines the sampling volume. To ensure that no more than a few molecules are likely to be found in the sampling volume, the concentration of the biomolecule must be less than $\sim 1 \times 10^{-6}$ M. Because the number of photons emitted in single-molecule spectroscopy is small, it is important to ensure that spurious photons, which could originate from scattered laser light or from Raman scattering outside of sampling volume, do not reach the detector.

Fluorescence spectroscopy is well suited for single-molecule studies, because as discussed in Section 19.9, internal conversion ensures that the emitted photons have a lower frequency than the laser used to excite the molecule. Therefore, optical filters can be used to ensure that scattered laser light does not reach the detector. If the molecules being investigated are immobilized, they can also be imaged using the same experimental techniques. Figure 19.21 shows an image of individual Rhodamine B dye molecules tethered to a glass slide. The apparent size of the molecules is determined by the wavelength of the light used, and not by the actual molecular size. For visible light the apparent molecular size is ~ 300 nm, so that both small and very large molecules have the same apparent size.

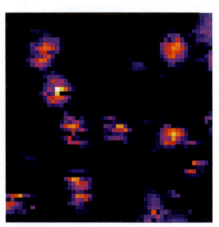

FIGURE 19.21
Microscope image of single Rhodamine B dye molecules on glass obtained using a confocal scanning microscope. The bright spots in the image correspond to fluorescence from single molecules. The image dimension is 5×5 μm.

19.13 Fluorescence Resonance Energy Transfer (FRET)

An electronically excited molecule can lose energy by either radiative or nonradiative events as discussed in Section 19.9. We refer to the molecule that loses energy as the **donor,** and the molecule that accepts the energy as the **acceptor.** If the emission spectrum of the donor overlaps the absorption spectrum of the acceptor as shown in Figure 19.22, then we refer to **resonance energy transfer,** meaning that the photon energy for fluorescence in the donor is equal to the photon energy for absorption in the acceptor as shown in Figure 19.23. Under resonance conditions, the energy transfer from the donor to the acceptor can occur with a high efficiency.

The probability for resonant energy transfer is strongly dependent on the distance between the two molecules. Theodore Förster showed that the rate at which resonant energy transfer occurs decreases as the sixth power of the donor–acceptor distance:

$$k_{ret} = \frac{1}{\tau_D^\circ}\left(\frac{R_0}{r}\right)^6 \qquad (19.11)$$

In Equation (19.11), τ_D° is the lifetime of the donor in its excited state and R_0, the critical Förster radius, is the distance at which the resonance transfer rate and the rate for

FIGURE 19.22

The emission spectrum of the excited donor occurs at a longer wavelength than the absorption as discussed in Section 19.9. If the emission spectrum of the donor overlaps the absorption spectrum of the acceptor, resonant energy transfer between the donor and acceptor can occur. Note the shift in the wavelength of the light emitted by the acceptor and the light absorbed by the donor. This shift allows the use of optical filters to detect acceptor emission in the presence of scattered light from the laser used to excite the donor.

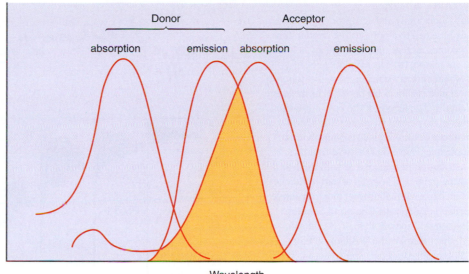

FIGURE 19.23

The energy levels of the ground-state and excited-state donor and acceptor are shown. Resonant energy transfer only occurs if the donor and acceptor match up in energy.

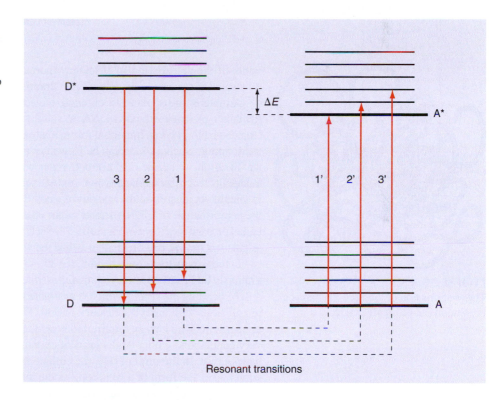

spontaneous decay of the excited-state donor are equal. Both of these quantities can be determined experimentally. The sensitive dependence of the resonance energy transfer on the donor–acceptance distance makes it possible to use **Fluorescence resonance energy transfer** (FRET) as a **spectroscopic ruler** to measure donor–acceptor distances in the 10- to 100-nm range as is described in Problem 19.15. Figure 19.24 illustrates how FRET can be used to determine the conformation of a biomolecule. Schuler *et al.* [*Proceedings of the National Academy of Sciences*, 102 (2005), 2754] attached dyes acting as donor and acceptor molecules to the ends of polyproline peptides of defined length containing between 6 and 40 proline residues. The donor absorbed a photon from a laser, and the efficiency with which the photon was transferred to the acceptor was measured. As Equation (19.11) shows, the efficiency falls off as the sixth power of the donor–acceptor distance. The results for a large number of measurements are shown in Figure 19.24b. A range of values for the efficiency is seen for each polypeptide. This is the case because the peptide is not a rigid rod, and a single molecule can have a variety of possible conformations, each of which has a different donor–acceptor distance. The

FIGURE 19.24

(a) Donor (left) and acceptor (right) dyes are attached to a polyproline peptide that becomes increasingly flexible as its length is increased. (b) The efficiency, E, of resonant energy transfer from the donor to acceptor is shown for peptides of different lengths. The length of the bars represents the relative event frequency for a large number of measurements on individual molecules. Note that the width of the distribution in E increases with the length of the peptide. The peak at zero efficiency is an experimental artifact due to inactive acceptors. (Source: Benjamin Schuler, Everett A. Lipman, Peter J. Steinbach, Michael Kumke and William A. Eaton, "Polyproline and the 'spectroscopic ruler' revisited with single-molecule fluorescence," *PNAS*, February 22, 2005, vol. 102, no. 8, 2754–2759, fig. 1)

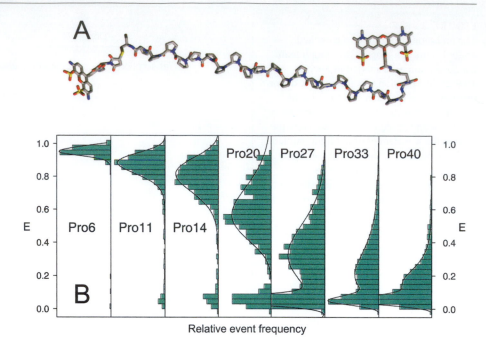

FIGURE 19.25

The conformation of long rod-like molecule in solution can be highly tangled.

width of the distribution in efficiency increases with the peptide length because more twists and turns can occur in a longer strand.

An interesting application of single-molecule FRET is in probing the conformational flexibility of single-stranded DNA in solution. The conformational flexibility of single-stranded DNA plays an important role in many DNA processes such a replication, repair, and transcription. Such a strand can be viewed as a flexible rod, approximately 2 nm in diameter. To put the dimensions of a strand in perspective, if it were a rubber tube of 1 cm in diameter, its length would be nearly 1 km. A single molecule of DNA can be as long as ~1 cm in length, yet must fit in the nucleus of a cell that is typically ~1 μm in diameter. To do so, the conformation of a DNA strand might take the form of a very long piece of spaghetti coiled on itself as shown in Figure 19.25. Such a complex conformation is best described by a statistical model, one of which is called the worm-like chain model.

In the **worm-like chain model,** the strand takes the form of a flexible rod that is continuously and randomly curved in all possible directions. However, there is an energetic cost of bending the strand, which is available to the strand through the energy transferred in collisions with other species in solution. This energy depends linearly on the absolute temperature. The energy required to bend the rod depends on the radius of curvature; a very gentle bend with a large radius of curvature requires much less energy than a sharp hairpin turn. In the limit of 0 K, the collisional energy transfer approaches zero, and the strand takes the form of a rigid rod. As the temperature increases, fluctuations in the radius of curvature increasingly occur along the rod. This behavior is described in the worm-like chain model by the **persistence length,** which is the length you can travel along the rod in a straight line before the rod bends in a different direction. As the temperature increases from 0 K to room temperature, a worm-like chain changes in conformation from a rigid rod of infinite persistence length to the tangled mess depicted in Figure 19.25, which has a very small persistence length.

How well does the worm-like chain model describe single-stranded DNA? This was tested by M. C. Murphy *et al.,* who attached single-stranded DNAs to a rigid tether that was immobilized by bonding the biotin at the end of the tether to a streptavidin-coated quartz surface as shown in Figure 19.26. A donor fluorophore was attached to the free end of the strand and an acceptor was attached to the rigid end. The length of the strand was varied between 10 and 70 nucleotides corresponding to distances from ~60 to ~420 nm between the donor and acceptor. After measuring τ_D° and R_0, the efficiency of resonant energy transfer from the donor to the acceptor was measured as a function of the strand length in NaCl solution whose concentration ranged for 2.5×10^{-3} to 2 M. The results are shown in Figure 19.27, and compared with calculations in which the persistence length was used as a parameter.

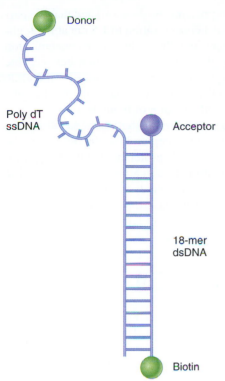

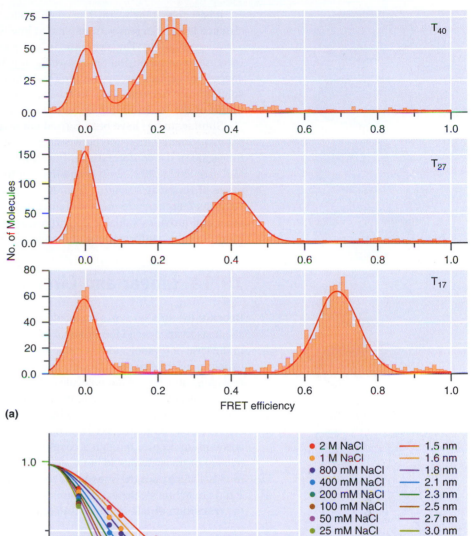

(a)

(b)

FIGURE 19.26

A donor and acceptor are attached to opposite ends of single strands of DNA. The strands are attached to a silica substrate. The function of the 18-mer dsDNA is to isolate the single strand from the substrate. (Source: Reprinted with permission from M. C. Murphy et al., *Biophysical J.,* 86 (2004), 2530.)

FIGURE 19.27

(a) The FRET efficiency is shown for Poly dT ssDNA of length 40, 27, and 17 nucleotides (top to bottom panel). The peak at zero efficiency is an experimental artifact due to inactive acceptors. (b) The FRET efficiency is shown as a function of N, the number of nucleotides for various concentrations of NaCl. The various curves are calculated curves using the persistence length as a parameter. The best fit curves and the corresponding persistence length are shown for each salt concentration. (Source: Reprinted with permission from M. C. Murphy et al., *Biophysical J.,* 86 (2004), 2530.)

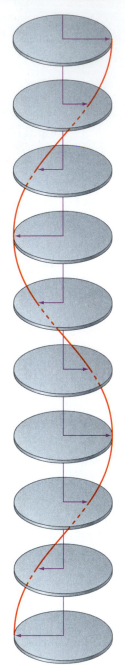

FIGURE 19.28
The arrows in successive images indicate the direction of the electric field vector as a function of time or distance. For linearly polarized light, the amplitude to the electric field vector changes periodically, but is confined to the plane of polarization.

FIGURE 19.29
The amide bonds in a polypeptide chain are shown. The transition dipole moment is shown for a $n \rightarrow \pi^*$ transition.

It is seen that the worm-like chain model represents the data well and that the persistence length decreases from 3 nm at low NaCl concentration to 1.5 nm at the highest concentration. The decrease in persistence length as the NaCl concentration increases can be attributed to a reduction in the repulsive interaction between the charged phosphate groups on the DNA through screening of the charge by the ionic solution (see Section 11.4). In this case, FRET measurements have provided a validation of the worm-like chain model for the conformation of biomolecules.

Similar studies have been carried out using electron transfer reactions rather than resonant energy transfer between the donor and acceptor in order to probe the timescale of the conformational fluctuations of a single protein molecule. Yang *et al.* [*Science*, 302 (2003), 262] found that conformational fluctuations occur over a wide range of timescales ranging from hundreds of microseconds to seconds. This result suggests that there are many different pathways that lead from one conformer to another and provides valuable data to researchers who model protein folding and other aspects of the conformational dynamics of biomolecules.

19.14 Linear and Circular Dichroism

Because the structure of a molecule is closely linked to its reactivity, it is a goal of chemists to understand the structure of a molecule of interest. This is a major challenge in the case of biomolecules because the larger the molecule, the more challenging it is to determine the structure. However, techniques are available to determine aspects of the molecular structure of biomolecules, although they do not give the positions of all atoms in the molecule. Linear and circular dichroism are particularly useful in giving information on the secondary structure of biomolecules.

As discussed in Section 18.1, light is a transverse electromagnetic wave that interacts with molecules through a coupling of the electric field, E, of the light to the permanent or transient dipole moment, μ, of the molecule. Both E and μ are vectors, and in classical physics the strength of the interaction is proportional to the scalar product $E\mu$. In quantum mechanics, the strength of the interaction is proportional to $E\mu^{fi}$, where the **transition dipole moment** is defined by

$$\mu^{fi} = \int \psi_f^*(\tau)\mu(\tau)\psi_i(\tau)d\tau \tag{19.12}$$

In Equation (19.12), τ is a shorthand symbol for the spatial coordinates x, y, z or r, θ, ϕ, and ψ_i and ψ_f refer to the initial and final states in the transition in which a photon is absorbed or emitted. The spatial orientation of μ^{fi} is determined by evaluating an integral such as Equation (19.12), which goes beyond the level of this text. In Figure 19.28, we show the orientation of μ^{fi} for the amide group, which is the building block for the backbone of a polypeptide for a given $\pi \rightarrow \pi^*$ transition.

Many biomolecules have a long rod-like shape and can be oriented by embedding them in a film and then stretching the film. For such a sample, the molecule and therefore μ^{fi} have a well-defined orientation in space. The electric field E can also be oriented in a plane with any desired orientation using a polarization filter, in which **linearly polarized light** is generated, as shown in Figure 19.29.

If the plane of polarization is varied with respect to the molecular orientation, the measured absorbance, A, will vary. It has a maximum value if E and μ^{fi} are parallel, and is zero if E and μ^{fi} are perpendicular. In **linear dichroism spectroscopy,** the variation of the absorbance with the orientation of plane polarized light is measured. It is useful because it allows the direction of μ^{fi} to be determined for an oriented molecule whose secondary structure is not known. One measures the absorbance with E parallel and perpendicular to the molecular axis. The difference $A_{\parallel} - A_{\perp}$ relative to the absorbance for randomly polarized light is the quantity of interest.

We illustrate the application of linear dichroism spectroscopy in determining the secondary structure of a polypeptide in the following discussion. The amide groups shown in Figure 19.28 interact with one another because of their close spacing, and that interaction can give rise to a splitting of the transition into two separate peaks. The orienta-

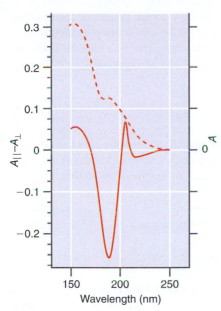

FIGURE 19.30

The normal isotropic absorbance, A and $A_{\parallel} - A_{\perp}$ are shown as a function of the wavelength for an oriented film of poly (g-ethyl-L-glutamate) in which it has the conformation of an α-helix. (Source: Reprinted with permission from J. Brahms et al., *Proc. National Academy of Sciences,* 60 (1968), 1130.)

tion of $\boldsymbol{\mu}^{fi}$ for each component depends on the polypeptide secondary structure. For the case of an α-helix, a transition near 208 nm with $\boldsymbol{\mu}^{fi}$ parallel to the helix axis and a transition near 190 nm with $\boldsymbol{\mu}^{fi}$ perpendicular to the helix axis is predicted from theory. Figure 19.30 shows the absorbance for randomly polarized light and $A_{\parallel} - A_{\perp}$ for a polypeptide. The data show that $A_{\parallel} < A_{\perp}$ for the transition near 190 nm and that $A_{\parallel} > A_{\perp}$ for the transition near 208 nm. This shows that the secondary structure of this polypeptide is an α-helix. Note that the absorption for randomly polarized light gives no structural information.

Because molecules must be oriented in space for linear dichroism spectroscopy, it cannot be used for biomolecules in solution except under conditions of a defined flow direction. For static solutions, **circular dichroism spectroscopy** is widely used to obtain secondary structural information. In this spectroscopy, **circularly polarized light,** which is depicted in Figure 19.31, is passed through the solution.

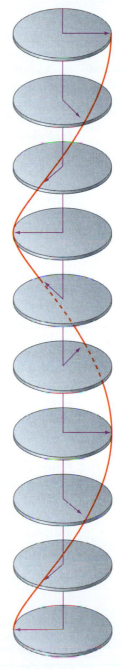

FIGURE 19.31

The arrows in successive images indicate the direction of the electric field vector as a function of time or distance. For circularly polarized light, the amplitude of the electric field vector is constant, but its plane of polarization undergoes a periodic variation.

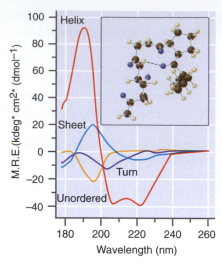

FIGURE 19.32

The rotation angle θ is shown as a function of wavelength for biomolecules having different secondary structures. Because the curves are distinctly different, circular dichroism spectra can be used to determine the secondary structure for optically active molecules. The inset shows the hydrogen bonding between different amide groups that generates different secondary structures. (Source: "Ideal CD spectra." from John T. Pelton, *Science*, vol. 291, no. 5511, pages 2175–2176 (16 March 2001). Reprinted with permission from AAAS.)

Biomolecules are optically active, meaning that they do not possess a center of inversion. For an optically active molecule, the absorption for circularly polarized light in which the direction of rotation is clockwise (*R*) differs from that in which the direction of rotation is counterclockwise (*L*). This difference in *A* can be expressed as a difference in the extinction coefficient ε:

$$\Delta A(\lambda) = A_L(\lambda) - A_R(\lambda) = [\varepsilon_L(\lambda) - \varepsilon_R(\lambda)]lc = \Delta\varepsilon lc \qquad (19.13)$$

In Equation (19.13), λ is the path length in the sample cell, and *c* is the concentration. In practice, the difference between $A_L(\lambda)$ and $A_R(\lambda)$ is usually expressed as the molar ellipticity, which is the shift in the phase angle θ between the components of the circularly polarized light in the form

$$\theta = 2.303 \times (A_L - A_R) \times 180/4\pi \text{ degrees} \qquad (19.14)$$

Circular dichroism can only be observed if $\Delta\varepsilon$ is nonzero, and is usually observed in the visible part of the light spectrum.

As in the case of linear dichroism, $\Delta A(\lambda)$ for a given transition is largely determined by the secondary structure, and is much less sensitive to other aspects of the conformation. A derivation of how $\Delta A(\lambda)$ depends on the secondary structure is beyond the level of this text, but it can be shown that common secondary structures such as the α-helix, the β-sheet, a single turn, and a random coil have a distinctly different $\varepsilon(\lambda)$ dependence as shown in Figure 19.32. In this range of wavelengths, the absorption corresponds to $\pi \rightarrow \pi^*$ transitions of the amide group.

The differences between the $\Delta\varepsilon(\lambda)$ curves are sufficient that the observed $\Delta\varepsilon(\lambda)$ curve obtained for a protein of unknown secondary structure in solution can be expressed in the form

$$\Delta\varepsilon_{observed}(\lambda) = \sum_i F_i \Delta\varepsilon_i(\lambda) \qquad (19.15)$$

where $\Delta\varepsilon(\lambda)$ is the curve corresponding to one of the secondary structures in Figure 19.32, and F_i is the fraction of the peptide chromophores in that particular secondary structure. A best fit of the data to Equation (19.15) using available software allows a determination of the F_i to be made.

Figure 19.33 shows the results of an application of circular dichroism in determining the secondary structure of α-synuclein bound to unilamellar phospholipid vesicles, which were used as a model for cell membranes. α-Synuclein is a small soluble protein of 140 to 143 amino acids that is found in high concentration in presynaptic nerve terminals. A mutation in this protein has been linked to Parkinson's disease and it is believed to be a precursor in the formation of extracellular plaques in Alzheimer's disease.

As can be seen by comparing the spectra in Figure 19.33 with those of Figure 19.32, the conformation of α-synuclein in solution is that of a random coil. However, on binding to unilamellar phospholipid vesicles, the circular dichroism spectrum is dramatically changed and is characteristic of an α-helix. These results show that the binding of α-synuclein requires a conformational change. This conformational change can be understood from the known sequence of amino acids in the protein. By forming an α-helix, the polar and nonpolar groups in the protein are shifted to opposite sides of the helix. This allows the polar groups to associate with the acidic phospholipids, leading to a stronger binding than would be the case for a random coil.

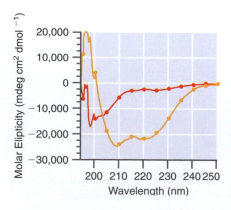

FIGURE 19.33

The molar ellipticity is shown as a function of the wavelength for α-synuclein in solution (yellow curve) and for α-synuclein bound to unilamellar phospholipid vesicles (red curve). (Source: Reprinted with permission from W. S. Davidson et al., *J. Biological Chemistry*, 273 (1998), 9443.)

Vocabulary

$n \rightarrow \pi^*$, $\pi \rightarrow \pi^*$, and $\sigma \rightarrow$
   σ^* transitions
acceptor
atomic absorption
   spectroscopy
atomic emission
   spectroscopy
Beer–Lambert law
Born–Oppenheimer
   approximation
chromophore
circular dichroism
   spectroscopy
circularly polarized light
coherent photon source

conformation
continuous energy spectrum
donor
Doppler broadening
Doppler effect
Doppler shift
fluorescence
Fluorescence resonance
   energy transfer (FRET)
Franck–Condon factor
Franck–Condon principle
incoherent photon source
integral absorption
   coefficient
internal conversion

intersystem crossing
laser
lasing transition
linear dichroism
   spectroscopy
linearly polarized light
molar extinction coefficient
nonradiative transition
optical resonator
persistence length
phosphorescence
population inversion
primary structure
pumping transition

radiative transition
resonance energy transfer
resonator modes
secondary structure
single-molecule spectroscopy
singlet state
spectroscopic ruler
spontaneous emission
stimulated emission
tertiary structure
transition dipole moment
triplet state
worm-like chain model

Questions on Concepts

Q19.1 What is the relationship between the Rydberg constant for H and for Li^{2+}?

Q19.2 Why is atomic absorption spectroscopy more sensitive in many applications than atomic emission spectroscopy?

Q19.3 Why does the Doppler effect lead to a shift in the wavelength of a star, but to a broadening of a transition in a gas?

Q19.4 How can the width of a laser line be less than that determined by the Doppler broadening?

Q19.5 The relative intensities of vibrational peaks in an electronic spectrum is determined by the Franck–Condon factors. How would the potential curve for the excited state in Figure 19.9 need to be shifted along the distance axis for the $n = 0 \rightarrow n' = 0$ transition to have the highest energy? The n term refers to the vibrational quantum number in the ground state, and n' refers to the vibrational quantum number in the excited state.

Q19.6 Make a sketch, like that in the inset of Figure 19.9, of what you might expect the electronic spectrum to look like for the ground and excited states shown here.

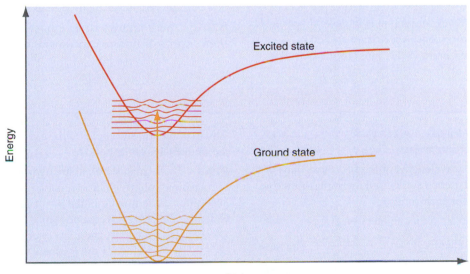

Q19.7   Make a sketch, like that in the inset of Figure 19.9, of what you might expect the electronic spectrum to look like for the ground and excited states shown here.

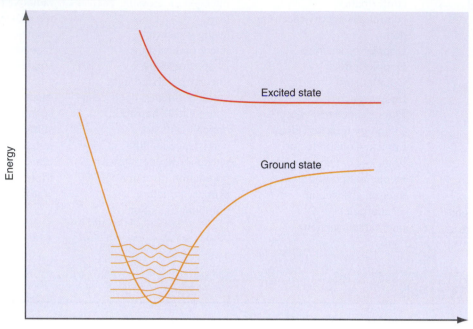

Q19.8   Explain why the fluorescence and absorption groups of peaks in Figure 19.17 are shifted and show mirror symmetry for idealized symmetrical ground-state and excited-state potentials.

Q19.9   What would the intensity versus frequency plot in Figure 19.17 look like if fluorescence were fast with respect to internal conversion?

Q19.10   The rate of fluorescence is higher than that for phosphorescence. Can you explain this fact?

Q19.11   Because internal conversion is in general very fast, the absorption and fluorescence spectra are shifted in frequency as shown in Figure 19.17. This shift is crucial in making fluorescence spectroscopy capable of detecting very small concentrations. Can you explain why?

Q19.12   What can be learned from single-molecule spectroscopy that cannot be learned from conventional spectroscopy?

Q19.13   What is the mechanism by which energy is transferred from the donor to the acceptor in FRET?

Q19.14   Why is spectroscopy based on circular dichroism more useful than linear dichroism in studying biomolecules?

Q19.15   What primary structural information can be obtained from circular dichroism experiments on biomolecules?

Problems

Problem numbers in **RED** indicate that the solution to the problem is given in the *Student Solutions Manual*.

P19.1   Calculate the wavelengths of the first three lines of the Lyman, Balmer, and Paschen series, and the series limit (the shortest wavelength) for each series.

P19.2   The energy levels for a one-electron atom or ion of nuclear charge Z are given by $E_n = -\dfrac{Z^2 \mu e^4}{32\pi^2 \varepsilon_0^2 \hbar^2 n^2}$. The masses of an electron, a proton, and a tritium (^{3}H or T) nucleus are given by 9.1094×10^{-31} kg, 1.6726×10^{-27} kg, and 5.0074×10^{-27} kg, respectively. Calculate the frequency of the $n = 1 \rightarrow n = 4$ transition in H and T to six significant figures.

Which of the transitions $1s \rightarrow 4s$, $1s \rightarrow 4p$, or $1s \rightarrow 4d$ could the frequencies correspond to?

P19.3   The Lyman series in the hydrogen atom corresponds to transitions, all of which originate from the $n = 1$ level in absorption, or terminate in the $n = 1$ level for emission. Calculate the energy, frequency (in s^{-1} and cm^{-1}), and wavelength of the least and most energetic transition in this series. Use the value for the Rydberg constant for hydrogen of 109,677 cm^{-1}.

P19.4   The absorption spectrum of the hydrogen atom shows lines at 5334, 7804, 9145, 9953, and 10,478 cm^{-1}. There are no lower frequency lines in the spectrum. Use the graphical methods discussed in Example Problem 11.1 to determine $n_{initial}$ and the ionization energy of the hydrogen atom in this state. Assume values for $n_{initial}$ of 1, 2, and 3.

P19.5   Calculate the transition dipole moment, $\mu_z^{mn} = \int \psi_m^*(\tau)\mu_z \psi_n(\tau)\, d\tau$, where $\mu_z = -er\cos\theta$ for a transition from the $1s$ level to the $2s$ level in H. Show that this transition is forbidden. The integration is over r, θ, and ϕ.

P19.6   Calculate the transition dipole moment, $\mu_z^{mn} = \int \psi_m^*(\tau)\mu_z \psi_n(\tau)\, d\tau$, where $\mu_z = -er\cos\theta$ for a transition from the $1s$ level to the $2p_z$ level in H. Show that this transition is allowed. The integration is over r, θ, and ϕ. Use

$$\psi_{210}(r,\theta,\phi) = \frac{1}{\sqrt{32\pi}}\left(\frac{1}{a_0}\right)^{3/2}\frac{r}{a_0}e^{-r/2a_0}\cos\theta$$ for the $2p_z$ wave function.

P19.7   The Doppler broadening in a gas can be expressed as $\Delta\nu = \frac{2\nu_0}{c}\sqrt{\frac{2RT\ln 2}{M}}$, where M is the molar mass. For a transition between states in sodium that arise from the $3p$ energy level, $\nu_0 = 5.0933 \times 10^{14}\ s^{-1}$. Calculate $\Delta\nu$ and $\Delta\nu/\nu_0$ at 500 K.

P19.8   An important biological application of absorption spectroscopy is the determination of the concentrations of solutions of nucleic acids. The $\pi \rightarrow \pi^*$ electronic transitions within the purine and pyrimidine bases of nucleic acids have absorption maxima near a wavelength of 260 nm. Assume the extinction coefficient of a nucleic acid is $1.00 \times 10^4\ M^{-1}cm^{-1}$ at 260 nm. If the concentration of a nucleic acid solution is $5.00 \times 10^{-4} M$, calculate the absorbance of this solution in a 1.00 cm cell at 260 nm.

P19.9   The absorbance of a nucleic acid solution at 260 nm is called A_{260}. The OD (i.e., optical density) is the amount of nucleic acid in a volume of 1.00 mL in a 1.00-cm path length cell for which $A_{260} = 1.00$. How many moles of nucleotide are contained in a 1.00 mL solution of a double-stranded nucleotide for which $A_{260} = 2.50$, assuming the extinction coefficient per nucleotide is $7.000 \times 10^3\ M^{-1}\ cm^{-1}$. Express this quantity in OD's. Assume a 1.00 cm path length.

P19.10   Because of interactions between transition dipoles of the constituent nucleotides, the extinction coefficient for a single strand polynucleotide is not simply the sum of the extinction coefficients for the individual nucleotides. These dipole-dipole interactions depend on $1/r^3$ where r is the distance between bases, so for the purpose of calculating the extinction coefficient for a single-stranded polynucleotide, only nearest neighbor interactions need be considered. For a hypothetical polynucleotide strand $GpCpUp\ldots ApG$ the extinction coefficient is

$$\varepsilon(GpCpUp\ldots ApG) = 2\,[\varepsilon(GpC) + \varepsilon(CpU) + \varepsilon(ApG)]$$
$$- [\varepsilon(Cp) + \varepsilon(Up) + \ldots \varepsilon(Ap)]$$

Interacting nucleotides pairs in a single strand are indicated by XpY, where p represents the phosphate group that joins the nucleotides X and Y. In the equation above $\varepsilon(ApG)$, $\varepsilon(ApC)$, etc., are extinction coefficients for component dinucleotide phosphates per mole of nucleotide. Hence they are counted twice, which accounts for the 2 in the expression above. To correct for this fact, the extinction coefficients of the individual nucleotides, except for the terminal nucleotides, are subtracted. The preceding equation gives good agreement with experimental extinction coefficients for DNA and RNA single strands.

Consider the tables of extinction coefficients per nucleotide ($M^{-1}\ cm^{-1}$) at 260 nm, 298K, and pH 7 for RNA nucleotides.

Ap	Cp	Gp	Up	ApA	ApC	ApG	ApU	CpA	CpC
15.34×10^3	7.60×10^3	12.16×10^3	10.21×10^3	13.65×10^3	10.67×10^3	12.79×10^3	12.14×10^3	10.67×10^3	7.52×10^3

CpG	CpU	GpA	GpC	GpG	GpU	UpA	UpC	UpG	UpU
9.39×10^3	8.37×10^3	12.92×10^3	9.19×10^3	11.43×10^3	10.96×10^3	12.52×10^3	8.90×10^3	10.40×10^3	10.11×10^3

   a. Calculate the extinction coefficient for the single strand RNA ApCpGpUpUpApGpU at 298K and pH 7.

   b. Calculate A_{260} for a 1.50×10^{-4} M solution of the RNA in part a for a path length of 1.00 cm.

   c. Repeat the calculations in part a and b for the single strand RNA GpCpUpUpApA. Assume a path length of 1.00 cm.

P19.11   Consider the tables of extinction coefficients per nucleotide for DNA nucleotides (units of $M^{-1}\ cm^{-1}$) at 260 nm, 298K, and pH 7.

Ap	Cp	Gp	Tp	ApA	ApC	ApG	ApT	CpA	CpC
15.34×10^3	7.60×10^3	12.16×10^3	8.70×10^3	13.65×10^3	10.67×10^3	12.79×10^3	11.42×10^3	10.67×10^3	7.52×10^3

CpG	CpT	GpA	GpC	GpG	GpT	TpA	TpC	TpG	TpT
9.39×10^3	7.66×10^3	12.92×10^3	9.19×10^3	11.43×10^3	10.22×10^3	11.78×10^3	8.15×10^3	9.70×10^3	8.61×10^3

   a. Calculate the extinction coefficient for the single-stranded DNA: ApCpGpTpApTpApG

   b. If the absorbance for a solution of the DNA in part a is $A_{260} = 1.2$ for a path length of 1.00 cm, calculate the concentration of DNA in the solution.

   c. Repeat the calculations in parts a and b for the single-stranded DNA GpCpApTpTpApApG. Assume a path length of 1.00 cm.

P19.12   For solutions composed of more than one absorbing species the absorbances at a given wavelength λ are additive. For two absorbing species M and N the absorbances of the individual species at λ are additive:

$$A_\lambda = A_\lambda^M + A_\lambda^N = \varepsilon_\lambda^M \ell c_M + \varepsilon_\lambda^N \ell c_N = \ell(\varepsilon_\lambda^M c_M + \varepsilon_\lambda^N c_N)$$

Therefore, the concentrations of the two species can be obtained by making absorbance measurements at two different wavelengths λ_1 and λ_2, and using known values of the four extinction coefficients $\varepsilon_{\lambda_1}^M, \varepsilon_{\lambda_2}^M, \varepsilon_{\lambda_1}^N, \varepsilon_{\lambda_2}^N$.

   a. Derive expressions for the concentrations c_M and c_N in terms of the extinction coefficients $\varepsilon_{\lambda_1}^M, \varepsilon_{\lambda_2}^M, \varepsilon_{\lambda_1}^N, \varepsilon_{\lambda_2}^N$, the path length l, and the absorbances A_{λ_1} and A_{λ_2}.

   b. For tyrosine (Y) the extinction coefficients at 240 and 280 nm are $\varepsilon_{240}^Y = 11,300 \text{ M}^{-1} \text{cm}^{-1}$ and $\varepsilon_{280}^Y = 1500 \text{ M}^{-1} \text{cm}^{-1}$. For tryptophan (W) the corresponding extinction coefficients are $\varepsilon_{240}^W = 1960 \text{ M}^{-1} \text{cm}^{-1}$ and $\varepsilon_{280}^W = 5380 \text{ M}^{-1} \text{cm}^{-1}$. A solution of tyrosine and tryptophane has absorbances of $A_{240} = 0.350$ and $A_{280} = 0.226$. Calculate the concentrations of tyrosine and tryptophan. Assume a path length of 1.00 cm.

P19.13   The extinction coefficient for a double-stranded nucleic acid depends quadratically on the base composition. The base composition for double-stranded DNA in turn is quantified by f_{AT}, the fraction of $A{\bullet}T$ base pairs. For a double-stranded DNA the extinction coefficient ε is given by

$$\varepsilon = \varepsilon_{AA} f_{AT}^2 + 2\varepsilon_{AG} f_{AT}(1 - f_{AT}) + \varepsilon_{GG}(1 - f_{AT})^2$$

where ε_{AA} is the extinction coefficient resulting from interacting $A{\bullet}T$ base pairs, ε_{GG} is the extinction coefficient resulting from interacting $G{\bullet}C$ base pairs, and ε_{AG} is the extinction coefficient resulting from interacting $A{\bullet}T$ and $G{\bullet}C$ base pairs.

   a. For double-stranded DNA:

   $\varepsilon_{AA} = 6300. \text{ M}^{-1} \text{cm}^{-1}$, $\varepsilon_{GG} = 7585 \text{ M}^{-1} \text{cm}^{-1}$, $\varepsilon_{AG} = 6100. \text{ M}^{-1} \text{cm}^{-1}$ at 260 nm.

   Calculate the extinction coefficient at 260 nm for DNA that is 50% $A{\bullet}T$.

   b. Calculate the concentration of a solution of double-stranded DNA that is 50% $A{\bullet}T$ if $A_{260} = 1.63$. Assume a path length of 1.00 cm.

   c. Suppose for a 1.25×10^{-4} M solution of a double-stranded DNA the absorbance at 260 nm is 0.81 for a path length of 1.00 cm. Calculate the fraction of $A{\bullet}T$ base pairs in the DNA.

P19.14   In double-stranded RNA, the base composition is quantified by f_{AU}, the number of $A{\bullet}U$ base pairs. An expression for the extinction coefficient of double-stranded RNA has the same form as given in Problem 19.13, where f_{AT} is replaced by f_{AU}. For double-stranded RNA $\varepsilon_{AA} = 6480 \text{ M}^{-1} \text{cm}^{-1}$, $\varepsilon_{GG} = 7010 \text{ M}^{-1} \text{cm}^{-1}$, $\varepsilon_{AG} = 6875 \text{ M}^{-1} \text{cm}^{-1}$, where the extinction coefficients ε_{AA}, ε_{GG}, and ε_{AG} reflect the interaction of, respectively, $A{\bullet}U$ base pairs with other $A{\bullet}U$ base pairs, $G{\bullet}C$ with other $G{\bullet}C$ base pairs, and $A{\bullet}U$ with $G{\bullet}C$. For a 1.55×10^{-4} M double-stranded RNA solution $A_{260} = 1.05$ in a 1.00-cm cell. Calculate f_{AU}. Assume a path length of 1.00 cm.

P19.15   As explained in Section 19.11, FRET can be used to determine distances between fluorescent chromophores in macromolcules, thus providing information on macromolecular conformation. The efficiency of energy transfer E_t in a FRET experiment is given by $E_t = \dfrac{R_0^6}{R_0^6 + r^6}$ where r is the distance between the donor (D) and acceptor (A) chromophores. The excitation transfer can be determined from fluorescent life times $E_t = 1 - \dfrac{\tau_{D+A}}{\tau_D}$ where τ_D is the fluorescent life time of the donor D in the absence of the acceptor A and τ_{D+A} is the life time of the donor in the presence of the acceptor.

   Consider a protein labeled with a donor naphthyl group and an acceptor dansyl group. The fluorescence lifetime for the naphthyl group in the protein is 23 ns. When dansyl is added to the protein the life time of the naphthyl group decreases to 18 ns. Calculate the distance r between the naphthyl and dansyl chromophores assuming the Förster radius $R_0 = 34$ Å.

Web-Based Simulations, Animations, and Problems

W19.1   The individual processes of absorption, spontaneous emission, and stimulated emission are simulated in a two-level system. The level of pumping needed to sustain lasing is ex-perimentally determined by comparing the population of the upper and lower levels in the lasing transition.

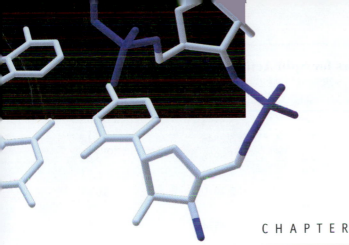

20

Nuclear Magnetic Resonance Spectroscopy

Although the nuclear magnetic moment interacts only weakly with an external magnetic field, this interaction provides a very sensitive probe of the local electron distribution in a molecule. An NMR spectrum can distinguish between chemically inequivalent nuclei such as ^1H at different sites in a molecule. NMR can also be used as a nondestructive imaging technique that is widely used in medicine and in the study of biological polymers. Pulsed NMR and multidimensional Fourier transform techniques provide a powerful combination to determine the structure of large molecules of biological interest.

20.1 Nuclear Spins in External Fields

Nuclear magnetic resonance (NMR) refers to the resonant absorption of radio-frequency energy by nuclear spins in a static magnetic field. The word *resonance* implies that radio-frequency energy will be absorbed by nuclear spins if the frequency ν of the energy fulfills the condition

$$h\nu = \Delta E \tag{20.1}$$

where ΔE refers to the energy difference between quantized nuclear spin angular momentum states. The spin magnetic dipole moment for the proton is given by

$$\vec{\mu} = g_N \frac{e\hbar}{2m_{proton}}\vec{I} = g_N\beta_N\vec{I} = \gamma\hbar\vec{I} \tag{20.2}$$

The vector $\vec{I}$ is the spin angular momentum of the proton. The symbol e stands for the unit of elementary nuclear charge, 1.60×10^{-19} C, and m_{proton} is the mass of a proton. Because the mass of the proton m_{proton} is greater than the mass of the electron m_e by a factor of about 2000, the **nuclear magnetic moment** is much smaller than the electron magnetic moment for the same I. The quantity $\beta_N = e\hbar/2m_{proton}$, which has the value $5.051 - 10^{-27}$ Joules Tesla^{-1}, is called the **nuclear magneton,** and γ is called the **magnetogyric ratio.** The **nuclear factor** g_N, which is a dimensionless number, is characteristic of a particular nucleus. Values of these quantities for some atomic nuclei are shown in Table 20.1.

The energy of interaction between a given nuclear magnetic moment and an external static magnetic field is quantized because the z component of the angular momentum is quantized:

$$E = -\vec{\mu}\cdot\vec{B}_0 = -\gamma\hbar\vec{I}\cdot\vec{B}_0 = -\gamma\hbar m_z B_0 \tag{20.3}$$

TABLE 20.1   Parameters for Spin Active Nuclei

Nucleus	Isotopic Abundance (%)	Spin	Nuclear factor g_N	Magnetogyric Ratio ($\gamma/10^7$ rad T^{-1} s^{-1})
^{1}H	99.985	1/2	5.5854	26.75
^{19}F	100	1/2	5.2546	25.18
^{17}O	0.037	5/2	−1.8928	−3.63
^{13}C	1.108	1/2	1.4042	6.73
^{31}P	100	1/2	2.2610	10.84
^{15}N	0.37	1/2		−2.71
^{2}H	0.015	1	0.8574	4.11
^{14}N	99.63	1	0.4036	1.93

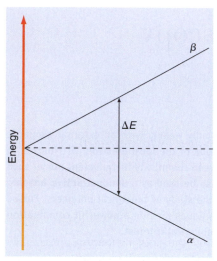

FIGURE 20.1
Energy of a nuclear spin of quantum number $\frac{1}{2}$ as a function of the magnetic field.

where we associate the z direction with the direction of the external magnetic field. For a nucleus with spin angular momentum I, the quantum number m_z varies integrally from $-I$ to $+I$. For a proton $I = \frac{1}{2}$ there are two possible values for m_z, that is, $m_z = \pm\frac{1}{2}$, leading to two energy states for a spin half particle in a magnetic field (see Figure 20.1). For a deuteron, which has spin $I = 1$, $m_z = 0, \pm 1$ and there are three energy states.

EXAMPLE PROBLEM 20.1

a. Calculate the two possible energies of the ^{1}H nuclear spin in a uniform magnetic field of 5.50 T.

b. Calculate the energy ΔE absorbed in making a transition from the α to the β state. If a transition is made between these levels by the absorption of electromagnetic radiation, what region of the spectrum is used?

Solution

a. The two energies are given by

$$E = \pm\frac{1}{2} g_N \beta_N B$$

$$= \pm\frac{1}{2} \times 5.5854 \times 5.051 \times 10^{-27} \text{ J/T} \times 5.50 \text{ T}$$

$$= \pm 7.76 \times 10^{-26} \text{ J}$$

b. The energy difference is given by

$$\Delta E = 2(7.76 \times 10^{-26} \text{ J})$$

$$= 1.55 \times 10^{-25} \text{ J}$$

$$\nu = \frac{\Delta E}{h} = \frac{1.55 \times 10^{-25}}{6.626 \times 10^{-34}} = 2.34 \times 10^8 \text{ s}^{-1}.$$

This is in the range of frequencies called *radio-frequencies*.

Consider a large number of protons in a magnetic field. Because the energies of nuclear magnetic moments oriented parallel and antiparallel to the magnetic field are unequal, there will be more nuclear spins in the lower energy level, and this fact will result in a **macroscopic magnetic moment** or magnetization *M*:

$$\vec{M} = \sum_i \vec{\mu}_i \qquad (20.4)$$

Although the z components of the nuclear magnetic moments will be oriented preferentially parallel to the z axis, their transverse components are oriented randomly as shown

FIGURE 20.2

Precession of a nuclear spin about the magnetic field direction for α spins. The right side of the figure shows that the magnetization vector $\vec{M}$ resulting from summing the individual spin magnetic moments is oriented parallel to the magnetic field. It has no transverse component.

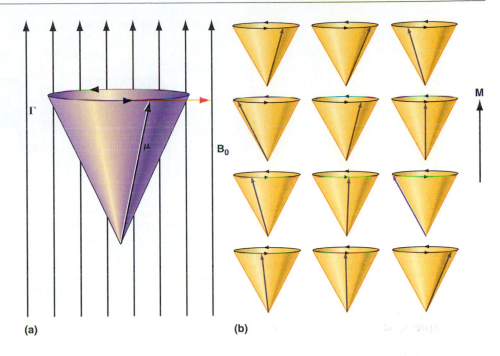

(a)   (b)

in Figure 20.2. Therefore, at equilibrium the magnetization is parallel to the magnetic field and the x and y components of the magnetization are zero: $M_{eq} = M_z$.

The nuclear dipoles are dynamic. This is because the magnetic field exerts a torque Γ on a nuclear magnetic dipole moment given by

$$\Gamma = \vec{\mu} \times \vec{B}_0 \tag{20.5}$$

The cross product in Equation (20.5) indicates that the torque is perpendicular to the plane defined by the magnetic field and the magnetic dipole moment. Because $\vec{\Gamma} = d\vec{I}/dt$, the torque results in a **precession** of the magnetic moment as shown in Figure 20.3:

$$\gamma\vec{\Gamma} = \frac{\gamma d\vec{I}}{dt} = \frac{d\vec{\mu}}{dt} = \gamma\vec{\mu} \times \vec{B}_0 \tag{20.6}$$

Note that the product of the magnetogyric ratio and the magnetic field γB has units of radians per second. We can therefore write

$$\nu_0 = \frac{\omega_0}{2\pi} = \frac{\gamma B_0}{2\pi} \tag{20.7}$$

where ν_0 is the frequency of precession of the magnetic moment around the magnetic field. Equation (20.6) is called the *Larmor equation* and states that if the field B is independent of time, the spin magnetic dipole moment will precess at a constant frequency ν_0, the **Larmor frequency,** about the magnetic field B.

We have thus far considered only the effect of a static magnetic field on a system of spin $\frac{1}{2}$ nuclei. Transitions between the energy levels can be accomplished by the application of a resonant radio-frequency (rf) field. This is accomplished technically as shown in Figure 20.4.

A constant current passing through a long solenoidal coil generates a static magnetic field $\vec{B}_0$ directed along the axis of the solendoid, which is designated the z axis. Around the sample containing spin 1/2 nuclei is wrapped a smaller solenoid whose axis is perpendicular to the direction of the static magnetic field. Designate this direction the y axis, as shown in Figure 20.4. An alternating field with a frequency in the rf domain is generated by the *rf solenoid*. The rf field is linearly polarized along the y axis and has the form

$$\vec{B}_1(t) = \vec{B}_y \cos \omega t = B_1 \hat{j} \cos \omega t \tag{20.8a}$$

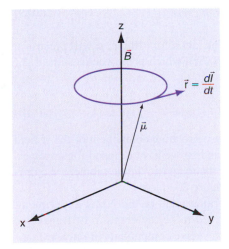

FIGURE 20.3

The effect of a static field $\vec{B}$ is to apply a torque $\vec{\Gamma}$ to the nuclear magnetic moment, which causes a precesstion of the moments about the static magnetic field $\vec{\Gamma} = d\vec{I}/dt = \vec{\mu} \times \vec{B}_0$. The precessional frequency is called the *Larmor frequency* $\nu_0 = \gamma\vec{B}_0/2\pi$.

FIGURE 20.4
Schematic picture of the NMR experiment showing the static field, and the rf field coil.

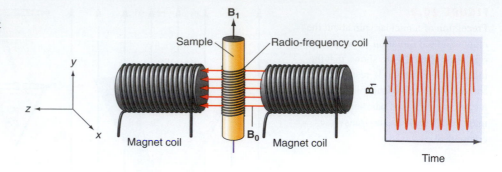

where $\hat{j}$ is the unit vector pointing in the y direction. The effect of a linearly polarized field can be represented mathematically as the superposition of two circularly polarized fields (see Figure 20.5):

$$\vec{B}_1^{cc} = B_1\left(\hat{i} \cos \omega t + \hat{j} \sin \omega t\right)$$

$$\vec{B}_1^{c} = B_1\left(\hat{i} \cos \omega t - \hat{j} \sin \omega t\right) \tag{20.8b}$$

We now alter Equation (20.6) to include the presence of $B_1^{cc}(t)$ only:

$$\frac{d\vec{\mu}}{dt} = \gamma\vec{\mu} \times \left(\vec{B}_0 + \vec{B}_1^{cc}(t)\right) \tag{20.9}$$

When viewed from a fixed frame of reference, the laboratory frame, $\vec{\mu}$ will precess at the Larmor frequency about an effective field $\vec{B}_{eff}(t)$, which is the vector sum of $\vec{B}_0$ and $\vec{B}_1^{cc}(t)$:

$$\vec{B}_{eff}(t) = \vec{B}_0 + \vec{B}_1^{cc}(t) = B_0\hat{k} + B_1\left(\hat{i} \cos \omega_1 t + \hat{j} \sin \omega_1 t\right) \tag{20.10}$$

where we understand that only the counterclockwise rotating component of the rf field is relevant to inducing nuclear spin transitions, and so we have dropped the cc superscript and replaced ω with ω_1. The precession of the magnetic dipole moment vector is now about a time-dependent field, as shown in Figure 20.6, left.

Figure 20.6, left, represents a laboratory frame view of the NMR experiment, where the effect of a resonant rf field is observed to be the creation of transverse magnetization. Recall that at equilibrium the transverse components of nuclear magnetic dipoles are randomly oriented, so the net magnetization is parallel to the static magnetic field. As a result of applying a resonant rf field, transverse components of nuclear magnetic dipoles are no longer randomly oriented, and transverse magnetization is observed as a precessional motion about an effective time-dependent field.

The NMR experiment is more easily viewed in a frame that rotates at the frequency of the rf field. One can imagine a **rotating frame,** in which the point of observation is no longer attached to a fixed laboratory frame, but rather is attached to the rotating rf field. In this frame of reference the observer sees an effective field

$$\vec{B}_{eff,R} = B_1\hat{j}' + \Delta B\hat{k}' = B_1\hat{i}' + \left(B_0 - \frac{\omega_1}{\gamma}\right)\hat{k}' \tag{20.11}$$

where $\hat{i}'$, $\hat{j}'$, $\hat{k}'$ are the rotating frame x, y, and z unit vectors and where $\Delta B = B_0 - \dfrac{\omega_1}{\gamma} = \dfrac{\omega_1}{\gamma}$. This view of the NMR experiment is depicted in Figure 20.6, right. A radio-frequency field linearly polarized along the x axis in the laboratory frame is viewed in the rotating frame as a static field along the rotating frame x axis. When the rf field is resonant with the NMR transition frequency (the Larmor frequency ω_0), that is, when $\omega_1 = \omega_0$, then $\Delta B = 0$, and the effective field is along the x axis of the

FIGURE 20.5

The expression for the field $B_1^c(t)$ is identical to that for $B_1^{cc}(t)$ except the frequency ω is replaced by $-\omega$, so $B_1^{cc}(t)$ is said to be rotating counterclockwise while $B_1^c(t)$ rotates clockwise. Of these two circularly polarized fields, only $B_1^{cc}(t)$ rotates in the same sense as the magnetic moments and for this reason only $B_1^{cc}(t)$ can induce transitions between nuclear spin states.

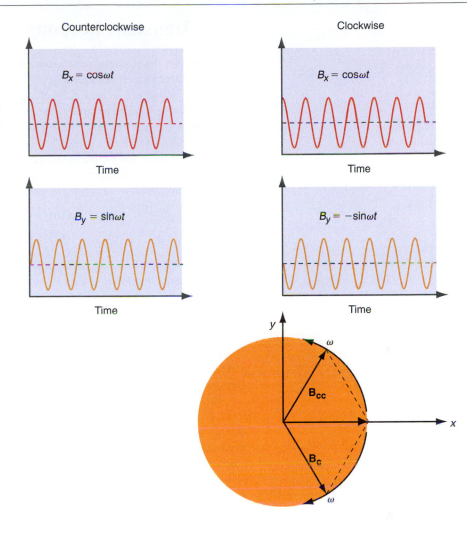

FIGURE 20.6

The NMR experiment as viewed from the laboratory and the rotating frame of reference.

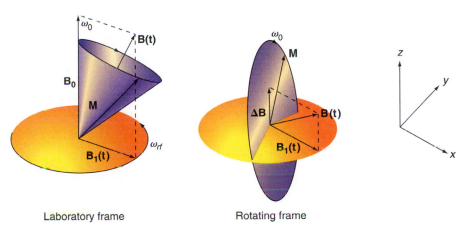

rotating frame and the magnetization precesses around the rotating frame x axis. This can be expressed as the rotating frame torque equation:

$$\frac{d\vec{\mu}_R}{dt} = \gamma\vec{\mu}_R \times B_1\hat{i}' \tag{20.12}$$

where the subscript R signifies that the magnetic dipole moment vector is viewed from the rotating frame.

20.2 Transient Response and Relaxation

At equilibrium the sum magnetic moment $\vec{M}_0 = \sum \mu_i$ is stationary and aligned along the z axis. Suppose an rf field is suddenly applied to the spin system for a time t, and then is suddenly turned off. If the time required for the field to turn on and off is very short compared to the irradiation time t, the rf "pulse" has a rectangular envelope as shown in Figure 20.7. If the frequency of this rf field is matched to the Larmor frequency, the effective field in the rotating frame is along the x axis of the rotating frame, and $\vec{M}$ begins to precess in the y-z plane. This can be seen as follows.

During the pulse the magnetization changes according to the rotating frame equation [see Equation (20.12)]:

$$\frac{d\vec{M}(t)}{dt} = \gamma\vec{M}(t) \times B_1\hat{i}' \tag{20.13}$$

where $\vec{M}(t) = (M_x(t), M_y(t), M_z(t))$ and $\vec{B}_1 = B_1\hat{i}' = (B_x, B_y, B_z) = (B_1, 0, 0)$. Solving this differential equation (see P20.23) we find that

$$\vec{M}(t) = (0, M_0 \sin \gamma B_1 t, -M_0 \cos \gamma B_1 t) \tag{20.14}$$

The physical interpretation of Equation (20.14) is that of a **magnetization vector** precessing about the x axis of the rotating frame, where the angle of precession is $\gamma B_1 t$. The pulse length t can be chosen so that $\vec{M}$ rotates 90°, that is, $t = \pi/2\gamma B_1$ after which time it lies in the x-y plane. This is called a **$\pi/2$ pulse.**

The transverse magnetization that exists after a $\pi/2$ pulse is aligned parallel to the y axis of the rotating frame. This magnetization will precess in the x-y plane around the static magnetic field. If a rf field linearly polarized along the axis of a solenoid is generated by applying an alternating voltage to the solenoid, magnetization precessing at the Larmor frequency will induce a sinusoidal voltage with a period equal to the inverse of the Larmor frequency. In fact, the experimentally observed nuclear induction signal is not an undamped sinusoid, but rather decays as shown at the bottom of Figure 20.8. This signal is called a **free induction decay (FID).**

FIGURE 20.7

RF pulse timing and the effect on $\vec{M}$ as viewed from the rotating frame. At time a, $\vec{M}$ points along the z axis. As the $\pi/2$ pulse is applied, $\vec{M}$ precesses in the x-y plane and points along the y axis as shown in c. After the pulse is turned off, $\vec{M}$ relaxes to its initial orientation along the z axis. The z component increases with the relaxation time T_1. Simultaneously, the x-y component of $\vec{M}$ decays with relaxation time T_2 as the individual spins dephase.

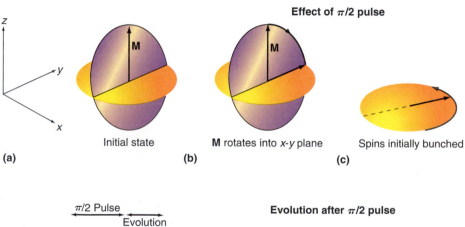

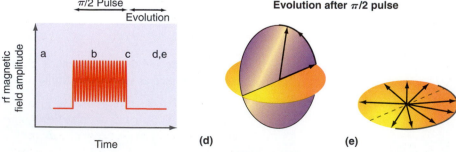

During evolution, M rotates toward z axis and the individual spins in the x-y plane dephase

FIGURE 20.8

Evolution of the magnetization vector M in three dimensions and M_{x-y} as a function of time. The variation of M_{x-y} with time leads to an exponentially decaying induced rf voltage in the detector coil.

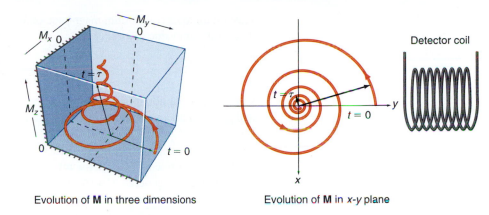

Evolution of **M** in three dimensions   Evolution of **M** in x-y plane

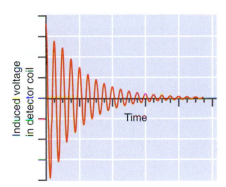

That the transverse magnetization decays is not surprising if one realizes that at equilibrium the transverse components of spin magnetic dipole moments were distributed randomly. Application of an rf field perturbs the system from equilibrium, creating transverse magnetization. The spin system will eventually return to equilibrium, once the rf field is turned off, and this return to equilibrium will involve (1) randomization of the transverse components of the spin magnetic dipole moments and (2) return of the populations of dipole moments in the spin-up α state and spin-down β states to their equilibrium values.

The fate of the transverse magnetization is conveniently viewed from the rotating frame where the x and y components M_x and M_y are observed to decay according to the rate laws:

$$\frac{dM_x}{dt} = -\frac{M_x}{T_2}, \frac{dM_y}{dt} = -\frac{M_y}{T_2} \tag{20.15}$$

which results in an exponential decay of transverse magnetization:

$$M_x(t) = M_x e^{-t/T_2}, M_y(t) = M_y e^{-t/T_2} \tag{20.16}$$

Transverse relaxation denotes the decay of transverse magnetization. T_2 is called the transverse relaxation time, and is the amount of time required for M_x and M_y to decay by a factor of $1/e$ where e ≈ 2.718 (i.e. lne = 1). When nuclear spin dipoles precess at slightly different frequencies, they get out of phase producing the observed decay of the FID. Transverse decay occurs as a result of a number of mechanisms. The magnetic field that a spin dipole moment μ generates at a neighboring spin at a distance r is proportional to $\mu(1 - 3\cos^2\theta)/r^3$ where θ is the angle between the internuclear vector and the magnetic field. In liquids, where molecules are constantly in motion translationally and rotationally, the field generated by a spin dipole moment at a neighboring spin fluctuates as the distance r and orientation θ between the spins change. These fluctuating local fields randomize the phase of the transverse components of spin magnetic dipoles.

Spin systems must also have the populations of spin-up and spin-down restored to their equilibrium values. This means that the longitudinal component of the magnetization M_z

must return to its equilibrium magnitude of M_0. Magnetization M_z is observed to return to its equilibrium value according to

$$\frac{dM_z}{dt} = -\frac{M_z - M_0}{T_1} \qquad (20.17)$$

where T_1 is the amount of time required for the difference $M_z - M_0$ to be reduced by $1/e$, and is called the longitudinal relaxation time. From Equation (20.17) it follows that

$$M_z = M_0(1 - e^{-t/T_1}) \qquad (20.18)$$

Like transverse relaxation, the relaxation of longitudinal magnetization results from fluctuating fields generated by the movements of surrounding spin dipole moments. But the nature of molecular motions that effect transverse and longitudinal relaxation differs in ways that can make T_2 less than T_1. To understand this fact, assume that a spin dipole moment experiences a torque from fluctuating field $\vec{b}$ generated by the motions of a neighboring spin. Viewed in the rotating frame, the torque equations are as follows:

$$\frac{dM_x}{dt} = \gamma(\vec{M} \times \vec{b})_x = \gamma(M_y b_z - M_z b_y)$$

$$\frac{dM_y}{dt} = \gamma(\vec{M} \times \vec{b})_y = \gamma(M_z b_x - M_x b_z)$$

$$\frac{dM_z}{dt} = \gamma(\vec{M} \times \vec{b})_z = \gamma(M_x b_y - M_y b_x) \qquad (20.19)$$

Note that M_x and M_y experience torques from all three components of $\vec{b}$, whereas M_z interacts with b_x and b_y only. Because the frame rotates at the Larmor frequency, the field components b_x and b_y must be fluctuating at frequencies comparable to the Larmor frequency when viewed from the lab frame. Because the rotating frame rotates around a static z field component, b_z must fluctuate very slowly relative to the Larmor frequency. Therefore, M_x and M_y decay as a result of fast and slow motions, whereas M_z decays as a result of fast motions only. For these reasons T_2 is generally shorter than T_1. For example, in solids T_2 can be on the order of 10^{-3}–10^{-5} seconds while T_1 can be many seconds, minutes, or even hours. In liquids composed of small internally rigid molecules that tumble rapidly, T_2 can approach but never exceed T_1.

20.3 The Spin–Echo

From the preceding discussion, we have learned that the decay of transverse magnetization has its origins in fluctuations of local magnetic fields produced by the motions of neighboring nuclear magnetic dipoles, and in static gradients of the external magnetic field. These two effects are summarized as:

$$\frac{1}{T_2^*} = \frac{1}{T_2} + \gamma|\Delta B| \qquad (20.20)$$

where $1/T_2^*$ is the observed rate of decay of transverse magnetization, $1/T_2$ is the rate of decay due to local field fluctuations, and $\gamma|\Delta B|$ is the **dephasing** rate due to a field gradient of magnitude $|\Delta B|$. Although local field fluctuations and external field gradients both cause the decay of transverse magnetization, they differ in one important respect. Local field fluctuations due to random motions of nuclear spins are random functions of time and their effects are irreversible. But if the external field varies across the sample and nuclear spin dipoles do not sample locations with differing fields randomly in time, it is possible to reverse the decay of transverse magnetization with a phenomenon called the *spin–echo*.

In 1950 Erwin Hahn reported that if a 90-degree pulse of rf radiation is followed at a time τ by a second rf pulse, a nuclear induction signal reappears at a time 2τ (see

Figure 20.9, bottom). This is a remarkable observation because according to Equations (20.15) and (20.16) transverse magnetization should decay exponentially after a 90-degree pulse as the system returns to equilibrium. Hahn's **spin–echo** experiment seems to indicate that the decay of transverse magnetization is somehow reversible. As explained in the caption of Figure 20.9, what is reversed in a spin–echo experiment is the decay of transverse magnetization due to a distribution of static magnetic fields across the sample.

Consider two spin dipole moments that exist in different locations in a sample. As a result of a static gradient of the external magnetic field, the fields at the two spin dipole moments differ by ΔB. This means that the precession frequencies differ by a factor of $\gamma |\Delta B|$ and after a time τ, the positions of the transverse components of the spin dipoles differ by $\theta = \gamma \Delta B \tau$. If a *180 degree* pulse (also called a π pulse) is applied at a time τ, the resulting torque rotates the dipoles whose rotational motion is subsequently reversed. The reversal of precessional motion causes the magnetization vectors to converge at a time τ after the π pulse or 2τ after the initial $\pi/2$ pulse. This convergent motion causes a reversal of the decay of transverse magnetization and the result is a spin–echo.

However, the spin–echo experiment cannot reverse dephasing due to motions that randomly modulate spin precession frequencies. Therefore, the spin–echo intensity will not be equal exactly to the intensity of transverse magnetization that directly follows a $\pi/2$ pulse. At a time 2τ following the initial $\pi/2$ pulse, the amplitude of the spin–echo is

$$I(2\tau) = I(0)e^{-2\tau/T_2} \qquad (20.21)$$

FIGURE 20.9

Schematic representation of the spin–echo experiment. The $\pi/2$ pulse applied along the y axis rotates $\vec{M}$ into the x-y plane. After an evolution time τ in which free induction decay occurs, a π pulse is applied along the x axis. This leads to a reversal of the precession direction of the spins, and the fanning out process resulting from field inhomogeneities and chemical shifts is reversed. An echo will be observed in the detector coil at time 2τ and successive echoes will be observed at further integral multiples of τ. The amplitude of the successive echoes decreases with time because of transverse spin relaxation.

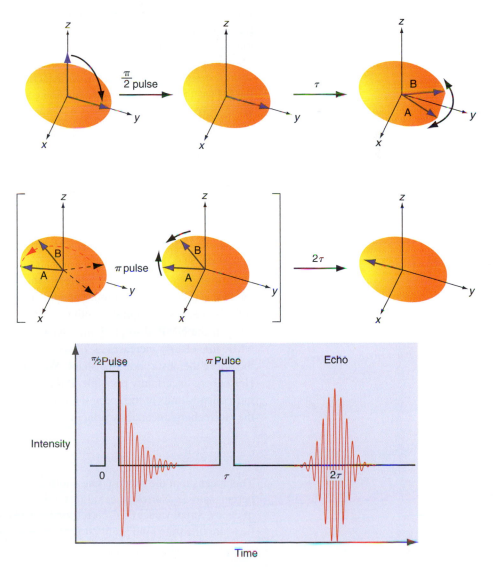

where $I(0)$ is the initial amplitude of the transverse magnetization that follows a 90-degree pulse. This means that T_2 can be measured with a spin–echo experiment by plotting the echo amplitude as a function of time.

20.4 Fourier Transform NMR Spectroscopy

A spectrum is a graph of the intensity of absorbed radiation as a function of frequency. In Chapters 18 and 19 we saw how vibrational, rotational, and electronic spectra consist of numerous transitions that represent various quantized vibrational and rotational motions of molecular species. For NMR spectroscopy we have thus far considered only the time domain response of a system of nuclear spin dipoles, yet there is a particularly simple relationship between the time domain response or FID signal obtained by pulsed NMR and the NMR absorption spectrum.

The motion of the transverse magnetization in the x-y plane of the rotating frame is periodic. Because the FID is induced by this periodic motion, the FID is similarly represented by a periodic function $f(t)$. Although we have portrayed FIDs as decaying cosine functions, in reality FIDs are represented by combinations of cosine and sine functions. For the simple case of a FID produced by an ensemble of nuclear dipole moments all precessing at the same frequency ω_0, we might write:

$$f(t) = \cos(\omega_0)te^{-t/T_2} \qquad (20.22)$$

But if a FID were represented simply as a cosine function, which is an even function, that is, $\cos\omega t = \cos(-\omega)t$ we would have no idea if the FID were produced by transverse magnetization rotating clockwise (ω) or counterclockwise ($-\omega$). NMR spectrometers always detect the direction or phase of precession as well as the magnitude of the precession frequency. Phase-sensitive detection represents each frequency component of the FID as the sum of a cosine and sine component, so instead of the form in Equation (20.23) we write the FID as

$$f(t) = (\cos \omega_0 t + i \sin \omega_0 t)e^{-t/T_2} = e^{(i\omega_0 - 1/T_2)t} \qquad (20.23)$$

To obtain the NMR spectrum $F(\omega)$ from the FID $f(t)$, we perform a **Fourier transform** of $f(t)$ given in Equation (20.23), which has the integral form:

$$F(\omega) = \int_{o}^{+\infty} f(t)e^{-i\omega t}\, dt \qquad (20.24)$$

Figure 20.10 shows that an absorption spectrum is obtained by Fourier transformation of the FID. For every precessional frequency component contained in the FID, the absorption spectrum displays a line at the precession frequency, where the intensity of the line is related to the number of spins possessing that precession frequency. But the FID also decays in amplitude with time, which, according to Equation (20.23), has form e^{-t/T_2}. If the NMR absorption line shape is plotted as a function of T_2, the width of the NMR line clearly increases as T_2 decreases. In other words, the faster the FID decays, the wider the absorption line shape. We can understand this line shape trend using the Heisenberg uncertainty principle. If the lifetime of the coherent state is Δt and the width in frequency units of the spectral line corresponding to the transition is $\Delta \nu$, according to the Heisenberg uncertainty principle

$$\Delta E\, \Delta t \approx h \ \text{ or } \ \Delta \nu \approx \frac{1}{\Delta t} \qquad (20.25)$$

This means that the spectral line width $\Delta \nu$ is inversely related to the lifetime of the coherent state Δt. Narrow spectral features correspond to large values of T_2 and slow rates of decay. Small values of T_2 correspond to fast rates of decay and large line widths. These facts are illustrated in Figure 20.11.

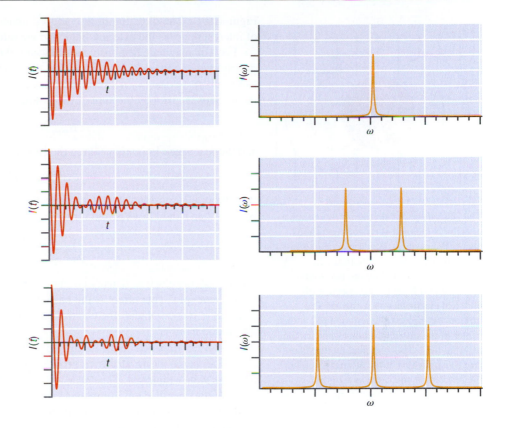

FIGURE 20.10

Juxtaposition of three free induction decays containing one (top), two (center), and three (bottom) frequency components with the respective absorption spectra obtained by Fourier transformation.

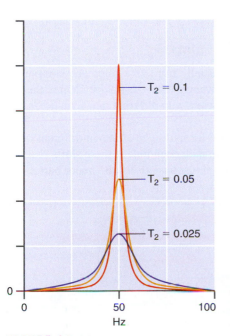

FIGURE 20.11

The NMR absorption line shape plotted as a function of T_2. The full-width at half-maximum is $\Delta\nu_{1/2} = 1/\pi T_2$.

Fourier transform NMR spectroscopy makes possible most of the modern applications of NMR spectroscopy to the study of large, complex macromolecules, which we will describe later in this chapter. For his work in developing Fourier transform NMR and for his contributions to the development of multidimensional NMR (see Section 20.6) Professor Richard Ernst of the Swiss Federal institute of Technology was awarded the Nobel Prize in Chemistry in 1991.

20.5 NMR Spectra: Chemical Shifts and Spin–Spin Couplings

Two important aspects of NMR make it very useful for obtaining additional chemical information at the molecular level. The first is that the magnetic field responsible for Larmor precession is not the applied external field, but rather the sum of the applied field and the local field. As we will see, the local field is influenced by the electron distribution on the atom of interest as well as by the electron distribution on nearby atoms. This difference between the external and induced magnetic fields is the origin of the **chemical shift.** The second important aspect is that individual magnetic dipoles interact with one another. This leads to a splitting of the energy levels of a multiple-spin system and the appearance of multiplet spectra in NMR. As will be discussed in Sections 20.5.2 and 20.5.3, the multiplet structure of a NMR resonance absorption gives direct structural information about the molecule.

20.5.1 The Chemical Shift

When an atom is placed in a magnetic field, a circulation current is induced by the electron charge around the nucleus that generates a secondary magnetic field, as shown in

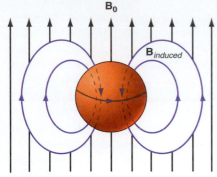

FIGURE 20.12

The shaded spherical volume represents a negatively charged classical continuous charge distribution. When placed in a magnetic field, the distribution will circulate as indicated by the horizontal orbit, viewed from the perspective of classical electromagnetic theory. The motion will induce a magnetic field at the center of the distribution that opposes the external field. This classical picture is not strictly applicable at the atomic level, but the outcome is the same as a rigorous quantum mechanical treatment.

Figure 20.12. The direction of the induced magnetic field at the position of the nucleus of interest opposes the external field, and we refer to a **diamagnetic response.**

For a diamagnetic response, the induced field at the nucleus of interest is opposite in direction and is linearly proportional to the external field in magnitude. Therefore, we can write $\vec{B}_{induced} = -\sigma\vec{B}_0$, which defines the **shielding constant** σ. The total field at the nucleus is given by the sum of the external and induced fields,

$$\vec{B}_{total} = (1 - \sigma)\vec{B}_0 \tag{20.26}$$

and the resonance frequency taking the shielding into account is given by

$$\omega_0 = 2\pi\nu_0 = \gamma B_0(1 - \sigma) \tag{20.27}$$

Because $\sigma > 0$ for a diamagnetic response, the resonance frequency of a nucleus in an atom is lower than would be expected for the bare nucleus. The frequency shift is given by

$$\nu = \frac{\gamma B}{2\pi} - \frac{\gamma B(1 - \sigma)}{2\pi} = \frac{\sigma\gamma B}{2\pi} \tag{20.28}$$

We see that the electron density around the nucleus reduces the resonance frequency of the nuclear spin. This effect is the basis for the chemical shift in NMR. The shielding constant σ increases with the electron density around the nucleus and, therefore, with the atomic number.

Because σ depends on the electron density around the nuclear spin of interest, it will change as neighboring atoms or groups either withdraw or increase electron density from the atom of interest. This leads to a shift in frequency that makes NMR a sensitive probe of the chemical environment around a nucleus. For ^{1}H, σ is typically in the range of 10^{-5} to 10^{-6}, so that the change in the resonance frequency due to the chemical shift is quite small. It is convenient to define a dimensionless quantity δ to characterize this frequency shift, with δ defined relative to a reference compound by

$$\delta = 10^6 = \frac{(\nu - \nu_{ref})}{\nu_{ref}} = 10^6 \frac{\gamma B(\sigma_{ref} - \sigma)}{\gamma B(1 - \sigma_{ref})} \approx 10^6(\sigma_{ref} - \sigma) \tag{20.29}$$

For ^{1}H NMR, tetramethylsilane, $(CH_3)_4Si$, is usually used as a reference compound. Defining the chemical shift in this way has the advantage that δ is independent of the frequency, so that all measurements using spectrometers with different magnetic fields will give the same value of δ.

Figure 20.13 shows the observed ranges of δ for hydrogen atoms in different types of chemical compounds. Although a quantitative understanding of these shifts requires the consideration of many factors, two factors are responsible for the major part of the

FIGURE 20.13

Chemical shifts δ as defined by Equation (20.29) for ^{1}H in different classes of chemical compounds. Extensive compilations of chemical shifts are available in the chemical literature.

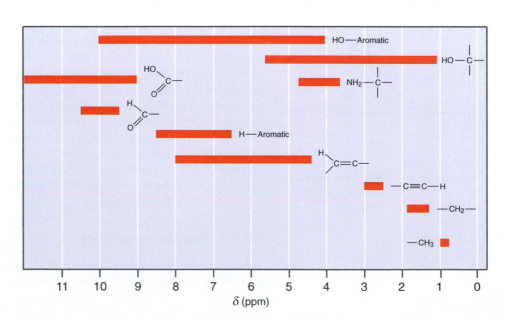

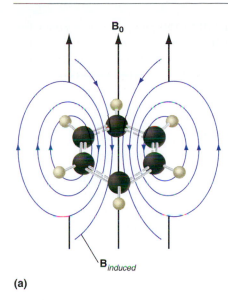

(a)

(b)

FIGURE 20.14

(a) The induced magnetic field generated by a circulating ring current in benzene. Note that within the ring of the molecule, the induced field is opposed to the external field while outside in the ring the induced field is along the applied field.
(b) 18-Annulene provides a confirmation of this model.

chemical shift: the electronegativity of the neighboring group and the induced magnetic field of the neighboring group at the position of the nucleus of interest.

If a neighboring group is more electronegative than hydrogen, it will withdraw electron density from the region around the ^1H nucleus. Therefore, the nucleus is less shielded, and the NMR resonance frequency will appear at a larger value of δ. For example, the chemical shift for ^1H in the methyl halides follows the sequence $CH_3I < CH_3Br < CH_3Cl < CH_3F$. The range of this effect is limited to about three or four bond lengths as can be shown by considering the chemical shifts in 1-chlorobutane. In this molecule, δ for the ^1H on the CH_2 group closest to the Cl is almost 3 ppm larger than the ^1H on the terminal CH_3 group, which has nearly the same δ that it has in propane.

Because σ is small for the H atom, the magnetic field at a ^1H nucleus is often dominated by the local induced magnetic fields from neighboring atoms or groups of atoms. It is helpful to think of the neighboring group as a magnetic dipole $\vec{\mu}$ whose strength and direction are determined by the magnitude and sign of its shielding constant σ. Groups containing delocalized electrons such as aromatic groups and other groups containing multiple bonds give rise to large values of $\vec{\mu}$ because the delocalized electrons give rise to a ring current as is illustrated in Figure 20.14a. This model predicts that the chemical shift of the interior and exterior ^1H atoms attached to an aromatic system should be in the opposite direction. In fact, δ for an exterior ^1H of 18-annulene is $+9.3$ and that of an interior ^1H is -3.0.

Although ^1H is the most utilized nuclear probe in NMR, the proton has a very small shielding constant, because it has only one electron orbiting around the nucleus. By comparison, σ for ^{13}C and ^{31}P are factors of 15 and 54 greater, respectively. According to Table 20.1, in contrast to the abundant stable spin 1/2 hydrogen isotope ^1H, the spin 1/2 isotope of carbon, ^{13}C, is only about 1.1% abundant. Despite the low sensitivity that arises from the low natural abundance of ^{13}C and from the fact that the magnetic moment of the ^{13}C nucleus is only about 1/4 that of the proton, ^{13}C NMR is a widely used technique for structural elucidation of organic molecular species.

The large chemical shift dispersion of ^{13}C versus ^1H is due to the extensive electron system around the ^{13}C nucleus. The chemical shift dispersion of ^{13}C is much greater than ^1H measured both in parts per million (ppm) and in frequency units (see Figure 20.15).

For example, according to Figure 20.13, methyl protons have a chemical shift range of 0.6 to 1.0 ppm, whereas aromatic protons resonate from 6.5 to 8.5 ppm. The Larmor frequency of protons in an 11.75-T field is 500 MHz. This means that in frequency units, a typical methyl proton will resonate about 2750 Hz from a typical aromatic proton. From Figure 20.15 a methyl ^{13}C nucleus resonates in a range of 5 to 20 ppm, whereas

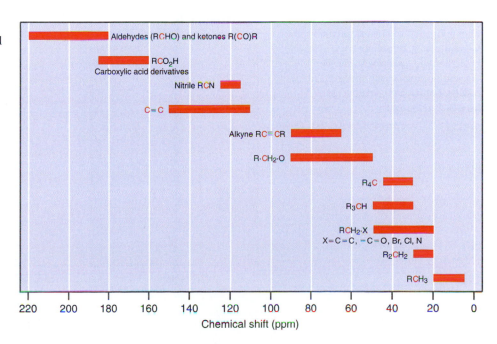

FIGURE 20.15
Typical chemical shifts for ^{13}C.

aromatic ^{13}C nuclei resonate in the range from 110 to 150 ppm. The ^{13}C Larmor frequency in an 11.75-T field is 125 MHz, so a typical methyl ^{13}C nucleus will resonate about 12,500 Hz away from a typical aromatic ^{13}C nucleus. Assuming comparable spectral line widths for ^{1}H and ^{13}C in a given molecule, the ^{13}C spectrum will be better resolved than the proton NMR spectrum.

20.5.2 Multiplet Splitting of NMR Peaks Arises through Spin–Spin Coupling

What might one expect the spectrum of a molecule of ethanol, with three different types of chemically equivalent hydrogens, to look like? A good guess is that the chemically nonequivalent protons resonate in three separate frequency ranges, one corresponding to the methyl group, another to the methylene group, and the third to the OH group. The OH proton is most strongly deshielded (largest δ) because it is directly bound to the electronegative oxygen atom. It is found near 5 ppm. As the methylene group gets closer to the electronegative OH group, the protons are more deshielded and appear at larger values of δ (near 3.5 ppm) than the methyl protons, which are found near 1 ppm. Furthermore, we expect that the areas of the peaks have the ratio $CH_3:CH_2:OH = 3:2:1$ because the proton NMR signal is proportional to the number of spins.

A simulated NMR spectrum for ethanol is shown in Figure 20.16. A very important feature that has not been discussed is also shown in this figure: the individual peaks are split into **multiplets.** At low temperature and in the absence of acidic protons, the OH proton resonance is a triplet, whereas the CH_3 proton resonance is a triplet and the CH_2 resonance is an octet. At higher temperature, a change in the NMR spectrum is observed. The OH proton resonance is a singlet, the CH_3 proton resonance is a triplet, and the CH_2 resonance is a quartet.

Multiplet splittings in NMR spectra arise as a result of spin–spin interactions between different nuclei. The multiplet splittings observed in the NMR spectra of molecules in liquid phases arise as a result of indirect interactions between the spin magnetic dipole moments of neighboring nuclei. These dipole–dipole interactions are indirect in the sense that they are mediated by intervening bonding electrons.

FIGURE 20.16
Simulated NMR spectrum for ethanol. The top panel shows the multiplet structure at room temperature. The lower panel shows the multiplet structure observed at lower temperature in acid-free water. The different portions of the spectrum are not to scale, but have the relative areas discussed in the text.

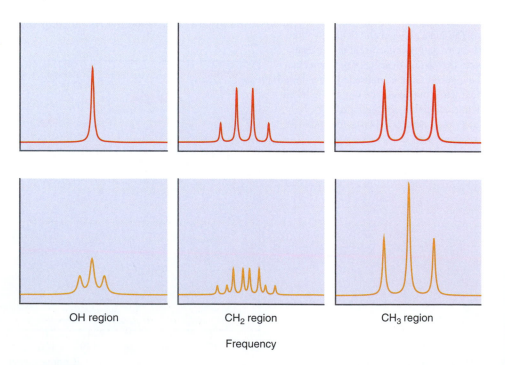

OH region   CH$_2$ region   CH$_3$ region

Frequency

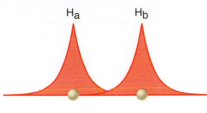

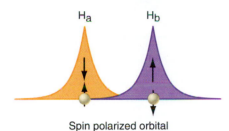

FIGURE 20.17
Schematic illustration of how spin-polarized orbitals couple nuclear spins even though they are highly shielded from one another through the electron density. The upper (lower) arrows in the right part of the figure indicate the electron (nuclear) spin.

An antiparallel orientation of the nuclear and electron spins is favored energetically over a parallel orientation. Therefore, the electrons around a spin down (i.e. nucleus whose spin magnetic dipole moment is anti-parallel to B_o) nucleus are more likely to be spin up (parallel to B_o) α than spin down. In a molecular orbital connecting two nuclei of nonzero spin, the β electrons around atom H_a are pushed toward atom H_b because of the electron sharing resulting from the chemical bond. This effect is called **spin polarization** (Figure 20.17). Nucleus H_b will now be slightly more stable if it has spin up rather than down spin, because this generates an antiparallel arrangement of nuclear and electron spins on this atom. A well-shielded nuclear spin senses the spin orientation of its neighbors through the interaction between the nuclear spin and the electrons. Because this is a very weak interaction and other factors favor molecular orbitals without spin polarization, the degree of spin polarization is very small. However, these very weak interactions are sufficient to account for multiplet splittings between protons ranging from a few hertz to a few tens of hertz.

Consider the case of two interacting spins. For the common case in which the chemical shift difference is much greater than the magnitude of the scalar coupling interaction, the wave functions associated with the four spin states are simple products of the single spin functions α and β, where α indicates spin up and β is spin down.

$$\psi_1 = \alpha(1)\alpha(2)$$
$$\psi_2 = \beta(1)\alpha(2)$$
$$\psi_3 = \alpha(1)\beta(2)$$
$$\psi_4 = \beta(1)\beta(2) \tag{20.30}$$

The spin energy that takes the chemical shifts and the scalar coupling into account is

$$E = -\gamma B_0 \hbar (1 - \sigma_1) m_{z_1} - \gamma B_0 \hbar (1 - \sigma_2) m_{z_2} + h J_{12} m_{z_1} m_{z_2}. \tag{20.31}$$

The first two terms in Equation (20.31) represent the chemical shift interactions of the isolated spins. The third term is the spin interaction energy. In this equation, J_{12}, which has the units s^{-1}, is called the indirect or scalar coupling constant and is a measure of the strength of interaction between the individual magnetic moments.

Using the fact that for an α spin state $m_z = 1/2$ and for a β spin state $m_z = -1/2$, the spin energies can be determined (Figure 20.18). NMR transitions follow the selection rule $\Delta m = \pm 1$, and given this fact, four transition frequencies will occur. As you will

FIGURE 20.18
The energy levels for two noninteracting spins and the allowed transitions between these levels are shown on the left. The same information is shown on the right for interacting spins in the weak coupling limit. The splitting between levels 2 and 3 and the energy shifts of all four levels for interacting spins are greatly magnified to emphasize the spin–spin interactions.

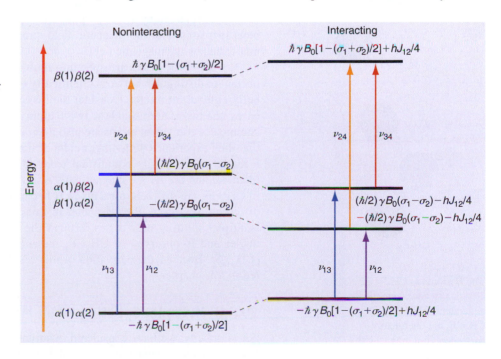

FIGURE 20.19
Splitting of a two interacting spin system into doublets for two values of B. The spacing within the doublet is independent of the magnetic field strength, but the spacing of the doublets increases linearly with B.

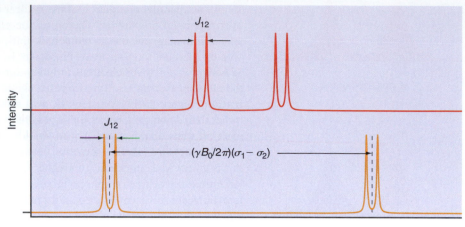

prove in the end-of-chapter problems, the frequencies of the allowed transitions including the **spin–spin coupling** are

$$\nu_{12} = \frac{E_2 - E_1}{h} = \frac{\gamma B(1 - \sigma_1)}{2\pi} - \frac{J_{12}}{2}$$

$$\nu_{34} = \frac{E_4 - E_3}{h} = \frac{\gamma B(1 - \sigma_1)}{2\pi} + \frac{J_{12}}{2}$$

$$\nu_{13} = \frac{E_3 - E_1}{h} = \frac{\gamma B(1 - \sigma_2)}{2\pi} - \frac{J_{12}}{2}$$

$$\nu_{24} = \frac{E_4 - E_2}{h} = \frac{\gamma B(1 - \sigma_2)}{2\pi} + \frac{J_{12}}{2} \tag{20.32}$$

What has this calculation shown? First, a given energy level is shifted to a higher energy if both spins are of the same orientation, and to a lower energy if the orientations are different. Second, spin–spin interactions result in the appearance of multiplet splittings in NMR spectra, as shown in Figure 20.19. The two peaks that would appear in the absence of spin–spin interactions are now split into a doublet in which the two components are separated by J_{12}. Note that, although the separation in frequency of the doublets increases with the magnetic field strength, the splitting within each doublet is unaffected by the magnetic field.

Not all NMR peaks are split into multiplets. To understand this result, it is important to distinguish between **chemically equivalent nuclei** as opposed to **magnetically equivalent nuclei**. Consider the two molecules shown in Figure 20.20. In both cases, the two H atoms and the two F atoms are chemically equivalent. The nuclei of chemically equivalent atoms are also magnetically equivalent only if the interactions that they experience with other nuclei of nonzero spin are identical. Because the two F nuclei in CH_2F_2 are equidistant from each H atom, the two H—F couplings are identical and the ^1H are magnetically equivalent. However, the two H—F couplings in CH_2CF_2 are different for a given H because the spacing between the two nuclei is different. Therefore, the ^1H nuclei in this molecule are magnetically inequivalent. Multiplet splitting only arises through the interaction of magnetically inequivalent nuclei and is observed in CH_2CF_2, but not in CH_2F_2 or the reference compound $(CH_3)_4Si$. Because the derivation of this result is somewhat lengthy, it will be omitted in this chapter.

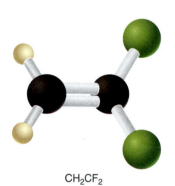

CH_2CF_2

CH_2F_2

FIGURE 20.20
The H atoms in CH_2F_2 are chemically and magnetically equivalent. The H atoms in CH_2CF_2 are chemically equivalent, but magnetically inequivalent.

20.5.3 Multiplet Splitting When More than Two Spins Interact

For simplicity, we have considered quantitatively only the case of two coupled spins. However, many organic molecules will have more than two inequivalent protons that are close enough to one another to generate multiplet splittings. In this section, we consider several different coupling schemes. The strength of the scalar coupling constant J falls off rapidly with the number of bonds separating the proton pair. Experimentally it is found that generally only those protons within three or four bond lengths of the nucleus of interest have a sufficiently strong interaction to generate peak splitting. In strongly coupled systems such as those with conjugated bonds, the coupling can still be strong when the spins are further apart.

To illustrate the effect of spin–spin interactions in generating multiplet splittings, we consider the coupling of three distinct spin 1/2 nuclei that we label A, M, and X. There will be two different coupling constants, J_{AM} and J_{AX} with $J_{AM} > J_{AX}$. We can think of the effect of these couplings by turning on the couplings individually as indicated in Figure 20.21. The result is that each of the lines in the doublet that arises from turning on the interaction J_{AM} is again split into a second doublet when the interaction J_{AX} is turned on as also shown in Figure 20.21.

A special case occurs when A and M are identical so that $J_{AM} = J_{AX}$. The middle two lines for the AMX case now lie at the same frequency giving rise to the AX_2 pattern shown in Figure 20.22. Because the two lines lie at the same frequency, the resulting spectrum is a triplet with the intensity ratio 1:2:1.

FIGURE 20.21

Coupling scheme and expected NMR spectrum for three coupled spins with different coupling constants J_{AX} and J_{AM}.

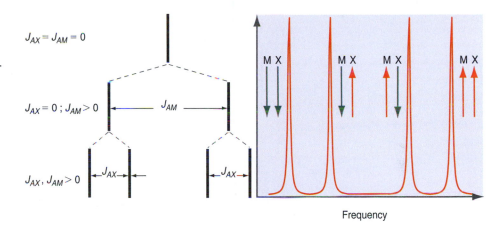

FIGURE 20.22

Coupling scheme and expected NMR spectrum for three coupled spins with only one coupling constant J_{AX}.

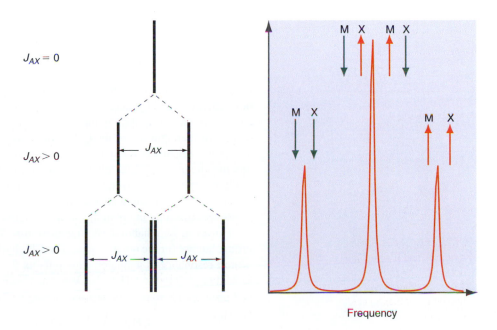

EXAMPLE PROBLEM 20.2

Using the same reasoning as was applied to the AX_2 case, predict the NMR spectrum for an AX_3 spin system.

Solution

Turning on each of the interactions in sequence, we obtain the following diagram:

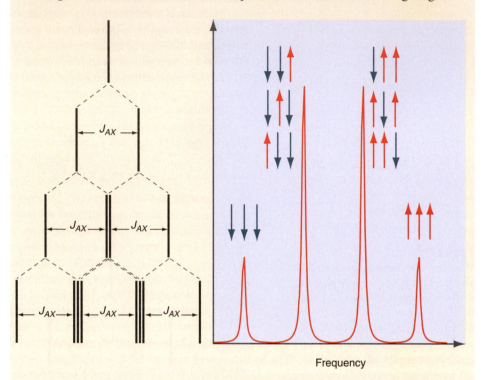

Frequency

The end result is a quartet with the intensity ratios 1:3:3:1. These results can be generalized to the rule that if a ^1H nucleus has n equivalent ^1H neighbors, its NMR spectral line will be split into $n + 1$ peaks. The relative intensity of these peaks is given by the coefficients in the expansion of $(1 + x)^n$, the binomial expression.

Given the results of the last two sections, we are (almost) at the point of being able to understand the fine structure in the NMR spectrum of ethanol shown earlier in Figure 20.16. As discussed earlier, the resonance near 5 ppm can be attributed to the OH proton, the resonance near 3.5 ppm can be attributed to the CH_2 protons, and the resonance near 1 ppm can be attributed to the CH_3 protons. This is consistent with the integrated intensities of the peaks, which from low to high δ have the ratio 1:2:3. We now consider the multiplet splitting. Invoking the guideline that spins located more than three bonds away will not generate peak splitting, we conclude that the CH_3 resonance will be a triplet because it is split by the two CH_2 protons. The OH proton is too distant to generate a further splitting of the CH_3 group. We conclude that the CH_2 resonance will be an octet (two pairs of quartets) because it will split by the three equivalent CH_3 protons and the OH proton. We would predict that the OH resonance should be a triplet because it will be split by the two equivalent CH_2 protons. In fact, this is exactly what is observed for the NMR spectrum of ethanol at low temperatures. This example shows the power of NMR spectroscopy in obtaining structural information at the molecular level.

However, for ethanol at room temperature, these predictions are correct for the CH_3 group, but not for the other groups. The CH_2 hydrogen resonance is a quartet and the OH proton resonance is a singlet. This tells us that there is something that we have overlooked regarding the OH group. What has been overlooked is the rapid exchange of the OH proton with water. The proton exchange reaction for ethanol is

$$CH_3CH_2OH + H_2O \rightleftharpoons CH_3CH_2O^- + H_3O^+ \qquad (20.33)$$

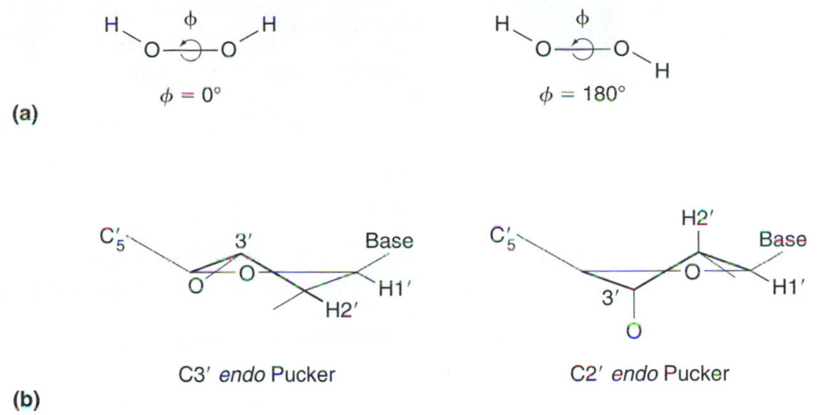

FIGURE 20.23
(a) The dihedral angle ϕ for a H—C—C—H moiety varies from 0 degrees in the *cis* conformation to 180 degrees in a *trans* conformation. (b) A five-membered furanose ring in a nucleoside is thought to adopt one of two low-energy puckering conformations where the dihedral angle between the 1′ and 2′ protons is 90 degrees in the C3′ *endo* conformation and 180 degrees in the C2′ *endo* conformation.

(a)

(b)

C3′ *endo* Pucker   C2′ *endo* Pucker

The OH proton on ethanol actually exchanges between two inequivalent environments: the OH group of ethanol and the hydronium ion in bulk water. We would expect to observe two distinct lines for the exchanging proton This effect is referred to as **motional broadening.** For a significantly faster exchange, only a single sharp peak will be observed. This is the case for ethanol at room temperature. For this reason, the portion of the ethanol NMR spectrum shown in Figure 20.16 corresponding to the OH proton is a singlet rather than a triplet. However, at low temperatures and under acid-free conditions, the exchange rate can be sufficiently reduced so that the exchange can be ignored. In this case, the OH ^{1}H signal is a triplet. We now understand why the 300 K CH_2 hydrogen resonance in ethanol is a quartet rather than an octet and why the OH hydrogen resonance is a singlet rather than a triplet.

20.5.4 Magnitude of Scalar Coupling Constants

The magnitude of scalar coupling constants depends on a number of factors. Scalar coupling constants are larger for nuclei with larger magnetogyric ratios. In general, the magnitude of the scalar coupling constant diminishes as the number of bonds between the interacting nuclei increases. An aspect of proton–proton scalar couplings widely exploited in biostructural studies is the dependence of the coupling constant on dihedral angle. Martin Karplus first proposed that the vicinal proton coupling constant in a H—C—C—H moiety is dependent on the bond dihedral angle (see Figure 20.23a) according to

$$^3J_{HH} = A + B \cos \phi + C \cos 2\phi \tag{20.34}$$

where for the H—C—C—H moiety in the *cis* conformation the dihedral angle is defined as $\phi = 0°$ and in the *trans* conformation as $\phi = 180°$. The superscript three indicates a three-bond coupling. Based on valence bond calculations made assuming a C—C bond length of 1.543 Å and sp^3 hybridization, Karplus found that the parameters $A = 4.22$ Hz, $B = -0.5$ Hz, and $C = 4.5$ Hz compared well on a qualitative basis to three-bond scalar couplings measured in ethanic (i.e., $XH_2C—CH_2Y$) molecules.

In large biological molecules where quantum mechanical calculations are not feasible, Karplus equations are derived empirically from model compounds. For example, $^3J_{HH}$ values in furanose rings in nucleosides fit the empirical equation $^3J_{HH} = A + B \cos \phi + C \cos^2 \phi$ where $A = 0.0$ Hz, $B = 0.8$ Hz, and $C = 10.2$ Hz.

EXAMPLE PROBLEM 20.3

Furanose rings in nucleosides are five-membered rings that are thought to adopt one of two low-energy puckering conformations: C3′ *endo* and C2′ *endo* (see Figure 20.23b). In the C3′ *endo* conformation, the dihedral angle for the H1′ H2′ vicinal pair is about $\phi \approx \pi/2$, whereas for C2′ *endo*, $\phi \approx \pi$. Calculate $^3J_{1'-2'}$ for the C3′ *endo* and C2′ *endo* conformations.

Solution

For C3′ *endo*

$$^3J_{1'-2'} = A + B\cos\phi + C\cos^2\phi$$
$$= 0.0\,\text{Hz} + 0.8\,\text{Hz} \times \cos\tfrac{\pi}{2} + 10.2\,\text{Hz} \times \cos^2\tfrac{\pi}{2} = 0.0$$

For C2′ *endo*

$$^3J_{1'-2'} = A + B\cos\phi + C\cos^2\phi = 0.0\,\text{Hz} + 0.8\,\text{Hz} \times \cos\pi + 10.2\,\text{Hz} \times \cos^2\pi$$
$$= -0.8\,\text{Hz} + 10.2\,\text{Hz} = 9.4\,\text{Hz}$$

Experimental measurements show that for C3′ *endo* $^3J_{1'-2'}$ is near zero and for C2′ *endo* $^3J_{1'-2'}$ is about 9 Hz.

20.6 Multidimensional NMR

20.6.1 General Scheme of a Two-Dimensional NMR Experiment

Using a combination of ^{1}H and ^{13}C NMR, the structures of small molecules in solution can be determined. However, NMR-based structural analysis requires that NMR line shapes and multiplets be well resolved. Although this may be the case for small molecules in solution, for large biomolecules (e.g., proteins, nucleic acids), the density of peaks in the NMR spectrum is very high. Figure 20.24 shows the proton NMR spectrum of U1A, a small RNA-binding protein with a molecular weight of about 10 kDa. Although the spectrum displays some resolution of proton types—methyl group protons range from 0 to 1.0 ppm, aliphatic protons from amino acid side chains range from approximately 1 to 4 ppm, amide and aromatic protons appear in the range from 6 to 8.5 ppm—the NMR spectral lines are not sufficiently resolved to enable assignment of individual NMR lines to particular protons located in particular amino acids in the protein's primary sequence. Multidimensional NMR improves the resolution of NMR spectra by separating the over-lapped spectra of chemically inequivalent spins into multiple frequency dimensions.

FIGURE 20.24

One-dimensional (1D) ^{1}H NMR spectrum of the protein U1A (molecular weight ~ 10 kDa) in aqueous solution. The large number of overlapping broad peaks precludes a detailed structural determination on the basis of the 1D spectrum. (Source: VIA Spectrum provided by G. Vavani)

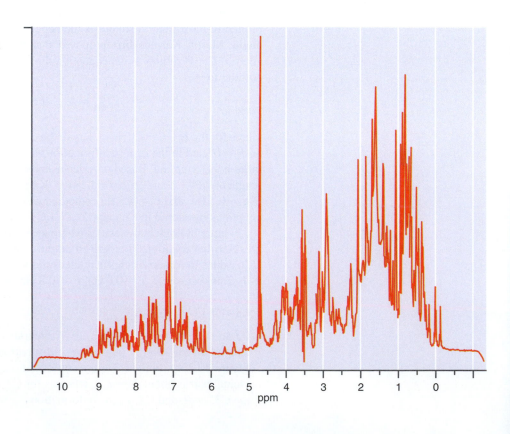

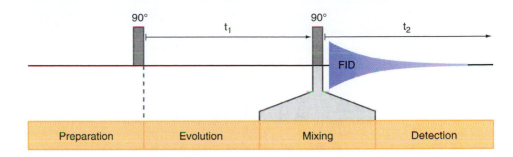

FIGURE 20.25

General scheme for a two-dimensional NMR experiment.

The simplest multidimensional NMR experiment is a **two-dimensional NMR (2D NMR)** experiment. The general experimental scheme for a 2D NMR experiment is shown in Figure 20.25. A 2D NMR **pulse sequence** is composed of four parts. The preparation period corresponds to the part of the pulse sequence that acts on the initial z magnetization. For most of the 2D NMR experiments we will discuss, the preparation period is simply a 90-degree pulse that rotates the initial z magnetization into the x-y plane. The t_2 period is the detection period during which the FID is observed and recorded. In 2D NMR, the FID acquired during t_2, is observed as a function of the length of an evolution period t_1. During the evolution period of length t_1, spin magnetization precesses at the rotating frame frequencies, which include the chemical shift frequencies and scalar coupling frequencies. The time domain signal in a 2D NMR experiment is now a matrix $f(t_1, t_2)$.

A key feature of a 2D NMR experiment is the mixing period. During the mixing period, pulsed irradiations correlate the spin precession frequencies in t_1 with precession frequencies in t_2. The nature of the correlation is determined by the type of irradiation applied during the mixing period, and is best viewed by way of the two-dimensional NMR spectrum. In analogy to one-dimensional Fourier transform NMR, a two-dimensional Fourier transform of $f(t_1, t_2)$ yields the two-dimensional spectrum $F(\omega_1, \omega_2)$:

$$F(\omega_1, \omega_2) = \int_0^{+\infty} dt_2 e^{-i\omega_2 t_2} \int_0^{+\infty} dt_1 f(t_1, t_2) e^{-i\omega_1 t_1} \tag{20.35}$$

A simple type of two-dimensional NMR pulse sequence is shown in Figure 20.25, top, where the mixing period contains just a second 90-degree pulse. A mixing irradiation of this type correlates spins precessing at frequency ω_1 during the evolution period with spins precessing at frequency ω_2 during the detection period, if the two classes of spins are scalar coupled. Suppose we have two coupled spin 1/2 nuclei that we label A and B. Spin A has chemical shift ν_A and spin B has chemical shift ν_B. Because A and B are coupled, the 1D NMR spectrum will be a pair of doublets centered at ν_A and ν_B. These 1D NMR frequencies appear as diagonal peaks aligned along the dashed line in Figure 20.26. If the mixing period consists of a single $\pi/2$, pulse the chemical shift frequencies will be correlated through the scalar coupling. These scalar coupling correlations manifest as cross-peak multiplets in the 2D NMR spectrum centered at (ν_A, ν_B) and (ν_B, ν_A). This chemical shift correlation spectroscopy is given the acronym **COSY.**

COSY spectra have been obtained for large numbers of biologically relevant compounds. As shown in Figure 20.27, proteins are polymers composed of amino acids connected via peptide bonds joining the amide nitrogen and the carboxyl carbon of adjacent amino acids. Except for proline, which has an imino proton, all amino acids in a protein have a single amide proton located at the peptide bond. Except for glycine with two protons attached to the alpha carbon atom (αC), all amino acids have a single proton attached to the alpha carbon. Amino acids are otherwise identified by their side chains, indicated by the letter R in Figure 20.27. COSY identifies amino acids by virtue of the unique pattern of cross peaks created between the amide proton, the alpha proton, and the protons of the side chain group R. Because the amide and alpha protons of different amino acids are not scalar coupled, there are no COSY cross peaks between protons belonging to different amino acids.

FIGURE 20.26
Two-dimensional NMR spectrum of two coupled spins A and B obtained using the COSY experiment shown in Figure 20.25.

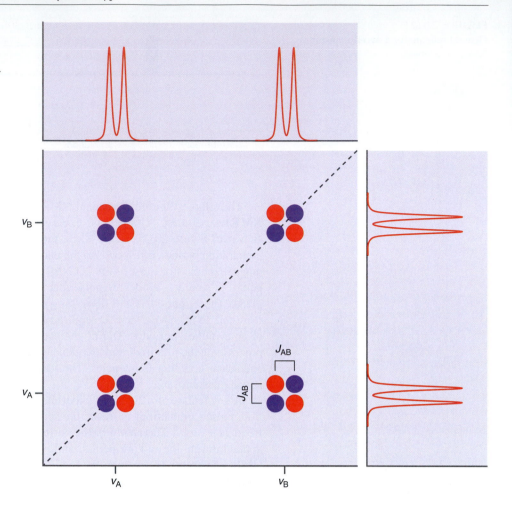

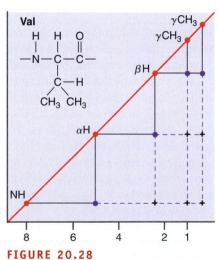

FIGURE 20.28
Schematic of the COSY spectra of valine. Cross-peak multiplets normally observed in COSY spectra are indicated by open circles. Cross peaks observed only in TOCSY spectra are indicated by crosses. COSY connectivities are solid lines. TOCSY connectivities are dotted lines. The frequency axis is in ppm (Source: Adapted from K. Wuetrich, *NMR of Proteins and Nucleic Acids,* John Wiley & Sons, New York, 1986.)

FIGURE 20.27
Schematic of a tripeptide segment derived from a polypeptide chain. R^i indicates the amino acid side chain associated with the amino acid that occupies the ith position in the primary sequence. The rectangles delimit the regions to which are confined connectivities established by proton COSY experiments.

We can use the average chemical shifts observed for amino acid protons in a large number of polypeptides (see Table 20.2 in Appendix B) to sketch the COSY spectrum for any amino acid as it might be observed in a polypeptide chain. For example, in addition to the amide proton (NH) and the alpha proton (αH), the amino acid valine has a single proton (βH) attached to the beta carbon (βC), and attached to βC are two magnetically inequivalent methyl groups (γCH_3) (Figure 20.28). As shown in Table 20.2 in Appendix B, in polypeptides the average NH chemical shift for valine is 8.44 ppm, for the αH the average chemical shift is 4.18 ppm, for βH the average chemical shift is 2.13 ppm, and the two methyl groups have average chemical shifts of 0.97 and 0.94 ppm. Using this information, the COSY spectrum of the valine located in a polypeptide chain is shown in Figure 20.28, where COSY cross-peak multiplets are shown as solid circles. The sequence of cross peaks traces the bonding connectivity in valine. The amide and alpha protons are separated by three bonds and have a large enough scalar coupling to produce detectable COSY cross peaks. The alpha and beta protons likewise have a detectable COSY cross peaks and the beta proton shows cross peaks to both methyl groups. The scalar couplings between more distant protons (e.g., the amide and beta) are generally too small to produce detectable COSY cross peaks.

EXAMPLE PROBLEM 20.4

Using the information in Table 20.2 (Appendix B), sketch the proton COSY spectrum of threonine as it would appear in a polypeptide chain. Exclude the hydroxyl proton, which exchanges too rapidly with solvent to be observed. Trace out the sequential connectivity.

Solution

The threonine side chain consists of a single beta proton (4.22 ppm) and a gamma methyl group (1.23 ppm). Also according to Table 20.2, the amide proton resonates at 8.25 ppm and the alpha proton resonates at 4.35 ppm. The COSY spectrum has the following appearance:

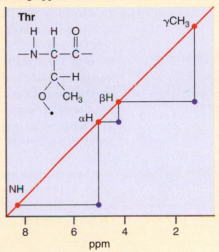

Example Problem 20.4 and the data in Table 20.2 (Appendix B) illustrate an important feature of proton NMR spectra. Some protons in amino acids are labile in the sense that they exchange with protons in the aqueous solvent. For example, if an amino acid in polypeptide is dissolved in D_2O, the following exchange reaction will occur if the amino acid has a labile amide proton:

$$-\underset{\underset{H}{|}}{N}-\underset{\underset{H}{|}}{C}-\underset{\underset{O}{\|}}{C}- + D_2O \longrightarrow -\underset{\underset{D}{|}}{N}-\underset{\underset{H}{|}}{C}-\underset{\underset{O}{\|}}{C}- + HDO$$

If the amide proton is replaced by a deuteron as a result of the exchange shown here, it will not appear in the NMR spectrum. Although amino protons in monomeric amino acids are labile, hydrogen bonding in the secondary structures of proteins, as well as solvent inaccessibility due to tertiary folding, slows the rate of amide exchange to the point that many amide protons will be observed in the NMR spectrum of a folded protein dissolved in D_2O. Hydroxyl protons are even more labile and are rarely observed in COSY spectra of proteins.

For large proteins the large numbers of amino acids may produce overlapping COSY cross peaks, obscuring the unique patterns of cross peaks that are signatures for amino acid types. In this case COSY data are augmented by an additional two-dimensional experiment called **TOCSY**. TOCSY stands for total correlation spectroscopy. In TOCSY, cross peaks between protons separated by more than three bonds may be observed by a sequence of pulses applied during the mixing period. We will not discuss exactly how

TOCSY works, but the results are of interest. For example, if side chain COSY cross peaks to the γ methyl groups for a particular valine in a protein cannot be resolved from other side chain COSY cross peaks arising from other amino acids, TOCSY data may enable the identification of the valine side chain from the $\alpha-\gamma$ cross peaks, represented by the crosses in Figure 20.28.

20.6.2 Cross Relaxation and the Nuclear Overhauser Effect

We have shown that the COSY experiment correlates chemical shifts of scalar-coupled spins and thus traces through-bond spin–spin connectivities. A second type of 2D NMR experiment correlates spins via a relaxation mechanism called the nuclear Overhauser effect (NOE). This 2D NMR experiment is called nuclear Overhauser effect spectroscopy (**NOESY**). Before describing the NOESY experiment, we discuss the physical origins of the NOE.

The simplest model for the nuclear Overhauser effect is a pair of nuclear spin dipoles undergoing a direct through-space interaction. Figure 20.29a shows the energy-level diagram for two coupled spin 1/2 nuclei I and S. Assume that at equilibrium the population of the $\alpha\alpha$ state is $N + \Delta$, the populations of the $\alpha\beta$ and $\beta\alpha$ states are each about equal to N, and the population of the $\beta\beta$ state is $N - \Delta$. If a weak rf field is applied to the spins so that only the S spins are affected (see Figure 20.29b), transitions will occur from the $\alpha\alpha$ state to $\alpha\beta$ and from $\beta\alpha$ to $\beta\beta$. If the irradiation continues for a period long compared to the spin lattice relaxation time of the S spin (i.e., T_{1S}), the S spin population of $\alpha\alpha$ will become equal to the S spin population of $\alpha\beta$, and similarly the S spin population of $\beta\alpha$ will become equal to $\beta\beta$, as shown in Figure 20.29c. This situation is called saturation. At equilibrium the S longitudinal magnetization S_Z assumes a value $S_Z = S_0$, but when saturated the S spin population difference between the $\alpha\alpha$ and $\alpha\beta$ levels is zero (and similarly for the $\beta\alpha$ and $\beta\beta$ levels), so the S spin longitudinal magnetization S_Z is zero, that is, $S_Z = 0$.

Continuous rf irradiation keeps the S spin populations in a state of saturation. Perturbed S spin populations can affect I spin populations through a process called *cross relaxation*. The magnetic dipole moment of an S spin generates a field that couples to the I spin magnetic moment. If the "dipolar" field at spin I generated by the magnetic dipole of spin S fluctuates as a result of molecular motion, the I spin can "flip" between α and β states. This fact provides a mechanism for the perturbed S spin populations to "cross relax" to the I spins.

When the I spin populations are perturbed from equilibrium, they relax to equilibrium according to the rate equation:

$$\frac{dM_Z^I}{dt} = -\rho_I(M_Z^I - M_0^I) - \sigma_{IS}(M_Z^S - M_0^S) \tag{20.36}$$

FIGURE 20.29

Selective saturation of spin S equalizes the populations of S spins in spin-up and spin-down orientations. Cross relaxation to the I spins perturbs the I spin populations. The I and S spin population differences are probed by observing the intensity of magnetization following a 90-degree pulse.

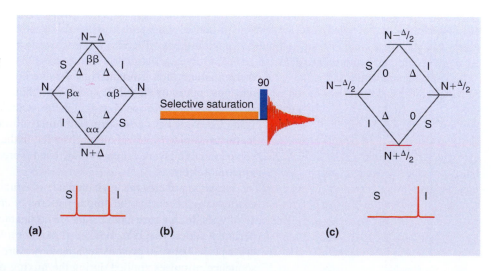

The first term on the right side of Equation (20.36) corresponds to the spin–lattice relaxation of the I spins, discussed in Section 20.2 where the rate of spin–lattice relaxation is ρ_I. The second term corresponds to the cross relaxation between the I and S spin systems. The cross relaxation rate is σ_{IS}. If the S spins are irradiated for a very long time the S spin-up and spin-down populations equalize and so $M_Z^S = 0$. Because of continuous cross relaxation to the saturated S spins, the I spin magnetization eventually ceases to change $\left(\dfrac{dM_Z^I}{dt} = 0\right)$, a condition called steady state. At steady state the I spin magnetization has a value $M_Z^{I,ss}$, which deviates from its equilibrium value according to (see Problem P20.20):

$$\frac{M_Z^{I,ss}}{M_0^I} = 1 + \frac{\sigma_{IS}M_0^S}{\rho_I M_0^I} = 1 + \frac{\sigma_{IS}\gamma_S}{\rho_I\gamma_I} = 1 + \eta_{IS} \tag{20.37}$$

Equation (20.37) describes the steady-state NOE experiment. Equation (20.37) says that in a steady-state NOE experiment, the I spin magnetization deviates from its equilibrium value by a factor of $\eta_{IS} = \dfrac{\sigma_{IS}\gamma_S}{\rho_I\gamma_I}$.

The value of η_{IS} can be positive or negative depending on the rate of molecular tumbling relative to the Larmor frequency. From Figure 20.30 cross relaxation occurs in two ways. For example, an S spin in a β state can flip an I spin in an α state. This mutual spin "flip-flop" causes a transition between the $\alpha\beta$ and the $\beta\alpha$ states. Because the $\alpha\beta$ and $\beta\alpha$ states correspond to $m = 0$, the change in the quantum number is $\Delta m = 0$. This is called a zero quantum transition and it occurs at a transition rate W_0. A second pathway involves a transition directly from the $\beta\beta$ state to the $\alpha\alpha$ state. This transition is called a double quantum transition because $\Delta m = 2$, and it occurs at a rate W_2. The I spin transitions between $\alpha\beta$ and $\beta\beta$ and between $\alpha\alpha$ and $\beta\alpha$ occur at a rate W_{1I}. The cross relaxation rate between spins I and S actually depends on the difference between W_0 and W_2, that is,

$$\sigma_{IS} = W_2 - W_0 \tag{20.38}$$

Spin–lattice relaxation of the I spins proceeds via zero and double quantum transition rates as well as by the two single quantum transitions. Therefore, the rate of spin lattice relaxation is $\rho_I = W_0 + 2W_{1I} + W_2$. The NOE enhancement factor is therefore

$$\eta_{IS} = \frac{\gamma_S}{\gamma_I}\frac{W_2 - W_0}{W_0 + 2W_{1I} + W_2} \tag{20.39}$$

From Equation (20.39) the size and sign of η_{IS} depends on the relative values of the double and zero quantum transition rates W_2 and W_0. These rates in turn depend on the rate

FIGURE 20.30
The sign of the NOE depends on the rate of molecular tumbling relative to the Larmor frequency. For small molecules where the rate of tumbling is fast compared to the Larmor frequency, spins relax via the double quantum transition at a rate W_2. This produces a positive NOE (left). Slow tumbling macromolecules will relax via zero quantum transitions at a rate W_0 and will display a negative NOE (right). The arrow in the spectrum indicates the intensity of the C spin line without cross relaxation.

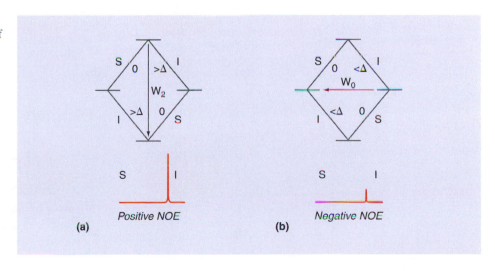

of molecular motions compared to the Larmor frequency. The timescale of molecular motions is quantified by the correlation time τ_c. The faster the rate of molecular motions, the shorter the correlation time.

It can be shown (see Problem P20.8) that for small molecules tumbling at rates higher than the Larmor frequency ν_0 (i.e., $1/\tau_c \gg \nu_0$), η_{IS} is a positive number with a theoretical maximum value of 0.5. In cases where $\eta_{IS} > 0$, the NOE is said to be positive (see Figure 20.30a). A biologically interesting limit is the case where a very large molecule such as a protein or a nucleic acid tumbles at a rate that is slow compared to the Larmor frequency, that is, $1/\tau_c \ll \omega_0$. The energy-level difference between the $\alpha\beta$ and $\beta\alpha$ states is much smaller in comparison to the difference between the $\alpha\alpha$ and $\beta\beta$ states, so cross relaxation via zero quantum transitions proceeding at a rate W_0 is the favored mechanism. Assuming purely dipolar relaxation, a detailed analysis shows that in the slow tumbling limit the transition rates are approximately

$$W_0 \approx q\tau_c \gg W_{1I}, W_{1S}, W_2 \tag{20.40}$$

where $q = \dfrac{\hbar^2}{10}\left(\dfrac{\mu_0}{4\pi}\right)^2 \dfrac{\gamma_I^2 \gamma_S^2}{r_{IS}^6}$ and r_{IS} is the distance between spin I and spin S and $\mu_0 = 4\pi \times 10^{-7}$ T mA^{-1} is the permeability constant. According to Equations (20.39) and (20.40), in the slow tumbling limit $\eta_{IS} \approx -1$. For $\eta < 1$ the NOE is said to be negative.

20.6.3 NOE Correlation Spectroscopy (NOESY)

The NOE effect can be exploited in the context of a two-dimensional NMR experiment to obtain through-space correlations. As shown in Figure 20.31, the NOESY experiment consists of three 90-degree pulses. The progress of a NOESY experiment can be followed

FIGURE 20.31

The NOESY pulse sequence and a vector picture of cross relaxation between two spin $\frac{1}{2}$ nuclei during a NOESY experiment.

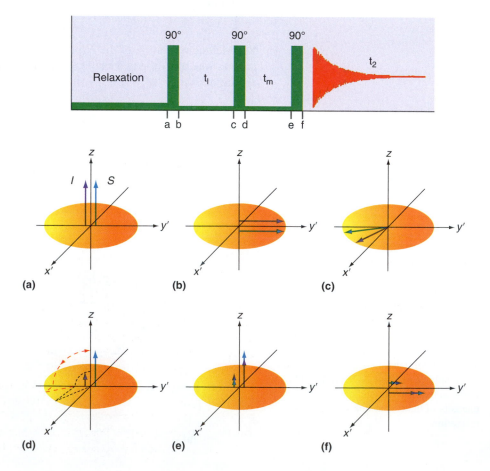

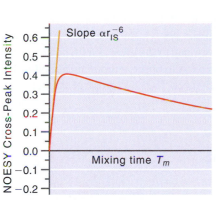

FIGURE 20.32
The buildup of a NOESY cross peak between two cross-relaxing spins I and S in the slow molecular tumbling limit. The slope of the NOE buildup curve extrapolated to zero mixing time is proportional to the inverse sixth power of the internuclear distance.

for a spin pair I-S. For simplicity assume there is no scalar coupling between I and S. Preceding the first pulse is a relaxation period, during which the I and S spin systems reach their equilibrium populations, and magnetizations I_0 and S_0 are established. In Figure 20.31a, I_0 and S_0 are indicated by dark and open arrows, respectively. The first 90-degree pulse creates **transverse magnetization** (see Figure 20.31b). During the evolution time t_1 transverse magnetizations from the I and S spin systems precess in the x-y plane as explained earlier. At the conclusion of the evolution period, I spin vectors and S spin vectors have accrued different angles because of chemical shift differences (i.e., $\delta_I \neq \delta_S$), as illustrated in Figure 20.31c. At the end of the evolution period, a second 90-degree pulse returns part of the transverse magnetization back to the z axis. As Figure 20.31d shows, the degree to which transverse magnetization that is converted to longitudinal magnetization by the second 90-degree pulse depends on the angle of precession at the end of the evolution period. Therefore, spins I and S with different precession frequencies will have different energy-level populations established by the second 90-degree pulse. Although neither spin I or S is saturated, they nevertheless have nonequilibrium populations after the second 90-degree pulse, so during the mixing time of length τ_m, spins I and S cross relax as explained earlier. The length of the mixing time must be selected with care to allow sufficient time for cross relaxation to occur. At the conclusion of the mixing period, a third 90-degree pulse returns the magnetization to the x-y plane. As with the COSY experiment, the FID is repeatedly detected as a function of t_1.

Note that in the NOESY experiment cross relaxation occurs because spin magnetizations are inverted to different degrees by the second 90-degree pulse as a result of precession during t_1. During the mixing period the NOE amplitude is perturbed by the combined effects of cross relaxation and spin lattice relaxation. This type of NOE is called a transient NOE.

If the NOESY cross-peak intensity is monitored as a function of the mixing time, at very short mixing times the second term in Equation (20.36) dominates and the slope extrapolated to $\tau_m = 0$ is linear and proportional to the cross relaxation rate σ_{IS}. In the slow tumbling limit σ_{IS} is proportional in turn to W_0, which according to Equation (20.40) is proportional to the inverse sixth power of the internuclear distance r_{IS}^{-6}. As the mixing time continues to increase, the effects of spin–lattice relaxation will cause the NOESY cross peak to eventually diminish, as shown in Figure 20.32.

20.7 Solution NMR Studies of Biomolecules

20.7.1 Determination of Protein Secondary and Tertiary Structure with NOESY

Professor Kurt Wüthrich and coworkers at the Swiss Federal Institute of Technology in Zurich pioneered the use of 2D NMR techniques to determine the solution structures of proteins. For his achievements in this field, Professor Wüthrich received the Nobel Prize in Chemistry in 2002. Wüthrich and coworkers used COSY to identify amino acid spin systems, and NOESY to determine internuclear distances, as described in the preceding section. Today, the ability to enrich proteins with ^{13}C and ^{15}N makes possible the use of heteronuclear correlation NMR experiments to identify spin systems, so COSY is used less often for this purpose. However, the NOESY experiment is still the primary tool for structure determination because by measuring internuclear distances, the secondary structure as well as the folding of proteins can be determined. We briefly consider next how this is done.

NOESY experiments can measure internuclear distances to a maximum limit of 5 to Å. Figure 20.33 shows proton–proton distances between neighboring amino acids in a protein that can be measured by NOESY. Many of these distances are sensitive to secondary structure. For example, the distance between amide protons d_{NN} varies as a function of secondary structure. Two very important secondary structural motifs in

FIGURE 20.33
Distances between protons on adjacent amino acids that can be measured by NOESY. Many of these distances are sensitive to secondary structure.

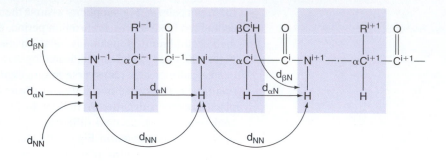

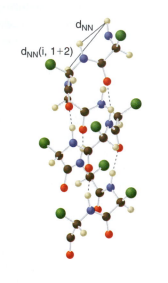

(a)

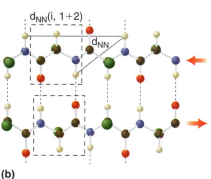

(b)

FIGURE 20.34
The proximity of amide protons in two secondary structural motifs. (a) The periodic structure of the α-helix brings both the adjacent and once removed amide protons into proximity. (b) The extended structure of the β-sheet causes adjacent amide pairs to be farther removed than adjacent amide pairs in the α-helix. Amide proton pairs on adjacent amino acids are too distant in the β-sheet to produce observable NOESY cross peaks.

proteins are α-helices and antiparallel β-sheets (see Figure 20.34). In an α-helix d_{NN} is typically 2.8 Å, but in an antiparallel β-sheet, the polypeptide chain is in a more extended conformation and the d_{NN} distance is typically 4.3 Å. On the other hand, $d_{\alpha N}$ is shorter in the β-sheet than in the α-helix (2.2 versus 3.5 Å). Because the buildup of a NOESY peak between two amide protons is proportional to the inverse sixth power of the d_{NN}, other things being equal, in an α-helix one expects to observe more intense NOESY cross peaks between amide protons than in a β-sheet. NOESY cross peaks between

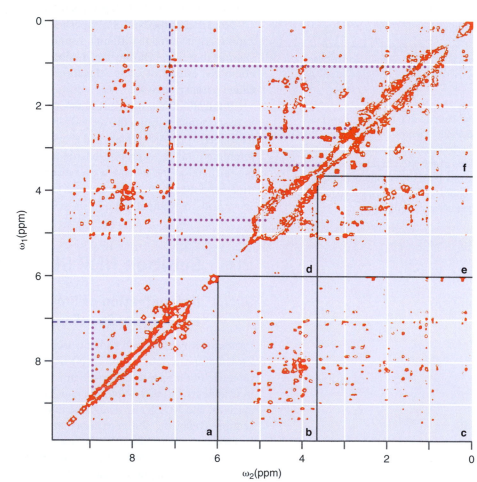

FIGURE 20.35
The proton NOESY spectrum of the proteinase inhibitor BUSI II. The spectrum below the diagonal is divided into six regions a through f that denote cross peaks between distinct proton types (see text). Above the diagonal, the vertical dotted line identifies cross peaks, between an amide proton at 7.2 ppm to six other types of protons, indicated by horizontal dotted lines. The amide proton at 7.2 ppm also has an NOE cross peak to another proton at 8.9 ppm. (Source: Wüthrich, K. *NMR of Proteins and Nucleic Acids,* John Wiley and Sons, New York, NY 1986)

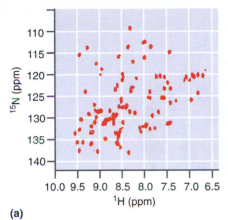

(a)

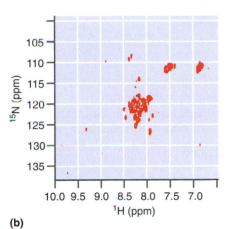

(b)

FIGURE 20.36

$^{15}N-{}^{1}H$ HSQC spectra show the correlation of ^{15}N amide chemical shifts with the chemical shifts of their attached amide protons. The HSQC spectrum of a folded protein is shown at the right in part (a). The unfolded protein is shown at the right in part (b).

amide and alpha protons on adjacent amino acids would be expected to be more intense than amide–amide NOESY cross peaks in β-sheet secondary structures.

Figure 20.35 is the NOESY spectrum of the proteinase inhibitor BUSI II. The spectrum below the diagonal is divided into six regions a through f, each region denoting cross peaks between different types of protons. For example, region a contains cross peaks between amide protons, so these cross peaks give information about d_{NN} distances. Region b contains cross peaks between amide protons and alpha protons, thus giving estimates of $d_{\alpha N}$. Region c contains cross peaks between amide protons and protons in aliphatic and aromatic side chains. Often NOESY cross peaks are observed between protons that are not located in adjacent amino acids. These *long-range NOEs* are indicative of folding of the protein structure that bring protons to within 5 to 6 Å that would otherwise be much farther apart if the protein were not folded.

20.7.2 Heteronuclear Correlation Spectroscopy

The applicability of NMR for determining the structure of proteins and protein complexes was greatly extended by the development of multidimensional NMR experiments. Isotopic enrichment of proteins and nucleic acids with ^{13}C and ^{15}N spins enables use of a large number of heteronuclear correlation NMR experiments that are used for assigning NMR spectra and for monitoring physical changes such as the unfolding of protein structures or formation of molecular complexes. A simple heteronuclear 2D NMR experiment that has been widely used in the study of protein structure and is the basis for many of the powerful multidimensional NMR experiments in use today is heteronuclear single quantum correlation spectroscopy or **HSQC.** HSQC is the heteronuclear analogue of COSY in the sense that while COSY correlates the chemical shift frequencies observed in the proton NMR spectrum with one another, thus tracing out through bond scalar coupling connectivities between protons, HSQC correlates the proton chemical shift frequencies with either ^{15}N or ^{13}C chemical shift frequencies via heteronuclear scalar couplings.

As applied to ^{15}N-enriched proteins, the $^{15}N-{}^{1}H$ HSQC correlates amide proton chemical shift precession frequencies with amide ^{15}N chemical shift precession frequencies. A typical $^{15}N-{}^{1}H$ HSQC spectrum of a protein is shown in Figure 20.36a. The HSQC spectrum is like a COSY spectrum except one axis labels the amide proton chemical shift and the other axis labels the amide ^{15}N chemical shifts. The tertiary structure of a folded protein maximizes chemical shift dispersion by local exposure of different amide sites to ring currents from aromatic rings in neighboring amino acids, hydrogen bonding in secondary structures, etc. Therefore, the amide proton and amide ^{15}N chemical shifts will show dispersion in a folded protein. If the protein is unfolded, chemical shift dispersions decreases at amide sites and the HSQC spectrum will be poorly resolved as shown in Figure 20.36b. Because of its convenience (many proteins can be easily enriched with ^{15}N) and high sensitivity, HSQC is widely used as a diagnostic test for protein folding.

Another important application of HSQC is monitoring the formation of biomolecular complexes. For example, when a drug or nucleic acid binds to a protein, structural changes induced by the formation of the molecular complex or ring currents from the drug induce changes in the chemical shifts of amide protons and amide ^{15}N spins that can be easily monitored using HSQC. Complexes between nucleic acids and proteins commonly occur in the course of cellular processes. Interactions between proteins and DNA are necessary for transcription. Storage of chromosomal material involves interactions between DNA and proteins called histones. Complexes are also formed between RNA and proteins.

The structure of an RNA-binding protein U1A has been studied extensively by multinuclear NMR methods. The production of the U1A protein is self-regulated by a negative feedback system. When excess U1A is produced, the protein binds to an RNA molecule called UTR, which sets into motion a series of chemical events that eventually inhibit further production of U1A. The complex formed between U1A and UTR RNA

FIGURE 20.37
The structure of the complex formed between UTR RNA (shown in blue) and the U1A protein (shown in green). (Source: Frederic H.-T. Allain, Peter W.A. Howe, David Neuhaus, Gabriele Varani, "Structural Basis of the RNA-binding Specificity of Human U1A Protein," *EMBO Journal,* Vol. 16, no.18, pgs 5794–5774, 1997)

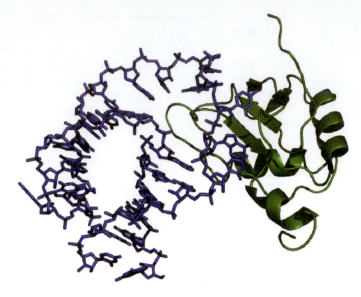

(Figure 20.37) has been studied intensely by NMR and X-ray crystallography in order to understand the structural basis for how this complex is stabilized.

When UTR RNA and U1A form a complex structural change in the U1A protein itself, together with changes in the local electronic environment of various nuclei in the protein due to close approaches of purine and pyrimidine rings from the RNA, result in extensive changes in the chemical shifts in the HSQC spectrum of U1A. Complex formation between RNA and U1A can be monitored by acquiring the ^{15}N–^{1}H HSQC spectrum for U1A before and after it has been bound to RNA. These data are shown in Figure 20.38. If each of the cross peaks in the HSQC spectrum is assigned to specific ^{15}N–^{1}H pairs in the protein, the degree of environmental perturbation at various sites in the protein due to binding of the RNA can be monitored.

In proteins that can be enriched with ^{15}N and ^{13}C, three- and even four-dimensional NMR experiments based on the type of heteronuclear correlations created in the

FIGURE 20.38
The TOCSY spectra of the aromatic protons of UTR RNA, shown in blue, when the RNA is free (left) and complexed (right) to U1A. The ^{15}N–^{1}H HSQC spectra of U1A are shown in red for the free protein (left) and the protein complexed to the RNA (right). (Source: Data provided by G. Varani)

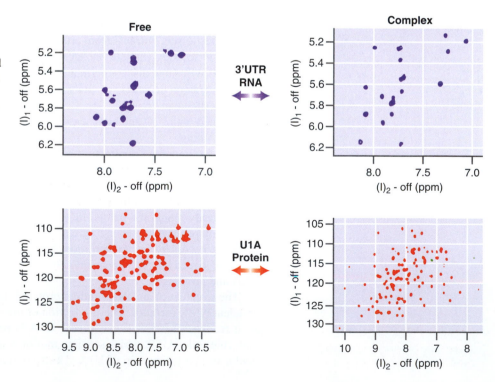

^{15}N–^{1}H HSQC experiment are now widely used to study and obtain solution structures of proteins, nucleic acids, and macromolecular complexes. Just as 2D NMR experiments like COSY and NOESY extended the size of molecules that could be structurally studied to small proteins composed of up to 100 amino acids, three- and four-dimensional NMR experiments have extended the protein size limit of NMR to hundreds of amino acids.

20.8 Solid-State NMR and Biomolecular Structure

20.8.1 The Chemical Shift Anisotropy and the Direct Dipolar Interaction

Nuclear magnetic resonance is not confined to studying molecules in solution, but has also been used to study molecules in the gas phase and in the solid state. Solid-state NMR is used to study macromolecular systems that are too large to be dissolved in solution or possess properties that make solution study impractical. For example, solid-state NMR has been used to study fibrillar amyloid peptides that are associated with diseases such as Alzheimer's syndrome and integral membrane proteins immobilized in lipid bilayers.

In solution, NMR molecules tumble rapidly and isotropically. Typical line widths observed in the proton NMR spectrum of a small molecule in solution tumbling isotropically at rates greater than the Larmor frequency may be less than 1 Hz. By comparison, line widths of tens of thousands of hertz may appear in the proton NMR spectrum of a polycrystalline solid where molecular tumbling is absent.

To understand why molecular tumbling narrows NMR lines, consider the chemical shift of a spin 1/2 nucleus in a liquid versus a solid. In general, the chemical shift is partly dependent on the orientation of the molecular framework relative to the direction of the magnetic field, that is,

$$\nu(\theta, \varphi) = \nu_{iso} + \nu_{aniso}(\theta, \varphi) \qquad (20.41)$$

where ν_{iso} is the isotropic chemical shift observed for molecules in solution, θ and φ are the angles that specify the orientation between the magnetic field vector and a frame of reference fixed to the molecule. The component of the chemical shift that depends on the angles θ and φ is called the anisotropic chemical shift $\nu_{aniso}(\theta, \varphi)$. The anisotropic chemical shift is quantified by three numbers called principal values: σ_1, σ_2, and σ_3. These three numbers can be physically represented relative to a frame fixed to a molecule's bonding framework as the semi-axes of an ellipsoid (see Figure 20.39a). The isotropic chemical shift observed for a nuclear spin in a liquid is the average of the three principal values:

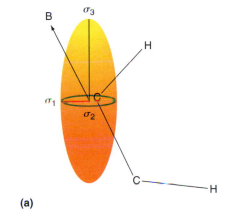

$$\nu_{iso} = -\nu_0 \sigma_{iso} = -\nu_0 \frac{\sigma_1 + \sigma_2 + \sigma_3}{3} \qquad (20.42)$$

The two angles θ and φ orient the magnetic field vector $\vec{B}$ in the reference frame formed by these mutually orthogonal semi-axes (see Figure 20.39b), and determine the anisotropic chemical shift for a molecule with a specific orientation relative to the magnetic field, as given by Equation (20.43):

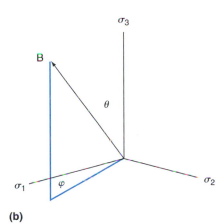

(b)

FIGURE 20.39

The principal values of the chemical shift represented as the semi-axes of an ellipsoid. The angles θ and ϕ orient the magnetic field vector $\vec{B}$ in the coordinate system formed by these semi-axes.

$$\nu(\theta,\varphi) = \nu_{iso} + \nu_{aniso}(\theta,\varphi) = -\nu_0(\sigma_{iso} + \sigma_{aniso})$$

$$= -\nu_0\left(\sigma_{iso} + \delta_{aniso}\left[\tfrac{1}{2}(3\cos^2\theta - 1) + \tfrac{1}{2}\eta\sin^2\theta\cos 2\varphi\right]\right) \qquad (20.43)$$

where the **chemical shift anisotropy** $\delta_{aniso} = \sigma_3 - \sigma_{iso}$ and the chemical shift asymmetry is $\eta = \dfrac{\sigma_1 - \sigma_2}{\delta}$.

In dilute solutions where molecules tumble isotropically, the angles θ and φ are changing rapidly and the chemical shift is averaged over all orientations as shown

FIGURE 20.40
(a) Molecules tumble isotropically at a rate fast compared to the Larmor frequency, causing the chemical shift to be averaged to its isotropic value σ_{iso}. The liquid NMR spectrum consists of a narrow line at σ_{iso}. (b) If the molecules are located in a perfect crystal such that they are oriented identically relative to the magnetic field, the line is observed at the anisotropic chemical shift σ_{aniso}, which is dependent on the orientation of the crystal relative to the field. (c) In a polycrystalline solid a random distribution of orientations occurs, producing a broad chemical shift powder pattern. (Source: M.H. Levitt "Spin Dynamics: Basics of Nuclear Magnetic Resonance." John Wiley & Sons. New York, NY 2001)

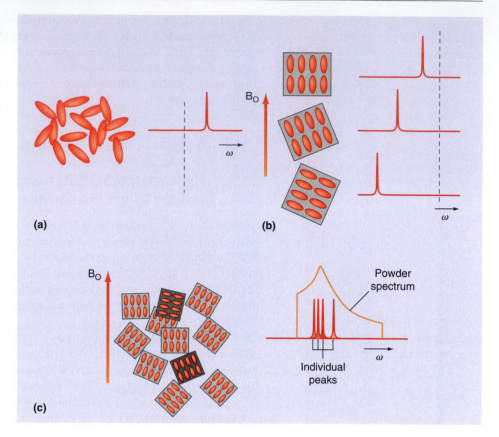

in Figure 20.40a. The anisotropic chemical shift is averaged to zero by rapid molecular tumbling and the observed chemical shift is simply ν_{iso} (see Problem P20.28).

In a single crystal where the ellipsoids are all oriented the same, the NMR line will be narrow, but the chemical shft will depend on the orientation relative to the magnetic field, as shown in Figure 20.40b and Example Problem 20.5.

EXAMPLE PROBLEM 20.5

A spin 1/2 nucleus has a chemical shift defined by the following principal values: $\sigma_1 = 243$ ppm, $\sigma_2 = 184$ ppm, and $\sigma_3 = 107$ ppm.

a. Calculate the isotropic chemical shift σ_{iso}.

b. Calculate the chemical shift for the following values of (θ, φ): $(0,0)$, $(\pi/2,0)$, $(\pi/2,\pi/2)$.

Solution

a. $\sigma_{iso} = \dfrac{\sigma_1 + \sigma_2 + \sigma_3}{3} = \dfrac{243 \text{ ppm} + 184 \text{ ppm} + 107 \text{ ppm}}{3} = 178 \text{ ppm}$

b. For $(\theta, \varphi) = (0,0)$ the magnetic field is parallel to σ_3:

$$\sigma_{iso} + \sigma_{aniso} = \sigma_{iso} + \delta_{aniso}\left[\tfrac{1}{2}\left(3\cos^2\theta - 1\right) + \tfrac{1}{2}\eta\sin^2\theta\cos 2\varphi\right]$$

$$= \sigma_{iso} + \delta_{aniso}\left[\tfrac{1}{2}\left(3\cos^2 0 - 1\right) + \tfrac{1}{2}\eta\sin^2 0\cos 0\right]$$

$$= \sigma_{iso} + \delta_{aniso} = \sigma_3 = 107 \text{ ppm}$$

For $(\theta,\varphi) = (\pi/2,0)$ the magnetic field is parallel to σ_1:

$$\sigma_{iso} + \sigma_{aniso} = \sigma_{iso} + \delta_{aniso}\left[\tfrac{1}{2}(3\cos^2\theta - 1) + \tfrac{1}{2}\eta\sin^2\theta\cos 2\varphi\right]$$

$$= \sigma_{iso} + \delta_{aniso}\left[\tfrac{1}{2}\left(3\cos^2\tfrac{\pi}{2} - 1\right) + \tfrac{1}{2}\eta\sin^2\tfrac{\pi}{2}\cos 0\right]$$

$$= \sigma_{iso} - \frac{\sigma_3 - \sigma_{iso}}{2} + \frac{\sigma_1 - \sigma_2}{2} = \sigma_1 = 243 \text{ ppm}$$

By analogy with the prior two calculations for $(\theta,\varphi) = (\pi/2, \pi/2)$ the magnetic field is parallel to σ_2 and the chemical shift will be 184 ppm.

Example Problem 20.5 shows how the chemical shift in a single crystal depends on the orientation of the crystal relative to the field. Therefore, in a polycrystalline sample, where all values of θ and φ are possible, the NMR spectrum will, as shown in Figure 20.40c, consist of a superposition of transitions covering the full range of chemical shifts possible. This spectrum will span a range of frequencies from σ_3 to σ_1 and this will be more than 130 ppm in width for the case shown in Example Problem 20.5 and much broader than the line width observed when the molecule is in solution.

In addition to the anisotropic chemical shift, two nuclear dipole moments can interact directly through space with an interaction that is proportional to the inverse cube of the internuclear distance:

$$\omega_{ij}^{dd} = 2\pi\nu_{ij}^{dd} - \frac{\mu_0}{8\pi}\frac{\gamma_i\gamma_j\hbar}{r_{ij}^3}\left(3\cos^2\theta_{ij} - 1\right) = \frac{b_{ij}}{2}\left(3\cos^2\theta_{ij} - 1\right) \quad (20.44)$$

In this equation, r_{ij} is the distance between the dipoles in meters and θ_{ij} is the angle between the magnetic field direction and the vector connecting the dipoles. The constant

$$b_{ij} = -\frac{\mu_0}{4\pi}\frac{\gamma_i\gamma_j\hbar}{r_{ij}^3} \quad (20.45)$$

is called the dipolar **coupling constant** and has units of rad s^{-1}. The constant $\mu_0 = 4\pi \times 10^{-7}$ T mA^{-1} is the permeability and is required whenever we desire the dipolar coupling in MKS units. The magnitude of the direct dipolar interaction is much larger than the scalar coupling observed in solution NMR spectra, as Example Problem 20.6 shows, and, like the chemical shift anisotropy, the direct dipolar interaction broadens the NMR spectrum of a polycrystalline solid. Also like the chemical shift anisotropy, the direct dipolar interaction is averaged to zero in solution by rapid, isotropic molecular tumbling (see Problem P20.28).

EXAMPLE PROBLEM 20.6

Calculate the dipolar coupling constant for two protons separated by 2.00 Å. Give your answer in units of hertz.

Solution

Consult Table 20.1 for the proton magnetogyric ratio.

$$b_{ij} = -\frac{\mu_0}{4\pi}\frac{\gamma_i\gamma_j\hbar}{r_{ij}^3} = -\frac{1.00 \times 10^{-7} \text{ T mA}^{-1}}{(2.00 \times 10^{-10} \text{ m})^3}$$

$$\times \frac{(267.5 \times 10^6 \text{ T}^{-1}\text{ s}^{-1})^2 \times 6.626 \times 10^{-34} \text{ J s}^{-1}}{2\pi}$$

$$= 9.44 \times 10^4 \text{ T}^{-1}\text{ A}^{-1}\text{ m}^{-2}\text{ J s}^{-1} = 9.44 \times 10^4 \text{ s}^{-1}$$

$$\times \left(\frac{\text{A m}}{\text{N}}\right) \times \text{A}^{-1}\text{ Nm}^{-1} = 9.44 \times 10^4 \text{ s}^{-1}$$

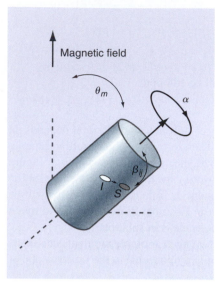

FIGURE 20.41

In magic angle spinning, the sample is rapidly spun about its goniometer axis, which is tilted $\theta_m = 54.74°$ with respect to the static magnetic field. The angles α and β orient a specific I-S internuclear vector in the frame of the goniometer.

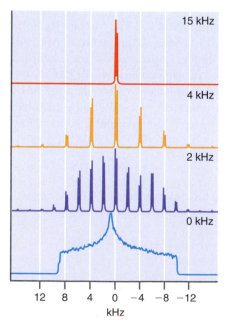

FIGURE 20.42

A ^{13}C NMR spectrum displayed as a function of spinning rate of a polycrystalline sample in which the unit cell contains a molecule with two chemically inequivalent ^{13}C-labeled carbonyl groups (see text).

20.8.2 Magic Angle Sample Spinning and Dipolar Recoupling

A technique called magic angle sample spinning (MASS) removes the spectral broadening imposed by anisotropic magnetic interactions in solid samples, and recovers a high-resolution NMR spectrum similar to that observed for solutions. Figure 20.41 schematizes how a MASS experiment is done. A sample of a polycrystalline or amorphous powder is confined to a cylindrical rotor and is spun rapidly around the axis of a goniometer. If the sample spinning rate is greater than the magnitude of the dipolar coupling, the dipolar interaction is averaged:

$$\overline{\nu_{ij}^{dd}} \propto b_{ij}\left\langle 3\cos^2\theta_{ij} - 1\right\rangle \propto b_{ij}(3\cos^2\theta_m - 1)(3\cos^2\beta_{ij} - 1) \quad (20.46)$$

A similar equation is obtained for the chemical shift. In Equation (20.46), θ_m is the angle that the sample rotation axis makes with the external magnetic field B_0 and β_{ij} is the angle between the internuclear vector and the goniometer axis. If we now choose to make $\theta_m = 54.74°$, that is, the **magic angle**, then $3\cos^2\theta_M - 1 = 0$. The dipolar coupling is reduced to zero and the chemical shift equals the isotropic chemical shift:

$$\overline{\nu_{ij}^{dd}} = 0 \quad \text{and} \quad \overline{\sigma} = \sigma_{iso} \quad (20.47)$$

An example of how a broad solid-state NMR spectrum can be transformed into a sharp spectrum through **magic angle spinning** is shown in Figure 20.42. In this series the NMR spectrum of two dipolar coupled carbonyl ^{13}C spins is detected as a function of the rate of sample rotation. The two ^{13}C spins are separated by 3 Å, which yields a dipolar coupling constant of about 300 Hz. The spectrum is mainly broadened by the very large chemical shift anisotropies of the carbonyl ^{13}C spins. As the sample rotation increases from zero, the powder pattern breaks up into side bands, which are caused by the modulation of the NMR signal by the periodic motion of the rotor. Note that the side bands are separated from each other by the spinning frequency and they diminish in amplitude as the spinning rate increases. At very high spinning rates the side bands disappear altogether and we observe only two lines at the isotropic chemical shifts of the two carbonyl spins.

By removing anisotropic magnetic interactions, MASS yields well-resolved spectra for ^{13}C, ^{15}N, and other spin 1/2 nuclei, but structural information from the direct dipolar interaction is lost. A modification of the MASS experiment called dipolar recoupling can recover the dipolar interaction and information on internuclear distances. In a dipolar recoupling experiment, rf pulses are applied synchronously with the mechanical sample spinning. The rf pulses act to average the spin angular momentum. Because the spin angular momentum is changing at a rate identical to the change of physical orientation by the sample rotation, the net averaging of the dipolar coupling by MASS is upset and a residual or "recoupled" dipolar coupling can be detected. It is also possible to achieve this while completely averaging the chemical shift anisotropy to zero, so high resolution is achieved by MASS without the loss of dipolar coupling information.

Dipolar couplings between homonuclear spins (e.g., ^{13}C and ^{13}C) or heteronuclear spins (e.g., ^{13}C and ^{15}N) can be recoupled. A widely used heteronuclear dipolar recoupling experiment is called rotational echo double resonance (**REDOR**). Figure 20.43 shows a REDOR-derived structural model of the drug vancomycin complexed to the

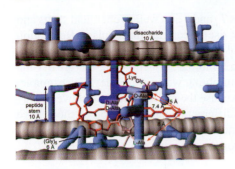

FIGURE 20.43
REDOR-derived structural model of fluorinated vancomycin derivative bound to the mature peptidoglycan of *S. aureus*. The glycan strands are represented as gray cylinders and amino acids at blue pipes. Rounded ends of the blue pipes are terminal D-alanines. The carbon framework of vancomycin is in red, and the ^{19}F spin is the green sphere at the right. (Source: Sung Joon Kim, Lynette Cegelski, Daniel R. Studelska, Robert D. O'Connor, Anil K. Mehta, and Jacob Schaefer, "Rotational-Echo Double Resonance Characterization of Vancomycin Binding Sites in Staphylococcus aureus," *Biochemistry* 2002, 41, 5667–6977, American Chemical Society)

bacterial cell wall of *Staphylococcus aureus*. Vancomycin is a powerful antibiotic reserved for the treatment of serious gram-positive bacterial infections. Clinical use of vancomycin has increased greatly in the United States because it is one of the few antimicrobial agents capable of killing methicillin-resistant staphylococci bacteria including *S. aureus*. Vancomycin kills bacteria by binding to a component of the bacterial cell wall called the peptidoglycan, a high molecular weight biopolymer consisting of amino acids and sugars that gives structure and rigidity to the bacterial cell wall and protects the bacterium from osmotic stress. Once vancomycin binds to the peptidoglycan, synthesis of the bacterial cell wall ceases and the bacterium dies.

Unlike solution NMR, in which biomolecular structures are derived proton–proton distances measured with the NOE, REDOR produces structural models from internuclear distances obtained from dipolar couplings between heteronuclear spin pairs such as ^{13}C—^{31}P, ^{13}C—^{15}N, and ^{15}N—^{19}F. Figure 20.43 shows a structural model of a fluorobiphenyl derivative of vancomycin complexed to the terminal D-alanine of a peptidoglycan that has had ^{13}C enriched glycine and alanine incorporated into its structure. Distances of up to 8 Å between the ^{19}F on the vancomycin and ^{13}C spins in amino acids within the peptidoglycan were measured using ^{13}C—^{19}F REDOR and elucidated how this drug binds and interacts with bacterial cell wall components.

Recently, dipolar recoupling pulse sequences have been incorporated into 2D NMR experiments, with a view to obtaining high-resolution structures of proteins only obtainable until recently by solution NMR and X-ray crystallography. This is an active area of current research.

20.9 NMR Imaging

One of the most important applications of NMR spectroscopy is its use in imaging the interior of solids. In the health sciences, **NMR imaging** has proven to be the most powerful and least invasive technique for obtaining information on soft tissue such as internal organs in humans. How is the spatial resolution needed for imaging obtained using NMR? For imaging, a **magnetic field gradient** is superimposed onto the constant magnetic field normally used in NMR. In this way, the resonance frequency of a given spin depends not only on the identity of the spin (i.e., ^{1}H or ^{13}C) but also on the local magnetic field, which is determined by the location of the spin relative to the poles of the magnet. Figure 20.44 illustrates how the addition of a field to the constant magnetic field allows the spatial mapping of spins to be carried out. Imagine a sphere and a cube containing ^{1}H$_2$O immersed in a background that contains no spin active nuclei. In the absence of the field gradient, all spins in the structures resonate at the same frequency, giving rise to a single NMR peak. However, with the field gradient present, each volume element of the structure along the gradient has a different resonance frequency. The intensity of the NMR peak at each frequency is proportional to the total number of spins in the volume. A plot of the NMR peak intensity versus field strength will give a projection of the volume of the structures along the gradient direction. If a number of scans corresponding to different directions of the gradient are obtained, the three-dimensional structure of the specimen can be reconstructed, provided that the scans cover a range of at least 180°.

The particular usefulness of NMR for imaging biological samples relies on the different properties that can be used to create contrast in an image. In X-ray radiography, the image contrast is determined by the differences in electron density in various parts of the structure. Because carbon has a lower atomic number than oxygen, it will not scatter X-rays as strongly as oxygen. Therefore, fatty tissue will appear lighter in a transmission image than tissues with a high density of water. However, this difference in scattering power is small and often gives insufficient contrast. To obtain a higher contrast, material that strongly scatters X-rays is injected or ingested. For NMR spectroscopy, several different properties can be utilized to provide image contrast without adding foreign substances.

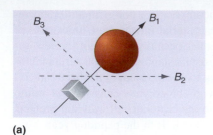

(a)

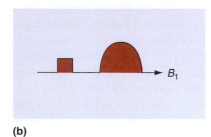

(b)

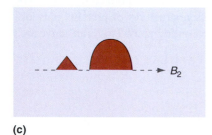

(c)

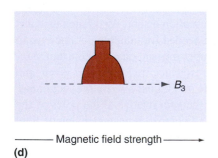

———— Magnetic field strength ————➤

(d)

FIGURE 20.44

(a) Two structures are shown along with the three gradient directions indicated along which NMR spectra will be taken. In each case, spins within a thin volume element slice along the gradient resonate at the same frequency. This leads to a spectrum that is a projection of the volume onto the gradient axis. Image reconstruction techniques originally developed for X-rays can be used to determine the three-dimensional structure. (b–d) NMR spectra taken along the directions B_1, B_2, and B_3 indicated in part (a).

Such properties include the relaxation times T_1 and T_2, as well as chemical shifts and flow rates. The relaxation time offers the most useful contrast mechanism. Relaxation times T_1 and T_2 for water can vary in biological tissues from 0.1 second to several seconds. The more strongly bound the water is to a biological membrane, the greater the change in its relaxation time relative to freely tumbling water molecules. For example, the brain can be imagined with high contrast because the relaxation times between gray matter, white matter, and spinal fluid are quite different. Data acquisition methods have been developed to enhance the signal amplitude for a particular range of relaxation times, enabling the contrast to be optimized for the problem of interest. Figure 20.45 shows an NMR image of a human brain.

Chemical shift imaging can be used to localize metabolic processes and to follow signal transmission in the brain through chemical changes that occur at nerve synapses. One variation of flow imaging is based on the fact that it takes times of several T_1 for the local magnetization to achieve its equilibrium value. If, for instance, blood flows into the region under investigation on shorter timescales, it will not have the full magnetization of the spins that have been exposed to the field for much longer times. The ^1H_2O in the blood will therefore resonate at a different frequency than the surrounding spins.

NMR imaging also has many applications in materials science, such as the measurement of the chemical cross-link density in polymers, the appearance of heterogeneities in elastomers such as rubber through vulcanization or aging, and the diffusion of solvents into polymers. Voids and defects in ceramics and the porosity of ceramics can be detected by nondestructive NMR imaging.

20.10 Electron Spin Resonance Spectroscopy

The electron is a spin 1/2 particle whose spin magnetic dipole moment, in analogy to the expression for the nuclear spin magnetic dipole moment given in Equation (20.2), is given by:

$$\vec{\mu}_e = g_e \frac{e\hbar}{2m_e} \vec{S} = g_e \beta_B \vec{S}$$

where for a free electron, $g_e = 2.0023193$ and the electron mass $m_e = 9.11 \times 10^{-31}$ kg. The parameter $\beta_B = e\hbar/2m_e$ is the Bohr magneton and has the value $\beta_B = 9.2740 \times 10^{-24}$ J T^{-1}. The Zeeman energy that results from the interaction between an electron spin magnetic dipole moment and an external field is

$$E = \mu_e \cdot B_0 = g_e \beta_B m_z B_0$$

where $m_z = \pm 1/2$. Again in analogy to NMR, if an electron spin system is irradiated with a resonant field such that the frequency of the field ν is related to the Zeeman energy difference ΔE by

$$\nu = \frac{\Delta E}{h} = \frac{g_e \beta_B B_0}{h}$$

a transition between the spin-up α state and spin-down β state of the electron will be observed. Note that because of its smaller mass, the electron has a spin magnetic dipole moment over 600 times that of the proton. This fact has great impact on the manner in which *electron spin resonance spectroscopy* (ESR; also called *electron paramagnetic resonance*, EPR) is conducted, as Example Problem 20.7 shows.

EXAMPLE PROBLEM 20.7

Calculate the Zeeman energy change ΔE that occurs in making a transition from the α to the β state of an electron spin in a 0.375-T magnetic field. If a transition is made between these levels by the absorption of electromagnetic radiation, what region of the electromagnetic spectrum is used?

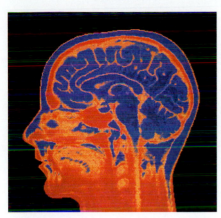

FIGURE 20.45
This figure shows an NMR image taken of a human brain. The section shown is obtained from a noninvasive scan of the patient's head. The contrast has its origin in the dependence of the relaxation time on the strength of binding of the water molecule to different biological tissues.

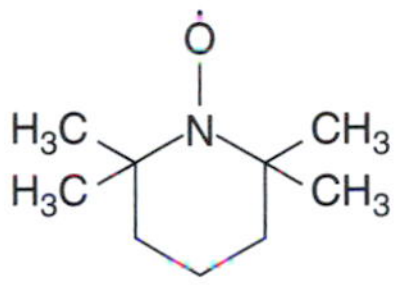

FIGURE 20.46
The stable nitroxide radical TEMPO.

TEMPO

Solution

The energy difference is given by

$$\Delta E = g_e\beta_B B_0 = 2.0023193 \times 9.2740 \times 10^{-24}\ \mathrm{J\ T^{-1}} \times 0.375\ \mathrm{T}$$

$$= 6.9636 \times 10^{-24}\ \mathrm{J}$$

$$\nu = \frac{\Delta E}{h} = \frac{6.9636 \times 10^{-24}\ \mathrm{J}}{6.6261 \times 10^{-34}\ \mathrm{J\ s}} = 1.0509 \times 10^{10}\ \mathrm{s^{-1}}$$

The electron Larmor frequency is about 10 GHz. This is in the range of frequencies called microwave frequencies. Because of its larger spin magnetic dipole moment, ESR is usually conducted at lower magnetic fields than NMR, although the electron Larmor frequencies are much higher than NMR Larmor frequencies obtained at much higher magnetic field strengths.

In addition to the large difference in spin magnetic dipole moments, the resonances of electrons and nuclei differ in another important respect. For a system to absorb microwaves due to electron spin resonance, there must occur in the system unpaired electrons. If this is not the case, paired electrons will possess opposite spin, the system will not display net electron spin, there will be no resonant absorption of microwave power by the sample, and no ESR signal will be observed. ESR can be performed on paramagnetic molecules such as O_2 with triplet ground states, proteins containing the paramagnetic ions of some transition metals such as Mn^{2+}, Fe^{3+}, and Cu^{2+}, and photochemical intermediates in the triplet state, all of which display the ESR phenomenon when irradiated with microwaves. In addition, molecules containing stable nitroxide radicals such as 2,2,6,6-tetramethyl–1-piperidinyloxyl (TEMPO; see Figure 20.46) display ESR transitions. Nitroxide radical derivatives can be tethered to lipids, nucleic acids, and proteins in order to allow probing by ESR the dynamics and structure of otherwise ESR-silent biomolecules.

To investigate the appearance of an ESR spectrum, we must consider the influence of surrounding nuclear spins. If nuclear spins are located around the electron spin, the Zeeman transition energy is shifted by interactions between the spin magnetic dipole moment of the electron and the nuclear spins. For an electron–nuclear spin pair located in a molecule tumbling isotropically in solution, the energy is

$$E = g_e\beta_B B_0 m_{Z,e} + A m_{Z,e} m_{Z,N}$$

where A is called the hyperfine coupling constant:

$$A = \frac{8\pi}{3}\, g_N\beta_N g_e\beta_e \rho(0)$$

and where $\rho(0)$ is the unpaired electron density at the nucleus. The hyperfine coupling constant is enormous compared to the scalar couplings and even the direct dipolar coupling observed between nuclei. For example, for an electron interacting with a hydrogen spin, $A/h = 1420$ MHz. Hyperfine couplings are frequently expressed in units of gauss (G), in which case the hyperfine coupling between a proton and an electron is about 508 G.

Because of the large bandwidth of the ESR spectrum, the microwave frequency is usually held constant and the magnetic field is swept through the range of the ESR spectrum. Weak, periodic field modulation is applied, which produces a spectrum that is the first derivative of the absorption spectrum (see Figure 20.47). Otherwise the appearance of ESR spectra for electrons coupled to nuclear spins can be deduced using the same rules outlined for nuclear scalar couplings (see Section 20.5.3). For an electron coupled to N spin I nuclei, the ESR transition is split into a multiplet with $2NI + 1$ components. For a spin 1/2 nucleus this is $N + 1$ components with relative intensities given by the coefficients of the binomial expansion $(1 + x)^N$, (see Figure 20.47a). For nitroxide radicals the electron spin interacts with the ^{14}N nucleus, a spin 1 nucleus that splits the ESR transition into a triplet (see Figure 20.47b). More distant protons pro-

FIGURE 20.47

(a) The ESR spectrum of an electron coupled equally to four spin $\frac{1}{2}$ nuclei. The red arrow denotes the size of the hyperfine coupling. (b) The ESR spectrum of an electron coupled to a single ^{14}N nucleus and coupled weakly to two spin $\frac{1}{2}$ nuclei. The ^{14}N nucleus is spin 1 and splits the ESR transition into a triplet. The red arrow denotes the ^{14}N hyperfine coupling. Due to rapid molecular tumbling, the components of the triplet are narrowed and the weaker hyperfine couplings are not observed.

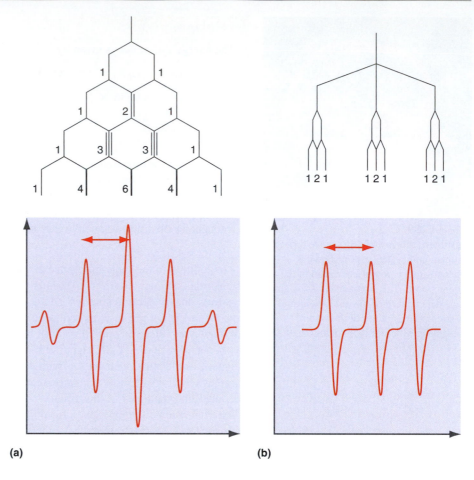

(a)   (b)

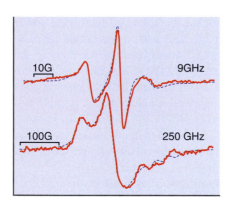

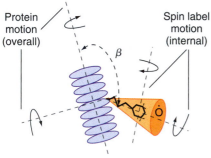

FIGURE 20.48

A multifrequency ESR study of the motions of T4-lysozyme. A nitroxide spin label is tethered to residue 44. The relevant molecular motions are schematized below the spectra. The ESR spectrum at 9 GHz is sensitive to slow, overall protein motions, whereas at 250 GHz, the slow motions appear "frozen out" and the spectrum is sensitive to faster local motions within the polymer. Dotted lines are the simulated line shapes based on the author's model for protein (Source: P.P. Borbat, A.J. Costa-Filho, K.A. Emle, J.K. Moscicki & J.H. Freed "Electron Spin Resonance in the Study of Membrane and Proteins" *Science*, 2001, 241: 55502, 266–269)

duce much weaker hyperfine splittings. Rapid molecular tumbling narrows the lines leaving the ^{14}N triplet (see Figure 20.47b).

An extremely valuable aspect of ESR is its sensitivity to molecular motions. Studying the molecular motions of a biopolymer is complicated by the fact that several motions may exist: the slow overall tumbling of the polymer as a whole is accompanied by local motions of amino acid sidechains and segments of the backbone, which occur at much higher rates. By tethering a nitroxide spin label to a segment of the T4-lysozyme protein in solution, ESR studies are conducted at multiple field strengths to distinguish overall motions of the polymer from local motions (Figure 20.48). At low magnetic field where the electron Larmor frequency is 9 GHz, the ESR spectrum is sensitive to slow, overall motions of the protein. At higher magnetic fields where the electron Larmor frequency is 250 GHz, the spectral line shape is sensitive to higher frequency local motions occurring within the polymer.

ESR has in recent years been extended to the time domain, with electron spin echo techniques, two-dimensional ESR, and other innovations enabling the study of the structure and dynamics of membrane proteins, nucleic acids, and other challenging biological systems.

For Further Reading

Bax, A., *Two Dimensional Nuclear Magnetic Resonance in Liquids.* Reidel Publishing, Dordrecht, Holland, 1982.

Farrar, T. C., *Introduction to Pulse NMR Spectroscopy.* Farragut Press, Chicago, IL, 1997.

Wüthrich, K., *NMR of Proteins and Nucleic Acids.* John Wiley & Sons, New York, 1986.

Vocabulary

$\pi/2$ pulse
chemical shift
chemical shift anisotropy
chemical shift imaging
chemically equivalent nuclei
COSY
coupling constant
dephasing
diamagnetic response
Fourier transform
Fourier transform NMR
   spectroscopy

free induction decay (FID)
HSQC
Larmor frequency
macroscopic magnetic
   moment
magic angle
magic angle spinning
magnetically equivalent
   nuclei
magnetic field gradient
magnetization vector

magnetogyric ratio
motional broadening
multiplet
multiplet splittings
NMR imaging
NOESY
nuclear factor
nuclear magnetic moment
nuclear magneton
precession
pulse sequence

REDOR
rotating frame
shielding constant
spin–echo technique
spin polarization
spin–spin coupling
transverse magnetization
transverse relaxation
two-dimensional NMR
   (2D NMR)

Questions on Concepts

Q20.1 Redraw Figure 20.2 for β spins. What is the direction of precession for the spins and for the macroscopic magnetic moment?

Q20.2 Explain why two magnetic fields, a static field and a radio-frequency field, are needed to carry out NMR experiments. Why must the two field directions be perpendicular?

Q20.3 Why do magnetic field inhomogeneities of only a few parts per million pose difficulties in NMR experiments?

Q20.4 Why is it useful to define the chemical shift relative to a reference compound as follows?

$$\delta = 10^6 \frac{(\nu - \nu_{ref})}{\nu_{ref}}$$

Q20.5 Order the molecules CH_3I, CH_3Br, CH_3Cl, and CH_3F in terms of increasing chemical shift for ^1H. Explain your answer.

Q20.6 Why do neighboring groups lead to a net-induced magnetic field at a given spin in a molecule in the solid state, but not for the same molecule in solution?

Q20.7 Explain the difference in the mechanism that gives rise to through-space dipole–dipole coupling and through-bond coupling.

Q20.8 Why is the multiplet splitting for coupled spins independent of the static magnetic field?

Q20.9 Why can the signal loss resulting from spin dephasing caused by magnetic field inhomogeneities and chemical shift be recovered in the spin–echo experiment?

Q20.10 Why is the measurement time in NMR experiments reduced by using Fourier transform techniques?

Q20.11 Why does NMR lead to a higher contrast in the medical imaging of soft tissues than X-ray techniques?

Problems

Problem numbers in **RED** indicate that the solution to the problem is given in the *Student Solutions Manual*.

P20.1 For a fixed frequency of the radio frequency field, ^1H, ^{13}C, and ^{31}P will be in resonance at different values of the static magnetic field. Calculate the value of $\mathbf{B}_0$ for these nuclei to be in resonance if the radio-frequency field has a frequency of 250 MHz.

P20.2 Using the information in Table 20.1, calculate the three Zeeman energies for a deuteron (^2H) in a magnetic field of 5.5 T. Calculate ΔE and the deuterium Larmor frequency in this field.

P20.3 A 250-MHz ^1H spectrum of a compound shows two peaks. The frequency of one peak is 510 Hz higher than that of the reference compound (tetramethylsilane), and the second peak is at a frequency 280 Hz lower than that of the reference compound. What chemical shift should be assigned to these two peaks?

P20.4   Assume the FID has the form $f(t) = (\cos \omega_0 t + i \sin \omega_0 t)e^{-t/T_2} = e^{(i\omega_0 - 1/T_2)t}$. Calculate the Fourier transform

$$F(\omega) = \int_0^{+\infty} f(t)e^{-i\omega t}\, dt.$$ The real part of the Fourier transform is

the absorption line shape. Show that the full-width at half-maximum of the absorption line shape is given by $\Delta\nu_{1/2} = 1/\pi T_2$.

P20.5   Show that there are four possible transitions between the energy levels of two interacting spins and that the frequencies are given by

$$\nu_{12} = \frac{\gamma B(1 - \sigma_1)}{2\pi} - \frac{J_{12}}{2}$$

$$\nu_{34} = \frac{\gamma B(1 - \sigma_1)}{2\pi} + \frac{J_{12}}{2}$$

$$\nu_{13} = \frac{\gamma B(1 - \sigma_2)}{2\pi} - \frac{J_{12}}{2}$$

$$\nu_{24} = \frac{\gamma B(1 - \sigma_2)}{2\pi} + \frac{J_{12}}{2}$$

P20.6   The nuclear spin operators can be represented as 2×2 matrices and α and β can be represented as column vectors in the form

$$\alpha = \begin{pmatrix} 1 \\ 0 \end{pmatrix} \text{ and } \beta = \begin{pmatrix} 0 \\ 1 \end{pmatrix}$$

Given that

$$\hat{I}_x = \frac{\hbar}{2}\begin{pmatrix} 0 & 1 \\ 1 & 0 \end{pmatrix},\ \hat{I}_y = \frac{\hbar}{2}\begin{pmatrix} 0 & -i \\ i & 0 \end{pmatrix},\ \hat{I}_z = \frac{\hbar}{2}\begin{pmatrix} 1 & 0 \\ 0 & -1 \end{pmatrix}$$

and

$$\hat{I}^2 = \left(\frac{\hbar}{2}\right)^2 \begin{pmatrix} 3 & 0 \\ 0 & 3 \end{pmatrix}$$

show that

$$\hat{I}^2\alpha = \frac{1}{2}\left(\frac{1}{2} + 1\right)\hbar^2\alpha,\ \hat{I}_z\alpha = +\frac{1}{2}\hbar\alpha,\ \hat{I}^2\beta = \frac{1}{2}\left(\frac{1}{2} + 1\right)\hbar^2\beta,$$

$$\text{and } \hat{I}_z\beta = -\frac{1}{2}\hbar\beta$$

P20.7   In liquids, the amplitude of the spin echo is limited by irreversible dephasing that results when a molecule diffuses through an inhomogeneous magnet field during the duration of time between the application of the $\pi/2$ pulse and the echo time 2τ. In the presence of molecular diffusion through a magnetic field gradient ΔB, the proton spin echo amplitude is given by

$$I(2\tau) = I(0)\exp\left[-\left(\frac{2\tau}{T_2}\right) - \frac{2\gamma^2\Delta B^2 D\tau^3}{3}\right]$$

where D is the coefficient of diffusion. Because of the τ^3 dependence, molecular diffusion strongly affects measurements of T_2. Suppose $T_2 = 2$ s, $D = 1.00 \times 10^{-9}$ m^2 s^{-1}, and $\Delta B = 0.10$ T m^{-1}. To what fraction will the proton spin echo intensity be reduced at $\tau = 0.01$ s?

P20.8   In the short correlation time limit, a molecule tumbles at a rate faster than the Larmor frequency such that $1/\tau_c >> \omega_0$, where τ_c is the correlation time. If the relaxation is purely due to dipole–dipole interactions, the transition rates are defined as

$$W_0 = q\tau_c$$

$$W_{1I} = W_{1S} = \frac{3q\tau_c}{2}$$

$$W_2 = 6q\tau_c$$

where $q = \dfrac{\hbar^2}{10}\left(\dfrac{\mu_0}{4\pi}\right)^2 \dfrac{\gamma_I^2\gamma_S^2}{r_{IS}^6}$ and r_{IS} is the distance between spin I and spin S. Show that in the short correlation time limit, the theoretical value for the NOE enhancement factor η_{IS} is 0.5.

P20.9   Predict the number of chemically shifted ^{1}H peaks and the multiplet splitting of each peak that you would observe for bromoethane. Justify your answer.

P20.10   Predict the number of chemically shifted ^{1}H peaks and the multiplet splitting of each peak that you would observe for 1-chloropropane. Justify your answer.

P20.11   Predict the number of chemically shifted ^{1}H peaks and the multiplet splitting of each peak that you would observe for 1,1,2-trichloroethane. Justify your answer.

P20.12   Predict the number of chemically shifted ^{1}H peaks and the multiplet splitting of each peak that you would observe for 1,1,1,2-tetrachloroethane. Justify your answer.

P20.13   Predict the number of chemically shifted ^{1}H peaks and the multiplet splitting of each peak that you would observe for 1,1,2,2-tetrachloroethane. Justify your answer.

P20.14   Predict the number of chemically shifted ^{1}H peaks and the multiplet splitting of each peak that you would observe for nitromethane. Justify your answer.

P20.15   Predict the number of chemically shifted ^{1}H peaks and the multiplet splitting of each peak that you would observe for nitroethane. Justify your answer.

P20.16   Sketch the positions of the diagonal and cross peaks in a COSY spectrum of an AMX spin system. Assume $J_{AM} = J_{MX} = J_{AX} \neq 0$.

P20.17   Using the information in Table 20.2 in Appendix B, determine the identity of the amino acid with the COSY/TOCSY spectrum shown here. Explain your reasoning. Assume the amino acid is monomeric and is dissolved in D$_2$O.

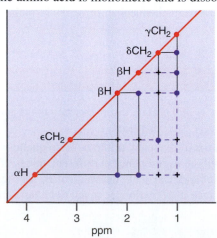

P20.18   Determine which of the COSY/TOCSY spectra shown here is that of leucine and which is that of isoleucine. Explain your answer.

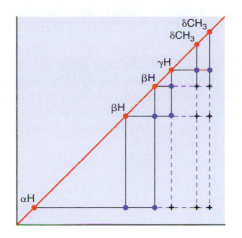

P20.19   The COSY spectrum obtained for aspartic acid is shown here. Based on the number of proton cross peaks that you observe, is the aspartic acid in monomeric form and dissolved in D_2O or is the aspartic acid incorporated into a structured protein domain and dissolved in D_2O? Explain your answer.

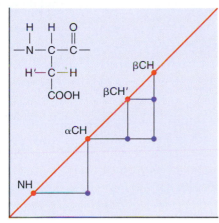

P20.20   Derive Equation (20.40), which states that for cross-relaxing spins I and S, with the S spins saturated and the I spins in steady state, the ratio of the steady-state I magnetization to its equilibrium value is given by $\dfrac{M_z^{I,ss}}{M_0^I} = 1 + \eta_{IS}$ where the NOE enhancement factor is $\eta_{IS} = \dfrac{\sigma_{IS}\gamma_S}{\rho_I\gamma_I}$.

P20.21   Assume two protons in a rigid macromolecule are separated by 3.00×10^{-10} m. Assume the molecule tumbles with a correlation time of 20 ns. Estimate the rate of cross relaxation between these two spins

P20.22   NOE experiments are often displayed in difference mode. For two cross-relaxing protons I and S, this means that the NOE spectrum obtained by selectively saturating the spin S and observing the NOE at the I spin is subtracted from a control spectrum where the S spin is not irradiated. Assume the I-S spin pair is in a rapidly tumbling molecule. A truncated NOE experiment results in a NOE enhancement factor of $\eta_{IS} = 0.4$. Sketch the I-S proton NOE difference spectrum. What

is the sign of the I spin peak versus the S spin peak? Repeat the exercise for two protons I and S in a slowly tumbling macromolecule where in a truncated NOE experiment $\eta_{IS} = -0.4$. Compare the relative signs of the two peaks in the difference NOE spectra to the signs of NOESY cross peaks versus diagonal peaks for the rapid and slow molecular tumbling cases.

P20.23   Using the rotating frame equation $\dfrac{d\vec{M}}{dt} = \gamma \vec{M} \times B_1 \hat{i}'$, show that the magnetization will precess about the rf field that is linearly polarized along the x axis in the rotating frame, that is, $\vec{B}_1 = B_1 \hat{i}' = (B_1, 0, 0)$. *Hint:* Expand the x, y, and z torque equations using the relationship $\dfrac{dM_i}{dt} = \gamma(\vec{M} \times \vec{B})_i = \gamma(M_j B_k - M_k B_j)$ for i, j, and k equal to x, y, and z.

P20.24   Assume a tetrapeptide fragment from a protein dissolved in H_2O with the primary sequence Ala-Gly-Val-Ile. Using the information in Table 20.3, Appendix B, sketch the NOESY cross peak region for the amide protons if the tetrapeptide is in an α helical secondary structure. Repeat the exercise if the same peptide is in a β-sheet structure.

P20.25   Assume four contiguous amino acids in a ^{13}C and ^{15}N enriched protein with the following chemical shift information (ppm):

$^1H_i^N$	7.2	8.1	7.5	8.3
$^{15}N_i$	116.2	120.5	123.7	118.3

Sketch the HSQC spectrum for this tetrapeptide.

P20.26   Suppose the following tetrapeptide sequence is observed in the α-helix of a protein: Phe-Val-Ala-Val. The two valine (Val) NH protons resonate at 8.40 and 8.34 ppm, the αH protons resonate at 4.18 and 4.14 ppm, the βH protons resonate at 2.10 and 2.14 ppm, and the four γ methyl groups resonate at 0.90, 0.93, 0.96, and 0.98 ppm. Assume the proton line widths are 4 Hz. The proton Larmor frequency is 600 MHz. Using a combination of proton NOESY, COSY, and TOCSY experiments, explain how these resonances can be assigned to the two different valines. Assume the phenylalanine (Phe) and the alanine (Ala) protons resonate at the average chemical shifts listed in Table 20.2 in Appendix B.

P20.27   Calculate the dipolar coupling constant between a ^{13}C spin and a ^{19}F spin separated by 5 Å. Give your answer in radians per second. Calculate the dipolar splitting of the ^{13}C line by the dipolar coupling to this ^{19}F spin if the angle between the internuclear vector and the magnetic field is $\theta = \pi/3$.

P20.28   Prove that the anisotropic chemical shift and the direct dipolar coupling, averaged over all angles θ and φ are both equal to zero.

P20.29   Consider a ^{13}C spin with principal values $\sigma_1 = -81$ ppm, $\sigma_2 = 4$ ppm, and $\sigma_3 = 78$ ppm and with a Larmor frequency of 125 MHz.

a. Calculate the chemical shift anisotropy, the chemical shift asymmetry, and the isotropic chemical shift

b. Calculate the chemical shift if the magnetic field is oriented at $\theta = \pi/4$ and $\varphi = 0$. Give your answer in rad s^{-1}.

P20.30   Calculate the ratio of the initial slopes of the NOE buildup curve for two amide proton pairs on adjacent amino acids, if one pair is located in an α-helix of a protein, and the other pair is located in a β-sheet of the same protein. Assume the amide proton pairs cross relax as isolated spin pairs.

Web-Based Simulations, Animations, and Problems

W20.1   Suppose a flexible molecule alternates between two structural forms. As the molecule "jumps" between the two forms the electronic environment of a proton in this molecule changes between two discrete values. We can portray this process as a jump of the proton between two structural sites with different chemical shifts. In this problem we will explore the proton NMR line shape as a function of the difference between the chemical shifts and the jump rate between the structural forms.

W20.2   The rules outline in section 20.5 for the multiplet structure in the spectra of coupled nuclear spins only apply if the difference between the chemical shifts of the coupled spins is much greater than the scalar coupling constant. In this problem we explore the structure of the NMR spectrum of two coupled spins as a function of chemical shift difference.

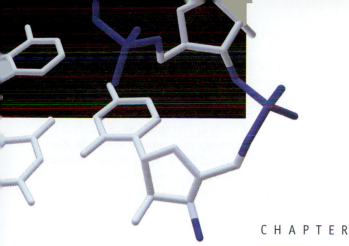

21

The Structure of Biomolecules at the Nanometer Scale: X-Ray Diffraction and Atomic Force Microscopy

Nanometer-scale structural probes are particularly useful in establishing a relationship linking the structure of a molecule to its chemical reactivity. X-ray crystallography, which relies on the diffraction of X-rays by a crystalline structure, allows the structure of a molecule to be imaged at the atomic level. However, a necessary condition is that the molecule forms macroscopic crystals. Atomic force microscopy (AFM) is a direct imaging probe that has a lower resolution than X-ray crystallography. However, AFM can be used under physiological conditions, and can be used to observe biochemical processes in real time, to carry out molecular recognition imaging, and for nanodissection. X-ray diffraction and AFM are complementary tools for carrying out structural studies in systems of biochemical interest.

21.1 Unit Cells and Bravais Lattices

Because the structure of a molecule is intimately linked to its function, chemists make every effort to determine the structure of molecules of interest. X-ray diffraction is one of the major tools that we have at our disposal for this purpose. Using this technique, the structure of molecules can be determined provided that they form a regular arrangement in three-dimensional space over macroscopic length scales. Before discussing X-ray diffraction, we discuss the most important features of crystal structures, namely, unit cells and Bravais lattices.

We first consider a simple crystalline structure that is exhibited by many elements in the periodic table. Figure 21.1 shows the face-centered cubic structure of nickel, which is the most stable phase of nickel from very low temperatures up to the melting point of 1728 K.

Any crystalline solid consists of a regular array of identical units, which may be one or several atoms, a molecule, or several molecules. The unit that is regularly repeated in space is called the **motif,** and the array of equivalent points in three dimensions that defines the spatial relation between the units is called the **lattice.** It is useful to think of a macroscopic crystal in terms of a repeat unit having molecular dimensions. However, the sample depicted in Figure 21.1 is not a particularly useful unit from which a sample of arbitrary size can be generated, because it is much larger than necessary. The choice of the repeating unit, which is called the **unit cell,** is not unique as is illustrated in Figure 21.2 for a two-dimensional lattice.

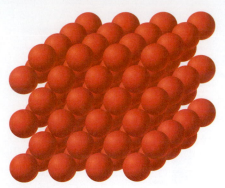

FIGURE 21.1

A portion of a crystal of nickel is shown that is large compared to atomic dimensions, but small compared to a macroscopic crystal.

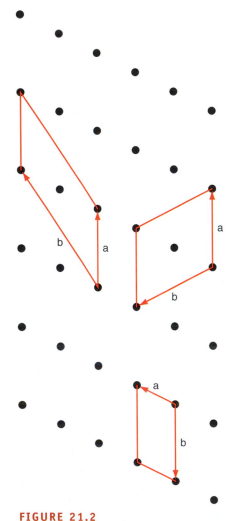

FIGURE 21.2

Three different choices for a unit cell in a two-dimensional lattice are shown. Which are primitive unit cells?

Any of the three unit cells in Figure 21.2 will reproduce the crystal by translation along the two axes indicated for each unit cell. The **primitive unit cell** has atoms located only at the corners of the unit cell. Although is it always possible to find a primitive unit cell, it is often more useful to choose a nonprimitive cell as shown in Figure 21.3 for a face-centered cubic lattice. The smallest unit cell for the face-centered cubic lattice that also has the symmetry of a cube is shown in Figure 21.3. The atoms at the corners of each face-centered cubic unit cell are shared by eight neighboring unit cells, and the atoms at the center of each face are shared by two adjacent unit cells. Taking this sharing into account, each unit cell contains four atoms. We could have chosen a primitive unit cell that contains only one atom. However, the faces of this unit cell are no longer mutually perpendicular and it does not have the full symmetry of the cube, which makes it less convenient to work with. As the earlier nickel example shows, the unit cell may contain several atoms even for a monatomic solid. The unit cell for the crystalline form of a biomolecule may contain thousands of atoms.

Although it is not immediately obvious, it can be shown that all crystalline solids can be described by only 7 fundamental lattices, each of which can be depicted as a periodic array of unit cells. These unit cells, and their equivalent points (indicated by circles), are shown in Figure 21.4. Primitive (P) unit cells have a single equivalent point in the cell. Several of the fundamental lattices have additional equivalent points at the center of the basic unit (I) or at the center of each face of the basic unit (F), or at the center of some faces of the basic unit (C) giving rise to the 14 lattices shown in Figure 21.4. They are called the **Bravais lattices** after the 19th-century French physicist August Bravais. The Bravais lattices do not represent atoms or molecules. They are the "anchoring points" for the lattice motif, which may consist of thousands of atoms.

The seven fundamental lattices differ in their symmetry, which places constraints on the relative length of the three unit cell vectors, a, b, and c, and the relative values for the three angles between the axes. By convention, $a \geq b \geq c$, and the angles α, β, and γ are opposite the unit cell vectors **a**, **b**, and **c**, respectively. For the triclinic lattice, $a \neq b \neq c$ and $\alpha \neq \beta \neq \gamma$. The monoclinic lattice is of higher symmetry. Although $a \neq b \neq c$, two of the angles are equal: $\alpha = \gamma = 90°$ and $\beta > 90°$. For the orthorhombic lattice, $a \neq b \neq c$ and all angles are equal to 90°. For the tetragonal lattice, $a = b \neq c$ and $\alpha = \beta = \gamma = 90°$. For the trigonal lattice, $a = b = c$, and $\alpha = \beta = \gamma \neq 90°$. For the hexagonal lattice, $a = b = c$, and $\alpha = \beta = 90°$; $\gamma = 120°$. For the cubic lattice, $a = b = c$ and $\alpha = \beta = \gamma = 90°$.

Figure 21.5 demonstrates how the presence of symmetry elements such as axes of rotation and mirror planes set requirements on the relative length of the three axes and the value of the three angles for the specific case of a tetragonal lattice. The tetragonal lattice has two kinds of symmetry elements, a fourfold rotation axis and three mutually perpendicular mirror planes as shown in Figure 21.5. A fourfold rotation axis has the property that a rotation of 90° about the indicated axis will leave the unit cell of the lattice in a position that is indistinguishable from its original position. The same is true for rotations of 180°, 270°, and 360°. The original lattice and that generated by the symmetry operation are only indistinguishable if the lengths of the top face, a and b, are equal (see Figure 21.4) and all angles are equal to 90°. Convince yourself that this statement is true.

We next consider the symmetry described by reflection through a mirror plane. A lattice contains a mirror plane if reflection of the part of the unit cell on one side of a plane through the plane generates the part of the unit cell on the other side of the plane. Convince yourself that the mirror planes only exist if a and b are equal and all angles are equal to 90°. The tetragonal lattice has the four symmetry elements described earlier. By comparison, the triclinic lattice has no rotation axes or mirror planes.

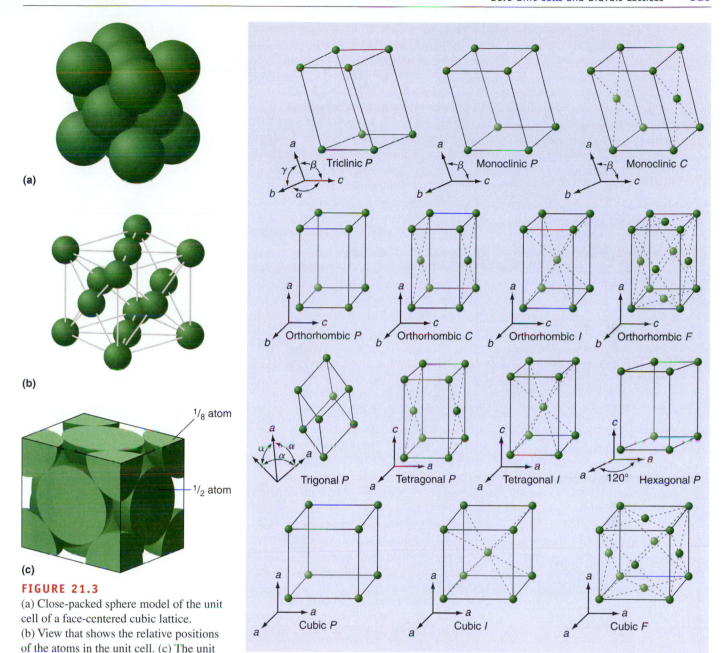

(a)

(b)

1/8 atom

1/2 atom

(c)

FIGURE 21.3
(a) Close-packed sphere model of the unit cell of a face-centered cubic lattice.
(b) View that shows the relative positions of the atoms in the unit cell. (c) The unit cell contains four atoms as the atoms at the corners and on the faces are shared among adjoining unit cells.

FIGURE 21.4
The 14 Bravais lattices include 7 fundamental lattices, some of which can have atoms at the center of the cell, or at the center of some or all faces of the cell.

It is important to realize that the spheres depicted in the Bravais lattices of Figure 21.4 indicate equivalent positions in the basic unit rather than atoms or molecules. As shown in Figure 21.6, the unit cell for the crystal structure of a particular molecule is generated by combining the appropriate Bravais lattice with a lattice motif placed at all equivalent positions in the Bravais lattice. The symmetry of the crystal structure generated by combining the indicated lattice motif with the tetragonal lattice may be lower than that of the lattice alone. Does the crystal structure shown in Figure 21.6 have a four-fold rotation axis? Does it have the three mutually perpendicular mirror planes shown in Figure 21.5?

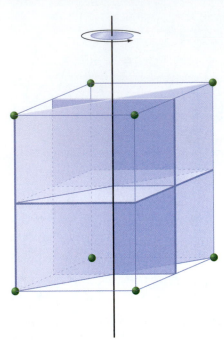

FIGURE 21.5

The unit cell of a tetragonal lattice has a fourfold rotation axis and only two unique mirror planes, because one of the mirror planes containing the rotation axis is produced from the other by rotation through 90°. What additional mirror plane(s) is (are) produced by applying the rotations to the indicated planes?

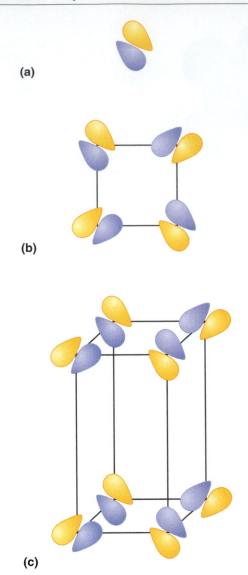

(a)

(b)

(c)

FIGURE 21.6

The unit cell is formed by combining the Bravais lattice with (a) a lattice motif. The lattice is combined with the lattice motif to form the crystal structure. (b) Top and (c) side views of the crystal structure are shown. The lattice motif lies in the plane shown in (b).

21.2 Lattice Planes and Miller Indices

To introduce the concept of lattice planes, we next consider the unit cell of the three cubic lattices in more detail. Each unit cell has a number of equivalent positions, and the number of these positions depends on whether the unit cell is primitive, body centered, or face centered. The equivalent positions can be expressed in terms of the axial lengths. Taking the origin of the coordinate system to be at the bottom left and front corner of the unit cell, the coordinates of this position can be written a $(0a, 0b, 0c)$ or more compactly, $(0,0,0)$.

EXAMPLE PROBLEM 21.1

What are the coordinates of the points for the primitive cubic unit cell shown in Figure 21.4? Only points that are not generated from other points by translation using the unit cell vectors **a, b,** and **c** are called equivalent points. What are the equivalent points for the unit cell of the primitive cubic Bravais lattice?

Solution

The eight points in the primitive cubic unit cell shown in Figure 21.4 are (0,0,0), (1,0,0), (0,1,0), (0,0,1), (1,1,0), (1,0,1), (0,1,1), and (1,1,1). The set of all positive and negative integral multiples of these coordinates generates the infinite lattice. All of these points are generated from (0,0,0) by addition of one or more of the unit cell vectors. Therefore, the primitive cubic unit cell has only one equivalent point, namely, (0,0,0).

Another way to view the structure of a macroscopic three-dimensional crystal is to construct it by stacking a set of equally spaced identical lattice planes parallel to one another where the equivalent positions of the Bravais lattice lie in the planes. To illustrate this alternate model of a crystal structure, we consider a three-dimensional crystal, made up of planes perpendicular to the plane of the page, as shown in Figure 21.7. Note that there is no unique set of planes that defines the crystal structure; for an infinitely large crystal, an infinite number of choices are possible, all of which can represent the same lattice equally well. As we will see in the next section, depicting a crystal by lattice planes is particularly useful in understanding X-ray diffraction. Figure 21.8 shows some of the sets of planes that can be used to represent an infinite three-dimensional cubic lattice.

We next discuss the nomenclature used to describe lattice planes. In naming a set of parallel planes, we consider the plane that is closest to the origin of the lattice. If the intersection of the plane with the three axes **a, b,** and **c** of length a, b, and c is given by a', b', and c', which are generally fractional multiples of the corresponding unit cell vector, then the plane is designated by the three **Miller indices** h, k, and l defined by

$$h = \frac{a}{a'}, \quad k = \frac{b}{b'}, \quad \text{and} \quad l = \frac{c}{c'} \tag{21.1}$$

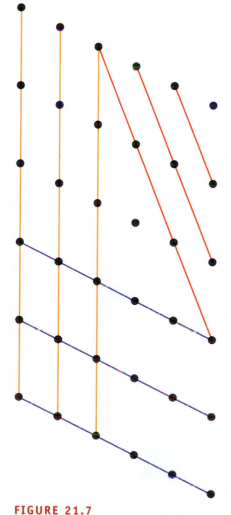

FIGURE 21.7
Three different sets of planes are shown that reproduce all points of the lattice.

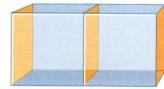
100 planes

110 planes

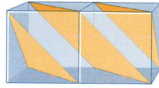
111 planes

FIGURE 21.8
A cubic lattice can be represented by a set of lattice planes, some of which are shown.

The exception to the previous statement is if the plane contains the unit cell vector. For this case, the intersection is at ∞. For example the (100) plane (no commas between the indices) intersects the **a** axis at the distance a, and the **b** and **c** axes at ∞. Therefore, the Miller indices for this plane are $h = a/a = 1$, $k = b/\infty = 0$, and $l = c/\infty = 0$. The other members of the set of $(h\,k\,l)$ planes are obtained from the first by translation along the unit cell axes. The distance between adjacent members of the set of $(h\,k\,l)$ planes, d, is determined by the Bravais lattice. It can be shown that for a cubic lattice of axis length a, distance d is given by

$$\frac{1}{d^2} = \frac{h^2 + k^2 + l^2}{a^2} \tag{21.2}$$

The integers h, k, and l can be positive or negative. If they are negative, the minus sign is indicated as a horizontal line above the integer as in the $(\bar{1}10)$ or $(1\bar{2}1)$ planes.

EXAMPLE PROBLEM 21.2

The crystalline structure of copper has a face-centered cubic unit cell and the cell length is $a = 0.361$ nm. Make a drawing showing the orientation of the (110) planes relative to the unit cell and calculate the distance between adjacent (110) planes.

Solution

The unit cell is indicated in the following figure, where the blue atom is at the point with the coordinates 0, 0, 0.

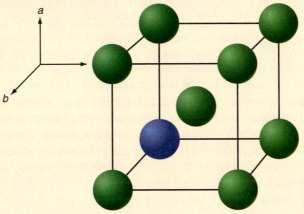

The position of the (110) plane in the unit cell is indicated here:

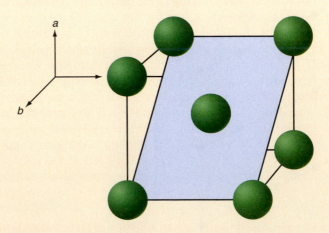

The distance between adjacent members in the set of planes described by the indices $h\,k\,l$ is

$$\frac{1}{d^2} = \frac{h^2 + k^2 + l^2}{a^2}$$

$$d = \sqrt{\frac{a^2}{h^2 + k^2 + l^2}} = \sqrt{\frac{(0.361 \times 10^{-9}\,\text{m})^2}{1^2 + 1^2}} = 0.255 \times 10^{-9}\,\text{m}$$

21.3 The Von Laue and Bragg Equations for X-Ray Diffraction

With this background on crystal structures, we next discuss the diffraction of X-rays from a crystal. If X-rays are incident on a single atom, they are scattered and the atom acts as a source of waves with spherical wavefronts. If X-rays are incident on a regular array of atoms as is the case for a crystal, the waves scattered from neighboring atoms will interfere constructively or destructively with one another as shown in Figure 21.7. Just as for diffraction from the double slit discussed in Section 12.6, constructive interference is observed if the difference in the path length of the spherical wave emanating from adjacent atoms is an integral multiple of the X-ray wavelength λ, which is typically in the range 0.07 to 0.15 nm.

To simplify the analysis of diffraction, we first consider a one-dimensional crystal that is periodic along the **c** axis as shown in Figure 21.9. For the X-rays incident perpendicular to the crystal, the condition for constructive interference is

$$l\lambda = c \cos \gamma \tag{21.3}$$

where l is an integer. If the X-rays are incident with an angle γ_0, the condition for constructive interference becomes

$$l\lambda = c(\cos \gamma - \cos \gamma_0) \tag{21.4}$$

as can be seen from Figure 21.10.

For a three-dimensional crystal, we can draw similar figures for the **a** and **b** axes, which generate the additional following conditions for constructive interference.

$$h\lambda = a(\cos \alpha - \cos \alpha_0) \tag{21.5}$$

$$k\lambda = b(\cos \beta - \cos \beta_0) \tag{21.6}$$

FIGURE 21.9
Diffraction is shown from a one-dimensional crystal. Constructive interference is observed between the waves scattered from the individual atoms if the path length differences are an integral multiple of the wavelength. The X-ray beam is incident perpendicular to the line of atoms.

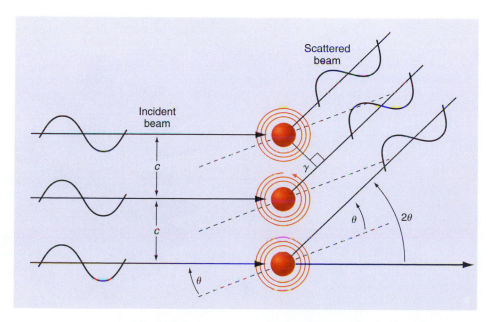

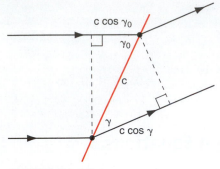

FIGURE 21.10
If the beam is incident at an angle other than 90°, the path difference before scattering must be included in the constructive interference condition.

In these equations, h, k, and l are integers. Equations (21.4) through (21.6), which are known as the **von Laue equations,** each describe a set of cones. For a one-dimensional crystal the set of cones, one for each value of l, is shown in Figure 21.11a.

EXAMPLE PROBLEM 21.3

Calculate the half-angles for the Laue cones for diffraction of an X-ray beam of wavelength 0.070926 nm from a one-dimensional crystal of periodicity 0.300 nm along the **c** direction. The angle $\gamma_0 = \pi/2$. Make a drawing showing the crystal direction, the X-ray beam direction, and the Laue cones labeled with their l values.

Solution

From Equation (21.4),

$$l\lambda = c(\cos \gamma - \cos \gamma_0)$$

$$\cos \gamma = \frac{l\lambda}{c} + \cos \gamma_0$$

$$= \frac{l \times 0.070926 \times 10^{-9}\,m}{0.300 \times 10^{-9}\,m} + \cos \frac{\pi}{2} = \frac{l \times 0.070926 \times 10^{-9}\,m}{0.300 \times 10^{-9}\,m}$$

Only those l values for which $-1 \le \cos \gamma \le 1$ are allowed. Calculating $\cos \gamma$ for various values of l gives the following results:

l	$\cos \gamma$	γ (degrees)
0	0	90
1	0.236	76.3
2	0.473	61.8
3	0.709	44.8
4	0.946	19.0
5	1.182	—
6	1.419	—
−1	−0.236	103.7
−2	−0.473	118.2
−3	−0.709	135.2
−4	−0.946	161.0
−5	−1.182	—
−6	−1.419	—

It is seen that only the Laue cones corresponding to l values of 0 and ±1, ±2, ±3, and ±4 are observed. The cones are shown below.

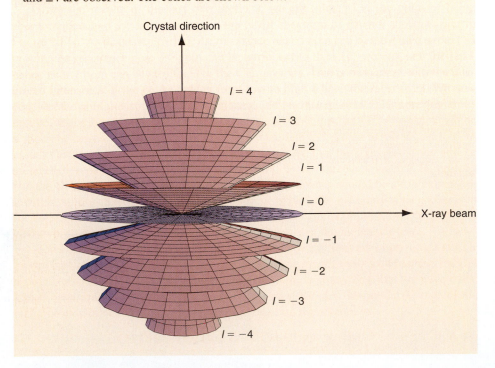

For an X-ray detector that moves on a spherical surface centered at the crystal, constructive interference is observed only along equally spaced circles such as the lines of latitude on a globe as shown in Figure 21.12. A diffraction pattern can be imaged by using photographic film or a detector. As Figure 21.11 shows, the diffraction pattern for a one-dimensional crystal consists of a series of equally spaced lines.

We next increase the number of dimensions to two and assume that the lattice is primitive cubic. For a two-dimensional crystal in the x-z plane, Equations (21.4) and (21.5) must be satisfied simultaneously. The solution corresponds to the intersection of two sets of cones, the axes of which are rotated by 90° as shown in Figure 21.12. The lines of intersection describe a set of lines normal to the spherical detector surface that intersect it in a regular array of points as shown in Figure 21.13. Adding the third dimension to the crystal along the y axis means that Equations (21.4) through (21.6) must be simultaneously satisfied for constructive interference, which is geometrically equivalent to the intersection of three sets of mutually perpendicular cones. This condition gives a discrete set of points (diffraction spots) along the lines that define constructive interference for the two-dimensional crystal.

In seeing the effect of changes in the lattice constant or the X-ray wavelength on the angle at which constructive interference occurs, it is more useful to use the angle $2\theta = \pi/2 - \gamma$, which is the angle through which the X-ray is scattered rather than γ. As the wavelength of the X-rays is increased for diffraction from a given crystal, the pattern of lines shown in Figure 21.11 expands. This can be seen by making the substitution $\gamma = (\pi/2) - 2\theta$ and writing Equation (21.3) in the form

$$\cos\left(\frac{\pi}{2} - 2\theta\right) = \sin 2\theta = \frac{l\lambda}{c} \tag{21.7}$$

using the identity $\cos(x + y) = \cos x \cos y - \sin x \sin y$.

For a two-dimensional crystal, the diffraction spots appear for all wavelengths because the conditions for constructive interference are lines along which the two sets of cones intersect. The pattern of diffraction spots expands as the wavelength increases. However, for a three-dimensional crystal, the diffraction spots are observed only under special conditions. This is the case because the conditions for constructive interference are points along the lines of constructive interference for the two-dimensional case. Consequently, the pattern of diffraction spots for the three-dimensional crystal expands as the wavelengths increases, but the intensity of the diffraction spots is different from zero only for the special case in which Equations (21.4) through (21.6) are all satisfied.

Because structural determination using X-ray diffraction requires that the intensity of a large number of diffraction peaks be measured under the same conditions, an alternate manner of recording a diffraction pattern must be used. By appropriately rotating the crystal during the exposure to X-rays (the *precession method*, the *precission method,* or the *Weisenberg method*), it is possible to obtain a diffraction pattern in which constructive interference conditions are obtained for all values of two of h, k, and l, while the third value is held constant. The measurements are repeated for other values of the third index. A precession pattern showing the h, k diffraction features with $l = 0$ is shown in Figure 21.14. We return to this figure later and assign indices to the individual diffraction spots.

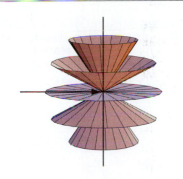

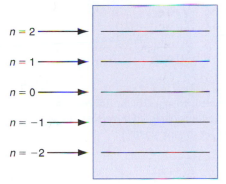

FIGURE 21.11
Consider the case in which the X-ray beam is incident perpendicular to the line of atoms. (Upper) The constructive interference condition is satisfied if the scattered beam lies on the surface of one of the depicted cones. (Lower) The cones intersect the spherical detector surface centered at the crystal in a set of parallel lines, one for each diffraction order.

$n = 2$
$n = 1$
$n = 0$
$n = -1$
$n = -2$

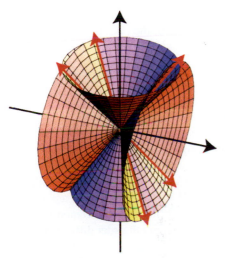

FIGURE 21.12
The constructive interference condition for a two-dimensional crystal is satisfied if the scattered beam lies on the intersection of the two sets of depicted cones. The black arrows are aligned along the mutually perpendicular rows of scattering centers in the plane of the two-dimensional crystal and five of the eight lines of intersection of the two cones are shown in red. To simplify the presentation, not all of the cones are shown.

FIGURE 21.13
The diffraction pattern of a two-dimensional crystal, the axes of which are perpendicular, is a rectangular array of points that we call diffraction spots.

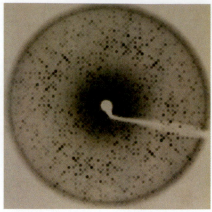

FIGURE 21.14
A diffraction image taken using the precession method is shown for the metalloprotein hemerythrin. All of the diffraction spots can be assigned indices h and k and $l = 0$. (Source: Courtesy of Ronald Stenkamp and Larry Sieker.)

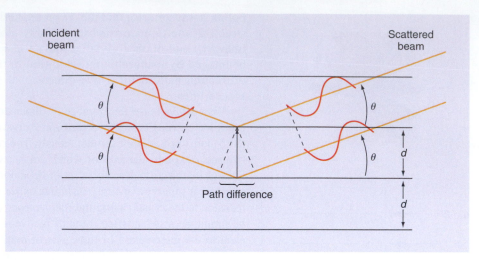

FIGURE 21.15
Diffraction from a set of parallel crystal planes is depicted. To simplify the drawing, the lattice motif is not shown. Diffraction or constructive interference is observed if the path difference between adjacent planes is an integral multiple of the X-ray wavelength.

An alternate and very useful way of describing diffraction of X-rays from a crystal in terms of scattering from the (hkl) lattice planes was formulated by W. L. Bragg. We can add up the scattering that occurs from each atom in the crystal, ensuring that interferences are taken into account, and assign it to the plane to which the lattice motif is assigned as shown in Figure 21.15 for a monatomic solid. For a given set of (hkl) planes perpendicular to the plane of the page with a characteristic spacing of d, the condition for constructive interference becomes

$$n\lambda = 2d \sin \theta \qquad (21.8)$$

where n is an integer. Equation (21.8) shows that the scattering angle θ decreases as d increases and increases as λ increases.

Combining Equation 21.8 with the relationship between d and the indices h, k, and l, we obtain a relation between the angle θ and the indices of the (hkl) plane from which the scattering occurs. For the specific case of a cubic crystal,

$$\sin^2 \theta = \frac{n^2 \lambda^2}{4a^2} (h^2 + k^2 + l^2) \qquad (21.9)$$

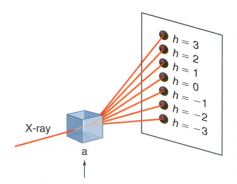

$h = 3$
$h = 2$
$h = 1$
$h = 0$
$h = -1$
$h = -2$
$h = -3$

X-ray

a

FIGURE 21.16
The diffraction pattern from the set of (100) planes is shown. Each spot corresponds to a different diffraction order.

The usefulness of the Bragg formulation is that Equation (21.9) provides a way to assign the spots observed in a diffraction pattern to particular lattice planes. There is a one-to-one association between the set of (hkl) planes and a single diffraction spot which we label (hkl). For example, the set of ($h00$) planes gives rise to the diffraction pattern shown in Figure 21.16 in which one spot is observed for each value of h. Note that the diffraction angle depends on the product n^2h^2 in Equation (21.9) rather than on n and h individually. Therefore, a spot labeled (hh nk nl), for which there is a common factor n between the indices, should be thought of as originating from nth-order diffraction from the (hkl) planes.

21.4 The Unit Cell Parameters Can Be Determined from a Diffraction Pattern

We next discuss how the molecular structure of the substance that has been crystallized is determined in an X-ray diffraction experiment. To do so, the information in the diffraction pattern must be extracted. There are two distinct sets of information in a diffraction

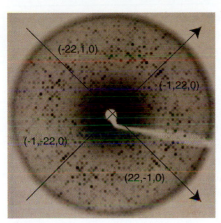

FIGURE 21.17
The diffraction pattern of Figure 21.14 with the indices of four planes giving rise to the diffraction spots is shown. On the basis of this indexing, the unit cell for hemerythrin has the dimensions $a = b =$ 8.66 nm. The diffraction pattern obtained by rotating the crystal by 90° gives $c =$ 8.08 nm. (Source: Courtesy of Ronald Stenkamp and Larry Sieker.)

FIGURE 21.18
Hemerythrin consists of eight identical subunits, each of which is made up of a polypeptide containing 113 amino acids, two iron atoms (red spheres), and a single oxygen atom bridging the iron atoms. The polypeptide folds into four α-helices that pack in an antiparallel arrangement forming the tertiary structure of the subunit. (Source: Courtesy of Ronald Stenkamp.)

pattern: the distribution of the diffraction spots in the pattern, and the relative intensity of the diffraction spots. In this section, we first discuss what can be learned from the distribution of the diffraction spots without a quantitative measurement of the intensities, and in the next section, we discuss what can be learned from the relative intensities.

We next determine the principal directions for the hemerythrin diffraction pattern of Figure 21.14. Hemerythrin is a protein used for oxygen transport or storage in marine worms. Amino acid residues in the protein bind to a dinuclear iron center bridged by a single oxygen atom. This complex reversibly binds a molecule of dioxygen, just as does the heme group in hemoglobin. The intensity pattern shows a fourfold rotation symmetry and therefore the principal axes of the diffraction pattern are perpendicular to one another, which is characteristic of a tetragonal lattice. We choose two perpendicular axes that pass through the center of the diffraction pattern and give a periodic unit in the diffraction pattern that is primitive. This square repeat unit is indicated in black at the center of Figure 21.17.

The observed spacing between adjacent diffraction spots along each of the three principal axes of the diffraction pattern can be used to obtain two of the unit cell lengths a, b, and c. Rotating the crystal by 90° will give an image analogous to Figure 21.17 showing the ($h0l$) or ($0kl$) diffraction spots, from which the third unit cell length can be obtained. Because the diffraction angle is inversely related to the lattice parameter [see Equation (21.8)], the principal axis along which the spacing of the spots is the shortest gives information on the longest of a, b, and c, which by convention is a. The measured angles between the three principal axes of the diffraction pattern can be used to determine α, β, and γ. If the diffraction pattern shows a fourfold rotation symmetry, the crystal lattice corresponds to one of the nine Bravais lattices in which $\alpha = \beta = \gamma = 90°$. For other lattice symmetries, the angles between the principal axes of the diffraction pattern are related to but are not equal to the angles α, β, and γ, and the spacing between the diffraction spots is not simply related to a, b, and c. Note that a knowledge of the unit cell parameters gives no information on the structure of the molecule of interest. The information on the molecular structure is obtained from the intensities of the diffraction spots. The structure of hemerythrin derived from the intensity of the diffraction spots shown in Figure 21.17 is shown in Figure 21.18.

21.5 The Electron Distribution in the Unit Cell Can Be Calculated from the Structure Factor

In general, a unit cell contains more than one atom, and may contain hundreds or even thousands of atoms for a biomolecule. As we will show in this section, although the incident X-ray beam is scattered from all the atoms in the crystal, only the scattering from an individual unit cell determines the relative intensities of the diffraction spots. The combined scattering from all unit cells in the macroscopic crystal is important because it amplifies the scattered signal from an individual unit cell sufficiently that it can be measured. In this section, we calculate the **structure factor,** which is the combined amplitude of the superposition of waves scattered by all atoms in a unit cell into the hkl diffraction spot.

Assume initially that the unit cell contains two atoms and consider the scattering from the (001) planes of this lattice, which are spaced a distance c apart. As shown in Figure 21.19, we choose the origin of the coordinate system so that one atom is at the origin. The second atom lies in a plane that is physically separated from, but parallel to, the (001) plane at a distance zc, where $0 < z < 1$. This situation corresponds to a lattice with a motif that extends beyond the planes containing the lattice points. We refer to the plane containing the first and second atoms as the first and intermediate planes, respectively. For first-order diffraction from the set of (001) planes, the path difference for waves scattered from adjacent planes of the set is λ.

As the sine function with which we represent the X-ray wave goes through one period in the length λ, the path difference between adjacent (001) planes corresponds to a

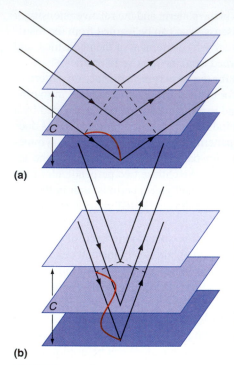

(a)

(b)

FIGURE 21.19

The top and bottom planes in (a) and (b) are (001) planes, and the middle plane is located somewhere between adjacent (001) planes. (a) At an angle corresponding to in-phase scattering of the two (001) planes for $n = 1$, the difference in path length is one wavelength, so that the difference in phase angle is 2π. (b) At an angle corresponding to in-phase scattering of the two (001) planes for $n = 2$, the difference in path length is two wavelengths, so that the difference in phase angle is 4π.

difference in phase, ϕ, of 2π in units of radians. Because only differences in phase are important, we can assign a phase of zero to scattering from the first (001) plane, and with this choice, the phase for scattering from the second (001) plane is 2π. What phase is associated with scattering from the intermediate plane containing the second atom? As can be seen from the following argument, the phase is given by

$$\phi = 2\pi z \qquad (21.10)$$

The preceding equation can be made plausible by considering a few cases. If the intermediate plane were at $z = 0$, it would have the phase $\phi = 0$ corresponding to that plane. If it were at $z = c$, it would have the phase $\phi = 2\pi$ corresponding to that plane. If it were located midway between the two planes (001) at $z = c/2$, the second plane would be exactly out of phase with the two (001) planes, corresponding to $\phi = \pi$. Equation (21.10) predicts exactly this behavior.

We next consider second-order diffraction from the (001) planes. For this case, the phase associated with the second (001) plane is $\phi = 4\pi z$ because the path difference between this plane and the first (001) plane is 2λ. With these two examples, we can generalize Equation (21.10) to give the phase corresponding to the lth-order diffraction from the second (001) plane, $\phi = l \times 2\pi z$. This argument can be extended to account for the jth atom at the position (x, y, z), where $0 < x < a$, $0 < b < y$, and $0 < z < c$, and an arbitrary diffraction feature with the indices (hkl). In this case, the phase of the X-ray scattered from the jth atom, where j is a running index for all atoms in the unit cell, is given by

$$\phi_{hkl}(j) = 2\pi (hx_j + ky_j + lz_j) \qquad (21.11)$$

Equation (21.11) gives the phase associated with a single atom for the hkl diffraction peak. It is important to note that the phase angle for each of the atomic scattering factors depends only on the position of the atom in the unit cell as shown by Equation (21.11).

The next step in calculating the structure factor is to combine the amplitudes of the X-ray waves scattered from each atom in the unit cell to give the amplitude of the resultant wave at the detector, taking the relative phases of each atom into account. It is convenient to give the value zero to the phase of the wave scattered from the atom at the origin measured at the detector. Doing so amounts to a shift in the phase of the waves scattered from each atom in the unit cell by the same amount, which does not affect their interference. We designate the amplitude of the wave scattered from the atom at the origin measured at the detector by f_1 and add in the amplitudes of all the outgoing X-ray waves from one unit cell, whereby it is essential to take their relative phases into account. The resulting amplitude F_{hkl} is given by

$$F_{hkl} = f_1 + f_2 e^{-2\pi i(hx_2 + ky_2 + lz_2)} + f_3 e^{-2\pi i(hx_3 + ky_3 + lz_3)} + \cdots$$

$$= \sum_{j=1}^{N} f_j e^{-2\pi i(hx_j + ky_j + lz_j)} = |F_{hkl}| e^{-i\delta_{hkl}} \qquad (21.12)$$

where N is the number of the atoms in the unit cell, and δ_{hkl} is the phase angle corresponding to the structure factor F_{hkl}. Because each of the terms in Equation (21.12) is a complex number, the sum is best depicted as vector addition in the complex plane as shown in Figure 21.20. Although the phase for a single atom depends only on the position of that atom in the unit cell, no analogous statement can be made for δ, the phase angle of F_{hkl}.

The scattering factor f_j is a measure of the ability of the jth atom to scatter X-rays and is proportional to the number of electrons in that atom. It has a maximum value equal to the number of electrons in the atom, and is a function of the variable $\sin \theta/\lambda$. The dependence of the scattering factor of carbon on $\sin \theta/\lambda$ is shown in Figure 21.21.

Equation (21.12) defines the structure factor F_{hkl}, which plays a central role in the formulation of the theory of X-ray diffraction. To this point, we have assumed that X-rays are scattered from atoms modeled as point masses. A more realistic model is that X-rays are scattered by electrons, and that the electron density, ρ, in the unit cell is a continuous function of the coordinates x, y, and z. Rather than summing over atoms, F_{hkl} is obtained by integrating over the unit cell as shown in Equation (21.13):

$$F_{hkl} = \iiint \rho(x, y, z)e^{-2\pi i(hx+ky+lz)} \, dx \, dy \, dz \qquad (21.13)$$

Because the electron density has the periodicity of the crystal lattice, it is convenient to write it as a Fourier series (see Appendix A, Math Supplement, for a discussion of Fourier series) as shown in Equation (21.14):

$$\rho(x,y,z) = \sum_{\alpha} \sum_{\beta} \sum_{\gamma} C_{\alpha\beta\gamma} e^{-2\pi i(\alpha x+\beta y+\gamma z)} \qquad (21.14)$$

In Equation (21.14), α β, and γ are integers that take on values between $-\infty$ and ∞ in the summation. To determine the as yet undefined coefficients $C_{\alpha\beta\gamma}$, we substitute the expression for $\rho(x,y,z)$ of Equation (21.14) into Equation (21.13). The result is

$$F_{hkl} = \iiint \sum_{\alpha} \sum_{\beta} \sum_{\gamma} C_{\alpha\beta\gamma} e^{-2\pi i[(\alpha+h)x+(\beta+k)y+(\gamma+l)z]} \, dx \, dy \, dz \qquad (21.15)$$

This complicated-looking equation can be simplified without a calculation by realizing that the individual terms in the summation cancel each other on integration over the unit cell except for the single term for which the exponent is zero, namely, $\alpha = -h$, $\beta = -k$, and $\gamma = -l$. This is the case because for every wave of the form $e^{-2\pi i[(\alpha+h)x+(\beta+k)y+(\gamma+l)z]}$, there is another wave that is exactly out of phase with it, so that these two waves sum to zero. Therefore, Equation (21.15) can be simplified to

$$F_{hkl} = \iiint C_{\underset{h\,k\,l}{---}} \, dx \, dy \, dz = C_{\underset{h\,k\,l}{---}} V \qquad (21.16)$$

where the horizontal lines over h, k, and l indicate that the integers are negative, and V is the volume of the unit cell. Equation (21.16) gives an expression for the $C_{\alpha\beta\gamma}$ in terms of the F_{hkl}, which can be substituted in Equation (21.14). The result is

$$\rho(x,y,z) = \frac{1}{V} \sum_{h} \sum_{k} \sum_{l} F_{hkl} \, e^{2\pi i(hx+ky+lz)} \qquad (21.17)$$

Because each of the summations is over all integers between $-\infty$ and ∞, we can use the indices h, k, and l rather than $-h$, $-k$, and $-l$ in the summations. Equation (21.17) shows that $\rho(x,y,z)$ can be calculated if we can experimentally determine the F_{hkl} for a large number of diffraction spots.

The goal of X-ray crystallography is the inverse to the problem just discussed, namely, to determine the positions of atoms in the unit cell from a diffraction experiment. This is done by determining $\rho(x,y,z)$, which has sharp maxima at the atomic positions. Equation (21.17) provides a way to reach this goal. Unfortunately, what is measured in an X-ray diffraction experiment is not the amplitude of the diffracted wave, F_{hkl}, but its intensity, I_{hkl}, where the two quantities are related by Equation (21.18):

$$I_{hkl} = |F_{hkl}|^2 \qquad (21.18)$$

Because F_{hkl} is a complex number that can be represented in the form $|F_{hkl}|e^{-i\phi}$ where ϕ is the phase,

$$I_{hkl} = F_{hkl}^* \times F_{hkl} = ZF_{hkl}|e^{i\delta} \times ZF_{hkl}|e^{-i\delta} = ZF_{hkl}|^2 \qquad (21.19)$$

Equation (21.19) shows that the phase information that is crucial in superposing the scattering from different volume elements of the electron density in the unit cell is lost if only the intensity can be measured.

To illustrate how essential the phase information is in calculating $\rho(x,y,z)$ from an X-ray diffraction experiment, we carry out a calculation to determine the F_h for a one-dimensional crystal with a periodicity of 1.00 nm in which one carbon atom is placed at

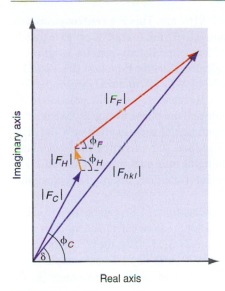

FIGURE 21.20

The scattering factor F_{hkl} can be represented as a vector in the complex number plane. It arises from vector addition of the individual scattering factors of the atoms in the unit cell, each of which has a magnitude proportional to its atomic number, and a phase determined only by the position of the atom in the unit cell. The superposition is shown for C, H, and F atoms.

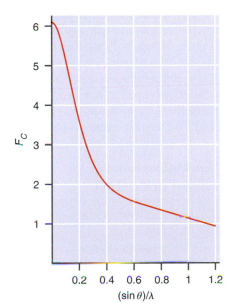

FIGURE 21.21

The magnitude of the scattering factor for atomic carbon is shown as a function of $(\sin \theta)/\lambda$.

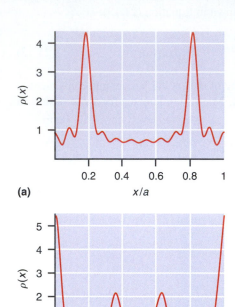

(a)

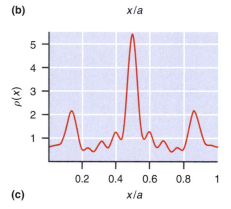

(b)

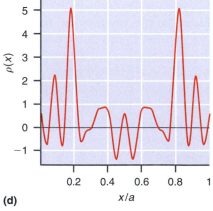

(c)

(d)

FIGURE 21.22

The electron density function calculated from the F_h in Table 21.1 is shown (a) for the correct phases, (b) for all phases set equal to 0, (c) for the phases set alternately equal to 0 and π, and (d) for the correct phases but with all f_C set equal to 6.

0.1833 nm and a second carbon atom is placed at 0.8167 nm. This is a **centrosymmetric unit cell,** because in a unit cell of length 1.00. nm, 0.8167 nm is equivalent to -0.1833 nm, making the two atoms at -0.1833 and 0.1833 nm equidistant from the origin. The X-ray wavelength is 0.15418 nm. We then calculate $\rho(x)$ using Equation (21.17). This calculation is based on an example in Section 8.10 of *X-Ray Structural Determination: A Practical Guide* by G. H. Stout and L. H. Jensen. Because we know the position of the atoms, this calculation gives us both the magnitude and the phase for the F_h. We next alter the phases for the F_h in order to mimic the experimental situation that only the magnitude of the F_h can be obtained from experiment, and show that the $\rho(x)$ calculated using the incorrect phases no longer gives meaningful information on the positions of the atoms. These calculations show that the phase information is essential in calculating structures from diffraction experiments.

Using the given values for d and λ, Bragg's law [Equation (21.8)] can be used to calculate $\sin\theta$ for the allowed values of h. The scattering factor f_C can be calculated for each value of h using the dependence on $(\sin\theta)/\lambda$ shown in Figure 21.21. Finally, the F_h can be calculated using Equation (21.12). The results are summarized in Table 21.1, and sample calculations are shown in Example Problem 21.4. Because the unit cell is centrosymmetric, the phase factors, $e^{-i\phi}$, for the F_h are all either $+1$ or -1, corresponding to phase angles of 0 or π, respectively. As Equation (21.22) shows, the function $\rho(x)$ can be calculated from the F_h and is shown in Figure 21.22a. The calculated electron density $\rho(x)$ is sharply peaked at $x = 0.1833$ nm and $x = 0.8167$ nm where the atoms were placed. The oscillations in $\rho(x)$ and the width of the peaks representing the atoms have their origin in the limited number of terms in the Fourier series (see the discussion of Figure A.7 in Appendix A, Math Supplement).

Because the phases cannot be determined from experiment, we simulate this case by guessing their values. Figure 21.22b shows the calculated function $\rho(x)$ if all phase factors are set equal to $+1$ (all the ϕ are 0), and Figure 21.22c shows $\rho(x)$ if the phase factors are alternately $+1$ and -1 (the ϕ are alternately 0 and π). For all $\phi = 0$, the calculated $\rho(x)$ erroneously suggests that there is an atom at $x = 0$, and two atoms with fewer electrons at $x = 0.35$ nm and $x = 0.65$ nm. For ϕ alternately 0 and π, one strong and two weak peaks are observed in $\rho(x)$ in the 100-nm period, incorrectly suggesting that there are three atoms in the unit cell, one of which has a larger number of electrons than the other two. Figure 21.22d shows the calculated function $\rho(x)$ using the correct phase factors, but setting all the scattering factors equal to $+6$. This is equivalent to saying that the positions of the atoms in the unit cell are known, but their identity is not known. It is seen that the peaks in $\rho(x)$ are found at the correct positions, although additional positive and negative peaks are found between the correct peaks. Negative peaks are unphysical because the electron density only has positive values. What can be concluded from Figure 21.22? We see that unless the phases are

TABLE 21.1 **Calculations for f_C and F_h for Diffraction from the One-Dimensional Lattice Discussed in the Text**

h	$\sin\theta$	f_C	F_h
0	0	6.00	12.0
±1	0.077	5.63	4.58
±2	0.154	4.89	−6.54
±3	0.231	4.23	−8.04
±4	0.308	3.65	−0.769
±5	0.385	3.15	5.44
±6	0.463	2.72	4.39
±7	0.504	2.35	−0.968
±8	0.617	2.04	−3.99
±9	0.694	1.79	−2.11
±10	0.771	1.59	1.58
±11	0.848	1.43	2.85
±12	0.925	1.31	0.818

known, the structure cannot be determined from diffraction data. We also see that knowing the correct phases is more important than knowing the correct scattering factors, because $\rho(x)$ in Figure 21.22d is more nearly correct than in Figures 21.22b and 21.22c.

EXAMPLE PROBLEM 21.4

Calculate F_h for $h = 0$, $+3$ and -3 for diffraction from the one-dimensional lattice discussed in the text. Take into account the dependence of f_C on the scattering angle shown in Figure 21.21. Assume that the dependence of f_C on the variable $\sin\theta/\lambda$ can be approximated by $f_C\left(\dfrac{\sin\theta}{\lambda}\right) = -1.961 + 17.47e^{-\sin\theta/\lambda}$
$- 9.051e^{-(\sin\theta/\lambda)^2}$ for $\sin\theta/\lambda > 0$ and $f_C = 6.00$ for $\sin\theta/\lambda = 0$ In this equation, λ is expressed in angstroms ($1\ \text{Å} = 10^{-10}$m).

Solution

We first calculate f_C for $h = 0$, $+3$ and -3. For $h = 0$, $\sin\theta/\lambda = 0$, and $f_C = 6.00$. For $h = +3$ and -3, the value of $\sin\theta/\lambda$ is the same. Therefore, f_C has the same value for $h = +3$ and -3.

$$\sin\theta/\lambda = 0.231/1.5418 = 0.1498$$

$$f_C\left(\frac{\sin\theta}{\lambda}\right) = -1.961 + 17.47e^{-0.1498} - 9.051e^{-(0.1498)^2} = 4.23$$

We next calculate F_h for $h = 0$:

$$F_h = f_C(e^{-2\pi i(hx_1)} + e^{-2\pi i(hx_2)})$$

$$F_0 = 6.00\left(e^{-2\pi i(hx_1)} + e^{-2\pi i(hx_2)}\right) = 6.00\left(e^{-0} + e^{-0}\right) = 12.00$$

For $h = +3$:

$$F_3 = f_C\left(e^{-2\pi i(hx_1)} + e^{-2\pi i(hx_2)}\right)$$

$$F_3 = 4.23\left(e^{-2\pi i 3 \times 0.1833} + e^{-2\pi i 3 \times 0.8167}\right)$$

$$= 4.23[\cos(2\pi \times 3 \times 0.1833) + i\sin(2\pi \times 3 \times 0.1833)$$

$$+ \cos(2\pi \times 3 \times 0.8167) + i\sin(2\pi \times 3 \times 0.8167)]$$

$$= 4.23[0.951 + 0.308i + 0.951 - 0.308i] = -8.04$$

The contribution of the sine terms has been canceled; therefore, F_h is a real number. This is the case because the two C atoms are equidistant from the origin. For F_h to be real, the phase angle must be either zero in which case $e^{-i\phi} = e^0 = 1$, making F_h positive, or π, in which case $e^{-i\phi} = e^{-i\pi} = -1$, making F_h negative. Because the sine terms do not contribute to F_h and $\cos\theta = \cos(-\theta)$, F_h also has the value -8.04 for $h = -3$.

21.6 Solutions to the Phase Problem

Although the phases δ_{hkl} corresponding to each F_{hkl} cannot be determined directly from experiment, several methods have been developed to extract them from the diffraction intensities. We first discuss the **direct methods**, developed by a number of scientists including Herbert Hauptman and Jerome Karle, who were awarded the 1985 Nobel Prize in chemistry for their work.

Hauptman and Karle realized that the phase information was contained in the measured intensities if a realistic physical model for the electron distribution in a unit cell is imposed. For example, one constraint imposed by reality is that the electron density must be positive everywhere, which severely restricts the values of the possible phases. This constraint also leads to inequality relationships between the known intensities and the desired phases, and further inequality relationships arise through the symmetry present in the crystal. A second constraint on the electron density is imposed by the fact that molecules consist of atoms. Therefore, the electron density function must reach a maximum value at the position of atoms and fall to small values between atoms. It turns out that these constraints are, *in principle,* sufficient to allow the phases to be calculated from measured intensities. Because the number of unknown phases is smaller than the number of values of I_{hkl} that can be measured, the problem is mathematically overdetermined. Although the solution of the problem is mathematically difficult, algorithms have been developed that start with a random distribution of atoms in the unit cell and converge to a well-defined structure using only the measured intensities for moderately sized molecules. Algorithms utilizing these constraints have been shown to converge for molecules containing more than 1000 nonhydrogen atoms. The reference to nonhydrogen atoms arises because hydrogen atoms contain only one electron. Therefore, they scatter very weakly and contribute much less to $|F_{hkl}|$ and δ than atoms with a larger atomic number as can be seen in Figure 21.20.

A second method for solving the phase problem is **multiple isomorphous replacement (MIR).** The principle behind MIR is that atoms with a large atomic number such as Br or U contribute much more to F_{hkl} and δ_{hkl} than do atoms with a small atomic number such as C, N, or O. This is the case because the magnitude of the scattering factor is proportional to the atomic number. Therefore, a sum such as

$$F_{hkl} = f_1 + f_2 e^{-2\pi i(hx_2+ky_2+lz_2)} + f_3 e^{-2\pi i(hx_3+ky_3+lz_3)} + \cdots \qquad (21.20)$$

is substantially changed by insertion of the heavy atom. Because the intensity of a diffraction feature is proportional to the square of the structure factor, a Br atom contributes more to the intensity of a diffraction feature than an H or a C atom by factors of $(35/1)^2 = 1255$ and $(35/6)^2 = 34.8$, respectively. Figure 21.23 shows how the vector addition of the individual contributions to the structure factor is changed if Br is substituted for the F atom in Figure 21.20. It is seen that the structure factor is largely determined by the scattering factor of the heavy atom.

In MIR, diffraction intensities are determined before and after a heavy atom is inserted at a location in the molecule that is fairly well known. This allows the phase of the heavy atom to be calculated. For example, a methyl group can often be replaced by a Br atom without changing the reactivity of a protein. It is also possible to add a heavy metal such as Pt in the form of a salt, which complexes with a sulfur atom in the biomolecule rather than replacing it. Crucial to the success of the method is that the replacement is isomorphous, meaning that the structure of the molecule is essentially unchanged by the insertion of the heavy atom. The "multiple" in MIR refers to the fact that several such heavy atom substitutions must be made to avoid an ambiguity in the derived phases.

A third method for solving the phase problem, the use of which has increased dramatically in recent years, is called **multiwavelength anomalous dispersion (MAD).** In MAD, the diffraction intensities are measured at several different wavelengths or photon energies, one of which corresponds to an electronic transition in an atom in the molecule. At this energy, there is an abrupt change in the scattering factor for the atom, which can be used to determine the phases in much the same way as in MIR. Nearly all atoms with atomic numbers greater than that of Ca are suitable for MAD, and many biomolecules naturally contain one or more of these atoms. It is also possible to incorporate suitable atoms into the biomolecule to use MAD. For example, selenium is a strong anomalous scatterer and can be substituted for sulfur to allow the position of the sulfur atom to be determined. Because of the requirement that the wavelength be variable, MAD experiments must be carried out at synchrotron sources, which have increasingly set aside dedicated beam lines for X-ray diffraction studies of biomolecules.

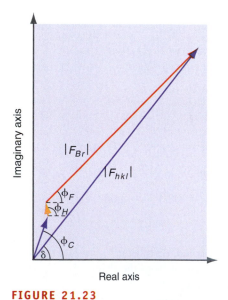

FIGURE 21.23

Figure 21.20 is redrawn after a Br atom is substituted in place of the F atom. It is seen that $|F_{hkl}|$ is largely determined by the Br atom.

21.7 Structure Determinations Are Crucial to Understanding Biochemical Processes

In this section, we demonstrate the power of X-ray crystallography in understanding photosynthesis. Photosynthetic organisms capture the energy of sunlight and harness it to synthesize ATP. Through evolution, this complex process has been optimized to the degree that a significant fraction of the photons incident on a photosynthetic organism are captured by light-harvesting units. About 5% of the energy incident in the form of sunlight is effectively used by photosynthetic organisms to produce molecules such as ATP. Although the necessity of sunlight for plant growth was known at the earliest stages of mankind, the mechanism by which light is converted to ATP has only become clear in the past 50 years. As quoted by Robert Huber, Ludwig Boltzmann, one of the greatest scientists of the 19th century, wrote in 1886: "Before reaching the temperature of the Earth, the energy radiated by the sun can assume improbable transitions forms. It is therefore possible to utilize the temperature drop between the sun and the Earth to perform work, as can be done for the temperature drop between steam and water. . . . To make the most of this transition, green plants spread the enormous surface of their leaves, and in an unknown way, force the energy of the sun to carry out chemical syntheses before it cools down to the temperature of the Earth's surface. These chemical syntheses are complete mysteries to us. . . ." In part due to scientific studies using X-ray crystallography, photosynthesis is no longer a mystery.

Although photosynthesis does not follow an identical pathway in different organisms, the pathways are so similar that what has been learned by studying bacterial photosynthesis, which are the most studied systems, can be transferred to other photosynthetic organisms with little change.

The essential structural elements in light harvesting are schematically shown in Figure 21.24. Each of the labeled units is a protein complex that spans the cell membrane. Electron transfer reactions initiated by light absorption occur at the reaction center (RC), which is surrounded by a concentric, cylindrically shaped, light-harvesting unit labeled LH1. A large number of additional light-harvesting units labeled LH2 absorb photons and transfer the energy to one another and to the same reaction center through LH1.

How is the light energy used by photosynthetic organisms to drive ATP synthesis? In this section, we focus on the mechanism by which energy is funneled to the reaction center, and in Section 26.5, the kinetics of photosynthetic reactions are discussed. As discussed in Chapter 19, light absorption occurs through a molecular electronic excitation as indicated schematically in Figure 21.25a. An important feature of photosynthesis is the high efficiency with which the energy of the initially absorbed photon is funneled from the initial light-harvesting molecule through a cascade of other light-harvesting molecules to the reaction center. How is this energy transfer accomplished? Intermolecular energy transfer can occur in two fundamentally different ways, namely, energy transfer and electron transfer. As shown in Figure 21.25b, resonant energy transfer can occur between two molecules according to the Förster mechanism, as discussed in Section 19.13, if the energy levels are appropriately aligned. This is the mechanism by which energy captured in LH1 and LH2 is funneled to the reaction center. Schematically, this process can be written as $A^* + B \rightarrow A + B^*$ where * designates an excited species. In a second mechanism, absorption of a photon leads to an ionization reaction, and the electron can be passed from one molecule to another by electron transfer, as shown in Figure 21.25c. Electron transfer reactions are used in the reaction center to synthesize ATP.

However, other mechanisms can lead to energy loss along the $A^* + B \rightarrow A + B^*$ pathway, which would lead to a decrease in the efficiency of photosynthesis. For example, the excited state can fluoresce back to the ground state with emission of a photon, or lose energy nonradiatively in which case the energy is converted to heat. Neither of these processes leads to ATP synthesis and they are therefore parasitic processes. One of the remarkable features of photosynthetic organisms is that their structure has evolved in such a way as to inhibit these parasitic processes as we will see later.

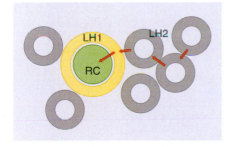

FIGURE 21.24

The transmembrane units involved in purple bacterial photosynthesis are shown looking down on the membrane surface. Light is captured by two light-harvesting systems, designated LH1 and LH2. The harvested energy is transferred to the reaction center (RC) where ATP synthesis is initiated. The orange arrows indicate energy transfer from LH2 to LH1, and from LH1 to the reaction system.

FIGURE 21.25
(a) Light absorption in LH1 and LH2
occurs through electronic excitations.
(b) Energy can be transferred from one
light-harvesting molecule to another
through resonant energy transfer.
(c) Energy can also be transferred from
one molecule to another through an
electron transfer (redox) reaction.

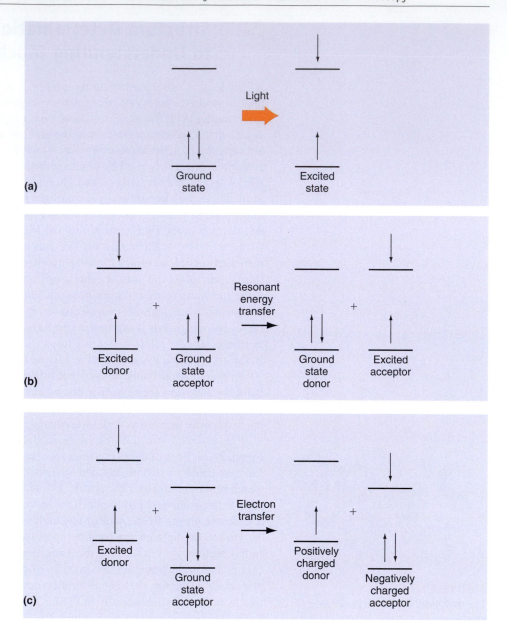

We next discuss light harvesting by the LH2 complex. A light-harvesting complex must contain photopigment molecules that absorb strongly in the visible part of the light spectrum. The structures of commonly occurring photopigments are shown in Figure 21.26. Chlorophylls *a* and *b* or bacteriochlorophyll *a* and *b* play a primary role as light absorbers in most photosynthetic systems, and other photopigments such as phycoerythrin and phycocyanin are used as light harvesters in cyanobacteria and red algae. Green plants contain auxiliary photopigments β-carotene and lutein in addition to chlorophylls *a* and *b*. All of these molecules have conjugated π-electron networks that shift the maximum in the absorption spectrum from higher energies to the visible range of the light spectrum as discussed in Section 14.7. The absorption spectrum of these molecules is shown together with the spectral distribution of sunlight in Figure 21.27. It is seen that the combination of chlorophylls *a* and *b* with other photopigments can result in a light-harvesting complex that fully covers the solar spectrum.

The structure of the LH2 system shown in Figure 21.28, which was not known until it was determined by X-ray diffraction, has made it possible to understand the high efficiency with which light energy is transferred to the LH1 complex and to the reaction center. The system consists of two concentric cylindrical units with 16 membrane-spanning

FIGURE 21.26

(a) Chlorophylls *a* and *b* and bacteriochlorophyll are the primary light absorbers in photosynthesis. The shading indicates the conjugated π-electron network. (b) β-carotene and (c) phycocyamin are examples of additional pigments that also absorb sunlight. (Source: Matthews, van Holde & Ahern, Fig. 17.7)

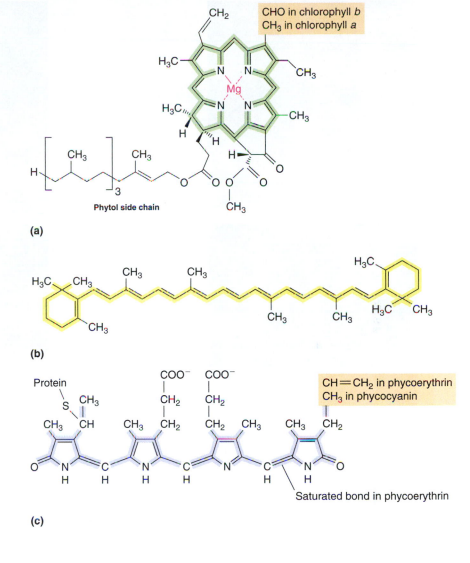

(a)

(b)

(c)

Key

— Chlorophyll *a* (green)
— Chlorophyll *b* (light blue)
— β-Carotene (yellow)
— Phycoerythrin (red)
— Phycocyanin (dark blue)

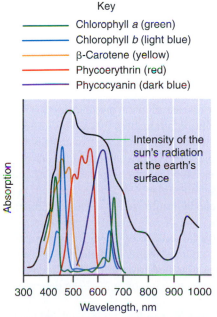

FIGURE 21.27

The colored curves show the absorption spectrum of the light absorbers shown in Figure 21.26. The black curve shows the wavelength distribution in sunlight. (Source: Matthews, van Holde & Ahern, Fig. 17.8)

α-helices that provide the structural scaffolding for the photopigments. The α-helices occur as dimer pairs that connect the outer and inner cylindrical units.

The arrangement of the bacteriochlorophyll *a* (Bchl-*a*) molecules in the LH2 unit is shown in Figure 21.29. The outer ring of eight Bchl-*a* molecules shown in Figure 21.29b has its absorption maximum near 800 nm, and the plane of the molecules is tilted by 38° with respect to the membrane surface. The orientation is determined through the interactions between the Bchl-*a* molecules and the residues of the α-helices. The inner ring of 16 Bchl-*a* molecules leads to a closer packing, shifting the maximum absorption to near 850 nm. The orientation of the Bchl-*a* molecules is perpendicular to the membrane surface. This complex but highly symmetric structure enhances the efficiency of the light-harvesting unit as an antenna. Having two different orientations relative to the membrane increases the overall absorption efficiency, and the different degree of interactions between the Bchl-*a* molecules in the inner and outer array broadens the spectral range over which absorption occurs.

The close spacing of the Bchl-*a* molecules to one another and to the lycopene photopigment ensures a rapid transfer of the energy from one light-harvesting molecule to another. In this way, parasitic processes such as fluorescence or nonradiative processes that dissipate the energy in the form of heat are avoided. Note also that the wavelength for maximum absorption of the inner ring of Bchl-*a* molecules is red shifted to lower energies than the outer ring Bchl-*a* molecules. This ensures a one-way transfer of energy from the outer to the inner ring, and enhances the transfer of energy to the reaction center. This example shows the power of X-ray crystallography in providing an understanding of biochemical processes.

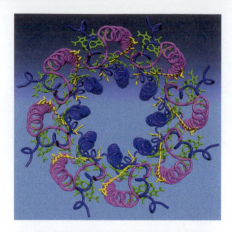

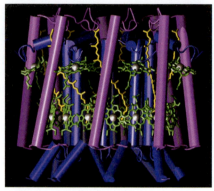

FIGURE 21.28

The LH2 complex of *Rhodospirillum molischianum* is shown in (a) a top view and (b) a side view. The α and β membrane-spanning helices are shown in blue and magenta, respectively. The Bchl-*a* molecules are shown in green and their phytol tails have been omitted for clarity. The additional lycopene photopigments are shown in yellow. In (b), the membrane-spanning α-helices are shown as rods. (Source: Courtesy of Koepke, J., Hu, X. C., Muenke, C., Schulten, K. & Michel, H., "The crystal structure of the light-harvesting complex II (B800-850) from Rhodospirillum molischianum." Structure, 4, 581–597 (1996), fig. 2)

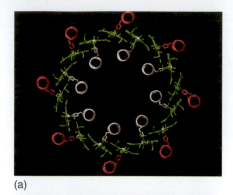

(a)

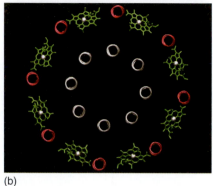

(b)

FIGURE 21.29

The orientation of the Bchl-*a* molecules relative to the LH2 unit is shown. (a) Sixteen Bchl-*a* molecules are situated in the space between the outer and inner cylindrical arrays. The plane of these Bchl-*a* molecules is oriented perpendicular to the membrane surface. (b) Eight Bchl-*a* molecules are located in the outer cylindrical array. The plane of these Bchl-*a* molecules is tilted by 38° with respect to the membrane surface. (Source: Courtesy of Koepke, J., Hu, X. C., Muenke, C., Schulten, K. & Michel, H., "The crystal structure of the light-harvesting complex II (B800-850) from Rhodospirillum molischianum." Structure, 4, 581–597 (1996), fig. 2)

The structures of the photosynthetic complex discussed in the previous section were determined at a resolution of 0.24 nm. The resolution attainable in X-ray crystallography is limited primarily by structural disorder in the crystal. There are two contributions to structural disorder, static and dynamic. Static disorder refers to slightly different structures in different unit cells of the crystal and arises from limitations in the ability to grow perfect crystals. Dynamic disorder arises because molecular vibrations are present at all temperatures including 0 K. Dynamic disorder can be minimized by measuring diffraction intensities at cryogenic temperatures. In the past decade, significantly more high-resolution X-ray structures have been obtained for biomolecules through a combination of the use of high-intensity synchrotron X-ray sources, better techniques for crystal growth, and better techniques to refine structures obtained from diffraction intensities. For example, the structure of the human aldose-reductase inhibitor has been determined at a resolution of 0.066 nm.

Higher resolution allows the electron distribution in bonds to be imaged and the location of H atoms to be determined. H atoms scatter X-rays weakly because they have only one electron, Locating active H atoms allows insights to be gained about proton transfer reactions that determine enzyme activity. An example is the enzyme cholesterol oxidase, a bifunctional pH-sensitive flavoenzyme that catalyzes the oxidation and isomerization of 3β-hydrosteroids [see A. Y. Lyubimov *et al.*, *Nature Chemical Biology*, 2 (2006), 259–264]. In the pH range of 4.5 to 7.5, the oxidation activity of the enzyme is

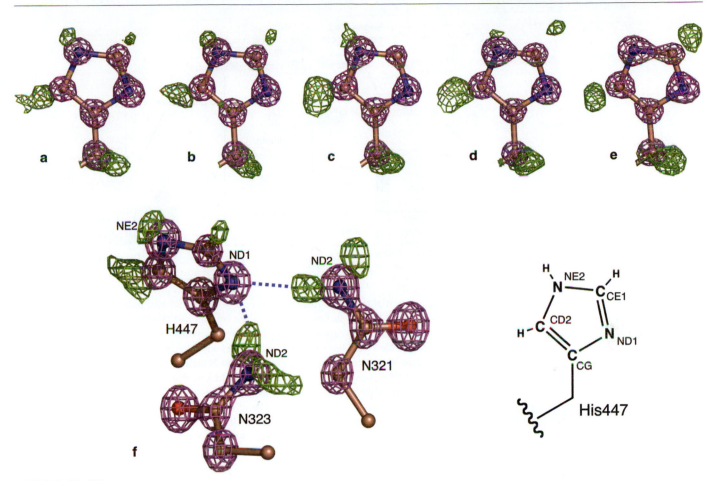

FIGURE 21.30
The protonation state of His447 is shown for various pH values. Note changes in the region labeled NE2 as the pH changes takes on the values (a) 5.2, (b) 5.8, (c) 7.3, and (d) 9.0. Part (f) shows that the negative charge left on the imidazole ring by removal of the H on NE2 is stabilized by hydrogen bonds formed to the nitrogen ND1. (Source: Reprinted by permission from Macmillan Publishers Ltd.: Atomic resoultion crystallography reveals how changes in pH shape the protein microenvironment, by Artem Y. Lyubimov, Paula I. Lario, Ibrahim Moustafa and Alice Vrielink, *Nature Chemical Biology,* vol. 2, issue 5, page 260, copyright 2006, fig. 4)

independent of pH. However, for pH values greater than 7.5, the rate of the oxidation reaction is reduced by a factor of ~2000. The reason for this marked decrease in catalytic activity is associated with the removal of a hydrogen on a histidine denoted His447 located near the β-OH group on the hydrosteroid that is oxidized. The evidence for this conclusion is shown in Figure 21.30.

The green regions in Figure 21.30 indicate the electron density associated with H atoms. As the pH increases from 4.5 to 9.0, a marked decrease in the electron density is observed at the site corresponding to the H at site NE2 (see Figure 21.30f). The removal of this hydrogen leaves a negative charge on the imidazole ring that is stabilized by neighboring hydrogen bond formation as indicated in Figure 21.30f.

21.8 The Atomic Force Microscope

The **atomic force microscope (AFM)** is a versatile tool that can measure the topography of a surface with subnanometer lateral resolution and can also measure forces between a surface and a scanned tip in the piconewton range. As a tool for structural determination, it complements X-ray diffraction. The resolution is lower as discussed later in this chapter, but the technique has the important advantage that the sample can be in a buffered solution under physiological conditions as it is being examined.

The German physicist Gerd Binnig is a co-inventor of the AFM as well as the STM (see Chapter 14). As shown in Figure 21.31, a tip attached to a flexible cantilever is

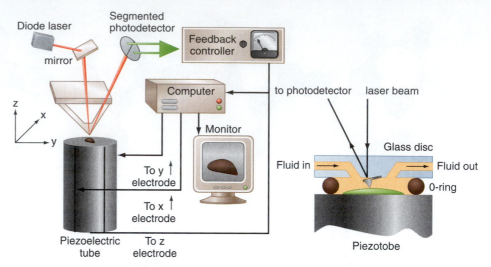

FIGURE 21.31

Schematic diagram of an atomic force microscope. (a) A tip mounted on a microfabricated cantilever is scanned over a surface in the *x-y* plane by applying dc voltages to a segmented piezoelectric tube. If the tip experiences an attractive or repulsive force from the surface, the cantilever is deflected from its horizontal position. As a result, the laser light reflected from the back of the cantilever onto a segmented photodetector is differently distributed on the segments, giving rise to a difference current, which is the input to a feedback controller. The controller changes the length of the piezoelectric tube in such a way to keep the cantilever deflection constant as the tip scans across the surface. Therefore, the surface image obtained corresponds to a constant force that can be varied using the feedback circuit. (b) The AFM can be modified to allow measurements in a liquid or controlled atmosphere using an O-ring seal mounted on the piezoelectric tube.

FIGURE 21.32

Image of a cantilever obtained with a scanning electron microscope. The arrow indicates the position of the tip. (Based on work by the NTUF)

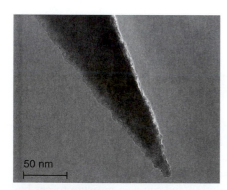

FIGURE 21.33

Electron microscope images of a tip typically used in atomic force microscopy. A radius of curvature of 10 to 20 nm can be routinely obtained. (Source: Courtesy of NanoWorld AG)

scanned over the surface of a sample using the same feedback circuitry as for an STM. The tip and cantilever shown in Figure 21.31 are microfabricated from Si_3N_4 and Si, respectively, and the deflection of the cantilever from its horizontal position is given by

$$x = -\frac{F}{k} \qquad (21.21)$$

where F is the force exerted on the cantilever shown in Figure 21.32, and k is its spring constant, which can have values in the range from 0.01 to 10 Nm^{-1} depending on the application. The tip shown in Figure 21.33 has a radius of curvature of $\sim$10 to 20 nm, and the force of interaction between the tip and the surface is primarily determined by those atoms on the tip closest to the surface. The deflection of the cantilever is measured using a laser similar to that in a CD player. The light reflected from the back of the cantilever is incident on a segmented photodetector, and the deflection of the cantilever can be determined by comparing the signal from the segments of the photodetector. The feedback circuit keeps the cantilever deflection, and therefore the tip–surface force, constant as the tip is scanned across the surface. Whereas in an STM, an image corresponds to a surface contour at constant tunneling current, in an AFM, an image corresponds to a surface contour at constant force. Image acquisition is sufficiently fast that many kinetic processes of biological interest can be imaged in real time.

The physical origins of the tip–surface force can be understood from Figure 21.34. As the tip approaches the surface from large distances along the dashed path, the initial force is zero. However, as the tip–surface distance, d, decreases, the tip is attracted to the surface by van der Waals forces, which vary as $1/d^6$. As d decreases further, the cantilever is increasingly attracted by the surface, until the tip snaps into contact with the surface. The cantilever is deflected in the opposite direction as it is brought closer to the surface, resulting in a linear increase in the force exerted by the tip on the surface with distance

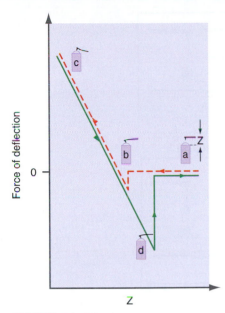

that is determined by the spring constant of the cantilever. Note that in the distance range in which snap-in occurs, the tip–surface force is attractive, whereas for smaller distances, the force becomes repulsive. As the cantilever is pulled away from the surface, a hysteresis is observed because the tip has been pushed into the surface in the repulsive region. The tip snaps away from the surface as the force exerted by the cantilever equals the adhesion force of the tip to the surface.

21.9 Measuring Adhesion Forces between Cells and Molecular Recognition Imaging Using the Atomic Force Microscope

As the forces between individual molecules and cells play an important role in biochemical processes, the ability to measure these forces can provide new scientific insights. For example, the adhesion of cells is an essential process in the embryonic development of multicellular organisms. Because the AFM is capable of measuring very small forces with high spatial resolution, it is possible to bring two cells into contact, and to measure the force needed to separate the cells again, which, by definition, is the adhesion force. Figure 21.35a shows how such a measurement can be carried out. Cells are strongly attached to a suitable flat substrate and to an AFM cantilever by appropriate chemical linkages. By moving the cantilever relative to the substrate under an optical microscope, the two cells are brought into contact. Subsequently, the cantilever is pulled away from the surface, and the force required to abruptly increase z corresponds to the adhesion force (see Figure 21.34). Figure 21.35b shows the cantilever with the attached cell before it is brought into contact with an immobilized cell on the substrate.

Benoit and coworkers [*Nature Cell Biology,* 2 (2000), 313–317] have studied adhesion between cells of the eukaryote *D. discoideum,* which is known to have a single type of glycoprotein adhesion molecule. It is known that the adhesion forces are dependent on the presence of Ca^{2+} in the solution. Figure 21.36 shows the result of the experiment described in Figure 21.35a. As the cantilever is retracted, the force on the cantilever sinitially increases. With a further increase in distance, the force falls, remains constant, and falls again. Each of these steps is due to the rupture of a single bond between the cells, and the number of glycogen molecules responsible for cell adhesion can be determined from such curves. If the Ca^{2+} is scavenged by adding EDTA to the solution, the adhesion force drops essentially to zero.

FIGURE 21.34
In an approach diagram, the cantilever deflection or force is shown as a function of the distance between the cantilever base and the sample surface. At large distances (a), the attractive force between the tip and surface is insufficient to cause a downward cantilever deflection. At a distance corresponding to (b), the attractive force is strong enough to deflect the cantilever until the tip and surface are in contact. At smaller distances where the force is repulsive, the cantilever is deflected away from the surface and the applied force increases linearly with the deflection.

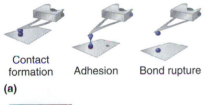

Contact   Adhesion   Bond rupture
formation

(a)

(b)

FIGURE 21.35
(a) The left panel shows contact formation between the cells on the substrate and the cantilever. (b) As z is increased (middle panel in part a), the contact area is stretched and the bond is ruptured if the deflection force is sufficiently high (right panel in part a). The black bar corresponds to a distance of 20 μm. (Source: Reprinted by permission from Macmillan Publishers Ltd.: Discrete interactions in cell adhesion measured by single-molecule force spectroscopy, by Martin Benoit, Daniela Gabriel, Gunther gerisch, Hermann E. Gaub, *Nature Cell Biology,* vol.2, issue 6, page 313-317, copyright 2008, fig. 1.)

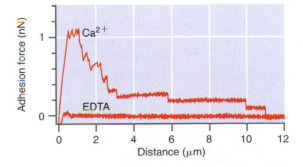

FIGURE 21.36
The measurement of adhesion forces for *D. discoideum* cells is depicted. Contact between the cells immobilized on the substrate and cantilever was maintained for 20 s at a force of 150 pN. The cells were pulled apart as a rate of 1.5 μm s^{-1}. (Source: Reprinted by permission from Macmillan Publishers Ltd.: Discrete interactions in cell adhesion measured by single-molecule force spectroscopy, by Martin Benoit, Daniela Gabriel, Gunther gerisch, Hermann E. Gaub, *Nature Cell Biology,* vol.2, issue 6, page 313-317, copyright 2008, fig. 2a)

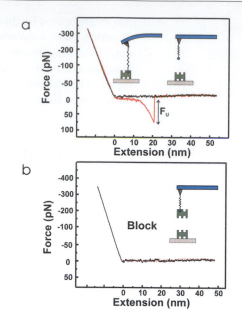

FIGURE 21.37

(a) A strong unbinding event is observed as the tip is withdrawn after having been brought into contact with the surface (red trace). (b) If the biotin on the tip is saturated with avidin added to the solution, the unbinding event is not seen. (Source: Courtesy of Dr. Ferry Kienberger, University of Linz, Institute of Biophysics, "Molecular Recognition Imaging and Force Spectroscopy of Single Biomolecules," *Accounts of Chemical Research,* 2006, Vol. 39, No. 1, pgs 29–36. Reprinted with permission from the American Chemical Society, fig. 2a-b)

Recognition image

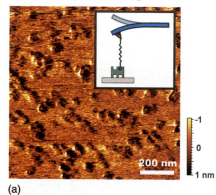

(a)

Recognition image (blocked)

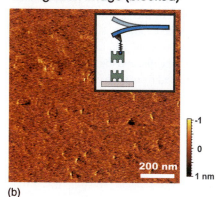

(b)

FIGURE 21.38

(a) Scanning a functionalized tip over a surface with binding sites gives an image that reflects both variations in local force and variations in height. (b) If the biotin on the tip is bound by avidin added to the solution, only variations in the surface height are seen. (Source: Courtesy of Dr. Ferry Kienberger, University of Linz, Institute of Biophysics, "Molecular Recognition Imaging and Force Spectroscopy of Single Biomolecules," *Accounts of Chemical Research,* 2006, Vol. 39, No. 1, pgs 29–36. Reprinted with permission from the American Chemical Society)

The experiment just described measures the attraction between whole cells. However, the high spatial resolution of the AFM makes it possible to functionalize the tip so that it experiences strongly attractive forces when scanned over a molecule of interest and weak forces otherwise. The principle of this measurement is shown in Figure 21.37. The tip was functionalized with a small molecule (biotin) through an extendible tether, and the surface was covered with avidin, a protein that binds biotin very tightly. As the tip approaches the surface, the biotin and avidin interact strongly. The shape of the approach curve (black trace) is as in Figure 21.34, but the snap-in is too weak to be seen. However, as the tip is pulled back, a strong unbinding event corresponding to an adhesion force of ~80 pN is observed. If the experiment is repeated after adding avidin to the solution so that the binding capacity of the tip is saturated, no unbinding event is observed.

If the tip is scanned over the surface, different images are obtained, depending on whether the biotin on the surface was saturated with avidin in solution. For the chemically active tip, strong deflections are observed as the tip is scanned over the surface as seen in Figure 21.38a. These regions correspond to biotin molecules. If the tip is made chemically inactive, the scan is largely featureless as seen in Figure 21.38b, and the remaining cantilever deflections are due to the surface topography. A comparison of the two images makes it possible to distinguish surface topography from the presence of chemically active molecules. These measurements suggest that in the future it will be possible to scan over membrane surfaces and to measure not only the surface topography, but also the distribution of molecules or functional groups of interest.

21.10 Nanodissection Using the Atomic Force Microscope

By increasing the force on the tip and repeatedly scanning along a line on the surface of a biological sample, the tip is driven into the surface, resulting in a cut whose depth can

(a)

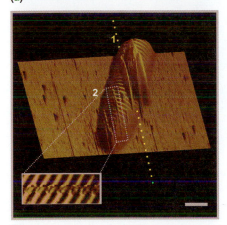

(b)

(c)

FIGURE 21.39

(a) The intact collagen fibril of ~210 nm diameter is seen prior to nanodissection. (b) The line labeled 1 shows an incision that is ~140 nm deep. The line labeled 2, also shown magnified in the box, is ~2 nm deep. (c) A three-dimensional height image of the dissected area is seen. The cross-section shows parallel-aligned fibrillar subcomponents. The scale bar is 1 μm in length. (Source: Courtesy of Professor M. Cynthia Goh, Department of Chemistry, University of Toronto, "AFM Nanodissection Reveals Internal Structural Details of Single Collagen Fibril," *Nano Letters 2004*, Vol. 4, No. 1, pgs 129–132, American Chemical Society)

be controlled on the nanometer scale. C. K. Wen and M. C. Goh have used this method to carry out a **nanodissection** of a single collagen fibril in order to determine its internal architecture.

Figure 21.39 shows stages in the nanodissection process. The intact fibril with a diameter of ~210 nm is shown in Figure 21.39a. Repeated scans can lead to deep or superficial cuts, depending on the applied force as shown in Figure 21.39b. The incised area can be peeled back by using the tip to apply a compression force on the area that is to be removed. The partially dissected fibril is shown in Figure 21.39c, showing parallel aligned fibrillar subcomponents. Because the compression force has only been applied to the area removed, the remaining material is essentially undamaged, allowing delicate samples to be nanodissected.

21.11 The Atomic Force Microscope as a Probe of Surface Structure

Because an AFM image corresponds to a surface scan at constant force, it always contains information on both the topography of the surface and the variation on the tip–surface force over the surface. The topography of the surface can be best measured by scanning the surface in the repulsive force range under conditions no strong variations occur in the force as the tip is scanned over the surface. Two ways have been developed to measure surface topography. In the **contact mode** shown in Figure 21.40b, the tip is dragged directly over the surface. This method is suitable for rigid samples such as ionic crystals, but can lead to the deformation of biomolecules attached to a surface because they are inherently soft. A more suitable technique for biomolecular surfaces is the **intermittent contact mode** in which the tip–surface distance is oscillated about a mean value as the tip is scanned across the surface. As the tip is in contact with the surface only for brief periods and forces parallel to the surface are minimized, much less deformation of the surface occurs. These two methods are depicted in Figure 21.40.

An example of a high-resolution intermittent mode image of DNA tethered to a mica surface is shown in Figure 21.41. Note that the helical pitch distance of 3.4 nm is clearly resolved. A primary requirement of using the AFM for structural studies is that the molecule of interest must be firmly attached to a flat surface. In this case, a mica surface was coated with a cationic bilayer. The positive charges were strongly attracted to the phosphate groups on the DNA, leading to strong binding of the DNA to the surface.

The resolution of the AFM for structural studies is determined by two factors: the tip–surface feature convolution and the degree to which the sample is deformed by the force used to obtain the image. Figure 21.42 illustrates the tip–surface convolution, and Figure 21.40b shows how sample deformation can falsify the surface structure, thereby limiting the resolution. The resolution of the AFM in the scanning plane can vary between 0.1 nm for a flat hard sample up to 100 nm for a highly corrugated soft sample. Typically, forces between 10 and 100 pN are used to image surfaces.

In a further example of structural studies, the AFM has been used in combination with nanodissection to clarify unresolved issues about the structure of the photosynthetic light-harvesting systems. The X-ray structural determinations discussed in Section 21.7 were carried out by removing the LH2 complexes from their site in the membrane and crystallizing them apart from other complexes with which they are associated. The same techniques were used to determine the structures of the light-harvesting system LH1 and the reaction center. As a result, X-ray diffraction measurements are not able to give detailed information on how the LH2, LH1components, and the reaction center fit together. By contrast, AFM measurements can be carried out on intact membranes under physiological conditions. Therefore, despite their limited resolution, AFM measurements are well suited to determining the hierarchical structure of the photosynthetic apparatus. This has been demonstrated by Scheuring *et al.* [*Proc. National Academy of Sciences*, 100 (2003), 1690–1693] in an AFM study of the photosynthetic apparatus of the purple bacterium *Blastochloris viridis.* This bacterium has a single light-harvesting center,

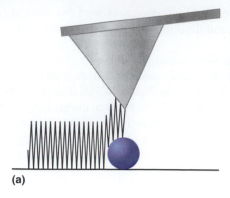

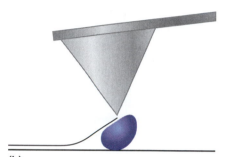

(a)

(b)

FIGURE 21.40

(a) In the intermittent contact mode, the cantilever is vibrated at its resonance frequency and touches the surface for a small fraction of the time needed for a scan. (b) The contact mode is suitable for hard surfaces, but can lead to reversible or irreversible deformation of a soft sample as shown.

FIGURE 21.41

High-resolution image of DNA obtained in the intermittent contact mode. The DNA was immobilized by binding it to cationic bilayers on mica. (Source: Courtesy of Dr. Zhifeng Shao, University of Virginia, Jianxun Mou, Daniel M. Czajkowsky, Yiyi Zhang, Zhifeng, "High-resolution atomic-force microscopy of DNA: the pitch of the double helix," *FEBS Letters* 371, 279–282, 1985)

LH1, that surrounds the reaction center. The reaction center structure including the cytochrome subunit determined using X-ray diffraction is shown in Figure 21.43. We will not be concerned with the structural details of the reaction center here. Note, however, the large globular cytochrome subunit with four hemes that protrudes on the P side of the membrane and covers the reaction center.

Figure 21.44 shows an AFM image of a portion of the membrane surface of *B. viridis.* It shows a cluster of structures. Scanning over larger portions of the membrane surface shows that the structures have a tendency to agglomerate, rather than to spread uniformly over the membrane surface. In these images, the resolution perpendicular and parallel to the membrane surface is 0.1 and 1 nm, respectively. By repeatedly scanning over the membrane surface, Scheuring *et al.* were able to remove the cytochrome subunit by nanodissection from many of the structures, enabling them to image the LH1 light-harvesting system and the reaction center as shown in Figure 21.44a. They observed three types of photosynthetic units after nanodissection. The structures labeled 3

FIGURE 21.42

The apparent size of a scanned object in the scanning plane is a convolution of the tip shape with the object shape. The top five images show the positions that a cantilever will take while scanning over the surface feature shown in red. The bottom image shows the apparent shape obtained from the scan.

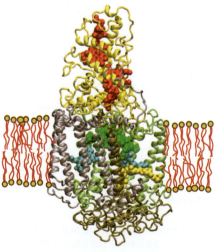

FIGURE 21.43

The photosynthetic reaction center in the membrane of the purple bacterium. The four hemes in the cytochrome subunit are shown in red. The LH1 light-harvesting system that surrounds the reaction center is not shown. (Figure provided by William Parson)

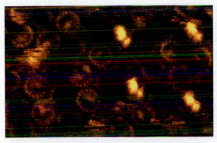

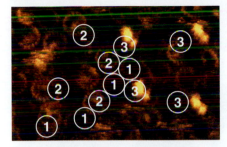

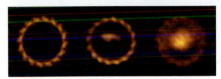

(a)

(b)

(c)

FIGURE 21.44

(a, b) A scan is shown over a cluster of LH1 complexes after a portion of the surface has been removed through nanodissection. The regions labeled 1, 2, and 3 correspond to the LH1 complex without a reaction center, the LH1 complex containing a reaction center without the cytochrome subunit, and the intact LH1–reaction center complex including the cytochrome subunit, respectively. (c) Higher resolution images are shown of 1, 2, and 3 obtained by averaging over similar units in part (a). (Source: "Nanodissection and high-resolution imaging of the Rhodopseudomonas viridis photosynthetic core complex in native membranes by AFM," by Simon Scheuring, Jerome Seguin, Sergio Marco, Daniel Levy, Bruno Robert, and Jean-Louis Rigaud, *PNAS,* Copyright © 2003, vol. 100, No. 4, National Academy of Sciences, U.S.A.)

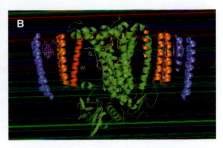

FIGURE 21.45

Schematic model of the RC–LH1 core complex in *Rhodopseudomonas palustris* obtained from X-ray diffractions studies. The views are (A) perpendicular and (B) parallel to the membrane plane. The most important elements of LH1 are the Bchl proto absorbers, shown in red, and the transmembrane helical proteins, shown in green and light blue. Note the gap in LH1 to the left of the helical protein labeled W. (Source: Aleksander W. Roszak, Tina D. Howard, June Southall, Alastair T. Gardiner, Christopher J. Law, Neil W. Isaacs, Richard J. Cogdell, "Crystal Structure of the RC-LH1 Core Complex from Rhodopseudomonas palustris," *Science* Vol. 32, 12 December 2003, Fig 4)

in Figure 21.44b protrude most strongly into the solution and correspond to the reaction center with the intact cytochrome subunit. Another subset of images labeled 1 shows a ring-like structure with a hollow core, corresponding to the LH1 complex without a reaction center inside. The third subset of images labeled 2 show the LH1 complex containing a reaction center without the cytochrome subunit. Averaging over all units of the same type to improve the signal-to-noise ratio leads to the three different structures shown in Figure 21.44c. The image of the LH1 complex shows 16 similar subunits in an elliptical arrangement.

A recent X-ray diffraction study by Roszak *et al.* [*Science,* 302 (2003), 1969] gives more detailed information on the structure of the LH1 complex and its spatial orientation with respect to the reaction center than is possible with AFM. The model shown in Figure 21.45 indicates that the LH1 complex consists of 15 identical units, each of which is believed to contain a pair of transmembrane α-helices, two Bchl molecules, and one or two carotenoids. However, complete enclosure of the reaction center by LH1 is prevented by a single transmembrane protein (labeled W in Figure 21.45a). The break in LH1 is believed to provide a portal through which the secondary electron acceptor ubiquinone can diffuse and transfer electrons to cytochrome b/c_1.

21.12 Observing Biochemical Processes in Real Time Using the Atomic Force Microscope

Because an AFM can be operated under a physiological environment with the sample in a buffered solution containing the components essential for the process, biochemical processes can be studied in real time. We consider the example of transcription, in which RNA is synthesized from a DNA template. This process is essential in DNA replication and is catalyzed by the enzyme RNA polymerase (RNAP). To carry out this process, RNAP must detect and transcribe distinct genes contained in long portions of DNA. DNA templates contain regions called promoter sites that specifically bind RNAP and thereby initiate the process of transcription. The rate at which the RNAP locates the promoter region is an important factor in the overall rate of transcription.

A number of studies that were carried out using standard biochemical techniques suggested that the high rate of transcription observed in specific systems occurred by a process in which the RNAP was initially bound to the DNA from the solution phase, and subsequently traveled along the DNA by diffusion, which is equivalent to one-dimensional diffusion. These studies were carried out on ensembles of molecules in solution, and therefore gave only indirect evidence for this process. Guthold *et al.* [*Biophysical Journal*, 77 (1999), 2284–2294] carried out a time-lapse experiment in which RNA polymerase was attached to a 1001 base-pair DNA segment without a promoter region. The motion of the RNAP along the DNA was tracked and its position was compared with the expected behavior for one-dimensional diffusion. The results are seen in Figure 21.46. The results agree well with predictions for the one-dimensional diffusion model.

2288   Biophysical Journal   Volume 77   October 1999

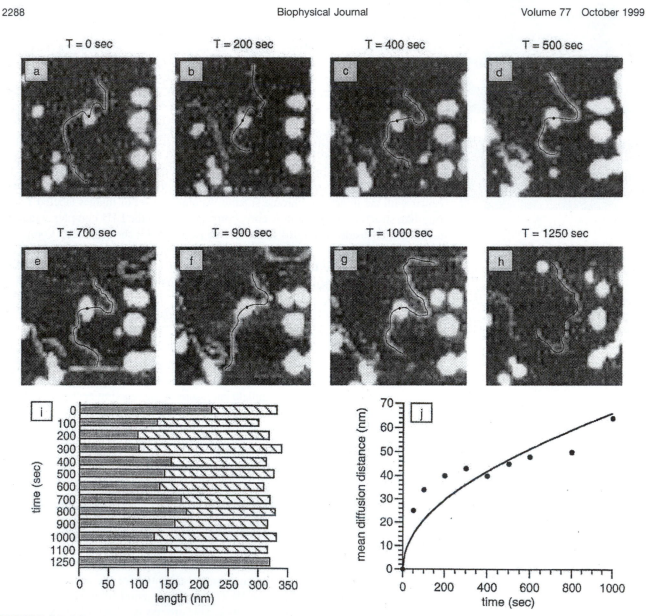

FIGURE 21.46

(a–h) Images are shown of the diffusion of nonspecific binary complexes of RNAP along the 1001 base-pair DNA. The elapsed time is shown above each image. The DNA is indicated by a line and the RNAP position is indicated by a dot. (i) Bar graph showing the apparent lengths of the DNA to the left and to the right of the RNAP at the indicated times. (j) The mean diffusion distance is plotted as a function of time. The solid curve shows what would be predicted for one-dimensional diffusion of the RNAP along the DNA. (Source: Martin Guthold, Xingshu Zhu, Claudio Rivetti, Guoliang Yang, Neil H. Thomson, Sandor Kasas, Helen G. Hansma, Bettye Smith, Paul K. Kansma and Carlos Bustamante, "Direct Observaton of One-dimensional Diffusion and Transcription by Escherichia coli RNA Polymerase," *Biophysical Journal*, Volume 77, October 1999, pgs 2284–2294)

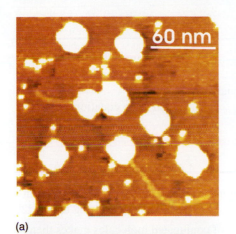

(a)

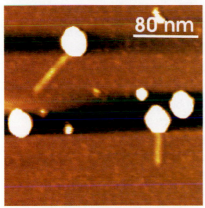

(b)

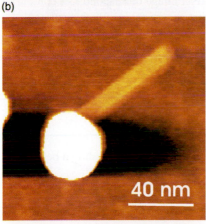

(c)

An additional example of the use of AFM to image biochemical processes is shown in Figure 21.47, in which RNA release from a virus is depicted. Human rhinoviruses are a cause of the common cold. The virus is a small roughly spherical particle having a radius of ~25 nm with an RNA genome enclosed within the protein coating, or capsid, that forms the outer surface of the virus. The virus can bind to cells and is brought into the cell via receptor-mediated endocytosis, after which the RNA is released. The release is triggered by a decrease in pH. In the initial stages of release, short segments of the RNA are observed to protrude from the virus capsid, as seen in Figure 21.47. These 40- to 330-nm-long segments can be straight or bent, and correspond to intermediates in the release of the RNA from the virus capsid. Isolated RNA molecules observed after complete release have a length of ~1000 nm.

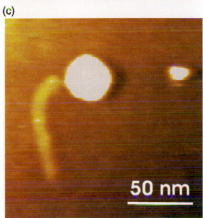

(d)

FIGURE 21.47
Partially released RNA molecules are seen as they are being released from the virus capsid. (a) Bent strands emerge from the ~31-nm-diameter virus capsids. Smaller white dots are debris of the virus capsid. (b, c) Straight RNA segments are shown emerging from the capsid. (d) A bent segment at higher magnification than the other parts. The apparent width of the RNA is due to a convolution of the true width with the ~20-nm tip radius. (Source: Ferry Kienberger, Rong Zhu, Rosita Moser, Dieter Blaas, and Peter Hinterdorfer, "Monitoring RNA Release from Human Rhinovirus by Dynamic Force Microscopy," *Journal of Virology*, April 2004, Pgs. 3203–3209, Vol. 78, No. 7)

Vocabulary

atomic force microscope (AFM)	direct methods	multiple isomorphous replacement (MIR)	primitive unit cell
Bravais lattices	intermittent contact mode		structure factor
centrosymmetric unit cell	lattice	multiwavelength anomalous	unit cell
contact mode	Miller indices	dispersion (MAD)	von Laue equations
	motif	nanodissection	

Questions on Concepts

Q21.1   Distinguish between the terms *crystal, lattice,* and *lattice motif.*

Q21.2   Explain why the Laue conditions for constructive interference can be depicted as cones.

Q21.3   As you are observing the diffraction pattern from a two-dimensional crystal, the X-ray photon energy is continuously increased. What changes will you observe in the diffraction pattern?

Q21.4   For which of the planes 100 and 111 will the diffraction angle θ be greater? Explain your answer.

Q21.5   Explain how the method of multiple isomorphous replacement is useful in solving the phase problem.

Q21.6   Diffraction is observed from two crystals, each of which has a tetragonal lattice with the same unit cell lengths. In one case, the lattice motif is a single atom, and in the second it is a triatomic molecule. Will the positions of the observed diffraction spots be different? Explain.

Q21.7   Which is more important to know in obtaining a structural model from diffraction data, $|F_{hkl}|$ or δ, the phase corresponding to $|F_{hkl}|$? Why?

Q21.8   Through what mechanism is energy transferred from one photoabsorber to another in the light-harvesting unit LH2?

Q21.9   What property of conjugated molecules makes them useful as photoabsorbers in photosynthesis?

Q21.10   What makes AFM particularly suitable for observing biochemical processes in real time?

Q21.11   Explain what makes the cantilever in an AFM "snap in" as the surface is approached.

Q21.12   How can the AFM be used to map out the spatial distribution of receptor molecules on a membrane surface?

Q21.13   Explain how nanodissection can be carried out using the AFM. Why is it useful?

Q21.14   What is the difference between the contact and noncontact modes of acquiring images in AFM?

Q21.15   Why does an AFM image correspond to a contour of constant force? Under what condition does such a surface correspond to a topographical image?

Problems

Problem numbers in **RED** indicate that the solution to the problem is given in the *Student Solutions Manual.*

P21.1   Taking sharing between neighboring cells into account, how many lattice points are contained in unit cells of the following Bravais lattices?

   a. monoclinic C

   b. orthorhombic I

   c. orthorhombic F

P21.2   Taking sharing between neighboring cells into account, how many lattice points are contained in unit cells of the following Bravais lattices?

   a. tetragonal I

   b. cubic F

   c. orthorhombic C

P21.3   What are the coordinates of the equivalent points in the unit cell of the cubic I Bravais lattice?

P21.4   What are the coordinates of all points shown in the unit cell of the orthorhombic F Bravais lattice in Figure 21.4? Which of these are the equivalent points?

P21.5   Using the tetragonal lattice unit cell shown below, draw in the following planes:

   a. 001   **b.** 101   **c.** 112   **d.** 111

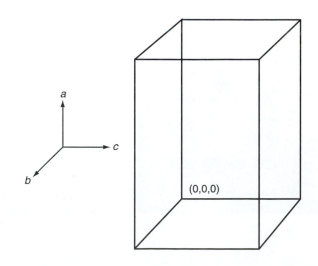

P21.6 Show by making a vector diagram such as that shown in Figure 21.23 that $I_{hkl} = I_{\overline{hkl}}$.

P21.7 If a unit cell contains more than one atom in a symmetric position, the intensity of some of the diffraction peaks may be zero for all angles of incidence and for all X-ray wavelengths. In this problem, you will calculate the systematic absences for the cubic body-centered lattice for a single identical atom at each equivalent position.

    a. Calculate F_{hkl} for a plane w with the indices hkl using the coordinates of the equivalent positions.

    b. For what conditions on the sum $h + k + l$ is $F_{hkl} = 0$?

P21.8 Carry out the preceding calculation for a C-centered orthorhombic cell with a single identical atom at each equivalent position.

    a. How many equivalent positions are there in the C-centered orthorhombic cell and what are their positions?

    b. For what conditions on the sum $h + k$ is $F_{hkl} = 0$?

P21.9 Aluminum crystals have a face-centered cubic structure with $a = 40.4$ nm. Calculate the density of aluminum from this information in units of kg m^{-3} and g cm^{-3}.

P21.10 What fraction of the volume in a face-centered cubic cell is occupied by atoms? To solve this problem, calculate the radius of atoms in a face of the lattice that just touch one another.

P21.11 The crystal radius of an atom can be calculated from the density of the crystalline material assuming that the atoms are hard spheres that just touch one another. The density of calcium metal is 1.55 g cm^{-3}. Calcium crystallizes in the face-centered cubic lattice. Calculate the atomic radius of Ca.

P21.12 Calculate the half-angles for the Laue cones for diffraction of an X-ray beam of wavelength 0.070926 nm from a one-dimensional crystal of periodicity 0.300 nm along the **c** direction. The angle $\gamma_0 = \pi/4$. Make a drawing showing the crystal direction, the X-ray beam direction, and the Laue cones labeled with their l values.

P21.13 Aluminum has a face-centered cubic structure with $a = 0.404$ nm. Calculate the angle θ at which diffraction is observed for (a) first-order and (b) second-order diffraction from the 112 planes. The X-ray wavelength is 0.15418 nm.

P21.14 The spacing, d, between planes of indices hkl for an orthorhombic unit cell is given by $1/d^2 = h^2/a^2 + k^2/b^2 + l^2/c^2$. Calculate the spacing between adjacent members of the (a) 011 and (b) 231 planes for unit cell lengths given by $a = 512$ pm, $b = 498$ pm, and $c = 622$ pm.

P21.15 Calculate F_h for $h = 0, +1$ and -6 for diffraction from the one-dimensional lattice discussed in Section 21.5. Assume that the dependence of f_C on the variable $\sin\theta/\lambda$ can be approximated by

$$f_C\left(\frac{\sin\theta}{\lambda}\right) = -1.961 + 17.47e^{-\sin\theta/\lambda} - 9.051e^{-(\sin\theta/\lambda)^2}$$

for $\sin\theta/\lambda > 0$ and $f_C = 6.00$ for $\sin\theta/\lambda = 0$. In this equation, λ is expressed in angstroms (1 Å $= 10^{-10}$ m).

P21.16 It is not possible in an X-ray diffraction experiment to measure the intensities of all of the allowed diffraction spots. In this problem, you will test how necessary it is to know all of the F_h in Table 21.1 and Example Problem 21.4 in order to determine the positions of the atoms in the units cell. Using the data in Table 21.1, calculate the electron density distribution for the following cases:

    a. Omit F_h for $h = 0$ and ±1.

    b. Omit F_h for $h = 0, \pm1, \pm2$, and ±3.

    c. Omit F_h for $h = 0, \pm1, \pm2, \pm3, \pm4$, and ±6.

    d. Omit F_h for $h = \pm11$ and ±12.

    e. Omit F_h for $h = \pm9, \pm10, \pm11$, and ±12.

    f. Omit F_h for $h = \pm7, \pm8, \pm9, \pm10, \pm11$, and ±12.

For which of these cases is it still possible to reliably determine the positions of the atoms in the unit cell? Explain your results.

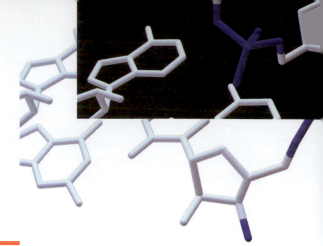

CHAPTER

The Boltzmann Distribution

Employing statistical concepts, one can determine the most probable distribution of energy in a chemical system. This distribution, referred to as the *Boltzmann distribution,* represents the most probable configuration of energy for a molecular system at equilibrium and also gives rise to the thermodynamic properties of the system. In this chapter, the Boltzmann distribution is derived using concepts from probability theory, and is applied to some elementary examples to demonstrate how the distribution of energy in a molecular system depends on both the available energy levels and the energy-level spacings that characterize the system. The concepts outlined here provide the framework required to apply statistical thermodynamics to molecular systems.

22.1 Microstates and Configurations

Statistical thermodynamics provides the bridge between the microscopic perspective of quantum mechanics and the macroscopic perspective of thermodynamics. The connection between these two limiting perspectives is constructed using ideas from probability theory and applying these ideas to chemical systems. Although extending probability theory to chemistry may appear to be a daunting task, the sheer size of chemical systems makes the application of statistical concepts straightforward. To begin, we introduce two key concepts from probability theory: permutations and configurations. These two concepts are illustrated by a simple experiment in which a single coin is tossed four times. The possible outcomes for this experiment are presented in Figure 22.1.

Figure 22.1 illustrates that there are five possible outcomes for this experiment: from no heads to all heads. Each outcome corresponds to a configuration, and each configuration has a number of specific orderings of the four coin flips, referred to as *permutations.* Which of these experimental outcomes is most likely? The simple answer to this question is that the most probable outcome is the one with the most possible ways to achieve that outcome. In the coin toss example, the configuration "2 Head" has the greatest number of ways to achieve this configuration; therefore, this configuration is the most likely experimental outcome. In the language of probability theory, the configuration with the largest number of corresponding permutations will be the most likely outcome. The probability, P_E, that a given configuration will be the outcome is given by:

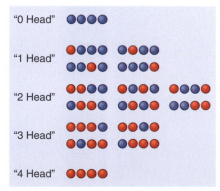

FIGURE 22.1
Possible configurations and permutations for an experiment involving the flipping of a coin four times. Blue indicates tails and red indicates heads.

$$P_E = \frac{E}{N} \tag{22.1}$$

where E is the number of permutations associated with the configuration of interest, and N is the total number of possible permutations (determined by summing over all configurations). This relatively simple equation provides the key idea for this entire chapter: *the most likely outcome for a trial is given by the configuration with the greatest number of associated permutations.* The application of this idea to macroscopic molecular systems is aided by the fact that for systems containing a large number of units, one configuration will have vastly more associated permutations than any other configuration. As such, this configuration will be the only one that is observed to an appreciable extent.

The connection of probability theory to chemistry is achieved by considering the distribution of energy in a system. Specifically, we are interested in developing a formalism that is capable of identifying the most likely configuration or distribution of energy in a chemical system. To begin, consider a simple "molecular" system consisting of three quantum harmonic oscillators that share a total of three quanta of energy. The energy levels for the oscillators are given by:

$$E_n = h\nu\left(n + \frac{1}{2}\right) \text{ for } n = 0, 1, 2, \ldots, \text{infinity} \tag{22.2}$$

In this equation, h is Planck's constant, ν is the oscillator frequency, and n is the quantum number associated with a given energy level of the oscillator. This quantum number can assume integer values starting from 0 and increasing to infinity; therefore, the energy levels of the quantum harmonic oscillator consist of a ladder or "manifold" of equally spaced levels. The lowest level ($n = 0$) has an energy of $h\nu/2$ referred to as the zero point energy. A modified version of the harmonic oscillator is employed here in which the ground-state energy ($n = 0$) is zero so that the level energies are given by

$$E_n = h\nu n \text{ for } n = 0, 1, 2, \ldots, \text{infinity} \tag{22.3}$$

When considering molecular systems in the upcoming chapters, the relative difference in energy levels will prove to be the relevant quantity of interest, and a similar modification will be employed. The final important point to note is that the oscillators are distinguishable. That is, each oscillator can be identified. This assumption must be relaxed when describing molecular systems such as an ideal gas because atomic or molecular motion prohibits identification of each particle. The extension of this approach to indistinguishable particles is readily accomplished; therefore, starting with the distinguishable case will not prove problematic.

In this example system, the three oscillators are of equal frequency such that the energy levels for all three are identical. Three quanta of energy are placed into this system, and the question becomes "What is the most probable distribution of energy?" From the preceding discussion, the most likely distribution of energy corresponds to the configuration with the largest number of associated permutations. How can the concepts of configuration and permutations be connected to oscillators and energy? Two central definitions are introduced here that will prove useful in making this connection: a **configuration** is a general arrangement of total energy available to the system, and a **microstate** is a specific arrangement of energy that describes the energy contained by each individual oscillator. This definition of a configuration is equivalent to the definition we have been employing in this chapter, and microstates are equivalent to permutations as previously defined. To determine the most likely configuration of energy in the example system, we can simply count all of the possible microstates and arrange them with respect to their corresponding configurations.

In the first configuration depicted in Figure 22.2, each oscillator has one quantum of energy so that all oscillators populate the $n = 1$ energy level. Only one permutation is associated with this configuration. In terms of the nomenclature just introduced, there is only one microstate corresponding to this energy configuration. In the next

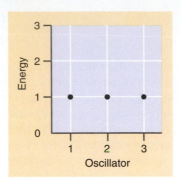

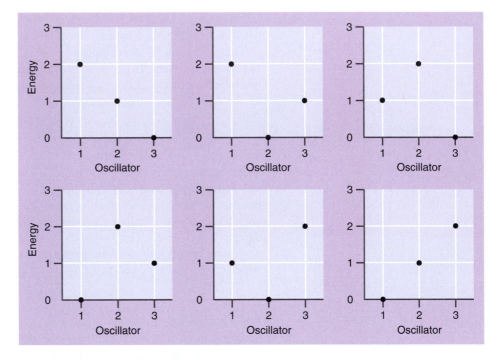

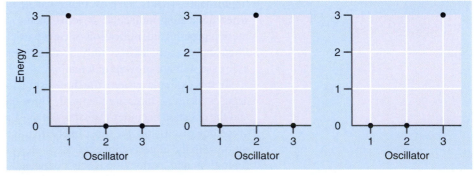

FIGURE 22.2
Configurations and associated microstates involving the distribution of three quanta of energy over three distinguishable oscillators.

configuration illustrated in the figure, one oscillator contains two quanta of energy, a second contains one quantum of energy, and a third contains no energy. Six potential arrangements correspond to this general distribution of energy; that is, six microstates correspond to this configuration. The last configuration depicted is one in which all three quanta of energy reside on a single oscillator. Because there are three choices for which oscillator will have all three quanta of energy, there are three corresponding microstates for this configuration. It is important to note that the total energy of all of the arrangements just mentioned is the same and that the only difference is the distribution of the energy over the oscillators.

Which configuration of energy would we expect to observe? Just like the coin toss example, the energy configuration that has the largest number of microstates is the most likely to be observed. In this example, that configuration is the second one discussed, or the "2, 1, 0" configuration. If all microstates depicted have an equal probability of being observed, the probability of observing the 2,1,0 configuration is simply the number of microstates associated with this configuration divided by the total number of microstates available, or

$$P_E = \frac{E}{N} = \frac{6}{6 + 3 + 1} = \frac{6}{10} = 0.6$$

Note that although this example involves a "molecular" system, the concepts encountered come directly from probability theory. Whether tossing a coin or distributing energy among distinguishable oscillators, the ideas are the same.

22.1.1 Counting Microstates and Weight

The three-oscillator example provides an approach for finding the most probable configuration of energy for a chemical system: determine all of the possible configurations of energy and corresponding microstates and identify the configuration with the greatest number of microstates. Clearly, this would be an extremely laborious task for a chemical system of interesting size. Fortunately, there are ways to obtain a quantitative count of all of the microstates associated with a given configuration without actually "counting" them. The total number of possible permutations given N objects is $N!$. For the most probable 2,1,0 configuration described earlier, there are three objects of interest (i.e., three oscillators) such that $N! = 3! = 6$. This is exactly the same number of microstates associated with this configuration. But what of the other configurations? Consider the 3,0,0 configuration in which one oscillator has all three quanta of energy. In assigning quanta of energy to this system to construct each microstate, there are three choices of where to place the three quanta of energy, and two remaining choices for zero quanta. However, this latter choice is redundant in that it does not matter which oscillator receives zero quanta first. The two conceptually different arrangements correspond to exactly the same microstate and, thus, are indistinguishable. To determine the number of microstates associated with such distributions of energy, the total number of possible permutations is divided by a factor that corrects for overcounting, which for the 3,0,0 configuration is accomplished as follows:

$$\text{Number of microstates} = \frac{3!}{2!} = 3$$

This expression is simply the probability expression for the number of permutations available using a subgroup from an overall group of size N. Therefore, if no two oscillators reside in the same energy level, then the total number of microstates available is given by $N!$, where N is the number of oscillators. However, if two or more oscillators occupy the same energy state (including the zero-energy state), then we need to correct for the overcounting of identical permutations. The total number of microstates associated with a given configuration of energy is referred to as the **weight** of the configuration, W, which is given by

$$W = \frac{N!}{a_0! \, a_1! \, a_2! \ldots a_n!} = \frac{N!}{\displaystyle\prod_n a_n!} \tag{22.4}$$

In Equation (22.4), W is the weight of the configuration of interest, N is the number of units over which energy is distributed, and the a_n terms represent the number of units occupying the nth energy level. The a_n quantities are referred to as **occupation numbers** because they describe how many units occupy a given energy level. For example, in the

3,0,0 configuration presented in Figure 22.2, $a_0 = 2$, $a_3 = 1$, and all other $a_n = 0$ (with $0! = 1$). The denominator in our expression for weight is evaluated by taking the product of the factorial of the occupation numbers ($a_n!$), with this product denoted by the Π symbol (which is analogous to the Σ symbol denoting summation). Equation (22.4) is not limited to our specific example, but is a general relationship that applies to any collection of distinguishable units for which only one state is available at a given energy level. The situation in which multiple states exist at a given energy level is discussed later in this chapter.

EXAMPLE PROBLEM 22.1

What is the weight associated with the configuration corresponding to observing 40 heads after flipping a coin 100 times? How does this weight compare to that of the most probable outcome?

Solution

Using the expression for weight of Equation (22.4), the coin flip can be envisioned as a system in which two states can be populated: heads or tails. In addition, the number of distinguishable units is 100, the number of coin tosses. Using these definitions,

$$W = \frac{N!}{a_H! \, a_T!} = \frac{100!}{40! 60!} = 1.37 \times 10^{28}$$

The most probable outcome corresponds to the configuration where 50 heads are observed such that

$$W = \frac{N!}{a_H! \, a_T!} = \frac{100!}{50! 50!} = 1.01 \times 10^{29}$$

22.1.2 Stirling's Approximation

When calculating values such a $N!$, if N is sufficiently small the factorial can be evaluated on a calculator. However, this approach to evaluating factorial quantities is limited to relatively small numbers. For example, 100! is equal to 9.3×10^{157}, which is an extremely large number, and beyond the range of many calculators. Furthermore, we are interested in extending the probability concepts we have developed up to chemical systems for which $N \sim 10^{23}$! The factorial of such a large number is simply beyond the computational ability of most calculators.

Fortunately, approximation methods are available that will allow us to calculate the factorial of large numbers. The most famous of these methods is known as **Stirling's approximation,** which provides a simple method by which to calculate the natural log of $N!$. A simplified version of this approximation is

$$\ln N! = N \ln N - N \tag{22.5}$$

Equation (22.5) is readily derived as follows:

$$\ln(N!) = \ln[(N)(N-1)(N-2)\ldots(2)(1)]$$

$$= \ln(N) + \ln(N-1) + \ln(N-2) + \cdots + \ln(2) + \ln(1)$$

$$= \sum_{n=1}^{N} \ln(n) \approx \int_1^N \ln(n)dn$$

$$= N \ln N - N - (1 \ln 1 - 1) \approx N \ln N - N \tag{22.6}$$

In this derivation, the summation over n is replaced by integration, an acceptable approximation when N is large. The final result is obtained by evaluating the integral over the limits indicated. Note that the main assumption inherent in this approximation is that N is a large number. The central concern in applying Stirling's approximation is whether

N is sufficiently large to justify its application. The following example problem illustrates this point.

EXAMPLE PROBLEM 22.2

Evaluate $\ln(N!)$ for $N = 10, 50$, and 100 using a calculator, and compare the result to that obtained using Stirling's approximation.

Solution

For $N = 10$ using a calculator we can determine that $N! = 3.63 \times 10^6$ and $\ln(N!)$ $= 15.1$. Using Stirling's approximation:

$$\ln(N!) = N \ln N - N = 10 \ln(10) - 10 = 13.0$$

This value represents a 13.9% error relative to the exact result, a substantial difference. The same procedure for $N = 50$ and 100 results in the following:

N	ln(N!) Calculated	ln(N!) Stirling	Error (%)
50	148.5	145.6	2.0
100	363.7	360.5	0.9

Example Problem 22.2 demonstrates that there are significant differences between the exact and approximate results even for $N = 100$. The example also demonstrates that the magnitude of this error decreases as N increases. For the chemical systems encountered in subsequent chapters, N will be $\sim 10^{22}$, many orders of magnitude larger than the values studied in this example. Therefore, for our purposes Stirling's approximation represents an elegant and sufficiently accurate method by which to evaluate the factorial of large quantities.

22.1.3 The Dominant Configuration

The three-oscillator example from the preceding section illustrates a few key ideas that will guide us toward our development of the Boltzmann distribution. Specifically, weight is the total number of microstates corresponding to a given configuration. The probability of observing a configuration is given by the weight of that configuration divided by the total weight:

$$P_i = \frac{W_i}{W_1 + W_2 + \cdots + W_N} = \frac{W_i}{\displaystyle\sum_{j=1}^{N} W_j} \tag{22.7}$$

where P_i is the probability of observing configuration i, W_i is the weight associated with this configuration, and the denominator represents the sum of weights for all possible configurations. Equation (22.7) predicts that the configuration with the largest weight will have the greatest probability of being observed. The configuration with the largest weight is referred to as the **dominant configuration.**

Given the definition of the dominant configuration, the question arises as to how dominant this configuration is relative to other configurations. A conceptual answer to this question is provided by an experiment in which a coin is tossed 10 times. How probable is the outcome of four heads relative to five heads? Although the outcome of $n_H = 5$ has the largest weight, the weight of $n_H = 4$ is of comparable value. Therefore, observing the $n_H = 4$ configuration would not be at all surprising. But what if the coin were flipped 100 times? How likely would the outcome of $n_H = 40$ be relative to $n_H = 50$? This question was answered in Example Problem 22.1, and the weight of $n_H = 50$ was significantly greater than that of $n_H = 40$. As the number of tosses increases, the probability of observing heads 50% of the time becomes greater relative to any other outcome. In other words, the configuration associated with observing 50% heads should become the more dominant configuration as the number of coin tosses increases. An illustration of this expectation is presented in Figure 22.3 where the relative weights associated with

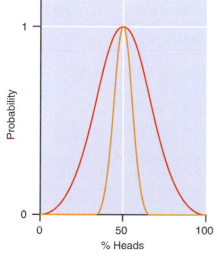

FIGURE 22.3
Comparison of relative probability (probability/maximum probability) for outcomes of a coin-flip trial in which the number of tosses is 10 (red line) and 100 (yellow line). The *x* axis is the percentage of tosses that are heads. Notice that all trials have a maximum probability at 50% heads; however, as the number of tosses increases, the probability distribution becomes more centered about this value as evidenced by the decrease in distribution width.

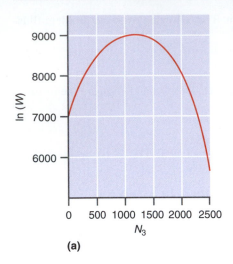

(a)

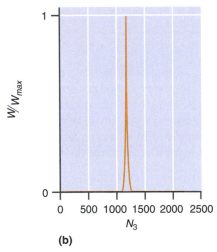

(b)

FIGURE 22.4
Illustration of the dominant configuration for a system consisting of 10,000 particles with each particle having three energy levels at energies of 0, ε, and 2ε as discussed in Example Problem 22.3. The number of particles populating the higher energy level is N_3, and the energy configurations are characterized by the population in this level. (a) Variation in the natural log of the weight, $\ln(W)$, for energy configurations as a function of N_3, demonstrating that $\ln(W)$ has a maximum at $N_3 \approx 1200$. (b) Variation in the weight associated with a given configuration to that of the dominant configuration. The weight is sharply peaked around $N_3 \approx 1200$ corresponding to the dominant configuration of energy.

observing heads a certain percentage of the time after tossing a coin 10 and 100 times are presented. The figure demonstrates that after 10 tosses, the relative probability of observing something other than five heads is still appreciable. However, as the number of coin tosses increases, the probability distribution narrows such that the final result of 50% heads becomes dominant. Taking this argument to sizes associated with molecular assemblies, imagine performing Avogadro's number of coin tosses! Given the trend illustrated in Figure 22.3, one would expect the probability to be sharply peaked at 50% heads. In other words, as the number of experiments in the trial increases, the 50% heads configuration will not only become the most probable, but the weight associated with this configuration will become so large that the probability of observing another outcome is minuscule. Indeed, the most probable configuration evolves into the dominant configuration as the size of the system increases.

EXAMPLE PROBLEM 22.3

Consider a collection of 10,000 particles with each particle capable of populating one of three energy levels having energies 0, ε, and 2ε with a total available energy of 5000ε. Under the constraint that the total number of particles and total energy be constant, determine the dominant configuration.

Solution

With constant total energy, only one of the energy-level populations is independent. Treating the number of particles in the highest energy level (N_3) as the independent variable, the number of particles in the intermediate (N_2) and lowest (N_1) energy levels is given by

$$N_2 = 5000 - 2N_3$$

$$N_1 = 10{,}000 - N_2 - N_3$$

Because the number of particles in a given energy level must be greater than or equal to 0, the preceding equations demonstrate that N_3 can range from 0 to 2500. Given the size of the state populations, it is more convenient to calculate the natural log of the weight associated with each configuration of energy as a function of N_3 by using Equation (22.4) in the following form:

$$\ln W = \ln\left(\frac{N!}{N_1!\,N_2!\,N_3!}\right) = \ln N! - \ln N_1! - \ln N_2! - \ln N_3!$$

Each term here can be readily evaluated using Stirling's approximation. The results of this calculation are presented in Figure 22.4. This figure demonstrates that $\ln(W)$ has a maximum value at $N_3 \approx 1200$ (or 1162 to be precise). The dominance of this configuration is also illustrated in Figure 22.4, where the weight of the configurations corresponding to the allowed values of N_3 are compared with that of the dominant configuration. Even for this relatively simple system having only 10,000 particles, the weight is sharply peaked at a configuration corresponding to the dominant configuration.

22.2 Derivation of the Boltzmann Distribution

As the size of the system increases, a single configuration will have such a large weight relative to the other configurations that only this configuration will be observed. In this limit, it becomes pointless to define all possible configurations, and only the outcome associated with this dominant configuration is of interest. Therefore, a method is needed to identify directly the configuration of interest. Figures 22.3 and 22.4 suggest that the dominant configuration can be determined as follows. The dominant configuration has the largest associated weight, and any change in outcome corresponding to a different configuration will be reflected by a reduction in weight. Therefore, the dominant con-

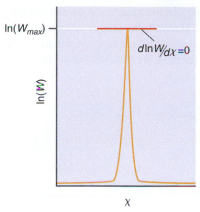

FIGURE 22.5

Mathematical definition of the dominant configuration. The change in the natural log of weight, $\ln(W)$, as a function of configuration index χ is presented. If we determine the change in $\ln(W)$ as a function of configuration index, the change will equal zero at the maximum of the curve corresponding to the location of the dominant configuration.

figuration can be identified by locating the peak of the curve corresponding to weight as a function of some **configurational index,** denoted by χ. Because W will be large for molecular systems, it is more convenient to work with $\ln W$, such that the search criterion for the dominant configuration becomes

$$\frac{d \ln W}{d\chi} = 0 \tag{22.8}$$

This expression is a mathematical definition of the dominant configuration, and it states that if a configuration space is searched by monitoring the change in $\ln W$ as a function of configurational index, a maximum will be observed that corresponds to the dominant configuration. A graphical description of the search criterion is presented in Figure 22.5.

The distribution of energy associated with the dominant configuration is known as the **Boltzmann distribution.** We begin our derivation of this distribution by taking the natural log of the weight using Equation (22.4) and applying Stirling's approximation:

$$\ln W = \ln N! - \ln \prod_n a_n!$$

$$= N \ln N - \sum_n a_n \ln a_n \tag{22.9}$$

The criterion for the dominant configuration requires differentiation of $\ln W$ by some relevant configurational index, but what is this index? We are interested in the distribution of energy among a collection of molecules, or the number of molecules that resides in a given energy level. Because the number of molecules residing in a given energy level is the occupation number, a_n, the occupation number provides a relevant configurational index. Recognizing this, differentiation of $\ln W$ with respect to a_n yields the following:

$$\frac{d \ln W}{da_n} = \frac{dN}{da_n} \ln N + N \frac{d \ln N}{da_n} - \sum_n \frac{d(a_n \ln a_n)}{da_n} \tag{22.10}$$

To evaluate the partial differentials on the right-hand side of Equation (22.10), the following relationships are used. (A discussion of partial derivatives is contained in Appendix A, Math Supplement.) First, the sum over occupation numbers is equal to the number of objects denoted by N:

$$N = \sum_n a_n \tag{22.11}$$

This last equation makes intuitive sense. The objects in our collection must be in one of the available energy levels; therefore, summation over the occupation numbers is equivalent to counting all objects. With this definition, the partial derivative of N with respect to a_n is simply one. Second, the following mathematical relationship is employed:

$$\frac{d \ln x}{dx} = \frac{1}{x}$$

such that

$$\frac{d \ln W}{da_n} = \ln N + N\left(\frac{1}{N}\right) - (\ln a_n + 1) = -\ln\left(\frac{a_n}{N}\right) \tag{22.12}$$

Comparison of Equation (22.12) and the search criteria for the dominant configuration suggests that if $\ln(a_n/N) = 0$ our search is complete. However, this assumes that the occupation numbers are independent, yet the sum of the occupation numbers must equal N dictating that reduction of one occupation number must be balanced by an increase in another. In other words, one object in the collection is free to gain or lose energy, but a corresponding amount of energy must be lost or gained elsewhere in the system. We can ensure that both the number of objects and total available energy are constant by requiring that

$$\sum_n da_n = 0 \quad \text{and} \quad \sum_n \varepsilon_n da_n = 0 \tag{22.13}$$

The da_n terms denote change in occupation number a_n. In the expression on the right, ε_n is the energy associated with a specific energy level. Because the conservation of N and energy has not been required to this point, these conditions are now included in the derivation by introducing Lagrange multipliers, α and β, as weights for the corresponding constraints into the differential as follows:

$$d \ln W = 0 = \sum_n -\ln\left(\frac{a_n}{N}\right)da_n + \alpha \sum_n da_n - \beta \sum_n \varepsilon_n \, da_n$$

$$= \sum_n \left(-\ln\left(\frac{a_n}{N}\right) + \alpha - \beta\varepsilon_n\right)da_n \tag{22.14}$$

The Lagrange method of undetermined multipliers is described in the Math Supplement (Appendix A). Formally, this technique allows for maximization of a function that is dependent on many variables that are constrained among themselves. The key step in this approach is to determine the identity of α and β by noting that in Equation (22.14) the equality is only satisfied when the collection of terms in parentheses is equal to zero:

$$0 = -\ln\left(\frac{a_n}{N}\right) + \alpha - \beta\varepsilon_n$$

$$\ln\left(\frac{a_n}{N}\right) = \alpha - \beta\varepsilon_n$$

$$\frac{a_n}{N} = e^\alpha e^{-\beta\varepsilon_n}$$

$$a_n = Ne^\alpha e^{-\beta\varepsilon_n} \tag{22.15}$$

At this juncture the Lagrange multipliers can be defined. First, α is defined by summing both sides of the preceding equality over all energy levels. Recognizing that $\Sigma a_n = N$,

$$N = \sum_n a_n = Ne^\alpha \sum_n e^{-\beta\varepsilon_n}$$

$$1 = e^\alpha \sum_n e^{-\beta\varepsilon_n}$$

$$e^\alpha = \frac{1}{\sum_n e^{-\beta\varepsilon_n}} \tag{22.16}$$

This last equality is a central result in statistical mechanics. The denominator in Equation (22.14) is referred to as the **partition function, q,** and is defined as follows:

$$q = \sum_n e^{-\beta\varepsilon_n} \tag{22.17}$$

The partition function represents the sum over all terms that describes the probability associated with the variable of interest, in this case ε_n, or the energy of level n. Using the partition function with Equation (22.13), the probability of occupying a given energy level, p_n, becomes

$$p_n = \frac{a_n}{N} = e^\alpha e^{-\beta\varepsilon_n} = \frac{e^{-\beta\varepsilon_n}}{q} \tag{22.18}$$

Equation (22.18) is the final result of interest. It quantitatively describes the probability of occupying a given energy for the dominant configuration of energy. This well-known and important result is referred to as the *Boltzmann distribution*.

How one applies Equation (22.18) to a molecular system is still unclear since we have yet to evaluate β, the second of our Lagrange multipliers. However, a conceptual

FIGURE 22.6
Two example oscillators. In case 1, the energy spacing is β^{-1}, and the energy spacing is $\beta^{-1}/2$ in case 2.

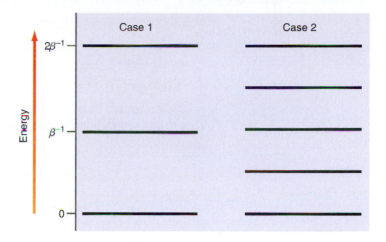

discussion can provide some insight into how this distribution describes molecular systems. Imagine a collection of harmonic oscillators as in the previous example, but instead of writing down all microstates and identifying the dominant configuration, the dominant configuration is instead given by the Boltzmann distribution law of Equation (22.18). This law establishes that the probability of observing an oscillator in a given energy level is dependent on level energy (ε_n) as $\exp(-\beta\varepsilon_n)$. Because the exponent must be unitless, β must have units of inverse energy.[1] Recall that the energy levels of the harmonic oscillator (neglecting zero point energy) are $\varepsilon_n = nh\nu$ for $n = 0, 1, 2, \ldots, \infty$. Therefore, our conceptual example will employ oscillators where $h\nu = \beta^{-1}$ as illustrated in Figure 22.6.

Using this value for the energy spacings, the exponential terms in the Boltzmann distribution are easily evaluated:

$$e^{-\beta\varepsilon_n} = e^{-\beta(n/\beta)} = e^{-n} \qquad (22.19)$$

The partition function is evaluated by performing the summation over the energy levels:

$$q = \sum_{n=0}^{\infty} e^{-n} = 1 + e^{-1} + e^{-2} + \cdots$$

$$= \frac{1}{1 - e^{-1}} = 1.58 \qquad (22.20)$$

In the last step of this example, we have used the following series expression (where $x < 1$):

$$\frac{1}{1 - x} = 1 + x + x^2 + \cdots$$

With the partition function, the probability of an oscillator occupying the first three levels ($n = 0, 1,$ and 2) is

$$p_0 = \frac{e^{-\beta\varepsilon_0}}{q} = \frac{e^{-0}}{1.58} = 0.633$$

$$p_1 = \frac{e^{-\beta\varepsilon_1}}{q} = \frac{e^{-1}}{1.58} = 0.233$$

$$p_2 = \frac{e^{-\beta\varepsilon_2}}{q} = \frac{e^{-2}}{1.58} = 0.086 \qquad (22.21)$$

---

[1]This simple unit analysis points to an important result: that β is related to energy. As will be shown, $1/\beta$ provides a measure of the energy available to the system.

EXAMPLE PROBLEM 22.4

For the example just discussed, what is the probability of finding an oscillator in energy levels $n \geq 3$?

Solution

The Boltzmann distribution is a normalized probability distribution. As such, the sum of all probabilities equals unity:

$$p_{total} = 1 = \sum_{n=0}^{\infty} p_n$$

$$1 = p_0 + p_1 + p_2 + \sum_{n=3}^{\infty} p_n$$

$$1 - (p_0 + p_1 + p_2) = 0.048 = \sum_{n=3}^{\infty} p_n$$

In other words, only 4.8% of the oscillators in our collection will be found in levels $n \geq 3$.

We continue with our conceptual example by asking the following question: "How will the probability of occupying a given level vary with a change in energy separation between levels?" In the first example, the energy spacings were equal to β^{-1}. A reduction in energy-level spacings to half this value requires that $h\nu = \beta^{-1}/2$. It is important to note that β has not changed relative to the previous example; only the separation in energy levels has changed. With this new energy separation, the exponential terms in the Boltzmann distribution become

$$e^{-\beta\varepsilon_n} = e^{-\beta(n/2\beta)} = e^{-n/2} \tag{22.22}$$

Substituting this equation into the expression for the partition function,

$$q = \sum_{n=0}^{\infty} e^{-n/2} = 1 + e^{-1/2} + e^{-1} + \cdots + e^{-\infty}$$

$$= \frac{1}{1 - e^{-1/2}} = 2.54 \tag{22.23}$$

Using this value for the partition function, the probability of occupying the first three levels ($n = 0, 1,$ and 2) corresponding to this new spacing is

$$p_0 = \frac{e^{-\beta\varepsilon_0}}{q} = \frac{e^{-0}}{2.54} = 0.394$$

$$p_1 = \frac{e^{-\beta\varepsilon_1}}{q} = \frac{e^{-1/2}}{2.54} = 0.239$$

$$p_2 = \frac{e^{-\beta\varepsilon_2}}{q} = \frac{e^{-1}}{2.54} = 0.145 \tag{22.24}$$

Comparison with the previous system probabilities in Equation (22.21) illustrates some interesting results. First, with a decrease in energy-level spacings, the probability of occupying the lowest energy level ($n = 0$) decreases, whereas the probability of occupying the higher energy levels increases. Reflecting this change in probabilities, the value of the partition function has also increased. Since the partition function represents the sum of the probability terms over all energy levels, an increase in the magnitude of the partition function reflects an increase in the probability of occupying higher energy levels. That is, the partition function provides a measure of the number of energy levels that are occupied for a given value of β.

EXAMPLE PROBLEM 22.5

For the preceding example with decreased energy-level spacings presented, what is the probability of finding an oscillator in energy states $n \geq 3$?

Solution

The calculation from the previous example is used to find that

$$\sum_{n=3}^{\infty} p_n = 1 - (p_0 + p_1 + p_2) = 0.222$$

Consistent with the discussion, the probability of occupying higher energy levels has increased substantially with a reduction in level spacings.

22.2.1 Degeneracy

To this point we have assumed that only one state is present at a given energy level; however, when discussing atomic and molecular systems more than a single state may be present for a given energy. The existence of multiple states at a given energy level is referred to as **degeneracy.** Degeneracy is incorporated into the expression for the partition function as follows:

$$q = \sum_n g_n e^{-\beta \varepsilon_n} \tag{22.25}$$

Equation (22.25) is identical to the previous definition of q from Equation (22.17) with the exception that the term g_n has been included. This term represents the number of states present at a given energy level, or the degeneracy of the level. The corresponding expression for the probability of occupying energy level ε_i is

$$p_i = \frac{g_i e^{-\beta \varepsilon_i}}{q} \tag{22.26}$$

where g_i is the degeneracy of the level with energy-level ε_i, and q is as defined in Equation (22.25).

How does degeneracy influence probability? Consider Figure 22.7 in which a system with single states at energy 0 and β^{-1} is shown with a similar system in which two states are present at energy β^{-1}. The partition function for the first system is

$$q_{system1} = \sum_n g_n e^{-\beta \varepsilon_n} = 1 + e^{-1} = 1.37 \tag{22.27}$$

FIGURE 22.7
Illustration of degeneracy. In system 1, one state is present at energies 0 and β^{-1}. In system 2, the energy spacing is the same, but at energy β^{-1} two states are present such that the degeneracy at this energy is two.

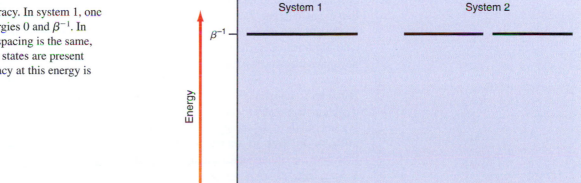

For the second system, q is

$$q_{system\,2} = \sum_n g_n e^{-\beta\varepsilon_n} = 1 + 2e^{-1} = 1.74 \qquad (22.28)$$

The corresponding probability of occupying a state at energy β^{-1} for the two systems is given by

$$p_{system\,1} = \frac{g_i e^{-\beta\varepsilon_i}}{q} = \frac{e^{-1}}{1.34} = 0.27$$

$$p_{system\,2} = \frac{g_i e^{-\beta\varepsilon_i}}{q} = \frac{2e^{-1}}{1.74} = 0.42 \qquad (22.29)$$

Notice that the probability of occupying a state at energy β^{-1} is greater for system 2, the system with degeneracy. This increase reflects the fact that two states are now available for population at this energy. However, this increase is not simply twice that of the nondegenerate case (system 1) because the value of the partition function also changes due to degeneracy.

22.3 Physical Meaning of the Boltzmann Distribution Law

How does one know other configurations exist if the dominant configuration is all one expects to see for a system at equilibrium? Furthermore, are the other nondominant configurations of the system of no importance? Modern experiments are capable of displacing systems from equilibrium and monitoring the system as it relaxes back toward equilibrium. Therefore, the capability exists to experimentally prepare a nondominant configuration so that these configurations can be studied. An illuminating, conceptual answer to this question is provided by the following logical arguments. First, consider the central postulate of statistical mechanics:

> Every possible microstate of an isolated assembly of units occurs with equal probability.

How does one know that this postulate is true? For example, imagine a collection of 100 oscillators having 100 quanta of energy. The total number of microstates available to this system is on the order of 10^{200}, which is an extremely large number. Now, imagine performing an experiment in which the energy content of each oscillator is measured such that the corresponding microstate can be established. Also assume that a measurement can be performed every 10^{-9} s (1 nanosecond) such that microstates can be measured at a rate of 10^9 microstates per second. Even with such a rapid determination of microstates, it would take us 10^{191} s to count every possible microstate, a period of time that is much larger than the age of the universe! In other words, the central postulate cannot be verified experimentally. However, we will operate under the assumption that the central postulate is true because statistical mechanical descriptions of chemical systems have provided successful and accurate descriptions of macroscopic systems.

Even if the validity of the central postulate is assumed, the question of its meaning remains. To gain insight into this question, consider a large or macroscopic collection of distinguishable and identical oscillators. Furthermore, the collection is isolated, resulting in both the total energy and the number of oscillators being constant. Finally, the oscillators are free to exchange energy such that any configuration of energy (and, therefore, any microstate) can be achieved. The system is set free to evolve, and the following features are observed:

1. All microstates are equally probable; however, one has the greatest probability of observing a microstate associated with the dominant configuration.

2. As demonstrated in the previous section, configurations having a significant number of microstates will be only infinitesimally different from the dominant configuration. The macroscopic properties of the system will be identical to that of the dominant configuration. Therefore, with overwhelming probability, one will observe a macroscopic state of the system characterized by the dominant configuration.

3. Continued monitoring of the system will result in the observation of macroscopic properties of the systems that appear unchanging, although energy is still being exchanged between the oscillators in our assembly. This macroscopic state of the system is called the **equilibrium state.**

4. Given items 1 through 3, the equilibrium state of the system is characterized by the dominant configuration.

This logical progression brings us to an important conclusion: *the Boltzmann distribution law describes the energy distribution associated with a chemical system at equilibrium.* In terms of probability, the fact that all microstates have equal probability of being observed does not translate into an equal probability of observing all configurations. As illustrated in Section 22.2, the vast majority of microstates correspond to the Boltzmann distribution, thereby dictating that the most probable configuration that will be observed is the one characterized by the Boltzmann distribution.

22.4 The Definition of β

Use of the Boltzmann distribution requires a definition for β, preferably one in which this quantity is defined in terms of measurable system variables. Such a definition can be derived by considering the variation in weight, W, as a function of total energy contained by an assembly of units, E. To begin, imagine an assembly of 10 oscillators having only three quanta of total energy. In this situation, the majority of the oscillators occupy the lowest energy states, and the weight corresponding to the dominant configuration should be small. However, as energy is deposited into the system, the oscillators will occupy higher energy states and the denominator in Equation (22.4) will be reduced, resulting in an increase in W. Therefore, one would expect E and W to be correlated.

The relationship between E and W can be determined by taking the natural log of Equation (22.4):

$$\ln W = \ln N! - \ln \prod_n a_n!$$

$$= \ln N! - \sum_n \ln a_n! \qquad (22.30)$$

Interest revolves around the change in W with respect to E, a relationship that requires the total differential of W:

$$d \ln W = - \sum_n d \ln a_n!$$

$$= - \sum_n \ln a_n \, da_n \qquad (22.31)$$

The result provided by Equation (22.31) was derived using Stirling's approximation to evaluate $\ln(a_n!)$ and recognizing that $\sum_n da_n = 0$. Simplification of Equation (22.31) is accomplished using the Boltzmann relationship to define the ratio between the occupation number for an arbitrary energy level, ε_n, versus the lowest or ground energy level ($\varepsilon_0 = 0$):

$$\frac{a_n}{a_0} = \frac{\dfrac{Ne^{-\beta\varepsilon_n}}{q}}{\dfrac{Ne^{-\beta\varepsilon_0}}{q}} = e^{-\beta\varepsilon_n} \qquad (22.32)$$

$$\ln a_n = \ln a_0 - \beta\varepsilon_n \qquad (22.33)$$

In the preceding steps, the partition function, q, and N are the same and simply cancel. Taking this expression for $\ln(a_n)$ and substituting into Equation (22.31) yields

$$d \ln W = -\sum_{n} (\ln a_0 - \beta \varepsilon_n) da_n$$

$$= -\ln a_0 \sum_{n} da_n + \beta \sum_{n} \varepsilon_n da_n \qquad (22.34)$$

The first summation in Equation (22.34) represents the total change in occupation numbers, and it is equal to the change in the total number of oscillators in the system. Because the system is closed with respect to the number of oscillators, $dN = 0$ and the first summation is also equal to zero. The second term represents the change in total energy of the system (dE) accompanying the deposition of energy into the system:

$$\sum_{n} \varepsilon_n da_n = dE \qquad (22.35)$$

With this last equality, the relationship between β, weight, and total energy is finally derived:

$$d \ln W = \beta dE \qquad (22.36)$$

This last equality is quite remarkable and provides significant insight into the physical meaning of β. We began by recognizing that weight increases in proportion with the energy available to the system, and β is simply the proportionality constant in this relationship. Unit analysis of Equation (22.36) also demonstrates that β must have units of inverse energy as inferred previously.

Associating β with measurable system variables is the last step in deriving a full definition of the Boltzmann distribution. This step can be accomplished through the following conceptual experiment. Imagine two separate systems of distinguishable units at equilibrium having associated weights W_x and W_y. Next, these assemblies are brought into thermal contact, and the composite system is allowed to evolve toward equilibrium, as illustrated in Figure 22.8. Furthermore, the composite system is isolated from the surroundings such that the total energy available to the composite system is the sum of energy contained in the individual assemblies. The total weight of the combined system immediately after establishing thermal contact is the product of W_x and W_y. If the two systems are initially at different equilibrium conditions, the instantaneous composite system weight will be less than the weight of the composite system at equilibrium. Since the composite weight will increase as equilibrium is approached

$$d(W_x \cdot W_y) \geq 0 \qquad (22.37)$$

This inequality can be simplified by applying the chain rule for differentiation (see the Math Supplement, Appendix A):

$$W_y dW_x + W_x dW_y \geq 0$$

$$\frac{dW_x}{W_x} + \frac{dW_y}{W_y} \geq 0 \qquad (22.38)$$

$$d \ln W_x + d \ln W_y \geq 0$$

Substitution of Equation (22.36) into the last expression of Equation (22.38) results in

$$\beta_x dE_x + \beta_y dE_y \geq 0 \qquad (22.39)$$

where β_x and β_y are the corresponding β values associated with the initial assemblies x and y. Correspondingly, dE_x and dE_y refer to the change in total energy for the individual assemblies. Because the composite system is isolated from the surroundings, any change in energy for assembly x must be offset by a corresponding change in assembly y:

$$dE_x + dE_y = 0$$

$$dE_x = -dE_y \qquad (22.40)$$

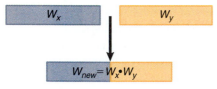

FIGURE 22.8

Two assemblies of distinguishable units, denoted x and y, are brought into thermal contact.

Now, if dE_x is positive, then by Equation (22.39)

$$\beta_x \geq \beta_y \qquad (22.41)$$

Can the preceding result be interpreted in terms of system variables? This question can be answered by considering the following. If dE_x is positive, energy flows into assembly x from assembly y. Thermodynamics dictates that because temperature is a measure of internal kinetic energy, and an increase in the energy will be accompanied by an increase in the temperature of assembly x. A corresponding decrease in the temperature of assembly y will occur. Therefore, before equilibrium is established, thermodynamic considerations dictate that

$$T_y \geq T_x \qquad (22.42)$$

In order for Equations (22.41) and (22.42) to be true, β must be inversely related to T. Furthermore, from unit analysis of Equation (22.36), we know that β must have units of inverse energy. This requirement is met by including a proportionality constant in the relationship between β and T, resulting in the final expression for β:

$$\beta = \frac{1}{kT} \qquad (22.43)$$

The constant in Equation (22.43), k, is referred to as **Boltzmann's constant** and has a numerical value of 1.381×10^{-23} J K^{-1}. The product of k and Avogadro's number is equal to R, the ideal gas constant (8.314 J mol^{-1} K^{-1}). Although the joule is the SI unit for energy, much of the information regarding molecular energy levels is derived from spectroscopic measurements. These spectroscopic quantities are generally expressed in units of wavenumbers (cm^{-1}). The **wavenumber** is simply the number of waves in an electromagnetic field per centimeter. Conversion from wavenumbers to joules is performed by multiplying the quantity in wavenumbers by Planck's constant, h, and the speed of light, c. In Example Problem 22.6, the vibrational energy levels for I$_2$ are given by the vibrational frequency of the oscillator, $\tilde{\nu} = 208$ cm^{-1} = 208 cm^{-1}. Using this spectroscopic information, the vibrational level energies in joules are

$$E_n = nhc\tilde{\nu} = n(6.626 \times 10^{-34} \text{ J s})(3 \times 10^{10} \text{ cm s}^{-1})(208 \text{ cm}^{-1})$$

$$= n(4.13 \times 10^{-21} \text{ J}) \qquad (22.44)$$

At times the conversion from wavenumbers to joules will prove inconvenient. In such cases, Boltzmann's constant can be expressed in units of wavenumbers instead of joules where $k = 0.695$ cm^{-1} K^{-1}. In this case, the spectroscopic quantities in wavenumbers can be used directly when evaluating partition functions and other statistical-mechanical expressions.

EXAMPLE PROBLEM 22.6

The vibrational frequency of I$_2$ is 208 cm^{-1}. What is the probability of I$_2$ populating the $n = 2$ vibrational level if the molecular temperature is 298 K?

Solution

Molecular vibrational energy levels can be modeled as harmonic oscillators; therefore, this problem can be solved employing a strategy identical to the one just presented. To evaluate the partition function q, the "trick" used earlier was to write the partition function as a series and use the equivalent series expression:

$$q = \sum_n e^{-\beta\varepsilon_n} = 1 + e^{-\beta hc\tilde{\nu}} + e^{-2\beta hc\tilde{\nu}} + e^{-3\beta hc\tilde{\nu}} + \cdots$$

$$= \frac{1}{1 - e^{-\beta hc\tilde{\nu}}}$$

Since $\tilde{\nu} = 208$ cm^{-1} and $T = 298$ K, the partition function is

$$q = \frac{1}{1 - e^{-\beta hc\tilde{\nu}}}$$

$$= \frac{1}{1 - \exp\left[-\left(\dfrac{(6.626 \times 10^{-34}\,\text{J s})(3.00 \times 10^{10}\,\text{cm s}^{-1})(208\,\text{cm}^{-1})}{(1.38 \times 10^{-23}\,\text{J K}^{-1})(298\,\text{K})}\right)\right]}$$

$$= \frac{1}{1 - e^{-1}} = 1.58$$

This result is then used to evaluate the probability of occupying the second vibrational state ($n = 2$) as follows:

$$p_2 = \frac{e^{-2\beta hc\tilde{\nu}}}{q}$$

$$= \frac{\exp\left[-2\left(\dfrac{(6.626 \times 10^{-34}\,\text{J s})(3.00 \times 10^{10}\,\text{cm s}^{-1})(208\,\text{cm}^{-1})}{(1.38 \times 10^{-23}\,\text{J K}^{-1})(298\,\text{K})}\right)\right]}{1.58} = 0.086$$

The last result in Example Problem 22.6 should look familiar. An identical example was worked earlier in this chapter where the energy-level spacings were equal to β^{-1} (case 1 in Figure 22.6) and the probability of populating states for which $n = 0$, 1, and 2 was determined. This previous example in combination with the molecular example just presented illustrates that the exponential term $\beta\varepsilon_n$ in the Boltzmann distribution and partition function can be thought of as a comparative term that describes the ratio of the energy needed to populate a given energy level versus the thermal energy available to the system, as quantified by kT. Energy levels that are significantly higher in energy than kT are not likely to be populated, whereas the opposite is true for energy levels that are small relative to kT.

Example Problem 22.6 is reminiscent of the development presented in Chapter 18 in which the use of the Boltzmann distribution to predict the relative population in vibrational and rotational states and the effect of these populations on vibrational and rotational transition intensities was presented. In addition, the role of the Boltzmann distribution in nuclear magnetic resonance spectroscopy (NMR) was outlined in Chapter 20 and is further explored in the following Example Problem.

the Chemistry place

22.1 Exploring the partition function of a harmonic oscillator

EXAMPLE PROBLEM 22.7

In NMR spectroscopy, energy separation between spin states is created by placing nuclei in a magnetic field. Protons have two possible spin states: $+1/2$ and $-1/2$. The energy separation between these two states, ΔE, is dependent on the strength of the magnetic field, and is given by

$$\Delta E = g_N \beta_N B = (2.82 \times 10^{-26}\,\text{J T}^{-1})B$$

where B is the magnetic field strength in tesla (T). Early NMR spectrometers employed magnetic field strengths of approximately 1.45 T. What is the ratio of the population between the two spin states given this magnetic field strength and $T = 298$ K?

Solution

Using the Boltzmann distribution, the occupation number for energy levels is given by

$$a_n = \frac{Ne^{-\beta\varepsilon_n}}{q}$$

where N is the number of particles, ε_n is the energy associated with the level of interest, and q is the partition function. Using the preceding equation, the ratio of occupation numbers is given by

$$\frac{a_{1/2}}{a_{-1/2}} = \frac{\dfrac{Ne^{-\beta\varepsilon_{1/2}}}{q}}{\dfrac{Ne^{-\beta\varepsilon_{-1/2}}}{q}} = e^{-\beta(\varepsilon_{1/2}-\varepsilon_{-1/2})} = e^{-\beta\Delta E}$$

Substituting for ΔE and β (and taking care that units cancel), the ratio of occupation numbers is given by

$$\frac{a_{1/2}}{a_{-1/2}} = e^{-\beta\Delta E} = \exp\left[\frac{-(2.82\times10^{-26}\text{ J T}^{-1})(1.45\text{ T})}{(1.38\times10^{-23}\text{ J K}^{-1})(298\text{ K})}\right]$$

$$= e^{-(9.94\times10^6)} = 0.999990$$

In other words, in this system the energy spacing is significantly smaller than the energy available (kT) such that the higher energy spin state is populated to a significant extent, and is nearly equal in population to that of the lower energy state.

For Further Reading

Chandler, D., *Introduction to Modern Statistical Mechanics*. Oxford, New York, 1987.

Hill, T., *Statistical Mechanics. Principles and Selected Applications*. Dover, New York, 1956.

McQuarrie, D., *Statistical Mechanics*. Harper & Row, New York, 1973.

Nash, L. K., "On the Boltzmann Distribution Law." *J. Chemical Education* 59 (1982), 824.

Nash, L. K., *Elements of Statistical Thermodynamics*. Addison-Wesley, San Francisco, 1972.

Noggle, J. H., *Physical Chemistry*. HarperCollins, New York, 1996.

Vocabulary

Boltzmann distribution	degeneracy	microstate	Stirling's approximation
Boltzmann's constant	dominant configuration	occupation number	wavenumber
configuration	equilibrium state	partition function	weight
configurational index			

Questions on Concepts

Q22.1 What is the difference between a configuration and a microstate?

Q22.2 How does one calculate the number of microstates associated with a given configuration?

Q22.3 What is Stirling's approximation, and when is it valid?

Q22.4 What is an occupation number? How is this number used to describe energy distributions?

Q22.5 Explain the significance of the Boltzmann distribution. What does this distribution describe?

Q22.6 What is degeneracy? Can you conceptually relate the expression for the partition function without degeneracy to that with degeneracy?

Q22.7 How is β related to temperature? What are the units of kT?

Problems

Problem numbers in **RED** indicate that the solution to the problem is given in the *Student Solutions Manual*.

P22.1

  a. What is the possible number of microstates associated with tossing a coin N times and having it come up H times heads and T times tails?

  b. For a series of 1000 tosses, what is the total number of microstates associated with 50% heads and 50% tails?

  c. How less probable is the outcome that the coin will land 40% heads and 60% tails?

P22.2

  a. Realizing that the most probable outcome from a series of N coin tosses is $N/2$ heads and $N/2$ tails, what is the expression for W_{max} corresponding to this outcome?

b. Given your answer for part (a), derive the following relationship between the weight for an outcome other than the most probable and W_{max}:

$$\log\left(\frac{W}{W_{max}}\right) = -H \log\left(\frac{H}{N/2}\right) - T \log\left(\frac{T}{N/2}\right)$$

c. We can define the deviation of a given outcome from the most probable outcome using a "deviation index," $\alpha = (H - T)/N$. Show that the number of heads or tails can be expressed as $H = (N/2)(1 + \alpha)$ and $T = (N/2)(1 - \alpha)$.

d. Finally, demonstrate that $W/W_{max} = e^{-N\alpha^2}$.

P22.3 Consider the case of 10 oscillators and eight quanta of energy. Determine the dominant configuration of energy for this system by identifying energy configurations and calculating the corresponding weights. What is the probability of observing the dominant configuration?

P22.4 Determine the weight associated with the following card hands:

a. Having any five cards

b. Having five cards of the same suit (known as a "flush")

P22.5 Suppose that you draw a card from a standard deck of 52 cards. What is the probability of drawing:

a. an ace of any suit?

b. the ace of spades?

c. How would your answers change if you were allowed to draw three times, replacing the card drawn back into the deck after each draw?

P22.6 You are dealt a hand consisting of 5 cards from a standard deck of 52 cards. Determine the probability of obtaining the following hands:

a. flush (five cards of the same suit)

b. a king, queen, jack, ten, and ace of the same suit (a "royal flush")

P22.7 A pair of standard dice are rolled. What is the probability of observing the following:

a. The sum of the dice is equal to 7.

b. The sum of the dice is equal to 9.

c. The sum of the dice is less than or equal to 7.

P22.8 Radio station call letters consist of four letters (for example, KUOW).

a. How many different station call letters are possible using the 26 letters in the English alphabet?

b. Stations west of the Mississippi River must use the letter K as the first call letter. Given this requirement, how many different station call letters are possible if repetition is allowed for any of the remaining letters?

c. How many different station call letters are possible if repetition is not allowed for any of the letters?

P22.9 Four bases (A, C, T, and G) appear in DNA. Assume that the appearance of each base in a DNA sequence is random.

a. What is the probability of observing the sequence AAGACATGCA?

b. What is the probability of finding the sequence GGGGGAAAAA?

c. How do your answers to parts (a) and (b) change if the probability of observing A is twice that of the probabilities used in parts (a) and (b) of this question when the preceding base is G?

P22.10 Imagine an experiment in which you flip a coin four times. Furthermore, the coin is balanced fairly such that the probability of landing heads or tails is equivalent. After tossing the coin 10 times, what is the probability of observing:

a. no heads?

b. two heads?

c. five heads?

d. eight heads?

P22.11 Imagine performing the coin-flip experiment of Problem P22.10, but instead of using a fair coin, a weighted coin is employed for which the probability of landing heads is twofold greater than landing tails. After tossing the coin 10 times, what is the probability of observing:

a. no heads?

b. two heads?

c. five heads?

d. eight heads?

P22.12 For a two-level system, the weight of a given energy distribution can be expressed in terms of the number of systems, N, and the number of systems occupying the excited state, n_1. What is the expression for weight in terms of these quantities?

P22.13 The probability of occupying a given excited state, p_i, is given by $p_i = n_i/N = e^{-\beta\varepsilon_i}/q$, where n_i is the occupation number for the state of interest, N is the number of particles, and ε_i is the energy of the level of interest. Demonstrate that the preceding expression is independent of the definition of energy for the lowest state.

P22.14 Barometric pressure can be understood using the Boltzmann distribution. The potential energy associated with being a given height above the Earth's surface is mgh, where m is the mass of the particle of interest, g is the acceleration due to gravity, and h is height. Using this definition of the potential energy, derive the following expression for pressure: $P = P_0 e^{-mgh/kT}$. Assuming that the temperature remains at 298 K, what would you expect the relative pressures of N_2 and O_2 to be at the tropopause, the boundary between the troposphere and stratosphere roughly 11 km above the Earth's surface? At the Earth's surface, the composition of air is roughly 78% N_2, 21% O_2, and the remaining 1% is other gases.

P22.15 Consider the following energy-level diagrams:

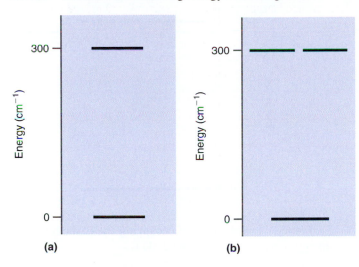

(a)   (b)

a. At what temperature will the probability of occupying the second energy level be 0.15 for the states depicted in part (a) of the figure?

b. Perform the corresponding calculation for the states depicted in part (b) of the figure. Before beginning the calculation, do you expect the temperature to be higher or lower than that determined in part (a) of this problem? Why?

P22.16 Consider the following energy-level diagrams, modified from Problem P22.15 by the addition of another excited state with energy of 600 cm^{-1}:

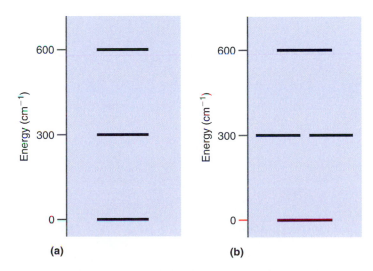

(a)   (b)

a. At what temperature will the probability of occupying the second energy level be 0.15 for the states depicted in part (a) of the figure?

b. Perform the corresponding calculation for the states depicted in part (b) of the figure.

(*Hint:* You may find this problem easier to solve numerically using a spreadsheet program such as Excel.)

P22.17 Consider the following sets of populations for four equally spaced energy levels:

ε/k (K)	Set A	Set B	Set C
300	5	3	4
200	7	9	8
100	15	17	16
0	33	31	32

a. Demonstrate that the sets have the same energy.

b. Determine which of the sets is the most probable.

c. For the most probable set, is the distribution of energy consistent with a Boltzmann distribution?

P22.18 A set of 13 particles occupies states with energies of 0, 100, and 200 cm^{-1}. Calculate the total energy and number of microstates for the following energy configurations:

a. $a_0 = 8$, $a_1 = 5$, and $a_2 = 0$

b. $a_0 = 9$, $a_1 = 3$, and $a_2 = 1$

c. $a_0 = 10$, $a_1 = 1$, and $a_2 = 2$

Do any of these configurations correspond to the Boltzmann distribution?

P22.19 For a set of nondegenerate levels with energy ε/k = 0, 100, and 200 K, calculate the probability of occupying each state when $T = 50$, 500, and 5000 K. As the temperature continues to increase, the probabilities will reach a limiting value. What is this limiting value?

P22.20 Consider a collection of molecules where each molecule has two nondegenerate energy levels that are separated by 6000 cm^{-1}. Measurement of the level populations demonstrates that there are eight times more molecules in the ground state than in the upper state. What is the temperature of the collection?

P22.21 The ^{13}C nucleus is a spin 1/2 particle as is a proton. However, the energy splitting for a given field strength is roughly 1/4 of that for a proton. Using a 1.45-T magnet as in Example Problem 22.7, what is the ratio of populations in the excited and ground spin states for ^{13}C at 298 K?

P22.22 ^{14}N is a spin 1 particle such that the energy levels are at 0 and $\pm \gamma B\hbar$, where γ is the magnetogyric ratio and B is the strength of the magnetic field. In a 4.8-T field, the energy splitting between any two spin states expressed as the resonance frequency is 14.45 MHz. Determine the occupation numbers for the three spin states at 298 K.

P22.23 The vibrational frequency of I_2 is 208 cm^{-1}. At what temperature will the population in the first excited state be half that of the ground state?

P22.24 The vibrational frequency of Cl_2 is 525 cm^{-1}. Will the temperature be higher or lower relative to I_2 (see Problem P22.6) at which the population in the first excited vibrational state is half that of the ground state? What is this temperature?

P22.25 Determine the partition function for the vibrational degrees of freedom of Cl_2 ($\tilde{\nu} = 525$ cm^{-1}) and calculate the probability of occupying the first excited vibrational level at 300 and 1000 K. Determine the temperature at which identical probabilities will be observed for F_2 ($\tilde{\nu} = 917$ cm^{-1}).

P22.26 A two-level system is characterized by an energy separation of 1.3×10^{-18} J. At what temperature will the population of the ground state be five times greater than that of the excited state?

P22.27 The lowest two electronic energy levels of the molecule NO are illustrated here:

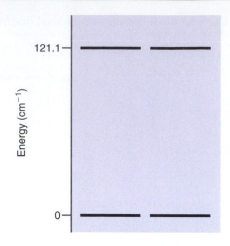

Determine the probability of occupying one of the higher energy states at 100, 500, and 2000 K.

Web-Based Simulations, Animations, and Problems

W22.1 In this simulation the behavior of the partition function for a harmonic oscillator with temperature and oscillator frequency is explored. The variation in q with temperature for an oscillator where $\tilde{\nu} = 1000$ cm^{-1} is studied, and variation in the individual level contributions to the partition function is studied. In addition, the variation in level contributions at fixed temperature, but as the oscillator frequency is varied, is depicted. This simulation provides insight into the elements of the partition function and the variation of this function with temperature and energy-level spacings.

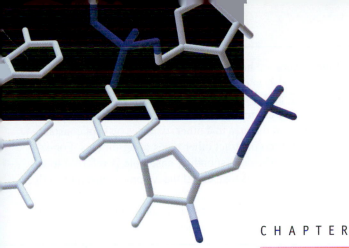

CHAPTER

23

Statistical Thermodynamics

The relationship between the microscopic description of individual molecules and the macroscopic properties of a collection of molecules is a central concept in statistical mechanics. In this chapter, the relationship between the partition function derived from the energy levels of an individual molecule and the partition function describing an ensemble of molecules is established. We demonstrate that the molecular partition function can be decomposed into the product of partition functions for each energetic degree of freedom, and the functional form of these partition functions is derived. With these partition functions and the application of statistical thermodynamics, expressions for fundamental thermodynamic quantities such as internal energy, entropy, and Gibbs free energy can be obtained. As will be shown, the statistical perspective is not only capable of reproducing the thermodynamic properties of matter, it also provides critical insight into the microscopic details behind these properties.

23.1 The Canonical Ensemble

An **ensemble** is defined as a collection of identical units or replicas of a system. For example, a mole of water can be envisioned as an ensemble with Avogadro's number of identical units of water molecules. The ensemble provides a theoretical concept by which the microscopic properties of matter can be related to the corresponding thermodynamic system properties as expressed in the following postulate:

> The average value for a property of the ensemble corresponds to the time-averaged value for the corresponding macroscopic property of the system.

What does this postulate mean? Imagine the individual units of the ensemble sampling the available energy space; the energy content of each unit is measured at a single time, and the measured unit energies are used to determine the average energy for the ensemble. According to the postulate, this energy will be equivalent to the average energy of the ensemble as measured over time. This idea, first formulated by J. W. Gibbs in the late 1800s, lies at the heart of statistical thermodynamics.

To connect the ensemble average values and thermodynamic properties of the ensemble, we begin by imagining a collection of identical copies of the system as illustrated

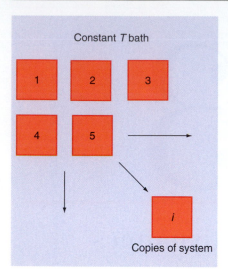

FIGURE 23.1
The canonical ensemble is comprised of a collection of identical systems having fixed temperature, volume, and number of particles. The units are embedded in a constant T bath. The arrows indicate that an infinite number of copies of the system comprises the ensemble.

in Figure 23.1. These copies of the system are held fixed in space such that they are distinguishable. The volume, V, temperature, T, and number of particles in each system, N, are constant. An ensemble in which V, T, and N are constant is referred to as a **canonical ensemble.** The term *canonical* means "by common practice" because this is the ensemble employed unless the problem of interest dictates that other variables be kept constant. Note that other quantities can be constant to construct other types of ensembles. For example, if N, V, and energy are held fixed, the corresponding ensemble is referred to as *microcanonical.* However, for the purposes of this text, use of the canonical ensemble will prove sufficient.

In the canonical ensemble, each ensemble member is embedded in a temperature bath such that the total ensemble energy is constant. Furthermore, the walls that define the volume of the units can conduct heat, allowing for energy exchange with the surroundings. The challenge is to link the statistical development presented in the previous chapter to a similar statistical description for this ensemble. We begin by considering the total energy of the ensemble, E_c, which is given by

$$E_c = \sum_i a_{(c)i} E_i \tag{23.1}$$

In Equation (23.1), the terms $a_{(c)i}$ are the occupation numbers corresponding to the number of ensemble members having energy E_i. Proceeding exactly as in the previous chapter, the weight (W_c) associated with a specific configuration of energy among the N_c members of the ensemble is given by

$$W_c = \frac{N_c!}{\prod_i a_{(c)i}!} \tag{23.2}$$

This relationship can be used to derive the probability of finding an ensemble unit at energy E_i:

$$p(E_i) = \frac{W_i e^{-\beta E_i}}{Q} \tag{23.3}$$

Equation (23.3) looks very similar to the probability expression derived in Chapter 22. In this equation, W_i can be thought of as the number of states present at a given energy E_i. The quantity Q in Equation (23.3) is referred to as the **canonical partition function** and is defined as follows:

$$Q = \sum_i e^{-\beta E_i} \tag{23.4}$$

In Equation (23.4), the summation is over all energy levels. The probability defined in Equation (23.3) is dependent on two factors: W_i, or the number of states present at a given energy that will increase with energy, and a Boltzmann term, $e^{-\beta E_i}/Q$, that describes the probability of an ensemble unit having energy E_i that decreases exponentially with energy. The generic behavior of each term with energy is depicted in Figure 23.2. The product of these terms will reach a maximum corresponding to the average ensemble energy. The figure illustrates that an individual unit of the ensemble will have an energy that is equal to or extremely close to the average energy, and that units having energy far from this value will be exceedingly rare. We know this to be the case from experience. Imagine a swimming pool filled with water divided up into 1-liter units. If the thermometer at the side of the pool indicates that the water temperature is 18°C, someone diving into the pool will not be worried that the liter of water immediately under his or her head will spontaneously freeze. That is, the temperature measured in one part of the pool is sufficient to characterize the temperature of the water in any part of the pool. Figure 23.2 provides an illustration of the statistical aspects underlying this expectation.

The vast majority of systems in the ensemble will have energy <E>; therefore, the thermodynamic properties of the unit are representative of the thermodynamic proper-

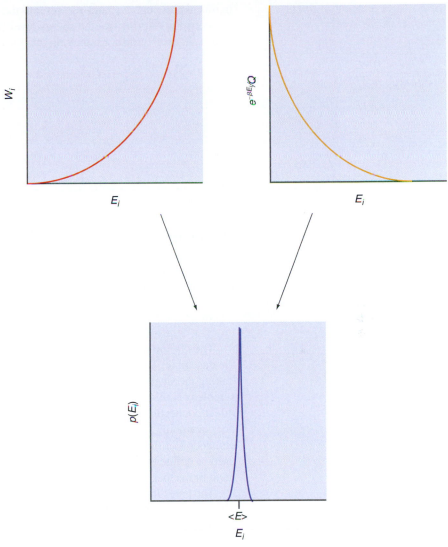

FIGURE 23.2
For the canonical ensemble, the probability of a member of the ensemble having a given energy is dependent on the product of W_i, the number of states present at a given energy, and the Boltzmann distribution function for the ensemble. The product of these two factors results in a probability distribution that is peaked about the average energy, $<E>$.

ties of the ensemble, demonstrating the link between the microscopic unit and the macroscopic ensemble. To make this connection mathematically exact, the canonical partition function, Q, must be related to the partition function describing the individual members of the ensemble, q.

23.2 Relating Q to q for an Ideal Gas

In relating the canonical partition function, Q, to the partition function describing the members of the ensemble, q, our discussion is limited to systems consisting of independent "ideal" particles in which the interactions between particles is negligible (for example, an ideal gas). The relationship between Q and q is derived by considering an ensemble made up of two distinguishable units, A and B, as illustrated in Figure 23.3. For this simple ensemble, the partition function is

$$Q = \sum_n e^{-\beta E_n} = \sum_n e^{-\beta(\varepsilon_{A_n} + \varepsilon_{B_n})} \tag{23.5}$$

FIGURE 23.3
A two-unit ensemble. In this ensemble, the two units, A and B, are distinguishable.

In this expression, ε_{A_n} and ε_{B_n} refer to the energy levels associated with unit A and B, respectively. Assume that the energy levels are quantized such that they can be indexed as 0, 1, 2, and so forth. Employing this idea, Equation (23.5) becomes

$$Q = \sum_n e^{-\beta(\varepsilon_A + \varepsilon_B)} = e^{-\beta(\varepsilon_{A_0} + \varepsilon_{B_0})} + e^{-\beta(\varepsilon_{A_0} + \varepsilon_{B_1})} + e^{-\beta(\varepsilon_{A_0} + \varepsilon_{B_2})} + \cdots$$

$$+ e^{-\beta(\varepsilon_{A_1} + \varepsilon_{B_0})} + e^{-\beta(\varepsilon_{A_1} + \varepsilon_{B_1})} + e^{-\beta(\varepsilon_{A_1} + \varepsilon_{B_2})} + \cdots$$

$$+ e^{-\beta(\varepsilon_{A_2} + \varepsilon_{B_0})} + e^{-\beta(\varepsilon_{A_2} + \varepsilon_{B_1})} + e^{-\beta(\varepsilon_{A_2} + \varepsilon_{B_2})} + \cdots$$

$$= (e^{-\beta\varepsilon_{A_0}} + e^{-\beta\varepsilon_{A_1}} + e^{-\beta\varepsilon_{A_2}} + \cdots)(e^{-\beta\varepsilon_{B_0}} + e^{-\beta\varepsilon_{B_1}} + e^{-\beta\varepsilon_{B_2}} + \cdots)$$

$$= (q_A)(q_B)$$

$$= q^2$$

The last step in the derivation is accomplished by recognizing that the ensemble units are identical such that the partition functions are also identical. Extending the preceding result to a system with N distinguishable units, the canonical partition function is found to be simply the product of unit partition functions

$$Q = q^N \text{ for } N \text{ distinguishable units} \qquad (23.6)$$

Thus far no mention has been made of the size of the identical systems comprising the ensemble. The systems can be as small as desired, including just a single molecule. Taking the single-molecule limit, a remarkable conclusion is reached: the canonical ensemble is nothing more than the product of the molecular partition functions. This is the direct connection between the microscopic and macroscopic perspectives that we have been searching for. The quantized energy levels of the molecular (or atomic) system are embedded in the **molecular partition function, q,** and this partition function can be used to define the partition function for the ensemble, Q. Finally, Q can be directly related to the thermodynamic properties of the ensemble.

The preceding derivation assumed that the ensemble members were distinguishable. This might be the case for a collection of molecules coupled to a surface where they cannot move, but what happens to this derivation when the ensemble is in the gaseous state? Clearly, the translational motion of the gas molecules will make identification of each individual molecule impossible. Therefore, how does Equation (23.6) change if the units are indistinguishable? A simple counting example will help to answer this question. Consider three distinguishable oscillators (A, B, and C) with three total quanta of energy as described in the previous chapter. The dominant configuration of energy was with the oscillators in three separate energy states, denoted "2, 1, 0." The energy states relative to the oscillators can be arranged in six different ways:

A	B	C
2	1	0
2	0	1
1	2	0
0	2	1
1	0	2
0	1	2

However, if the three oscillators are indistinguishable, there is no difference among the arrangements listed. In effect, there is only one arrangement of energy that should be counted. To correct for overcounting in the case of indistinguishable particles, the total number of permutations is divided by $N!$ where N is the number of particles. Extending this logic to a molecular ensemble dictates that the canonical partition function for indistinguishable particles have the following form:

$$Q = \frac{q^N}{N!} \text{ for } N \text{ indistinguishable units} \qquad (23.7)$$

Equation (23.7) is correct in the limit for which the number of energy levels available is significantly greater than the number of particles. This discussion of statistical mechanics is limited to systems for which this is true, and the validity of this statement is demonstrated later in this chapter. It is also important to keep in mind that Equation (23.7) is limited to ideal systems of noninteracting particles such as an ideal gas.

23.3 Molecular Energy Levels

The relationship between the canonical and molecular partition functions provides the link between the microscopic and macroscopic descriptions of the system. The molecular partition function can be evaluated by considering molecular energy levels. For polyatomic molecules, there are four **energetic degrees of freedom** to consider in constructing the molecular partition function:

1. Translation
2. Rotation
3. Vibration
4. Electronic

Assuming the energetic degrees of freedom are not coupled, the total molecular partition function that includes all of these degrees of freedom can be decomposed into a product of partition functions corresponding to each degree of freedom. Let ε_{Total} represent the energy associated with a given molecular energy level. This energy will depend on the translational, rotational, vibrational, and electronic level energies as follows:

$$\varepsilon_{Total} = \varepsilon_T + \varepsilon_R + \varepsilon_V + \varepsilon_E \tag{23.8}$$

Recall that the molecular partition function is obtained by summing over molecular energy levels. Using the expression for the total energy and substituting into the expression for the partition function, the following expression is obtained:

$$q_{Total} = \sum g_{Total} e^{-\beta \varepsilon_{Total}}$$

$$= \sum \left(g_T g_R g_V g_E \right) e^{-\beta(\varepsilon_T + \varepsilon_R + \varepsilon_V + \varepsilon_E)}$$

$$= \sum \left(g_T e^{-\beta \varepsilon_T} \right) \left(g_R e^{-\beta \varepsilon_R} \right) \left(g_V e^{-\beta \varepsilon_V} \right) \left(g_E e^{-\beta \varepsilon_E} \right)$$

$$= q_T q_R q_V q_E \tag{23.9}$$

This relationship demonstrates that the total molecular partition function is simply the product of partition functions for each molecular energetic degree of freedom. Using this definition for the molecular partition function, the final relationships of interest are

$$Q_{Total} = q_{Total}^N \ (\text{distinguishable}) \tag{23.10}$$

$$Q_{Total} = \frac{1}{N!} q_{Total}^N \ (\text{indistinguishable}) \tag{23.11}$$

All that remains to derive are partition functions for each energetic degree of freedom, a task that is accomplished in the remainder of this chapter.

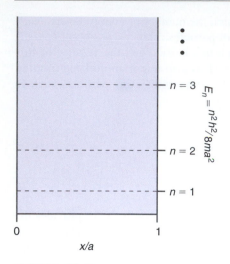

FIGURE 23.4
Particle-in-a-box model for translational energy levels.

23.4 Translational Partition Function

Translational energy levels correspond to the translational motion of atoms or molecules in a container of volume V. Rather than work directly in three dimensions, a one-dimensional model is first employed and then later extended to three dimensions. From quantum mechanics, the energy levels of a molecule confined to a box were described by the "particle-in-a-box" model as illustrated in Figure 23.4. In this figure, a particle with mass m is free to move in the domain $0 \leq x \leq a$, where a is the length of the box. Using the expression for the energy levels provided in Figure 23.4, the partition function for translational energy in one dimension becomes

$$q_{T,1D} = \sum_{n=1}^{\infty} e^{-\beta n^2 h^2 / 8ma^2} \tag{23.12}$$

Notice that the summation consists of an infinite number of terms. Furthermore, a closed-form expression for this series does not exist such that it appears one must evaluate the sum directly. However, a way around this apparently impossible task becomes evident when the spacing between energy translational energy states is considered, as illustrated in the following example.

EXAMPLE PROBLEM 23.1

What is the difference in energy between the $n = 2$ and $n = 1$ states for molecular oxygen constrained by a one-dimensional box having a length of 1 cm?

Solution

The energy difference is obtained by using the expression for the one-dimensional particle-in-a-box model as follows:

$$\Delta E = E_2 - E_1 = 3E_1 = \frac{3h^2}{8ma^2}$$

The mass of an O_2 molecule is 5.31×10^{-26} kg such that

$$\Delta E = \frac{3(6.626 \times 10^{-34} \text{ J s})^2}{8(5.31 \times 10^{-26} \text{ kg})(0.01 \text{ m})^2}$$

$$= 3.10 \times 10^{-38} \text{ J}$$

Converting to units of cm^{-1}:

$$\Delta E = \frac{3.10 \times 10^{-38} \text{ J}}{hc} = 1.57 \times 10^{-15} \text{ cm}^{-1}$$

At 298 K, the amount of thermal energy available as given by the product of Boltzmann's constant and temperature, kT, is 208 cm^{-1}. Clearly, the spacings between translational energy levels are extremely small relative to kT at room temperature.

Because numerous translational energy levels are accessible at room temperature, the summation in Equation (23.12) can be replaced by integration with negligible error:

$$q_T = \sum e^{-\beta \alpha n^2} \approx \int_0^{\infty} e^{-\beta \alpha n^2} \, dn \tag{23.13}$$

In this expression, the following substitution was made to keep the collection of constant terms compact:

$$\alpha = \frac{h^2}{8ma^2} \tag{23.14}$$

The integral in Equation (23.13) is readily evaluated (see Appendix A, Math Supplement):

$$q_T \approx \int_0^\infty e^{-\beta\alpha n^2} dn = \frac{1}{2}\sqrt{\frac{\pi}{\beta\alpha}} \qquad (23.15)$$

Substituting for α, the **translational partition function** in one dimension becomes

$$q_{T,1D} = \left(\frac{2\pi m}{h^2\beta}\right)^{1/2} a \qquad (23.16)$$

This expression can be simplified by defining the **thermal de Broglie wavelength,** or simply the **thermal wavelength, Λ,** as follows:

$$\Lambda = \left(\frac{h^2\beta}{2\pi m}\right)^{1/2} \qquad (23.17)$$

such that

$$q_{T,1D} = \frac{a}{\Lambda} = (2\pi mkT)^{1/2}\frac{a}{h} \qquad (23.18)$$

Referring to Λ as the thermal wavelength reflects the fact that the average momentum of a gas particle, p, is equal to $(mkT)^{1/2}$. Therefore, Λ is essentially h/p, or the de Broglie wavelength of the particle. Extension of the one-dimensional result to three dimensions is straightforward. The translational degrees of freedom are considered separable; therefore, the three-dimensional translational partition function is the product of one-dimensional partition functions for each dimension:

$$q_{T,3D} = q_{T_x} q_{T_y} q_{T_z}$$
$$= \left(\frac{a_x}{\Lambda}\right)\left(\frac{a_y}{\Lambda}\right)\left(\frac{a_z}{\Lambda}\right)$$
$$= \left(\frac{1}{\Lambda}\right)^3 V$$

$$q_{T,3D} = \frac{V}{\Lambda^3} = (2\pi mkT)^{3/2}\frac{V}{h^3} \qquad (23.19)$$

where V is volume and Λ is the thermal wavelength [Equation (23.17)]. Notice that the translational partition is a function of both V and T. Recall the discussion from the previous chapter in which the partition function was described conceptually as providing a measure of the number of energy states available to the system at a given temperature. The increase in q_T with volume reflects the fact that as volume is increased, the translational energy-level spacings decrease such that more states are available for population at a given T. Given the small energy spacings between translational energy levels relative to kT at room temperature, we might expect that at room temperature a significant number of translational energy states are accessible. The following example provides a test of this expectation.

EXAMPLE PROBLEM 23.2

What is the translational partition function for Ar confined to a volume of 1 L at 298 K?

Solution

Evaluation of the translational partition function is dependent on determining the thermal wavelength [Equation (23.17)]:

$$\Lambda = \left(\frac{h^2\beta}{2\pi m}\right)^{1/2} = \frac{h}{(2\pi mkT)^{1/2}}$$

The mass of Ar is 6.63×10^{-26} kg. Using this value for m, the thermal wavelength becomes

$$\Lambda = \frac{6.626 \times 10^{-34} \text{ J s}}{[2\pi(6.63 \times 10^{-26} \text{ kg})(1.38 \times 10^{-23} \text{ J K}^{-1})(298 \text{ K})]^{1/2}}$$

$$= 1.60 \times 10^{-11} \text{ m}$$

The units of volume must be such that the partition function is unitless. Therefore, conversion of volume to units of cubic meters (m^3) is performed as follows:

$$V = 1 \text{ L} = 1000 \text{ mL} = 1000 \text{ cm}^3 \left(\frac{1\text{m}}{100 \text{ cm}}\right)^3 = 0.001 \text{ m}^3$$

The partition function is simply the volume divided by the thermal wavelength cubed:

$$q_{T,3D} = \frac{V}{\Lambda^3} = \frac{0.001 \text{ m}^3}{(1.60 \times 10^{-11} \text{ m})^3} = 2.44 \times 10^{29}$$

The magnitude of the translational partition function determined in Example Problem 23.2 illustrates that a vast number of translational energy states are available at room temperature. In fact, the number of accessible states is roughly 10^6 times larger than Avogadro's number, illustrating that the assumption that many more states are available relative to units in the ensemble (Section 23.2) is reasonable.

23.5 Rotational Partition Function

A **diatomic** molecule consists of two atoms joined by a chemical bond as illustrated in Figure 23.5. In treating rotational motion of diatomic molecules, the rigid-rotor approximation is employed in which the bond length is assumed to remain constant during rotational motion and effects such as centrifugal distortion are neglected.

In deriving the rotational partition function for a diatomic, an approach similar to that used in deriving the translational partition function is employed. Within the rigid-rotor approximation, the quantum mechanical description of rotational energy levels for diatomic molecules dictates that the energy of a given rotational state, E_J, is dependent on the rotational quantum number, J, as follows:

$$E_J = BJ(J+1) \text{ for } J = 0, 1, 2, \ldots \qquad (23.20)$$

where J is the quantum number corresponding to rotational energy level and can take on integer values beginning with zero. The quantity B is the **rotational constant** and is given by

$$B = \frac{h}{8\pi^2 cI} \qquad (23.21)$$

where I is the moment of inertia, which is equal to

$$I = \mu r^2 \qquad (23.22)$$

In the expression for the moment of inertia, r is the distance separating the two atomic centers and μ is the reduced mass, which for a diatomic consisting of atoms having masses m_1 and m_2 is equal to

$$\mu = \frac{m_1 m_2}{m_1 + m_2} \qquad (23.23)$$

Because diatomic molecules differ depending on the masses of atoms in the molecule and the bond length, the value of the rotational constant is molecule dependent. Using the preceding expression for the rotational energy, the rotational partition function can be constructed by simply substituting into the general form of the molecular partition function:

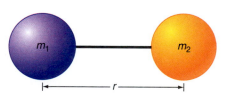

FIGURE 23.5

Schematic representation of a diatomic molecule, consisting of two masses (m_1 and m_2) joined by a chemical bond with the separation of atomic centers equal to the bond length, r.

$$q_R = \sum_J g_J e^{-\beta hcBJ(J+1)} \qquad (23.24)$$

In this expression, the energies of the levels included in the summation are given by $hcBJ(J + 1)$. However, notice that the expression for the rotational partition function contains an addition term, g_J, that represents the number of rotational states present at a given energy level, or the degeneracy of the rotational energy level. To determine the degeneracy, refer to Chapter 18 and the discussion of the rigid rotor and the time-independent Schrödinger equation:

$$H\psi = E\psi \qquad (23.25)$$

For the rigid rotor, the Hamiltonian (H) is proportional to the square of the total angular momentum given by the operator $\hat{l}^2$. The eigenstates of this operator are the spherical harmonics with the following eigenvalues:

$$\hat{l}^2\psi = \hat{l}^2 Y_{l,m}(\theta,\phi) = \frac{h^2}{4\pi^2}l(l + 1)Y_{l,m}(\theta,\phi) \qquad (23.26)$$

In this expression, l is a quantum number corresponding to total angular momentum, and it ranges from $0, 1, 2, \ldots$, to infinity. The spherical harmonics are also eigenfunctions of the $\hat{l}_z$ operator corresponding to the z component of the angular momentum. The corresponding eigenvalues employing the $\hat{l}_z$ operator are given by

$$\hat{P}_Z Y_{l,m}(\theta,\phi) = \frac{h}{2\pi}m Y_{l,m}(\theta,\phi) \qquad (23.27)$$

Possible values for the quantum number m in Equation (23.27) are dictated by the quantum number l:

$$m = -l \ldots 0 \ldots l \text{ or } (2l + 1) \qquad (23.28)$$

Thus, the degeneracy of the rotational energy levels originates from the quantum number m because all values of m corresponding to a given quantum number l will have the same total angular momentum and, therefore, the same energy. Using the value of $(2l + 1)$ for the degeneracy, the rotational partition function is

$$q_R = \sum_J (2J + 1)e^{-\beta hcBJ(J+1)} \qquad (23.29)$$

As written, evaluation of Equation (23.29) involves summation over all rotational states. A similar issue was encountered when the expression for the translational partition function was evaluated. The spacings between translational levels were very small relative to kT such that the partition function could be evaluated by integration rather than discrete summation. Are the rotational energy-level spacings also small relative to kT such that integration can be performed instead of summation?

To answer this question, consider the energy-level spacings for the rigid rotor presented in Figure 23.6. The energy of a given rotational state (in units of the rotational constant B) is presented as a function of the rotational quantum number, J. The energy-level spacings are multiples of B. The value of B will vary depending on the molecule of interest, with representative values provided in Table 23.1. Inspection of the table reveals a few interesting trends. First, the rotational constant depends on the atomic mass, with an increase in atomic mass resulting in a reduction in the rotational constant. Second, the values for B are quite different; therefore, any comparison of rotational state energies to kT will depend on the diatomic of interest. For example, at 298 K, $kT = 208 \text{ cm}^{-1}$, which is roughly equal to the energy of the $J = 75$ level of I_2. For this species, the energy-level spacings are clearly much smaller than kT and integration of the partition function is appropriate. However, for H_2 the $J = 2$ energy level is greater than kT so that integration would be inappropriate, and evaluation of the partition function by direct summation must be performed. In the remainder of this chapter, we assume that integration of the rotational partition function is appropriate unless stated otherwise.

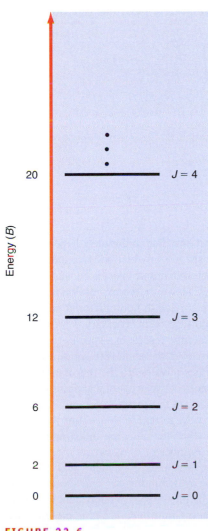

FIGURE 23.6

Rotational energy levels as a function of the rotational quantum number J. The energy of a given rotational state is equal to $BJ(J + 1)$.

TABLE 23.1 **Rotational Constants for Some Representative Diatomic Molecules**

Molecule	B (cm^{-1})	Molecule	B (cm^{-1})
H^{35}Cl	10.595	H$_2$	60.853
H^{37}Cl	10.578	^{14}N^{16}O	1.7046
D^{35}Cl	5.447	^{127}I^{127}I	0.03735

Source: G. Herzberg, *Molecular Spectra and Molecular Structure, Volume 1: Spectra of Diatomic Molecules*, Krieger Publishing, Melbourne, FL, 1989.

With the assumption that the rotational energy-level spacings are small relative to kT, evaluation of the rotational partition function is performed with integration over the rotational states:

$$q_R = \int_0^\infty \left(2J + 1\right) e^{-\beta hcBJ(J+1)} dJ \tag{23.30}$$

Evaluation of the preceding integral is simplified by recognizing the following:

$$\frac{d}{dJ} e^{-\beta hcBJ(J+1)} = -\beta hcB \left(2J + 1\right) e^{-\beta hcBJ(J+1)}$$

Using this relationship, the expression for the **rotational partition function** can be rewritten and the result evaluated as follows:

$$q_R = \int_0^\infty \left(2J + 1\right) e^{-\beta hcBJ(J+1)} dJ = \int_0^\infty \frac{-1}{\beta hcB} \frac{d}{dJ} e^{-\beta hcBJ(J+1)} dJ$$

$$= \frac{-1}{\beta hcB} e^{-\beta hcBJ(J+1)} \Big|_0^\infty = \frac{1}{\beta hcB}$$

$$q_R = \frac{1}{\beta hcB} = \frac{kT}{hcB} \tag{23.31}$$

23.5.1 The Symmetry Number

The expression for the rotational partition function of a diatomic molecule provided in the previous section is correct for heterodiatomic species in which the two atoms comprising the diatomic are not equivalent. HCl is a heterodiatomic species because the two atoms in the diatomic, H and Cl, are not equivalent. However, the expression for the rotational partition function must be modified when applied to homodiatomic molecules such as N$_2$. A simple illustration of why such a modification is necessary is presented in Figure 23.7. In the figure, rotation of the heterodiatomic results in a species that is distinguishable from the molecule before rotation. However, the same 180° rotation applied to a homodiatomic results in a configuration that is equivalent to the prerotation form. This difference in behavior is similar to the differences between canonical partition functions for distinguishable and indistinguishable units. In the partition function case, the result for the distinguishable case was divided by $N!$ to take into account the "overcounting" of nonunique microstates encountered when the units are indistinguishable. In a similar spirit, for homodiatomic species the number of classical rotational states (i.e., distinguishable rotational configurations) is overcounted by a factor of 2.

To correct our rotational partition function for overcounting, we can simply divide the expression for the rotational partition function by the number of equivalent rotational configurations. This factor is known as the **symmetry number,** σ, and is incorporated into the partition function as follows:

$$q_R = \frac{1}{\sigma \beta hcB} = \frac{kT}{\sigma hcB} \tag{23.32}$$

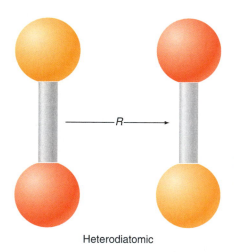

Heterodiatomic

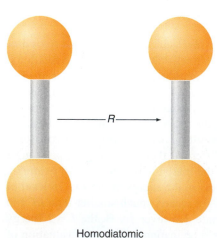

Homodiatomic

FIGURE 23.7
A 180° rotation of heterodiatomic and homodiatomic molecules.

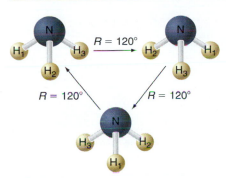

FIGURE 23.8
Rotational configurations for NH_3.

The concept of a symmetry number can be extended to molecules other than diatomics. For example, consider a trigonal pyramidal molecule such as NH_3 as illustrated in Figure 23.8. Imagine that performing a 120° rotation about an axis through the nitrogen atom and the center of the triangle made by the three hydrogens. The resulting configuration would be exactly equivalent to the previous configuration before rotation. Furthermore, a second 120° rotation would produce a third configuration. A final 120° would result in the initial prerotation configuration. Therefore, NH_3 has three equivalent rotational configurations; therefore, $\sigma = 3$.

EXAMPLE PROBLEM 23.3

What is the symmetry number for methane (CH_4)?

Solution

To determine the number of equivalent rotational configurations, we will proceed in a fashion similar to that employed for NH_3. The tetrahedral structure of methane is shown in the following figure:

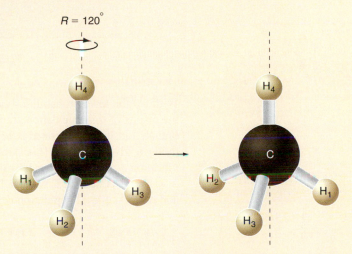

Similar to NH_3, three equivalent configurations can be generated by 120° rotation about the axis depicted by the dashed line in the figure. Furthermore, we can draw four such axes of rotation aligned with each of the four C—H bonds. Therefore, there are 12 total rotational configurations for CH_4 corresponding to $\sigma = 12$.

23.5.2 Rotational Level Populations and Spectroscopy

In Chapter 18, the relationship between the populations in various rotational energy levels and the rotational-vibrational infrared absorption intensities were described. With the rotational partition function, we are in a position to explore this relationship in detail. The probability of occupying a given rotational energy level, p_J, is given by

$$P_J = \frac{g_J e^{-\beta hcBJ(J+1)}}{q_R} = \frac{(2J+1)e^{-\beta hcBJ(J+1)}}{q_R} \tag{23.33}$$

$H^{35}Cl$ where $B = 10.595 \text{ cm}^{-1}$ was previously employed to illustrate the relationship between p_J and absorption intensity. At 300 K the rotational partition function for $H^{35}Cl$ is

$$q_R = \frac{1}{\sigma \beta hcB}$$

$$= \frac{kT}{\sigma hcB} = \frac{\left(1.38 \times 10^{-23} \text{J K}^{-1}\right)\left(300 \text{ K}\right)}{\left(1\right)\left(6.626 \times 10^{-34} \text{J s}\right)\left(3.00 \times 10^{10} \text{cm s}^{-1}\right)\left(10.595 \text{ cm}^{-1}\right)}$$

$$= 19.7 \tag{23.34}$$

FIGURE 23.9
Probability of occupying a rotational
energy level, p_J, as a function of rotational
quantum number J for H^{35}Cl at 300 K.

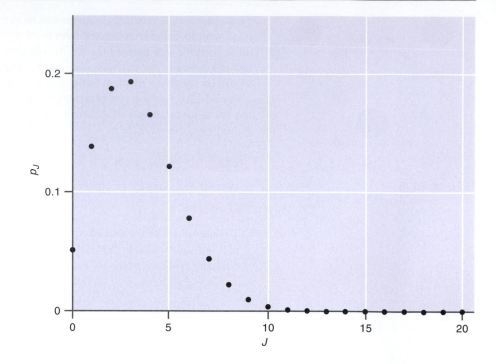

FIGURE 23.9
Probability of occupying a rotational
energy level, p_J, as a function of rotational
quantum number J for H^{35}Cl at 300 K.

With q_R, the level probabilities can be readily determined using Equation (23.33), and the results of this calculation are presented in Figure 23.9. The intensity of the P and R branch transitions illustrated in Figure 19.16 are proportional to the probability of occupying a given J level. This dependence is reflected by the evolution in rotational-vibrational transition intensity as a function of J. The transition moment demonstrates modest J dependence as well such that the correspondence between transition intensity and rotational level population is not exact.

EXAMPLE PROBLEM 23.4

In a rotational spectrum of HBr ($B = 8.46$ cm^{-1}), the maximum intensity is observed for the $J = 4$ to 5 transition. At what temperature was the spectrum obtained?

Solution

The information provided for this problem dictates that the $J = 4$ rotational energy level was the most populated at the temperature at which the spectrum was taken. To determine the temperature, we first determine the change in occupation number for the rotational energy level a_J versus J as follows:

$$a_J = \frac{N(2J + 1)e^{-\beta hcBJ(J+1)}}{q_R} = \frac{N(2J + 1)e^{-\beta hcBJ(J+1)}}{\left(\dfrac{1}{\beta hcB}\right)}$$

$$= N\beta hcB(2J + 1)e^{-\beta hcBJ(J+1)}$$

Next, we take the derivative of a_J with respect to J and set the derivative equal to zero to find the maximum of the function:

$$\frac{da_J}{dJ} = 0 = \frac{d}{dJ}N\beta hcB(2J + 1)e^{-\beta hcBJ(J+1)}$$

$$0 = \frac{d}{dJ}(2J + 1)e^{-\beta hcBJ(J+1)}$$

$$0 = 2e^{-\beta hcBJ(J+1)} - \beta hcB(2J + 1)^2 e^{-\beta hcBJ(J+1)}$$

$$0 = 2 - \beta hcB(2J + 1)^2$$

$$2 = \beta hcB(2J + 1)^2 = \frac{hcB}{kT}(2J + 1)^2$$

$$T = \frac{(2J + 1)^2 hcB}{2k}$$

Substitution of $J = 4$ into the preceding expression results in the following temperature at which the spectrum was obtained:

$$T = \frac{(2J + 1)^2 hcB}{2k}$$

$$= \frac{(2(4) + 1)^2(6.626 \times 10^{-34}\,\text{J s})(3.00 \times 10^{10}\,\text{cm s}^{-1})(8.46\,\text{cm}^{-1})}{2(1.38 \times 10^{-23}\,\text{J K}^{-1})}$$

$$= 494\,\text{K}$$

23.5.3 The Rotational Temperature

Whether the rotational partition function should be evaluated by direct summation or integration is entirely dependent on the size of the rotational energy spacings relative to the amount of thermal energy available (kT). This comparison is facilitated through the introduction of the **rotational temperature,** Θ_R, defined as the rotational constant divided by Boltzmann's constant:

$$\Theta_R = \frac{hcB}{k} \tag{23.35}$$

Unit analysis of Equation (23.35) dictates that Θ_R has units of temperature. We can rewrite the expression for the rotational partition function in terms of the rotational temperature as follows:

$$q_R = \frac{1}{\sigma \beta hcB} = \frac{kT}{\sigma hcB} = \frac{T}{\sigma \Theta_R} \tag{23.36}$$

A second application of the rotational temperature is as a comparative metric to the temperature at which the partition function is being evaluated. For temperatures where $T/\Theta_R \geq 10$, the integrated form of the rotational partition function can be used instead of performing a summation to evaluate q_R. The integrated form of the partition function is referred to as the **high-temperature** or **high-T limit** because it is applicable when kT is significantly greater than the rotational energy spacings. The following example illustrates the use of the rotational temperature in deciding which functional form of the rotational partition function to use.

EXAMPLE PROBLEM 23.5

Evaluate the rotational partition functions for I_2 at $T = 100$ K.

Solution

Because $T = 100$ K, it is important to ask how kT compares to the rotational energy-level spacings. Using Table 23.1, $B(I_2) = 0.0374$ cm^{-1} corresponding to rotational temperatures of

$$\Theta_R(I_2) = \frac{hcB}{k} = \frac{(6.626 \times 10^{-34}\,\text{J s})(3.00 \times 10^{10}\,\text{cm s}^{-1})(0.0374\,\text{cm}^{-1})}{1.38 \times 10^{-23}\,\text{J K}^{-1}} = 0.054\,\text{K}$$

Comparison of these rotational temperatures to 100 K indicates that the high-temperature expression for the rotational partition function is valid for I_2:

$$q_R(I_2) = \frac{T}{\sigma \Theta_R} = \frac{100\,\text{K}}{(2)(0.054\,\text{K})} = 926$$

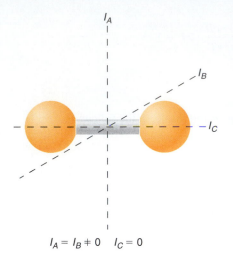

$I_A = I_B \neq 0 \quad I_C = 0$

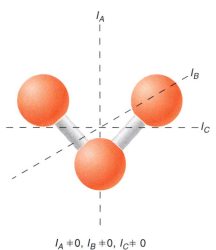

$I_A \neq 0, \; I_B \neq 0, \; I_C \neq 0$

FIGURE 23.10
Moments of inertia for diatomic and nonlinear polyatomic molecules. Note that in the case of the diatomic, $I_C = 0$ in the limit that the atomic masses are considered to be point masses that reside along the axis connecting the two atomic centers. Each moment of inertia will have a corresponding rotational constant.

23.5.4 Rotational Partition Function: Polyatomics

In the diatomic systems described in the preceding section, there are two nonvanishing moments of inertia as illustrated in Figure 23.10. For **polyatomic** molecules (more than two atoms) the situation can become more complex.

If the polyatomic system is linear, there are again only two nonvanishing moments of inertia such that a linear polyatomic molecule can be treated using the same formalism as diatomic molecules. However, if the polyatomic molecule is not linear, then there are three nonvanishing moments of inertia. Therefore, the partition function that describes the rotational energy levels must take into account rotation about all three axes. Derivation of this partition function is not trivial; therefore, the result without derivation is stated here:

$$q_R = \frac{\sqrt{\pi}}{\sigma} \left(\frac{1}{\beta h c B_A} \right)^{1/2} \left(\frac{1}{\beta h c B_B} \right)^{1/2} \left(\frac{1}{\beta h c B_C} \right)^{1/2} \qquad (23.37)$$

The subscript on B in Equation (23.37) indicates the corresponding moment of inertia as illustrated in Figure 23.10, and σ is the symmetry number as discussed earlier. In addition, the assumption is made that the polyatomic is "rigid" during rotational motion. The development of the rotational partition function for diatomic systems provides some intuition into the origin of this partition function. One can envision each moment of inertia contributing $(\beta h c B)^{-1/2}$ to the overall partition function. In the case of diatomics or linear polyatomics, the two nonvanishing moments of inertia are equivalent such that the product of the contribution from each moment results in the expression for the diatomic derived earlier. For the nonlinear polyatomic system, the partition function is the product of the contribution from each of the moments of inertia, which may or may not be equivalent as indicated by the subscripts on the corresponding rotational constants in the partition function presented earlier.

> **EXAMPLE PROBLEM 23.6**
>
> Evaluate the rotational partition functions for the following species at 298 K. You can assume that the high-temperature expression is valid.
>
> **a.** OCS ($B = 1.48$ cm^{-1})
>
> **b.** ONCl ($B_A = 2.84$ cm^{-1}, $B_B = 0.191$ cm^{-1}, $B_C = 0.179$ cm^{-1})
>
> **c.** CH$_2$O ($B_A = 9.40$ cm^{-1}, $B_B = 1.29$ cm^{-1}, $B_C = 1.13$ cm^{-1})
>
> **Solution**
>
> **a.** OCS is a linear molecule as indicated by the single rotational constant. In addition, the molecule is asymmetric such that $\sigma = 1$. Using the rotational constant, the rotational partition function is
>
> $$q_R = \frac{1}{\sigma \beta h c B} = \frac{kT}{hcB}$$
>
> $$= \frac{(1.38 \times 10^{-23} \, \text{J K}^{-1})(298 \, \text{k})}{(6.626 \times 10^{-34} \, \text{J s})(3.00 \times 10^{10} \, \text{cm s}^{-1})(1.48 \, \text{cm}^{-1})} = 140$$
>
> **b.** ONCl is a nonlinear polyatomic. It is asymmetric such that $\sigma = 1$, and the partition function becomes:
>
> $$q_R = \frac{\sqrt{\pi}}{\sigma} \left(\frac{1}{\beta h c B_A} \right)^{1/2} \left(\frac{1}{\beta h c B_B} \right)^{1/2} \left(\frac{1}{\beta h c B_C} \right)^{1/2}$$
>
> $$= \sqrt{\pi} \left(\frac{kT}{hc} \right)^{3/2} \left(\frac{1}{B_A} \right)^{1/2} \left(\frac{1}{B_B} \right)^{1/2} \left(\frac{1}{B_C} \right)^{1/2}$$

$$= \sqrt{\pi}\left(\frac{\left(1.38 \times 10^{-23}\text{J K}^{-1}\right)\left(298\text{ K}\right)}{\left(6.626 \times 10^{-34}\text{J s}\right)\left(3.00 \times 10^{10}\text{cm s}^{-1}\right)}\right)^{3/2}\left(\frac{1}{2.84\text{ cm}^{-1}}\right)^{1/2}\left(\frac{1}{0.191\text{ cm}^{-1}}\right)^{1/2}$$

$$\times \left(\frac{1}{0.179\text{ cm}^{-1}}\right)^{1/2}$$

$$= 16{,}940$$

c. CH_2O is a nonlinear polyatomic. However, the symmetry of this molecule is such that $\sigma = 2$. With this value for the symmetry number, the rotational partition function becomes:

$$q_R = \frac{\sqrt{\pi}}{\sigma}\left(\frac{1}{\beta h c B_A}\right)^{1/2}\left(\frac{1}{\beta h c B_B}\right)^{1/2}\left(\frac{1}{\beta h c B_C}\right)^{1/2}$$

$$= \sqrt{\frac{\pi}{2}}\left(\frac{kT}{hc}\right)^{3/2}\left(\frac{1}{B_A}\right)^{1/2}\left(\frac{1}{B_B}\right)^{1/2}\left(\frac{1}{B_C}\right)^{1/2}$$

$$= \sqrt{\frac{\pi}{2}}\left(\frac{\left(1.38 \times 10^{-23}\text{J K}^{-1}\right)\left(298\text{ K}\right)}{\left(6.626 \times 10^{-34}\text{J s}\right)\left(3.00 \times 10^{10}\text{cm s}^{-1}\right)}\right)^{3/2}\left(\frac{1}{9.40\text{ cm}^{-1}}\right)^{1/2}\left(\frac{1}{1.29\text{ cm}^{-1}}\right)^{1/2}$$

$$\times \left(\frac{1}{1.13\text{ cm}^{-1}}\right)^{1/2}$$

$$= 711$$

Note that the values for all three partition functions indicate that a substantial number of rotational states are populated at room temperature.

23.6 Vibrational Partition Function

The quantum mechanical model for vibrational degrees of freedom is the harmonic oscillator. In this model, each vibrational degree of freedom is characterized by a quadratic potential as illustrated in Figure 23.11. The energy levels of the harmonic oscillator are as follows:

$$E_n = hc\tilde{\nu}\left(n + \frac{1}{2}\right) \tag{23.38}$$

This equation demonstrates that the energy of a given level, E_n, is dependent on the quantum number n, which can take on integer values beginning with zero ($n = 0, 1, 2, \ldots$). The frequency of the oscillator, or vibrational frequency, is given by $\tilde{\nu}$ in units of cm^{-1}. Note that the energy of the $n = 0$ level is not zero, but $hc\tilde{\nu}/2$. This residual energy is known as the zero point energy and was discussed in detail during the quantum mechanical development of the harmonic oscillator. The expression for E_n provided in Equation (23.38) can be used to construct the vibrational partition function as follows:

$$q_V = \sum_{n=0}^{\infty} e^{-\beta E_n}$$

$$= \sum_{n=0}^{\infty} e^{-\beta h c\tilde{\nu}\left(n+\frac{1}{2}\right)}$$

$$= e^{-\beta h c\tilde{\nu}/2}\sum_{n=0}^{\infty} e^{-\beta h c\tilde{\nu}n} \tag{23.39}$$

The sum can be rewritten using the series identity:

$$\frac{1}{1 - e^{-\alpha x}} = \sum_{n=0}^{\infty} e^{-n\alpha x} \tag{23.40}$$

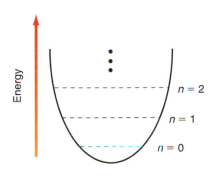

$$E_n = hc\tilde{\nu}(n + \tfrac{1}{2})$$

FIGURE 23.11
The harmonic oscillator model. Each vibrational degree of freedom is characterized by a quadratic potential. The energy levels corresponding to this potential are evenly spaced.

With this substitution, we arrive at the following expression for the **vibrational partition function:**

$$q_V = \frac{e^{-\beta h c \tilde{\nu}/2}}{1 - e^{-\beta h c \tilde{\nu}}} \quad \text{(with zero point energy)} \tag{23.41}$$

Although this expression is correct as written, at times it is advantageous to redefine the vibrational energy levels such that $E_0 = 0$. In other words, the energy of all levels is decreased by an amount equal to the zero point energy. Why would this be an advantageous thing to do? Consider the calculation of the probability of occupying a given vibrational energy level, p_n, as follows:

$$p_n = \frac{e^{-\beta E_n}}{q_V} = \frac{e^{-\beta h c \tilde{\nu}(n + \frac{1}{2})}}{\dfrac{e^{-\beta h c \tilde{\nu}/2}}{1 - e^{-\beta h c \tilde{\nu}}}} = \frac{e^{-\beta h c \tilde{\nu}/2} e^{-\beta h c \tilde{\nu} n}}{\dfrac{e^{-\beta h c \tilde{\nu}/2}}{1 - e^{-\beta h c \tilde{\nu}}}} = e^{-\beta h c \tilde{\nu} n}(1 - e^{-\beta h c \tilde{\nu}}) \tag{23.42}$$

Notice that in Equation (23.42) the zero point energy contributions for both the energy level and the partition function cancel. Therefore, the relevant energy for determining p_n is not the absolute energy of a given level, but the *relative* energy of the level. Given this, one can simply eliminate the zero point energy, resulting in the following expression for the vibrational partition function:

$$q_V = \frac{1}{1 - e^{-\beta h c \tilde{\nu}}} \quad \text{(without zero point energy)} \tag{23.43}$$

It is important to be consistent in including or not including zero point energy. Once a decision has been made to include or ignore zero point energy, the approach taken must be consistently applied.

EXAMPLE PROBLEM

At what temperature will the vibrational partition function for I_2 ($\tilde{\nu} = 208 \text{ cm}^{-1}$) be greatest: 298 or 1000 K?

Solution

Because the partition function is a measure of the number of states that are accessible given the amount of energy available (kT), we would expect the partition function to be greater for $T = 1000$ K relative to $T = 298$ K. We can confirm this expectation by numerically evaluating the vibrational partition function at these two temperatures:

$$(q_V)_{298 \text{ K}} = \frac{1}{1 - e^{-\beta h c \tilde{\nu}}} = \frac{1}{1 - e^{-h c \tilde{\nu}/kT}}$$

$$= \frac{1}{1 - \exp\left[-\dfrac{(6.626 \times 10^{-34} \text{ J s})(3.00 \times 10^{10} \text{ cm s}^{-1})(208 \text{ cm}^{-1})}{(1.38 \times 10^{-23} \text{ J K}^{-1})(298 \text{ K})} \right]} = 1.58$$

$$(q_V)_{1000 \text{ K}} = \frac{1}{1 - e^{-\beta h c \tilde{\nu}}}$$

$$= \frac{1}{1 - \exp\left[-\dfrac{(6.626 \times 10^{-34} \text{ J s})(3.00 \times 10^{10} \text{ cm s}^{-1})(208 \text{ cm}^{-1})}{(1.38 \times 10^{-23} \text{ J K}^{-1})(1000 \text{ K})} \right]} = 3.86$$

Consistent with our expectation, the partition function increases with temperature, indicating that more states are accessible at elevated temperatures.

23.6.1 Beyond Diatomics: Multidimensional q_V

The expression for the vibrational partition function derived in the preceding subsection is for a single vibrational degree of freedom and is sufficient for diatomic molecules.

However, triatomics and larger molecules (collectively referred to as *polyatomics*) require a different form for the partition function that takes into account all vibrational degrees of freedom. To define the vibrational partition function for polyatomics, we first need to know how many vibrational degrees of freedom there will be. A polyatomic molecule consisting of N atoms has $3N$ total degrees of freedom corresponding to three Cartesian degrees of freedom for each atom. The atoms are connected by chemical bonds; therefore, the atoms are not free to move independently of each other. First, the entire molecule can translate through space; therefore, three of the $3N$ total degrees of freedom correspond to translational motion of the entire molecule. Next, a rotational degree of freedom will exist for each nonvanishing moment of inertia. As discussed in the section on rotational motion, linear polyatomics have two rotational degrees of freedom because there are two nonvanishing moments of inertia, and nonlinear polyatomic molecules have three rotational degrees of freedom. The remaining degrees of freedom are vibrational such that the number of vibrational degrees of freedom are

$$\text{Linear polyatomics } 3N - 5 \tag{23.44}$$

$$\text{Nonlinear polyatomics } 3N - 6 \tag{23.45}$$

Note that a diatomic molecule can be viewed as linear polyatomic with $N = 2$, and the preceding expressions dictate that there is only one vibrational degree of freedom [$3(2) - 5 = 1$] as stated earlier.

The final step in deriving the partition function for a polyatomic system is to recognize that within the harmonic approximation, the vibrational degrees of freedom are separable, and each vibration can be treated as a separate energetic degree of freedom. In Section 23.3, various forms of molecular energy were shown to be separable so that the total molecular partition function is simply the sum of the partition functions for each energetic degree of freedom. Similar logic applies to vibrational degrees of freedom in which the total vibrational partition function is simply the product of vibrational partition functions for each vibrational degree of freedom:

$$(q_V)_{Total} = \prod_{i=1}^{3N-5 \text{ or } 3N-6} (q_V)_i \tag{23.46}$$

In Equation (23.46), the total vibrational partition function is equal to the product of vibrational partition functions for each vibrational mode (denoted by the subscript i). There will be $3N - 5$ or $3N - 6$ mode-specific partition functions depending on the geometry of the molecule.

EXAMPLE PROBLEM 23.8

The triatomic chlorine dioxide (OClO) has three vibrational modes of frequency: 450, 945, and 1100 cm^{-1}. What is the value of the vibrational partition function for $T = 298$ K?

Solution

The total vibrational partition function is simply the product of partition functions for each vibrational degree of freedom. Setting the zero point energy equal to zero, we find that

$$q_{450} = \frac{1}{1 - e^{-\beta hc(450 \text{ cm}^{-1})}}$$

$$= \frac{1}{1 - \exp\left[-\dfrac{(6.626 \times 10^{-34} \text{ J s})(3.00 \times 10^{10} \text{ cm s}^{-1})(450 \text{ cm}^{-1})}{(1.38 \times 10^{-23} \text{ J s})(298 \text{ K})}\right]}$$

$$= 1.12$$

$$q_{945} = \frac{1}{1 - e^{-\beta hc(945 \text{ cm}^{-1})}}$$

$$= \cfrac{1}{1 - \exp\left[-\cfrac{(6.626 \times 10^{-34}\,\text{J s})(3.00 \times 10^{10}\,\text{cm s}^{-1})(945\ \text{cm}^{-1})}{(1.38 \times 10^{-23}\,\text{J s})(298\,\text{K})}\right]}$$

$$= 1.01$$

$$q_{1100} = \frac{1}{1 - e^{-\beta hc(1100\ \text{cm}^{-1})}}$$

$$= \cfrac{1}{1 - \exp\left[-\cfrac{(6.626 \times 10^{-34}\,\text{J s})(3.00 \times 10^{10}\,\text{cm s}^{-1})(1100\ \text{cm}^{-1})}{(1.38 \times 10^{-23}\,\text{J s})(298\,\text{K})}\right]} = 1.00$$

$$(q_V)_{Total} = \prod_{i=1}^{3N-6} (q_V)_i = (q_{450})(q_{950})(q_{1100}) = (1.12)(1.01)(1.00) = 1.13$$

Note that the total vibrational partition function is close to unity. This is consistent with the fact that the vibrational energy spacings for all modes are significantly greater than kT such that few states other than $n = 0$ are populated.

23.6.2 High-Temperature Approximation to q_V

Similar to the development of rotations, the **vibrational temperature** (Θ_V) is defined as the frequency of a given vibrational degree of freedom divided by k:

$$\Theta_V = \frac{hc\tilde{\nu}}{k} \tag{23.47}$$

Unit analysis of Equation (23.47) dictates that Θ_V will have units of temperature (K). We can incorporate this term into our expression for the vibrational partition function as follows:

$$q_V = \frac{1}{1 - e^{-\beta hc\tilde{\nu}}} = \frac{1}{1 - e^{-hc\tilde{\nu}/kT}} = \frac{1}{1 - e^{-\Theta_V/T}} \tag{23.48}$$

The utility of this form of the partition function is that the relationship between vibrational energy and temperature becomes transparent. Specifically, as T becomes large relative to Θ_V, the exponent becomes smaller and the exponential term approaches one. The denominator in Equation (23.48) will decrease such that the vibrational partition function will increase. If the temperature becomes sufficiently large relative to Θ_V, q_V can be reduced to a simpler form. The Math Supplement (Appendix A) provides the following series expression for e^{-x}:

$$e^{-x} = 1 - x + \frac{x^2}{2} - \cdots$$

For the vibrational partition function in Equation (23.48), $x = -\Theta_V/T$. When $T \gg \Theta_V$, x becomes sufficiently small that only the first two terms can be included in the series expression for e^{-x} because higher order terms are negligible. Substituting into the expression for the vibrational partition function:

$$q_V = \frac{1}{1 - e^{-\Theta_V/T}} = \frac{1}{1 - \left(1 - \dfrac{\Theta_V}{T}\right)} = \frac{T}{\Theta_V} \tag{23.49}$$

This result is the high-temperature (or high-T) limit for the vibrational partition function:

$$q_V = \frac{T}{\Theta_V} \quad \text{(high-T limit)} \tag{23.50}$$

When is Equation (23.50) appropriate for evaluating q_V as opposed to the exact expression? The answer to this question depends on both the vibrational frequency of interest and the temperature. When $T \geq 10\Theta_V$, the fractional difference between the high-T and exact results is sufficiently small that the high-T result for q_V can be used. For the majority of molecules, this temperature will be extremely high, as shown in Example Problem 23.9.

EXAMPLE PROBLEM 23.9

At what temperature is the high-T limit for q_V appropriate for F_2 ($\tilde{\nu} = 917 \text{ cm}^{-1}$)?

Solution

The high-T limit is applicable when $T = 10\Theta_V$. The vibrational temperature for F_2 is

$$\Theta_V = \frac{hc\tilde{\nu}}{k} = \frac{(6.626 \times 10^{-34} \text{ J s})(3.00 \times 10^{10} \text{ cm s}^{-1})(917 \text{ cm}^{-1})}{1.38 \times 10^{-23} \text{ J K}^{-1}} = 1319 \text{ K}$$

Therefore, the high-T limit is applicable when T = ~13,000 K. To make sure this is indeed the case, we can compare the value for q_V determined by both the full expression for the partition function and the high-T approximation:

$$q_v = \frac{1}{1 - e^{-\Theta_V/T}} = \frac{1}{1 - e^{-1319 \text{ K}/13,000 \text{ K}}} = 10.4$$

$$= \frac{T}{\Theta_V} = \frac{13,000 \text{ K}}{1319 \text{ K}} = 9.9$$

Comparison of the two methods for evaluating the partition function demonstrates that the high-T limit expression provides a legitimate estimate of the partition function at this temperature. However, the temperature at which this is true is exceedingly high.

23.6.3 Degeneracy and q_V

The total vibrational partition function for a polyatomic molecule is the product of the partition functions for each vibrational degree of freedom. What if two or more of these vibrational degrees of freedom have the same frequency such that the energy levels are degenerate? It is important to keep in mind that there will always be $3N - 5$ or $3N - 6$ vibrational degrees of freedom depending on geometry. In the case of degeneracy, two or more of these degrees of freedom will have the same vibrational energy spacings, or the same vibrational temperature. The total partition function is still the product of partition functions for each vibrational degree of freedom; however, the degenerate vibrational modes have identical partition functions. There are two ways to incorporate the effects of vibrational degeneracy into the expression for the vibrational partition function. First, one can simply use the existing form of the partition function and keep track of all degrees of freedom irrespective of frequency. A second method is to rewrite the total partition function as a product of partition functions corresponding to a given vibrational frequency and include degeneracy at a given frequency, resulting in the corresponding partition function being raised to the power of the degeneracy:

$$(q_V)_{Total} = \prod_{i=1}^{n'} (q_V)_i^{g_i} \tag{23.51}$$

where n' is the total number of unique vibrational frequencies indexed by i. It is important to note that n' is *not* the number of vibrational degrees of freedom! Carbon dioxide serves as a classic example of vibrational degeneracy, as Example Problem 23.10 illustrates.

EXAMPLE PROBLEM 23.10

CO_2 has the following vibrational degrees of freedom: 1388, 667.4 (doubly degenerate), and 2349 cm^{-1}. What is the total vibrational partition function for this molecule at 1000 K?

Solution

Evaluation of the partition function can be performed by calculating the individual vibrational partition functions for each unique frequency, then taking the product of these partition functions raised to the power of the degeneracy at a given frequency:

$$(q_V)_{1388} = \cfrac{1}{1 - \exp\left[-\cfrac{(6.626 \times 10^{-34}\,\text{J s})(3.00 \times 10^{10}\,\text{cm s}^{-1})(1388\,\text{cm}^{-1})}{(1.38 \times 10^{-23}\,\text{J K}^{-1})(1000\,\text{K})}\right]} = 1.16$$

$$(q_V)_{667.4} = \cfrac{1}{1 - \exp\left[-\cfrac{(6.626 \times 10^{-34}\,\text{J s})(3.00 \times 10^{10}\,\text{cm s}^{-1})(667.4\,\text{cm}^{-1})}{(1.38 \times 10^{-23}\,\text{J K}^{-1})(1000\,\text{K})}\right]} = 1.62$$

$$(q_V)_{2349} = \cfrac{1}{1 - \exp\left[-\cfrac{(6.626 \times 10^{-34}\,\text{J s})(3.00 \times 10^{10}\,\text{cm s}^{-1})(2349\,\text{cm}^{-1})}{(1.38 \times 10^{-23}\,\text{J K}^{-1})(1000\,\text{K})}\right]} = 1.04$$

$$(q_V)_{Total} = \prod_{i=1}^{n'} (q_V)_i^{g_i} = (q_V)_{1388}(q_V)_{667.4}^2(1.04)_{2349}$$

$$= (1.16)(1.62)^2(1.04) = 3.17$$

the **Chemistry place**

Variation of q_T, q_R, and q_V with Temperature

23.7 The Equipartition Theorem

In the previous sections regarding rotations and vibrations, equivalence of the high-T and exact expressions for q_R and q_V, respectively, was observed when the temperature was sufficiently large that the thermal energy available to the system was significantly greater than the energy-level spacings. At these elevated temperatures, the quantum nature of the energy levels becomes unimportant and a classical description of the energetics is all that is needed.

The definition of partition function involves summation over quantized energy levels, and one might assume that there is a corresponding classical expression for the partition function in which a classical description of the system energetics is employed. Indeed there is such an expression; however, its derivation is beyond the scope of this text, so we simply state the result here. The expression for the three-dimensional partition function for a molecule consisting of N atoms is

$$q_{classical} = \frac{1}{h^{3N}}\int \cdots \int e^{-\beta H}\, dp^{3N}\, dx^{3N} \tag{23.52}$$

In the expression for the partition function, the terms p and x represent the momentum and position coordinates for each particle, respectively, with three Cartesian dimensions available for each term. The integral is multiplied by h^{-3N}, which has units of (momentum $\times$ distance)$^{-3N}$ such that the partition function is unitless.

What does the term $e^{-\beta H}$ represent in $q_{classical}$? The H represents the classical Hamiltonian and, like the quantum Hamiltonian, is the sum of a system's kinetic and potential energy. Therefore, $e^{-\beta H}$ is equivalent to $e^{-\beta \varepsilon}$ in our quantum expression for the molecular partition function. Consider the Hamiltonian for a classical one-dimensional harmonic oscillator with reduced mass μ and force constant k:

$$H = \frac{p^2}{2\mu} + \frac{1}{2}kx^2 \tag{23.53}$$

Using this Hamiltonian, the corresponding classical partition function for the one-dimensional harmonic oscillator is

$$q_{classical} = \frac{1}{h}\int dp \int dx\, e^{-\beta\left(\frac{p^2}{2\mu} + \frac{1}{2}kx^2\right)} = \frac{T}{\Theta_V} \tag{23.54}$$

This result is in agreement with the high-T approximation to q_V derived using the quantum partition function of Equation (23.50). This example illustrates the applicability of classical statistical mechanics to molecular systems when the temperature is sufficiently high such that summation over the quantum states can be replaced by integration. Under these temperature conditions, knowledge of the quantum details of the system is not necessary because, when evaluating Equation (23.54), nothing was implied regarding the quantization of the harmonic oscillator energy levels.

The applicability of classical statistical mechanics to molecular systems at high temperature finds application in an interesting theorem known as the **equipartition theorem.** This theorem states that any term in the classical Hamiltonian that is quadratic with respect to momentum or position (i.e., p^2 or x^2) will contribute $kT/2$ to the average energy. For example, the Hamiltonian for the one-dimensional harmonic oscillator [Equation (23.53)] has both a p^2 and x^2 term such that the average energy for the oscillator by equipartition should be kT (or NkT for a collection of N harmonic oscillators). In the next chapter, the equipartition result will be directly compared to the average energy determined using quantum statistical mechanics. At present, it is important to recognize that the concept of equipartition is a consequence of classical mechanics because, for a given energetic degree of freedom, the change in energy associated with passing from one energy level to the other must be significantly less than kT. As discussed earlier, this is true for translational and rotational degrees of freedom, but is not the case for vibrational degrees of freedom except at relatively high temperatures.

23.8 Electronic Partition Function

Electronic energy levels correspond to the various arrangements of electrons in an atom or molecule. The hydrogen atom provides an excellent example of an atomic system where the orbital energies are given by

$$E_n = \frac{-m_e e^4}{8\varepsilon_0^2 h^2 n^2} = -109{,}737 \text{ cm}^{-1} \frac{1}{n^2} \quad (n = 1, 2, 3, \dots) \qquad (23.55)$$

This expression demonstrates that the energy of a given orbital in the hydrogen atom is dependent on the quantum number n. In addition, each orbital has a degeneracy of $2n^2$. Using Equation (23.55), the energy levels for the electron in the hydrogen atom can be determined as illustrated in Figure 23.12.

From the perspective of statistical mechanics, the energy levels of the hydrogen atom represent the energy levels for the electronic energetic degree of freedom, with the corresponding partition function derived by summing over the energy levels. However, rather than use the absolute energies as determined in the quantum mechanical solution to the hydrogen atom problem, we will adjust the energy levels such that the energy associated with the $n = 1$ orbital is zero, similar to the adjustment of the ground-state energy of the harmonic oscillator to zero by elimination of the zero point energy. With this redefinition of the orbital energies, the electronic partition for the hydrogen atom becomes

$$q_E = \sum_{n=1}^{\infty} g_n e^{-\beta hc E_n} = 2e^{-\beta hc E_1} + 8e^{-\beta hc E_2} + 18e^{-\beta hc E_3} + \cdots$$

$$= 2e^{-\beta hc(0 \text{ cm}^{-1})} + 8e^{-\beta hc(82{,}303 \text{ cm}^{-1})} + 18e^{-\beta hc(97{,}544 \text{ cm}^{-1})} + \cdots$$

$$= 2 + 8e^{-\beta hc(82{,}303 \text{ cm}^{-1})} + 18e^{-\beta hc(97{,}544 \text{ cm}^{-1})} + \cdots \qquad (23.56)$$

The magnitude of the terms in the partition function corresponding to $n \geq 2$ will depend on the temperature at which the partition function is being evaluated. However, note that these energies are quite large. Consider defining the "electronic temperature," or Θ_E, in exactly the same way as the rotational and vibrational temperatures were defined:

$$\Theta_E = \frac{hc E_n}{k} = \frac{E_n}{0.695 \text{ cm}^{-1} \text{K}^{-1}} \qquad (23.57)$$

FIGURE 23.12

Orbital energies for the hydrogen atom. (a) The orbital energies as dictated by solving the Schrödinger equation for the hydrogen atom. (b) The energy levels shifted by the addition of 109,737 cm^{-1} of energy such that the lowest orbital energy is 0.

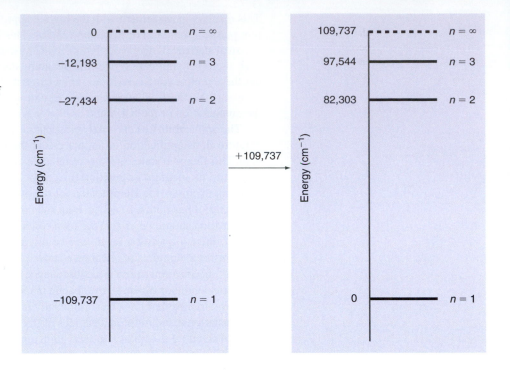

With the definition of $E_1 = 0$, the energy of the $n = 2$ orbital, E_2, is 82,303 cm^{-1} corresponding to $\Theta_E = 118,421$ K! This is an extremely high temperature, and this simple calculation illustrates the primary difference between electronic energy levels and the other energetic degrees of freedom discussed thus far. Electronic degrees of freedom are generally characterized by level spacings that are quite large relative to kT. Therefore, only the ground electronic state is populated to a significant extent (although exceptions are known, as presented in the problems at the end of this chapter). Applying this conceptual picture to the hydrogen atom, the terms in the partition function corresponding to $n \geq 2$ should be quite small at 298 K. For example, the term for the $n = 2$ state is as follows:

$$e^{-\beta hcE_2} = \exp\left[\frac{-(6.626 \times 10^{-34}\text{J s})(3.00 \times 10^{10}\text{cm s}^{-1})(82,303 \text{ cm}^{-1})}{(1.38 \times 10^{-23}\text{J K}^{-1})(298 \text{ K})}\right]$$

$$= e^{-397.3} \approx 0$$

Terms corresponding to higher energy orbitals will also be extremely small such that the electronic partition function for the hydrogen atom is ~2 at 298 K. In general, the contribution of each state to the partition function must be considered, resulting in the following expression for the **electronic partition function:**

$$q_E = \sum_n g_n e^{-\beta hcE_n} \tag{23.58}$$

In the expression for the electronic partition function, the exponential term for each energy level is multiplied by the degeneracy of the level, g_n. If the energy-level spacings are very large compared to kT, then $q_E \approx g_0$, or the degeneracy of the ground state.

Although the presentation of q_E has focused on atomic systems, similar logic applies to molecular systems. Molecular electronic energy levels were first introduced in the discussion of molecular orbital (MO) theory. In MO theory, linear combinations of atomic orbitals are used to construct a new set of electronic orbitals known as molecular orbitals. The molecular orbitals differ in energy, and the electronic configuration of the molecule is determined by placing spin-paired electrons into the orbitals starting with the lowest energy orbital. The highest energy occupied molecular orbital was designated as the HOMO. The molecular orbital energy-level diagram for butadiene is presented in Figure 23.13.

Figure 23.13 presents both the lowest and next highest energy electronic energy states for butadiene, with the difference in energies corresponding to the promotion of

FIGURE 23.13
Depiction of the molecular orbitals for butadiene. The highest occupied molecular orbital (HOMO) is indicated by the yellow rectangle, and the lowest unoccupied molecular orbital (LUMO) is indicated by the pink rectangle. The lowest energy electron configuration is shown on the left, and the next highest energy configuration is shown on the right, corresponding to the promotion of an electron from the HOMO to the LUMO.

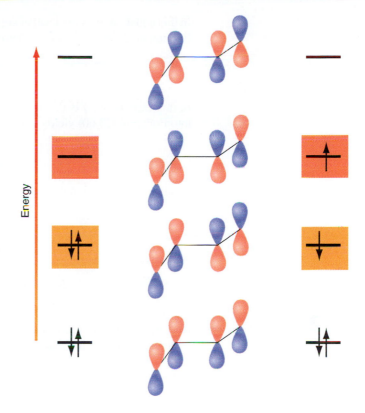

an electron from the HOMO to the lowest unoccupied molecular orbital (LUMO). The separation in energy between these two states corresponds to the amount of energy it takes to excite the electron. The wavelength of the lowest energy electronic transition of butadiene is ~220 nm, demonstrating that the separation between the HOMO and LUMO is ~45,000 cm^{-1}, significantly greater than kT at 298 K. As such, only the lowest energy electronic state contributes to the electronic partition function, and $q_E = 1$ for butadiene because the degeneracy of the lowest energy level is one. Typically, the first electronic excited level of molecules will reside 5000 to 50,000 cm^{-1} higher in energy than the lowest level such that at room temperature only this lowest level is considered in evaluating the partition function:

$$q_E = \sum_{n=0} g_n e^{-\beta E_n} \approx g_0 \tag{23.59}$$

In the absence of degeneracy of the ground electronic state, the electronic partition function will equal one.

23.9 Statistical Thermodynamics

With the central concepts of statistical mechanics in hand, we are now in a position to connect these concepts to thermodynamics. Specifically, quantities such as internal energy, entropy, and the Gibbs energy can be expressed as functions of the canonical partition function. These relationships are developed and explored in the following subsections.

23.9.1 Internal Energy

Returning to the canonical ensemble (Section 23.1), the **average energy** content of an ensemble unit, $\langle \varepsilon \rangle$, is equal to the **total energy** of the ensemble, E, divided by the number of units in the ensemble, N:

$$\langle \varepsilon \rangle = \frac{E}{N} = \frac{\sum_n \varepsilon_n a_n}{N} = \sum_n \varepsilon_n \frac{a_n}{N} \tag{23.60}$$

In this equation, ε_n is the level energy and a_n is the occupation number for the level. The Boltzmann distribution for a series of nondegenerate energy levels is

$$\frac{a_n}{N} = \frac{e^{-\beta\varepsilon_n}}{q} \tag{23.61}$$

In this expression, q is the molecular partition function. Substituting Equation (23.61) into Equation (23.60) yields

$$\langle \varepsilon \rangle = \sum_n \varepsilon_n \frac{a_n}{N} = \frac{1}{q}\sum_n \varepsilon_n e^{-\beta\varepsilon_n} \tag{23.62}$$

Finally, the summation can be rewritten as $-dq/d\beta$ so that:

$$\langle \varepsilon \rangle = \frac{-1}{q}\left(\frac{dq}{d\beta}\right) = -\left(\frac{d\ln q}{d\beta}\right) \tag{23.63}$$

$$E = N\langle \varepsilon \rangle = \frac{-N}{q}\left(\frac{dq}{d\beta}\right) = -N\left(\frac{d\ln q}{d\beta}\right) \tag{23.64}$$

At times, Equations (23.63) and (23.64) are easier to evaluate by taking the derivative with respect to T rather than β:

$$\langle \varepsilon \rangle = kT^2\left(\frac{d\ln q}{dT}\right) \tag{23.65}$$

$$E = NkT^2\left(\frac{d\ln q}{dT}\right) \tag{23.66}$$

Equation (23.66) allows one to calculate the ensemble energy, but to what thermodynamic quantity is this energy related? Recall that we are interested in the canonical ensemble in which N, V, and T are held constant. Because V is constant, there can be no P-V-type work, and by the first law of thermodynamics any change in internal energy must occur by heat flow, q_V. Using the first law, the change in heat is related to the change in system **internal energy** at constant volume by

$$U - U_0 = q_V \tag{23.67}$$

In Equation (23.67), q_V is heat, not the molecular partition function. The energy as expressed by Equation (23.67) is the difference in internal energy at some finite temperature to that at 0 K. If there is residual, internal energy present at 0 K, it must be included to determine the overall energy of the system. However, by convention U_0 is generally set to zero. For example, in the two-level example described earlier, the internal energy will be zero at 0 K since the energy of the ground state is defined as zero.

The second important relationship to establish is the one between the internal energy and the canonical partition function. Fortunately, we have already encountered this relationship. For an ensemble of indistinguishable noninteracting particles, the canonical partition function is given by

$$Q = \frac{q^N}{N!} \tag{23.68}$$

In Equation (23.68), q is the molecular partition function. Taking the natural log of Equation (23.68):

$$\ln Q = \ln\left(\frac{q^N}{N!}\right) = N\ln q - \ln N! \tag{23.69}$$

Finally, taking the derivative of Equation (23.69) with respect to β and recognizing that $\ln(N!)$ is constant in the canonical ensemble,

$$\frac{d \ln Q}{d\beta} = \frac{d}{d\beta}\left(N \ln q\right) - \frac{d}{d\beta}\left(\ln N!\right)$$

$$= N\frac{d \ln q}{d\beta} \qquad (23.70)$$

The last relationship is simply the total energy; therefore, the relationship between the canonical partition function and total internal energy, U, is simply

$$U = -\left(\frac{d \ln Q}{d\beta}\right)_V \qquad (23.71)$$

EXAMPLE PROBLEM 23.11

For an ensemble consisting of a mole of particles having two energy levels separated by $h\nu = 1.00 \times 10^{-20}$ J, at what temperature will the internal energy of this system equal 1.00 kJ?

Solution

Using the expression for total energy and recognizing that $N = nN_a$,

$$U = -\left(\frac{d \ln Q}{d\beta}\right)_V = -nN_A\left(\frac{d \ln q}{d\beta}\right)_V$$

$$U = -nN_A\left(\frac{d}{d\beta}\ln q\right)_V = -\frac{nN_A}{q}\left(\frac{dq}{d\beta}\right)_V$$

$$\frac{U}{nN_A} = \frac{-1}{\left(1 + e^{-\beta h\nu}\right)}\left(\frac{d}{d\beta}\left(1 + e^{-\beta h\nu}\right)\right)_V$$

$$= \frac{h\nu e^{-\beta h\nu}}{1 + e^{-\beta h\nu}} = \frac{h\nu}{e^{\beta h\nu} + 1}$$

$$\frac{nN_A h\nu}{U} - 1 = e^{\beta h\nu}$$

$$\ln\left(\frac{nN_A h\nu}{U} - 1\right) = \beta h\nu = \frac{h\nu}{kT}$$

$$T = \frac{h\nu}{k \ln\left(\frac{nN_A h\nu}{U} - 1\right)}$$

$$= \frac{1.00 \times 10^{-20}\text{ J}}{\left(1.38 \times 10^{-23}\text{J K}^{-1}\right)\ln\left(\frac{\left(1\text{ mol}\right)\left(6.022 \times 10^{-23}\text{ mol}^{-1}\right)\left(1.00 \times 10^{-20}\text{ J}\right)}{\left(1.00 \times 10^{3}\text{J}\right)} - 1\right)}$$

$$= 449\text{ K}$$

23.9.2 Energy and Molecular Energetic Degrees of Freedom

The total molecular partition function (q_{Total}) is equal to the product of partition functions for each individual energetic degree of freedom:

$$q_{Total} = q_T q_R q_V q_E \qquad (23.72)$$

In this expression, the subscripts T, R, V, and E refer to translational, rotational, vibrational, and electronic energetic degrees of freedom, respectively. In a similar fashion, the internal energy can be decomposed into the contributions from each energetic degree of freedom:

$$U = -\left(\frac{d \ln Q}{d\beta}\right)_V = -N\left(\frac{d \ln q}{d\beta}\right)_V$$

$$= -N\left(\frac{d \ln(q_T q_R q_V q_E)}{d\beta}\right)_V$$

$$= -N\left(\frac{d}{d\beta}(\ln q_T + \ln q_R + \ln q_V + \ln q_E)\right)_V$$

$$= -N\left[\left(\frac{d \ln q_T}{d\beta}\right)_V + \left(\frac{d \ln q_R}{d\beta}\right)_V + \left(\frac{d \ln q_V}{d\beta}\right)_V + \left(\frac{d \ln q_E}{d\beta}\right)_V\right]$$

$$= U_T + U_R + U_V + U_E \tag{23.73}$$

The last line in this expression demonstrates a very intuitive result—that the total internal energy is simply the sum of contributions from each molecular energetic degree of freedom. This result also illustrates the connection between the macroscopic property of the ensemble (internal energy) and the microscopic details of the units themselves (molecular energy levels). To relate the total internal energy to the energetic degrees of freedom, expressions for the energy contribution from each energetic degree of freedom (U_T, U_R, and so on) are needed.

Translations.   The contribution to the system internal energy from translational motion is

$$U_T = \frac{-N}{q_T}\left(\frac{dq_T}{d\beta}\right)_V \tag{23.74}$$

In Equation (23.74), q_T is the translational partition function, which in three dimensions is given by

$$q_T = \frac{V}{\Lambda^3} \quad \text{with} \quad \Lambda^3 = \left(\frac{h^2\beta}{2\pi m}\right)^{3/2} \tag{23.75}$$

With this partition function, the translational contribution to the internal energy becomes

$$U_T = \frac{-N}{q_T}\left(\frac{dq_T}{d\beta}\right)_V = \frac{-N\Lambda^3}{V}\left(\frac{d}{d\beta}\frac{V}{\Lambda^3}\right)_V = \frac{3}{2}NkT = \frac{3}{2}nRT \tag{23.76}$$

Equation (23.76) should look familiar. Recall that the internal energy of an ideal monatomic gas is $nC_v \Delta T = 3/2(nRT)$ with $T_{initial} = 0$ K, identical to the result just obtained. The convergence between the thermodynamic and statistical mechanical descriptions of monatomic gas systems is remarkable. It is also important to note that the starting point in deriving the preceding relationship was a quantum mechanical description of translational motion and the partition function.

It is also interesting to note that the contribution of translational motion to the internal energy is equal to that predicted by the equipartition theorem (Section 23.7). The equipartition theorem states that any term in the classical Hamiltonian that is quadratic with respect to momentum or position will contribute $kT/2$ to the energy. The Hamiltonian corresponding to three-dimensional translational motion of an ideal monatomic gas is

$$H_{trans} = \frac{1}{2m}\left(p_x^2 + p_y^2 + p_z^2\right) \tag{23.77}$$

Each p^2 term in Equation (23.77) will contribute $1/2\ kT$ to the energy by equipartition such that the total contribution will be $3/2\ kT$, or $3/2\ RT$ for a mole of particles that is identical to the result derived using the quantum mechanical description of translational motion. This agreement is not surprising given the small energy gap between translational energy levels such that classical behavior is expected.

Rotations.   Within the rigid-rotor approximation, the rotational partition function for a diatomic molecule in the high-temperature limit is given by

$$q_R = \frac{1}{\sigma \beta hcB} \tag{23.78}$$

With this partition function, the contribution to the internal energy from rotational motion is

$$U_R = \frac{-N}{q_R} \left(\frac{dq_R}{d\beta} \right)_V = NkT = nRT \tag{23.79}$$

Recall that the expression for the rotational partition function employed here is for a diatomic in which the rotational temperature, Θ_R, is much less than kT. In this limit, a significant number of rotational states are accessible. If Θ_R is not small relative to kT, then full evaluation of the sum form of the rotational partition function is required. In the high-temperature limit, the rotational energy can be thought of as containing contributions of 1/2 kT from each nonvanishing moment of inertia. This partitioning of energy is analogous to the case of translational energy discussed earlier within the context of the equipartition theorem. The concept of equipartition can be used to extend the result for U_R obtained for a diatomic molecule to linear and nonlinear polyatomic molecules. Because each nonvanishing moment of inertia will provide 1/2 kT to the rotational energy equipartition, we can state that

$$U_R = nRT \text{ (diatomic and linear polyatomic)} \tag{23.80}$$

$$U_R = \frac{3}{2} nRT \text{ (nonlinear polyatomic)} \tag{23.81}$$

Vibrations. Unlike translational and rotational degrees of freedom, vibrational energy-level spacings are typically greater than kT such that the equipartition theorem is generally not applicable to this energetic degree of freedom. Fortunately, within the harmonic oscillator model, the regular energy-level spacings provide for a relatively simple expression for the vibrational partition function:

$$q_V = \left(1 - e^{-\beta hc\tilde{\nu}} \right)^{-1} \tag{23.82}$$

The term $\tilde{\nu}$ represents vibrational frequency in units of cm^{-1}. Using this partition function, the vibrational contribution to the average energy is

$$U_V = \frac{-N}{q_V} \left(\frac{dq_V}{d\beta} \right)_V = \frac{Nhc\tilde{\nu}}{e^{\beta hc\tilde{\nu}} - 1} \tag{23.83}$$

The temperature dependence of U_V/Nhc for a vibrational degree of freedom with $\tilde{\nu} = 1000$ cm^{-1} is presented in Figure 23.14. First, note that at lowest temperatures U_V is zero. At low temperatures, $kT \ll hc\tilde{\nu}$ where the available thermal energy is insufficient to populate the first excited vibrational state to an appreciable extent. However, for temperatures ≥ 1000 K the average energy increases linearly with temperature, identical to the behavior observed for translational and rotational energy. This observation suggests that a high-temperature expression for U_V also exists.

To derive the high-temperature limit expression for the vibrational energy, the exponential term in q_V is written using the following series expression:

$$e^x = 1 + x + \frac{x^2}{2!} + \cdots$$

Because $x = \beta hc\tilde{\nu} = hc\tilde{\nu}kT$, when $kT \gg hc\tilde{\nu}$, the series is approximately equal to 1 + x, yielding

$$U_V = \frac{Nhc\tilde{\nu}}{e^{\beta hc\tilde{\nu}} - 1} = \frac{Nhc\tilde{\nu}}{\left(1 + \beta hc\tilde{\nu} \right) - 1}$$

$$= \frac{N}{\beta}$$

$$U_V = NkT = nRT \tag{23.84}$$

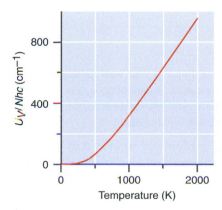

FIGURE 23.14
The variation in average vibrational energy as a function of temperature where $\tilde{\nu} = 1000$ cm^{-1}.

When the high-temperature limit is applicable, the vibrational contribution to the internal energy is nRT, identical to the prediction of the equipartition theorem. Note that this is the contribution for a single vibrational degree of freedom, and that the overall contribution will be the sum over all vibrational degrees of freedom. Furthermore, the applicability of the high-temperature approximation is dependent on the details of the vibrational energy spectrum such that while the high-temperature approximation may be applicable for some low-energy vibrations, it may not be applicable to higher energy vibrations.

Electronic. Because electronic energy-level spacings are generally quite large compared to kT, the partition function is simply equal to the ground-state degeneracy. Since the degeneracy is a constant, the derivative of this quantity with respect to β must be zero and

$$U_E = 0 \tag{23.85}$$

Exceptions to this result do exist. In particular, for systems in which electronic energy levels are comparable to kT, it is necessary to evaluate the full partition function.

23.9.3 Heat Capacity

As discussed in Chapter 2, the thermodynamic definition of the **heat capacity** at constant volume (C_V) is

$$C_V = \left(\frac{\partial U}{\partial T}\right)_V = -k\beta^2 \left(\frac{\partial U}{\partial \beta}\right)_V \tag{23.86}$$

Because the internal energy can be decomposed into contributions from each energetic degree of freedom, the heat capacity can also be decomposed in a similar fashion. In the previous section, the average internal energy contribution from each energetic degree of freedom was determined. Correspondingly, the heat capacity is given by the derivative of the internal energy with respect to temperature for a given energetic degree of freedom.

Translations. The contribution of translational motion to the average molecular energy is

$$U_T = \frac{3}{2} NkT \tag{23.87}$$

The translational contribution to C_V is readily determined by taking the derivative of the preceding expression with respect to temperature:

$$(C_V)_T = \left(\frac{dU_T}{dT}\right)_V = \frac{3}{2} Nk \tag{23.88}$$

This result demonstrates that the contribution of translations to the overall constant volume heat capacity is simply a constant, with no temperature dependence. The constant value of the translational contribution to C_V down to very low temperatures is a consequence of the dense manifold of closely spaced translational states.

Rotations. Assuming the rotational degrees of freedom can be treated in the high-temperature limit, the internal energy is dependent on molecular geometry:

$$U_R = NkT \text{ (linear)} \tag{23.89}$$

$$U_R = \frac{3}{2} NkT \text{ (nonlinear)} \tag{23.90}$$

Using these two equations, the rotational contribution to C_V becomes

$$(C_V)_R = Nk \text{ (linear)} \tag{23.91}$$

$$(C_V)_R = \frac{3}{2} Nk \text{ (nonlinear)} \tag{23.92}$$

Vibrational. In contrast to translations and rotations, the high-temperature limit is generally not applicable to the vibrational degrees of freedom. Therefore, the exact functional form of the energy must be evaluated to determine the vibrational contribution to C_V. The derivation is somewhat more involved, but still straightforward. First, recall that the vibrational contribution to U is [Equation (23.83)]

$$U_V = \frac{Nhc\tilde{\nu}}{e^{\beta hc\tilde{\nu}} - 1} \qquad (23.93)$$

Given this equation, the vibrational contribution to the constant volume heat capacity is

$$(C_V)_{Vib} = \left(\frac{dU_V}{dT}\right)_V = Nk\beta^2(hc\tilde{\nu})^2 \frac{e^{\beta hc\tilde{\nu}}}{\left(e^{\beta hc\tilde{\nu}} - 1\right)^2} \qquad (23.94)$$

The heat capacity can also be cast in terms of the vibrational temperature Θ_V as follows

$$(C_V)_{Vib} = Nk\left(\frac{\Theta_V}{T}\right)^2 \frac{e^{\Theta_V/T}}{\left(e^{\Theta_V/T} - 1\right)^2} \qquad (23.95)$$

A polyatomic molecule has $3N - 6$ or $3N - 5$ vibrational degrees of freedom. Each vibrational degree of freedom contributes to the overall vibrational constant volume heat capacity such that

$$(C_V)_{Vib, Total} = \sum_{k=1}^{3N-6 \text{ or } 3N-5} (C_V)_{Vib, k} \qquad (23.96)$$

How does the vibrational heat capacity change as a function of T? The contribution of a given vibrational degree of freedom depends on the spacing between vibrational levels relative to kT. Therefore, at lowest temperatures we expect the vibrational contribution to C_V to be zero. As the temperature increases, the lowest energy vibrational modes have spacings comparable to kT such that these modes will contribute to the heat capacity. Finally, the highest energy vibrational modes contribute only at high temperature. In summary, we expect the vibrational contribution to C_V to demonstrate a significant temperature dependence.

Figure 23.15 presents the vibrational contribution to C_V as a function of temperature for a nonlinear triatomic molecule having three nondegenerate vibrational modes with $\tilde{\nu}$ equal to 100, 1000, and 3000 cm^{-1}. As expected, the lowest frequency mode contributes to the heat capacity at lowest temperatures, reaching a constant value of Nk for temperatures greatly in excess of Θ_V for a given vibration. As the temperature increases, each successively higher energy vibrational mode begins to contribute to the heat capacity and reaches the same limiting value of k at highest temperatures. Finally, the total vibrational heat capacity, which is simply the sum of the contributions from the individual modes, approaches a limiting value of $3Nk$.

The behavior illustrated in Figure 23.15 demonstrates that for temperatures where $kT \gg hc\tilde{\nu}$, the heat capacity approaches a constant value, suggesting that, like translations and rotations, there is a high-temperature limit for the vibrational contribution to C_V. Recall that the high-temperature approximation for the average energy for an individual vibrational degree of freedom is

$$U_V = \frac{N}{\beta} = NkT \qquad (23.97)$$

Differentiation of this expression with respect to temperature reveals that the vibrational contribution to the C_V is indeed equal to Nk per vibrational mode in the high-temperature limit as expected from the equipartition theorem.

Electronic. Because the partition function for the energetic degree of freedom is generally equal to the ground-state degeneracy, the resulting average energy is zero. Therefore, there is no contribution to the constant volume heat capacity from these degrees of freedom. However, for systems with electronic excited states that are comparable to kT, the contribution to C_V from electronic degrees of freedom can be finite and must be determined using the summation form of the partition function.

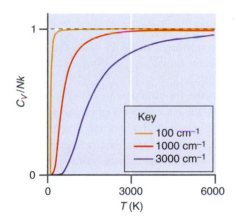

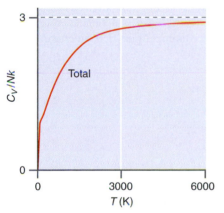

FIGURE 23.15
Evolution in the vibrational contribution to C_V as a function of temperature. Calculations are for a molecule with three vibrational degrees of freedom as indicated in the top panel. Contribution for each vibrational mode (top) and the total vibrational contribution (bottom) are shown.

Temperature Dependence of C_V for Vibrations

23.9.4 Entropy

Entropy is perhaps the most misunderstood thermodynamic property of matter. Introductory chemistry courses generally describe the entropic driving force for a reaction as an inherent desire for the system to increase its "randomness" or "disorder." With statistical mechanics, we will see that the tendency of an isolated system to evolve toward the state of maximum entropy is a direct consequence of statistics. To illustrate this point, recall that a system approaches equilibrium by achieving the configuration of energy with the maximum weight. In other words, the configuration of energy that will be observed at equilibrium corresponds to W_{max}. The tendency of a system to maximize W and entropy, S, suggests that a relationship exists between these quantities. Boltzmann expressed this relationship in following equation, known as **Boltzmann's formula:**

$$S = k \ln W \tag{23.98}$$

This relationship states that entropy is directly proportional to $\ln(W)$, with the Boltzmann constant serving as the proportionality constant. Equation (23.98) makes it clear that a maximum in W will correspond to a maximum in S.

The definition of entropy provided by the Boltzmann formula provides insight into the underpinnings of entropy, but how does one calculate the entropy for a molecular system? The answer to this question requires a bit more work. Because the partition function provides a measure of energy state accessibility, it can be assumed that a relationship between the partition function and entropy exists. Beginning with the Boltzmann formula, the following relationship between entropy and the canonical partition function can be derived:

$$S = \frac{U}{T} + k \ln Q = \left(\frac{d}{dT}(kT \ln Q) \right)_V \tag{23.99}$$

Applying this expression to an ideal monatomic gas, assuming that the electronic partition function is unity (i.e., the ground electronic energy level is nondegenerate), only translational degrees of freedom contribute to the canonical partition function. Therefore, the entropy can be written as:

$$
\begin{aligned}
S &= \left(\frac{d}{dT}(kT \ln Q) \right)_V = \frac{U}{T} + k \ln Q \\[2mm]
&= \frac{1}{T}\left(\frac{3}{2}NkT \right) + k \ln \frac{q_{trans}^N}{N!} \\[2mm]
&= \frac{3}{2}Nk + Nk \ln q_{trans} - k(N \ln N - N) \\[2mm]
&= \frac{5}{2}Nk + Nk \ln q_{trans} - Nk \ln N \\[2mm]
&= \frac{5}{2}Nk + Nk \ln \frac{V}{\Lambda^3} - Nk \ln N \\[2mm]
&= \frac{5}{2}Nk + Nk \ln V - Nk \ln \Lambda^3 - Nk \ln N \\[2mm]
&= \frac{5}{2}Nk + Nk \ln V - Nk \ln\left(\frac{h^2}{2\pi mkT} \right)^{3/2} - Nk \ln N \\[2mm]
&= \frac{5}{2}Nk + Nk \ln V + \frac{3}{2}Nk \ln T - Nk \ln\left(\frac{N^{2/3}h^2}{2\pi mk} \right)^{3/2} \\[2mm]
&= \frac{5}{2}nR + nR \ln V + \frac{3}{2}nR \ln T - nR \ln\left(\frac{n^{2/3}N_A^{2/3}h^2}{2\pi mk} \right)^{3/2}
\end{aligned}
\tag{23.100}
$$

The final line of Equation (23.100) is a version of the **Sackur–Tetrode equation,** which can be written in the more compact form:

$$S = nR \ln \left[\frac{RTe^{5/2}}{\Lambda^3 N_A P} \right] \quad \text{where } \Lambda^3 = \left(\frac{h^2}{2\pi mkT} \right)^{3/2} \tag{23.101}$$

The Sackur–Tetrode equation reproduces many of the classical thermodynamics properties of ideal monatomic gases encountered previously. For example, consider the isothermal expansion of an ideal monatomic gas from an initial volume V_1 to a final volume V_2. Inspection of the expanded form of the Sackur–Tetrode equation [Equation (23.100)] demonstrates that all of the terms in this expression are unchanged except for the second term involving volume such that

$$\Delta S = S_{final} - S_{initial} = nR \ln \frac{V_2}{V_1} \tag{23.102}$$

This is the same result obtained from classical thermodynamics. What if the entropy change were initiated by isochoric ($\Delta V = 0$) heating? Using the difference in temperature between initial (T_1) and final (T_2) states, Equation (23.100) yields

$$\Delta S = S_{final} - S_{initial} = \frac{3}{2} nR \ln \frac{T_2}{T_1} = nC_V \ln \frac{T_2}{T_1} \tag{23.103}$$

Recognizing that $C_V = 3/2R$ for an ideal monatomic gas, we again arrive at a result first encountered in thermodynamics.

Does the Sackur–Tetrode equation provide any information not available from thermodynamics? Indeed, note the first and fourth terms in Equation (23.100). These terms are simply constants, with the latter varying with the atomic mass. Classical thermodynamics is entirely incapable of explaining the origin of these terms, and only through empirical studies could the presence of these terms be determined. However, their contribution to entropy appears naturally (and elegantly) when using the statistical perspective.

EXAMPLE PROBLEM 23.12

Determine the standard molar entropy of Ne and Kr under standard thermodynamics conditions.

Solution

Beginning with the expression for entropy derived in the text:

$$S = \frac{5}{2} R + R \ln \left(\frac{V}{\Lambda^3} \right) - R \ln N_A$$

$$= \frac{5}{2} R + R \ln \left(\frac{V}{\Lambda^3} \right) - 54.75\, R$$

$$= R \ln \left(\frac{V}{\Lambda^3} \right) - 52.25\, R$$

The conventional standard state is defined by $T = 298$ K and $V_m = 24.4$ l (0.0244 m^3). The thermal wavelength for Ne is

$$\Lambda = \left(\frac{h^2}{2\pi mkT} \right)^{1/2}$$

$$= \left(\frac{(6.626 \times 10^{-34}\ \text{J s})^2}{2\pi \left(\dfrac{0.02018\ \text{kg mol}^{-1}}{N_A} \right)(1.38 \times 10^{-23}\ \text{J K}^{-1})(298\ \text{K})} \right)^{1/2}$$

$$= 2.25 \times 10^{-11}\ \text{m}$$

Using this value for the thermal wavelength, the entropy becomes

$$S = R \ln\left(\frac{0.0244 \text{ m}^2}{(2.25 \times 10^{-11} \text{ m})^3}\right) - 52.25 \, R$$

$$= 69.83R - 52.25R = 17.59R = 146 \text{ J mol}^{-1} \text{K}^{-1}$$

The experimental value is 146.48 J/mol K. Rather than determining the entropy of Kr directly, it is easier to determine the difference in entropy relative to Ne:

$$\Delta S = S_{Kr} - S_{Ne} = S = R \ln\left(\frac{V}{\Lambda_{Kr}^3}\right) - R \ln\left(\frac{V}{\Lambda_{Ne}^3}\right)$$

$$= R \ln\left(\frac{\Lambda_{Ne}}{\Lambda_{Kr}}\right)^3$$

$$= 3R \ln\left(\frac{\Lambda_{Ne}}{\Lambda_{Kr}}\right)$$

$$= 3R \ln\left(\frac{m_{Kr}}{m_{Ne}}\right)^{1/2}$$

$$= \frac{3}{2} R \ln\left(\frac{m_{Kr}}{m_{Ne}}\right) = \frac{3}{2} R \ln(4.15)$$

$$= 17.7 \text{ J mol}^{-1} \text{K}^{-1}$$

Using this difference, the standard molar entropy of Kr becomes

$$S_{Kr} = \Delta S + S_{Ne} = 164 \text{ J mol}^{-1} \text{K}^{-1}$$

The experimental value is 163.89 J mol^{-1} K^{-1} in excellent agreement with the calculated value.

As illustrated by the preceding example problem, when the entropy calculated using statistical mechanics is compared to experiment, good agreement is observed for a variety of atomic and molecular systems. However, for many molecular systems, this agreement is less than ideal. A famous example of such a system is carbon monoxide, for which the calculated entropy at thermodynamic standard temperature and pressure is 197.9 J mol^{-1} K^{-1} and the experimental value is only 193.3 J mol^{-1} K^{-1}. In this and other systems, the calculated entropy is always greater than that observed experimentally. The reason for the systematic discrepancy between calculated and experimental entropies for such systems is **residual entropy,** or entropy associated with molecular orientation in the molecular crystal at low temperature. Using CO as an example, the weak electric dipole moment of the molecule dictates that dipole–dipole interactions do not play a dominant role in determining the orientation of one CO molecule relative to neighboring molecules in a crystal. Therefore, each CO can assume one of two orientations as illustrated in Figure 23.16. The solid corresponding to the possible orientations of CO will have an inherent randomness to it. Because each CO molecule can assume one of two possible orientations, the entropy associated with this orientational disorder is

$$S = k \ln W = k \ln 2^N = Nk \ln 2 = nR \ln 2 \tag{23.104}$$

FIGURE 23.16

The origin of residual entropy for CO. Each CO molecule in the solid can have one of two possible orientations as illustrated by the central CO. Each CO will have two possible directions such that the total number of arrangements possible is 2^N where N is the number of CO molecules.

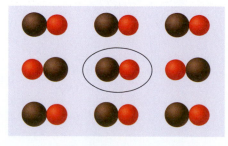
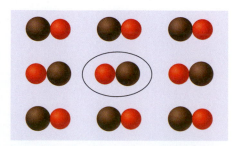

In Equation (23.104), W is the total number of CO arrangements possible, and it is equal to 2^N where N is the number of CO molecules. For a system consisting of 1 mol of CO, the residual entropy is predicted to be $R \ln 2$ or $5.76 \text{ J mol}^{-1} \text{ K}^{-1}$, roughly equal to the difference between the experimental and calculated entropy values.

Finally, note that the concept of residual entropy sheds light on the origin of the third law of thermodynamics. As discussed in Chapter 5, the third law states that the entropy of a pure and crystalline substance is zero at 0 K. By "pure and crystalline," the third law means that the system must be pure with respect to both composition (i.e., a single component) and orientation in the solid at 0 K. For such a pure system, $W = 1$ and correspondingly $S = 0$ by Equation (23.104). Therefore, the definition of zero entropy provided by the third law is a natural consequence of the statistical nature of matter.

EXAMPLE PROBLEM 23.13

The Van der Waals radii of H and F are similar such that steric effects on molecular ordering in the crystal are minimal. Do you expect the residual molar entropies for crystalline 1,2-difluorobenzene and 1,4-difluorobenzene to be the same?

Solution

The structures of 1,2-difluorobenzene and 1,4-difluorobenzene are shown here:

1,2-Difluorobenzene   1,4-Difluorobenzene

In crystalline 1,2-difluorobenzene, there are six possible arrangements that can be visualized by rotation of the molecule about the C_6 symmetry axis of the molecule. Therefore, $W = 6^N$ and

$$S = K \ln W = k \ln 6^{N_A} = N_A k \ln 6 = R \ln 6$$

Similarly, for 1,4-difluorobenzene, there are three possible arrangements such that $W = 3^N$ and $S = R \ln 3$. The residual molar entropies are expected to differ for molecular crystals involving these species.

23.9.5 Enthalpy, Helmholtz Energy, and Gibbs Energy

The relationship between the canonical partition function and other thermodynamic quantities can be derived using the following familiar thermodynamics relationships for enthalpy, H, Helmholtz energy, A, and Gibbs energy, G:

$$H = U + PV \tag{23.105}$$

$$A = U - TS \tag{23.106}$$

$$G = H - TS \tag{23.107}$$

Using these expressions, relationships between the canonical partition function and the previous thermodynamic quantities can be derived. For example, the thermodynamic definition of A from Equation (23.106) provides for the following relationship between A and the canonical partition function:

$$A = U - TS$$

$$= U - T\left(\frac{U}{T} + k \ln Q\right)$$

$$A = -kT \ln Q \tag{23.108}$$

Using the thermodynamic definition of enthalpy from Equation (23.105):

$$H = U + PV$$

$$= \left(\frac{-\partial}{\partial \beta} \ln Q\right)_V + V\left(\frac{-\partial A}{\partial V}\right)_T$$

$$= \left(\frac{-\partial}{\partial \beta} \ln Q\right)_V + V\left(\frac{-\partial}{dV}(-kT \ln Q)\right)_T$$

$$= kT^2\left(\frac{-\partial}{\partial T} \ln Q\right)_V + VkT\left(\frac{\partial}{\partial V} \ln Q\right)_T$$

$$H = T\left[kT\left(\frac{\partial}{\partial T} \ln Q\right)_V + Vk\left(\frac{\partial}{\partial V} \ln Q\right)_T\right] \qquad (23.109)$$

Although Equation (23.109) is correct, it clearly requires a bit of work to implement. Yet, the statistical perspective has shown that one can relate enthalpy to microscopic molecular details through the partition function. When calculating enthalpy, it is sometimes easier to use a combination of thermodynamic and statistical perspectives, as the following example demonstrates.

EXAMPLE PROBLEM 23.14

What is the enthalpy of 1 mol of an ideal monatomic gas?

Solution

One approach to this problem is to start with the expression for the canonical partition function in terms of the molecular partition function for an ideal monatomic gas and evaluate the result. However, a more efficient approach is to begin with the thermodynamic definition of enthalpy:

$$H = U + PV$$

Recall that the translational contribution to U is 3/2RT, and this is the only degree of freedom operative for the monatomic gas. In addition, we can apply the ideal gas law (because we have now demonstrated its validity from a statistical perspective) such that the enthalpy is simply

$$H = U + PV$$

$$= \frac{3}{2}RT + RT$$

$$= \frac{5}{2}RT$$

The interested reader is encouraged to obtain this result through a full evaluation of the statistical expression for enthalpy.

Perhaps the most important state function to emerge from thermodynamics is the **Gibbs energy.** Using this quantity, one can determine if a chemical reaction will occur spontaneously. The statistical expression for the Gibbs energy is also derived starting with the thermodynamic definition of this quantity [Equation (23.107)]:

$$G = A + PV$$

$$= -kT \ln Q + NkT$$

$$= -kT \ln\left(\frac{q^N}{N!}\right) + NkT$$

$$= -kT \ln q^N + kT \ln N! + NkT$$

$$G = -NkT \ln\left(\frac{q}{N}\right) = -nRT \ln\left(\frac{q}{N}\right) \tag{23.110}$$

This relationship is extremely important because it provides insight into the origin of the Gibbs energy. At constant temperature the nRT prefactor in the expression for G is equivalent for all species; therefore, differences in the Gibbs energy must be due to the partition function. Because the Gibbs energy is proportional to $-\ln(q)$, an increase in the value for the partition function will result in a lower Gibbs energy. The partition function quantifies the number of states that are accessible at a given temperature; therefore, the statistical perspective dictates that species with a comparatively greater number of accessible energy states will have a lower Gibbs energy. This relationship will have profound consequences when discussing chemical equilibria in the next section.

EXAMPLE PROBLEM 23.15

Calculate the Gibbs energy for 1 mol of Ar at 298.15 K and standard pressure (10^5 Pa) assuming that the gas demonstrates ideal behavior.

Solution

Argon is a monatomic gas; therefore, $q = q_{trans}$. Using Equation (23.110),

$$G° = -nRT \ln\left(\frac{q}{N}\right) = -nRT \ln\left(\frac{V}{N\Lambda^3}\right)$$

$$= -nRT \ln\left(\frac{kT}{P\Lambda^3}\right)$$

The superscript on G indicates standard thermodynamic conditions of 298.15 K and 1 bar. In the last step, the ideal gas law was used to express V in terms of P, and the relationships $N = nN_A$ and $R = N_A k$ were employed. The units of pressure must be Pa = J m^{-3}. Solving for the thermal wavelength term, Λ^3, we get

$$\Lambda^3 = \left(\frac{h^2}{2\pi mkT}\right)^{3/2}$$

$$= \left(\frac{(6.626 \times 10^{-34} \text{ J s})^2}{2\pi\left(\frac{0.040 \text{ kg mol}^{-1}}{6.022 \times 10^{23} \text{ mol}^{-1}}\right)(1.38 \times 10^{-23} \text{ J K}^{-1})(298 \text{ K})}\right)^{3/2}$$

$$= 4.09 \times 10^{-33} \text{ m}^3$$

With this result, $G°$ becomes

$$G° = -nRT \ln\left(\frac{kT}{P\Lambda^3}\right) = -(1 \text{ mol})(8.314 \text{ J mol}^{-1}\text{K}^{-1})$$

$$\times (298 \text{ K}) \ln\left(\frac{(1.38 \times 10^{-23} \text{ J K}^{-1})(298 \text{ K})}{(10^5 \text{ Pa})(4.09 \times 10^{-33} \text{ m}^3)}\right)$$

$$= -3.99 \times 10^4 \text{ J} = -39.9 \text{ kJ}$$

23.10 Chemical Equilibrium

Consider the following generic reaction:

$$a\text{A} + b\text{B} \Leftrightarrow c\text{C} + d\text{D} \tag{23.111}$$

The change in Gibbs energy for this reaction is related to the Gibbs energy for the associated species as follows:

$$\Delta G^\circ = cG^\circ{}_C + dG^\circ{}_D - aG^\circ{}_A - bG^\circ{}_B \tag{23.112}$$

In this expression, the superscript indicates standard thermodynamic state. In addition, the equilibrium constant K is given by

$$\Delta G^\circ = -RT \ln K \tag{23.113}$$

In the previous section of this chapter, the Gibbs energy was related to the molecular partition function. Therefore, it should be possible to define ΔG° and K in terms of partition functions for the various species involved. This relationship is obtained by combining the expression for ΔG° given in Equation (23.112) with the expression for G given in Equation (23.110):

$$\Delta G^\circ = c\left(-RT \ln\left(\frac{q^\circ_C}{N}\right)\right) + d\left(-RT \ln\left(\frac{q^\circ_D}{N}\right)\right) - a\left(-RT \ln\left(\frac{q^\circ_A}{N}\right)\right)$$

$$-b\left(-RT \ln\left(\frac{q^\circ_B}{N}\right)\right)$$

$$= -RT \ln\left(\frac{\left(\frac{q^\circ_C}{N}\right)^c\left(\frac{q^\circ_D}{N}\right)^d}{\left(\frac{q^\circ_A}{N}\right)^a\left(\frac{q^\circ_B}{N}\right)^b}\right) \tag{23.114}$$

Comparison of the preceding relationship to the thermodynamics definition of ΔG° demonstrates that the equilibrium constant can be defined as follows:

$$K_P = \frac{\left(\frac{q^\circ_C}{N}\right)^c\left(\frac{q^\circ_D}{N}\right)^d}{\left(\frac{q^\circ_A}{N}\right)^a\left(\frac{q^\circ_B}{N}\right)^b} \tag{23.115}$$

Although Equation (23.115) is correct as written, there is one final detail to consider. Specifically, imagine taking the preceding relationship to $T = 0$ K such that only the lowest energy states along all energetic degrees of freedom are populated. The translational and rotational ground states for all species are equivalent; however, the vibrational and electronic ground states are not. Figure 23.17 illustrates the origin of this discrepancy. The figure illustrates the ground *vibronic* (vibrational and electronic) potential for a diatomic molecule. The presence of a bond between the two atoms in the molecule lowers the energy of the molecule relative to the separated atomic fragments. Because the energy of the atomic fragments is defined as zero, the ground vibrational state is

FIGURE 23.17

The ground-state potential energy curve for a diatomic molecule. The lowest three vibrational levels are indicated.

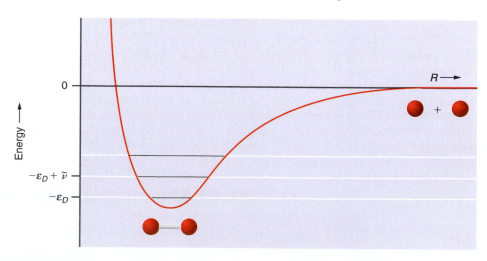

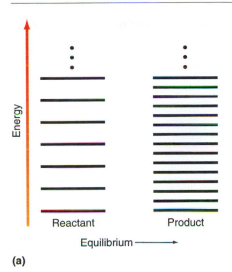

(a)

(b)

FIGURE 23.18
The statistical interpretation of equilibrium. (a) Reactant and product species having equal ground-state energies are depicted. However, the energy spacings of the product are less than the reactant such that more product states are available at a given temperature. Therefore, equilibrium will lie with the product. (b) Reactant and product species having equal state spacings are depicted. In this case, the product states are higher in energy than those of the reactant such that equilibrium lies with the reactant.

lower than zero by an amount equal to the **dissociation energy** of the molecule, ε_D. Furthermore, different molecules will have different values for the dissociation energy such that this offset from zero will be molecule specific.

Establishing a common reference state for vibrational and electronic degrees of freedom is accomplished as follows. First, differences in ε_D can be incorporated into the vibrational part of the problem such that the electronic partition function remains identical to our earlier definition. Turning to the vibrational problem, the general expression for the vibrational partition function can be written incorporating ε_D as follows:

$$q'_{vib} = \sum_n e^{-\beta \varepsilon_n} = e^{-\beta(-\varepsilon_D)} + e^{-\beta(-\varepsilon_D + \tilde{\nu})} + e^{-\beta(-\varepsilon_D + 2\tilde{\nu})} + \cdots$$

$$= e^{-\beta \varepsilon_D}(1 + e^{-\beta \tilde{\nu}} + e^{-2\beta \tilde{\nu}} + \cdots)$$

$$= e^{-\beta \varepsilon_D} q_{vib} \qquad (23.116)$$

In other words, the corrected form for the vibrational partition function is simply the product of our original q_{vib} (without zero point energy) times a factor that corrects for the offset in energy due to dissociation. With this correction factor in place, the ground vibrational states are all set to zero. Therefore, the final expression for the equilibrium constant is

$$K_P = \frac{\left(\dfrac{q^\circ_C}{N}\right)^c \left(\dfrac{q^\circ_D}{N}\right)^d}{\left(\dfrac{q^\circ_A}{N}\right)^a \left(\dfrac{q^\circ_B}{N}\right)^b} e^{\beta(c\varepsilon_C + d\varepsilon_D - a\varepsilon_A - b\varepsilon_B)} = \frac{\left(\dfrac{q^\circ_C}{N}\right)^c \left(\dfrac{q^\circ_D}{N}\right)^d}{\left(\dfrac{q^\circ_A}{N}\right)^a \left(\dfrac{q^\circ_B}{N}\right)^b} e^{-\beta \Delta \varepsilon} \qquad (23.117)$$

The dissociation energies needed to evaluate Equation (23.117) are readily obtained using spectroscopic techniques.

What insight does this expression for the equilibrium constant provide? Equation (23.117) can be viewed as consisting of two parts. The first part is the ratio of partition functions for products and reactants. Because the partition functions quantify the number of energy states available, this ratio dictates that equilibrium will favor those species with the greatest number of available energy states at a given temperature. The second half is the dissociation energy. This term dictates that equilibrium will favor those species with the lowest energy states. This behavior is illustrated in Figure 23.18. Figure 23.18a shows a reactant and product species having equal ground-state energies. The only difference between the two is that the product species has more accessible energy states at a given temperature than the reactant. Another way to envision this relationship is that the partition function describing the product will be greater than that for the reactant at the same temperature such that, at equilibrium, products will be favored ($K > 1$). In Figure 23.18b, the reactant and product energy spacings are equivalent; however, the product states lie higher in energy than those of the reactant. In this case, equilibrium will lie with the reactant ($K < 1$).

EXAMPLE PROBLEM 23.16

What is the general form of the equilibrium constant for the dissociation of a diatomic molecule?

Solution

The dissociation reaction is

$$X_2(g) \rightleftarrows 2X(g)$$

We first need to derive the partition functions that describe the reactants and products. The products are monatomic species such that only translations and electronic degrees of freedom are relevant. The partition function is then

$$q^\circ_X = q^\circ_T q_E = \left(\frac{V^\circ}{\Lambda^3_X}\right) g_o$$

The superscripts indicate standard thermodynamic conditions. With these conditions, $V° = RT/P°$ (for 1 mol) such that

$$q_X° = q_T°q_E = \left(\frac{RT}{\Lambda_X^3 P°}\right)g_o$$

and

$$\frac{q_X°}{N_A} = \frac{g_o RT}{N_A \Lambda_X^3 P°}$$

The partition function for X_2 will be equivalent to that for X, with the addition of rotational and vibrational degrees of freedom:

$$\frac{q_{X_2}°}{N_A} = \frac{g_o RT}{N_A \Lambda_{X_2}^3 P°} q_R q_V$$

Using the preceding expressions, the equilibrium constant for the dissociation of a diatomic becomes

$$K_P = \frac{\left(\dfrac{q_X°}{N_A}\right)^2}{\left(\dfrac{q_{X_2}°}{N_A}\right)} e^{-\beta\varepsilon_D} = \frac{\left(\dfrac{g_{o,X} RT}{N_A \Lambda_X^3 P°}\right)^2}{\left(\dfrac{g_{o,X_2} RT}{N_A \Lambda_{X_2}^3 P°}\right) q_R q_V} e^{-\beta\varepsilon_D}$$

$$= \left(\frac{g_{o,X}^2}{g_{o,X_2}}\right)\left(\frac{RT}{N_A P°}\right)\left(\frac{\Lambda_{X_2}^3}{\Lambda_X^6}\right)\frac{1}{q_R q_V} e^{-\beta\varepsilon_D}$$

For a specific example, we can use the preceding expression to predict K_P for the dissociation of I_2 at 298 K given the following parameters:

$$g_{o,I} = 4 \quad \text{and} \quad g_{o,I_2} = 1$$

$$\Lambda_I = 3.20 \times 10^{-12}\text{m} \quad \text{and} \quad \Lambda_{I_2} = 2.26 \times 10^{-12}\text{m}$$

$$q_R = 2773$$

$$q_V = 1.58$$

$$\varepsilon_D = 12{,}461 \text{ cm}^{-1}$$

With these values, K_P becomes

$$K_P = \left(\frac{g_{o,I}^2}{g_{o,I_2}}\right)\left(\frac{RT}{N_A P°}\right)\left(\frac{\Lambda_{I_2}^3}{\Lambda_I^6}\right)\frac{1}{q_R q_V} e^{-\beta\varepsilon_D}$$

$$= (16)\left(4.06 \times 10^{-26}\text{m}^3\right)\left(\frac{(2.26 \times 10^{-12}\text{m})^3}{(3.20 \times 10^{-12}\text{m})^6}\right)\frac{1}{(2773)(1.58)} \exp\left[\frac{-hc(12461 \text{ cm}^{-1})}{kT}\right]$$

$$= (1.59 \times 10^6)\exp\left[\frac{-(6.626 \times 10^{-34}\text{ J s})(3.00 \times 10^{10}\text{cm s}^{-1})(12461 \text{ cm}^{-1})}{(1.38 \times 10^{-23}\text{J K}^{-1})(298 \text{ K})}\right]$$

$$= 1.10 \times 10^{-20}$$

Using tabulated values for ΔG provided in the back of the text and the relationship $\Delta G = -RT \ln K$, a value of $K = 6 \times 10^{-22}$ is obtained for this reaction.

23.11 **The Helix-Coil Transition**

From molecular spectroscopy it is known that polypeptides can adopt a helical structure known as an α-helix. An illustration of a right-handed α-helix is presented in Figure 23.19. The structure is largely the result of hydrogen bonding between the hydrogen on the amine (N—H) of an amino-acid residue in the polypeptide to the carbonyl oxygen (C=O) of another residue located four residues away from the amine group. A current area of extensive research is protein folding, or understanding how a polypeptide sequence folds from a random orientation of amino acid residues into three-dimensional structures such as an α-helix.

By modifying the pH, temperature, or solvent conditions, it is possible to initiate a structural change in a polypeptide, taking it from an ordered conformational state to a less ordered state (or vice versa). For example, Figure 23.20 presents an experimental study of poly-L-glutamic acid in which the optical activity of the sample was measured as a function of pH. A dramatic evolution in optical activity is observed at ~pH = 6 consistent with a conformational change in the polypeptide from a random coil at high pH to an α-helix at low pH. For pH ranges above the pK_a of glutamic acid, the residues become charged and the α-helical structure that results in the residues being in proximity is unfavored relative to a random coil conformation. Notice that the conformational transition occurs over a relatively limited pH range. This implies that the transition occurs cooperatively with conversion from coil to helix occurring without a significant population of intermediate structural states.

Why is the helix-coil transition for poly-L-glutamic acid a cooperative process? To gain some insight into this process, first consider the thermodynamic factors involved in forming an α-helix from a random coil. The formation of the first helical segment requires a significant reduction in entropy. However, once the first turn of the helix is complete, hydrogen bonding to the next residues in the helix turn occurs that is enthalpically favorable. Therefore, once the structural transition is initiated, the overall change in polypeptide conformation can proceed in a facile fashion.

Using the concepts of statistical mechanics, we can model the cooperative behavior of the helix-coil transition. We begin by considering a model polypeptide containing four segments, with each segment existing in one of two conformational states, h (for helix) or c (for coil). We ignore symmetry such that configurations such as hccc an ccch are treated as being distinct. As we have discussed in this chapter, the partition function describing the configurations will have associated electronic, vibrational, rotational, and translational energy states; however, the details of these states are ignored in this model. As we will see, this gross simplification will turn out to be inconsequential with regard to the predictive ability of the model.

Each total configuration is described by a partition function associated with the number of residues that assume the h configuration. For example, q_0 corresponding to the cccc configuration, q_1 to the hccc configuration, and so forth with the subscript indicating the number of h-configuration residues. Next, there are four possible locations of the h-configuration residue in hccc; therefore, there are four arrangements corresponding to the q_1 partition function. Using these two ideas, the total partition function expressed as the sum of partition functions for the possible structural configuration is written as follows:

$$q = q_0 + 4q_1 + 6q_2 + 4q_3 + q_4 \tag{23.118}$$

The ratio of the partition function for a given configuration to the cccc configuration is equal to the microscopic equilibrium constant for the configuration of interest versus the cccc configuration, or $K_i = q_i/q_0$. Using this identity the total partition function is rewritten as

$$q = q_0 (1 + 4K_1 + 6K_2 + 4K_3 + K_4) \tag{23.119}$$

The probability of residing in a configuration (p_i) is equal to the values for the partition function corresponding to that configuration divided by the total partition function. For

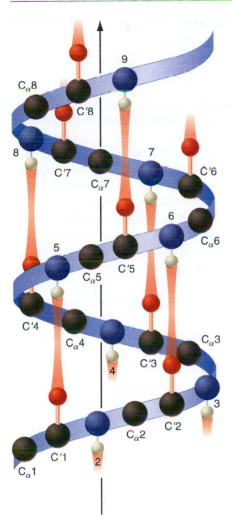

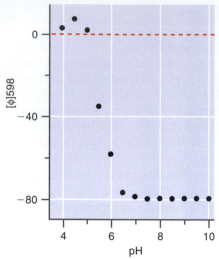

FIGURE 23.20

Optical activity of poly-L-glutamic acid as a function of pH. Data correspond to the specific rotation measured at 598 nm. (Adapted from P. Doty *et al., J. Polymer Science*, 23 (1957), 851.)

example, the probability of the four-segment polypeptide residing in the 2-h configuration is

$$p_2 = \frac{6q_2}{q} = \frac{6(K_2 q_0)}{q_0(1 + 4K_1 + 6K_2 + 4K_3 + K_4)} = \frac{6K_2}{(1 + 4K_1 + 6K_2 + 4K_3 + K_4)}$$

Notice that q_0 cancels out in the preceding expression; therefore, the value of p_i is independent of the value of q_0. Therefore, a common simplification is made by setting q_0 equal to unity.

The microscopic equilibrium constant is related to the statistical weight of a specific configuration relative to that of the reference configuration. Building on this idea, K_1 is assigned the statistical weight s, which will be a variable in the conformational model. Next, we assume that the free energy associated with changing a segment from the c to h configuration is independent of segmentation location so that ΔG associated with the 2-h configuration is twice that of the 1-h configuration. The relationship between ΔG and K establishes that the weight of K_2 is the square of K_1, or s^2. Proceeding with this logic, the total partition function is written as

$$q = 1 + 4s + 6s^2 + 4s^3 + s^4 = 1 + \sum_{i=1}^{4} \Omega_i s^i \qquad (23.120)$$

In this expression, Ω_i corresponds to the number of ways of arranging i h-configuration segments in the four-unit chain.

Using this expression for the partition function, we are now in a position to model the helix-coil transition. This model is known as the **zipper model.** In addition to assuming that each segment in the polypeptide exists in either the c or h configuration, this model also assumes that all helical segments are arranged contiguously so that an arrangement such as ....ccchhccccc.... is allowed but a ....ccchchcchhccc... arrangement is not. In this model, the statistical weights associated with h-containing are determined as follows:

1. The statistical weight associated with a configuration having an additional h-configuration segment at the end of an existing helical sequence is s:

$$\frac{(.......cchhhhh\underline{h}ccc.......)}{(.......cchhhh\underline{c}ccc.......)} = s$$

2. The statistical weight associated with initiating or nucleating an h segment is σs:

$$\frac{(.......ccccc\underline{h}ccc.......)}{(.......ccccc\underline{c}ccc.......)} = \sigma s$$

From the preceding discussion, nucleation is an entropically unfavorable process; therefore, $\sigma \ll 1$. Using these rules the configuration ccchhhccc would contribute σs^4 to the total partition function. For a segment of N units, the total partition function is

$$q = 1 + \sum_{i=1}^{N} \Omega_i \sigma s^i \qquad (23.121)$$

In this expression, Ω_i is the number of ways to arrange i h-configuration segments in a polypeptide containing N segments, equal to

$$\Omega_i = (N - i + 1) \qquad (23.122)$$

Substituting this result into q yields:

$$q = 1 + \frac{\sigma s^2 (s^N + N s^{-1} - (N + 1))}{(s - 1)^2} \qquad (23.123)$$

FIGURE 23.21
Predictions of the zipper model for the helix-coil transition of a polypeptide. As the statistical weight of the helical conformer is increased, the helical form of the peptide is favored.

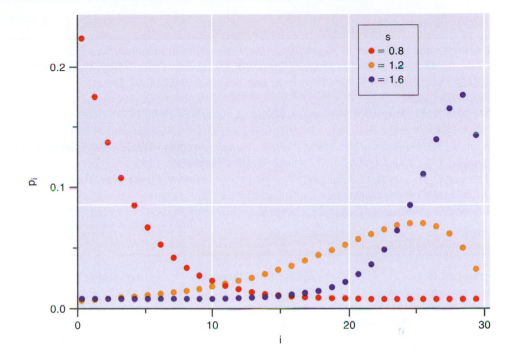

With the total partition function, the probability that the polypeptide contains i helical units is simply

$$p_i = \frac{(N - i + 1)\sigma s^i}{q} \qquad (23.124)$$

Notice that both q and p_i depend only on σ and s. Figure 23.21 presents the variation of p_i with s predicted using the zipper model for a 30-segment polypeptide ($N = 30$) where $\sigma = 10^{-3}$. The figure illustrates some interesting predictions to emerge from this model. First, if the statistical weight favors helical conformations (that is, $s > 1$), the helical segments in the polypeptide are quite long and extend to nearly the entire length of the polypeptide. That is, once nucleation of a helical segment occurs, the majority of the polypeptide undergoes a structural transition to an α-helix. This prediction is consistent with the cooperativity of the helix-coil transition observed for poly-L-glutamic acid. If the statistical weight favors the coil conformations (that is, $s < 1$), then the polypeptide exists largely as a random coil. Finally, the value of σ controls the breadth of the conformational distribution for a given value of s. As σ decreases, longer coil or helical segments will be favored for $s < 1$ and $s > 1$, respectively, reflecting the fact that a decrease in σ corresponds to a greater barrier to helix initiation. As such, longer helical segments are required to overcome the barrier.

The zipper model provides significant insight into the underlying factors that influence the helix-coil transition in polypeptides. However, its applicability is limited to smaller polypeptides. In longer polypeptides multiple helical segments can exist separated by random-coil regions, an occurrence that is not allowed in the zipper model. However, more complex models have been developed to take this behavior into account, and these models are explored in the end-of-chapter problems.

For Further Reading

Cantor, C. and Schimmel, P. *Biophysical Chemistry, Part III*, Freeman, New York, 1980.

Chandler, D., *Introduction to Modern Statistical Mechanics.* Oxford, New York, 1987.

Hill, T., *Statistical Mechanics. Principles and Selected Applications.* Dover, New York, 1956.

McQuarrie, D., *Statistical Mechanics.* Harper & Row, New York, 1973.

Nash, L. K., "On the Boltzmann Distribution Law." *J. Chemical Education* 59 (1982), 824.

Nash, L. K., *Elements of Statistical Thermodynamics.* Addison-Wesley, San Francisco, 1972.

Noggle, J. H., *Physical Chemistry.* HarperCollins, New York, 1996.

Townes, C. H., and A. L. Schallow, *Microwave Spectroscopy.* Dover, New York, 1975. (This book contains an excellent appendix of spectroscopic constants.)

Widom, B., *Statistical Mechanics.* Cambridge University Press, Cambridge, 2002.

Vocabulary

average energy

Boltzmann's formula

canonical ensemble

canonical partition function

diatomic

dissociation energy

electronic partition function

energetic degrees of freedom

ensemble

equipartition theorem

Gibbs energy

heat capacity

high-temperature (high-T) limit

internal energy

molecular partition function

polyatomic

residual entropy

rotational constant

rotational partition function

rotational temperature

Sackur–Tetrode equation

symmetry number

thermal de Broglie wavelength

thermal wavelength

total energy

translational partition function

vibrational partition function

vibrational temperature

zipper model

Questions on Concepts

Q23.1   What is the canonical ensemble? What properties are held constant in this ensemble?

Q23.2   What is the relationship between Q and q? How does this relationship differ if the particles of interest are distinguishable versus indistinguishable?

Q23.3   List the atomic and/or molecular energetic degrees of freedom discussed in this chapter. For each energetic degree of freedom, briefly summarize the corresponding quantum mechanical model.

Q23.4   For which energetic degrees of freedom are the spacings between energy levels small relative to kT at room temperature?

Q23.5   For the translational and rotational degrees of freedom, evaluation of the partition function involved replacement of the summation by integration. Why could integration be performed? How does this relate back to the discussion of probability distributions of discrete variables treated as continuous?

Q23.6   What is the high-T approximation for rotations and vibrations? For which degree of freedom do you expect this approximation to be generally valid at room temperature?

Q23.7   State the equipartition theorem. Why is this theorem inherently classical?

Q23.8   Why is the electronic partition function generally equal to the degeneracy of the ground electronic state?

Q23.9   What is q_{Total}, and how is it constructed using the partition functions for each energetic degree of freedom discussed in this chapter?

Q23.10   Why is it possible to set the energy of the ground vibrational and electronic energy level to zero?

Q23.11   List the energetic degrees of freedom for which the contribution to the internal energy determined by statistical mechanics is equal to the prediction of the equipartition theorem at 289 K.

Q23.12   Write down the contribution to the constant volume heat capacity from translations and rotations for an ideal monatomic, diatomic, and nonlinear polyatomic gas, assuming that the high-temperature limit is appropriate for the rotational degrees of freedom.

Q23.13   When are rotational degrees of freedom expected to contribute R or $3/2R$ (linear and nonlinear, respectively) to the molar constant volume heat capacity? When will a vibrational degree of freedom contribute R to the molar heat capacity?

Q23.14   Why do electronic degrees of freedom generally not contribute to the constant volume heat capacity?

Q23.15   What is the Boltzmann formula, and how can it be used to predict residual entropy?

Q23.16   How does the Boltzmann formula provide an understanding of the third law of thermodynamics?

Q23.17   What are the assumptions of the zipper model for helix-coil transitions in polypeptides? What is meant by cooperativity when describing the helix-coil conformational change?

Problems

Problem numbers in **RED** indicate that the solution to the problem is given in the *Student Solutions Manual*.

P23.1   Evaluate the translational partition function for H_2 confined to a volume of 100 cm^3 at 298 K. Perform the same calculation for N_2 under identical conditions (*Hint:* Do you need to reevaluate the full expression for q_T?).

P23.2   Evaluate the translational partition function for Ar confined to a volume of 1000 cm^3 at 298 K. At what temperature will the translational partition function of Ne be identical to that of Ar at 298 K confined to the same volume?

P23.3   At what temperature are there Avogadro's number of translational states available for O_2 confined to a volume of 1000 cm^3?

P23.4   Imagine gaseous Ar at 298 K confined to move in a two-dimensional plane of area 1.00 cm^2. What is the value of the translational partition function?

P23.5   For N_2 at 77.3 K, 1 atm, in a 1-cm^3 container, calculate the translational partition function and ratio of this partition function to the number of N_2 molecules present under these conditions.

P23.6   What is the symmetry number for the following molecules?

   a. $^{35}Cl^{37}Cl$

   b. $^{35}Cl^{35}Cl$

   c. H_2O

   d. C_6H_6

   e. CH_2Cl_2

P23.7   Which species will have the largest rotational partition function: H_2, HD, or D_2? Which of these species will have the largest translational partition function assuming that volume and temperature are identical? When evaluating the rotational partition functions, you can assume that the high-temperature limit is valid.

P23.8   For which of the following diatomic molecules is the high-temperature expression for the rotational partition function valid if $T = 40$ K?

   a. DBr ($B = 4.24$ cm^{-1})

   b. DI ($B = 3.25$ cm^{-1})

   c. CsI ($B = 0.0236$ cm^{-1})

   d. F^{35}Cl ($B = 0.516$ cm^{-1})

P23.9   Calculate the rotational partition function for SO$_2$ at 298 K where $B_A = 2.03$ cm^{-1}, $B_B = 0.344$ cm^{-1}, and $B_C = 0.293$ cm^{-1}.

P23.10   Calculate the rotational partition function for ClNO at 500 K where $B_A = 2.84$ cm^{-1}, $B_B = 0.187$ cm^{-1}, and $B_C = 0.175$ cm^{-1}.

P23.11

   a. In the rotational spectrum of H^{35}Cl ($I = 2.65 \times 10^{-47}$ kg m^2), the transition corresponding to the $J = 4$ to $J = 5$ transition is the most intense. At what temperature was the spectrum obtained?

   b. At 1000 K, which rotational transition of H^{35}Cl would you expect to demonstrate the greatest intensity?

   c. Would you expect the answers for parts (a) and (b) to change if the spectrum were of H^{37}Cl?

P23.12

   a. Calculate the percent population of the first 10 rotational energy levels for HBr ($B = 8.46$ cm^{-1}) at 298 K.

   b. Repeat this calculation for HF assuming that the bond length of this molecule is identical to that of HBr.

P23.13   In general, the high-temperature limit for the rotational partition function is appropriate for almost all molecules at temperatures above the boiling point. Hydrogen is an exception to this generality because the moment of inertia is small due to the small mass of H. Given this, other molecules with H may also represent exceptions to this general rule. For example, methane (CH$_4$) has relatively modest moments of inertia ($I_A = I_B = I_C = 5.31 \times 10^{-40}$ g cm^2) and has a relatively low boiling point of $T = 112$ K.

   a. Determine B_A, B_B, and B_C for this molecule.

   b. Use the answer from part (a) to determine the rotational partition function. Is the high-temperature limit valid?

P23.14   Calculate the vibrational partition function for H^{35}Cl at 300 and 3000 K. What fraction of molecules will be in the ground vibrational state at these temperatures?

P23.15   For IF ($\tilde{\nu} = 610$ cm^{-1}) calculate the vibrational partition function and populations in the first three vibrational energy levels for $T = 300$ and 3000 K. Repeat this calculation for IBr ($\tilde{\nu} = 269$ cm^{-1}). Compare the probabilities for IF and IBr. Can you explain the differences between the probabilities of these molecules?

P23.16   Evaluate the vibrational partition function for H_2O at 2000 K where the vibrational frequencies are 1615, 3694, and 3802 cm^{-1}.

P23.17   Evaluate the vibrational partition function for SO$_2$ at 298 K where the vibrational frequencies are 519, 1151, and 1361 cm^{-1}.

P23.18   Evaluate the vibrational partition function for NH$_3$ at 1000 K for which the vibrational frequencies are 950, 1627.5 (doubly degenerate), 3335, and 3414 cm^{-1} (doubly degenerate). Are there any modes that you can disregard in this calculation? Why or why not?

P23.19   In deriving the vibrational partition function, a mathematical expression for the series expression for the partition function was employed. However, what if one performed integration instead of summation to evaluate the partition function? Evaluate the following expression for the vibrational partition function:

$$q_v = \sum_{n=0}^{\infty} e^{-\beta hcn\tilde{\nu}} \approx \int_0^{\infty} e^{-\beta hcn\tilde{\nu}} dn$$

Under what conditions would you expect the resulting expression for q_V to be applicable?

P23.20   You have in your possession the first vibrational spectrum of a new diatomic molecule, X_2, obtained at 1000 K. From the spectrum you determine that the fraction of molecules occupying a given vibrational energy state n is as follows:

n	0	1	2	3	>3
Fraction	0.352	0.184	0.0963	0.050	0.318

What are the vibrational energy spacings for X_2?

P23.21

a. In this chapter, the assumption was made that the harmonic oscillator model is valid such that anharmonicity can be neglected. However, anharmonicity can be included in the expression for vibrational energies. The energy levels for an anharmonic oscillator are given by

$$E_n = hc\tilde{\nu}\left(n + \frac{1}{2}\right) - hc\tilde{\chi}\tilde{\nu}\left(n + \frac{1}{2}\right)^2 + \cdots$$

Neglecting zero point energy, the energy levels become $E_n = hc\tilde{\nu}n - hc\tilde{\chi}\tilde{\nu}n^2 + \cdots$. Using the preceding expression, demonstrate that the vibrational partition function for the anharmonic oscillator is

$$q_{V,anharmonic} = q_{V,harm}(1 + \beta hc\tilde{\chi}\tilde{\nu} q_{V,harm}^2(e^{-2\beta hc\tilde{\nu}} + e^{-\beta hc\tilde{\nu}}))$$

In deriving the preceding result, the following series relationship will prove useful:

$$\sum_{n=0}^{\infty} n^2 x^n = \frac{x^2 + x}{(1-x)^3}$$

b. For H_2, $\tilde{\nu} = 4401.2 \text{ cm}^{-1}$ and $\tilde{\chi}\tilde{\nu} = 121.3 \text{ cm}^{-1}$. Use the result from part (a) to determine the percent error in q_V if anharmonicity is ignored.

P23.22   Consider a particle free to translate in one dimension. The classical Hamiltonian is $H = p^2/2m$.

a. Determine $q_{classical}$ for this system. To what quantum system should you compare it in order to determine the equivalence of the classical and quantum statistical mechanical treatments?

b. Derive $q_{classical}$ for a system with translational motion in three dimensions for which $H = (p_x^2 + p_y^2 + p_z^2)/2m$.

P23.23   Evaluate the electronic partition function for atomic Fe at 298 K given the following energy levels. Terms in parenthesis are divided by $2m$.

Level (n)	Energy (cm^{-1})	Degeneracy
0	0	9
1	415.9	7
2	704.0	5
3	888.1	3
4	978.1	1

P23.24   Determine the total molecular partition function for I_2, confined to a volume of 1000 cm^3 at 298 K. Other information you will find useful: $B = 0.0374 \text{ cm}^{-1}$, $\tilde{\nu} = 208 \text{ cm}^{-1}$, and the ground electronic state is nondegenerate.

P23.25   Determine the total molecular partition function for gaseous H_2O at 1000 K confined to a volume of 1 cm^3. The rotational constants for water are $B_A = 27.8 \text{ cm}^{-1}$, $B_B = 14.5 \text{ cm}^{-1}$, and $B_C = 9.95 \text{ cm}^{-1}$. The vibrational frequencies are 1615, 3694, and 3802 cm^{-1}. The ground electronic state is nondegenerate.

P23.26   What is the contribution to the internal energy from translations for an ideal monatomic gas confined to move on a surface? What is the expected contribution from the equipartition theorem?

P23.27   Consider the following table of diatomic molecules and associated rotational constants:

Molecule	B (cm^{-1})	$\tilde{\nu}$(cm^{-1})
$H^{35}Cl$	10.59	2886
$^{12}C^{16}O$	1.93	2170
^{39}KI	0.061	200
CsI	0.024	120

a. Calculate the rotational temperature for each molecule.

b. Assuming that these species remain gaseous at 100 K, for which species is the equipartition theorem prediction for the rotational contribution to the internal energy appropriate?

c. Calculate the vibrational temperature for each molecule.

d. If these species were to remain gaseous at 1000 K, for which species is the equipartition theorem prediction for the vibrational contribution to the internal energy appropriate?

P23.28   Consider an ensemble of units in which the first excited electronic state at energy ε_1 is m_1-fold degenerate, and the energy of the ground state is m_o-fold degenerate with energy ε_0.

a. Demonstrate that if $\varepsilon_0 = 0$, the expression for the electronic partition function is

$$q_E = m_o\left(1 + \frac{m_1}{m_o} e^{-\varepsilon_1/kT}\right)$$

b. Determine the expression for the internal energy U of a ensemble of N such units. What is the limiting value of U as the temperature approaches zero and infinity?

P23.29   Calculate the internal energy of He, Ne, and Ar under standard thermodynamic conditions. Do you need to redo the entire calculation for each species?

P23.30 The ground state of O_2 is ${}^3\Sigma_g^-$. When O_2 is electronically excited, emission from the excited state (${}^1\Delta_g$) to the ground state is observed at 1263 nm. Calculate q_E and determine the electronic contribution to U for a mole of O_2 at 500 K.

P23.31 Determine the vibrational contribution to C_V for HCl ($\tilde{\nu} = 2886$ cm^{-1}) over a temperature range from 500 to 5000 K in 500-K intervals and plot your result. At what temperature do you expect to reach the high-temperature limit for the vibrational contribution to C_V?

P23.32 Determine the vibrational contribution to C_V for HCN where $\tilde{\nu}_1 = 2041$ cm^{-1}, $\tilde{\nu}_2 = 712$ cm^{-1} (doubly degenerate), and $\tilde{\nu}_3 = 3369$ cm^{-1} at $T = 298, 500$, and 1000 K.

P23.33 The speed of sound is given by the relationship

$$c_{sound} = \left(\frac{\dfrac{C_P}{C_V} RT}{M} \right)^{1/2}$$

where C_p is the constant pressure heat capacity (equal to $C_V + R$), R is the ideal gas constant, T is temperature, and M is molar mass.

a. What is the expression for the speed of sound for an ideal monatomic gas?

b. What is the expression for the speed of sound of an ideal diatomic gas?

c. What is the speed of sound in air at 298 K, assuming that air is mostly made up of nitrogen ($B = 2.00$ cm^{-1} and $\tilde{\nu} = 2359$ cm^{-1})?

P23.34 Inspection of the thermodynamic tables in the back of the text reveals that many molecules have quite similar constant volume heat capacities.

a. The value of $C_{V,m}$ for Ar(g) at standard temperature and pressure is 12.48 J mol^{-1} K^{-1}, identical to gaseous He(g). Using statistical mechanics, demonstrate why this equivalence is expected.

b. The value of $C_{V,m}$ for N$_2$(g) is 20.81 J mol^{-1} K^{-1}. Is this value expected given your answer to part (a)? For N$_2$, $\tilde{\nu} = 2359$ cm^{-1} and $B = 2.00$ cm^{-1}.

P23.35 Determine the molar entropy of N$_2$ ($\tilde{\nu} = 2359$ cm^{-1} and $B = 2.00$ cm^{-1}, $g_0 = 1$) and the entropy when $P = 1$ atm, but $T = 2500$ K.

P23.36 Determine the standard molar entropy of H^{35}Cl at 298 K where $B = 10.58$ cm^{-1}, $\tilde{\nu} = 2886$ cm^{-1}, and the ground-state electronic level degeneracy is one.

P23.37 Derive the expression for the standard molar entropy of a monatomic gas restricted to two-dimensional translational motion. (*Hint:* You are deriving the two-dimensional version of the Sackur–Tetrode equation.)

P23.38 The molecule NO has a ground electronic level that is doubly degenerate, and a first excited level at 121.1 cm^{-1} that is also twofold degenerate. Determine the contribution of electronic degrees of freedom to the standard molar entropy of

NO. Compare your result to $R \ln(4)$. What is the significance of this comparison?

P23.39 Determine the residual molar entropies for molecular crystals of the following:

a. ^{35}Cl^{37}Cl

b. CFCl$_3$

c. CF$_2$Cl$_2$

d. CO$_2$

P23.40 Determine the standard Gibbs energy for ^{35}Cl^{35}Cl where $\tilde{\nu} = 560$ cm^{-1}, $B = 0.244$ cm^{-1}, and the ground electronic state is nondegenerate.

P23.41 Determine the rotational and vibrational contributions to the standard Gibbs energy for N$_2$O (NNO), a linear triatomic molecule where $B = 0.419$ cm^{-1}, $\tilde{\nu}_1 = 1285$ cm^{-1}, $\tilde{\nu}_2 = 589$ cm^{-1} (doubly degenerate), and $\tilde{\nu}_3 = 2224$ cm^{-1}.

P23.42 Determine the equilibrium constant for the dissociation of sodium at 298 K: Na$_2$(g) $\rightleftarrows$ 2Na(g). For Na$_2$, $B = 0.155$ cm^{-1}, $\tilde{\nu} = 159$ cm^{-1}, the dissociation energy is 70.4 kJ/mol, and the ground-state electronic degeneracy for Na is 2.

P23.43 The isotope-exchange reaction for Cl$_2$ is as follows: ^{35}Cl^{35}Cl + ^{37}Cl^{37}Cl $\rightleftarrows$ 2^{37}Cl^{35}Cl. The equilibrium constant for this reaction is ~4. Furthermore, the equilibrium constant for similar isotope-exchange reactions is also close to this value. Demonstrate why this would be so.

P23.44 In deriving the zipper model, the statement was made that if the statistical weight of the hccc configuration were s, then the weight of the hhcc configuration would be s^2. Using the relationship between ΔG, K, and s, justify this statement mathematically.

P23.45 Determine the statistical weights of the following sequences using the parameters of the zipper model:

a. cchhhhcc

b. cchccccc

c. cchhhccHhc

P23.46 The average length of a helical segment in a polypeptide is given by

$$\langle i \rangle = \sum_l i p_i$$

Using the expression for p_i from the zipper model, derive the expression for $\langle i \rangle$. For $s = 1.6$, $\sigma = 10^{-3}$, and $N = 30$, what is $\langle i \rangle$?

P23.47 Using the expression for p_i obtained from the zipper model, what is the most probable helical length for $s = 1.5$, $\sigma = 10^{-3}$, and $N = 30$. The most straightforward way to solve this problem is to numerically evaluate p_i for all values of i, but can you think of a way to derive an expression for the maximum of p_i versus i?

P23.48 One of the limitations of the zipper model is that helical segments must be contiguous; however, in longer polypeptides more than one helical segment can be formed. Helix formation in longer polypeptides can be modeled using the Zimm–Bragg model. In this model, the conformational state of the polypeptide is described by the fractional helicity (Θ), defined as the average of amino-acid residues that exist in helical regions of the polypeptide divided by the total number of residues:

$$\Theta = 0.5\left(1 + \frac{(s-1) + s\sigma}{[(s-1)^2 + 4s\sigma]^{1/2}}\right)$$

In this expression, s and σ have identical definitions to those used in the zipper model.

a. For what value of s does this model predict half the amino-acid residues to exist in helical regions?

b. Assuming $\sigma = 0.01$, generate a plot of Θ versus s in the range $0 \leq s \leq 2$.

c. Repeat part (b), but for $\sigma = 0.0001$. Comment on the differences between the plots, and in particular why a decrease in σ should affect the sharpness of the helix-coil transition.

Web-Based Simulations, Animations, and Problems

W23.1 In this web-based simulation, the variation in q_T, q_R, and q_V with temperature is investigated for three diatomic molecules: HF, $H^{35}Cl$, and ^{35}ClF. Comparisons of q_T and q_R are performed to illustrate the mass and temperature dependence of these partition functions. Also, the expected dependence of q_T, q_R, and q_V on temperature in the high-temperature limit is investigated.

W23.2 In this simulation, the temperature dependence of C_V for vibrational degrees of freedom is investigated for diatomic and polyatomic molecules with and without mode degeneracy. Comparisons of exact values to those expected for the high-temperature limit are performed.

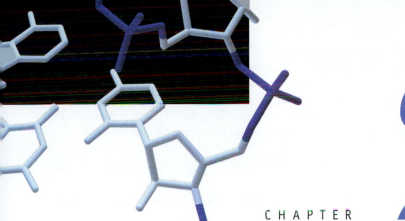

Transport Phenomena

How will a system respond when it is not at equilibrium? The first steps toward answering this question are provided in this chapter. The study of system relaxation toward equilibrium is known as *dynamics*. In this chapter, transport phenomena involving the evolution of a system's physical properties such as mass or energy are described. All transport phenomena are connected by one central idea: the rate of change for a system's physical property is dependent on the spatial gradient of the property. In this chapter, this underlying idea is first described as a general concept, then applied to mass (diffusion), linear momentum (viscosity), and charge transport. The timescale for mass transport is discussed and approached from both the macroscopic and microscopic perspectives. It is important to note that although the various transport phenomena outlined here look different, the underlying concepts describing these phenomena have a common origin.

24.1 What Is Transport?

To this point we have been concerned with describing system properties at equilibrium; however, what happens when an external perturbation results in the property of the system (for example, mass or energy) being shifted away from equilibrium? Once the external perturbation is removed, the system will evolve to reestablish the equilibrium distribution of the property. **Transport phenomena** involve the evolution of a system property in response to a nonequilibrium distribution of the property. In order for a system property to be transported, a spatial distribution of the property must exist that is different from that at equilibrium. For example, consider a collection of gas particles in which the equilibrium spatial distribution of the particles corresponds to a homogeneous particle number density throughout the container. What would happen if the particle number density were greater on one side of the container than the other? The expectation is that gas particle translation will result in the reestablishment of a homogeneous number density throughout the container. That is, the system evolves to reestablish a distribution of the system property that is consistent with equilibrium.

A central concept in transport phenomena is **flux,** defined as a quantity transferred through a given area per unit of time. Flux will occur when a spatial imbalance or gradient exists for a system property, and the flux will act in opposition to this gradient. In the example just discussed, imagine dividing the container in two parts with a partition

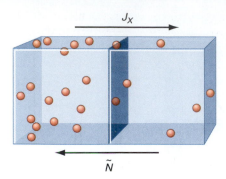

FIGURE 24.1
Illustration of flux. The flux J_x of gas particles is in opposition to the gradient in particle number density $\tilde{N}$.

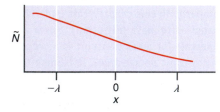

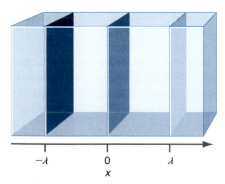

FIGURE 24.2
Model used to describe gas diffusion. The gradient in number density $\tilde{N}$ results in particles diffusing from $-x$ to $+x$. The plane located at $x = 0$ is where the flux of particles in response to the gradient is calculated (the *flux plane*). Two planes, located one mean free path distance away ($\pm\lambda$), are considered with particles traveling from either of these planes to the flux plane. The total flux through the flux plane is equal to the difference in flux from the planes located at $\pm\lambda$.

and counting the number of particles that move from one side of the container to the other side, as illustrated in Figure 24.1. The flux in this case is equal to the number of particles that move through the partition per unit time. A central theoretical underpinning of all transport processes is the relationship between a flux and the spatial gradient in a system property that gives rise to the corresponding flux. The most basic relationship between flux and the spatial gradient in transported property is as follows:

$$J_x = -\alpha\frac{d\,(\text{property})}{dx} \tag{24.1}$$

In Equation (24.1), J_x is the flux expressed in units of property area^{-1} time^{-1}. The derivative in Equation (24.1) represents the spatial gradient of the quantity of interest (mass, energy, etc.). The linear relationship between flux and the property spatial gradient is reasonable when the displacement away from equilibrium is modest. The limit of modest displacement of the system property away from equilibrium is assumed in the remainder of this chapter. The negative sign in Equation (24.1) indicates that the flux occurs in the opposite direction of the gradient; therefore, flux will result in a reduction of the gradient if external action is not taken to maintain the gradient. Correspondingly, if the gradient is externally maintained at a constant value, the flux will also remain constant. Again, consider Figure 24.1, which presents a graphical example of the relationship between gradient and flux. The gas density is greatest on the left-hand side of the container such that the particle density increases as one goes from the right side of the container to the left. According to Equation (24.1), particle flux occurs in opposition to the number density gradient in an attempt to make the particle density spatially homogeneous. The final quantity of interest in Equation (24.1) is the factor α. Mathematically, this quantity serves as the proportionality constant between the gradient and flux and is referred to as the **transport coefficient**. Although the derivations for each transport property will look different, it is important to note that all originate from Equation (24.1). That is, the underlying principle behind all transport phenomena is the relationship between flux and gradient.

24.2 Mass Transport: Diffusion

Diffusion is the process by which particle density evolves in response to a spatial gradient in concentration. With respect to thermodynamics, this spatial gradient represents a gradient in chemical potential, and the system will relax toward equilibrium by eliminating this gradient. The first case to be considered is diffusion in an ideal gas. Diffusion in liquids is treated later in this chapter.

Consider a gradient in gas particle number density, $\tilde{N}$, as depicted in Figure 24.2. According to Equation (24.1), there will be a flux of gas particles in opposition to the gradient. The flux is determined by quantifying the flow of particles per unit time through an imaginary plane located at $x = 0$ with area A. We will refer to this plane as the *flux plane*. Two other planes are located one mean free path $\pm\lambda$ away on either side of the flux plane, and the net flux arises from particles traveling from either of these planes to the flux plane. The mean free path is defined as the distance a particle travels on average between collisions. Using gas kinetic theory, this quantity can be quantitatively defined (see the Further Reading section at the end of the chapter), but only the conceptual definition is needed at present.

Figure 24.2 demonstrates that a gradient in particle number density exists in the x direction such that:

$$\frac{d\tilde{N}}{dx} \neq 0$$

If the gradient in $\tilde{N}$ were equal to zero, flux J_x would also equal zero by Equation (24.1). However, this does not mean that particles are now stationary. Instead, $J_x = 0$ indicates that the flow of particles through the flux plane from left to right is exactly balanced by the flow of particles from right to left. Therefore, the flux expressed in Equation (24.1) represents the net flux, or sum of flux in each direction through the flux plane.

Equation (24.1) provides the relationship between the flux and the spatial gradient in $\tilde{N}$. Solution of this **mass transport** problem involves determining the proportionality constant α. This quantity is referred to as the diffusion coefficient for mass transport. To determine this constant, consider the particle number density at $\pm\lambda$:

$$\tilde{N}(-\lambda) = \tilde{N}(0) - \lambda\left(\frac{d\tilde{N}}{dx}\right)_{x=0} \tag{24.2}$$

$$\tilde{N}(\lambda) = \tilde{N}(0) + \lambda\left(\frac{d\tilde{N}}{dx}\right)_{x=0} \tag{24.3}$$

These expressions state that the value of $\tilde{N}$ away from $x = 0$ is equal to the value of $\tilde{N}$ at $x = 0$ plus a second term representing the change in concentration as one moves toward the planes at $\pm\lambda$. Formally, Equations (24.2) and (24.3) are derived from a Taylor series expansion of the number density with respect to distance, and only the first two terms in the expansion are kept, consistent with being sufficiently small such that higher order terms of the expansion can be neglected. Effusion is the movement of gas molecules through an aperture whose size is small compared to the mean free path of the molecules. The diffusion process can be viewed as an effusion of particles through a plane or aperture of area A. The number of particles, N, striking a given area per unit time is equal to

$$\frac{dN}{dt} = J_x \times A \tag{24.4}$$

Consider the flux of particles traveling from the plane at $-\lambda$ to the flux plane (Figure 24.2). Because we are interested in the number of particles striking the flux plane per unit time, we only want to count particles that are traveling toward the flux plane. To determine this quantity, we need to describe the particles in terms of their velocity (that is, consider the direction the particles are traveling as well as their speed), and include only those particles that are traveling toward the wall. To accomplish this task, we first use the one-dimensional velocity distribution for a collection of gas particles, $f(v_x)$, derived from gas kinetic theory:

$$f(v_x) = \left(\frac{m}{2\pi kT}\right)^{1/2} e^{-mv_x^2/2kT} \tag{24.5}$$

In Equation (24.5), m is the mass of an individual gas particle, T is temperature, and k is Boltzmann's constant. Using this expression, the flux is defined as the number density of the particles times the average velocity of particles traveling toward the flux plane:

$$\begin{aligned} J_x &= \tilde{N}\int_0^\infty v_x f(v_x)\,dv_x \\ &= \tilde{N}\int_0^\infty v_x\left(\frac{m}{2\pi kT}\right)^{1/2} e^{-mv_x^2/2kT}\,dv_x \\ &= \tilde{N}\left(\frac{kT}{2\pi m}\right)^{1/2} \\ &= \frac{\tilde{N}}{4} v_{ave} \end{aligned} \tag{24.6}$$

Notice that the integral is restricted to those molecules traveling in the $+x$ direction, ensuring that only those particles traveling toward the flux plane are included. In addition, the last step in Equation (24.6) is performed to express the flux in terms of the average speed of the gas particles, also defined using gas kinetic theory:

$$v_{ave} = \left(\frac{8kT}{\pi m}\right)^{1/2} \tag{24.7}$$

Using this expression for J_x and substituting into Equations (24.2) and (24.3) yields:

$$J_{-\lambda,0} = \frac{1}{4}v_{ave}\tilde{N}(-\lambda) = \frac{1}{4}v_{ave}\left[\tilde{N}(0) - \lambda\left(\frac{d\tilde{N}}{dx}\right)_{x=0}\right] \tag{24.8}$$

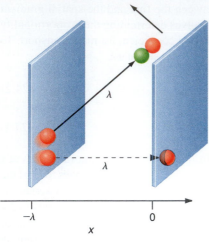

FIGURE 24.3
Particle trajectories aligned with the x axis (dashed line) result in the particle traveling between planes without collision. However, trajectories not aligned with the x axis (solid line) result in the particle not reaching the flux plane before a collision with another particle occurs. This collision may result in the particle being directed away from the flux plane.

$$J_{\lambda,0} = \frac{1}{4}\nu_{ave}\tilde{N}(\lambda) = \frac{1}{4}\nu_{ave}\left[\tilde{N}(0) + \lambda\left(\frac{d\tilde{N}}{dx}\right)_{x=0}\right] \tag{24.9}$$

The total flux through the flux plane is simply the difference in flux from the planes at $\pm\lambda$:

$$J_{Total} = J_{-\lambda,0} - J_{\lambda,0} = \frac{1}{4}\nu_{ave}\left(-2\lambda\left(\frac{d\tilde{N}}{dx}\right)_{x=0}\right)$$

$$= -\frac{1}{2}\nu_{ave}\lambda\left(\frac{d\tilde{N}}{dx}\right)_{x=0} \tag{24.10}$$

One correction remains before the derivation is complete. We have assumed that the particles move from the planes located at $\pm\lambda$ to the flux plane directly along the x axis. However, Figure 24.3 illustrates a subtle point: What if the particle trajectory is not aligned with the x axis? In this case the particle will not reach the flux plane after traveling one mean free path such that collisions with other particles can occur, resulting in postcollision particle trajectories away from the flux plane. Inclusion of these trajectories requires one to take the orientational average of the mean free path, and this averaging results in a reduction in total flux as expressed by Equation (24.10) by a factor of 2/3. With this averaging, the total flux becomes

$$J_{Total} = -\frac{1}{3}\nu_{ave}\lambda\left(\frac{d\tilde{N}}{dx}\right)_{x=0} \tag{24.11}$$

Equation (24.11) is identical in form to Equation (24.1), with the diffusion proportionality constant, or simply the **diffusion coefficient,** defined as follows:

$$D = \frac{1}{3}\nu_{ave}\lambda \tag{24.12}$$

Because ν_{ave} has units of m s^{-1} and λ has units of meters, the diffusion coefficient has units of m^2 s^{-1}. Quantitative evaluation of the preceding expression can be performed using the definition for λ obtained from gas kinetic theory for a homogeneous collection of gas particles:

$$\lambda = \frac{kT}{\sqrt{2}P\sigma} \tag{24.13}$$

In Equation (24.13), P is the pressure of the gas (in Pa) and σ is the collisional cross-section of the gas particles. With the diffusion coefficient, Equation (24.11) becomes:

$$J_{Total} = -D\left(\frac{d\tilde{N}}{dx}\right)_{x=0} \tag{24.14}$$

Equation (24.11) is referred to as **Fick's first law** of diffusion. It is important to note that the diffusion coefficient is defined using parameters derived from gas kinetic theory, namely, the average speed of the gas and the mean free path. Example Problem 24.1 illustrates the dependence of the diffusion coefficient on these parameters.

EXAMPLE PROBLEM 24.1

Determine the diffusion coefficient for Ar ($\sigma = 3.6 \times 10^{-19}$ m^2) at 298 K and a pressure of 1.00 atm (101,325 Pa).

Solution

Using Equation (24.12), the diffusion coefficient is calculated as follows:

$$D_{Ar} = \frac{1}{3}\nu_{ave,Ar}\lambda_{Ar}$$

$$= \frac{1}{3}\left(\frac{8kT}{\pi m}\right)^{1/2}\left(\frac{RT}{\sqrt{2}P\sigma_{Ar}}\right)$$

$$= \frac{1}{3}\left(\frac{8(1.38 \times 10^{-23}\ \text{J K}^{-1})298\ \text{K}}{\pi\left(0.040\ \text{kg mol}^{-1}\Big/N_A\right)}\right)^{1/2}\left(\frac{(1.38 \times 10^{-23}\ \text{J K}^{-1})298\ \text{K}}{\sqrt{2}(101{,}325\ \text{Pa})(3.6 \times 10^{-19}\text{m}^2)}\right)$$

$$= \frac{1}{3}(397\ \text{m s}^{-1})(7.98 \times 10^{-8}\ \text{m})$$

$$= 1.06 \times 10^{-5}\ \text{m}^2\ \text{s}^{-1}$$

The diffusion constant for Ar is representative of many gaseous molecules where D is typically on the order of $10^{-5}\ \text{m}^2\ \text{s}^{-1}$.

One criticism of this development is that parameters such as average velocity and the mean free path are derived using an equilibrium distribution, yet these concepts are now being applied in a nonequilibrium context when discussing transport phenomena. This development is performed under the assumption that the displacement of the system away from equilibrium is modest; therefore, equilibrium-based quantities remain relevant. That said, transport phenomena can be described using nonequilibrium distributions; however, the mathematical complexity of this approach is beyond the scope of this text.

With the expression for the diffusion coefficient in hand [Equation (24.12)], the relationship between this quantity and the details of the gas particles is clear. This relationship suggests that transport properties such as diffusion can be used to determine particle parameters such as effective size as described by the collisional cross-section. Example Problem 24.2 illustrates the connection between the diffusion coefficient and particle size.

EXAMPLE PROBLEM 24.2

Under identical temperature and pressure conditions, the diffusion coefficient of He is roughly four times larger than that of Ar. Determine the ratio of the collisional cross-sections.

Solution

Using Equation (24.12), the ratio of diffusion coefficients (after canceling the 1/3 constant term) can be written in terms of the average speed and mean free path as follows:

$$\frac{D_{He}}{D_{Ar}} = 4 = \frac{\nu_{ave,He}\lambda_{He}}{\nu_{ave,Ar}\lambda_{Ar}}$$

$$= \frac{\left(\dfrac{8kT}{\pi m_{He}}\right)^{1/2}\left(\dfrac{kT}{\sqrt{2}P_{He}\sigma_{He}}\right)}{\left(\dfrac{8kT}{\pi m_{Ar}}\right)^{1/2}\left(\dfrac{kT}{\sqrt{2}P_{Ar}\sigma_{Ar}}\right)}$$

$$= \left(\frac{m_{Ar}}{m_{He}}\right)^{1/2}\left(\frac{\sigma_{Ar}}{\sigma_{He}}\right)$$

$$\left(\frac{\sigma_{He}}{\sigma_{Ar}}\right) = \frac{1}{4}\left(\frac{m_{Ar}}{m_{He}}\right)^{1/2} = \frac{1}{4}\left(\frac{M_{Ar}}{M_{He}}\right)^{1/2} = \frac{1}{4}\left(\frac{39.9\ \text{g mol}^{-1}}{4.00\ \text{g mol}^{-1}}\right)^{1/2} = 0.79$$

24.3 The Time Evolution of a Concentration Gradient

As illustrated in the previous section, the existence of a concentration gradient results in particle diffusion. What is the timescale for diffusion, and how far can a particle diffuse in a given amount of time? These questions are addressed by the diffusion equation,

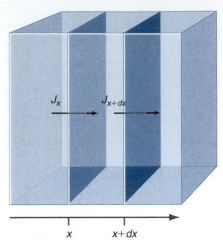

FIGURE 24.4
Depiction of flux through two separate planes. If $J_x = J_{x+dx}$ then the concentration between the planes will not change. However, if the fluxes are unequal, then the concentration will change with time.

which can be derived as follows. Beginning with Fick's first law, the particle flux is given by

$$J_x = -D\left(\frac{d\tilde{N}(x)}{dx}\right) \quad (24.15)$$

The quantity J_x in Equation (24.15) is the flux through a plane located at x as illustrated in Figure 24.4. The flux at location $x + dx$ can also be written using Fick's first law:

$$J_{x+dx} = -D\left(\frac{d\tilde{N}(x + dx)}{dx}\right) \quad (24.16)$$

The particle density at $(x + dx)$ is related to the corresponding value at x as follows:

$$\tilde{N}(x + dx) = \tilde{N}(x) + dx\left(\frac{d\tilde{N}(x)}{dx}\right) \quad (24.17)$$

This equation is derived by keeping the first two terms in the Taylor series expansion of number density with distance equivalent to the procedure described earlier for obtaining Equations (24.2) and (24.3). Substituting Equation (24.17) into Equation (24.16), the flux through the plane at $(x + dx)$ becomes

$$J_{x+dx} = -D\left(\frac{d\tilde{N}(x)}{dx} + \left(\frac{d^2\tilde{N}(x)}{dx^2}\right)dx\right) \quad (24.18)$$

Consider the space between the two flux planes illustrated in Figure 24.4. The change in particle density in this region of space depends on the difference in flux through the two planes. If the fluxes are identical, the particle density will remain constant in time. However, a difference in flux will result in an evolution in particle number density as a function of time. The flux is equal to the number of particles that pass through a given area per unit time. If the area of the two flux planes is equivalent, then the difference in flux is directly proportional to the difference in number density. This relationship between the time dependence of the number density and the difference in flux is expressed as follows:

$$\frac{d\tilde{N}(x,t)}{dt} = \frac{d(J_x - J_{x+dx})}{dx}$$

$$= \frac{d}{dx}\left[-D\left(\frac{d\tilde{N}(x,t)}{dx}\right) - \left(-D\left(\left(\frac{d\tilde{N}(x,t)}{dx}\right) + \left(\frac{d^2\tilde{N}(x,t)}{dx^2}\right)dx\right)\right)\right]$$

$$= D\frac{d^2\tilde{N}(x,t)}{dx^2} \quad (24.19)$$

Equation (24.19) is called the **diffusion equation,** and it is also known as **Fick's second law of diffusion.** Notice that $\tilde{N}$ depends on both position x and time t. Equation (24.19) demonstrates that the time evolution of the concentration gradient is proportional to the second derivative of the spatial gradient in concentration. That is, the greater the "curvature" of the concentration gradient, the faster the relaxation will proceed. Equation (24.19) is a differential equation that can be solved using standard techniques and a set of initial conditions (see the Crank entry in the Further Reading section at the end of the chapter) resulting in the following expression for $\tilde{N}(x,t)$:

$$\tilde{N}(x,t) = \frac{N_0}{2A(\pi Dt)^{1/2}}e^{-x^2/4Dt} \quad (24.20)$$

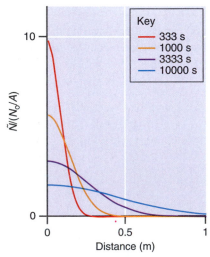

Key
— 333 s
— 1000 s
— 3333 s
— 10000 s

$\tilde{N}(N_0/A)$

Distance (m)

FIGURE 24.5
The spatial variation in particle number density, $\tilde{N}(x,t)$, as a function of time. The number density is defined with respect to N_0/A, the number of particles confined to a plane located at $x = 0$ of area A. In this example, $D = 10^{-5}$ m^2 s^{-1}, a typical value for a gas at 1 atm and 298 K (see Example Problem 24.1). The corresponding diffusion time for a given concentration profile is indicated.

In this expression, N_0 represents the initial number of molecules confined to a plane at $t = 0$, A is the area of this plane, x is distance away from the plane, and D is the diffusion coefficient. Equation (24.20) can be viewed as a distribution function that describes the probability of finding a particle at time $= t$ at a plane located a distance x away from the initial plane at $t = 0$. An example of the spatial variation in $\tilde{N}$ versus time is provided in Figure 24.5 for a species with $D = 10^{-5}$ m^2 s^{-1}. The figure demonstrates that with an increase in time, $\tilde{N}$ increases at distances farther away from the initial plane (located at 0 m in Figure 24.5).

Instead of describing the entire distribution, it is more convenient to use a metric or benchmark value that provides a measure of $\tilde{N}(x,t)$. The primary metric employed to describe $\tilde{N}(x,t)$ is the root-mean-square (rms) displacement, determined as follows:

$$x_{rms} = \langle x^2 \rangle^{1/2} = \left[\frac{A}{N_0} \int_{-\infty}^{\infty} x^2 \tilde{N}(x,t)\, dx \right]^{1/2}$$

$$= \left[\frac{A}{N_0} \int_{-\infty}^{\infty} x^2 \frac{N_0}{A2(\pi Dt)^{1/2}} e^{-x^2/4Dt}\, dx \right]^{1/2}$$

$$= \left[\frac{1}{2(\pi Dt)^{1/2}} \int_{-\infty}^{\infty} x^2 e^{-x^2/4Dt}\, dx \right]^{1/2}$$

$$x_{rms} = \sqrt{2Dt} \qquad (24.21)$$

Notice that the rms displacement increases as the square root of both the diffusion coefficient and time. Equation (24.21) represents the rms displacement in a single dimension. For diffusion in three dimensions, the corresponding term, r_{rms}, can be determined using the Pythagorean theorem under the assumption that diffusion is equivalent in all three dimensions:

$$r_{rms} = \sqrt{6Dt} \qquad (24.22)$$

The diffusion relationships derived in this section and the previous section involved gases and employed concepts from gas kinetic theory. However, these relationships are also applicable to diffusion in solution, as demonstrated later in this chapter.

Time-Evolution of a Concentration Gradient

EXAMPLE PROBLEM 24.3

Determine x_{rms} for a particle where $D = 10^{-5}$ m^2 s^{-1} for diffusion times of 1000 and 10,000 s.

Solution

Employing Equation (24.21):

$$x_{rms,1000s} = \sqrt{2Dt} = \sqrt{2(10^{-5}\ \text{m}^2\ \text{s}^{-1})(1000\ \text{s})} = 0.141\ \text{m}$$

$$x_{rms,10,000s} = \sqrt{2Dt} = \sqrt{2(10^{-5}\ \text{m}^2\ \text{s}^{-1})(10{,}000\ \text{s})} = 0.447\ \text{m}$$

The diffusion coefficient employed in this example is equivalent to that used in Figure 24.5, and the rms displacements determined here can be compared to the spatial variation in $\tilde{N}(x,t)$ depicted in the figure to provide a feeling for the x_{rms} distance versus the overall distribution of particle diffusion distances versus time.

24.4 Supplemental: Statistical View of Diffusion

In deriving Fick's first law of diffusion, a gas particle was envisioned to move a distance equal to the mean free path before colliding with another particle. After this collision, memory of the initial direction of motion is lost and the particle is free to move in the same or a new direction until the next collision occurs. This conceptual picture of particle motion is mathematically described by the statistical approach to diffusion. In the statistical approach, illustrated in Figure 24.6, particle diffusion is also modeled as a series of discrete displacements or steps, with the direction of one step being uncorrelated with that of the previous step. That is, once the particle has taken a step, the direction of the next step is random. A series of such steps is referred to as a **random walk.**

In the previous section we determined the probability of finding a particle at a distance x away from the origin after a certain amount of time. The statistical model of diffusion can be connected directly to this idea using the random walk model. Consider a particle undergoing a random walk along a single dimension x such that the particle moves one step

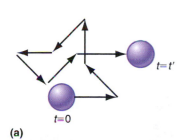

$t=t'$

$t=0$

(a)

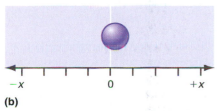

$-x$ 0 $+x$

(b)

FIGURE 24.6

(a) Illustration of a random walk. Diffusion of the particle is modeled as a series of discrete steps (each arrow); the length of each step is the same, but the direction of the step is random. (b) Illustration of the one-dimensional random walk model.

in either the $+x$ or $-x$ direction (Figure 24.6). After a certain number of steps, the particle will have taken Δ total steps with Δ_- steps in the $-x$ direction and Δ_+ steps in the $+x$ direction. The probability that the particle will have traveled a distance X from the origin is related to the weight associated with that distance, as given by the following expression:

$$W = \frac{\Delta!}{\Delta_+!\Delta_-!} = \frac{\Delta!}{\Delta_+!(\Delta - \Delta_+)!} \tag{24.23}$$

Equation (24.23) is identical to the weight associated with observing a certain number of heads after tossing a coin Δ times, as discussed in Chapter 22. This similarity is not a coincidence—the one-dimensional random walk model is very much like tossing a coin. Each outcome of a coin toss is independent of the previous outcome, and only one of two outcomes is possible per toss. Evaluation of Equation (24.23) requires an expression for Δ_+ in terms of Δ. This relationship can be derived by recognizing that X is equal to the difference in the number of steps in the $+x$ and $-x$ direction:

$$X = \Delta_+ - \Delta_- = \Delta_+ - (\Delta - \Delta_+) = 2\Delta_+ - \Delta$$

$$\frac{X + \Delta}{2} = \Delta_+$$

With this definition for Δ_+, the expression for W becomes

$$W = \frac{\Delta!}{\left(\dfrac{\Delta + X}{2}\right)!\left(\Delta - \dfrac{\Delta + X}{2}\right)!} = \frac{\Delta!}{\left(\dfrac{\Delta + X}{2}\right)!\left(\dfrac{\Delta - X}{2}\right)!} \tag{24.24}$$

The probability of the particle being a distance X away from the origin is given by the weight associated with this distance divided by the total weight, 2^Δ, such that

$$P = \frac{W}{W_{Total}} = \frac{W}{2^\Delta} \propto e^{-X^2/2\Delta} \tag{24.25}$$

The final proportionality can be derived by evaluation of Equation (24.24) using Stirling's approximation. Recall from the solution to the diffusion equation in the previous section that the distance a particle diffuses away from the origin was also proportional to an exponential term:

$$\tilde{N}(x,t) \propto e^{-X^2/4Dt} \tag{24.26}$$

In Equation (24.26), x is the actual diffusion distance, D is the diffusion coefficient, and t is time. For these two pictures of diffusion to converge on the same physical result, the exponents must be equivalent such that

$$\frac{x^2}{4Dt} = \frac{X^2}{2\Delta} \tag{24.27}$$

At this point, the random walk parameters X and Δ must be expressed in terms of the actual quantities of diffusion distance x and total diffusion time t. The total number of random walk steps is expressed as the total diffusion time, t, divided by the time per random walk step, τ:

$$\Delta = \frac{t}{\tau} \tag{24.28}$$

In addition, x can be related to the random walk displacement X by using a proportionality constant, x_o, that represents the average distance in physical space a particle traverses between collisions such that

$$x = X x_o \tag{24.29}$$

With these definitions for Δ and X, substitution into Equation (24.27) results in the following definition of D:

$$D = \frac{x_o^2}{2\tau} \tag{24.30}$$

Equation (24.30) is the **Einstein–Smoluchowski equation.** The importance of this equation is that it relates a macroscopic quantity, D, to microscopic aspects of the diffusion as described by the random walk model. For reactions in solution, x_o is generally taken to be the particle diameter. Using this definition and the experimental value for D, the timescale associated with each random walk event can be determined.

EXAMPLE PROBLEM 24.4

The diffusion coefficient of liquid benzene is $2.2 \times 10^{-5} \ cm^2 \ s^{-1}$. Liquids typically demonstrate diffusion coefficients on the order of $10^{-5} \ cm^2 \ s^{-1}$ under standard conditions. Given an estimated molecular diameter of 0.3 nm, what is the timescale for a random walk?

Solution

Rearranging Equation (24.30), the time per random walk step is

$$\tau = \frac{x_o^2}{2D} = \frac{(0.3 \times 10^{-9} \ m)^2}{2(2.2 \times 10^{-9} \ m^2 \ s^{-1})} = 2.0 \times 10^{-11} \ s$$

This is an extremely short time, only 20 ps! This example illustrates that, on average, the diffusional motion of a benzene molecule in the liquid phase is characterized by short-range translational motion between frequent collisions with neighboring molecules.

The Einstein–Smoluchowski equation can also be related to gas diffusion as described previously. If we equate x_o with the mean free path λ and define the time per step as the average time it takes a gas particle to translate one mean free path λ/v_{ave}, then the diffusion coefficient is given by

$$D = \frac{\lambda^2}{2\left(\dfrac{\lambda}{v_{ave}}\right)} = \frac{1}{2}\lambda v_{ave} \tag{24.31}$$

This is exactly the same expression for D derived from gas kinetic theory in the absence of the 2/3 correction for particle trajectories as discussed earlier. That is, the statistical and kinetic theory viewpoints of diffusion provide equivalent descriptions of gas diffusion.

24.5 Properties of Rigid Macromolecules in Solution

24.5.1 Spherical Molecules

Many experimental techniques including velocity sedimentation, dynamic light scattering, nuclear magnetic resonance, and electron spin resonance experiments are sensitive to the translational and/or rotational diffusion of macromolecules. As we will see, by quantifying these diffusive motions, information about the structures of biological macromolecules can be obtained. The concept of random motion resulting in particle diffusion was evidenced in the famous microscopy experiments of Robert Brown performed in 1827. [An excellent account of this work can be found in *The Microscope* 40 (1992), 235–241.] In these experiments, Brown took ~5-μm-diameter pollen grains suspended in water and, by using a microscope, he was able to see that the particles were "very evidently in motion." After performing experiments to show that the motion was not from convection or evaporation, Brown concluded that the motion was associated with the particle itself. This apparently random motion of the pollen grain is referred to as **Brownian motion,** and this motion is actually the diffusion of a large particle in solution.

Consider Figure 24.7 where a particle of mass m is embedded in a liquid having viscosity η. The motion of the particle is driven by collisions with liquid particles, which will provide a time-varying force, $F(t)$. We decompose $F(t)$ into its directional components

FIGURE 24.7
Illustration of Brownian motion. A spherical particle with radius r is embedded in a liquid of viscosity η. The particle undergoes collisions with solvent molecules (the red dot in the right-hand figure) resulting in a time-varying force, $F(t)$, that initiates particle motion. Particle motion is opposed by a frictional force that is dependent on the solvent viscosity.

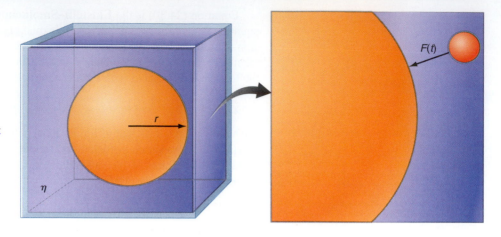

and focus on the component in the x direction, $F_x(t)$. Motion of the particle in the x direction will result in a frictional force due to the liquid's viscosity:

$$F_{fr,x} = -fv_x = -f\left(\frac{dx}{dt}\right) \tag{24.32}$$

The negative sign in Equation (24.32) indicates that the frictional force is in opposition to the direction of motion. In Equation (24.32), f is referred to as the friction coefficient and is dependent on both the geometry of the particle and the viscosity of the fluid. The total force on the particle is simply the sum of the collisional and frictional forces:

$$F_{total,x} = F_x(t) + F_{fr,x}$$

$$m\left(\frac{d^2x}{dt^2}\right) = F_x(t) - f\left(\frac{dx}{dt}\right) \tag{24.33}$$

This differential equation was studied by Einstein in 1905, who demonstrated that when averaged over numerous fluid–particle collisions the average square displacement of the particle in a specific amount of time t is given by

$$\langle x^2 \rangle = \frac{2kTt}{f} \tag{24.34}$$

where k is Boltzmann's constant and T is temperature. If the particle undergoing diffusion is spherical, the frictional coefficient is given by

$$f = 6\pi\eta r \tag{24.35}$$

such that

$$\langle x^2 \rangle = 2\left(\frac{kT}{6\pi\eta r}\right)t \tag{24.36}$$

Comparison of this result to the value of $\langle x^2 \rangle$ determined by the diffusion equation [Equation (24.21)] dictates that the term in parentheses is the diffusion coefficient, D, given by

$$D = \frac{kT}{f} = \frac{kT}{6\pi\eta r} \tag{24.37}$$

Equation (24.37) is the **Stokes–Einstein equation** for diffusion of a spherical particle. This equation states that the diffusion coefficient is dependent on the viscosity of the medium, the size of the particle, and the temperature. This form of the Stokes–Einstein equation is applicable to diffusion in solution when the radius of the diffusing particle is significantly greater than the radius of a liquid particle. It is also applicable when the solvent surrounding the particle interacts strongly with the particle such that solvent immediately adjacent to the particle moves with the particle, referred to as the **stick boundary condition.** For particle diffusion in a fluid where the solvent is similar in size to the particle, or when solvent molecules interact weakly with the particle and thus "slip" over its surface (the **slip boundary condition**), experimental data demonstrate that Equation (24.37) should be modified as follows:

$$D = \frac{kT}{4\pi\eta r} \qquad (24.38)$$

Equation (24.38) can also be used for **self-diffusion,** in which the diffusion of a single solvent molecule in the solvent itself occurs. Note that the friction coefficient for the slip boundary condition ($4\pi\eta r$) is less than the coefficient when stick boundary conditions apply ($6\pi\eta r$).

EXAMPLE PROBLEM 24.5

Hemoglobin is a protein responsible for oxygen transport. The diffusion coefficient of human hemoglobin in water at 298 K and for $\eta = 0.891 \times 10^{-3}$ kg m^{-1} s^{-1} (or $\eta = 0.891$ centipoise of 1 cP) is 6.9×10^{-11} m^2 s^{-1}. Assuming this protein can be approximated as spherical, what is the radius of hemoglobin?

Solution

Rearranging Equation (24.36) and paying close attention to units, the radius is

$$r = \frac{kT}{6\pi\eta D} = \frac{(1.38 \times 10^{-23}\, \text{J K}^{-1})(298\, \text{K})}{6\pi(0.891 \times 10^{-3}\, \text{kg m}^{-1}\, \text{s}^{-1})(6.9 \times 10^{-11}\, \text{m}^2\, \text{s}^{-1})}$$

$$= 3.55\, \text{nm}$$

X-ray crystallographic studies of hemoglobin have shown that this globular protein can be approximated as a sphere having a radius of 2.75 nm.

In Example Problem 24.5 the radius for hemoglobin determined from the diffusion coefficient is larger than the radius deduced from the crystallographic measurment. Because the diffusion coefficient is associated with the dynamics of particle motion in water, the radius calculated using Equation (24.37) is a called the *hydrodynamic radius*. The difference between the hydrodynamic radius and the radius of an unhydrated particle is due to water molecules that strongly associate with the particle and contribute to the volume. The dependence of the friction coefficient on the hydrodynamic radius can be used to determine the number of water molecules bound to the particle. To develop this connection, we first consider the specific volumes of the components of the macromolecular solution. The specific volume of component i of a macromolcular solution, $\overline{V}_i$, is defined as the volume per gram of component i in the solution. For a two-component solution consisting of water ($i = 1$) and a single macromolecular solute ($i = 2$), the total volume (V_T) is

$$V_T = g_1\overline{V}_1 + g_2\overline{V}_2 \qquad (24.39)$$

In Equation (24.39), g_1 and g_2 are the grams of water and macromolecule, respectively. Consider the case where δ_1 grams of water hydrate each gram of macromolecule. In this case Equation (24.39) can be rewritten as:

$$V_T = (g_1 - \delta_1 g_2)\overline{V}_1 + g_2(\overline{V}_2 + \delta_1\overline{V}_1) \qquad (24.40)$$

The second term in Equation (24.40) is the total macromolecular particle volume in solution including the waters of hydration. In Equation (24.40) we have assumed that the specific volumes of bound and bulk water are equivalent. Because the volume of a sphere is related to the cube of the sphere radius, and the frictional coefficient is directly proportional to the radius, the solute volume in Equation (24.40) is directly related to the frictional coefficient as follows:

$$f^3 = (6\pi\eta)^3 r^3 = (6\pi\eta)^3 \times \frac{3V_h}{4\pi} \qquad (24.41)$$

In Equation (24.41) V_h is the volume of a single hydrated spherical particle determined by the second term in Equation (24.40):

$$V_h = \frac{M_2}{N_0} \times (\overline{V}_2 + \delta_1 \overline{V}_1) \qquad (24.42)$$

For proteins that do not contain large amounts of material besides amino acids, $\overline{V}_2$ ranges from 0.69 to 0.75 mL g^{-1}.

EXAMPLE PROBLEM 24.6

Using the information in Example Problem 24.5, calculate the number of water molecules that hydrate a hemoglobin molecule where $\overline{V}_2 = 0.75$ mL g^{-1}. The molecular weight of hemoglobin is 64.6 kg mol^{-1}.

Solution

We first equate the volume of a sphere of radius 3.55 nm with the expression for the molecular volume given in Equation (24.42): $\frac{4\pi r^3}{3} = v_h = \frac{M_2}{N_0} \times (\overline{V}_2 + \delta_1 \overline{V}_1)$.
A analogous expression for the unhydrated protein is

$$\frac{4\pi r_0^3}{3} = \frac{M_2}{N_0} \overline{V}_2$$

Then taking the ratio

$$\left(\frac{r}{r_0}\right)^3 = \frac{\overline{V}_2 + \delta_1 \overline{V}_1}{\overline{V}_2}$$

Solving for δ_1:

$$\delta_1 = \frac{1}{\overline{V}_1} \times \left(\left(\frac{r}{r_0}\right)^3 \overline{V}_2 - \overline{V}_2 \right)$$

$$= \frac{1}{1.00 \text{ L g}^{-1}}$$

$$\times \left(\left(\frac{3.55 \text{ nm}}{2.75 \text{ nm}}\right)^3 \times 0.75 \text{ mL g}^{-1} - 0.75 \text{ m L g}^{-1} \right)$$

$$= 0.86$$

The grams of water hydrating a gram of hemoglobin can now be multiplied by the ratio of the molecular weight of hemoglobin to the molecular weight of water to obtain the number of water molecules hydrating a hemoglobin molecule:

$$\text{Number of water molecules per protein} = \delta_1 \times \frac{M_2}{M_1} = 0.84 \times \frac{64.6 \text{ kg mol}^{-1}}{0.018 \text{ kg mol}^{-1}} = 3014$$

Hydrodynamic measurements yield δ_1 values for structured proteins ranging from 0.24 to 0.86.

Under stick boundary conditions the solvent may also be disturbed by the rotational motion of a sphere. In this case the rotational coefficient of friction is given by

$$f_{rot} = 6\eta V_h = 6\eta \times \frac{4\pi}{3} r^3 = 8\pi\eta r^3 \qquad (24.43)$$

where V_h is the volume of the hydrated sphere. Under slip boundary conditions, the surrounding solvent is not disturbed by the rotational motion of a sphere so the rotational coefficient of friction is zero. From now on we will designate the coefficient of friction associated with translation of a particle through a fluid as f_{tr}, and the coefficient of friction associated with the rotational motion of a particle in a fluid as f_{rot}.

EXAMPLE PROBLEM 24.7

Using the information given in Example Problem 24.5, calculate the coefficients of translational and rotational friction for hemoglobin. Assume stick boundary conditions. Also calculate also the coefficients of rotational and translational diffusion. Assume $T = 298$ K.

Solution

For the coefficient of translational friction using Equation (24.35):

$$f_{trans} = 6\pi\eta r = 6\pi \times (0.891 \times 10^{-3} \text{ kg m}^{-1}\text{s}^{-1}) \times (3.55 \times 10^{-9} \text{ m}) = 5.96 \times 10^{-11} \text{ kg s}^{-1})$$

Then

$$D_{trans} = \frac{k_B T}{f_{trans}} = \frac{1.38 \times 10^{-23} \text{ J K}^{-1} \times 298 \text{ K}}{5.96 \times 10^{-11} \text{ kg s}^{-1}} = 6.90 \times 10^{-11} \text{ m}^2\text{ s}^{-1}$$

For the coefficient of rotational friction, applying Equation (24.43):

$$f_{rot} = 8\pi\eta r^3$$
$$= 8\pi \times 0.891 \times 10^{-3} \text{ kg m}^{-1}\text{s}^{-1} \times (3.55 \times 10^{-9} \text{m})^3 = 1.00 \times 10^{-27} \text{ kg m}^2\text{ s}^{-1}$$

Then

$$D_{rot} = \frac{k_B T}{f_{rot}} = \frac{1.38 \times 10^{-23} \text{ J K}^{-1} \times 298 \text{ K}}{1.00 \times 10^{-27} \text{ kg m}^2\text{s}^{-1}} = 4.11 \times 10^6 \text{ s}^{-1}$$

Note the difference in units between translations and rotations friction coefficients and diffusion coefficients.

24.5.2 Nonspherical Molecules

We have thus far treated the frictional properties of macromolecules using the Stokes–Einstein equation, which assumes the macromolecule is spherical. But the shapes of many biological macromolecules are not well approximated as spheres. For example, high molecular weight DNA composed of many thousands of base pairs is a flexible linear polymer, and its solution properties are not well represented by the rigid sphere model used to treat small globular proteins. Intermediate to the extremes of the rigid sphere and the flexible linear polymer are nucleic acid oligomers, virus coats, small protein aggregates, and other compounds whose hydrodynamics properties can be approximated by rigid, nonspherical bodies.

Expressions for the frictional properties of nonspherical particles under stick boundary conditions have been determined for a number of idealized cases. For example, rigid, nonspherical particles may be described as prolate ellipsoids of revolution. A prolate ellipsoid of revolution is obtained by rotating an ellipse with major semi-axis of length a and minor semi-axis of length b around the major semi-axis as shown in Figure 24.8,

FIGURE 24.8

A prolate ellipsoid of revolution is obtained by rotating an ellipse with semimajor and semiminor axis lengths of a and b, respectively, around the major axis. An oblate ellipsoid is produced by rotating an ellipse around its minor semi-axis. For both ellipsoids the aspect ratio is $P = a/b$.

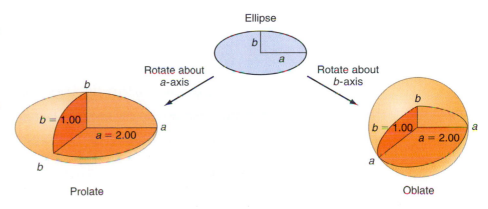

Ellipse

Rotate about a-axis

Rotate about b-axis

$b = 1.00$
$a = 2.00$

Prolate

$b = 1.00$
$a = 2.00$

Oblate

left. If the ellipse is rotated around the minor axis, the result is an oblate ellipsoid of revolution (see Figure 24.8, right). For ellipsoids, the ratio of the major and minor semi-axis lengths is referred to as the axial ratio, P, that is, $P = a/b$.

For a prolate ellipsoid with major and minor semi-axis lengths a and b, respectively, the coefficient of translational friction is

$$f_{tr} = f_{tr}^{\circ} \frac{P^{-1/3}\sqrt{P^2 - 1}}{\ln\left[P + \sqrt{P^2 - 1}\right]} \tag{24.44}$$

where $f_{tr}^{\circ} = 6\pi\eta r_e$ and r_e is the radius of a sphere whose volume is equal to the volume of a prolate ellipsoid $V_{pro} = \dfrac{4\pi}{3}ab^2$. The frictional properties of nonspherical molecules are frequently described in terms of the Perrin or shape factor. For a prolate ellipsoid the **Perrin factor** is defined as

$$F_{tr} = \frac{f_{tr}}{f_{tr}^{\circ}} = \frac{P^{-1/3}\sqrt{P^2 - 1}}{\ln\left[P + \sqrt{P^2 - 1}\right]} \tag{24.45}$$

Analogous expressions for the frictional properties of oblate ellipsoids will be given in the end-of-chapter problems. For a very long cylindrical particle with length L and diameter d, the Perrin factor is approximated as follows:

$$F_{tr} = \frac{f_{tr}}{f_{tr}^{\circ}} \approx \frac{(2/3)^{1/3} P^{2/3}}{\ln(P) + 0.312} \tag{24.45a}$$

In this expression the axial ratio is defined as $P = L/d$, and it is assumed in Equation (24.45a) that $L \gg d$ so that $P \gg 1$. Again, the term $f_{tr}^{\circ} = 6\pi\eta r_e$ is the coefficient of friction for a sphere with radius r_e for which the sphere is equal in volume to a rod of length L and diameter d, that is, $V_{rod} = \pi(d/2)^2 L = V_{sphere} = \dfrac{4}{3}\pi r_e^3$.

EXAMPLE PROBLEM 24.8

A hydrated DNA oligomer can be treated as a rod-like molecule of length 20.00 nm and diameter 2.00 nm. Calculate f_{tr}, f_{tr}°, and the Perrin factor. Assume $\eta = 0.891$ cP. Calculate also the translational diffusion coefficient at $T = 298$ K.

Solution

To obtain, f_{tr}°, r_e is first determined as follows. Equating the volume of the sphere $\left(V_{sphere} = \dfrac{4\pi}{3}r_e^3\right)$ to that of a rod $\left(V_{rod} = \pi\left(\dfrac{d}{2}\right)^2 L\right)$ with length L and diameter d yields:

$$\frac{4\pi}{3}r_e^3 = \pi\left(\frac{d}{2}\right)^2 L$$

Solving for r_e,

$$r_e = \left(\frac{3d^2 L}{16}\right)^{1/3} = \left(\frac{3 \times 20.00 \text{ nm} \times 4.00 \text{ nm}^2}{16}\right)^{1/3} = 2.47 \text{ nm}$$

Now f_{tr}° can be readily determined:

$$f_{tr}^{\circ} = 6\pi\eta r_e = 6\pi \times 0.891 \times 10^{-3} \text{ kg m}^{-1}\text{ s}^{-1} \times 2.47 \times 10^{-9} \text{ m} = 4.15 \times 10^{-11} \text{ kg s}^{-1}$$

The Perrin factor is obtained from Equation (24.45a):

$$F_{tr} = \frac{f_{tr}}{f_{tr}^{\circ}} \approx \frac{(2/3)^{1/3} P^{2/3}}{\ln(P) + 0.312} = \frac{(2/3)^{1/3} 10^{2/3}}{\ln(10) + 0.312} = \frac{0.874 \times 4.64}{2.62} = 1.55$$

where the axial ratio $P = 10$.

The coefficient of translational friction is

$$f_{tr} = f_{tr}^{\circ} F_{tr} = 4.15 \times 10^{-11} \text{ kg s}^{-1} \times 1.55 = 6.43 \times 10^{-11} \text{ kg s}^{-1}$$

The translational diffusion coefficient is

$$D_{tr} = \frac{k_B T}{f_{tr}} = \frac{1.38 \times 10^{-23} \text{ J K}^{-1} \times 298 \text{ K}}{6.43 \times 10^{-11} \text{ kg s}^{-1}} = 6.40 \times 10^{-11} \text{ m}^2 \text{ s}^{-1}.$$

Figure 24.9 presents the Perrin factors for a cylinder and a prolate ellipsoid as a function of axial ratio P. The Perrin factors for translational motions of rigid, nonspherical molecules such as prolate ellipsoids and cylinders are not strongly dependent on the axial ratio. For example, as P increases from 1 to 20, F_{tr} for both cylinders and prolate ellipsoids undergoes a modest increase from 1.0 to ~2.0.

24.6 Supplemental: Properties of Flexible Macromolecules in Solution

Our discussion of macromolecular friction has assumed that the molecule remains rigid and thus maintains its shape. However, many macromolecules (e.g., high molecular weight DNA) are flexible. To calculate transport properties, these polymeric materials are modeled as a linear array of rigid structural subunits connected by a series of bonds (see Figure 24.10a). The shape of the polymer is determined by rotations around the "bonds." The rotational state of the bond between the (i-1) and the (i) sub-units is designated by the torsion angle ϕ_i, with the overall shape of the polymer chain is determined by the set of torsion angles $\{\phi_i\}$.

The shape of the polymer can be is quantified by either of two parameters: the end-to-end distance and the radius of gyration. Suppose a polymer chain is composed of $n + 1$ repeating units connected by n bonds. Associated with each bond is a vector $\vec{l}_i$, oriented parallel to the bond axis. The end-to-end distance vector $\vec{r}$ is the sum of the n bond vectors: $\vec{r} = \sum_{i=1}^{n} \vec{l}_i$ as illustrated in Figure 24.10b. The magnitude of the end-to-end vector is $(\vec{r} \cdot \vec{r})^{1/2}$ where

$$\vec{r} \cdot \vec{r} = \left(\sum_{i=1}^{n} \vec{l}_i \right) \cdot \left(\sum_{j=1}^{n} \vec{l}_j \right) = \sum_{i=1}^{n} \sum_{j=1}^{n} \vec{l}_i \cdot \vec{l}_j = nl^2 + 2 \sum_{j>i} \vec{l}_i \cdot \vec{l}_i \quad (24.46)$$

The scalar product $\vec{r} \cdot \vec{r}$ in Equation (24.46) is composed of the scalar products of all possible combinations of bond vectors. There will be n inner products of bond vectors with themselves for which $\sum_{i=1}^{n} \vec{l}_i \cdot \vec{l}_i = \sum_{i=1}^{n} l^2 \cos \theta = nl^2$, which accounts for the first term on the right-hand side of Equation (24.46). Many shapes are possible for a linear polymer and any collection of polymers will display a distribution of shapes. Therefore, we can identify a mean-squared end-to-end distance :

$$\langle r^2 \rangle = nl^2 + 2 \sum_{j>i} \langle l_i \cdot l_j \rangle \quad (24.47)$$

where the average $\langle l_i \cdot l_j \rangle$ in Equation (24.47) is over all mutual orientations of the bond vectors i and j.

The simplest model used to describe biopolymers is a random coil. In a random coil, the direction of the bond between each repeating unit is random, and the shape of the polymer can be visualized as following a random walk as described in Section 24.4 for diffusive processes. This equivalence of descriptions makes the structural description of a polymer analogous to the description of particle diffusion. A consequence of the random

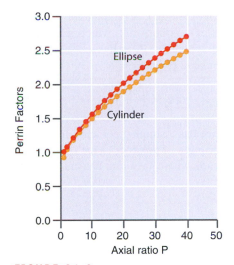

FIGURE 24.9
Perrin factors for translational motions of cylinders and prolate ellipsoids of revolution as functions of axial ratio P.

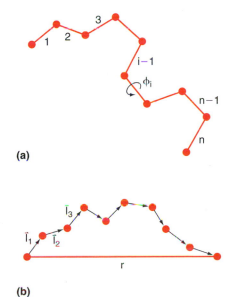

(a)

(b)

FIGURE 24.10
(a) A linear polymer composed of $n + 1$ subunits connected by bonds each of length l. The shape of the polymer is described by the set of bond torsion angles $\{\phi_i\}$. (b) The mean end-to-end distance of a linear polymer composed of n subunits is the vector sum of the $n - 1$ bond vectors l_i.

nature of bond directions in a **random coil polymer** is that the average $\langle l_i \cdot l_j \rangle$ in Equation (24.47) has the result

$$\langle \vec{li} \cdot \vec{lj} \rangle = \langle l^2 \cos \theta \rangle = \int_0^{2\pi} d\varphi \int_0^\pi d\theta l^2 \cos \theta \sin \theta = 0 \qquad (24.48)$$

which simplifies Equation (24.47) to

$$\langle r^2 \rangle = nl^2 \qquad (24.49)$$

Note that this is the same result we obtained for the mean-squared displacement of a randomly jumping particle that executes n jumps each of length l. A random coil polymer, therefore, occupies a volume described by an n-step random walk, which may be approximated by a spherical domain with radius $\langle r^2 \rangle^{1/2} = r_{rms} = l\sqrt{n}$. It can be shown that the coefficient of translational friction for a random coil polymer is

$$f_{tr} = 6\pi\eta \frac{3\sqrt{\pi}}{8} \left(\frac{nl^2}{6} \right)^{1/2} = 6\pi\eta \times 0.665 \times \left(\frac{nl^2}{6} \right)^{1/2} \qquad (24.50)$$

Although the mean-squared end-to-end distance has a simple geometric interpretation, it is not easily measured. An alternative description of polymer shape is the radius of gyration R_G. If the polymer is composed of n units where each unit has mass m_j, the radius of gyration is defined as

$$R_G^2 = \frac{\sum_{j=1}^n m_j d_j^2}{\sum_{j=1}^n m_j} \qquad (24.51)$$

where d_j is the distance between structural subunit j and the polymer's center of mass. If the polymer is flexible, the radius of gyration, like the mean-squared end-to-end vector, is an average, and if all the masses m_j are equal we get:

$$\langle R_G^2 \rangle = \frac{\sum_{j=1}^N m_j \langle d_j^2 \rangle}{\sum_{j=1}^N m_j} = \frac{m \sum_{j=1}^N \langle d_j^2 \rangle}{Nm} = \frac{1}{N} \sum_{j=1}^N \langle d_j^2 \rangle \qquad (24.52)$$

For a random coil polymer it can be shown that the radius of gyration and the mean-squared end-to-end distance have a particularly simple relationship:

$$\langle R_G^2 \rangle = \frac{\langle r^2 \rangle}{6} = \frac{nl^2}{6} \qquad (24.53)$$

Therefore, the coefficient of translational friction of a random coil polymer is very simply related to the radius of gyration:

$$f_{tr} = 6\pi\eta \times 0.665 \times \langle R_G^2 \rangle^{1/2} \qquad (24.54)$$

that is, the effective hydrodynamic radius of a random coil polymer is about two-thirds of the radius of gyration.

EXAMPLE PROBLEM 24.9

Suppose a high molecular weight DNA molecule is treated as a random coil polymer for which $n = 100,000$ and $l = 0.350$ nm. Calculate the mean-squared end-to-end distance, the radius of gyration, the coefficient of translational friction, and the diffusion coefficient. Assume $\eta = 0.891$ cP and $T = 298$ K.

Solution

The mean squared end-to-end distance is

$$\langle r^2 \rangle = nl^2 = 10^5 \times (0.350 \text{ nm})^2 = 10^5 \times (3.50 \times 10^{-10} \text{m})^2 = 1.23 \times 10^{-14} \text{ m}^2$$

and for a random coil polymer:

$$\langle R_G^2 \rangle = \frac{nl^2}{6} = \frac{1.23 \times 10^{-14} \text{m}^2}{6} = 2.05 \times 10^{-15} \text{ m}^2$$

The coefficient of friction is

$$f_{tr} = 6\pi\eta \times 0.665 \times \langle R_G^2 \rangle^{1/2}$$

$$= 6\pi \times 0.891 \times 10^{-3} \text{ kg m}^{-1}\text{ s}^{-1} \times 0.665 \times (2.05 \times 10^{-15} \text{ m}^2)^{1/2} = 5.05 \times 10^{-10} \text{ kg s}^{-1}$$

The coefficient of diffusion is

$$D_{tr} = \frac{k_B T}{f_{tr}} = \frac{1.38 \times 10^{-23} \text{ J K}^{-1} \times 298 \text{ K}}{5.05 \times 10^{-10} \text{ kg s}^{-1}} = 8.14 \times 10^{-12} \text{ m}^2 \text{ s}^{-1}$$

The random coil polymer model can only approximate the behavior of a real polymer because not all shapes adopted by a random coil polymer are physically allowed. Local geometric constraints imposed on chemical bonding networks restrict polymer shape. Polymer shape is also restricted by other factors including excluded volume and persistence length effects. Excluded volume effects refer to the fact that shapes of a random coil are not physically allowed if they result in a segment of a polymer chain entering a volume already occupied by another segment of the chain. Excluded volume effects are accounted for by the empirical correction:

$$\langle r^2 \rangle \approx l^2 n^{1+\varepsilon} \tag{24.55}$$

where the value of ε varies with the polymer. For example, for high molecular weight DNA $\varepsilon = 0.1$.

Persistence length effects arise from the fact that while biopolymers such as DNA demonstrate long-range flexibility, the polymer chain is locally stiff and will not bend as freely as a random coil polymer over short distances. A polymer chain that displays local stiffness but also shows long-range flexibility is called a worm-like chain. Two factors determine the shape of a worm-like chain: the contour length (L) and the persistence length (p). The contour length is the length of the polymer chain measured along the backbone of the chain. For DNA, the contour length is measured along the direction of the double helix. The persistence length is defined in terms of the average cosine of the angle θ formed between the tangent lines to the chain at its beginning and at its end. A detailed analysis shows that this average is for a worm-like chain:

$$\langle \cos \theta \rangle = e^{-L/p} \tag{24.56}$$

For double-stranded DNA the persistence length is about 50 nm. Using the concepts of contour length L and persistence length p the mean-squared end-to-end distance of a worm-like chain can be shown to be is

$$\langle r^2 \rangle = 2p(L - p + pe^{-L/p}) \tag{24.57}$$

Suppose the length of the chain is much smaller than the persistence length, that is, $L \ll p$. Then the exponent in Equation (24.57) can be expanded:

$$\langle r^2 \rangle \approx 2p\left(L - p + p\left(1 - \frac{L}{P} + \frac{1}{2}\left(\frac{L}{P}\right)^2 + \cdots\right)\right) \approx L^2 \tag{24.58}$$

which is the result expected for a rigid chain of length L. If the contour length is much greater than the persistence length, that is, $L \gg p$, then $pe^{-L/p} \approx 0$ and

$$\langle r^2 \rangle \approx 2p(L - p) \approx 2pL \tag{24.59}$$

By definition, the contour length $L = nl$, so for a random coil polymer, $\langle r^2 \rangle = nl^2 = lL$. According to Equation (24.59), twice the persistence length $2p$ replaces the bond length l in the expression for the mean-squared end-to-end distance of a worm-like chain.

24.7 Velocity and Equilibrium Sedimentation

Velocity **sedimentation** measures frictional coefficients for macromolecules and so provides information on macromolecular shape, aggregation state, and so forth. In a velocity sedimentation experiment, a tube containing a macromolecular solution is spun at high angular velocity around an axis. The resulting solute transport will continue until equilibrium is reestablished in the centrifugal field. A number of physical properties affect the distribution of solute prior to equilibrium being established in the centrifugal field. In addition to the specific volume of the solute and solvent, both of which are generally known, sedimentation profiles are affected by the solute molecular weight and the solute particle coefficient of translational friction. Because sedimentation experiments detect two quantities, molecular weight and coefficient of friction, and both of these quantities might be unknowns, sedimentation data are generally combined with additional experimental data. For example, sedimentation data can be used with diffusion data to determine the molecular weights of macromolecules.

If the molecular weight of the macromolecule is known, sedimentation experiments can yield the coefficient of friction directly, with this value allowing for a determination of the macromolecular radius using the Stokes–Einstein equation. If the hydrodynamic radius of a macromolecule is already known, sedimentation can yield information on the state of molecular aggregation. Finally, once equilibrium is reached in the centrifugal field, transport ceases and further centrifugation enables an unambiguous measure of the molecular weight.

Figure 24.11 depicts a molecule with mass m undergoing sedimentation in a liquid of density ρ under the influence of the Earth's gravitational field. Three distinct forces are acting on the particle:

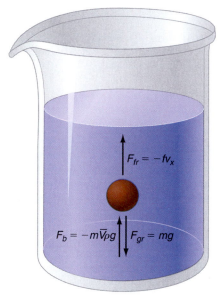

FIGURE 24.11

Illustration of the forces involved in sedimentation of a particle; F_{fr} is the frictional force, F_{gr} is the gravitational force, and F_b is the buoyant force.

1. The frictional force: $F_{fr} = -f v_x$,

2. The gravitational force: $F_{gr} = mg$, and

3. The buoyant force: $F_b = -m \overline{V}_2 \rho_1 g$.

In the expression for the buoyant force, $\overline{V}_2$ is the **specific volume** of the solute, equal to the change in solution volume per mass of solute with units of $cm^3 \, g^{-1}$.

Imagine placing the particle at the top of the solution and then letting it fall. Initially, the downward velocity (v_x) is zero, but the particle will accelerate and the velocity will increase. Eventually, a particle velocity will be reached where the frictional and buoyant forces are balanced by the gravitational force. This velocity is known as the terminal velocity, and when this velocity is reached the particle acceleration is zero. Using Newton's second law,

$$F_{total} = ma = F_{fr} + F_{gr} + F_b$$

$$0 = -f v_{x,ter} + mg - m\overline{V}_2 rg$$

$$v_{x,ter} = \frac{mg(1 - \overline{V}_2 \rho_1)}{f} \tag{24.60}$$

$$\overline{s} = \frac{v_{x,ter}}{g} = \frac{m(1 - \overline{V}_2 \rho_1)}{f}$$

The **sedimentation coefficient,** $\overline{s}$, is defined as the terminal velocity divided by the acceleration due to gravity. Sedimentation coefficients are generally reported in the units of **Svedbergs** (S) with $1 \, S = 10^{-13}$ s.

Sedimentation is generally not performed using acceleration due to the Earth's gravity. Instead, acceleration of the particle is accomplished using a **centrifuge,** with ultracentrifuges capable of producing accelerations on the order of 10^5 times the acceleration due to gravity. In centrifugal sedimentation, the acceleration is equal to $\omega^2 x$ where ω is the angular velocity (rad s^{-1}) and x is the distance of the particle from the center of rotation. During centrifugal sedimentation, the particles will also reach a terminal velocity that depends on the acceleration due to centrifugation, and the sedimentation coefficient is expressed as

$$\overline{s} = \frac{v_{x,ter}}{\omega^2 x} = \frac{m(1 - \overline{V}_2 \rho_1)}{f} \qquad (24.61)$$

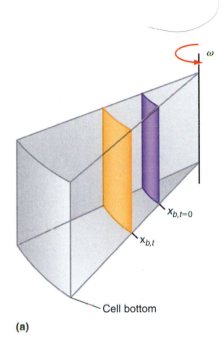

EXAMPLE PROBLEM 24.10

The sedimentation coefficient of lysozyme ($M = 14,100$ g mol^{-1}) in water at 20°C is 1.91×10^{-13} s, and the specific volume is 0.703 cm^3 g^{-1}. The density of water at this temperature is 0.998 g cm^{-3} and $\eta = 1.002$ cP. Assuming lysozyme is spherical, what is the radius of this protein?

Solution

The frictional coefficient is dependent on molecular radius. Rearranging Equation (24.61) to isolate f, we obtain

$$f = \frac{m(1 - \overline{V}_2 \rho_1)}{\overline{s}} = \frac{\dfrac{(14,100 \text{ g mol}^{-1})}{(6.022 \times 10^{23} \text{ mol}^{-1})}(1 - (0.703 \text{ mL g}^{-1})(0.998 \text{ g mL}^{-1}))}{1.91 \times 10^{-13} \text{ s}}$$

$$= 3.66 \times 10^{-8} \text{ g s}^{-1}$$

The frictional coefficient is related to the radius for a spherical particle by Equation (24.35) such that

$$r = \frac{f}{6\pi\eta} = \frac{3.66 \times 10^{-8} \text{ g s}^{-1}}{6\pi(1.002 \text{ g m}^{-1} \text{ s}^{-1})} = 1.94 \times 10^{-9} \text{ m} = 1.93 \text{ nm}$$

One method for measurement of macromolecular sedimentation coefficients is by centrifugation, as illustrated in Figure 24.12a. In this process, an initially homogeneous solution of macromolecules is placed in a centrifuge and spun. Sedimentation occurs, resulting in regions of the sample farther away from the axis of rotation experiencing an increase in macromolecule concentrations, and a corresponding reduction in concentration for the sample regions closest to the axis of rotation. A boundary between these two concentration layers will be established, and this boundary will move away from the axis of rotation with time in the centrifuge. If we define the x_b as the midpoint of the **boundary layer** (Figure 24.12b), the following relationship exists between the location of the boundary layer and centrifugation time:

$$\overline{s} = \frac{v_{x,ter}}{\omega^2 x} = \frac{\dfrac{dx_b}{dt}}{\omega^2 x_b}$$

$$\omega^2 \overline{s} \int_0^t dt = \int_{x_{b,t=0}}^{x_{b,t}} \frac{dx_b}{x_b} \qquad (24.62)$$

$$\omega^2 \overline{s} t = \ln\left(\frac{x_{b,t}}{x_{b,t=0}}\right)$$

Equation (24.62) suggests that a plot of $\ln(x_b/x_{b,t=0})$ versus time will yield a straight line with slope equal to ω^2 times the sedimentation coefficient. The determination of sedimentation coefficients by boundary centrifugation is illustrated in Example Problem 24.11.

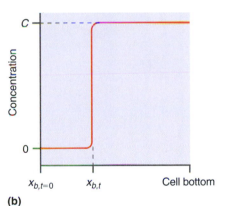

(a)

(b)

FIGURE 24.12
Determination of sedimentation coefficient by centrifugation. (a) Schematic drawing of the centrifuge cell, which is rotating with angular velocity ϖ. The blue plane at $x_{b,t=0}$ is the location of the solution meniscus before centrifugation. As the sample is centrifuged, a boundary between the solution with increased molecular concentration versus the solvent is produced. This boundary is represented by the yellow plane at $x_{b,t}$. (b) As centrifugation proceeds, the boundary layer will move toward the cell bottom. A plot of $\ln(x_{b,t}/x_{b,t=0})$ versus time will yield a straight line with slope equal to ω^2 times the sedimentation coefficient.

EXAMPLE PROBLEM 24.11

The sedimentation coefficient of lysozyme is determined by centrifugation at 55,000 rpm in water at 20°C. The following data were obtained regarding the location of the boundary layer as a function of time:

Time (min)	x_b (cm)
0	6.00
30	6.07
60	6.14
90	6.21
120	6.28
150	6.35

Using these data, determine the sedimentation coefficient of lysozyme in water at 20°C.

Solution

First, we transform the data to determine $\ln(x_b/x_{b,t=0})$ as a function of time:

Time (min)	x_b (cm)	$(x_b/x_{b,t=0})$	$\ln(x_b/x_{b,t=0})$
0	6.00	1	0
30	6.07	1.01	0.00995
60	6.14	1.02	0.01980
90	6.21	1.03	0.02956
120	6.28	1.04	0.03922
150	6.35	1.05	0.04879

The plot of $\ln(x_b/x_{b,t=0})$ versus time is shown here:

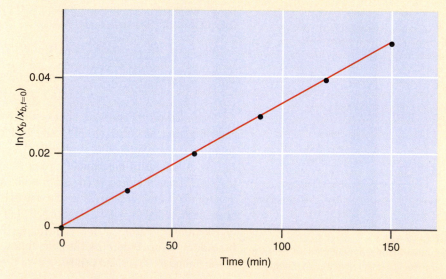

The slope of the line in the preceding plot is 3.75×10^{-4} min^{-1}, which is equal to ω^2 times the sedimentation coefficient:

$$3.75 \times 10^{-4}\ \text{min}^{-1} = 6.25 \times 10^{-6}\ \text{s}^{-1} = \omega^2 \overline{s}$$

$$\overline{s} = \frac{6.25 \times 10^{-6}\ \text{s}^{-1}}{\omega^2} = \frac{6.25 \times 10^{-6}\ \text{s}^{-1}}{((55,000\ \text{rev min}^{-1})(2\pi\ \text{rad rev}^{-1})(0.0167\ \text{min s}^{-1}))^2}$$

$$= 1.88 \times 10^{-13}\ \text{s} = 1.88\ \text{S}$$

Finally with knowledge of the sedimentation coefficient and diffusion coefficient, the molecular weight of a macromolecule can be determined. Equation (24.61) can be rearranged to isolate the frictional coefficient:

$$f = \frac{m(1 - \overline{V}_2\rho_1)}{\overline{s}} \qquad (24.63)$$

The frictional coefficient can also be expressed in terms of the diffusion coefficient $f = kT/D$ [Equation (24.37)]: putting the expression for f in terms of D into Equation (24.63), the mass of the molecule is given by

$$m = \frac{kT\overline{s}}{D(1 - \overline{V}_2\rho_1)} \quad \text{or} \quad M = \frac{RT\overline{s}}{D(1 - \overline{V}_2\rho_1)} \qquad (24.64)$$

where m is the mass of a single macromolecule and M is the molar mass. Equation (24.64) demonstrates that with the sedimentation and diffusion coefficients, as well as the specific volume of the molecule, the molecular weight of a macromolecule can be determined.

Sedimentation coefficients are usually reported for aqueous solutions of macromolecules at $T = 20°C = 293$ K, indicated by the subscript "20,w" as follows:

$$\overline{s}_{20,w} = \frac{m(1 - \overline{V}_2\rho_1)_{20,w}}{f_{20,w}} = \frac{m(1 - \overline{V}_2\rho_1)_{20,w}}{6\pi\eta_{20,w}R} \qquad (24.65)$$

Sedimentation coefficients obtained under other conditions are left unsubscripted. Assuming the molecular radii do not change with solvent or temperature:

$$\frac{\overline{s}_{20,w}}{\overline{s}} = \frac{m(1 - \overline{V}_2\rho_1)_{20,w}}{6\pi\eta_{20,w}R} \frac{6\pi\eta R}{m(1 - \overline{V}_2\rho_1)} = \frac{(1 - \overline{V}_2\rho_1)_{20,w}}{(1 - \overline{V}_2\rho_1)} \frac{\eta}{\eta_{20,w}} \qquad (24.66)$$

Therefore, a sedimentation constant from nonstandard conditions can be readily converted to standard conditions by:

$$\overline{s}_{20,w} = \overline{s} \frac{(1 - \overline{V}_2\rho_1)_{20,w}}{(1 - \overline{V}_2\rho_1)} \frac{\eta}{\eta_{20,w}} \qquad (24.66a)$$

Equation (24.65) demonstrates that $\overline{s}_{20,w}$ is proportional to the ratio $m/f_{20,w}$. For example, the dependence of $\overline{s}_{20,w}$ on molecular weight observed for DNA can be understood by considering this ratio for various forms of $f_{20,w}$. Figure 24.13 presents the variation in $\overline{s}_{20,w}$ on DNA molecular weight, and illustrates that modest variation in $\overline{s}_{20,w}$ occurs for small DNA masses, and a more pronounced increase occurs for larger DNAs. This trend in the molecular weight dependence of $\overline{s}_{20,w}$ can be derived from the dependence of the frictional coefficient $f_{20,w}$ on molecular weight. Low molecular weight DNA has the frictional properties of a rigid cylinder. Figure 24.9 illustrates that the translational friction coefficient changes almost linearly as a function of P for $P > 15$. Therefore, for short rigid cylinders the frictional coefficient is proportional to m so that $\overline{s}_{20,w} \propto m/f_{20,w} \approx m/m$, and so $\overline{s}_{20,w}$ for DNA has a weak dependence on mass in the low mass limit. In contrast, high molecular weight DNA demonstrates random coil behavior. From Equation (24.50) for a random coil polymer, $f_{20,w} \propto (nl^2)^{1/2} = (lL)^{1/2}$. The DNA mass will be proportional to the contour length, which in turn is proportional to the mass m, so in the high mass limit for DNA $s_{20,w} \propto m/m^{1/2} = m^{1/2}$. We expect the sedimentation coefficient of DNA to change more rapidly as a function of m in the high mass limit than the low mass limit, and this is exactly the trend observed in Figure 24.13.

Sedimentation experiments can also be performed using the band sedimentation technique. Instead of starting with a uniform distribution of macromolecular solute in the solution (see Figure 24.14a), a band of macromolecular solution is placed over the solvent. The band propagates through the solvent with time as shown in Figure 24.14b. The broadening of the band with time is due to diffusive motion of the solute particles. To avoid band spreading, the solution below the macromolecular solute band is composed of layers of a solution containing a solute like sucrose, with more concentrated layers occurring at the bottom of the centrifuge tube. According to Equation (24.60), the sedimentation velocity $v_{x,ter} \propto 1 - \overline{V}_2\rho_1$. As the leading edge of the solute band progresses toward the bottom of the centrifuge tube during sedimentation, it encounters a higher solution

FIGURE 24.13
Dependence of the sedimentation coefficient and intrinsic viscosity of solutions of DNA on molecular weight. (Source: Reprinted with permission from J. Eigner and P. Doty, "The Native, Denatured, and Renatured States of DNA," *J. Molecular Biology* 12 (1965), 549–580.)

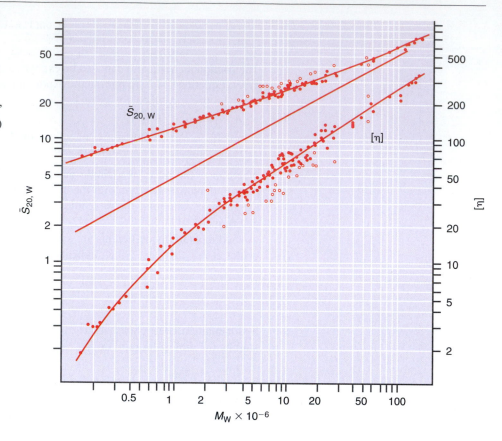

FIGURE 24.14
Comparison of (a) boundary and (b) **band sedimentation**. In boundary sedimentation the solution is uniform in solute concentration prior to centrifugation. The boundary spreads during the course of the sedimentation experiment as a result of diffusion. In band centrifugation a layer of macromolecular solution initially overlays the solvent. The band also progressively broadens during sedimentation due to diffusion, but this effect can be minimized using a sucrose gradient.

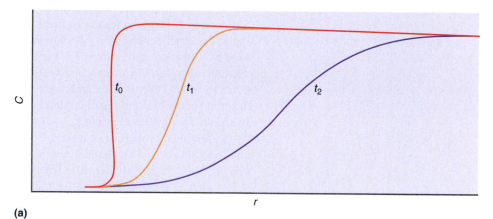

(a)

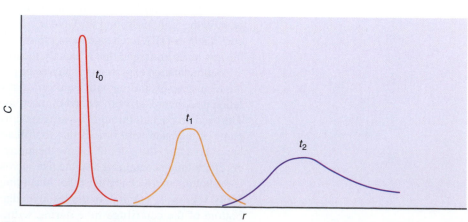

(b)

density ρ_1 due to the increasing sucrose concentration. The leading edge of the band will therefore slow down. The trailing edge of the band meanwhile encounters a lower ρ_1 corresponding to lower sucrose concentration and it therefore moves faster compared to the leading edge. The overall effect of the sucrose gradient is to counteract diffusive band spreading so the band stays narrow. Band sedimentation is used when solutions contain more than one macromolecular component.

A macromolecular solution subjected to a centrifugal field will quickly attain a steady-state condition during which transport of solute mass occurs at constant velocity. However, after a long enough time, the system will attain equilibrium, at which point net transport will cease. The condition of equilibrium in a centrifugal field requires that the system's centrifugal-chemical potential, the sum of the chemical and centrifugal potentials, be minimized; that is, its derivative with respect to r, the distance from the rotation axis, must be zero. For convenience we designate the total potential of the system μ_{total}:

$$\mu_{total} = \mu_{chemical} + \mu_{centrifugal} = \mu_2^0 + kT \ln a_2(r) + U(r) \qquad (24.67)$$

where the subscript "2" indicates the chemical potential and activity of the macromolecular solute being centrifuged, and $U(r)$ is the centrifugal potential at position r in the centrifuge tube. The condition for equilibrium is that the derivative of the centrifugal-chemical potential is zero so that

$$\frac{d\mu_{total}}{dr} = \frac{d}{dr}(\mu_{chemical} + \mu_{centrifugal}) = \frac{d\mu_2^0}{dr} + kT\frac{d(\ln a_2(r))}{dr} + \frac{dU(r)}{dr} = 0 \qquad (24.68)$$

To simplify Equation (24.68), we note that $d\mu_2^0/dr = 0$. Next, the dilute solution limit is assumed such that the solute concentration $c_2(r)$ can be directly substituted for solute activity $a_2(r)$. Finally, the derivative of the centrifugal potential is related to the centrifugal force, corrected for buoyancy effects by:

$$-\frac{dU(r)}{dr} = F_b + F_{centrifugal} = m(1 - \overline{V}_2\rho_1)\omega^2 r \qquad (24.69)$$

Substituting these relationships into Equation (24.68):

$$k_BT\frac{d(\ln c_2(r))}{dr} - m(1 - \overline{V}_2\rho_1)\omega^2 r = 0 \qquad (24.70)$$

Integrating both sides of Equation (24.70) between r and r_0,

$$\ln\left(\frac{c_2(r)}{c_2(r_0)}\right) = \frac{\omega^2 m(1 - \overline{V}_2\rho)}{2kT}(r^2 - r_0^2) = \frac{\omega^2 M(1 - \overline{V}_2\rho)}{2RT}(r^2 - r_0^2) \qquad (24.71)$$

where M is the molar mass of the solute. Equation (24.71) demonstrates that when a solution reaches equilibrium in a centrifugal field, a concentration gradient will be generated from which macromolecular masses can be determined.

EXAMPLE PROBLEM 24.12

To determine the molecular weight of carbonyl hemoglobin, T. Svedberg [*Z. Physik. Chem.* 121 (1926), 65] measured the sedimentation equilibrium in an ultracentrifuge. The results of Svedberg's measurements on a solution containing originally 1.00 g of carbonyl hemoglobin per 100.0 mL are shown in the following table. The temperature was $T = 293.3$ K, the rotation rate was 8708 min^{-1}, the specific volume of carbonyl hemoglobin was $\overline{V}_2 = 0.749$ mL g^{-1}, and the solution density was $\rho = 0.9988$ g mL^{-1}. Calculate the average molecular weight of carbonyl hemoglobin.

r(cm)	r_0(cm)	$c_2(r)$ (percent)	$c_2(r_0)$ (percent)
4.61	4.56	1.220	1.061
4.51	4.46	0.930	0.832
4.36	4.31	0.639	0.564

Solution

The equation relating concentration c to distance r in an equilibrium sedimentation is

$$\ln\left(\frac{c_2(r)}{c_2(r_0)}\right) = \frac{M(1 - \overline{V}_2\rho_1)\omega^2(r^2 - r_0^2)}{2RT}$$

Rearranging this expression to isolate the molecular weight of carbonyl hemoglobin yields:

$$M = \frac{2RT}{(1 - \overline{V}_2\rho)\omega^2(r_2^2 - r_1^2)}\ln\left(\frac{c_2}{c_1}\right)$$

Evaluating the angular velocity, we obtain

$$\omega = 2\pi\nu = 2\pi \times 8708 \text{ min}^{-1} \times \frac{1 \text{ min}}{60 \text{ s}} = 912 \text{ s}^{-1}$$

Finally, using the data from the preceding table, the molecular weight of carbonyl hemoglobin can be calculated:

$$M = \frac{2 \times 8.31 \text{ J K}^{-1} \text{ mol}^{-1} \times 293 \text{ K}}{(1 - 0.9988 \times 0.749) \times 912^2 \text{ s}^{-2} \times (4.61^2 - 4.56^2)\text{cm}^2 \times 1.00 \times 10^{-4} \text{ m}^2 \text{ cm}^2}$$

$$\times \ln\left(\frac{1.220}{1.061}\right)$$

$$= 70.7 \text{ kg mol}^{-1}$$

A second type of equilibrium centrifugation has proven very useful in the study of nucleic acids. Suppose a solution of a macromolecule (e.g., DNA) also contains sucrose or a salt such as CsCl. Initially the salt and the DNA have uniform concentrations. However, once centrifugation has commenced, the CsCl concentration quickly reaches equilibrium according to Equation (24.71), and the density of the solution will vary as a function of r, or the distance from the spinning axis. Suppose at r' the solution has a density $\rho(r') = 1/\overline{V}_2$, where $\overline{V}_2$ is the specific volume of the macromolecule. Expanding the density of the solution around the point r',

$$\rho(r) \approx \frac{1}{\overline{V}_2} + (r - r')\frac{d\rho}{dr} + \cdots \tag{24.72}$$

Next, substituting Equation (24.72) into (24.71) and truncating to first order:

$$\frac{d \ln c_2}{dr} \approx \frac{M\left(1 - \overline{V}_2\left(\frac{1}{\overline{V}_2} + (r - r')\frac{d\rho}{dr}\right)\right)\omega^2 r}{2RT} \tag{24.73}$$

Finally, integrating both sides of Equation (24.73) we obtain the final result:

$$\ln\left(\frac{c_2(r)}{c_2(r')}\right) = -\frac{\omega^2 M\overline{V}_2}{2RT}\frac{d\rho}{dr}r'(r - r')^2 \tag{24.74}$$

The meaning of Equation (24.74) becomes clear when the logarithm is removed. The solute concentration profile is a Gaussian curve:

$$c_2(r) = c_2(r')\exp\left(-\frac{\omega^2 M_2\overline{V}_2}{2RT}\frac{d\rho}{dr}r'(r - r')^2\right) = c_2(r')\exp\left(-\frac{(r - r')^2}{2\sigma^2}\right) \tag{24.75}$$

with the squared deviation being defined as follows:

$$\sigma^2 = \frac{RT}{\omega^2 r' M_2\overline{V}_2(d\rho/dr)} \tag{24.76}$$

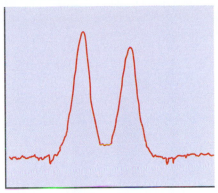

FIGURE 24.15
The resolution of ^{14}N and ^{15}N DNA by ultracentrifugation in a CsCl gradient. The result obtained by centrifuging ^{14}N and ^{15}N bacterial lysates for 24 hours at 44,770 rpm. (Source: Reprinted with permission from M. Meselson and F.W. Stahl "The Replication of DNA in *Escherichia coli*," *Proc. Natl. Acad. Sci.* USA 44 (1958), 671–682.)

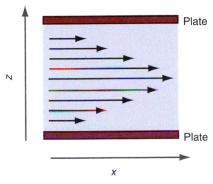

FIGURE 24.16
Cross-section of a fluid flowing between two plates. The fluid is indicated by the blue area between the plates, with arrow lengths representing the speed of the fluid.

When DNA is centrifuged to equilibrium in a salt or sucrose gradient, the resulting concentration profile will be Gaussian or bell-shaped centered at the r' where $\rho(r') = 1/\overline{V}_2$. This effect can be understood on a qualitative basis. At $r < r'$ the density of the solution is less than $1/\overline{V}_2$, the buoyant force has minimal effect, and the DNA "sinks" to the bottom of the tube, pulled "downward" by the centrifugal force. At $r > r'$ the density of the solution is greater than $1/\overline{V}_2$, and the DNA "floats" upward toward the top of the centrifuge tube as a result of the buoyant force. Note that the width of the concentration spatial profile increases as the salt gradient $d\rho/dr$ decreases, and decreases with an increase in molecular weight. This technique has very high resolution with respect to molecular mass. Figure 24.15 demonstrates the mass resolution attainable when DNA is centrifuged in a CsCl gradient. *Escherichia coli* bacteria were grown on a medium where the sole nitrogen source was ^{15}NH$_4$Cl. After several generations of bacterial growth, the nitrogen source in the medium was abruptly changed to ^{14}NH$_4$Cl. The resulting cells were lysed and centrifuged in a CsCl gradient. Figure 24.15 shows the resolution of two DNA bands, one consisting of ^{15}N-labeled DNA and the other consisting of DNA with only ^{14}N.

24.8 Viscosity of Gases, Liquids, and Solutions

The phenomenon of viscosity is the manifestation of the transport of linear momentum. Practical experience provides an intuitive guide with respect to this area of transport. Consider the flow of a gas through a pipe under pressure. Some gases will flow more easily than others, and the property that characterizes resistance of flow is **viscosity**, represented by the symbol η (lowercase Greek eta). What does viscosity have to do with linear momentum? Figure 24.16 provides a cutaway view of a gas flowing between two plates. It can be shown experimentally that the velocity of the gas, v_x, is greatest midway between the plates and decreases as the gas approaches either plate with $v_x = 0$ at the fluid–plate boundary. Therefore, a gradient in v_x exists along the coordinate orthogonal to the direction of flow (z in Figure 24.16). Because the linear momentum in the x direction is mv_x, a gradient in linear momentum must also exist.

We assume that the gas flow is **laminar flow,** meaning that the gas can be decomposed into layers of constant speed as illustrated in Figure 24.16. This regime will exist for most gases and some liquids provided the flow rate is not too high (see Problem P24.8 at the end of the chapter). At high flow rates, the **turbulent flow** regime is reached where the layers are intermixed such that a clear dissection of the gas in terms of layers of the same speed cannot be performed. The discussion presented here is limited to conditions of laminar flow.

The analysis of linear momentum transport proceeds in direct analogy to diffusion and thermal conductivity. As illustrated in Figure 24.16, a gradient in linear momentum exists in the z direction; therefore, planes of similar linear momentum are defined parallel to the direction of fluid flow as illustrated in Figure 24.17. The transfer of linear momentum occurs by a particle from one momentum layer colliding with the flux plane and thereby transferring its momentum to the adjacent layer.

To derive the relationship between flux and the gradient in linear momentum, we proceed in a fashion analogous to that used to derive diffusion (Section 24.2). First, the linear momentum, p, at $\pm\lambda$ is given by

$$p(-\lambda) = p(0) - \lambda\left(\frac{dp}{dz}\right)_{z=0} \tag{24.77}$$

$$p(\lambda) = p(0) + \lambda\left(\frac{dp}{dz}\right)_{z=0} \tag{24.78}$$

Proceeding just as before for diffusion, the flux in linear momentum from each plane located at $\pm\lambda$ to the flux plane is

$$J_{-\lambda,0} = \frac{1}{4}v_{ave}\tilde{N}p(-\lambda) \tag{24.79}$$

FIGURE 24.17
Parameterization of the box model used to
derive viscosity. Planes of identical particle
velocity (v_x) are given by the blue planes,
with the magnitude of velocity given by
the arrows. The gradient in linear
momentum will result in momentum
transfer from regions of high momentum
(darker blue plane) to regions of lower
momentum (indicated by the light blue
plane). The plane located at $z = 0$ is the
location at which the flux of linear
momentum in response to the gradient is
determined.

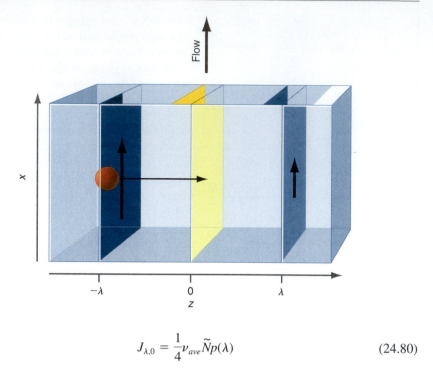

FIGURE 24.17
Parameterization of the box model used to
derive viscosity. Planes of identical particle
velocity (v_x) are given by the blue planes,
with the magnitude of velocity given by
the arrows. The gradient in linear
momentum will result in momentum
transfer from regions of high momentum
(darker blue plane) to regions of lower
momentum (indicated by the light blue
plane). The plane located at $z = 0$ is the
location at which the flux of linear
momentum in response to the gradient is
determined.

$$J_{\lambda,0} = \frac{1}{4}v_{ave}\tilde{N}p(\lambda) \tag{24.80}$$

The total flux is the difference between Equations (24.79) and (24.80):

$$J_{Total} = \frac{1}{4}v_{ave}\tilde{N}\left(-2\lambda\left(\frac{dp}{dz}\right)_{z=0}\right)$$

$$= -\frac{1}{2}v_{ave}\tilde{N}\lambda\left(\frac{d(mv_x)}{dz}\right)_{z=0}$$

$$= -\frac{1}{2}v_{ave}\tilde{N}\lambda m\left(\frac{dv_x}{dz}\right)_{z=0} \tag{24.81}$$

Finally, Equation (24.81) is multiplied by 2/3 as a result of orientational averaging of
the particle trajectories, resulting in the final expression for the total flux in linear
momentum:

$$J_{Total} = -\frac{1}{3}v_{ave}\tilde{N}\lambda m\left(\frac{dv_x}{dz}\right)_0 \tag{24.82}$$

Comparison of Equation (24.1) and Equation (24.82) indicates that the proportionality
constant between flux and the gradient in velocity is defined as

$$\eta = \frac{1}{3}v_{ave}\tilde{N}\lambda m \tag{24.83}$$

Equation (24.83) represents the viscosity of the gas, η, given in terms of parameters derived from gas kinetic theory. The units of viscosity are the poise (P), or 0.1 kg m^{-1} s^{-1}.
Notice the quantity 0.1 in the conversion to SI units. Viscosities are generally reported
in $\mu P(10^{-6}\ P)$ for gases and cP ($10^{-2}\ P$) for liquids.

EXAMPLE PROBLEM 24.13

The viscosity of Ar is 227 μP at 300 K and 1 atm. What is the collisional cross-
section of Ar assuming ideal gas behavior?

Solution

Because the collisional cross-section is related to the mean free path, Equation
(24.83) is first rearranged as follows:

$$\lambda = \frac{3\eta}{v_{ave}\tilde{N}m}$$

Next, we evaluate each term separately, then use these terms to calculate the mean free path:

$$v_{ave} = \left(\frac{8RT}{\pi M}\right)^{1/2} = 399 \text{ m s}^{-1}$$

$$\tilde{N} = \frac{PN_A}{RT} = 2.45 \times 10^{25} \text{ m}^{-3}$$

$$m = \frac{M}{N_A} = \frac{0.040 \text{ kg mol}^{-1}}{6.022 \times 10^{23} \text{ mol}^{-1}} = 6.64 \times 10^{-26} \text{ kg}$$

$$\lambda = \frac{3\eta}{v_{ave}\tilde{N}m} = \frac{3(227 \times 10^{-6}\text{P})}{(399 \text{ m s}^{-1})(2.45 \times 10^{25} \text{ m}^{-3})(6.64 \times 10^{-26}\text{ kg})}$$

$$= \frac{3(227 \times 10^{-7}\text{ kg m}^{-1}\text{ s}^{-1})}{(399 \text{ m s}^{-1})(2.45 \times 10^{25} \text{ m}^{-3})(6.64 \times 10^{-26}\text{ kg})} = 1.05 \times 10^{-7} \text{ m}$$

Note the conversion of poise to SI units in the last step. Using the definition of the mean free path, the collisional cross-section can be determined:

$$\sigma = \frac{1}{\sqrt{2}\tilde{N}\lambda} = \frac{1}{\sqrt{2}(2.45 \times 10^{25} \text{ m}^{-3})(1.05 \times 10^{-7} \text{ m})} = 2.75 \times 10^{-19} \text{ m}^2$$

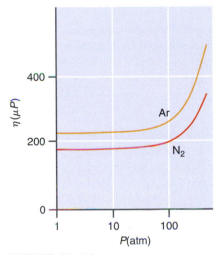

FIGURE 24.18
Pressure dependence of η for gaseous N_2 and Ar at 300 K.

In gases, viscosity is dependent on v_{ave}, $\tilde{N}$, and λ, quantities that are both temperature and pressure dependent. Both $\tilde{N}$ and λ demonstrate pressure dependence, but the product of $\tilde{N}$ and λ contains no net $\tilde{N}$ dependence; therefore, η is predicted to be independent of pressure. Figure 24.18 presents the pressure dependence of η for N_2 and Ar at 300 K with η demonstrating little pressure dependence until $\sim P = 50$ atm. At elevated pressures, intermolecular interactions become important, and the increased interaction between particles gives rise to a substantial increase in η for $P > 50$ atm.

With respect to the temperature dependence of η, the v_{ave} term in Equation (24.83) dictates that η should increase as $T^{1/2}$. This result is perhaps a bit surprising because liquids demonstrate a decrease in η with an increase in temperature. However, the predicted increase in η with temperature for a gas is born out by experiment as illustrated in Figure 24.19 where the variation in η with temperature for N_2 and Ar at 1 atm is presented. Also shown is the predicted $T^{1/2}$ dependence. The prediction of increased viscosity with temperature is a remarkable confirmation of gas kinetic theory. However, comparison of the experimental and predicted temperature dependence demonstrates that η increases more rapidly than predicted due to the presence of intermolecular interactions that are neglected in the hard-sphere model, similar to the discussion of the temperature dependence of κ provided earlier. The increase in η with temperature for a gas is consistent with the increase in velocity accompanying a rise in temperature, and corresponding increase in momentum flux.

Unlike gases, the dependence of fluid viscosity on temperature is difficult to explain theoretically. The primary reason for this is that, on average, the distance between particles in a liquid is small, and intermolecular interactions become important when describing particle interactions. As a general rule, the stronger the intermolecular interactions in a liquid, the greater the viscosity of the liquid. With respect to temperature, liquid viscosities increase as the temperature decreases. The reason for this behavior is that as the temperature is reduced, the kinetic energy of the particles is also reduced, and the particles have less kinetic energy to overcome the potential energy arising from intermolecular interactions, resulting in the fluid being more resistant to flow.

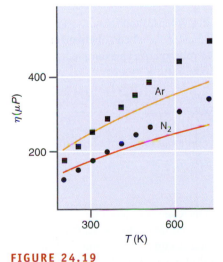

FIGURE 24.19
Temperature dependence of η for N_2 and Ar at 1 atm. Experimental values are given by the squares and circles, and the predicted $T^{1/2}$ dependence is given as the solid line.

Introducing macromolecular particles into a fluid solvent introduces yet another level of complexity. The presence of particles in the solvent distorts the laminar flow

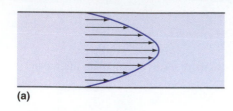

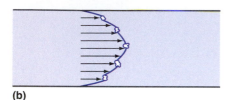

(a)

(b)

FIGURE 24.20

(a) Laminar flow lines for a fluid moving between two plates. (b) Laminar flow in the presence of solute particles.

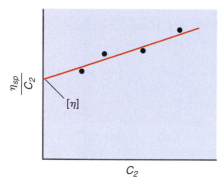

FIGURE 24.21

The intrinsic viscosity is equivalent to the first viral coefficient in the power series expansion of the solution viscosity. See Equation (24.86).

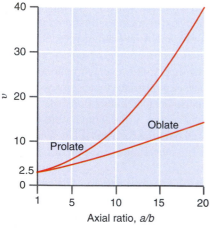

FIGURE 24.22

Dependence of the Simha factor ν on axial ratio $P = a/b$ for prolate and oblate ellipsoids. (Source: Reprinted with permission from G.M. Barrow, *Physical Chemistry for the Life Sciences,* 2nd ed., Figure 8.18, McGraw Hill, New York, 2000.)

lines, as shown in Figure 24.20. The overall effect of the presence of solute particles is to increase the viscosity of the solution, but the dependence of viscosity on solute concentration is complicated. Generally the viscosity of a macromolecular solution is expressed as a power series in the solute concentration:

$$\eta = \eta_0(1 + k_1 c_2 + k_2 c_2^2 + \cdots) \tag{24.84}$$

where η_0 is the viscosity of the pure solvent and c_2 is the solute concentration. Equation (24.84) is called a virial expansion, and the constants k_1, k_2, and so on, are called virial coefficients. The first virial coefficient k_1 reflects the effect exerted by individual solute molecules on the viscosity. The second virial coefficient k_2 reflects the effect exerted by pairs of solute molecules on the viscosity, and so on. Equation (24.84) indicates that viscosity has a complicated dependence on molecular properties that involves progressively larger clusters of molecules as solute concentration increases. Of primary interest is the first viral coefficient k_1, because it reflects the effect of individual solute properties on viscosity. Rearranging Equation (24.84),

$$\frac{\eta_{sp}}{c_2} = \frac{\eta - \eta_0}{\eta_0 c_2} = k_1 + k_2 c_2 + \cdots \tag{24.85}$$

In Equation (24.85), η_{sp} is called the specific viscosity. The right-hand side of Equation (24.85) is still dependent on a hierarchy of molecular properties and interactions. However, in the infinitely dilute limit the effect of k_1 can be isolated:

$$[\eta] = \lim_{c_2 \to 0} \frac{\eta_{sp}}{c_2} = k_1 \tag{24.86}$$

The quantity $[\eta]$ is referred to as the intrinsic viscosity, which is measured by extrapolating a plot of η_{sp}/c_2 versus c_2 to the $c_2 = 0$ limit as illustrated in Figure 24.21.

The interpretation of k_1 in terms of molecular properties was first accomplished by Einstein. Beginning with the power series expansion of the specific viscosity,

$$\eta_{sp} = \frac{\eta - \eta_0}{\eta_0} = k_1 c_2 + \cdots = \nu V_h c_2 + \cdots \tag{24.87}$$

Einstein found that $k_1 = \nu V_h$ where V_h is the volume per gram of the hydrated polymer. The term ν is called the asymmetry or **Simha factor** and is dependent on molecular shape. In general, $V_h = \overline{V}_2 + \delta_1 \overline{V}_1$ so that in the dilute limit:

$$[\eta] = \nu V_h = \nu(\overline{V}_2 + \delta_1 \overline{V}_1) \tag{24.88}$$

Einstein found that for a rigid sphere the Simha factor $\nu = 5/2$, and for nonspherical particles $\nu > 5/2$. As with friction coefficients, Simha factors have been obtained for idealized cases including prolate and oblate ellipsoids of rotation. As an example, for prolate ellipsoids the Simha factor ν is dependent on the axial ratio $P = a/b$:

$$\nu = \frac{P^2}{15(\ln(2P) - \frac{3}{2})} + \frac{P^2}{5(\ln(2P) - \frac{1}{2})} + \frac{14}{15} \tag{24.89}$$

The dependence of the Simha factor for ellipsoids on axial ratio is more easily viewed graphically, as shown in Figure 24.22.

EXAMPLE PROBLEM 24.14

Treated as a rod-like molecule, a 200-base-pair DNA segment has a length of 68.0nm and a diameter of 2.00 nm. Calculate the axial ratio and the Simha factor, treating the DNA as a prolate ellipsoid with a volume equal to the volume of the cylinder, with the assumption that $L = 2a$. In addition, calculate the intrinsic viscosity for DNA with $\delta_1 \approx 1.5$ and $\overline{V}_2 = 0.51$ mL g^{-1}.

Solution

The first step in solving this problem is to establish the relationship between the ratios L/d and a/b. Because the cylinder and prolate ellipsoid have equal volumes,

$$\pi L \left(\frac{d}{2}\right)^2 = \frac{4}{3}\pi a b^2$$

Substituting $L = 2a$ into the preceding expression, we obtain

$$\frac{L}{d} = \sqrt{\frac{3}{2}}\frac{a}{b}$$

Therefore,

$$P = \frac{a}{b} = \frac{L}{d}\sqrt{\frac{2}{3}} = \frac{680}{20}\sqrt{\frac{2}{3}} = 27.8$$

$$\nu = \frac{P^2}{15(\ln(2P) - \frac{3}{2})} + \frac{P^2}{5(\ln(2P) - \frac{1}{2})} + \frac{14}{15}$$

$$= \frac{(27.8)^2}{15(\ln(2 \times 27.8) - \frac{3}{2})} + \frac{(27.8)^2}{5(\ln(2 \times 27.8) - \frac{1}{2})} + \frac{14}{15} = 62.5$$

Finally, the intrinsic viscosity is

$$[\eta] = \nu(\overline{V}_2 + \delta_1 \overline{V}_1) = 62.5 \times (0.51 \text{ mL g}^{-1} + 1.5 \times 1.00 \text{ mL g}^{-1}) = 131 \text{ mL g}^{-1}$$

Note that if the DNA were spherical then $\nu = 2.5$ and the intrinsic viscosity would be reduced by a factor of $62.5/2.5 = 25$.

Intrinsic viscosity is extremely sensitive to molecular shape, although Equation (24.88) shows that a change in $[\eta]$ may in principle be due either to a shape change reflected in a change in ν, or a change in hydration reflected in a change in δ_1. Intrinsic viscosity is also extremely sensitive to protein denaturation as shown by the data in Table 24.1. When a protein undergoes denaturation, a transition in molecular shape from a compact globular structure, for example, to a random coil structure occurs. This change in shape results in a significant increase in intrinsic viscosity.

The structural dependence of viscosity is also evident in the dependence of the intrinsic viscosity on molecular weight (M). For spherical molecules $[\eta] = \frac{5}{2}V_h = \frac{10\pi}{3}\frac{r_e^3 N_A}{M}$.

However, because the sphere volume is itself proportional to M, the dependence of $[\eta]$ on M for globular proteins is weak as shown in Table 24.1. For rigid rods it can be shown that $[\eta] \propto M^{1.8}$, and as also shown in Table 24.1, the intrinsic viscosity of rod-like proteins

TABLE 24.1   Intrinsic Viscosities for Globular and Random Coil Proteins

Shape	Protein	Molecular Weight (g mol^{-1})	Intrinsic Viscosity (mL g^{-1})
Globular	Ribonuclease	13,680	3.4
	Serum albumin	67,500	3.7
	Bushy stunt virus	10,700,000	3.4
Random coil	Insulin	2,970	6.1
	Ribonuclease	13,680	16.0
	Serum albumin	67,500	52
Rod	Fibrinogen	330,000	27
	Myosin	440,000	217

demonstrates a stronger dependence on M than occurs for globular proteins. Finally, for a random coil we associate the volume V_h with a sphere of radius on the order of the radius of gyration: $V_h \approx \dfrac{4\pi}{3} \dfrac{N_A}{M} R_G^3$. We have shown that for a random coil polymer $R_G \propto M^{1/2}$; therefore, $[\eta] \propto M^{1/2}$. In Figure 24.13 the variation in the slope of the plot of the intrinsic viscosity $[\eta]$ of DNA versus molecular weight M reflects the transition from rigid rod behavior at low molecular weights to random coil behavior at higher molecular weights.

24.9 Viscometry

Viscosity is a measure of a fluid's resistance to flow; therefore, it is not surprising that the viscosity of gases and liquids is measured using flow. Viscosity is typically measured by monitoring the flow of a fluid through a tube with the underlying idea that the greater the viscosity, the smaller the flow through the tube. The following equation was derived by Poiseuille to describe flow of a liquid through a round tube under conditions of laminar flow:

$$\frac{\Delta V}{\Delta t} = \frac{\pi r^4}{8\eta} \left(\frac{P_2 - P_1}{x_2 - x_1} \right) \tag{24.90}$$

This equation is referred to as **Poiseuille's law.** In Equation (24.90), $\Delta V / \Delta t$ represents the volume of fluid, ΔV, that passes through the tube in a specific amount of time, Δt; r is the radius of the tube through which the fluid flows; η is the fluid viscosity; and the factor in parentheses represents the macroscopic pressure gradient over the tube length. Notice that the fluid flow rate is dependent on the radius of the tube and is also inversely proportional to fluid viscosity. As anticipated, the more viscous the fluid, the smaller the flow rate. The flow of an ideal gas through a tube is given by

$$\frac{\Delta V}{\Delta t} = \frac{\pi r^4}{16\eta L P_0} (P_2^2 - P_1^2) \tag{24.91}$$

where L is the length of the tube, P_2 and P_1 are the pressures at the entrance and exit of the tube, respectively, and P_0 is the pressure at which the volume is measured (and is equal to P_1 if the volume is measured at the end of the tube).

EXAMPLE PROBLEM 24.15

Gas cylinders of CO_2 are sold in terms of weight of CO_2. A cylinder contains 50 lb (22.7 kg) of CO_2. How long can this cylinder be used in an experiment that requires flowing CO_2 at 293 K ($\eta = 146 \ \mu P$) through a 1.00-m-long tube (diameter = 0.75 mm) with an input pressure of 1.05 atm and output pressure of 1.00 atm? The flow is measured at the tube output.

Solution

Using Equation (24.91), the gas flow rate $\Delta V / \Delta t$ is

$$\frac{\Delta V}{\Delta t} = \frac{\pi r^4}{16\eta L P_0} (P_2^2 - P_1^2)$$

$$= \frac{\pi (0.375 \times 10^{-3} \, \text{m})^4}{16(1.46 \times 10^{-5} \, \text{kg m}^{-1} \, \text{s}^{-1})(1.00 \, \text{m})(101{,}325 \, \text{Pa})}$$

$$\times \left((106{,}391 \, \text{Pa})^2 - (101{,}325 \, \text{Pa})^2 \right)$$

$$= 2.76 \times 10^{-6} \, \text{m}^3 \, \text{s}^{-1}$$

Converting the CO_2 contained in the cylinder to the volume occupied at 298 K and 1 atm pressure, we get

$$n_{CO_2} = 22.7 \text{ kg}\left(\frac{1}{0.044 \text{ kg mol}^{-1}}\right) = 516 \text{ mol}$$

$$V = \frac{nRT}{P} = 1.24 \times 10^4 \text{ L}\left(\frac{10^{-3} \text{ m}^3}{\text{L}}\right) = 12.4 \text{ m}^3$$

Given the effective volume of CO_2 contained in the cylinder, the duration over which the cylinder can be used is

$$\frac{12.4 \text{ m}^3}{2.76 \times 10^{-6} \text{ m}^3 \text{ s}^{-1}} = 4.49 \times 10^6 \text{ s}$$

This time corresponds to roughly 52 days.

Poiseuille's law is also used to describe the relationship between blood vessel geometry and blood pressure, where $\Delta P = P_2 - P_1$ is the pressure drop from one point x_1 in the blood vessel to another point x_2 and where the length of a blood vessel $L = x_2 - x_1$. The shorter and wider the blood vessel the less resistance to flow.

EXAMPLE PROBLEM 24.16

Using Poiseuille's law, calculate the pressure drop from one end of the aorta ($r = 0.01$ m) to a point 1 cm away ($L = 0.01$ m) using the fact that blood flows at a rate of $Q = \Delta V/\Delta t = 0.08$ L s^{-1} and the viscosity of blood is 0.04 P at $T = 310$ K.

Solution

Subsitute the definitions above in Equation (24.90):

$$\Delta P = \frac{8\eta L Q}{\pi r^4}$$

$$= \frac{8 \times 0.004 \text{ kg m}^{-1}\text{s}^{-1} \times 0.010 \text{ m} \times 8.00 \times 10^{-2} \text{Ls}^{-1} \times 1.00 \times 10^{-3} \text{m}^3\text{L}^{-1}}{\pi \times (0.01 \text{ m})^4}$$

$$= 0.80 \text{ kg m}^{-1}\text{s}^{-2}$$

$$= 0.80 \text{ Pa} \times \frac{1.00 \text{ atm}}{101325 \text{ Pa}} = 7.9 \times 10^{-6} \text{ atm} \times \frac{760 \text{ mmHg}}{1.00 \text{ atm}} = 0.006 \text{ mmHg}$$

This very small pressure drop is expected since the diameter of the aorta is large. Blood pressure is measured in terms of systolic pressure, or the maximum pressure at the peak of the heart pulse, and the diastolic pressure, or the minimum pressure between pulses. In a healthy adult the systolic pressure is 120 mmHg (0.158 atm) and the diastolic pressure is 80 mmHg (0.105 atm). Therefore, the average blood pressure is roughly 100 mmHg. Note that these pressures correspond to the excess pressure over atmospheric (760 mmHg). Thus, the average systolic and diastolic pressures are actually 880 mmHg (1.158 atm) and 840 mmHg (1.105 atm), respectively. As blood enters smaller diameter vessels, the pressure drops increase until the venous system is reached, where the blood pressure is barely 10 mmHg. The pressure drop ΔP is obviously a strong function of r, the radius of the vessel in question. Because of the inverse fourth-power dependence of ΔP on vessel radius r, a slight decrease of the blood vessel radius results in a large pressure drop, and blood flow Q must be maintained by a higher blood pressure. This strong dependence of blood pressure on vessel radius is the cause of much of the heart disease that occurs in humans.

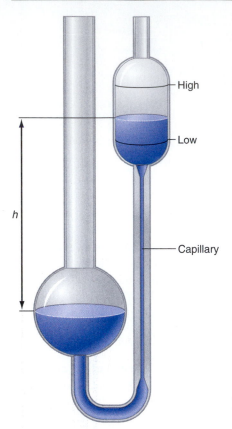

FIGURE 24.23
An Ostwald viscometer. The time for a volume of fluid (ΔV) to flow from the "High" level mark to the "Low" level mark is measured and then used to determine the viscosity of the fluid by means of Equation (24.91).

A convenient tool for measuring the viscosity of liquids is an **Ostwald viscometer** (Figure 24.23). To determine the viscosity, one measures the time it takes for the liquid level to fall from the "High" level mark to the "Low" level mark, and the fluid flows through a thin capillary that ensures laminar flow. The pressure driving the liquid through the capillary is ρgh, where ρ is the fluid density, g is the acceleration due to gravity, and h is the difference in liquid levels in the two sections of the viscometer, as illustrated in the figure. Because the height difference will evolve as the fluid flows, h represents an average height. Because ρgh is the pressure generating flow, Equation (24.90) can be rewritten as follows:

$$\frac{\Delta V}{\Delta t} = \frac{\pi r^4}{8\eta} \frac{\rho gh}{(x_2 - x_1)} = \frac{\pi r^4}{8\eta} \frac{\rho gh}{l} \tag{24.92}$$

where l is the length of the capillary. This equation can be rearranged:

$$\eta = \left(\frac{\pi r^4}{8} \frac{gh}{\Delta Vl} \right) \rho \Delta t = A\rho \Delta t \tag{24.93}$$

In Equation (24.93), A is known as the viscometer constant and is dependent on the geometry of the viscometer. All of the viscometer parameters can be determined through careful measurement of the viscometer dimensions; however, the viscometer constant is generally determined by calibration using a fluid of known density and viscosity.

24.10 Electrophoresis

In diffusion, molecular mass is transported as a result of a concentration gradient. We have also learned that molecular mass can be transported by application of a centrifugal field. Electrophoresis refers to transport that occurs when a charged molecule is placed in an electricfield. When charged molecules are placed in an electric field, they will migrate toward the oppositely charged electrode. Because electrophoresis is performed as an electrolytic process, anions migrate to the positive electrode and cations to the negative electrode. Suppose a macromolecule with total charge Zq (where Z is an integer and q is the charge on an electron, or 1.6×10^{-19} C) is placed in a static electric field of magnitude E. The electric force resulting from the presence of the charged molecule in the field is

$$F_{elec} = ZqE \tag{24.94}$$

The molecule will initially accelerate after the field is initially applied, but the presence of a frictional force opposing the electrical force quickly brings about a steady state defined by

$$fv = ZqE \tag{24.95}$$

The steady-state transport velocity v divided by the electric field E is defined as the electrophoretic mobility, μ:

$$\mu = \frac{v}{E} = \frac{Zq}{f} = \frac{Zq}{6\pi\eta r} \tag{24.96}$$

where the final equality assumes a charged sphere of radius r.

EXAMPLE PROBLEM 24.17

A globular protein has a positive charge Zq where $Z = 5$ and a radius $r = 3.55$ nm. Calculate the steady-state velocity corresponding to an electric field $E = 1.00$ V cm^{-1} with $\eta = 0.891$ cP. At this velocity, how far will the protein move in 1 hour? Toward which electrode will the protein migrate?

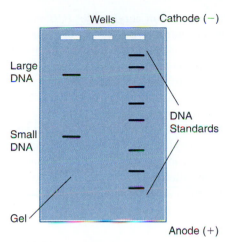

FIGURE 24.24

Schematic of a gel electrophoresis experiment performed on two DNA molecules. The DNAs are initially confined to wells at the top end of the gel and subsequently migrate downward toward the anode. Molecular weight is determined by comparison to known standards. Note that electrode conventions in electrophoresis follow the conventions of an electrolytic experiment, in which the anode is designated as the positive electrode to which anions will migrate, and the negative electrode is the cathode to which cations will migrate.

Solution

$$v = \frac{ZqE}{6\pi\eta r} = \frac{5 \times 1.6 \times 10^{-19}\ \mathrm{C} \times 1.00\ \mathrm{V\ cm^{-1}} \times 100\ \mathrm{cm\ m^{-1}}}{6\pi(0.891 \times 10^{-3}\ \mathrm{kg\ m^{-1}\ s^{-1}})(3.55 \times 10^{-9}\ \mathrm{m})}$$

$$= 1.34 \times 10^{-6}\ \mathrm{m\ s^{-1}}$$

The distance moved in 1 hour is $d = v \times t = 1.34 \times 10^{-6}\ \mathrm{m\ s^{-1}} \times 1.00\mathrm{h} \times 3600\ \mathrm{s\ h^{-1}} = 4.82 \times 10^{-3}\ \mathrm{m}$. Because Zq is a positive number, the protein is cationic and will migrate toward the negatively charged electrode.

In reality, it is more difficult to quantify the electrophoretic mobility than is suggested by Equation (24.96). For example, the binding of counter ions to the macromolecule diminishes the charge to less than Zq. In addition, an ionic atmosphere exists around the charged macromolecule (see Section 9.4) that reduces the electric field experienced by the charged species. Both of these effects will reduce the mobility from the value expected using the analysis given earlier. In addition, the movement of counter ions toward the opposite electrode sets up a charge current that opposes the motion of the charged macromolecule (i.e., the **electrophoretic effect**). Although theoretical treatments aimed at accounting for the ion atmosphere and counter-ion effects have been published, electrophoretic mobilities have only been calculated for special cases such as a uniform sphere of charge, a poor model for a charged nonspherical protein or a nucleic acid.

In spite of these issues, electrophoresis is a powerful tool for separating and characterizing proteins and nucleic acids. **Moving boundary electrophoresis** is essentially the analogue of velocity sedimentation (see Figure 24.14) in that a charged macromolecular front moves through a solvent in response to an electric field. Similar to velocity sedimentation, spreading of the front occurs as a result of diffusion and also, like sedimentation, diffusive spreading can be partly mediated by initially confining the charged macromolecule to a narrow zone (i.e., **zonal electrophoresis**) and conducting the experiment in a density gradient.

Equation (24.96) indicates a dependence of the electrophoretic mobility in a moving boundary experiment on the molecular weight of the charged macromolecule. However, in the case of charged random coil polymers (e.g., high molecular weight DNA), the frictional and electrical forces on each monomer almost cancel, making the electrophoresis of high molecular weight DNA in free solution essentially insensitive to molecular weight.

Electrophoresis experiments aimed at separating DNA based on molecular weight are performed using a gel support, which introduces sensitivity to molecular size. In **gel electrophoresis,** a gel is cast into a thin slab and then immersed in a buffer that maintains the pH at a constant value during the experiment. At one end of the gel are small wells for initially confining the charged macromolecule as illustrated in Figure 24.24. When the electric field is applied, the charged macromolecule migrates to the oppositely charged electrode at the other side of the gel slab. For example, in the case of DNA, which is a polyanion, migration will occur toward the anode. Once sufficient time has passed for the components to separate, the gel is treated with a fluorescent dye, such as ethidium bromide for DNA, and the bands are visualized under a fluorescent lamp. Alternatively, the DNA can be radiolabeled with ^{32}P and the bands visualized by **autoradiography.**

Because of the complications that arise in quantifying the electrophoretic mobility described earlier, molecular weight determination in electrophoretic experiments is accomplished using molecules of known molecular weight and homologous structure as standards. For example, electrophoretic mobility of unknown nucleic acids and nucleic acid standards are graphed as a function of the logarithm of molecular weight, and the molecular weight of the unknowns must be determined by interpolation, as shown in Figure 24.25.

Two types of gels are commonly used in DNA electrophoresis: **agarose** and **polyacrylamide.** Agarose is a polysaccharide extracted from seaweed. Agarose gels can separate a wide range of DNA molecular sizes but have low resolution. By varying the concentration of agarose, fragments of DNA between 200 and 50,000 base pairs can be separated. Therefore, the primary function of agarose gel electrophoresis is in

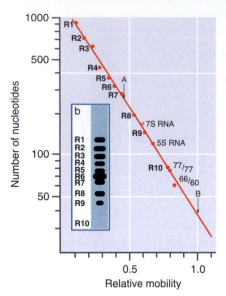

FIGURE 24.25

(a) Relative mobilities of denatured DNA fragments (closed circles marked R1–R10) by polyacrylamide gel electrophoresis. Open triangles are synthetic DNA standards. Open circles are RNA standards. Mobilities are reported relative to bromophenol blue (marked B). (Source: Reprinted with permission from T. Maniatis, A. Jeffrey, and H. van deSande "Chain Length Determination of Small Double-Stranded and Single-Stranded DNA Molecules by Polyacrylamide Gel Electrophoresis," *Biochemistry* 14(17) (1975), 3787–3794.)

"sizing" DNA. Polyacrylamide is a cross-linked polymer of acrylamide: $H_2C = CH - (CO) - NH_2$. The chain length of the polymer is controlled by varying the concentration of acrylamide. Polyacrylamide gels can separate small DNAs of less than 500 base pairs. Although polyacrylamide gel electrophoresis has a smaller range than agarose gel electrophoresis, polyacrylamide gel electrophoresis has extremely high resolution. Under appropriate conditions DNAs differing in size by only a few base pairs can be separated by polyacrylamide gel electrophoresis.

The separation of DNAs by size occurs by virtue of the pores that form in the gel. Agarose has relatively large pores (ca. 100 nm), whereas the high resolution and narrow range of polyacrylamide is due to smaller pore sizes (1 to 2 nm). DNA is envisioned to move through tubes in the gel in a worm-like fashion called **reptation.** According to this description, if a DNA of contour length L is exposed to an electric field in the x direction, the mobility of DNA is given by

$$\mu = \frac{\langle h_x^2 \rangle Zq}{L^2 f} \tag{24.97}$$

In Equation (24.97), h_x is the component of the end-to-end vector of the DNA in the direction of the electric field. The DNA is assumed to be flexible so the mean-squared x component $\langle h_x^2 \rangle$ appears in Equation (24.97). See Figure 24.26. If the DNA is treated as a random coil, then the mean-squared x component of the end-to-end vector $\langle h_x^2 \rangle$ is proportional to the contour length L. Because Z and the frictional coefficient are both also proportional to L, the mobility is overall inversely proportional to L. This prediction of reptation theory has been confirmed by experiment.

However, when the DNA exceeds ~20,000 base pairs, the electrophoretic mobility becomes insensitive to L. This is because high molecular weight DNA has a random coil conformation that is much larger than the pore size. To move through the gel the DNA must stretch out to fit through the pores in the gel, and under these conditions the friction coefficient (f) is proportional to L as is Z, such that the mobility becomes $\mu = Zq/f$ (i.e., independent of L). To circumvent this difficulty, pulse field gradient electrophoresis (PFGE) is used in which the direction of the electric field is alternated. In response to this alternation, the DNA must reorient and does so in a time that is dependent on its length L.

Polyacrylamide gel electrophoresis is also used to separate mixtures of proteins and to determine the isoelectric point (pI) of proteins or, . . . To determine protein molecular weight, polyacrylamide gel electrophoresis is performed in the presence of sodium dodecylsulfate (SDS). SDS is a protein denaturant so in the presence of SDS structural effects are removed from the mobility because proteins are converted by SDS to unfolded forms that differ in molecular weight. The isoelectric point is the pH at which the protein is uncharged. To determine the pI, the electrophoretic mobility is observed as a function of pH as shown in Figure 24.26. The pI is indicated by the pH at which the protein shows zero electrophoretic mobility.

FIGURE 24.26

The electrophoretic mobility of the protein β-lactoglobulin as a function of pH. The isoelectric point occurs at pH = 5.1. For pH < pI the protein is cationic and migrates toward the negatively charged cathode. For pH > pI the protein is negatively charged and migrates toward the anode. At pH = pI the protein is uncharged and has zero electrophoretic mobility. (Source: Reprinted with permission from K.O. Pearson, *Biochemistry J.* 30 (1936), 1961.)

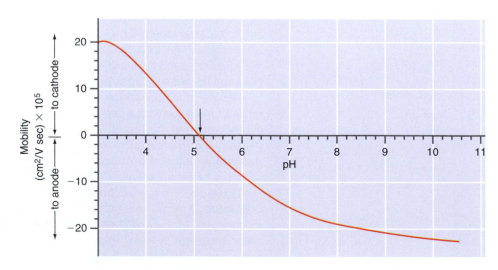

For Further Reading

Bird, R. B., W. E. Stewart, and E. N. Lightfoot, *Transport Phenomena.* Wiley, New York, 1960.

Bloomfield, V. A., D. M. Crothers, and I. Tinoco, *Nucleic Acids: Structure, Properties, and Functions.* University Science Books, Sausalito, CA, 2000.

Cantor, C. R., and P. R. Schimmel, *Biophysical Chemistry. Part II: Techniques for the Study of Biological Structure and Function.* W H. Freeman, San Francisco, 1980.

Castellan, G. W., *Physical Chemistry.* Addison-Wesley, Reading, MA, 1983.

Crank, J., *The Mathematics of Diffusion.* Clarendon Press, Oxford, 1975.

Eisenberg, D., and D. Crothers, *Physical Chemistry with Applications to the Life Sciences,* Benjamin-Cummings, Menlo Park, CA, 1979.

Hirschfelder, J. O., C. F. Curtiss, and R. B. Bird, *The Molecular Theory of Gases and Liquids.* Wiley, New York, 1954.

Reid, R. C., J. M. Prausnitz, and T. K. Sherwood, *The Properties of Gases and Liquids.* McGraw-Hill, New York, 1977.

Van Holde, K. E., W. C. Johnson, and P.-S. Ho, *Principles of Physical Biochemistry.* Prentice Hall, Upper Saddle River, NJ, 1998.

Vargaftik, N. B., *Tables on the Thermophysical Properties of Liquids and Gases.* Wiley, New York, 1975.

Welty, J. R., C. E. Wicks, and R. E. Wilson, *Fundamentals of Momentum, Heat, and Mass Transfer.* Wiley, New York, 1969.

Vocabulary

agarose
autoradiography
band sedimentation
boundary layer
Brownian motion
centrifuge
diffusion
diffusion coefficient
diffusion equation
Einstein–Smoluchowski equation

electrophoretic effect
Fick's first law
Fick's second law of diffusion
flux
gel electrophoresis
laminar flow
mass transport
moving boundary electrophoresis
Ostwald viscometer

Perrin factor
Poiseuille's law
polyacrylamide
random coil polymer
random walk
reptation
sedimentation
sedimentation coefficient
self-diffusion
Simha factor
slip boundary conditions

specific volume
stick boundary conditions
Stokes–Einstein equation
Svedbergs
transport coefficient
transport phenomena
turbulent flow
viscosity
zonal electrophoresis

Questions on Concepts

Q24.1 What is the general relationship between the spatial gradient in a system property and the flux of that property?

Q24.2 What is the expression for the diffusion coefficient, D, in terms of gas kinetic theory parameters? How is D expected to vary with an increase in molecular mass or collisional cross-section?

Q24.3 Particles are confined to a plane and then allowed to diffuse. How does the number density vary with distance away from the initial plane?

Q24.4 How does the root-mean-square diffusion distance vary with the diffusion coefficient? How does this quantity vary with time?

Q24.5 Explain the advantages of zonal sedimentation.

Q24.6 Explain the dependence of the sedimentation coefficient on molecular weight for small DNA oligomers versus high molecular weight DNA.

Q24.7 In describing viscosity, what system quantity was transported? What is the expression for viscosity in terms of particle parameters derived from gas kinetic theory?

Q24.8 What observable is used to measure the viscosity of a gas or liquid?

Q24.9 What is Brownian motion?

Q24.10 In the Stokes–Einstein equation, which describes particle diffusion for a spherical particle, how does the diffusion coefficient depend on fluid viscosity and particle size?

Q24.11 Explain the variation of the sedimentation coefficient for DNA as a function of molecular weight, shown in Figure 24.13

Q24.12 Explain the relationship between the viscosity, the specific viscosity and the intrinsic viscosity of a macromolecular solution.

Q24.13 Discuss the uses of agarose and polyacrylamide gel electrophoresis in DNA studies.

Problems

Problem numbers in **RED** indicate that the solution to the problem is given in the *Student Solutions Manual*.

P24.1 The diffusion coefficient for CO_2 at 273 K and 1 atm is 1.00×10^{-5} m^2 s^{-1}. Estimate the collisional cross-section of CO_2 given this diffusion coefficient.

P24.2

a. The diffusion coefficient for Xe at 273 K and 1 atm is 0.5×10^{-5} m^2 s^{-1}. What is the collisional cross-section of Xe?

b. The diffusion coefficient of N_2 is threefold greater than that of Xe under the same pressure and temperature conditions. What is the collisional cross-section of N_2?

P24.3

a. The diffusion coefficient of sucrose in water at 298 K is 0.522×10^{-9} m^2 s^{-1}. Determine the time it will take a sucrose molecule on average to diffuse an rms distance of 1 mm.

b. If the molecular diameter of sucrose is taken to be 0.8 nm, what is the time per random walk step?

P24.4

a. The diffusion coefficient of the protein lysozyme (MW = 14.1 kg/mol) is 0.104×10^{-5} cm^2 s^{-1}. How long will it take this protein to diffuse an rms distance of 1 μm? Model the diffusion as a three-dimensional process.

b. You are about to perform a microscopy experiment in which you will monitor the fluorescence from a single lysozyme molecule. The spatial resolution of the microscope is 1 μm. You intend to monitor the diffusion using a camera that is capable of one image every 60 s. Is the imaging rate of the camera sufficient to detect the diffusion of a single lysozyme protein over a length of 1 μm?

c. Assume that in the microscopy experiment of part (b) you use a thin layer of water such that diffusion is constrained to two dimensions. How long will it take a protein to diffuse an rms distance of 1 μm under these conditions?

P24.5 A solution consisting of 1 g of sucrose in 10 mL of water is poured into a 1-L graduated cylinder with a radius of 2.5 cm. Then the cylinder is filled with pure water.

a. The diffusion of sucrose can be considered diffusion in one dimension. Derive an expression for the average distance of diffusion, x_{ave}.

b. Determine x_{ave} and x_{rms} for sucrose for time periods of 1 s, 1 min, and 1 h.

P24.6 Myoglobin is a protein that participates in oxygen transport. For myoglobin in water at 20°C, $\bar{s} = 2.04 \times 10^{-13}$ s, $D = 11.3 \times 10^{-11}$ m^2 s^{-1}, and $\bar{V} = 0.740$ cm^3 g^{-1}. The density of water is 0.998 g cm^3 and the viscosity is 1.002 cP at this temperature.

a. Using the information provided, estimate the size of myoglobin.

b. What is the molecular weight of myoglobin?

P24.7 You are interested in purifying a sample containing the protein alcohol dehydrogenase obtained from horse liver;

however, the sample also contains a second protein, catalase. These two proteins have the following transport properties:

	Catalase	Alcohol Dehydrogenase
$\bar{s}$ (s)	11.3×10^{-13}	4.88×10^{-13}
D (m^2 s^{-1})	4.1×10^{-11}	6.5×10^{-11}
$\bar{V}$ (cm^3 g^{-1})	0.715	0.751

a. Determine the molecular weight of catalase and alcohol dehydrogenase.

b. You have access to a centrifuge that can provide angular velocities up to 35,000 rpm. For the species you expect to travel the greatest distance in the centrifuge tube, determine the time it will take to centrifuge until a 3-cm displacement of the boundary layer occurs relative to the initial 5-cm location of the boundary layer relative to the centrifuge axis.

c. To separate the proteins, you need a separation of at least 1.5 cm between the boundary layers associated with each protein. Using your answer to part (b), will it be possible to separate the proteins by centrifugation?

P24.8 Boundary centrifugation is performed at an angular velocity of 40,000 rpm to determine the sedimentation coefficient of cytochrome c ($M = 13,400$ g mol^{-1}) in water at 20°C ($\rho = 0.998$ g cm^3, $\eta = 1.002$ cP). The following data are obtained on the position of the boundary layer as a function of time:

Time (h)	x_b (cm)
0	4.00
2.5	4.11
5.2	4.23
12.3	4.57
19.1	4.91

a. What is the sedimentation coefficient for cytochrome c under these conditions?

b. The specific volume of cytochrome c is 0.728 cm^3 g^{-1}. Estimate the size of cytochrome c.

P24.9 For a one-dimensional random walk, determine the probability that the particle will have moved six steps in either the $+x$ or $-x$ direction after 10, 20, and 100 steps.

P24.10 In the early 1990s, fusion involving hydrogen dissolved in palladium at room temperature, or *cold fusion*, was proposed as a new source of energy. This process relies on the diffusion of H_2 into palladium. The diffusion of hydrogen gas through a 0.005-cm-thick piece of palladium foil with a cross-section of 0.750 cm^2 is measured. On one side of the foil, a volume of gas maintained at 298 K and 1 atm is applied, while a vacuum is applied to the other side of the foil. After 24 h, the volume of hydrogen has decreased by 15.2 cm^3. What is the diffusion coefficient of hydrogen gas in palladium?

P24.11 The viscosity of Cl_2 at 293 K and 1 atm is 132 μP. Determine the collisional cross-section of this molecule based on the viscosity.

P24.12

a. The viscosity of O_2 at 293 K and 1 atm is 204 μP. What is the expected flow rate through a tube having a radius of 2 mm, length of 10 cm, input pressure of 765 torr, output pressure of 760 torr, with the flow measured at the output end of the tube?

b. If Ar were used in the apparatus ($\eta = 223$ μP) of part (a), what would be the expected flow rate? Can you determine the flow rate without evaluating Poiseuille's equation?

P24.13   The Reynolds' number (Re) is defined as $Re = \rho \langle v_x \rangle d / \eta$ where ρ and η are the fluid density and viscosity, respectively; d is the diameter of the tube through which the fluid is flowing; and $\langle v_x \rangle$ is the average velocity. Laminar flow occurs when Re < 2000, the limit in which the equations for gas viscosity were derived in this chapter. Turbulent flow occurs when Re > 2000. For the following species, determine the maximum value of $\langle v_x \rangle$ for which laminar flow will occur:

a. Ne at 293 K ($\eta = 313$ μP, $\rho =$ that of an ideal gas) through a 2-mm-diameter pipe

b. Liquid water at 293 K ($\eta = 0.891$ cP, $\rho = 0.998$ g mL^{-1}) through a 2-mm-diameter pipe

P24.14   The viscosity of H_2 at 273 K at 1 atm is 84 μP. Determine the viscosities of D_2 and HD.

P24.15   An Ostwald viscometer is calibrated using water at 20°C ($\eta = 1.0015$ cP, $\rho = 0.998$ g mL^{-1}). It takes 15 s for the fluid to fall from the upper to the lower level of the viscometer. A second liquid is then placed in the viscometer and it takes 37 s for the fluid to fall between the levels. Finally, 100 mL of the second liquid weighs 76.5 g. What is the viscosity of the liquid?

P24.16   How long will it take to pass 200 mL of H_2 at 273 K through a 10-cm-long capillary tube of 0.25 mm if the gas input and output pressures are 1.05 and 1.00 atm, respectively?

P24.17

a. Derive the general relationship between the diffusion coefficient and viscosity for a gas.

b. Given that the viscosity of Ar is 223 μP at 293 K and 1 atm, what is the diffusion coefficient?

P24.18   As mentioned in the text, the viscosity of liquids decreases with increasing temperature. The empirical equation $\eta(T) = Ae^{E/RT}$ provides the relationship between viscosity and temperature for a liquid. In this equation, A and E are constants, with E being referred to as the activation energy for flow.

a. How can one use the equation provided to determine A and E given a series of viscosity versus temperature measurements?

b. Use your answer in part (a) to determine A and E for liquid benzene given the following data:

T (°C)	η (cP)
5	0.826
40	0.492
80	0.318
120	0.219
160	0.156

P24.19   Assume a rod-like DNA 200 base pairs long has a length $L = 680$ Å and diameter $d = 20$ Å. Calculate the axial ratio a/b if the DNA is treated as a prolate ellipsoid with volume equal to the rod.

P24.20   Calculate the Simha factor ν of the DNA in Problem P24.19 where the DNA is treated as a prolate ellipsoid. Also calculate the intrinsic viscosity of the DNA assuming the $\overline{V}_2$ for DNA is 0.51 mL g^{-1} and 0.30 g of water hydrate each gram of DNA.

P24.21   A more exact formulation of the frictional coefficient of a cylinder is $F_{tr} = \dfrac{f_{tr}}{f_{tr}^\circ} \approx \dfrac{(2/3)^{1/3} P^{2/3}}{\ln(P) + \gamma}$ where $\gamma = 0.312 + \dfrac{0.565}{P} + \dfrac{0.100}{P^2}$ is intended to account for the abrupt ends of the cylinder. A hydrated DNA oligomer can be treated as a rod-like molecule of length 20.00 nm and diameter 2.00 nm. Calculate f_{tr}, f_{tr}°, and the Perrin factor. Assume $\eta = 0.891$ cP. Also calculate the translational diffusion coefficient at $T = 298$ K. Compare this result to the result obtained in Example Problem 24.8.

P24.22   A hydrated DNA oligomer can be treated as a prolate ellipsoid whose major semi-axis length is $a = 10.00$ nm and minor semi-axis length $b = 1.00$ nm. Calculate f_{tr}, f_{tr}°, and the Perrin factor. Assume $\eta = 0.891$ cP. Calculate also the translational diffusion coefficient at $T = 298$ K.

P24.23   An isolated proton spin pair is embedded in a rigid, globular protein. If the rate of rotational diffusion is greater than the Larmor frequency and if the spins relax through their dipole–dipole interaction, the spin lattice relaxation time T_1 is related to the coefficient of rotational diffusion by

$$\frac{1}{T_1} \approx \frac{1}{4D_{rot}} \left(\frac{\mu_0}{4\pi}\right)^2 \frac{\gamma^4 \hbar^2}{r^6}$$

Where the permeability constant $\mu_0 = 4\pi \times 10^{-7}$ T mA^{-1} and r is the internuclear distance. Assume the protons are separated by 0.400 nm. If the observed spin lattice relaxation time is $T_1 = 0.400$ s, calculate the coefficient of rotational diffusion and the molecular radius. Assume $\eta = 0.891$ cP and $T = 298$ K. Hint: See Example Problem 20.6.

P24.24   In the random coil model, the direction of each bond is perfectly random; however, real biopolymers depart from this ideal behavior due to any number of short-range and/or long-range geometric constraints. For example, many polymers are composed of bonds of fixed length but due to the sp^3 hybridization of the bonds of backbone carbon atoms, the bond vectors are at an angle θ. At a given step, the direction of the chain is now defined by a bond that rotates through by an angle φ on the surface of a cone with half-angle θ (see following figure). In this case the mean-squared end-to-end distance is $\langle r^2 \rangle = nl^2 \dfrac{1 + \cos\theta}{1 - \cos\theta}$

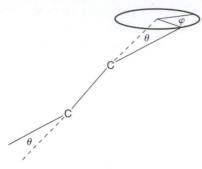

A polymer composed of a polymethylene chain where bonds are of fixed length l, but are at a fixed angle θ. The direction of the polymer chain is now indicated by the excursion of a bond by an angle φ on the surface of a cone of half-angle θ.

For polyethylene chains $\theta = 70°32'$. Calculate the mean-squared end-to-end distance in terms of the number of monomer units n and the effective bond length l. How much does this result differ from the random coil approximation?

P24.25 In reality, polymethylene chains cannot rotate freely about the C—C bond. This means that not all values of the bond angle φ in Problem 24.24 are energetically equivalent. This impacts the mean-squared end-to-end distance as follows:

$$\langle r^2 \rangle = nl^2 \, \frac{1 + \cos \theta}{1 - \cos \theta} \times \frac{1 + \langle \cos \varphi \rangle}{1 - \langle \cos \varphi \rangle} \text{ where } \langle \cos \varphi \rangle$$

$$= \frac{\displaystyle\int_0^{2\pi} d\varphi \cos \varphi e^{-\beta U(\varphi)}}{\displaystyle\int_0^{2\pi} d\varphi e^{-\beta U(\varphi)}}$$

where $U(\varphi)$ is the energy associated with a bond angle φ and $\beta = \dfrac{1}{k_B T}$. Suppose for a polymethylene chain $U(\varphi) = \dfrac{U_0}{2}(1 + \cos \varphi)$. Assume the high temperature limit where $k_B T \gg U(\varphi)$, and where $e^{-\beta U(\varphi)} \approx 1 - \beta U(\varphi)$. Also assume as before $\cos \theta = \frac{1}{3}$. Calculate the mean-squared end-to-end distance.

P24.26 The flow of blood in the capillaries is regulated in part by constriction of small blood vessels, called arterioles.

   a. Assume an arteriole has a radius 0.005 cm and a length of 0.05 cm. Assume the pressure drop across this arteriole is 50 mmHg. Calculate the flow Q in cubic meters per second through this arteriole. Assume the viscosity of blood is 0.04 P. Assume flow of blood in the arterioles is governed by Poiseuille's law. Note 0.0075 mmHg = 1 Nt/m² = 1 Pa.

   b. By what fraction is the radius of the arteriole reduced if the flow is reduced by 40%?

   c. To what radius must the arteriole dilate to increase the blood flow by 50%?

P24.27 A 200-base-pair fragment of DNA is 680 Å long and 20 Å in diameter. It has a molecular weight of 135,000 g mol⁻¹.

   a. Calculate the frictional coefficient of this DNA fragment, assuming it can be treated as a rod of length L = 680 Å and diameter d = 20 Å.

   b. Calculate the sedimentation coefficient of the DNA assuming $\overline{V}_2$ of the DNA is 0.51 mL g⁻¹.

P24.28 For the purpose of calculating the frictional coefficient f, a rod-like, linear polymer can be more accurately treated as a string of N touching beads, with each spherical bead having a diameter δ. The rod is therefore of length $N\delta$.

For such a polymer the frictional coefficient is $f = \dfrac{3\pi \eta N \delta}{\ln N}$ where η is the viscosity of the solvent, assumed here to be water.

   a. Calculate the bead diameter and number of beads required to have the same total volume and length as the DNA described in Problem P24.27.

   b. Calculate the frictional coefficient and the sedimentation coefficient of the DNA described in part (a).

   c. The vertical distance between adjacent base pairs is about 3.4 Å in B form DNA. How large is each sphere compared to a base pair in B form DNA?

P24.29 Chromatin is the complex of DNA and proteins (mostly histones) found in all eukaryotic cells. The fundamental repeating unit of chromatin is the nucleosome particle. The properties of the DNA and the protein in nucleosome particles were determined using a combination of velocity sedimentation, dynamic light scattering, and gel electrophoresis.

   • Dynamic light scattering studies determined the diffusion coefficient of the nucleosome particle in solution at 293 K to be 4.37×10^{-11} m² s⁻¹.

   • In the same solution velocity sedimentation performed at 18,100 revolutions per minute, obtained the following data:

Time (min)	Boundary Position r (cm)
0	4.460
80	4.593
160	4.713
240	4.844

   • Gel electrophoresis showed that the DNA molecule associated with a single nucleosome protein complex is 200 base pairs in length.

   a. What is the molecular weight of the nucleosome particle? Assume the solution has a density of 1.02 g cm⁻³ and the specific volume of the nucleosome particle is 0.66 cm³g⁻¹.

b. Assuming the nucleosome particle is spherical, calculate its Stokes radius. Assume the viscosity of the solution is $0.01 \text{ g cm}^{-1} \text{ s}^{-1}$.

c. Assuming each base pair in DNA is separated from the adjacent base pairs by about 3.4×10^{-10} m, approximately how long is a piece of DNA 200 base pairs in length? Based on this result, and the result from part (b), comment on how tightly packed the DNA is in the nucleosome.

d. Other evidence suggests that the protein component of the nucleosome is composed of a complex of eight protein molecules. Assuming that a single DNA base pair weighs about 660 g mol^{-1}, estimate the total weight of the protein in the nucleosome and the weight of each component protein.

e. How are the proteins and DNA packed in the nucleosome? To answer this question, assume the eight proteins form an unhydrated, spherical complex with specific volume $0.74 \text{ cm}^3 \text{ g}^{-1}$. Calculate the radius of this hypothetical protein sphere. Assume the 200-base-pair DNA behaves as a random coil polymer. Calculate the rms end-to-end distance. Comparing the protein sphere radius with the rms end-to-end distance for the DNA, is most of the DNA packed outside or inside the protein core of the nucleosome? Explain.

P24.30 T4 is a bacterial virus. The virus has a head group that is roughly spherical, and which contains DNA. The virus head particles were found to have the following properties at $T = 293$ K and in water solution: the sedimentation coefficient $\bar{s} = 1025$ S, the diffusion coefficient $D = 3.6 \times 10^{-8} \text{ cm}^2 \text{ s}^{-1}$, and the gram-specific volume for the unhydrated head group is 0.605 mL g^{-1}. Calculate the following:

a. The molar weight of the head group.

b. The volume of the unhydrated head group.

c. The frictional coefficient of the hydrated head group.

d. From your result in part (c), calculate the radius and volume of the hydrated head group. Assume the viscosity of water is 0.01 P.

e. From your results in parts (a), (b), and (d), calculate the number of grams of water that hydrate each gram of head group.

P24.31 Suppose the intrinsic viscosity $[\eta]$ of a globular protein solution is $4.35 \text{ cm}^3 \text{g}^{-1}$. The *unhydrated* protein has a partial specific volume of $\bar{V}_2 = 0.74 \text{ cm}^3 \text{ g}^{-1}$. Calculate the number of grams of water that hydrate each gram of protein in the solution. Assume the specific volume of water is $1.00 \text{ cm}^3 \text{ g}^{-1}$ and that the protein can be modeled as a hydrated sphere.

P24.32 A rod-shaped protein has a molecular weight of 20,000 and a sedimentation coefficient of 1.74 S. At high concentrations, the protein dimerizes and has a sedimentation coefficient of 2.36 S. Determine whether the dimer is formed by a side-by-side attachment of the monomers, a linear end-to-end attachment of the monomers, or a bent end-to-end attachment of the monomers. *Hint:* Assume the side-by-side and linear end-to-end geometries can be modeled as prolate ellipsoids with different aspect ratios.

P24.33 Bovin serum albumin (BSA) has a molecular weight of $M = 66.500 \text{ kg mol}^{-1}$ and a specific volume of $\bar{V}_2 = 0.734 \text{ cm}^3 \text{ g}^{-1}$.

a. In a velocity centrifugation experiment, the sedimentation coefficient $\bar{s}_{20,w}$ is determined to be 4.31 S. Determine the friction coefficient f of BSA. Assume the density of the solution is 1.00 g cm^{-3}.

b. Assuming that BSA is spherically shaped in solution, calculate the effective radius. Assume a viscosity of $\eta = 0.010$ P.

c. Assuming that BSA is a hydrated sphere in solution, calculate the number of water molecules of hydration associated with each BSA molecule. Assume the specific volume of water is $1 \text{ cm}^3 \text{ g}^{-1}$.

P24.34 The molecular weight of carbonyl hemoglobin was determined by T. Svedberg [*Z. Physik. Chem.* 127 (1927) 51] using velocity sedimentation. The sample was composed of 0.96 g of carbonyl hemoglobin dissolved in 100 mL of water, which was spun at 39,300 rpm at a temperature of 303 K. After 30.0 min the concentration boundary advanced by 0.0740 cm from a mean position of 4.525 cm relative to the spinning axis. Calculate the sedimentation coefficient. Assuming the diffusion coefficient of carbonyl hemoglobin at $T = 303$ K is $7.00 \times 10^{-11} \text{ m}^2 \text{ s}^{-1}$, calculate the molecular weight of carbonyl hemoglobin. Assume $\bar{V}_2 = 0.755 \text{ mL g}^{-1}$ and $\rho = 0.9469 \text{ g mL}^{-1}$.

Web-Based Simulations, Animations, and Problems

W24.1 In this problem, concentration time dependence in one dimension is depicted as predicted using Fick's second law of diffusion. Specifically, variation of particle number density as a function of distance with time and diffusion constant D is investigated. Comparisons of the full distribution to x_{rms} are performed to illustrate the behavior of x_{rms} with time and D.

W24.2 In this problem, the dependence of $\bar{s}_{20,w}$ for DNA as a function of molecular weight is investigated as a function of the molecular mass of the DNA. Specifically, in the low molecular weight range ($M < 10^5 \text{ g mol}^{-1}$), DNA can be considered a rigid rod. For DNA greater than this molecular

weight, the DNA chain will be considered flexible and the impact of polymer flexibility on $\bar{s}_{20,w}$ will be investigated.

W24.3 The dependence of the translational friction coefficient on the aspect ratio P for non-spherical molecules has been considered for the case of rods and for ellipsoids of revolution (i.e. prolate and oblate). Although frictional effects arise from the rotation of rigid bodies, we only considered the coefficient of rotational friction for a sphere subjected to stick boundary conditions. Here, we augment the discussion in the text by investigating the dependence of the rotational friction coefficient on P for cylinders and ellipsoids of revolution.

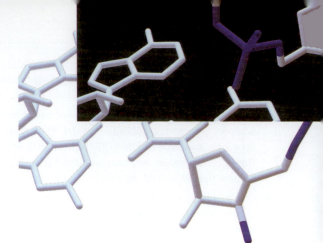

25 CHAPTER

Elementary Chemical Kinetics

In the next two chapters, the evolution of a system toward equilibrium with respect to chemical composition is discussed. The central question of interest is perhaps the first question one asks about any chemical reaction: just how do the reactants become products? In chemical kinetics, this question is answered by determining the timescale and mechanism of a chemical reaction. The basic tools of this field are presented here. First, reaction rates and methods for their determination are discussed. Next, the concept of reaction mechanisms in terms of elementary reaction steps is presented. Approaches to describing elementary reaction steps, including integrated rate law expressions and numerical methods, are outlined. Basic reaction types, such as unidirectional, sequential, and parallel reactions, are described. The possibility of back or reverse reactions is introduced, and these reactions are used to establish the kinetic definition of equilibrium. Finally, reaction dynamics are described. The ideas presented here provide a kinetic "toolkit" that will be employed to describe complex biological reactions in the following chapter.

25.1 Introduction to Kinetics

In the previous chapter, the evolution of a system's physical properties toward equilibrium was discussed. In these transport phenomena, the system undergoes relaxation without a change in chemical composition. Transport phenomena are sometimes referred to as physical kinetics to indicate that the physical properties of the system are evolving and not the system composition. In this chapter and the next, we focus on chemical kinetics where the composition of the system evolves with time.

Chemical kinetics involves the study of the rates and mechanisms of chemical reactions. This area bridges an important gap in our discussion of chemical reactions. Thermodynamic descriptions of chemical reactions involved the Gibbs or Helmholtz energy for a reaction and the corresponding equilibrium constant. These quantities are sufficient to predict the reactant and product concentrations at equilibrium, but are of little use in determining the timescale over which the reaction occurs. That is, thermodynamics may dictate that a reaction is spontaneous, but does it occur in 10^{-15} s (a femtosecond, the timescale for the fastest chemical processes known to date) or 10^{15} s (the age of the universe)? The answer to this question lies in the domain of chemical kinetics.

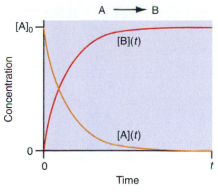

FIGURE 25.1

Concentration as a function of time for the conversion of reactant A into product B. The concentration of A at $t = 0$ is $[A]_0$, and the concentration of B is zero. As the reaction proceeds, the loss of A results in the production of B.

In the course of a chemical reaction, concentrations will change with time as "reactants" become "products." Figure 25.1 presents possibly the first chemical reaction you were introduced to in your introductory chemistry class: the conversion of reactant A into product B. The figure illustrates that, as the reaction proceeds, a decrease in reactant concentration and a corresponding increase in product concentration is observed. One way to describe this process is to define the rate of concentration change with time, a quantity that is referred to as the reaction rate.

The central ideal behind chemical kinetics is this: by monitoring the rate at which chemical reactions occur and determining the dependence of this rate on system parameters such as temperature, concentration, and pressure, we can gain insight into the mechanism of the reaction. Experimental chemical kinetics includes the development of techniques that allow for the study of chemical reactions including the measurement and analysis of chemical reaction dynamics. In addition to experiments, a substantial amount of theoretical work has been performed to understand reaction mechanisms and the underlying physics that govern the rates of chemical transformations. The synergy between experiment and theoretical chemical kinetics has provided for dramatic advances in this field. Finally, the importance of chemical kinetics is evidenced by its application in nearly every area of chemistry. Found in areas such as enzyme catalysis, materials processing, and atmospheric chemistry, chemical kinetics is clearly an important area of physical chemistry.

25.2 Reaction Rates

Consider the following "generic" chemical reaction:

$$aA + bB + \cdots \rightarrow cC + dD + \cdots \tag{25.1}$$

In Equation (25.1), uppercase letters indicate a chemical species and lowercase letters represent the stoichiometric coefficient for the species in the balanced reaction. The species on the left-hand and right-hand sides of the arrow are referred to as *reactants* and *products,* respectively. The number of moles of a species during any point of the reaction is given by

$$n_i = n_i^o + \nu_i \xi \tag{25.2}$$

where n_i is the number of moles of species i at any given time during the reaction, n_i^o is the number of moles of species i present before initiation of the reaction, ν_i is related to the stoichiometric coefficient of species i, and ξ represents the advancement of the reaction and is equal to zero at the start of the reaction. The advancement variable allows us to quantify the rate of the reaction with respect to all species, irrespective of stoichiometry (see later discussion). For the reaction depicted in Equation (25.1), reactants will be consumed and products formed during the reaction. To ensure that this behavior is reflected in Equation (25.2), ν_i is set equal to -1 times the stoichiometric coefficient for reactants, and is set equal to the stoichiometric coefficient for products.

The time evolution of the reactant and product concentrations is quantified by differentiating both sides of Equation (25.2) with respect to time:

$$\frac{dn_i}{dt} = \nu_i \frac{d\xi}{dt} \tag{25.3}$$

The **reaction rate** is defined as the change in the advancement of the reaction with time:

$$\text{Rate} = \frac{d\xi}{dt} \tag{25.4}$$

With this definition, the rate of reaction with respect to the change in the number of moles of a given species with time is

$$\text{Rate} = \frac{1}{\nu_i} \frac{dn_i}{dt} \tag{25.5}$$

As an example of how the rate of reaction is defined relative to the change in moles of reactant or product with time, consider the following reaction:

$$4NO_2(g) + O_2(g) \longrightarrow 2N_2O_5(g) \tag{25.6}$$

The rate of reaction can be expressed with respect to any species in Equation (25.6):

$$\text{Rate} = -\frac{1}{4}\frac{dn_{NO_2}}{dt} = -\frac{dn_{O_2}}{dt} = \frac{1}{2}\frac{dn_{N_2O_5}}{dt} \tag{25.7}$$

Notice the sign convention of the coefficient with respect to reactants and products: negative for reactants and positive for products. Also, notice that the rate of reaction can be defined with respect to both reactants and products. In our example, 4 mol of NO_2 react with 1 mol of O_2 to produce 2 mol of N_2O_5 product. Therefore, the **rate of conversion** of NO_2 will be four times greater than the rate of O_2 conversion. Although the conversion rates are different, the reaction rate defined with respect to either species will be the same. Furthermore, because both NO_2 and O_2 are reactants, the change in the moles of these species with respect to time is negative. However, by using a negative stoichiometric coefficient, the reaction rate defined with respect to the reactants is still a positive quantity.

In applying Equation (25.5) to define a rate of reaction, a set of stoichiometric coefficients must be employed; however, these coefficients are not unique. For example, if we multiply both sides of Equation (25.6) by a factor of 2, the expression for the rate of conversion must also change. Generally, one decides on a given set of coefficients for a balanced reaction and uses these coefficients consistently throughout a given kinetics problem.

In our present definition, the rate of reaction as written is an extensive property; therefore, it will depend on the system size. The rate can be made intensive by dividing Equation (25.5) by the volume of the system:

$$R = \frac{\text{Rate}}{V} = \frac{1}{V}\left(\frac{1}{\nu_i}\frac{dn_i}{dt}\right) = \frac{1}{\nu_i}\frac{d[i]}{dt} \tag{25.8}$$

In Equation (25.8), R is the intensive reaction rate. The last equality in Equation (25.8) is performed recognizing that moles of species i per unit volume is simply the molarity of species i, or $[i]$. Equation (25.8) is the definition for the rate of reaction at constant volume. For species in solution, the application of Equation (25.8) in defining the rate of reaction is clear, but it can also be used for gases, as Example Problem 25.1 illustrates.

EXAMPLE PROBLEM 25.1

The decomposition of acetaldehyde is given by the following balanced reaction:

$$CH_3COH(g) \longrightarrow CH_4(g) + CO(g)$$

Define the rate of reaction with respect to the pressure of the reactant.

Solution

Beginning with Equation (25.2) and focusing on the acetaldehyde reactant, we obtain

$$n_{CH_3COH} = n^o_{CH_3COH} - \xi$$

Using the ideal gas law, the pressure of acetaldehyde is expressed as

$$P_{CH_3COH} = \frac{n_{CH_3COH}}{V}RT = [CH_3COH]RT$$

Therefore, the pressure is related to the concentration by the quantity RT. Substituting this result into Equation (25.8) with $\nu_i = 1$ yields

$$R = \frac{\text{Rate}}{V} = -\frac{1}{\nu_{CH_3COH}}\frac{d[CH_3COH]}{dt}$$

$$= -\frac{1}{RT}\frac{dP_{CH_3COH}}{dt}$$

25.3 Rate Laws

We begin our discussion of rate laws with a few important definitions. The rate of a reaction will generally depend on the temperature, pressure, and concentrations of species involved in the reaction. In addition, the rate may depend on the phase or phases in which the reaction occurs. Homogeneous reactions occur in a single phase, whereas heterogeneous reactions involve more than one phase. Reactions that involve a surface are classic examples of heterogeneous reactions. We will limit our initial discussion to homogeneous reactions, with heterogeneous reactivity discussed in Chapter 26. For the majority of homogeneous reactions, an empirical relationship between reactant concentrations and the rate of a chemical reaction can be written. This relationship is known as a **rate law,** and for the reaction shown in Equation (25.1) it is written as

$$R = k[A]^\alpha [B]^\beta \ldots \tag{25.9}$$

where [A] is the concentration of reactant A, [B] is the concentration of reactant B, and so forth. The constant α is known as the **reaction order** with respect to species A, β the reaction order with respect to species B, and so forth. The overall reaction order is equal to the sum of the individual reaction orders ($\alpha + \beta + \ldots$). Finally, the constant k is referred to as the **rate constant** for the reaction. The rate constant is independent of concentration, but dependent on pressure and temperature, as discussed later in Section 25.9.

The reaction order dictates the concentration dependence of the reaction rate. The reaction order may be integer, zero, or fractional. *It cannot be overemphasized that reaction orders have no relation to stoichiometric coefficients, and they are determined by experiment.* For example, reconsider the reaction of nitrogen dioxide with molecular oxygen [Equation (25.6)]:

$$4NO_2(g) + O_2(g) \longrightarrow 2N_2O_5(g)$$

The experimentally determined rate law expression for this reaction is

$$R = k[NO_2]^2[O_2]$$

That is, the reaction is second order with respect to NO_2, first order with respect to O_2, and third order overall. Notice that the reaction orders are not equal to the stoichiometric coefficients. *All rate laws must be determined experimentally with respect to each reactant,* and there is no insight to be gained by considering the stoichiometry of the reaction.

In the rate law expression of Equation (25.9), the rate constant serves as the proportionality constant between the concentration of the various species and the reaction rate. Inspection of Equation (25.8) demonstrates that the reaction rate will *always* have units of concentration per unit here, or time^{-1}. Therefore, the units of k must change with respect to the overall order of the reaction to ensure that the reaction rate has the correct units. The relationship between the rate law expression, order, and the units of k is presented in Table 25.1.

25.3.1 Measuring Reaction Rates

With the definitions for the reaction rate and rate law provided by Equations (25.8) and (25.9), the question of how one measures the rate of reaction becomes important. To illustrate this point, consider the following reaction:

$$A \xrightarrow{k} B \tag{25.10}$$

The rate of this reaction in terms of [A] is given by

$$R = -\frac{d[A]}{dt} \tag{25.11}$$

Furthermore, suppose experiments demonstrate that the reaction is first order in A, first order overall, and $k = 40 \text{ s}^{-1}$ so that

$$R = k[A] = (40 \text{ s}^{-1})[A] \tag{25.12}$$

TABLE 25.1 **Relationship between Rate Law, Order, and the Rate Constant, Σ**

Rate Law	Order	Units* of k
Rate = k	Zero	M s^{-1}
Rate = $k[A]$	First order with respect to A	s^{-1}
	First order overall	
Rate = $k[A]^2$	Second order with respect to A	M^{-1} s^{-1}
	Second order overall	
Rate = $k[A][B]$	First order with respect to A	M^{-1} s^{-1}
	First order with respect to B	
	Second order overall	
Rate = $k[A][B][C]$	First order with respect to A	M^{-2} s^{-1}
	First order with respect to B	
	First order with respect to C	
	Third order overall	

*In the units of k, M represents mol L^{-1} or moles per liter.

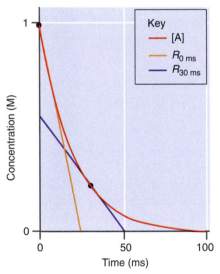

FIGURE 25.2

Measurement of the reaction rate. The concentration of reactant A as a function of time is presented. The rate R is equal to the slope of the tangent of this curve. This slope depends on the time at which the tangent is determined.

Equation (25.11) states that the rate of the reaction is equal to the negative of the time derivative of [A]. Imagine that we perform an experiment in which [A] is measured as a function of time as shown in Figure 25.2. The derivative in Equation (25.11) is simply the slope of the tangent for the concentration curve at a specific time. Therefore, the reaction rate will depend on the time at which the rate is determined. Figure 25.2 presents a measurement of the rate at two time points, $t = 0$ ms (1 ms = 10^{-3} s) and $t = 30$ ms. At $t = 0$ ms, the reaction rate is given by the negative slope of the line corresponding to the change in [A] with time, per Equation (25.11).

The tangent determined 30 ms into the reaction is presented as the blue line, and the tangent at $t = 0$ is presented as the yellow line:

$$R_{t=0} = -\frac{d[A]}{dt} = 40 \text{ M s}^{-1}$$

However, when measured at 30 ms the rate is

$$R_{t=30 \text{ ms}} = -\frac{d[A]}{dt} = 12 \text{ M s}^{-1}$$

Notice that the reaction rate is decreasing with time. This behavior is a direct consequence of the change of [A] as a function of time, as expected from the rate law of Equation (25.12). Specifically, at $t = 0$,

$$R_{t=0} = 40 \text{ s}^{-1}[A]_{t=0} = 40 \text{ s}^{-1} (1 \text{ M}) = 40 \text{ M s}^{-1}$$

However, by $t = 30$ ms the concentration of A has decreased to 0.3 M so that the rate is

$$R_{t=30 \text{ ms}} = 40 \text{ s}^{-1}[A]_{t=30 \text{ ms}} = 40 \text{ s}^{-1} (0.3 \text{ M}) = 12 \text{ M s}^{-1}$$

This difference in rates brings to the forefront an important issue in kinetics: how does one define a reaction rate if the rate changes with time? One convention is to define the rate before the reactant concentrations have undergone any substantial change from their initial values. The reaction rate obtained under such conditions is known as the **initial rate.** The initial rate in the previous example is that determined at $t = 0$. In the remainder of our discussion of kinetics, the rate of reaction is taken to be synonymous with initial rate. However, the rate constant is independent of concentration; therefore, if the rate constant, concentrations, and order dependence of the reaction rate are known, the reaction rate can be determined at any time.

25.3.2 Determining Reaction Orders

Consider the following reaction:

$$A + B \xrightarrow{k} C \tag{25.13}$$

The rate law expression for this reaction is

$$R = k[A]^\alpha[B]^\beta \tag{25.14}$$

How can one determine the order of the reaction with respect to A and B? First, note that the measurement of the rate under a single set of concentrations for A and B will not by itself provide a measure of α and β because one would have only one equation [Equation (25.14)], but two unknown quantities. Therefore, the determination of reaction order will involve the measurement of the reaction rate under various concentration conditions. This approach assumes that one can vary the reactant concentrations, but for some reactions this is not possible. In this case, another approach must be taken to determine the order dependence of the reaction, and this approach will be discussed shortly. For now, assume that the reactant concentrations can be varied. The question then becomes "What set of concentrations should be used to determine the reaction rate?" One answer to this question is known as the **isolation method.** In this approach, the reaction is performed with all species but one in excess. Under these conditions, only the concentration of one species will vary to a significant extent during the reaction. For example, consider the example A + B reaction shown in Equation (25.13). Imagine performing the experiment where the initial concentration of A is 1.00 M and the concentration of B is 0.01 M. The rate of the reaction will be zero when all of reactant B has been used; however, the concentration of A will have been reduced to 0.99 M, only a slight reduction from the initial concentration. This simple example demonstrates that the concentration of species present in excess will be essentially constant with time. This time independence simplifies the reaction rate expression because the reaction rate will depend only on the concentration of the nonexcess species. In our example reaction where A is in excess, Equation (25.14) simplifies to

$$R = k'[B]^\beta \tag{25.15}$$

In Equation (25.15), k' is the product of the original rate constant and $[A]^\alpha$, both of which are time independent. Isolation results in the dependence of the reaction rate on [B] exclusively, and the reaction order with respect to B is determined by measuring the reaction rate as [B] is varied. Of course, the isolation method could just as easily be applied to determine α by performing measurements with B in excess.

A second strategy employed to determine reaction rates is referred to as the **method of initial rates.** In this approach, the concentration of a single reactant is changed while holding all other concentrations constant, and the initial rate of the reaction determined. The variation in the initial rate as a function of concentration is then analyzed to determine the order of the reaction with respect to the reactant that is varied. Consider the reaction depicted by Equation (25.13). To determine the order of the reaction for each reactant, the reaction rate is measured as [A] is varied and the concentration of B is held constant. The reaction rates at two different values of [A] are then analyzed to determine the order of the reaction with respect to [A] as follows:

$$\frac{R_1}{R_2} = \frac{k[A]_1^\alpha[B]_0^\beta}{k[A]_2^\alpha[B]_0^\beta} = \left(\frac{[A]_1}{[A]_2}\right)^\alpha$$

$$\ln\left(\frac{R_1}{R_2}\right) = \alpha \ln\left(\frac{[A]_1}{[A]_2}\right) \tag{25.16}$$

Notice that [B] and k are constant in each measurement; therefore, they cancel when one evaluates the ratio of the measured reaction rates. Using Equation (25.16), the order of the reaction with respect to A is readily determined. A similar experiment to determine β can be performed where [A] is held constant and the dependence of the reaction rate on [B] is measured.

EXAMPLE PROBLEM 25.2

Using the following data for the reaction illustrated in Equation (25.13), determine the order of the reaction with respect to A and B, and the rate constant for the reaction:

[A] (M)	[B] (M)	Initial Rate (M s^{-1})
2.30×10^{-4}	3.10×10^{-5}	5.25×10^{-4}
4.60×10^{-4}	6.20×10^{-5}	4.20×10^{-3}
9.20×10^{-4}	6.20×10^{-5}	1.70×10^{-2}

Solution

Using the last two entries in the table, the order of the reaction with respect to A is

$$\ln\left(\frac{R_1}{R_2}\right) = \alpha \ln\left(\frac{[A]_1}{[A]_2}\right)$$

$$\ln\left(\frac{4.20 \times 10^{-3}}{1.70 \times 10^{-2}}\right) = \alpha \ln\left(\frac{4.60 \times 10^{-4}}{9.20 \times 10^{-4}}\right)$$

$$-1.398 = \alpha(-0.693)$$

$$2 = \alpha$$

Using this result and the first two entries in the table, the order of the reaction with respect to B is given by

$$\frac{R_1}{R_2} = \frac{k[A]_1^2[B]_1^\beta}{k[A]_2^2[B]_2^\beta} = \frac{[A]_1^2[B]_1^\beta}{[A]_2^2[B]_2^\beta}$$

$$\left(\frac{5.25 \times 10^{-4}}{4.20 \times 10^{-3}}\right) = \left(\frac{2.30 \times 10^{-4}}{4.60 \times 10^{-4}}\right)^2\left(\frac{3.10 \times 10^{-5}}{6.20 \times 10^{-5}}\right)^\beta$$

$$0.500 = (0.500)^\beta$$

$$1 = \beta$$

Therefore, the reaction is second order in A, first order in B, and third order overall. Using any row from the table, the rate constant is readily determined:

$$R = k[A]^2[B]$$

$$5.2 \times 10^{-4} \text{ M s}^{-1} = k(2.3 \times 10^{-4} \text{ M})^2(3.1 \times 10^{-5} \text{ M})$$

$$3.17 \times 10^8 \text{ M}^{-2} \text{ s}^{-1} = k$$

Having determined k, the overall rate law is

$$R = (3.17 \times 10^8 \text{ M}^{-2} \text{ s}^{-1})[A]^2[B]$$

The remaining question to address is how one experimentally determines the rate of a chemical reaction. Measurement techniques are usually separated into one of two categories: chemical and physical. As the name implies, **chemical methods** in kinetic studies rely on chemical processing to determine the progress of a reaction with respect to time. In this method, a chemical reaction is initiated, and samples are removed from the reaction and manipulated such that the reaction in the sample is terminated. Termination of the reaction is accomplished by rapidly cooling the sample or by adding a chemical species that depletes one of the reactants. After stopping the reaction, the sample contents are analyzed. By performing this analysis on a series of samples removed from the original reaction container as a function of time after initiation of the reaction, the kinetics of the reaction can be determined. Chemical methods are generally cumbersome to use and are limited to reactions that occur on slow timescales.

The majority of modern kinetics experiments involve **physical methods.** In these methods, a physical property of the system is monitored as the reaction proceeds. For some reactions, the system pressure or volume provides a convenient physical property for monitoring the progress of a reaction. For example, consider the thermal decomposition of PCl_5:

$$PCl_5(g) \longrightarrow PCl_3(g) + Cl_2(g)$$

As the reaction proceeds, for every gaseous PCl_5 molecule that decays, two gaseous product molecules are formed. Therefore, the total system pressure will increase as the reaction proceeds. Measurement of this pressure increase as a function of time provides information on the reaction kinetics.

More complex physical methods involve techniques that are capable of monitoring the concentration of an individual species as a function of time. Many of the spectroscopic techniques described in this text are extremely useful for such measurements. For example, electronic absorption measurements can be performed in which the concentration of a species is monitored using the electronic absorption of a molecule and the Beer–Lambert law. Vibrational spectroscopic measurements using infrared absorption and Raman scattering can be employed to monitor vibrational transitions of reactants or products providing information on their consumption or production. Finally, NMR spectroscopy is a useful technique for following the reaction kinetics of complex systems.

The challenge in chemical kinetics is to perform measurements with sufficient time resolution to monitor the chemistry of interest. If the reaction is slow (seconds or longer), then the chemical methods just described can be used to monitor the kinetics. However, many chemical reactions occur on timescales as short as picoseconds (10^{-12} s) and femtoseconds (10^{-15} s). Reactions occurring on these short timescales are most easily studied using physical methods.

For reactions that occur on timescales as short at 1 ms (10^{-3} s), **stopped-flow techniques** provide a convenient method by which to measure solution-phase reactions. These techniques are exceptionally popular for biochemical studies. A stopped-flow experiment is illustrated in Figure 25.3. Two reactants (A and B) are held in reservoirs connected to a syringe pump. The reaction is initiated by depressing the reactant syringes, and the reactants are mixed at the junction indicated in the figure. The reaction is monitored by observing the change in absorbance of the reaction mixture as a function of time. The temporal resolutions of stopped-flow techniques are generally limited by the time it takes for the reactants to mix.

Reactions that can be triggered by light are studied using **flash photolysis techniques.** In flash photolysis, the sample is exposed to a temporal pulse of light that initiates the reaction. Light pulses as short as 10 femtoseconds (10 fs = 10^{-14} s) in the visible region of the electromagnetic spectrum are available such that reaction dynamics on this extremely short or ultrafast timescale can be studied. For reference, a 3000-cm^{-1} vibrational mode has a period of roughly 10 fs. Therefore, reactions can be initiated on the same timescale as vibrational molecular motion, and this capability has opened up many exciting fields in chemical kinetics. This capability has been used to determine the ultrafast reaction kinetics associated with vision, photosynthesis, atmospheric processes, and charge-carrier dynamics in semiconductors. Recent references to some of this work are included in the For Further Reading section at the end of this chapter. Short optical pulses can be used to perform vibrational spectroscopic measurements (infrared absorption or

FIGURE 25.3
Schematic of a stopped-flow experiment. Two reactants are rapidly introduced into the mixing chamber by syringes. After mixing, the reaction kinetics are monitored by observing the change in sample concentration versus time, in this example by measuring the absorption of light as a function of time after mixing.

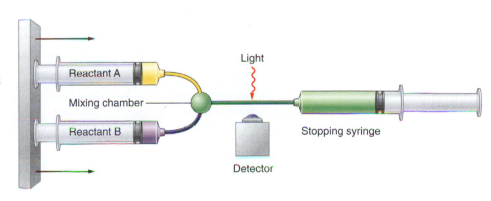

Raman) on the 100-fs timescale. Finally, NMR techniques, as well as optical absorption and vibrational spectroscopy, can be used to study reactions that occur on the microsecond (10^{-6} s) and longer timescales.

Another approach to studying chemical kinetics is that of **perturbation-relaxation methods.** In this approach, a chemical system initially at equilibrium is perturbed such that the system is no longer at equilibrium. By following the relaxation of the system back toward equilibrium, the rate constants for the reaction can be determined. Any system variable that affects the position of the equilibrium such as pressure, pH, or temperature can be used in a perturbation-relaxation experiment. Temperature perturbation or T-jump experiments are the most common type of perturbation experiment and are described in detail later in this chapter (see Section 25.11).

In summary, the measurement technique chosen for reaction rate determination will depend on both the specifics of the reaction as well as the timescale over which the reaction occurs. In any event, the determination of reaction rates is an experimental exercise and must be accomplished through careful measurements involving well-designed experiments.

25.4 Reaction Mechanisms

As discussed in the previous section, the order of a reaction with respect to a given reactant is not determined by the stoichiometry of the reaction. The reason for the inequivalence of the stoichiometric coefficient and reaction order is that the balanced chemical reaction provides no information with respect to the mechanism of the chemical reaction. A **reaction mechanism** is defined as the collection of individual kinetic processes or elementary steps involved in the transformation of reactants into products. The rate law expression for a chemical reaction, including the order of the reaction, is entirely dependent on the reaction mechanism. In contrast, the Gibbs energy for a reaction is dependent on the equilibrium concentration of reactants and products. Just as the study of concentrations as a function of reaction conditions provides information on the thermodynamics of the reaction, the study of reaction rates as a function of reaction conditions provides information on the reaction mechanism.

All reaction mechanisms consist of a series of **elementary reaction steps,** or a chemical process that occurs in a single step. The **molecularity** of a reaction step is the stoichiometric quantity of reactants involved in the step. For example, unimolecular reactions involve a single reactant species. An example of a unimolecular reaction step is the decomposition of a diatomic molecule into its atomic fragments:

$$I_2 \xrightarrow{k_d} 2I \tag{25.17}$$

Although Equation (25.17) is referred to as a unimolecular reaction, enthalpy changes accompanying the reaction generally involve the transfer of this heat through collisions with other, neighboring molecules. The role of collisional energy exchange with surrounding molecules will figure prominently in the discussion of unimolecular dissociation reactions in the following chapter, but these energy-exchange processes are suppressed in this discussion. Bimolecular reaction steps involve the interaction of two reactants. For example, the reaction of nitric oxide with ozone is a biomolecular reaction:

$$NO + O_3 \xrightarrow{k_r} NO_2 + O_2 \tag{25.18}$$

The importance of elementary reaction steps is that the corresponding rate law expression for the reaction can be written based on the molecularity of the reaction. For the unimolecular reaction, the rate law expression is that of a first-order reaction. For the unimolecular decomposition of I_2 presented in Equation (25.17), the rate law expression for this elementary step is

$$R = -\frac{d[I_2]}{dt} = k_d[I_2] \tag{25.19}$$

Likewise, the rate law expression for the bimolecular reaction of NO and O_3 [Equation (25.18)] is

$$R = -\frac{d[NO]}{dt} = k_r[NO][O_3] \tag{25.20}$$

Comparison of the rate law expressions with their corresponding reactions demonstrates that the order of the reaction is equal to the stoichiometric coefficient. For elementary reactions, the order of the reaction can be inferred from the molecularity of the reaction. Keep in mind that *the equivalence of order and molecularity is only true for elementary reaction steps.*

A common problem in kinetics is identifying which of a variety of proposed reaction mechanisms is the "correct" mechanism. The design of kinetic experiments to differentiate between proposed mechanisms is quite challenging. Due to the complexity of many reactions, it is often difficult to experimentally differentiate between several potential mechanisms. A general rule of kinetics is that although it may be possible to rule out a proposed mechanism, it is never possible to prove unequivocally that a given mechanism is correct. The following example illustrates the origins of this rule. Consider this reaction:

$$A \longrightarrow P \tag{25.21}$$

As written, the reaction is a simple first-order transformation of reactant A into product P, and it may occur through a single elementary step. However, what if the reaction were to occur through two elementary steps as follows:

$$A \xrightarrow{k_1} I$$
$$I \xrightarrow{k_2} P \tag{25.22}$$

In this mechanism, the decay of reactant A results in the formation of an intermediate species, I, that undergoes subsequent decay to produce the reaction product, P. One way to validate this mechanism is to observe the formation of the intermediate species. However, if the rate of the second reaction step is fast compared to the rate of the first step, the concentration of [I] will be quite small such that detection of the intermediate may be difficult. As will be seen later, in this limit the product formation kinetics will be consistent with the single elementary step mechanism, and verification of the two-step mechanism is not possible. It is usually assumed that the simplest mechanism consistent with the experimentally determined order dependence is correct until proven otherwise. In this example, a simple single-step mechanism would be considered "correct" until a clever chemist discovered a set of reaction conditions that demonstrates the reaction must occur by a sequential mechanism.

In order for a reaction mechanism to be valid, the order of the reaction predicted by the mechanism must be in agreement with the experimentally determined rate law. In evaluating a reaction mechanism, one must express the mechanism in terms of elementary reaction steps. The remainder of this chapter involves an investigation of various elementary reaction processes and derivations of the rate law expressions for these elementary reactions. The techniques developed in this chapter can be readily employed in the evaluation of complex kinetic problems, as illustrated in Chapter 26.

25.5 Integrated Rate Law Expressions

The rate law determination methods described in the previous section assume that one has a substantial amount of control over the reaction. Specifically, application of the initial-rate method requires that the reactant concentrations be controlled and mixed in any proportion desired. In addition, this method requires that the rate of reaction be measured immediately after initiation of the reaction. Unfortunately, many reactions cannot be studied by this technique due to the instability of the reactants involved, or the timescale of the reaction of interest. In this case, other approaches must be employed.

One approach is to assume that the reaction occurs with a given order dependence and then determine how the concentrations of reactants and products will vary as a function of time. The predictions of the model are compared to experiment to determine if the model provides an appropriate description of the reaction kinetics. **Integrated rate law expressions** provide the predicted temporal evolution in reactant and product concentrations for reactions having an assumed order dependence. In this section these expressions are derived. For many elementary reactions, integrated rate law expressions can be derived, and some of those cases are considered in this section. However, more complex reactions may be difficult to approach using this technique, and one must resort to numerical methods to evaluate the kinetic behavior associated with a given reaction mechanism. Numerical techniques are discussed in Section 25.6.

25.5.1 First-Order Reactions

Consider the following elementary reaction step where reactant A decays, resulting in the formation of product P:

$$A \xrightarrow{k} P \tag{25.23}$$

If the reaction is first order with respect to [A], the corresponding rate law expression is

$$R = k[A] \tag{25.24}$$

where k is the rate constant for the reaction. The reaction rate can also be written in terms of the time derivative of [A]:

$$R = -\frac{d[A]}{dt} \tag{25.25}$$

Because the reaction rates given by Equations (25.24) and (25.25) are the same, we can write

$$\frac{d[A]}{dt} = -k[A] \tag{25.26}$$

Equation (25.26) is known as a differential rate expression. It relates the time derivative of A to the rate constant and concentration dependence of the reaction. It is also a standard differential equation that can be integrated as follows:

$$\int_{[A]_0}^{[A]} \frac{d[A]}{[A]} = \int_0^t -k \, dt$$

$$\ln\left(\frac{[A]}{[A]_0}\right) = -kt$$

$$[A] = [A]_0 e^{-kt} \tag{25.27a}$$

The limits of integration employed in obtaining Equation (25.27a) correspond to the initial concentration of reactant when the reaction is initiated ([A] = [A]$_0$ at $t = 0$) and the concentration of reactant at a given time after the reaction has started. If only the reactant is present at $t = 0$, the sum of reactant and product concentrations at any time must be equal to [A]$_0$. Using this idea, the concentration of product with time for this **first-order reaction** is

$$[P] + [A] = [A]_0$$
$$[P] = [A]_0 - [A]$$

$$[P] = [A]_0 (1 - e^{-kt}) \tag{25.27b}$$

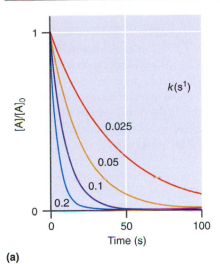

(a)

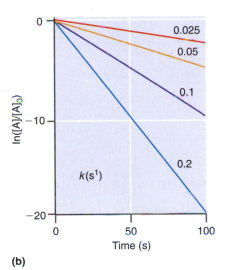

(b)

FIGURE 25.4
Reactant concentration as a function of time for a first-order chemical reaction as given by Equation (25.27). (a) Plots of [A] as a function of time for various rate constants k. The rate constant of a given curve is provided in the figure. (b) The natural log of reactant concentration as a function of time for a first-order chemical reaction as given by Equation (25.28).

Equation (25.27a) demonstrates that for a first-order reaction, the concentration of A will undergo exponential decay with time. A graphically convenient version of Equation (25.27a) for comparison to experiment is obtained by taking the natural log of the equation:

$$\ln[A] = \ln[A]_0 - kt \qquad (25.28)$$

Equation (25.28) predicts that for a first-order reaction, a plot of the natural log of the reactant concentration versus time will be a straight line of slope $-k$ and y intercept equal to the natural log of the initial concentration. Figure 25.4 provides a comparison of the concentration dependences predicted by Equations (25.27) and Equation (25.28) for first-order reactions. It is important to note that the comparison of experimental data to an integrated rate law expression requires that the variation in concentration with time be accurately known over a wide range of reaction times to determine if the reaction indeed follows a certain order dependence.

25.5.2 Half-Life and First-Order Reactions

The time it takes for the reactant concentration to decrease to one-half of its initial value is called the **half-life** of the reaction and is denoted as $t_{1/2}$. For a first-order reaction, substitution of the definition for $t_{1/2}$ into Equation (25.28) results in the following:

$$-kt_{1/2} = \ln\left(\frac{[A]_0/2}{[A]_0}\right) = -\ln 2 \qquad (25.29)$$

$$t_{1/2} = \frac{\ln 2}{k}$$

Notice that the half-life for a first-order reaction is independent of the initial concentration, and only the rate constant of the reaction influences $t_{1/2}$.

EXAMPLE PROBLEM 25.3

The decomposition of N_2O_5 is an important process in tropospheric chemistry. The half-life for the first-order decomposition of this compound is 2.05×10^4 s. How long will it take for a sample of N_2O_5 to decay to 60% of its initial value?

Solution

Using Equation (25.29), the rate constant for the decay reaction is determined using the half-life as follows:

$$k = \frac{\ln 2}{t_{1/2}} = \frac{\ln 2}{2.05 \times 10^4 \text{ s}} = 3.38 \times 10^{-5} \text{ s}^{-1}$$

The time at which the sample has decayed to 60% of its initial value is then determined using Equation (25.27a):

$$[N_2O_5] = 0.6[N_2O_5]_0 = [N_2O_5]_0 e^{-(3.38 \times 10^{-5} \text{ s}^{-1})t}$$

$$0.6 = e^{-(3.38 \times 10^{-5} \text{ s}^{-1})t}$$

$$\frac{-\ln(0.6)}{3.38 \times 10^{-5} \text{ s}^{-1}} = t = 1.51 \times 10^4 \text{ s}$$

Radioactive decay of unstable nuclear isotopes is an important example of a first-order process. The decay rate is usually stated as the half-life. Example Problem 25.4 demonstrates the use of radioactive decay in determining the age of a carbon-containing material.

Carbon-14 is a radioactive nucleus with a half-life of 5760 years. Living matter exchanges carbon with its surroundings (for example, through CO_2) so that a constant level of ^{14}C is maintained, corresponding to 15.3 decay events per minute. Once living matter has died, carbon contained in the matter is not exchanged with the surroundings, and the amount of ^{14}C that remains in the dead material decreases with time due to radioactive decay. Consider a piece of fossilized wood that demonstrates 2.4 ^{14}C decay events per minute. How old is the wood?

Solution

The ratio of decay events yields the amount of ^{14}C present currently versus the amount that was present when the tree died:

$$\frac{[^{14}C]}{[^{14}C]_0} = \frac{2.40 \text{ min}^{-1}}{15.3 \text{ min}^{-1}} = 0.157$$

The rate constant for isotope decay is related to the half-life as follows:

$$k = \frac{\ln 2}{t_{1/2}} = \frac{\ln 2}{5760 \text{ years}} = \frac{\ln 2}{1.82 \times 10^{11} \text{ s}} = 3.81 \times 10^{-12} \text{ s}^{-1}$$

With the rate constant and ratio of isotope concentrations, the age of the fossilized wood is readily determined:

$$\frac{[^{14}C]}{[^{14}C]_0} = e^{-kt}$$

$$\ln\left(\frac{[^{14}C]}{[^{14}C]_0}\right) = -kt$$

$$-\frac{1}{k}\ln\left(\frac{[^{14}C]}{[^{14}C]_0}\right) = -\frac{1}{3.81 \times 10^{-12} \text{ s}}\ln(0.157) = t$$

$$4.86 \times 10^{11} \text{ s} = t$$

This time corresponds to an age of roughly 15,400 years.

25.5.3 Second-Order Reaction (Type I)

Consider the following reaction, which is second order with respect to the reactant A:

$$2A \xrightarrow{k} P \qquad (25.30)$$

Second-order reactions involving a single reactant species are referred to as **type I.** Another reaction that is second order overall involves two reactants, A and B, with a rate law that is first order with respect to each reactant. Such reactions are referred to as **second-order reactions of type II.** We focus first on the type I case. For this reaction, the corresponding rate law expression is

$$R = k[A]^2 \qquad (25.31)$$

The rate as expressed as the derivative of reactant concentration is

$$R = -\frac{1}{2}\frac{d[A]}{dt} \qquad (25.32)$$

The rates in the preceding two expressions are equivalent such that

$$-\frac{d[A]}{dt} = 2k[A]^2 \qquad (25.33)$$

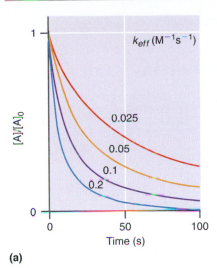

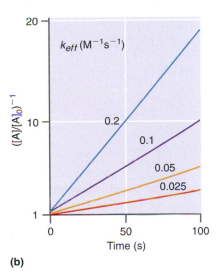

FIGURE 25.5

Reactant concentration as a function of time for a type I second-order chemical reaction. (a) Plots of [A] as a function of time for various rate constants. The rate constant of a given curve is provided in the figure. (b) The inverse of reactant concentration as a function of time as given by Equation (25.34).

Generally, the quantity $2k$ is written as an effective rate constant, denoted as k_{eff}. With this substitution, integration of Equation (25.33) yields

$$-\int_{[A]_0}^{[A]} \frac{d[A]}{[A]^2} = \int_0^t k_{eff}\, dt$$

$$\frac{1}{[A]} - \frac{1}{[A]_0} = k_{eff}\, t \qquad (25.34)$$

$$\frac{1}{[A]} = \frac{1}{[A]_0} + k_{eff}\, t$$

Equation (25.34) demonstrates that for a second-order reaction, a plot of the inverse of reactant concentration versus time will result in a straight line having a slope of k_{eff} and y intercept of $1/[A]_0$. Figure 25.5 presents a comparison between [A] versus time for a second-order reaction and $1/[A]$ versus time. The linear behavior predicted by Equation (25.34) is evident.

25.5.4 Half-Life and Reactions of Second Order (Type I)

Recall that the definition of half-life is when the concentration of a reactant is half of its initial value. With this definition, the half-life for a type I second-order reaction is

$$t_{1/2} = \frac{1}{k_{eff}[A]_0} \qquad (25.35)$$

In contrast to first-order reactions, the half-life for a second-order reaction is dependent on the initial concentration of reactant, with an increase in initial concentration resulting in a decrease in $t_{1/2}$. This behavior is consistent with a first-order reaction occurring through a unimolecular process, whereas the second-order reaction involves a bimolecular process in which the concentration dependence of the reaction rate is anticipated.

25.5.5 Second-Order Reaction (Type II)

Second-order reactions of type II involves two different reactants, A and B, as follows:

$$A + B \xrightarrow{k} P \qquad (25.36)$$

Assuming that the reaction is first order in both A and B, the reaction rate is

$$R = k[A][B] \qquad (25.37)$$

In addition, the rate with respect to the time derivative of the reactant concentrations is

$$R = -\frac{d[A]}{dt} = -\frac{d[B]}{dt} \qquad (25.38)$$

Notice that the loss rate for the reactants is equal such that

$$[A]_0 - [A] = [B]_0 - [B]$$

$$[B]_0 - [A]_0 + [A] = [B] \qquad (25.39)$$

$$\Delta + [A] = [B]$$

Equation (25.39) provides a definition for [B] in terms of [A] and the difference in initial concentration, $[B]_0 - [A]_0$, denoted as Δ. With this definition, the integrated rate law expression can be solved as follows. First, setting Equations (25.37) and (25.38) equal, the following expression is obtained:

$$\frac{d[A]}{dt} = -k[A][B] = -k[A](\Delta + [A])$$

$$\int_{[A]_0}^{[A]} \frac{d[A]}{[A](\Delta + [A])} = -\int_0^t k\, dt \tag{25.40}$$

Next, solution to the integral involving [A] is given by

$$\int \frac{dx}{x(c + x)} = -\frac{1}{c} \ln\left(\frac{c + x}{x}\right)$$

Using this solution to the integral, the integrated rate law expression becomes

$$-\frac{1}{\Delta} \ln\left(\frac{\Delta + [A]}{[A]}\right)\Big|_{[A]_0}^{[A]} = -kt$$

$$\frac{1}{\Delta}\left[\ln\left(\frac{\Delta + [A]}{[A]}\right) - \ln\left(\frac{\Delta + [A]_0}{[A]_0}\right)\right] = kt$$

$$\frac{1}{\Delta}\left[\ln\left(\frac{[B]}{[A]}\right) - \ln\left(\frac{[B]_0}{[A]_0}\right)\right] = kt$$

$$\frac{1}{[B]_0 - [A]_0} \ln\left(\frac{[B]/[B]_0}{[A]/[A]_0}\right) = kt \tag{25.41}$$

Equation (25.41) is not applicable in the case for which the initial concentrations are equivalent, that is, when $[B]_0 = [A]_0$. For this specific case, the concentrations of [A] and [B] reduce to the expression for a second-order reaction of type I with $k_{eff} = k$. The time evolution in reactant concentrations depends on the amount of each reactant present. Finally, the concept of half-life does not apply to second-order reactions of type II. Unless the reactants are mixed in stoichiometric proportions (1:1 for the case discussed in this section), the concentrations of both species will not be 1/2 their initial concentrations at the identical time.

25.6 Supplemental: Numerical Approaches

For the simple reactions outlined in the preceding section, an integrated rate law expression can be readily determined. However, there is a wide variety of kinetic problems for which an integrated rate law expression cannot be obtained. How can one compare a kinetic model with experiment in the absence of an integrated rate law? In such cases, numerical methods provide another approach by which to determine the time evolution in concentrations predicted by a kinetic model. To illustrate this approach, consider the following first-order reaction:

$$A \xrightarrow{k} P \tag{25.42}$$

The differential rate expression for this reaction is

$$\frac{d[A]}{dt} = -k[A] \tag{25.43}$$

The time derivative corresponds to the change in [A] for a time duration that is infinitesimally small. Using this idea, we can state that for a finite time duration, Δt, the change in [A] is given by

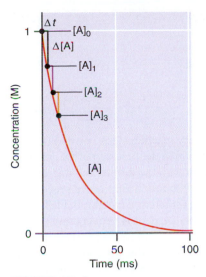

FIGURE 25.6
Schematic representation of the numerical evaluation of a rate law.

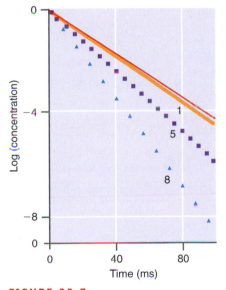

FIGURE 25.7
Comparison of the numerical approximation method to the integrated rate law expression for a first-order reaction. The rate constant for the reaction is 0.1 m s^{-1}. The time evolution in reactant concentration determined by the integrated rate law expression of Equation (25.27a) is shown as the solid red line. Comparison to three numerical approximations is given, and the size of the time step (in milliseconds) employed for each approximation is indicated. Notice the improvement in the numerical approximation as the time step is decreased.

$$\frac{\Delta[A]}{\Delta t} = -k[A]$$

$$\Delta[A] = -\Delta t(k[A]) \qquad (25.44)$$

In Equation (25.44), [A] is the concentration of [A] at a specific time. Therefore, we can use this equation to determine the change in the concentration of A, or $\Delta[A]$, over a time period Δt and then use this concentration change to determine the concentration at the end of the time period. This new concentration can be used to determine the subsequent change in [A] over the next time period, and this process is continued until the reaction is complete. Mathematically,

$$[A]_{t+\Delta t} = [A]_t + \Delta[A]$$

$$= [A]_t + \Delta t(-k[A]_t)$$

$$= [A]_t - k\Delta t[A]_t \qquad (25.45)$$

In Equation (25.45), $[A]_t$ is the concentration at the beginning of the time interval, and $[A]_{t+\Delta t}$ is the concentration at the end of the time interval. This process is illustrated in Figure 25.6. In the figure, the initial concentration is used to determine $\Delta[A]$ over the time interval Δt. The concentration at this next time point, $[A]_1$, is used to determine $\Delta[A]$ over the next time interval, resulting in concentration $[A]_2$. This process is continued until the entire concentration profile is evaluated.

The specific example discussed here is representative of the general approach to numerically integrating differential equations, known as **Euler's method.** Application of Euler's method requires some knowledge of the timescale of interest, and then selection of a time interval, Δt, that is sufficiently small to capture the evolution in concentration. Figure 25.7 presents a comparison of the reactant concentration determined using the integrated rate law expression for a first-order reaction to that determined numerically for three different choices for Δt. The figure illustrates that the accuracy of this method is highly dependent on an appropriate choice for Δt. In practice, convergence of the numerical model is demonstrated by reducing Δt and observing that the predicted evolution in concentrations does not change.

The numerical method can be applied to any kinetic process for which differential rate expressions can be prescribed. Euler's method provides the most straightforward way by which to predict how reactant and product concentrations will vary for a specific kinetic scheme. However, this method uses "brute force" in that a sufficiently small time step must be chosen to accurately capture the slope of the concentration, and the time steps may be quite small, requiring a large number of iterations in order to reproduce the full time course of the reaction. As such, Euler's method can be computationally demanding. More elegant approaches, such as the Runge–Kutta method, exist that allow for larger time steps to be performed in numerical evaluations, and the interested reader is encouraged to investigate these approaches.

25.7 Sequential First-Order Reactions

Many chemical reactions occur in a series of steps in which reactants are transformed into products through multiple sequential elementary reaction steps. For example, consider the following **sequential reaction** scheme:

$$A \xrightarrow{k_A} I \xrightarrow{k_I} P \qquad (25.46)$$

In this scheme, the reactant A decays to form intermediate I, and this **intermediate** species undergoes subsequent decay resulting in the formation of product P. The sequential reaction scheme illustrated in Equation (25.46) involves a series of elementary first-order reactions. Recognizing this, the differential rate expressions for each species can be written as follows:

$$\frac{d[A]}{dt} = -k_A[A] \qquad (25.47)$$

$$\frac{d[I]}{dt} = k_A[A] - k_I[I] \qquad (25.48)$$

$$\frac{d[P]}{dt} = k_I[I] \qquad (25.49)$$

These expressions follow naturally from the elementary reaction steps in which a given species participates. For example, the decay of A occurs in the first step of the reaction. The decay is a standard first-order process, consistent with the differential rate expression in Equation (25.47). The formation of product P is also a first-order process per Equation (25.49). The expression of Equation (25.48) for intermediate I reflects the fact that I is involved in both elementary reaction steps, the decay of A ($k_A[A]$), and the formation of P ($-k_I[I]$). Correspondingly, the differential rate expression for [I] is the sum of the rates associated with these two reaction steps. To determine the concentrations of each species as a function of time, we begin with Equation (25.47), which can be readily integrated given a set of initial concentrations. Let only the reactant A be present at $t = 0$ such that

$$[A]_0 \neq 0 \quad [I]_0 = 0 \quad [P]_0 = 0 \qquad (25.50)$$

With these initial conditions, the expression for [A] is exactly that derived previously:

$$[A] = [A]_0 e^{-k_A t} \qquad (25.51)$$

The expression for [A] given by Equation (25.51) can be substituted into the differential rate expression for I resulting in

$$\frac{d[I]}{dt} = k_A[A] - k_I[I]$$

$$= k_A[A]_0 e^{-k_A t} - k_I[I] \qquad (25.52)$$

Equation (25.52) is a differential equation that when solved yields the following expression for [I]:

$$[I] = \frac{k_A}{k_I - k_A}(e^{-k_A t} - e^{-k_I t})[A]_0 \qquad (25.53)$$

Finally, the expression for [P] is readily determined using the initial conditions of the reaction, with the initial concentration of A, $[A]_0$, equal to the sum of all concentrations for $t > 0$:

$$[A]_0 = [A] + [I] + [P]$$

$$[P] = [A]_0 - [A] - [I] \qquad (25.54)$$

Substituting Equations (25.51) and (25.53) into Equation (25.54) results in the following expression for [P]:

$$[P] = \left(\frac{k_A e^{-k_I t} - k_I e^{-k_A t}}{k_I - k_A} + 1\right)[A]_0 \qquad (25.55)$$

Although the expressions for [I] and [P] look complicated, the temporal evolution in concentration predicted by these equations is intuitive as shown in Figure 25.8. Figure 25.8a presents the evolution in concentration when $k_A = 2k_I$. Notice that A undergoes exponential decay resulting in the production of I. The intermediate in turn undergoes subsequent decay to form the product. The temporal evolution of [I] is extremely

FIGURE 25.8

Concentration profiles for a sequential reaction in which the reactant (A, blue line) forms an intermediate (I, yellow line) that undergoes subsequent decay to form the product (P, red line) where (a) $k_A = 2k_I$ $= 0.1$ s^{-1} and (b) $k_A = 8k_I = 0.4$ s^{-1}. Notice that both the maximal amount of I in addition to the time for the maximum is changed relative to the first panel. (c) $k_A = 0.025k_I = 0.0125$ s^{-1}. In this case, very little intermediate is formed, and the maximum in [I] is delayed relative to the first two examples.

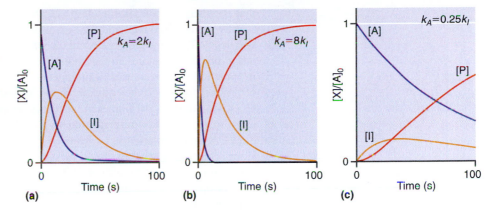

(a)   (b)   (c)

25.1 Sequential Kinetics

dependent on the relative rate constants for the production, k_A, and decay, k_I. Figure 25.8b presents the case where $k_A \gg k_I$. Here, the maximum intermediate concentration is greater than in the first case. The opposite limit is illustrated in Figure 25.8c, where $k_A < k_I$ and the maximum in intermediate concentration is significantly reduced. This behavior is consistent with intuition: if the intermediate undergoes decay at a faster rate than the rate at which it is being formed, then the intermediate concentration will be small. Of course, the opposite logic holds as evidenced by the $k_A \gg k_I$ example presented in the Figure 25.8b.

25.7.1 Maximum Intermediate Concentration

Inspection of Figure 25.8 demonstrates that the time at which the concentration of the intermediate species will be at a maximum depends on the rate constants for its production and decay. Can we predict when [I] will be at a maximum? The maximum intermediate concentration has been reached when the derivative of [I] with respect to time is equal to zero:

$$\left(\frac{d[I]}{dt}\right)_{t=t_{max}} = 0 \tag{25.56}$$

Using the expression for [I] given in Equation (25.53) in the preceding equation, the time at which [I] is at a maximum, t_{max}, is

$$t_{max} = \frac{1}{k_A - k_I}\ln\left(\frac{k_A}{k_I}\right) \tag{25.57}$$

EXAMPLE PROBLEM 25.5

Determine the time at which [I] is at a maximum for $k_A = 2k_I = 0.1$ s^{-1}.

Solution

This is the first example illustrated in Figure 25.8 where $k_A = 0.1$ s^{-1} and $k_I = 0.05$ s^{-1}. Using these rate constants and Equation (25.57), t_{max} is determined as follows:

$$t_{max} = \frac{1}{k_A - k_I}\ln\left(\frac{k_A}{k_I}\right) = \frac{1}{0.1\ \text{s}^{-1} - 0.05\ \text{s}^{-1}}\ln\left(\frac{0.1\ \text{s}^{-1}}{0.05\ \text{s}^{-1}}\right) = 13.9\ \text{s}$$

25.7.2 Rate-Determining Steps

In the preceding subsection, the rate of product formation in a sequential reaction was found to depend on the timescale for production and decay of the intermediate species. Two limiting situations can be envisioned at this point. The first limit is where the rate constant for intermediate decay is much greater than the rate constant for production,

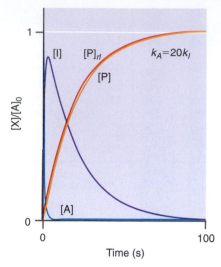

(a)

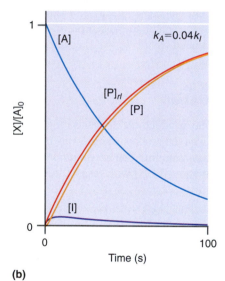

(b)

FIGURE 25.9
Rate-limiting step behavior in sequential reactions. (a) $k_A = 20k_I = 1 \text{ s}^{-1}$ such that the rate-limiting step is the decay of intermediate I. In this case, the reduction in [I] is reflected by the appearance of [P]. The time evolution of [P] predicted by the sequential mechanism is given by the yellow line, and the corresponding evolution assuming rate-limiting step behavior, $[P]_{rl}$, is given by the red curve. (b) The opposite case from part (a) in which $k_A = 0.04k_I = 0.02 \text{ s}^{-1}$ such that the rate-limiting step is the decay of reactant A.

that is, where $k_I \gg k_A$ in Equation (25.46). In this limit, any intermediate formed will rapidly go on to product, and the rate of product formation depends on the rate of reactant decay. The opposite limit occurs when the rate constant for intermediate production is significantly greater than the intermediate decay rate constant, that is, where $k_A \gg k_I$ in Equation (25.46). In this limit, reactants quickly produce intermediates, but the rate of product formation depends on the rate of intermediate decay. These two limits give rise to one of the most important approximations made in the analysis of kinetic problems, that of the **rate-determining step** or the *rate-limiting step*. The central idea behind this approximation is as follows: if one step in the sequential reaction is much slower than any other step, this slow step will control the rate of product formation and is therefore the rate-determining step.

Consider the sequential reaction illustrated in Equation (25.46) when $k_A \gg k_I$. In this limit, the kinetic step corresponding to the decay of intermediate I is the rate-limiting step. Because $k_A \gg k_I$, $e^{-k_A t} \ll e^{-k_I t}$ and the expression for [P] of Equation (25.55) becomes

$$\lim_{k_A \to \infty} [P] = \lim_{k_A \to \infty} \left(\left(\frac{k_A e^{-k_I t} - k_I e^{-k_A t}}{k_I - k_A} + 1 \right) [A]_0 \right) = (1 - e^{-k_I t})[A]_0 \quad (25.58)$$

The time dependence of [P] when k_I is the rate-limiting step is identical to that predicted for first-order decay of I resulting in product formation. The other limit occurs when $k_I \gg k_A$, where $e^{-k_I t} \ll e^{-k_A t}$, and the expression for [P] becomes

$$\lim_{k_I \to \infty} [P] = \lim_{k_I \to \infty} \left(\left(\frac{k_A e^{-k_I t} - k_I e^{-k_A t}}{k_I - k_A} + 1 \right) [A]_0 \right) = (1 - e^{-k_A t})[A]_0 \quad (25.59)$$

In this limit, the time dependence of [P] is identical to that predicted for the first-order decay of the reactant A, resulting in product formation.

When is the rate-determining step approximation appropriate? For the two-step reaction under consideration, 20-fold differences between rate constants are sufficient to ensure that the smaller rate constant will be rate determining. Figure 25.9 presents a comparison for [P] determined using the exact result from Equation (25.55) and the rate-limited prediction of Equations (25.58) and (25.59), for the case where $k_A = 20k_I = 1 \text{ s}^{-1}$ and where $k_A = 0.04k_I = 0.02 \text{ s}^{-1}$. In Figure 25.9a, decay of the intermediate is the rate-limiting step in product formation. Notice the rapid reactant decay, resulting in an appreciable intermediate concentration, with the subsequent decay of the intermediate reflected by a corresponding increase in [P]. The similarity of the exact and rate-limiting curves for [P] demonstrates the validity of the rate-limiting approximation for this ratio of rate constants. The opposite limit is presented in Figure 25.9b. In this case, decay of the reactant is the rate-limiting step in product formation. When reactant decay is the rate-limiting step, very little intermediate is produced. In this case, the loss of [A] is mirrored by an increase in [P]. Again, the agreement between the exact and rate-limiting descriptions of [P] demonstrates the validity of the rate-limiting approximation when a substantial difference in rate constants for intermediate production and decay exists.

25.7.3 The Steady-State Approximation

Consider the following sequential reaction scheme:

$$A \xrightarrow{k_A} I_1 \xrightarrow{k_1} I_2 \xrightarrow{k_2} P \quad (25.60)$$

In this reaction, product formation results from the formation and decay of two intermediate species, I_1 and I_2. The differential rate expressions for this scheme are as follows:

$$\frac{d[A]}{dt} = -k_A[A] \quad (25.61)$$

$$\frac{d[I_1]}{dt} = k_A[A] - k_1[I_1] \quad (25.62)$$

$$\frac{d[I_2]}{dt} = k_1[I_1] - k_2[I_2] \quad (25.63)$$

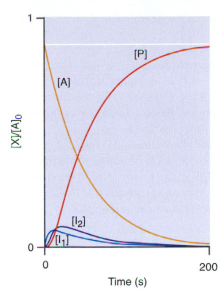

FIGURE 25.10

Concentrations determined by numerical evaluation of the sequential reaction scheme presented in Equation (25.46) where $k_A = 0.02 \text{ s}^{-1}$ and $k_1 = k_2 = 0.2 \text{ s}^{-1}$.

$$\frac{d[P]}{dt} = k_2[I_2] \tag{25.64}$$

A determination of the time-dependent concentrations for the species involved in this reaction by integration of the differential rate expressions is not trivial; therefore, how can the concentrations be determined? One approach is to use Euler's method to determine numerically the concentrations as a function of time. The result of this approach for $k_A = 0.02 \text{ s}^{-1}$ and $k_1 = k_2 = 0.2 \text{ s}^{-1}$ is presented in Figure 25.10. Notice that the relative magnitude of the rate constants results in only modest intermediate concentrations.

Inspection of Figure 25.10 illustrates that in addition to the modest intermediate concentrations, $[I_1]$ and $[I_2]$ change very little with time such that the time derivative of these concentrations can be set approximately equal to zero:

$$\frac{d[I]}{dt} = 0 \tag{25.65}$$

Equation (25.65) is known as the **steady-state approximation.** This approximation is used to evaluate the differential rate expressions by simply setting the time derivative of all intermediates to zero. This approximation is particularly good when the decay rate of the intermediate is greater than the rate of production so that the intermediates are present at very small concentrations during the reaction (as in the case illustrated in Figure 25.10). Applying the steady-state approximation to I_1 in our example reaction results in the following expression for $[I_1]$:

$$\frac{d[I_1]_{ss}}{dt} = 0 = k_A[A] - k_1[I_1]_{ss}$$

$$[I_1]_{ss} = \frac{k_A}{k_1}[A] = \frac{k_A}{k_1}[A]_0 e^{-k_A t} \tag{25.66}$$

where the subscript ss indicates that the concentration is that predicted using the steady-state approximation. The final equality in Equation (25.66) results from integration of the differential rate expression for $[A]$ with the initial conditions that $[A]_0 \neq 0$ and all other initial concentrations are zero. The corresponding expression for $[I_2]$ under the steady-state approximation is

$$\frac{d[I_2]_{ss}}{dt} = 0 = k_1[I_1]_{ss} - k_2[I_2]_{ss}$$

$$[I_2]_{ss} = \frac{k_1}{k_2}[I_1]_{ss} = \frac{k_A}{k_2}[A]_0 e^{-k_A t} \tag{25.67}$$

Finally, the differential expression for P is

$$\frac{d[P]_{ss}}{dt} = k_2[I_2] - k_A[A]_0 e^{-k_A t} \tag{25.68}$$

Integration of Equation (25.68) results in the now familiar expression for $[P]$:

$$[P]_{ss} = [A]_0(1 - e^{-k_A t}) \tag{25.69}$$

Equation (25.69) demonstrates that within the steady-state approximation, $[P]$ is predicted to demonstrate appearance kinetics consistent with the first-order decay of A.

When is the steady-state approximation valid? The approximation requires that the concentration of intermediate be constant as a function of time. Consider the concentration of the first intermediate under the steady-state approximation. The time derivative of $[I_1]_{ss}$ is

$$\frac{d[I_1]_{ss}}{dt} = \frac{d}{dt}\left(\frac{k_A}{k_1}[A]_0 e^{-k_A t}\right) = -\frac{k_A^2}{k_1}[A]_0 e^{-k_A t} \tag{25.70}$$

The steady-state approximation is valid when Equation (25.70) is equal to zero, which is true when $k_1 \gg k_A^2[A]_0$. In other words, k_1 must be sufficiently large such that $[I_1]$ is

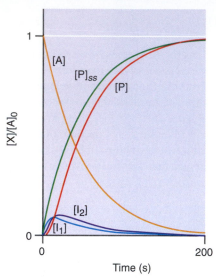

FIGURE 25.11

Comparison of the numerical and steady-state concentration profiles for the sequential reaction scheme presented in Equation (25.46) where $k_A = 0.02$ s^{-1} and $k_1 = k_2 = 0.2$ s^{-1}. Curves corresponding to the steady-state approximation are indicated by the subscript ss.

small at all times. Similar logic applies to I_2 for which the steady-state approximation is valid when $k_2 \gg k_A^2[A]_0$.

Figure 25.11 presents a comparison between the numerically determined concentrations and those predicted using the steady-state approximation for the two-intermediate sequential reaction where $k_A = 0.02$ s^{-1} and $k_1 = k_2 = 0.2$ s^{-1}. Notice that even for these conditions where the steady-state approximation is expected to be valid, the discrepancy between [P] determined by numerical evaluation versus the steady-state approximation value, $[P]_{ss}$, is evident.

For the examples presented here, the steady-state approximation is relatively easy to implement; however, for many reactions the approximation of constant intermediate concentration with time is not appropriate. In addition, the steady-state approximation is difficult to implement if the intermediate concentrations are not isolated to one or two of the differential rate expressions derived from the mechanism of interest.

EXAMPLE PROBLEM 25.6

Consider the following sequential reaction scheme:

$$A \xrightarrow{k_A} I \xrightarrow{k_I} P$$

Assuming that only reactant A is present at $t = 0$, what is the expected time dependence of [P] using the steady-state approximation?

Solution

The differential rate expressions for this reaction were provided in Equations (25.47), (25.48), and (25.49):

$$\frac{d[A]}{dt} = -k_A[A]$$

$$\frac{d[I]}{dt} = k_A[A] - k_I[I]$$

$$\frac{d[P]}{dt} = k_I[I]$$

Applying the steady-state approximation to the differential rate expression for I and substituting in the integrated expression for [A] of Equation (25.51) yield

$$\frac{d[I]}{dt} = 0 = k_A[A] - k_I[I]$$

$$\frac{k_A}{k_I}[A] = \frac{k_A}{k_I}[A]_0 e^{-k_A t} = [I]$$

Substituting the preceding expression for [I] into the differential rate expression for the product and integrating yield

$$\frac{d[P]}{dt} = k_I[I] = \frac{k_A}{k_I}(k_I[A]_0 e^{-k_A t})$$

$$\int_0^{[P]} d[P] = k_A[A]_0 \int_0^t e^{-k_A t}$$

$$[P] = k_A[A]_0 \left[\frac{1}{k_A}(1 - e^{-k_A t}) \right]$$

$$[P] = [A]_0(1 - e^{-k_A t})$$

This expression for [P] is identical to that derived in the limit that the decay of A is the rate-limiting step in the sequential reaction [Equation (25.59)].

25.8 Parallel Reactions

In the reactions discussed thus far, reactant decay results in the production of only a single species. However, in many instances a single reactant can become a variety of products. Such reactions are referred to as **parallel reactions**. Consider the following reaction in which the reactant A can form one of two products, B or C:

$$\begin{array}{c} \qquad\qquad B \\ k_B \nearrow \\ A \\ k_C \searrow \\ \qquad\qquad C \end{array} \qquad (25.71)$$

The differential rate expressions for the reactant and products are

$$\frac{d[A]}{dt} = -k_B[A] - k_C[A] = -(k_B + k_C)[A] \qquad (25.72)$$

$$\frac{d[B]}{dt} = k_B[A] \qquad (25.73)$$

$$\frac{d[C]}{dt} = k_C[A] \qquad (25.74)$$

Integration of the preceding expression involving [A] with the initial conditions $[A]_0 \neq 0$ and $[B] = [C] = 0$ yields

$$[A] = [A]_0 e^{-(k_B + k_C)t} \qquad (25.75)$$

The product concentrations can be determined by substituting the expression for [A] into the differential rate expressions and integrating, which results in

$$[B] = \frac{k_B}{k_B + k_C}[A]_0(1 - e^{-(k_B + k_C)t}) \qquad (25.76)$$

$$[C] = \frac{k_C}{k_B + k_C}[A]_0(1 - e^{-(k_B + k_C)t}) \qquad (25.77)$$

Figure 25.12 provides an illustration of the reactant and product concentrations for this branching reaction where $k_B = 2k_C = 0.1 \text{ s}^{-1}$. A few general trends demonstrated by branching reactions are evident in the figure. First, notice that the decay of A occurs with an apparent rate constant equal to $k_B + k_C$, the sum of rate constants for each reaction branch. Second, the ratio of product concentrations is independent of time. That is, at any time point the ratio [B]/[C] is identical. This behavior is consistent with Equations (25.76) and (25.77) where this ratio of product concentrations is predicted to be

$$\frac{[B]}{[C]} = \frac{k_B}{k_C} \qquad (25.78)$$

Equation (25.78) is a very interesting result. The equation states that as the rate constant for one of the reaction branches increases relative to the other, the greater the final concentration of the corresponding product will be. Furthermore, there is no time dependence in Equation (25.78); therefore, the product ratio remains constant with time.

Equation (25.78) demonstrates that the extent of product formation in a parallel reaction is dependent on the rate constants. Another way to view this behavior is with respect to probability; the larger the rate constant for a given process, the more likely that product will be formed. The **yield, Φ,** is defined as the probability that a given product will be formed by decay of the reactant:

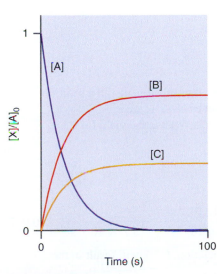

FIGURE 25.12

Concentrations for a parallel reaction where $k_B = 2k_C = 0.1 \text{ s}^{-1}$.

$$\Phi_i = \frac{k_i}{\displaystyle\sum_n k_n} \tag{25.79}$$

In Equation (25.79), k_i is the rate constant for the reaction leading to formation of the product of interest indicated by the subscript i. The denominator is the sum over all rate constants for the reaction branches. The total yield is the sum of the yields for forming each product, and it is normalized such that

$$\sum_i \Phi_i = 1 \tag{25.80}$$

In the example reaction depicted in Figure 25.12 where $k_B = 2k_C$, the yield for the formation of product C is

$$\Phi_C = \frac{k_C}{k_B + k_C} = \frac{k_C}{(2k_C) + k_C} = \frac{1}{3} \tag{25.81}$$

Because there are only two branches in this reaction, $\Phi_B = 2/3$. Inspection of Figure 25.12 reveals that $[B] = 2[C]$, which is consistent with the calculated yields.

the Chemistry place

25.2 Parallel Reactions

EXAMPLE PROBLEM 25.7

In acidic conditions, benzyl penicillin (BP) undergoes the following parallel reaction:

In the molecular structures, R_1 and R_2 indicate alkyl substituents. In a solution where pH = 3, the rate constants for the processes at 22°C are $k_1 = 7.0 \times 10^{-4}$ s^{-1}, $k_2 = 4.1 \times 10^{-3}$ s^{-1}, and $k_3 = 5.7 \times 10^{-3}$ s^{-1}. What is the yield for P_1 formation?

Solution

Using Equation (25.79),

$$\Phi_{P_1} = \frac{k_1}{k_1 + k_2 + k_3} = \frac{7.0 \times 10^{-4}\,s^{-1}}{7.0 \times 10^{-4}\,s^{-1} + 4.1 \times 10^{-3}\,s^{-1} + 5.7 \times 10^{-3}\,s^{-1}} = 0.067$$

Of the BP that undergoes acid-catalyzed dissociation, 6.7% will result in the formation of P_1.

25.9 Temperature Dependence of Rate Constants

As mentioned at the beginning of this chapter, rate constants k are generally temperature-dependent quantities. Experimentally, it is observed that for many reactions a plot of $\ln(k)$ versus T^{-1} demonstrates linear or close to linear behavior. The following empirical relationship between temperature and k, first proposed by Arrhenius in the late 1800s, is known as the **Arrhenius expression:**

$$k = Ae^{-E_a/RT} \tag{25.82}$$

In Equation (25.82), the constant A is referred to as the **frequency factor** or **Arrhenius preexponential factor,** and E_a is the **activation energy** for the reaction. The units of the preexponential factor are identical to those of the rate constant and will vary depending on the order of the reaction. The activation energy is in units of energy mol^{-1} (for example kJ mol^{-1}). The natural log of Equation (25.82) results in the following expression:

$$\ln(k) = \ln(A) - \frac{E_a}{R}\frac{1}{T} \tag{25.83}$$

Equation (25.83) predicts that a plot of $\ln(k)$ versus T^{-1} will yield a straight line with slope equal to $-E_a/R$ and y intercept equal to $\ln(A)$. Example Problem 25.8 provides an example of the application of Equation (25.83) to determine the Arrhenius parameters for a reaction.

EXAMPLE PROBLEM 25.8

The temperature dependence of the acid-catalyzed hydrolysis of penicillin (illustrated in Example Problem 25.7) is investigated, and the dependence of k_1 on temperature is given in the following table. What is the activation energy and Arrhenius preexponential factor for this branch of the hydrolysis reaction?

Temperature (°C)	k_1 (s^{-1})
22.2	7.0×10^{-4}
27.2	9.8×10^{-4}
33.7	1.6×10^{-3}
38.0	2.0×10^{-3}

Solution

A plot of $\ln(k_1)$ versus T^{-1} is shown here:

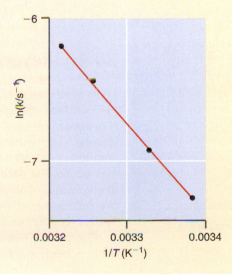

The data are indicated by the points, and the solid line corresponds to the linear least-squares fit to the data. The equation for the line is

$$\ln(k) = (-6306.3\text{ K})\frac{1}{T} + 14.1$$

As shown in Equation (25.83), the slope of the line is equal to $-E_a/R$ such that

$$6306.3\text{ K} = \frac{E_a}{R} \Rightarrow \text{E}_\text{a} = 52{,}400\text{ J mol}^{-1} = 52.4\text{ kJ mol}^{-1}$$

The y intercept is equal to $\ln(A)$ such that

$$A = e^{14.1} = 1.33 \times 10^{6}\text{ s}^{-1}$$

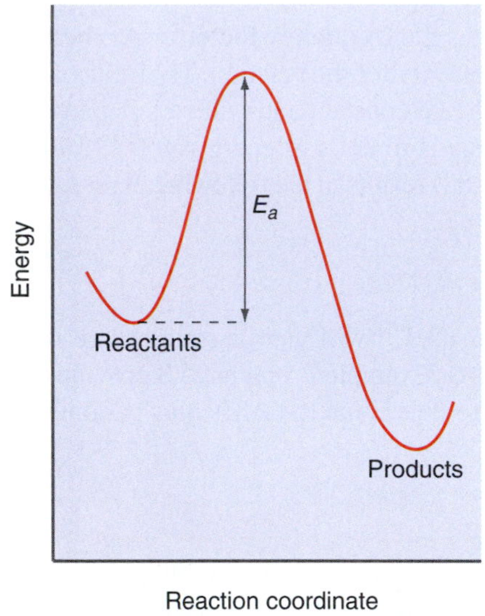

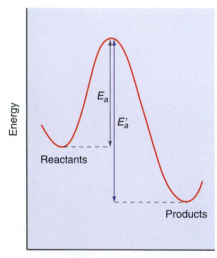

FIGURE 25.13
A schematic drawing of the energy profile for a chemical reaction. Reactants must acquire sufficient energy to overcome the activation energy, E_a, for the reaction. The reaction coordinate represents the bonding and geometry changes that occur in the transformation of reactants into products.

The origin of the energy term in the Arrhenius expression can be understood as follows. The activation energy corresponds to the energy needed for the chemical reaction to occur. Conceptually, we envision a chemical reaction as occurring along an energy profile as illustrated in Figure 25.13. If the reactants have an energy greater than the activation energy, the reaction can proceed. The exponential dependence on the activation energy is consistent with Boltzmann statistics, with $\exp(-E_a/RT)$ representing the fraction of molecules with sufficient kinetic energy to undergo reaction. As the activation energy increases, the fraction of molecules with sufficient energy to react will decrease as will the reaction rate.

Not all chemical reactions demonstrate Arrhenius behavior. Specifically, the inherent assumption in Equation (25.83) is that both E_a and A are temperature-independent quantities. However, there are many reactions for which a plot of $\ln(k)$ versus T^{-1} does not yield a straight line, consistent with the temperature dependence of one or both of the Arrhenius parameters. Modern theories of reaction rates predict that the rate constant will demonstrate the following behavior:

$$k = aT^{m}e^{-E'/RT}$$

where a and E' are temperature-independent quantities, and m can assume values such as 1, 1/2, and $-1/2$ depending on the details of the theory used to predict the rate constant. For example, in the upcoming section on activated complex theory (Section 25.14), a value of $m = 1$ is predicted. With this value for m, a plot of $\ln(k/T)$ versus T^{-1} should yield a straight line with slope equal to $-E'/R$ and the y intercept equal to $\ln(a)$. Although the limitations of the Arrhenius expression are well known, this relationship still provides an adequate description of the temperature dependence of reaction rate constants for a wide variety of reactions.

25.10 Reversible Reactions and Equilibrium

FIGURE 25.14
Reaction coordinate demonstrating the activation energy for reactants to form products, E_a, and the back reaction in which products form reactants, E'_a.

In the kinetic models discussed in earlier sections, it was assumed that once reactants form products, the opposite or "back" reaction does not occur. However, the **reaction coordinate** presented in Figure 25.14 suggests that, depending on the energetics of the reaction, such reactions can indeed occur. Specifically, the figure illustrates that reactants form products if they have sufficient energy to overcome the activation energy for the reaction. But what if the reaction coordinate is viewed from the product's perspective? Can the coordinate be followed in reverse, with products returning to reactants by overcoming the activation energy barrier from the product side, E'_a, of the coordinate? Such **reversible reactions** are discussed in this section.

Consider the following reaction in which the forward reaction is first order in A, and the back reaction is first order in B:

$$\text{A} \underset{k_B}{\overset{k_A}{\rightleftarrows}} \text{B} \qquad (25.84)$$

The forward and back rate constants are k_A and k_B, respectively. Integrated rate law expressions can be obtained for this reaction starting with the differential rate expressions for the reactant and product:

$$\frac{d[A]}{dt} = -k_A[A] + k_B[B] \tag{25.85}$$

$$\frac{d[B]}{dt} = k_A[A] - k_B[B] \tag{25.86}$$

Equation (25.85) should be contrasted with the differential rate expression for first-order reactant decay given in Equation (25.26). Reactant decay is included through the $-k_A[A]$ term similar to first-order decay discussed earlier; however, a second term involving the formation of reactant by product decay, $k_B[B]$, is now included. The initial conditions are identical to those employed in previous sections. Only reactant is present at $t = 0$, and the concentration of reactant and product for $t > 0$ must be equal to the initial concentration of reactant:

$$[A]_0 = [A] + [B] \tag{25.87}$$

With these initial conditions, Equation (25.59) can be integrated as follows:

$$\frac{d[A]}{dt} = -k_A[A] + k_B[B]$$

$$= -k_A[A] + k_B([A]_0 - [A])$$

$$= -[A](k_A + k_B) + k_B[A]_0$$

$$\int_{[A]_0}^{[A]} \frac{d[A]}{[A](k_A + k_B) - k_B[A]_0} = -\int_0^t dt \tag{25.88}$$

Equation (25.88) can be evaluated using the following standard integral:

$$\int \frac{dx}{(a + bx)} = \frac{1}{b} \ln(a + bx)$$

Using this relationship with the initial conditions specified earlier, the concentrations of reactant and products are

$$[A] = [A]_0 \frac{k_B + k_A e^{-(k_A + k_B)t}}{k_A + k_B} \tag{25.89}$$

$$[B] = [A]_0 \left(1 - \frac{k_B + k_A e^{-(k_A + k_B)t}}{k_A + k_B} \right) \tag{25.90}$$

Figure 25.15 presents the time dependence of [A] and [B] for the case where $k_A = 2k_B = 0.06 \text{ s}^{-1}$. Note that [A] undergoes exponential decay with an apparent rate constant equal to $k_A + k_B$, and [B] appears exponentially with an equivalent rate constant. If the back reaction were not present, [A] would be expected to decay to zero; however, the existence of the back reaction results in both [A] and [B] being nonzero at long times. The concentration of reactant and product at long times is defined as the equilibrium concentration. The equilibrium concentrations are equal to the limit of Equations (25.89) and (25.90) as time goes to infinity:

$$[A]_{eq} = \lim_{t \to \infty} [A] = [A]_0 \frac{k_B}{k_A + k_B} \tag{25.91}$$

$$[B]_{eq} = \lim_{t \to \infty} [B] = [A]_0 \left(1 - \frac{k_B}{k_A + k_B} \right) \tag{25.92}$$

Equations (25.91) and (25.92) demonstrate that the reactant and product concentrations reach a constant or equilibrium value that depends on the relative size of the forward and back reaction rate constants.

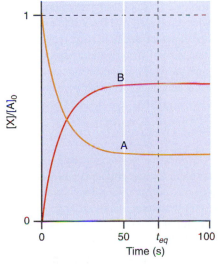

FIGURE 25.15

Time-dependent concentrations in which both forward and back reactions exist between reactant A and product B. In this example, $k_A = 2k_B = 0.06 \text{ s}^{-1}$. Note that the concentrations reach a constant value at longer times ($t \geq t_{eq}$) at which point the reaction reaches equilibrium.

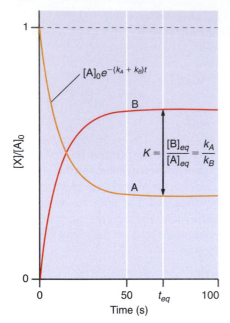

FIGURE 25.16

Methodology for determining forward and back rate constants. The apparent rate constant for reactant decay is equal to the sum of forward, k_A, and back, k_B, rate constants. The equilibrium constant is equal to k_A/k_B. These two measurements provide a system of two equations and two unknowns that can be readily evaluated to produce k_A and k_B.

Theoretically, one must wait an infinite amount of time before equilibrium is reached. In practice, there will be a time after which the reactant and product concentrations are sufficiently close to equilibrium, and the change in these concentrations with time is so modest, that approximating the system as having reached equilibrium is reasonable. This time is indicated by t_{eq} in Figure 25.15, where inspection of the figure demonstrates that the concentrations are near their equilibrium values for times after t_{eq}. After equilibrium has been established, the time independence of the reactant and product concentrations can be expressed as

$$\frac{d[A]_{eq}}{dt} = \frac{d[B]_{eq}}{dt} = 0 \qquad (25.93)$$

The subscripts in Equation (25.93) indicate that equality applies only after equilibrium has been established. A common misconception is that Equation (25.93) states that at equilibrium the forward and back reaction rates are zero. Instead, at equilibrium the forward and back reaction rates are equal, but not zero, such that the macroscopic concentration of reactant or product does not evolve with time. That is, the forward and back reactions still occur, but they occur with equal rates at equilibrium. Using Equation (25.93) in combination with the differential rate expressions for the reactant [Equation (25.85)], we arrive at what is hopefully a familiar relationship:

$$\frac{d[A]_{eq}}{dt} = \frac{d[B]_{eq}}{dt} = 0 = -k_A[A]_{eq} + k_B[B]_{eq}$$

$$\frac{k_A}{k_B} = \frac{[B]_{eq}}{[A]_{eq}} = K_c \qquad (25.94)$$

In this equation, K_c is the equilibrium constant defined in terms of concentration. This quantity is identical to that first encountered in thermodynamics (Chapter 6) and statistical mechanics (Chapter 23). We now have a definition of equilibrium from the kinetic perspective; therefore, Equation (25.94) is a remarkable result in which the concept of equilibrium as described by these three different perspectives is connected into one deceptively simple equation. From the kinetic standpoint, K_c is related to the ratio of forward and back rate constants for the reaction. The greater the forward rate constant relative to that for the back reaction, the more equilibrium will favor products over reactants.

Figure 25.16 illustrates the methodology by which forward and back rate constants can be determined. Specifically, measurement of the reactant decay kinetics (or equivalently the product formation kinetics) provides a measure of the apparent rate constant, $k_A + k_B$. The measurement of K_c, or the reactant and product concentrations at equilibrium, provides a measure of the ratio of the forward and backward rate constants. Together, these measurements represent a system of two equations and two unknowns that can be readily solved to determine k_A and k_B.

EXAMPLE PROBLEM 25.9

Proflavin has been used as a bacterial disinfectant and topical antiseptic; however, its use has been limited because it is also a mutagenic compound arising from its ability to intercalate between nucleic-acid base pairs in DNA. Although the monomeric form of proflavin can intercalate, the dimer cannot. The monomer and dimer forms of proflavin exist in equilibrium in solution.

Proflavin

The monomer/dimer equilibrium of proflavin was studied by Turner et al. [*Nature* 239 (1972), 215], who determined that the dimerization rate constant was $k_1 = 8 \times 10^8$ M^{-1} s^{-1}, and the rate constant for dissociation of the dimer back to monomers was $k_{-1} = 2.0 \times 10^6$ s^{-1}. In addition, the apparent rate constant for the relaxation of this system toward equilibrium is equal to $k_{-1} + 4k_1[M]_{eq}$. If 0.100 mM of monomer is present at equilibrium, what is the equilibrium constant for the dimerization reaction, and how long does it take for equilibrium to be achieved?

Solution

The dimerization reaction can be written as follows:

$$2M \underset{k_{-1}}{\overset{k_1}{\rightleftharpoons}} M_2$$

where $k_1 = 8 \times 10^8$ M^{-1} s^{-1} and $k_2 = 2.0 \times 10^6$ s^{-1}. The differential rate expression for the monomer is

$$-\frac{1}{2}\frac{d[M]}{dt} = -k_1[M]^2 + k_{-1}[M_2]$$

$$\frac{d[M]}{dt} = 2k_1[M]^2 - 2k_{-1}[M_2]$$

At equilibrium, the change in [M] as a function of time is zero; therefore,

$$\frac{d[M]}{dt} = 0 = 2k_1[M]^2 - 2k_{-1}[M_2]$$

$$\frac{[M_2]}{[M]^2} = \frac{k_1}{k_{-1}} = \frac{8 \times 10^8 \text{ M}^{-1}\text{s}^{-1}}{2 \times 10^6 \text{ s}^{-1}} = 400 \text{ M}^{-1} = K$$

The apparent time constant for the relaxation is:

$$k_{app} = k_{-1} + 4k_1[M]_{eq} = 2.0 \times 10^6 \text{ s}^{-1} + 4(8.0 \times 10^8 \text{ M}^{-1}\text{s}^{-1})(1 \times 10^{-4} \text{ M})$$

$$= 2.3 \times 10^6 \text{ s}^{-1}$$

The inverse of k_{app} is the time constant for the relaxation toward equilibrium, or 43 μs for this example. The relaxation can be considered complete after five time constants, or approximately 215 μs.

25.11 Perturbation-Relaxation Methods

The previous section demonstrated that for reactions with appreciable forward and backward rate constants, concentrations approaching those at equilibrium will be established at some later time after initiation of the reaction. The forward and backward rate constants for such reactions can be determined by monitoring the evolution in reactant or product concentrations as equilibrium is approached, and by measuring the concentrations at equilibrium. But what if the initial conditions for the reaction cannot be controlled? For example, what if it is impossible to sequester the reactants such that initiation of the reaction at a specified time is impossible? In such situations, application of the methodology described in the preceding section to determine forward and back rate constants is not possible. However, if one can perturb the system by changing temperature, pressure, or concentration, the system will no longer be at equilibrium and will evolve until a new equilibrium is established. If the perturbation occurs on a timescale that is rapid compared to the system relaxation, the kinetics of the relaxation can be monitored and related to the forward and back rate constants. This is the conceptual idea behind perturbation methods and their application to chemical kinetics.

There are many perturbation techniques; however, the focus here is on **temperature jump** (or T-jump) methods to illustrate the type of information available using perturbation techniques. Consider again the following reaction in which both the forward and back reactions are first order:

$$A \underset{k_B}{\overset{k_A}{\rightleftarrows}} B \tag{25.95}$$

Next, a rapid change in temperature occurs such that the forward and back rate constants are altered in accord with the Arrhenius expression of Equation (25.82), and a new equilibrium is established:

$$A \underset{k_B^+}{\overset{k_A^+}{\rightleftarrows}} B \tag{25.96}$$

The superscript + in this expression indicates that the rate constants correspond to the conditions after the temperature jump. Following the temperature jump, the concentrations of reactants and products will evolve until the new equilibrium concentrations are reached. At the new equilibrium, the differential rate expression for the reactant is equal to zero so that

$$\frac{d[A]_{eq}}{dt} = 0 = -k_A^+[A]_{eq} + k_B^+[B]_{eq}$$

$$k_A^+[A]_{eq} = k_B^+[B]_{eq} \tag{25.97}$$

The subscripts on the reactant and product concentrations represent the new equilibrium concentrations after the temperature jump. The evolution of reactant and product concentrations from the pre-temperature to post-temperature jump values can be expressed using a coefficient of reaction advancement (Section 25.2). Specifically, let the variable ξ represent the extent to which the pre-temperature jump concentration is shifted away from the concentration for the post-temperature jump equilibrium:

$$[A] - \xi = [A]_{eq} \tag{25.98}$$

$$[B] + \xi = [B]_{eq} \tag{25.99}$$

Immediately after the temperature jump, the concentrations will evolve until equilibrium is reached. Using this idea, the differential rate expression describing the extent of reaction advancement is as follows:

$$\frac{d\xi}{dt} = -k_A^+[A] + k_B^+[B]$$

Notice in this equation that the forward and back rate constants are the post-temperature jump values. Substitution of Equations (25.98) and (25.99) into the differential rate expression yields the following:

$$\frac{d\xi}{dt} = -k_A^+(\xi + [A]_{eq}) + k_B^+(-\xi + [B]_{eq})$$

$$= -k_A^+[A]_{eq} + k_B^+[B]_{eq} - \xi(k_A^+ + k_B^+)$$

$$= -\xi(k_A^+ + k_B^+) \tag{25.100}$$

In the final step of the preceding equation, derivation was performed recognizing that at equilibrium the first two terms cancel in accord with Equation (25.97). The relaxation time, τ, is defined as follows:

$$\tau = (k_A^+ + k_B^+)^{-1} \tag{25.101}$$

Employing the relaxation time, Equation (25.100) is readily evaluated:

$$\frac{d\xi}{dt} = -\frac{\xi}{\tau}$$

$$\int_{\xi_0}^{\xi} \frac{d\xi}{\xi} = -\frac{1}{\tau}\int_0^\tau dt$$

$$\xi = \xi_0 e^{-t/\tau} \tag{25.102}$$

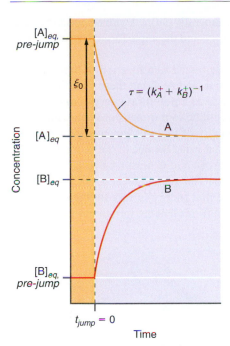

FIGURE 25.17

Example of a temperature-jump experiment for a reaction in which the forward and back rate processes are first order. The yellow and blue portions of the graph indicate times before and after the temperature jump, respectively. After the temperature jump, [A] decreases with a time constant related to the sum of the forward and back rate constants. The change between the pre-jump and post-jump equilibrium concentrations is given by ξ_0.

Equation (25.102) demonstrates that for this reaction, the concentrations will change exponentially, and that the relaxation time is the time it takes for the coefficient of reaction advancement to decay to e^{-1} of its initial value. The timescale for relaxation after the temperature jump is related to the sum of the forward and back rate constants. This information in combination with the equilibrium constant (given by measurement of $[A]_{eq}$ and $[B]_{eq}$) can be used to determine the individual values for the rate constants. Figure 25.17 presents a schematic of this process.

EXAMPLE PROBLEM 25.10

The temperature-induced structural changes in tRNA were reported by Cole and Crothers [*Biochemistry* 11 (1972), 4368]. In this experiment a temperature jump was performed from 25.0° up to 47.0°C and the changes in structure were measured by monitoring the absorbance changes in the ultraviolet. Following the T-jump, a slow relaxation with an apparent time constant of 7.2 ms was observed. With an increase in temperature, the tRNA undergoes relaxation to conformations having reduced secondary structure. Assuming that the equilibrium constant for these two conformations is 50 after the T-jump, determine the forward and back rate constants for the conformational change. Which of these rate constants dominates the observed relaxation?

Solution

Treating the relaxation of tRNA as a simple equilibrium between two species:

$$tRNA_{high-2°} \underset{k_{-1}}{\overset{k_1}{\rightleftarrows}} tRNA_{low-2°}$$

The equilibrium constant and apparent relaxation time constants are:

$$\tau = \frac{1}{k_1 + k_{-1}} = 7.2 \text{ ms}$$

$$K = \frac{k_1}{k_{-1}} = 50$$

Solving the preceding two equations yields the following values for k_{-1} and k_1:

$$k_{-1} = 2.72 \text{ s}^{-1}$$

$$k_1 = 136 \text{ s}^{-1}$$

Comparison of these rate constants demonstrates that the rate constant for structural relaxation dominates the observed relaxation. Temperature-jump experiments with faster time resolutions have shown that the structural relaxation of tRNA is complex, with numerous intermediates and timescales of relaxation [see E. B. Brauns and R. B. Dyer, *Biophys. J.* 89 (2005), 3523].

25.12 Potential Energy Surfaces

In the discussion of the Arrhenius equation, the reaction energetics were identified as an important factor determining the rate of a reaction. This connection between reaction kinetics and energetics is central to the concept of the potential energy surface. To illustrate this concept, consider the following bimolecular reaction:

$$AB + C \longrightarrow A + BC \qquad (25.103)$$

The diatomic species AB and BC are stable, but we will assume that the triatomic species ABC and the diatomic species AC are not formed during the course of the reaction. This reaction can be viewed as the interaction of three atoms, and the potential energy of this collection of atoms can be defined with respect to the relative positions in space. The

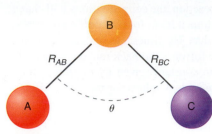

FIGURE 25.18
Definition of geometric coordinates for the AB + C ⟶ A + BC reaction.

geometric relationship between these species is generally defined with respect to the distance between two of the three atoms (R_{AB} and R_{BC}), and the angle formed between these two distances, as illustrated in Figure 25.18.

The potential energy of the system can be expressed as a function of these coordinates. The variation of the potential energy with a change along these coordinates can then be presented as a graph or surface referred to as a **potential energy surface.** Formally, for our example reaction, this surface would be four dimensional (the three geometric coordinates and energy). The dimensionality of the problem can be reduced by considering the energetics of the reaction at a fixed value for one of the geometric coordinates. In the example reaction, the centers of A, B, and C must be aligned during the reaction such that $\theta = 180°$. With this requirement, the potential energy is reduced to a three-dimensional problem, and plots of the potential energy surface for the case where $\theta = 180°$ are presented in Figure 25.19. The graphs represent the variation in energy with

FIGURE 25.19
Illustration of a potential energy surface for the AB + C reaction at a colinear geometry ($\theta = 180°$ in Figure 25.18). (a, b) Three-dimensional views of the surface. (c) Contour plot of the surface with contours of equipotential energy. The curved dashed line represents the path of a reactive event, corresponding to the reaction coordinate. The transition state for this coordinate is indicated by the symbol ‡. (d, e) Cross sections of the potential energy surface along the lines a′–a and b′–b, respectively. These two graphs correspond to the potential for two-body interactions of B with C, and A with B. (Source: Adapted from J.H. Noggle, Physical Chemistry, Harper Collins, New York, 1996).

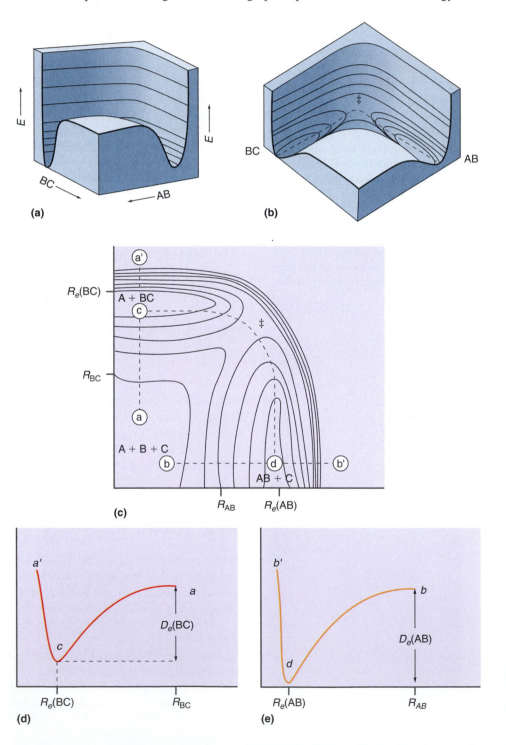

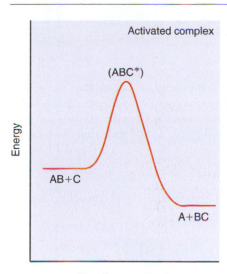

FIGURE 25.20
Reaction coordinates involving an activated complex and a reactive intermediate. The graph corresponds to the reaction coordinate derived from the dashed line between points c and d on the contour plot of Figure 25.19c. The maximum in energy along this coordinate corresponds to the transition state, and the species at this maximum is referred to as an activated complex.

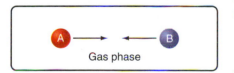

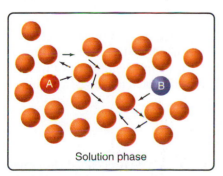

FIGURE 25.21
Top: Reactants A and B approach each other and collide in the gas phase. *Bottom*: In solution, the reactants undergo a series of collisions with the solvent. In this case, the approach of the reactants is dependent on the rate of reactant diffusion in solution.

displacement along R_{AB} and R_{BC} with the arrows indicating the direction of increased separation.

Figures 25.19a and b illustrate the three-dimensional potential energy surface and the two minima in this surface corresponding to the stable diatomic molecules AB and BC. A more convenient way to view the potential energy surface is to use a two-dimensional **contour plot,** as illustrated in Figure 25.19c. One can think of this plot as a view straight down onto the three-dimensional surface presented in Figure 25.19b. The lines on the contour plot connect regions of equal energy. On the lower left-hand region of the surface is a broad energetic plateau that corresponds to the energy when the three atoms are separated, or the dissociated state A + B + C. The pathway corresponding to the reaction of B + C to form BC is indicated by the dashed line between points a and a'. The cross section of the potential energy surface along this line is presented in Figure 25.19d, and this contour is simply the potential energy diagram for the diatomic molecule BC. The depth of the potential is equal to the dissociation energy of the diatomic, $D_e(BC)$, and the minimum along R_{BC} corresponds to the equilibrium bond length of the diatomic. Figure 25.19e presents the corresponding diagram for the diatomic AB, as indicated by the dashed line between points b and b' in Figure 25.19c.

The dashed line between points c and d in Figure 25.19c represents the system energy as C approaches AB and reacts to form BC and A under the constraint that $\theta = 180°$. This pathway represents the AB + C $\longrightarrow$ A + BC reaction. The maximum in energy along this pathway is referred to as the **transition state** and is indicated by the double dagger symbol, ‡. The variation in energy as one proceeds from reactants to products along this reactive pathway can be plotted to construct a reaction coordinate as presented in Figure 25.20. Note that the transition state corresponds to a maximum along this coordinate; therefore, the activated complex is not a stable species (i.e., an intermediate) along the reaction coordinate.

The discussion of potential energy surfaces just presented suggests that the kinetics and product yields will depend on the energy content of the reactants and the relative orientation of reactants. This sensitivity can be explored using techniques of crossed-molecular beams. In this approach, reactants with well-defined energies are seeded into a molecular beam that intersects another beam of reactants at well-defined beam geometries. The products formed in the reaction can be analyzed in terms of their energetics, spatial distribution of the products, and beam geometry. This experimental information is then used to construct a potential energy surface (following a substantial amount of analysis). Crossed-molecular beam techniques have provided much insight into the nature of reactive pathways, and detailed, introductory references to this important area of research are presented at the end of this chapter.

25.13 Diffusion-Controlled Reactions

For bimolecular chemical reactions in solution, the presence of solvent molecules can result in reaction dynamics that differ significantly from those in the gas phase. For example, activation energy and relative orientation were identified as being key factors in defining the magnitude of the rate constant. That is, when two reactant molecules collide, the reaction rate is dependent on the energy contained by the reactants and the collision geometry. Imagine this same reaction occurring in solution as illustrated in Figure 25.21. Since the average kinetic energy of the reactants is $3/2RT$, the average translational velocity is the same in either phase. However, in solution the presence of the solvent molecules results in a number of solvent–solute collisions before the reactants collide. Subsequently, the uninterrupted approach of the reactants characteristic of a gas-phase reaction is replaced by the reactants undergoing diffusion in solution until they encounter each other. In this case, the rate of diffusion can determine the rate of reaction.

The role of diffusion in solution-phase chemistry can be explored using the following mechanism:

$$A + B \xrightarrow{k_d} AB \tag{25.104}$$

$$AB \xrightarrow{k_r} A + B \qquad (25.105)$$

$$AB \xrightarrow{k_p} P \qquad (25.106)$$

In this mechanism, reactants A and B diffuse with rate constant k_d until they make contact and form the intermediate complex AB. Once this complex is formed, dissociation can occur to reform the separate reactants (with rate constant k_r) or the reaction can continue resulting in product formation (with rate constant k_p). The expression for the reaction rate consistent with this mechanism is

$$\text{Rate} = k_p[AB] \qquad (25.107)$$

AB is an intermediate; therefore, the steady-state approximation is employed to express the concentration of this species in terms of the reactants:

$$\frac{d[AB]}{dt} = 0 = k_d[A][B] - k_r[AB] - k_p[AB]$$

$$[AB] = \frac{k_d[A][B]}{k_r + k_p} \qquad (25.108)$$

Using this expression for [AB], the expression for the reaction rate becomes:

$$R = \frac{k_p k_d}{k_r + k_p}[A][B] \qquad (25.109)$$

if the rate constant for product formation is much greater than the decay of the intermediate complex to re-form the reactants ($k_p \gg k_r$), the rate expression becomes:

$$R = k_d[A][B] \qquad (25.110)$$

This is the **diffusion-controlled limit,** in which diffusion for the reactants limits the rate of product formation. Building on the previous discussion of molecular diffusion in solution, the rate constant for diffusion can be related to the viscosity of the solvent as follows:

$$k_d = 4\pi N_A(r_A + r_B)D_{AB} \qquad (25.111)$$

In Equation (25.111), r_A and r_B are the radii of the reactants, and D_{AB} is the mutual diffusion coefficient equal to the sum of diffusion coefficients for the reactants ($D_{AB} = D_A + D_B$). To provide some insight into the magnitude of diffusion-controlled rate constants, consider the protonation of acetate (CH_3COO^-) in aqueous solution. The diffusion coefficient of CH_3COO^- is $1.1 \times 10^{-5} \, cm^2 \, s^{-1}$ and H^+ is $9.3 \times 10^{-5} \, cm^2 \, s^{-1}$, and $(r_A + r_B)$ is on the order of 5 Å. Assuming that this reaction is diffusion controlled, the rate constant for the reaction is $6.5 \times 10^{-11} \, cm^3 \, s^{-1}$ per reactant pair. Using N_A to convert this quantity to a molar value, $k_d = 3.9 \times 10 \, M^{-1} s^{-1}$. Experimentally, the rate constants for acid–base neutralization reactions are greater than the diffusion-controlled limit due to Coulombic attraction of the ions accelerating the collisional rate beyond the diffusion-controlled limit. Another prediction of the diffusion-controlled limit is that the rate of reaction is dependent on the viscosity of solution. Specifically, for a spherical particle in solution, the diffusion coefficient is related to the viscosity by the Stokes–Einstein equation:

$$D = \frac{kT}{6\pi\eta r}$$

Therefore, the reaction rate constant should decrease linearly with an increase in solvent viscosity.

The opposite limit for the reaction occurs when the rate constant for product formation is much smaller than the rate of complex dissociation. In this activation-controlled limit, the expression for the reaction rate [Equation (25.109)] is

$$\text{Rate} = \frac{k_p k_d}{k_r}[A][B] \qquad (25.112)$$

In this limit, the rate of the reaction depends on the energetics of the reaction contained in k_p.

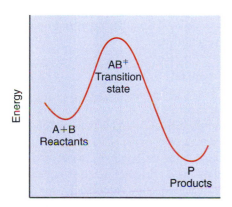

FIGURE 25.22
Illustration of transition state theory. Similar to reaction coordinates depicted previously, the reactants (A and B) and products (P) are separated by an energy barrier. The transition state is an activated reactant complex envisioned to exist at the free-energy maximum along the reaction coordinate.

25.14 Activated Complex Theory

The concept of equilibrium is central to a theoretical description of reaction rates developed principally by Henry Eyring in the 1930s. This theory, known as **activated complex theory** or **transition state theory**, provides a theoretical description of reaction rates. To illustrate the conceptual ideas behind activated complex theory, consider the following bimolecular reaction:

$$A + B \xrightarrow{k} P \tag{25.113}$$

Figure 25.22 illustrates the reaction coordinate for this process, where A and B react to form an activated complex that undergoes decay, resulting in product formation. The **activated complex** represents the system at the transition state. This complex is not stable and has a lifetime on the order of one or a few vibrational periods (~10^{-14} s). When this theory was first proposed, experiments were incapable of following reaction dynamics on such short timescales such that evidence for an activated complex corresponding to the transition state was not available. However, recent developments in experimental kinetics have allowed for the investigation of these transient species, and a few references to this work are provided in the For Further Reading section at the end of this chapter.

Activated complex theory involves a few major assumptions that are important to acknowledge before proceeding. The primary assumption is that an equilibrium exists between the reactants and the activated complex. It is also assumed that the reaction coordinate describing decomposition of the activated complex can be mapped onto a single energetic degree of freedom of the activated complex. For example, if product formation involves the breaking of a bond, then the vibrational degree of freedom corresponding to bond stretching is taken to be the reactive coordinate.

With these approximations in mind, we can take the kinetic methods derived earlier in this chapter and develop an expression for the rate of product formation. For the example bimolecular reaction from Equation (25.113), the kinetic mechanism corresponding to the activated complex model described earlier is

$$A + B \underset{k_{-1}}{\overset{k_1}{\rightleftharpoons}} AB^{\ddagger} \tag{25.114}$$

$$AB^{\ddagger} \xrightarrow{k_2} P \tag{25.115}$$

Equation (25.114) represents the equilibrium between reactants and the activated complex, and Equation (25.115) represents the decay of the activated complex to form product. With the assumption of an equilibrium between the reactants and the activated complex, the differential rate expression for one of the reactants (A in this case) is set equal to zero consistent with equilibrium, and an expression for $[AB^{\ddagger}]$ is obtained as follows:

$$\frac{d[A]}{dt} = 0 = -k_1[A][B] + k_{-1}[AB^{\ddagger}]$$

$$[AB^{\ddagger}] = \frac{k_1}{k_{-1}}[A][B] = \frac{K_c^{\ddagger}}{c^{\circ}}[A][B] \tag{25.116}$$

In Equation (25.116), $K_c^{\ddagger}$ is the equilibrium constant involving the reactants and the activated complex, and it can be expressed in terms of the molecular partition functions of these species as described in Chapter 23. In addition, c° is the standard state concentration (typically 1 M), which appears in the following definition for $K_c^{\ddagger}$:

$$K_c^{\ddagger} = \frac{[AB^{\ddagger}]/c^{\circ}}{([A]/c^{\circ})([B]/c^{\circ})} = \frac{[AB^{\ddagger}]c^{\circ}}{[A][B]}$$

The rate of the reaction is equal to the rate of product formation, which by Equation (25.115) is equal to

$$R = \frac{d[P]}{dt} = k_2[AB^{\ddagger}] \tag{25.117}$$

Substitution into Equation (25.117) of the expression for $[AB^{\ddagger}]$ provided in Equation (25.116) yields the following expression for the reaction rate:

$$R = \frac{d[P]}{dt} = \frac{k_2 K_c^{\ddagger}}{c^{\circ}}[A][B] \tag{25.118}$$

Further evaluation of the reaction rate expression requires that k_2 be defined. This rate constant is associated with the rate of activated complex decay. Imagine that product formation requires the dissociation of a bond in the activated complex. The activated complex is not stable; therefore, the dissociating bond must be relatively weak and the complex can dissociate with the initial motion along the corresponding bond-stretching coordinate. Therefore, k_2 is related to the vibrational frequency associated with bond stretching, ν. The rate constant is equal to ν only if every time an activated complex is formed, it dissociates, resulting in product formation. However, it is possible that the activated complex will instead revert back to reactants. If the reverse reaction can occur, then only a fraction of the activated complexes that are formed will continue along the reaction coordinate and result in product formation. To account for this possibility, a term referred to as the *transmission coefficient, κ,* is included in the definition of k_2:

$$k_2 = \kappa\nu \tag{25.119}$$

With this definition of k_2, the reaction rate becomes

$$R = \frac{\kappa\nu K_c^{\ddagger}}{c^{\circ}}[A][B] \tag{25.120}$$

As stated earlier, one can express $K_c^{\ddagger}$ in terms of the partition function of reactants and the activated complex using the techniques outlined in Chapter 23. In addition, the partition function for the activated complex can decompose into a product of partition functions corresponding to the reactive coordinate and the remaining energetic degrees of freedom. Removing the partition function for the reactive coordinate from the expression of $K_c^{\ddagger}$ yields

$$K_c^{\ddagger} = q_{rc}\overline{K}_c^{\ddagger} = \frac{k_B T}{h\nu}\overline{K}_c^{\ddagger} \tag{25.121}$$

where q_{rc} is the partition function associated with the reactive coordinate, $\overline{K}_c^{\ddagger}$ is the remainder of the original equilibrium constant in the absence of q_{rc}, and k_B is Boltzmann's constant. The final equality in Equation (25.121) is made by recognizing that the reactive vibrational coordinate corresponds to a weak bond for which $h\nu \ll kT$, and the high-temperature approximation for q_{rc} is valid. Substituting Equation (25.121) into Equation (25.120) yields the following expression for k_2:

$$k_2 = \kappa \frac{k_B T}{hc^\circ} \overline{K_c^\ddagger} \qquad (25.122)$$

Equation (25.122) is the central result of activated complex theory, and it provides a connection between the rate constant for product formation and the molecular parameters for species involved in the reaction. Evaluation of this rate expression requires that one determine $\overline{K_c^\ddagger}$, which is related to the partition functions of the activated complex and reactants (Chapter 23). The partition functions for the reactants can be readily determined using the techniques discussed in Chapters 23; however, the partition function for the activated complex requires some thought.

The translational partition function for the complex can also be determined using the techniques described earlier, but determination of the rotational and vibrational partition functions requires some knowledge of the structure of the activated complex. The determination of the vibrational partition function is further complicated by the requirement that one of the vibrational degrees of freedom be designated as the reactive coordinate; however, identification of this coordinate may be far from trivial for an activated complex with more than one weak bond. At times, computational techniques, can be used to provide insight into the structure of the activated complex and assist in determination of the partition function for this species. With these complications acknowledged, Equation (25.122) represents an important theoretical accomplishment in chemical reaction kinetics.

Note that the presentation of activated complex theory provided here is a very rudimentary description of this field. Work continues to the present day to advance and refine this theory, and references are provided at the end of this chapter to review articles describing the significant advances in this field.

We end this discussion by connecting the results of activated complex theory to earlier thermodynamics descriptions of chemical reactions. Recall from thermodynamics that equilibrium constant $K_c^\ddagger$ is related to the corresponding change in Gibbs energy using the following thermodynamic definition:

$$\Delta G^\ddagger = -RT \ln K_c^\ddagger \qquad (25.123)$$

In this definition, $\Delta G^\ddagger$ is the difference in Gibbs energy between the transition state and the reactants. With this definition for $K_c^\ddagger$, k_2 becomes (with $\kappa = 1$ for convenience)

$$k_2 = \frac{k_B T}{hc^\circ} e^{-\Delta G^\ddagger/RT} \qquad (25.124)$$

In addition, $\Delta G^\ddagger$ can be related to the corresponding changes in enthalpy and entropy using

$$\Delta G^\ddagger = \Delta H^\ddagger - T\Delta S^\ddagger \qquad (25.125)$$

Substituting Equation (25.125) into Equation (25.124) yields

$$k_2 = \frac{k_B T}{hc^\circ} e^{\Delta S^\ddagger/R} e^{-\Delta H^\ddagger/RT} \qquad (25.126)$$

Equation (25.126) is known as the **Eyring equation.** Notice that the temperature dependence of the reaction rate constant predicted by transition state theory is different than that assumed by the Arrhenius expression of Equation (25.82). In particular, the preexponential term in the Eyring equation demonstrates temperature dependence as opposed to the assumed temperature independence of the corresponding term in the Arrhenius expression. However, both the Eyring equation and the Arrhenius expression provide an expression for the temperature dependence of rate constants; therefore, one might expect that the parameters in the Eyring equation ($\Delta H^\ddagger$ and $\Delta S^\ddagger$) can be related to corresponding parameters in the Arrhenius expression (E_a and A). To derive this relationship, we begin with a modification of Equation (25.82) where the Arrhenius activation energy is written as

$$E_a = RT^2\left(\frac{d \ln k}{dT}\right) \tag{25.127}$$

Substituting for k the expression for k_2 given in Equation (25.122) yields

$$E_a = RT^2\left(\frac{d}{dT}\ln\left(\frac{kT}{hc^\circ}\overline{K_c^{\ddagger}}\right)\right) = RT + RT^2\left(\frac{d \ln \overline{K_c^{\ddagger}}}{dT}\right)$$

From thermodynamics (Chapter 6), the temperature derivative of $\ln(K_c)$ is equal to $\Delta U/RT^2$. Applying this definition to the previous equation results in the following:

$$E_a = RT + \Delta U^{\ddagger}$$

We also make use of the thermodynamic definition of enthalpy, $H = U + PV$, to write

$$\Delta U^{\ddagger} = \Delta H^{\ddagger} - \Delta(PV)^{\ddagger} \tag{25.128}$$

In Equation (25.128), the $\Delta(PV)^{\ddagger}$ term is related to the difference in the product PV with respect to the activated complex and reactants. For a solution-phase reaction, P is constant and the change in V is negligible such that $\Delta U^{\ddagger} \cong \Delta H^{\ddagger}$ and the activation energy in terms of $\Delta H^{\ddagger}$ becomes

$$E_a = \Delta H^{\ddagger} + RT \text{ (solutions)} \tag{25.129}$$

Comparison of this result with Equation (25.126) demonstrates that the Arrhenius preexponential factor in this case is

$$A = \frac{ek_BT}{hc^\circ} e^{\Delta S^{\ddagger}/R} \text{ (solutions, bimolecular)} \tag{25.130}$$

For solution-phase unimolecular reactions, $\Delta U^{\ddagger} \cong \Delta H^{\ddagger}$, and the activation energy for a unimolecular solution-phase reaction is identical to Equation (25.129). All that changes relative to the bimolecular case is the Arrhenius preexponential factor, resulting in

$$A = \frac{ek_BT}{h} e^{\Delta S^{\ddagger}/R} \text{ (solutions, unimolecular)} \tag{25.131}$$

For a gas-phase reaction, $\Delta(PV)^{\ddagger}$ in Equation (25.128) is proportional to the difference in the number of moles between the transition state and reactants. For a unimolecular ($\Delta n^{\ddagger} = 0$), bimolecular ($\Delta n^{\ddagger} = -1$), and trimolecular ($\Delta n^{\ddagger} = -2$) reaction, E_a and A are given by

$$gas,uni \quad E_a = \Delta H^{\ddagger} + RT \quad A = \frac{ek_BT}{h} e^{\Delta S^{\ddagger}/R} \tag{25.132}$$

$$gas,bi \quad E_a = \Delta H^{\ddagger} + 2RT \quad A = \frac{e^2 k_BT}{hc^\circ} e^{\Delta S^{\ddagger}/R} \tag{25.133}$$

$$gas,tri \quad E_a = \Delta H^{\ddagger} + 3RT \quad A = \frac{e^3 k_BT}{h(c^\circ)^2} e^{\Delta S^{\ddagger}/R} \tag{25.134}$$

Notice now that both the Arrhenius activation energy and preexponential terms are expected to demonstrate temperature dependence. If $\Delta H^{\ddagger} \gg RT$, then the temperature dependence of E_a will be modest. Also notice that if the enthalpy of the transition state is lower than that of the reactants, then the reaction rate may become faster as temperature is decreased! However, the entropy difference between the transition state and reactants is also important in determining the rate. If this entropy difference is positive and the activation energy is near zero, then the reaction rate is determined by entropic rather than enthalpic factors.

EXAMPLE PROBLEM 25.12

The thermal decomposition reaction of nitrosyl halides is important in tropospheric chemistry. For example, consider the decomposition of NOCl:

$$2NOCl(g) \longrightarrow 2NO(g) + Cl_2(g)$$

The Arrhenius parameters for this reaction are $A = 1.00 \times 10^{13} \, M^{-1} \, s^{-1}$ and $E_a = 104 \, kJ \, mol^{-1}$. Calculate $\Delta H^{\ddagger}$ and $\Delta S^{\ddagger}$ for this reaction with $T = 300$ K.

Solution

This is a bimolecular reaction such that

$$\Delta H^{\ddagger} = E_a - 2RT = 104 \, kJ \, mol^{-1} - 2(8.314 \, J \, mol^{-1} \, K^{-1})(300 \, K)$$

$$= 104 \, kJ \, mol^{-1} - (4.99 \times 10^3 \, J \, mol^{-1})\left(\frac{1 \, kJ}{1000 \, J}\right) = 99.00 \, kJ \, mol^{-1}$$

$$\Delta S^{\ddagger} = R \ln\left(\frac{Ahc^{\circ}}{e^2 kT}\right)$$

$$= (8.314 \, J \, mol^{-1} \, K^{-1}) \ln\left(\frac{(1.00 \times 10^{13} \, M^{-1} \, s^{-1})(6.626 \times 10^{-34} \, J \, s)(1 \, M)}{e^2(1.38 \times 10^{-23} \, J \, K^{-1})(300 \, K)}\right)$$

$$= -12.7 \, J \, mol^{-1} \, K^{-1}$$

One of the utilities of this calculation is that the sign and magnitude of $\Delta S^{\ddagger}$ provide information on the structure of the activated complex at the transition state relative to the reactants. The negative value in this example illustrates that the activated complex has a lower entropy (or is more ordered) than the reactants. This observation is consistent with a mechanism in which the two NOCl reactants form a complex that eventually decays to produce NO and Cl.

For Further Reading

Brooks, P. R., "Spectroscopy of Transition Region Species," *Chemical Reviews* 87 (1987), 167.

Callender, R. H., R. B. Dyer, R. Blimanshin, and W. H. Woodruff, "Fast Events in Protein Folding: The Time Evolution of a Primary Process," *Annual Review of Physical Chemistry* 49 (1998), 173.

Castellan, G. W., *Physical Chemistry*. Addison-Wesley, Reading, MA, 1983.

Eyring, H., S. H. Lin, and S. M. Lin, *Basic Chemical Kinetics*. Wiley, New York, 1980.

Frost, A. A., and R. G. Pearson, *Kinetics and Mechanism*. Wiley, New York, 1961.

Hammes, G. G., *Thermodynamics and Kinetics for the Biological Sciences*. Wiley, New York, 2000.

Laidler, K. J., *Chemical Kinetics*. Harper & Row, New York, 1987.

Martin, J.-L., and M. H. Vos, "Femtosecond Biology," *Annual Review of Biophysical and Biomolecular Structure* 21 (1992), 1999.

Pannetier, G., and P. Souchay, *Chemical Kinetics*. Elsevier, Amsterdam, 1967.

Schoenlein, R. W., L. A. Peteanu, R. A. Mathies, and C. V. Shank, "The First Step in Vision: Femtosecond Isomerization of Rhodopsin," *Science* 254 (1991), 412.

Steinfeld, J. I., J. S. Francisco, and W. L. Hase, *Chemical Kinetics and Dynamics*. Prentice-Hall, Upper Saddle River, NJ, 1999.

Truhlar, D. G., W. L. Hase, and J. T. Hynes, "Current Status in Transition State Theory," *Journal of Physical Chemistry* 87 (1983), 2642.

Vos, M. H., F. Rappaport, J.-C. Lambry, J. Breton, and J.-L. Martin, "Visualization of Coherent Nuclear Motion in a Membrane Protein by Femtosecond Spectroscopy," *Nature* 363 (1993), 320.

Zewail, H., "Laser Femtochemistry," *Science* 242 (1988), 1645.

Vocabulary

activated complex
activated complex theory
activation energy
Arrhenius expression

Arrhenius preexponential factor
chemical kinetics
chemical methods

contour plot
diffusion-controlled limit
diffusion-controlled reaction
elementary reaction step

Euler's method
Eyring equation
first-order reaction
flash photolysis techniques

frequency factor	parallel reaction	reaction coordinate	sequential reaction
half-life	perturbation-relaxation	reaction mechanism	steady-state approximation
initial rate	methods	reaction order	stopped-flow techniques
integrated rate law	physical methods	reaction rate	temperature jump
expression	potential energy surface	reversible reaction	transition state
intermediate	rate constant	second-order reaction	transition state theory
isolation method	rate-determining step	(type I)	yield
method of initial rates	rate law	second-order reaction	
molecularity	rate of conversion	(type II)	

Questions on Concepts

Q25.1   Why is the stoichiometry of a reaction generally not sufficient to determine reaction order?

Q25.2   What is an elementary chemical step, and how is one used in kinetics?

Q25.3   What is the difference between chemical and physical methods for studying chemical kinetics?

Q25.4   What is the method of initial rates, and why is it used in chemical kinetics studies?

Q25.5   What is a rate law expression, and how is it determined?

Q25.6   What is the difference between a first-order reaction and a second-order reaction?

Q25.7   What is a half-life? Is the half-life for a first-order reaction dependent on concentration?

Q25.8   In a sequential reaction, what is an intermediate?

Q25.9   What is meant by the rate-determining step in a sequential reaction?

Q25.10   What is the steady-state approximation, and when is this approximation employed?

Q25.11   In a parallel reaction in which two products can be formed from the same reactant, what determines the extent to which one product will be formed over another?

Q25.12   What is the kinetic definition of equilibrium?

Q25.13   In a temperature-jump experiment, why does a change in temperature result in a corresponding change in equilibrium?

Q25.14   What is a transition state? How is the concept of a transition state used in activated complex theory?

Q25.15   What is meant by a diffusion-controlled reaction? What is a typical rate constant for a diffusion-controlled reaction in aqueous solution?

Q25.16   What is the relationship between the parameters in the Arrhenius equation and in the Eyring equation?

Problems

Problem numbers in **RED** indicate that the solution to the problem is given in the *Student Solutions Manual*.

P25.1   Express the rate of reaction with respect to each species in the following reactions:

   a. $2NO(g) + O_2(g) \longrightarrow N_2O_4(g)$

   b. $H_2(g) + I_2(g) \longrightarrow 2HI(g)$

   c. $ClO(g) + BrO(g) \longrightarrow ClO_2(g) + Br(g)$

P25.2   Consider the first-order decomposition of cyclobutane at 438°C at constant volume:

$$C_4H_8(g) \longrightarrow 2C_2H_4(g)$$

   a. Express the rate of the reaction in terms of the change in total pressure as a function of time.

   b. The rate constant for the reaction is $2.48 \times 10^{-4} \text{ s}^{-1}$. What is the half-life?

   c. After initiation of the reaction, how long will it take for the initial pressure of C_4H_8 to drop to 90% of its initial value?

P25.3   As discussed in the text, the total system pressure can be used to monitor the progress of a chemical reaction. Consider the following reaction: $SO_2Cl_2(g) \longrightarrow SO_2(g) +$

$Cl_2(g)$. The reaction is initiated, and the following data are obtained:

Time (h)	0	3	6	9	12	15
P_{Total} (kPa)	11.07	14.79	17.26	18.90	19.99	20.71

   a. Is the reaction first or second order with respect to SO_2Cl_2?

   b. What is the rate constant for this reaction?

P25.4   Consider the following reaction involving bromophenol blue (BPB) and OH^-: $BPB(aq) + OH^-(aq) \longrightarrow BPBOH^-(aq)$. The concentration of BPB can be monitored by following the absorption of this species and using the Beer–Lambert law. In this law, absorption, A, and concentration are linearly related.

   a. Express the reaction rate in terms of the change in absorbance as a function of time.

   b. Let A_o be the absorbance due to BPB at the beginning of the reaction. Assuming that the reaction is first order with respect to both reactants, how is the absorbance of BPB expected to change with time?

c. Given your answer to part (b), what plot would you construct to determine the rate constant for the reaction?

P25.5 For the following rate expressions, state the order of the reaction with respect to each species, the total order of the reaction, and the units of the rate constant, k:

a. Rate = $k[ClO][BrO]$

b. Rate = $k[NO]^2[O_2]$

c. Rate = $k\dfrac{[HI]^2[O_2]}{[H^+]^{1/2}}$

P25.6 What is the overall order of the reaction corresponding to the following rate constants?

a. $k = 1.63 \times 10^{-4} \, M^{-1} \, s^{-1}$

b. $k = 1.63 \times 10^{-4} \, M^{-2} \, s^{-1}$

c. $k = 1.63 \times 10^{-4} \, M^{-1/2} \, s-1$

P25.7 The reaction rate as a function of initial reactant pressures was investigated for the reaction $2NO(g) + 2H_2(g)$ $\longrightarrow N_2(g) + 2H_2O(g)$, and the following data were obtained:

Run	$P_o \, H_2$ (kPa)	$P_o \, NO$ (kPa)	Rate (kPa s^{-1})
1	53.3	40.0	0.137
2	53.3	20.3	0.033
3	38.5	53.3	0.213
4	19.6	53.3	0.105

What is the rate law expression for this reaction?

P25.8 The disaccharide lactose can be decomposed into its constituent sugars galactose and glucose. This decomposition can be accomplished through acid-based hydrolysis, or by the enzyme lactase. Lactose intolerance in humans is due to the lack of lactase production by cells in the small intestine. However, the stomach is an acidic environment; therefore, one might expect lactose hydrolysis to still be an efficient process. The following data were obtained on the rate of lactose decomposition as a function of acid and lactose concentration. Using this information, determine the rate law expression for the acid-based hydrolysis of lactose.

Initial Rate (M^{-1} s^{-1})	[lactose]$_0$ (M^{-1})	[H$^+$] (M^{-1})
0.00116	0.01	0.001
0.00232	0.02	0.001
0.00464	0.01	0.004

P25.9 You are given the following data for the decomposition of acetaldehyde:

Initial Concentration (M)	9.72×10^{-3}	4.56×10^{-3}
Half-Life (s)	328	572

Determine the order of the reaction and the rate constant for the reaction.

P25.10 Consider the schematic reaction $A \xrightarrow{k} P$.

a. If the reaction is one-half order with respect to [A], what is the integrated rate law expression for this reaction?

b. What plot would you construct to determine the rate constant k for the reaction?

c. What would be the half-life for this reaction? Will it depend on initial concentration of the reactant?

P25.11 A certain reaction is first order, and 540 s after initiation of the reaction, 32.5% of the reactant remains.

a. What is the rate constant for this reaction?

b. At what time after initiation of the reaction will 10% of the reactant remain?

P25.12 The half-life of ^{238}U is 4.5×10^9 years. How many disintegrations occur in 1 min for a 10-mg sample of this element?

P25.13 You are performing an experiment using ^3H (half-life $= 4.5 \times 10^3$ days) labeled phenylalanine in which the five aromatic hydrogens are labeled. To perform the experiment, the initial activity cannot be lower than 10% of the initial activity when the sample was received. How long after receiving the sample can you wait before performing the experiment?

P25.14 A convenient source of gamma rays for radiation therapy is ^{60}Co, which undergoes the following decay process:
$$^{60}_{27}Co \xrightarrow{k} {}^{60}_{28}Ni + \beta^- + \gamma.$$ The half-life of ^{60}Co is 1.9×10^3 days.

a. What is the rate constant for the decay process?

b. How long will it take for a sample of ^{60}Co to decay to half of its original concentration?

P25.15 The growth of a bacterial colony can be modeled as a first-order process in which the probability of cell division is linear with respect to time such that $dN/N = \zeta dt$, where dN is the number of cells that divide in the time interval dt, and ζ is a constant.

a. Use the preceding expression to show that the number of cells in the colony is given by $N = N_0 e^{\zeta t}$, where N is the number of cell colonies and N_0 is the number of colonies present at $t = 0$.

b. The generation time is the amount of time it takes for the number of cells to double. Using the answer to part (a), derive an expression for the generation time.

c. In milk at $37°C$, the bacteria *Lactobacillus acidophilus* has a generation time of about 75 min. Construct a plot of the acidophilus concentration as a function of time for time intervals of 15, 30, 45, 60, 90, 120, and 150 min after a colony of size N_0 is introduced into a container of milk.

P25.16 Show that the ratio of the half-life to the three-quarter life, $t_{1/2}/t_{3/4}$, for a reaction that is nth order ($n > 1$) in reactant A can be written as a function of n alone (that is, there is no concentration dependence in the ratio).

P25.17 Given the following kinetic scheme and associated rate constants, determine the concentration profiles of all species using Euler's method. Assume that the reaction is initiated with only the reactant A present at an initial concentration of 1 M. To perform this calculation, you may want to use a spreadsheet program such as Excel.

$$A \xrightarrow{k=1.5\times10^{-3} \, s^{-1}} B$$
$$A \xrightarrow{k=2.5\times10^{-3} \, s^{-1}} C \xrightarrow{k=1.8\times10^{-3} \, s^{-1}} D$$

P25.18 For the sequential reaction $A \xrightarrow{k_A} B \xrightarrow{k_B} C$, the rate constants are $k_A = 5 \times 10^6$ s^{-1} and $k_B = 3 \times 10^6$ s^{-1}. Determine the time at which [B] is at a maximum.

P25.19 For the sequential reaction $A \xrightarrow{k_A} B \xrightarrow{k_B} C$, $k_A = 1.00 \times 10^{-3}$ s^{-1}. Using a computer spreadsheet program such as Excel, plot the concentration of each species for cases where $k_B = 10k_A$, $k_B = 1.5k_A$, and $k_B = 0.1k_A$. Assume that only the reactant is present when the reaction is initiated.

P25.20 (Challenging) For the sequential reaction in Problem P25.19, plot the concentration of each species for the case where $k_B = k_A$. Can you use the analytical expression for [B] in this case?

P25.21 For a type II second-order reaction, the reaction is 60% complete in 60 seconds when $[A]_0 = 0.1$ M and $[B]_0 = 0.5$ M.

a. What is the rate constant for this reaction?

b. Will the time for the reaction to reach 60% completion change if the initial reactant concentrations are decreased by a factor of 2?

P25.22 Bacteriorhodopsin is a protein found in *Halobacterium halobium* that converts light energy into a transmembrane proton gradient that is used for ATP synthesis. After light is absorbed by the protein, the following initial reaction sequence occurs:

$$Br \xrightarrow{k_1 = 2.0 \times 10^{12} s^{-1}} J \xrightarrow{k_2 = 3.3 \times 10^{11} s^{-1}} K$$

a. At what time will the maximum concentration of the intermediate J occur?

b. Construct plots of the concentration of each species versus time.

P25.23 Bananas are somewhat radioactive due to the presence of substantial amounts of potassium. Potassium-40 decays by two different paths:

$$^{40}_{19}K \rightarrow \, ^{40}_{20}Ca + \beta \quad (89.3\%)$$

$$^{40}_{19}K \rightarrow \, ^{40}_{18}Ar + \beta^+ \quad (10.7\%)$$

The half-life for potassium decay is 1.3×10^9 years. Determine the rate constants for the individual channels.

P25.24 In the stratosphere, the rate constant for the conversion of ozone to molecular oxygen by atomic chlorine is $Cl + O_3 \longrightarrow ClO + O_2$ $[k = (1.7 \times 10^{10}$ M^{-1} s$^{-1})e^{-260\,K/T}]$.

a. What is the rate of this reaction at 20 km where [Cl] = 5×10^{-17} M, $[O_3] = 8 \times 10^{-9}$ M, and $T = 220$ K?

b. The actual concentrations at 45 km are [Cl] = 3×10^{-15} M and $[O_3] = 8 \times 10^{-11}$ M. What is the rate of the reaction at this altitude where $T = 270$ K?

c. (Optional) Given the concentrations in part (a), what would you expect the concentrations at 45 km to be assuming that the gravity represents the operative force defining the potential energy?

P25.25 An experiment is performed on the parallel reaction illustrated in Equation 25.71:

Two things are determined: (1) The yield for B at a given temperature is found to be 0.3 and (2) the rate constants are described well by an Arrhenius expression with the activation to B and C formation being 27 and 34 kJ mol^{-1}, respectively, and with identical preexponential factors. Demonstrate that these two statements are inconsistent with each other.

P25.26 A standard "rule of thumb" for thermally activated reactions is that the reaction rate doubles for every 10 K increase in temperature. Is this statement true independent of the activation energy (assuming that the activation energy is positive and independent of temperature)?

P25.27 Calculate the ratio of rate constants for two thermal reactions that have the same Arrhenius preexponential term, but with activation energies that differ by 1, 10, and 30 kJ/mol.

P25.28 The rate constant for the reaction of molecular hydrogen with molecular iodine is 2.45×10^{-4} M^{-1} s^{-1} at 302°C and 0.950 M^{-1} s^{-1} at 508°C.

a. Calculate the activation energy and Arrhenius preexponential factor for this reaction.

b. What is the value of the rate constant at 400°C?

P25.29 The melting of double-strand DNA into two single strands can be initiated using temperature-jump methods. Derive the expression for the T-jump relaxation time for the following equilibrium involving double-strand (DS) and single-strand (SS) DNA:

$$DS \underset{k_r}{\overset{k_f}{\rightleftarrows}} 2SS$$

P25.30 Consider the reaction

$$A + B \underset{k'}{\overset{k}{\rightleftarrows}} P$$

A temperature-jump experiment is performed where the relaxation time constant is measured to be 310 μs, resulting in an equilibrium where $K_{eq} = 0.7$ with $[P]_{eq} = 0.2$ M. What are k and k'? (Watch the units!)

P25.31 In the limit where the diffusion coefficients and radii of two reactants are equivalent, demonstrate that the rate constant for a diffusion–controlled reaction can be written as follows:

$$k_d = \frac{8RT}{3\eta}$$

P25.32 In the following chapter, enzyme catalysis reactions will be extensively reviewed. The first step in these reactions involves the binding of a reactant molecule (referred to as a substrate) to a binding site on the enzyme. If this binding is extremely efficient (that is, equilibrium strongly favors the enzyme–substrate complex over separate enzyme and substrate) and the formation of product rapid, then the rate of catalysis could be diffusion limited. Estimate the expected rate constant for a diffusion-controlled reaction using typical values for an enzyme ($D = 1 \times 10^{-7}$ cm^2 s^{-1} and $r = 40$ Å) and a small molecular substrate ($D = 1 \times 10^{-5}$ cm^2 s^{-1} and $r = 5$ Å).

P25.33 Imidazole is a common molecular species in biological chemistry. For example, it constitutes the side chain of the amino acid histidine. Imidazole can be protonated in solution as follows:

The rate constant for the protonation reaction is 5.5×10^{10} $M^{-1} s^{-1}$. Assuming that the reaction is diffusion controlled, estimate the diffusion coefficient of imidazole when $D(H^+) = 9.31 \times 10^{-5}$ cm^2 s^{-1}, $r(H^+) \sim 1.0$ Å and r(imidazole) = 6 Å. Use this information to predict the rate of deprotonation of imidazole by OH^- ($D = 5.30 \times 10^{-5}$ cm^2 s^{-1} and $r = \sim1.5$ Å).

P25.34 Catalase is an enzyme that promotes the conversion of hydrogen peroxide (H_2O_2) into water and oxygen. The diffusion constant and radius for catalase are 6.0×10^{-7} cm^2 s^{-1} and 51.2 Å, respectively. For hydrogen peroxide the corresponding values are 1.5×10^{-5} cm^2 s^{-1} and $r \sim 2.0$ Å. The experimentally determined rate constant for the conversion of hydrogen peroxide by catalase is 5×10^6 M^{-1} s^{-1}. Is this a diffusion-controlled reaction?

P25.35 The unimolecular decomposition of urea in aqueous solution is measured at two different temperatures and the following data are observed:

Trial Number	Temperature (°C)	k (s^{-1})
1	60.0	1.2×10^{-7}
2	71.5	4.40×10^{-7}

a. Determine the Arrhenius parameters for this reaction.

b. Using these parameters, determine $\Delta H^{\ddagger}$ and $\Delta S^{\ddagger}$ as described by the Eyring equation.

P25.36 The gas-phase decomposition of ethyl bromide is a first-order reaction, occurring with a rate constant that demonstrates the following dependence on temperature:

Trial Number	Temperature (K)	k (s^{-1})
1	800	0.036
2	900	1.410

a. Determine the Arrhenius parameters for this reaction.

b. Using these parameters, determine $\Delta H^{\ddagger}$ and $\Delta S^{\ddagger}$ as described by the Eyring equation.

P25.37 Hydrogen abstraction from hydrocarbons by atomic chlorine is a mechanism for Cl loss in the atmosphere. Consider the reaction of Cl with ethane:

$$C_2H_6(g) + Cl(g) \longrightarrow C_2H_5(g) + HCl$$

This reaction was studied in the laboratory, and the following data were obtained:

T (K)	$10^{-10} k$ (M^{-1}s^{-1})
270	3.43
370	3.77
470	3.99
570	4.13
670	4.23

a. Determine the Arrhenius parameters for this reaction.

b. At the tropopause (the boundary between the troposphere and stratosphere located approximately 11 km above the surface of the Earth), the temperature is roughly 220 K. What do you expect the rate constant to be at this temperature?

c. Using the Arrhenius parameters obtained in part (a), determine the Eyring parameters $\Delta H^{\ddagger}$ and $\Delta S^{\ddagger}$ for this reaction.

P25.38 Reactions involving hydroxyl radical (OH) are extremely important in atmospheric chemistry. The reaction of hydroxyl radical with molecular hydrogen is as follows:

$$OH \cdot (g) + H_2(g) \longrightarrow H_2O(g) + H \cdot (g)$$

Determine the Eyring parameters $\Delta H^{\ddagger}$ and $\Delta S^{\ddagger}$ for this reaction where $A = 8 \times 10^{13}$ M^{-1} s^{-1} and $E_a = 42$ kJ mol^{-1}.

P25.39 Chlorine monoxide (ClO) demonstrates three bimolecular self-reactions:

$$Rxn_1: ClO \cdot (g) + ClO \cdot (g) \xrightarrow{k_1} Cl_2(g) + O_2(g)$$

$$Rxn_2: ClO \cdot (g) + ClO \cdot (g) \xrightarrow{k_2} Cl \cdot (g) + ClOO \cdot (g)$$

$$Rxn_3: ClO \cdot (g) + ClO \cdot (g) \xrightarrow{k_3} Cl \cdot (g) + OClO \cdot (g)$$

The following table provides the Arrhenius parameters for this reaction:

	A (M^{-1} s^{-1})	E_a (kJ/mol)
Rxn_1	6.08×10^8	13.2
Rxn_2	1.79×10^{10}	20.4
Rxn_3	2.11×10^8	11.4

a. For which reaction is $\Delta H^{\ddagger}$ greatest and by how much relative to the next closest reaction?

b. For which reaction is $\Delta S^{\ddagger}$ the smallest and by how much relative to the next closest reaction?

Web-Based Simulations, Animations, and Problems

W25.1 In this problem, concentration profiles as a function of rate constant are explored for the following sequential reaction scheme:

$$A \xrightarrow{k_a} B \xrightarrow{k_b} C$$

Students vary the rate constants k_a and k_b and explore the following behavior:

a. The variation in concentrations as the rate constants are varied,

b. Comparison of the maximum intermediate concentration time determined by simulation and through computation, and

c. Visualization of the conditions under which the rate-limiting step approximation is valid.

W25.2 In this simulation, the kinetic behavior of the following parallel reaction is studied:

$$A \xrightarrow{k_b} B$$

$$A \xrightarrow{k_c} C$$

The variation in concentrations as a function of k_b and k_c is studied. In addition, the product yields for the reaction determined based on the relative values of k_b and k_c are compared to the simulation result.

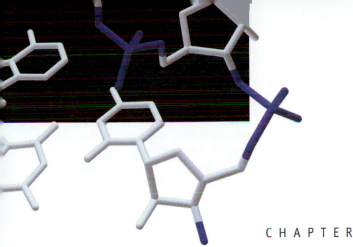

CHAPTER

26

Complex Biological Reactions

In this chapter, the kinetic techniques developed in the previous chapter are applied to complex biological reactions. Reaction mechanisms and their use in predicting reaction rate law expressions are explored. The preequilibrium approximation is presented and used in the evaluation of catalytic reactions. The kinetics of biological catalysts, or enzymes, are described using the Michaelis–Menten model, and the role of inhibitors in enzyme catalysis is discussed. An introduction to photochemistry and photobiology including the process of vision and light harvesting in photosynthetic pigments is presented. The chapter concludes with a discussion of electron transfer reactions, and the application of Marcus theory in describing the rate of electron transfer. The unifying theme behind these apparently different topics is that all of the reaction mechanisms for these phenomena can be developed using the techniques of elementary chemical kinetics. Seemingly complex reactions can be decomposed into a series of well-defined kinetic steps, thereby providing substantial insight into the underlying chemical reaction dynamics.

26.1 Reaction Mechanisms and Rate Laws

Reaction mechanisms are defined as the collection of individual kinetic processes or steps involved in the transformation of reactants into products. The rate law expression for a chemical reaction, including the order of the reaction, is entirely dependent on the reaction mechanism. For a reaction mechanism to be valid, the order of the reaction predicted by the mechanism must agree with the experimentally determined order. Consider the following reaction:

$$2N_2O_5 \longrightarrow 4NO_2 + O_2 \qquad (26.1)$$

One possible mechanism for this reaction is a single step consisting of a bimolecular collision between two N_2O_5 molecules. A reaction mechanism that consists of a single elementary step is known as a **simple reaction.** The rate law predicted by this mechanism is second order with respect to N_2O_5. However, the experimentally determined rate law for this reaction is first order in N_2O_5, not second order. Therefore, the single-step mechanism cannot be correct. To explain the observed order dependence of the reaction rate, the following mechanism was proposed:

$$2\left\{N_2O_5 \underset{k_{-1}}{\overset{k_1}{\rightleftarrows}} NO_2 + NO_3\right\} \qquad (26.2)$$

701

$$NO_2 + NO_3 \xrightarrow{k_2} NO_2 + O_2 + NO \qquad (26.3)$$

$$NO + NO_3 \xrightarrow{k_3} 2NO_2 \qquad (26.4)$$

This mechanism is an example of a **complex reaction,** defined as a reaction that occurs in two or more elementary steps. In this mechanism, the first step, Equation (26.2), represents an equilibrium between N_2O_5 and NO_2/NO_3. In the second step, Equation (26.3), the bimolecular reaction of NO_2 and NO_3 results in the dissociation of NO_3 to product NO and O_2. In the final step of the reaction, Equation (26.4), NO and NO_3 undergo a bimolecular reaction to produce $2NO_2$.

In addition to the reactant (N_2O_5) and overall reaction products (NO_2 and O_2), two other species appear in the mechanism (NO and NO_3) that are not in the overall reaction of Equation (26.1). These species are referred to as **reaction intermediates.** Reaction intermediates that are formed in one step of the mechanism must be consumed in a subsequent step. Given this requirement, step 1 of the reaction must occur twice in order to balance the NO_3 that appears in steps 2 and 3. Therefore, we have multiplied this reaction by two in Equation (26.2) to emphasize that step 1 must occur twice for every occurrence of steps 2 and 3. The number of times a given step occurs in a reaction mechanism is referred to as the **stoichiometric number.** In the mechanism under discussion, step 1 has a stoichiometric number of two, whereas the other two steps have stoichiometric numbers of one. With correct stoichiometric numbers, the sum of the elementary reaction steps will produce an overall reaction that is stoichiometrically equivalent to the reaction of interest.

For a reaction mechanism to be considered valid, the mechanism must be consistent with the experimentally determined rate law. Using the mechanism depicted by Equations (26.2) through (26.4), the rate of the reaction is

$$R = -\frac{1}{2}\frac{d[N_2O_5]}{dt} = \frac{1}{2}(k_1[N_2O_5] - k_{-1}[NO_2][NO_3]) \qquad (26.5)$$

Notice that the stoichiometric number of the reaction is not included in the differential rate expression. Equation (26.5) corresponds to the loss of N_2O_5 by unimolecular decay and production by the bimolecular reaction of NO_2 with NO_3. As discussed earlier, NO and NO_3 are reaction intermediates. Writing the differential rate expression for these species and applying the steady-state approximation to the concentrations of both intermediates (Section 25.7) yields

$$\frac{d[NO]}{dt} = 0 = k_2[NO_2][NO_3] - k_3[NO][NO_3] \qquad (26.6)$$

$$\frac{d[NO_3]}{dt} = 0 = k_1[N_2O_5] - k_{-1}[NO_2][NO_3] - k_2[NO_2][NO_3] - k_3[NO][NO_3] \quad (26.7)$$

Equation (26.6) can be rewritten to produce the following expression for [NO]:

$$[NO] = \frac{k_2[NO_2]}{k_3} \qquad (26.8)$$

Substituting this result into Equation (26.7) yields

$$0 = k_1[N_2O_5] - k_{-1}[NO_2][NO_3] - k_2[NO_2][NO_3] - k_3\left(\frac{k_2[NO_2]}{k_3}\right)[NO_3]$$

$$0 = k_1[N_2O_5] - k_{-1}[NO_2][NO_3] - 2k_2[NO_2][NO_3]$$

$$\frac{k_1[N_2O_5]}{k_{-1} + 2k_2} = [NO_2][NO_3] \qquad (26.9)$$

Substituting Equation (26.9) into Equation (26.5) results in the following predicted rate law expression for this mechanism:

$$R = \frac{1}{2}(k_1[N_2O_5] - k_{-1}[NO_2][NO_3])$$

$$= \frac{1}{2}\left(k_1[N_2O_5] - k_{-1}\left(\frac{k_1[N_2O_5]}{k_{-1} + 2k_2}\right)\right)$$

$$= \frac{k_1k_2}{k_{-1} + 2k_2}[N_2O_5] = k_{eff}[N_2O_5] \tag{26.10}$$

In Equation (26.10), the collection of rate constants multiplying $[N_2O_5]$ has been renamed k_{eff}. Equation (26.10) demonstrates that the mechanism is consistent with the experimentally observed first-order dependence on $[N_2O_5]$. However, as discussed in Chapter 25, the consistency of a reaction mechanism with the experimental order dependence of the reaction is not proof that the mechanism is absolutely correct, but instead demonstrates that the mechanism is consistent with the experimentally determined order dependence. It is quite possible that an alternative mechanism could be constructed that is also consistent with the experimental order dependence.

The example just presented illustrates how reaction mechanisms are used to explain the order dependence of the reaction rate. A theme that reoccurs throughout this chapter is the relation between reaction mechanisms and elementary reaction steps. As we will see, the mechanisms for many complex reactions can be decomposed into a series of elementary steps, and the techniques developed in the previous chapter can be readily employed in the evaluation of these complex kinetic problems.

26.2 The Preequilibrium Approximation

The preequilibrium approximation is a central concept employed in the evaluation of reaction mechanisms. This approximation is used when equilibrium among a subset of species is established before product formation occurs. In this section, the preequilibrium approximation is defined. This approximation will prove to be extremely useful in subsequent sections.

26.2.1 General Solution

Consider the following reaction:

$$A + B \underset{k_r}{\overset{k_f}{\rightleftharpoons}} I \overset{k_p}{\longrightarrow} P \tag{26.11}$$

In Equation (26.11), forward and back rate constants link the reactants A and B with an intermediate species, I. Decay of I results in the formation of product, P. If the forward and back reactions involving the reactants and intermediate are more rapid than the decay of the intermediate to form products, then the reaction of Equation (26.11) can be envisioned as occurring in two distinct steps:

1. First, equilibrium between the reactants and the intermediate is maintained during the course of the reaction.

2. The intermediate undergoes decay to form product.

This description of events is referred to as the **preequilibrium approximation.** This approximation is appropriate when the rate of the backward reaction of I to form reactants is much greater than the rate of product formation, a condition that occurs when $k_r \gg k_p$. The application of the preequilibrium approximation in evaluating reaction mechanisms containing equilibrium steps is performed as follows. The differential rate expression for the product is

$$\frac{d[P]}{dt} = k_p[I] \tag{26.12}$$

In Equation (26.12), [I] can be rewritten by recognizing that this species is in equilibrium with the reactants; therefore,

$$\frac{[I]}{[A][B]} = \frac{k_f}{k_r} = K_c \tag{26.13}$$

$$[I] = K_c[A][B] \tag{26.14}$$

In the preceding equations, K_c is the equilibrium constant expressed in terms of reactant and product concentrations. Substituting the definition of [I] provided by Equation (26.14) into Equation (26.12), the differential rate expression for the product becomes

$$\frac{d[P]}{dt} = k_p[I] = k_p K_c[A][B] = k_{eff}[A][B] \tag{26.15}$$

Equation (26.15) demonstrates that with the preequilibrium approximation, the predicted rate law is second order overall and first order with respect to both reactants (A and B). Finally, the rate constant for product formation is not simply k_p, but is instead the product of this rate constant with the equilibrium constant, which is in turn equal to the ratio of the forward and back rate constants.

26.2.2 A Preequilibrium Example

The atmospheric reaction of NO and O_2 to form product NO_2 provides an example in which the preequilibrium approximation provides insight into the mechanism of NO_2 formation. The specific reaction of interest is

$$2NO(g) + O_2(g) \longrightarrow 2NO_2(g) \tag{26.16}$$

One possible mechanism for this reaction is that of a single elementary step corresponding to a trimolecular reaction of two NO molecules and one O_2 molecule. The experimental rate law for this reaction is second order in NO and first order in O_2, consistent with this mechanism. However, this mechanism was further evaluated by measuring the temperature dependence of the reaction rate. If correct, raising the temperature will increase the number of collisions, and the reaction rate should increase. However, as the temperature is increased, a *reduction* in the reaction rate is observed, proving that the trimolecular-collisional mechanism is incorrect. In contrast, these experimental observations were explained using the following mechanism:

$$2NO \underset{k_r}{\overset{k_f}{\rightleftharpoons}} N_2O_2 \tag{26.17}$$

$$N_2O_2 + O_2 \xrightarrow{k_p} 2NO_2 \tag{26.18}$$

In the first step of this mechanism, Equation (26.17), an equilibrium between NO and the dimer, N_2O_2, is established rapidly compared to the rate of product formation. In the second step, Equation (26.18), a bimolecular reaction involving the dimer and O_2 results in the production of the NO_2 product. The stoichiometric number for each step is one. To evaluate this mechanism, the preequilibrium approximation is applied to step 1 of the mechanism, and the concentration of N_2O_2 is expressed as

$$[N_2O_2] = \frac{k_f}{k_r}[NO]^2 = K_c[NO]^2 \tag{26.19}$$

Using the second step of the mechanism, the reaction rate is written as

$$R = \frac{1}{2}\frac{d[NO_2]}{dt} = k_p[N_2O_2][O_2] \tag{26.20}$$

Substitution of Equation (26.19) into the above rate expression yields

$$R = k_p[N_2O_2][O_2] = k_p k_c[NO]^2[O_2]$$
$$= k_{eff}[NO]^2[O_2] \qquad (26.21)$$

The rate law predicted by this mechanism is second order in NO and first order in O_2, consistent with experiment. Furthermore, the preequilibrium approximation provides an explanation for the temperature dependence of product formation. Specifically, the formation of N_2O_2 is an exothermic process such that an increase in temperature shifts the equilibrium between NO and N_2O_2 toward NO. As such, there is less N_2O_2 to react with O_2, and this is reflected by a reduction in the rate of NO_2 formation with increased temperature.

26.3 Catalysis

A **catalyst** is a substance that participates in chemical reactions by increasing the rate of reaction, yet the catalyst itself remains intact after the reaction is complete. The general function of a catalyst is to provide an additional mechanism by which reactants are converted to products. The presence of a new reaction mechanism involving the catalyst results in a second reaction coordinate that connects reactants and products. The activation energy along this second reaction coordinate will be lower in comparison to the reaction coordinate for the uncatalyzed reaction; therefore, the overall reaction rate will increase. For example, consider Figure 26.1, in which a reaction involving the conversion of reactant A to product B with and without a catalyst is depicted. In the absence of a catalyst, the rate of product formation is given by rate $= r_0$. In the presence of the catalyst, a second pathway is created, and the reaction rate is now the sum of the original rate plus the rate for the catalyzed reaction, or $r_0 + r_c$.

An analogy for a catalyzed reaction is found in the electrical circuits depicted in Figure 26.1. In the "catalyzed" electrical circuit, a second, parallel pathway for current flow has been added, allowing for increased total current when compared to the "uncatalyzed" circuit. By analogy, the addition of the second, parallel pathway is equivalent to the alternative reaction mechanism involving the catalyst.

To be effective, a catalyst must combine with one or more of the reactants or with an intermediate species involved in the reaction. After the reaction has taken place, the catalyst is freed, and can combine with another reactant or intermediate in a subsequent reaction. The catalyst is not consumed during the reaction so that a small amount of catalyst can participate in numerous reactions. The simplest mechanism describing a catalytic process is as follows:

$$S + C \underset{k_{-1}}{\overset{k_1}{\rightleftharpoons}} SC \qquad (26.22)$$

$$SC \overset{k_2}{\longrightarrow} P + C \qquad (26.23)$$

where S represents the reactant or substrate, C is the catalyst, and P is the product. The **substrate–catalyst complex** is represented by SC and is an intermediate species in this mechanism. The differential rate expression for product formation is

$$\frac{d[P]}{dt} = k_2[SC] \qquad (26.24)$$

Because SC is an intermediate, we write the differential rate expression for this species and apply the steady-state approximation:

$$\frac{d[SC]}{dt} = k_1[S][C] - k_{-1}[SC] - k_2[SC] = 0$$

$$[SC] = \frac{k_1[S][C]}{k_{-1} + k_2} = \frac{[S][C]}{K_m} \qquad (26.25)$$

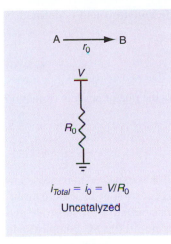

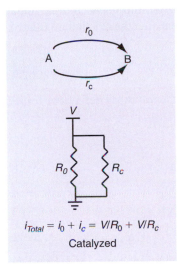

FIGURE 26.1

Illustration of catalysis. In the uncatalyzed reaction, the rate of reaction is given by r_0. In the catalyzed case, a new pathway is created by the presence of the catalyst with corresponding rate r_c. The total rate of reaction for the catalyzed case is $r_0 + r_c$. The analogous electrical circuits are also presented for comparison.

In Equation (26.25), K_m is referred to as the **composite constant** and is defined as follows:

$$K_m = \frac{k_{-1} + k_2}{k_1} \tag{26.26}$$

Substituting the expression for [SC] into Equation (26.24), the rate of product formation becomes

$$\frac{d[P]}{dt} = \frac{k_2[S][C]}{K_m} \tag{26.27}$$

Equation (26.27) illustrates that the rate of product formation is expected to increase linearly with both substrate and catalyst concentrations. This equation is difficult to evaluate over the entire course of the reaction because the concentrations of substrate and catalyst given in Equation (26.27) correspond to species not in the SC complex, and these concentrations can be quite difficult to measure. A more convenient measurement is to determine how much substrate and catalyst are present at the beginning of the reaction. Conservation of mass dictates the following relationship between these initial concentrations and the concentrations of all species present after the reaction is initiated:

$$[S]_0 = [S] + [SC] + [P] \tag{26.28}$$

$$[C]_0 = [C] + [SC] \tag{26.29}$$

Rearrangement of Equations (26.28) and (26.29) yields the following definitions for [S] and [C]:

$$[S] = [S]_0 - [SC] - [P] \tag{26.30}$$

$$[C] = [C]_0 - [SC] \tag{26.31}$$

Substituting these expressions into Equation (26.25) yields

$$K_m[SC] = [S][C] = ([S]_0 - [SC] - [P])([C]_0 - [SC])$$

$$0 = [[C]_0([S]_0 - [P])] - [SC]([S]_0 + [C]_0 - [P] + K_m) + [SC]^2 \tag{26.32}$$

Equation (26.32) can be evaluated as a quadratic equation to determine [SC]. However, two assumptions are generally employed at this point to simplify matters. First, through control of the initial substrate and catalyst concentrations, conditions can be employed such that [SC] is small. Therefore, the $[SC]^2$ term in Equation (26.32) can be neglected. Second, we confine ourselves to early stages of the reaction when little product has been formed; therefore, terms involving [P] can also be neglected. With these two approximations, Equation (26.32) is readily evaluated, providing the following expression for [SC]:

$$[SC] = \frac{[S]_0[C]_0}{[S]_0 + [C]_0 + K_m} \tag{26.33}$$

Substituting Equation (26.33) into Equation (26.24), the rate of the reaction becomes

$$R_0 = \frac{d[P]}{dt} = \frac{k_2[S]_0[C]_0}{[S]_0 + [C]_0 + K_m} \tag{26.34}$$

In Equation (26.34), the subscript on the rate indicates that this expression applies to the early-time or initial reaction rate. We next consider two limiting cases of Equation (26.34).

26.3.1 Case 1: $[C]_0 \ll [S]_0$

In the $[C]_0 \ll [S]_0$ case, the most common case in catalysis, much more substrate is present in comparison to catalyst. In this limit $[C]_0$ can be neglected in the denominator of Equation (26.34) and the rate becomes

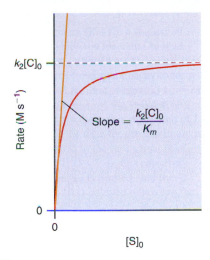

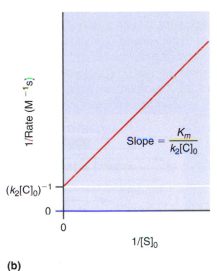

(a)

(b)

FIGURE 26.2
Illustration of the variation in the reaction rate with substrate concentration under Case 1 conditions as described in the text. (a) Plot of the initial reaction rate with respect to substrate concentration [Equation (26.35)]. At low substrate concentrations, the reaction rate increases linearly with substrate concentration. At high substrate concentrations, a maximum reaction rate of $k_2[C]_0$ is reached. (b) Reciprocal plot where the inverse of the reaction rate is plotted with respect to the inverse of substrate concentration [Equation (26.36)]. The y intercept of this line is equal to the inverse of the maximum reaction rate, or $(k_2[C]_0)^{-1}$. The slope of the line is equal to $K_m(k_2[C]_0)^{-1}$; therefore, with the slope and y intercept, K_m can be determined.

$$R_0 = \frac{k_2[S]_0[C]_0}{[S]_0 + K_m} \tag{26.35}$$

For substrate concentrations where $[S]_0 < K_m$, the reaction rate should increase linearly with substrate concentration, with a slope equal to $k_2[C]_0/K_m$. Although parameters such as k_2 and K_m can be obtained by comparing experimental reaction rates to Equation (26.35), another approach is generally taken. Specifically, inverting Equation (26.35) provides the following relationship between the reaction rate and initial substrate concentration:

$$\frac{1}{R_0} = \left(\frac{K_m}{k_2[C]_0}\right)\frac{1}{[S]_0} + \frac{1}{k_2[C]_0} \tag{26.36}$$

Equation (26.36) demonstrates that a plot of the inverse of the initial reaction rate versus $[S]_0^{-1}$, referred to as a **reciprocal plot,** should yield a straight line. The y intercept and slope of this line provide a measure of K_m and k_2, assuming $[C]_0$ is known.

At elevated concentrations of substrate where $[S]_0 \gg K_m$, the denominator in Equation (26.49) can be approximated as $[S]_0$, resulting in the following expression for the reaction rate:

$$R_0 = k_2[C]_0 = R_{max} \tag{26.37}$$

In other words, the rate of reaction will reach a limiting value where the rate becomes zero order in substrate concentration. In this limit, the reaction rate can only be enhanced by increasing the amount of catalyst. An illustration of the variation in the reaction rate with initial substrate concentration predicted by Equations (26.35) and (26.36) is provided in Figure 26.2.

26.3.2 Case 2: $[C]_0 \gg [S]_0$

In the $[C]_0 \gg [S]_0$ limit, Equation (26.34) becomes

$$R_0 = \frac{k_2[S]_0[C]_0}{[C]_0 + K_m} \tag{26.38}$$

In this concentration limit, the reaction rate is first order in $[S]_0$, but can be first or zero order in $[C]_0$ depending on the size of $[C]_0$ relative to K_m. In catalysis studies, this limit is generally avoided because the insight to be gained regarding the rate constants for the various reaction steps are more easily evaluated for the previously discussed Case 1. In addition, good catalysts can be expensive; therefore, employing excess catalyst in a reaction is not cost effective.

26.4 Enzyme Kinetics

Enzymes are biological molecules that serve as catalysts in a wide variety of chemical reactions. Enzymes are noted for their reaction specificity, with nature having developed specific catalysts to facilitate the vast majority of biological reactions required for organism survival. An illustration of an enzyme with associated substrate is presented in Figure 26.3. The figure presents a space-filling model derived from a crystal structure of phospholipase A_2 (white) containing a bound substrate analogue (red). This enzyme catalyzes the hydrolysis of esters in phospholipids. The substrate analogue contains a stable phosphonate group in place of the enzyme-susceptible ester. The substrate analogue is resistant to enzymatic hydrolysis so that it does not suffer chemical breakdown during the structure determination process. With reactive substrate, ester hydrolysis occurs and the products of the reaction are released from the enzyme, resulting in the regeneration of the free enzyme.

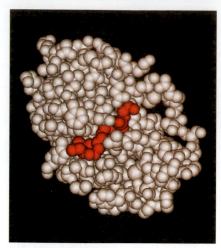

FIGURE 26.3

Space-filling model of the enzyme phospholipase A_2 (white) containing a bound substrate analogue (red). The substrate analogue contains a stable phosphonate group in place of the enzyme-susceptible ester; therefore, the substrate analogue is resistant to enzymatic hydrolysis and the enzyme–substrate complex remains stable in the complex during the X-ray diffraction structure determination process (Source: Structural data from Scott, White, Browning, Rosa, Gelb, and Sigler, *Science* 5034 (1991), 1007.)

26.4.1 Michaelis–Menten Model

The mechanism of phospholipase A_2 catalysis can be described using the **Michaelis–Menten mechanism** of enzyme activity illustrated in Figure 26.4. The figure depicts the "lock-and-key" model for enzyme reactivity in which the substrate is bound to the active site of the enzyme where the reaction is catalyzed. The enzyme and substrate form the enzyme–substrate complex, which dissociates into product and uncomplexed enzyme. The interactions involved in the creation of the enzyme–substrate complex are enzyme specific. For example, the active site may bind the substrate in more than one location, thereby creating geometric strain that promotes product formation. The enzyme may orient the substrate so that the reaction geometry is optimized. In summary, the details of enzyme-mediated chemistry are highly dependent on the reaction of interest. Rather than an exhaustive presentation of enzyme kinetics, our motivation here is to describe enzyme kinetics within the general framework of catalyzed reactions.

A schematic description of the mechanism illustrated in Figure 26.4 is as follows:

$$E + S \underset{k_{-1}}{\overset{k_1}{\rightleftarrows}} ES \overset{k_2}{\longrightarrow} E + P \tag{26.39}$$

In this mechanism, E is enzyme, S is substrate, ES is the complex, and P is product. Comparison of the mechanism of Equation (26.39) to the general catalytic mechanism described earlier in Equations (26.22) and (26.23) demonstrates that this mechanism is identical to the general catalysis mechanism except that the catalyst C is now the enzyme E. In the limit where the initial substrate concentration is substantially greater than that of the enzyme ($[S]_0 \gg [E]_0$ or Case 1 conditions as described previously), the rate of product formation is given by

$$R_0 = \frac{k_2 [S]_0 [E]_0}{[S]_0 + K_m} \tag{26.40}$$

The composite constant, K_m, in Equation (26.40) is referred to as the **Michaelis constant** in enzyme kinetics, and Equation (26.40) is referred to as the **Michaelis–Menten rate law.** When $[S]_0 \gg K_m$, the Michaelis constant can be neglected, resulting in the following expression for the rate:

$$R_0 = k_2 [E]_0 = R_{max} \tag{26.41}$$

With this definition, Equation (26.40) becomes:

$$R_0 = \frac{R_{max} [S]_0}{[S]_0 + K_m} \tag{26.42}$$

Equation (26.42) demonstrates that the rate of product formation will plateau at some maximum value equal to the product of initial enzyme concentration and k_2, the rate constant for product formation, consistent with the behavior depicted in Figure 26.2. A reciprocal plot of the reaction rate can also be constructed by inverting Equation (26.42), which results in the **Lineweaver–Burk equation:**

$$\frac{1}{R_0} = \frac{1}{R_{max}} + \frac{K_m}{R_{max}} \frac{1}{[S]_0} \tag{26.43}$$

For the Michaelis–Menten mechanism to be consistent with experiment, a plot of the inverse of the initial rate with respect to $[S]^{-1}_0$ should yield a straight line from which the y intercept and slope can be used to determine the maximum reaction rate and the Michaelis constant. This reciprocal plot is referred to as a **Lineweaver–Burk plot.** In addition, because $[E]_0$ is readily determined experimentally, the maximum rate can be used to determine k_2, referred to as the **turnover number** of the enzyme [Equation (26.41)]. The turnover number is the maximum number of substrate molecules per unit time that can be converted into product, with most enzymes demonstrating turnover numbers between 1 and 10^5 s^{-1} under physiological conditions.

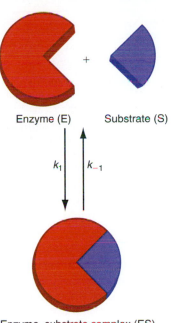

Enzyme (E)   Substrate (S)

k_1 k_{-1}

Enzyme–substrate complex (ES)

k_2

Product (P)

FIGURE 26.4
Schematic of enzyme catalysis.

EXAMPLE PROBLEM 26.1

DeVoe and Kistiakowsky [*J. Am. Chem. Soc.* 83 (1961), 274] studied the kinetics of CO_2 hydration catalyzed by the enzyme carbonic anhydrase:

$$CO_2 + H_2O \rightleftharpoons HCO_3^- + H^+$$

In this reaction, CO_2 is converted to bicarbonate ion. Bicarbonate is transported in the bloodstream and converted back to CO_2 in the lungs, a reaction that is also catalyzed by carbonic anhydrase. The following initial reaction rates for the hydration reaction were obtained for an initial enzyme concentration of 2.3 nM and temperature of 0.5°C:

Rate (M s^{-1})	[CO_2] (mM)
2.78×10^{-5}	1.25
5.00×10^{-5}	2.5
8.33×10^{-5}	5.0
1.67×10^{-4}	20.0

Determine K_m and k_2 for the enzyme at this temperature.

Solution

The Lineweaver–Burk plot of the $rate^{-1}$ versus $[CO_2]^{-1}$ is shown here:

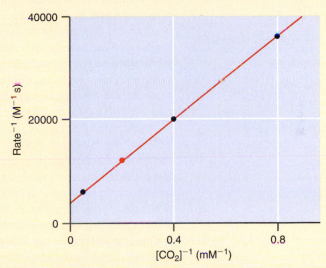

The y intercept for the best fit line to the data is 4000 M^{-1} s corresponding to R_{max} = 2.5×10^{-4} M s^{-1}. Using this value and $[E]_0$ = 2.3 nM, k_2 is

$$k_2 = \frac{R_{max}}{[E]_0} = \frac{2.5 \times 10^{-4} \text{ M s}^{-1}}{2.3 \times 10^{-9} \text{ M}} = 1.1 \times 10^5 \text{ s}^{-1}$$

Notice that the units of k_2, the turnover number, are consistent with a first-order process, in agreement with the Michaelis–Menten mechanism. The slope of the best fit line is 40 s such that, per Equation (26.43), K_m is given by

$$K_m = slope \times R_{max} = (40 \text{ s})(2.5 \times 10^{-4} \text{ M s}^{-1})$$

$$= 10 \text{ mM}$$

In addition to the Lineweaver–Burk plot, K_m can be estimated if the maximum rate is known. Specifically, if the initial rate is equal to one-half the maximum rate, Equation (26.40) reduces to

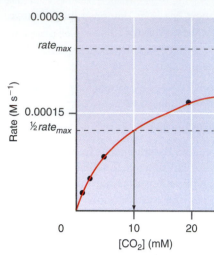

FIGURE 26.5

Determination of K_m for the carbonic-anhydrase catalyzed hydration of CO_2. The substrate concentration at which the rate of reaction is equal to half that of the maximum rate is equal to K_m.

26.1 Michendis-Menter Enzyme Kinetics

$$R_0 = \frac{k_2[S]_0[E]_0}{[S]_0 + \check{K}_m} = \frac{R_{max}[S]_0}{[S]_0 + K_m}$$

$$\frac{R_{max}}{2} = \frac{R_{max}[S]_0}{[S]_0 + K_m}$$

$$[S]_0 + K_m = 2[S]_0$$

$$K_m = [S]_0 \qquad (26.44)$$

Equation (26.44) demonstrates that when the initial rate is half the maximum rate, K_m is equal to the initial substrate concentration. Therefore, K_m can be determined by viewing a substrate saturation curve, as illustrated in Figure 26.5 for the carbonic-anhydrase catalyzed hydration of CO_2 discussed in Example Problem 26.1. The figure demonstrates that the initial rate is equal to half the maximum rate when $[S]_0 = 10$ mM. Therefore, the value of K_m determined in this relatively simple approach is in excellent agreement with that determined from the Lineweaver–Burk plot. Notice in Figure 26.5 that the maximum rate depicted was that employed using the Lineweaver–Burk analysis as shown in the example problem. When employing this method to determine K_m, the high-substrate-concentration limit must be carefully explored to ensure that the reaction rate is indeed at a maximum.

26.4.2 Enzyme Inhibition

Enzyme activity can be altered by the introduction of species other than the substrate, a process known as *enzyme inhibition*. Two limits of inhibition can be envisioned: reversible and irreversible. In irreversible inhibition, the substrate is covalently bound to the enzyme and permanently alters enzyme functionality. The other limit is reversible inhibition where enzyme activity is altered through a noncovalent binding of the inhibitor to the enzyme. Once the inhibitor unbinds from the enzyme, the enzyme activity is recovered. The function of many drugs is based on irreversible enzyme inhibition; therefore, we will explore this type of inhibition in this section.

Perhaps the first type of irreversible inhibition one could imagine is one in which the inhibitor structurally resembles the substrate (Figure 26.6a). The structural similarity of the inhibitor and substrate is such that both species can occupy the enzyme active site, and if the inhibitor is bound, conversion of the substrate to product cannot occur. Such molecules are referred to as **competitive inhibitors.** The phosphonated substrate bound to phospholipase A_2 in Figure 26.3 is an example of a competitive inhibitor because the substrate undergoes competitive binding with active substrates at the enzyme active site, but once bound it does not undergo hydrolytic cleavage. A second type of reversible inhibition involves **noncompetitive inhibitors,** or substrates that bind at a site other than the enzyme active site, yet still reduce the rate of product formation (Figure 26.6b). For example, binding of the noncompetitive inhibitor could alter the enzyme structure so that the catalytic efficiency of the enzyme is reduced. Both types of enzyme inhibition are described next.

Competitive Inhibition.   Competitive inhibition can be described using the following mechanism:

$$E + S \underset{k_{-1}}{\overset{k_1}{\rightleftharpoons}} ES \qquad (26.45)$$

$$ES \xrightarrow{k_2} E + P \qquad (26.46)$$

$$E + I \underset{k_{-3}}{\overset{k_3}{\rightleftharpoons}} EI \qquad (26.47)$$

In this mechanism, I is the inhibitor, EI is the enzyme–inhibitor complex, and the other species are identical to those employed in the standard enzyme kinetic scheme of Equa-

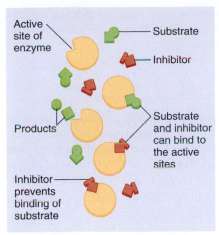

Active site of enzyme

Substrate

Inhibitor

Products

Substrate and inhibitor can bind to the active sites

Inhibitor prevents binding of substrate

(a)

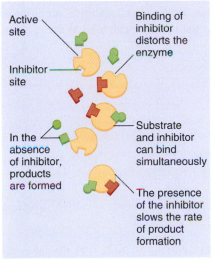

Active site

Binding of inhibitor distorts the enzyme

Inhibitor site

In the absence of inhibitor, products are formed

Substrate and inhibitor can bind simultaneously

The presence of the inhibitor slows the rate of product formation

(b)

FIGURE 26.6
(a) Competitive and (b) noncompetitive enzyme inhibition.

tion (26.39). Notice in this mechanism that if the enzyme is bound to the inhibitor, it cannot bind the substrate consistent with competitive inhibition where the substrate and inhibitor bind to the same site on the enzyme. How does the rate of reaction differ from the noninhibited case discussed earlier? To answer this question, we first define the initial enzyme concentration:

$$[E]_0 = [E] + [EI] + [ES] \tag{26.48}$$

Next, assuming that k_1, k_{-1}, k_3, and $k_{-3} \gg k_2$ and using Equations (26.45) and (26.47), we write:

$$K_s = \frac{[E][S]}{[ES]} \approx K_m \tag{26.49}$$

$$K_i = \frac{[E][I]}{[EI]} \tag{26.50}$$

In Equation (26.49), the equilibrium constant describing the enzyme and substrate is equivalent to K_m when $k_{-1} \gg k_2$. With these relationships, Equation (26.48) can be written as

$$
\begin{aligned}
[E]_0 &= \frac{K_m[ES]}{[S]} + \frac{[E][I]}{K_i} + [ES] \\
&= \frac{K_m[ES]}{[S]} + \left(\frac{K_m[ES]}{[S]}\right)\frac{[I]}{K_i} + [ES] \\
&= [ES]\left(\frac{K_m}{[S]} + \frac{K_m[I]}{[S]K_i} + 1\right)
\end{aligned}
\tag{26.51}
$$

Solving Equation (26.51) for [ES] yields

$$[ES] = \frac{[E]_0}{1 + \dfrac{K_m}{[S]} + \dfrac{K_m[I]}{[S]K_i}} \tag{26.52}$$

Finally, the rate of product formation is given by

$$R_0 = \frac{d[P]}{dt} = k_2[ES] = \frac{k_2[E]_0}{1 + \dfrac{K_m}{[S]} + \dfrac{K_m[I]}{[S]K_i}} = \frac{k_2[S][E]_0}{[S] + K_m\left(1 + \dfrac{[I]}{K_i}\right)}$$

$$R_0 \cong \frac{k_2[S]_0[E]_0}{[S]_0 + K_m\left(1 + \dfrac{[I]}{K_i}\right)} \tag{26.53}$$

In Equation (26.53), the assumption that [ES] and [P] ≪ [S] has been employed so that $[S] = [S]_0$, consistent with the previous treatment of uninhibited catalysis. Comparison of Equation (26.53) to the corresponding expression for the uninhibited case of Equation (26.40) illustrates that with competitive inhibition, a new apparent Michaelis constant can be defined:

$$K_m^* = K_m\left(1 + \frac{[I]}{K_i}\right) \tag{26.54}$$

Notice that K_m^* reduces to K_m in the absence of inhibitor ([I] = 0). Next, using the definition of maximum reaction rate defined earlier in Equation (26.41), the reaction rate in the case of competitive inhibition can be written as

$$R_0 \cong \frac{R_{max}[S]_0}{[S]_0 + K_m^*} \tag{26.55}$$

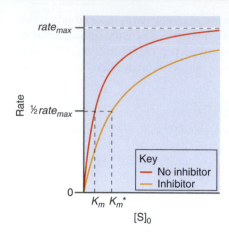

(a)

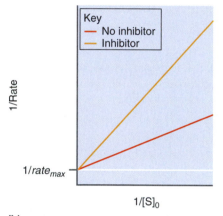

(b)

FIGURE 26.7
Comparison of enzymatic reaction rates in the presence and absence of a competitive inhibitor. (a) Plot of rate versus initial substrate concentration. The location of K_m and K_m^* is indicated. (b) Reciprocal plots ($1/Rate$ versus $1/[S]_0$). Notice that $1/rate_{max}$ is identical in the presence and absence of a competitive inhibitor.

In the presence of inhibitor, $K_m^* > K_m$, and more substrate is required to reach half the maximum rate in comparison to the uninhibited case. The effect of inhibition can also be observed in a Lineweaver–Burk plot of the following form:

$$\frac{1}{R_0} = \frac{1}{R_{max}} + \frac{K_m^*}{R_{max}} \frac{1}{[S]_0} \tag{26.56}$$

Because $K_m^* > K_m$, the slope of the Lineweaver–Burk plot will be greater with inhibitor compared to the slope without inhibitor. Figure 26.7 presents an illustration of this effect.

Competitive inhibition has been used in drug design for antiviral, antibacterial, and antitumor applications. These drugs are molecules that serve as competitive inhibitors for enzymes required for viral, bacterial, or cellular replication. For example, sulfanilamide (Figure 26.8) is a powerful antibacterial drug. This compound is similar to p-aminobenzoic acid, the substrate for the enzyme dihydropteroate synthetase that participates in the production of folate. When present, the enzyme in bacteria cannot produce folate, and the bacteria die. However, humans can obtain folate from other sources; therefore, sulfanilamide is not toxic.

Noncompetitive Inhibition. The reaction mechanism for noncompetitive inhibition is more complex than the competitive case because the inhibitor can bind to the enzyme and enzyme–substrate complex:

$$E + S \underset{k_{-1}}{\overset{k_1}{\rightleftharpoons}} ES \tag{26.57}$$

$$ES \overset{k_2}{\longrightarrow} E + P \tag{26.58}$$

$$EI \underset{k_{-3}}{\overset{k_3}{\rightleftharpoons}} E + I \tag{26.59}$$

$$ES + I \underset{k_{-4}}{\overset{k_4}{\rightleftharpoons}} EIS \tag{26.60}$$

Notice that in this mechanism the enzyme can bind either the inhibitor (EI), the substrate (ES), or both (EIS). Derivation of the reaction rate proceeds exactly as that for competitive inhibition. The initial enzyme concentration is given by

$$[E]_0 = [E] + [EI] + [ES] + [EIS] \tag{26.61}$$

Next, the preequilibrium approximation is applied to reactions Equations (26.57), (26.59), and (26.60), yielding

$$K_s = \frac{[E][S]}{[ES]} \approx K_m \tag{26.62}$$

$$K_i = \frac{[E][I]}{[EI]} \tag{26.63}$$

$$K_{is} = \frac{[ES][I]}{[EIS]} \tag{26.64}$$

In Equation (26.62), the equilibrium constant describing the enzyme and substrate is equivalent to K_m when $k_{-1} \gg k_2$. With these relationships, Equation (26.61) can be written as

$$\begin{aligned}
[E]_0 &= \frac{K_m[ES]}{[S]} + \frac{[E][I]}{K_i} + [ES] + \frac{[ES][I]}{K_{is}} \\
&= \frac{K_m[ES]}{[S]} + \left(\frac{K_m[ES]}{[S]}\right)\frac{[I]}{K_i} + [ES] + \frac{[ES][I]}{K_{is}} \\
&= [ES]\left(\frac{K_m}{[S]} + \frac{K_m[I]}{[S]K_i} + 1 + \frac{[I]}{K_{is}}\right) \tag{26.65}
\end{aligned}$$

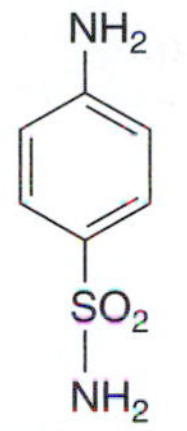

Sulfanilamide

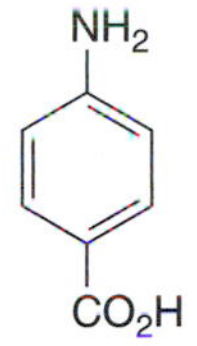

p-Aminobenzoic acid

FIGURE 26.8
Structural comparison of the antibacterial drug sulfanilamide, a competitive inhibitor of the enzyme dihydropteroate synthetase, and the active substrate, p-aminobenzoic acid. The change in functional group from $-CO_2H$ to $-SO_2NH_2$ is such that sulfanilamide cannot be used by bacteria to synthesize folate, and the bacterium starves.

Solving Equation (26.65) for [ES] yields

$$[ES] = \frac{[E]_0}{1 + \dfrac{[I]}{K_{is}} + \dfrac{K_m}{[S]} + \dfrac{K_m[I]}{[S]K_i}} \quad (26.66)$$

Finally, the rate of product formation is given by

$$R_0 = \frac{d[P]}{dt} = k_2[ES] = \frac{k_2[S][E]_0}{[S]\left(1 + \dfrac{[I]}{K_{is}}\right) + K_m\left(1 + \dfrac{[I]}{K_i}\right)}$$

$$\cong \frac{k_2[S]_0[E]_0}{[S]_0\left(1 + \dfrac{[I]}{K_{is}}\right) + K_m\left(1 + \dfrac{[I]}{K_i}\right)}$$

$$= \frac{R_{max}[S]_0}{[S]_0\left(1 + \dfrac{[I]}{K_{is}}\right) + K_m\left(1 + \dfrac{[I]}{K_i}\right)} \quad (26.67)$$

The effect of noncompetitive inhibition on enzyme activity can be quantified by using the preceding expression to construct a Lineweaver–Burk plot of the following form:

$$\frac{1}{R_0} = \frac{1}{R_{max}}\left(1 + \frac{[I]}{K_{is}}\right) + \frac{K_m}{R_{max}}\left(1 + \frac{[I]}{K_i}\right)\frac{1}{[S]_0} \quad (26.68)$$

Notice in this case that both the rate and *y* intercept of the Lineweaver–Burk plot are altered relative to the uninhibited case. Figure 26.9 presents a comparison of the reaction rates for uninhibited and noncompetitive inhibition reactions. Notice that the apparent Michaelis constant is not affected by the presence of the inhibitor, but instead the maximum rate of the reaction is reduced. This is exactly opposite of the competitive inhibition case where the apparent Michaelis constant was altered, but the maximum rate remains the same. This behavior is also reflected in the reciprocal plots where the change in maximum rate is reflected by a change in the value for the *y* intercept. In addition, notice that the plots in the presence and absence of inhibitor intersect the *y* axis at the same point consistent with identical values for the apparent Michaelis constant. Comparison of the competitive and noncompetitive cases demonstrates that evaluation of a reciprocal plot of the rate allows one to determine which of these inhibition mechanisms is operative.

EXAMPLE PROBLEM 26.2

Dihydrofolate reductase (DHFR) is an enzyme that reduces 7,8-dihydrofolate (DHF) to 5,6,7,8-tetrahydrofolate, a step in the biosynthesis of thymidine. Methotrexate (MTX), a structural analogue of dihydrofolate, inhibits this enzyme. Partial structures for the folates and MTX are shown in the following figure:

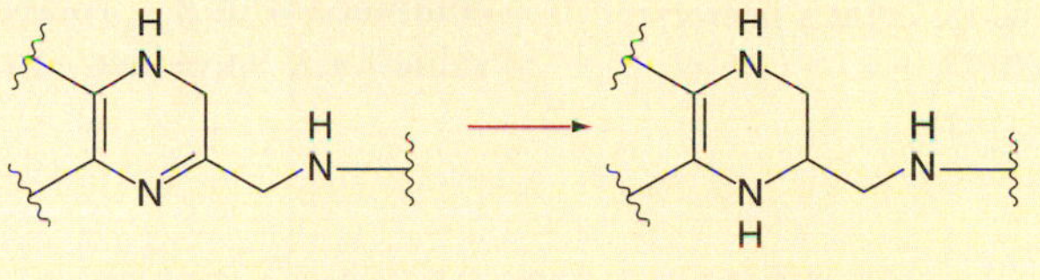

Dihydrofolate (DHF)   **Tetrahydrofolate**

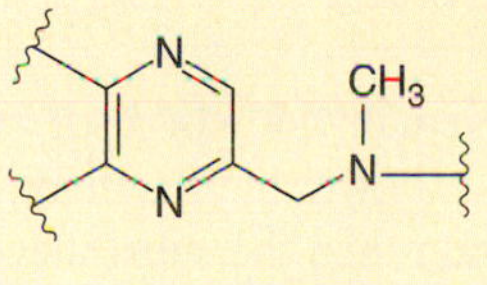

Methotrexate (MTX)

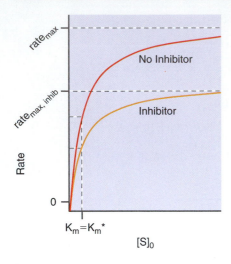

(a)

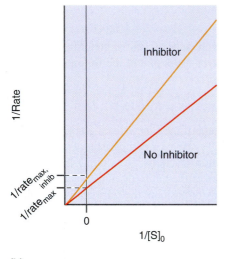

(b)

FIGURE 26.9

Comparison of enzymatic reaction rates in the presence and absence of a noncompetitive inhibitor. (a) Plot of rate versus initial substrate concentration. Notice that the apparent Michaelis constant is unaffected by the addition of inhibitor; however, the maximum rate is reduced. (b) Reciprocal plots (1/*Rate* versus 1/[S]$_0$). Notice that the difference in reaction rates is reflected by a change in the *y* intercept.

MTX has been used in cancer therapies because the inhibition of DHFR restricts the thymidine production required for cell division. In the absence of thymidine, the cancerous cells cannot multiply. Using the following data of DHFR activity in the presence of MTX, determine the inhibition mechanism of MTX and the maximum reaction rate. Finally, if K_m for DHFR is 0.100 μM, what is K_i?

[DHF] (mM)	Rate w/50 nM MTX (mM s^{-1})	Rate w/100 nM MTX (mM s^{-1})	Rate w/200 nM MTX (mM s^{-1})
3	7.38	5.56	3.72
6	8.84	7.38	5.55
9	9.46	8.29	6.65
12	9.79	8.84	7.38

Solution

Differentiation between inhibition mechanisms is readily accomplished by inspection of reciprocal plots using the data from the preceding table. A plot of 1/*Rate* versus 1/[DHF] for the three MTX concentrations is shown here:

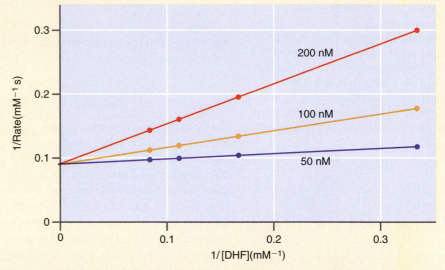

The variation in slope and common *y* intercept for the reciprocal plots as a function of MTX concentration demonstrates that MTX is a competitive inhibitor. Best fit by a straight line to the 200-nM data corresponds to a *y* intercept of 0.091 mM^{-1} s and slope of 0.535 s. The reciprocal plot equation for competitive inhibition is

$$\frac{1}{R_0} = \frac{1}{R_{max}} + \frac{K_m^*}{R_{max}}\frac{1}{[S]_0}$$

Inspection of this equation shows that the *y* intercept is equal to the inverse of the maximum rate; therefore, $R_{max} = 11.0$ *m*M s^{-1}. The slope from the straight-line fit combined with R_{max} establishes that $K_m^* = 5.89$ mM. With this value and the value for K_m provided earlier, K_i is determined as follows:

$$K_m^* = K_m\left(1 + \frac{[I]}{K_i}\right)$$

$$5890\ \mu M = 0.100\ \mu M\left(1 + \frac{200\ nM}{K_i}\right)$$

$$0.00340\ nM = 3.40\ pM = K_i$$

The small value of K_i demonstrates that MTX binds strongly to DHFR. Studies have shown that the binding affinity of MTX to DHFT is roughly 1000 times greater than the substrate affinity [see, for example, Appleman *et al.*, *J. Biol. Chem.* 263 (1988), 10,304].

26.4.3 Homogeneous and Heterogeneous Catalysis

A **homogeneous catalyst** is a catalyst that exists in the same phase as the species involved in the reaction, and heterogeneous catalysts exist in a different phase. Enzymes serve as an example of a homogeneous catalyst; they exist in solution and catalyze reactions that occur in solution. Another example of gas-phase catalysis is the catalytic depletion of stratospheric ozone by atomic chlorine. In the mid-1970s, F. Sherwood Rowland and Mario Molina proposed that Cl atoms catalyze the decomposition of stratospheric ozone by the following mechanism:

$$Cl + O_3 \xrightarrow{k_1} ClO + O_2 \tag{26.69}$$

$$ClO + O \xrightarrow{k_2} Cl + O_2 \tag{26.70}$$

$$O_3 + O \longrightarrow 2O_2 \tag{26.71}$$

In this mechanism, Cl reacts with ozone to produce chlorine monoxide (ClO) and molecular oxygen. The ClO undergoes a second reaction with atomic oxygen, resulting in the reformation of Cl and the product of O_2. The sum of these reactions leads to the net conversion of O_3 and O to $2O_2$. Notice that the Cl is not consumed in the net reaction.

The catalytic efficiency of Cl can be determined using standard techniques in kinetics. The experimentally determined rate law expression for the uncatalyzed reaction of Equation (26.71) is

$$R_{vc} = k_{vc}[O][O_3] \tag{26.72}$$

The stratospheric temperature where this reaction occurs is roughly 220 K, at which temperature k_{nc} has a value of $3.30 \times 10^5 \ M^{-1} \ s^{-1}$. For the Cl catalyzed decomposition of ozone, the rate constants at this temperature are $k_1 = 1.56 \times 10^{10} \ M^{-1} \ s^{-1}$ and $k_2 = 2.44 \times 10^{10} \ M^{-1} \ s^{-1}$. To employ these rates in determining the overall rate of reaction, the rate law expression for the catalytic mechanism must be determined. Notice that both Cl and ClO are intermediates in this mechanism. Applying the steady-state approximation, the concentration of intermediates is taken to be a constant such that

$$[X] = [Cl] + [ClO] \tag{26.73}$$

where [X] is defined as the sum of reaction intermediate concentrations, a definition that will prove useful in deriving the rate law. In addition, the steady-state approximation is applied in evaluating the differential rate expression for [Cl] as follows:

$$\frac{d[Cl]}{dt} = 0 = -k_1[Cl][O_3] + k_2[ClO][O]$$

$$k_1[Cl][O_3] = k_2[ClO][O]$$

$$\frac{k_1[Cl][O_3]}{k_2[O]} = [ClO] \tag{26.74}$$

Substituting Equation (26.74) into Equation (26.73) yields the following expression for [Cl]:

$$[Cl] = \frac{k_2[X][O]}{k_1[O_3] + k_2[O]} \tag{26.75}$$

Using Equation (26.76), the rate law expression for the catalytic mechanism is determined as follows:

$$R_{cat} = -\frac{d[O_3]}{dt} = k_1[Cl][O_3] = \frac{k_1 k_2[X][O][O_3]}{k_1[O_3] + k_2[O]} \tag{26.76}$$

The composition of the stratosphere is such that $[O_3] \gg [O]$. Taken in combination with the numerical values for k_1 and k_2 presented earlier, the $k_2[O]$ term in the denominator of Equation (26.76) can be neglected, and the rate law expression for the catalyzed reaction becomes

$$R_{cat} = k_2[X][O] \tag{26.77}$$

The ratio of catalyzed to uncatalyzed reaction rates is

$$\frac{R_{cat}}{R_{uc}} = \frac{k_2[X]}{k_{nc}[O_3]} \tag{26.78}$$

In the stratosphere $[O_3]$ is roughly 10^3 greater than $[X]$, and Equation (26.78) becomes

$$\frac{R_{cat}}{R_{uc}} = \frac{k_2}{k_{nc}} \times 10^{-3} = \frac{2.44 \times 10^{10} \, M^{-1} \, s^{-1}}{3.30 \times 10^5 \, M^{-1} \, s^{-1}} \times 10^{-3} \approx 74$$

Therefore, through Cl-mediated catalysis, the rate of O_3 loss is roughly two orders of magnitude greater than the loss through the bimolecular reaction of O_3 and O directly.

Where does stratospheric Cl come from? Rowland and Molina proposed that a major source of Cl was from the photolysis of chlorofluorocarbons such as $CFCl_3$ and CF_2Cl_2, anthropogenic compounds that were common refrigerants at the time. These molecules are extremely robust, and when released into the atmosphere, they readily survive transport through the troposphere and into the stratosphere. Once in the stratosphere, these molecules can absorb a photon of light with sufficient energy to dissociate the C—Cl bond, and Cl is produced. This proposal served as the impetus for understanding the details of stratospheric ozone depletion, and led to the Montreal Protocol in which the vast majority of nations agreed to phase out the industrial use of chlorofluorocarbons.

Heterogeneous catalysts are extremely important in industrial chemistry. The majority of industrial catalysts are solids. For example, the synthesis of NH_3 from N_2 and H_2 is catalyzed using Fe. This is an example of heterogeneous catalysis because the reactant products are in the gas phase, but the catalyst is a solid. An important step in reactions involving solid catalysis is the adsorption of one or more of the reactants to the solid surface. There are two modes of surface interaction. First, a substrate can absorb to the surface without a change in bonding, a process known as **physisorption.** In contrast, the adsorption can occur through bonding, a process known as **chemisorption.** In both cases, a dynamic equilibrium exists between the free and surface-adsorbed species or adsorbate, and information regarding the kinetics of surface adsorption and desorption can be obtained by studying this equilibrium as a function of reactant pressure over the surface of the catalyst. Models for surface binding in heterogeneous catalysis were explored earlier in this book when discussing equilibrium surface binding.

26.5 Photochemistry and Photobiology

Photochemical processes involve the initiation of a chemical reaction through the absorption of a photon by an atom or molecule. In these reactions, photons can be thought of as reactants, and initiation of the reaction occurs when the photon is absorbed. Photochemical reactions are important in a wide variety of areas. The primary event in vision involves the absorption of a photon by the visual pigment rhodopsin. Photosynthesis involves the conversion of light energy into chemical energy by plants and bacteria. Finally, numerous photochemical reactions occur in the atmosphere (e.g., ozone production and decomposition) that are critical to life on Earth. As illustrated by these examples, photochemical reactions are an extremely important area of chemistry, and are explored in this section.

26.5.1 Photophysical Processes

When a molecule absorbs a photon of light, the energy contained in the photon is transferred to the molecule. The amount of energy contained by a photon is given by the Planck equation:

$$E_{photon} = h\nu = \frac{hc}{\lambda} \tag{26.79}$$

In Equation (26.79), h is Planck's constant (6.626×10^{-34} J s), c is the speed of light in a vacuum (3.00×10^8 m s^{-1}), ν is the frequency of light, and λ is the corresponding wavelength of light. A mole of photons is referred to as an Einstein, and the energy contained by an Einstein of photons is Avogadro's number times E_{photon}. The intensity of light is generally stated as energy per unit area per unit time. Because one joule per second is a watt, a typical intensity unit is W cm^{-2}.

The simplest photochemical process is the absorption of a photon by a reactant resulting in product formation:

$$A \xrightarrow{h\nu} P \tag{26.80}$$

The rate of reactant photoexcitation is given by

$$\text{Rate} = -\frac{d[A]}{dt} = \frac{I_{abs}1000}{l} \tag{26.81}$$

In Equation (26.81), I_{abs} is the intensity of absorbed light in units of Einstein cm^{-2} s^{-1}, l is the path length of the sample in centimeters, and 1000 represents the conversion from cubic centimeters to liters such that the rate has appropriate units of M s^{-1}. In Equation (26.81), it is assumed that reactant excitation occurs through the absorption of a single photon. According to the Beer–Lambert law, the intensity of light transmitted through a sample (I_{trans}) is given by

$$I_{trans} = I_0 \, 10^{-\varepsilon l[A]} \tag{26.82}$$

where I_0 is the intensity of incident radiation, ε is the **molar absorptivity** of species A, and $[A]$ is the concentration of reactant. Recall that the molar absorptivity will vary with excitation wavelength. Because $I_{abs} = I_0 - I_{trans}$,

$$I_{abs} = I_0(1 - 10^{-\varepsilon l[A]}) \tag{26.83}$$

The series expansion of the exponential term in Equation (26.83) is

$$10^{-\varepsilon l[A]} = 1 - 2.303\varepsilon l[A] + \frac{(2.303\varepsilon l[A])^2}{2!} - \cdots \tag{26.84}$$

If the concentration of reactant is kept small, only the first two terms in Equation (26.84) are appreciable, and substitution into Equation (26.83) yields

$$I_{abs} = I_0(2.303)\varepsilon l[A] \tag{26.85}$$

Substitution of Equation (26.85) into the rate expression for reactant photoexcitation of Equation (26.81) and integration yield the following expression for $[A]$:

$$[A] = [A]_0 \, e^{-I_0(2303)\varepsilon t} = [A]_0 \, e^{-kt} \tag{26.86}$$

Equation (26.86) demonstrates that the absorption of light will result in the decay of reactant concentration consistent with first-order kinetic behavior. Most photochemical reactions are first order in reactant concentration such that Equation (26.86) describes the evolution in reactant concentration for the majority of photochemical processes. At times it is more useful to discuss photochemical processes with respect to the number of molecules as opposed to concentration. This is precisely the limit one encounters when considering the **photochemistry** of individual molecules as presented later. In this case, Equation (26.81) becomes

$$-\frac{dA}{dt} = I_0 \frac{2303\varepsilon}{N_A} A \tag{26.87}$$

where A represents the number of molecules of reactant and N_A is Avogadro's number. Integrating Equation (26.87), we obtain

$$A = A_0 e^{-I_0(2303\varepsilon/N_A)t} = A_0 e^{-I_0\sigma_A t} \tag{26.88}$$

where σ_A is known as the **absorption cross-section,** and the rate constant for excitation, k_a, is equal to $I_0\sigma_A$ with I_0 in units of photons cm^{-2} s^{-1}.

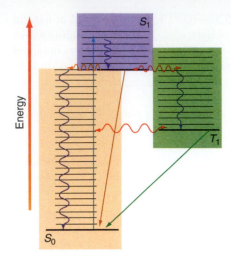

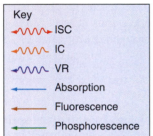

A Jablonski diagram depicting various photophysical processes, where S_0 is the ground electronic singlet state, S_1 is the first excited singlet state, and T_1 is the first excited triplet state. Radiative processes are indicated by the straight lines. The nonradiative processes of intersystem crossing (ISC), internal conversion (IC), and vibrational relaxation (VR) are indicated by the wavy lines.

The absorption of light may occur when the photon energy is equal to the energy difference between two energy states of the molecule. A schematic of the processes that occur following photon absorption resulting in an electronic energy-level transition (or "electronic transition") is given in Figure 26.10. Such diagrams are referred to as **Jablonski diagrams** after Aleksander Jablonski, a Polish physicist who developed these diagrams for describing kinetic processes initiated by electronic transitions. In a Jablonski diagram, the vertical axis represents increasing energy. The electronic states depicted are the ground-state singlet, S_0, first excited singlet, S_1, and triplet, T_1. In the singlet states, the electrons are spin paired such that the spin multiplicity is one (i.e., a "singlet"), and in the triplet state two electrons are unpaired such that the spin multiplicity is three (a "triplet"). The subscripts indicate the energy ordering of the states. Because triplets are generally formed by electronic excitation, the lowest energy triplet state is labeled T_1 as opposed to T_0 (the lowest energy spin configuration of molecular oxygen is a triplet, a famous exception to this generality). Finally, the lowest vibrational level for each electronic state is indicated by dark horizontal lines, with higher vibrational levels indicated by the lighter horizontal lines. In addition, a manifold of rotational states will exist for each vibrational level; however, the rotational energy levels have been suppressed for clarity in Figure 26.10.

The solid and wavy lines in Figure 26.10 represent a variety of processes that couple the electronic states. These processes, including the absorption of light and subsequent energetic relaxation pathways, are referred to as **photophysical processes** because the structure of the molecule remains unchanged. In fact, many of the processes of interest in "photochemistry" do not involve photochemical transformation of the reactant at all, but are instead photophysical in nature. The absorption of light decreases the population in the lowest energy singlet state, S_0, referred to as a *depletion*. Correspondingly, the population in the first excited singlet, S_1, is increased. The absorption transition depicted in Figure 26.10 is to a higher vibrational level in S_1, with the probability of transition to a specific vibrational level determined by the Franck–Condon factor (Chapter 19) between the lowest energy vibrational level in S_0 and the vibrational states in S_1.

After populating S_1, thermal equilibration of the vibrational energy will occur, a process referred to as *vibrational relaxation*. Vibrational relaxation is extremely rapid (~100 fs), and when complete the vibrational state population in S_1 will be governed by the Boltzmann distribution. The vibrational energy-level spacings are assumed to be sufficiently large such that only the lowest vibrational level of S_1 is populated to a significant extent after equilibration. Decay of S_1 resulting in repopulation of S_0 can occur through one of three paths:

1. *Path 1*: Loss of excess electronic energy through the emission of a photon. Such processes are referred to as **radiative transitions.** The process by which photons are emitted in the radiative transitions between S_1 and S_0 is referred to as **fluorescence.** This process is equivalent to the spontaneous emission.

2. *Path 2*: **Intersystem crossing** (ISC in Figure 26.10) resulting in population of T_1. This process involves a change in spin state, a process that is forbidden by quantum mechanics. As such, intersystem crossing is significantly slower than vibrational relaxation, but is competitive with fluorescence in systems where the triplet state is populated to a significant extent. Following intersystem crossing, vibrational relaxation in the triplet vibrational manifold occurs, resulting in population of the lowest energy vibrational level. From this level, a second radiative transition can occur where S_0 is populated and the excess energy is released as a photon. This process is referred to as **phosphorescence.** Because the $T_1 - S_0$ transition also involves a change in spin, it is also forbidden by spin selection rules. Therefore, the rate for this process is slow and phosphorescence occurs over longer timescales (~10^{-6} s to seconds) as compared to fluorescence (~10^{-9} s).

3. *Path 3*: Rather than undergoing a radiative transition, decay from S_1 to a high vibrational level of S_0 can occur followed by rapid vibrational relaxation. This process is referred to as **internal conversion** or nonradiative decay. Nonradiative decay can also occur through the triplet state by crossing to S_0 followed by vibrational relaxation.

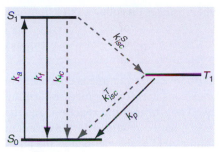

FIGURE 26.11
Kinetic description of photophysical processes. Rate constants are indicated for absorption (k_a), fluorescence (k_f), internal conversion (k_{ic}), intersystem crossing from S_1 to T_1 (k_{isc}^S), intersystem crossing from T_1 to S_0 (k_{isc}^T), and phosphorescence (k_p).

From the viewpoint of kinetics, the absorption of light and subsequent relaxation processes can be viewed as a collection of reactions with corresponding rates. Figure 26.11 presents a modified version of the Jablonski diagram that focuses on these processes and corresponding rate constants. The individual processes, reactions, and notation for the reaction rates are provided in Table 26.1.

26.5.2 Fluorescence and Fluorescence Quenching

The photophysical processes outlined in Table 26.1 are present for any molecular system. To study excited state lifetimes, another photophysical process is introduced: **collisional quenching.** In this process, a collision occurs between a species, Q, and a molecule populating an excited electronic state. The result of the collision is the removal of energy from the molecule with the accompanying conversion of the molecule from S_1 to S_0:

$$S_1 + Q \xrightarrow{k_q} S_0 + Q \qquad (26.89)$$

The rate expression for this process is

$$R_q = k_q[S_1][Q] \qquad (26.90)$$

By studying the rate of collisional quenching as a function of [Q], it is possible to determine the k_f. To demonstrate this procedure, we begin by recognizing that in the kinetic scheme illustrated in Figure 26.11, S_1 can be considered an intermediate species. Under constant illumination, the concentration of this intermediate will not change. Therefore, we can write the differential rate expression for S_1 and apply the steady-state approximation:

$$\frac{d[S_1]}{dt} = 0 = k_a[S_0] - k_f[S_1] - k_{ic}[S_1] - k_{isc}^S[S_1] - k_q[S_1][Q] \quad (26.91)$$

The **fluorescence lifetime,** τ_f, is defined as

$$\frac{1}{\tau_f} = k_f + k_{ic} + k_{isc}^S + k_q[Q] \qquad (26.92)$$

Using this definition of τ_f, Equation (26.91) becomes

$$\frac{d[S_1]}{dt} = 0 = k_a[S_0] - \frac{[S_1]}{\tau_f} \qquad (26.93)$$

Equation (26.93) is readily solved for $[S_1]$:

$$[S_1] = k_a[S_0]\tau_f \qquad (26.94)$$

The fluorescence intensity, I_f, depends on the rate of fluorescence given by

$$I_f = k_f[S_1] \qquad (26.95)$$

Substituting Equation (26.94) into Equation (26.95) results in

$$I_f = k_a[S_0]k_f\tau_f \qquad (26.96)$$

TABLE 26.1   Photophysical Reactions and Corresponding Rate Expressions

Process	Reaction	Rate
Absorption/excitation	$S_0 + h\nu \longrightarrow S_1$	$k_a[S_0] \ (k_a = I_0\sigma_A)$
Fluorescence	$S_1 \longrightarrow S_0 + h\nu$	$k_f[S_1]$
Internal conversion	$S_1 \longrightarrow S_0$	$k_{ic}[S_1]$
Intersystem crossing	$S_1 \longrightarrow T_1$	$k_{isc}^s[S_1]$
Phosphorescence	$T_1 \longrightarrow S_0 + h\nu$	$k_p[T_1]$
Intersystem crossing	$T_1 \longrightarrow S_0$	$k_{isc}^T[T_1]$

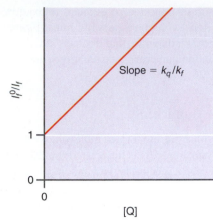

FIGURE 26.12

A Stern–Volmer plot. Intensity of fluorescence as a function of quencher concentration is plotted relative to the intensity in the absence of quencher. The slope of the line provides a measure of the quenching rate constant relative to the rate constant for fluorescence.

Inspection of the last two factors in Equation (26.96) illustrates the following relationship:

$$k_f \tau_f = \frac{k_f}{k_f + k_{ic} + k_{ics}^S + k_q[Q]} = \Phi_f \qquad (26.97)$$

The product of the fluorescence rate constant and fluorescence lifetime is equivalent to the radiative rate constant divided by the sum of rate constants for all processes leading to the decay of S_1. In effect, S_1 decay can be viewed as a **branching reaction,** and the ratio of rate constants contained in Equation (26.97) can be rewritten as the quantum yield for fluorescence, Φ_f. The fluorescence quantum yield is also defined as the number of photons emitted as fluorescence divided by the number of photons absorbed. Comparison of this definition to Equation (26.97) demonstrates that the fluorescence quantum yield will be large for molecules in which k_f is significantly greater than other rate constants corresponding to S_1 decay. Inverting Equation (26.96) and using the definition of τ_f, the following expression is obtained:

$$\frac{1}{I_f} = \frac{1}{k_a[S_0]}\left(1 + \frac{k_{ic} + k_{ics}^S}{k_f}\right) + \frac{k_q[Q]}{k_a[S_0]k_f} \qquad (26.98)$$

In fluorescence quenching experiments, fluorescence intensity is measured as a function of [Q]. Measurements are generally performed by referencing to the fluorescence intensity observed in the absence of quencher, I_f^0, such that

$$\frac{I_f^0}{I_f} = 1 + \frac{k_q}{k_f}[Q] \qquad (26.99)$$

Equation (26.99) reveals that a plot of the fluorescence intensity ratio as a function of [Q] will yield a straight line, with slope equal to k_q/k_f. Such plots are referred to as **Stern–Volmer plots,** an example of which is shown in Figure 26.12.

26.5.3 Measurement of τ_f

In the development presented in the preceding subsection, it was assumed that the system of interest was subjected to continuous irradiation so that the steady-state approximation could be applied to $[S_1]$. However, it is often more convenient to photoexcite the system with a temporally short burst of photons, or pulse of light. If the temporal duration of the pulse is short compared to the rate of S_1 decay, the decay of this state can be measured directly by monitoring the fluorescence intensity as a function of time. Optical pulses as short as 4 fs (4×10^{-15} s) can be produced that provide excitation on a timescale that is significantly shorter than the decay time of S_1.

After excitation by a temporally short optical pulse, the concentration of molecules in $[S_1]$ will be finite. In addition, the rate constant for excitation is zero because $I_0 = 0$; therefore, the differential rate expression for S_1 becomes

$$\frac{d[S_1]}{dt} = -k_f[S_1] - k_{ic}[S_1] - k_{isc}^S[S_1] - k_q[Q][S_1]$$

$$\frac{d[S_1]}{dt} = -\frac{[S_1]}{\tau_f} \qquad (26.100)$$

Equation (26.100) can be solved for $[S_1]$ resulting in

$$[S_1] = [S_1]_0 e^{-t/\tau_f} \qquad (26.101)$$

Because the fluorescence intensity is linearly proportional to $[S_1]$ per Equation (26.95), Equation (26.101) predicts that the fluorescence intensity will undergo exponential decay with time constant τ_f. In the limit where $k_f \gg k_{ic}$ and $k_f \gg k_{isc}^S$, τ_f can be approximated as follows:

$$\lim_{k_f \gg k_{ic}, k_{isc}^S} \tau_f = \frac{1}{k_f + k_q[Q]} \qquad (26.102)$$

In this limit, measurement of the fluorescence lifetime at a known quencher concentration combined with the slope from a Stern–Volmer plot is sufficient to uniquely determine k_f and k_q. Taking the reciprocal of Equation (26.102), we obtain

$$\frac{1}{\tau_f} = k_f + k_q[Q] \tag{26.103}$$

Equation (26.103) demonstrates that a plot of $(\tau_f)^{-1}$ versus [Q] will yield a straight line with y intercept equal to k_f and slope equal to k_q.

EXAMPLE PROBLEM 26.3

Thomaz and Stevens (in *Molecular Luminescence*, Lim, 1969) studied the fluorescence quenching of pyrene in solution. Using the following information, determine k_f and k_q for pyrene in the presence of the quencher Br_6C_6:

$[Br_6C_6]$ (M)	τ_f (s)
0.0005	2.66×10^{-7}
0.001	1.87×10^{-7}
0.002	1.17×10^{-7}
0.003	8.50×10^{-8}
0.005	5.51×10^{-8}

Solution

Using Equation (26.103), a plot of $(\tau_f)^{-1}$ versus [Q] for this system is as follows:

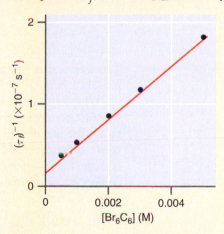

The best fit to the data by a straight line corresponds to a slope of 3.00×10^9 s^{-1}, which is equal to k_q by Equation (26.103), and a y intercept of 1.98×10^6 s^{-1}, which is equal to k_f.

26.5.4 Fluorescence Resonance Energy Transfer

Another fluorescence quenching technique involves the transfer of excitation from one chromophore to another, thereby reducing the excited-state population of the initially photoexcited chromophore and correspondingly the fluorescence from this chromophore. This process, known as fluorescence resonance energy transfer, or FRET, has been extensively used to measure the structure and dynamics of many biological systems. The following mechanism can be employed to describe energy transfer between donor (D) and acceptor (A) chromophores:

$$D \xrightarrow{h\nu} D* \tag{26.104}$$

$$D* \xrightarrow{k_f} D \tag{26.105}$$

$$D^* \xrightarrow{k_{nr}} D \tag{26.106}$$

$$D^* + A \xrightarrow{k_{fret}} D + A^* \tag{26.107}$$

$$A^* \xrightarrow{k_f} A \tag{26.108}$$

In the preceding scheme, the donor is initially photoexcited, resulting in population of the first excited singlet state. Decay from this state can occur through fluorescence, nonradiative decay (representing the sum of internal conversion and intersystem crossing), and resonant energy transfer to A, resulting in this species populating the first excited singlet state (A^*). Decay of the acceptor excited state occurs through fluorescence of this species. Proceeding as in the previous section, in the absence of A the mechanism provides the following expression for the fluorescence quantum yield:

$$\Phi_f = \frac{k_f}{k_f + k_{nr}} \tag{26.109}$$

FRET experiments are generally performed with high-fluorescence quantum yield donors where $k_f \gg k_{nr}$. In the presence of A, the expression for the fluorescence quantum yield becomes

$$\Phi_{fw/FRET} = \frac{k_f}{k_f + k_{nr} + k_{FRET}} \tag{26.110}$$

The efficiency of excitation transfer is related to the ratio of the fluorescence quantum yields as follows:

$$Eff = 1 - \frac{\Phi_{fw/FRET}}{\Phi_f} \tag{26.111}$$

This expression illustrates that as k_{FRET} becomes greater than k_f, the efficiency approaches unity.

What factors influence FRET efficiency? The theory for resonance energy transfer was first developed by T. Förster in the late 1950s. The central ideas inherent in Förster theory are that the efficiency of resonance energy transfer is dependent on the distance between the donor and acceptor, that the absorption band of the acceptor should overlap with the fluorescence band of the donor (that is, $S_0 - S_1$ energy gaps of the donor and acceptors are comparable), and that the relative orientation of the donor and acceptor pair will influence the efficiency of transfer. The distance dependence of the transfer efficiency predicted by Förster theory is

$$Eff = \frac{r_0^6}{r_0^6 + r^6} \tag{26.112}$$

In Equation (26.112), r is the separation distance between donor and acceptor, and r_0 is a pair-dependent quantity that defines the distance at which the transfer efficiency is 0.5. The value of r_0 depends on the spectral overlap of the donor fluorescence and acceptor absorption as well as the relative orientation between donor and acceptor:

$$r_0(\text{Å}) = 8.79 \times 10^{-5} \left(\frac{\kappa^2 \, J \Phi_f}{n^4} \right)^{1/6} \tag{26.113}$$

In this expression, κ depends on the relative orientation of the transition dipole moments, and it is equal to 0 if they are perpendicular, 2 if they are parallel, and 1/3 for random orientation. Because κ can be difficult to determine, the random-orientation value for this quantity is generally assumed. Also in the expression for r_0, Φ_f is the fluorescence quantum yield of the donor, n is the refractive index of the medium in which the trans-

TABLE 26.2   **Values of r_0 for FRET Pairs**

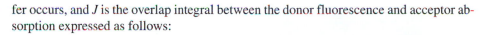

Donor*	Acceptor	r_0 (Å)
EDANS	DABCYL	33
Pyrene	Coumarin	39
Dansyl	Octadecylrhodamine	43
IAEDANS	Fluorescein	46
Fluorescein	Tetramethylrhodamine	55

*EDANS = 5-((2-aminoethyl)amino)naphthalene-1-sulfonic acid; IAEDANS = 5-(((2-iodoacetyl)amino)ethyl) amino)naphthalene-1-sulfonic acid; and DABCYL = 4-((4-(dimethylamino)phenyl)azo)benzoic acid, succinimidyl ester.

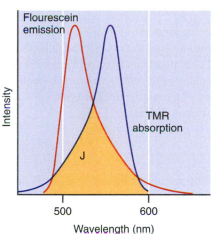

FIGURE 26.13
Illustration of the overlap integral ($J(\lambda)$) for the TMR FRET donor–acceptor pair.

fer occurs, and J is the overlap integral between the donor fluorescence and acceptor absorption expressed as follows:

$$J = \int \varepsilon_A(\lambda) F_D(\lambda) \lambda^4 \, d\lambda \qquad (26.114)$$

In this expression, ε_A is the extinction coefficient of the acceptor which defines the absorption spectrum of this species, F_D the fluorescence spectrum of the donor, and the integral is performed over all wavelengths. The value of this integral, and correspondingly the value for r_0, will vary as a function of donor–acceptor pair. To illustrate the connection between donor emission, acceptor absorption, and the overlap integral, Figure 26.13 presents the emission and absorption for the fluorescein/tetramethylrhodamine (TMR) FRET pair, and provides an illustration of J for this FRET pair. Table 26.2 lists values of r_0 for various FRET pairs.

When using FRET to measure distance, it is critical to choose a donor–acceptor pair whose r_0 is close to the length scale of interest. For distances where $r \gg r_0$, the FRET efficiency will be 0 so that the fluorescence quantum yield of the donor will be largely unaffected by the presence of the acceptor. In the other limit where $r_0 \gg r$, the quenching of the donor fluorescence from energy transfer will be extremely efficient and little emission from the donor will be observed.

EXAMPLE PROBLEM 26.4

You are designing a FRET experiment to determine the structural change introduced by a substrate binding to an enzyme. Using site-specific mutagenesis, you have constructed a mutant form of the enzyme that possesses a single tyrosine and tryptophan residue, and these residues are separated by 11 Å. You would like to determine if the distance between these residues changes with substrate binding. The fluorescence of tyrosine overlaps with the tryptophan absorption; therefore, these two amino acids form a FRET pair for which $r_0 = 9$ Å determined using the absorption and emission spectra in combination with Equation (26.113). Calculate the FRET efficiency at 11-Å separation and how much this distance must increase in order for the efficiency to decrease by 20%, the experimental detection limit.

Solution

Using the initial separation distance and r_0, the efficiency is determined as follows:

$$Eff = \frac{r_0^6}{r_0^6 + r^6} = \frac{(9 \text{ Å})^6}{(9 \text{ Å})^6 + (11 \text{ Å})^6} = 0.23$$

The detection limit corresponds to $Eff = 0.18$. Solving for r yields:

$$Eff = 0.18 = \frac{r_0^6}{r_0^6 + r^6} = \frac{(9 \text{ Å})^6}{(9 \text{ Å})^6 + r^6}$$

$$\frac{(9\,\mathring{A})^6}{0.18} = (9\,\mathring{A})^6 + r^6$$

$$2.42 \times 10^6\,\mathring{A}^6 = r^6$$

$$11.6\,\mathring{A} = r$$

Notice that for this FRET pair the modification of the tyrosine-tryptophan separation accompanying substrate binding can be measured for a relatively limited change in r. This example illustrates the importance of choosing FRET pairs having r_0 values that are close to the length scale of interest.

An important application of resonant energy transfer involves light harvesting in photosynthetic pigments. The absorption of light is accomplished by light harvesting by antenna pigments contained in the thylakoid membranes of the chloroplast, the photosynthetic organelles of green plants and algae. These pigments have evolved such that light primarily in the visible and near-infrared regions of the electromagnetic spectrum is absorbed corresponding to electronic transitions of the pigments. The photosynthetic pigments and their role in photosynthesis are described in Chapter 21.

26.5.5 Single-Molecule Fluorescence

Equation (26.101) describes how the population of molecules in S_1 will evolve as a function of time, and the fluorescence intensity is predicted to demonstrate exponential decay. This predicted behavior is for a collection, or ensemble, of molecules; however, recent spectroscopic techniques and advances in light detection have allowed for the detection of fluorescence from a single molecule. Figure 26.14 presents an image of single molecules obtained using a confocal scanning microscope. In a confocal microscope, the excitation source and image occur at identical focal distances such that fluorescence from sample areas not directly in focus can be rejected. Using this technique in combination with laser excitation and efficient detectors, it is possible to observe the fluorescence from a single molecule. In Figure 26.14, the bright features represent fluorescence from single molecules. The spatial dimension of these features is determined by the diameter of the light beam at the sample (~300 nm).

What does the fluorescence from a single molecule look like as a function of time? Instead of a population of molecules in S_1 being responsible for the emission, the fluorescence is derived from a single molecule. Figure 26.15 presents the observed fluorescence intensity from a single molecule with continuous photoexcitation. Fluorescence is observed after the light field is turned on, and the molecule cycles between S_0 and S_1 due to photoexcitation and subsequent relaxation via fluorescence. This regime of constant

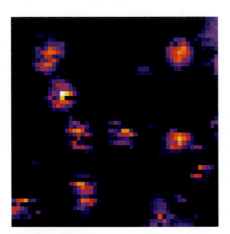

FIGURE 26.14
Microscope image of single Rhodamine B dye molecules on glass. Image was obtained using a confocal scanning microscope with the bright spots in the image corresponding to molecular fluorescence. The image dimension is 5 μm by 5 μm.

FIGURE 26.15
Fluorescence from a single Rhodamine B dye molecule. Steady illumination of the single molecule occurs at t_{on} resulting in fluorescence, I_f. The fluorescence continues until decay of the S_1 state leads to population of a nonfluorescent state. At the end of the time axis, a brief period of fluorescence is observed corresponding to decay of the nonfluorescent state to populate S_0 followed by photoexcitation resulting in the population of S_1 and fluorescence. However, this second period of fluorescence ends abruptly due to photodestruction of the molecule as evidenced by the absence of fluorescence after the decay event (t_{pd}).

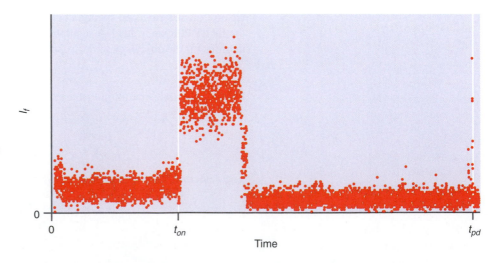

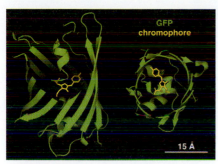

FIGURE 26.16

Green fluorescent protein. The protein is a β-barrel structure with the fluorescent chromophore located in the center of the structure. The chromophore is produced by post-transcriptional oxidation of the Ser65-Tyr66-Gly67 amino-acid sequence in the protein. (Source: Courtesy of Tom Groothuis, www.mekentosj.com/science/fret/gfp.html.)

fluorescence intensity continues until the fluorescence abruptly stops. At this point, depopulation of the S_1 state results in the production of T_1 or some other state of the molecule that does not fluoresce. Eventually, fluorescence is again observed at later times corresponding to the eventual recovery of S_0 by relaxation from these other states, with excitation resulting in the repopulation of S_1 followed by fluorescence. This pattern continues until a catastrophic event occurs in which the structural integrity of the molecules is lost. This catastrophic event is referred to as *photodestruction,* and it results in irreversible photochemical conversion of the molecule to another, nonemissive species.

Clearly, the fluorescence behavior evident in Figure 26.15 is dramatically different from the behavior predicted for an ensemble of molecules. Current interest in this field involves the application of single-molecule techniques to elucidate behavior that is not reflected by the ensemble. Such studies are extremely useful for isolating molecular dynamics from an ensemble of molecules having inherently inhomogeneous behavior. In addition, molecules can be studied in isolation of the bulk, thereby providing a window into the connection between molecular and ensemble behavior.

The fluorescence from individual probe molecules has been recently used to monitor a variety of biological processes. One extremely useful probe is green fluorescent protein (GFP), derived from the jellyfish *Aequorea victoria,* shown in Figure 26.16. The utility of this protein arises from the posttranscriptional formation of a *p*-hydroxybenzylidene-imidazolinone chromophore through cyclization and subsequent oxidation of the Ser65-Tyr66-Gly67 amino-acid sequence. Irradiation of the chromophore with near-ultraviolet light results in the production of green fluorescence (λ_{max} = 508 nm) with high quantum efficiency. These photophysical characteristics have resulted in GFP being employed as a genetically encodable spectroscopic label because the gene that expresses GFP can be incorporated into a gene of interest, and GFP will be formed after transcription. This ability has allowed GFP to be used as a marker for gene and protein expression.

One example of the utility of GFP in biophysical studies is found in the work of Xie and coworkers who have used a derivative of GFP (specifically, yellow fluorescent protein or YFP) to measure gene expression in *Escherichia coli* cells. Figure 26.17 presents a series of images of *E. coli* cells growing as a function of time. The images correspond with the differential interference contrast (DIC) image that provides information on cell shape, and fluorescence images that record the production and location of the expressed membrane protein with YFP. By studying the number of expressed proteins and the times at which expression occurs at the single-molecule level, the researchers are able to address fundamental questions regarding the kinetics and mechanism of gene transcription.

26.5.6 Photochemical Processes

As discussed earlier, photochemical processes are distinct from photophysical processes in that the absorption of a photon results in chemical transformation of the reactant. For a photochemical process that occurs through the first excited singlet state, S_1, a photochemical reaction can be viewed kinetically as another reaction

FIGURE 26.17

Images of gene expression during *E, coli* cell growth as a function of growth time. The images correspond to a combination of differential interference contrast and fluorescence from yellow-fluorescent protein. The DIC images provide information on cellular shape, and the fluorescence images correspond to the bright spots in the images. These spots provide information on the frequency and spatial location of the expressed membrane-protein during cell growth. (Source: Ji Yu, Jie Xiao, Xiaojia Ren, Kaiqin Lao, X. Sunney Xie, "Probing Gene Expression in Live Cells, One Protein Molecule at a Time," Science Vol. 311, 17 March 2006, Fig. 3a)

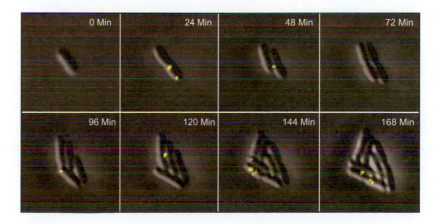

branch resulting in decay of S_1. The corresponding expression for the rate corresponding to this photochemical reaction branch is

$$R_{photochem.} = k_{photo}^S [S_1] \tag{26.115}$$

where k_{photo} is the rate constant for the photochemical reaction. For photochemical processes occurring through T_1, a rate expression similar to Equation (26.115) can be constructed as follows:

$$R_{photochem.} = k_{photo}^T [T_1] \tag{26.116}$$

The absorption of a photon can also provide sufficient energy to initiate a chemical reaction. However, given the range of photophysical processes that occurs, absorption of a photon is not sufficient to guarantee that the photochemical reaction will occur. The extent of photochemistry is quantified by the overall **quantum yield, ϕ**, which is defined as the number of reactant molecules consumed in photochemical processes per photon absorbed. The overall quantum yield can be greater than one, as demonstrated by the photoinitiated decomposition of HI that proceeds by the following mechanism:

$$HI + h\nu \longrightarrow H\bullet + I\bullet \tag{26.117}$$

$$H\bullet + HI \longrightarrow H_2 + I\bullet \tag{26.118}$$

$$I\bullet + I\bullet \longrightarrow I_2 \tag{26.119}$$

In this mechanism, absorption of a photon results in the loss of two HI molecules such that $\phi = 2$. In general, the overall quantum yield can be determined experimentally by comparing the molecules of reactant lost to the number of photons absorbed, as illustrated in Example Problem 26.5.

EXAMPLE PROBLEM 26.5

The reactant 1,3 cyclohexadiene can be photochemically converted to *cis*-hexatriene. In an experiment, 2.5 mmol of cyclohexadiene are converted to *cis*-hexatriene when irradiated with 100 W of 280-nm light for 27 s. All of the light is absorbed by the sample. What is the overall quantum yield for this photochemical process?

Solution

First, the total photon energy absorbed by the sample, E_{abs}, is

$$E_{abs} = (power)\Delta t = (100 \text{ J s}^{-1})(27 \text{ s}) = 2.7 \times 10^3 \text{ J}$$

Next, the photon energy at 280 nm is

$$E_{ph} = \frac{hc}{\lambda} = \frac{(6.626 \times 10^{-34} \text{ J s})(3 \times 10^8 \text{ m s}^{-1})}{2.80 \times 10^{-7} \text{ m}} = 7.10 \times 10^{-19} \text{ J}$$

The total number of photons absorbed by the sample is therefore

$$\frac{E_{abs}}{E_{ph}} = \frac{2.7 \times 10^3 \text{ J}}{7.10 \times 10^{-19} \text{ J photon}^{-1}} = 3.80 \times 10^{21} \text{ photons}$$

Dividing this result by Avogadro's number results in 6.31×10^{-3} Einsteins or moles of photons. Therefore, the overall quantum yield is

$$\phi = \frac{moles_{react}}{moles_{photon}} = \frac{2.50 \times 10^{-3} \text{ mol}}{6.31 \times 10^{-3} \text{ mol}} = 0.396 \approx 0.4$$

An important photochemical process involves molecular isomerization where absorbed photon energy results in a conformational change around a molecular bond. Photoisomerization represents the primary event in vision as illustrated in Figure 26.18. Photons enter the eye and are focused onto the retina at which rod cells containing rhodopsin are located. Rhodopsin is a protein that consists of seven α-helical segments

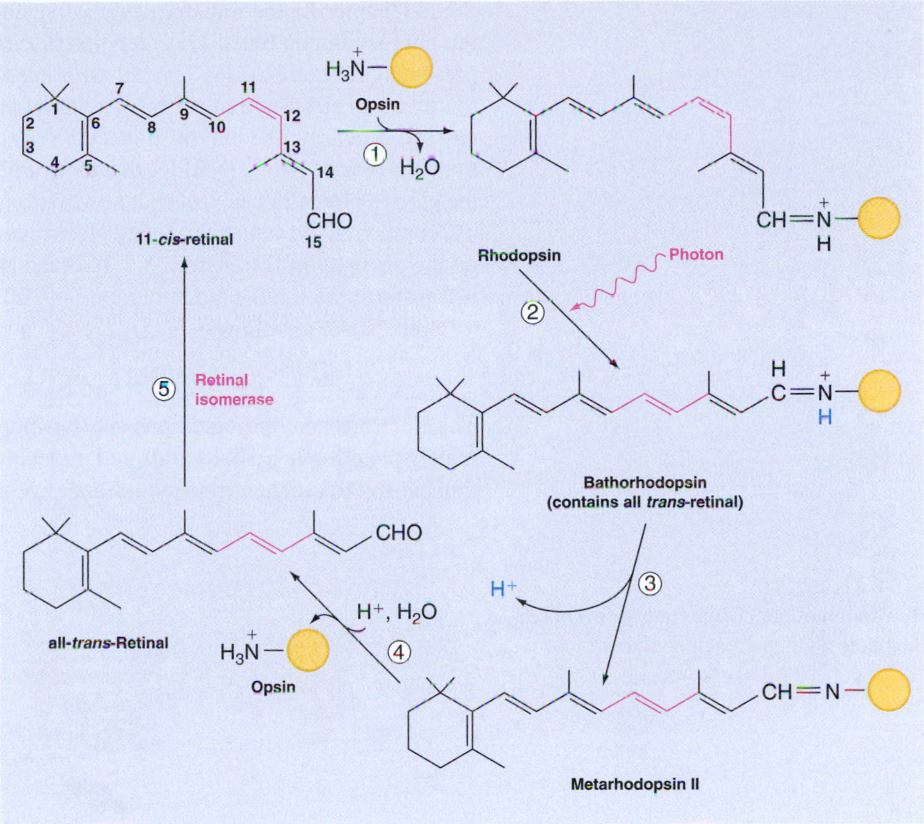

FIGURE 26.18

Transformations of retinal involved in vision. The primary event in vision is the absorption of a photon by 11-*cis*-retinal, resulting in photoisomerization to form the all-*trans* conformation.

to which the visual pigment retinal is attached. Retinal absorbs the photon energy and undergoes an isomerization to form an all-*trans* configuration referred to as bathorhodopsin. The photoisomerization of 11-*cis*-retinal to all-*trans* occurs in approximately 200 fs (200×10^{-15} s!) with a photochemical quantum yield of 0.67. Subsequent conformation changes in the surrounding protein occur and the chromophore deprotonates transferring its proton to the surrounding protein matrix, resulting in the formation of the intermediate species metarhodopsin II, a process that occurs in ~0.5 ms. The formation of metarhodopsin II initiates a series of reactions that ultimately results in an electrical impulse down the optic nerve.

26.6 Electron Transfer

Electron transfer reactions involve the exchange an electron between two chemical species. These reactions are ubiquitous in biological chemistry. For example, photosynthesis involves electron transfer in biological energy transduction. In plants, the net reaction from photosynthesis is the conversion of CO_2 and H_2O to form carbohydrates and molecular oxygen:

$$6CO_2(g) + 6H_2O(l) \longrightarrow C_6H_{12}O_6(s) + 6O_2(g) \qquad \Delta G° = 2870 \text{ kJ}$$

Photosynthesis also occurs in certain bacteria. For example, in green sulfur bacteria H_2S serves as the reactant rather than water:

$$CO_2(g) + 2H_2S(g) \longrightarrow CH_2O(g) + H_2O(l) + 2S(rh) \qquad \Delta G° = 88 \text{ kJ}$$

In these reactions, the carbon in CO_2 is reduced and H_2O/H_2S is oxidized such that net reactions involve the electron transfer. Although photosynthesis is based on a series of coupled reactions rather than the single-step reaction shown earlier, the net reactions do illustrate the importance of electron transfer in this process.

In Chapter 21 the transfer of radiative energy from the light-harvesting complex to the photosynthetic reaction center was discussed. A schematic of the reaction center is presented in Figure 26.19. The radiative energy transferred to the reaction center is used to initiate an electron transfer from a pair of chlorophyll molecules (known as the special pair) to a nearby pheophytin molecule (structurally similar to chlorophyl) in approximately 3 ps (3×10^{-12} s). Notice that there are two arms of the reaction center along which the electron transfer can proceed; however, electron transfer only occurs down the active or A branch of the center. A second electron transfer process occurs in which the electron on the pheophytin is transferred to A-branch quinone in 200 ps. Ultimately, the electron is transferred to the B-branch quinone in 100 μs. The electrons on the quinone are used to "split" water as follows:

$$2H_2O \longrightarrow O_2 + 4H^+ + 4e^-$$

The electrons and hydrogen ions are transported in a series of sequential reactions, ultimately resulting in carbohydrate and molecular oxygen production. The net effect of this transport is to create a transmembrane proton gradient that is used by the cell to drive

FIGURE 26.19

Illustration of the charge transfer events in the photosynthetic reaction center.

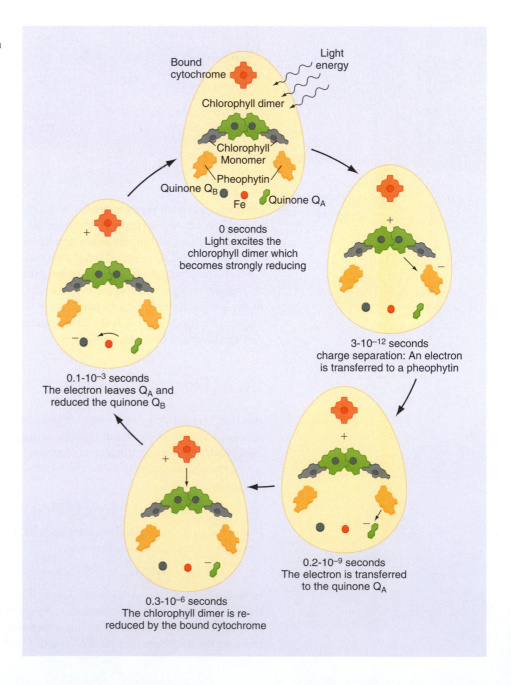

ATP synthesis. Given the importance of electron transfer in photosynthesis and other areas of biological chemistry discussed earlier in the text, in this section we investigate simple kinetic models for charge transfer, and explore Marcus theory for describing electron transfer processes.

26.6.1 Kinetic Model of Electron Transfer

Electron transfer involves the exchange of an electron from a donor molecule (D) to an acceptor molecule (A) resulting in donor oxidation (D^+) and reduction of the acceptor (A^-):

$$D + A \rightleftharpoons D^+A^- \tag{26.120}$$

The reaction mechanism employed to describe bimolecular electron transfer in solution is as follows. First, the donor and acceptor form the donor–acceptor complex by diffusing in solution until these species come in contact. Formation of the donor–acceptor complex is modeled as a reversible process; therefore,

$$D + A \underset{k_{d'}}{\overset{k_d}{\rightleftharpoons}} DA \tag{26.121}$$

In the next step of the mechanism, electron transfer from the donor to the acceptor occurs in the complex. In addition, the back-electron transfer process is also possible corresponding to reformation of the donor-acceptor complex:

$$DA \underset{k_{b-et}}{\overset{k_{et}}{\rightleftharpoons}} D^+A^- \tag{26.122}$$

In the final step of the mechanism, the post-electron transfer complex separates to form an isolated oxidized donor and reduced acceptor:

$$D^+A^- \overset{k_{sep}}{\longrightarrow} D^+ + A^- \tag{26.123}$$

The expression for the rate of reaction consistent with this mechanism is derived as follows. The last step of the mechanism results in the formation of products; therefore, the reaction rate can be written as:

$$R = k_{sep}[D^+A^-] \tag{26.124}$$

Writing the differential-rate expression for D^+A^- and applying the steady-state approximation results in the following expression for $[D^+A^-]$:

$$\frac{d[D^+A^-]}{dt} = 0 = k_{et}[DA] - k_{b-et}[D^+A^-] - k_{sep}[D^+A^-]$$

$$[D^+A^-] = \frac{k_{et}[DA]}{k_{b-et} + k_{sep}} \tag{26.125}$$

The donor–acceptor complex (DA) is also an intermediate; therefore, writing the differential rate expression for this species and applying the steady-state approximation yields

$$\frac{d[DA]}{dt} = 0 = k_d[D][A] - k_{d'}[DA] - k_{et}[DA] + k_{b-et}[D^+A^-]$$

$$[DA] = \frac{k_d[D][A] + k_{b-et}[D^+A^-]}{k_{d'} + k_{et}} \tag{26.126}$$

Substituting these results into the expression for $[D^+A^-]$ results in the following:

$$[D^+A^-] = \frac{k_{et}[DA]}{k_{b-et} + k_{sep}} = \left(\frac{k_{et}}{k_{b-et} + k_{sep}}\right)\left(\frac{k_d[D][A] + k_{b-et}[D^+A^-]}{k_{d'} + k_{et}}\right)$$

$$[D^+A^-] = \frac{k_{et}\,k_d}{k_{b-et}\,k_{d'} + k_{sep}\,k_{d'} + k_{sep}\,k_{et}}[D][A] \tag{26.127}$$

Using this expression for $[D^+A^-]$, the reaction rate becomes:

$$R = \frac{k_{sep}k_{et}k_d}{k_{b-et}k_{d'} + k_{sep}k_{d'} + k_{sep}k_{et}}[D][A] \tag{26.128}$$

The reaction is predicted to be first order with respect to both donor and acceptor concentrations (as expected), and the apparent microscopic rate constant is a composite of the microscopic rate constants for the various steps in the mechanism. If we assume that dissociation of the post-electron transfer complex is rapid compared to back-electron transfer ($k_{sep} \gg k_{b-et}$), then the reaction rate becomes:

$$R = \frac{k_{et}k_d}{k_{d'} + k_{et}}[D][A] \tag{26.129}$$

Finally, if the rate constant for electron transfer is sufficiently large relative to dissociation of the donor–acceptor complex ($k_{et} \gg k_{d'}$), then formation of the donor acceptor becomes the rate-limiting step and electron transfer becomes a diffusion-controlled reaction as described in the previous chapter. In the opposite limit ($k_{et} \ll k_{d'}$), electron transfer becomes the rate-limiting step and the expression for the reaction rate becomes:

$$R_{k_{d'} \gg k_{et}} = k_{et}K_{d,d'}[D][A] = k_{exp}[D][A] \tag{26.130}$$

In the preceding expression, $K_{d,d'}$ is the equilibrium constant for the $D + A \rightleftharpoons DA$ step of the reaction mechanism described earlier, and k_{exp} is the experimentally measured rate constant. In this limit, the rate of electron transfer is determined by the rate constant for the transfer process itself. This later limit is also representative of systems in which the donor and acceptor are already in contact (for example, by covalently linking the donor and acceptor) and in many biological systems in which the electron donor and acceptor are held at a fixed distance by the surrounding protein matrix.

26.6.2 Marcus Theory

What factors determine the magnitude of the observed rate constant for electron transfer? The kinetic model for electron transfer suggests two important factors in determining k_{exp}. First, the model requires that donor and acceptor be in proximity before electron transfer can occur; therefore, the rate constant should depend on separation distance. This distance dependence is expressed as

$$k_{exp} \propto e^{-\beta r} \tag{26.131}$$

where β is a constant that varies as a function of the system of interest and the medium which electron transfer occurs, and r is the donor–acceptor separation distance.

The second factor that influences the observed rate constant is thermodynamic in origin. If the charge transfer occurs over an energy barrier, we can refer back to transition state theory described in Chapter 25 and describe this barrier as the difference in free energy between the donor–acceptor complex and the activated complex corresponding to the free-energy maximum along the reaction Therefore, the experimentally observed rate constant for electron transfer depends on the Gibbs energy of activation, $\Delta G^{\ddagger}$, as follows:

$$k_{exp} \propto e^{-\Delta G^{\ddagger}/kT} \tag{26.132}$$

Therefore, the rate constant for electron transfer can be written as

$$k_{exp} \propto e^{-\beta r}\, e^{-\Delta G^{\ddagger}/kT} \tag{26.133}$$

Rudolph Marcus received the Nobel Prize in Chemistry in 1992 for his contributions to defining the preceding relationship and exploring the chemical factors that are important in determining the value of $\Delta G^{\ddagger}$. Specifically, Marcus noted that after electron transfer both the solvent and solute will undergo relaxation. For the solutes, the transfer of an electron will result in a change in bond order and the nuclei will subsequently relax. In addition, the solvent will rearrange in response to the change in charge distribution accompanying the formation of D^+ and A^-. Taking these factors into account, Marcus defined the Gibbs energy of activation in terms of the standard change in Gibbs energy accompanying the charge transfer ($D + A \rightleftharpoons D^+A^-$) and the change in Gibbs energy accompanying relaxation of the solvent and solute, referred to as the **reorganization energy** or λ:

$$\Delta G^{\ddagger} = \frac{(\Delta G° + \lambda)^2}{4\lambda} \qquad (26.134)$$

The reorganization energy can be thought of as the Gibbs energy accompanying rearrangement of bonds in the donor and acceptor as well as solvent rearrangement of accompanying evolution along the electron-transfer reaction coordinate. The preceding expression demonstrates that the Gibbs energy of activation will be at a minimum when $-\Delta G° = \lambda$. If the separation distance of the donor and acceptor are fixed, then the experimentally observed rate constant will be greatest at this value of $\Delta G°$. A sketch of $\ln(k_{exp})$ versus $-\Delta G°$ is presented in Figure 26.20. The maximum in the rate constant occurs when $-\Delta G° = \lambda$, but notice that the rate constant is predicted to decrease as the charge-transfer state decreases in free energy relative to the neutral state. This is the so-called Marcus inverted regime of electron transfer.

The connection between the rate constant for reaction and the Gibbs energy for the reactants and products is presented in Figure 26.21. Two curves are shown, one representing the neutral donor and acceptor, and the other representing the donor cation and acceptor anion formed as a result of electron transfer. The potential energy surfaces representing donor and acceptor configurations are parabolic, consistent with the quadratic dependence of energy with displacement along nuclear vibrational coordinates [i.e., $V(x) = 1/2kx^2$ where k is the force constant of the bond and x is displacement]. The electron transfer occurs at the point along the reaction coordinate where the two curves cross so that energy is conserved in the transfer. In the normal regime, the value $\Delta G°$ is such that there is a barrier to electron transfer that must be overcome ($\Delta G^{\dagger}$). As the Gibbs energy of the charge transfer state is reduced relative to that of the reactant state, the thermodynamic driving force for the reaction increases, and when $-\Delta G° = \lambda$ the electron transfer is predicted to be a barrierless process corresponding to a maximum in the rate constant for electron transfer. A further decrease in the Gibbs energy of the products relative to the reactants results in an increased barrier to electron transfer and corresponding reduction in the reaction rate constant, corresponding to the Marcus inverted regime.

A significant amount of experimental research has been performed to test the validity of the Marcus theory, and in particular to obtain evidence for the inverted regime. One successful demonstration of the inverted regime is illustrated is Figure 26.22. In

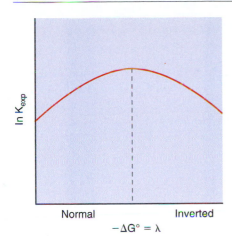

FIGURE 26.20

Illustration of the Marcus normal and inverted regimes. The experimentally measured rate constant is predicted to increase as the driving force for the reaction increases, corresponding to the normal regime, and be at a maximum when $-\Delta G° = \lambda$. As the driving force for the reaction increases beyond this point, the rate constant should decrease corresponding to the inverted regime.

FIGURE 26.21

Plots of the Gibbs energy versus the reaction coordinate for electron transfer. Three different values for $\Delta G°$ are presented corresponding to the normal, maximum, and inverted Marcus regimes as illustrated in Figure 26.22.

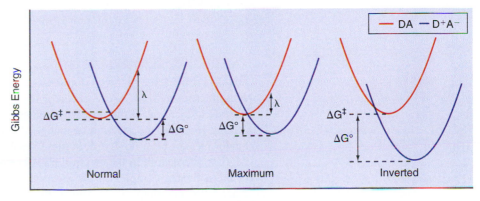

Reaction Coordinate

FIGURE 26.22
Experimental verification of the Marcus inverted regime in electron transfer. In this experiment, the electron transfer rates between the biphenyl donor and the various acceptors was measured in 2-methyl-tetrahydrofuran at 296 K. Notice that the rate constant reaches a maximum and then decreases as $-\Delta G°$ increases, corresponding to an increase in driving force for the reaction. (Source: Reprinted with permission from Miller *et al.*, *J. Am. Chem. Soc.* 106 (1984), 3047.)

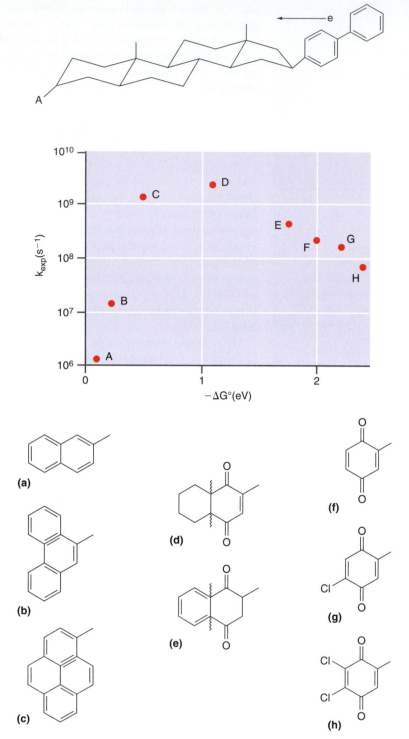

this experiment, the biphenyl donor was connected to a variety of acceptors through a cyclohexane scaffold. As the acceptor is changed, $\Delta G°$ is altered. The data clearly illustrate a decrease in the observed rate constant for electron transfer with increased driving force for the reaction beyond ~1.1 eV. For other systems inverted-regime behavior has not been observed. In these systems, the participation of solute vibrational degrees of freedom in the electron transfer process decreases the barrier to formation of the activated complex in the inverted regime; therefore, the rate constant remains close to the maximum value even as the driving force for the reaction increases.

EXAMPLE PROBLEM 26.6

The experimental results presented earlier demonstrate that a maximum in the rate constant for electron transfer ($\sim 2 \times 10^9$ s^{-1}) occurs when $-G° = -1.20$ eV. Given this observation, estimate the rate constant for electron transfer when 2-napthoquinoyl is employed as the acceptor for which $-G° = -1.93$ eV.

Solution

Because the rate constant for electron transfer is predicted to be at a maximum when $-\Delta G° = \lambda$, therefore, $\lambda = 1.20$ eV. Using this information, the barrier to electron transfer is determined as follows:

$$\Delta G^{\ddagger} = \frac{(\Delta G° + \lambda)^2}{4\lambda} = \frac{(-1.93 \text{ eV} + 1.20 \text{ eV})^2}{4(1.20 \text{ eV})} = 0.111 \text{ eV}$$

With the barrier to electron transfer, the rate constant is estimated as follows:

$$\frac{k_{1.93 \text{ eV}}}{k_{max}} = e^{-\Delta G^{\ddagger}/kT} = e^{\left(-0.111 \text{ eV} \times \frac{1.60 \times 10^{-19} \text{J}}{\text{eV}} \middle/ (1.38 \times 10^{-23} \text{ J K}^{-1})(296 \text{ K}) \right)} = e^{-4.35} = 0.0129$$

$$k_{1.93 \text{ eV}} = k_{max}(0.0129) = (2 \times 10^9 \text{ s}^{-1})(0.0129) = 2.58 \times 10^7 \text{ s}^{-1}$$

For Further Reading

CRC Handbook of Photochemistry and Photophysics. CRC Press, Boca Raton, FL, 1990.

Eyring, H., S. H. Lin, and S. M. Lin, *Basic Chemical Kinetics*. Wiley, New York, 1980.

Fersht, A., *Enzyme Structure and Mechanism*. W. H. Freeman, New York, 1985.

Fersht, A., *Structure and Mechanism in Protein Science*. W. H. Freeman, New York, 1999.

Hammes, G. G., *Thermodynamics and Kinetics for the Biological Sciences*. Wiley, New York, 2000.

Laidler, K. J., *Chemical Kinetics*. Harper & Row, New York, 1987.

Lim, E. G., *Molecular Luminescence*. Benjamin Cummings, Menlo Park, CA, 1969.

Noggle, J. H., *Physical Chemistry*. HarperCollins, New York, 1996.

Pannetier, G., and P. Souchay, *Chemical Kinetics*. Elsevier, Amsterdam, 1967.

Robinson, P. J., and K. A. Holbrook. *Unimolecular Reactions*. Wiley, New York, 1972.

Simons, J. P., *Photochemistry and Spectroscopy*. Wiley, New York, 1971.

Turro, N. J., *Modern Molecular Photochemistry*. Benjamin Cummings, Menlo Park, CA, 1978.

Vocabulary

absorption cross-section
branching reaction
catalyst
chemisorption
collisional quenching
competitive inhibitor
complex reaction
composite constant
enzymes
fluorescence
fluorescence lifetime

heterogeneous catalyst
homogeneous catalyst
internal conversion
intersystem crossing
Jablonski diagram
Lineweaver–Burk equation
Lineweaver–Burk plot
Michaelis constant
Michaelis–Menten
    mechanism
Michaelis–Menten rate law

molar absorptivity
noncompetitive inhibitor
phosphorescence
photochemistry
photophysical processes
physisorption
preequilibrium
    approximation
quantum yield
radiative transitions
reaction intermediate

reaction mechanism
reciprocal plot
reorganization energy
simple reaction
Stern–Volmer plots
stoichiometric number
substrate–catalyst complex
turnover number

Questions on Concepts

Q26.1 How is a simple reaction different from a complex reaction?

Q26.2 For a reaction mechanism to be considered correct, what property must it demonstrate?

Q26.3 What is a reaction intermediate? Can an intermediate be present in the rate law expression for the overall reaction?

Q26.4 What is the preequilibrium approximation, and under what conditions is it considered valid?

Q26.5 How is a catalyst defined, and how does such a species increase the reaction rate?

Q26.6 What is an enzyme? What is the general mechanism describing enzyme catalysis?

Q26.7 What is the Michaelis–Menten rate law? What is the maximum reaction rate predicated by this rate law?

Q26.8 How is the standard enzyme kinetic scheme modified to incorporate competitive inhibition? What plot is used to establish competitive inhibition and to determine the kinetic parameters associated with inhibition?

Q26.9 How can one tell the difference between competitive and noncompetitive enzyme inhibition

Q26.10 What is photochemistry? How does one calculate the energy of a photon?

Q26.11 What depopulation pathways occur from the first excited singlet state? For the first excited triplet state?

Q26.12 What is the expected variation in excited state lifetime with quencher concentration in a Stern–Volmer plot?

Q26.13 What is the distance dependence for fluorescence resonance energy transfer predicted by the Förster model?

Q26.14 What two factors influence the electron transfer rate constant according to Marcus theory?

Problems

Problem numbers in **RED** indicate that the solution to the problem is given in the *Student Solutions Manual*.

P26.1 A proposed mechanism for the formation of N_2O_5 from NO_2 and O_3 is

$$NO_2 + O_3 \xrightarrow{k_1} NO_3 + O_2$$

$$NO_2 + NO_3 + M \xrightarrow{k_2} N_2O_5 + M$$

Determine the rate law expression for the production of N_2O_5 given this mechanism.

P26.2 The Rice–Herzfeld mechanism for the thermal decomposition of acetaldehyde (CH_3CHO) is

$$CH_3CHO \xrightarrow{k_1} CH_3 \cdot + CHO \cdot$$

$$CH_3 \cdot + CH_3CHO \xrightarrow{k_2} CH_4 + CH_2CHO \cdot$$

$$CH_2CHO \cdot \xrightarrow{k_3} CO + CH_3 \cdot$$

$$CH_3 \cdot + CH_3 \cdot \xrightarrow{k_4} C_2H_6$$

Using the steady-state approximation, determine the rate of methane (CH_4) formation.

P26.3 Consider the following mechanism for ozone decomposition:

$$O_3 \underset{k_{-1}}{\overset{k_1}{\rightleftharpoons}} O_2 + O$$

$$O_3 + O \xrightarrow{k_2} 2O_2$$

a. Derive the rate law expression for the loss of O_3.

b. Under what conditions will the rate law expression for O_3 decomposition be first order with respect to O_3?

P26.4 The hydrogen-bromine reaction corresponds to the production of HBr from H_2 and Br_2 as follows: $H_2 + Br_2 \rightleftharpoons 2HBr$. This reaction is famous for its complex rate law, determined by Bodenstein and Lind in 1906:

$$\frac{d[HBr]}{dt} = \frac{k[H_2][Br_2]^{1/2}}{1 + \dfrac{m[HBr]}{[Br_2]}}$$

where k and m are constants. It took 13 years for the correct mechanism of this reaction to be proposed, and this feat was accomplished simultaneously by Christiansen, Herzfeld, and Polyani. The mechanism is as follows:

$$Br_2 \underset{k_{-1}}{\overset{k_1}{\rightleftharpoons}} 2Br \cdot$$

$$Br \cdot + H_2 \xrightarrow{k_2} HBr + H \cdot$$

$$H \cdot + Br_2 \xrightarrow{k_3} HBr + Br \cdot$$

$$HBr + H \cdot \xrightarrow{k_4} H_2 + Br \cdot$$

Construct the rate law expression for the hydrogen-bromine reaction by performing the following steps:

a. Write down the differential rate expression for [HBr].

b. Write down the differential rate expressions for [Br] and [H].

c. Because Br and H are reaction intermediates, apply the steady-state approximation to the result of part (b).

d. Add the two equations from part (c) to determine [Br] in terms of [Br_2].

e. Substitute the expression for [Br] back into the equation for [H] derived in part (c) and solve for [H].

f. Substitute the expressions for [Br] and [H] determined in part (e) into the differential rate expression for [HBr] to derive the rate law expression for the reaction.

P26.5

a. For the hydrogen-bromine reaction presented in Problem P26.4 imagine initiating the reaction with only Br_2 and

H_2 present. Demonstrate that the rate law expression at $t = 0$ reduces to

$$\left(\frac{d[HBr]}{dt}\right)_{t=0} = 2k_2\left(\frac{k_1}{k_5}\right)^{1/2}[H_2]_0[Br_2]_0^{1/2}$$

b. The activation energies for the rate constants are as follows:

Rate Constant	ΔE_a(kJ/mol)
k_1	192
k_2	0
k_5	74

What is the overall activation energy for this reaction?

c. How much will the rate of the reaction change if the temperature is increased to 400 K from 298 K?

P26.6 For the reaction I^- (aq) + OCl^- $(aq)^+$ OI^- (aq) + $Cl^-(aq)$ occurring in aqueous solution, the following mechanism has been proposed:

$$OCl^- + H_2O \underset{k_{-1}}{\overset{k_1}{\rightleftharpoons}} HOCl + OH^-$$

$$I^- + HOCl \overset{k_2}{\longrightarrow} HOI + Cl^-$$

$$HOI + OH^- \overset{k_3}{\longrightarrow} H_2O + OI^-$$

a. Derive the rate law expression for this reaction based on this mechanism. (*Hint:* $[OH^-]$ should appear in the rate law.)

b. The initial rate of reaction was studied as a function of concentration by Chia and Connick [*J. Physical Chemistry* 63 (1959), 1518], and the following data were obtained:

$[I^-]_0$ (M)	$[OCl^-]_0$ (M)	$[OH^-]_0$ (M)	Initial Rate (M s^{-1})
2.0×10^{-3}	1.5×10^{-3}	1.00	1.8×10^{-4}
4.0×10^{-3}	1.5×10^{-3}	1.00	3.6×10^{-4}
2.0×10^{-3}	3.0×10^{-3}	2.00	1.8×10^{-4}
4.0×10^{-3}	3.0×10^{-3}	1.00	7.2×10^{-4}

Is the predicted rate law expression derived from the mechanism consistent with these data?

P26.7 Using the preequilibrium approximation, derive the predicted rate law expression for the following mechanism:

$$A_2 \underset{k_{-1}}{\overset{k_1}{\rightleftharpoons}} 2A$$

$$A + B \overset{k_2}{\longrightarrow} P$$

P26.8 Consider the following mechanism that describes the formation of product P:

$$A \underset{k_{-1}}{\overset{k_1}{\rightleftharpoons}} B \underset{k_{-2}}{\overset{k_2}{\rightleftharpoons}} C$$

$$B \overset{k_3}{\longrightarrow} P$$

If only the species A is present at $t = 0$, what is the expression for the concentration of P as a function of time? You can apply the preequilibrium approximation in deriving your answer.

P26.9 The enzyme fumarase catalyzes the hydrolysis of fumarate: Fumarate + $H_2O \rightarrow$ L-malate. The turnover number for this enzyme is 2.5×10^3 s^{-1}, and the Michaelis constant is 4.2×10^{-6} M. What is the rate of fumarate conversion if the initial enzyme concentration is 1×10^{-6} M and the fumarate concentration is 2×10^{-4} M?

P26.10 The enzyme catalase catalyzes the decomposition of hydrogen peroxide. The following data are obtained regarding the rate of reaction as a function of substrate concentration:

$[H_2O_2]_0$ (M)	0.001	0.002	0.005
Initial Rate (M s^{-1})	1.38×10^{-3}	2.67×10^{-3}	6.00×10^{-3}

The concentration of catalase is 3.5×10^{-9} M. Use these data to determine R_{max}, K_m, and the turnover number for this enzyme.

P26.11 Protein tyrosine phosphatases (PTPases) are a general class of enzymes that are involved in a variety of disease processes including diabetes and obesity. In a study by Z.-Y. Zhang and coworkers [*J. Medicinal Chemistry* 43 (2000), 146], computational techniques were used to identify potential competitive inhibitors of a specific PTPase known as PTP1B. The structure of one of the identified potential competitive inhibitors is shown here:

PTP1B inhibitor

The reaction rate was determined in the presence and absence of inhibitor, I, and revealed the following initial reaction rates as a function of substrate concentration:

[S] (μM)	Rate$_0$ (μM s^{-1}), [I] = 0	Rate$_0$ (μM s^{-1}) [I] = 200 μM
0.299	0.071	0.018
0.500	0.100	0.030
0.820	0.143	0.042
1.22	0.250	0.070
1.75	0.286	0.105
2.85	0.333	0.159
5.00	0.400	0.200
5.88	0.500	0.250

a. Determine K_m and R_{max} for PTP1B.

b. Demonstrate that the inhibition is competitive, and determine K_i.

P26.12 The rate of reaction can be determined by measuring the change in optical rotation of the sample as a function of time if a reactant or product is chiral. This technique is especially useful for kinetic studies of enzyme catalysis involving sugars. For example, the enzyme invertase catalyzes the hydrolysis of sucrose, an optically active sugar. The initial reaction rates as a function of sucrose concentration are as follows:

[Sucrose]$_0$ (M)	Rate (M s^{-1})
0.029	0.182
0.059	0.266
0.088	0.310
0.117	0.330
0.175	0.372
0.234	0.371

Use these data to determine the Michaelis constant and the turnover number for invertase.

P26.13 Peptide bond hydrolysis is performed by a family of enzymes known as serine proteases. The name is derived from a highly conserved serine residue in these enzymes that is critical for enzyme function. One member of this enzyme class is chymotrypsin, which preferentially cleaves proteins at residues sites with hydrophobic side chains such as phenylalanine, leucine, and tyrosine. For example, *N*-benzoyl-tyrosylamide (NBT) and *N*-acetyl-tyrosylamide (NAT) are cleaved by chymotrypsin:

N-benzoyl-tyroslyamide (NBT)   *N*-acetyl-tyroslyamide (NAT)

a. The cleavage of NBT by chymotrypsin was studied and the following reaction rates were measured as a function of substrate concentration:

[NBT] (mM)	1.00	2.00	4.00	6.00	8.00
$Rate_0$ (mM s^{-1})	0.040	0.062	0.082	0.099	0.107

Use these data to determine K_m and R_{max} for chymotrypsin with NBT as the substrate.

b. The cleavage of NAT is also studied and the following reaction rates versus substrate concentration were measured:

[NAT] (mM)	1.00	2.00	4.00	6.00	8.00
$Rate_0$ (mM s^{-1})	0.004	0.008	0.016	0.022	0.028

Use these data to determine K_m and R_{max} for chymotrypsin with NAT as the substrate.

c. Compare the K_m values for chymotrypsin in the presence of NBT versus NAT. What does this difference in K_m imply about the relative binding efficiencies of these substrates?

P26.14 The enzyme glycogen synthase kinase 3β (GSK–3β) plays a central role in Alzheimer's disease. The onset of Alzheimer's disease is accompanied by the production of highly phosphorylated forms of a protein referred to as τ. GSK–3β contributes to the hyperphosphorylation of τ such that inhibiting the activity of this enzyme represents a pathway for the development of an Alzheimer's drug. A compound known as Ro 31-8220 is a competitive inhibitor of GSK-3β. The following data were obtained for the rate of GSK-3β activity in the presence and absence of Ro 31-8220 [A. Martinez *et al.*, *J. Medicinal Chemistry* 45 (2002), 1292]:

[S] (μM)	$Rate_0$ (μM s^{-1}), [I] = 0	$Rate_0$ (μM s^{-1}) [I] = 200 μM
66.7	4.17×10^{-8}	3.33×10^{-8}
40.0	3.97×10^{-8}	2.98×10^{-8}
20.0	3.62×10^{-8}	2.38×10^{-8}
13.3	3.27×10^{-8}	1.81×10^{-8}
10.0	2.98×10^{-8}	1.39×10^{-8}
6.67	2.31×10^{-8}	1.04×10^{-8}

Determine K_m and R_{max} for GSK–3β and, using the data with the inhibitor, determine K_m^* and K_i.

P26.15 The steady-state kinetics of an enzyme are studied in the absence and presence of an inhibitor (I). The following initial rates of reaction were measured as a function of substrate (S) and initiator concentration:

[S]$_0$ (mM)	$Rate_0$ (mM s^{-1}) [I] = 0 μM	$Rate_0$ (mM s^{-1}) [I] = 2 mM	$Rate_0$ (mM s^{-1}) [I] = 5 mM
1.20	5.25	1.65	0.810
2.60	8.25	2.72	1.37
3.31	9.22	3.12	1.59
6.23	11.5	4.11	2.10
10.3	13.0	4.80	2.45

a. Determine K_m and R_{max} for the uninhibited enzyme.

b. Determine the inhibition mechanism and determine R_{max} for the two different inhibitor concentrations.

P26.16 The enzyme glutamate dehydrogenase catalyzes the conversion of glutamate to α-ketoglutarate:

Glutamate α-Ketoglutarate

a. At pH = 7 and 25°C in the presence of 25 μM NAD(P)$^+$, the following reaction rates as a function of glutamate concentration were observed:

Rate (μM s^{-1})	16.6	10.9	6.66	4.76
[Glutamate] (μM)	40.0	11.1	5.02	3.01

Determine K_m and R_{max} for glutamate under these conditions.

b. NAD(P)$^+$ acts as a coenzyme in this reaction; therefore, the catalytic efficiency of glutamate dehydrogenase is dependent on the concentration of this species. This concentration dependence was investigated by Engel and Dalziel [*Biochem. J.* 116 (1969), 621], where in the presence of 9.18 μM NAD(P)$^+$ at pH = 7 and 25°C the following reaction rates as a function of glutamate concentration were observed:

Rate (μM s^{-1})	8.33	4.54	2.33	1.47
[Glutamate] (μM)	40.0	11.1	5.02	3.01

Determine K_m and R_{max} for glutamate dehydrogenase under these conditions, and compare these values to those determined in part (a). What is the effect of decreased NAD(P)$^+$ concentration on the enzyme activity?

P26.17 In the Michaelis–Menten mechanism, it is assumed that the formation of product from the enzyme–substrate complex is irreversible. However, consider the following modified version in which the product formation step is reversible:

$$E + S \underset{k_{-1}}{\overset{k_1}{\rightleftharpoons}} ES \underset{k_{-2}}{\overset{k_2}{\rightleftharpoons}} E + P$$

Derive the expression for the Michaelis constant for this mechanism in the limit where $[S]_0 \gg [E]_0$.

P26.18 In addition to competitive and noncompetitive enzyme inhibition, inhibition can also be *un*competitive. This occurs when binding of the inhibitor can occur only after the substrate is bound. The mechanism for this type of inhibition is:

$$E + S \underset{k_{-1}}{\overset{k_1}{\rightleftharpoons}} ES$$

$$ES \overset{k_2}{\longrightarrow} E + P$$

$$ES + I \underset{k_{-4}}{\overset{k_4}{\rightleftharpoons}} EIS$$

a. Derive the expression for the reaction rate. Is K_m, R_{max}, or both of these quantities affected?

b. Using your expression for the reaction rate, what is the equation for the corresponding reciprocal plot?

c. How does the reciprocal plot for uncompetitive inhibition deviate from those of competitive and noncompetitive inhibition? That is, could you still use reciprocal plots at varying inhibitor concentrations to differentiate between these inhibition mechanisms?

P26.19 Reciprocal plots provide a relatively straightforward way to determine if an enzyme demonstrates Michaelis–Menten kinetics and to determine the corresponding kinetic parameters. However, the slope determined from these plots can require significant extrapolation to regions corresponding to low substrate concentrations. An alternative to the reciprocal plot is the Eadie–Hofstee plot where the reaction rate is plotted versus the rate divided by the substrate concentration and the data are fit to a straight line.

a. Beginning with the general expression for the reaction rate given by the Michaelis–Menten mechanism:

$$R_0 = \frac{R_{max}[S]_0}{[S]_0 + K_m}$$

Rearrange this equation to construct the following expression, which is the basis for the Eadie–Hofstee plot:

$$R_0 = R_{max} - K_m\left(\frac{R_0}{[S]_0}\right)$$

b. Using an Eadie–Hofstee plot, determine R_{max} and K_m for hydrolysis of sugar by the enzyme invertase using the following data:

[Sucrose]$_0$ (M)	Rate (M s^{-1})
0.029	0.182
0.059	0.266
0.088	0.310
0.117	0.330
0.175	0.372

P26.20 Sunburn is caused primarily by sunlight in what is known as the UVB band, or the wavelength range from 290 to 320 nm. The minimum dose of radiation needed to create a sunburn (erythema) is known as a MED (minimum erythema

dose). The MED for a person of average resistance to burning is 50 mJ cm^{-2}.

a. Determine the number of 290-nm photons corresponding to the MED, assuming each photon is absorbed. Repeat this calculation for 320-nm photons.

b. At 20° latitude, the solar flux in the UVB band at the surface of the earth is 1.45 mW cm^{-2}. Assuming that each photon is absorbed, how long would a person with unprotected skin be able to stand in the sun before acquiring one MED?

P26.21 A likely mechanism for the photolysis of acetaldehyde is

$$CH_3CHO + h\nu \longrightarrow CH_3\cdot + CHO\cdot$$

$$CH_3\cdot + CH_3CHO \overset{k_1}{\longrightarrow} CH_4 + CH_3CO\cdot$$

$$CH_3CO\cdot \overset{k_2}{\longrightarrow} CO + CH_3\cdot$$

$$CH_3\cdot + CH_3\cdot \overset{k_3}{\longrightarrow} C_2H_6$$

Derive the rate law expression for the formation of CO based on this mechanism.

P26.22 If $\tau_f = 1 \times 10^{-10}$ s and $k_{ic} = 5 \times 10^8$ s^{-1}, what is ϕ_f? Assume that the rate constants for intersystem crossing and quenching are sufficiently small that these processes can be neglected.

P26.23 The quantum yield for CO production in the photolysis of gaseous acetone is unity for wavelengths between 250 and 320 nm. After 20 min of irradiation at 313 nm, 18.4 cm^3 of CO (measured at 1008 Pa and 22°C) is produced. Calculate the number of photons absorbed and the absorbed intensity in J s^{-1}.

P26.24 If 10% of the energy of a 100-W incandescent bulb is in the form of visible light having an average wavelength of 600 nm, how many quanta of light are emitted per second from the lightbulb?

P26.25 For phenanthrene, the measured lifetime of the triplet state, τ_p, is 3.3 s, the fluorescence quantum yield is 0.12, and the phosphorescence quantum yield is 0.13 in an alcohol-ether glass at 77 K. Assume that no quenching and no internal conversion from the singlet state occurs. Determine k_p, k_{isc}^T and k_{isc}^S/k_r.

P26.26 In this problem you will investigate the parameters involved in a single-molecule fluorescence experiment. Specifically, the incident photon power needed to see a single molecule with a reasonable signal-to-noise ratio will be determined.

a. Rhodamine dye molecules are typically employed in such experiments because their fluorescence quantum yields are large. What is the fluorescence quantum yield for Rhodamine B (a specific rhodamine dye) where $k_r = 1 \times 10^9$ s^{-1} and $k_{ic} = 1 \times 10^8$ s^{-1}? You can ignore intersystem crossing and quenching in deriving this answer.

b. If care is taken in selecting the collection optics and detector for the experiment, a detection efficiency of

10% can be readily achieved. Furthermore, detector dark noise usually limits these experiments, and dark noise on the order of 10 counts s^{-1} is typical. If we require a signal-to-noise ratio of 10:1, then we will need to detect 100 counts s^{-1}. Given the detection efficiency, a total emission rate of 1000 fluorescence photons s^{-1} is required. Using the fluorescence quantum yield and a molar extinction coefficient for Rhodamine B of ~40,000 M^{-1} cm^{-1}, what is the intensity of light needed in this experiment in terms of photons cm^{-2} s^{-1}?

c. The smallest diameter focused spot one can obtain in a microscope using conventional refractive optics is one-half the wavelength of incident light. Studies of Rhodamine B generally employ 532-nm light such that the focused-spot diameter is ~270 nm. Using this diameter, what incident power in watts is required for this experiment? Don't be surprised if this value is relatively modest.

P26.27 A central issue in the design of aircraft is improving the lift of aircraft wings. To assist in the design of more efficient wings, wind-tunnel tests are performed in which the pressures at various parts of the wing are measured generally using only a few localized pressure sensors. Recently, pressure-sensitive paints have been developed to provide a more detailed view of wing pressure. In these paints, a luminescent molecule is dispersed into an oxygen-permeable paint and the aircraft wing is painted. The wing is placed into an airfoil, and luminescence from the paint is measured. The variation in O_2 pressure is measured by monitoring the luminescence intensity, with lower intensity demonstrating areas of higher O_2 pressure due to quenching.

a. The use of platinum octaethylprophyrin (PtOEP) as an oxygen sensor in pressure-sensitive paints was described by Gouterman and coworkers [Review of Scientific Instruments 61 (1990), 3340]. In this work, the following relationship between luminescence intensity and pressure was derived: $I_0/I = A + B(P/P_0)$ where I_0 is the fluorescence intensity at ambient pressure P_0, and I is the fluorescence intensity at an arbitrary pressure P. Determine coefficients A and B in the preceding expression using the Stern–Volmer equation: $k_{total} = 1/\tau_l = k_r + k_q[Q]$. In this equation τ_l is the luminescence lifetime, k_r is the luminescent rate constant, and k_q is the quenching rate constant. In addition, the luminescent intensity ratio is equal to the ratio of luminescence quantum yields at ambient pressure, Φ_0, and an arbitrary pressure, Φ:

$$\Phi_0/\Phi = I_0/I.$$

b. Using the following calibration data of the intensity ratio versus pressure observed for PtOEP, determine A and B:

I_0/I	P/P_0	I_0/I	P/P_0
1.0	1.0	0.65	0.46
0.9	0.86	0.61	0.40
0.87	0.80	0.55	0.34
0.83	0.75	0.50	0.28
0.77	0.65	0.46	0.20
0.70	0.53	0.35	0.10

c. At an ambient pressure of 1 atm, $I_0 = 50,000$ (arbitrary units) and 40,000 at the front and back of the wing. The wind tunnel is turned on to a speed of Mach 0.36 and the measured luminescence intensity is 65,000 and 45,000 at the respective locations. What is the pressure differential between the front and back of the wing?

P26.28 Oxygen sensing is important in biological studies of many systems. The variation in oxygen content of sapwood trees was measured by del Hierro and coworkers [J. Experimental Biology 53 (2002), 559] by monitoring the luminescence intensity of $[Ru(dpp)_3]^{2+}$ immobilized in a sol-gel that coats the end of an optical fiber implanted into the tree. As the oxygen content of the tree increases, the luminescence from the ruthenium complex is quenched. The quenching of $[Ru(dpp)_3]^{2+}$ by O_2 was measured by Bright and coworkers [Applied Spectroscopy 52 (1998), 750] and the following data were obtained:

I_0/I	% O_2
3.6	12
4.8	20
7.8	47
12.2	100

a. Construct a Stern–Volmer plot using the data supplied in the table. For $[Ru(dpp)_3]^{2+}$ and $k_r = 1.77 \times 10^5$ s^{-1}, what is k_q?

b. Comparison of the Stern–Volmer prediction to the quenching data led the authors to suggest that some of the $[Ru(dpp)_3]^{2+}$ molecules are located in sol-gel environments that are not equally accessible to O_2. What led the authors to this suggestion?

P26.29 The pyrene/coumarin FRET pair ($r_0 = 39$ Å) is used to study the fluctuations in enzyme structure during the course of a reaction. Computational studies suggest that the pair will be separated by 35 Å in one conformation, and 46 Å in a second configuration. What is the expected difference in FRET efficiency between these two conformational states?

P26.30 In a FRET experiment designed to monitor conformational changes in T4 lysozyme, the fluorescence intensity fluctuates between 5000 and 10,000 counts per section. Assuming that 7500 counts represents a FRET efficiency of 0.5, what is the change in FRET pair separation distance during the reaction? For the tetramethylrhodamine/Texas red FRET pair employed, $r_0 = 50$ Å.

P26.31 One complication when using FRET is that if fluctuations in the local environment can affect the S_0–S_1 energy gap for the donor or acceptor. Explain how this fluctuation would impact a FRET experiment.

P26.32 In Marcus theory for electron transfer, the reorganization energy is partitioned into solvent and solute contributions. Modeling the solvent as a dielectric continuum, the solvent reorganization energy is given by

$$\lambda_{sol} = \frac{(\Delta e)^2}{4\pi\varepsilon_0}\left(\frac{1}{d_1} + \frac{1}{d_2} - \frac{1}{r}\right)\left(\frac{1}{n^2} - \frac{1}{\varepsilon}\right)$$

Where Δe is the amount of charge transferred, d_1 and d_2 are the ionic diameters of ionic products, r is the separation distance of the reactants, n^2 is the index of refraction of the surrounding medium, and ε is the dielectric constant of the medium. In addition, $(4\pi\varepsilon_0)^{-1} = 8.99 \times 10^9$ J m C^{-2}.

a. For an electron transfer in water ($n = 1.33$ and $\varepsilon = 80$), for which the ionic diameters of both species are 6 Å and the separation distance is 15 Å, what is the expected solvent reorganization energy?

b. Redo the above calculation for the the same reaction occurring in a protein. The dielectric constant of a protein is dependent on sequence, structure, and the amount on included water; however, a dielectric constant of 4 is generally assumed consistent with a hydrophobic environment.

P26.33 An experiment is performed in which the rate constant for electron transfer is measured as a function of distance by attaching an electron donor an acceptor to pieces of DNA of varying length. The measured rate constant for electron transfer as a function of separation distance is as follows:

k_{exp} (s^{-1})	2.10×10^8	2.01×10^7	2.07×10^5	204
Distance (Å)	14	17	23	32

a. Determine the value for β that defines the distance dependence of the electron transfer rate constant.

b. It has been proposed that DNA can serve as an electron "π-way" that facilitates electron transfer over long distances. Using the rate constant at 17 Å presented earlier, what value of β would result in the rate of electron transfer decreasing by only a factor of 10 at a separation distance of 23 Å?

Web-Based Simulations, Animations, and Problems

W26.1 In this problem, Michaelis–Menten enzyme kinetics are investigated, specifically, the variation in reaction rate with substrate concentration for three enzymes having significantly different Michaelis–Menten kinetic parameters. Students will investigate how the maximum reaction rate and overall kinetics depend on K_m and turnover number. Finally, hand calculations of enzyme kinetic parameters performed by the students are compared to the results obtained by simulation.

APPENDIX A

Math Supplement

A.1 Working with Complex Numbers and Complex Functions

Imaginary numbers can be written in the form

$$z = a + ib \tag{A.1}$$

where a and b are real numbers and $i = \sqrt{-1}$. It is useful to represent complex numbers in the complex plane shown in Figure A.1. The vertical and horizontal axes correspond to the imaginary and real parts of z, respectively.

In the representation shown in Figure A.1, a complex number corresponds to a point in the complex plane. Note the similarity to the polar coordinate system. Because of this analogy, a complex number can be represented either as the pair (a,b), or by the magnitude of the radius vector r and the angle θ. From Figure A.1, it can be seen that

$$r = \sqrt{a^2 + b^2} \text{ and } \theta = \cos^{-1}\frac{a}{r} = \sin^{-1}\frac{b}{r} = \tan^{-1}\frac{b}{a} \tag{A.2}$$

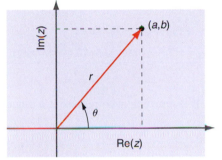

FIGURE A.1

Using the relations between a, b, and r, as well as the Euler relation, $e^{i\theta} = \cos\theta + i\sin\theta$, a complex number can be represented in either of two equivalent ways:

$$a + ib = r\cos\theta + r\sin\theta = re^{i\theta} = \sqrt{a^2 + b^2}\exp[i\tan^{-1}(b/a)] \tag{A.3}$$

If a complex number is represented in one way, it can easily be converted to the other way. For example, we express the complex number $6 - 7i$ in the form $re^{i\theta}$. The magnitude of the radius vector r is given by $\sqrt{6^2 + 7^2} = \sqrt{85}$. The phase is given by $\tan\theta = (-7/6)$ or $\theta = \tan^{-1}(-7/6)$. Therefore, we can write $6 - 7i$ as $\sqrt{85}\exp[i\tan^{-1}(-7/6)]$.

In a second example, we convert the complex number $2e^{i\pi/2}$, which is in the $re^{i\theta}$ notation, to the $a + ib$ notation. Using the relation $e^{i\alpha} = \exp(i\alpha) = \cos\alpha + i\sin\alpha$, we can write $2e^{i\pi/2}$ as

$$2\left(\cos\frac{\pi}{2} + i\sin\frac{\pi}{2}\right) = 2(0 + i) = 2i$$

The complex conjugate of a complex number z is designated by z^* and is obtained by changing the sign of i, wherever it appears in the complex number. For example, if $z = (3 - \sqrt{5}i)e^{i\sqrt{2}\phi}$, then $z^* = (3 + \sqrt{5}i)e^{-i\sqrt{2}\phi}$. The magnitude of a complex number is defined by $\sqrt{zz^*}$ and is always a real number. This is the case for the previous example:

$$zz^* = (3 - \sqrt{5}i)e^{i\sqrt{2}\phi}(3 + \sqrt{5}i)e^{-i\sqrt{2}\phi} = (3 - \sqrt{5}i)(3 + \sqrt{5}i)e^{i\sqrt{2}\phi - i\sqrt{2}\phi} = 14 \tag{A.4}$$

Note also that $zz^* = a^2 + b^2$.

FIGURE A.2

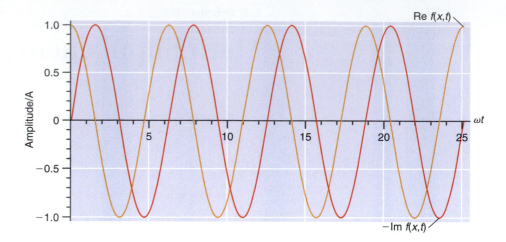

Complex numbers can be added, multiplied, and divided just like real numbers. A few examples follow:

$$(3 + \sqrt{2}i) + (1 - \sqrt{3}i) = [4 + (\sqrt{2} - \sqrt{3})i]$$

$$(3 + \sqrt{2}i)(1 - \sqrt{3}i) = 3 - 3\sqrt{3}i + \sqrt{2}i - \sqrt{6}i^2 = (3 + \sqrt{6}) + (\sqrt{2} - 3\sqrt{3})i$$

$$\frac{(3 + \sqrt{2}i)}{(1 - \sqrt{3}i)} = \frac{(3 + \sqrt{2}i)}{(1 - \sqrt{3}i)}\frac{(1 + \sqrt{3}i)}{(1 + \sqrt{3}i)} = \frac{3 + 3\sqrt{3}i + \sqrt{2}i + \sqrt{6}i^2}{4} = \frac{(3 - \sqrt{6}) + (3\sqrt{3} + \sqrt{2})i}{4}$$

Functions can depend on a complex variable. It is convenient to represent a plane traveling wave usually written in the form

$$\psi(x,t) = A\sin(kx - \omega t) \tag{A.5}$$

in the complex form

$$Ae^{i(kx - \omega t)} = A\cos(kx - \omega t) - iA\sin(kx - \omega t) \tag{A.6}$$

Note that

$$\psi(x,t) = -\text{Im}\,Ae^{i(kx - \omega t)} \tag{A.7}$$

The reason for working with the complex form rather than the real form of a function is that calculations such as differentiation and integration can be carried out more easily. Waves in classical physics have real amplitudes, because their amplitudes are linked directly to observables. For example, the amplitude of a sound wave is the local pressure that arises from the expansion or compression of the medium through which the wave passes. However, in quantum mechanics, observables are related to $|\psi(x,t)|^2$ rather than $\psi(x,t)$. Because $|\psi(x,t)|^2$ is always real, $\psi(x,t)$ can be complex, and the observables associated with the wave function are still real.

For the complex function $f(x,t) = Ae^{i(kx - \omega t)}$, $zz^* = \psi(x,t)\psi^*(x,t) = Ae^{i(kx - \omega t)}$ $A^*e^{-i(kx - \omega t)} = AA^*$, so that the magnitude of the function is a constant and does not depend on t or x. As Figure A.2 shows, the real and imaginary parts of $Ae^{i(kx - \omega t)}$ depend differently on the variables x and t; they are phase shifted by $\pi/2$. The figure shows the amplitudes of the real and imaginary parts as a function of ωt for $x = 0$.

A.2 Differential Calculus

A.2.1 The First Derivative of a Function

The derivative of a function has as its physical interpretation the slope of the function evaluated at the position of interest. For example, the slope of the function $y = x^2$ at the point $x = 1.5$ is indicated by the line tangent to the curve shown in Figure A.3.

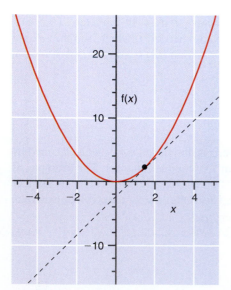

FIGURE A.3

Mathematically, the first derivative of a function $f(x)$ is denoted $f'(x)$ or $df(x)/dx$. It is defined by

$$\frac{df(x)}{dx} = \lim_{h \to 0} \frac{f(x + h) - f(x)}{h} \tag{A.8}$$

For the function of interest,

$$\frac{df(x)}{dx} = \lim_{h \to 0} \frac{(x + h)^2 - (x)^2}{h} = \lim_{h \to 0} \frac{2hx + h^2}{h} = \lim_{h \to 0} 2x + h = 2x \tag{A.9}$$

To define $df(x)/dx$ over an interval in x, $f(x)$ must be continuous over the interval.

Based on this example, $df(x)/dx$ can be calculated if $f(x)$ is known. Several useful rules for differentiating commonly encountered functions are listed next:

$$\frac{d(ax^n)}{dx} = anx^{n-1}, \text{ where } a \text{ is a constant and } n > 0 \tag{A.10}$$

For example, $d(\sqrt{3}x^{4/3})/dx = (4/3)\sqrt{3}x^{1/3}$

$$\frac{d(ae^{bx})}{dx} = abe^{bx}, \text{ where } a \text{ and } b \text{ are constants} \tag{A.11}$$

For example, $d(5e^{3\sqrt{2}x})/dx = 15\sqrt{2}e^{3\sqrt{2}x}$

$$\frac{d(ae^{bx})}{dx} = abe^{bx}, \text{ where } a \text{ and } b \text{ are constants}$$

$$\frac{d(a \sin x)}{dx} = a \cos x, \text{ where } a \text{ is a constant} \tag{A.12}$$

$$\frac{d(a \cos x)}{dx} = -a \sin x, \text{ where } a \text{ is a constant}$$

Two rules are used to evaluate the derivative of a function that is itself the sum or product of two functions. The first is as follows:

$$\frac{d[f(x) + g(x)]}{dx} = \frac{df(x)}{dx} + \frac{dg(x)}{dx} \tag{A.13}$$

$$\frac{d(x^3 + \sin x)}{dx} = \frac{dx^3}{dx} + \frac{d \sin x}{dx} = 3x^2 + \cos x$$

The second rule is:

$$\frac{d[f(x)g(x)]}{dx} = g(x)\frac{df(x)}{dx} + f(x)\frac{dg(x)}{dx} \tag{A.14}$$

For example,

$$\frac{d[\sin(x)\cos(x)]}{dx} = \cos(x)\frac{d \sin(x)}{dx} + \sin(x)\frac{d \cos(x)}{dx}$$

$$= \cos^2 x - \sin^2 x$$

A.2.2 The Reciprocal Rule and the Quotient Rule

How is the first derivative calculated if the function to be differentiated does not have a simple form such as those listed in the preceding section? In many cases, the derivative can be found by using the product and quotient rules stated here:

$$\frac{d\left(\dfrac{1}{f(x)}\right)}{dx} = -\frac{1}{[f(x)]^2}\frac{df(x)}{dx} \tag{A.15}$$

For example,

$$\frac{d\left(\frac{1}{\sin x}\right)}{dx} = -\frac{1}{\sin^2 x}\frac{d\sin x}{dx} = \frac{-\cos x}{\sin^2 x}$$

$$\frac{d\left[\frac{f(x)}{g(x)}\right]}{dx} = \frac{g(x)\frac{df(x)}{dx} - f(x)\frac{dg(x)}{dx}}{[g(x)]^2} \tag{A.16}$$

For example,

$$\frac{d\left(\frac{x^2}{\sin x}\right)}{dx} = \frac{2x\sin x - x^2\cos x}{\sin^2 x}$$

A.2.3 The Chain Rule

In this section, we deal with the differentiation of more complicated functions. Suppose that $y = f(u)$ and $u = g(x)$. From the previous section, we know how to calculate $df(u)/du$. How do we calculate $df(u)/dx$? The answer to this question is stated as the chain rule:

$$\frac{df(u)}{dx} = \frac{df(u)}{du}\frac{du}{dx} \tag{A.17}$$

Several examples illustrating the chain rule follow:

$$\frac{d\sin(3x)}{dx} = \frac{d\sin(3x)}{d(3x)}\frac{d(3x)}{dx} = 3\cos(3x)$$

$$\frac{d\ln(x^2)}{dx} = \frac{d\ln(x^2)}{d(x^2)}\frac{d(x^2)}{dx} = \frac{2x}{x^2} = \frac{2}{x}$$

$$\frac{d\left(x + \frac{1}{x}\right)^{-4}}{dx} = \frac{d\left(x + \frac{1}{x}\right)^{-4}}{d\left(x + \frac{1}{x}\right)}\frac{d\left(x + \frac{1}{x}\right)}{dx} = -4\left(x + \frac{1}{x}\right)^{-5}\left(1 - \frac{1}{x^2}\right)$$

$$\frac{d\exp(ax^2)}{dx} = \frac{d\exp(ax^2)}{d(ax^2)}\frac{d(ax^2)}{dx} = 2ax\exp(ax^2), \quad \text{where } a \text{ is a constant}$$

A.2.4 Higher Order Derivatives: Maxima, Minima, and Inflection Points

A function $f(x)$ can have higher order derivatives in addition to the first derivative. The second derivative of a function is the slope of a graph of the slope of the function versus the variable. Mathematically,

$$\frac{d^2 f(x)}{dx^2} = \frac{d}{dx}\left(\frac{df(x)}{dx}\right) \tag{A.18}$$

For example,

$$\frac{d^2\exp(ax^2)}{dx^2} = \frac{d}{dx}\left[\frac{d\exp(ax^2)}{dx}\right] = \frac{d[2ax\exp(ax^2)]}{dx}$$

$$= 2a\exp(ax^2) + 4a^2x^2\exp(ax^2), \quad \text{where } a \text{ is a constant}$$

The second derivative is useful in identifying where a function has its minimum or maximum value within a range of the variable, as shown next.

Because the first derivative is zero at a local maximum or minimum, $df(u)/dx = 0$ at the values x_{max} and x_{min}. Consider the function $f(x) = x^3 - 5x$ shown in Figure A.4 over the range $-2.75 \leq x \leq 2.75$.

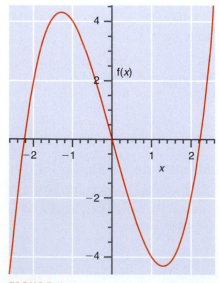

FIGURE A.4

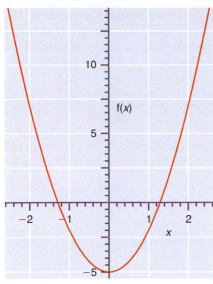

FIGURE A.5

By taking the derivative of this function and setting it equal to zero, we find the minima and maxima of this function in the range

$$\frac{d(x^3 - 5x)}{dx} = 3x^2 - 5 = 0, \text{ which has the solutions } x = \pm\sqrt{\frac{5}{3}} = 1.291$$

The maxima and minima can also be determined by graphing the derivative and finding the zero crossings as shown in Figure A.5.

Graphing the function clearly shows that the function has one maximum and one minimum in the range specified. What criterion can be used to distinguish between these extrema if the function is not graphed? The sign of the second derivative, evaluated at the point for which the first derivative is zero, can be used to distinguish between a maximum and a minimum:

$$\frac{d^2 f(x)}{dx^2} = \frac{d}{dx}\left[\frac{df(x)}{dx}\right] < 0 \text{ for a maximum} \tag{A.19}$$

$$\frac{d^2 f(x)}{dx^2} = \frac{d}{dx}\left[\frac{df(x)}{dx}\right] > 0 \text{ for a minimum}$$

We return to the function graphed earlier and calculate the second derivative:

$$\frac{d^2(x^3 - 5x)}{dx^2} = \frac{d}{dx}\left[\frac{d(x^3 - 5x)}{dx}\right] = \frac{d(3x^2 - 5)}{dx} = 6x$$

By evaluating

$$\frac{d^2 f(x)}{dx^2} \text{ at } x = \pm\sqrt{\frac{5}{3}} = \pm1.291$$

we see that $x = 1.291$ corresponds to the minimum, and $x = -1.291$ corresponds to the maximum.

If a function has an inflection point in the interval of interest, then

$$\frac{df(x)}{dx} = 0 \quad \text{and} \quad \frac{d^2 f(x)}{dx^2} = 0 \tag{A.20}$$

An example for an inflection point is $x = 0$ for $f(x) = x^3$. A graph of this function in the interval $-2 \le x \le 2$ is shown in Figure A.6. As you can verify,

$$\frac{dx^3}{dx} = 3x^2 = 0 \text{ at } x = 0 \quad \text{and} \quad \frac{d^2(x^3)}{dx^2} = 6x = 0 \text{ at } x = 0$$

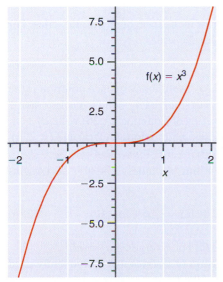

FIGURE A.6

A.2.5 Maximizing a Function Subject to a Constraint

A frequently encountered problem is that of maximizing a function relative to a constraint. We first outline how to carry out a constrained maximization, and subsequently apply the method to maximizing the volume of a cylinder while minimizing its area. The theoretical framework for solving this problem originated with the French mathematician Lagrange, and the method is known as Lagrange's method of undetermined multipliers. We wish to maximize the function $f(x,y)$ subject to the constraint that $\phi(x,y) - C = 0$, where C is a constant. For example, you may want to maximize the area, A, of a rectangle while minimizing its circumference, C. In this case, $f(x, y) = A(x, y) = xy$ and $\phi(x, y) = C(x, y) = 2(x + y)$, where x and y are the length and width of the rectangle. The total differentials of these functions are given by Equation (A.21):

$$df = \left(\frac{\partial f}{\partial x}\right)_y dx + \left(\frac{\partial f}{\partial y}\right)_x dy = 0 \quad \text{and} \quad d\phi = \left(\frac{\partial \phi}{\partial x}\right)_y dx + \left(\frac{\partial \phi}{\partial y}\right)_x dy = 0 \tag{A.21}$$

If x and y were independent variables (that is, no constraining relationship exists), the maximization problem would be identical to those dealt with earlier. However, because

$d\phi = 0$ also needs to be satisfied, x and y are not independent variables. In this case, Lagrange found that the appropriate function to minimize is $f - \lambda\phi$, where λ is an undetermined multiplier. He showed that each of the expressions in the square brackets in the differential given by Equation (A.22) can be maximized independently. A separate multiplier is required for each constraint:

$$df = \left[\left(\frac{\partial f}{\partial x}\right)_y - \lambda\left(\frac{\partial \phi}{\partial x}\right)_y\right]dx + \left[\left(\frac{\partial f}{\partial y}\right)_x - \lambda\left(\frac{\partial \phi}{\partial y}\right)_x\right]dy \qquad \text{(A.22)}$$

We next use this method to maximize the volume, V, of a cylindrical can subject to the constraint that its exterior area, A, be minimized. The functions f and ϕ are given by

$$V = f(r,h) = \pi r^2 h \quad \text{and} \quad A = \phi(r,h) = 2\pi r^2 + 2\pi rh \qquad \text{(A.23)}$$

Calculating the partial derivatives and using Equation (A.22), we have

$$\left(\frac{\partial f(r,h)}{r}\right)_h = 2\pi rh \quad \left(\frac{\partial f(r,h)}{h}\right)_r = \pi r^2$$

$$\left(\frac{\partial \phi(r,h)}{r}\right)_h = 4\pi r + 2\pi h \quad \left(\frac{\partial \phi(r,h)}{h}\right)_r = 2\pi r \qquad \text{(A.24)}$$

$$(2\pi rh - \lambda[4\pi r + 2\pi h])dr = 0 \quad \text{and} \quad (\pi r^2 - \lambda 2\pi r)dh = 0$$

Eliminating λ from these two equations gives

$$\frac{2\pi rh}{4\pi r + 2\pi h} = \frac{\pi r^2}{2\pi r} \qquad \text{(A.25)}$$

Solving for h in terms of r gives the result $h = 2r$. Note that there is no need to determine the value of multiplier λ. Perhaps you have noticed that beverage cans do not follow this relationship between r and h. Can you think of factors other than minimizing the amount of metal used in the can that might be important in this case?

A.3 Series Expansions of Functions

A.3.1 Convergent Infinite Series

Physical chemists often express functions of interest in the form of an infinite series. For this application, the series must converge. Consider the series

$$a_0 + a_1 x + a_2 x^2 + a_3 x^3 + \ldots a_n x^n + \ldots \qquad \text{(A.26)}$$

How can we determine if such a series converges? A useful convergence criterion is the ratio test. If the absolute ratio of successive terms (designated u_{n-1} and u_n) is less than 1 as $n \to \infty$, the series converges. We consider the series of Equation (A.26) with (a) $a_n = n!$ and (b) $a_n = 1/n!$, and apply the ratio test as shown in Equations (A.27a) and (A.27b).

$$(a) \lim_{n\to\infty}\left|\frac{u_n}{u_{n-1}}\right| = \left|\frac{n!x^n}{(n-1)!x^{n-1}}\right| = \lim_{n\to\infty} > 1 \text{ unless } x = 0 \qquad \text{(A.27a)}$$

$$(b) \lim_{n\to\infty}\left|\frac{u_n}{u_{n-1}}\right| = \left|\frac{x^n/n!}{x^{n-1}/(n-1)!}\right| = \lim_{n\to\infty}\left|\frac{x}{n}\right| < 1 \text{ for all } x \qquad \text{(A.27b)}$$

We see that the infinite series converges if $a_n = 1/n!$ but diverges if $a_n = n!$.

The power series is a particularly important form of a series that is frequently used to fit experimental data to a functional form. It has the form

$$a_0 + a_1 x + a_2 x^2 + a_3 x^3 + a_1 x + a_4 x^4 + \cdots = \sum_{n=0}^{\infty} a_n x^n \qquad \text{(A.28)}$$

Fitting a data set to a series with a large number of terms is impractical, and to be useful, the series should contain as few terms as possible to satisfy the desired accuracy. For example, the function $\sin x$ can be fit to a power series over the interval $0 \leq x \leq 1.5$ by the following truncated power series:

$$\sin x \approx -1.20835 \times 10^{-3} + 1.02102x - 0.0607398x^2 - 0.11779x^3$$

$$\sin x \approx -8.86688 \times 10^{-5} + 0.996755x + 0.0175769x^2 - 0.200644x^3 - 0.027618x^4 \quad \text{(A.29)}$$

The coefficients in Equation (A.29) have been determined using a least squares fitting routine. The first series includes terms in x up to x^3, and is accurate to within 2% over the interval. The second series includes terms up to x^4, and is accurate to within 0.1% over the interval. Including more terms will increase the accuracy further.

A special case of a power series is the geometric series, in which successive terms are related by a constant factor. An example of a geometric series and its sum is given in Equation (A.30). Using the ratio criterion of Equations (A.27), convince yourself that the following series converges for $|x| < 1$:

$$a(1 + x + x^2 + x^3 + \cdots) = \frac{a}{1 - x}, \quad \text{for } |x| < 1 \quad \text{(A.30)}$$

A.3.2 Representing Functions in the Form of Infinite Series

Assume that you have a function in the form $f(x)$ and wish to express it as a power series in x of the form

$$f(x) = a_0 + a_1 x + a_2 x^2 + a_3 x^3 + \ldots \quad \text{(A.31)}$$

To do so, we need a way to find the set of coefficients $(a_0, a_1, a_2, a_3, \ldots)$. How can this be done?

If the functional form $f(x)$ is known, the function can be expanded about a point of interest using the Taylor–Mclaurin expansion. In the vicinity of $x = a$, the function can be expanded in the series

$$f(x) = f(a) + \left(\frac{df(x)}{dx}\right)_{x=a}(x - a) + \frac{1}{2!}\left(\frac{d^2 f(x)}{dx^2}\right)_{x=a}(x - a)^2$$

$$+ \frac{1}{3!}\left(\frac{d^3 f(x)}{dx^3}\right)_{x=a}(x - a)^3 + \cdots \quad \text{(A.32)}$$

For example, consider the expansion of $f(x) = e^x$ about $x = 0$. Because $(d^n e^x / dx^n)_{x=0} = 1$ for all values of n, the Taylor–Mclaurin expansion for e^x about $x = 0$ is

$$f(x) = 1 + x + \frac{1}{2!}x^2 + \frac{1}{3!}x^3 + \cdots \quad \text{(A.33)}$$

Similarly, the Taylor–Mclaurin expansion for $\ln(1 + x)$ is found by evaluating the derivatives in turn:

$$\frac{d \ln(1 + x)}{dx} = \frac{1}{1 + x}$$

$$\frac{d^2 \ln(1 + x)}{dx^2} = \frac{d}{dx}\frac{1}{(1 + x)} = -\frac{1}{(1 + x)^2}$$

$$\frac{d^3 \ln(1 + x)}{dx^3} = -\frac{d}{dx}\frac{1}{(1 + x)^2} = \frac{2}{(1 + x)^3}$$

$$\frac{d^4 \ln(1 + x)}{dx^4} = \frac{d}{dx}\frac{2}{(1 + x)^3} = \frac{-6}{(1 + x)^4}$$

Each of these derivatives must be evaluated at $x = 0$.

Using these results, the Taylor–Mclaurin expansion for $\ln(1 + x)$ about $x = 0$ is

$$f(x) = x - \frac{x^2}{2!} + \frac{2x^3}{3!} - \frac{6x^4}{4!} + \cdots = x - \frac{x^2}{2} + \frac{x^3}{3} - \frac{x^4}{4} + \cdots \quad \text{(A.34)}$$

The number of terms that must be included to adequately represent the function depends on the value of x. For $-1 \ll x \ll 1$, the series converges rapidly and, to a very good approximation, we can truncate the Taylor–Mclaurin series after the first one or two terms involving the variable. For the two functions just considered, it is reasonable to write $e^x \approx 1 + x$ and $\ln(1 \pm x) \approx \pm x$ if $-1 \ll x \ll 1$.

A second widely used series is the Fourier sine and cosine series. This series can be used to expand functions that are periodic over an interval $-L \le x \le L$ by the series

$$f(x) = \frac{1}{2} b_0 + \sum_n b_n \cos \frac{n\pi x}{L} + \sum_n a_n \sin \frac{n\pi x}{L} \tag{A.35}$$

A Fourier series is an infinite series, and the coefficients a_n and b_n can be calculated using the equations

$$a_n = \frac{1}{L} \int_{-L}^{+L} f(x) \sin \frac{n\pi x}{L}\, dx \quad \text{and} \quad b_n = \frac{1}{L} \int_{-L}^{+L} f(x) \cos \frac{n\pi x}{L}\, dx \tag{A.36}$$

The usefulness of the Fourier series is that a function can often be approximated by a few terms, depending on the accuracy desired.

For functions that are either even or odd with respect to the variable x, only either the sine or the cosine terms will appear in the series. For even functions, $f(-x) = f(x)$, and for odd functions, $f(-x) = -f(x)$. Because $\sin(-x) = -\sin(x)$ and $\cos(-x) = \cos(x)$, all coefficients a_n are zero for an even function, and all coefficients b_n are zero for an odd function. Note that Equations (A.29) are not odd functions of x because the function was only fit over the interval $0 \le x \le 1.5$.

Whereas the coefficients for the Taylor–Mclaurin series can be readily calculated, those for the Fourier series require more effort. To avoid mathematical detail here, the Fourier coefficients a_n and b_n are not explicitly calculated for a model function. It is much easier to calculate the coefficients using a program such as *Mathematica* than to calculate them by hand. Our focus here is to show that periodic functions can be approximated to a reasonable degree by using the first few terms in a Fourier series, rather than to carry out the calculations.

To demonstrate the usefulness of expanding a function in a Fourier series, consider the function

$$f(x) = 1 \qquad \text{for } 0 \le x \le L$$

$$f(x) = -x \quad \text{for } -L \le x \le 0 \tag{A.37}$$

which is periodic in the interval $-L \le x \le L$, in a Fourier series. This function is a demanding function to expand in a Fourier series because the function is discontinuous at $x = 0$ and the slope is discontinuous at $x = 0$ and $x = 1$. The function and the approximate functions obtained by truncating the series at $n = 2$, $n = 5$, and $n = 10$ are shown in Figure A.7. The agreement between the truncated series and the function is reasonably good for $n = 10$. The oscillations seen near $x/L = 0$ are due to the discontinuity in the function. More terms in the series are required to obtain a good fit, because of the discontinuities in the function and its slope.

A.4 Integral Calculus

A.4.1 Definite and Indefinite Integrals

In many areas of physical chemistry, the property of interest is the integral of a function over an interval in the variable of interest. For example, the total probability of finding a particle within an interval $0 \le x \le a$ is the integral of the probability density $P(x)$ over the interval

$$P_{total} = \int_0^a P(x)\, dx \tag{A.38}$$

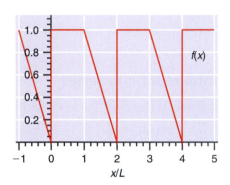

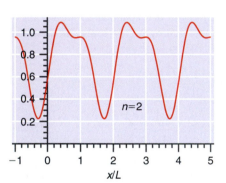

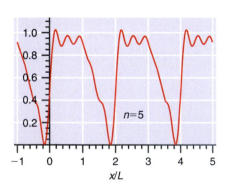

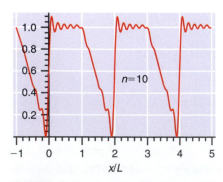

FIGURE A.7

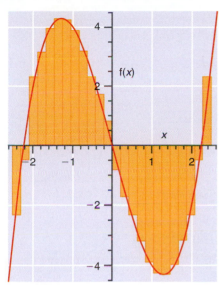

FIGURE A.8

Geometrically, the integral of a function over an integral is the area under the curve describing the function. For example, the integral $\int_{-2.3}^{2.3} (x^3 - 5x)\, dx$ is the sum of the areas of the individual rectangles in Figure A.8 in the limit within which the width of the rectangles approaches zero. If the rectangles lie below the zero line, the incremental area is negative; if the rectangles lie above the zero line, the incremental area is positive. In this case the total area is zero because the total negative area equals the total positive area. This is the case because $f(x)$ is an odd function of x.

The integral can also be understood as an antiderivative. From this point of view, the integral symbol is defined by the relation

$$f(x) = \int \frac{df(x)}{dx}\, dx \tag{A.39}$$

and the function that appears under the integral sign is called the integrand. Interpreting the integral in terms of area, we evaluate a definite integral, and the interval over which the integration occurs is specified. The interval is not specified for an indefinite integral.

The geometrical interpretation is often useful in obtaining an integral from experimental data when the functional form of the integrand is not known. For our purposes, the interpretation of the integral as an antiderivative is more useful. The value of the indefinite integral $\int (x^3 - 5x)\, dx$ is that function which, when differentiated, gives the integrand. Using the rules for differentiation discussed earlier, you can verify that

$$\int (x^3 - 5x)\, dx = \frac{x^4}{4} - \frac{5x^2}{2} + C \tag{A.40}$$

Note the constant that appears in the evaluation of every indefinite integral. By differentiating the function obtained upon integration, you should convince yourself that any constant will lead to the same integrand. In contrast, a definite integral has no constant of integration. If we evaluate the definite integral

$$\int_{-2.3}^{2.3} (x^3 - 5x)\, dx = \left(\frac{x^4}{4} - \frac{5x^2}{2} + C \right)_{x=2.3} - \left(\frac{x^4}{4} - \frac{5x^2}{2} + C \right)_{x=-2.3} \tag{A.41}$$

we see that the constant of integration cancels. Because the function obtained upon integration is an even function of x, $\int_{-2.3}^{2.3} (x^3 - 5x)\, dx = 0$, just as we saw in the geometric interpretation of the integral.

It is useful for the student of physical chemistry to commit the integrals listed next to memory, because they are encountered frequently. These integrals are directly related to the derivatives discussed in Section A.2:

$$\int df(x) = f(x) + C$$

$$\int x^n\, dx = \frac{x^{n+1}}{n+1} + C$$

$$\int \frac{dx}{x} = \ln x + C$$

$$\int e^{ax} = \frac{e^{ax}}{a} + C, \quad \text{where } a \text{ is a constant}$$

$$\int \sin x\, dx = -\cos x + C$$

$$\int \cos x\, dx = \sin x + C$$

However, the primary tool for the physical chemist in evaluating integrals is a good set of integral tables. The integrals that are most frequently used in elementary quantum mechanics are listed here; the first group lists indefinite integrals:

$$\int (\sin ax) \, dx = -\frac{1}{a} \cos ax + C$$

$$\int (\cos ax) \, dx = \frac{1}{a} \sin ax + C$$

$$\int (\sin^2 ax) \, dx = \frac{1}{2}x - \frac{1}{4a} \sin 2ax + C$$

$$\int (\cos^2 ax) \, dx = \frac{1}{2}x + \frac{1}{4a} \sin 2ax + C$$

$$\int (x^2 \sin^2 ax) \, dx = \frac{1}{6}x^3 - \left(\frac{1}{4a}x^2 - \frac{1}{8a^3}\right) \sin 2ax - \frac{1}{4a^2} x \cos 2ax + C$$

$$\int (x^2 \cos^2 ax) \, dx = \frac{1}{6}x^3 + \left(\frac{1}{4a}x^2 - \frac{1}{8a^3}\right) \sin 2ax + \frac{1}{4a^2} x \cos 2ax + C$$

$$\int x^m e^{ax} \, dx = \frac{x^m e^{ax}}{a} - \frac{m}{a} \int x^{m-1} e^{ax} \, dx + C$$

$$\int \frac{e^{ax}}{x^m} \, dx = -\frac{1}{m-1} \frac{e^{ax}}{x^{m-1}} + \frac{a}{m-1} \int \frac{e^{ax}}{x^{m-1}} \, dx + C$$

The following group lists definite integrals:

$$\int_0^a \sin\left(\frac{n\pi x}{a}\right) \times \sin\left(\frac{m\pi x}{a}\right) dx = \int_0^a \cos\left(\frac{n\pi x}{a}\right) \times \cos\left(\frac{m\pi x}{a}\right) dx = \frac{a}{2}\delta_{mn}$$

$$\int_0^a \left[\sin\left(\frac{n\pi x}{a}\right)\right] \times \left[\cos\left(\frac{n\pi x}{a}\right)\right] dx = 0$$

$$\int_0^\pi \sin^2 mx \, dx = \int_0^\pi \cos^2 mx \, dx = \frac{\pi}{2}$$

$$\int_0^\infty \frac{\sin x}{\sqrt{x}} \, dx = \int_0^\infty \frac{\cos x}{\sqrt{x}} \, dx = \sqrt{\frac{\pi}{2}}$$

$$\int_0^\infty x^n e^{-ax} \, dx = \frac{n!}{a^{n+1}} \quad (a > 0, \, n \text{ positive integer})$$

$$\int_0^\infty x^{2n} e^{-ax^2} \, dx = \frac{1 \cdot 3 \cdot 5 \cdots (2n-1)}{2^{n+1} a^n} \sqrt{\frac{\pi}{a}} \quad (a > 0, \, n \text{ positive integer})$$

$$\int_0^\infty x^{2n+1} e^{-ax^2} \, dx = \frac{n!}{2 \, a^{n+1}} \quad (a > 0, \, n \text{ positive integer})$$

$$\int_0^\infty e^{-ax^2} dx = \left(\frac{\pi}{4a}\right)^{1/2}$$

In the first integral above, $\delta_{mn} = 1$ if $m = n$, and 0 if $m \neq n$.

A.4.2 Multiple Integrals and Spherical Coordinates

In the previous section, integration with respect to a single variable was discussed. Often, however, integration occurs over two or three variables. For example, the wave functions for the particle in a two-dimensional box are given by

$$\psi_{n_x n_y}(x,y) = N \sin\frac{n_x \pi x}{a} \sin\frac{n_y \pi y}{b} \tag{A.42}$$

In normalizing a wave function, the integral of $|\psi_{n_x n_y}(x,y)|^2$ is required to equal one over the range $0 \le x \le a$ and $0 \le y \le b$. This requires solving the double integral

$$\int_0^b dy \int_0^a \left(N \sin\frac{n_x \pi x}{a} \sin\frac{n_y \pi y}{b} \right)^2 dx = 1 \tag{A.43}$$

to determine the normalization constant N. We sequentially integrate over the variables x and y or vice versa using the list of indefinite integrals from the previous section.

$$\int_0^b dy \int_0^a \left(N \sin\frac{n_x \pi x}{a} \sin\frac{n_y \pi y}{b} \right)^2 dx = \left[\frac{1}{2}x - \frac{a}{4n\pi}\sin\frac{2n_x\pi x}{a} \right]_{x=0}^{x=a} \times N^2 \int_0^b \left(\sin\frac{n_y \pi y}{b} \right)^2 dy$$

$$1 = \left[\frac{1}{2}a - \frac{a}{4n\pi}(\sin 2n_x\pi - 0) \right] \times N^2 \int_0^b \left(\sin\frac{n_y \pi y}{b} \right)^2 dy$$

$$1 = N^2 \left[\frac{1}{2}a - \frac{a}{4n\pi}(\sin 2n_x\pi - 0) \right] \times \left[\frac{1}{2}b - \frac{a}{4n\pi}(\sin 2n_y\pi - 0) \right] = \frac{N^2 ab}{4}$$

$$N = \frac{2}{\sqrt{ab}}$$

Convince yourself that the normalization constant for the wave functions of the three-dimensional particle in the box

$$\psi_{n_x n_y n_z}(x, y, z) = N \sin\frac{n_x \pi x}{a} \sin\frac{n_y \pi y}{b} \sin\frac{n_z \pi z}{c} \tag{A.44}$$

has the value $N = 2\sqrt{2}/\sqrt{abc}$.

Up to this point, we have considered functions of a single variable. This restricts us to dealing with a single spatial dimension. The extension to three independent variables becomes important in describing three-dimensional systems. The three-dimensional system of most importance to us is the atom. Closed-shell atoms are spherically symmetric, so we might expect atomic wave functions to be best described by spherical coordinates. Therefore, you should become familiar with integrations in this coordinate system. In transforming from spherical coordinates r, θ, and ϕ to Cartesian coordinates x, y, and z, the following relationships are used:

$$x = r \sin\theta \cos\phi$$

$$y = r \sin\theta \sin\phi \tag{A.45}$$

$$z = r \cos\theta$$

These relationships are depicted in Figure A.9. For small increments in the variables r, θ, and ϕ, the volume element depicted in this figure is a rectangular solid of volume

$$dV = (r \sin\theta \, d\phi)(dr)(rd\theta) = r^2 \sin\theta \, dr \, d\theta \, d\phi \tag{A.46}$$

Note in particular that the volume element in spherical coordinates is not $dr \, d\theta \, d\phi$ in analogy with the volume element $dx \, dy \, dz$ in Cartesian coordinates.

In transforming from Cartesian coordinates x, y, and z to the spherical coordinates r, θ, and ϕ, these relationships are used:

$$r = \sqrt{x^2 + y^2 + z^2} \quad \theta = \cos^{-1}\frac{z}{\sqrt{x^2 + y^2 + z^2}} \quad \text{and } \phi = \tan^{-1}\frac{y}{x} \tag{A.47}$$

What is the appropriate range of variables to integrate over all space in spherical coordinates? If we imagine the radius vector scanning over the range $0 \le \theta \le \pi$; $0 \le \phi \le 2\pi$, the whole angular space is scanned. If we combine this range of θ and ϕ with $0 \le r \le \infty$, all of the three-dimensional space is scanned. Note that $r = \sqrt{x^2 + y^2 + z^2}$ is always positive.

FIGURE A.9

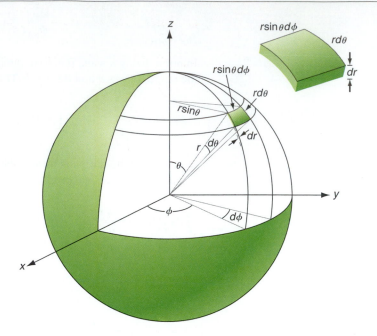

To illustrate the process of integration in spherical coordinates, we normalize the function $e^{-r} \cos \theta$ over the interval $0 \leq r \leq \infty;\ 0 \leq \theta \leq \pi;\ 0 \leq \phi \leq 2\pi$:

$$N^2 \int_0^{2\pi} d\phi \int_0^{\pi} \sin \theta\, d\theta \int_0^{\infty} (e^{-r} \cos \theta)^2 r^2\, dr = N^2 \int_0^{2\pi} d\phi \int_0^{\pi} \cos^2 \theta \sin \theta\, d\theta \int_0^{\infty} r^2 e^{-2r}\, dr = 1$$

It is most convenient to integrate first over ϕ, giving

$$2\pi N^2 \int_0^{\pi} \cos^2 \theta \sin \theta\, d\theta \int_0^{\infty} r^2 e^{-2r}\, dr = 1$$

We next integrate over θ, giving

$$2\pi N^2 \left[\frac{-\cos^3 \pi + \cos^3 0}{3} \right] \times \int_0^{\infty} r^2 e^{-2r}\, dr = \frac{4\pi N^2}{3} \int_0^{\infty} r^2 e^{-2r}\, dr = 1$$

We finally integrate over r using the standard integral

$$\int_0^{\infty} x^n e^{-ax}\, dx = \frac{n!}{a^{n+1}} (a > 0,\ n \text{ positive integer})$$

The result is

$$\frac{4\pi N^2}{3} \int_0^{\infty} r^2 e^{-2r}\, dr = \frac{4\pi N^2}{3} \frac{2!}{8} = 1 \quad \text{or} \quad N = \sqrt{\frac{3}{\pi}}$$

We conclude that the normalized wave function is $\sqrt{3/\pi}\ e^{-r} \cos \theta$.

A.5 Vectors

The use of vectors occurs frequently in physical chemistry. Consider circular motion of a particle at constant speed in two dimensions, as depicted in Figure A.10. The particle is moving in a counterclockwise direction on the ring-like orbit. At any instant in time,

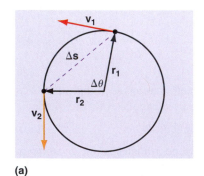

(a)

(b)

FIGURE A.10

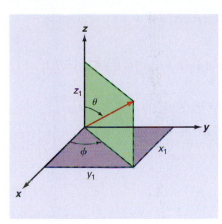

FIGURE A.11

its position, velocity, and acceleration can be measured. The two aspects to these measurements are the magnitude and the direction of each of these observables. Whereas a scalar quantity such as speed has only a magnitude, a vector has both a magnitude and a direction.

For the particular case under consideration, the position vectors $\mathbf{r}_1$ and $\mathbf{r}_2$ extend outward from the origin and terminate at the position of the particle. The velocities $\mathbf{v}_1$ and $\mathbf{v}_2$ are related to the position vector as $\mathbf{v} = \lim_{\Delta t \to 0} [\mathbf{r}(t + \Delta t) - \mathbf{r}(t)]/\Delta t$. Therefore, the velocity vector is perpendicular to the position vector. The acceleration vector is defined by $\mathbf{a} = \Delta \mathbf{v}/\Delta t = \lim_{\Delta t \to 0} [\mathbf{v}(t + \Delta t) - \mathbf{v}(t)]/\Delta t$. As we see in part (b) of Figure A.10, $\mathbf{a}$ is perpendicular to $\mathbf{v}$ and is antiparallel to $\mathbf{r}$. As this example of a relatively simple motion shows, vectors are needed to describe the situation properly by keeping track of both the magnitude and direction of each of the observables of interest. For this reason, it is important to be able to work with vectors.

In three-dimensional Cartesian coordinates, any vector can be written in the form

$$\mathbf{r} = x_1\mathbf{i} + y_1\mathbf{j} + z_1\mathbf{k} \tag{A.48}$$

where $\mathbf{i}$, $\mathbf{j}$, and $\mathbf{k}$ are the mutually perpendicular vectors of unit length along the x, y, and z axes, respectively, and x_1, y_1, and z_1 are numbers. The length of a vector is defined by

$$|\mathbf{r}| = \sqrt{x_1^2 + y_1^2 + z_1^2} \tag{A.49}$$

This vector is depicted in the three-dimensional coordinate system shown in Figure A.11.

By definition, the angle θ is measured from the z axis, and the angle ϕ is measured in the x-y plane from the x axis. The angles θ and ϕ are related to x_1, y_1, and z_1 by

$$\theta = \cos^{-1} \frac{z_1}{\sqrt{x_1^2 + y_1^2 + z_1^2}} \quad \text{and} \quad \phi = \tan^{-1} \frac{y_1}{x_1} \tag{A.50}$$

We next consider the addition and subtraction of two vectors. Two vectors $\mathbf{a} = x\mathbf{i} + y\mathbf{j} + z\mathbf{k}$ and $\mathbf{b} = x'\mathbf{i} + y'\mathbf{j} + z'\mathbf{k}$ can be added or subtracted according to following the equation:

$$\mathbf{a} \pm \mathbf{b} = (x \pm x')\mathbf{i} + (y \pm y')\mathbf{j} + (z + z')\mathbf{k} \tag{A.51}$$

The addition and subtraction of vectors can also be depicted graphically, as done in Figure A.12.

The multiplication of two vectors can occur in either of two forms. Scalar multiplication of $\mathbf{a}$ and $\mathbf{b}$, also called the dot product of $\mathbf{a}$ and $\mathbf{b}$, is defined by

$$\mathbf{a} \cdot \mathbf{b} = |\mathbf{a}||\mathbf{b}|\cos \alpha \tag{A.52}$$

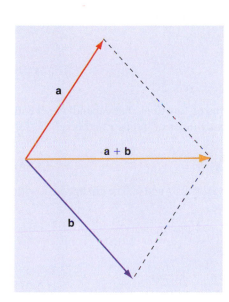

where α is the angle between the vectors. For $\mathbf{a} = 3\mathbf{i} + 1\mathbf{j} - 2\mathbf{k}$ and $\mathbf{b} = 2\mathbf{i} + -1\mathbf{j} + 4\mathbf{k}$, the vectors in the previous equation can be expanded in terms of their unit vectors:

$\mathbf{a} \cdot \mathbf{b} = (3\mathbf{i} + 1\mathbf{j} - 2\mathbf{k}) \cdot (2\mathbf{i} + -1\mathbf{j} + 4\mathbf{k})$

$= 3\mathbf{i} \cdot 2\mathbf{i} + 3\mathbf{i} \cdot (-1\mathbf{j}) + 3\mathbf{i} \cdot 4\mathbf{k} + 1\mathbf{j} \cdot 2\mathbf{i} + 1\mathbf{j} \cdot (-1\mathbf{j}) + 1\mathbf{j} \cdot 4\mathbf{k} - 2\mathbf{k} \cdot 2\mathbf{i} - 2\mathbf{k} \cdot (-1\mathbf{j}) - 2\mathbf{k} \cdot 4\mathbf{k}$

However, because $\mathbf{i}$, $\mathbf{j}$, and $\mathbf{k}$ are mutually perpendicular vectors of unit length, $\mathbf{i} \cdot \mathbf{i} = \mathbf{j} \cdot \mathbf{j} = \mathbf{k} \cdot \mathbf{k} = 1$ and $\mathbf{i} \cdot \mathbf{j} = \mathbf{i} \cdot \mathbf{k} = \mathbf{j} \cdot \mathbf{k} = 0$. Therefore, $\mathbf{a} \cdot \mathbf{b} = 3\mathbf{i} \cdot 2\mathbf{i} + 1\mathbf{j} \cdot (-1\mathbf{j}) - 2\mathbf{k} \cdot 4\mathbf{k} = -3$.

The other form in which vectors are multiplied is the vector product, also called the cross product. The vector multiplication of two vectors results in a vector, whereas the scalar multiplication of two vectors results in a scalar. The cross product is defined by

$$\mathbf{a} \times \mathbf{b} = \mathbf{c}|\mathbf{a}||\mathbf{b}|\sin \alpha \tag{A.53}$$

Note that $\mathbf{a} \times \mathbf{b} = -\mathbf{b} \times \mathbf{a}$ as shown in Figure A.13. By contrast, $\mathbf{a} \cdot \mathbf{b} = \mathbf{b} \cdot \mathbf{a}$.

In Equation (A.53), $\mathbf{c}$ is a vector of unit length that is perpendicular to the plane containing $\mathbf{a}$ and $\mathbf{b}$ and has a positive direction found by using the right-hand rule and α is the angle between $\mathbf{a}$ and $\mathbf{b}$.

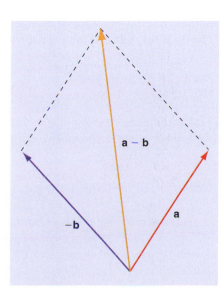

FIGURE A.12

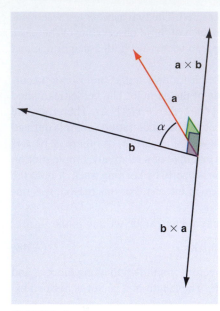

FIGURE A.13

The cross product between two three-dimensional vectors **a** and **b** is given by

$$\mathbf{a} \times \mathbf{b} = (a_x\mathbf{i} + a_y\mathbf{j} + a_z\mathbf{k}) \times (b_x\mathbf{i} + b_y\mathbf{j} + b_z\mathbf{k})$$

$$\begin{aligned} = a_x\mathbf{i} \times b_x\mathbf{i} + a_x\mathbf{i} \times b_y\mathbf{j} + a_x\mathbf{i} \times b_z\mathbf{k} + a_y\mathbf{j} \times b_x\mathbf{i} + a_y\mathbf{j} \times b_y\mathbf{j} + a_y\mathbf{j} \times b_z\mathbf{k} \\ + a_z\mathbf{k} \times b_x\mathbf{i} + a_z\mathbf{k} \times b_y\mathbf{j} + a_z\mathbf{k} \times b_z\mathbf{k} \end{aligned} \tag{A.54}$$

However, using the definition of the cross product in Equation (A.53),

$$\mathbf{i} \times \mathbf{i} = \mathbf{j} \times \mathbf{j} = \mathbf{k} \times \mathbf{k} = 0, \quad \mathbf{i} \times \mathbf{j} = \mathbf{k}, \quad \mathbf{i} \times \mathbf{k} = -\mathbf{j}$$
$$\mathbf{j} \times \mathbf{i} = -\mathbf{k}, \quad \mathbf{j} \times \mathbf{k} = \mathbf{i}, \quad \mathbf{k} \times \mathbf{i} = \mathbf{j}, \quad \mathbf{k} \times \mathbf{j} = -\mathbf{i}$$

Therefore, Equation (A.54) simplifies to

$$\mathbf{a} \times \mathbf{b} = (a_y b_z - a_z b_y)\mathbf{i} + (a_z b_x - a_x b_z)\mathbf{j} + (a_x b_y - a_y b_x)\mathbf{k} \tag{A.55}$$

As we will see in Section A.7, there is a simple way to calculate cross products using determinants.

The angular momentum $\mathbf{l} = \mathbf{r} \times \mathbf{p}$ where **p** is the linear momentum is of particular interest in quantum chemistry, because s, p, and d electrons are distinguished by their orbital angular momentum. For the example of the particle rotating on a ring depicted at the beginning of this section, the angular momentum vector is pointing upward in a direction perpendicular to the plane of the page. In analogy to Equation (A.55),

$$\mathbf{l} = \mathbf{r} \times \mathbf{p} = (yp_z - zp_y)\mathbf{i} + (zp_x - xp_z)\mathbf{j} + (xp_y - yp_x)\mathbf{k}$$

A.6 Partial Derivatives

In this section, we discuss the differential calculus of functions that depend on several independent variables. Consider the volume of a cylinder of radius r and height h, for which

$$V = f(r, h) = \pi r^2 h \tag{A.56}$$

where V can be written as a function of the two variables r and h. The change in V with a change in r or h is given by the partial derivatives

$$\left(\frac{\partial V}{\partial r}\right)_h = \lim_{\Delta r \to 0} \frac{V(r + \Delta r, h) - V(r, h)}{\Delta r} = 2\pi r h$$

$$\left(\frac{\partial V}{\partial h}\right)_r = \lim_{\Delta h \to 0} \frac{V(r, h + \Delta h) - V(r, h)}{\Delta h} = \pi r^2 \tag{A.57}$$

The subscript h in $(\partial V/\partial r)_h$ reminds us that h is being held constant in the differentiation. The partial derivatives in Equations (A.57) allow us to determine how a function changes when one of the variables changes. How does V change if the values of both variables change? In this case, V changes to $V + dV$ where

$$dV = \left(\frac{\partial V}{\partial r}\right)_h dr + \left(\frac{\partial V}{\partial h}\right)_r dh \tag{A.58}$$

These partial derivatives are useful in calculating the error in the function that results from errors in measurements of the individual variables. For example, the relative error in the volume of the cylinder is given by

$$\frac{dV}{V} = \frac{1}{V}\left[\left(\frac{\partial V}{\partial r}\right)_h dr + \left(\frac{\partial V}{\partial h}\right)_r dh\right] = \frac{1}{\pi r^2 h}[2\pi r h\, dr + \pi r^2\, dh] = \frac{2dr}{r} + \frac{dh}{h}$$

This equation shows that a given relative error in r generates twice the relative error in V as a relative error in h of the same size.

We can also take second or higher derivatives with respect to either variable. The mixed second partial derivatives are of particular interest. The mixed partial derivatives of V are given by

$$\left(\frac{\partial}{\partial h}\left(\frac{\partial V}{\partial r}\right)_h\right)_r = \left(\partial\left(\frac{\partial[\pi r^2 h]}{\partial r}\right)_h \middle/ \partial h\right)_r = \left(\frac{\partial[2\pi rh]}{\partial h}\right)_r = 2\pi r$$

$$\left(\frac{\partial}{\partial r}\left(\frac{\partial V}{\partial h}\right)_r\right)_h = \left(\partial\left(\frac{\partial[\pi r^2 h]}{\partial h}\right)_r \middle/ \partial r\right)_h = \left(\frac{\partial[\pi r^2]}{\partial r}\right)_h = 2\pi r \qquad (A.59)$$

For the specific case of V, the order in which the function is differentiated does not affect the outcome. Such a function is called a state function. Therefore, for any state function f of the variables x and y,

$$\left(\frac{\partial}{\partial y}\left(\frac{\partial f(x, y)}{\partial x}\right)_y\right)_x = \left(\frac{\partial}{\partial x}\left(\frac{\partial f(x, y)}{\partial y}\right)_x\right)_y \qquad (A.60)$$

Because Equation (A.60) is satisfied by all state functions f, it can be used to determine if a function f is a state function.

We demonstrate how to calculate the partial derivatives

$$\left(\frac{\partial f}{\partial x}\right)_y, \left(\frac{\partial f}{\partial y}\right)_x, \left(\frac{\partial^2 f}{\partial x^2}\right)_y, \left(\frac{\partial^2 f}{\partial y^2}\right)_x, \left(\partial\left(\frac{\partial f}{\partial x}\right)_y \middle/ \partial y\right)_x, \quad \text{and} \quad \left(\partial\left(\frac{\partial f}{\partial y}\right)_x \middle/ \partial x\right)_y$$

for the function $f(x, y) = ye^{ax} + xy \cos x + y \ln xy$, where a is a real constant:

$$\left(\frac{\partial f}{\partial x}\right)_y = aye^{ax} + \frac{y}{x} + y \cos x - xy \sin x, \left(\frac{\partial f}{\partial y}\right)_x = 1 + e^{ax} + x \cos x + \ln xy$$

$$\left(\frac{\partial^2 f}{\partial x^2}\right)_y = a^2 ye^{ax} - \frac{y}{x^2} - 2y \sin x - xy \cos x, \left(\frac{\partial^2 f}{\partial y^2}\right)_x = \frac{1}{y}$$

$$\left(\partial\left(\frac{\partial f}{\partial x}\right)_y \middle/ \partial y\right)_x = ae^{ax} + \frac{1}{x} + \cos x - x \sin x, \left(\partial\left(\frac{\partial f}{\partial y}\right)_x \middle/ \partial x\right)_y = ae^{ax} + \frac{1}{x} + \cos x - x \sin x$$

Because we have shown that

$$\left(\partial\left(\frac{\partial f}{\partial x}\right)_y \middle/ \partial y\right)_x = \left(\partial\left(\frac{\partial f}{\partial y}\right)_x \middle/ \partial x\right)_y$$

$f(x,y)$ is a state function of the variables x and y.

Whereas the partial derivatives tell us how the function changes if the value of one of the variables is changed, the total differential tells us how the function changes when all of the variables are changed simultaneously. The total differential of the function $f(x,y)$ is defined by

$$df = \left(\frac{\partial f}{\partial x}\right)_y dx + \left(\frac{\partial f}{\partial y}\right)_x dy \qquad (A.61)$$

The total differential of the function used earlier is calculated as follows:

$$df = \left(aye^{ax} + \frac{y}{x} + y \cos x - xy \sin x\right) dx + \left(1 + e^{ax} + x \cos x + \log xy\right) dy$$

Two other important results from multivariate differential calculus are used frequently. For a function $z = f(x,y)$, which can be rearranged to $x = g(y,z)$ or $y = h(x,z)$,

$$\left(\frac{\partial x}{\partial y}\right)_z = \frac{1}{\left(\dfrac{\partial y}{\partial x}\right)_z} \qquad (A.62)$$

The other important result that is used frequently is the cyclic rule:

$$\left(\frac{\partial x}{\partial y}\right)_z \left(\frac{\partial y}{\partial z}\right)_x \left(\frac{\partial z}{\partial x}\right)_y = -1 \qquad (A.63)$$

Consider an additional example of calculating partial derivatives for a function encountered in quantum mechanics. The Schrödinger equation for the hydrogen atom takes the form

$$-\frac{\hbar^2}{2\mu}\left[\frac{1}{r^2}\frac{\partial}{\partial r}\left(r^2\frac{\partial\psi(r,\theta,\phi)}{\partial r}\right)+\frac{1}{r^2\sin\theta}\frac{\partial}{\partial\theta}\left(\sin\theta\frac{\partial\psi(r,\theta,\phi)}{\partial\theta}\right)+\frac{1}{r^2\sin\theta}\frac{\partial^2\psi(r,\theta,\phi)}{\partial\phi^2}\right]$$

$$-\frac{e^2}{4\pi\varepsilon_0 r}\,\psi(r,\theta,\phi)=E\psi(r,\theta,\phi)$$

Note that each of the first three terms on the left side of the equation involves partial differentiation with respect to one of the variables r, θ, and ϕ, in turn. Two of the solutions to this differential equation are $(r/a_0)e^{-r/2a_0}\sin\theta e^{\pm i\phi}$. Each of these terms is evaluated separately to demonstrate how partial derivatives are taken in quantum mechanics. Although this is a more complex exercise than those presented earlier, it provides good practice in partial differentiation. For the first term, the partial derivative is taken with respect to r:

$$-\frac{\hbar^2}{2\mu}\frac{1}{\sqrt{64\pi}}\left(\frac{1}{a_0}\right)^{3/2}\left[\frac{1}{r^2}\frac{\partial}{\partial r}\left(r^2\frac{\partial\left(\dfrac{r}{a_0}e^{-r/2a_0}\sin\theta e^{\pm i\phi}\right)}{\partial r}\right)\right]$$

$$=-\frac{\hbar^2}{2\mu}\frac{1}{\sqrt{64\pi}}\left(\frac{1}{a_0}\right)^{3/2}\sin\theta e^{\pm i\phi}\left[\frac{1}{r^2}\frac{\partial}{\partial r}\left(r^2\frac{\partial\left(\dfrac{r}{a_0}e^{-r/2a_0}\right)}{\partial r}\right)\right]$$

$$=-\frac{\hbar^2}{2\mu}\frac{1}{\sqrt{64\pi}}\left(\frac{1}{a_0}\right)^{3/2}\sin\theta e^{\pm i\phi}\left[\frac{1}{r^2}\frac{\partial}{\partial r}\left(r^2\left(\frac{1}{a_0}e^{-r/2a_0}-(r/2a_0^2)e^{-r/2a_0}\right)\right)\right]$$

$$=-\frac{\hbar^2}{2\mu}\frac{1}{\sqrt{64\pi}}\left(\frac{1}{a_0}\right)^{3/2}\sin\theta e^{\pm i\phi}\left[\frac{1}{r^2}\left(\begin{array}{c}-r^2\dfrac{e^{-r/2a_0}}{a_0^2}+r^3\dfrac{e^{-r/2a_0}}{4a_0^3}+2r\dfrac{e^{-r/2a_0}}{a_0}\\[2mm]-2r^2\dfrac{e^{-r/2a_0}}{a_0^2}\end{array}\right)\right]$$

$$=-\frac{\hbar^2}{2\mu}\frac{1}{\sqrt{64\pi}}\left(\frac{1}{a_0}\right)^{3/2}\sin\theta e^{\pm i\phi}e^{-r/2a_0}\frac{(8a_0^2-8a_0r+r^2)}{4a_0^3 r}$$

Partial differentiation with respect to θ is easier, because the terms that depend on r and ϕ are constant:

$$-\frac{\hbar^2}{2\mu}\frac{1}{\sqrt{64\pi}}\left(\frac{1}{a_0}\right)^{3/2}\left[\frac{1}{r^2\sin\theta}\frac{\partial}{\partial\theta}\left(\sin\theta\frac{\partial\left(\dfrac{r}{a_0}e^{-r/2a_0}\sin\theta e^{\pm i\phi}\right)}{\partial\theta}\right)\right]$$

$$=-\frac{\hbar^2}{2\mu}\frac{1}{\sqrt{6\pi}}\left(\frac{1}{a_0}\right)^{3/2}\frac{r}{a_0}e^{-r/2a_0}e^{\pm i\phi}\left[\frac{1}{r^2\sin\theta}\frac{\partial}{\partial\theta}\left(\sin\theta\frac{\partial(\sin\theta)}{\partial\theta}\right)\right]$$

$$=-\frac{\hbar^2}{2\mu}\frac{1}{\sqrt{64\pi}}\left(\frac{1}{a_0}\right)^{3/2}\frac{r}{a_0}e^{-r/2a_0}e^{\pm i\phi}\left[\frac{1}{r^2\sin\theta}\frac{\partial}{\partial\theta}(\sin\theta\cos\theta)\right]$$

$$=-\frac{\hbar^2}{2\mu}\frac{1}{\sqrt{64\pi}}\left(\frac{1}{a_0}\right)^{3/2}\frac{r}{a_0}e^{-r/2a_0}e^{\pm i\phi}\left[\frac{1}{r^2\sin\theta}(\cos^2\theta-\sin^2\theta)\right]$$

Partial differentiation with respect to ϕ is also not difficult, because the terms that depend on r and θ are constant:

$$-\frac{\hbar^2}{2\mu}\frac{1}{\sqrt{64\pi}}\left(\frac{1}{a_0}\right)^{3/2}\left[\frac{1}{r^2\sin\theta}\frac{\partial^2\dfrac{r}{a_0}e^{-r/2a_0}\sin\theta\,e^{\pm i\phi}}{\partial\phi^2}\right]$$

$$= -\frac{\hbar^2}{2\mu} \frac{1}{\sqrt{64\pi}} \left(\frac{1}{a_0}\right)^{3/2} \frac{r}{a_0} e^{-r/2a_0} \frac{1}{r^2} \left[\frac{\partial^2 e^{\pm i\phi}}{\partial \phi^2}\right]$$

$$= -\frac{\hbar^2}{2\mu} \frac{1}{\sqrt{64\pi}} \left(\frac{1}{a_0}\right)^{3/2} \frac{r}{a_0} e^{-r/2a_0} \frac{1}{r^2} \left[e^{\pm i\phi}\right]$$

A.7 Working with Determinants

A determinant of nth order is a square $n \times n$ array of numbers symbolically enclosed by vertical lines. A fifth-order determinant is shown here with the conventional indexing of the elements of the array:

$$\begin{vmatrix} a_{11} & a_{12} & a_{13} & a_{14} & a_{15} \\ a_{21} & a_{22} & a_{23} & a_{24} & a_{25} \\ a_{31} & a_{32} & a_{33} & a_{34} & a_{35} \\ a_{41} & a_{42} & a_{43} & a_{44} & a_{45} \\ a_{51} & a_{52} & a_{53} & a_{54} & a_{55} \end{vmatrix} \tag{A.64}$$

A 2×2 determinant has a value that is defined in Equation (A.65). It is obtained by multiplying the elements in the diagonal connected by a line with a negative slope and subtracting from this the product of the elements in the diagonal connected by a line with a positive slope.

$$\begin{vmatrix} a_{11} & a_{12} \\ a_{21} & a_{22} \end{vmatrix} = a_{11}a_{22} - a_{12}a_{21} \tag{A.65}$$

The value of a higher order determinant is obtained by expanding the determinant in terms of determinants of lower order. This is done using the method of cofactors. We illustrate the use of the method of cofactors by reducing a 3×3 determinant to a sum of 2×2 determinants. Any row or column can be used in the reduction process. We use the first row of the determinant in the reduction. The recipe is spelled out in this equation:

$$\begin{vmatrix} a_{11} & a_{12} & a_{13} \\ a_{21} & a_{22} & a_{23} \\ a_{31} & a_{32} & a_{33} \end{vmatrix} = (-1)^{1+1} a_{11} \begin{vmatrix} a_{22} & a_{23} \\ a_{32} & a_{33} \end{vmatrix} + (-1)^{1+2} a_{12} \begin{vmatrix} a_{21} & a_{23} \\ a_{31} & a_{33} \end{vmatrix}$$

$$+ (-1)^{1+3} a_{13} \begin{vmatrix} a_{21} & a_{22} \\ a_{31} & a_{32} \end{vmatrix} \tag{A.66}$$

$$= a_{11} \begin{vmatrix} a_{22} & a_{23} \\ a_{32} & a_{33} \end{vmatrix} - a_{12} \begin{vmatrix} a_{21} & a_{23} \\ a_{31} & a_{33} \end{vmatrix} + a_{13} \begin{vmatrix} a_{21} & a_{22} \\ a_{31} & a_{32} \end{vmatrix}$$

Each term in the sum results from the product of one of the three elements of the first row, $(-1)^{m+n}$, where m and n are the indices of the row and column designating the element, respectively, and the 2×2 determinant obtained by omitting the entire row and column to which the element used in the reduction belongs. The product $(-1)^{m+n}$ and the 2×2 determinant are called the cofactor of the element used in the reduction. For example, the value of the following 3×3 determinant is found using the cofactors of the second row:

$$\begin{vmatrix} 1 & 3 & 4 \\ 2 & -1 & 6 \\ -1 & 7 & 5 \end{vmatrix} = (-1)^{2+1} 2 \begin{vmatrix} 3 & 4 \\ 7 & 5 \end{vmatrix} + (-1)^{2+2} (-1) \begin{vmatrix} 1 & 4 \\ -1 & 5 \end{vmatrix} + (-1)^{2+3} 6 \begin{vmatrix} 1 & 3 \\ -1 & 7 \end{vmatrix}$$

$$= -1 \times 2 \times (-13) + 1 \times (-1) \times 9 + (-1) \times 6 \times 10 = -43$$

If the initial determinant is of a higher order than 3, multiple sequential reductions as outlined earlier will reduce it in order by one in each step until a sum of 2×2 determinants is obtained.

The main usefulness for determinants is in solving a system of linear equations. Such a system of equations is obtained in evaluating the energies of a set of molecular orbitals obtained by combining a set of atomic orbitals. Before illustrating this method, we list some important properties of determinants that we will need in solving a set of simultaneous equations.

Property I   The value of a determinant is not altered if each row in turn is made into a column or vice versa as long as the original order is kept. By this we mean that the nth row becomes the nth column. This property can be illustrated using 2×2 and 3×3 determinants:

$$\begin{vmatrix} 2 & 1 \\ 3 & -1 \end{vmatrix} = \begin{vmatrix} 2 & 3 \\ 1 & -1 \end{vmatrix} = -5 \text{ and } \begin{vmatrix} 1 & 3 & 4 \\ 2 & -1 & 6 \\ -1 & 7 & 5 \end{vmatrix} = \begin{vmatrix} 1 & 2 & -1 \\ 3 & -1 & 7 \\ 4 & 6 & 5 \end{vmatrix} = -43$$

Property II   If any two rows or columns are interchanged, the sign of the value of the determinant is changed. For example,

$$\begin{vmatrix} 2 & 1 \\ 3 & -1 \end{vmatrix} = -5, \text{ but } \begin{vmatrix} 1 & 2 \\ -1 & 3 \end{vmatrix} = +5 \text{ and } \begin{vmatrix} 1 & 3 & 4 \\ 2 & -1 & 6 \\ -1 & 7 & 5 \end{vmatrix} = -43, \text{ but } \begin{vmatrix} 2 & -1 & 6 \\ 1 & 3 & 4 \\ -1 & 7 & 5 \end{vmatrix} = +43$$

Property III   If two rows or columns of a determinant are identical, the value of the determinant is zero. For example,

$$\begin{vmatrix} 2 & 1 \\ 2 & 1 \end{vmatrix} = 2 - 2 = 0 \text{ and } \begin{vmatrix} 1 & 1 & 4 \\ 2 & 2 & 6 \\ -1 & -1 & 5 \end{vmatrix} = (-1)^{2+1} 2 \begin{vmatrix} 1 & 4 \\ -1 & 5 \end{vmatrix} + (-1)^{2+2} 2 \begin{vmatrix} 1 & 4 \\ -1 & 5 \end{vmatrix}$$

$$+ (-1)^{2+3} 6 \begin{vmatrix} 1 & 1 \\ -1 & -1 \end{vmatrix}$$

$$= -1 \times 2 \times 9 + 1 \times 2 \times 9 + (-1) \times 6 \times 0 = 0$$

Property IV   If each element of a row or column is multiplied by a constant, the value of the determinant is multiplied by that constant. For example,

$$\begin{vmatrix} 2 & 1 \\ 3 & -1 \end{vmatrix} = -5 \text{ and } \begin{vmatrix} 8 & 4 \\ 3 & -1 \end{vmatrix} = -20 \text{ and } \begin{vmatrix} 1 & 2 & -1 \\ 3 & -1 & 7 \\ 4 & 6 & 5 \end{vmatrix} = -43 \text{ and } \begin{vmatrix} 1 & 3\sqrt{2} & 4 \\ 2 & -\sqrt{2} & 6 \\ -1 & 7\sqrt{2} & 5 \end{vmatrix} = -43\sqrt{2}$$

Property V   The value of a determinant is unchanged if a row or column multiplied by an arbitrary number is added to another row or column. For example,

$$\begin{vmatrix} 2 & 1 \\ 3 & -1 \end{vmatrix} = \begin{vmatrix} 2+1 & 1 \\ 3-1 & -1 \end{vmatrix} = \begin{vmatrix} 2 & 1 \\ 3 & -1 \end{vmatrix} = -5 \text{ and}$$

$$\begin{vmatrix} 1 & 3 & 4 \\ 2 & -1 & 6 \\ -1 & 7 & 5 \end{vmatrix} = \begin{vmatrix} 1 & 3 & 4 \\ 2-1 & -1+7 & 6+5 \\ -1 & 7 & 5 \end{vmatrix} = \begin{vmatrix} 1 & 3 & 4 \\ 2 & -1 & 6 \\ -1 & 7 & 5 \end{vmatrix} = -43$$

How are determinants useful? This question can be answered by illustrating how determinants can be used to solve a set of linear equations:

$$x + y + z = 10$$

$$3x + 4y - z = 12 \qquad \text{(A.67)}$$

$$-x + 2y + 5z = 26$$

This set of equations is solved by first constructing the 3×3 determinant that is the array of the coefficients of x, y, and z:

$$\mathbf{D}_{coefficients} = \begin{vmatrix} 1 & 1 & 1 \\ 3 & 4 & -1 \\ -1 & 2 & 5 \end{vmatrix} \tag{A.68}$$

Now imagine that we multiply the first column by x. This changes the value of the determinant as stated in Property IV:

$$\begin{vmatrix} 1x & 1 & 1 \\ 3x & 4 & -1 \\ -1x & 2 & 5 \end{vmatrix} = x\mathbf{D}_{coefficients} \tag{A.69}$$

We next add to the first column of $x\mathbf{D}_{coefficients}$ the second column of $\mathbf{D}_{coefficients}$ multiplied by y, and the third column multiplied by z. According to Properties IV and V, the value of the determinant is unchanged. Therefore,

$$\mathbf{D}_{c1}\begin{vmatrix} 1 & 1 & 1 \\ 3 & 4 & -1 \\ -1 & 2 & 5 \end{vmatrix} = \begin{vmatrix} x+y+z & 1 & 1 \\ 3x+4y-z & 4 & -1 \\ -x+2y+5z & 2 & 5 \end{vmatrix} = \begin{vmatrix} 10 & 1 & 1 \\ 12 & 4 & -1 \\ 26 & 2 & 5 \end{vmatrix} = x\mathbf{D}_{coefficients} \tag{A.70}$$

To obtain the third determinant in the previous equation, the individual equations in Equation (A.67) are used to substitute the constants for the algebraic expression in the preceding determinants. From the previous equation, we conclude that

$$x = \frac{\mathbf{D}_{c1}}{\mathbf{D}_{coefficients}} = \frac{\begin{vmatrix} 10 & 1 & 1 \\ 12 & 4 & -1 \\ 26 & 2 & 5 \end{vmatrix}}{\begin{vmatrix} 1 & 1 & 1 \\ 3 & 4 & -1 \\ -1 & 2 & 5 \end{vmatrix}} = 3$$

To determine y and z, the exact same procedure can be followed, but we substitute instead in columns 2 and 3, respectively. The first step in each case is to multiply all elements of the second (third) row by $y(z)$. If we do so, we obtain the determinants $\mathbf{D}_{c2}$ and $\mathbf{D}_{c3}$:

$$\mathbf{D}_{c2}\begin{vmatrix} 1 & 10 & 1 \\ 3 & 12 & -1 \\ -1 & 26 & 5 \end{vmatrix} \quad \text{and} \quad \mathbf{D}_{c3} = \begin{vmatrix} 1 & 1 & 10 \\ 3 & 4 & 12 \\ -1 & 2 & 26 \end{vmatrix}$$

and we conclude that

$$y = \frac{\mathbf{D}_{c2}}{\mathbf{D}_{coefficients}} = \frac{\begin{vmatrix} 1 & 10 & 1 \\ 3 & 12 & -1 \\ -1 & 26 & 5 \end{vmatrix}}{\begin{vmatrix} 1 & 1 & 1 \\ 3 & 4 & -1 \\ -1 & 2 & 5 \end{vmatrix}} = 2 \quad \text{and} \quad z = \frac{\mathbf{D}_{c3}}{\mathbf{D}_{coefficients}} = \frac{\begin{vmatrix} 1 & 1 & 10 \\ 3 & 4 & 12 \\ -1 & 2 & 26 \end{vmatrix}}{\begin{vmatrix} 1 & 1 & 1 \\ 3 & 4 & -1 \\ -1 & 2 & 5 \end{vmatrix}} = 5$$

This method of solving a set of simultaneous linear equations is known as Cramer's method.

If the constants in the set of equations are all zero, as in Equations (A.71a) and (A.71b),

$$x + y + z = 0$$

$$3x + 4y - z = 0 \tag{A.71a}$$

$$-x + 2y + 5z = 0$$

$$3x - y + 2x = 0$$

$$-x + y - z = 0 \tag{A.71b}$$

$$(1 + \sqrt{2})x + (1 - \sqrt{2})y + \sqrt{2}z = 0$$

the determinants D_{c1}, D_{c2}, and D_{c3} all have the value zero. An obvious set of solutions is $x = 0$, $y = 0$, and $z = 0$. For most problems in physics and chemistry, this set of solutions is not physically meaningful and is referred to as the set of trivial solutions. A set of nontrivial solutions only exists if the equation $D_{coefficients} = 0$ is satisfied. There is no nontrivial solution to the set of Equations (A.71a) because $D_{coefficients} \neq 0$. There is a set of nontrivial solutions to the set of Equations (A.71b), because $D_{coefficients} = 0$ in this case.

Determinants offer a convenient way to calculate the cross product of two vectors, as discussed in Section A.5. The following recipe is used:

$$\mathbf{a} \times \mathbf{b} = \begin{vmatrix} \mathbf{i} & \mathbf{j} & \mathbf{k} \\ a_x & a_y & a_z \\ b_x & b_y & b_z \end{vmatrix} = \mathbf{i} \begin{vmatrix} a_y & a_z \\ b_y & b_z \end{vmatrix} - \mathbf{j} \begin{vmatrix} a_x & a_z \\ b_x & b_z \end{vmatrix} + \mathbf{k} \begin{vmatrix} a_x & a_y \\ b_x & b_y \end{vmatrix}$$

$$= (a_y b_z - a_z b_y)\mathbf{i} + (a_z b_x - a_x b_z)\mathbf{j} + (a_x b_y - a_y b_x)\mathbf{k} \tag{A.72}$$

Note that by referring to Property II, you can show that $\mathbf{b} \times \mathbf{a} = -\mathbf{a} \times \mathbf{b}$.

A.8 Working with Matrices

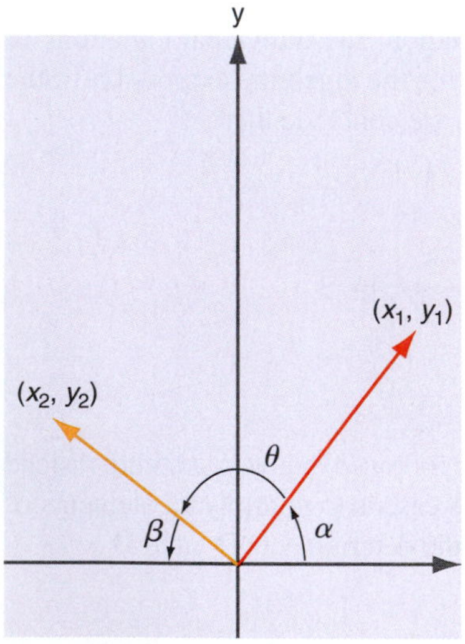

Physical chemists find widespread use for matrices. Matrices can be used to represent symmetry operations in the application of group theory to problems concerning molecular symmetry. They can also be used to obtain the energies of molecular orbitals formed through the linear combination of atomic orbitals. We next illustrate the use of matrices for representing the rotation operation that is frequently encountered in molecular symmetry considerations.

Consider the rotation of a three-dimensional vector about the z axis. Because the z component of the vector is unaffected by this operation, we need only consider the effect of the rotation operation on the two-dimensional vector formed by the projection of the three-dimensional vector on the x-y plane. The transformation can be represented by $(x_1, y_1, z_1) \rightarrow (x_2, y_2, z_1)$. The effect of the operation on the x and y components of the vector is shown in Figure A.14.

Next, relationships are derived among (x_1, y_1, z_1), (x_2, y_2, z_1), the magnitude of the radius vector r, and the angles α and β, based on the preceding figure. The magnitude of the radius vector r is

$$r = \sqrt{x_1^2 + y_1^2 + z_1^2} = \sqrt{x_2^2 + y_2^2 + z_1^2} \tag{A.73}$$

Although the values of x and y change in the rotation, r is unaffected by this operation. The relationships between x, y, r, α, and β are given by

$$\theta = 180° - \alpha - \beta$$

$$x_1 = r \cos \alpha, \qquad y_1 = r \sin \alpha$$

$$x_2 = -r \cos \beta, \qquad y_1 = r \sin \beta \tag{A.74}$$

In the following discussion, these identities are used:

$$\cos(\alpha \pm \beta) = \cos \alpha \cos \beta \mp \sin \alpha \sin \beta$$

$$\sin(\alpha \pm \beta) = \sin \alpha \cos \beta \pm \cos \alpha \sin \beta \tag{A.75}$$

From Figure A.14, the following relationship between x_2 and x_1 and y_1 can be derived using the identities of Equation (A.75):

$$x_2 = -r \cos \beta = -r \cos(180° - \alpha - \theta)$$

$$= r \sin 180° \sin(-\theta - \alpha) - r \cos 180° \cos(-\theta - \alpha)$$

$$= r \cos(-\theta - \alpha) = r \cos(\theta + \alpha) = r \cos \theta \cos \alpha - r \sin \theta \sin \alpha$$

$$= x_1 \cos \theta - y_1 \sin \theta \tag{A.76}$$

FIGURE A.14

Using the same procedure, the following relationship between y_2 and x_1 and y_1 can be derived:

$$y_2 = x_1 \sin \theta + y_1 \cos \theta \tag{A.77}$$

Next, these results are combined to write the following equations relating x_2, y_2, and z_2 to x_1, y_1, and z_1:

$$x_2 = x_1 \cos \theta - y_1 \sin \theta$$
$$y_2 = x_1 \sin \theta + y_1 \cos \theta$$
$$z_2 = 0x_1 + 0y_1 + z_1 \tag{A.78}$$

At this point, the concept of a matrix can be introduced. An $n \times m$ matrix is an array of numbers, functions, or operators that can undergo mathematical operations such as addition and multiplication with one another. The operation of interest to us in considering rotation about the z axis is matrix multiplication. We illustrate how matrices, which are designated in bold script, such as $\mathbf{A}$, are multiplied using 2×2 matrices as an example.

$$\mathbf{AB} = \begin{pmatrix} a_{11} & a_{12} \\ a_{21} & a_{22} \end{pmatrix} \begin{pmatrix} b_{11} & b_{12} \\ b_{21} & b_{22} \end{pmatrix} = \begin{pmatrix} a_{11}b_{11} + a_{12}b_{21} & a_{11}b_{12} + a_{12}b_{22} \\ a_{21}b_{11} + a_{22}b_{21} & a_{21}b_{12} + a_{22}b_{22} \end{pmatrix} \tag{A.79}$$

Using numerical examples,

$$\begin{pmatrix} 2 & 1 \\ -3 & 4 \end{pmatrix} \begin{pmatrix} 1 & 6 \\ 2 & -1 \end{pmatrix} = \begin{pmatrix} 4 & 11 \\ 5 & -22 \end{pmatrix} \quad \text{and} \quad \begin{pmatrix} 1 & 6 \\ 2 & -1 \end{pmatrix} \begin{pmatrix} 1 \\ -1 \end{pmatrix} = \begin{pmatrix} -5 \\ 3 \end{pmatrix}$$

Now consider the initial and final coordinates (x_1, y_1, z_1) and (x_2, y_2, z_1) as 3×1 matrices (x_1, y_1, z_1) and (x_2, y_2, z_1). In that case, the set of simultaneous equations of Equation (A.78) can be written as

$$\begin{pmatrix} x_2 \\ y_2 \\ z_2 \end{pmatrix} = \begin{pmatrix} \cos \theta & -\sin \theta & 0 \\ \sin \theta & \cos \theta & 0 \\ 0 & 0 & 1 \end{pmatrix} \begin{pmatrix} x_1 \\ y_1 \\ z_1 \end{pmatrix} \tag{A.80}$$

We see that we can represent the operator for rotation about the z axis, R_z, as the following 3×3 matrix:

$$\mathbf{R}_z = \begin{pmatrix} \cos \theta & -\sin \theta & 0 \\ \sin \theta & \cos \theta & 0 \\ 0 & 0 & 1 \end{pmatrix} \tag{A.81}$$

The rotation operator for $180°$ and $120°$ rotation can be obtained by evaluating the sine and cosine functions at the appropriate values of θ. These rotation operators have the form

$$\begin{pmatrix} -1 & 0 & 0 \\ 0 & -1 & 0 \\ 0 & 0 & 1 \end{pmatrix} \quad \text{and} \quad \begin{pmatrix} 1/2 & -\sqrt{3}/2 & 0 \\ \sqrt{3}/2 & 1/2 & 0 \\ 0 & 0 & 1 \end{pmatrix}, \quad \text{respectively} \tag{A.82}$$

One special matrix, the identity matrix designated $\mathbf{I}$, deserves additional mention. The identity matrix corresponds to an operation in which nothing is changed. The matrix that corresponds to the transformation $(x_1, y_1, z_1) \rightarrow (x_2, y_2, z_2)$ expressed in equation form as $x_2 = x_1 + 0y_1 + 0z_1$

$$y_2 = 0x_1 + y_1 + 0z_1$$
$$z_2 = 0x_1 + 0y_1 + z_1 \tag{A.83}$$

is the identity matrix

$$\mathbf{I} = \begin{pmatrix} 1 & 0 & 0 \\ 0 & 1 & 0 \\ 0 & 0 & 1 \end{pmatrix}$$

The identity matrix is an example of a diagonal matrix. It has this name because only the diagonal elements are nonzero. In the identity matrix of order $n \times n$, all diagonal elements have the value one.

The operation that results from the sequential operation of two individual operations represented by matrices $\mathbf{A}$ and $\mathbf{B}$ is the products of the matrices: $\mathbf{C} = \mathbf{AB}$. An interesting case illustrating this relationship is counterclockwise rotation through an angle θ followed by clockwise rotation through the same angle, which corresponds to rotation by $-\theta$. Because $\cos(-\theta) = \cos\theta$ and $\sin\theta = -\sin\theta$, the rotation matrix for $-\theta$ must be

$$\mathbf{R}_{-z} = \begin{pmatrix} \cos\theta & \sin\theta & 0 \\ -\sin\theta & \cos\theta & 0 \\ 0 & 0 & 1 \end{pmatrix} \tag{A.84}$$

Because the sequential operations leave the vector unchanged, it must be the case that $\mathbf{R}_z\mathbf{R}_{-z} = \mathbf{R}_{-z}\mathbf{R}_z = \mathbf{I}$. We verify here that the first of these relations is obeyed:

$$\mathbf{R}_z\mathbf{R}_{-z} = \begin{pmatrix} \cos\theta & -\sin\theta & 0 \\ \sin\theta & \cos\theta & 0 \\ 0 & 0 & 1 \end{pmatrix} \begin{pmatrix} \cos\theta & \sin\theta & 0 \\ -\sin\theta & \cos\theta & 0 \\ 0 & 0 & 1 \end{pmatrix} \tag{A.85}$$

$$= \begin{pmatrix} \cos^2\theta + \sin^2\theta + 0 & \sin\theta\cos\theta - \sin\theta\cos\theta + 0 & 0 \\ \sin\theta\cos\theta - \sin\theta\cos\theta + 0 & \cos^2\theta + \sin^2\theta + 0 & 0 \\ 0 & 0 & 1 \end{pmatrix} = \begin{pmatrix} 1 & 0 & 0 \\ 0 & 1 & 0 \\ 0 & 0 & 1 \end{pmatrix}$$

Any matrix $\mathbf{B}$ that satisfies the relationship $\mathbf{AB} = \mathbf{BA} = \mathbf{I}$ is called the inverse matrix of $\mathbf{A}$ and is designated $\mathbf{A}^{-1}$. Inverse matrices play an important role in finding the energies of a set of molecular orbitals that is a linear combination of atomic orbitals.

Sources of Data

The most extensive databases for thermodynamic data (and the abbreviations listed with the tables) are as follows:

HCP Lide, D. R., Ed., *Handbook of Chemistry and Physics,* 83rd ed., CRC Press, Boca Raton, FL, 2002.

NIST Chemistry Webbook Linstrom, P. J., and W. G. Mallard, Eds., *NIST Chemistry Webbook: NIST Standard Reference Database Number 69,* National Institute of Standards and Technology, Gaithersburg, MD; retrieved from http://webbook.nist.gov.

Additional data sources used in the tables include the following:

AS Alberty, R. A., and R. S. Silbey, *Physical Chemistry,* John Wiley & Sons, New York, 1992.

Bard Bard, A. J., R. Parsons, and J. Jordan, *Standard Potentials in Aqueous Solution,* Marcel Dekker, New York, 1985.

DAL Blachnik, R., Ed., *D'Ans Lax Taschenbuch für Chemiker und Physiker,* 4th ed., Springer, Berlin, 1998.

HCAA Tsuzuki, T., and H. Hunt, "Heats of Combustion. VI The Heats of Combustion of Some Amino Acids," *J. Phys. Chem.* 61 (1957), 1668.

HCFE Huffman, H. M., E. L. Ellis, and S. W. Fox, "Thermal Data VI. The Heats of Combustion and Free Energies of Seven Organic Compounds Containing Nitrogen," *J. Am. Chem. Soc.* 58 (1936), 1728–1733.

HP Benenson, W., J. W. Harris, H. Stocker, and H. Lutz, *Handbook of Physics,* Springer, New York, 2002.

HTTD Lide, D. R., Ed., *CRC Handbook of Thermophysical and Thermochemical Data,* CRC Press, Boca Raton, FL, 1994.

TDOC Pedley, J. B., R. D. Naylor, and S. P. Kirby, *Thermochemical Data of Organic Compounds,* Chapman and Hall, London, 1977.

TDPS Barin, I., *Thermochemical Data of Pure Substances,* VCH Press, Weinheim, 1989.

TPBS R. C. Wilhoit, "Thermodynamic Properties of Biochemical Substances," Chap. 2 in H. D. Brown, Ed., *Biochemical Microcalorimetry,* Academic Press, New York, 1969.

Wüthrich Wüthrich, K. *NMR of Proteins and Nucleic Acids,* John Wiley & Sons, New York, 1986.

TABLE 1.3   van der Waals and Redlich–Kwong Parameters for Selected Gases

Substance	Formula	van der Waals a (dm^6 bar mol^{-2})	b (dm^3 mol^{-1})	Redlich–Kwong a (dm^6 bar mol^{-2} K$^{1/2}$)	b (dm^3 mol^{-1})
Ammonia	NH_3	4.225	0.0371	86.12	0.02572
Argon	Ar	1.355	0.0320	16.86	0.02219
Benzene	C_6H_6	18.82	0.1193	452.0	0.08271
Bromine	Br_2	9.75	0.0591	236.5	0.04085
Carbon dioxide	CO_2	3.658	0.0429	64.63	0.02971
Carbon monoxide	CO	1.472	0.0395	17.20	0.02739
Ethane	C_2H_6	5.580	0.0651	98.79	0.04514
Ethanol	C_2H_5OH	12.56	0.0871	287.7	0.06021
Ethene	C_2H_4	4.612	0.0582	78.51	0.04034
Ethyne	C_2H_2	4.533	0.05240	80.65	0.03632
Fluorine	F_2	1.171	0.0290	14.17	0.01993
Hydrogen	H_2	0.2452	0.0265	1.427	0.01837
Methane	CH_4	2.303	0.0431	32.20	0.02985
Methanol	CH_3OH	9.476	0.0659	217.1	0.04561
Nitrogen	N_2	1.370	0.0387	15.55	0.02675
Oxygen	O_2	1.382	0.0319	17.40	0.02208
Pentane	C_5H_{12}	19.09	0.1448	419.2	0.1004
Propane	C_3H_8	9.39	0.0905	182.9	0.06271
Pyridine	C_5H_5N	19.77	0.1137	498.8	0.07877
Tetrachloromethane	CCl_4	20.01	0.1281	473.2	0.08793
Water	H_2O	5.537	0.0305	142.6	0.02113
Xenon	Xe	4.192	0.0516	72.30	0.03574

Source: Calculated from critical constants.

TABLE 2.2 **Physical Properties of Selected Elements**

Densities are shown for nongaseous elements under standard conditions.

Substance	Atomic Weight	Melting Point (K)	Boiling Point (K)	$\rho°$ (kg m^{-3})	$C°_{P,m}$ (J K^{-1}mol^{-1})	Oxidation States
Aluminum	26.982	933.47	2792.15	2698.9	25.4	3
Argon	39.948	83.79 tp* (69 kPa)	87.30	—	20.79	
Barium	137.33	1000.15	2170.15	3620	28.07	2
Boron	10.811	2348.15	4273.15	2340	11.1	3
Bromine	79.904	265.95	331.95	3103	36.05	1, 3, 4, 5, 6
Calcium	40.078	1115.15	1757.15	1540	25.9	2
Carbon	12.011	4713.15 (12.4 GPa) 4762.15 tp (10.3 MPa)	4098.15 (graphite)	3513 (diamond) 2250 (graphite)	6.113 (diamond) 8.527 (graphite)	2, 4
Cesium	132.91	301.65	944.15	1930	32.20	1
Chlorine	35.453	171.65	239.11	—	33.95	1, 3, 4, 5, 6, 7
Copper	63.546	1357.77	2835.15	8960	24.4	1, 2
Fluorine	18.998	53.48 tp	85.03	—	31.30	1
Gold	196.97	1337.33	3129.15	19320	25.42	1, 3
Helium	4.0026	0.95	4.22	—	20.79	
Hydrogen	1.0079	13.81	20.28	—	28.84	1
Iodine	126.90	386.85	457.55	4933	54.44	1, 3, 5, 7
Iron	55.845	1811.15	3134.15	7874	25.10	2, 3
Krypton	83.80	115.77 tp (73.2 kPa)	119.93	—	20.79	
Lead	207.2	600.61	2022.15	11343	26.44	2, 4
Lithium	6.941	453.65	1615.15	534	24.77	1
Magnesium	24.305	923.15	1363.15	1740	24.89	2
Manganese	54.938	1519.15	2334.15	7300	26.3	2, 3, 4, 6, 7
Mercury	200.59	234.31	629.88	13534	27.98	1, 2
Molybdenum	95.94	2896.15	4912.15	10222	23.90	2, 3, 4, 5, 6
Neon	20.180	24.54 tp (43 kPa)	27.07	—	20.79	
Nickel	58.693	1728.15	3186	8902	26.07	2, 3
Nitrogen	14.007	63.15	77.36	—	29.12	1, 2, 3, 4, 5
Oxygen	15.999	54.36	90.20	—	29.38	1, 2
Palladium	106.42	1828.05	3236.15	11995	25.98	2, 4
Phosphorus (white)	30.974	317.3	553.65	1823	23.84	3, 5
Platinum	195.08	2041.55	4098.15	21500	25.85	2, 4, 6
Potassium	39.098	336.65	1032.15	890	29.58	1
Rhenium	186.21	3459.15	5869.15	20800	25.31	2, 4, 5, 6, 7
Rhodium	102.91	2237.15	3968.15	12410	24.98	2, 3, 4
Ruthenium	101.07	2607.15	4423.15	12100	24.04	3, 4, 5, 6, 8
Silicon	28.086	1687.15	3538.15	2330	20.00	4
Silver	107.87	1234.93	2435.15	10500	25.35	1
Sodium	22.990	370.95	1156.15	971	28.24	1
Sulfur	32.066	388.36	717.75	1819	22.76	2, 4, 6
Tin	118.71	505.08	2879	7310	26.99	2, 4
Titanium	47.867	1941.15	3560.15	4506	25.05	2, 3, 4
Vanadium	50.942	2183.15	3680.15	6000	24.89	2, 3, 4, 5
Xenon	131.29	161.36 tp (81.6 kPa)	165.03	—	20.79	2, 4, 6, 8
Zinc	65.39	692.68	1180.15	7135	25.40	2

* triple point

Sources: HCP and DAL.

TABLE 2.3   Physical Properties of Selected Compounds

Densities are shown for nongaseous compounds under standard conditions.

Formula	Name	Molecular Weight	Melting Point (K)	Boiling Point (K)	Density ρ^0 (kg m^{-3})	Heat Capacity $C^0_{P,m}$ (J K^{-1}mol^{-1})
CO(g)	Carbon monoxide	28.01	68.13	81.65	—	29.1
COCl$_2$(g)	Phosgene	98.92	145.4	281	—	
CO$_2$(g)	Carbon dioxide	44.01	216.6 tp	194.75	—	37.1
D$_2$O(l)	Deuterium oxide	20.03	277	374.6	1108	
HCl(g)	Hydrogen chloride	36.46	158.98	188.15	—	29.1
HF(g)	Hydrogen fluoride	20.01	189.8	293.15	—	
H$_2$O	Water	18.02	273.15	373.15	998(l)	75.3(l)
					20°C	36.2(s)
					917(s) 0°C	
H$_2$O$_2$(l)	Hydrogen peroxide	34.01	272.72	423.35	1440	43.1
H$_2$SO$_4$(l)	Sulfuric acid	98.08	283.46	610.15	1800	
KBr(s)	Potassium bromide	119.00	1007.15	1708.15	2740	52.3
KCl(s)	Potassium chloride	74.55	1044.15		1988	51.3
KI(s)	Potassium iodide	166.0	954.15	1596.15	3120	52.9
NaCl(s)	Sodium chloride	58.44	1073.85	1738.15	2170	50.5
NH$_3$(g)	Ammonia	17.03	195.42	239.82	—	35.1
SO$_2$(g)	Sulfur dioxide	64.06	197.65	263.10	—	39.9
CCl$_4$(l)	Carbon tetrachloride	153.82	250.3	349.8	1594	131.3
CH$_4$(g)	Methane	16.04	90.75	111.65	—	35.7
HCOOH(l)	Formic acid	46.03	281.45	374.15	1220	99.04
CH$_3$OH(l)	Methanol	32.04	175.55	337.75	791.4	81.1
CH$_3$CHO(l)	Acetaldehyde	44.05	150.15	293.25	783.4	89.0
CH$_3$COOH(l)	Acetic acid	60.05	289.6	391.2	1044.6	123.1
CH$_3$COCH$_3$(l)	Acetone	58.08	178.5	329.2	789.9	125.45
C$_2$H$_5$OH(l)	Ethanol	46.07	158.8	351.5	789.3	112.3
C$_3$H$_7$OH(l)	1-Propanol	60.10	147.05	370.35	799.8	156.5
C$_4$H$_{11}$OH(l)	1-Butanol	74.12	183.35	390.85	809.8	176.9
C$_5$H$_5$NH$_2$(l)	Pyridine	79.10	231.55	388.35	981.9	193
C$_5$H$_{12}$(l)	Pentane	72.15	143.45	309.15	626.2	167.2
C$_5$H$_{11}$OH(l)	1-Pentanol	88.15	194.25	411.05	814.4	207.5
C$_6$H$_{12}$(l)	Cyclohexane	84.16	279.6	353.9	773.9	156.0
C$_6$H$_5$CHO(l)	Benzaldehyde	106.12	247.15	452.15	1041.5	172.0
C$_6$H$_5$COOH(s)	Benzoic acid	122.12	395.55	522.35	1265.9	147.8
C$_6$H$_5$CH$_3$(l)	Toluene	92.14	178.2	383.8	866.9	157.1
C$_6$H$_5$NH$_2$(l)	Aniline	93.13	267	457	1021.7	194.1
C$_6$H$_5$OH(s)	Phenol	94.11	314.05	454.95	1057.6	127.2
C$_6$H$_6$(l)	Benzene	78.11	278.6	353.3	876.5	135.7
1,2-(CH$_3$)$_2$C$_6$H$_5$(l)	o-Xylene	106.17	248	417.6	880.2	187.7
C$_8$H$_{18}$(l)	Octane	114.23	216.35	398.75	698.6	254.7

Sources: HCP and TDOC.

TABLE 2.4   Molar Heat Capacity, $C_{P,m}$, of Gases in the Range 298–800 K

Given by

$$C_{P,m} \text{ (J K}^{-1} \text{ mol}^{-1}) = A(1) + A(2)\frac{T}{K} + A(3)\frac{T^2}{K^2} + A(4)\frac{T^3}{K^3}$$

Note that $C_{P,m}$ for solids and liquids at 298.15 K is listed in Tables 2.1 and 2.2.

Name	Formula	C°_P (298.15 K) (J K^{-1}mol^{-1})	A(1)	A(2)	A(3)	A(4)
All monatomic gases	He, Ne, Ar, Xe, O, H, among others	20.79	20.79			
Bromine	Br_2	36.05	30.11	0.03353	-5.5009×10^{-5}	3.1711×10^{-8}
Chlorine	Cl_2	33.95	22.85	0.06543	-1.2517×10^{-4}	1.1484×10^{-7}
Carbon monoxide	CO	29.14	31.08	-0.01452	3.1415×10^{-5}	-1.4973×10^{-8}
Carbon dioxide	CO_2	37.14	18.86	0.07937	-6.7834×10^{-5}	2.4426×10^{-8}
Fluorine	F_2	31.30	23.06	0.03742	-3.6836×10^{-5}	1.351×10^{-8}
Hydrogen	H_2	28.84	22.66	0.04381	-1.0835×10^{-4}	1.1710×10^{-7}
Water	H_2O	33.59	33.80	-0.00795	2.8228×10^{-5}	-1.3115×10^{-8}
Hydrogen bromide	HBr	29.13	29.72	-0.00416	7.3177×10^{-6}	
Hydrogen chloride	HCl	29.14	29.81	-0.00412	6.2231×10^{-6}	
Hydrogen fluoride	HF	29.14	28.94	0.00152	-4.0674×10^{-6}	$\times 3.8970 \times 10^{-9}$
Ammonia	NH_3	35.62	29.29	0.01103	4.2446×10^{-5}	-2.7706×10^{-8}
Nitrogen	N_2	29.13	30.81	-0.01187	2.3968×10^{-5}	-1.0176×10^{-8}
	NO	29.86	33.58	-0.02593	5.3326×10^{-5}	-2.7744×10^{-8}
	NO_2	37.18	32.06	-0.00984	1.3807×10^{-4}	-1.8157×10^{-7}
Oxygen	O_2	29.38	32.83	-0.03633	1.1532×10^{-4}	-1.2194×10^{-7}
Sulfur dioxide	SO_2	39.83	26.07	0.05417	-2.6774×10^{-5}	
Methane	CH_4	35.67	30.65	-0.01739	1.3903×10^{-4}	8.1395×10^{-8}
Methanol	CH_3OH	44.07	26.53	0.03703	9.451×10^{-5}	-7.2006×10^{-8}
Ethyne	C_2H_2	44.05	10.82	0.15889	-1.8447×10^{-4}	8.5291×10^{-8}
Ethene	C_2H_4	42.86	8.39	0.12453	-2.5224×10^{-5}	-1.5679×10^{-8}
Ethane	C_2H_6	52.38	6.82	0.16840	-5.2347×10^{-5}	
Propane	C_3H_8	73.52	0.56	0.27559	-1.0355×10^{-4}	
Butane	C_4H_{10}	101.01	172.02	-1.08574	4.4887×10^{-3}	-6.5539×10^{-6}
Pentane	C_5H_{12}	120.11	2.02	0.44729	-1.7174×10^{-4}	
Benzene	C_6H_6	82.39	-46.48	0.53735	-3.8303×10^{-4}	1.0184×10^{-7}
Hexane	C_6H_{12}	142.13	-13.27	0.61995	-3.5408×10^{-4}	7.6704×10^{-8}

Sources: HCP and HTTD.

TABLE 2.5 **Molar Heat Capacity, $C_{P,m}$, of Solids**

$$C_{P,m} \text{ (J K}^{-1}\text{ mol}^{-1}) = A(1) + A(2)\frac{T}{K} + A(3)\frac{T^2}{K^2} + A(4)\frac{T^3}{K^3} + A(5)\frac{T^4}{K^4}$$

Formula	Name	$C_{P,m}$ (298.15 K) (J K^{-1} mol^{-1})	A(1)	A(2)	A(3)	A(4)	A(5)	Range (K)
Ag	Silver	25.35	26.12	−0.0110	3.826×10^{-5}	3.750×10^{-8}	1.396×10^{-11}	290–800
Al	Aluminum	24.4	6.56	0.1153	-2.460×10^{-4}	1.941×10^{-7}		200–450
Au	Gold	25.4	34.97	−0.0768	2.117×10^{-4}	-2.350×10^{-7}	9.500×10^{-11}	290–800
CsCl	Cesium chloride	52.5	43.38	0.0467	-8.973×10^{-5}	1.421×10^{-7}	-8.237×10^{-11}	200–600
CuSO$_4$	Copper sulfate	98.9	−13.81	0.7036	-1.636×10^{-3}	2.176×10^{-6}	-1.182×10^{-9}	200–600
Fe	Iron	25.1	−10.99	0.3353	-1.238×10^{-3}	2.163×10^{-6}	-1.407×10^{-9}	200–450
NaCl	Sodium chloride	50.5	25.19	0.1973	-6.011×10^{-4}	8.815×10^{-7}	-4.765×10^{-10}	200–600
Si	Silicon	20.0	−6.25	0.1681	-3.437×10^{-4}	2.494×10^{-7}	6.667×10^{-12}	200–450
C (graphite)	C	8.5	−12.19	0.1126	-1.947×10^{-4}	1.919×10^{-7}	-7.800×10^{-11}	290–600
C$_6$H$_5$OH	Phenol	127.4	−5.97	1.0380	-6.467×10^{-3}	2.304×10^{-5}	-2.658×10^{-8}	100–314
C$_{10}$H$_8$	Naphthalene	165.7	−6.16	1.0383	-5.355×10^{-3}	1.891×10^{-5}	-2.053×10^{-8}	100–353
C$_{14}$H$_{10}$	Anthracene	210.5	11.10	0.5816	2.790×10^{-4}			100–488

Sources: HCP and HTTD.

TABLE 4.1   Thermodynamic Data for Inorganic Compounds at 298.15 K

Substance	ΔH_f° (kJ mol^{-1})	ΔG_f° (kJ mol^{-1})	S° (J mol^{-1} K^{-1})	$C_{P,m}^\circ$ (J mol^{-1} K^{-1})	Atomic or Molecular Weight (amu)
Aluminum					
Al(s)	0	0	28.3	24.4	26.98
Al$_2$O$_3$(s)	-1675.7	-1582.3	50.9	79.0	101.96
Al^{3+}(aq)	-538.4	-485.0	-325		26.98
Antimony					
Sb(s)	0	0	45.7	25.2	121.75
Argon					
Ar(g)	0	0	154.8	20.8	39.95
Barium					
Ba(s)	0	0	62.5	28.1	137.34
BaO(s)	-548.0	-520.3	72.1	47.3	153.34
BaCO$_3$(s)	-1216.3	-1137.6	112.1	85.4	197.35
BaCl$_2$(s)	-856.6	-810.3	123.7	75.1	208.25
BaSO$_4$(s)	-1473.2	-1362.3	132.2	101.8	233.40
Ba^{2+}(aq)	-537.6	-560.8	9.6		137.34
Bromine					
Br$_2$(l)	0	0	152.2	75.7	159.82
Br$_2$(g)	30.9	3.1	245.5	36.0	159.82
Br(g)	111.9	82.4	175.0	20.8	79.91
HBr(g)	-36.3	-53.4	198.7	29.1	90.92
Br$^-$(aq)	-121.6	-104.0	82.4		79.91
Calcium					
Ca(s)	0	0	41.6	25.9	40.08
CaCO$_3$(s) calcite	-1206.9	-1128.8	92.9	83.5	100.09
CaCl$_2$(s)	-795.4	-748.8	104.6	72.9	110.99
CaO(s)	-634.9	-603.3	38.1	42.0	56.08
CaSO$_4$(s)	-1434.5	-1322.0	106.5	99.7	136.15
Ca^{2+}(aq)	-542.8	-553.6	-53.1		40.08
Carbon					
Graphite(s)	0	0	5.74	8.52	12.011
Diamond(s)	1.89	2.90	2.38	6.12	12.011
C(g)	716.7	671.2	158.1	20.8	12.011
CO(g)	-110.5	-137.2	197.7	29.1	28.011
CO$_2$(g)	-393.5	-394.4	213.8	37.1	44.010
HCN(g)	135.5	124.7	201.8	35.9	27.03
CN$^-$(aq)	150.6	172.4	94.1		26.02
HCO$_3^-$(aq)	-692.0	-586.8	91.2		61.02
CO$_3^{2-}$(aq)	-675.2	-527.8	-50.0		60.01
Chlorine					
Cl$_2$(g)	0	0	223.1	33.9	70.91
Cl(g)	121.3	105.7	165.2	21.8	35.45
HCl(g)	-92.3	-95.3	186.9	29.1	36.46
ClO$_2$(g)	104.6	105.1	256.8	45.6	67.45
ClO$_4^-$(aq)	-128.1	-8.52	184.0		99.45
Cl$^-$(aq)	-167.2	-131.2	56.5		35.45
Copper					
Cu(s)	0	0	33.2	24.4	63.54
CuCl$_2$(s)	-220.1	-175.7	108.1	71.9	134.55
CuO(s)	-157.3	-129.7	42.6	42.3	79.54

TABLE 4.1 Thermodynamic Data for Inorganic Compounds at 298.15 K *(continued)*

Substance	ΔH_f°(kJ mol^{-1})	ΔG_f°(kJ mol^{-1})	S° (J mol^{-1} K^{-1})	$C_{P,m}^\circ$ (J mol^{-1} K^{-1})	Atomic or Molecular Weight (amu)
Cu$_2$O(s)	−168.6	−146.0	93.1	63.6	143.08
CuSO$_4$(s)	−771.4	−662.2	109.2	98.5	159.62
Cu$^+$(aq)	71.7	50.0	40.6		63.54
Cu^{2+}(aq)	64.8	65.5	−99.6		63.54
Deuterium					
D$_2$(g)	0	0	145.0	29.2	4.028
HD(g)	0.32	−1.46	143.8	29.2	3.022
D$_2$O(g)	−249.2	−234.5	198.3	34.3	20.028
D$_2$O(l)	−294.6	−243.4	75.94	84.4	20.028
HDO(g)	−246.3	−234.5	199.4	33.8	19.022
HDO(l)	−289.9	−241.9	79.3		19.022
Fluorine					
F$_2$(g)	0	0	202.8	31.3	38.00
F(g)	79.4	62.3	158.8	22.7	19.00
HF(g)	−273.3	−275.4	173.8	29.1	20.01
F$^-$(aq)	−332.6	−278.8	−13.8		19.00
Gold					
Au(s)	0	0	47.4	25.4	196.97
Au(g)	366.1	326.3	180.5	20.8	197.97
Hydrogen					
H$_2$(g)	0	0	130.7	28.8	2.016
H(g)	218.0	203.3	114.7	20.8	1.008
OH(g)	39.0	34.2	183.7	29.9	17.01
H$_2$O(g)	−241.8	−228.6	188.8	33.6	18.015
H$_2$O(l)	−285.8	−237.1	70.0	75.3	18.015
H$_2$O(s)			48.0	36.2 (273 K)	18.015
H$_2$O$_2$(g)	−136.3	−105.6	232.7	43.1	34.015
H$^+$(aq)	0	0	0		1.008
OH$^-$(aq)	−230.0	−157.24	−10.9		17.01
Iodine					
I$_2$(s)	0	0	116.1	54.4	253.80
I$_2$(g)	62.4	19.3	260.7	36.9	253.80
I(g)	106.8	70.2	180.8	20.8	126.90
I$^-$(aq)	−55.2	−51.6	111.3		126.90
Iron					
Fe(s)	0	0	27.3	25.1	55.85
Fe(g)	416.3	370.7	180.5	25.7	55.85
Fe$_2$O$_3$(s)	−824.2	−742.2	87.4	103.9	159.69
Fe$_3$O$_4$(s)	−1118.4	−1015.4	146.4	150.7	231.54
FeSO$_4$(s)	−928.4	−820.8	107.5	100.6	151.92
Fe^{2+}(aq)	−89.1	−78.9	−137.7		55.85
Fe^{3+}(aq)	−48.5	−4.7	−315.9		55.85
Lead					
Pb(s)	0	0	64.8	26.4	207.19
Pb(g)	195.2	162.2	175.4	20.8	207.19
PbO$_2$(s)	−277.4	−217.3	68.6	64.6	239.19
PbSO$_4$(s)	−920.0	−813.20	148.5	86.4	303.25
Pb^{2+}(aq)	0.92	−24.4	18.5		207.19

TABLE 4.1   Thermodynamic Data for Inorganic Compounds at 298.15 K *(continued)*

Substance	ΔH_f°(kJ mol^{-1})	ΔG_f°(kJ mol^{-1})	S° (J mol^{-1} K^{-1})	$C_{P,m}^\circ$ (J mol^{-1} K^{-1})	Atomic or Molecular Weight (amu)
Lithium					
Li(s)	0	0	29.1	24.8	6.94
Li(g)	159.3	126.6	138.8	20.8	6.94
LiH(s)	−90.5	−68.3	20.0	27.9	7.94
LiH(g)	140.6	117.8	170.9	29.7	7.94
Li$^+$(aq)	−278.5	−293.3	13.4		6.94
Magnesium					
Mg(s)	0	0	32.7	24.9	24.31
Mg(g)	147.1	112.5	148.6	20.8	24.31
MgO(s)	−601.6	−569.3	27.0	37.2	40.31
MgSO$_4$(s)	−1284.9	−1170.6	91.6	96.5	120.38
MgCl$_2$(s)	−641.3	−591.8	89.6	71.4	95.22
MgCO$_3$(s)	−1095.8	−1012.2	65.7	75.5	84.32
Mg^{2+}(aq)	−466.9	−454.8	−138.1		24.31
Manganese					
Mn(s)	0	0	32.0	26.3	54.94
Mn(g)	280.7	238.5	173.7	20.8	54.94
MnO$_2$(s)	−520.0	−465.1	53.1	54.1	86.94
Mn^{2+}(aq)	−220.8	−228.1	−73.6		54.94
MnO$_4^-$(aq)	−541.4	−447.2	191.2		118.94
Mercury					
Hg(l)	0	0	75.9	28.0	200.59
Hg(g)	61.4	31.8	175.0	20.8	200.59
Hg$_2$Cl$_2$(s)	−265.4	−210.7	191.6	101.9	472.09
Hg^{2+}(aq)	170.2	164.4	−36.2		401.18
Hg$_2^{2+}$(aq)	166.9	153.5	65.7		401.18
Nickel					
Ni(s)	0	0	29.9	26.1	58.71
Ni(g)	429.7	384.5	182.2	23.4	58.71
NiCl$_2$(s)	−305.3	−259.0	97.7	71.7	129.62
NiO(s)	−239.7	−211.5	38.0	44.3	74.71
NiSO$_4$(s)	−872.9	−759.7	92.0	138.0	154.77
Ni^{2+}(aq)	−54.0	−45.6	−128.9		58.71
Nitrogen					
N$_2$(g)	0	0	191.6	29.1	28.013
N(g)	472.7	455.5	153.3	20.8	14.007
NH$_3$(g)	−45.9	−16.5	192.8	35.1	17.03
NO(g)	91.3	87.6	210.8	29.9	30.01
N$_2$O(g)	81.6	103.7	220.0	38.6	44.01
NO$_2$(g)	33.2	51.3	240.1	37.2	46.01
NOCl(g)	51.7	66.1	261.7	44.7	65.46
N$_2$O$_4$(g)	11.1	99.8	304.4	79.2	92.01
N$_2$O$_4$(l)	−19.5	97.5	209.2	142.7	92.01
HNO$_3$(l)	−174.1	−80.7	155.6	109.9	63.01
HNO$_3$(g)	−133.9	−73.5	266.9	54.1	63.01
NO$_3^-$(aq)	−207.4	−111.3	146.4		62.01
NH$_4^+$(aq)	−132.5	−79.3	113.4		18.04

TABLE 4.1   **Thermodynamic Data for Inorganic Compounds at 298.15 K** *(continued)*

Substance	ΔH_f° (kJ mol^{-1})	ΔG_f° (kJ mol^{-1})	S° (J mol^{-1} K^{-1})	$C_{P,m}^\circ$ (J mol^{-1} K^{-1})	Atomic or Molecular Weight (amu)
Oxygen					
$O_2(g)$	0	0	205.2	29.4	31.999
$O(g)$	249.2	231.7	161.1	21.9	15.999
$O_3(g)$	142.7	163.2	238.9	39.2	47.998
$OH(g)$	39.0	34.22	183.7	29.9	17.01
$OH^-(aq)$	−230.0	−157.2	−10.9		17.01
Phosphorus					
$P(s)$ white	0	0	41.1	23.8	30.97
$P(s)$ red	−17.6	−12.1	22.8	21.2	30.97
$P_4(g)$	58.9	24.4	280.0	67.2	123.90
$PCl_5(g)$	−374.9	−305.0	364.6	112.8	208.24
$PH_3(g)$	5.4	13.5	210.2	37.1	34.00
$H_3PO_4(l)$	−1271.7	−1123.6	150.8	145.0	94.97
$PO_4^{3-}(aq)$	−1277.4	−1018.7	−220.5		91.97
$HPO_4^{2-}(aq)$	−1299.0	−1089.2	−33.5		92.97
$H_2PO_4^-(aq)$	−1302.6	−1130.2	92.5		93.97
Potassium					
$K(s)$	0	0	64.7	29.6	39.10
$K(g)$	89.0	60.5	160.3	20.8	39.10
$KCl(s)$	−436.5	−408.5	82.6	51.3	74.56
$K_2O(s)$	−361.5	−322.8	102.0	77.4	94.20
$K_2SO_4(s)$	−1437.8	−1321.4	175.6	131.5	174.27
$K^+(aq)$	−252.4	−283.3	102.5		39.10
Silicon					
$Si(s)$	0	0	18.8	20.0	28.09
$Si(g)$	450.0	405.5	168.0	22.3	28.09
$SiCl_4(g)$	−662.7	−622.8	330.9	90.3	169.70
$SiO_2(quartz)$	−910.7	−856.3	41.5	44.4	60.09
Silver					
$Ag(s)$	0	0	42.6	25.4	107.87
$Ag(g)$	284.9	246.0	173.0	20.8	107.87
$AgCl(s)$	−127.0	−109.8	96.3	50.8	143.32
$AgNO_2(s)$	−44.4	19.8	140.6	93.0	153.88
$AgNO_3(s)$	−124.4	−33.4	140.9	93.1	169.87
$Ag_2SO_4(s)$	−715.9	−618.4	200.4	131.4	311.80
$Ag^+(aq)$	105.6	77.1	72.7		107.87
Sodium					
$Na(s)$	0	0	51.3	28.2	22.99
$Na(g)$	107.5	77.0	153.7	20.8	22.99
$NaCl(s)$	−411.2	−384.1	72.1	50.5	58.44
$NaOH(s)$	−425.8	−379.7	64.4	59.5	40.00
$Na_2SO_4(s)$	−1387.1	−1270.2	149.6	128.2	142.04
$Na^+(aq)$	−240.1	−261.9	59.0		22.99
Sulfur					
$S(rhombic)$	0	0	32.1	22.6	32.06
$SF_6(g)$	−1220.5	−1116.5	291.5	97.3	146.07
$H_2S(g)$	−20.6	−33.4	205.8	34.2	34.09
$SO_2(g)$	−296.8	−300.1	248.2	39.9	64.06
$SO_3(g)$	−395.7	−371.1	256.8	50.7	80.06

TABLE 4.1 Thermodynamic Data for Inorganic Compounds at 298.15 K *(continued)*

Substance	ΔH_f° (kJ mol^{-1})	ΔG_f° (kJ mol^{-1})	S° (J mol^{-1} K^{-1})	$C_{P,m}^\circ$ (J mol^{-1} K^{-1})	Atomic or Molecular Weight (amu)
$SO_3^{2-}(aq)$	−635.5	−486.6	−29.3		80.06
$SO_4^{2-}(aq)$	−909.3	−744.5	20.1		96.06
Tin					
$Sn(white)$	0	0	51.2	27.0	118.69
$Sn(g)$	301.2	266.2	168.5	21.3	118.69
$SnO_2(s)$	−577.6	−515.8	49.0	52.6	150.69
$Sn^{2+}(aq)$	−8.9	−27.2	−16.7		118.69
Titanium					
$Ti(s)$	0	0	30.7	25.0	47.87
$Ti(g)$	473.0	428.4	180.3	24.4	47.87
$TiCl_4(l)$	−804.2	−737.2	252.4	145.2	189.69
$TiO_2(s)$	−944.0	−888.8	50.6	55.0	79.88
Xenon					
$Xe(g)$	0	0	169.7	20.8	131.30
$XeF_4(s)$	−261.5	−123	146	118	207.29
Zinc					
$Zn(s)$	0	0	41.6	25.4	65.37
$ZnCl_2(s)$	−415.1	−369.4	111.5	71.3	136.28
$ZnO(s)$	−350.5	−320.5	43.7	40.3	81.37
$ZnSO_4(s)$	−982.8	−871.5	110.5	99.2	161.43
$Zn^{2+}(aq)$	−153.9	−147.1	−112.1		65.37

Sources: HCP, HTTD, and TDPS.

TABLE 4.2 Thermodynamic Data for Selected Organic Compounds at 298.15 K

Substance	Formula	Molecular Weight	ΔH_f° (kJ mol^{-1})	$\Delta H_{combustion}^\circ$ (kJ mol^{-1})	ΔG_f° (kJ mol^{-1})	S° (J mol^{-1} K^{-1})	C_P° (J mol^{-1} K^{-1})
Carbon (graphite)	C	12.011	0	−393.5	0	5.74	8.52
Carbon (diamond)	C	12.011	1.89	−395.4	2.90	2.38	6.12
Carbon monoxide	CO	28.01	−110.5	−283.0	−137.2	197.7	29.1
Carbon dioxide	CO_2	44.01	−393.5		−394.4	213.8	37.1
Acetaldehyde (*l*)	C_2H_4O	44.05	−192.2	−1166.9	−127.6	160.3	89.0
Acetic acid (*l*)	$C_2H_4O_2$	60.05	−484.3	−874.2	−389.9	159.8	124.3
Acetone (*l*)	C_3H_6O	58.08	−248.4	−1790	−155.2	199.8	126.3
L-Alanine (*s*)	$C_3H_7O_2N$	89.09	−562.8	−1575	−370.2	129.2	122.3
L-Arginine (*s*)	$C_6H_{14}O_2N_4$	174.20	−621.7	−3480	−656.9	250.6	233.5
L-Asparagine (*s*)	$C_4H_8O_3N_2$	132.10	−790.4	−1925	−531.0	194.5	160.7
L-Aspartic acid (*s*)	$C_4H_7O_4N$	69.11	−972.5	−1599	−729.4	170.1	155.3
Benzene (*l*)	C_6H_6	78.12	49.1	−3268	124.5	173.4	136.0
Benzene (*g*)	C_6H_6	78.12	82.9	−3268	129.7	269.2	82.4
Benzoic acid (*s*)	$C_7H_6O_2$	122.13	−385.2	−3227	−245.5	167.6	146.8
1,3-Butadiene (*g*)	C_4H_6	54.09	110.0	−2541			79.8
n-Butane (*g*)	C_4H_{10}	58.13	−125.7	−2878	−17.0	310.2	97.5
1-Butene (*g*)	C_4H_8	56.11	−0.63	−2718	71.1	305.7	85.7
Carbon disulfide (*g*)	CS_2	76.14	116.9	−1112	66.8	238.0	45.7
Carbon tetrachloride (*l*)	CCl_4	153.82	−128.2	−360	−62.5	214.4	133.9
Carbon tetrachloride (*g*)	CCl_4	153.82	−95.7		−58.2	309.7	83.4

TABLE 4.2   **Thermodynamic Data for Selected Organic Compounds at 298.15 K** *(continued)*

Substance	Formula	Molecular Weight	ΔH_f° (kJ mol^{-1})	$\Delta H_{combustion}^\circ$ (kJ mol^{-1})	ΔG_f° (kJ mol^{-1})	S° (J mol^{-1} K^{-1})	C_P° (J mol^{-1} K^{-1})
Cyclohexane (*l*)	C_6H_{12}	84.16	−156.4	−3920	26.8	204.5	154.9
Cyclopentane (*l*)	C_5H_{10}	70.13	−105.1	−3291	38.8	204.5	128.8
Cyclopropane (*g*)	C_3H_6	42.08	53.3	−2091	104.5	237.5	55.6
L-Cysteine (*s*)	$C_3H_7O_2NS$	121.15	−532.6	−1877	−342.7	169.9	173.2
Dimethyl Ether (*g*)	C_2H_6O	131.6	−184.1	−1460	−112.6	266.4	64.4
Ethane (*g*)	C_2H_6	30.07	−84.0	−1561	−32.0	229.2	52.5
Ethanol (*l*)	C_2H_6O	46.07	−277.6	−1367	−174.8	160.7	112.3
Ethanol (*g*)	C_2H_6O	46.07	−234.8	−1367	−167.9	281.6	65.6
Ethene (*g*)	C_2H_4	28.05	52.4	−1411	68.4	219.3	42.9
Ethyne (*g*)	C_2H_2	26.04	227.4	−1310	209.2	200.9	44
Formaldehyde (*g*)	CH_2O	30.03	−108.6	−571	−102.5	218.8	35.4
Formic acid (*l*)	CH_2O_2	46.03	−425.0	−255	−361.4	129.0	99.0
Formic acid (*g*)	CH_2O_2	46.03	−378.7	−256	−351.0	248.7	45.2
α-D-Glucose (*s*)	$C_6H_{12}O_6$	180.16	−1273.1	−2805	−910.6	209.2	219.2
L-Glutamic acid (*s*)	$C_5H_9O_4N$	147.13	−1009	−2245	−731.0	188.2	175.2
L-Glutamine (*s*)	$C_5H_{10}O_3N_2$	146.15	−825.9	−2570	−532.2	195.1	183.8
Glycerol (*l*)	$C_3H_8O_3$	92.09	−670.70	−1661	−479.5	204.6	216.7
Glycine (*s*)	$C_2H_5O_2N$	75.07	−537.2	−965.2	−377.7	103.5	99.2
Glycylglycine (*s*)	$C_4H_8O_3N_2$	132.12	−746.0	−1969	−491.5	190.0	163.6
n-Hexane (*l*)	C_6H_{14}	86.18	−198.7	−4163	−4.0	296.0	195.6
Hydrogen cyanide (*l*)	HCN	27.03	108.9		125.0	112.8	70.6
Hydrogen cyanide (*g*)	HCN	27.03	135.5		124.7	201.8	35.9
DL-Lactic acid (*l*)	$C_3H_6O_3$	90.08	−673.6	−1364	−518.8	192.1	211.3
β-Lactose (*s*)	$C_{12}H_{22}O_{11}$	342.30	−2237	−5652	−1567	386.2	410.5
L-Leucine (*s*)	$C_6H_{13}O_2N$	131.18	−646.9	−3566	−356.5	209.6	208.4
Methane (*g*)	CH_4	16.04	−74.6	−891	−50.5	186.3	35.7
Methanol (*l*)	CH_4O	32.04	−239.2	−726	−166.6	126.8	81.1
Methanol (*g*)	CH_4O	32.04	−201.0	−764	−162.3	239.9	44.1
L-Methionine (*s*)	$C_5H_{11}O_2NS$	149.21	−761.1	−3174	−508.4	231.5	290.2
Oxalic acid (*g*)	$C_2H_2O_4$	90.04	−731.8	−246	−662.7	320.6	86.2
Palmitic acid (*s*)	$C_{16}H_{32}O_2$	256.43	−890.77	−10,035	−315.1	455.2	460.7
n-Pentane (*g*)	C_5H_{12}	72.15	−146.9	−3509	−8.2	349.1	120.1
Phenol (*s*)	C_6H_6O	94.11	−165.1	−3054	−50.2	144.0	127.4
Propane (*g*)	C_3H_8	44.10	−103.8	−2219	−23.4	270.3	73.6
Propene (*g*)	C_3H_6	42.08	20.0	−2058	62.7	266.9	64.0
Propyne (*g*)	C_3H_4	40.07	184.9	−2058	194.5	248.2	60.7
Pyridine (*l*)	C_5H_5N	79.10	100.2	−2782		177.9	132.7
Sucrose (*s*)	$C_{12}H_{22}O_{11}$	342.3	−2226.1	−5643	−1544.6	360.2	424.3
Thiophene (*l*)	C_4H_4S	84.14	80.2	−2829		181.2	123.8
Toluene (*g*)	C_7H_8	92.14	50.5	−3910	122.3	320.8	104
Urea (*s*)	CH_4N_2O	60.06	−333.1	2635	2197.4	104.3	92.8

Sources: HCP, HCAA, HCFE, HTTD, TDPS, TDOC, and TPBS.

TABLE 7.1   Triple Point Temperatures and Pressures of Selected Substances

Formula	Name	T_{tp} (K)	P_{tp} (Pa)
Ar	Argon	83.806	68950
Br_2	Bromine	280.4	5879
Cl_2	Chlorine	172.17	1392
HCl	Hydrogen chloride	158.8	
H_2	Hydrogen	13.8	7042
H_2O	Water	273.16	611.73
H_2S	Hydrogen sulfide	187.67	23180
NH_3	Ammonia	195.41	6077
Kr	Krypton	115.8	72920
NO	Nitrogen oxide	109.54	21916
O_2	Oxygen	54.36	146.33
SO_3	Sulfur trioxide	289.94	21130
Xe	Xenon	161.4	81590
CH_4	Methane	90.694	11696
CO	Carbon monoxide	68.15	15420
CO_2	Carbon dioxide	216.58	518500
C_3H_6	Propene	87.89	9.50×10^{-4}

Sources: HCP, HTTP, and DAL.

TABLE 7.2   Melting and Boiling Temperatures and Enthalpies of Transition at 1 atm Pressure

Substance	Name	mp (K)	ΔH_{fusion} (kJ mol^{-1}) at T_m	bp (K)	$\Delta H_{vaporization}$ (kJ mol^{-1}) at T_b
Ar	Argon	83.8	1.12	87.3	6.43
Cl_2	Chlorine	171.6	6.41	239.18	20.41
Fe	Iron	1811	13.81	3023	349.5
H_2	Hydrogen	13.81	0.12	20.4	0.90
H_2O	Water	273.15	6.010	373.15	40.65
He	Helium	0.95	0.021	4.22	0.083
I_2	Iodine	386.8	14.73	457.5	41.57
N_2	Nitrogen	63.5	0.71	77.5	5.57
Na	Sodium	370.87	2.60	1156	98.0
NO	Nitric oxide	109.5	2.3	121.41	13.83
O_2	Oxygen	54.36	0.44	90.7	6.82
SO_2	Sulfur dioxide	197.6		263.1	24.94
Si	Silicon	1687	50.21	2628	359
W	Tungsten	3695	52.31	5933	422.6
Xe	Xenon	161.4	1.81	165.11	12.62
CCl_4	Carbon tetrachloride	250	3.28	349.8	29.82
CH_4	Methane	90.68	0.94	111.65	8.19
CH_3OH	Methanol	175.47	3.18	337.7	35.21
CO	Carbon monoxide	68	0.83	81.6	6.04
C_2H_4	Ethene			169.38	13.53
C_2H_6	Ethane	90.3	2.86	184.5	14.69
C_2H_5OH	Ethanol	159.0	5.02	351.44	38.56
C_3H_8	Propane	85.46	3.53	231.08	19.04
C_5H_5N	Pyridine			388.38	35.09
C_6H_6	Benzene	278.68	9.95	353.24	30.72
C_6H_5OH	Phenol	314.0	11.3	455.02	45.69
$C_6H_5CH_3$	Toluene	178.16	6.85	383.78	33.18
$C_{10}H_8$	Naphthalene	353.3	17.87	491.14	43.18

Sources: Data from HCP, HTTD, and DAL.

TABLE 7.3   Vapor Pressure and Boiling Temperature of Liquids

$$\ln \frac{P(T)}{\text{Pa}} = A(1) - \frac{A(2)}{\dfrac{T}{K} + A(3)} \qquad T_b(P) = \frac{A(2)}{A(1) - \ln \dfrac{P}{\text{Pa}}} - A(3)$$

Molecular Formula	Name	T_b (K)	A(1)	A(2)	A(3)	$10^{-3}P(298.15\,K)$ (Pa)	Range (K)
Ar	Argon	87.28	22.946	1.0325×10^3	3.130	—	73–90
Br_2	Bromine	331.9	20.729	2.5782×10^3	−51.77	28.72	268–354
HF	Hydrogen fluoride	292.65	22.893	3.6178×10^3	25.627	122.90	273–303
H_2O	Water	373.15	23.195	3.8140×10^3	−46.290		353–393
SO_2	Sulfur dioxide	263.12	21.661	2.3024×10^3	−35.960		195–280
CCl_4	Tetrachloromethane	349.79	20.738	2.7923×10^3	−46.6667	15.28	287–350
$CHCl_3$	Trichloromethane	334.33	20.907	2.6961×10^3	−46.926	26.24	263–335
HCN	Hydrogen cyanide	298.81	22.226	3.0606×10^3	−12.773	98.84	257–316
CH_3OH	Methanol	337.70	23.593	3.6971×10^3	−31.317	16.94	275–338
CS_2	Carbon disulfide	319.38	20.801	2.6524×10^3	−33.40	48.17	255–320
C_2H_5OH	Ethanol	351.45	23.58	3.6745×10^3	−46.702	7.87	293–366
C_3H_6	Propene	225.46	20.613	1.8152×10^3	−25.705	1156.6	166–226
C_3H_8	Propane	231.08	20.558	1.8513×10^3	−26.110	948.10	95–370
C_4H_9Br	1-Bromobutane	374.75	17.076	1.5848×10^3	−11.188	5.26	195–300
C_4H_9Cl	1-Chlorobutane	351.58	20.612	2.6881×10^3	−55.725	13.68	256–352
$C_5H_{11}OH$	1-Pentanol	411.133	20.729	2.5418×10^3	−134.93	0.29	410–514
C_6H_5Cl	Chlorobenzene	404.837	20.964	3.2969×10^3	−55.515	1.57	335–405
C_6H_5I	Iodobenzene	461.48	21.088	3.8136×10^3	−62.654	0.13	298–462
C_6H_6	Benzene	353.24	20.767	2.7738×10^3	−53.08	12.69	294–378
C_6H_{14}	Hexane	341.886	20.749	2.7081×10^3	−48.251	20.17	286–343
C_6H_5CHO	Benzaldehyde	451.90	21.213	3.7271×10^3	−67.156	0.17	311–481
$C_6H_5CH_3$	Toluene	383.78	21.600	3.6266×10^3	−23.778	3.80	360–580
$C_{10}H_8$	Naphthalene	491.16	21.100	4.0526×10^3	−67.866	0.01	353–453
$C_{14}H_{10}$	Anthracene	614.0	21.965	5.8733×10^3	−51.394		496–615

Sources: HCP and HTTP.

TABLE 7.4   Sublimation Pressure of Solids

$$\ln \frac{P(T)}{\text{Pa}} = A(1) - \frac{A(2)}{\dfrac{T}{K} + A(3)}$$

Molecular Formula	Name	A(1)	A(2)	A(3)	Range (K)
CCl_4	Tetrachloromethane	17.613	1.6431×10^3	−95.250	232–250
C_6H_{14}	Hexane	31.224	4.8186×10^3	−23.150	168–178
C_6H_5COOH	Benzoic acid	14.870	4.7196×10^3		293–314
$C_{10}H_8$	Naphthalene	31.143	8.5750×10^3		270–305
$C_{14}H_{10}$	Anthracene	31.620	1.1378×10^4		353–400

Sources: HCP and HTTP.

TABLE 9.2 Dielectric Constants, ε_r, of Selected Liquids

Substance	Dielectric Constant	Substance	Dielectric Constant
Acetic acid	6.2	Heptane	1.9
Acetone	21.0	Isopropyl alcohol	20.2
Benzaldehyde	17.8	Methanol	33.0
Benzene	2.3	Nitrobenzene	35.6
Carbon tetrachloride	2.2	o-Xylene	2.6
Cyclohexane	2.0	Phenol	12.4
Ethanol	25.3	Toluene	2.4
Glycerol	42.5	Water (273 K)	88.0
1-Hexanol	13.0	Water (373 K)	55.3

Source: HCP.

TABLE 9.3 Mean Activity Coefficients in Terms of Molalities at 298 K

Substance	0.1m	0.2m	0.3m	0.4m	0.5m	0.6m	0.7m	0.8m	0.9m	1.0m
$AgNO_3$	0.734	0.657	0.606	0.567	0.536	0.509	0.485	0.464	0.446	0.429
$BaCl_2$	0.500	0.444	0.419	0.405	0.397	0.391	0.391	0.391	0.392	0.395
$CaCl_2$	0.518	0.472	0.455	0.448	0.448	0.453	0.460	0.470	0.484	0.500
$CuCl_2$	0.508	0.455	0.429	0.417	0.411	0.409	0.409	0.410	0.413	0.417
$CuSO_4$	0.150	0.104	0.0829	0.0704	0.0620	0.0559	0.0512	0.0475	0.0446	0.0423
HCl	0.796	0.767	0.756	0.755	0.757	0.763	0.772	0.783	0.795	0.809
HNO_3	0.791	0.754	0.735	0.725	0.720	0.717	0.717	0.718	0.721	0.724
H_2SO_4	0.2655	0.2090	0.1826		0.1557		0.1417			0.1316
KCl	0.770	0.718	0.688	0.666	0.649	0.637	0.626	0.618	0.610	0.604
KOH	0.798	0.760	0.742	0.734	0.732	0.733	0.736	0.742	0.749	0.756
$MgCl_2$	0.529	0.489	0.477	0.475	0.481	0.491	0.506	0.522	0.544	0.570
$MgSO_4$	0.150	0.107	0.0874	0.0756	0.0675	0.0616	0.0571	0.0536	0.0508	0.0485
NaCl	0.778	0.735	0.710	0.693	0.681	0.673	0.667	0.662	0.659	0.657
NaOH	0.766	0.727	0.708	0.697	0.690	0.685	0.681	0.679	0.678	0.678
$ZnSO_4$	0.150	0.140	0.0835	0.0714	0.0630	0.0569	0.0523	0.0487	0.0458	0.0435

Source: HCP.

TABLE 9.5 **Standard Reduction Potentials in Alphabetical Order**

Reaction	$E°$ (V)	Reaction	$E°$ (V)
$Ag^+ + e^- \rightarrow Ag$	0.7996	$HClO + H^+ + e^- \rightarrow 1/2Cl_2 + H_2O$	1.611
$Ag^{2+} + e^- \rightarrow Ag^+$	1.980	$HClO_2 + 3H^+ + 3e^- \rightarrow 1/2Cl_2 + 2H_2O$	1.628
$AgBr + e^- \rightarrow Ag + Br^-$	0.07133	$HO_2 + H^+ + e^- \rightarrow H_2O_2$	1.495
$AgCl + e^- \rightarrow Ag + Cl^-$	0.22233	$HO_2 + H_2O + 2e^- \rightarrow 3OH^-$	0.878
$AgCN + e^- \rightarrow Ag + CN^-$	−0.017	$2H_2O + 2e^- \rightarrow H_2 + 2OH^-$	−0.8277
$AgF + e^- \rightarrow Ag + F^-$	0.779	$H_2O_2 + 2H^+ + 2e^- \rightarrow 2H_2O$	1.776
$Ag_4[Fe(CN)_6] + 4e^- \rightarrow 4Ag + [Fe(CN)_6]^{4-}$	0.1478	$H_3PO_4 + 2H^+ + 2e^- \rightarrow H_3PO_3 + H_2O$	−0.276
$AgI + e^- \rightarrow Ag + I^-$	−0.15224	$Hg^{2+} + 2e^- \rightarrow Hg$	0.851
$AgNO_2 + e^- \rightarrow Ag + NO_2^-$	0.564	$Hg_2^{2+} + 2e^- \rightarrow 2Hg$	0.7973
$Al^{3+} + 3e^- \rightarrow Al$	−1.662	$Hg_2Cl_2 + 2e^- \rightarrow 2Hg + 2Cl^-$	0.26808
$Au^+ + e^- \rightarrow Au$	1.692	$Hg_2SO_4 + 2e^- \rightarrow 2Hg + SO_4^{2-}$	0.6125
$Au^{3+} + 2e^- \rightarrow Au^+$	1.401	$I_2 + 2e^- \rightarrow 2I^-$	0.5355
$Au^{3+} + 3e^- \rightarrow Au$	1.498	$I_3^- + 2e^- \rightarrow 3I^-$	0.536
$AuBr_2 + e^- \rightarrow Au + 2Br^-$	0.959	$In^+ + e^- \rightarrow In$	−0.14
$AuCl_4 + 3e^- \rightarrow Au + 4Cl^-$	1.002	$In^{2+} + e^- \rightarrow In^+$	−0.40
$Ba^{2+} + 2e^- \rightarrow Ba$	−2.912	$In^{3+} + 3e^- \rightarrow In$	−0.3382
$Be^{2+} + 2e^- \rightarrow Be$	−1.847	$K^+ + e^- \rightarrow K$	−2.931
$Bi^{3+} + 3e^- \rightarrow Bi$	0.20	$Li^+ + e^- \rightarrow Li$	−3.0401
$Br_2(aq) + 2e^- \rightarrow 2Br^-$	1.0873	$Mg^{2+} + 2e^- \rightarrow Mg$	−2.372
$BrO^- + H_2O + 2e^- \rightarrow Br^- + 2OH^-$	0.761	$Mg(OH)_2 + 2e^- \rightarrow Mg + 2OH^-$	−2.690
$Ca^+ + e^- \rightarrow Ca$	−3.80	$Mn^{2+} + 2e^- \rightarrow Mn$	−1.185
$Ca^{2+} + 2e^- \rightarrow Ca$	−2.868	$Mn^{3+} + e^- \rightarrow Mn^{2+}$	1.5415
$Cd^{2+} + 2e^- \rightarrow Cd$	−0.4030	$MnO_2 + 4H^+ + 2e^- \rightarrow Mn^{2+} + 2H_2O$	1.224
$Cd(OH)_2 + 2e^- \rightarrow Cd + 2OH^-$	−0.809	$MnO_4^- + 4H^+ + 3e^- \rightarrow MnO_2 + 2H_2O$	1.679
$CdSO_4 + 2e^- \rightarrow Cd + SO_4^{2-}$	−0.246	$MnO_4^{2-} + 2H_2O + 2e^- \rightarrow MnO_2 + 4OH^-$	0.595
$Ce^{3+} + 3e^- \rightarrow Ce$	−2.483	$MnO_4^- + 8H^+ + 5e^- \rightarrow Mn^{2+} + 4H_2O$	1.507
$Ce^{4+} + e^- \rightarrow Ce^{3+}$	1.61	$MnO_4^- + e^- \rightarrow MnO_4^{2-}$	0.558
$Cl_2(g) + 2e^- \rightarrow 2Cl^-$	1.35827	$2NO + 2H^+ + 2e^- \rightarrow N_2O + H_2O$	1.591
$ClO_4^- + 2H^+ + 2e^- \rightarrow ClO_3^- + H_2O$	1.189	$HNO_2 + H^+ + e^- \rightarrow NO + H_2O$	0.983
$ClO^- + H_2O + 2e^- \rightarrow Cl^- + 2OH^-$	.81	$NO_2 + H_2O + 3e^- \rightarrow NO + 2OH^-$	−0.46
$ClO_4^- + H_2O + 2e^- \rightarrow ClO_3^- + 2OH^-$	0.36	$NO_3^- + 4H^+ + 3e^- \rightarrow NO + 2H_2O$	0.957
$Co^{2+} + 2e^- \rightarrow Co$	−0.28	$NO_3^- + 2H^+ + e^- \rightarrow NO_2^- + H_2O$	0.835
$Co^{3+} + e^- \rightarrow Co^{2+}$ (2 mol / l H_2SO_4)	1.83	$NO_3^- + H_2O + 2e^- \rightarrow NO_2^- + 2OH^-$	0.10
$Cr^{2+} + 2e^- \rightarrow Cr$	−0.913	$Na^+ + e^- \rightarrow Na$	−2.71
$Cr^{3+} + e^- \rightarrow Cr^{2+}$	−0.407	$Ni^{2+} + 2e^- \rightarrow Ni$	−0.257
$Cr^{3+} + 3e^- \rightarrow Cr$	−0.744	$NiO_2 + 2H_2O + 2e^- \rightarrow Ni(OH)_2 + 2OH^-$	0.49
$Cr_2O_7^{2-} + 14H^+ + 6e^- \rightarrow 2Cr^{3+} + 7H_2O$	1.232	$Ni(OH)_2 + 2e^- \rightarrow Ni + 2OH^-$	−0.72
$Cs^+ + e^- \rightarrow Cs$	−2.92	$NiO_2 + 4H^+ + 2e^- \rightarrow Ni^{2+} + 2H_2O$	1.678
$Cu^+ + e^- \rightarrow Cu$	0.521	$NiOOH + H_2O + e^- \rightarrow Ni(OH)_2 + OH^-$	+0.52
$Cu^{2+} + e^- \rightarrow Cu^+$	0.153	$O_2 + e^- \rightarrow O_2^-$	−0.56
$Cu(OH)_2 + 2e^- \rightarrow Cu + 2OH^-$	−0.222	$O_2 + 2H^+ + 2e^- \rightarrow H_2O_2$	0.695
$F_2 + 2H^+ + 2e^- \rightarrow 2HF$	3.053	$O_2 + 4H^+ + 4e^- \rightarrow 2H_2O$	1.229
$F_2 + 2e^- \rightarrow 2F^-$	2.866		
$Fe^{2+} + 2e^- \rightarrow Fe$	−0.447	$O_2 + 2H_2O + 2e^- \rightarrow H_2O_2 + 2OH^-$	−0.146
$Fe^{3+} + 3e^- \rightarrow Fe$	−0.030	$O_2 + 2H_2O + 4e^- \rightarrow 4OH^-$	0.401
$Fe^{3+} + e^- \rightarrow Fe^{2+}$	0.771	$O_2 + H_2O + 2e^- \rightarrow HO_2^- + OH^-$	−0.076
$[Fe(CN)_6]^{3-} + e^- \rightarrow [Fe(CN)_6]^{4-}$	0.358	$O_3 + 2H^+ + 2e^- \rightarrow O_2 + H_2O$	2.076
$2H^+ + 2e^- \rightarrow H_2$	0	$O_3 + H_2O + 2e^- \rightarrow O_2 + 2OH^-$	1.24
$HBrO + H^+ + e^- \rightarrow 1/2Br_2 + H_2O$	1.574	$[PtCl_4]^{2-} + 2e^- \rightarrow Pt + 4Cl^-$	0.755

TABLE 9.5   **Standard Reduction Potentials in Alphabetical Order** *(continued)*

Reaction	E° (V)	Reaction	E° (V)
$[PtCl_6]^{2-} + 2e^- \rightarrow [PtCl_4]^{2-} + 2Cl^-$	0.68	$S + 2H^+ + 2e^- \rightarrow H_2S(aq)$	0.142
$Pt(OH)_2 + 2e^- \rightarrow Pt + 2OH^-$	0.14	$S_2O_6^{2-} + 4H^+ + 2e^- \rightarrow 2H_2SO_3$	0.564
$Rb^+ + e^- \rightarrow Rb$	-2.98	$S_2O_6^{2-} + 2e^- + 2\,H^+ \rightarrow 2HSO_3^-$	0.464
$Re^{3+} + 3e^- \rightarrow Re$	0.300	$S_2O_8^{2-} + 2e^- \rightarrow 2SO_4^{2-}$	2.010
$S + 2e^- \rightarrow S^{2-}$	-0.47627	$2H_2SO_3 + H^+ + 2e^- \rightarrow H_2SO_4^- + 2H_2O$	-0.056
$Pb^{2+} + 2e^- \rightarrow Pb$	-0.1262	$H_2SO_3 + 4H^+ + 4e^- \rightarrow S + 3H_2O$	0.449
$Pb^{4+} + 2e^- \rightarrow Pb^{2+}$	1.67	$Sn^{2+} + 2e^- \rightarrow Sn$	-0.1375
$PbBr_2 + 2e^- \rightarrow Pb + 2Br^-$	-0.284	$Sn^{4+} + 2e^- \rightarrow Sn^{2+}$	0.151
$PbCl_2 + 2e^- \rightarrow Pb + 2Cl^-$	-0.2675	$Ti^{2+} + 2e^- \rightarrow Ti$	-1.630
$PbO + H_2O + 2e^- \rightarrow Pb + 2OH^-$	-0.580	$Ti^{3+} + 2e^- \rightarrow Ti^{2+}$	-0.368
$PbO_2 + 4H^+ + 2e^- \rightarrow Pb^{2+} + 2H_2O$	1.455	$TiO_2 + 4H^+ + 2e^- \rightarrow Ti^{2+} + 2H_2O$	-0.502
$PbO_2 + SO_4^{2-} + 4H^+ + 2e^- \rightarrow PbSO_4 + 2H_2O$	1.6913	$Zn^{2+} + 2e^- \rightarrow Zn$	-0.7618
$PbSO_4 + 2e^- \rightarrow Pb + SO_4^{2-}$	-0.3505	$ZnO_2^{2-} + 2H_2O + 2e^- \rightarrow Zn + 4OH^-$	-1.215
$Pd^{2+} + 2e^- \rightarrow Pd$	0.951	$Zr(OH)_2 + H_2O + 4e^- \rightarrow Zr + 4OH^-$	-2.36
$Pt^{2+} + 2e^- \rightarrow Pt$	1.118		

Source: HCP and Bard.

TABLE 9.6   Standard Reduction Potentials Ordered by Reduction Potential

Reaction	$E°$ (V)	Reaction	$E°$ (V)
$Ca^+ + e^- \rightarrow Ca$	-3.80	$Fe^{3+} + 3e^- \rightarrow Fe$	-0.030
$Li^+ + e^- \rightarrow Li$	-3.0401	$AgCN + e^- \rightarrow Ag + CN^-$	-0.017
$Rb^+ + e^- \rightarrow Rb$	-2.98	$2H^+ + 2e^- \rightarrow H_2$	0
$K^+ + e^- \rightarrow K$	-2.931	$AgBr + e^- \rightarrow Ag + Br^-$	0.07133
$Cs^+ + e^- \rightarrow Cs$	-2.92	$NO_3^- + H_2O + 2e^- \rightarrow NO_2^- + 2OH^-$	0.10
$Ba^{2+} + 2e^- \rightarrow Ba$	-2.912	$Pt(OH)_2 + 2e^- \rightarrow Pt + 2OH^-$	0.14
$Ca^{2+} + 2e^- \rightarrow Ca$	-2.868	$S + 2H^+ + 2e^- \rightarrow H_2S(aq)$	0.142
$Na^+ + e^- \rightarrow Na$	-2.71	$Ag_4[Fe(CN)_6] + 4e^- \rightarrow 4Ag + [Fe(CN)_6]^{4-}$	0.1478
$Mg(OH)_2 + 2e^- \rightarrow Mg + 2OH^-$	-2.690	$Sn^{4+} + 2e^- \rightarrow Sn^{2+}$	0.151
$Ce^{3+} + 3e^- \rightarrow Ce$	-2.483	$Cu^{2+} + e^- \rightarrow Cu^+$	0.153
$Mg^{2+} + 2e^- \rightarrow Mg$	-2.372	$Bi^{3+} + 3e^- \rightarrow Bi$	0.20
$Zr(OH)_2 + H_2O + 4e^- \rightarrow Zr + 4OH^-$	-2.36	$AgCl + e^- \rightarrow Ag + Cl^-$	0.22233
$Be^{2+} + 2e^- \rightarrow Be$	-1.847	$Hg_2Cl_2 + 2e^- \rightarrow 2Hg + 2Cl^-$	0.26808
$Al^{3+} + 3e^- \rightarrow Al$	-1.662	$Re^{3+} + 3e^- \rightarrow Re$	0.300
$Ti^{2+} + 2e^- \rightarrow Ti$	-1.630	$[Fe(CN)_6]^{3-} + e^- \rightarrow [Fe(CN)_6]^{4-}$	0.358
$ZnO_2^{2-} + 2H_2O + 2e^- \rightarrow Zn + 4OH^-$	-1.215	$ClO_4^- + H_2O + 2e^- \rightarrow ClO_3^- + 2OH^-$	0.36
$Mn^{2+} + 2e^- \rightarrow Mn$	-1.185	$O_2 + 2H_2O + 4e^- \rightarrow 4OH^-$	0.401
$Cr^{2+} + 2e^- \rightarrow Cr$	-0.913	$H_2SO_3 + 4H^+ + 4e^- \rightarrow S + 3H_2O$	0.449
$2H_2O + 2e^- \rightarrow H_2 + 2OH^-$	-0.8277	$S_2O_6^{2-} + 2e^- + 2H^+ \rightarrow 2HSO_3^-$	0.464
$Cd(OH)_2 + 2e^- \rightarrow Cd + 2OH^-$	-0.809	$NiO_2 + 2H_2O + 2e^- \rightarrow Ni(OH)_2 + 2OH^-$	0.49
$Zn^{2+} + 2e^- \rightarrow Zn$	-0.7618	$NiOOH + H_2O + e^- \rightarrow Ni(OH)_2 + OH^-$	0.52
$Cr^{3+} + 3e^- \rightarrow Cr$	-0.744	$Cu^+ + e^- \rightarrow Cu$	0.521
$Ni(OH)_2 + 2e^- \rightarrow Ni + 2OH^-$	-0.72	$I_2 + 2e^- \rightarrow 2I^-$	0.5355
$PbO + H_2O + 2e^- \rightarrow Pb + 2OH^-$	-0.580	$I_3^- + 2e^- \rightarrow 3I^-$	0.536
$O_2 + e^- \rightarrow O_2^-$	-0.56	$MnO_4^- + e^- \rightarrow MnO_4^{2-}$	0.558
$TiO_2 + 4H^+ + 2e^- \rightarrow Ti^{2+} + 2H_2O$	-0.502	$AgNO_2 + e^- \rightarrow Ag + NO_2^-$	0.564
$S + 2e^- \rightarrow S^{2-}$	-0.47627	$S_2O_6^{2-} + 4H^+ + 2e^- \rightarrow 2H_2SO_3$	0.564
$NO_2 + H_2O + 3e^- \rightarrow NO + 2OH^-$	-0.46	$MnO_4^{2-} + 2H_2O + 2e^- \rightarrow MnO_2 + 4OH^-$	0.595
$Fe^{2+} + 2e^- \rightarrow Fe$	-0.447	$Hg_2SO_4 + 2e^- \rightarrow 2Hg + SO_4^{2-}$	0.6125
$Cr^{3+} + e^- \rightarrow Cr^{2+}$	-0.407	$[PtCl_6]^{2-} + 2e^- \rightarrow [PtCl_4]^{2-} + 2Cl^-$	0.68
$Cd^{2+} + 2e^- \rightarrow Cd$	-0.4030	$O_2 + 2H^+ + 2e^- \rightarrow H_2O_2$	0.695
$In^{2+} + e^- \rightarrow In^+$	-0.40	$[PtCl_4]^{2-} + 2e^- \rightarrow Pt + 4Cl^-$	0.755
$Ti^{3+} + 2e^- \rightarrow Ti^{2+}$	-0.368	$BrO^- + H_2O + 2e^- \rightarrow Br^- + 2OH^-$	0.761
$PbSO_4 + 2e^- \rightarrow Pb + SO_4^{2-}$	-0.3505	$Fe^{3+} + e^- \rightarrow Fe^{2+}$	0.771
$In^{3+} + 3e^- \rightarrow In$	-0.3382	$AgF + e^- \rightarrow Ag + F^-$	0.779
$PbBr_2 + 2e^- \rightarrow Pb + 2Br^-$	-0.284	$Hg_2^{2+} + 2e^- \rightarrow 2Hg$	0.7973
$Co^{2+} + 2e^- \rightarrow Co$	-0.28	$Ag^+ + e^- \rightarrow Ag$	0.7996
$H_3PO_4 + 2H^+ + 2e^- \rightarrow H_3PO_3 + H_2O$	-0.276	$ClO^- + H_2O + 2e^- \rightarrow Cl^- + 2OH^-$	0.81
$PbCl_2 + 2e^- \rightarrow Pb + 2Cl^-$	-0.2675	$NO_3^- + 2H^+ + e^- \rightarrow NO_2^- + H_2O$	0.835
$Ni^{2+} + 2e^- \rightarrow Ni$	-0.257	$Hg^{2+} + 2e^- \rightarrow Hg$	0.851
$CdSO_4 + 2e^- \rightarrow Cd + SO_4^{2-}$	-0.246	$HO_2 + H_2O + 2e^- \rightarrow 3OH^-$	0.878
$Cu(OH)_2 + 2e^- \rightarrow Cu + 2OH^-$	-0.222	$Pd^{2+} + 2e^- \rightarrow Pd$	0.951
$AgI + e^- \rightarrow Ag + I^-$	-0.15224	$NO_3^- + 4H^+ + 3e^- \rightarrow NO + 2H_2O$	0.957
$O_2 + 2H_2O + 2e^- \rightarrow H_2O_2 + 2OH^-$	-0.146	$AuBr_2 + e^- \rightarrow Au + 2Br^-$	0.959
$In^+ + e^- \rightarrow In$	-0.14	$HNO_2 + H^+ + e^- \rightarrow NO + H_2O$	0.983
$Sn^{2+} + 2e^- \rightarrow Sn$	-0.1375	$AuCl_4 + 3e^- \rightarrow Au + 4Cl^-$	1.002
$Pb^{2+} + 2e^- \rightarrow Pb$	-0.1262	$Br_2(aq) + 2e^- \rightarrow 2Br^-$	1.0873
$O_2 + H_2O + 2e^- \rightarrow HO_2^- + OH^-$	-0.076	$Pt^{2+} + 2e^- \rightarrow Pt$	1.118
$2H_2SO_3 + H^+ + 2e^- \rightarrow H_2SO_4^- + 2H_2O$	-0.056	$ClO_4^- + 2H^+ + 2e^- \rightarrow ClO_3^- + H_2O$	1.189

TABLE 9.6   **Standard Reduction Potentials Ordered by Reduction Potential** *(continued)*

Reaction	E^0 (V)	Reaction	$E°$ (V)
$MnO_2 + 4H^+ + 2e^- \rightarrow Mn^{2+} + 2H_2O$	1.224	$HClO + H^+ + e^- \rightarrow 1/2Cl_2 + H_2O$	1.611
$O_2 + 4H^+ + 4e^- \rightarrow 2H_2O$	1.229	$HClO_2 + 3H^+ + 3e^- \rightarrow 1/2Cl_2 + 2H_2O$	1.628
$Cr_2O_7^{2-} + 14H^+ + 6e^- \rightarrow 2Cr^{3+} + 7H_2O$	1.232	$Pb^{4+} + 2e^- \rightarrow Pb^{2+}$	1.67
$O_3 + H_2O + 2e^- \rightarrow O_2 + 2OH^-$	1.24	$NiO_2 + 4H^+ + 2e^- \rightarrow Ni^{2+} + 2H_2O$	1.678
$Cl_2(g) + 2e^- \rightarrow 2Cl^-$	1.35827	$MnO_4^- + 4H^+ + 3e^- \rightarrow MnO_2 + 2H_2O$	1.679
$Au^{3+} + 2e^- \rightarrow Au^+$	1.401	$PbO_2 + SO_4^{2-} + 4H^+ + 2e^- \rightarrow PbSO_4 + 2H_2O$	1.6913
$PbO_2 + 4H^+ + 2e^- \rightarrow Pb^{2+} + 2H_2O$	1.455	$Au^+ + e^- \rightarrow Au$	1.692
$HO_2 + H^+ + e^- \rightarrow H_2O_2$	1.495	$H_2O_2 + 2H^+ + 2e^- \rightarrow 2H_2O$	1.776
$Au^{3+} + 3e^- \rightarrow Au$	1.498	$Co^{3+} + e^- \rightarrow Co^{2+}$ (2 mol / l H_2SO_4)	1.83
$MnO_4^- + 8H^+ + 5e^- \rightarrow Mn^{2+} + 4H_2O$	1.507	$Ag^{2+} + e^- \rightarrow Ag^+$	1.980
$Mn^{3+} + e^- \rightarrow Mn^{2+}$	1.5415	$S_2O_8^{2-} + 2e^- \rightarrow 2SO_4^{2-}$	2.010
$HBrO + H^+ + e^- \rightarrow 1/2Br_2 + H_2O$	1.574	$O_3 + 2H^+ + 2e^- \rightarrow O_2 + H_2O$	2.076
$2NO + 2H^+ + 2e^- \rightarrow N_2O + H_2O$	1.591	$F_2 + 2e^- \rightarrow 2F^-$	2.866
$Ce^{4+} + e^- \rightarrow Ce^{3+}$	1.61	$F_2 + 2H^+ + 2e^- \rightarrow 2HF$	3.053

Sources: HCP and Bard.

TABLE 20.2   **Chemical Shifts for Protons in the 20 Common Amino Acids**[a]

NH indicates the amide proton. Hα indicates alpha protons, and Hβ indicates beta protons.

Residue	NH	αH	βH	Others	
Gly	8.39	3.97			
Ala	8.25	4.35	1.39		
Val	8.44	4.18	2.13	γCH$_3$	0.97, 0.94
Ile	8.19	4.23	1.90	γCH$_2$	1.48, 1.19
				γCH$_3$	0.95
				δCH$_3$	0.89
Leu	8.42	4.38	1.65, 1.65	γH	1.64
				δCH$_3$	0.94, 0.90
Pro[b]		4.44	2.28, 2.02	γCH$_2$	2.03, 2.03
				δCH$_2$	3.68, 3.65
Ser	8.38	4.50	3.88, 3.88		
Thr	8.24	4.35	4.22	γCH$_3$	1.23
Asp	8.41	4.76	2.84, 2.75		
Glu	8.37	4.29	2.09, 1.97	γCH$_2$	2.31, 2.28
Lys	8.41	4.36	1.85, 1.76	γCH$_2$	1.45, 1.45
				δCH$_2$	1.70, 1.70
				ϵCH$_2$	3.02, 3.02
				ϵNH$_3$	7.52
Arg	8.27	4.38	1.89, 1.79	γCH$_2$	1.70, 1.70
				δCH$_2$	3.32, 3.32
				NH	7.17, 6.62
Asn	8.75	4.75	2.83, 2.75	γNH$_2$	7.59, 6.91
Gln	8.41	4.37	2.13, 2.01	γCH$_2$	2.38, 2.38
				δNH$_2$	6.87, 7.59
Met	8.42	4.52	2.15, 2.01	γCH$_2$	2.64, 2.64
				ϵCH$_3$	2.13
Cys	8.31	4.69	3.28, 2.96		
Trp	8.09	4.70	3.32, 3.19	2H	7.24
				4H	7.65
				5H	7.17
				6H	7.24
				7H	7.50
				NH	10.22
Phe	8.23	4.66	3.22, 2.99	2,6H	7.30
				3,5H	7.39
				4H	7.34
Tyr	8.18	4.60	3.13, 2.92	2,6H	7.15
				3,5H	6.86
His	8.41	4.63	3.26, 3.20	2H	8.12
				4H	7.14

[a]Data for the nonterminal residues X in tetrapeptides GGXA, pH 7.0, 35°C [from Bundi and Wüthrich (1979a), except that more precise data were obtained for Leu, Pro, Lys, Arg, Met, and Phe using new measurements at 500 MHz].

[b]Data for *trans*-Pro.

Source: Wüthrich, K. *NMR of Proteins and Nucleic Acids,* John Wiley and Sons, New York, 1986.

Answers to Selected End of Chapter Problems

Numerical answers to problems are included here. Complete solutions to selected problems can be found in the *Student Solution's Manual*.

Chapter 1

P1.1 723 K

P1.2 a. $P_{H_2} = 6.24 \times 10^5$ Pa; $P_{O_2} = 3.90 \times 10^4$ Pa
$P_{total} = 6.57 \times 10^5$ Pa
mol % H_2 = 94.1%; mol % O_2 = 5.9%
b. $P_{N_2} = 4.45 \times 10^4$ Pa; $P_{O_2} = 3.90 \times 10^4$ Pa
$P_{total} = 8.35 \times 10^4$ Pa
mol % N_2 = 53.3%; mol % O_2 = 46.7%
c. $P_{NH_3} = 7.32 \times 10^4$ Pa; $P_{CH_4} = 7.77 \times 10^4$ Pa
$P_{total} = 1.51 \times 10^5$ Pa
mol % NH_3 = 48.5%; mol % O_2 = 51.5%

P1.3 $N_{O_2} = 1.04 \times 10^{21}$ s^{-1}
$N_{mitochondria} = 800 \times 10^{12}$
$\dfrac{N_{O_2}}{N_{mitochondria}} = 1.30 \times 10^6 \text{s}^{-1}$

P1.4 $n_{O_2} = 1.97 \times 10^{-2}$ mol min^{-1}

P1.5 0.08200 atm mol^{-1} °C^{-1}; 280.2°C

P1.7 4.84×10^5 Pa

P1.8 Mass output = 21.9 g h^{-1}

P1.9 Net loss = 17.6 g h^{-1}

P1.10 1.26×10^3 L

P1.11 $V_{air} = 6.67 \times 10^6$ L

P1.12 26.8 L

P1.13 $N_{O_2} = 5.38 \times 10^{21}$
$N_{hemoglobin} = 1.34 \times 10^{21}$

P1.14 $N_{myoglobin} = 5.38 \times 10^{21}$

P1.15 a. 68.6%, 18.5%, 0.9%, and 12.0% for $N_2, O_2, Ar,$ and H_2O
b. 12.2 L
c. 0.992

P1.16 a. 2.88×10^{-2} bar
b. $x_{O_2} = 0.0179, x_{N_2} = 0.803$
$x_{CO} = 0.178$
$P_{O_2} = 5.16 \times 10^{-4}$ bar, $P_{N_2} = 2.31 \times 10^{-2}$ bar
$P_{CO} = 5.10 \times 10^{-3}$ bar

P1.17 $\dfrac{n_{O_2}}{n_{CO}} = 250.0$

P1.18 $N_{breaths} = 1.00 \times 10^4$

P1.19   1.50 L

P1.20   Mass glucose = 3.66 g

P1.21 $x_{CO_2} = 0.176, x_{H_2O} = 0.235, x_{O_2} = 0.588$

P1.22 $V = 0.25$ L

P1.23   59.9%

P1.24   158 amu

P1.25 $x_{O_2} = 0.20; x_{H_2} = 0.80$

P1.26   Mass product = 16.3 g

P1.27   17.3 bar

P1.28   41.6 bar

P1.29   54

P1.30 $x_{CO_2} = 0.028 \; x_{N_2} = 0.972$

P1.31 8.34×10^4 Pa

P1.32 $x_{H_2}^\circ = 0.103; x_{O_2}^\circ = 0.897$

P1.33 $V_{O_2} = 7.47$ L

P1.34 $V = 1.02 \times 10^{-2}$ L

Chapter 2

P2.1   a. -4.00×10^3 J
        b. -8.22×10^3 J

P2.2 $\Delta T = 0.43$ K

P2.3 $w = -2.03 \times 10^4$ J; $\Delta U = 0$ and $\Delta H = 0$
        $q = 2.03 \times 10^4$ J

P2.4   312 K

P2.5   Heat = 7.09×10^3 J kg^{-1} h^{-1}

P2.6 $\Delta T = 5.1$ K

P2.7 $\Delta U = -935$ J; $\Delta H = q_P = -1.56 \times 10^3$ J
        $w = 624$ J

P2.8 -28.6×10^3 J, -15.1×10^3 J, 0

P2.9   31 g

P2.10 $w = 0.61$ J

P2.11 $\Delta H = 63.0 \times 10^3$ J; $\Delta U = 60.5 \times 10^3$ J

P2.12 $w = -1.87 \times 10^3$ J; 0.944 bar

P2.13 $q = 0; \Delta U = w = -1200$ J
        $\Delta H = -2.00 \times 10^3$ J

P2.14 $\Delta T = 0.070$ K

P2.16   235 K

P2.17   18.0 m

P2.18   Candle burns 4.97×10^3 J min^{-1}

P2.19   a. -21.1×10^3 J; b. -9.34×10^3 J

P2.20   step 1: 0; step 2: -23.0×10^3 J
        step 3: 9.00×10^3 J; cycle -14.0×10^3 J

P2.21 4.25×10^3 J; 0.69 m

P2.22 110.5×10^3 Pa; 107.8×10^3 Pa

P2.23 -379 J

P2.24 $w = 0$; $\Delta U = q = 5.72 \times 10^3$ J
        $\Delta H = 8.01 \times 10^3$ J

P2.25   a. $P_2 = 0.500 \times 10^6$ Pa
            $w = -1.69 \times 10^3$ J
            $\Delta U = 0$ and $\Delta H = 0$
            $q = -w = 1.69 \times 10^3$ J
        b. $P_2 = 6.02 \times 10^5$ Pa; $\Delta U = 748$ J
            $w = 0$; $q = 748$ J
            $\Delta H = 1.25 \times 10^3$ J
        Overall:
            $q = 2.44 \times 10^3$ J
            $w = -1.69 \times 10^3$ J
            $\Delta U = 748$ J
            $\Delta H = 1.25 \times 10^3$ J

P2.26   749 K; $q = 0$;
        $w = \Delta U = 5.62 \times 10^3$ J
        $\Delta H = 9.37 \times 10^3$ J

P2.27 $q = 0$;
        $w = \Delta U = 463$ J; $\Delta H = 771$ J

P2.28   a. $\Delta U = \Delta H = 0$;
            $w = -q = -1.25 \times 10^3$ J
        b. $w = 0$;
            $q = \Delta U = 854$ J; $\Delta H = 1.42 \times 10^3$ J
        For the overall process, $w = -1.25 \times 10^3$ J, $q = 2.02 \times 10^3$ J, $\Delta U = 854$ J, and
        $\Delta H = 1.42 \times 10^3$ J

P2.29   a.   667 K; $w = 9.30 \times 10^3$ J
        b. 3.80×10^3 K; 97.5×10^3 J

P2.30   299 K

P2.31 $w = 2.25 \times 10^{-19}$ J

P2.32 $k_B T = 4.28 \times 10^{-21}$ J. The bending energy is 53 times the thermal energy.

P2.33 7.26×10^{-21} J

P2.35 -8.99×10^3 J

P2.36   a.   188 K b. 217 K

P2.37   a. 2.55×108 Nm^{-2}
        b. $w = 1.18 \times 10^{-20}$ J
        c. 4.28×10^{-21} J

P2.38 $M = 0.25$ kg

P2.39   c. 1.27×10^3 kg, 2.54×10^3 kg

P2.40 $q = 0$;
        $\Delta U = w = -2.43 \times 10^3$ J
        $\Delta H = -4.05 \times 10^3$ J

P2.41   472 K

P2.42 $q = 0$;
        $\Delta U = w = -1.43 \times 10^3$ J; $\Delta H = -2.39 \times 10^3$ J

P2.43   a. -5.54×10^3 J
        b. -5.52×10^3 J; -0.4%

P2.44 a. $w = -496$ J; ΔU and $\Delta H = 0$
$q = -w = 496$ J
  b. $\Delta U = -623$ J
$w = 0$; $q = \Delta U = -623$ J
$\Delta H = -1.04 \times 10^3$ J
$\Delta U_{total} = 623$ J
$w_{total} = -496$ J
$q_{total} = -127$ J
$\Delta H_{total} = -1.04 \times 10^3$ J

P2.45 Change in contact area $= 4.50 \times 10^{-10}$ m². The work ratio is the ratio of the contact areas $= 2.79$.

Chapter 3
P3.5 77.8 bar

P3.9 $\Delta H = q = 6.67 \times 10^4$ J
$\Delta U = 5.61 \times 10^4$ J; $w = -1.06 \times 10^4$ J

P3.15 306 K

P3.16 304 K

P3.17 345 K

P3.18 $q = \Delta H = 4.35 \times 10^4$ J
$w = -3.74 \times 10^3$ J
$\Delta U = 3.98 \times 10^4$ J

P3.19 3.06×10^3 J, 0

Chapter 4
P4.1 a. -1816 kJ mol^{-1}; -1814 kJ mol^{-1}
  b. -116.2 kJ mol^{-1}; -113.7 kJ mol^{-1}
  c. 62.6 kJ mol^{-1}; 52.7 kJ mol^{-1}
  d. -111.6 kJ mol^{-1}; -111.6 kJ mol^{-1}
  e. 205.9 kJ mol^{-1}; 200.9 kJ mol^{-1}
  f. -172.8 kJ mol^{-1}; -167.8 kJ mol^{-1}

P4.2 $\Delta H_{combustion}^\circ = -3268$ kJ mol^{-1}
$\Delta U_{reaction}^\circ = -3264$ kJ mol^{-1}; 0.0122

P4.3 49.6 kJ mol^{-1}

P4.4 10.41 kJ mol^{-1}, -1.54%

P4.5 -59.8 kJ mol^{-1}

P4.6 -266.3 kJ mol^{-1}; -824.2 kJ mol^{-1}

P4.7 91.6 kJ mol^{-1}

P4.8 -1810 kJ mol^{-1}

P4.9 -20.6 kJ mol^{-1}; -178.2 kJ mol^{-1}

P4.10 415.8 kJ mol^{-1}; 1.2%

P4.11 -134 kJ mol^{-1}; $\approx 0\%$

P4.12 -180.0 kJ mol^{-1}

P4.13 -91.96 kJ mol^{-1}

P4.14 132.86 kJ mol^{-1}

P4.15 -812.2 kJ mol^{-1}

P4.16 a. -73.0 kJ mol^{-1};
  b. -804 kJ mol^{-1}

P4.17 a. 428.22 kJ mol^{-1}; 425.74 kJ mol^{-1}
         b. 926.98 kJ mol^{-1}; 922.02 kJ mol^{-1}
         c. 498.76 kJ mol^{-1}; 498.28 kJ mol^{-1}

P4.18 Si–F: 596 kJ mol^{-1}; 593 kJ mol^{-1}
         Si–Cl: 398 kJ mol^{-1}; 396 kJ mol^{-1}
         C–F: 489 kJ mol^{-1}; 487 kJ mol^{-1}
         N–F: 279 kJ mol^{-1}; 276 kJ mol^{-1}
         O–F: 215 kJ mol^{-1}; 213 kJ mol^{-1}
         H–F: 568 kJ mol^{-1}; 565 kJ mol^{-1}

P4.19 a. 416 kJ mol^{-1}; 413 kJ mol^{-1}
         b. 329 kJ mol^{-1}; 329 kJ mol^{-1}
         c. 589 kJ mol^{-1}; 588 kJ mol^{-1}

P4.20 $\Delta U_f - 757$ kJ mol^{-1}; $\Delta H_f^\circ = -756$ kJ mol^{-1}

P4.21 5.16×10^3 J $^\circ$C^{-1}

P4.22 -2.86×10^3J mol^{-1}, 16%

P4.23 $\Delta H_{330K} = -1336.8$ kJ mol^{-1}

P4.24 $\Delta H_{330K} = 2793.7$ kJ mol^{-1}

P4.25 $\Delta H_{land} = 2.368 \times 10^{18}$ J

P4.26 $\Delta H_{ocean} = 5.86 \times 10^{17}$ J

P4.27 7.39×10^{-4}

P4.28 $\Delta H = -68.9$ kJ mol^{-1}, $w = -4.96$ kJ,
         $\Delta U = -73.9$ kJ

P4.29 a. $\Delta H_{reaction} = 5639$ kJ mol^{-1}
         b. Stored energy $= 9.16 \times 10^4$ W hectare^{-1}
         c. 0.916% of the radiation can be stored in the form of carbohydrates.

P4.30 a. $\Delta H_{298K} = -114.9$ kJ mol^{-1}
         b. $\Delta H_{310K} = -114.5$ kJ mol^{-1}

P4.31 $T_m = 304$ K. From the graph the intrinsic excess heat capacity δC_P^{int} is about
         0.42 J K^{-1} g^{-1}. The transition excess heat capacity δC_P^{trs} is about 0.83 J K^{-1} g^{-1}. The
         excess heat capacity is 1.25 J K^{-1}g^{-1}.

P4.32 $\Delta H_{den} = M\Delta h_{den} \approx 217$ kJ mol^{-1}

P4.33 Energy per pound per mile for the auto is 3.12 kJ lb^{-1} mi^{-1}. Therefore, the auto
         consumes 1.56 times as much area per pound per mile as the walker.

P4.34 Heat per mole of O_2 is 357 kJ.

P4.35 The camper needs 49 grams of rations.

P4.36 For 10 grams: $q = -149.1$ kJ, $\Delta U = -150.3$ kJ, and $W = -1.24$ kJ.

P4.37 $\Delta H = -816.7$ kJ mol^{-1}

Chapter 5
P5.2 a. $V_c = 29.6$ L; $V_d = 10.4$ L
        b. $w_{ab} = -7.62 \times 10^3$ J
            $w_{bc} = -5.61 \times 10^3$ J
            $w_{cd} = 3.68 \times 10^3$ J
            $w_{da} = 5.61 \times 10^3$ J
            $w_{total} = -3.94 \times 10^3$ J
         c. 0.515; 1.94 kJ

P5.3 $a \rightarrow b$: $\Delta U = \Delta H = 0$; $q = -w = 7.62 \times 10^3$ J
$b \rightarrow c$: $\Delta U = w = -5.61 \times 10^3$ J; $q = 0$
$\Delta H = -9.35 \times 10^3$ J
$c \rightarrow d$: $\Delta U = \Delta H = 0$; $q = -w = 3.68 \times 10^3$ J
$d \rightarrow a$: $\Delta U = w = 5.61 \times 10^3$ J; $q = 0$
$\Delta H = 9.35 \times 10^3$ J
$q_{total} = 3.94 \times 10^3$ J $= -w_{total}$
$\Delta U_{total} = \Delta H_{total} = 0$

P5.4 $a \rightarrow b$: $\Delta S = -\Delta S_{surroundings} = 8.73$ J K^{-1}
$\Delta S_{total} = 0$
$b \rightarrow c$: $\Delta S = -\Delta S_{surroundings} = 0$; $\Delta S_{total} = 0$
$c \rightarrow d$: $\Delta S = -\Delta S_{surroundings} = -8.70$ J K^{-1}
$\Delta S_{total} = 0$
$d \rightarrow a$: $\Delta S = -\Delta S_{surroundings} = 0$. $\Delta S_{total} = 0$ to within the round-off error.
For the cycle, $\Delta S = \Delta S_{surroundings} = \Delta S_{total} = 0$ to within the round-off error.

P5.5 a. 17.6 J K^{-1}; b. 10.6 J K^{-1}

P5.6 16.8 J K^{-1}

P5.7 a. $w = -1.25 \times 10^3$ J; $\Delta U = 1.87 \times 10^3$ J
$q = \Delta H = 3.12 \times 10^3$ J; $\Delta S = 8.43$ J K^{-1}
b. $w = 0$; $\Delta U = q = 1.87 \times 10^3$ J
$\Delta H = 3.12 \times 10^3$ J; $\Delta S = 5.06$ J K^{-1}
c. $\Delta U = \Delta H = 0$; $w_{reversible} = -q = -1.73 \times 10^3$ J
$\Delta S = 5.76$ J K^{-1}

P5.8 a. $\Delta S_{surroundings} = -6.93$ J K^{-1}
$\Delta S_{total} = 1.50$ J K^{-1}; spontaneous
b. $\Delta S_{surroundings} = -4.16$ J K^{-1}
$\Delta S_{total} = 0.90$ J K^{-1}; spontaneous
c. $\Delta S_{surroundings} = -5.76$ J K^{-1}
$\Delta S_{total} = 0$; not spontaneous

P5.9 $\Delta S = 4.13 \times 10^4$ J K^{-1}

P5.10 $\Delta S_{water} = -41.09$ J K^{-1}
$\Delta S_{surr} = 42.73$ J K^{-1}
$\Delta S_{univ} = \Delta S_{water} + \Delta S_{surr} = 1.64$ J K^{-1}

P5.11 a. 1.03 J K^{-1} mol^{-1}
b. 3.14 J K^{-1} mol^{-1}
c. $\Delta S_{transition} = 8.24$ J K^{-1} mol^{-1}
$\Delta S_{fusion} = 25.12$ J K^{-1} mol^{-1}

P5.12 a. 23.49 J K^{-1}
b. 154.4 J K^{-1}

P5.13 a. $q = 0$ $\Delta U = w = -935$ J
$\Delta H = -1.31 \times 10^3$ J; $\Delta S = 0$
b. $q = 0$
$\Delta U = w = -748$ J
$\Delta H = -1.05 \times 10^3$ J
$\Delta S = 1.24$ J K^{-1}
c. $w = 0$; $\Delta U = \Delta H = 0$; $q = 0$; $\Delta S = 5.76$ J K^{-1}

P5.14 a. 100.8 J mol^{-1} K^{-1}
b. 18.94 × 10^3 J mol^{-1}

P5.15 $\Delta H_m = 2.84 \times 10^3$ J mol^{-1}
$\Delta S_m = 8.90$ J K^{-1} mol^{-1}

P5.16 $\Delta U = 18.5$ J; $w = -2.73 \times 10^3$ J
$\Delta H = 32.1$ J; $q \approx 2.73 \times 10^3$ J
$\Delta S = 9.10$ J K^{-1}

P5.17 21.88 J K^{-1} mol^{-1}

P5.18 $\Delta S = \Delta S_{reaction} = 53.00 \text{ J K}^{-1} \text{ mol}^{-1}$
$\Delta H_{reaction} = 42.6 \text{ kJ mol}^{-1}$
$\Delta S_{surr} = -142.95 \text{ J K}^{-1} \text{ mol}^{-1}$
$\Delta S_{universe} = \Delta S + \Delta S_{surr} = -89.95 \text{ J K}^{-1} \text{ mol}^{-1}$

P5.19 $\Delta S_{310} = 176.42 \text{ J K}^{-1} \text{ mol}^{-1}$
$\Delta S_{surr,310} = 240.86 \text{ J K}^{-1} \text{ mol}^{-1}$
$\Delta S_{universe} = 417.29 \text{ J K}^{-1} \text{ mol}^{-1}$

P5.20 $\Delta S_{sys,330} = -379.01 \text{ J K}^{-1} \text{ mol}^{-1}$
$\Delta S_{surr,330} = -8.38 \times 10^3 \text{ J K}^{-1} \text{ mol}^{-1}$
$\Delta S_{universe} = \Delta S_{sys,330} + \Delta S_{surr,330} = -8.76 \times 10^3 \text{ J K}^{-1} \text{ mol}^{-1}$

P5.21 $\Delta S = 44.30 \text{ J K}^{-1} \text{ mol}^{-1}$

P5.22 $\Delta U - 4.36 \times 10^3 \text{ J}; \Delta H = -7.27 \times 10^3 \text{ J}$
$\Delta S = -30.7 \text{ J K}^{-1}$

P5.23 18.2 J K^{-1}

P5.24 a. $q = 0$
  $\Delta U = w = -5.21 \times 10^3 \text{ J}; \Delta H = -8.68 \times 10^3 \text{ J}$
  $\Delta S = 0; \Delta S_{surroundings} = 0; \Delta S_{total} = 0$
b. $w = 0$
  $\Delta U = q = 5.21 \times 10^3 \text{ J}; \Delta H = 8.68 \times 10^3 \text{ J}$
  $\Delta S = 14.5 \text{ J K}^{-1}$
  $\Delta S_{surroundings} = -17.4 \text{ J K}^{-1}; \Delta S_{total} = -2.90 \text{ J K}^{-1}$
c. $\Delta H = \Delta U = 0 \ w = -q = 6.48 \times 10^3 \text{ J}$
  $\Delta S = -14.5 \text{ J K}^{-1}$
  $\Delta S_{surroundings} = 21.6 \text{ J K}^{-1}; \Delta S_{total} = 7.1 \text{ J K}^{-1}$
For the cycle,
$w_{total} = 1.27 \times 10^3 \text{ J}; q_{total} = -1.27 \times 10^3 \text{ J}$
$\Delta U_{total} = 0; \Delta H_{total} = 0; \Delta S_{total} = 0$
$\Delta S_{surroundings} = 0; \Delta S_{total} = 4.20 \text{ J K}^{-1}$

P5.25 30.7 J K^{-1}

P5.26 9.0 J K^{-1}

P5.27 $\Delta S = 44.19 \text{ J K}^{-1} \text{mol}^{-1}$

P5.28 $0.564, 0.744$
$\Delta S_{total} = 0.2 \text{ J K}^{-1}$

P5.29 $\Delta S = -21.7 \text{ J K}^{-1}; \Delta S_{surroundings} = 21.9 \text{ J K}^{-1}$

P5.30 a. $\Delta S_{total} = \Delta S + \Delta S_{surroundings} = 0 + 0 = 0;$ not spontaneous
b. $\Delta S_{total} = \Delta S + \Delta S_{surroundings} = 1.24 \text{ J K}^{-1} + 0 = 1.24 \text{ J K};$ spontaneous

P5.31 6.25 m^2

P5.32 640 J s^{-1}

P5.33 $4.5 \times 10^2 \text{ g}$

P5.34 a. $\Delta S_{surroundings} = 0; \Delta S = 0; \Delta S_{total} = 0;$ not spontaneous
b. $\Delta S = 27.7 \text{ J K}^{-1}; \Delta S_{surroundings} = 0$
  $\Delta S_{total} = 27.17 \text{ J K}^{-1};$ spontaneous
c. $\Delta S = 27.7 \text{ J K}^{-1}$
  $\Delta S_{surroundings} = -27.7 \text{ J K}^{-1}$
  $\Delta S_{total} = 0;$ not spontaneous

P5.36 $S_{310} = 135.28 \text{ J K}^{-1} \text{ mol}^{-1}$

P5.37 $S(T = 300 \text{ K}) = 108.9 \text{ J K}^{-1} \text{ mol}^{-1}$

P5.38 $3.24 \times 10^8 \text{ J}$

P5.39 2.5

P5.40   a. 0.627
            b. 0.398
            c. 110.7 tons

P5.41   30.69 J K^{-1} mol^{-1}

P5.42 $\Delta S_{den} \approx 717$ J K^{-1} mol^{-1}

P5.43 $\Delta S_{den} = 714$ J K^{-1} mol^{-1}

P5.44 $S(T = 300$ K$) = 131.5$ J K^{-1} mol^{-1}

P5.45 $\Delta S_{340} = 1880$ J K^{-1} mol^{-1}
            $\Delta H_{310} = 389.0$ kJ mol^{-1}
            $\Delta S_{310} = 1110$ J K^{-1} mol^{-1}

P5.46   Efficiency $= 0.071$

Chapter 6

P6.1 -40.96 kJ g^{-1}, -117.6 kJ g^{-1}

P6.2 5.30×10^3 J

P6.3 -22.1×10^3 J

P6.4   a. -9.97×10^3 J, -9.97×10^3 J
            b. same as part (a).

P6.5   216

P6.6   52.8 J; 11.4×10^3 J; -218.5×10^3 J mol^{-1}

P6.7 -257.2×10^3 J mol^{-1}; -226.8×10^3 J mol^{-1}

P6.8 $\Delta G^{\circ}_{combustion} = -818.6 \times 10^3$ J mol^{-1}
            $\Delta A^{\circ}_{combustion} = -813.6 \times 10^3$ J mol^{-1}

P6.9   167.9 g of glucose required

P6.11   a. 0.1408
            b. 2.00×10^{-18}
            c. 101 kJ mol^{-1}

P6.12 $K = 6.68$

P6.13 $K = 5.8 \times 10^{-49}$

P6.14   a. 0.379; 1.284
            b. $\Delta H^{\circ}_{reaction} = 56.8 \times 10^3$ J mol^{-1}; $\Delta G^{\circ}_{reaction}$ (298.15 K) $= 35.0 \times 10^3$ J mol^{-1}

P6.15   a. 1.40; $\Delta G^{\circ}_{reaction} = -2.80 \times 10^3$ J mol^{-1}
            b. -29.7 kJ mol^{-1}

P6.16   a. $\Delta H^{\circ}_{reaction} = -19.0$ kJ mol^{-1}
                $\Delta G^{\circ}_{reaction}$ (600° C) $= 765$ J mol^{-1}
                $\Delta S^{\circ}_{reaction}$ (600° C) $= -22.6$ J K^{-1} mol^{-1}
            b. $x_{co_2} = 0.47$ $x_{co} = 0.53$

P6.17   a. K_P (700 K) $= 3.85$
                K_P (800 K) $= 1.56$
            b. $\Delta H^{\circ}_{reaction} = -42.1$ kJ mol^{-1}
                $\Delta G^{\circ}_{reaction}$ (700 K) $= -7.81$ kJ mol^{-1}
                $\Delta G^{\circ}_{reaction}$ (800 K) $= -2.91$ kJ mol^{-1}
                $\Delta S^{\circ}_{reaction}$ (700 K) $= 60.1$ J mol^{-1} K^{-1}
                $\Delta S^{\circ}_{reaction}$ (800 K) $= 52.6$ J mol^{-1} K^{-1}
            c. -27.5 kJ mol^{-1}

P6.18   a. $\Delta G_{den} = 24.0$ kJ mol^{-1}
            b. $K = 7.0 \times 10^{-5}$
            c. The protein is structurally stable at pH $= 2$ and $T = 303$ K.

P6.19 $T = 320$ K

P6.21 $\Delta G_{den} = -49.2$ kJ mol^{-1}. ΔV is taken as constant over 1000 bar.

P6.22 $T_m = 338.1$ K at $P = 1000$ bar

P6.23 $\Delta G_{reaction}$ (298 K) $= -126.3$ kJ mol^{-1}
$\Delta G_{reaction}$ (310 K) $= -128.4$ kJ mol^{-1}

P6.24 $\Delta G_{den} = 14.1$ kJ mol^{-1}

P6.25 a. $\dfrac{x_F}{x_G} = 2.025 \times 10^{-4}$

$\dfrac{x_E}{x_G} = 4.581 \times 10^{-7}; \dfrac{x_D}{x_G} = 2.486 \times 10^{-5}$

$\dfrac{x_C}{x_G} = 4.109 \times 10^{-6}; \dfrac{x_B}{x_G} = 1.497 \times 10^{-6}$

$\dfrac{x_A}{x_G} = 9.803 \times 10^{-8}$

b. $F\ 2.025 \times 10^{-2}\%; E\ 4.581 \times 10^{-5}\%$
$D\ 2.486 \times 10^{-3}\%; C\ 4.109 \times 10^{-4}\%$
$B\ 1.497 \times 10^{-4}\%; A\ 9.803 \times 10^{-6}\%$

P6.26 a. 3.78×10^{-5} bar
b. 6.20×10^{-5} bar

P6.28 -65.2×10^3 J mol^{-1}

P6.29 -18.6×10^3 J; 62.5 J K^{-1}

P6.30 a. -34.4 kJ
b. -47.3 kJ
c. -12.9 kJ

P6.31 $4; -32.6 \times 10^3$ J mol^{-1}

P6.32 468 K; 1.03×10^4

P6.33 9.95×10^5

P6.34 371

P6.35 4.68×10^{-2}

P6.36 1456 K; 9.12 Torr

P6.37 a. 1.11×10^{-2}
b. 1.76 mol of $N_2(g)$, 5.28 mol of $H_2(g)$, and 0.48 mol of $NH_3(g)$

P6.38 a. 5.13×10^{-35}
1.03×10^{-34}

P6.39 a. 8.68×10^{-2}; 0.045
b. 2.2%

P6.40 a. 3.31×10^{-3}
b. 0.0139 bar

P6.41 b. 0.55
d. 0.72

P6.42 $T = 298$ K
$\Delta G^\circ_{reaction} = -1362.6$ kJ mol^{-1}
$T = 310$ K $\Delta G_{reaction} = -1363.5$ kJ mol^{-1}

P6.43 a. 3700 bar

P6.44 $\Delta G^\circ_{duplex} = -31.0$ kJ mol^{-1} and $K = 1.54 \times 10^5$

Chapter 7

P7.1 a. 110 J mol^{-1}
b. 594 J mol^{-1}

P7.5 $T_{b, \, normal} = 271.8 \text{ K}; T_{b, \, standard} = 269.6 \text{ K}$

P7.6 354.4 K

P7.7 $6.17 \times 10^3 \text{ Pa}$

P7.8 $30.58 \text{ kJ mol}^{-1}$

P7.9 $22.88 \text{ kJ mol}^{-1}$

P7.10 $20.32 \text{ kJ mol}^{-1}$

P7.11 $25.28 \text{ kJ mol}^{-1}$

P7.12 $50.99 \text{ kJ mol}^{-1}$

P7.13 a. $\Delta H_m^{vaporization} = 32.1 \times 10^3 \text{ J mol}^{-1}$
$\Delta H_m^{sublimation} = 37.4 \times 10^3 \text{ J mol}^{-1}$
b. $5.3 \times 10^3 \text{ J mol}^{-1}$
c. $349.5 \text{ K}; 91.8 \text{ J mol}^{-1} \text{ K}^{-1}$
d. $T_{tp} = 264 \text{ K}; P_{tp} = 2.84 \times 10^3 \text{ Pa}$

P7.14 a. 720 bar
b. $2.2 \times 10^2 \text{ bar}$
c. $-1.5°\text{C}$

P7.15 a. 56.22 Torr
b. 52.65 Torr

P7.16 a. 4.66 bar
b. 4.10 bar

P7.17 $8.2 \, °\text{C}$

P7.18 $T_m = 355 \text{ K}$

P7.19 $\Delta H_{den} = 230. \text{ kJ mol}^{-1}$

P7.20 269 Pa

P7.21 a. 335.9 K
b. $38.19 \text{ kJ mol}^{-1}$ at 298 K and
$37.20 \text{ kJ mol}^{-1}$ at 335.9 K

P7.23 $\Delta H° = -90.7 \text{ kJ mol}^{-1}$
$\Delta S° = -253 \text{ J K}^{-1} \text{ mol}^{-1}$

P7.25 a. $\Delta H_{sublimation} = 16.92 \times 10^3 \text{ J mol}^{-1}$
$\Delta H_{vaporization} = 14.43 \times 10^3 \text{ J mol}^{-1}$
b. $2.49 \times 10^3 \text{ J mol}^{-1}$
c. $T_{tp} = 73.62 \text{ K}; P_{tp} = 5.36 \times 10^{-3} \text{ Torr}$

P7.26 $H° = -90.7 \text{ kJ mol}^{-1}$
$\Delta S° = -241 \text{ J K}^{-1} \text{ mol}^{-1}$

P7.27 $\Gamma = 4.8 \times 10^8 \text{ mol m}^{-2}$

P7.28 $K_{335} = 74$
$\Delta G°_{335} = -12 \text{ kJ mol}^{-1}$

P7.29 $38.4 \text{ J K}^{-1} \text{ mol}^{-1}; 16.4 \times 10^3 \text{ J mol}^{-1}$

P7.30 0.061%

P7.31 $K_{310 \text{ K}} = 1.24 \times 10^{-2}$
$\Delta G°_{310 \text{ K}} = 11.3 \text{ kJ mol}^{-1}$

P7.32 $467.7 \text{ K}; 2.513 \times 10^5 \text{ Pa}$

P7.33 $7.806 \times 10^4 \text{ Pa}$

P7.34 $\Delta H_{sublimation} = 231.7 \text{ kJ mol}^{-1}$
$\Delta H_{vaporization} = 206.5 \text{ kJ mol}^{-1}$
$\Delta H_{fusion} = 25.2 \text{ kJ mol}^{-1}$
1398 K; 128 Torr

P7.35 $\Delta H^{\circ}_{sublimation} = 32.6 \text{ kJ mol}^{-1}$
$\Delta H^{\circ}_{vaporization} = 26.9 \text{ kJ mol}^{-1}$
$\Delta H^{\circ}_{fusion} = 5.6 \text{ kJ mol}^{-1}$
240.3 K; 402 Torr

P7.36   142 K; 2984 Torr
$\Delta H_{sublimation} = 10.07 \times 10^{3} \text{ J mol}^{-1}$
$\Delta H_{vaporization} = 9.38 \times 10^{3} \text{ J mol}^{-1}$
$\Delta H_{fusion} = 0.69 \times 10^{3} \text{ J mol}^{-1}$

P7.37 8.5 kJ mol^{-1}

P7.38 -0.72 K at 100 bar and -3.62 K at 500 bar

P7.39   1.95 atm

P7.40 $9.60 \times 10^{5} \text{ Pa}$

P7.41   1.068

P7.42 $6.66 \times 10^{4} \text{ Pa}$

P7.44 $P = 26.4 \text{ Torr}$

P7.45 $\Gamma_{max} = 5.1 \times 10^{-6} \text{ mol m}^{-2}$

P7.46 $\Delta G = 915 \text{ kJ mol}^{-1}$

P7.50 $w_{rev} = -2.2 \times 10^{-10} \text{ J}$

Chapter 8

P8.1   121 Torr

P8.2   0.116 bar

P8.3   0.272

P8.4 $P_{A}^{*} = 0.623 \text{ bar}; P_{B}^{*} = 1.414 \text{ bar}$

P8.5   a. $P_A = 63.0 \text{ Torr}; P_B = 25.4 \text{ Torr}$
b. $P_A = 74.9 \text{ Torr}; P_B = 18.2 \text{ Torr}$

P8.6 $x_{bromo} = 0.67; y_{bromo} = 0.44$

P8.7   a.   2651 Torr
b.   0.525
c. $Z_{chloro} = 0.614$

P8.8   0.301

P8.9   a.   25.0 Torr, 0.50
b. $Z_{EB} = (1 - Z_{EC}) = 0.387$

P8.10   a.   0.560
b.   0.884

P8.11   0.337

P8.13   a.   for ethanol
$a_1 = 0.9504; \gamma_1 = 1.055;$ for isooctane
$a_2 = 1.411; \gamma_2 = 14.20$
b.   121.8 Torr

P8.16   413 Torr

P8.17   61.9 Torr

P8.18 -4.2 cm^{3}

P8.19 33.5 g mol^{-1}

P8.20 1.86 K kg mol^{-1}

P8.21 $M = 37.6$ g mol^{-1}; $\Delta T_f = -1.26$ K

$$\frac{P_{benzene}}{P^*_{benzene}} = 0.981$$

$\pi = 5.37 \times 10^5$ Pa

P8.22 2.37 m; 2.32×10^4 Pa

P8.23 1400 kg mol^{-1}

P8.24 $C_{O_2} = 3.9$ mg L^{-1}; $C_{He} = 1.4$ mg L^{-1}

P8.25 $h = 0.315$ m

P8.26 $c = 0.299$ mol L^{-1}

P8.28 57.8 cm^3 mol^{-1}

P8.29 0.327 mol

P8.30 -0.034 L

P8.31 $a_A = 0.569$; $\gamma_A = 2.00$

$a_B = 0.986$; $\gamma_B = 1.38$

P8.32 $a^R_{CS_2} = 0.8723$; $\gamma^R_{CS_2} = 1.208$

$a^H_{CS_2} = 0.2223$; $\gamma^H_{CS_2} = 0.3079$

P8.33   c.   For 6-methylpurine:

$C_M = 6.6 \times 10^{-2}$ M; $C_D = 6.7 \times 10^{-2}$ M; $K = 15$

For cytidine:

$C_M = 0.13M$; $C_D = 3.5 \times 10^{-2}$ M;$K = 2.1$

P8.34 $\gamma = 0.285$

P8.35 $\Delta T_f = 1.0 \times 10^{-5}$K

P8.36   Assuming 100 g of water, $\pi = 106$ Torr

P8.37 $\pi = 8.0$ atm

P8.38 $\pi = 5.33$ atm

P8.39 7.14×10^{-3} g; 2.67×10^{-3} g

P8.40 $M = 1.00 \times 10^3$ g mol^{-1}

P8.41 $M = 69.0 \times 10^3$ g mol^{-1}

P8.42   Assuming $T = 298$ K, $M = 210$ g mol^{-1}

P8.43 $\pi = 4.32 \times 10^{-3}$ atm

P8.44 $P = 7.0$ bar; $C_{O_2} = 1.6 \times 10^{-3}$ M; $C_{N_2} = 3.4 \times 10^{-3}$ M

P8.45 $\pi = 27.0$ bars; $h = 0.28$ km

Chapter 9

P9.1 $\Delta H^\circ_{reaction} = -65.4$ kJ mol^{-1}

$\Delta G^\circ_{reaction} = -55.7$ kJ mol^{-1}

P9.2 $\Delta H^\circ_{reaction} = 17$ kJ mol^{-1}

$\Delta G^\circ_{reaction} = 16.5$ kJ mol^{-1}

P9.3 -32.9 J K^{-1} mol^{-1}

P9.4 1.1 J K^{-1} mol^{-1}

P9.5 $\Delta G^\circ_{solvation} = -379$ kJ mol^{-1}

P9.6   a. 5.0×10^{-4} mol kg^{-1}

b. 7.9×10^{-4} mol kg^{-1}

c. 5.0×10^{-4} mol kg^{-1}

P9.10 $0.0285 \text{ mol kg}^{-1}$

P9.11 0.0111

P9.12 $0.238 \text{ mol kg}^{-1}$; 0.0393

P9.13 43.0 nm

P9.14 304 nm

P9.15 0.736

P9.16 a. 0.92;
b. 0.77;
c. 0.52

P9.17 $I = 0.1500 \text{ mol kg}^{-1}$
$\gamma_\pm = 0.2559$
$a_\pm = 0.0146$

P9.18 $I = 0.0750 \text{ mol kg}^{-1}$
$\gamma_\pm = 0.523$
$a_\pm = 0.0209$

P9.19 $I = 0.325 \text{ mol kg}^{-1}$
$\gamma_\pm = 0.069$
$a_\pm = 0.0068$

P9.20 a. $1.07 \times 10^{-5} \text{ mol L}^{-1}$
b. $1.21 \times 10^{-5} \text{ mol kg}^{-1}$

P9.21 a. 49%
b. 40%

P9.22 a. 13.6%
b. 14.8%

P9.23 a. 6.89%
b. 8.08%

P9.24 a. 0.0770
b. 0.0422
c. 0.0840

P9.25 a. $0.0794 \text{ mol kg}^{-1}$
b. $0.0500 \text{ mol kg}^{-1}$
c. $0.0500 \text{ mol kg}^{-1}$
d. $0.1140 \text{ mol kg}^{-1}$

P9.26 a. $0.150 \text{ mol kg}^{-1}$
b. $0.0500 \text{ mol kg}^{-1}$
c. $0.200 \text{ mol kg}^{-1}$
d. $0.300 \text{ mol kg}^{-1}$

P9.27 Using limiting law $0.100m$
1.32%; $1.00m$ 0.453%; no ionic interactions $0.100m$ 1.31%; $1.00m$ 0.418%

P9.28 $I = 0.13m$

P9.29 a. 2.91%
b. 2.02%
c. 17.7%

P9.31 $210.4 \text{ kJ mol}^{-1}$; 1.21×10^{-37}

P9.32 $713.2 \text{ kJ mol}^{-1}$; 9.06×10^{124}

P9.33 8.28×10^{-84}; -1.22869 V

P9.34 $-103.8 \text{ kJ mol}^{-1}$

P9.35 $-131.2 \text{ kJ mol}^{-1}$

P9.36   a. 1.30×10^8
         b. 6.67×10^{-56}

P9.37 $\Delta G^{\circ\prime} = -438 \text{ kJ mol}^{-1}$

P9.39 $\dfrac{c_{ethanol}}{c_{acetaldehyde}} = 55.6$

P9.40   a. 1.52×10^{-82}
         b. 3.34×10^{13}

P9.41 4.16×10^{-4}

P9.42   a. 2.65×10^6
         b. $-36.7 \text{ kJ mol}^{-1}$

P9.43 -0.913 V

P9.44 4.90×10^{-13}

P9.45 $\Delta G^{\circ}_{reaction} = -212.3 \text{ kJ mol}^{-1}$
         $\Delta S^{\circ}_{reaction} = -12.5 \text{ J K}^{-1}$
         $\Delta H^{\circ}_{reaction} = -216.0 \text{ kJ mol}^{-1}$

P9.46 $2.38 \text{ V}; 1.81 \times 10^{-4} \text{ V K}^{-1}$

P9.47   a. 1.0122 V
         b. $1.0050 \text{ V } 0.72\%$ or ~0 with correct number of significant figures

P9.49   a. 1.110 V
         b. 0.626
         c. 1.106 V

P9.51 $-131.1 \text{ kJ mol}^{-1}$

P9.53 1.75×10^{-12}

P9.54 $K' = 2.5 \times 10^4; \Delta G^{\circ\prime} = -25 \text{ kJ mol}^{-1}$
         $K = 2.5 \times 10^{11}; \Delta G^{\circ} = -65 \text{ kJ mol}^{-1}$

P9.55 $E^{\circ} = 0.337 \text{ V}; E^{\circ\prime} = 0.130 \text{ V}$

P9.56 $Q = 2.73 \times 10^{-4}; \Delta G = -70.7 \text{ kJ mol}^{-1}$

P9.57 $Q = 2.2 \times 10^{-6}; \Delta G = -12.8 \text{ kJ mol}^{-1}$
         $E = 0.066 \text{ V}$

P9.58 $\pi = 25 \text{ Pa}$

P9.60 $c^L_- = 4.5 \times 10^{-2} \text{ M} ; c^L_+ = 6.5 \times 10^{-2} \text{M}$
         $c^R_- = c^R_+ = 5.5 \times 10^{-2} \text{ M}$
         $r_D = 1.2; 0_D = -4.7 \text{ mV}$

P9.61 $\pi = 25 \text{Pa}$

P9.64 $c^L_- = 4.60 \times 10^{-3} \text{ M}; c^L_+ = 7.60 \times 10^{-3} \text{ M}$
         $c^R_- = c^R_+ = 5.9 \times 10^{-3} \text{ M}$
         $\Delta pH = pH^R - pH^L = 0.11$

Chapter 10

P10.1   a. $-56.82 \text{ kJ mol}^{-1}$
         b. $-56.60 \text{ kJ mol}^{-1}$

P10.2 $K = 846$

P10.3 $-33.3 \text{ kJ mol}^{-1}$

P10.4   a. 14.4 kJ mol^{-1}
         b. $-22.7 \text{ kJ mol}^{-1}$
         c. $K = 3.2 \times 10^6$

P10.5 2.20 kJ

P10.6   10.0 kJ

P10.7   a. $\Delta\tilde{\mu} = -RT \ln 10 \times \Delta pH + zF\Delta\phi$
       b. 20.5 kJ mol^{-1}

P10.8   0.082 V

P10.9   a. $1.45 \times 10^{-4} M$
       b. 24.16 kJ mol^{-1}
       c. 11

P10.10   a. 13.55 kJ
       b. 0.97 kJ
       c. 42.59 kJ

P10.11   ATP/ADP = 1.09

P10.12   a. -55.2 kJ mol^{-1}
       b. 0.78

P10.13 $K' = 1.38 \times 10^5$
       $K = 1.38 \times 10^{12}$
       $\Delta G^\circ = -72.0$ kJ mol^{-1}

P10.14 -21.78 kJ mol^{-1}

P10.15 $K_{obs} = 9.359 \times 10^6$
       $\Delta G^\circ_{obs} = -39.792$ kJ mol^{-1}
       $\Delta S^\circ_{obs} = -60.41$ J K^{-1} mol^{-1}

P10.16 $\chi_{ATP} = 2.60 \times 10^{-3}$
       $\chi_{ADP} = 3.27 \times 10^{-3}$
       $\chi_P = 5.20 \times 10^{-3}$

P10.17 $\chi_{ATP} = 8.39 \times 10^{-2}$
       $\chi_{ADP} = 0.359$
       $\chi_P = 0.596$

P10.18   Proof not given.

P10.19   94.0 mol

P10.20   7.00 mol

Chapter 11

P11.1 -0.120V

P11.3 $E^\circ - E^{\circ\prime} = 7x\dfrac{RT}{nF}\ln 10$

P11.5 $K' = 10.1, \Delta G^{\circ\prime} = -5.73$ kJ mol^{-1}
       $K = 1.01 \times 10^8, \Delta G^\circ = -45.6$ kJ mol^{-1}

P11.6 $E^\circ = 0.236$V, $E^{\circ\prime} = 0.03$V

P11.7 -18.8 kJ mol^{-1}

P11.8 $E^{\circ\prime} = -0.14$V, $\Delta G^{\circ\prime} = 27.$kJ mol^{-1},
       $K' = 2.8 \times 10^{-5}$

P11.9   69.kJ mol^{-1}

P11.10 $\dfrac{c_{Pyr}}{c_{Lact}} = 7.9 \times 10^{14}$

P11.11   a. $K_{293} = 1.0 \times 10^4, N = 4$
       b. $\Delta H^\circ = -31.$kJ mol^{-1}
       c. $\Delta G^\circ_{293} = -22.$kJ mol^{-1}
       d. $\Delta S^\circ_{293} = -31.$J mol^{-1} K^{-1}

P11.12   For cooperative, $\theta = 0.98$. For independent, $\theta = 0.79$.

P11.13 $\theta = 0.9957$ without salt

P11.14 Reduces binding by a factor of 5.

P11.16 $2.4 \text{ kJ K}^{-1} \text{mol}^{-1}$

Chapter 12

P12.1 6.40×10^{-6} m, 1.92×10^{-6} m, and 6.39×10^{-7} m for 450 K, 1500 K, and 4500 K, respectively

P12.2 1.26×10^{-10} m for He at 100 K and 5.65×10^{-11} m for He at 500 K. For Ar, 4.00×10^{-11} m and 1.79×10^{-11} m at 100 K and 500 K, respectively

P12.3 40 K for He and 4.0 K for Ar

P12.4 $4.85 \times 10^{6} \text{ m s}^{-1}$

P12.5 864 m s^{-1}, $3.87 \times 10^{3} \text{ m s}^{-1}$, $8.65 \times 10^{3} \text{ m s}^{-1}$, and $2.73 \times 10^{5} \text{ m s}^{-1}$ for 10^{4} nm, 500 nm, 100 nm, and 0.1 nm, respectively; 958 K, 1.92×10^{4} K, 9.60×10^{4} K, and 9.56×10^{7} K for 1000 nm, 500 nm, 100 nm, and 0.1 nm, respectively

P12.6 a. $1.0 \times 10^{7} \text{ J s}^{-1}$
b. 5.0×10^{17}

P12.7 6.31×10^{19}

P12.8 $4.23 \times 10^{18} \text{ s}^{-1}$

P12.9 4.51 m s^{-1}

P12.10 0.707 m s^{-1}

P12.11 $h \approx 7.0 \times 10^{-34}$ J s; $\phi \approx 4.0 \times 10^{-19}$ J or 2.5 eV

P12.12 At 800 K, $3.10 \times 10^{-4} \text{ J m}^{-3}$; at 4000 K, 0.194 J m^{-3}

P12.13 $1.78 \times 10^{5} \text{ J s}^{-1}$, 0.0800 m

P12.14 4.52×10^{26} W

P12.15 $\tilde{\nu} = 109{,}677 \text{ cm}^{-1}$, $27{,}419 \text{ cm}^{-1}$, and $12{,}186 \text{ cm}^{-1}$ and $E_{max} = 2.18 \times 10^{-18}$ J, 5.45×10^{-19} J, and 2.42×10^{-19} J for the Lyman, Balmer, and Paschen series, respectively

P12.16 4.78 cm

P12.17 150.4 V

P12.18 4.91×10^{15} electrons, $E = 2.77 \times 10^{-19}$ J
$v = 7.80 \times 10^{5} \text{ m s}^{-1}$

P12.19 $\nu \geq 1.37 \times 10^{15} \text{ s}^{-1}$ $v = 9.59 \times 10^{5} \text{ m s}^{-1}$

P12.20 $\lambda = 1.24$ nm, $n = 9.36 \times 10^{13} \text{ s}^{-1}$

P12.21 $\Delta v_{H_2} = 1.98 \text{ m s}^{-1}$, $\dfrac{\Delta v}{v} = 1.03 \times 10^{-3}$

Chapter 13

P13.1 $\dfrac{n_5}{n_1}(100 \text{ K}) = 0.086$ $\dfrac{n_5}{n_1}(650 \text{ K}) = 2.67$

$\dfrac{n_{10}}{n_1}(100 \text{ K}) = 5.2 \times 10^{-7}$ $\dfrac{n_{10}}{n_1}(650 \text{ K}) = 0.757$

P13.2 For $n_2/n_1 = 0.150$, $T = 115$ K; for $n_2/n_1 = 0.999$, $T = 2.17 \times 10^{5}$ K

P13.3 $x = 0.79$ m $t_0 = -5.5 \times 10^{-4}$ s

P13.7 a. $2\sqrt{5} \exp(0.352 i\pi)$
b. $6 \exp(0)$

c. $\dfrac{\sqrt{10}}{4} \exp(0.398 i\pi)$

d. $\dfrac{\sqrt{13}}{2} \exp(0.392 i\pi)$

P13.8   a. $2i$

b. $-2\sqrt{5}i$

c. -1

d. $\dfrac{3}{5 + \sqrt{3}}(1+i)$

P13.12 $N = \dfrac{1}{\sqrt{2\pi}}$

P13.13 a. $N = \dfrac{2}{\sqrt{ab}}$

b. $N = \sqrt{\dfrac{6}{\pi a^3}}$

P13.15 a. $r = \sqrt{14}, \theta = 1.30$ radians

$\phi = 0.588$ radians

b. $x = -2.5, y = 2.5, z = \dfrac{5}{\sqrt{2}}$

Chapter 14

P14.6   a. 0.059

b. 0.0010

P14.7   a. 20

b. 20

c. 42.9

P14.9 $\sqrt{\dfrac{30}{a^3}}; \dfrac{a}{2}; \dfrac{2a^2}{7}$

P14.13 $N = \sqrt{\dfrac{4}{ab}}$

P14.16 $N = \sqrt{\dfrac{8}{abc}}$

P14.17 a. 2

b. 3

P14.18 a. $\alpha = 6.86 \times 10^{10}$

b. 1.80×10^{-31} J

c. 4.44×10^{-7}

P14.19 4.92×10^{-21} J; 0.073

P14.21 $\dfrac{\Delta\rho_{total}(x)}{\langle\rho_{total}(x)\rangle} = 0.089$

$\dfrac{\Delta\rho_{n=11}(x)}{\langle\Delta\rho_{total}(x)\rangle} = 2.0$

P14.22 $\lambda = 119$ nm

P14.23 $\lambda = 368$ nm

P14.24 $T_{Si} = 901$ K; $T_C = 4427$ K

P14.25 1.3×10^9 A/m^2

P14.26 4.10×10^{-20} J, 1.24×10^{14} s^{-1}

P14.33 516 kg s^{-2}; 0.525 kg

P14.34 $E_0 = 5.28 \times 10^{-32}$ J,
$E_0/kT = 1.27 \times 10^{-11}, 7.27 \times 10^{-15}$ m s^{-1}

P14.35 For $n = 0, 1,$ and 2, 5.97×10^{-2}, 0.103, and 0.134, respectively

P14.36 8.963×10^{13} s^{-1}

P14.37 $2299 \text{ N m}^{-1}, 4.28 \times 10^{-3} \text{ m}$

P14.38 $2.97 \times 10^{-20} \text{ J}, 997 \text{ m s}^{-1}$
$|v|/|v_{rms}| = 2.19$

Chapter 15

P15.1 $1.1 \times 10^{-13}; 0.999456$

P15.9 c. $0.920, 0.0620, 2.77 \times 10^{-3}$

P15.10 $1.1 \times 10^{-3}, 0.32, 0.99$

P15.13 $1.5a_0$

P15.15 $(3/4)(a_0)^2$

P15.16 $5a_0$

P15.17 $\langle r \rangle_H = (3/2)a_0; \langle r \rangle_{He^+} = (3/4)a_0;$
$\langle r \rangle_{Li^{2+}} = (1/2)a_0; \langle r \rangle_{Be^{3+}} = (3/8)a_0$

P15.18 $I_H = 13.60 \text{ eV}; I_{He^+} = 54.42 \text{ eV};$
$I_{Li^{2+}} = 122.4 \text{ eV}; I_{Be^{3+}} = 217.7 \text{ eV}$

P15.20 1.26

P15.22 $0, a_0^2$

P15.25 $r = 4a_0$

P15.26 $(3/2)a_0, a_0$

P15.27 364.508 nm

P15.28 Most energetic: $109, 678; 27419 \text{ cm}^{-1}; 12186 \text{ cm}^{-1}$

Least energetic: $822583, 15233.0, 5331.56 \text{ cm}^{-1}$

P15.29 $26, 0.0201 \text{ eV}$

Chapter 18

P18.1 $6.51 \times 10^{13} \text{ s}^{-1}, 1.54 \times 10^{-14} \text{ s}, 2.16 \times 10^{-20} \text{ J}, 0.134 \text{ eV}$

P18.2 0.935

P18.3 For H_2 at 300 and 1000 K, $n_1/n_0 = 6.81 \times 10^{-10}$ and 1.78×10^{-3}
For H_2 at 300 and 1000 K, $n_2/n_0 = 4.65 \times 10^{-19}$ and 3.07×10^{-6}
For Br_2 at 300 and 1000 K, $n_1/n_0 = 0.212$ and 0.628
For Br_2 at 300 and 1000 K, $n_2/n_0 = 4.50 \times 10^{-2}$ and 0.394

P18.4 $E_0 = 2.942 \times 10^{-20} \text{ J}, E_1 = 8.647 \times 10^{-20} \text{ J}$
$E_2 = 1.411 \times 10^{-19} \text{ J}, E_3 = 1.934 \times 10^{-19} \text{ J}$
$\nu_{0 \to 1} = 8.61 \times 10^{13} \text{ s}^{-1}, \nu_{0 \to 2} = 1.69 \times 10^{14} \text{ s}^{-1}$
$\nu_{0 \to 3} = 2.47 \times 10^{14} \text{ s}^{-1}$
$\text{Error}(\nu_{0 \to 2}), (\nu_{0 \to 3}) = -1.89\%, -4.57\%$

P18.5 12

P18.7 $1.738 \times 10^{-18} \text{ J}, 1.717 \times 10^{-18} \text{ J or } 1.034 \times 10^3 \text{ kJ mol}^1$

P18.8 $1.133 \times 10^{-20} \text{ J}$

P18.10 $1.28 \times 10^6 \text{ cm}, 5.61 \times 10^2 \text{ cm}$

P18.11 25%

P18.15 $1.5 \text{ cm}, 0.35 \text{ cm}$

P18.16 $1.742035 \times 10^{-10} \text{ m}$

P18.17 $9.1707 \times 10^{-11} \text{ m}$

P18.18 $4.657 \times 10^{-48} \text{ kg m}^2, 2.39 \times 10^{-21} \text{ J}$

P18.19 $1.938 \times 10^{-46} \text{ kg m}^2, 5.743 \times 10^{-23} \text{ J}$

P18.20 1.424×10^{-10} m, 1.443×10^{-10} m

P18.23 a. $k = 1595$ N m^{-1}
   b. $\tilde{v} = 1391$ cm^{-1}

P18.24 $k = 1903$ N m^{-1}; $\tilde{v} = 1641$ cm^{-1}

P18.26 $\tilde{v}(^{18}O_2) = 751.1$ cm^{-1}

Chapter 19

P19.1 Lyman:
   82258 cm^{-1} $\lambda = 121.6$ nm;
   97491 cm^{-1} $\lambda = 102.6$ nm;
   102823 cm^{-1} $\lambda = 97.3$ nm;
   109677 cm^{-1} $\lambda = 91.2$ nm

   Balmer:
   15233 cm^{-1} $\lambda = 656.5$ nm;
   20565 cm^{-1} $\lambda = 486.3$ nm;
   23032 cm^{-1} $\lambda = 434.2$ nm;
   27419 cm^{-1} $\lambda = 364.7$ nm

   Paschen:
   5331.5 cm^{-1} $\lambda = 1876$ nm;
   7799.3 cm^{-1} $\lambda = 1282$ nm;
   9139.8 cm^{-1} $\lambda = 1094$ nm;
   12186.4 cm^{-1} $\lambda = 820.6$ nm

P19.2 3.08255×10^{15} s^{-1}; 3.08367×10^{15} s^{-1}

P19.3 $E_{max} = 2.178 \times 10^{-18}$ J; $\nu_{max} = 3.288 \times 10^{15}$ s^{-1}
   $\lambda_{max} = 91.18$ nm
   $E_{min} = 1.634 \times 10^{-18}$ J; $\nu_{min} = 2.466 \times 10^{15}$ s^{-1}
   $\lambda_{min} = 121.6$ nm

P19.6 $-\dfrac{16\sqrt{2}e}{81}a_0$

P19.7 1.70×10^{9} s^{-1}, 3.33×10^{-6}

P19.8 $A = 5.00$

P19.9 $n = 3.6 \times 10^{-7}$ mol; OD $= 2.5$

P19.10 a. $\varepsilon = 87.12 \times 10^{3}$ M^{-1} cm^{-1}
   b. $A_{260} = 13.1$;
   c. $\varepsilon = 64.32 \times 10^{3}$ M^{-1} cm^{-1}; $A_{260} = 9.65$

P19.11 a. $\varepsilon = 88.26 \times 10^{3}$ M^{-1} cm^{-1}
   b. $c = 1.4 \times 10^{-5}$ M;
   c. $\varepsilon = 85.20 \times 10^{3}$ M^{-1} cm^{-1}; $c = 1.4 \times 10^{-5}$ M

P19.12 a. $A_{\lambda_1} = \left(c_M \varepsilon_{\lambda_1}^M + c_N \varepsilon_{\lambda_1}^N\right)\ell$; $A_{\lambda_2} = \left(c_M \varepsilon_{\lambda_2}^M + c_N \varepsilon_{\lambda_2}^N\right)\ell$
   b. $c_Y = 2.5 \times 10^{-5}$ M; $c_W = 3.5 \times 10^{-5}$ M

P19.13 a. $\varepsilon = 6500$ M^{-1} cm^{-1}
   b. $c = 2.5 \times 10^{-4}$ M
   c. $f_{AT} = 0.53$

P19.14 $f_{AU} = 0.57$

P19.15 $r = 4.2$nm

Chapter 20

P20.1   5.87 T, 23.3 T, 14.5 T

P20.2 -2.3839×10^{-26} J, 0 J, 2.3839×10^{-26} J

P20.3 -2.04 ppm, 1.12 ppm

P20.7 $I(2\tau)/I(0) = 0.239$

P20.8 $\eta_{IS} = 0.5$

P20.9   Two sets of equivalent protons. A triplet and a quartet.

P20.10 Three sets of equivalent protons. Two triplets and a triplet of quartets.

P20.11 Two sets of equivalent protons. A triplet and a doubet.

P20.12 One singlet

P20.13 One singlet

P20.14 One singlet

P20.15 Two sets of equivalent protons. A triplet and a quartet.

P20.17 Lysine

P20.18 Left spectrum is leucine.

P20.21 $W_0 = 2.47 \times 10^{-12}\,\text{s}^{-1}$

P20.27 $D_{CF} = 1429.4\,\text{s}^{-1}$ At $\theta = \pi/3$,
            $\Delta\omega = 714.7\,\text{s}^{-1}$

P20.29 $\sigma_{iso} = 1/3$ ppm, $\sigma_{aniso} = -81.33$ ppm, $\eta = 0.91$

P20.30 3.81

Chapter 21

P21.9 2.70×10^3 kg m^{-3}

P21.10 74.1%

P21.11 0.197 nm

P21.13 a. 27.8°;
            b. 69.1°

P21.14 a. 389 pm;
            b. 136 pm

P21.15 12.0, 4.58, 4.39

Chapter 22

P22.1   a. exp(692)
            b. exp(673)

P22.3   0.25

P22.4   a. 2.60×10^6
            b. 5148

P22.5   a. 4/52
            b. 1/52
            c. 12/52 and 3/52, respectively.

P22.6   a. 0.002
            b. 1.52×10^6

P22.7   a. 1/6
            b. 1/9
            c. 21/36

P22.8 a. 4.57×10^5
           b. 1.76×10^4
           c. 3.59×10^5

P22.9 a. 9.54×10^{-7}
           b. 9.54×10^{-7}
           c. 1.27×10^{-6} and 3.77×10^{-7}

P22.10 a. 9.77×10^{-4}
           b. 0.044
           c. 0.246
           d. 0.044

P22.11 a. 1.69×10^{-5}
           b. 3.05×10^{-3}
           c. 0.137
           d. 0.195

P22.14 $P_{N_2} = 0.230$ atm

           $P_{O_2} = 0.052$ atm

P22.15 a. 248 K

           b. 178 K

P22.16 a. 254 K

           b. 179 K

P22.17 b. set C

P22.19 0.333

P22.20 4150 K

P22.21 0.999998

P22.22 $a_- = 0.333334$

           $a_0 = 0.333333$

           $a_+ = 0.333333$

P22.23 432 K

P22.24 1090 K

P22.25 @ 300 K, $p = 0.074$. F_2 equivalent at 524 K
           @ 1000 K, $p = 0.249$. F_2 equivalent at 1742 K

P22.26 5.85×10^4K

P22.27 @ 100 K, 0.149
           @ 500 K, 0.414
           @ 200 K, 0.479

Chapter 23

P23.1 $q_T(H_2) = 2.74 \times 10^{26}$
           $q_T(N_2) = 1.42 \times 10^{28}$

P23.2 $q_T(Ar) - 2.44 \times 10^{29}$, $T = 590$ K

P23.3 0.086 K

P23.4 3.91×10^{17}

P23.5 2.00×10^5

P23.6 a. 1
           b. 2
           c. 2
           d. 12
           e. 2

P23.7   rotational: HD, translational: D_2

P23.8   a.   no
            b.   no
            c.   yes
            d.   yes

P23.9 $q_R = 5832$

P23.10 $q_R = 3.78 \times 10^4$

P23.11 a.   616 K
            b. $J = 5$ to 6

P23.12

a.

J	p_J	J	p_J
0	0.041	5	0.132
1	0.113	6	0.095
2	0.160	7	0.062
3	0.175	8	0.037
4	0.167	9	0.019

b.

J	p_J	J	p_J
0	0.043	5	0.131
1	0.117	6	0.093
2	0.165	7	0.059
3	0.179	8	0.034
4	0.163	9	0.018

P23.13 $\Theta_R = 7.58$ K

P23.14 @300 K, $q = 1, p_0 = 1$
            @300 K, $q = 1.32, p_0 = 0.762$

P23.15 IF @300 K: $q = 1.06, p_0 = 0.943, p_1 = 0.051, p_2 = 0.003$
            IF @ 3000 K: $q = 3.94, p_0 = 0.254, p_1 = 0.189, p_2 = 0.141$
            IBr @ 300 K: $q = 1.38, p_0 = 0.725, p_1 = 0.199, p_2 = 0.054$
            IBr @ 3000 K: $q = 8.26, p_0 = 0.121, p_1 = 0.106, p_2 = 0.094$

P23.16 $q_v = 1.67$

P23.17 $q_v = 1.09$

P23.18 $q_v = 1.70$

P23.20 451 cm^{-1}

P23.23 $q_E = 10.2$

P23.24 $q = 1.71 \times 10^{34}$

P23.25 $q = 1.30 \times 10^{29}$

P23.26 nRT

P23.27

Molecule	θ_R (K)	High-T for R?	θ_V (K)	High-T for V?
H^{35}Cl	15.3	no	4153	no
^{12}C^{16}O	2.78	yes	3123	no
^{39}KI	0.088	yes	288	no
CsI	0.035	yes	173	no

P23.29 3.72 kJ mol^{-1}

P23.30 1.46×10^{-28} J, or 0.

P23.32 C_v values are as follows:

	298 K	500 K	1000 K
2041 cm^{-1}	0.042	0.811	4.24
712 cm^{-1}	3.37	5.93	7.62
3369 cm^{-1}	0.000	0.048	1.56
Total	6.78	12.7	21.0

P23.33   c. 352 m s^{-1}

P23.35 $260 \text{ J mol}^{-1} \text{ K}^{-1}$

P23.36 $186 \text{ J mol}^{-1} \text{ K}^{-1}$

P23.38 $11.2 \text{ J mol}^{-1} \text{ K}^{-1}$

P23.39   a. $R \ln 2$
       b. $R \ln 4$
       c. $R \ln 2$
       d.   0

P23.40 $-57.2 \text{ kJ mol}^{-1}$

P23.41 $G^\circ_{R, m} = -15.4 \text{ kJ mol}^{-1}$
       $G^\circ_{v, m} = -0.30 \text{ kJ mol}^{-1}$

P23.42 2.25×10^{-9}

P23.45   a. σs
       b. σs^4
       c.   0 (not allowed)

P23.46   26.7

P23.47   28.5

Chapter 24

P24.1 $3.18 \times 10^{-19} \text{ m}^2$

P24.2   a. $3.68 \times 10^{-19} \text{ m}^2$
       b. $2.65 \times 10^{-19} \text{ m}^2$

P24.3   a. 319 s
       b. $6.13 \times 10^{-10} \text{ s}$

P24.4   a. $1.60 \times 10^{-3} \text{ s}$
       c. $2.40 \times 10^{-3} \text{ s}$

P24.5   a. $x_{ave} = 2.58 \times 10^{-5} \text{ m}$

P24.6 $r = 1.89 \times 10^{-9} \text{ m}, M = 16.8 \text{ kg mol}^{-1}$

P24.7   a.   catalase, $M = 238 \text{ kg mol}^{-1}$; alcohol dehydrogenase, $M = 74.2 \text{ kg mol}^{-1}$
       b.   8.6 h
       c.   6.1 cm

P24.8   a. $s = 1.70 \times 10^{-13} \text{ s}$
       b. $r = 1.89 \times 10^{-9} \text{ m}$

P24.9 $P(10, \pm 6) = 0.0439$
       $P(20, \pm 6) = 0.0739$
       $P(100, \pm 6) = 0.0666$

P24.10 $1.17 \times 10^{-10} \text{ m}^2 \text{ s}^{-1}$

P24.11 $6.23 \times 10^{-19} \text{ m}^2$

P24.12 a. $2.05 \times 10^{-3} \, \text{m}^3 \, \text{s}^{-1}$
    b. $1.88 \times 10^{-3} \, \text{m}^3 \, \text{s}^{-1}$

P24.13 a. $37.3 \, \text{m s}^{-1}$
    b. $0.893 \, \text{m s}^{-1}$

P24.14 $\eta_{D_2} = 1.18 \times 10^2 \, \mu P$, $\eta_{HD} = 1.03 \times 10^2 \, \mu P$

P24.15 1.89 cP

P24.16 21.9 s

P24.17 b. $1.34 \times 10^{-5} \, \text{m}^2 \, \text{s}^{-1}$

P24.18 b. $E = 10.4 \, \text{kJ mol}^{-1}$; $A = 9.32 \times 10^{-3} \, \text{cP}$

P24.19 $P = a/b = 27.76$

P24.20 $\nu = 65.2$; $[\eta] = 52.8 \, \text{mL g}^{-1}$

P24.21 $F = 1.20$; $f_{tr,0} = 3.29 \times 10^{-11} \, \text{kg s}^{-1}$; $f_{tr} = 3.94 \times 10^{-11} \, \text{kg s}^{-1}$;
$D_{tr} = 1.04 \times 10^{-10} \, \text{m}^2 \, \text{s}^{-1}$

P24.22 $F = 1.54$ using Eq. (24.45);
$f_{tr,0} = 3.62 \times 10^{-11} \, \text{kgs}^{-1}$; $f_{tr} = 5.49 \times 10^{-11} \, \text{kgs}^{-1}$; $D_{tr} = 7.49 \times 10^{-11} \, \text{m}^2 \, \text{s}^{-1}$

P24.23 $3.46 \times 10^{-9} \, \text{m}$

P24.24 $\langle r^2 \rangle = 2nl^2$

P24.25 $\langle r^2 \rangle \approx 2nl^2 \left(1 - \dfrac{U_0}{4kT} \right)$

P24.26 a. $8 \times 10^{-9} \, \text{m}^3 \, \text{s}^{-1}$
    b. 0.88
    c. 0.006 cm

P24.27 a. $f = 1.5 \times 10^{-10} \, \text{kg s}^{-1}$
    b. $s = 7.3 \times 10^{-13} \, \text{s} = 7.3 \, \text{S}$

P24.28 a. $\delta = P24. \, \text{Å}, N = 28$
    b. $f = 1.7 \times 10^{-10} \, \text{kg s}^{-1}$, $s = 6.5 \times 10^{-13} \, \text{s} = 6.5 \, \text{S}$
    c. 7

P24.29 a. $270 \, \text{kg mol}^{-1}$
    b. 4.9 nm
    c. 68 nm; tight packing
    d. $140 \, \text{kg mol}^{-1}$; $18 \, \text{kg mol}^{-1}$
    e. $R_{protein} = 34.0 \, \text{Å}$; $R_{DNA} = 48.0 \, \text{Å}$

P24.30 a. $1.8 \times 10^5 \, \text{kg mol}^{-1}$
    b. $1.8 \times 10^{-22} \, \text{m}^3$
    c. $1.2 \times 10^{-9} \, \text{kg s}^{-1}$
    d. $r = 64 \times 10^{-9} \, \text{m}$; $V = 1.1 \times 10^{-21} \, \text{m}^3$

P24.31 $\delta_1 = 1.00$

P24.32 Partially overlapped or bent

P24.33 a. $6.9 \times 10^{-11} \, \text{kg s}^{-1}$
    b. 3.7 nm

P24.34 $67.2 \, \text{kg mol}^{-1}$

Chapter 25
P25.2 a. 2.79×10^3
    b. 425 s

P25.3 First order, $k = 3.9 \times 10^{-5} \, \text{s}^{-1}$

P25.5 a. First order wrt ClO
First order wrt BrO
Second order overall
Units of k: $s^{-1} M^{-1}$

b. Second order wrt NO
First order wrt O_2
Third order overall
Units of k: $s^{-1} M^{-2}$

c. Second order wrt HI
First order wrt O_2

$-\dfrac{1}{2}$ order wrt H^+
2.5 order over all
Units of k: $s^{-1} M^{-3/2}$

P25.6 a. second
b. third
c. 1.5

P25.7 Second order with respect to NO_2, first order with respect to H_2, $k = 6.43 \times 10^5 \, kPa^2 \, s^{-1}$

P25.8 First order with respect to both lactose and H^+

P25.9 Second order. $k = 0.317 \, M^{-1} \, s^{-1}$

P25.11 a. $k = 2.08 \times 10^{-3} \, s^{-1}$
b. $t = 1.11 \times 10^3 \, s$

P25.12 1.43×10^{24}

P25.13 1.50×10^4 days

P25.14 a. $k = 3.65 \times 10^{-4} \, day^{-1}$
b. 1.90×10^3 days

P25.18 $2.55 \times 10^{-7} \, s$

P25.21 a. $k = 0.0329 \, M^{-1} \, s^{-1}$
b. 120 s

P25.22 a. $1.08 \times 10^{-12} \, s$

P25.23 $4.76 \times 10^{-10} \, yr^{-1}$ and $k_2 = 5.70 \times 10^{-11} \, yr^{-1}$

P25.24 a. Rate $= 2.08 \times 10^{-15} \, M \, s^{-1}$
b. Rate $= 1.56 \times 10^{-15} \, M \, s^{-1}$
c. $[Cl] = 1.10 \times 10^{-18} \, M$, $[O_3] = 4.24 \times 10^{-11} \, M$

P25.28 a. $E_a = 1.50 \times 10^5 \, J \, mol^{-1}$, $A = 1.02 \times 10^{10} \, M^{-1} \, s^{-1}$
b. $k = 0.0234 \, M^{-1} \, s^{-1}$

P25.30 1845 s^{-1} and 1291 $M^- \, s^{-1}$

P25.32 $3.44 \times 10^{11} \, M^{-1} \, s^{-1}$

P25.33 $D = 1.08 \times 10^{-5} \, cm^2 \, s^{-1}$, $k = 3.62 \times 10^{10} \, M^{-1} \, s^{-1}$

P25.35 a. $E_a = 108 \, kJ \, mol^{-1}$, $A = 1.05 \times 10^{10} \, s^{-1}$
b. $\Delta S^{\ddagger} = -60.8 \, J \, mol \, K^{-1}$. $\Delta H^{\ddagger} = 105.2 \, kJ \, mol^{-1}$

P25.36 a. $E_a = 219 \, kJ \, mol^{-1}$, $A = 7.20 \times 10^{12} \, s^{-1}$
b. $\Delta S^{\ddagger} = -14.0 \, J \, mol \, K^{-1}$. $\Delta H^{\ddagger} = 212.4 \, kJ \, mol^{-1}$

P25.37 a. $E_a = 790 \, J \, mol^{-1}$, $A = 4.48 \times 10^{10} \, M^{-1} \, s^{-1}$
b. $k = 3.17 \times 10^{10} \, M^{-1} \, s^{-1}$
c. $\Delta S^{\ddagger} = -57.8 \, J \, mol \, K^{-1}$. $\Delta H^{\ddagger} = -2.87 \, kJ \, mol^{-1}$

P25.38 $\Delta S^{\ddagger} = 7.33 \, J \, mol \, K^{-1}$. $\Delta H^{\ddagger} = 37 \, kJ \, mol^{-1}$

Chapter 26

P26.5   a. $170\ kJ\ mol^{-1}$
             b. 3.98×10^7

P26.9 $2.45 \times 10^{-3}\ M\ s^{-1}$

P26.10 $R_{max} = 3.75 \times 10^{-2}\ M\ s^{-1}$, $K_m = 2.63 \times 10^{-2}\ M$, $k_2 = 1.08 \times 10^7\ s^{-1}$

P26.11   a. $2.5\ M$
             b. $2.71 \times 10^{-5}\ M$

P26.12 $0.0431\ M$

P26.13   a. $R_{max} = 0.137\ mM\ s^{-1}$, $K_m = 2.42\ mM$
             b. $R_{max} = 0.300\ mM\ s^{-1}$, $K_m = 28.5\ mM$

P26.14 $K_m = 6.49\ \mu M$, $R_{max} = 4.74 \times 10^{-8}\ \mu M\ s^{-1}$, $K_m{}^* = 24.9\ \mu M$, $K_i = 70.4\ \mu M$

P26.15   a. $R_{max} = 16.1\ mM\ s^{-1}$, $K_m = 2.48\ mM$
             b.   noncompetitive. R_{max} (2 mM) $= 6.37\ mM\ s^{-1}$, R_{max} (5 mM) $= 3.39\ mM\ s^{-1}$

P26.16   a. $R_{max} = 20.6\ \mu M\ s^{-1}$, $K_m = 10.1\ \mu M$
             b. $R_{max} = 15.5\ \mu M\ s^{-1}$, $K_m = 2.48\ mM$

P26.19 $R_{max} = 0.460\ M\ s^{-1}$, $K_m = 0.045\ M$

P26.20   a.   at 290 nm: 7.29×10^{16} photons cm^{-2}
                    at 320 nm: 8.05×10^{16} photons cm^{-2}
             b. $34.5\ s$

P26.22 0.95

P26.23 4.55×10^{18} photons, $I = 2.41 \times 10^{-3}\ J\ s^{-1}$

P26.24 3.02×10^{19} photons s^{-1}

P26.25 $\dfrac{k_{ISC}^{S}}{k_f} = 7.33$, $k_P = 3.88 \times 10^{-2}\ s^{-1}$, $k_{ics}^{T} = 0.260\ s^{-1}$

P26.26   a. 0.91
             b. 7.19×10^{18} photons $cm^{-1}\ s^{-1}$
             c. $1.53\ nW$

P26.27   a. $A = 0.312$, $B = 0.697$
             b. $-0.172\ atm$

P26.29 0.386

P26.30 $11.5\ \text{Å}$

P26.32   a. $3.40 \times 10^{-19}\ J$
             b. $1.94 \times 10^{-19}\ J$

P26.33   a. $\beta = 0.768\ \text{Å}^{-1}$
             b. $\beta = 0.384\ \text{Å}^{-1}$

Index

Masses and Natural Abundances for Selected Isotopes

Nuclide	Symbol	Mass (amu)	Percent Abundance
H	^{1}H	1.0078	99.985
	^{2}H	2.0140	0.015
He	^{3}He	3.0160	0.00013
	^{4}He	4.0026	100
Li	^{6}Li	6.0151	7.42
	^{7}Li	7.0160	92.58
B	^{10}B	10.0129	19.78
	^{11}B	11.0093	80.22
C	^{12}C	12 (exact)	98.89
	^{13}C	13.0034	1.11
N	^{14}N	14.0031	99.63
	^{15}N	15.0001	0.37
O	^{16}O	15.9949	99.76
	^{17}O	16.9991	0.037
	^{18}O	17.9992	0.204
F	^{19}F	18.9984	100
P	^{31}P	30.9738	100
S	^{32}S	31.9721	95.0
	^{33}S	32.9715	0.76
	^{34}S	33.9679	4.22
Cl	^{35}Cl	34.9688	75.53
	^{37}Cl	36.9651	24.4
Br	^{79}Br	78.9183	50.54
	^{81}Br	80.9163	49.46
I	^{127}I	126.9045	100